Features of the Companion Website

Hands-On Problem Solving Exercises: These inquiry-based exercises challenge you to think as a scientist and to analyze and interpret experimental data. Exercises include data manipulation questions based on real experiments, as well as problems involving simulations of model systems.

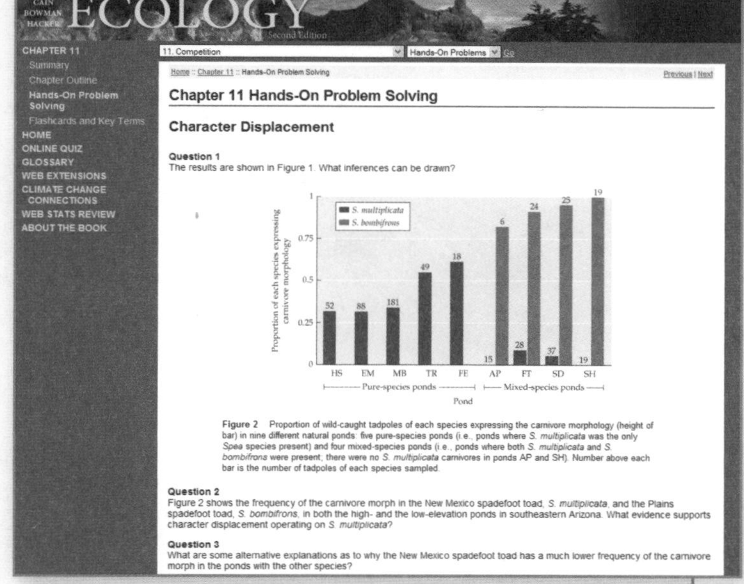

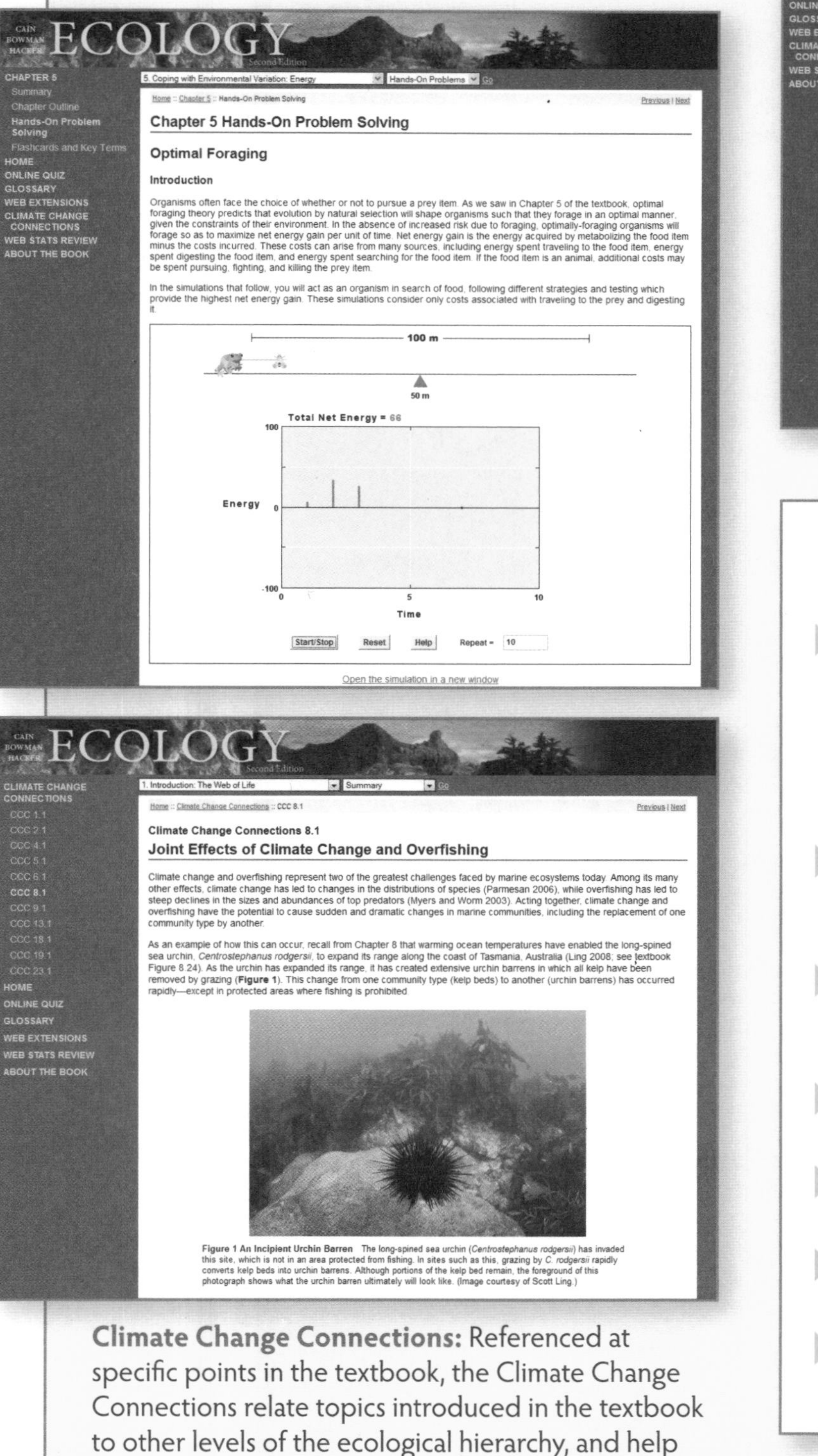

Climate Change Connections: Referenced at specific points in the textbook, the Climate Change Connections relate topics introduced in the textbook to other levels of the ecological hierarchy, and help you better understand ongoing climate change.

ADDITIONAL FEATURES OF THE Companion Website:

- **Online Quizzes:** Multiple-choice quizzes cover all the main topics presented in each chapter. Your instructor may assign these quizzes, or they may be made available to you as self-study tools. (Instructor registration is required for student access to the quizzes.)
- **Chapter Outlines and Summaries:** Concise overviews of the important concepts and topics covered in each chapter.
- **Flashcards & Key Terms:** Flashcard activities help you master the many new terms introduced in each chapter of the textbook.
- **Web Extensions:** Expanded and additional coverage of selected topics introduced in the textbook.
- **Web Stats Review:** A brief review of statistical methods and techniques mentioned in the textbook.
- **Glossary:** A complete online version of the glossary, for quick access to definitions of important terms.
- **Suggested Readings** for each chapter of the textbook.

Hands-On Problem Solving Exercises

The following Hands-On Problems are available on the Companion Website (sites.sinauer.com/ecology2e).

1. Mosquitoes, Drought, and Disease: This Web exercise explores the connections between mosquito populations, periodic drought, and the incidence of mosquito transmitted diseases. You will read a recent paper that shows that mosquito populations increase in size after droughts. You will then plot data on drought severity compared to incidence of disease, and discuss implications for human health.

2. Simulating Seasons: This Web exercise illustrates connections between the axial tilt of Earth and temperature variation. Seasonal patterns and the range of temperature variations result from the degree of axial tilt. You will use a simulation model of Earth to vary axial tilt and explore latitudinal and seasonal variations in temperature.

3. Changing Climate and Changing Tree Lines: This Web exercise explores connections between elevation of tree lines and climate patterns. You will read a paper that discusses factors determining upper tree lines and which types of tree lines are likely to advance with changes in temperature. You will then plot recent temperature changes in high elevation areas and discuss the probability of tree line advance there.

4. Thermal Adaptations in Urban Ants: This Web exercise demonstrates thermal adaptations in ants that live in cities. You will investigate whether adaptations to heat in urban leaf-cutter ants reduces tolerance to cold temperatures.

5. Optimal Foraging: This Web exercise illustrates patterns of movement predicted by optimal foraging. Foraging decisions are based on relative costs and benefits. You will manipulate the foraging decision rules of a predator to explore how distance to and size of prey influence foraging strategies and benefits.

6. Effects of Natural Selection and Genetic Drift: This Web exercise demonstrates how natural selection and genetic drift can alter the frequencies of alleles in populations. You will investigate the effects of manipulating population sizes (and thus the strength of genetic drift) and strengths of selection.

7. The Growth/Reproduction Trade-off: This Web exercise explores the trade-off that species must make between growth and reproduction. You will investigate the effects of manipulating the set point at which fish start allocating resources into reproduction rather than further growth, under different levels of predation.

8. Effort and Accuracy of Population Estimates: This Web exercise illustrates the relationship between the effort required to obtain population size estimates and their accuracy. Species and population characteristics influence the ease of obtaining population estimates and the accuracy of those estimates. You will choose a method of population estimation and manipulate the amount of effort, to explore the effects on estimate accuracy.

9. Density-Dependent and Density-Independent Factors: This Web exercise explores the effects of density-dependent and density-independent factors on populations. You will analyze data from populations of arctic ground squirrels to determine these effects on weaning success, over-wintering survival, and population growth rate.

10. Population Dynamics: This Web exercise explores how changes in population growth rates and the extent of delayed density dependence affect population dynamics. You will manipulate the values of these two factors, and run a simulation to determine their effects on a population.

11. Character Displacement: This Web exercise explores character displacement and phenotypic plasticity for a polyphenism in a spadefoot toad species. You will interpret data collected on different populations of toads and the results of common garden experiments to assess genetic effects.

12. Cascading Effects of Predators: This Web exercise explores the effect of predators through multiple trophic levels. You will read a recent paper on predator-driven cascades in marine systems. You will then use data from a system in which wolves are the top predator to test for a trophic cascade.

13. Dynamics of Disease: This Web exercise explores the dynamics of host–pathogen systems. You will manipulate traits of the interacting species to simulate various strategies of hosts and pathogens. You will also explore the spread or decline of pathogens based on the population size of vulnerable hosts, and discuss the ecological and evolutionary implications of parasitic interactions.

14. Population Dynamics of a Mutualism: This Web exercise explores the dynamics of a mutualism between cacti and moths. You will read a recent paper that presents a model of this mutualism and discusses the natural history of the two species. Using the model, you will explore the effects of starting size and proportion of the two species on population dynamics.

15. Measuring Marine Species Diversity: This Web exercise explores various methods of measuring species diversity. You will interpret data from a recent paper that measured marine species diversity, with emphasis on the benthic fauna of the continental shelf of Norway, using the Shannon index as well as other indices of diversity.

16. Soil Invertebrates and Succession: This Web exercise explores how invertebrates that live in the soil could affect patterns of succession. You will interpret data from a recent paper that shows the effects of soil invertebrates on the growth of early- and mid-succession plants.

17. Island Biogeography: This Web exercise explores the factors that determine the number of species that can live on different islands according to the theory of island biogeography. In a series of simulations, you will manipulate the size of an island and the distance from the island to the mainland to demonstrate how these factors affect the equilibrium number of species on that island.

18. Periodic Disturbance and Its Effect on Species: This Web exercise demonstrates how periodic disturbance can maintain species in a community that otherwise could not coexist. You will manipulate the frequency and intensity of disturbances to investigate this effect.

19. Drought Reduces Productivity across Europe: This Web exercise examines the effect of the 2003 European drought on primary productivity. You will interpret data from a recent paper that demonstrated that the drought did substantially reduce primary productivity at various sites in Europe.

20. Trophic Efficiency in a Coral Reef System: This Web exercise explores energy flow and efficiency of energy transfer in a coral reef community. You will read a recent paper that quantifies energy flows through multiple trophic levels in a community. Using data from the paper, you will calculate efficiencies of various steps in this system, and discuss the effects of trophic level on energy flow.

21. Dry Decomposition: This Web exercise explores how plant litter decomposes in a dry climate. You will interpret data from a recent paper demonstrating the factors responsible for litter decomposition in a semi-arid ecosystem in Patagonia.

22. Population Augmentation and Recovery of Endangered Species: This Web exercise explores the consequences of augmenting populations of endangered species with captive-raised individuals. You will read a recent paper on population augmentation in an endangered butterfly. Then, using a transition matrix model, you will explore the relative costs and benefits of population augmentation and habitat enhancement for an endangered fish, the June sucker.

23. Patch Movement: Crickets vs. Cybercrickets: This Web exercise explores how organisms move across patches in the landscape. You will interpret data from a recent paper that simulated the movement of virtual organisms between patches across landscapes with different levels of connectivity, and then compared the simulation results to results from manipulation studies. Do real crickets move like cybercrickets?

24. Nitrogen Cycle: Too Much or Too Little? This Web exercise explores global flows in reactive nitrogen from anthropogenic sources. You will read a recent paper on anthropogenic transformation of the global nitrogen cycle. You will then calculate gains and losses of nitrogen on a continental scale and discuss the potential effects on humans and the natural environment.

ECOLOGY
Second Edition

ECOLOGY

Second Edition

MICHAEL L. CAIN • Bowdoin College
WILLIAM D. BOWMAN • University of Colorado
SALLY D. HACKER • Oregon State University

Sinauer Associates, Inc. • Sunderland, Massachusetts

Cover photograph

Pride of Madeira (*Echium candicans*) blooms in Julia Pfeiffer Burns State Park, California. Photo © Tomas Kaspar/Alamy.

Ecology, Second Edition

Address inquiries and orders to
Sinauer Associates, Inc.
23 Plumtree Road
Sunderland, MA 01375 U.S.A.

FAX: 413-549-1118
E-mail: publish@sinauer.com
Internet: www.sinauer.com

Library of Congress Cataloging-in-Publication Data

Cain, Michael L. (Michael Lee), 1956-
Ecology / Michael L. Cain, William D. Bowman, Sally D. Hacker. -- 2nd ed.
p. cm.
ISBN 978-0-87893-445-4
1. Ecology. I. Bowman, William D. II. Hacker, Sally D. III. Title.
QH541.E31933 2011
577--dc22
2011000946

Printed in U.S.A.

5 4 3 2

For Debra and Hannah, with thanks and love.

MLC

For Jen, Gordon, and Miles and their unending patience,
and to my students for pushing me as much as I pushed them.

WDB

For my family and my students, whose gift of time has
made all the difference.

SDH

About the Authors

MICHAEL L. CAIN, having opted to change careers and focus full-time on writing, is currently affiliated with Bowdoin College. After receiving his Ph.D. in Ecology and Evolutionary Biology from Cornell University, he taught at New Mexico State University and the Rose-Hulman Institute of Technology. In addition to his work on this book, Dr. Cain is a coauthor of Campbell's *Biology*, Ninth Edition. He has instructed students across a wide range of subjects, including introductory biology, ecology, field ecology, evolution, botany, mathematical biology, and biostatistics. His research interests include: plant ecology; long-distance dispersal; ecological and evolutionary dynamics in hybrid zones; and search behavior in plants and animals.

WILLIAM D. BOWMAN is a Professor at the University of Colorado at Boulder, affiliated with the Department of Ecology and Evolutionary Biology and the Institute of Arctic and Alpine Research. He earned his Ph.D. from Duke University. Dr. Bowman has taught courses in introductory ecology, plant ecology, plant–soil interactions, and ecosystems ecology, and has directed undergraduate summer field research programs. He is coeditor of *Structure and Function of an Alpine Ecosystem, Niwot Ridge, Colorado* (Oxford University Press, 2001). His research focuses on plant ecology, biogeochemistry, and community dynamics, and has been supported by the National Science Foundation, the Environmental Protection Agency, the National Park Service, and the Andrew W. Mellon Foundation.

SALLY D. HACKER is a Professor at Oregon State University, Corvallis, where she has been a faculty member since 2004. As a community ecologist interested in natural and managed coastal, dune, and estuarine communities, Dr. Hacker's research explores the structures, functions, and services of communities under varying contexts of species interactions and physical conditions. She teaches courses in introductory ecology, community ecology, and marine biology. Dr. Hacker received her Ph.D. in 1996 from Brown University, where she conducted research on the role of positive interactions in communities. This work, conducted in salt marsh systems, has been widely cited and featured in a number of ecology textbooks.

Brief Contents

Preface

It was a joy to write this Second Edition because we love what we are writing about, ecology. Indeed, this is an exciting and challenging time to study ecology. New discoveries are pouring in, revealing factors that affect local communities and link ecosystems to one another across broad geographic areas. The progress in these and other areas of ecology could not come at a better time: Ecologists are increasingly being asked to apply their knowledge toward efforts to solve current environmental problems and prevent future ones.

Developments such as these fuel the excitement that grips the field of ecology—but they also mean that what we know about ecology is increasing very rapidly. The explosion of ecological information makes ecology a daunting subject, both to study and to teach. Students need to master a heady mix of abstract concepts, experimental reasoning, mathematical equations, and details about particular organisms and their habitats. For their part, instructors are faced with the challenge of conveying fundamental concepts, new discoveries, and the relevance and rigor of modern ecology—all in a manner that works well for students taking their first course in ecology. With these challenges in mind, the overarching goal for the Second Edition of *Ecology* was to enhance the book as a learning tool for students and as a teaching tool for professors. In setting out to achieve this goal, the book's two core principles guided our every step.

Core Principles of *Ecology*, Second Edition

This book is written for undergraduate students of ecology. We set out to introduce our readers to the beauty and importance of ecology, and to do so without boring them or overwhelming them with unnecessary detail. This is a tall order, and so when we began writing the Second Edition of *Ecology*, we kept our focus on the two core principles of this book: **"Teaching comes First!"** and **"Less is More!"**

Teaching truly does come first in *Ecology*—it motivates everything we did. The structure and content of our chapters is designed primarily to make them good tools for teaching. For example, to introduce the material covered and capture student interest, each chapter begins with a story (a "Case Study," as described more fully below) about an applied problem or interesting bit of natural history. Once students are drawn in by the Case Study, the "storyline" that begins there is maintained throughout the rest of the chapter. We use a narrative writing style to link the sections of the chapter to one another, thus helping students keep the big picture in mind. In addition, the sections of the chapter are organized around a small number of Key Concepts (also described more thoroughly below) that were carefully selected to summarize current knowledge and provide students with a clear overview of the subject at hand. Similarly, when designing the art, pedagogy came first: Many students are visual learners, so we worked very hard to ensure that each figure "tells a story" that can be understood on its own, without reference to the main body of the text.

As another way to help us achieve our primary goal of teaching students, we followed a "less is more" philosophy. We were guided by the principle that if we covered less material—but presented it clearly and well—students would learn more. Hence, our chapters are relatively short and they are built around a small number of Key Concepts (typically, three to five). We made these choices to prevent students from being overwhelmed by long, diffuse chapters, and to allow them to master the big ideas first, then fill in the details. In addition, as we worked on the drafts of our chapters, we put our "less is more" philosophy into action by asking each other whether the text served one of the following purposes:

- Does it help to explain an essential concept?
- Does it show how the process of ecological inquiry works?
- Does it motivate readers by focusing on a key ecological application or a fascinating piece of natural history?

This approach made for some tough choices as we strove to balance the addition of new material with cuts made from the First Edition, but it enabled us to focus on teaching students what is currently known about ecology without overwhelming them with excess information.

New Features of *Ecology*, Second Edition

In striving to make *Ecology* the best teaching tool possible, we updated, replaced, or cut sections of the text as appropriate, and added several new pedagogical features. These include:

Climate Change Connection Climate change has broad ecological effects with important implications for conservation and ecosystem services. Roughly one-half of the Second Edition chapters now include a major climate change example, followed immediately by a sentence directing students to additional content on the Companion Website. These web-based *Climate Change Connections* discuss how the example in the text connects to other levels of the ecological hierarchy, while enriching the student's understanding of ongoing climate change.

Ecological Toolkits A number of chapters include an *Ecological Toolkit*, a type of box that describes ecological "tools" such as experimental design, remote sensing, GIS, mark–recapture techniques, stable isotope analysis, DNA fingerprinting, and the calculation of species–area curves.

Figure Legend Questions Each chapter includes 3–6 *Figure Legend Questions* that are highlighted in color at the end of the legend. These questions encourage students to grapple with the figure and make sure they understand its content. The questions range from those that test whether students understand the axes or other simple aspects of the figure to those that ask students to develop or evaluate hypotheses.

In-Class Exercises For the Second Edition, a new type of inquiry exercise has been added to the Instructor's Resource Library: ready-to-go problems that take about 10 minutes to do and can be used in class or assigned as homework. One or more of these exercises has been added for each chapter.

Error Bars Where appropriate, error bars have been added to figures. To provide support for students related to this change, the Web Stats Review now includes new material on the relationship between the standard deviation and the standard error of the mean, as well as confidence intervals and linear regression.

Hallmark Features of *Ecology*

We've also revised and strengthened the following key pedagogical features of *Ecology*, introduced in the First Edition:

Pedagogical Excellence Students taking their first course in ecology are exposed to a great deal of material, on a conceptual as well as individual-systems level. To help them manage this vast amount of information, each chapter of *Ecology* is organized around a small number of Key Concepts that provide up-to-date summaries of fundamental ecological principles. All of these Key Concepts are listed on the book's back end papers.

Case Studies Each chapter opens with an interesting vignette—a *Case Study*. By presenting an engaging story or interesting application, the *Case Study* captures the reader's attention while introducing the topic of the chapter. Later, the reader is brought full circle with the corresponding "*Case Study Revisited*" section at chapter's end. Each *Case Study* relates naturally to multiple levels of the ecological hierarchy, thereby providing a nice lead-in to the *Connections in Nature* feature, described next.

Connections in Nature In most ecology textbooks, connections among levels of the ecological hierarchy are discussed briefly, perhaps only in the opening chapter. As a result, many opportunities are missed to highlight for students the fact that events in natural systems *really are* interconnected. To facilitate the ability of students to grasp how events in nature are interconnected, each chapter of *Ecology* closes with a section that discusses how the material covered in that chapter affects and is affected by interactions at other levels of the ecological hierarchy. Where appropriate, these interconnections are also emphasized in the main body of the text.

Ecological Inquiry Our understanding of ecology is constantly changing due to new observations and new results from ecological experiments and models. All chapters of the book emphasize the active, inquiry-based nature of what is known about ecology. In addition, *Ecology* includes hands-on interpretative and quantitative exercises, described next.

Hands-On Problem Solving Exercises This popular feature of the Companion Website asks students to manipulate data, explore mathematical aspects of ecology in more detail, interpret results from real experiments, and analyze simple model systems using simulations. The Second Edition includes 24 revised and new *Hands-On Problem Solving Exercises*, one for each chapter of the book. These inquiry exercises can be used in two important ways: portions of them can serve as ready-to-go, five- to ten-minute problems for in-class use (e.g., with "clickers" or to stimulate class discussion), or the entire exercise can be assigned as homework.

Ecological Applications In recent years, ecologists have increasingly focused their attention on applied issues. Similarly, many students taking introductory ecology are very interested in applied aspects of ecology. Thus, ecological applications (including conservation biology) receive great attention in this book. Discussions of applied topics are woven into each chapter, helping to capture and retain student interest.

Links to Evolution Evolution is a central unifying theme of all biology, and its connections with ecology are very strong. Yet, ecology textbooks typically present evolution almost as a separate subject. As an alternative to the standard approach, *Ecology*'s Chapter 6 is devoted to describing the joint effects of ecology and evolution. This chapter explores the ecology of evolution at both the population level and as documented in the sweeping history of life on Earth. Concepts or applications that relate to evolution are also described in many other chapters.

Art Program Many of *Ecology*'s illustrations feature "balloon captions," which tell a story that can be understood at a glance, without relying on the accompanying text. The art program is available as part of the Instructor's Resource Library (see Media and Supplements section).

Ecology Is a Work in Progress

This book, like the subject we write about, does not consist of a set of unchanging ideas and fixed bits of information. Instead, the book will develop and change over time as we respond to new discoveries and new ways of teaching. As we roll up our sleeves to begin working on the next edition, we would love to hear from you—what you like about the book, what you don't like, and any questions or suggestions you may have for how we can improve the book. You can reach us individually or as a group by sending an email message to ecology@sinauer.com, or by writing us at *Ecology*, Sinauer Associates, 23 Plumtree Road, Sunderland, MA 01375.

Acknowledgments

We'd like to thank some of the many people who helped us turn our ideas into a book in print. Special thanks to Mary Santelmann at Oregon State University, who read every chapter in page proofs and helped us catch lingering problems. We are grateful to all of our colleagues who generously critiqued the plan for the book or read one or more chapters in manuscript, who are listed on the following pages. Among the hundreds of people we contacted while researching this book, we also wish to thank the following individuals for their special efforts in providing guidance and generously sharing their time and expertise: Jocelyn Aycrigg, Jon Evans, Jenifer Hall-Bowman, John Jaenike, Justin Kitzes, Scott Ling, Nathan Stephenson, Debra VamVikites, and Ophelia Wang.

Finally, we would like to express our appreciation to some of our colleagues at Sinauer Associates, with whom we worked closely during the writing but especially during the book's production. Andy Sinauer supported the plan for the book from Day One. He enthusiastically participated in every phase along the way. Kathaleen Emerson and Laura Green did a terrific job guiding the book through its many stages of production. Norma Roche did a superb job of copyediting our manuscript, often encouraging us to expand certain topics or omit others to help us meet the goals outlined above. Elizabeth Morales provided the beautiful illustrations in the book and her queries always helped us sharpen the visual messages. David McIntyre, our photo editor, always managed to find exquisite images that enhance the information in the figures. Joan Gemme stayed with us through several rounds of page design. We love her final version that you see here as well as her elegant cover design. Susan McGlew coordinated the more than 100 reviews of the manuscript with great perseverance. Dean Scudder masterminded the entire marketing effort and Marie Scavotto produced the attractive brochure. Jason Dirks pulled together the impressive array of supplementary materials listed on page xiii. And Mark Belk (Brigham Young University), Charlene D'Avanzo (Hampshire College), Christiane I. M. Healey (University of Massachusetts, Amherst), Norman Johnson (University of Massachusetts, Amherst), and Amy McEuen (University of Illinois, Springfield) also demonstrated their great skill and breadth in writing the online and instructor resources.

MICHAEL L. CAIN
mcain@bowdoin.edu

WILLIAM D. BOWMAN
william.bowman@colorado.edu

SALLY D. HACKER
hackers@science.oregonstate.edu

February 2011

Reviewers

Reviewers for the Second Edition

David Ackerly, University of California, Berkeley
Stephano Allesina, University of Chicago
Robert Baldwin, Clemson University
Betsy Bancroft, Southern Utah University
Jeb Barrett, Virginia Polytechnic Institute and State University
Beatrix Beisner, University of Quebec at Montreal
Mark Belk, Brigham Young University
Kim Bjorgo-Thorne, West Virginia Wesleyan College
Steve Blumenshine, California State University, Fresno
Michael Booth, Principia College
Steve Brewer, University of Mississippi
Linda Brooke Stabler, University of Central Oklahoma
Kenneth Brown, Louisiana State University
Stephen Burton, Grand Valley State University
Peter Chabora, Queens College, CUNY
Gary Chang, Gonzaga University
Elsa Cleland, University of California, San Diego
Liane Cochran-Stafira, Saint Xavier University
Robert Colwell, University of Connecticut
William Crampton, University of Central Florida
Megan Dethier, University of Washington
John Ebersole, University of Massachusetts, Boston
Erle Ellis, University of Maryland, Baltimore County
Sally Entrekin, University of Central Arkansas
Jennifer Fox, Georgetown University
Kamal Gandhi, University of Georgia
Elise Granek, Portland State University
Martha Groom, University of Washington
Vladislav Gulis, Coastal Carolina University
Christiane Healey, University of Massachusetts, Amherst
Kevin Higgins, University of South Carolina
Randall Hughes, Florida State University
Art Johnson, Pennsylvania State University
Jerry Johnson, Brigham Young University
Piet Johnson, University of Colorado
Vedham Karpakakunjaram, University of Maryland
Michael Kinnison, University of Maine
Astrid Kodric-Brown, University of New Mexico
Jennifer Lau, Michigan State University
Stacey Lettini, Gwynedd-Mercy College
Gary Ling, University of California, Riverside
Scott Ling, University of Tasmania
Dale Lockwood, Colorado State University
Daniel Markewitz, University of Georgia
Scott Meiners, Eastern Illinois University
Thomas Miller, Florida State University
Gary Mittelbach, Kellogg Biological Station, Michigan State University
David Morgan, University of West Georgia
Shannon Murphy, George Washington University
Timothy Nuttle, Indiana University of Pennsylvania
Mike Palmer, Oklahoma State University
Kevin Pangle, The Ohio State University
Keith Pecor, The College of New Jersey
Jeff Podos, University of Massachusetts, Amherst
Andrea Previtalli, Cary Institute of Ecosystem Studies
Alysa Remsburg, Unity College
Jason Rohr, University of South Florida
Nathan Sanders, University of Tennessee, Knoxville
Mary Santelmann, Oregon State University
Tom Sarro, Mount Saint Mary College
Dov Sax, Brown University
Sam Scheiner
Tom Schoener, University of California, Davis
Catherine Searle, Oregon State University
Jonathan Shurin, University of California, San Diego
Christopher Steiner, Wayne State Univesity
Michael Toliver, Eureka College
Monica Turner, University of Wisconsin
Stuart Wooley, California State University, Stanislaus
Brenda Young, Daemen College
Richard Zimmerman, Old Dominion University

Reviewers for the First Edition

David Ackerly, Stanford University
Gregory H. Adler, University of Wisconsin, Oshkosh
Stuart Allison, Knox College
Kama Almasi, University of Wisconsin, Stevens Point

Peter Alpert, University of Massachusetts, Amherst
David Armstrong, University of Colorado
James Barron, Montana State University
Christopher Beck, Emory University
Mark C. Belk, Brigham Young University
Michael A. Bell, Stony Brook University
Eric Berlow, University of California, Merced
Charles Blem, Virginia Commonwealth University
Carl Bock, University of Colorado
Daniel Bolnick, University of Texas, Austin
April Bouton, Villanova University
Steve Brewer, University of Mississippi
David D. Briske, Texas A&M University
Judie Bronstein, University of Arizona
Romi Burks, Southwestern University
Aram Calhoun, University of Maine
Mary Anne Carletta, Georgetown College
Walter Carson, University of Pittsburgh
David D. Chalcraft, East Carolina University
Colin A. Chapman, University of Florida
Cory Cleveland, University of Montana
Rob Colwell, University of Connecticut
James Cronin, Louisiana State University
Todd Crowl, Utah State University
Anita Davelos Baines, University of Texas, Pan American
Andrew Derocher, University of Alberta
Megan Dethier, University of Washington
Jonathan Evans, University of the South
John Faaborg, University of Missouri
William F. Fagan, University of Maryland
Rick Gillis, University of Wisconsin, LaCrosse
Thomas J. Givnish, University of Wisconsin
Elise Granek, Portland State University
Martha Groom, University of Washington, Bothell
Jack Grubaugh, University of Memphis
Jessica Gurevitch, Stony Brook University
Nelson Hairston, Cornell University
Jenifer Hall-Bowman, University of Colorado
Jason Hamilton, Ithaca College
Christopher Harley, University of British Columbia
Bradford Hawkins, University of California, Irvine
Mike Heithaus, Florida International University
Kringen Henein, Carleton University, Ontario
Nat Holland, Rice University
Stephen Howard, Middle Tennessee State University
Vicki Jackson, Central Missouri State University
John Jaenike, University of Rochester
Pieter Johnson, University of Colorado
Timothy Kittel, University of Colorado
Jeff Klahn, University of Iowa
Tom Langen, Clarkson University
Jack R. Layne, Jr., Slippery Rock University
Jeff Leips, University of Maryland, Baltimore County
Svata Louda, University of Nebraska
Sheila Lyons-Sobaski, Albion College
Richard Mack, Washington State University
Lynn Mahaffy, University of Delaware
Michael Mazurkiewicz, University of Southern Maine
Andrew McCall, Denison University
Shannon McCauley, University of Michigan
Mark McPeek, Dartmouth College
Bruce Menge, Oregon State University
Thomas E. Miller, Florida State University
Sandra Mitchell, Western Wyoming College
Russell Monson, University of Colorado
Daniel Moon, University of North Florida
William F. Morris, Duke University
Kim Mouritsen, University of Aarhus
Courtney Murren, College of Charleston
Shahid Naeem, Columbia University
Jason Neff, University of Colorado
Scott Newbold, Colorado State University
Shawn Nordell, Saint Louis University
Timothy Nuttle, University of Pittsburgh
Christopher Paradise, Davidson College
Matthew Parris, University of Memphis
William D. Pearson, University of Louisville
Jan Pechenik, Tufts University
Karen Pfennig, University of North Carolina
David M. Post, Yale University
Joe Poston, Catawba College
Seth R. Reice, University of North Carolina
Heather Reynolds, Indiana University, Bloomington
Willem Roosenburg, Ohio University, Athens
Richard B. Root, Cornell University
Scott Ruhren, University of Rhode Island
Nathan Sanders, University of Tennessee
Dov Sax, Brown University
Maynard H. Schaus, Virginia Wesleyan College
Thomas Schoener, University of California, Davis
Janet Schwengber, SUNY Delhi
Erik P. Scully, Towson University
Dennis K. Shiozawa, Brigham Young University
Frederick Singer, Radford University
John J. Stachowicz, University of California, Davis
Cheryl Swift, Whittier College
Ethan Temeles, Amherst College
Michael Toliver, Eureka College
Bill Tonn, University of Alberta
Kathleen Treseder, University of Pennsylvania
Thomas Veblen, University of Colorado
Don Waller, University of Wisconsin
Carol Wessman, University of Colorado
Jake F. Weltzin, University of Tennessee
Jon Witman, Brown University

Media & Supplements to Accompany *Ecology*, Second Edition

For the Student

Companion Website (sites.sinauer.com/ecology2e)

The *Ecology* Companion Website offers students a wealth of study and review material, all available free of charge. Climate Change Connections and Web Extensions expand on the coverage of selected topics introduced in the textbook. Hands-On Problem Solving Exercises provide practical experience working with experimental data and interpreting results from simulations and models. The online quizzes (instructor registration required) are a great way for students to check their comprehension of the material covered in each chapter. And the flashcards encourage familiarity with the many new terms introduced in the ecology course.

The *Ecology*, Second Edition Companion Website includes:

- Chapter Outlines
- Chapter Summaries
- Hands-On Problem Solving Exercises
- Climate Change Connections
- Web Extensions
- Online Quizzes
- Flashcards & Key Terms
- Suggested Readings
- Web Stats Review
- Complete Glossary

(See the inside front cover for additional details.)

For the Instructor

(Available to qualified adopters)

Instructor's Resource Library

The *Ecology* Instructor's Resource Library includes a variety of resources to aid instructors in course planning, lecture development, and student assessment. The Resource Library includes:

- Figures & Tables: All of the line-art illustrations, photos, and tables from the textbook are provided as both high-resolution and low-resolution JPEGs, all optimized for use in lecture.
- PowerPoint Resources: Two different PowerPoint presentations are provided for each chapter of the textbook.
- Figures: All figures, photos, and tables from each chapter, with titles.
- Lecture: A complete lecture outline, including selected figures.
- In-Class Exercises: New for the Second Edition, these exercises provide instructors with ready-to-use problems and questions designed to be incorporated into the lecture.
- Hands-on Problem Solving Exercises: The exercises from the Companion Website are included in Microsoft Word format, with suggested answers.

All of the resources included in the Instructor's Resource Library are also available to instructors online, via the instructor's side of the Companion Website. (Instructor registration required.)

Test Bank

The *Ecology*, Second Edition Test Bank (included on the Instructor's Resource Library) includes a thorough set of multiple-choice questions for each chapter of the textbook. All important concepts are covered, and each question is referenced to a specific chapter heading, concept number, and page number. The Test Bank also includes key terms lists, for use in terminology quizzes, and all of the questions from the Companion Website online quizzes. The test bank is included in the Instructor's Resource Library in two formats:

- Microsoft Word
- Wimba Diploma (software included): Diploma is a powerful, easy-to-use exam creation program that lets you quickly assemble exams using any combination of publisher-provided questions and your own questions.

Online Quizzing

The Companion Website includes online quizzes that can be assigned or opened for use by students as self-quizzes. Quizzes can be customized with any combination of the default questions and an instructor's own questions. Quiz results are stored in the online grade book. (Note: Instructors must register in order for their students to be able to take the quizzes.)

eBook (ISBN 978-0-87893-579-6)

www.coursesmart.com

Ecology, Second Edition is available as an eBook via CourseSmart, at a substantial discount off the price of the printed textbook. The CourseSmart eBook reproduces the look of the printed book exactly, and includes convenient tools for searching the text, highlighting, and note-taking. The eBook is available in both downloadable and online formats, and supports the iPhone and iPad.

Contents

UNIT 1 Organisms and Their Environment

UNIT 2 Populations

UNIT 3 Interactions among Organisms

UNIT 4 Communities

UNIT 5 Ecosystems

UNIT 6 Applied and Large-Scale Ecology

ECOLOGY
Second Edition

CHAPTER

1 The Web of Life

KEY CONCEPTS

- **CONCEPT 1.1** Events in the natural world are interconnected.
- **CONCEPT 1.2** Ecology is the scientific study of interactions between organisms and their environment.
- **CONCEPT 1.3** Ecologists evaluate competing hypotheses about natural systems with observations, experiments, and models.

Deformity and Decline in Amphibian Populations: A Case Study

In August of 1995, a group of elementary and middle school students from Henderson, Minnesota, made a gruesome discovery as they caught leopard frogs (*Rana pipiens*) for a summer science project: 11 of the 22 frogs they found were severely deformed. Some of the frogs had missing or extra limbs, others had legs that were shortened or bent in odd directions, and still others had legs projecting from their stomachs or bony growths coming out of their backs (**Figure 1.1**). The students reported their findings to the Minnesota Pollution Control Agency, which investigated the pond and found that, overall, 30%–40% of the frogs there were deformed.

News of the students' discovery traveled fast, capturing public attention and spurring scientists to check for similar deformities in other parts of the country and in other amphibian species. It soon became apparent that the problem was widespread. In the United States, misshapen individuals were found in 46 states and in more than 60 species of frogs, salamanders, and toads. In some localities, more than 90% of the individuals were deformed. Deformed amphibians were also found in Europe, Asia, and Australia. Worldwide, it appeared that the frequency of amphibian deformities was on the rise.

Adding to the alarm caused by the gruesome deformities were observations, beginning in the late 1980s, of another disturbing trend: global amphibian populations seemed to be in decline. By 1993, over 500 populations of frogs and salamanders from around the world were reported to be decreasing in size or under threat of extinction. In some cases, entire species were in danger; across the globe, hundreds of species were extinct, missing, or threatened (**Figure 1.2**). Since 1980, at least 9 amphibian species have become extinct. An additional 113 species have not been seen since that time and are listed as "possibly extinct" (Vié 2009).

Species in other groups of organisms were also showing signs of decline, but scientists were especially worried about amphibians for three reasons. First, the decline appeared to have started recently across wide regions of the world. Second, some of the populations in decline were located in protected or pristine regions, seemingly far from the effects of human activities. Third, many scientists view amphibians as good "biological indicators" of environmental conditions for a number of reasons. Amphibians have permeable skin (through which pollutants and other molecules can pass), with no hair, scales, or feathers to protect them, and their eggs lack shells or other protective coverings. Furthermore, most amphibians spend part of their lives in water and part on land. As a result, they are exposed to a

FIGURE 1.1 Deformed Leopard Frog With its misshapen and extra leg, this individual shows one of the types of limb deformities that have become common in leopard frogs and other amphibian species.

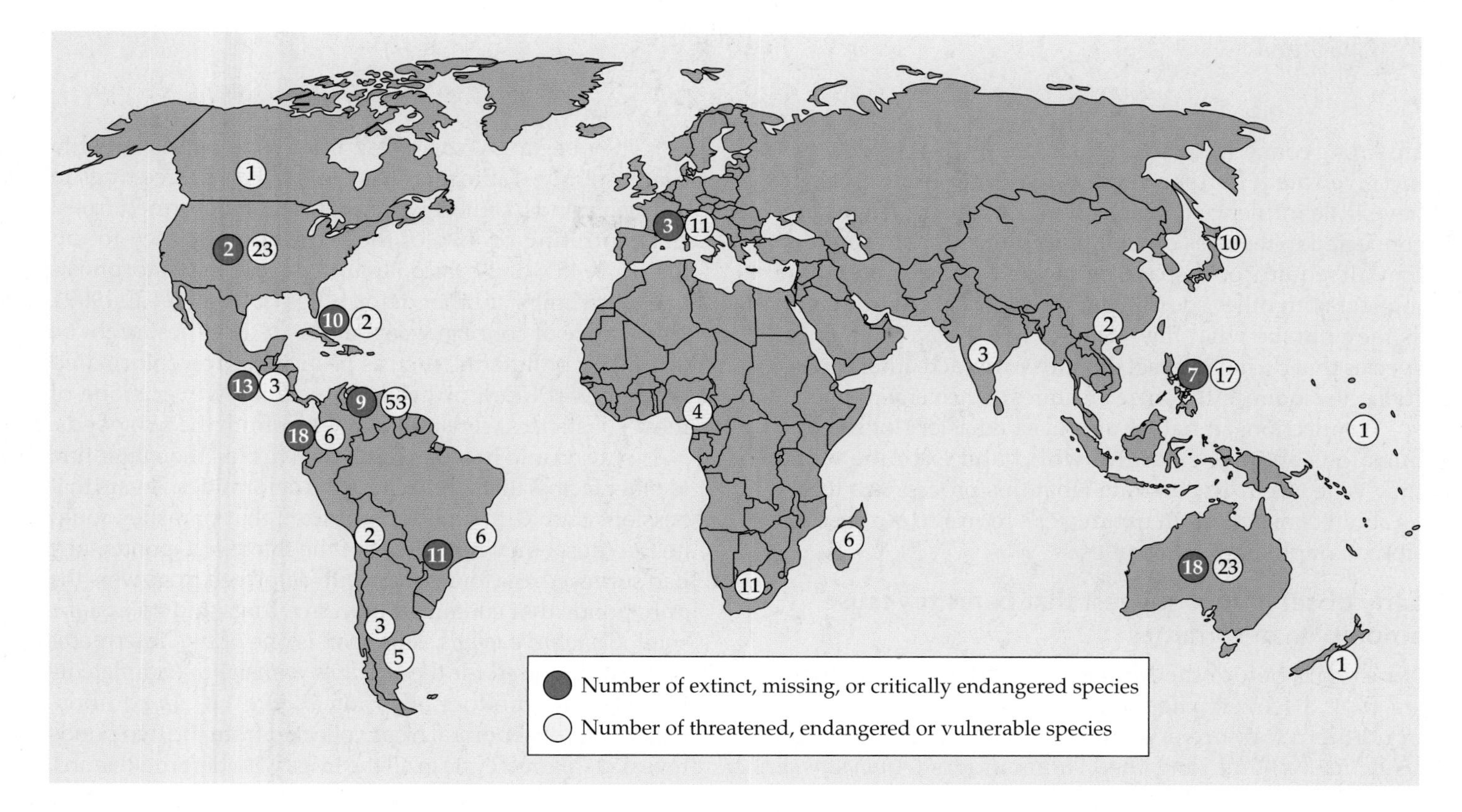

FIGURE 1.2 Amphibians in Decline In many regions of the world, amphibian species face increased risk of extinction.

wide range of potential threats, including water and air pollution as well as changes in temperature and in the amount of ultraviolet (UV) light in their environment. Many amphibians never travel far from their birthplace, so the decline of a local population is likely to indicate a deterioration of local environmental conditions.

Because amphibians worldwide were showing declining numbers and frequent deformities, scientists initially tried to find one or a few global causes that might explain these problems. However, as we'll see in this chapter, the story turned out to be more complicated than that: a single "smoking gun" has not emerged. What, then, has caused the global decline of amphibian populations?

Introduction

We humans have an enormous impact on our planet. Our activities have transformed nearly half of Earth's land surface and have altered the composition of the atmosphere, leading to global climate change. We have introduced many species to new regions, an action that can have severe negative effects on both native species and human economies. Even the oceans, seemingly so vast, show many signs of deterioration due to human activities, including declining fish stocks, the decline of once-spectacular coral reefs, and the formation of large "dead zones," regions where oxygen concentrations have dropped to levels that kill many species.

But people do not just affect the global environment, we are a part of it. Despite this, we have often taken actions that have affected our environment without considering how the natural systems of our environment work. Fortunately, we are beginning to realize that we must understand those systems so that we can anticipate the consequences of our actions and fix the problems we have already caused.

Our growing realization that we must understand how natural systems work brings us to the subject of this book. Natural systems are driven by the ways in which organisms interact with one another and with their physical environment. Thus, to understand how natural systems work, we must understand *ecology*, the scientific study of how organisms affect—and are affected by—other organisms and their environment.

In this chapter, we'll introduce the study of ecology and its relevance for humans. We'll begin by exploring a theme that runs throughout this book: connections in nature.

CONCEPT 1.1 Events in the natural world are interconnected.

Connections in Nature

From what you have read or observed about nature, can you think of examples that might illustrate the phrase *connections in nature*? In this book, we use that phrase to refer to the fact that events in the natural world can be linked or connected to one another. These connections occur as organisms interact with one another and with their physical environment. This does not necessarily mean that there are strong connections among all of the organisms that live in a given area. Two species may live in the same area but have little influence on each other. But all organisms are connected to features of their environment. For example, they all require food, space, and other resources, and they interact with other species and the physical environment as they pursue what they need to live. As a result, even species that do not interact directly with each other can be connected indirectly by shared features of their environment.

Connections in nature are revealed as ecologists ask questions about the natural world and examine what they've learned. To illustrate what this process can teach us about connections in nature, let's return to our discussion of amphibian deformities.

Early observations suggest that parasites cause amphibian deformities

Nine years before the Minnesota students made their startling discovery, Stephen Ruth was exploring ponds in northern California when he found Pacific tree frogs (*Hyla regilla*) and long-toed salamanders (*Ambystoma macrodactylum*) with extra limbs, missing limbs, and other deformities. He asked Stanley Sessions, an expert in amphibian limb development, to examine his specimens. Sessions found that the deformed amphibians all contained a parasite, now known to be *Ribeiroia ondatrae*, a trematode flatworm. Sessions and Ruth hypothesized that the parasite caused the deformities. As an initial test of this hypothesis, they implanted small glass beads near the developing limb buds of tadpoles. These beads were meant to mimic the effects of *Ribeiroia*, which often produces cysts close to the area where limbs form as a tadpole begins its metamorphosis into an adult frog. In a 1990 paper, they reported that the beads caused deformities similar to those found by Ruth.

A laboratory experiment tests the role of parasites

When Ruth found deformed amphibians in the mid-1980s, he assumed (quite reasonably) that they were an isolated, local phenomenon. By 1996, Pieter Johnson, then an undergraduate at Stanford University, had learned of the Minnesota students' findings and of the paper by Sessions and Ruth. Although Sessions and Ruth provided indirect evidence that *Ribeiroia* could cause amphibian deformities, they did not infect *H. regilla* or *A. macrodactylum* with *Ribeiroia* and show that deformities resulted. Furthermore, the two amphibian species they used in their experiments (the African clawed frog, *Xenopus laevis*, and the axolotl salamander, *A. mexicanum*) were not known to have limb deformities in nature. Building on the work done by Sessions and Ruth (1990), Johnson and his colleagues set out to provide a more direct test of whether *Ribeiroia* parasites can cause limb deformities in amphibians.

They began by surveying thirty-five ponds in Santa Clara County, California. They found Pacific tree frogs in thirteen ponds, four of which contained deformed frogs. Concentrating on two of these four ponds, they found that 15%–45% of the tadpoles undergoing metamorphosis had extra limbs or other deformities (Johnson et al. 1999). One source of concern was that the deformities might be caused by pollutants, such as pesticides, polychlorinated biphenyls (PCBs), or heavy metals. However, none of these substances were found in water from the two ponds.

Johnson and his colleagues then turned their attention to other factors that might cause the deformities. Aware that Sessions and Ruth had hypothesized that parasites could be the cause, they noted that of the thirty-five ponds they had surveyed, the four ponds with deformed frogs were the only ponds that contained both tree frogs and an aquatic snail, *Planorbella tenuis*. This snail is one of two intermediate hosts required for the *Ribeiroia* parasite to complete its life cycle and produce offspring (**Figure 1.3**). In addition, dissections of abnormal frogs collected from the two ponds revealed *Ribeiroia* cysts in all the frogs with deformed limbs.

Like the findings of Sessions and Ruth, the results described in the previous paragraph provided only indirect evidence that *Ribeiroia* caused deformities in Pacific tree frogs. Next, Johnson and his colleagues returned to the laboratory to perform a more rigorous test of that idea. They did this by using a standard scientific approach: they performed a **controlled experiment** in which an *experimental group* (that has the factor being tested) was compared with a *control group* (that lacks the factor being tested). Johnson et al. collected *H. regilla* eggs from a region not known to have frog deformities, brought the eggs into the laboratory, and placed the tadpoles that hatched from them in 1-liter containers (one tadpole per container). Each tadpole was then assigned at random to one of four *treatments*, in which 0 (the control group), 16, 32, or 48 *Ribeiroia* parasites were placed in its container; these numbers were selected to match parasite levels that had been observed in the field.

Johnson and his colleagues found that as the number of parasites increased, fewer of the tadpoles survived to metamorphosis, and more of the survivors had deformities (**Figure 1.4**). In the control group (with zero *Ribeiroia*), 88% of the tadpoles survived and none had deformities (Johnson et al. 1999). The link had been made: *Ribeiroia* could cause frog deformities. Furthermore, since exposure to *Ribeiroia* killed up to 60% of the tadpoles, the results also suggested that the parasites could contribute to amphibian declines.

A field experiment suggests that multiple factors influence frog deformities

A few years after Johnson and his colleagues published their research, scientists had shown that *Ribeiroia* parasites could cause limb deformities in other amphibian species, including western toads (*Bufo boreas*), wood frogs (*Rana sylvatica*), and

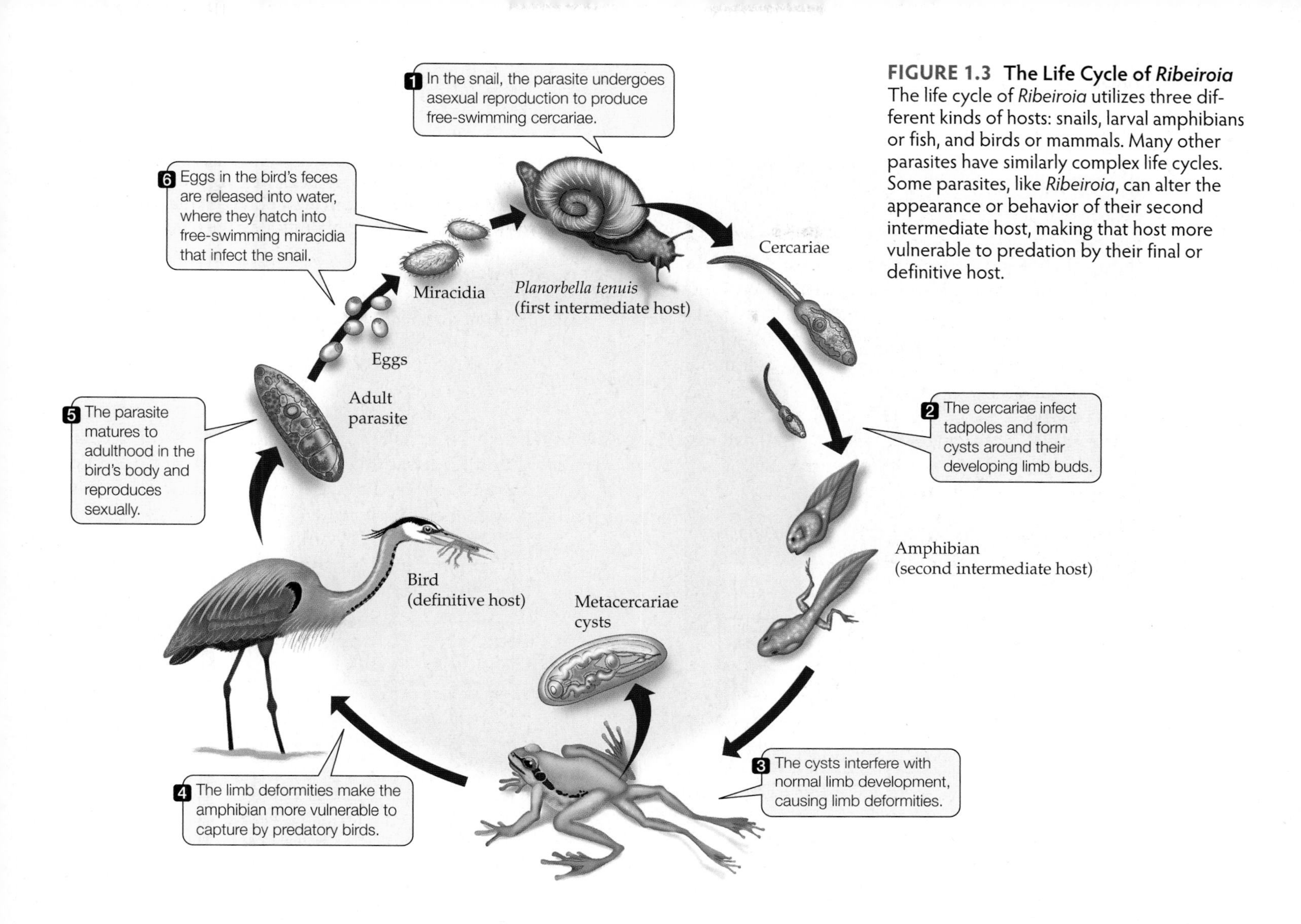

FIGURE 1.3 The Life Cycle of *Ribeiroia* The life cycle of *Ribeiroia* utilizes three different kinds of hosts: snails, larval amphibians or fish, and birds or mammals. Many other parasites have similarly complex life cycles. Some parasites, like *Ribeiroia*, can alter the appearance or behavior of their second intermediate host, making that host more vulnerable to predation by their final or definitive host.

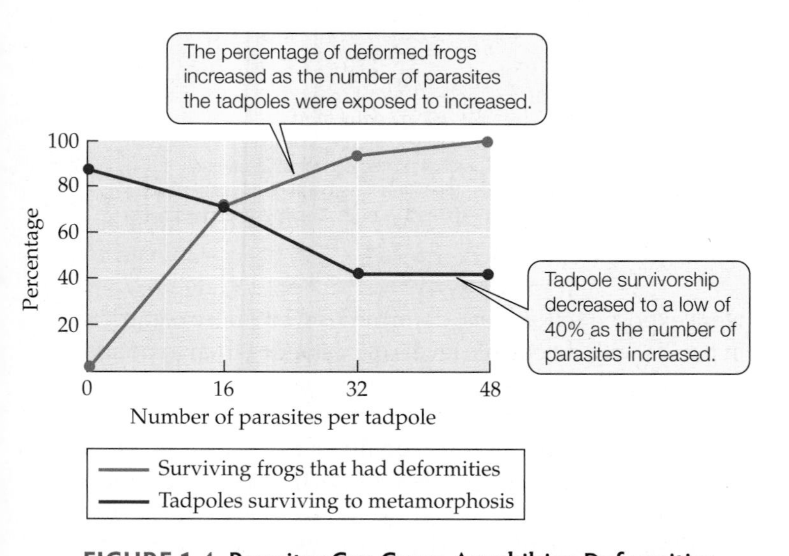

FIGURE 1.4 Parasites Can Cause Amphibian Deformities The graph shows the relationship between the number of *Ribeiroia* parasites that tadpoles were exposed to and their rates of survival and deformity. Initial numbers of tadpoles were 35 in the control group (0 parasites) and 45 in each of the other three treatments. (After Johnson et al. 1999.)

leopard frogs (*R. pipiens*, the species in which the Minnesota students had discovered deformities). While *Ribeiroia* was clearly important, some researchers suspected that other factors might also play a role. Pesticides, for example, were known to contaminate some of the ponds in which deformed frogs were found. To examine the possible joint effects of parasites and pesticides, Joseph Kiesecker conducted a field experiment in six ponds, all of which contained *Ribeiroia*, but only some of which contained pesticides (Kiesecker 2002).

Three of the ponds in Kiesecker's study were close to farm fields, and water tests indicated that each of these ponds contained detectable levels of pesticides. The other three ponds were not as close to farm fields, and none of them showed detectable levels of pesticides. In each of the six ponds, Kiesecker placed wood frog tadpoles in cages made with a mesh through which water could flow, but tadpoles could not escape. Six cages were placed in each pond; three of the cages had a mesh through which *Ribeiroia* parasites could pass, while the other three had a mesh too small for the parasites. Thus, in each pond, the tadpoles in three cages were exposed to the parasites, while the tadpoles in the other three cages were not.

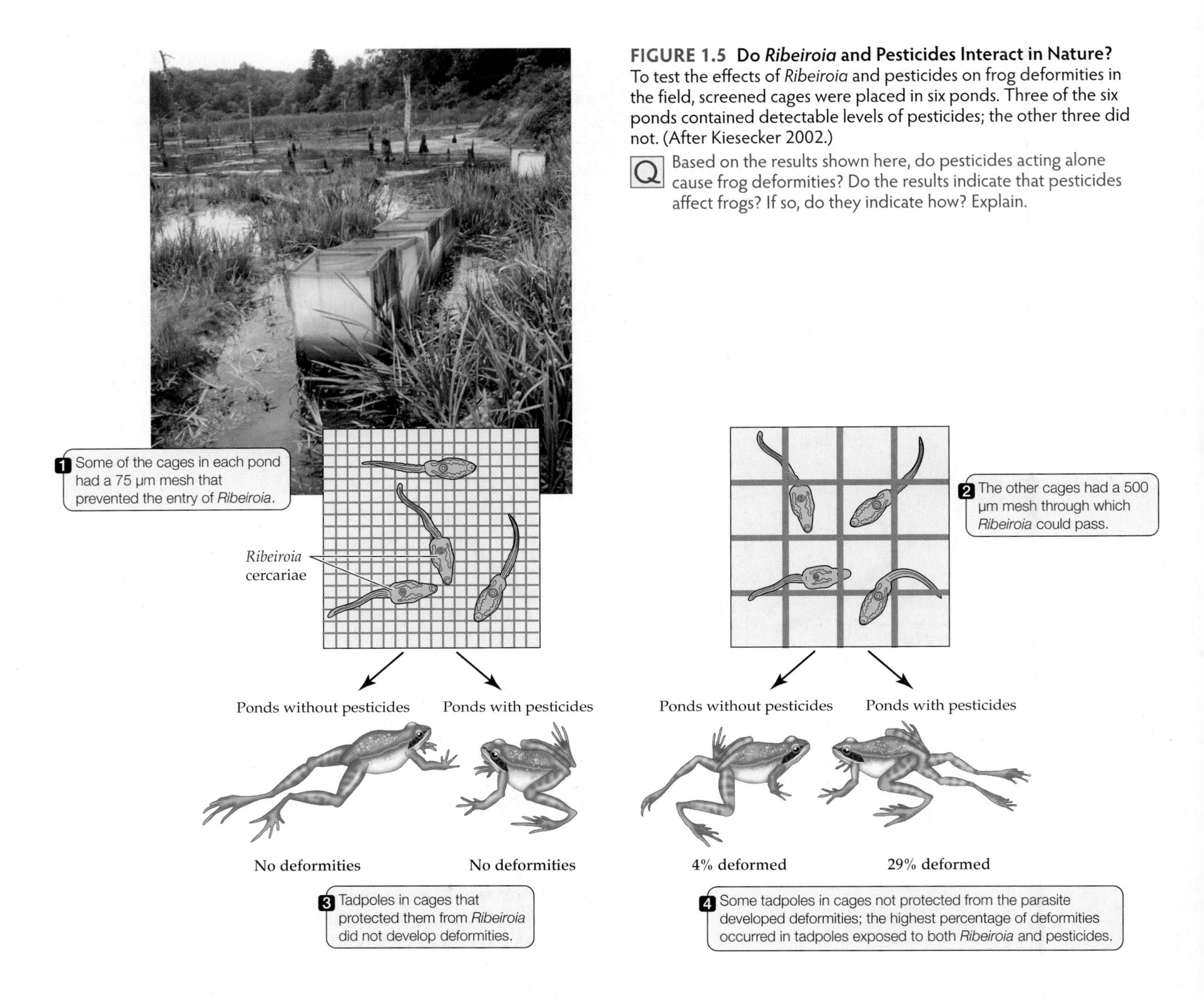

FIGURE 1.5 Do *Ribeiroia* and Pesticides Interact in Nature? To test the effects of *Ribeiroia* and pesticides on frog deformities in the field, screened cages were placed in six ponds. Three of the six ponds contained detectable levels of pesticides; the other three did not. (After Kiesecker 2002.)

Q Based on the results shown here, do pesticides acting alone cause frog deformities? Do the results indicate that pesticides affect frogs? If so, do they indicate how? Explain.

The results showed that *Ribeiroia* caused limb deformities in the field (**Figure 1.5**). No deformities were found in frogs raised in cages whose small mesh size (75 µm) prevented the entry of *Ribeiroia*, regardless of which pond the cages were in. Deformities were found in some of the frogs raised in cages whose larger mesh size (500 µm) allowed the entry of *Ribeiroia*. In addition, dissections revealed that every frog with a deformity was infected by *Ribeiroia*. However, a greater percentage of frogs had deformities in the ponds that contained pesticides than in the ponds that did not (29% vs. 4%). Overall, the results of this experiment indicated that (1) exposure to *Ribeiroia* was necessary for deformities to occur, and (2) when frogs were exposed to *Ribeiroia*, deformities were more common in ponds with detectable levels of pesticides than in ponds without detectable levels of pesticides.

Based on these results, Kiesecker hypothesized that pesticides might decrease the ability of frogs to resist infection by parasites. To test whether pesticides had such an effect, Kiesecker (2002) brought wood frog tadpoles into the laboratory, where he reared some in an environment with pesticides and others in an environment without pesticides, then exposed all of them to *Ribeiroia*. The tadpoles exposed to pesticides had fewer white blood cells (indicating a suppressed immune system) and a higher rate of *Ribeiroia* cyst formation (**Figure 1.6**). Together, Kiesecker's laboratory and

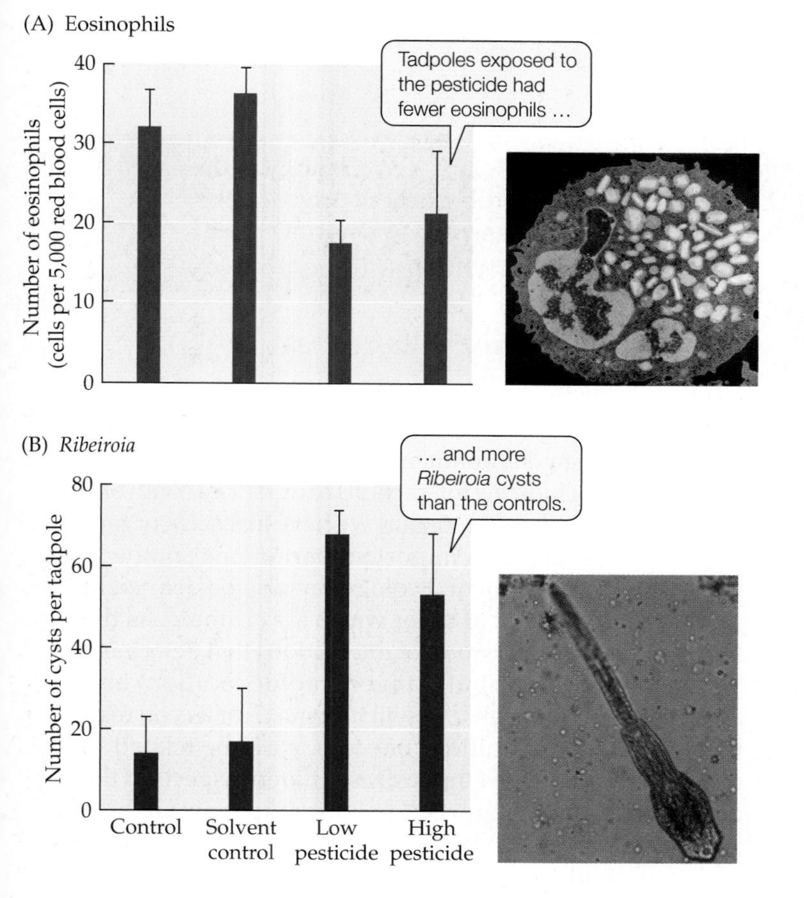

FIGURE 1.6 Pesticides May Weaken Tadpole Immune Systems In a laboratory experiment, wood frog (*Rana sylvatica*) tadpoles were exposed to low or high concentrations of the pesticide esfenvalerate and then exposed to 50 *Ribeiroia* parasites per tadpole. The tadpoles were then examined for (A) numbers of eosinophils (a type of white blood cell used in the immune response) and (B) numbers of *Ribeiroia* cysts. Two types of controls were used: one in which only parasites were added to the tadpoles' containers ("control"), and another in which both parasites and the solvent used to dissolve the pesticide were added ("solvent control"). Error bars show one SE of the mean (see **Web Stats Review**). (After Kiesecker 2002.)

Q What was the purpose of using two types of controls in this experiment?

field results suggested that pesticide exposure can affect the frequency with which parasites cause deformities in amphibian populations. This conclusion has since been supported by other studies. Field surveys and laboratory experiments in Rohr et al. (2008), for example, indicated that exposure to pesticides can increase the number of *Ribeiroia* infections and decrease survival rates in several frog species. As in Kiesecker's study, one reason for the increased number of parasitic infections appeared to be that the frogs' immune response was suppressed by the pesticide.

Connections in nature can lead to unanticipated side effects

As we have seen, the immediate cause of amphibian deformities is often infection by *Ribeiroia* parasites. But we also noted on p. 2 that amphibian deformities are occurring more often now than in the past. Why has the frequency of amphibian deformities increased?

One possible answer is suggested by the results of Kiesecker (2002) and Rohr et al. (2008): pesticides may decrease the ability of amphibians to ward off parasite attack, and hence deformities are more likely in environments that contain pesticides. The first synthetic pesticides were developed in the late 1930s, and their use has risen dramatically since that time. Thus, it is likely that amphibian exposure to pesticides has increased considerably in recent years, which may help to explain the recent rise in the frequency of amphibian deformities. If this scenario turns out to be correct, it illustrates how any action (such as increased pesticide use by people) can have unanticipated side effects (such as more frequent deformities in amphibians).

Other environmental changes may also contribute to the observed increase in amphibian deformities. For example, the addition of nutrients to natural or artificial ponds (used to store water for cattle or crops) can lead to increases in parasite infections and amphibian deformities (Johnson et al. 2007). Nutrients can enter a pond when rain or snowmelt washes fertilizers from an agricultural field into it. Fertilizer inputs often stimulate increased growth of algae, and the snails that harbor *Ribeiroia* parasites eat algae (to refresh your memory of the parasite's life cycle, see Figure 1.3). Thus, as the algae increase, so do the snail hosts of *Ribeiroia*. An increase in snails tends to increase the number of *Ribeiroia* found in the pond. Here, a chain of events that begins with increased fertilizer use by people ends with increased numbers of *Ribeiroia*, and hence increased numbers of deformed amphibians. As this and other results described in this section suggest, events in the natural world are connected, one to another.

We live in an ecological world

The fact that events in nature are interconnected means that when people alter one aspect of the environment, we can cause other changes that we do not intend or anticipate. For example, when we increased our use of pesticides and fertilizers, we did not intend to increase the frequency of deformities in frogs. Nevertheless, we seem to have done just that.

The indirect effects of human actions include more than bizarre deformities in frogs. For example, some changes we are making to our local and global environment appear to have increased human health risks. The damming of rivers in Africa has created favorable habitat for snails that harbor trematode parasites that cause schistosomiasis, thereby increasing the spread of an infection that can weaken or kill people. Globally, the past few decades have seen an increase in the appearance and spread of new diseases, such as AIDS, Lyme disease, hantavirus pulmonary syndrome, Ebola hemorrhagic fever, and West Nile virus.

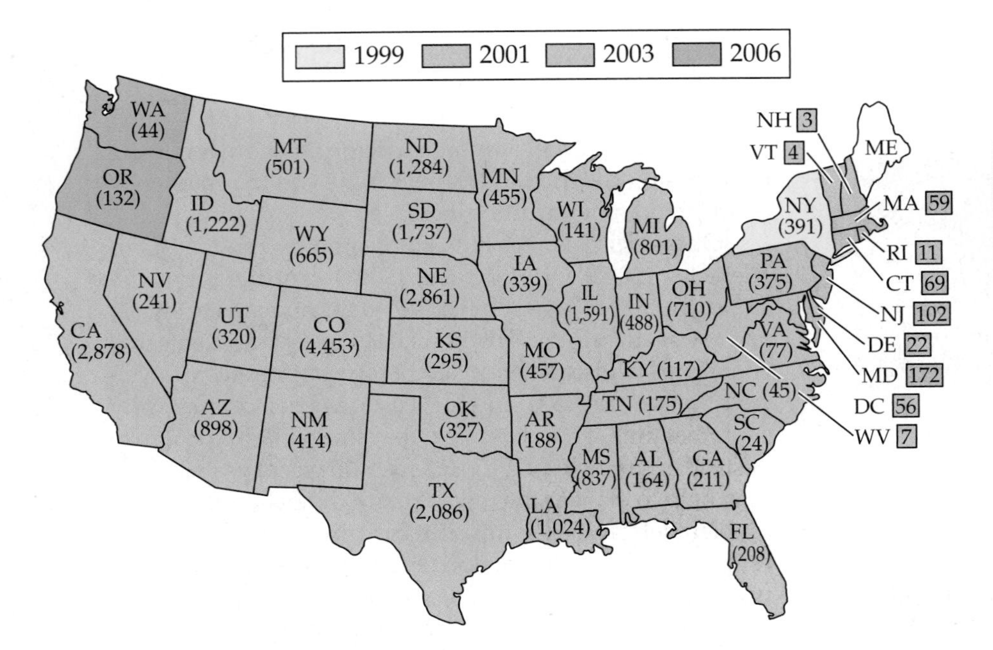

FIGURE 1.7 Rapid Spread of a Deadly Disease Within 7 years, West Nile virus (shaded areas) had spread from its North American point of entry (New York City) to all but one of the lower 48 states of the United States (Maine). Birds are a primary host for the West Nile virus, which may help to explain its rapid spread. Mosquitoes transmit the disease from birds and other animal hosts to people. Colors indicate disease spread over time; numbers in parentheses indicate the cumulative number of human cases by December 31, 2009. (Data from Centers for Disease Control and Prevention.)

Many public health experts think that the effects of human actions on the environment have contributed to the emergence of these and other new diseases (Weiss and McMichael 2004). For example, West Nile virus, which is transmitted by mosquitoes and infects birds and humans, is thought to have been introduced into North America by people in 1999 (**Figure 1.7**). Furthermore, the incidence of West Nile virus in humans is influenced by factors such as human population size, the extent of land development, the abundance and identity of mosquito and bird species, and variations in temperature and rainfall (Reisen et al. 2006; Landesman et al. 2007; Allan et al. 2009). Each of these factors can be affected by human actions, either directly (e.g., by urban or agricultural development) or indirectly (e.g., as a result of climate change; see Chapter 24).

If you live in a city, it can be easy to forget the extent to which everything you do depends on the natural world. Your house or apartment shelters you from the elements and keeps you warm in winter and cool in summer. Similarly, you obtain food from a grocery store, clothes from a shop or department store, water from a faucet. Ultimately, however, each of these items—and everything else you use or own—comes from or depends on the natural environment. No matter how far from the natural world our day-to-day activities take us, people, like all other organisms on Earth, are part of an interconnected web of life. Let's turn now to the study of these connections, the scientific discipline of *ecology*.

CONCEPT 1.2 Ecology is the scientific study of interactions between organisms and their environment.

Ecology

In this book, **ecology** is defined as the scientific study of interactions between organisms and their environment. This definition is meant to include the interactions of organisms with one another because, as we have just seen, organisms are an important part of one another's environment. Ecology can also be defined in a variety of other ways; for example, as the scientific study of interactions that determine the distribution (geographic location) and abundance of organisms. As will become clear as you read this book, these definitions of ecology can be related to one another, and each emphasizes different aspects of the discipline. A more important point for our purpose here is that the term "ecology," as used by ecologists, refers to a scientific endeavor.

We emphasize this point because "ecology" has other meanings in its public usage. People who are not scientists may assume that an "ecologist" is an environmental activist. Some ecologists are activists, but many are not. Furthermore, as a scientific discipline, ecology is related to—yet different from—other disciplines such as environmental science. Ecology is a branch of biology, while **environmental science** is an interdisciplinary field that incorporates concepts from the natural sciences (including ecology) and the social sciences (e.g., politics, economics, ethics). Compared with ecology, environmental science is focused more specifically on how people affect the environment and how we can address environmental problems. While an ecologist might examine pollution as one of several factors that influence the reproductive success of wetland plants, an environmental scientist might focus on how economic and political systems could be used to reduce pollution.

Public and professional ideas about ecology often differ

Surveys have shown that many members of the public think that (1) there is a "balance of nature," in which natural systems are stable and tend to return to an original, preferred state after a disturbance, and that (2) each species in nature has a distinct role to play in maintaining that balance. Such ideas about ecological systems can have moral or ethical implications for the people who hold them. For example, the view that each species has a distinct function can lead people to think that each species is important and irreplaceable, which in turn can cause people to feel that it is wrong to harm other species. As summarized by one interviewee in a survey on the meaning of ecology (Uddenberg et al. 1995, as quoted in Westoby 1997), "There is a certain balance in nature, and there is a place for all species. There is a reason for their existence and we are not free to exterminate them."

Public views on the balance of nature and the unique function of each species are not surprising, since these views were once held by many ecologists. Ecologists, however, now recognize that natural systems do not necessarily return to their original state after a disturbance, and they understand that seemingly random perturbations often play an important role in nature. In addition, as we will see in Unit 4, the current evidence suggests that the different species in an area often respond in different ways to changing conditions, a finding at odds with the idea that each species has a distinct function within a larger, tightly knit group of species. Therefore, unless they provide careful qualifications, few ecologists today speak of a balance of nature or the unique function of each species.

While early views about a balance of nature and the unique function of each species have not stood the test of time, other ideas have. In particular, early ecologists and modern ecologists would agree that events in nature are interconnected. As a result, a change in one part of an ecological system can alter other parts of that system, including those that govern life-supporting processes such as the purification and replenishment of air, water, and soil.

Connections in nature form the basis for the first of eight ecological maxims that we will mention in this book; namely, "You can never do just one thing" (see **Table 1.1** for the full list). This maxim is meant to suggest that all actions have multiple effects because events in nature are interconnected. Overall, although the natural world may not be as predictable or as tightly woven as early ecologists may have thought, species are connected to one another. For some people, the fact that events in nature are interconnected provides an ethical imperative to protect natural systems. A person who feels an ethical obligation to protect human life, for example, may also feel an ethical obligation to protect the natural systems on which human life depends.

TABLE 1.1

Some Ecological Maxims

1. You can never do just one thing.
   Organisms interact with one another and with their physical environment. As a result, events in nature are connected, and what affects one organism or place can affect others as well.
2. Everything goes somewhere.
   There is no "away" into which waste materials disappear.
3. No population can increase in size forever.
   There are limits to the growth and resource use of every population, including our own.
4. There is no free lunch.
   An organism's energy and resources are finite, and increasing inputs into one function (such as reproduction) results in a trade-off in which there is a loss for other functions (such as growth).
5. Evolution matters.
   Organisms evolve or change over time—it is a mistake to view them as static. Evolution is an ongoing process because organisms continually face new challenges from changes in both the living and nonliving components of their environment.
6. Time matters.
   Ecosystems change over time. When we look at the world as we know it, it is easy to forget how past events may have affected our present, and how our present actions may affect the future.
7. Space matters.
   Abiotic and biotic environmental conditions can change dramatically from one place to another, sometimes across very short distances. This variation matters because organisms are simultaneously influenced by processes acting at multiple spatial scales, from local to regional to global.
8. Life would be impossible without species interactions.
   Species depend on one another for energy, nutrients, and habitat.

The scale of an ecological study affects what can be learned from it

Whether they study individual organisms or the diversity of life on Earth—or anything in between—ecologists always draw boundaries around what they observe. An ecologist interested in frog deformities might ignore the birds that migrate above the study site, while an ecologist studying bird migrations might ignore the details of what occurs in the ponds below. It is not possible or desirable to study everything at once.

When they seek to answer a particular question, ecologists must select the most appropriate dimension, or **scale**, for collecting observations, including both time and space. Every ecological study addresses events on some scales but ignores events on other scales. A study on the activities of soil microorganisms, for example, might be conducted at a small spatial scale (e.g., measurements might be collected at centimeter to meter scales). For a study addressing how atmospheric pollutants affect the global climate, on the other hand, the scale of observation would be large indeed and might include Earth's entire atmosphere. Ecological studies also differ greatly in the time scales they cover. Some studies, such as those that document how leaves respond to momentary increases in the availability of sunlight, concern events on short time scales (seconds to hours). Others, such as studies that use fossil data to show how the species found in a given area have changed over time, address events on much longer time scales (centuries to millennia or longer).

FIGURE 1.8 An Ecological Hierarchy As suggested by this series of photographs, life in the Sonoran desert can be studied at a number of levels, from individuals to the biosphere. These levels are nested within one another, in the sense that each level is composed of groups of the entity found in the level below it.

Ecology is broad in scope

Ecologists study interactions in nature across many levels of biological organization. Some ecologists are interested in how particular genes or proteins enable organisms to respond to environmental challenges. Other ecologists study how hormones influence social interactions, or how specialized tissues or organ systems allow animals to cope with extreme environments. However, even among ecologists whose research is focused on lower levels of biological organization (e.g., from molecules to organ systems), ecological studies usually emphasize one or more of the following levels: individuals, populations, communities, ecosystems, landscapes, or the entire biosphere (**Figure 1.8**).

A **population** is a group of individuals of a single species that live in a particular area and interact with one another. Many of the central questions in ecology concern how and why the locations and abundances of populations change over time. To answer such questions, it is often helpful to understand the roles played by other species. Thus, many ecologists study nature at the level of the **community**, which is an association of interacting populations of different species that live in the same area. Communities can cover large or small areas, and they can differ greatly in terms of the numbers and types of species found within them (**Figure 1.9**).

Ecological studies at the population and community levels often examine not only the effects of the **biotic**, or living, components of a natural system, but also those of the **abiotic**, or physical, environment. For example, a population or community ecologist might ask whether features of the abiotic environment, such as temperature, precipitation, or nutrients, influence the fertility of individuals or the relative abundances of the different species found in a community. Other ecologists are particularly interested in how ecosystems work. An **ecosystem** is a community of organisms plus the physical environment in which they live. An ecologist studying ecosystems might want to know the rate at which a chemical (such as nitrogen from fertilizers) enters a particular community, as well as how the species living there affect what happens to the chemical once it enters the community. For example, ecosystem ecologists studying amphibian deformities might document the rates at which nitrogen enters ponds that do and do not contain deformed amphibians, or they might determine how the presence or absence of algae affects what happens to nitrogen once it has entered the ponds.

Across larger spatial regions, ecologists study **landscapes**, which are areas that vary substantially from one place to another, typically including multiple ecosystems. Finally, global patterns of air and water circulation (see Chapter 2) link the world's ecosystems into the **biosphere**, which consists of all living organisms on Earth plus the environments in which they live. The biosphere forms the highest

FIGURE 1.9 A Few of Earth's Many Communities Moving counterclockwise, these photographs show (A) a savanna in Kenya, where a male topi antelope stands on an ant mound to attract a mate; (B) a rainforest on Saint Lucia; (C) dunes in the Namib Desert, Namibia, an extremely dry desert that receives 2 to 85 mm (0.08 to 3.3 inches) of rain per year; (D) a shallow-water marine community off the coast of Indonesia.

level of biological organization. Over recent decades, as we will see in Unit 6, ecologists have acquired new tools that improve their ability to study the big picture: how the biosphere works. As just one example, ecologists can now use satellite data to answer questions such as, How do different ecosystems contribute to ongoing changes in the global concentration of carbon dioxide (CO_2) in the atmosphere (see Chapter 24)?

Some key terms are helpful for studying connections in nature

Whether we are discussing individuals, populations, communities, or ecosystems, all chapters of this book incorporate the principle that events in the natural world are interconnected. For example, in Unit 2, we will see how an explosion in the population size of an introduced species (the comb jelly, *Mnemiopsis*) altered the entire Black

TABLE 1.2

Key Terms for Studying Connections in Nature

Term	Definition
Adaptation	A feature of an organism that improves its ability to survive or reproduce in its environment
Natural selection	An evolutionary process in which individuals that possess particular characteristics survive or reproduce at a higher rate than other individuals because of those characteristics
Consumer	An organism that obtains its energy by eating other organisms or their remains
Producer	An organism that uses energy from an external source, such as the sun, to produce its own food without having to eat other organisms or their remains
Net primary production (NPP)	The amount of energy (per unit of time) that producers fix by photosynthesis or other means, minus the amount they use in cellular respiration
Nutrient cycle	The cyclic movement of a nutrient between organisms and the physical environment

Sea ecosystem. Because we stress connections in nature in every chapter, and hence may discuss ecosystems in a chapter about organisms, or vice versa, we describe here a handful of key terms that you will need to know as you begin your study of ecology. These terms are also summarized in **Table 1.2**.

A universal feature of living systems is that they change over time, or *evolve*. Depending on the questions or time scale of interest, **evolution** can be defined as (1) a change in the genetic characteristics of a population over time or as (2) *descent with modification*, the process by which organisms gradually accumulate differences from their ancestors. We will discuss evolution in the context of ecology more fully in Chapter 6, but here we define two key evolutionary terms: adaptation and natural selection.

An **adaptation** is a characteristic of an organism that improves its ability to survive or reproduce within its environment. Adaptations are of critical importance for understanding how organisms function and interact with one another. As we'll see in Chapter 6, although several mechanisms can cause evolutionary change, only natural selection can produce adaptations consistently. In the process of **natural selection**, individuals with particular characteristics tend to survive and reproduce at a higher rate than other individuals *because of those characteristics*. If the characteristics being selected for are heritable, then the offspring of individuals favored by natural selection will tend to have the same characteristics that gave their parents an advantage. As a result, the frequency of those characteristics in a population may increase over time. If that occurs, the population will have evolved.

Consider what happens within the body of a person taking an antibiotic. Some of the bacteria that live inside that person may possess genes that provide resistance to the antibiotic. Because of those genes, those bacteria will survive and reproduce at a higher rate than will nonresistant bacteria (**Figure 1.10**). Because the trait on which natural selection acts (antibiotic resistance) is heritable, the offspring of the resistant bacteria will tend to be resistant. As a result, the proportion of resistant bacteria in the person's body will increase over time, and the bacterial population will have evolved.

The remaining four key terms that we'll introduce here concern ecosystem processes. One way to look at how ecosystems work is to consider the

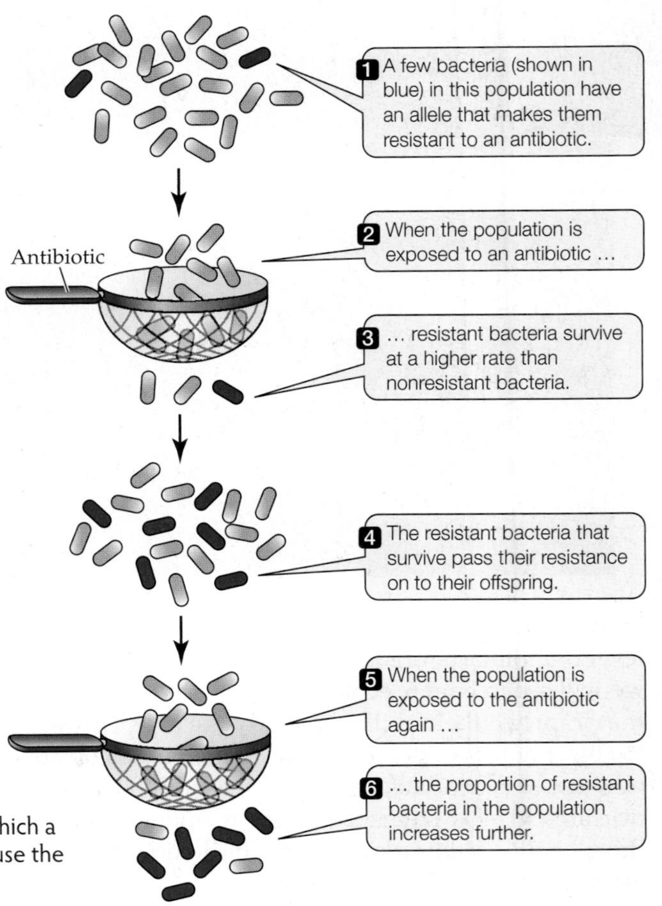

FIGURE 1.10 Natural Selection in Action As shown in this diagram, in which a sieve represents the selective effects of an antibiotic, natural selection can cause the frequency of antibiotic resistance in bacteria to increase over time.

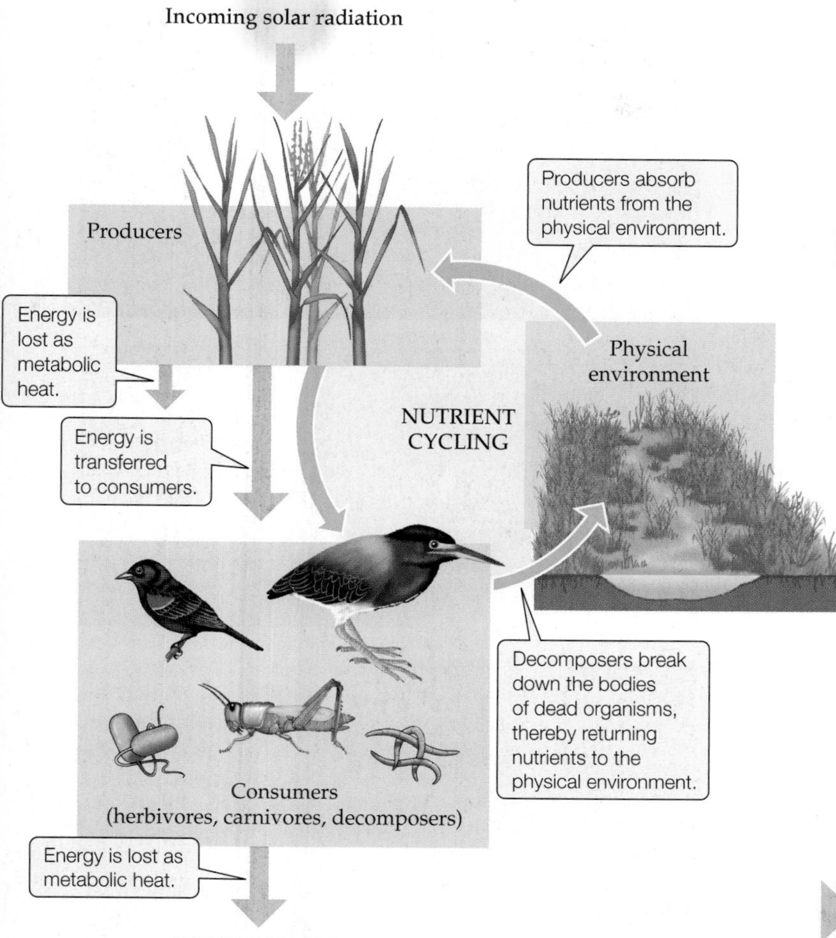

FIGURE 1.11 How Ecosystems Work Each time one organism eats another, a portion of the energy originally captured by a producer is lost as heat given off during the chemical breakdown of food by cellular respiration. As a result, energy flows through the ecosystem in a single direction and is not recycled. Nutrients such as carbon and nitrogen, on the other hand, cycle between organisms and the physical environment.

Q Describe the three main steps by which a nutrient cycles through an ecosystem.

movement of energy and materials through a community. Energy enters the community when an organism such as a plant or bacterium captures energy from an external source, such as the sun, and uses that energy to produce food. An organism that can produce its own food from an external energy source without having to eat other organisms or their remains is called a **producer** (such organisms are also called *primary producers* or *autotrophs*). An organism that obtains its energy by eating other organisms or their remains is called a **consumer**. Per unit of time, the amount of energy that producers capture by photosynthesis or other means, minus the amount they lose as heat in cellular respiration, is called **net primary production** (**NPP**). Changes in NPP can have large effects on ecosystem function, and NPP varies greatly from one ecosystem to another.

Each unit of energy captured by producers is eventually lost from the ecosystem as heat (**Figure 1.11**). As a result, energy moves through ecosystems in a single direction only—it cannot be recycled. Nutrients, however, are recycled from the physical environment to organisms and back again. The cyclic movement of nutrients such as nitrogen or phosphorus between organisms and the physical environment is referred to as a **nutrient cycle**. Life as we know it would cease if nutrients were not recycled because the molecules organisms need for their growth and reproduction would be much less readily available.

Whether they are concerned with adaptations or NPP, populations or ecosystems, the scientists who study ecological systems have not produced a fixed body of knowledge, engraved in stone. Instead, what we know about ecology changes constantly as ideas are tested and, if necessary, revised or discarded as new information emerges. As we will see in the next section, ecology, like all branches of science, is about answering questions and seeking to understand the underlying causes of natural phenomena.

CONCEPT 1.3 Ecologists evaluate competing hypotheses about natural systems with observations, experiments, and models.

Answering Ecological Questions

The studies of amphibian deformities that we discussed earlier in this chapter illustrate several ways in which ecologists seek to answer questions about the natural world. The study by Johnson and his colleagues (1999), for example, had two key components: observational studies in the field and a controlled experiment in the laboratory. In the observational part of their work, the researchers surveyed ponds, noted the species present, and observed that deformities were found only in ponds that contained both tree frogs and a snail that harbored the *Ribeiroia* parasite. These findings suggested that *Ribeiroia* might cause deformities, so Johnson and his colleagues performed a laboratory experiment to test whether that was the case (it was).

Kiesecker (2002) extended these results in two experiments, one performed in the field, the other in the laboratory. To examine the effects of pesticides on frog deformities, Kiesecker compared results from three ponds with pesticides with results from three ponds without detectable levels of pesticides. While this approach had the advantage of allowing the effects of *Ribeiroia* to be examined under different field conditions (in ponds with and without pesticides), Kiesecker could not control the conditions as precisely as he did in his laboratory experiment. The constraints of working in the field meant, for example, that he could not start out with six identical ponds, then add pesticides to three of them but not to the other three—an experiment that would test more directly whether pesticides were responsible for the results he obtained. As this example suggests, no single approach works best in all situations, so ecologists use a variety of methods when seeking to answer ecological questions.

Ecologists use experiments, observations, and models to answer ecological questions

In an ecological experiment, an investigator alters one or more features of the environment and observes the effect of that change on natural processes. When possible, such experiments include both a control group (which is not subjected to any alterations) and one or more experimental groups (see p. 4). When performing an experiment, ecologists have a range of types and scales to choose from, including laboratory studies, small-scale field studies that cover a few square meters, and large-scale field studies in which entire ecosystems, such as lakes or forests, are manipulated (**Figure 1.12**).

In some cases, however, it can be difficult or impossible to perform an appropriate experiment. For example, when ecologists are seeking to understand events that cover large geographic regions or occur over long periods, experiments can provide useful information, but they cannot provide convincing answers to the underlying questions of interest. Consider global warming. As we will see in Chapter 24, temperature data show that Earth's climate is warming, but the future magnitude and effects of global warming remain uncertain. We are not sure, for example, how the geographic ranges of different species will change as a result of the projected temperature increases. There is only one Earth, so of course even if we wanted to, we could not apply different levels of global warming to copies of the planet and then observe how species' ranges change over time in each of our experimental treatments.

Instead, we must approach such problems using a mixture of observational studies, experiments, and quantitative (mathematical or computer) models. Field observations reveal that many species have shifted their ranges poleward or up the sides of mountains in a manner that is consistent with the amount of global warming that has already occurred (Parmesan 2006). Field observations can also be used to summarize the environmental conditions under which species are currently found, and experiments can be used to examine the performance of species under different environmental conditions. To put all of this information together, results from observational studies and experiments can be used to help formulate quantitative models that predict how changing environmental conditions are likely to affect the geographic ranges of species, depending on how much the planet actually warms in the future.

(A)

(B)

(C)

FIGURE 1.12 Ecological Experiments Experiments in ecology range from (A) laboratory experiments to (B) small-scale field experiments conducted in natural or artificial environments to (C) large-scale experiments that alter major components of an ecosystem, as seen in this watershed.

The observation that global warming has already altered the geographic ranges of species brings us to a topic addressed in many chapters of this book: **climate change**. This term refers to a directional change in climate (such as global warming) that occurs over three decades or longer. As you'll read in later chapters, climate affects many aspects of ecology, such as the growth and survival of individuals, interactions between members of different species, and the relative abundances of species in ecological communities. These observations suggest that *changes* to climate may have far-reaching effects—and they do, as shown by changes that have already occurred in the physiology, survival, reproduction, or geographic range of hundreds of species (Parmesan 2006). (See **Climate Change Connection 1.1** for further information on the ecological effects of climate change.)

Experiments are designed and analyzed in consistent ways

When ecologists perform experiments, they often take the three additional steps described in **Ecological Toolkit 1.1**: they **replicate** (i.e., perform more than once) each treatment, including the control; they assign treatments at random; and they analyze the results using statistical methods.

An advantage of replication is that as the number of replicates increases, it becomes less likely that the results are due to a variable that was not measured or controlled in the study. Imagine that Kiesecker had performed his field experiment with only two ponds, one with detectable levels of pesticides and the other without. Suppose he had found that frog deformities were more common in the single pond that contained pesticides. While pesticides might have been responsible for this result, the two ponds could have differed in many other ways, too, one or more of which might have been the real cause of the deformities. By using three ponds with pesticides and three ponds without pesticides, Kiesecker made it less likely that each of the three ponds with pesticides also contained something else—some variable not controlled in his experiment—that increased the chance of frog deformities. In his experiment, Kiesecker accounted for the possible effects of some uncontrolled variables: he showed, for example, that the number of snails and the frequency of their infection by *Ribeiroia* were similar in all six ponds, thus making it unlikely that the ponds with pesticides had many more *Ribeiroia* than the ponds without pesticides.

Ecologists also seek to limit the effects of unmeasured variables by assigning treatments at random. Suppose an investigator wanted to test whether insects that eat plants decrease the number of seeds the plants produce. One way to test this idea would be to divide an area into a series of plots (see Ecological Toolkit 1.1), some of which would be sprayed regularly with an insecticide (the experimental plots) while others would be left alone (the control plots). The decision as to whether a particular plot would be sprayed (or not) would be made at random at the start of the experiment. The purpose of assigning treatments at random would be to make it less likely that plots that receive a particular treatment share other characteristics that might influence seed production, such as high or low levels of soil nutrients.

Finally, ecologists often use statistical analyses to determine whether their results are "significant." To understand why, let's turn again to Kiesecker's experiment. It would have been surprising if Kiesecker had found that rates of frog deformities in ponds with pesticides were exactly equal to those in ponds without pesticides. But how different would those rates have to be to show that the pesticides are having an effect? Since the results of different experimental treatments will rarely be identical, the investigator must ask whether an observed difference is great enough to be of biological importance. Statistical methods are often used as a standardized way to help make this decision. We describe general statistical principles and one statistical method, the *t*-test, in the **Web Stats Review**. There are many different types of statistical analyses; books such as those by Zar (2006), Sokal and Rohlf (1995), or Gotelli and Ellison (2004) provide examples of the best statistical methods to use under various circumstances.

What we know about ecology is always changing

The information in this book is not a static body of knowledge. Instead, like the natural world itself, our understanding of ecology is constantly changing. Like all scientists, ecologists observe nature and ask questions about how nature works. For example, when the existence of amphibian deformities became widely known in 1995, some scientists set out to answer a series of questions about those deformities. There were many things they wanted to know: How many species were afflicted by deformities? Did amphibian deformities occur in a few or many geographic regions? What caused the deformities?

The questions stimulated by the discovery of amphibian deformities illustrate the first in a series of four steps by which scientists can learn about the natural world. These four steps constitute the **scientific method**, which can be summarized as follows:

1. Observe nature and ask a question about those observations.
2. Use previous knowledge or intuition to develop possible answers to that question (**hypotheses**).
3. Evaluate competing hypotheses by performing experiments, gathering carefully selected observations, or analyzing results of quantitative models.
4. Use the results of those experiments, observations, or models to modify one or more of the hypotheses, to pose new questions, or to draw conclusions about the natural world.

This four-step process is iterative and self-correcting. New observations lead to new questions, which stimulate ecologists to formulate and test new ideas about how

1.1 Designing Ecological Experiments

A key step in any ecological experiment occurs well before it is performed: the experiment must be designed carefully. As described on p. 4, in a controlled experiment, an *experimental group*, which has the factor being tested, is compared with a *control group*, which does not. Different levels of the factor being tested are often referred to as different *treatments*. For example, in the experiment by Johnson et al. (1999) discussed on p. 4, the control group received a treatment of 0 parasites per container, while members of the experimental group were assigned to one of three other treatments (16, 32, or 48 parasites per container).

The design of many ecological experiments includes three additional steps: *replication*, *random assignment of treatments*, and *statistical analyses*. Replication and random assignment of treatments are used to reduce the chance that variables not under the control of the experimenter will unduly influence the results of the experiment. Once the experiment has been completed, statistical analyses are used to assess the extent to which the results from the different treatments differ from one another.

Several features of experimental design can be illustrated by the layout used in field studies performed by Richard B. Root and colleagues at Cornell University. In one such study, Carson and Root (2000) examined how herbivorous (plant-eating) insects affected a plant community dominated by the goldenrod *Solidago altissima*. Their first step was to define their research question: Do plant abundance, growth, or reproduction differ between insecticide-treated and control plots? To find out, they divided a field of goldenrods into the grid of 5 × 5 m plots shown in **Figure A**. The experiment ran for 10 years and used two treatments: a control, in which natural processes were left undisturbed in one set of plots, and an insect removal treatment, in which an insecticide was applied annually to reduce the numbers of herbivorous insects in another set of plots. Carson and Root selected 30 plots at random for use in the experiment; half of those plots were then selected at random to receive the insecticide treatment, while the remaining plots served as controls. Thus, there were 15 replicates for each treatment. Statistical analyses of the results indicated that the herbivorous insects had major effects on the plant community, as is also suggested by the photograph in **Figure B**.

(A)

Figure A Carson and Root's Field Experiment This aerial photograph shows the field divided (by mowing) into 112 5 × 5 m plots.

(B)

Figure B Carson and Root's Results A plot sprayed with insecticide (right) is shown surrounded by several control plots.

nature works. The results from such tests can lead to new knowledge, still more questions, or the abandonment of ideas that fail to explain the results. Although this four-step process is not followed exactly in all scientific studies, the back-and-forth between observations, questions, and results—potentially leading to a reevaluation of existing ideas—captures the essence of how science is done.

We've already seen some examples of how the process of scientific inquiry works: as answers to some questions about amphibian deformities were found, new questions arose, and new discoveries were made. Such ongoing discoveries occur in all fields of ecology, suggesting that our understanding of ecological processes is, and always will be, a work in progress.

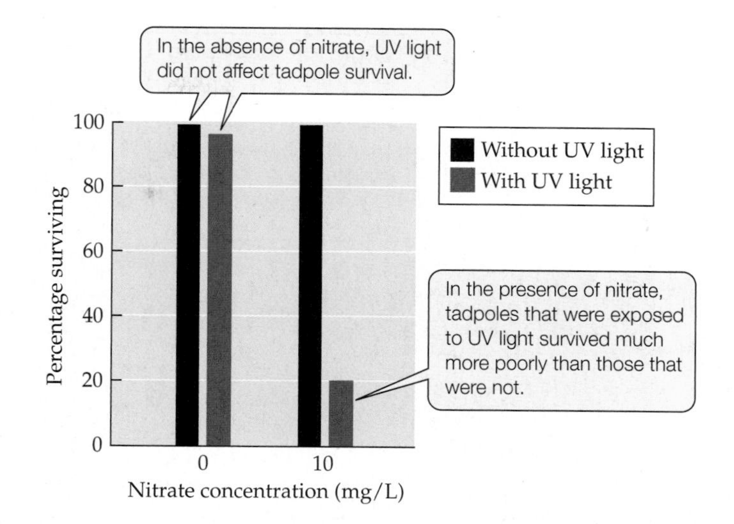

FIGURE 1.13 Joint Effects of Nitrate and UV Light on Tadpole Survival In a field experiment in the Cascade Range, nitrate and UV light had joint, but not individual, effects on the survivorship of Pacific tree frog (*Hyla regilla*) tadpoles. UV light levels were manipulated by raising tadpoles in tanks covered with clear plastic that either transmitted or blocked UV light. (After Hatch and Blaustein 2003.)

Q In the treatments in which tadpoles were exposed to UV light, why were the tanks covered with plastic that transmitted UV light instead of simply being left uncovered?

A CASE STUDY REVISITED

Deformity and Decline in Amphibian Populations

As we've seen in this chapter, amphibian deformities are often caused by parasites, but they can also be influenced by other factors, such as exposure to pesticides or fertilizers. With respect to population declines, studies have suggested that a range of factors can cause amphibian abundances to drop. Such factors include habitat loss, parasites and diseases, pollution, overexploitation, climate change, introduced species, and UV light.

A consensus has yet to be reached on the relative importance of these and other factors that affect amphibian declines. For example, Stuart et al. (2004) analyzed the results of studies on 435 amphibian species that have experienced rapid declines since 1980. Habitat loss was the primary cause of the decline of the largest number of species (183 species), followed by overexploitation (50 species). The cause of decline for the remaining 207 species was listed as "enigmatic": populations of these species were declining rapidly for reasons that were poorly understood. Skerrat et al. (2007) argued that many such enigmatic declines may be due to pathogens such as the chytrid *Batrachochytrium dendrobatidis*, a fungus that causes a lethal skin disease. This fungus has spread rapidly in recent years, leaving a host of devastated amphibian populations in its wake. However, other researchers have suggested that climate change is the underlying factor because it has altered conditions in ways that favor the growth and transmission of the fungus (Pounds et al. 2006). Note that these hypotheses are not mutually exclusive: indeed, Rohr and Raffel (2010) found that while disease spread often played a primary role in amphibian declines, climate change (specifically, changes in temperature variation) also played a key role.

Collectively, these and other studies of amphibian declines suggest that no single factor can explain most of them—a finding that differs from early views that a single, global factor might be responsible. Instead, the declines seem to be caused by complex factors that often act together and may vary from place to place. For example, Hatch and Blaustein (2003) conducted a field experiment to examine the effects of UV light and nitrate (a fertilizer) on tadpole survival. At a high-elevation site, they found that neither exposure to UV light nor exposure to nitrate, by themselves, affected the survival of Pacific tree frog tadpoles. Acting together, however, UV light and nitrate reduced the survival of the tadpoles considerably (**Figure 1.13**). In contrast, when they conducted a similar experiment at a low-elevation site, no such joint effects were found—perhaps because UV light levels were lower there (see Table 1 in Hatch and Blaustein 2003).

Recent discoveries about the effects of pesticides also suggest that the causes of amphibian decline are complex and variable. While pesticides may contribute to frog deformities (see pp. 6–7), many studies have failed to link pesticides to decreases in the size of amphibian populations. Many of these negative findings came from laboratory studies that held other factors constant and examined the effect of pesticides alone on amphibian growth or survival. Rick Relyea, of the University of Pittsburgh, repeated such experiments, but with an added twist: predators. In two of six amphibian species studied, pesticides became up to 46 times more lethal if the tadpoles sensed

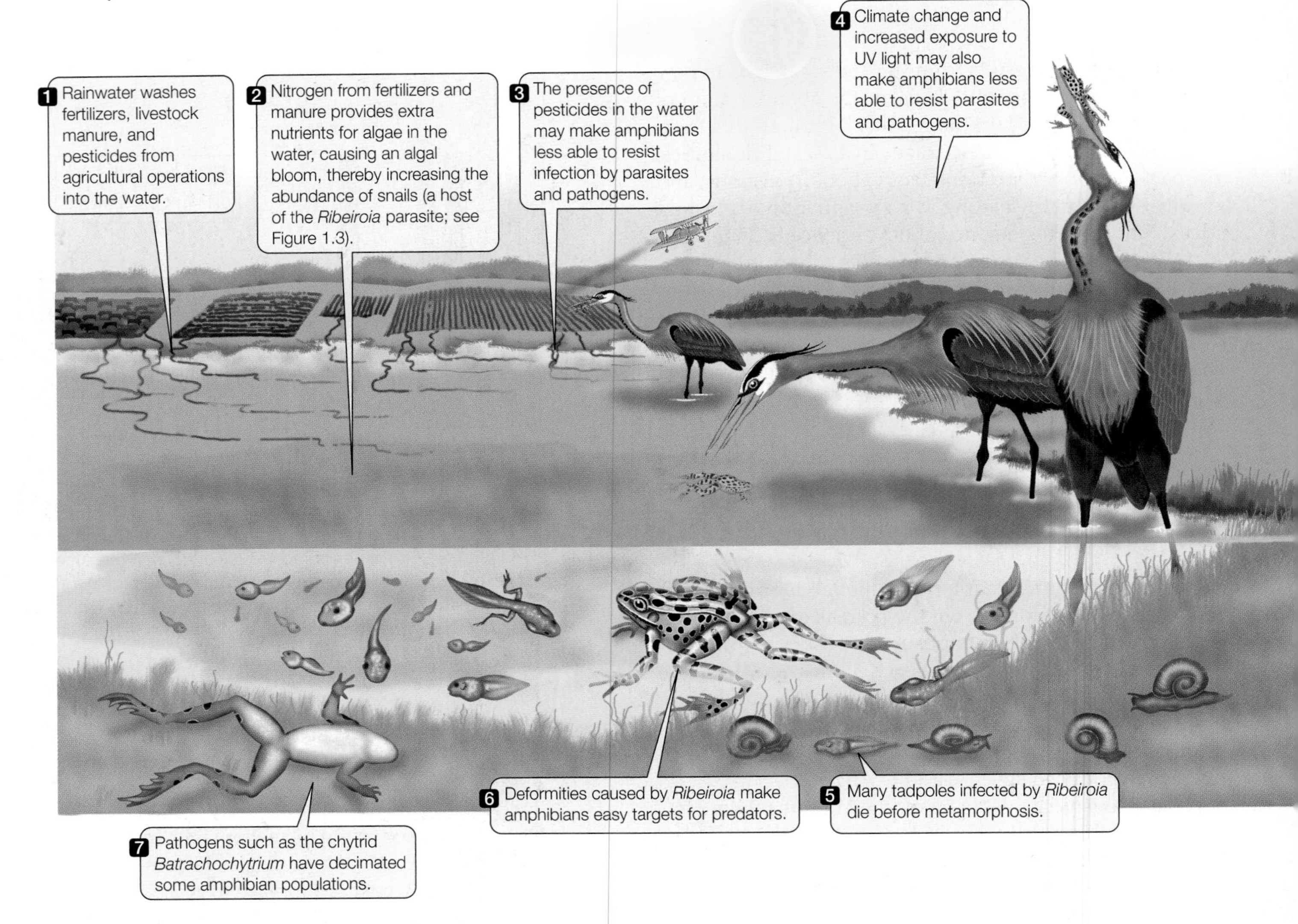

FIGURE 1.14 Complex Causation of Amphibian Deformities and Declines As we have seen, amphibian deformities can be caused by parasites such as *Ribeiroia*. However, other factors—many of them a result of human actions—may also be involved in amphibian deformities and declines. (After Blaustein and Johnson 2003.)

the presence of a predator (Relyea 2003). The predators were kept separate from the tadpoles by netting, but the tadpoles could smell them.

In Relyea's experiments, the ability of some tadpoles to cope with pesticides was reduced by stress caused by the presence of a predator. The mechanism by which this works is not known. In general, although we know that a broad set of factors can cause frog deformities and declines (**Figure 1.14**), relatively little is known about the extent to which these factors interact or how any such interactions exert their effects. In this and many other areas of ecology, we have learned enough to solve parts of the mystery, yet more remains to be discovered.

CONNECTIONS IN NATURE
Mission Impossible?

As we emphasized in the opening pages of this chapter, people have begun to realize that it is important for us to understand how nature works, if only to protect ourselves from inadvertently changing our environment in ways that cause us harm. Does the fact that the natural world is vast, complex, and interconnected mean that it is impossible to understand? Most ecologists do not think so. Our understanding of natural systems has improved greatly in the last hundred years. Ongoing efforts to understand how nature works are sure to be challenging, but such efforts are also enormously exciting and important. What we learn, and how we use that knowledge, will have a great impact on the current and future well-being of human societies. Whatever your career path, we hope this book will help you to understand the natural world in which you live, as well as how you affect—and are affected by—that world.

SUMMARY

CONCEPT 1.1 Events in the natural world are interconnected.

- Laboratory and field experiments on the effects of parasites on amphibian deformities illustrate how events in nature can be connected with one another.
- Because events in the natural world are interconnected, any action can have unanticipated side effects.
- People both depend on and affect the natural environment.

CONCEPT 1.2 Ecology is the scientific study of interactions between organisms and their environment.

- Ecology is a scientific discipline that is related to, but differs from, other disciplines such as environmental science.
- Public and professional ideas about ecology often differ.
- Ecology is broad in scope and encompasses studies at many levels of biological organization.
- All ecological studies address events on some spatial and temporal scales while ignoring events at other scales.

CONCEPT 1.3 Ecologists evaluate competing hypotheses about natural systems with observations, experiments, and models.

- In an ecological experiment, an investigator alters one or more features of the environment and observes the effect of that change on natural processes.
- Some features of the natural world are best investigated with a combination of field observations, experiments, and quantitative models.
- Experiments are designed and analyzed in consistent ways: typically, each treatment, including the control, is replicated; treatments are assigned at random; and statistical methods are used to analyze the results.
- The information in this book is not a static body of knowledge; what we know about ecology is always changing.

REVIEW QUESTIONS

1. Describe what the phrase "connections in nature" means, and explain how such connections can lead to unanticipated side effects. Illustrate your points with an example discussed in the chapter.
2. What is ecology, and what do ecologists study? If an ecologist studied the effects of a particular gene, describe how the emphasis of that researcher's work might differ from the emphasis of a geneticist or cell biologist.
3. How does the scientific method work? Include in your answer a description of a controlled experiment.

ON THE COMPANION WEBSITE
sites.sinauer.com/ecology2e

The website includes Chapter Outlines, Online Quizzes, Flashcards & Key Terms, Suggested Readings, a complete Glossary, and the Web Stats Review. In addition, the following resources are available for this chapter:

HANDS-ON PROBLEM SOLVING

Mosquitoes, Drought, and Disease

This Web exercise explores the connections between mosquito populations, periodic drought, and the incidence of mosquito-transmitted diseases. You will read a recent paper that shows that mosquito populations increase in size after droughts. You will then plot data on drought severity compared to incidence of disease, and discuss implications for human health.

CLIMATE CHANGE CONNECTIONS

1.1 Ecological Effects of Climate Change

UNIT 1
Organisms and their Environment

CHAPTER

2 The Physical Environment

KEY CONCEPTS

- **CONCEPT 2.1** Climate is the most fundamental component of the physical environment.
- **CONCEPT 2.2** Winds and ocean currents result from differences in solar radiation across Earth's surface.
- **CONCEPT 2.3** Large-scale atmospheric and oceanic circulation patterns establish global patterns of temperature and precipitation.
- **CONCEPT 2.4** Regional climates reflect the influence of the distribution of oceans and continents, mountains, and vegetation.
- **CONCEPT 2.5** Seasonal and long-term climatic variation are associated with changes in Earth's position relative to the sun.
- **CONCEPT 2.6** Salinity, acidity, and oxygen concentrations are major determinants of the chemical environment.

Climatic Variation and Salmon Abundance: A Case Study

Grizzly bears of the Pacific Northwest feast seasonally on the salmon that arrive in huge numbers to reproduce in the streams of the region (**Figure 2.1**). Salmon are *anadromous*; that is, they are born in freshwater streams, spend their adult lives in the ocean, and then return to spawn in the freshwater habitats where they were born. Grizzlies capitalize on the salmon's reproductive habits to gorge themselves on this rich food resource. These normally aggressive bears will forgo their usual territorial behavior and tolerate high densities of other bears while fishing for salmon.

It is not only bears that rely on salmon for food. Salmon have been an important part of the human economy of the Pacific Northwest for millennia. The fish were a staple of the diets of Native Americans in this region as well as a central part of their cultural and spiritual lives. Salmon are now fished commercially in the waters of the North Pacific Ocean, providing an important economic base for many coastal communities. Commercial salmon fishing is a risky venture, however. Successful reproduction for salmon depends on the health of the streams in which they spawn. The construction of dams, increases in stream sediments due to logging operations, water pollution, and overharvesting have all been blamed for declines in salmon populations, primarily from the California coast northward to British Columbia (Walters 1995). Despite efforts to mitigate this environmental degradation, the recovery of salmon stocks has been marginal at best in the southern portion of the region.

Researchers, environmental advocates, and government policy experts have focused primarily on the degradation of freshwater habitat as a cause for the declines in salmon. In 1994, however, Steven Hare and Robert Francis suggested that changes in the marine environment, where salmon spend the majority of their adult lives, could be contributing to the declines in salmon abundance. In particular, they noted that records of fish harvests covering more than a century indicated that multi-decadal periods of low or high fish production have occurred repeatedly, separated by abrupt changes in production rather than gradual transitions (**Figure 2.2**). In addition, Nathan Mantua and colleagues (Mantua et al. 1997) noted that periods of high salmon production in Alaska corresponded with periods of low salmon production at the southern end of the salmon range, particularly in Oregon and Washington. They found

FIGURE 2.1 A Seasonal Opportunity Grizzly bears congregate in the streams and rivers of Alaska to feed on salmon migrating upstream to reproduce. The size of the salmon run each year depends in part on physical conditions in the Pacific Ocean, many miles away.

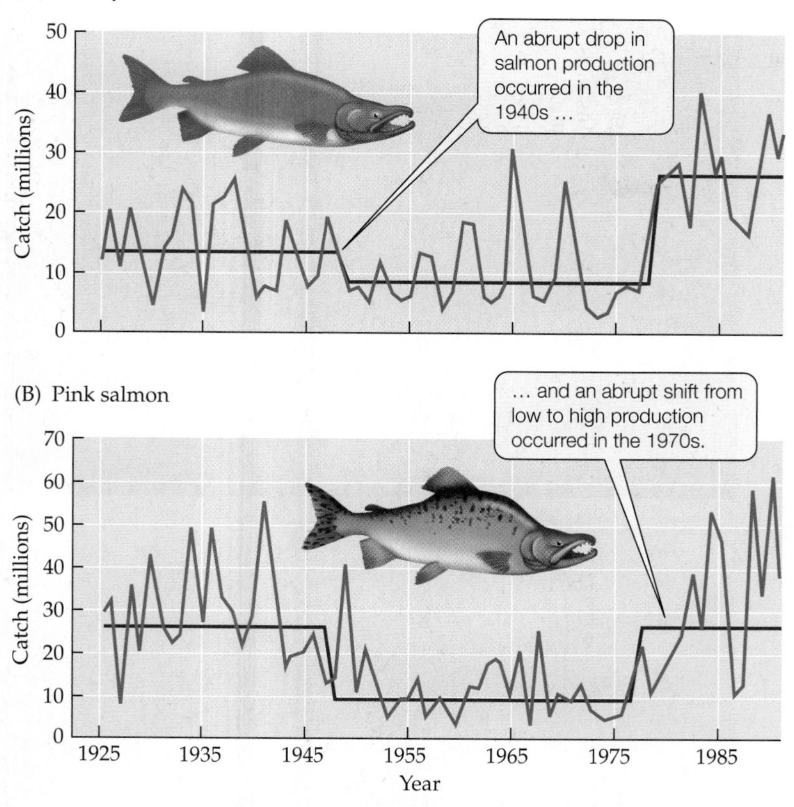

FIGURE 2.2 Changes in Salmon Harvests over Time Records of commercial harvests of (A) sockeye salmon and (B) pink salmon in Alaska over 65 years show abrupt drops and increases in production. Red lines represent annual catch; purple lines are a statistical fit to the data. (After Hare and Francis 1994.)

telling quotes in commercial fishing publications that told the same story: when the fishing was poor in Washington and Oregon, it was good in Alaska, and vice versa.

> Pacific Fisherman (published in 1915)
>
> Never before have the Bristol Bay (Alaska) salmon packers returned to port after the season's operations so early. [i.e., a bad year, with few fish to catch]
>
> The spring [chinook salmon] fishing season on the Columbia River [Washington and Oregon] closed at noon on August 25, and proved to be one of the best for some years.
>
> Pacific Fisherman (1939)
>
> The Bristol Bay Red [Alaska sockeye salmon] run was regarded as the greatest in history.
>
> The [chinook] catch this year is one of the lowest in the history of the Columbia [Washington].

Hare and Francis hypothesized that the abrupt shifts in salmon production were associated with long-term climatic variation in the North Pacific. The nature and cause(s) of these underlying climatic shifts, however, were unclear. Additional work by Mantua and colleagues found good correspondence between the multi-decadal shifts in salmon production and changes in sea surface temperatures in the North Pacific.

How widespread is this variation in climate and its effects on salmon and the associated marine ecosystem? As we will see at the end of this chapter, the research on variation in salmon production led to the discovery of an important long-term cyclic climatic pattern that affects a large area.

Introduction

The physical environment is the ultimate determinant of where organisms can live, the resources that are available to them, and the rate at which their populations can grow. Therefore, an understanding of the physical environment is key to understanding all ecological phenomena, from the outcome of interactions between bacteria and fungi in the soil to the exchange of carbon dioxide between the biosphere and the atmosphere.

The physical environment includes climate, which consists of long-term trends in temperature, wind, and precipitation. Radiation from the sun ultimately drives the climate system as well as biological energy production. Another aspect of the physical environment is the chemical environment, which includes salinity (concentrations of dissolved salts), acidity (pH), and concentrations of gases in the atmosphere and dissolved in water. Soil is an important component of the physical environment because it is a medium in which microorganisms, plants, and animals live. Soil also influences the availability of critical resources, particularly water and nutrients. This chapter will focus on climate and the chemical environment; we will cover soils, their development, and nutrient supply in Chapter 21.

This chapter will provide a framework for characterizing the physical environment, including its variability, at a variety of spatial and temporal scales. We will begin by exploring the processes that create the climatic patterns we see at global to regional scales.

CONCEPT 2.1 Climate is the most fundamental component of the physical environment.

Climate

In our daily lives, we are aware of the **weather** around us: the current temperature, humidity, precipitation, wind, and cloud cover. Weather is an important determinant of our behavior: what we wear, the activities we engage in, and our mode of transportation. **Climate** is the long-term description of weather at a given location, based on averages and variation measured over decades. Climatic variation includes the daily and seasonal cycles associated with changes in solar radiation as Earth rotates on its axis and orbits the sun. Climatic variation also includes changes over years or decades, such as large-scale cyclic weather patterns related to changes in the atmosphere and oceans (El Niño events, discussed later in this chapter, are one example). Longer-term climate change occurs as a result of changes in the intensity and distribution of solar radiation reaching Earth's surface. Earth's climate is currently changing due to changes in concentrations of gases such as carbon dioxide that are emitted into the atmosphere as a result of human activities. These gases absorb and radiate energy back to Earth's surface, creating a greenhouse effect.

Climate controls where and how organisms live

Where organisms live (their geographic distribution) and how they function are determined by climate. Temperature determines the rates of biochemical reactions and physiological activity for all organisms. Water supplied by precipitation is an essential resource for terrestrial organisms. Freshwater organisms are dependent on precipitation for the maintenance and quality of their habitats. Marine organisms depend on ocean currents that influence the temperature and chemistry of the waters they live in.

We often characterize climate—or any aspect of the physical environment—at a given location by the average conditions over time. For example, the average temperature at a location provides information on the long-term conditions organisms must face if they are to survive there. However, the distributions of organisms often reflect *extreme* conditions more than average conditions because extreme events are important determinants of mortality. Temperature and moisture extremes can affect long-lived organisms such as forest trees. For example, record high temperatures, along with a severe drought in 2000–2003, contributed to widespread mortality in large stands of piñon pines (*Pinus edulis*) in the southwestern United States (Breshears et al. 2005) (**Figure 2.3**). Even these long-lived plants could no longer survive in the region where they had existed for many decades. Thus, the physical environment must also be characterized by its *variability* over time, not just by average conditions, if we are to understand its ecological importance. The frequency and severity of extreme temperature events such as the one that led to the die-off in piñon pines are predicted to increase in association with global climate change (Jentsch et al. 2007). These events will increase the probability of large-scale mortality of vegetation (**Climate Change Connection 2.1**).

The *timing* of changes in the physical environment is also ecologically important. The seasonality of rainfall, for example, is an important determinant of water availability for terrestrial organisms. In regions with a "Mediterranean-type" climate, the majority of precipitation falls in winter. Although these regions receive more precipitation than most desert areas, they experience regular dry periods during the summer growing season. The lack of water during summer reduces the growth of plants and promotes the occurrence of fires. In contrast, some grasslands have the same *average annual temperature* (the average temperature measured over an entire year) and precipitation as these Mediterranean-type ecosystems, but the precipitation is spread evenly throughout the year.

Climate also influences the rates of abiotic processes that affect organisms. The rate at which rocks and soil are broken down to supply nutrients to plants and microorganisms, for example, is determined by climate. Climate can also influence the rates of periodic *disturbances*, such as fires, rockslides, and avalanches. These events kill organisms and disrupt biological communities, but subsequently create opportunities for the establishment and growth of new organisms and communities.

(A)

(B)

FIGURE 2.3 Widespread Mortality in Piñon Pines A historic drought from 2000 to 2003 killed large areas of piñon pines (*Pinus edulis*) throughout the southwestern United States. (A) Here, stands in the Jemez Mountains, New Mexico, begin to show substantial needle death due to water and temperature stress, combined with a bark beetle outbreak in October 2002. (B) By May 2004, most of the trees had died.

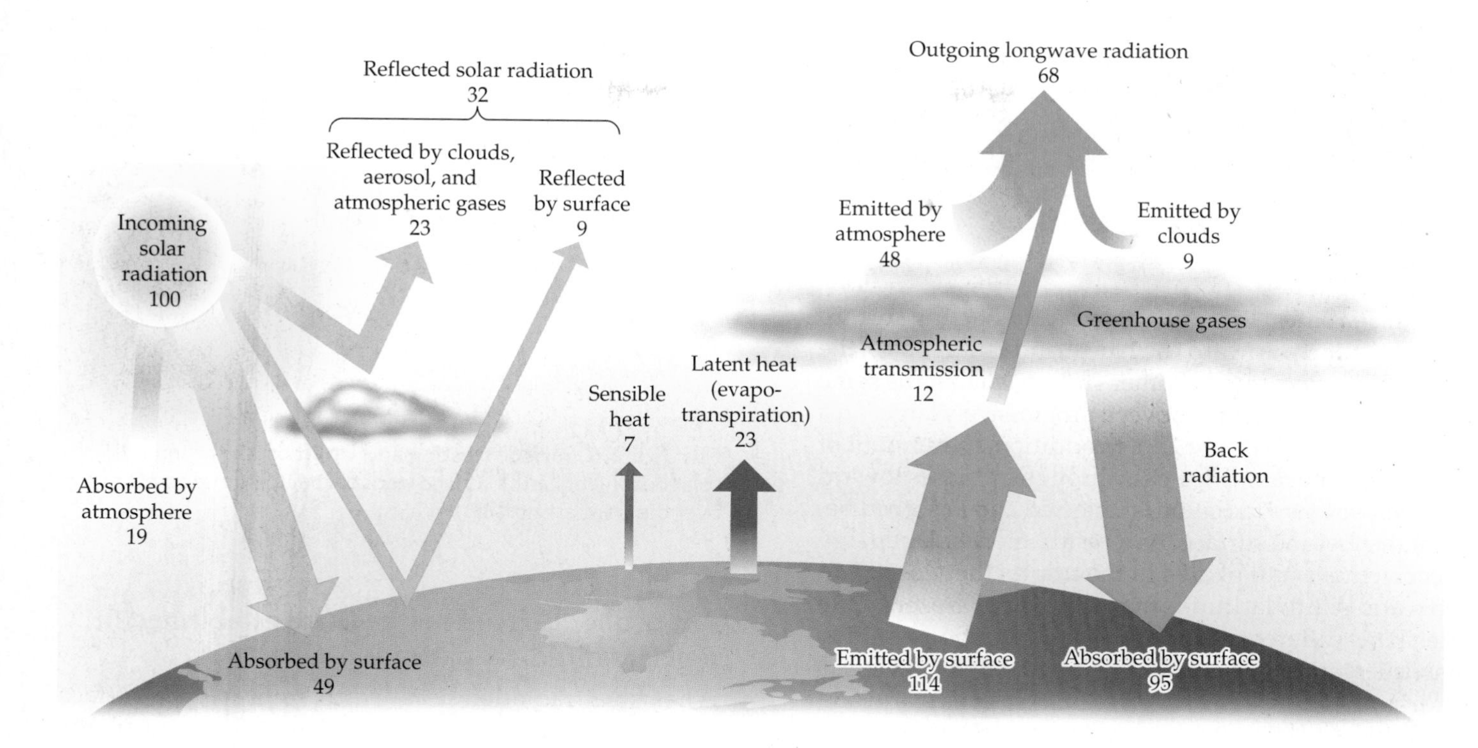

FIGURE 2.4 Earth's Energy Balance Average annual energy balance for Earth's surface, including gains from solar radiation and gains and losses due to emission of infrared radiation, latent heat flux, and sensible heat flux. The numbers are gains and losses of energy, given as percentages of the average annual incoming solar radiation at the top of Earth's atmosphere (342 W/m^2). (After Kiehl and Trenberth 1997.)

Q What component of Earth's energy balance would be influenced by an increase in greenhouse gases? What would the effect on Earth's energy balance be if there were an increase in atmospheric aerosols?

Global energy balance drives the climate system

The energy that drives the global climate system is ultimately derived from solar radiation. On average, the top of Earth's atmosphere receives 342 watts (W) of solar radiation per square meter each year. About a third of this solar radiation is reflected back out of the atmosphere by clouds, fine atmospheric particles called *aerosols*, and Earth's surface. Another fifth of the incoming solar radiation is absorbed by ozone, clouds, and water vapor in the atmosphere. The remaining half of the incoming solar radiation is absorbed by land and water at Earth's surface (**Figure 2.4**).

If Earth's temperature is to remain the same, these energy gains from solar radiation must be offset by energy losses. Much of the solar radiation absorbed by Earth's surface is emitted to the atmosphere as infrared radiation (also known as *longwave* radiation). Earth's surface also loses energy when water evaporates. The change in phase from liquid water to water vapor absorbs energy and is thus a cooling process. The heat loss due to evaporation is known as **latent heat flux**. Energy is also transferred through the exchange of kinetic energy by molecules in direct contact with one another (**conduction**) and by the movement of currents of air (wind) and water (**convection**). Energy transfer from the warm air immediately above Earth's surface to the cooler atmosphere by convection and conduction is known as **sensible heat flux**.

Earth's surface actually releases more energy than it receives by direct solar radiation (see Figure 2.4). However, the atmosphere absorbs and reradiates much of the infrared radiation emitted by Earth. The atmosphere contains several gases, known as *radiatively active gases* or **greenhouse gases**, that absorb and reradiate that infrared radiation. These gases include water vapor (H_2O), carbon dioxide (CO_2), methane (CH_4), and nitrous oxide (N_2O). Some of these greenhouse gases are produced through biological activity (e.g., CO_2, CH_4, N_2O), linking the biosphere to the climate system. Without these greenhouse gases, Earth's climate would be considerably cooler than it is (by approximately 33°C, or 59°F). As noted earlier, increases in atmospheric concentrations of greenhouse gases due to human activities are altering Earth's energy balance, changing the climate system, and causing global climate change (see Chapter 24).

Our discussion of Earth's energy balance has focused on average annual transfers of energy to and from Earth as a whole. But not every location on Earth receives the same amount of energy from the sun. Let's consider how these differences in solar radiation affect the circulation of Earth's atmosphere and ocean waters.

CONCEPT 2.2 Winds and ocean currents result from differences in solar radiation across Earth's surface.

Atmospheric and Oceanic Circulation

Near the equator, the sun's rays strike Earth's surface perpendicularly. Toward the North and South Poles, the angle of the sun's rays becomes steeper, so that the same amount of energy is spread over a progressively larger area of Earth's surface (**Figure 2.5**). In addition, the amount of atmosphere the rays must pass through increases toward the poles, so more radiation is reflected and absorbed before it reaches the surface. As a result, more solar energy is received per unit of area in the tropics (between 23.5° north and south latitude) than in regions closer to the poles. This differential input of solar radiation not only establishes latitudinal gradients in temperature, but is also the driving force for climate dynamics such as warm and cold fronts, and large storms (e.g., hurricanes). In addition, the movement of Earth around the sun, in combination with the tilt of Earth's axis of rotation, results in changes in the amount of solar radiation received at any location over the course of the year, as we'll see in the discussion of Concept 2.5. These changes are the cause of seasonal climatic variation: winter–spring–summer–fall changes at high latitudes and wet–dry shifts in tropical regions.

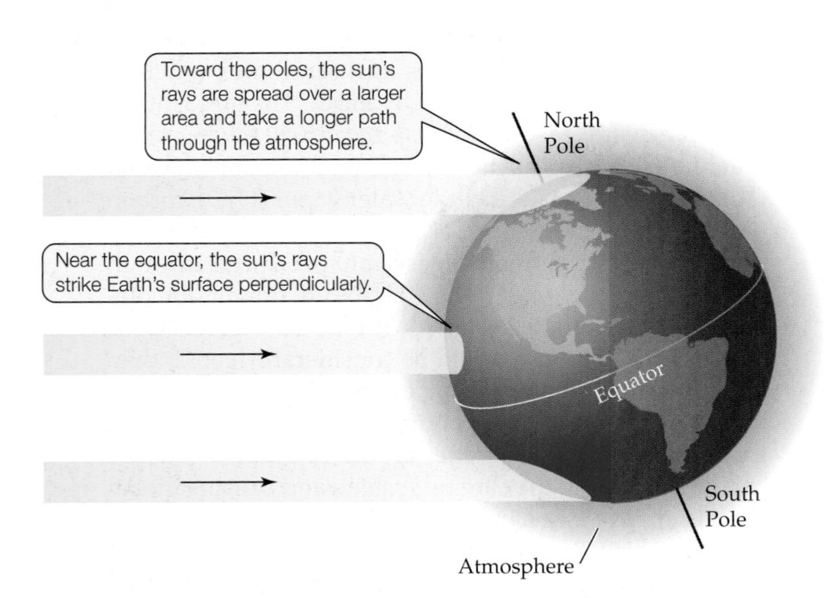

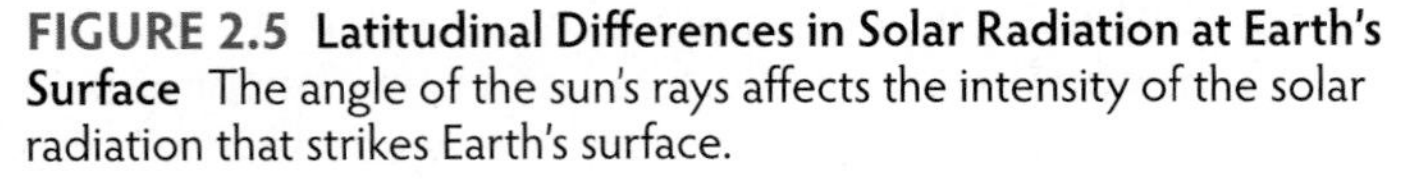

FIGURE 2.5 Latitudinal Differences in Solar Radiation at Earth's Surface The angle of the sun's rays affects the intensity of the solar radiation that strikes Earth's surface.

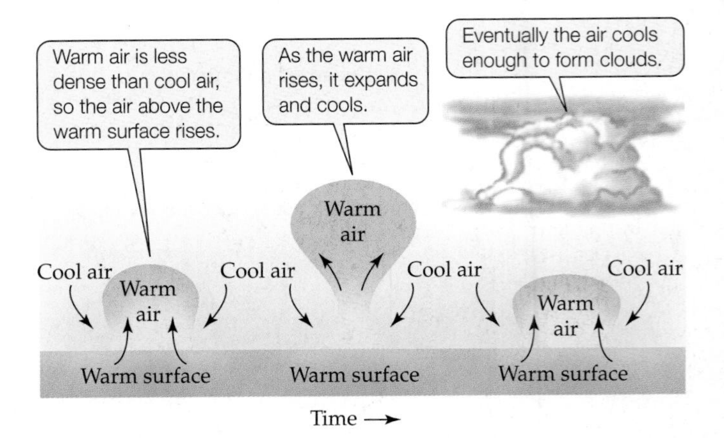

FIGURE 2.6 Surface Heating and Uplift of Air Differential solar heating of Earth's surface leads to the uplift of pockets of air over the warmest surfaces.

Atmospheric circulation cells are established in regular latitudinal patterns

When solar radiation heats Earth's surface, the surface warms and emits infrared radiation to the atmosphere, warming the air above it. As we have just seen, the heating of Earth's surface varies with latitude, and it can also vary with topography. Such differential warming creates pockets of warm air surrounded by cooler air. Warm air is less dense (has fewer molecules per unit of volume) than cool air, so as long as a pocket of air remains warmer than the surrounding air, it will rise (a process called **uplift**) (**Figure 2.6**). **Atmospheric pressure** results from the force exerted on a packet of air (or on Earth's surface) by the air molecules above it, so it decreases with increasing altitude. Thus, as the warm air rises higher, it expands. This expansion cools the rising air. Cool air cannot hold as much water vapor as warm air, so as the air continues to rise and cool, the water vapor contained within it begins to condense into droplets and form clouds.

The condensation of water into clouds is a warming process (another form of latent heat transfer), which may act to keep the pocket of air warmer than the surrounding atmosphere and enhance its uplift, despite its cooling due to expansion. You may have observed this process on a warm summer day as bubble-shaped cumulus clouds began to form thunderstorms. When there is substantial heating of Earth's surface and a progressively cooler atmosphere above the surface, the uplifted air will form clouds with wedge-shaped tops. The clouds reach to the boundary between the *troposphere*, the atmospheric layer above Earth's surface, and the *stratosphere*, the next atmospheric layer above the troposphere. This boundary is marked by a transition from progressively cooler temperatures in the troposphere to warmer temperatures in the stratosphere. Thus, the air pocket ceases to rise once it reaches the warmer temperatures of the stratosphere.

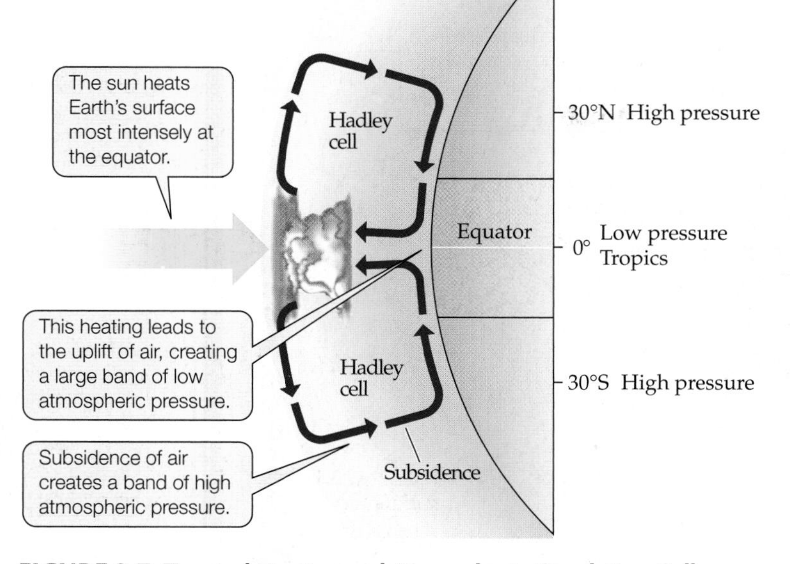

FIGURE 2.7 Tropical Heating and Atmospheric Circulation Cells The heating of Earth's surface in the tropics causes air to rise and release precipitation.

The tropics receive the most precipitation of any area on Earth because they receive the most solar radiation and thus experience the greatest amount of surface heating, uplift of air, and cloud formation. The uplift of air in the tropics creates a band of low atmospheric pressure relative to zones to the north and south. When air rising over the tropics reaches the boundary between the troposphere and stratosphere, it flows toward the poles (**Figure 2.7**). Eventually, this poleward-moving air cools as it exchanges heat with the surrounding air and meets cooler air moving from the poles toward the equator. Once the air reaches a temperature similar to that of the surrounding atmosphere, it descends toward Earth's surface, a process known as **subsidence**. This subsidence of air creates regions of high atmospheric pressure at latitudes 30° N and S, which inhibit the formation of clouds. It is not surprising, therefore, that the major deserts of the world are found at these latitudes.

The tropical uplift of air creates a large-scale pattern of atmospheric circulation in each hemisphere known as a **Hadley cell**, named after George Hadley, the eighteenth-century British meteorologist and physicist who first proposed its existence. Additional atmospheric circulation cells are formed at higher latitudes (**Figure 2.8**). The **polar cell**, as its name indicates, occurs at the North and South Poles. Cold, dense air subsides at the poles and moves toward the equator when it reaches Earth's surface. The descending air at the poles is replaced by air moving through the upper atmosphere from lower latitudes. Subsidence at the poles creates an area of high pressure, so the polar regions, despite the abundance of ice and snow on the ground, actually receive little precipitation, and are known as "polar deserts." An intermediate **Ferrell cell** (named after American meteorologist William Ferrell), exists at mid-latitudes between the Hadley and polar cells. The Ferrell cell is driven by the movement of the Hadley and polar cells and by exchange of energy between tropical and polar air masses in a region known as the *polar front*.

These three atmospheric circulation cells establish the major climatic zones on Earth. Between 30° N and S is the **tropical zone**. The **temperate zones** are between 30° and 60° N and S, and the **polar zones** are above 60° N and S (see Figure 2.8).

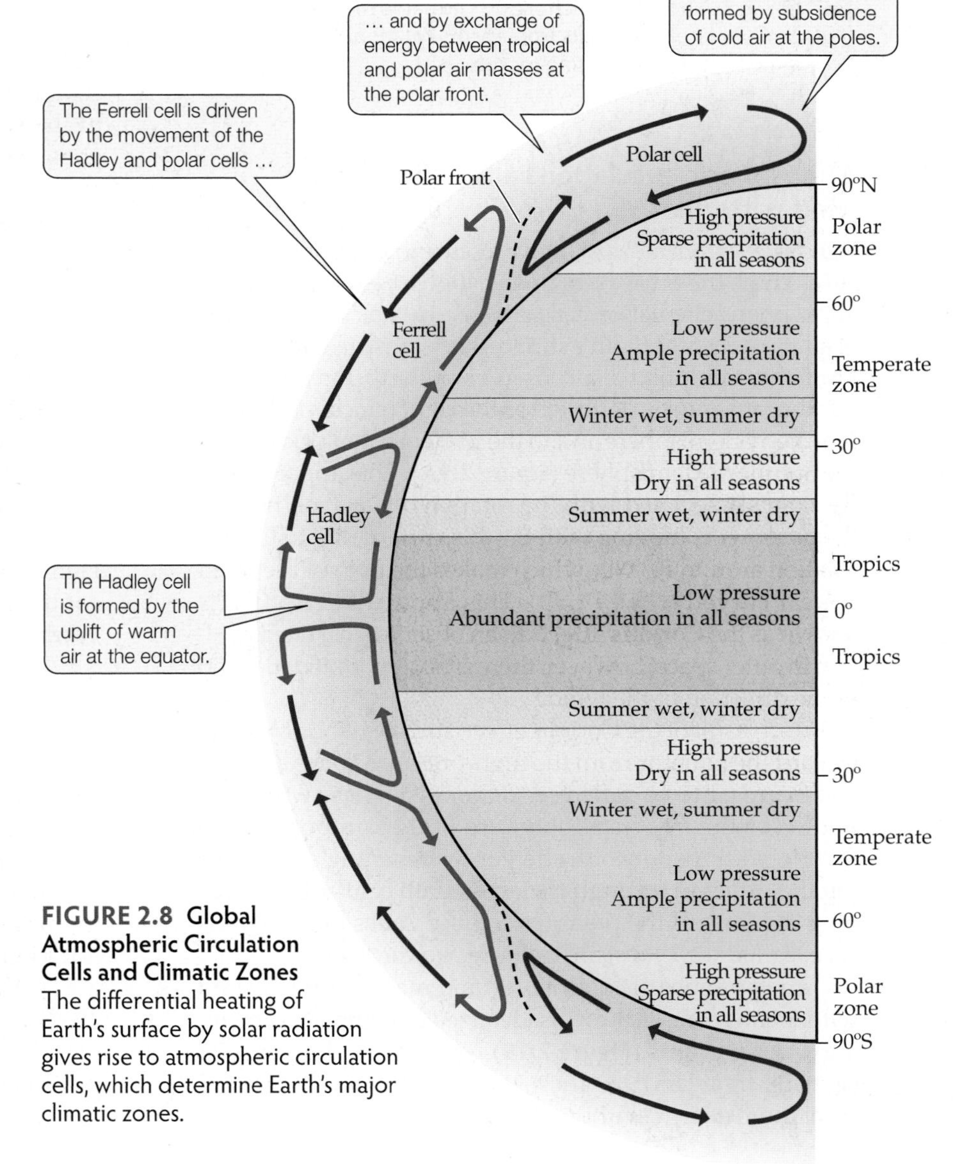

FIGURE 2.8 Global Atmospheric Circulation Cells and Climatic Zones The differential heating of Earth's surface by solar radiation gives rise to atmospheric circulation cells, which determine Earth's major climatic zones.

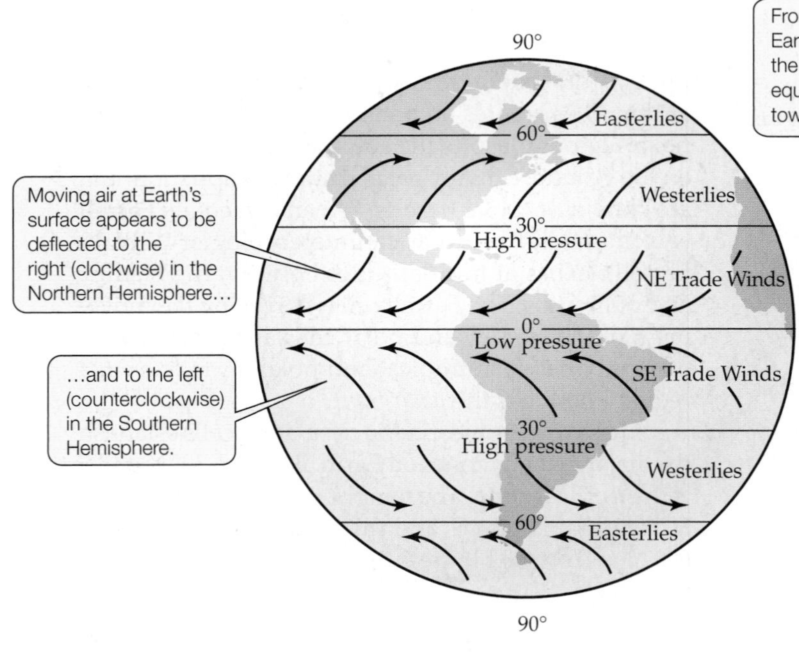

FIGURE 2.9 The Coriolis Effect on Global Wind Patterns (A) The Coriolis effect results from Earth's rotation. (B) Visualization of the Coriolis effect using rockets.

Atmospheric circulation cells create surface wind patterns

Winds flow from areas of high pressure to areas of low pressure. Thus, the areas of high and low pressure formed by atmospheric circulation cells give rise to consistent patterns of air movement at Earth's surface, known as *prevailing winds*. From the standpoint of an observer on Earth, the prevailing winds appear to be deflected to the right (clockwise) in the Northern Hemisphere and to the left (counterclockwise) in the Southern Hemisphere (**Figure 2.9A**). This apparent deflection is associated with the rotation of Earth on its axis: the observer is moving with Earth's surface due to Earth's rotation around its axis, which makes the path of the wind appear curved (**Figure 2.9B**). This apparent deflection is known as the **Coriolis effect**. To an observer in a fixed position in outer space, however, there is no apparent deflection in the direction of the wind.

As a result of the Coriolis effect, surface winds blowing toward the equator from the high-pressure zones at 30° N and S appear to be deflected to the west. These winds are known as the *trade winds* because of their importance to the global transport of trade goods in sailing ships during the fifteenth through the nineteenth centuries. Winds blowing toward the poles from those zones of high pressure, called *westerlies*, are deflected to the east.

The presence of continental land masses interspersed with oceans complicates this idealized depiction of prevailing wind patterns (**Figure 2.10**). Water has a higher **heat capacity** than land; in other words, water can absorb and store more energy without its temperature changing than land can. For this reason, solar radiation heats the land surface more than ocean water in summer, but the oceans retain more heat, and remains warmer in winter, than land at the same latitude. As a result, seasonal temperature changes over the oceans are smaller and less extreme than those on land. In summer, air over the oceans is cooler and denser than that over land. As a result, semipermanent zones of high pressure (*high-pressure cells*) form over the oceans, particularly around 30° N and S. In winter, the opposite situation exists: the air over the continents is cooler and denser than that over the oceans, so high-pressure cells develop in temperate zones over large continental areas. Because winds blow from areas of high pressure to areas of low pressure, these seasonal shifts in pressure cells influence the direction of the prevailing winds. The effect of land areas on the development of these semipermanent pressure cells is more pronounced in the Northern Hemisphere than in the Southern Hemisphere because continental land masses make up a larger proportion of Earth's surface there.

Ocean currents are driven by surface winds

Wind moving across the ocean surface creates a frictional drag that moves the surface water. As a result of the Coriolis effect, the water appears to move at an angle to the wind because from the perspective of an observer on Earth it is deflected to the right in the Northern Hemisphere and to the left in the Southern Hemisphere. For this reason, the pattern of ocean surface currents is similar to, but not identical to, the pattern of prevailing winds (**Figure 2.11**). The speed of ocean currents is usually only about 2%–3% of the wind speed. An average wind speed of 10 meters per second (22 miles per hour) would therefore produce an ocean current moving at 30 centimeters per second (0.7 miles per hour). In the North Atlantic Ocean,

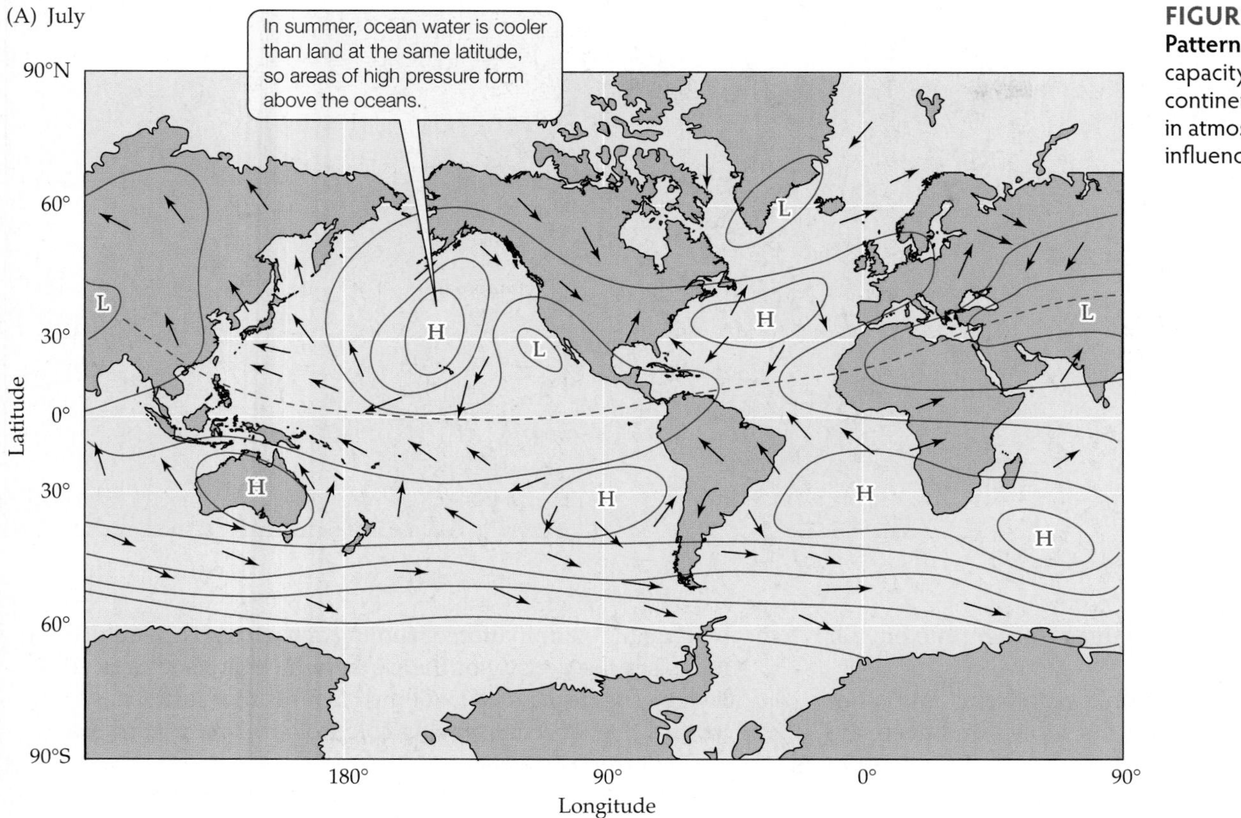

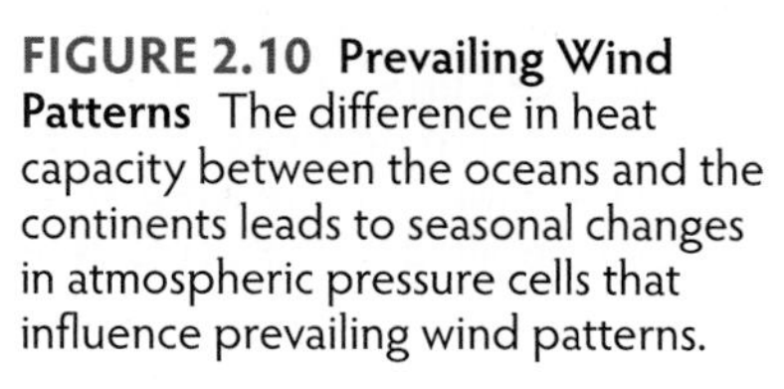

FIGURE 2.10 Prevailing Wind Patterns The difference in heat capacity between the oceans and the continents leads to seasonal changes in atmospheric pressure cells that influence prevailing wind patterns.

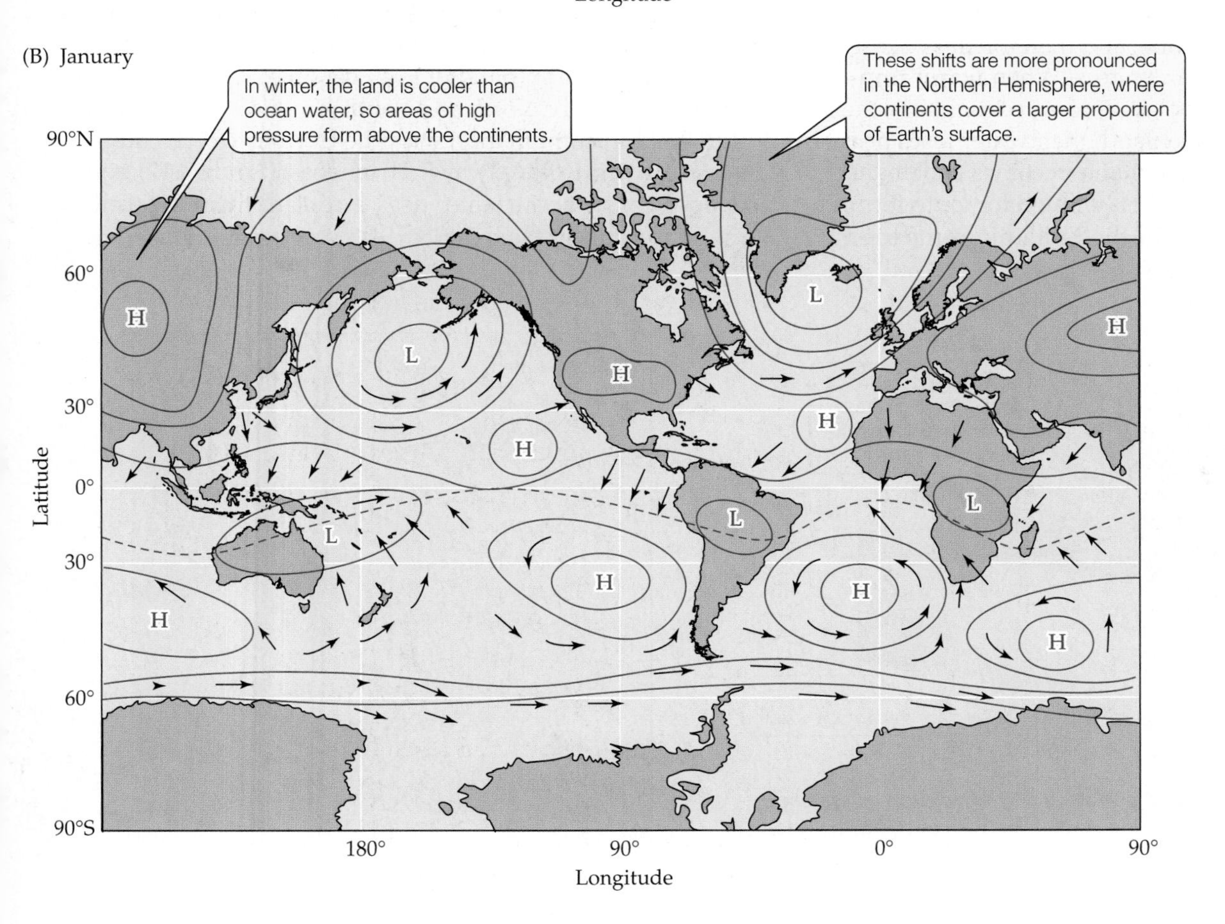

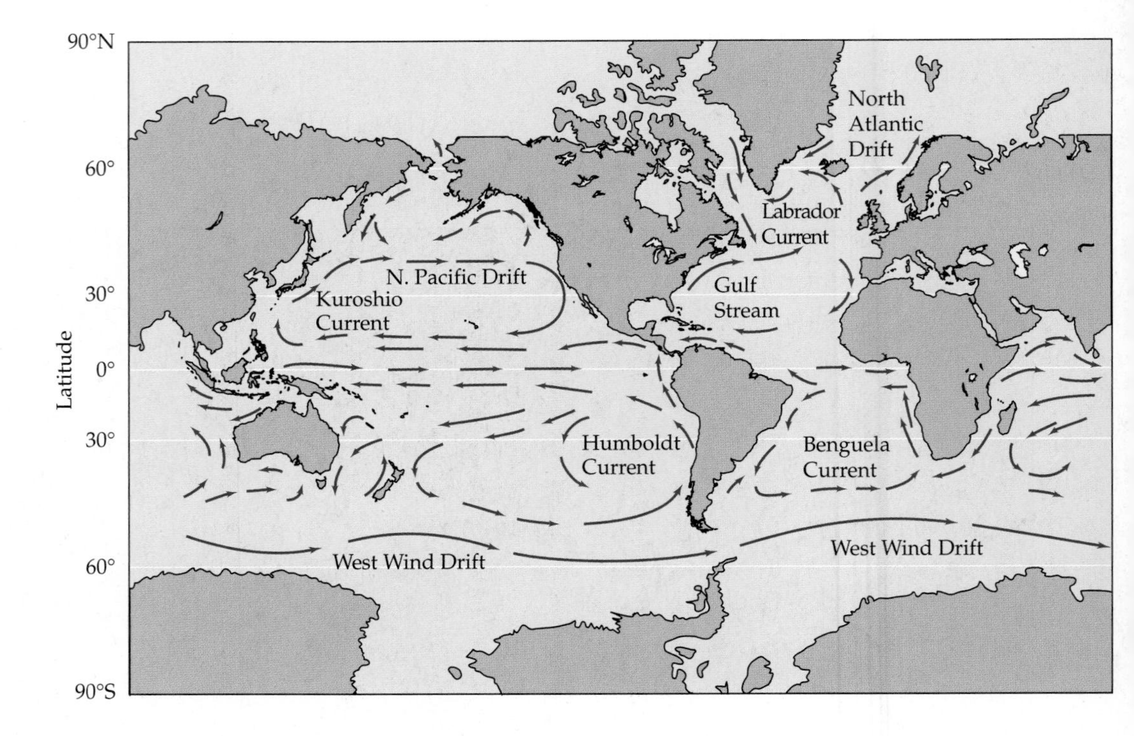

FIGURE 2.11 Global Ocean Surface Currents The major ocean surface currents are driven by the surface winds shown in Figure 2.9A, but modified by the Coriolis effect.

current velocities may be as high as 200 centimeters per second (4.5 miles per hour).

Like atmospheric circulation, oceanic circulation has a vertical dimension. Generally, the surface and deep layers of ocean water do not mix because of differences in their temperature and *salinity* (concentration of dissolved salts). The surface waters (those above 75–200 m, 250–600 feet) are warmer and less saline, and therefore less dense, than the deeper, cooler ocean waters. When warm tropical surface currents reach polar regions, particularly the coasts of Antarctica and Greenland, their water loses heat to the surrounding environment and becomes cooler and denser. The water eventually cools enough for ice to form, which increases the salinity of the remaining unfrozen water. This combination of cooling and increasing salinity increases the density of the water, which sinks to deeper layers. The dense downwelling currents that result move toward the equator, carrying cold polar water toward the warmer tropical oceans.

These deep ocean currents connect with surface currents again at zones of **upwelling**, where deep ocean water rises to the surface. Upwelling occurs where prevailing winds blow nearly parallel to a coastline, such as off the western coasts of North and South America. The force of the wind, in combination with the Coriolis effect, causes surface waters to flow away from the coast (**Figure 2.12**), and deeper, colder ocean water rises to replace them. Upwelling also occurs in the westward-flowing equatorial Pacific

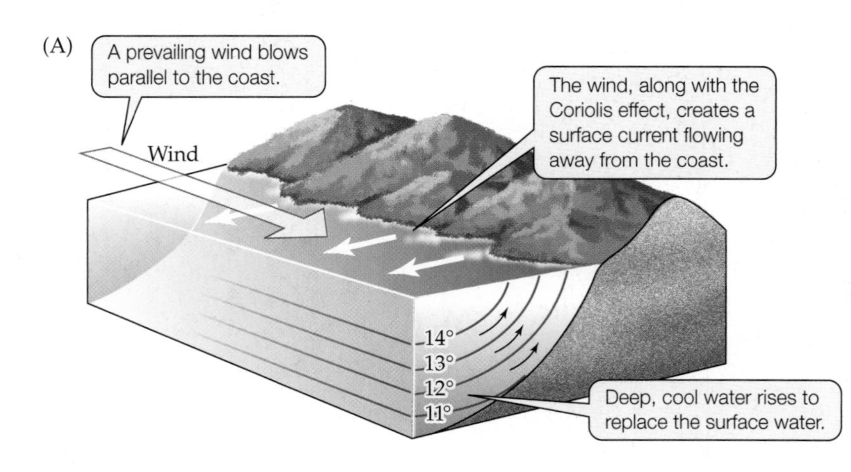

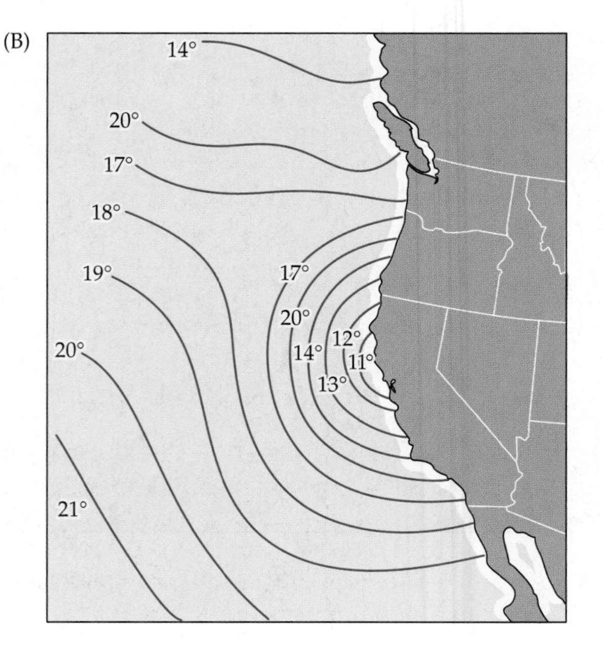

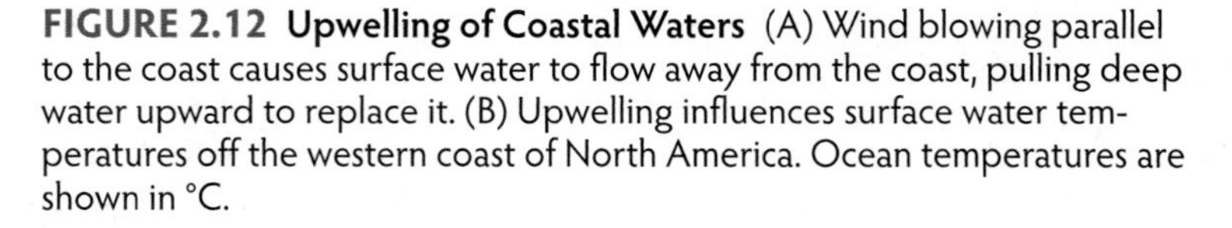

FIGURE 2.12 Upwelling of Coastal Waters (A) Wind blowing parallel to the coast causes surface water to flow away from the coast, pulling deep water upward to replace it. (B) Upwelling influences surface water temperatures off the western coast of North America. Ocean temperatures are shown in °C.

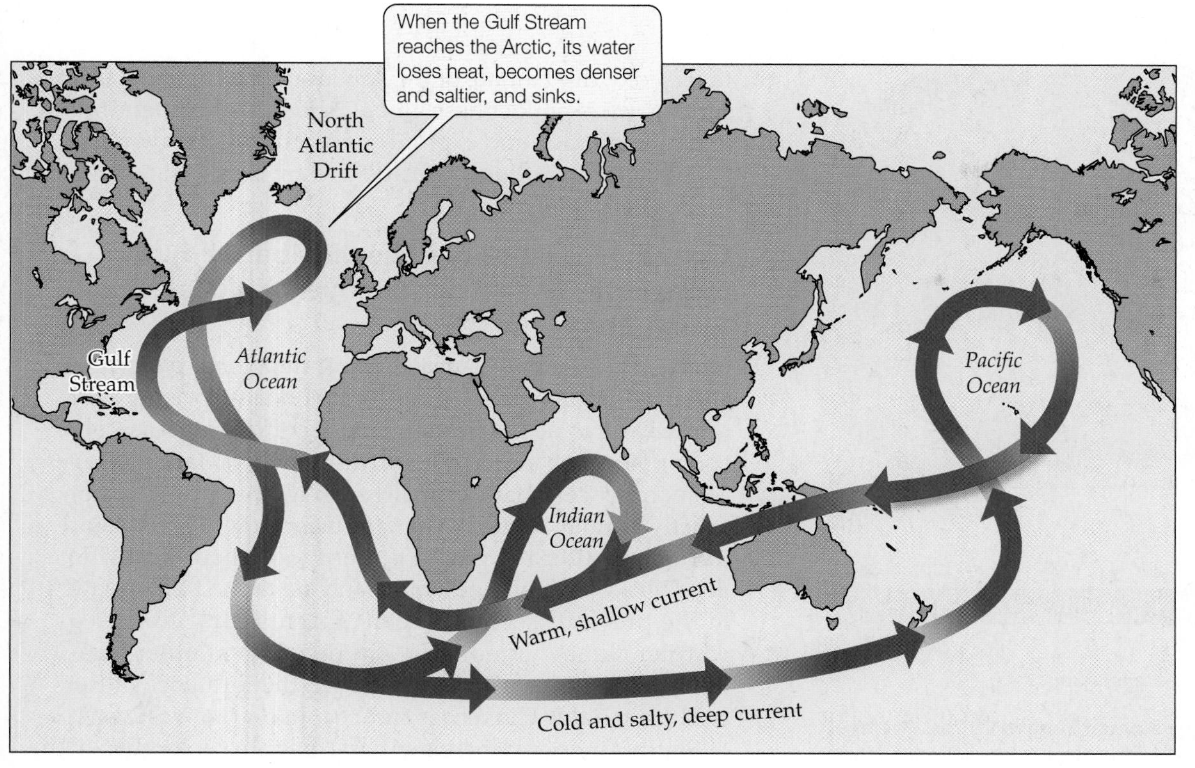

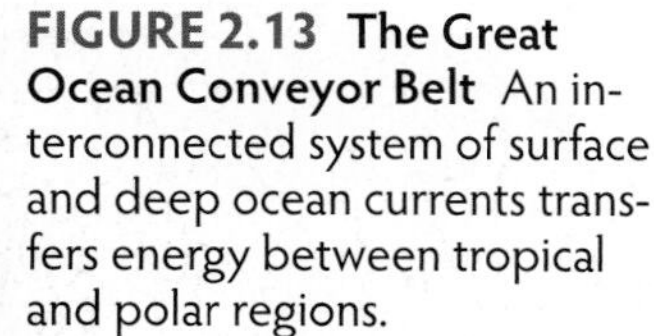

FIGURE 2.13 The Great Ocean Conveyor Belt An interconnected system of surface and deep ocean currents transfers energy between tropical and polar regions.

Ocean. As a result of the Coriolis effect, water just to the north and south of the equator is deflected slightly away from the equator, causing divergence of surface water and a zone of upwelling.

Upwelling has important consequences for the local climate, creating a cooler, moister environment. Upwelling also has a strong effect on biological activity in the surface waters. When organisms in the surface waters die, their bodies, and the nutrients they contain, fall into deeper water. Thus, nutrients tend to accumulate in deep water and in sediments at the ocean bottom. Upwelling brings those nutrients back to the *photic zone*, the layer of surface water where there is enough light to support photosynthesis. Upwelling zones are among the most productive open ocean ecosystems because these nutrients increase the growth of *phytoplankton* (small, free-floating algae and other photosynthetic organisms), which provide food for *zooplankton* (free-floating animals and protists), which in turn support the growth of their consumers.

Ocean currents influence the climates of the regions where they flow. For example, the Gulf Stream and North Atlantic Drift, a current system that flows from the tropical Atlantic northward to the North Atlantic (see Figure 2.11), contributes to warmer winters in Scandinavia relative to locations at the same latitude across the Atlantic in North America. In addition, winds blowing across the Atlantic pick up heat from the ocean, which also contributes to a warmer climate in northern Europe. Winter temperatures on the west coast of Scandinavia are approximately 15°C (22°F) warmer than those on the coast of Labrador. This temperature difference is reflected in the vegetation: deciduous forests are more common on the Scandinavian coast, while boreal forests of spruce and pine dominate the coast of Labrador. The Gulf Stream also keeps the North Atlantic ice free most of the winter, while sea ice forms at the same latitude off the North American coast.

Ocean currents are responsible for about 40% of the heat exchanged between the tropics and the polar regions; the remaining 60% is transferred by winds. Thus, ocean currents are sometimes referred to as the "heat pumps" or "thermal conveyers" of the planet. A large system of interconnected currents that links the Pacific, Indian, and Atlantic oceans, sometimes called the "great ocean conveyor belt," is an important means of transferring heat to the polar regions (**Figure 2.13**).

Now that we have seen how the differential heating of Earth's surface generates prevailing winds and ocean currents, let's examine the effects of these atmospheric and oceanic circulation patterns on Earth's climates, including global patterns of temperature and precipitation.

CONCEPT 2.3 Large-scale atmospheric and oceanic circulation patterns establish global patterns of temperature and precipitation.

Global Climatic Patterns

Earth's climates reflect a variety of temperature and precipitation regimes, from the warm, wet climate of the tropics

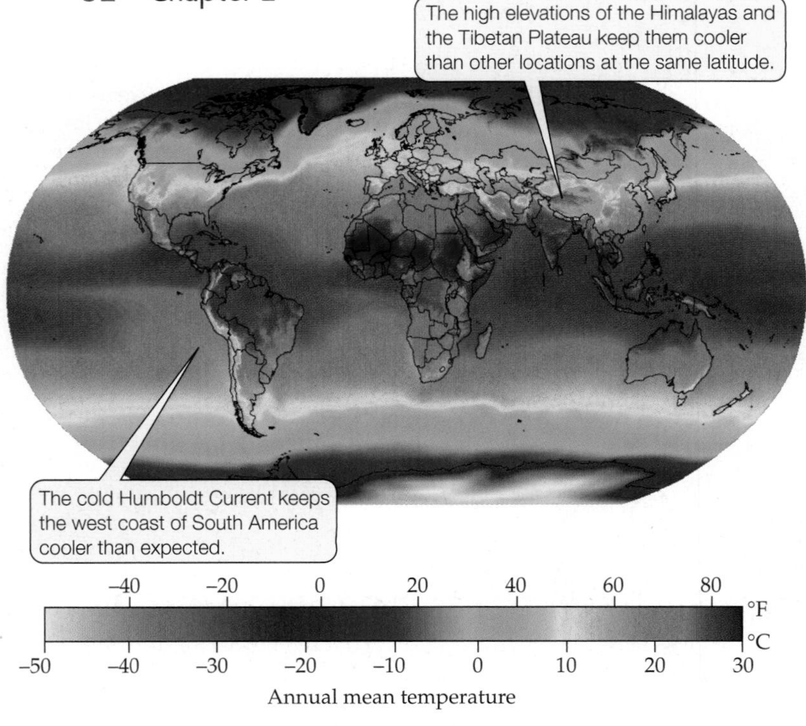

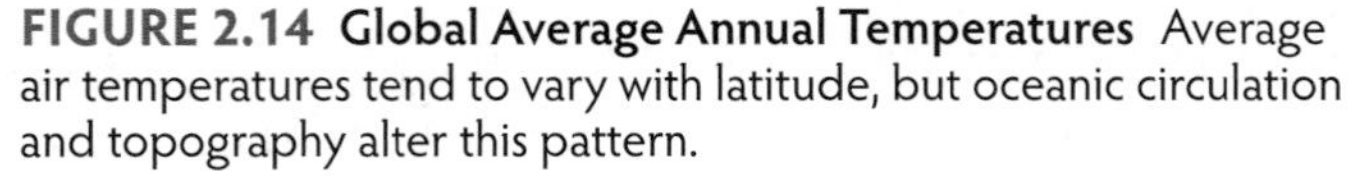

FIGURE 2.14 Global Average Annual Temperatures Average air temperatures tend to vary with latitude, but oceanic circulation and topography alter this pattern.

to the cold, dry climate of the Arctic and Antarctic. In this section, we examine these global patterns of temperature and precipitation and explore how both climatic averages and climatic variation are influenced by prevailing winds and ocean currents.

Oceanic circulation and the distribution and topography of continents influence global temperatures

Given what we have learned about the global pattern of solar radiation (see Figure 2.5), it is not surprising that temperatures at Earth's surface become progressively cooler from the equator toward the poles (**Figure 2.14**). Note, however, that the changes in temperature are not exactly parallel with the changes in latitude. Three major influences alter the global pattern: ocean currents, the distribution of land and water, and elevation. As we saw in the previous section, ocean currents contribute to a warmer climate in northern Europe than at other locations at the same latitude. Similarly, the influence of the cold Humboldt Current is noticeable on the west coast of South America, where temperatures are cooler than at similar latitudes elsewhere.

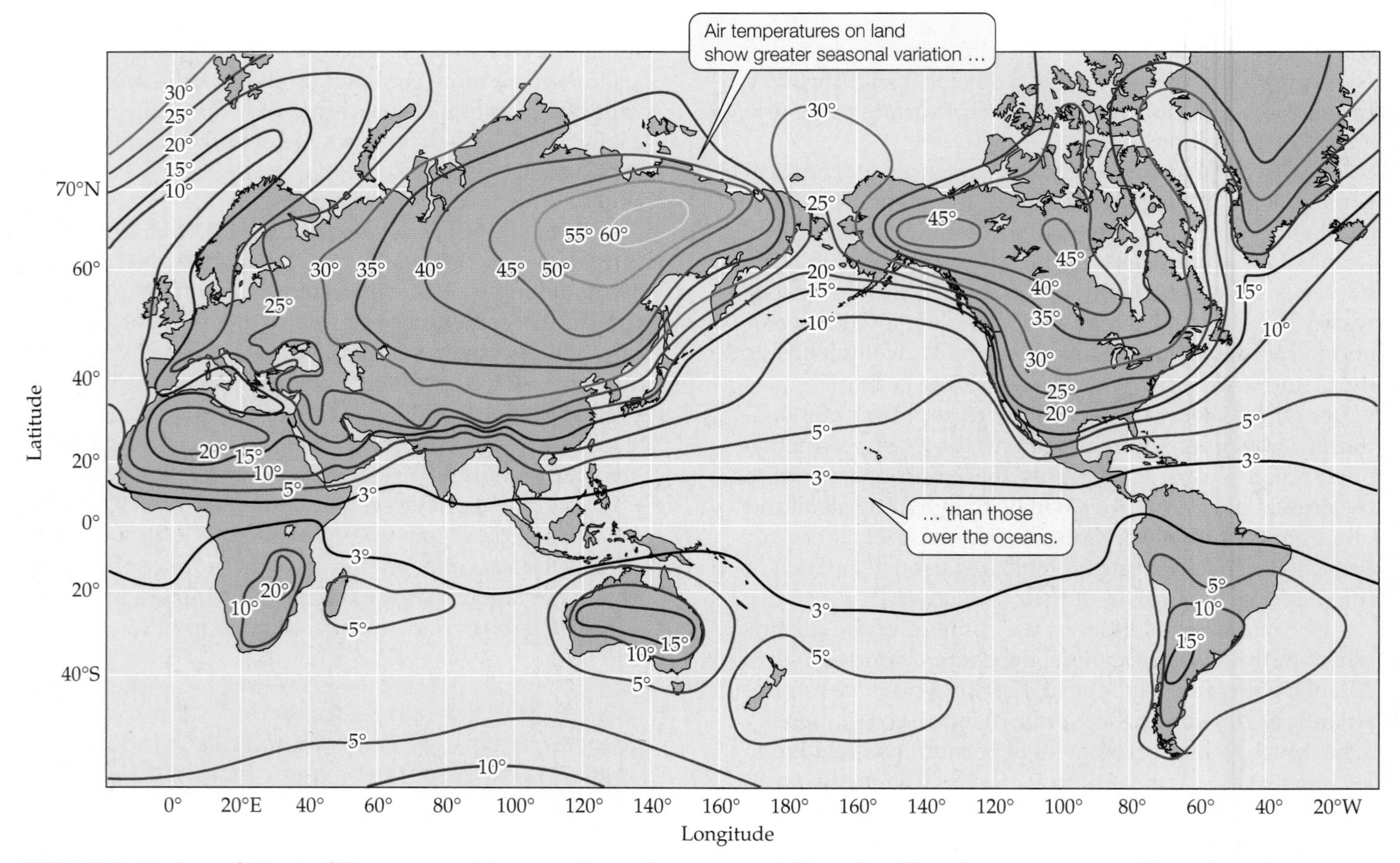

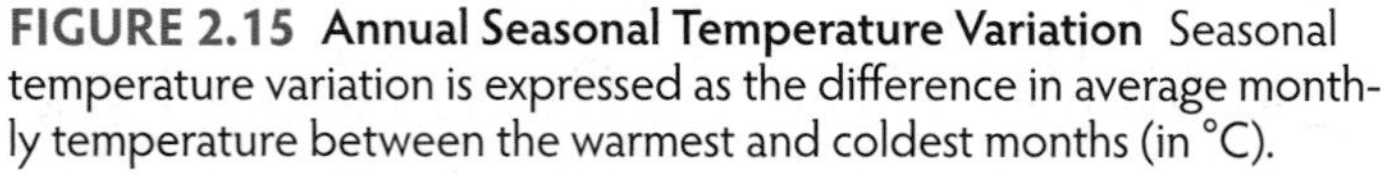

FIGURE 2.15 Annual Seasonal Temperature Variation Seasonal temperature variation is expressed as the difference in average monthly temperature between the warmest and coldest months (in °C).

Q What is the effect of continent size and latitude on the magnitude of seasonal temperature change?

The difference in heat capacity between the oceans and the continents is not reflected in the average annual temperatures shown in Figure 2.14. Why is this so? Because the annual temperature *variation* is not depicted in that figure. Air temperatures over land show greater seasonal variation, with warmer temperatures in summer and colder temperatures in winter, than those over the oceans (**Figure 2.15**).

Elevation above sea level has an important influence on continental temperatures. Note in Figure 2.14 the large difference in temperature between the Indian subcontinent and Asia. The sharp change in air temperature in this region is due to the influence of the Himalayas and the Tibetan Plateau. The change in elevation is extreme here, from about 150 m (500 feet) on the Ganges Plain in India to over 8,000 m (28,000 feet) in the highest peaks of the Himalayas in only 200 km (120 miles).

Why is it colder in mountains and highlands than in surrounding lowlands? Two factors contribute to the colder climates found at higher elevations. First, atmospheric pressure, and therefore the density of air, decreases with increasing elevation above sea level. As a result, there are fewer air molecules to absorb the infrared energy radiating from Earth's surface. Thus, even though the highlands may receive as much solar radiation as nearby lowlands, the heating of air by the ground surface is less effective because of the lower air density. Second, highlands exchange air more effectively with cooler air in the surrounding atmosphere. Because the atmosphere is warmed mainly by infrared radiation emitted by Earth's surface, the temperature of the atmosphere decreases with increasing distance from the ground. This decrease in temperature with increasing height above the surface is known as the **lapse rate**. In addition, wind velocity increases with increasing elevation because there is less friction with the ground surface. As a result, the decrease in air temperature with increasing elevation tends to follow the lapse rate.

Patterns of atmospheric pressure and topography influence precipitation

The locations of the Hadley, Ferrell, and polar circulation cells suggest that precipitation should be highest in the tropical latitudes between 23.5° N and S and in a band at about 60° N and S and lowest in zones around 30° N and S (see Figure 2.8). The African continent displays the pattern closest to this idealized precipitation distribution. However, there are substantial deviations from the expected latitudinal precipitation pattern in other areas, particularly in the Americas (**Figure 2.16**). These deviations

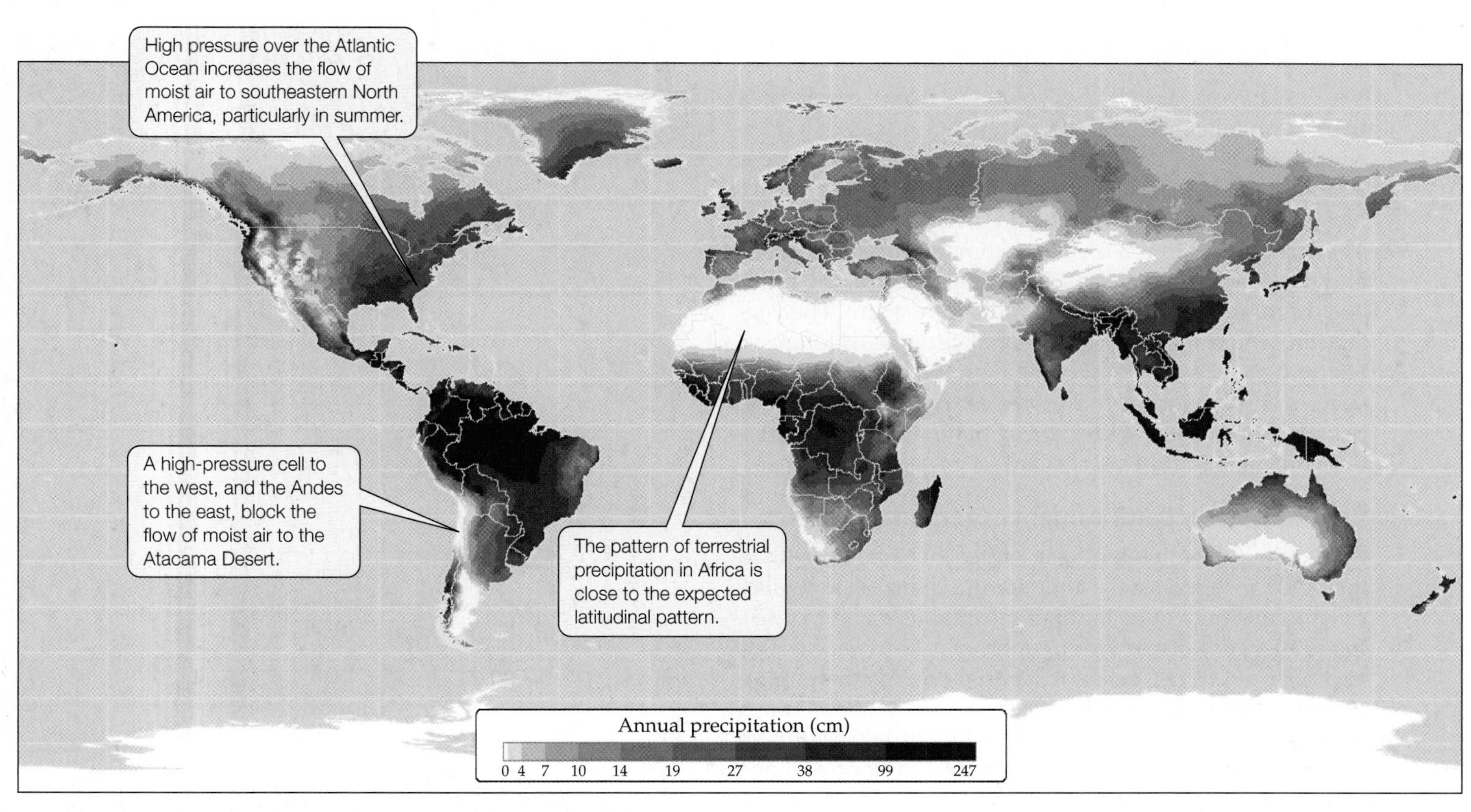

FIGURE 2.16 Average Annual Terrestrial Precipitation The latitudinal pattern of precipitation deviates from what would be expected based on atmospheric circulation patterns alone (see Figure 2.8).

are associated with the semipermanent high-pressure and low-pressure cells discussed earlier (see Figure 2.10) as well as with large mountain chains.

Pressure cells influence the movement of moist air from oceans to continents as well as cloud formation. For example, high pressure over the South Pacific Ocean decreases precipitation along the central west coast of South America. One of the driest deserts in the world, the Atacama, located along the Pacific coast of Chile, is associated with the presence of this high-pressure cell and with the blockage of air masses moving from the east by the Andes. In contrast, high pressure over the Atlantic Ocean increases the flow of moist air to southeastern North America, particularly in summer, increasing precipitation and supporting the development of a productive deciduous forest there.

Mountains also influence precipitation patterns by forcing air moving across them to rise, which enhances local precipitation. The effects of mountains, as well as those of oceans and vegetation, on regional climatic patterns are addressed in the next section.

CONCEPT 2.4 Regional climates reflect the influence of the distribution of oceans and continents, mountains, and vegetation.

Regional Climatic Influences

You may have noticed that as you travel from a coastal area to an inland location, the climate changes. This change in climate can be abrupt, particularly when you travel across a mountain chain. The daily variation in air temperature increases, humidity decreases, and precipitation decreases. These climatic differences result from the effects of oceans and continents on regional energy balance and the influence of mountains on air flow and temperature. The vegetation often reflects these regional climatic differences, testifying to the effects of climate on the distributions of species and biological communities. The vegetation also has important feedbacks to the climate through its influence on energy and water balance.

Proximity to oceans influences regional climates

As we have seen, water has a higher heat capacity than land. As a result, seasonal temperature changes are smaller over oceans than over continental areas (see Figure 2.15). In addition, oceans provide a source of moisture for clouds and precipitation. Coastal terrestrial regions that are influenced by an adjacent ocean have a **maritime climate**. Maritime climates are characterized by little variation in daily and seasonal temperatures, and they often have higher humidity than regions more distant from the coast. In contrast, areas in the middle of large continental land masses have a **continental climate**, which is characterized by much greater variation in daily and seasonal temperatures. Maritime climates occur in all climatic zones, from tropical to polar. In the temperate zones, the influence of oceans on coastal climates tends to be accentuated on west coasts in the Northern Hemisphere and on east coasts in the Southern Hemisphere because of the prevailing wind patterns. Continental climates are limited to mid- and high latitudes (primarily in the temperate zones), where large seasonal changes in solar radiation accentuate the effect of the low heat capacity of land masses.

The influence of land and water on climate can be exemplified by comparing the seasonal temperature variation in locations at similar latitudes and elevations in Siberia (**Figure 2.17**). Sangar, a town on the Lena River in the middle of the Asian continent, has a maximum aver-

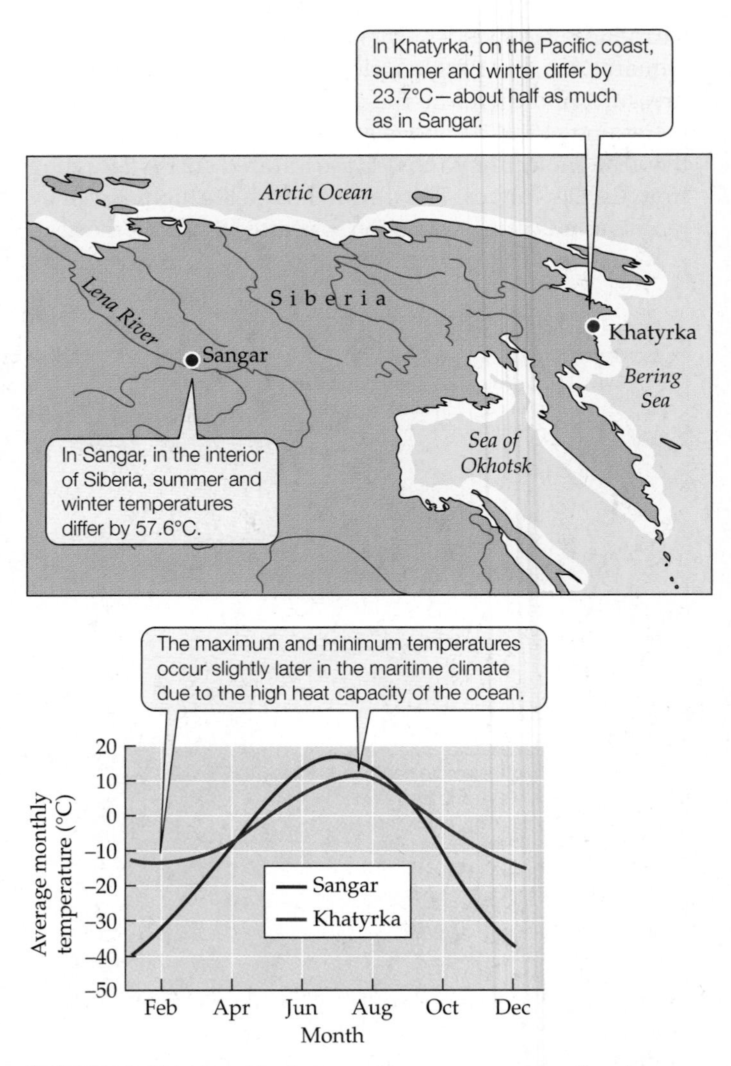

FIGURE 2.17 Monthly Average Temperatures in a Continental and a Maritime Climate The difference in seasonal temperature variation between two locations in Siberia at about the same latitude and elevation illustrates the effect of the high heat capacity of ocean water.

age monthly temperature of 17.9°C (64°F) (in July) and a minimum average monthly temperature of –39.7°C (–40°F) (in January), for an average seasonal variation of 57.6°C (114°F). Khatyrka, on the Pacific coast, has a maximum average monthly temperature of 9.5°C (49°F) (in July) and a minimum average monthly temperature of –14.2°C (6°F) (in February), for an average seasonal variation of 23.7°C (43°F). The seasonal temperature variation in continental Sangar is therefore more than double that in coastal Khatyrka, at nearly the same latitude and elevation. Note that the maximum and minimum temperatures occur slightly later in the year in the maritime climate (Khatyrka), another reflection of the high heat capacity of the ocean and its effect on local climate.

Mountains influence wind patterns and gradients in temperature and precipitation

The effects of mountains on climate are visually apparent in the elevational patterns of vegetation, particularly in arid regions. As we move up a mountain, grasslands may abruptly change to forests, and at higher elevations, forests may give way to alpine grasslands. These abrupt shifts in vegetation patterns reflect the rapid changes in climate that occur over short distances in mountains as temperatures decrease, precipitation increases, and wind speed increases with elevation. What causes these abrupt changes? The climates of mountains are the product of the effects of topography and elevation on air temperatures, the behavior of air masses, and their own generation of unique local wind patterns.

When air moves across Earth's surface and encounters a mountain range, the slopes of the mountains force it upward. This uplifted air cools as it rises, and water vapor condenses to form clouds and precipitation. As a result, the amount of precipitation increases with elevation. This enhancement of precipitation in mountains is particularly apparent in north–south-trending mountain ranges on the slopes that face into the prevailing wind (the *windward* slopes). In the temperate zones, the westerlies encounter the western slopes of these mountain ranges (such as the Andes and Sierra Nevada in the Americas) and lose most of their moisture there as precipitation before cresting over the summits. The loss of moisture, as well as the warming of the air as it moves down the eastern slopes, dries the air mass (**Figure 2.18A**). This **rain-shadow effect** results in lower precipitation and soil moisture on the slopes facing away from the prevailing wind (the *leeward* slopes) and higher precipitation and soil moisture on the windward slope. The rain-shadow effect influences the types and amounts of vegetation on mountain ranges: lush, productive plant communities tend to be found on the windward slopes and sparser, more drought-resistant vegetation on the leeward slopes (**Figure 2.18B**).

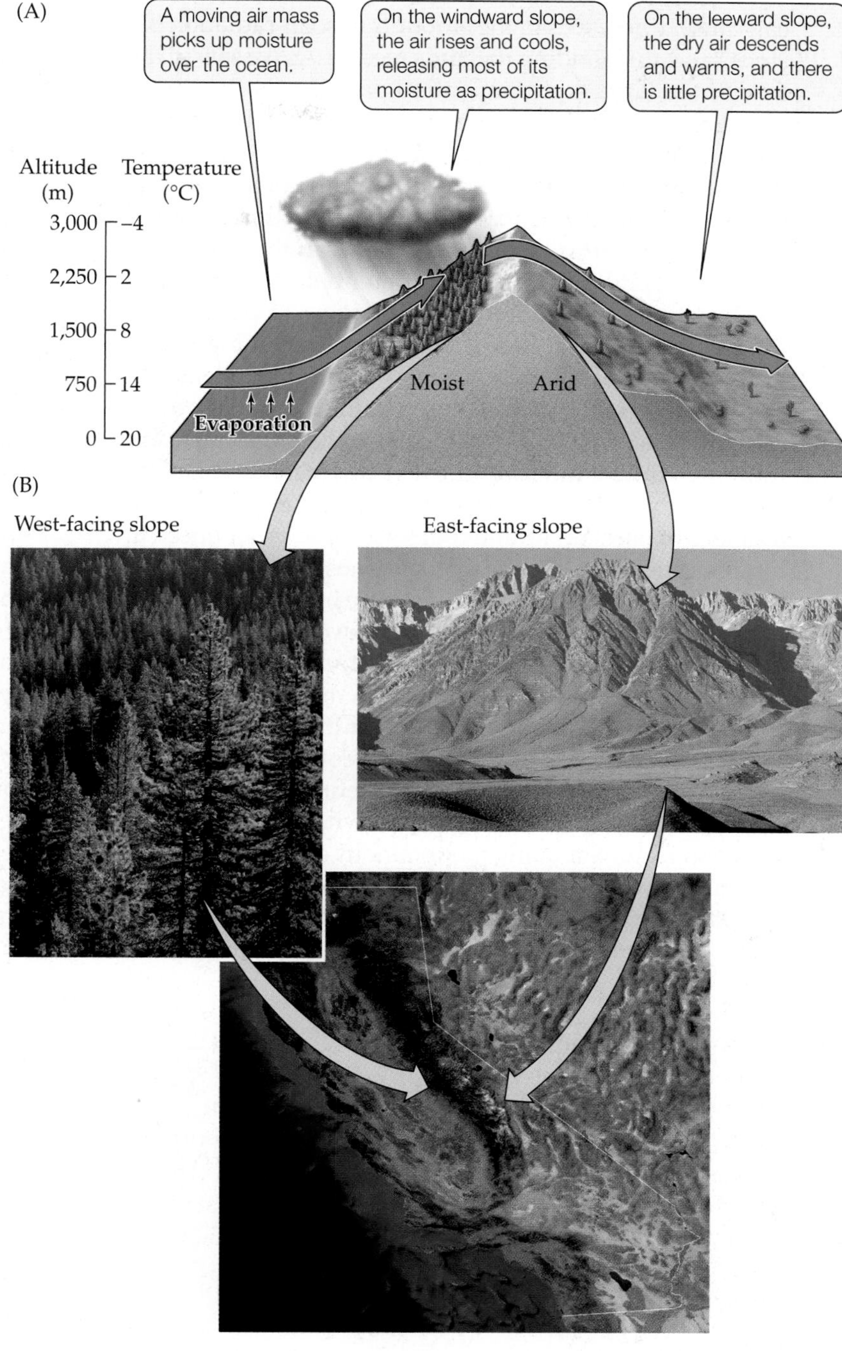

FIGURE 2.18 The Rain-Shadow Effect (A) Precipitation tends to be greater on the windward slope of a mountain range than on the leeward slope. (B) Vegetation on west-facing and east-facing slopes in the Sierra Nevada of California reflects the rain-shadow effect.

Mountains can also generate local wind and precipitation patterns. Differences in the direction that mountain slopes face (referred to as the slope exposure or *aspect*) can cause differences in the amounts of solar radiation the slopes and surrounding flatlands receive. As we saw in the case of the large-scale circulation patterns that generate Hadley cells (see Figure 2.6), differences in solar heating of the ground surface can cause the uplift of air parcels that are warmer than the surrounding air. In the morning, east-facing slopes receive more solar radiation from the rising sun and thus become warmer than the surrounding slopes and lowlands. This differential heating creates localized upslope winds in the mountains. Depending on the moisture content of the air and the prevailing winds at higher elevations, clouds may form on the eastern flanks of the mountains. These clouds can generate local thunderstorms that may move off the mountains and into surrounding lowlands, increasing local precipitation.

At night, the ground surface cools, and the air above it becomes denser. Nighttime cooling is more pronounced at high elevations because the thinner atmosphere absorbs and reradiates less energy and allows more heat to be lost from the ground surface. Cold, dense air behaves like water, flowing downslope and pooling in low-lying areas. As a result, valley bottoms are the coldest sites in mountainous areas during clear, calm nights. This *cold air drainage* influences vegetation distributions in temperate zones because of the higher frequency of subfreezing temperatures in low-lying areas. Daily upslope and nightly downslope winds are a common feature of many mountainous areas, particularly in summer when the input of solar radiation is highest.

At continental scales, mountains influence the movement, position, and behavior of air masses, and as a result, they influence temperature patterns in surrounding lowlands. Large mountain chains, or *cordilleras*, can act to channel the movement of air masses. The Rocky Mountains, for example, steer cold Arctic air through the central part of North America to their east and inhibit its movement through the intermountain basins to their west.

Vegetation affects climate via surface energy exchange

Climate determines where and how organisms can live, but organisms, in turn, influence the climate system. The amount and type of vegetation influences how the ground surface interacts with solar radiation and wind and how much water it loses to the atmosphere. The capacity of a land surface to reflect solar radiation, known as its **albedo**, is influenced by the presence and type of vegetation as well as by soil and topography. Light-colored surfaces have the highest albedo. A coniferous forest, for example, has a darker color, and thus a lower albedo, than most types of bare soil or grasslands, so it absorbs more solar energy. The texture of Earth's surface is also influenced by vegetation. A smooth surface, such as a continuous grassland, allows greater transfer of energy to the atmosphere by wind (sensible heat loss by convection) than a rough surface, such as mixed forest and grassland. Finally, vegetation affects the movement of water into the atmosphere. **Evapotranspiration**, the sum of water loss through **transpiration** (evaporation of water from inside a plant) and evaporation, increases with the area of leaves per unit of ground surface area. Evapotranspiration transfers en-

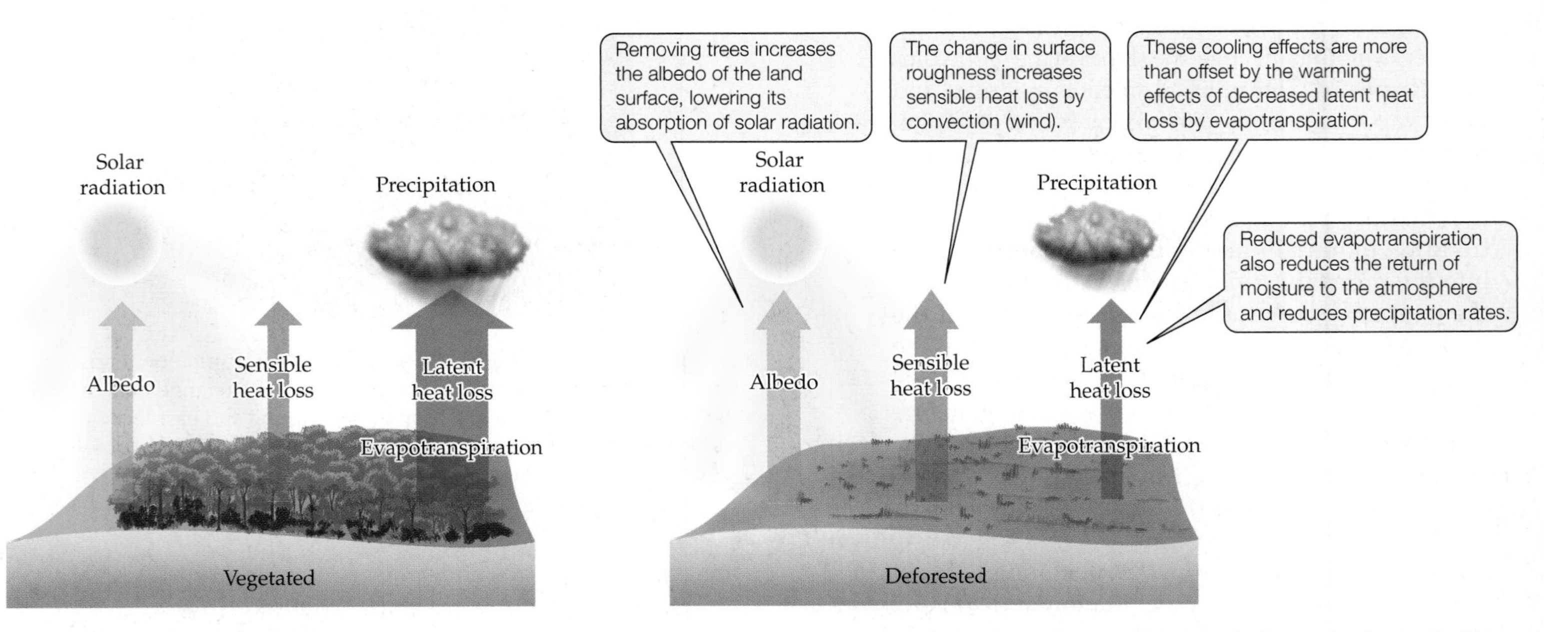

FIGURE 2.19 The Effects of Deforestation Illustrate the Influence of Vegetation on Climate The conversion of forest to pasture in the tropics results in a number of changes in energy exchange with the atmosphere. (After Foley et al. 2003.)

Q What changes in energy exchange with the atmosphere would you expect if an arid, light-colored grassland was converted to an irrigated dark green agricultural field?

ergy (as latent heat) as well as water into the atmosphere, thereby affecting air temperature and moisture.

What happens to climate when the type or amount of vegetation is altered (e.g., the conversion of forest to pasture)? This question is particularly important because of the current high rates of deforestation in the tropics: each year, between 6 and 7 million hectares of tropical forest are cleared worldwide (Wright 2005). Loss of the trees increases the albedo of the land surface as bare soil is exposed and the trees are partially replaced with lighter-colored grasses (**Figure 2.19**). The higher albedo decreases the absorption of solar radiation, resulting in less heating of the land surface. However, the lower heat gain from solar radiation is offset by lower evapotranspirative cooling (lower latent heat transfer) due to loss of leaf area (Foley et al. 2003). Lower evapotranspiration rates not only reduce surface cooling, but also lead to lower precipitation because less moisture is returned from the ground surface to the atmosphere. Thus, the outcome of tropical deforestation may be a warmer, drier regional climate (Foley et al. 2003). Widespread deforestation may lead to climate change that is significant enough to inhibit reforestation and may thus lead to long-term changes in tropical ecosystems.

In Chapter 24 we will return to the effects of human activities on climate, especially over the past two centuries. Human activities, however, are not the only cause of long-term climate change. We turn next to the natural climatic variation that has occurred throughout Earth's history.

CONCEPT 2.5 Seasonal and long-term climatic variation are associated with changes in Earth's position relative to the sun.

Climatic Variation over Time

As noted at the beginning of this chapter, understanding climatic variation is critical to understanding ecological phenomena such as the distribution of microorganisms, plants, and animals. Climatic variation at daily to multidecadal time scales determines the range of environmental conditions experienced by organisms as well as the availability of the resources and habitats they need to survive. Long-term climatic variation over hundreds and thousands of years influences the evolutionary history of organisms and the development of ecosystems. As we will see, the global climate has changed substantially over the course of Earth's history. In this section, we will review climatic variation from seasonal to 100,000-year time scales.

Seasonality results from the annual orbit of Earth around the sun

The amount of sunlight striking any point on Earth's surface varies as Earth makes its 365.25-day journey around the sun. Earth rotates on an axis that is tilted at an angle of 23.5° relative to the sun's direct rays (**Figure 2.20**). Thus, the angle and intensity of the rays striking any point on

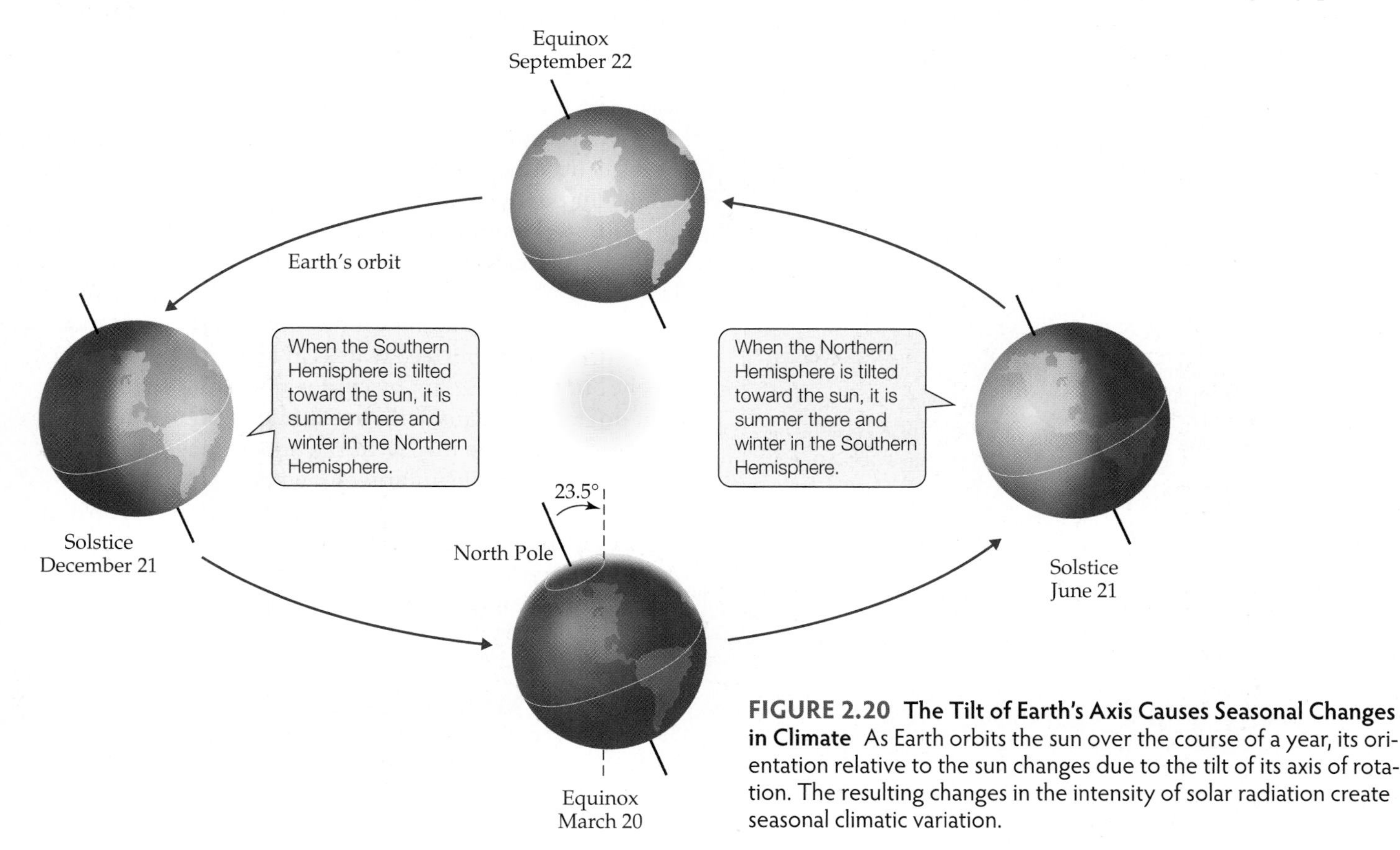

FIGURE 2.20 The Tilt of Earth's Axis Causes Seasonal Changes in Climate As Earth orbits the sun over the course of a year, its orientation relative to the sun changes due to the tilt of its axis of rotation. The resulting changes in the intensity of solar radiation create seasonal climatic variation.

Earth vary as Earth orbits the sun. This influence of the tilt of Earth's axis overrides the variation associated with seasonal changes in the distance between Earth and the sun due to Earth's slightly elliptical orbit. Earth is closest to the sun in January (at a point called the perihelion, 147 million miles) and farthest away in July (at the aphelion, 152 million miles). As we will see later, however, the effect of Earth–sun distance on climate is important over much longer time scales.

Seasonality in the tropics is marked primarily by changes in precipitation, rather than by changes in temperature, because the seasonal changes in solar radiation are relatively small there compared with those in the temperate and polar zones. These seasonal changes in precipitation are associated with the movement of the zone of maximum solar radiation, known as the **Intertropical Convergence Zone** or **ITCZ**, between the Northern and Southern hemispheres as Earth orbits the sun. The zone of maximum solar radiation is also the zone of maximum warm air uplift and precipitation, as we saw earlier in this chapter. Thus, as the ITCZ moves from 23.5° N in June to 23.5° S in December, it brings the wet season with it (**Figure 2.21**). For example, Tampico, Mexico (22°12′ N), reaches its maximum precipitation levels from July to October and has a dry season from November to April. In contrast, Vicosa, Brazil (20°45′ S), has a wet season from October to February and a dry season from April to August.

The temperate and polar zones experience pronounced changes in temperature associated with variation in solar radiation over the year. Summer occurs in the Northern Hemisphere in June, when that hemisphere is tilted toward the sun, while the Southern Hemisphere is oriented away from the sun and enters its winter. The difference in solar radiation, and thus the temperature variation, between summer and winter increases from the tropics toward the poles. The seasonal changes in the angle of the sun affect not only the intensity of solar radiation, but also the length of the day. Above 66.5° N and S latitudes, the sun does not set for several days, weeks, or even months in summer. At these same latitudes, the sun does not rise and warm the surface of Earth during portions of the winter. Because air temperatures regularly drop below freezing during winter in the temperate and polar zones, seasonality is an important determinant of biological activity and strongly influences the distributions of organisms in those zones.

Seasonal changes in aquatic environments are associated with changes in water temperature and density

Aquatic environments in the temperate and polar zones also experience seasonal changes in temperature, but as we have seen, they are not as extreme as those on land. Liquid water has the unique property of being most dense at 4°C. Ice has a lower density than liquid water and therefore forms on the surfaces of water bodies in winter. Because it has a higher albedo than open water, ice on the surface of lakes or polar oceans effectively prevents warming of the water below it.

Differences in water temperature with depth result in the **stratification**, or layering, of water in oceans and lakes. Stratification has important implications for aquatic organisms because it determines the movement of nutrients and oxygen. Surface waters in lakes and oceans mix freely, but are underlain by colder, denser layers of water that do not mix easily with the surface waters. In oceans, the surface waters mix with the subsurface layers only rarely—for example, in upwelling zones.

In temperate-zone lakes, seasonal changes in water temperature and density result in seasonal changes in stratification (**Figure 2.22**). In summer, the surface layer, or **epilimnion**, is the warmest and contains active popula-

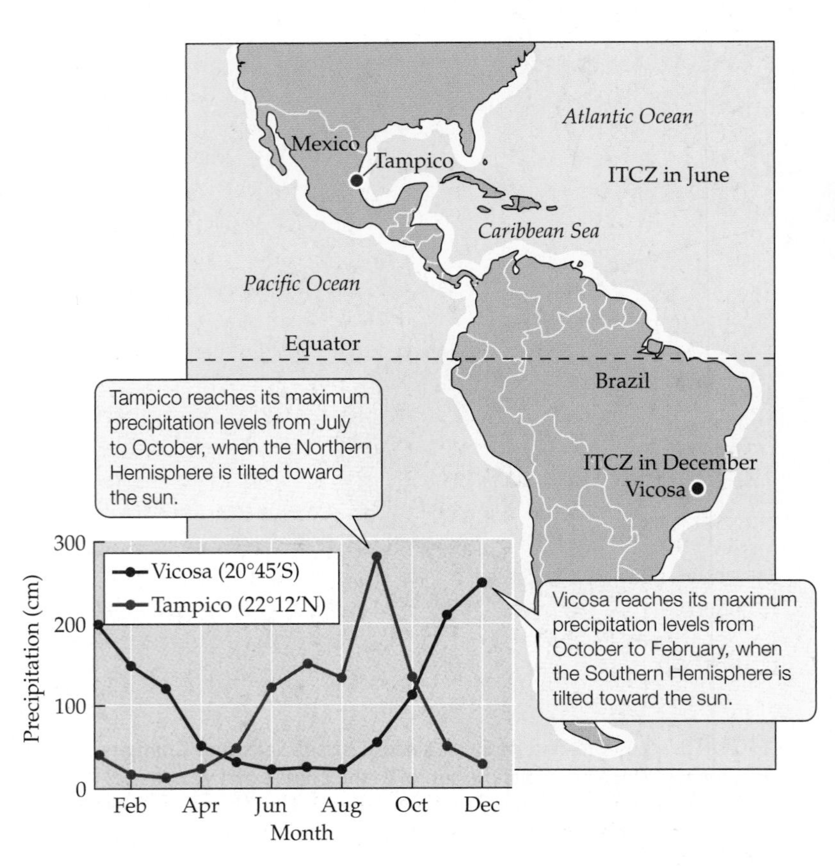

FIGURE 2.21 Wet and Dry Seasons and the ITCZ Seasonality of precipitation in the tropics is associated with movement of the Intertropical Convergence Zone (ITCZ) between the Northern and Southern hemispheres as Earth orbits the sun.

tions of phytoplankton and zooplankton. The epilimnion is underlain by a zone of rapid temperature decline, called the **thermocline**. Below the thermocline is a stable layer of the densest, coldest water in the lake, known as the **hypolimnion**. During the summer dead organisms from the epilimnion will drop to the hypolimnion and bottom (*benthic*) zone, carrying nutrients and energy away from the surface layers.

During the fall, the air above the water surface cools, and the lake loses heat to the atmosphere through emission of infrared radiation as well as latent (evaporation) and sensible (conduction and convection) heat transfers. As the epilimnion cools, its density increases until it is the same as that of the layers below it. Eventually, the water at all depths of the lake has the same temperature and density, and winds blowing on the surface lead to a mixing of epilimnion and hypolimnion, known as lake **turnover**.

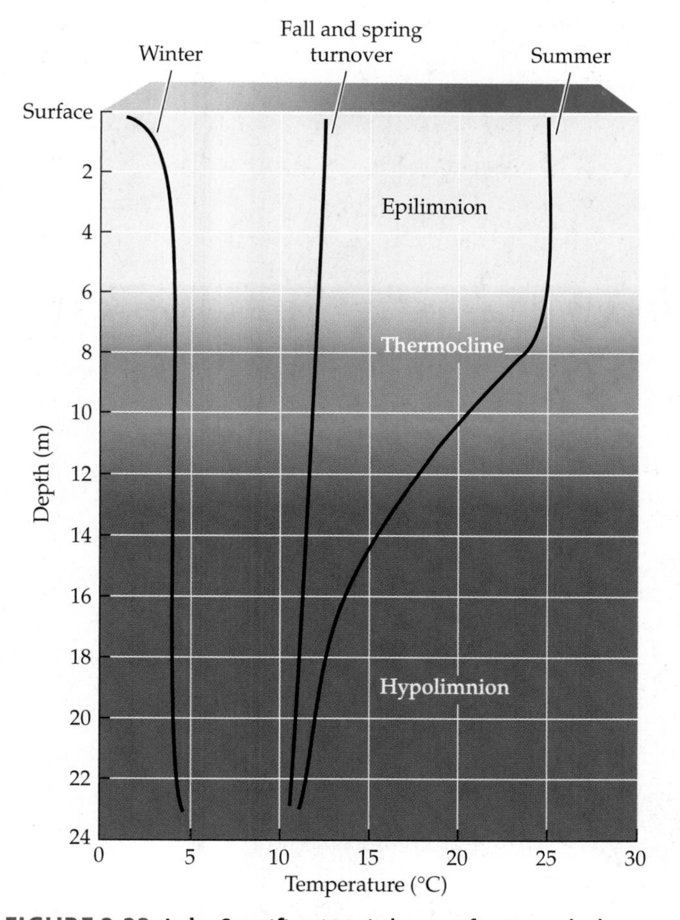

FIGURE 2.22 Lake Stratification Lake stratification, which is most apparent in summer in temperate and polar regions, results from the effects of temperature on water density. Seasonal changes in water temperature result in the turnover of water that mixes little during summer and winter.

Q Why would seasonal changes in lake stratification be unlikely to occur in tropical lakes?

This mixing is important for recycling of the nutrients that are lost from the epilimnion during summer. In addition, lake turnover moves oxygen into the hypolimnion and the sediments at the lake bottom. The replenishment of oxygen, which is used up by the respiration of aerobic bacteria during summer, increases biological activity in these deep lake zones. Turnover occurs again in spring when the surface ice melts and the lake water has a uniform density once again.

Climatic variation over years and decades results from changes in atmospheric pressure cells

Peruvian fisherman and other inhabitants of the west coast of South America have long been aware of times when the normally productive ocean waters hold few fish and the weather becomes extremely wet. They named these climatic episodes "El Niño," for the Christ child, because they usually started around Christmas. El Niño events are associated with a switch (or oscillation) in the positions of high-pressure and low-pressure cells over the equatorial Pacific, which leads to a weakening of the easterly trade winds that normally push warm water toward Southeast Asia. Climatologists refer to this oscillation and the climate changes associated with it as the **El Niño Southern Oscillation**, or **ENSO** for short. Its underlying causes are still not well understood. The frequency of ENSO is somewhat irregular, but it occurs at intervals of 3–8 years and generally lasts for about 18 months. During El Niño events, the upwelling of deep ocean water off the coast of South America ceases as the easterly winds weaken, or in some events, shift to westerly winds (**Figure 2.23A**). ENSO also includes **La Niña** events, which are stronger-than-average phases of the normal pattern, with high pressure off the coast of South America and low pressure in the western Pacific (**Figure 2.23B**). La Niña events usually follow El Niño events, but tend to be less frequent.

ENSO is associated with unusual climatic conditions, even at localities distant from the tropical Pacific, through its complex interactions with atmospheric circulation patterns (**Figure 2.23C,D**). El Niño events are associated with unusually dry conditions in Indonesia, other parts of Southeast Asia, and Australia. The likelihood of fires in the grasslands, shrublands, and forests of these areas increases as precipitation decreases and vegetation dries out. In contrast, in the southern United States and northern Mexico, El Niño events may increase precipitation, while the ensuing La Niña events bring drought conditions. The increase in plant growth associated with an El Niño event, followed by dry La Niña conditions, may intensify fires in the southwestern United States (Veblen et al. 2000).

Similar atmospheric pressure–ocean current oscillations occur in the North Atlantic Ocean. The **North Atlantic Oscillation** affects climatic variation in Europe,

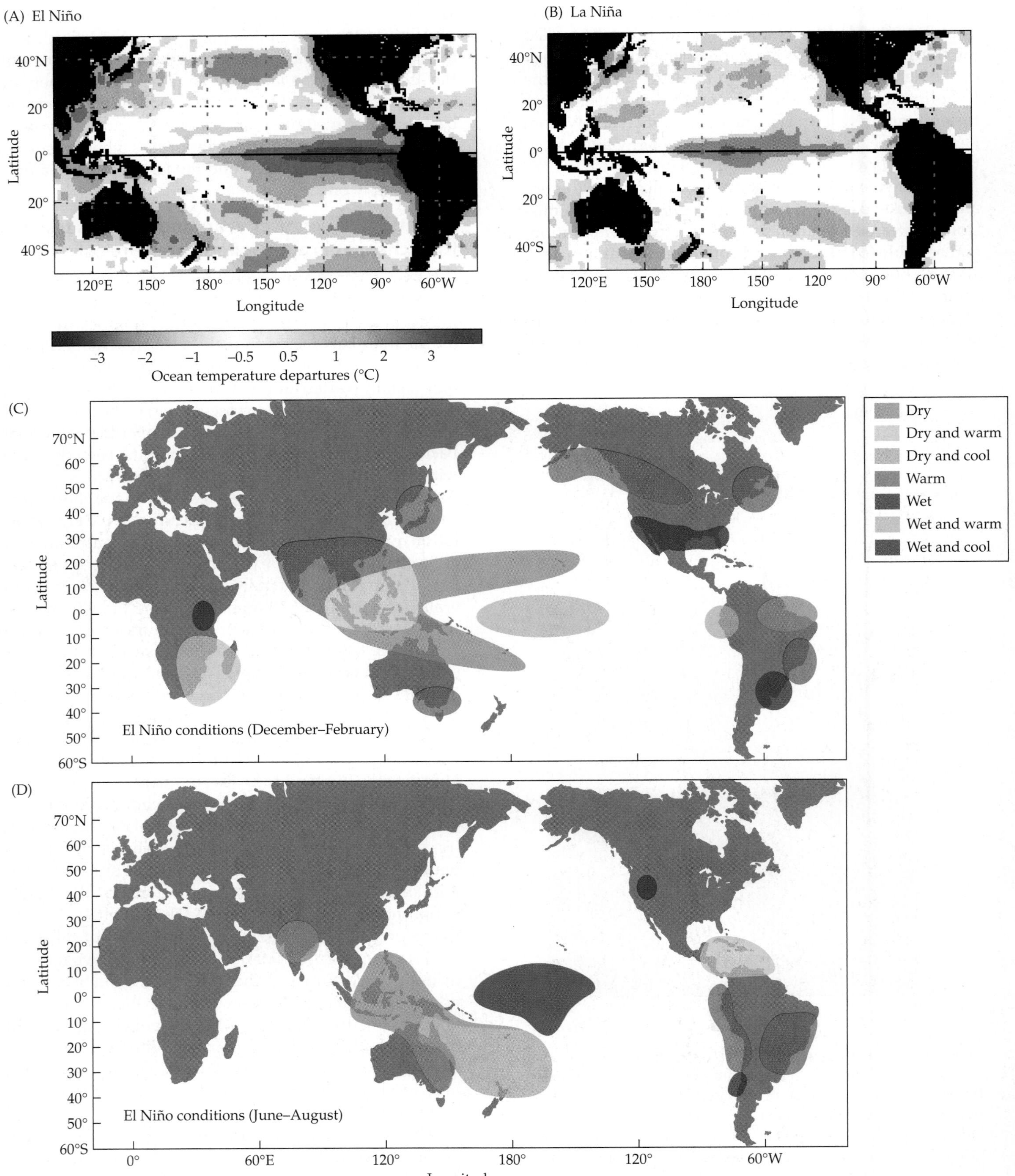
(A) El Niño
40°N
20°
0°
20°
40°S
Latitude
120°E
150°
180°
150°
120°
90°
60°W
Longitude
(B) La Niña
40°N
20°
0°
20°
40°S
Latitude
120°E
150°
180°
150°
120°
90°
60°W
Longitude
–3
–2
–1
–0.5
0.5
1
2
3
Ocean temperature departures (°C)
(C)
70°N
60°
50°
40°
30°
20°
10°
0°
10°
20°
30°
40°
50°
60°S
Latitude
El Niño conditions (December–February)
Dry
Dry and warm
Dry and cool
Warm
Wet
Wet and warm
Wet and cool
(D)
70°N
60°
50°
40°
30°
20°
10°
0°
10°
20°
30°
40°
50°
60°S
Latitude
El Niño conditions (June–August)
0°
60°E
120°
180°
120°
60°W
Longitude

FIGURE 2.23 The El Niño Southern Oscillation (ENSO) (A, B) Departures from long-term average ocean temperatures in equatorial waters of the Pacific under (A) El Niño conditions and (B) La Niña conditions. (C, D) El Niño events have widespread climatic effects that vary seasonally, altering temperature and precipitation patterns at a global scale.

in northern Asia, and on the east coast of North America. Another long-term oscillation in sea surface temperature and atmospheric pressure, known as the **Pacific Decadal Oscillation**, or **PDO**, has been discovered in the North Pacific. The PDO affects climate in ways similar to ENSO and can moderate or intensify the effects of ENSO. The effects of the PDO are felt primarily in northwestern North America, although southern parts of North America, Central America, Asia, and Australia may also be affected. We will return to the PDO at the end of this chapter.

Long-term climate change is associated with variation in Earth's orbital path

Antarctica today is almost completely covered with ice, and life is limited to its outer margins. Yet evidence from fossils indicates that there were once forests in interior parts of Antarctica and that dinosaurs once roamed the continent. Based on our knowledge of the biology of trees and dinosaurs, we can conclude that there must have been a warmer climate in the Antarctic past. What could have caused the climate to be so much warmer? The latitudinal position of Antarctica has shifted over geologic time (over hundreds of millions of years) due to *continental drift*, the movement of land masses across the surface of Earth as they float on the mantle beneath the crust. Continental drift could have resulted in a warmer or cooler climate as Antarctica moved closer to or farther away from the equator. However, the age of many of the Antarctic fossils, along with a reconstruction of the movements of the continent, indicates that Antarctica was in a latitudinal position similar to today's when the forests and dinosaurs occurred there (190–65 million years ago). Thus, Earth's climate must have once been much warmer than it is now.

Many lines of evidence indicate that Earth has experienced several episodes of warmer and cooler climates over the past 500 million years (**Figure 2.24**). During the course of Earth's history, the amount of radiation emitted by the sun has been gradually increasing, so that factor does not explain these irregular cycles of long-term global climate change. Hypotheses for the causes of these climatic shifts have focused on changes in the concentrations of greenhouse gases in the atmosphere. Warmer periods are associated with higher concentrations of greenhouse gases, while cooler periods are associated with lower concentrations of those gases.

Earth is currently in a cool climatic phase that has lasted about 3 million years. This phase has been characterized by regular periods of cooling, accompanied by the formation and advance of glaciers, followed by periods of warming, accompanied by glacial melting. The peaks of glacial advance are referred to as *glacial maxima*, and the periods of glacial melting and retreat are referred to *interglacial periods*. These glacial–interglacial cycles occur at frequencies of about 100,000 years. We are currently in an interglacial period; in the past, these periods have lasted

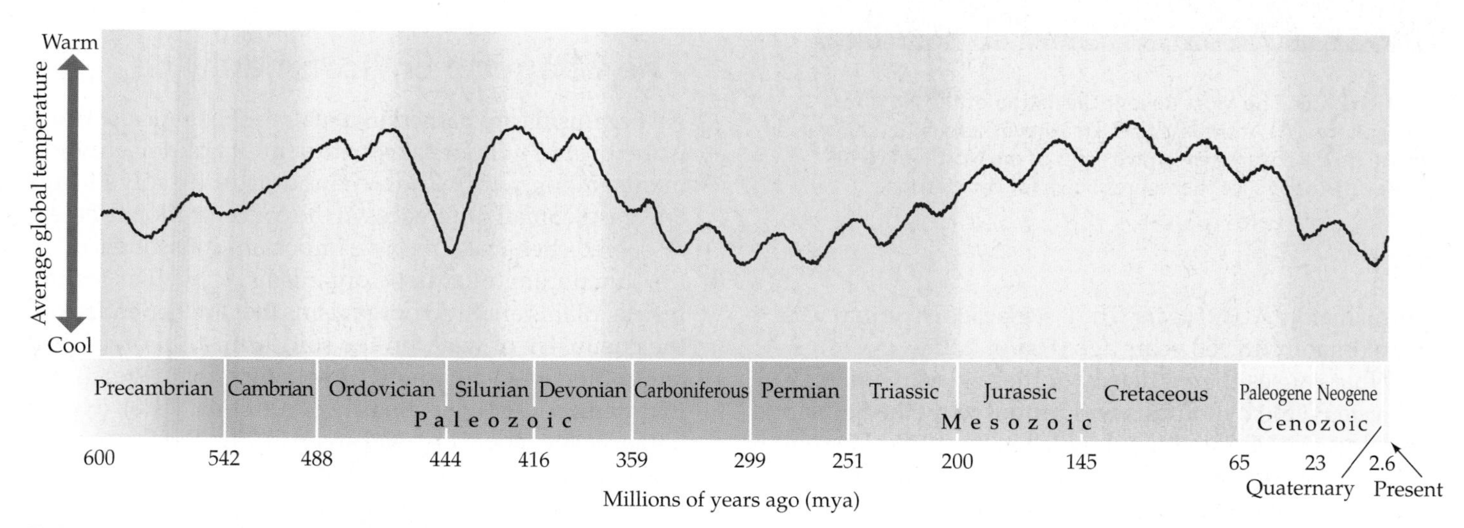

FIGURE 2.24 Long-Term Record of Average Global Temperature Global climatic patterns have alternated between warm and cool cycles for the past 500 million years.

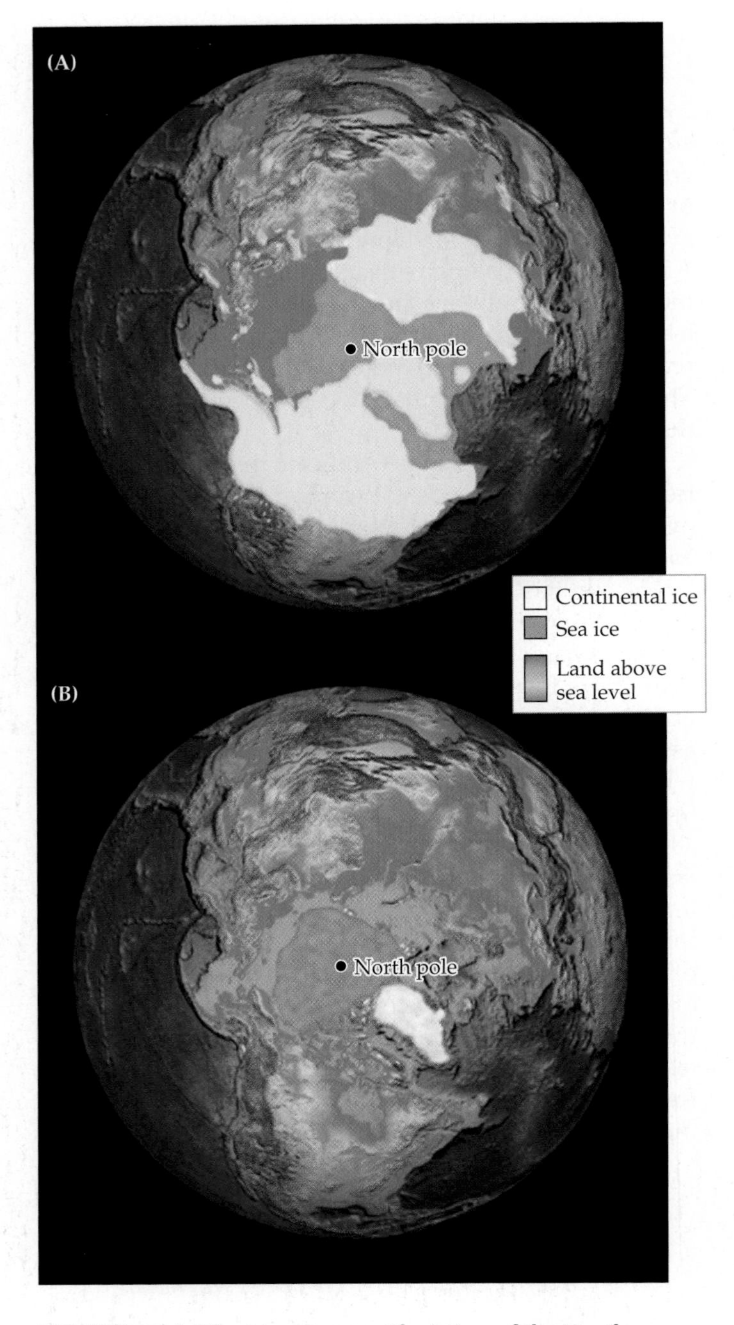

FIGURE 2.25 The Most Recent Glaciation of the Northern Hemisphere (A) At the last glacial maximum, about 18,000 years ago, ice sheets covered extensive areas of the Northern Hemisphere. (B) Today's ice sheets are shown for comparison.

approximately 23,000 years. The last glacial maximum was approximately 18,000 years ago (**Figure 2.25**).

What causes the regularity of these glacial–interglacial cycles? One hypothesis was proposed by the Serbian astrophysicist Milutin Milankovitch in the 1920s. He suggested that a combination of regular changes in the shape of Earth's orbit and the tilt of its axis, collectively known now as **Milankovitch cycles**, changed the intensity of solar radiation at high latitudes. Currently, the shape of Earth's orbit is nearly circular, but it shifts between a circular and a more elliptical shape in regular cycles of 100,000 years (**Figure 2.26A**). When Earth has a more elliptical orbit, the distance between Earth and the sun at the aphelion is greater, and thus the intensity of solar radiation is less, accentuating seasonal variation in climate. The angle of tilt of Earth's axis also changes, in regular cycles of about 41,000 years (**Figure 2.26B**); the greater the angle of tilt, the greater the seasonal variation in solar radiation at Earth's surface. Currently, the axis is tilted at an angle of 23.5°, which is near the middle of its range of variation (24.5°–22.1°). Finally, Earth's orientation relative to other celestial bodies changes in regular cycles of about 22,000 years. Today, the North Pole is oriented toward Polaris, the North Star, but that has not always been the case; there have been other "North Stars" in the past. These changes in Earth's orientation influence the timing of the seasons by determining the hemisphere that receives more solar radiation (**Figure 2.26C**).

Long-term climatic cycles correlate well with the Milankovitch cycles (Hays et al. 1976). The dominant glacial–interglacial cycle has a period of 100,000 years, corresponding to the changes in the shape of Earth's orbit. Indications of smaller climatic fluctuations with 41,000-year and 22,000-year cycles have also been observed.

These changes in climate over time have had profound effect on the distributions of organisms, as we will see in Chapter 3. But climate is not the only factor that determines where organisms can live. The chemical environment also plays an important role.

CONCEPT 2.6 Salinity, acidity, and oxygen concentrations are major determinants of the chemical environment.

The Chemical Environment

All organisms are bathed in a matrix of chemicals. Water is the primary chemical constituent of aquatic environments, along with variable amounts of dissolved salts and gases. Small differences in the concentrations of these dissolved chemicals can have important consequences for the functioning of aquatic organisms, as well as for terrestrial plants and microorganisms that are dependent on the chemistry of water in the soil. Terrestrial organisms are immersed in a gaseous atmosphere that is relatively invariant, consisting primarily of nitrogen (78%), oxygen (20%), water vapor (1%), and argon (0.9%). The atmosphere also contains trace gases, including the greenhouse gases that play a critical role in Earth's energy balance, and pollutants derived from human activities, which can have important effects on atmospheric chemistry. We will dis-

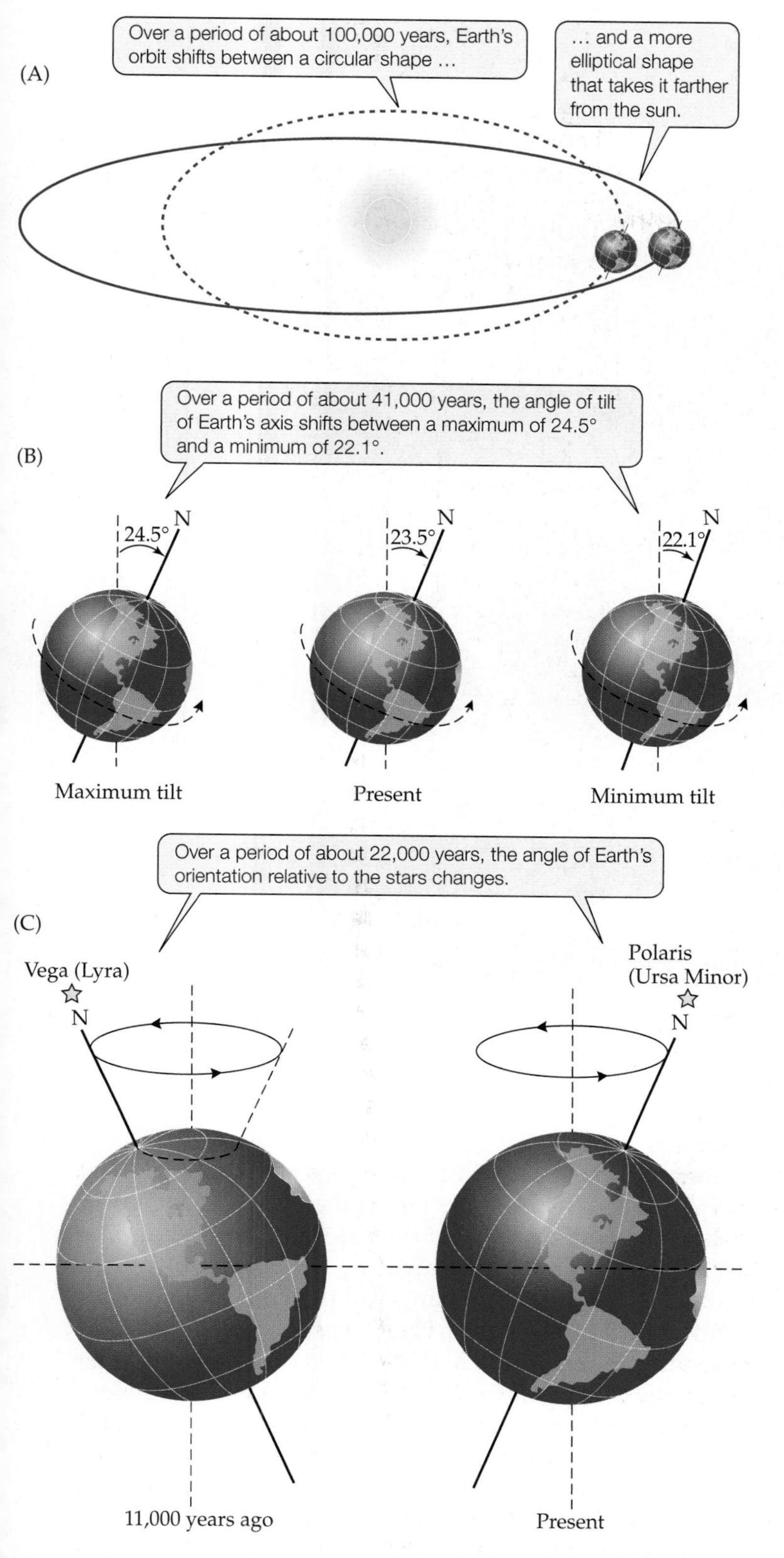

FIGURE 2.26 Milankovitch Cycles and Long-Term Climatic Variation Variations in (A) the shape of Earth's orbit around the sun, (B) the tilt of Earth's axis of rotation, and (C) the orientation of the axis to other celestial objects all affect the intensity and distribution of solar radiation striking Earth and correlate with glacial–interglacial cycles over the past 400,000 years.

Q What combination of conditions associated with these three cycles would promote the occurrence of a glacial period?

cuss the effects of air pollutants and greenhouse gases in Chapter 24. Here, we will review three chemical variables that influence biological function: salinity, acidity, and the availability of oxygen.

All waters contain dissolved salts

Salinity refers to the concentration of dissolved salts in water. *Salts* are ionic compounds composed of cations (positively charged ions) and anions (negatively charged ions) that disassociate when placed in water. Dissolved salts are important from a biological perspective because they influence properties of water that affect the ability of organisms to absorb it. The salts also have direct influences on organisms as nutrients (as we will see in Chapter 21) and can inhibit metabolic activity if their concentrations are too high or too low.

Although all waters contain dissolved salts, we often think about salinity in the context of ocean water because it is one of the saltiest solutions we come into contact with. Seventy percent of Earth's surface is ocean water, and the oceans contain 97% of the water on Earth. The salinity of the oceans varies between 33 and 37 parts per thousand; this variation is a result of evaporation, precipitation, and the freezing and melting of sea ice (**Figure 2.27**). The salinity of ocean surface waters is highest near the equator and lowest at high latitudes.

What are the salts that make water saline, and where do they come from? Ocean salts consist mainly of sodium, chloride, magnesium, calcium, sulfate, bicarbonate, and potassium. These salts come from gases emitted by volcanic eruptions early in Earth's history, when its crust was cooling, and from the gradual breakdown of minerals in the rocks that make up the crust. Most landlocked bodies of water become more saline over time, reflecting a balance between water inputs from precipitation, water losses due to evaporation, and inputs of salts. When these inland "seas" occur in arid areas, as do the Great Salt Lake and the Dead Sea, their salinity may exceed that of ocean water. The types of salts that contribute to their salinity vary, reflecting the chemistry of the minerals in the rocks that make up their basins.

High levels of salinity occur naturally in the water in soils adjacent to oceans, such as those in salt marshes. Soils may also become more saline in arid regions as water from deeper soil layers is brought to the surface by plant roots or through pumping of groundwater for irrigation. As this water evaporates, it leaves behind the salts it contains. If there is little precipitation to leach the salts to deeper soil layers, or if drainage of the water is impeded by impervious layers beneath the soil, high rates of evapotranspiration will result in a progressive buildup of salts at the soil surface. This process, known as **salinization**, occurs naturally in some desert soils, but may also occur

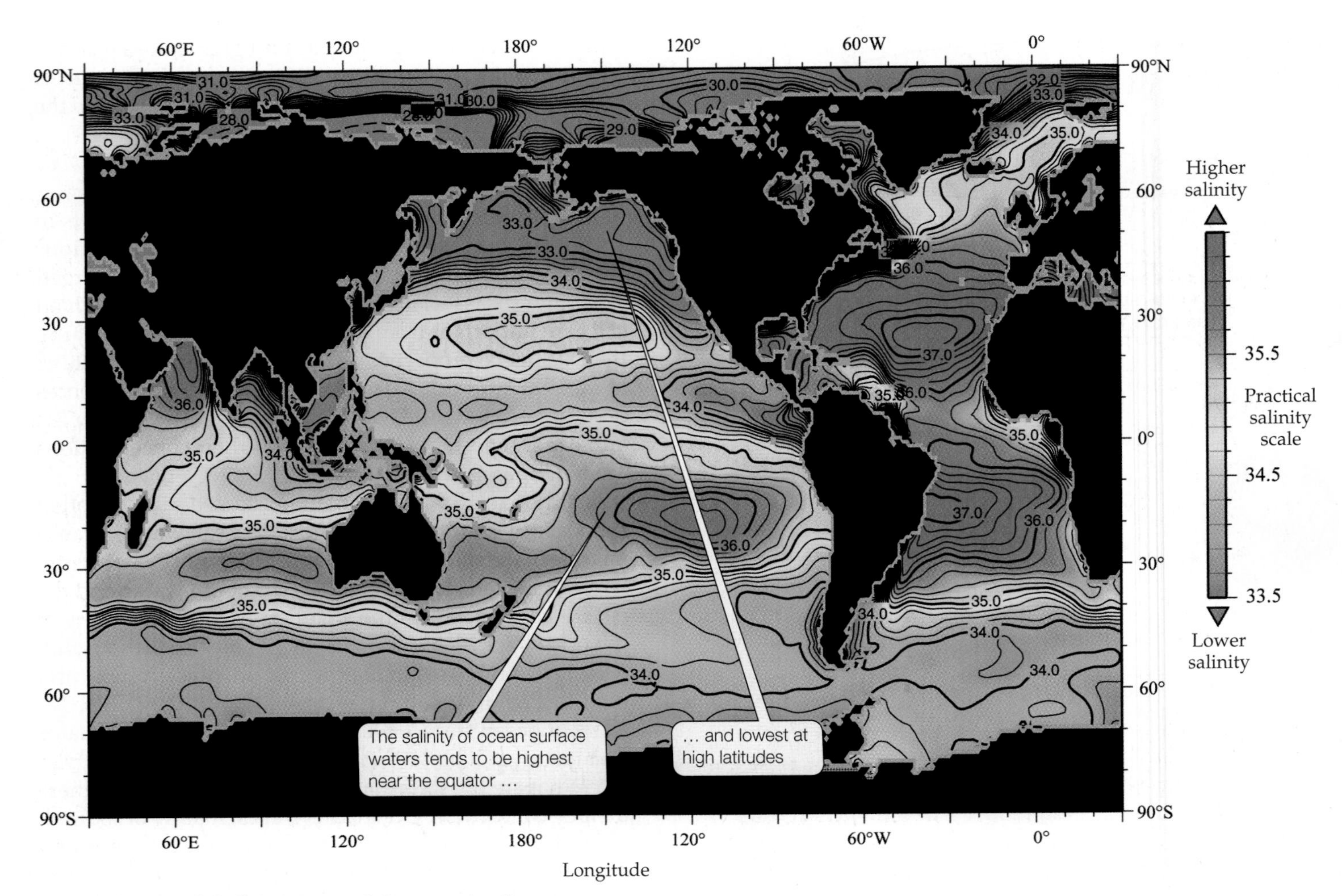

FIGURE 2.27 Global Variation in Salinity at the Ocean Surface Variations in the salinity of ocean surface water reflect the concentrating effect of evaporation, dilution by melting sea ice, and precipitation.

in irrigated agricultural soils of arid regions (**Figure 2.28**). Salinization contributed to agricultural decline in ancient Mesopotamia (modern-day Iraq) and is a problem today in California's Central Valley, Australia, and other regions.

Organisms are sensitive to the acidity of their environment

Acidity, and its converse, **alkalinity**, are measures of the ability of a solution to behave as an acid or a base, respectively. *Acids* are compounds that give up protons (H^+) to the water they are dissolved in. *Bases* take up protons or give up hydroxide ions (OH^-). Examples of common acids include the citric, tannic, and ascorbic acids found in fruits. Examples of common bases include sodium bicarbonate (baking soda) and other carbonate minerals in rock. Acidity and alkalinity are measured in pH, which is equal to the negative of the logarithm ($-\log_{10}$) of the concentration of H^+. Thus, one pH unit represents a tenfold change in the concentration of H^+. Pure water has a neutral pH of 7.0. Solutions with pH values higher than 7.0 are alkaline (basic), and solutions with pH values lower than 7.0 are acidic.

The pH values of water have important effects on organismal function. Changes in pH values can directly affect metabolic activity. The pH values of water also determine the chemistry and availability of nutrients, as we will see in Chapter 21. Organisms have a limited range of pH values that they can tolerate. Natural levels of alkalinity (when the pH of the environment exceeds 7) tend not to be as important as levels of acidity as a constraint on organismal function and distributions.

In the oceans, pH does not vary appreciably because the chemistry of seawater acts to buffer changes in pH. The salts in seawater can bind free protons and thereby minimize changes in pH. Thus, pH tends to be most important in terrestrial and freshwater ecosystems. However, increases in atmospheric CO_2 concentrations are expected to increase the acidity of the oceans appreciably during the next century, with potentially negative effects on marine ecosystems (Orr et al. 2005).

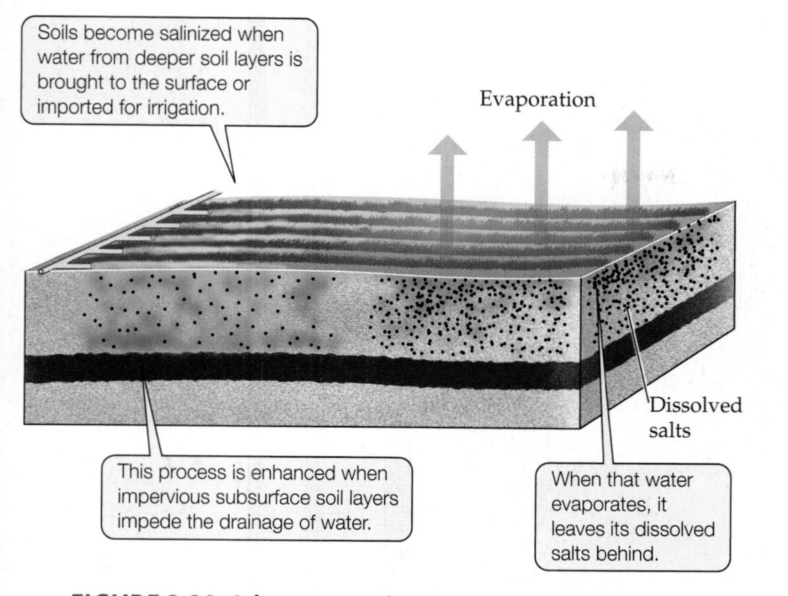

FIGURE 2.28 Salinization Salinization of soils is disrupting agricultural production in many areas, especially in arid regions.

On land, the pH of surface waters and soils varies naturally. What causes this variation? Water can become more acidic over time through the input of acidic compounds derived from several sources, most associated with soil development (which is covered in more detail in Chapter 21). Two of the main components of soil are mineral particles from the breakdown of rocks and organic matter from the decomposition of dead plants and other organisms. Some rock types, such as granites, generate acidic salts, while other rock types generate basic salts. Soils become more acidic as they age because the basic salts leach away more easily and because decomposition and leaching of plant matter adds organic acids to the soil. The emission of acidic pollutants into the atmosphere by the burning of fossil fuels, as well as overuse of agricultural fertilizers, can increase the acidity of soil and water. We will cover these sources of acidity in more detail in Chapter 24.

Oxygen concentrations vary with elevation, diffusion, and consumption

When life on Earth first evolved, there was no oxygen in the atmosphere, and oxygen was toxic to the earliest forms of life. Even today, there are many organisms that are intolerant of oxygen. However, with the exception of some archaea, bacteria, and fungi, most organisms require oxygen to carry out their metabolic processes and cannot survive in **hypoxic** (low-oxygen) conditions. Hypoxic conditions can also promote the formation of chemicals (e.g., hydrogen sulfide) that are toxic to many organisms. In addition, oxygen levels are important for chemical reactions that determine the availability of nutrients.

Oxygen concentrations in the atmosphere have been stable at about 21% for the past 65 million years, so most terrestrial environments have invariant oxygen concentrations. However, the availability of atmospheric oxygen decreases with elevation above sea level. As we have seen, the overall density of air decreases with elevation, so there are fewer molecules of oxygen in a given volume of air at higher elevations. We will discuss the repercussions of this variation for human health in Chapter 4.

Oxygen concentrations can vary substantially in aquatic environments and in soils. The rate of diffusion of oxygen into water is slow and may not keep pace with its consumption by organisms. Waves and currents mix oxygen from the atmosphere into ocean surface waters, so its concentration is usually stable there. Oxygen concentrations are low in the deep ocean and in marine sediments, where biological uptake is greater than replenishment from surface waters. The same holds true in deep lakes, lake sediments, and flooded soils (e.g., in wetlands). Oxygen concentrations are highest in freshwater ecosystems with moving water (streams and rivers) because mixing with the atmosphere is greatest there.

A CASE STUDY REVISITED

Climatic Variation and Salmon Abundance

The research of Steven Hare and Robert Francis on salmon production in the North Pacific contributed to the discovery of the Pacific Decadal Oscillation (PDO). As noted earlier, the PDO is a multi-decadal shift in sea surface temperature and atmospheric pressure cells. A review of existing records of sea surface temperatures during the twentieth century indicated that the PDO was associated with alternating 20–30-year periods of warm and cool temperatures in the North Pacific (**Figure 2.29**). The length of the phases of the PDO differentiate it from other climatic oscillations whose phases tend to be much shorter (e.g., 18 months–2 years for ENSO). The warm and cool phases of the PDO had influenced the marine ecosystems that Pacific salmon depended on, and had thus shifted salmon production north or south, depending on the phase.

The PDO has been linked to changes in the abundances and distributions of many marine organisms and, through its climatic effects, in the functioning of terrestrial ecosystems (Mantua and Hare 2002). Its effects have been found primarily in western North America and eastern Asia, but effects have also been reported in Australia. Thus, the influence of the PDO on climate extends throughout the Western Hemisphere. Evidence for the existence of climate changes associated with the PDO dates back to the 1850s, in the form of instrumental temperature records, and to the 1600s, in the form of information from corals and tree rings. The underlying mechanisms that cause the PDO are unclear, but its effect on climate is significant and widespread (**Table 2.1**).

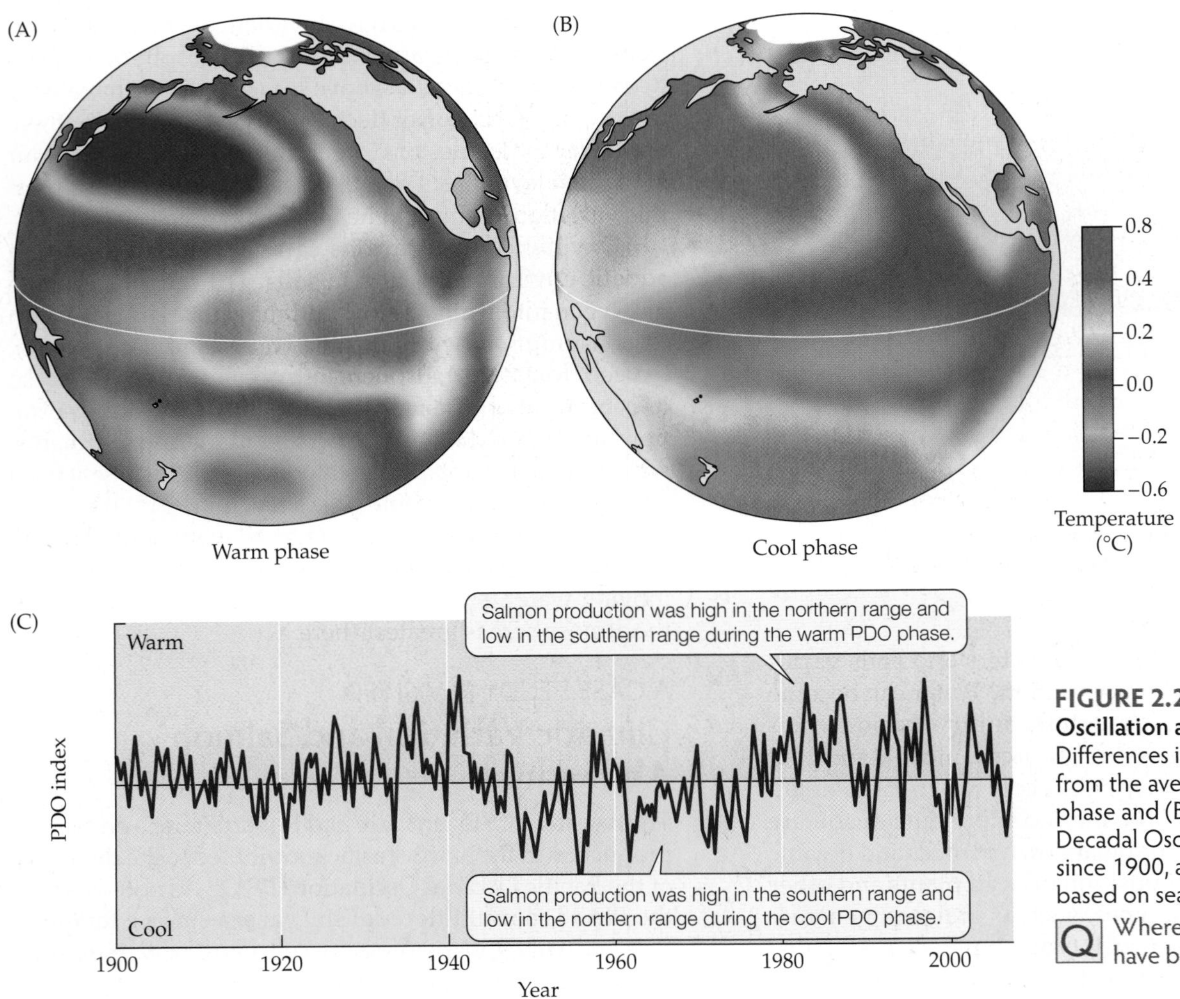

FIGURE 2.29 The Pacific Decadal Oscillation and Ocean Temperatures Differences in ocean surface temperatures from the average during (A) the warm phase and (B) the cool phase of the Pacific Decadal Oscillation (PDO). (C) PDO phases since 1900, as estimated using a PDO index based on sea surface temperatures.

Q Where would the best salmon fishing have been found in 1987?

TABLE 2.1

Summary of Climatic Effects of the Pacific Decadal Oscillation (PDO)

Climatic effect	Warm phase PDO	Cool phase PDO
Ocean surface temperature in the northeastern and tropical Pacific	Above average	Below average
October–March northwestern North American air temperature	Above average	Below average
October–March southeastern U.S. air temperature	Below average	Above average
October–March southern U.S./northern Mexico precipitation	Above average	Below average
October–March northwestern North American and Great Lakes precipitation	Below average	Above average
Northwestern North American spring snowpack and water year (October–September stream flow)	Below average	Above average
Winter and spring flood risk in the Pacific Northwest	Below average	Above average

Source: Mantua 2001.

CONNECTIONS IN NATURE
Climatic Variation and Ecology

Two aspects of the PDO are particularly important in the context of ecology. First, the realization that the PDO existed was driven initially by an attempt to understand variation in the size of an animal population. This observation underscores the relationship between physical conditions (the topic of this chapter), the functioning of individual organisms and their growth and reproduction (Chapters 4 and 5), and population and community processes (Units 2 and 4, respectively). This relationship is one of the central themes of ecology that will form a common thread throughout this book. Ultimately, the physical environment, including climate and the myriad factors, such as the PDO, that control it, determines whether an organism can exist in a given location (as we'll see in Chapter 3). Extremes in the physical environment, including those that are driven by climatic oscillations, play a critical role in our understanding of ecological phenomena.

Second, the time scale of the climatic variation associated with the PDO is long relative to the human life span. The abrupt changes in climate, and the associated ecological responses of the marine ecosystem, were therefore perceived by people as unusual events. Indeed, the phases of the PDO may be longer than the life spans of most of the organisms affected by it, limiting their ability to adapt to this climatic oscillation. As a result, from the perspective of an ecological community, the PDO represents a disturbance, an event that detrimentally affects some species' populations and disrupts the community.

Although we don't yet understand what causes it, the PDO has been a part of the climate system for at least the last 400 years, and a better understanding of its effects will help us place other climatic phenomena, including global climate change, in perspective.

SUMMARY

CONCEPT 2.1 Climate is the most fundamental characteristic of the physical environment.

- Weather refers to the current conditions of temperature, precipitation, humidity, wind, and cloud cover. Climate is the long-term average weather at a given location.
- Climate determines the geographic distribution and physiological functioning of organisms.
- The climate system is driven by the balance between energy gains from solar radiation and reradiation by the atmosphere and energy losses due to infrared radiation from Earth's surface, latent heat flux, and sensible heat flux.

CONCEPT 2.2 Winds and ocean currents result from differences in solar radiation across Earth's surface.

- Latitudinal differences in the intensity of solar radiation at Earth's surface establish atmospheric circulation cells.
- The Coriolis effect and the difference in heat capacity between the oceans and the continents act on atmospheric circulation cells to determine the pattern of prevailing winds at Earth's surface.
- Ocean currents are driven by surface winds and by differences in water temperature and salinity.
- Winds and ocean currents transfer energy from the tropics to higher latitudes.

CONCEPT 2.3 Large-scale atmospheric and oceanic circulation patterns establish global patterns of temperature and precipitation.

- Global temperature patterns are determined by latitudinal variation in solar radiation, but are also influenced by oceanic circulation patterns and by the distribution of continents.
- Temperature decreases as the elevation of the land surface increases.
- Global patterns of terrestrial precipitation are determined by atmospheric circulation cells, but are also influenced by semipermanent pressure cells.

CONCEPT 2.4 Regional climates reflect the influence of the distribution of oceans and continents, mountains, and vegetation.

- Seasonal variation in temperature is greater in the middle of a continent than on the coast because ocean water has a higher heat capacity than land.
- Mountains force air masses passing over them to rise and drop most of their moisture as precipitation, resulting in moister environments on windward slopes and drier environments on leeward slopes.
- Vegetation influences regional climates through its effects on energy exchange associated with albedo, evapotranspiration (latent heat transfer), and surface winds (sensible heat transfer).

SUMMARY (continued)

CONCEPT 2.5 Seasonal and long-term climatic variation are associated with changes in Earth's position relative to the sun.

- The tilt of Earth's axis as it orbits the sun causes seasonal changes in precipitation in tropical regions and in temperature and day length in temperate and polar regions.
- Temperature-induced differences in water density result in nonmixing layers of water in oceans and lakes. In temperate-zone lakes, these layers break down in fall and spring, allowing the movement of oxygen and nutrients.
- Variations in climate over years to decades are caused by cyclic changes in atmospheric pressure cells. These changes have widespread effects beyond the regions where the pressure cells are located.
- Long-term climatic cycles over hundreds and thousands of years are associated with changes in the shape of Earth's orbit, the angle of tilt of its axis, and Earth's orientation to other celestial bodies.

CONCEPT 2.6 Salinity, acidity, and oxygen concentrations are major determinants of the chemical environment.

- The salinity of Earth's waters, including water in soils, is determined by the balance between inputs of salts and gains (by precipitation) and losses (by evaporation) of water.
- The pH of soils and surface waters is determined by inputs of salts from the breakdown of rock minerals, organic acids from plants, and acidic pollutants.
- Oxygen concentrations are stable in most terrestrial ecosystems, but oxygen availability decreases as elevation increases. Concentrations of oxygen in aquatic ecosystems are low where its consumption by organisms exceeds its slow rate of diffusion into water.

REVIEW QUESTIONS

1. Why is the variability of physical conditions potentially more important than average conditions as a determinant of ecological patterns, such as species distributions?
2. Describe the factors that determine the major latitudinal climatic zones (the tropics, temperate zones, and polar zones).
3. How do the following features affect regional terrestrial climates?
   a. Mountains
   b. Oceans
4. Why would deserts be more prone to salinization from irrigation than areas with greater precipitation?

ON THE COMPANION WEBSITE
sites.sinauer.com/ecology2e

The website includes Chapter Outlines, Online Quizzes, Flashcards & Key Terms, Suggested Readings, a complete Glossary, and the Web Stats Review. In addition, the following resources are available for this chapter:

HANDS-ON PROBLEM SOLVING

Simulating Seasons

This Web exercise illustrates connections between the axial tilt of Earth and temperature variation. Seasonal patterns and the range of temperature variations result from the degree of axial tilt. You will use a simulation model of Earth to vary axial tilt and explore latitudinal and seasonal variations in temperature.

CLIMATE CHANGE CONNECTIONS

2.1 The Importance of Extreme Events to Ecological Responses to Climate Change

CHAPTER

3

The Biosphere

KEY CONCEPTS

CONCEPT 3.1 Terrestrial biomes are characterized by the growth forms of the dominant vegetation.

CONCEPT 3.2 Biological zones in freshwater ecosystems are associated with the velocity, depth, temperature, clarity, and chemistry of the water.

CONCEPT 3.3 Marine biological zones are determined by ocean depth, light availability, and the stability of the bottom substrate.

The American Serengeti—Twelve Centuries of Change in the Great Plains: A Case Study

Today, the region covering the central part of North America, known as the Great Plains, bears little resemblance to the Serengeti Plain of Africa. Biological diversity is very low in many parts of the current landscape, which contains large stands of uniform crop plants (which are often even genetically identical) and a few species of domesticated herbivores. In the Serengeti, on the other hand, some of the largest and most diverse herds of wild animals in the world roam a picturesque savanna (**Figure 3.1**). If not for a series of important environmental changes, however, the two ecosystems might superficially look very similar.

Temperate and high-latitude biological communities have been subjected to natural, long-term climate change, leading to latitudinal or elevational shifts in their positions and species composition. Eighteen thousand years ago, at the last glacial maximum of the Pleistocene epoch, ice sheets covered the northern portion of North America. Over the next 12,000 years, the climate warmed and the ice receded. Vegetation followed the retreating ice northward and colonized the newly exposed substrate. Grasslands in the center of the continent expanded into former spruce and aspen woodlands. These grasslands contained species of grasses, sedges, and low-growing herbaceous plants similar to those found in the natural grasslands that exist today.

The animal inhabitants of those earlier grasslands were, however, strikingly different from today's. A diverse collection of *megafauna* (animals larger than 45 kg, or 100 pounds) existed in North America, rivaling the diversity found today in the Serengeti (Martin 2005) (**Figure 3.2**). Thirteen thousand years ago—a relatively short time in an evolutionary context—North American herbivores included woolly mammoths and mastodons (relatives of elephants), as well as several species of horses, camels, and giant ground sloths. Predators included saber-toothed cats with 18 cm (7-inch) incisors, cheetahs, lions, and giant short-faced bears that were larger and faster than grizzly bears.

About 10,000–13,000 years ago, as the extensive grasslands of the Great Plains were developing, many of the large mammals of North America suddenly went extinct (Barnosky et al. 2004). The rapidity of the disappearance of approximately 28 genera (40–70 species) made this extinction unlike any previous extinction event during the previous 65 million years. Another unusual aspect of this extinction was that nearly all of the animals that went extinct belonged to the same group: large mammals. The causes of this extinction are a mystery to paleontologists

FIGURE 3.1 The Serengeti Plain of Africa Large, diverse herds of native animals migrate across the Serengeti in search of food and water.

FIGURE 3.2 Pleistocene Animals of the Great Plains The animals of the grasslands of central North America 13,000 years ago included woolly mammoths, horses, and giant bison. Many of these large mammals went extinct within a short time between 13,000 and 10,000 years ago.

(scientists who investigate the development of life on Earth through the study of fossils and other ancient remains).

Several hypotheses have been proposed to account for the disappearance of the North American megafauna. Changes in the climate during the extinction period were rapid and could have led to changes in habitat or food supply that would have negatively affected the animals. Another hypothesis that has generated substantial controversy suggests that the arrival of humans in North America may have hastened the demise of the animals (Martin 1984). When this hypothesis was first proposed, it was met with widespread skepticism, and the initial supporting evidence was considered weak. Although humans first appeared in the central part of North America about 14,000 years ago, it is unclear how hunters bearing stone and wooden tools could have driven so many species of large mammals to extinction. What evidence is there to support the hypothesis that humans were involved in this extinction event?

Introduction

Living things can be found in remarkable places. Birds such as ravens, lammergeyers (Eurasian vultures), and alpine choughs (crows) fly over the highest summits of the Himalayas, over 8,000 m above sea level. Fish (e.g., *Abyssobrotula galatheae*, the "fangtooth") can be found over 8,000 m below the ocean surface. Bacteria and archaea can be found almost everywhere on Earth, in hot sulfur springs at the extreme chemical and temperature limits for life, under glaciers, and on dust particles many kilometers above Earth's surface and kilometers deep in ocean sediments. However, most living things occur in a range of habitats that cover a thin veneer of Earth's surface, from the tops of trees to the surface soil layers in terrestrial environments and within 200 m of the surface of the oceans.

The **biosphere**, the zone of life on Earth, is sandwiched between the *lithosphere*, Earth's surface crust and upper mantle, and the *troposphere*, the lowest layer of the atmosphere. In this chapter, we provide an introduction to the amazing diversity of life on Earth. Biological communities can be studied at multiple scales of varying complexity, as we saw under Concept 1.2. Here, we will use **biomes** as the unit to introduce the diversity of terrestrial life on Earth. The diversity of aquatic life is not as easily categorized, but we will describe several freshwater and marine biological zones, which, like terrestrial biomes, reflect the physical conditions where they are found.

CONCEPT 3.1 Terrestrial biomes are characterized by the growth forms of the dominant vegetation.

Terrestrial Biomes

Biomes are large-scale biological communities shaped by the physical environment in which they are found. In particular, they reflect the climatic variation described in Chapter 2. Biomes are categorized by the most common forms of plants distributed across large geographic areas. The categorization of biomes does not take taxonomic relationships among organisms into account; instead, it relies on similarities in the morphological responses of organisms to the physical environment. Biomes occur as similar biotic assemblages on distant continents, indicating similar responses to climatic forces in different locations. In addition to providing a useful introduction to the diversity of life on Earth, the biome concept provides a convenient biological unit for modelers simulating the effects of environmental change on biological communities as well as the effects of vegetation on the climate system (see the discussion under Concept 2.4).

Terrestrial communities vary considerably, from the warm, wet tropics to the cold, dry polar regions. Tropical forests have multiple verdant layers, high growth rates, and tremendous species diversity. Lowland tropical forests in Borneo have an estimated 10,000 species of vascular plants, and most other tropical forest communities have about 5,000 species. In contrast, polar deserts have a scattered cover of tiny plants clinging to the ground, reflecting a harsh climate of high winds, low temperatures, and dry soils. High-latitude Arctic communities contain about 100 species of vascular plants. Tropical rainforest vegetation may reach over 75 m (250 feet) in height and contain over 400,000 kg of aboveground biomass in a single hectare (about 2.5 acres). Plants of polar deserts, on the other hand, rarely exceed 5 cm (2 inches) in height and contain less than 1,000 kg of surface biomass per hectare.

Terrestrial biomes are classified by the *growth form* (size and morphology) of the dominant plants (for example trees, shrubs, or grasses) (**Figure 3.3**). Characteristics of their leaves, such as *deciduousness* (seasonal shedding of leaves), thickness, and *succulence* (development of fleshy water storage tissues), may also be used. Why use plants rather than animals to categorize terrestrial biomes? Plants are immobile, so in order to occupy a site successfully for a long time, they must be able to cope with its environmental extremes as well as its biological pressures (such as competition for water, nutrients, and light). Plant growth forms are therefore good indicators of the physical environment, reflecting the climatic zones discussed in Chapter 2 as well as rates of disturbance (e.g., fire frequency). In addition, animals are a less visible component of most large landscapes, and their mobility allows them to avoid exposure to adverse environmental conditions. Microorganisms (archaea, bacteria, and fungi) are important components of biomes, and the composition of microbial communities reflects physical conditions in a manner similar to plant growth forms. However, their tiny size, as well as rapid temporal and spatial changes in their composition, makes them impractical for classifying biomes.

FIGURE 3.3 Plant Growth Forms The growth form of a plant is an evolutionary response to the environment, particularly climate and soil fertility.

Since their emergence from the oceans about 500 million years ago, plants have taken on a multitude of different forms in response to the selection pressures of the terrestrial environment (see Figure 3.3). These selection pressures include aridity, high and subfreezing temperatures, intense solar radiation, nutrient-poor soils, grazing by animals, and crowding by neighbors. Having deciduous leaves, for example, is one solution to seasonal exposure to subfreezing temperatures or extended dry periods. Trees and shrubs invest energy in woody tissues in order to increase their height and ability to capture sunlight and to protect their tissues from damage by wind or large amounts of snow. Perennial grasses, unlike most other plants, can grow from the bases of their leaves and keep their vegetative and reproductive buds below the soil surface, which facilitates their tolerance of grazing, fire, sub-freezing temperatures, and dry soils. Similar plant growth forms appear in similar climatic zones on different continents, even though the plants may not be genetically related. The evolution of similar growth forms among distantly related species in response to similar selection pressures is called **convergence**.

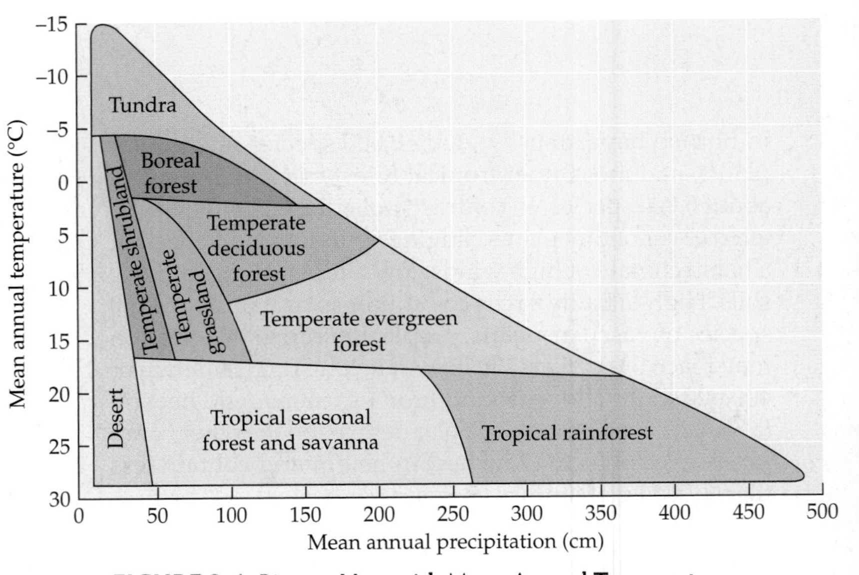

FIGURE 3.4 Biomes Vary with Mean Annual Temperature and Precipitation When plotted on a graph of temperature and precipitation, the nine major terrestrial biomes form a triangle.

Q What factor(s) might result in grasslands or shrublands "invading" climate space occupied by forest or savanna?

Terrestrial biomes reflect global patterns of precipitation and temperature

Chapter 2 described Earth's climatic zones and their association with the atmospheric and oceanic circulation patterns that result from the differential heating of Earth's surface by the sun. These climatic zones are major determinants of the distribution of terrestrial biomes.

The tropics (between 23.5° N and S) are characterized by high rainfall and warm, invariant temperatures. In the subtropical regions that border the tropics, rainfall becomes more seasonal, with pronounced dry and wet seasons. The major deserts of the world are associated with the zones of high pressure at about 30° N and S and with the rain-shadow effects of large mountain ranges. Subfreezing temperatures during winter are an important climatic feature of the temperate and polar zones. The amount of precipitation north and south of 40° varies depending on proximity to the ocean and the influence of mountain ranges (see Figure 2.16).

The locations of terrestrial biomes are correlated with these variations in temperature and precipitation. Temperature influences the distribution of plant growth forms directly through its effect on the physiological functioning of plants. Precipitation and temperature act in concert to influence the availability of water and its rate of loss by plants. Water availability and soil temperature are important in determining the supply of nutrients in the soil, which is also an important control on plant growth form.

The association between climatic variation and terrestrial biome distribution can be visualized using a graph of average annual precipitation and temperature (**Figure 3.4**). While these two factors predict biome distributions reasonably well, this approach fails to incorporate seasonal variation in temperature and precipitation. As we saw in the discussion under Concept 2.1, climatic extremes are sometimes more important in determining species distributions than mean annual conditions. For example, grasslands and shrublands have wider global distributions than Figure 3.4 would suggest, occurring in regions with relatively high average annual precipitation, but with regular dry periods (e.g., Mediterranean-type shrublands; grasslands at the margins of deciduous forests). In addition, factors such as soil texture and chemistry as well as proximity to mountains and large bodies of water can influence biome distribution.

The potential distributions of terrestrial biomes differ from their actual distributions due to human activities

The effects of land conversion and resource extraction by humans are increasingly apparent on the land surface. These human effects are collectively described as **land use change**. Human modification of terrestrial ecosystems began at least 10,000 years ago with the use of fire as a tool to clear forests and enhance the size of game populations. The greatest changes have occurred over the last 150 years, since the onset of mechanized agriculture and logging and an exponential increase in the human population (see Figure 9.2) (Harrison and Pearce 2001). Between 50% and 60% of Earth's land surface has been altered by human activities, primarily agriculture, forestry, and livestock grazing; a smaller amount (2%–3%) has been transformed by urban development and transportation corridors (Harrison and Pearce 2001, Sanderson et al. 2002). As a result of these human influences, the potential and the actual distributions of biomes are markedly different (**Figure 3.5**). Temperate biomes, particularly grasslands, have been transformed the most, although tropical and subtropical biomes are experiencing rapid change as well.

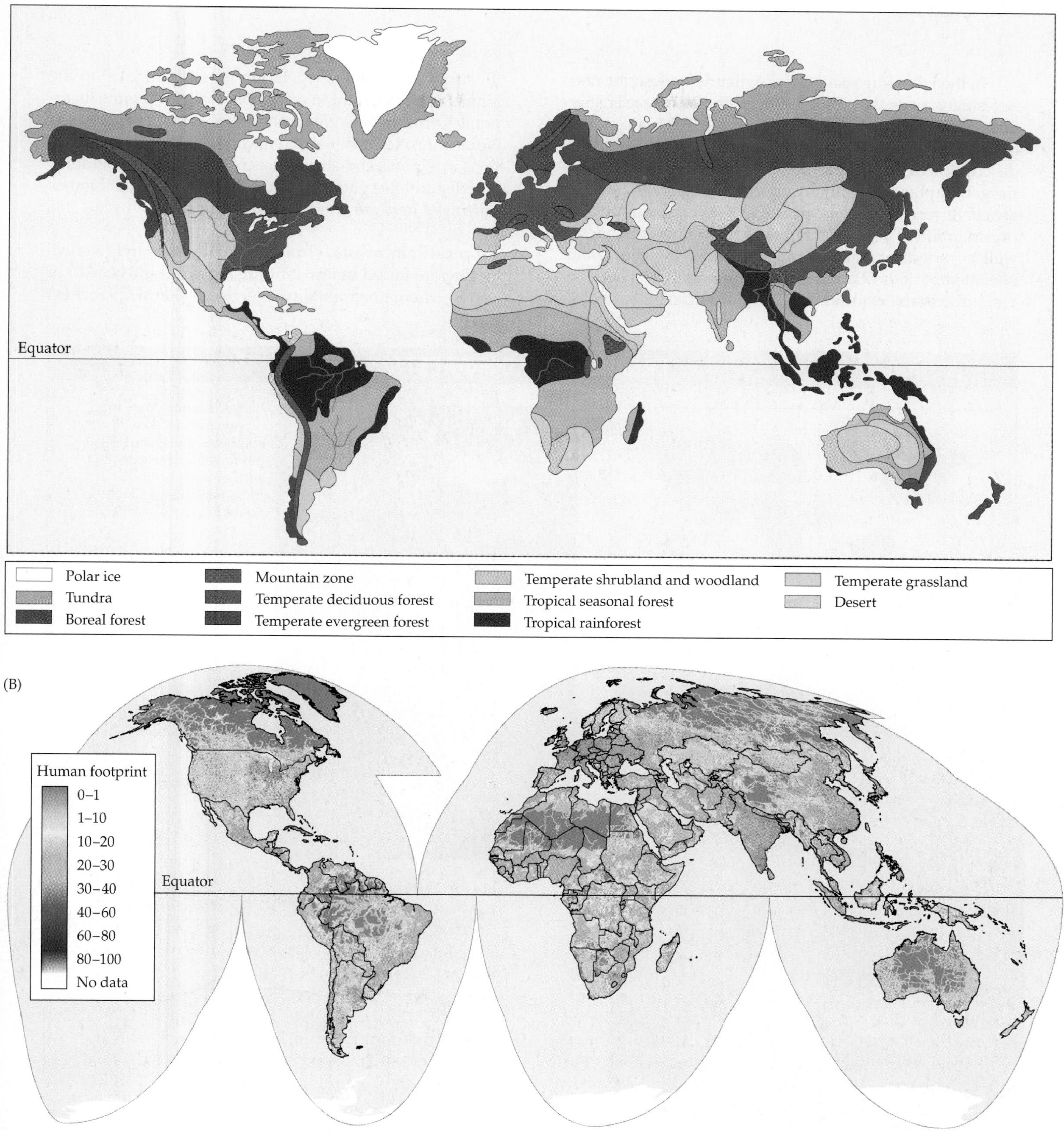

FIGURE 3.5 Global Biome Distributions Are Affected by Human Activities The potential distributions of biomes differ from their actual distributions because human activities have altered much of Earth's land surface. (A) The potential global distribution of biomes. (B) Alteration of terrestrial biomes by human activities. The colors and associated numbers represent a quantitative measure (100 = maximum) of the overall human impact on the environment based on geographic data describing human population size, land development, and resource use. (After Sanderson et al. 2002.)

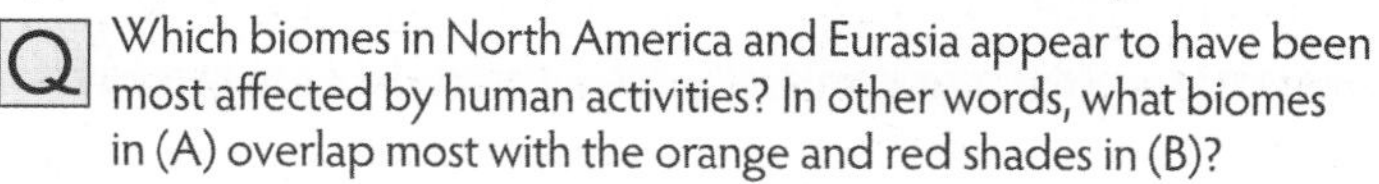
Which biomes in North America and Eurasia appear to have been most affected by human activities? In other words, what biomes in (A) overlap most with the orange and red shades in (B)?

In the following sections, we will briefly describe nine terrestrial biomes, their biological and physical characteristics, and the human activities that influence the actual amount of natural vegetation cover that remains in each biome. The description of each biome begins with a map of its potential geographic distribution and a *climate diagram* showing the characteristic seasonal patterns of air temperature and precipitation at a representative location in that biome as well as periods of water shortages (shaded in yellow) and extended periods of subfreezing temperatures (shaded in blue) (see **Web Extension 3.1**). In addition, sample photos illustrate some of the vegetation types that make up the biome. It is important to remember that each biome incorporates a mix of different communities. Boundaries between biomes are often gradual and may be complex due to variations in regional climatic influences, soil types, topography, and disturbance patterns. Thus, the boundaries of biomes portrayed here are only approximations.

Tropical Rainforests Tropical rainforests are aptly named, as they are found in low-latitude regions (between 10° N and S) where precipitation exceeds 2,000 mm (79 inches)

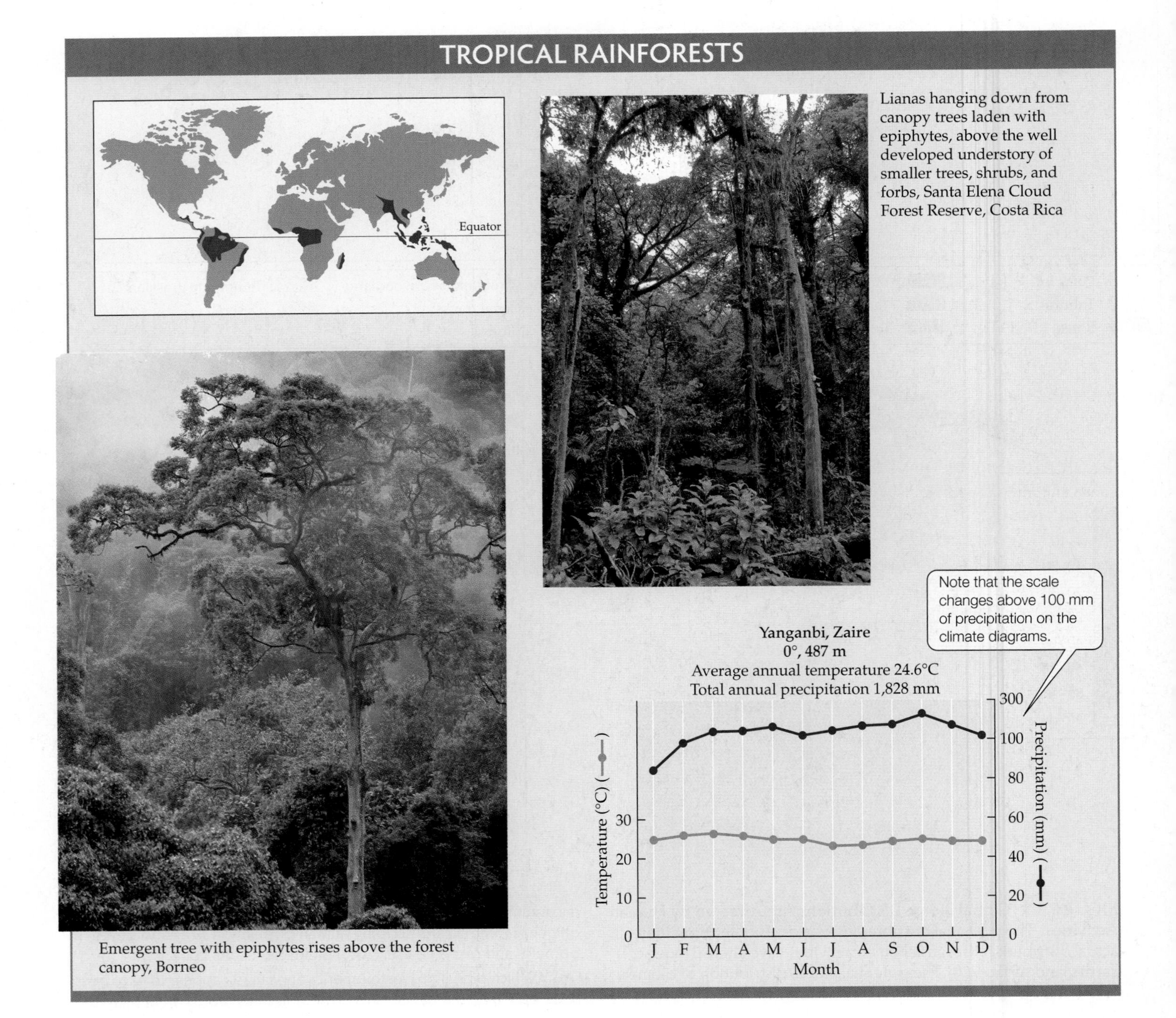

Lianas hanging down from canopy trees laden with epiphytes, above the well developed understory of smaller trees, shrubs, and forbs, Santa Elena Cloud Forest Reserve, Costa Rica

Emergent tree with epiphytes rises above the forest canopy, Borneo

annually. Tropical rainforests experience warm, seasonally invariant temperatures. The abundant precipitation may be spread evenly throughout the year or occur in one or two main peaks associated with the movement of the Intertropical Convergence Zone (ITCZ) (see Figure 2.21). Seasonal climatic rhythms are generally absent from this biome, and plants grow continuously throughout the year. Tropical rainforests contain a substantial amount of living plant biomass, as mentioned earlier, and they include the most productive ecosystems on Earth. They contain an estimated 50% of Earth's species in only about 11% of its terrestrial vegetation cover (Dirzo and Raven 2003). Tropical rainforests occur in Central and South America, Africa, Australia, and Southeast Asia.

The tropical rainforest biome is characterized by broad-leaved evergreen and deciduous trees. Light is a key environmental factor determining the vegetation structure of this biome. Climatic conditions favoring plant growth exert selection pressure either to grow tall above neighboring plants or to adjust physiologically to low light levels. As many as five layers of plants occur in tropical rainforests. *Emergent trees* rise above the majority of the other trees that make up the *canopy* of the forest. The canopy consists primarily of the leaves of evergreen trees, which form a continuous layer approximately 30 to 40 m above the ground. Below the canopy, plants that utilize trees for support and to elevate their leaves above the ground are found draped over or clinging to the canopy and emergent trees, including *lianas* (woody vines) and *epiphytes* (plants that grow on tree branches). *Understory* plants grow in the shade of the canopy, further reducing the light that finally reaches the forest floor. Shrubs and *forbs* (broad-leaved herbaceous plants) occupy the forest floor, where they must rely on light flecks that move across the forest floor during the day for photosynthesis.

Globally, tropical rainforests are disappearing rapidly due to logging and conversion of forests to pasture and croplands (**Figure 3.6**). Approximately half of the tropical rainforest biome has been altered by deforestation (Asner et al. 2009). Rainforests in Africa and Southeast Asia have been altered the most, and rates of deforestation continue to be greatest in these areas (Wright 2005). In some cases, rainforests have been replaced by disturbance-maintained pastures of forage grasses. In other cases, rainforest is regrowing, but the recovery of the previous rainforest structure is uncertain. Rainforest soils are often nutrient-poor, and recovery of nutrient supplies may take a very long time, hindering forest regrowth.

Tropical Seasonal Forests and Savannas As we move to the north and south of the wet tropics toward the Tropics of Capricorn (23.5° S) and Cancer (23.5° N), rainfall becomes more seasonal, with pronounced wet and dry seasons associated with shifts in the ITCZ. This region is marked by

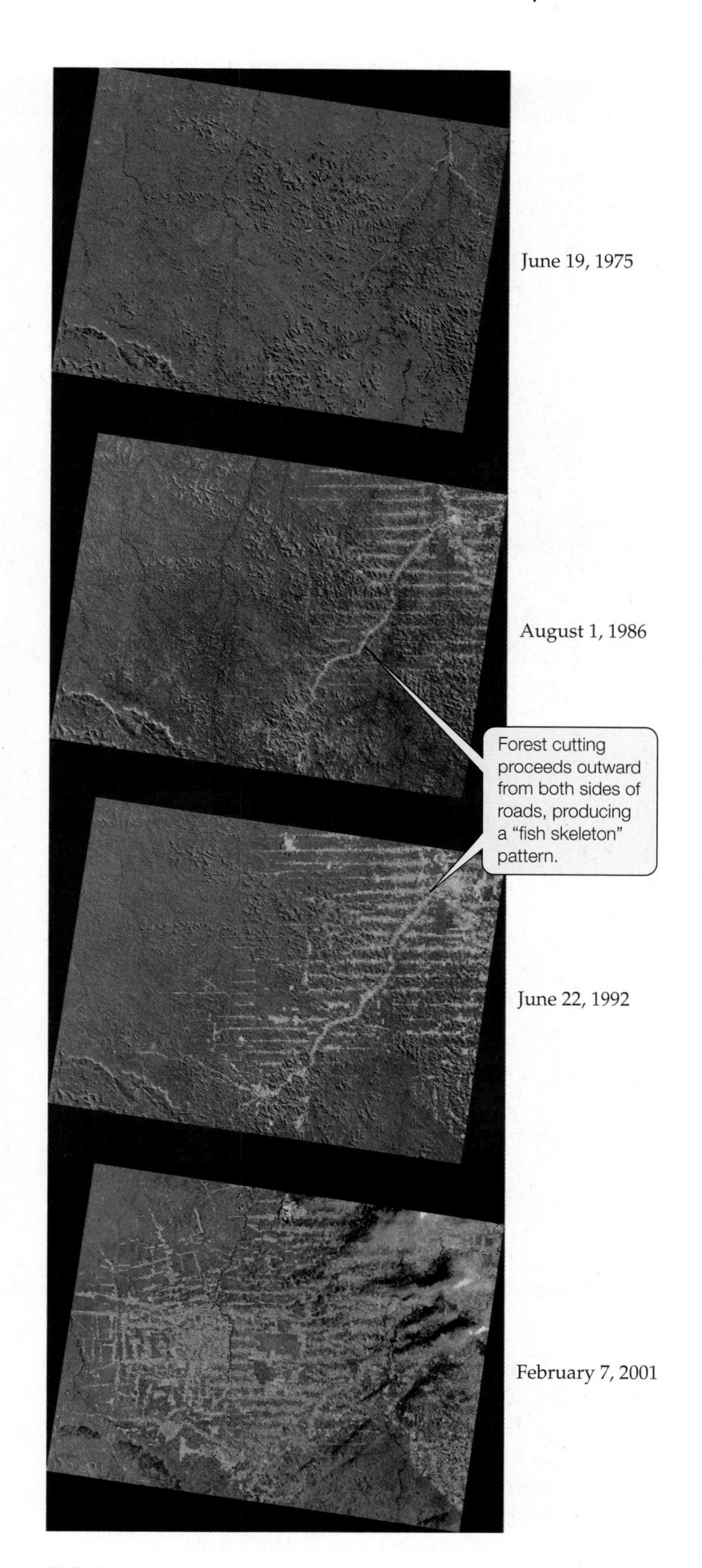

FIGURE 3.6 Tropical Deforestation These satellite photos show the extent of rainforest clearing in Rondônia, Brazil, over a 26-year period (1975–2001).

Baobab trees in dry season in Zambia

Semi-evergreen forest of Pijio trees (*Cavanillesia platanifolia*) during the dry season, Cerro Blanco, Ecuador

a large gradient in climate primarily associated with the seasonality of rainfall. The responses of vegetation to this climatic gradient include shorter stature, lower tree densities, and an increasing degree of drought deciduousness, with leaves dropping from the trees during the dry season. In addition, there is a greater abundance of grasses and shrubs and fewer trees relative to rainforests.

The tropical seasonal biome includes a complex of vegetation types, including *tropical dry forests*, *thorn woodlands*, and *tropical savannas*. The frequency of fires, which increases with the length of the dry season, influences the vegetation growth forms. Recurrent fires, sometimes set by humans, promote the establishment of **savannas**, communities dominated by grasses with intermixed trees and shrubs. In Africa, large herds of herbivores, such as wildebeests, zebras, elephants, and antelopes, also influence the balance between trees and grasses and act as an important force promoting the establishment of savannas. On the floodplains of the Orinoco River in South America, seasonal flooding contributes to the establishment of savannas, as trees are intolerant of long periods of soil saturation. Thorn woodlands (communities dominated by widely spaced trees and shrubs) get their name from the heavy armaments of thorns on the trees, which act as a deterrent to herbivores that would consume the vegetation. Thorn woodlands typically occur in regions with climates intermediate between tropical dry forests and savannas.

Tropical seasonal forests and savannas once covered an area greater than tropical rainforests. Increasing human demand for supplies of wood and agricultural land has resulted in rates of loss of tropical seasonal forests and savannas equal to or greater than those for tropical rainforests (Bullock et al. 1995). Less than half of this biome remains intact. Large increases in human populations in tropical dry forest regions have had a particularly large effect. Large tracts of tropical dry forest in Asia and Central and South America have been converted to cropland and pasture to meet the needs of growing human populations for food and earnings from agricultural goods exported to more developed countries.

Hot Deserts In contrast to the tropical ecosystems, deserts contain sparse populations of plants and animals, reflecting sustained periods of high temperatures and low water availability. The subtropical positions of hot deserts

HOT DESERTS

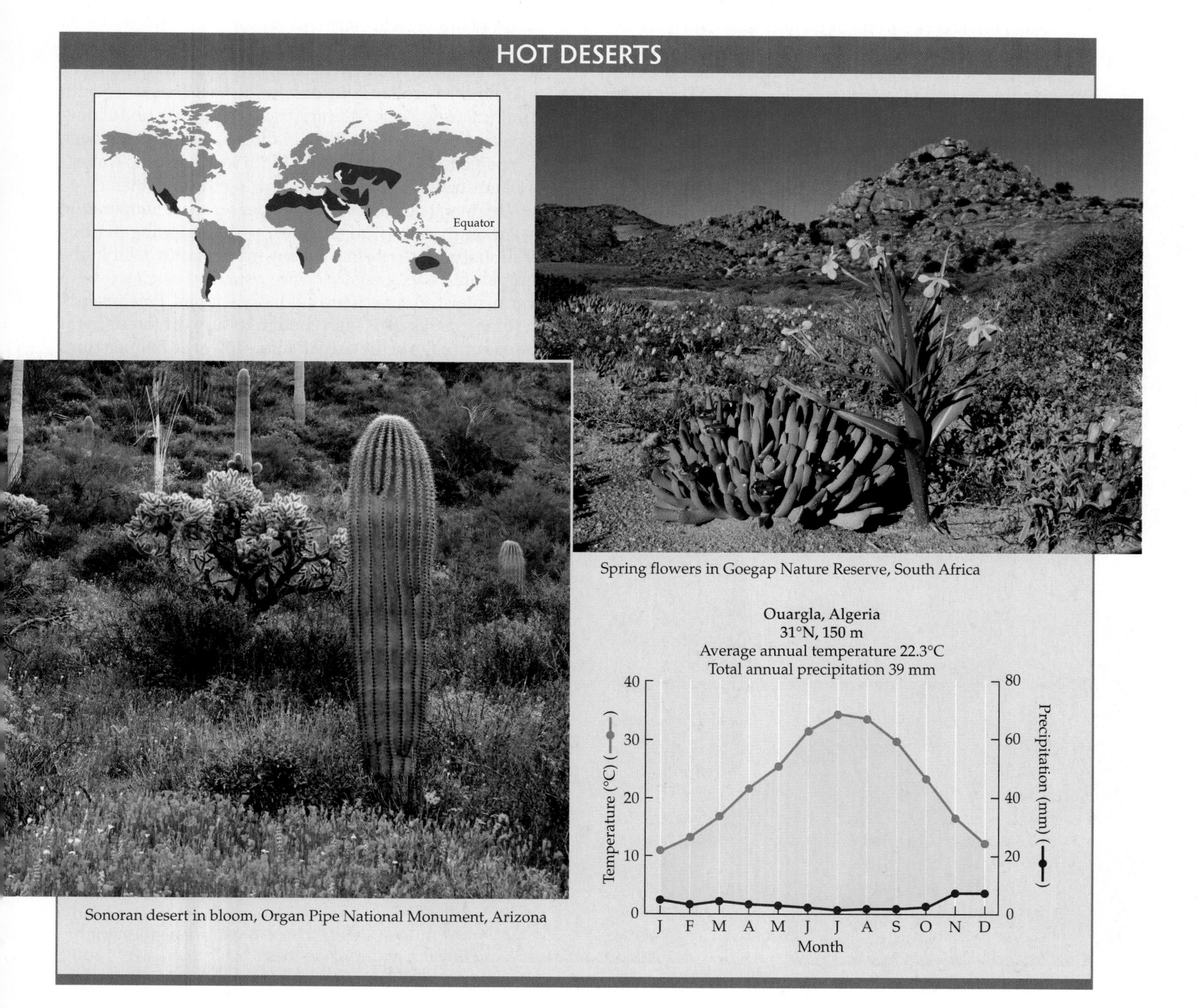

Spring flowers in Goegap Nature Reserve, South Africa

Sonoran desert in bloom, Organ Pipe National Monument, Arizona

generally correspond to the descending air of Hadley cells (see Figure 2.8). This descending air creates zones of high pressure around 30° N and S, which inhibit the formation of storms and their associated precipitation. Low precipitation levels, coupled with high temperatures and high rates of evapotranspiration, result in a limited supply of water for desert organisms. The major desert zones include the Sahara, the Arabian deserts, the Atacama Desert of Chile and Peru, and the Chihuahuan, Sonoran, and Mojave deserts of North America.

Low water availability is an important constraint on the abundance of desert plants as well as an important influence on their form and function. One of the best examples of convergence in plant form is the occurrence of stem succulence in desert plants. Stem succulence occurs in both the cacti of the Western Hemisphere and the euphorb family of the Eastern Hemisphere (**Figure 3.7**). Plants with succulent stems can store water in their tissues to help the plant continue to function during dry periods. Other plants of the desert biome include drought-deciduous shrubs and grasses. Some short-lived annual plants are active only after sufficient precipitation has fallen; these plants carry out their entire life cycle, from germination through flowering and seed production, in a few short weeks. Although the abundance of organisms may be low, species diversity can be high in some deserts. The Sonoran Desert, for example, has over 4,500 plant species, 1,200 bee species, and 500 bird species (Nabhan and Holdsworth 1998).

Humans have used deserts for livestock grazing and agriculture for centuries. Agricultural development in desert areas is dependent on irrigation, often using water that flows in from distant mountains or extracted from deep underground. Unfortunately, irrigated agriculture in deserts has repeatedly failed due to salinization (see p. 58). Livestock grazing in deserts is also a risky venture due to the unpredictable nature of the precipitation needed to support plant growth for herbivores. Long-term droughts in association with unsustainable grazing practices can result in loss of plant cover and soil erosion, a process known as **desertification**. Desertification is a concern in populated regions at the margins of deserts, such as the Sahel to the south of the Sahara in Africa.

Temperate Grasslands Large expanses of grasslands once occurred throughout North America and Eurasia (the Great Plains and the steppes of Central Asia) at latitudes between 30° N and 50° N. Southern Hemisphere grasslands (pampas) are found at similar latitudes on the east coasts of South America, New Zealand, and Africa. These vast, undulating expanses of grass-dominated landscape have often been compared to a terrestrial ocean, with wind-driven "waves" of plants bending to the gusts blowing through them.

Temperate climates have greater seasonal temperature variation than tropical climates, with increasing periods of subfreezing temperatures toward the poles. Within the temperate zone, grasslands are usually associated with warm, moist summers and cold, dry winters. Precipitation in some grasslands is high enough to support forests, as at the eastern edge of the Great Plains. However, frequent fires and grazing by large herbivores such as bison prevent the establishment of trees and thus maintain the dominance of grasses in these environments. The use of fire to man-

(A) Cactus

(B) Euphorb

FIGURE 3.7 Convergence in the Forms of Desert Plants (A) The blue candle cactus (*Myrtillocactus geometrizans*) is native to the Chihuahuan Desert of Mexico. (B) This desert species of *Euphorbia* has cactus-like characteristics. These plants both have succulent stems, water-conserving photosynthetic pathways, upright stems that minimize midday sun exposure, and spines that protect them from herbivores. The two plants are only distantly related, however, and these traits evolved independently in each species.

TEMPERATE GRASSLANDS

Sand Hills grasslands at Valentine National Wildlife Refuge, Nebraska, USA

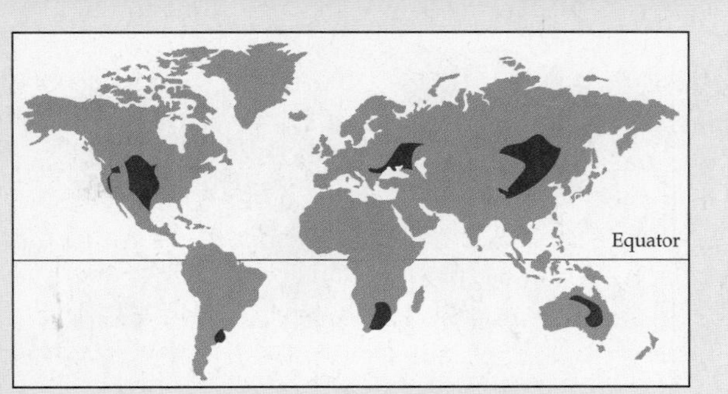

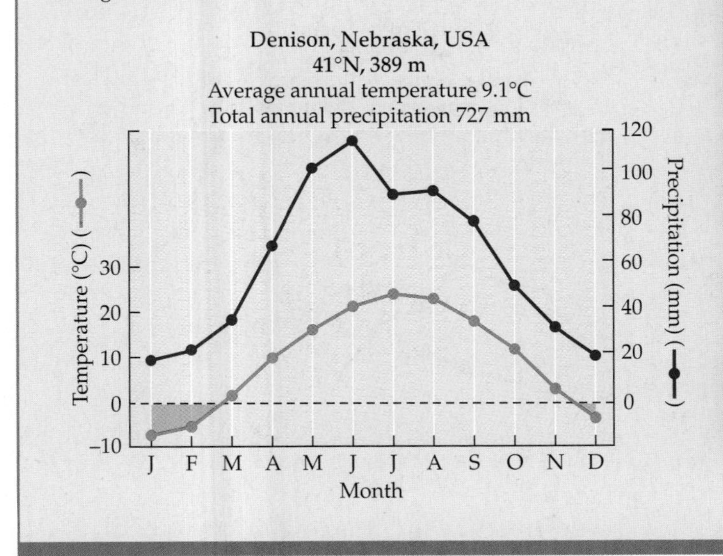

Grassland with chamomile flowers, Altai Plateau, Russia

age grasslands near the edges of forests was probably one of the first human activities with a widespread effect on a terrestrial biome.

The grasslands of the world have been a major focus for agricultural and pastoral development. In order to acquire enough water, grasses grow more roots than stems and leaves. The rich organic matter that accumulates in the soils as a result enhances their fertility, so grassland soils are particularly well suited for agricultural development. Most of the fertile grasslands of central North America and Eurasia have been converted to agriculture. The diversity of the crop species grown on these lands is far less than the diversity of the grasslands they replaced. In more arid grasslands, rates of grazing by domesticated animals can exceed the capacity of the plants to produce new tissues, and grassland degradation, including desertification, may occur. As in deserts, irrigation of some grassland soils has resulted in salinization, decreasing their fertility over time. In parts of Europe, cessation of centuries-old grazing practices has resulted in increased forest invasion into grasslands. This long legacy of grassland use for agriculture and grazing has made grasslands the most human-influenced biome on Earth.

Temperate Shrublands and Woodlands The seasonality of precipitation is an important control on the distribution of temperate biomes. Woodlands (characterized by an open canopy of short trees) and shrublands occur in regions with a winter rainy season. *Mediterranean-type climates*, which occur on the west coasts of the Americas,

TEMPERATE SHRUBLANDS AND WOODLANDS

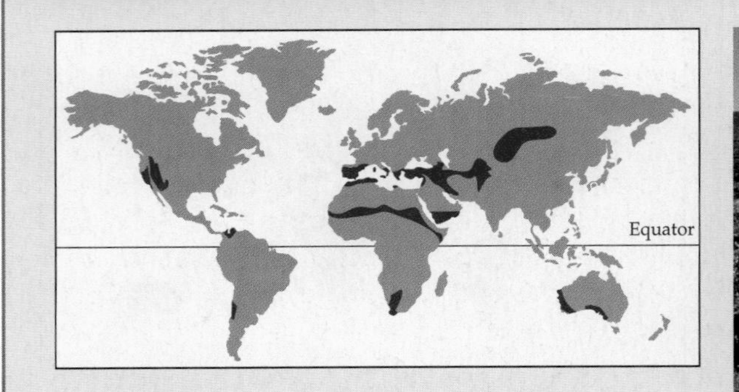

Shrubland on the coast of Western Australia

Gerona, Spain
41°N, 70 m
Average annual temperature 16.7°C
Total annual precipitation 747 mm

Temperature (°C)
Precipitation (mm)
Month
J F M A M J J A S O N D

Fynbos landscape with everlastings (*Helichrysum* sp.), Cape Peninsula National Park, South Africa

Africa, Australia, and Europe between 30° and 40° N and S, are an example of such a climate regime. As we saw in the discussion under Concept 2.1, these Mediterranean-type climates are characterized by asynchrony between precipitation and the summer *growing season* (the period of time with suitable temperatures to support growth). Precipitation falls primarily in winter, and hot, dry weather occurs throughout the late spring, summer, and fall. The vegetation of Mediterranean-type climates is characterized by evergreen shrubs and trees. Evergreen leaves allow plants to be active during cooler, wetter periods and also lower their nutrient requirements, since they do not have to develop new leaves every year. Many plants of Mediterranean-type climates have *sclerophyllous* leaves, which are tough, leathery, and stiff. These plants are well adapted to dry soils and may continue to photosynthesize and grow at reduced rates during the hot, dry summer. Their sclerophyllous leaves help to deter consumption by herbivores and prevent wilting as water is lost. Sclerophyllous shrublands are found in each of the zones characterized by a Mediterranean-type climate, including the *mallee* of Australia, the *fynbos* of South Africa, the *matorral* of Chile, the *maquis* around the Mediterranean Sea, and the *chaparral* of North America.

Fire is a common feature in Mediterranean-type shrublands, and as in some grasslands, may promote their persistence. Some of the shrubs recover after fires by resprouting from woody storage organs protected from the heat below the ground surface. Other shrubs produce seeds that germinate and grow quickly after a fire. Without regular fires at 30–40-year intervals, some temperate shrublands may be replaced by forests of oaks, pines, junipers, or eucalypts. Regular disturbance by fire, combined with the unique climate of temperate shrublands, is thought to promote high species diversity.

Shrublands and woodlands are also found in the continental interior of North America and Eurasia, associated with rain-shadow effects and seasonally cold climates. The Great Basin, for example, occupies the interior of North America between the Sierra Nevada and Cascade mountain ranges to the west and the Rocky Mountains to the east. Large expanses of sagebrush (*Artemesia tridentata*), saltbush (*Atriplex* spp.), creosote bush (*Larrea tridentata*), and piñon pine and juniper woodland occur throughout this area.

Humans have occupied temperate shrublands and woodlands for many thousands of years. While some have been converted to croplands and vineyards, their climates and nutrient-poor soils have limited the extent of agricultural and pastoral development. In the Mediterranean basin, agricultural development using irrigation was attempted but failed due to the infertile soils. Urban development has reduced the cover of shrublands in some regions (e.g., southern California). Increases in local human populations have increased the frequency of fires, which decreases the ability of shrubs to recover and may lead to their replacement by invasive annual grasses.

Temperate Deciduous Forests Deciduous leaves are a solution to the extended periods of freezing weather in the temperate zone. Leaves are more sensitive to freezing

Autumn foliage prior to leaf fall, Great Smoky Mountains National Park, North Carolina, USA

Beech forest in summer, Japan

than other plant tissues because of the high level of physiological activity associated with photosynthesis. Temperate deciduous forests occur in areas with enough rainfall to support tree growth (500–2,500 mm, or 20–100 inches, per year) and where soils are fertile enough to supply the nutrients lost when leaves are shed in the fall. Temperate deciduous forests are primarily limited to the Northern Hemisphere, as the Southern Hemisphere contains less land area and lacks extensive areas with the continental climates associated with the deciduous forest biome.

Deciduous forests occur at 30° to 50° N on the eastern and western edges of Eurasia and in eastern North America, extending inland to the continental interior before diminishing due to lack of rainfall and, in some cases, increased fire frequency. Similar species occur on each of these continents in this biome, reflecting a common biogeographic history (see Chapter 17). Oak, maple, and beech trees, for example, are components of this forest biome on each continent. The vertical structure of the forest includes canopy trees as well as shorter trees, shrubs, and forbs below the canopy. Species diversity is lower than in tropical forests but can include as many as 3,000 plant species (e.g., in eastern North America). Disturbances such as fire and outbreaks of herbivorous insects do not play a major role in determining the development and persistence of temperate deciduous forest vegetation, although they can be important in determining its boundaries, and periodic outbreaks of herbivores do occur (e.g., Gypsy moths, a non-native insect introduced to North America).

The temperate deciduous forest biome has been a focus for agricultural development for centuries. The fertile soils and climate are conducive to the growth of crops. Forest clearing for crop and wood production has therefore been widespread in this biome. Very little old-growth temperate deciduous forest remains on any continent. Since the early twentieth century, however, agriculture has gradually shifted away from temperate zone forests toward temperate grasslands and the tropics, particularly in the Americas. Abandonment of agricultural fields has resulted in reforestation in some parts of Europe and North America. However, the species composition of the second-growth forests often differs from what was present prior to agricultural development. Nutrient losses from soils due to long-term agricultural use is one reason for this difference. Another is the loss of some species due to introductions of invasive species. For example, the chestnut blight fungus, introduced from Asia, nearly wiped out the chestnut trees (*Castanea* spp.) of North America in the early twentieth century (see Chapter 13). As a result, oak species are more widespread than they were prior to agricultural development.

Temperate Evergreen Forests Evergreen forests span a wide range of environmental conditions in the temperate zone, from warm coastal zones to cool continental and maritime climates. Precipitation also varies substantially among these forests, from 500 to 4,000 mm (20–150 inches) per year. Some temperate evergreen forests with high levels of precipitation, which are typically located on west coasts at latitudes between 45° and 50°, are referred to as "temperate rainforests" (**Figure 3.8**). Temperate evergreen forests are commonly found on nutrient-poor soils, whose condition is in part related to the acidic nature of the leaves of the evergreen trees. Some evergreen forests are subject

FIGURE 3.8 Temperate Rainforest in Tasmania Rainforests occur in temperate zones with high precipitation (over 5,000 mm, or 200 inches) and relatively mild winter temperatures. Here, understory tree ferns and canopy trees of leatherwood and myrtle crowd together near Montezuma Falls in western Tasmania, Australia.

TEMPERATE EVERGREEN FORESTS

Araucaria (monkey puzzle tree) forest, Lanin National Park, Argentina

Grove of giant sequoias (*Sequoiadendron giganteum*), with Douglas fir (*Pseudotsuga menziesii*), Mariposa Grove, Yosemite National Park, California

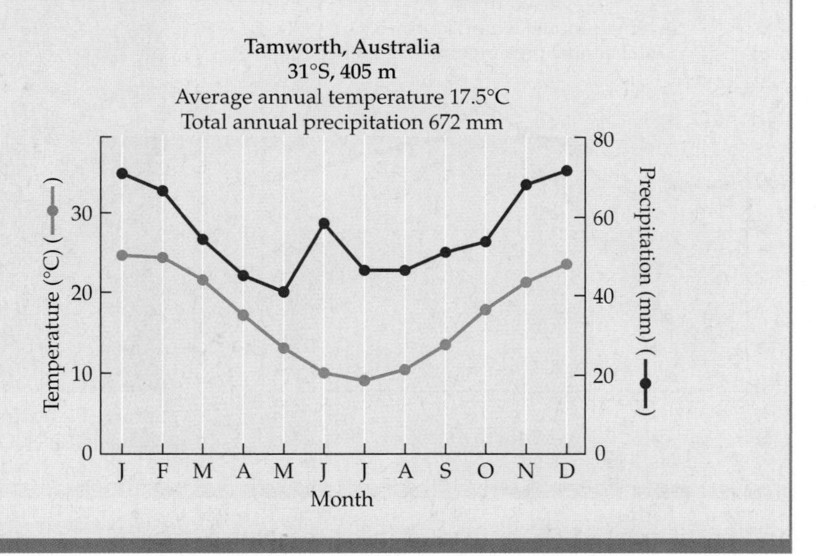

to regular fires at intervals of 30–60 years, which may promote their persistence.

Temperate evergreen forests are found in both the Northern and Southern hemispheres between 30° and 50° latitude. Their diversity is generally lower than that of deciduous and tropical forests. In the Northern Hemisphere, the tree species include needle-leaved conifers such as pines, junipers, and Douglas fir (*Pseudotsuga menziesii*). In the Southern Hemisphere, on the west coasts of Chile and Tasmania and in southeastern and southwestern Australia, there is greater tree diversity, including southern beeches (*Nothofagus* spp.), eucalypts, the Chilean cedar (*Austrocedrus*), and podocarps (see Chapter 17).

Evergreen trees provide a source of high-quality wood and pulp for paper production. The temperate evergreen forest biome has been subjected to extensive clearing, and little old-growth forest remains. Some forestry practices tend to promote sustainable use of these forests, although in some regions the planting of non-native species (such as Monterey pine in New Zealand), the uniform age and density of the trees, and losses of formerly dominant species have created forests that are ecologically very different from their pre-logging condition. The suppression of naturally occurring fires in western North America has increased the density of some forest stands, which has resulted in more intense fires when they do occur

BOREAL FORESTS

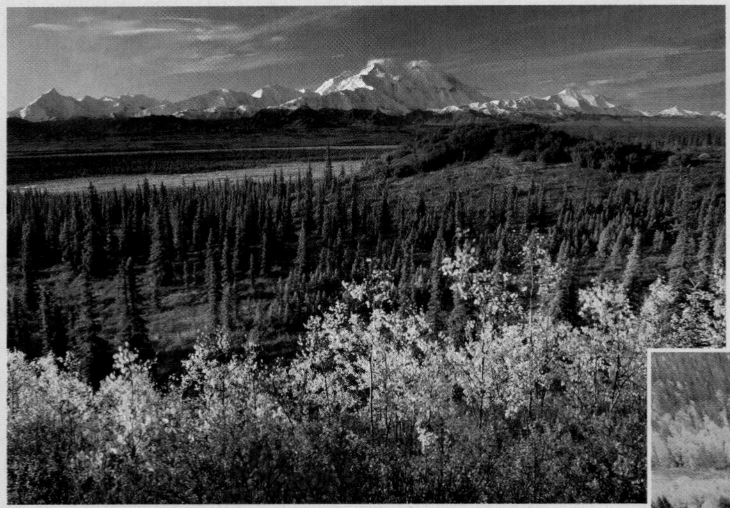

Spruce trees in autumn, Denali National Park, Alaska

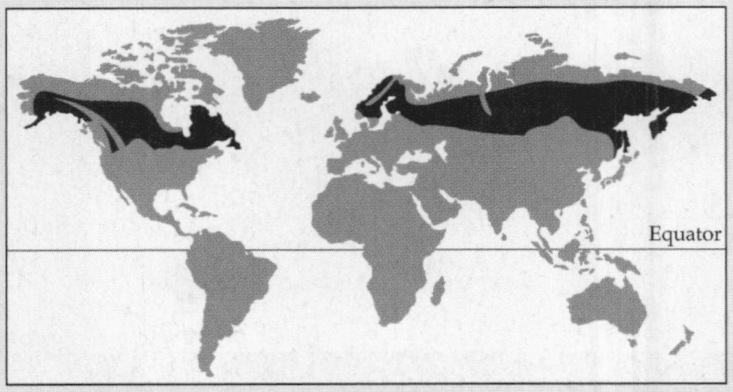
Equator

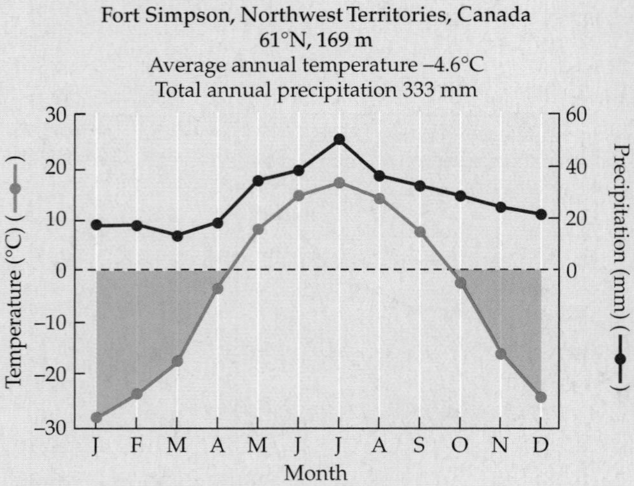
Fort Simpson, Northwest Territories, Canada
61°N, 169 m
Average annual temperature –4.6°C
Total annual precipitation 333 mm
Temperature (°C)
Precipitation (mm)
Month
J F M A M J J A S O N D
30
20
10
0
–10
–20
–30
60
40
20
0

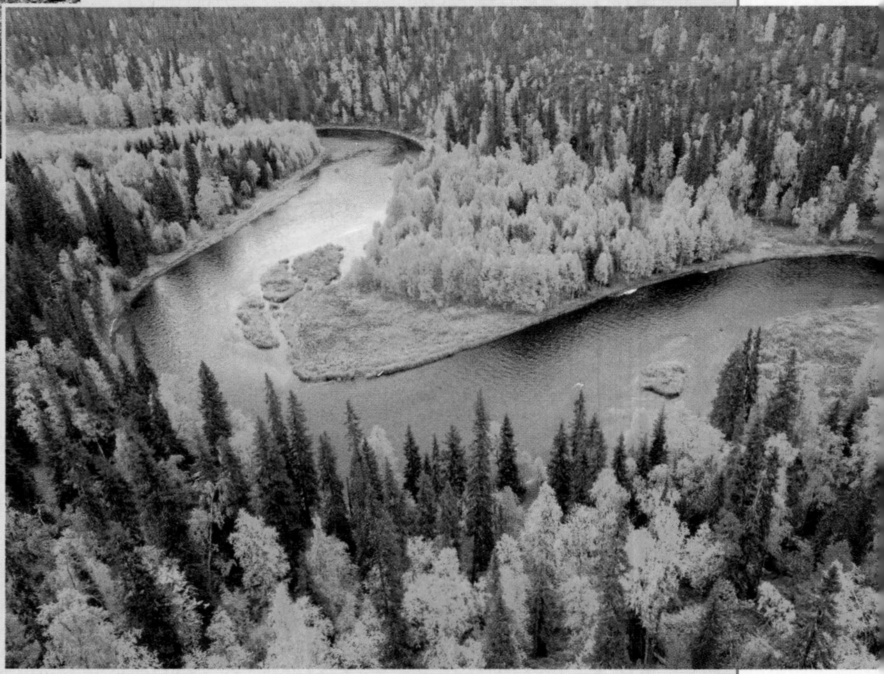

Spruce (*Picea abies*) and silver birch (*Betula verrucosa*) along the Kitkajoki River, Oulanka National Park, Finland

and has increased the spread of insect pests (e.g., bark beetles) and pathogens. In industrialized countries, the effects of air pollution have damaged some temperate evergreen forests (see Chapter 24) and made them more susceptible to other stresses.

Boreal Forests Above 50° N, the severity of winters increases. Minimum temperatures of –50°C (–58°F) are common in continental locations such as Siberia, and continuous subfreezing temperatures may last up to 6 months. The extreme weather in these subarctic regions is an important determinant of the vegetation structure. Not only must the plants cope with low air temperatures, but soils may regularly freeze, leading to the formation of **permafrost**, defined as a subsurface soil layer that remains frozen year-round for at least 3 years. Although precipitation is low, the permafrost impedes water drainage, so soils are moist to saturated.

The biome that occupies the zone between 50° and 65° N is the boreal (far northern) forest. This biome is also known as *taiga*, the Russian word for this northern forest. It is composed primarily of coniferous species, including spruces, pines, and larches (deciduous needle-leaved trees), but also includes extensive deciduous birch forests in maritime locations, particularly in Scandinavia. Conifers tend to resist damage from winter freezing better than angiosperm trees, despite maintaining green leaves year-round. Although the boreal forest is found only in the Northern Hemisphere, it is the largest biome in area and contains one-third of Earth's forested land.

FIGURE 3.9 Fire in the Boreal Forest Despite its cold climate, fire is an important part of the boreal forest environment.

The cold, wet condition of boreal forest soils limits the decomposition of plant material such as leaves, wood, and roots. Thus, the rate of plant growth exceeds the rate of decomposition, and the soils contain large amounts of organic matter. During extensive summer droughts, conditions are conducive to forest fires, which are set off by lightning. These fires may burn both the trees and the soil (**Figure 3.9**). Soil fires may continue to burn slowly for several years, even through the cold winters. In the absence of fire, forest growth enhances permafrost formation by lowering the amount of sunlight absorbed by the soil surface. In low-lying areas, soils become saturated, killing the trees and forming extensive peat bogs.

Boreal forests have been less affected by human activities than other forest biomes. Logging occurs in some regions, as does oil and gas development, including the mining of oil sands. These activities will pose an increasing threat to the boreal forest as demands for wood and energy increase. In addition, the large store of organic matter in the soil makes boreal forests an important component of the global carbon cycle. Climate warming may result in more rapid decomposition and thus higher rates of carbon release from boreal forest soils, enhancing atmospheric greenhouse gas concentrations and resulting in a positive feedback to global warming (see Chapter 24).

Tundra Trees cease to be the dominant vegetation beyond approximately 65° latitude. The tree line that marks the transition from boreal forest to tundra is associated with low growing-season temperatures, although the causes of this transition are complex and can also include other climatic and soil conditions. The tundra biome occurs primarily in the Arctic but can also be found on the edges of the Antarctic Peninsula and on a few islands in the Southern Ocean. The poleward decrease in temperature and precipitation across the tundra biome is associated with the zones of high pressure generated by the polar atmospheric circulation cells (see Figure 2.8).

The tundra biome is characterized by sedges, forbs, grasses, and low-growing shrubs such as heaths, willows, and birches. Lichens and mosses are also important components of this biome. Although the summer growing season is short, the days are long, with continuous periods of light for 1 to 2 months of the summer. The plants and lichens survive the long winter by going dormant, maintaining living tissues under the snow or soil, insulated from the cold air temperatures.

The tundra has several similarities to the boreal forest: temperatures are cold, precipitation is low, and permafrost is widespread. Despite the low precipitation, many tundra areas are wet, as the permafrost keeps the precipitation that does fall from percolating to deeper soil layers. Repeated freezing and thawing of surface soil layers over several decades results in sorting of soil materials according to their texture. This process forms polygons of soil at the surface with upraised rims and depressed centers (**Figure 3.10**). Where soils are coarser or permafrost does not develop, the soils may be dry, and plants must be able to cope with low water availability. These polar deserts are most common at the highest latitudes where tundra is found.

Human inhabitants of the tundra are scattered in sparse settlements. As a result, this biome contains some of the largest pristine regions on Earth. Herds of caribou and musk oxen and predators such as wolves and brown

FIGURE 3.10 Soil Polygons and Pingo Pingos are small hills found in the Arctic caused by an intrusion of water that freezes in the subsurface permafrost zone, thrusting the soil above it upward. On the periphery of the pingo, polygons can be seen; these result from freezing and thawing of soils, which pushes coarse soil materials toward the edges and finer soil to the middle of the polygons.

TUNDRA

Looking out to the Arctic plain at midnight from the northern edge of the Brooks Range, Alaska

Arctic tundra in early autumn color, Dovrefjell-Sunndalsfjella National Park, Norway

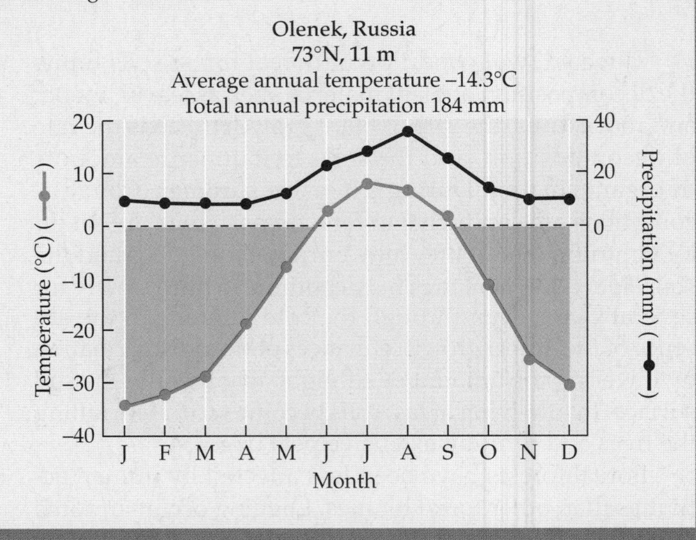

bears, which have been extirpated throughout much of their previous range in other biomes, inhabit the tundra, and many species of migratory birds nest there. The influence of human activities on the Arctic tundra is increasing, however. Exploration and development of energy resources has accelerated. A key to limiting the effects of energy development is preventing damage to the permafrost, which can cause long-term erosion. The Arctic has experienced climatic warming almost double the global average during the late twentieth and early twenty-first centuries. Increased losses of permafrost, catastrophic lake drainage, and reduced carbon storage in the soil have been linked to climate change.

Now that we've completed our tropics-to-tundra tour of terrestrial biomes, let's consider the influence of mountains on more local-scale patterning of biological communities. In some mountainous locations, elevational changes result in a smaller version of our latitudinal description of biomes.

Biological communities in mountains occur in elevational bands

Approximately one-fourth of Earth's land surface is mountainous. Mountains create elevational climatic gradients that change more rapidly over a given distance than those associated with changes in latitude. Temperatures in temperate continental mountain ranges, for example, decrease approximately 6.4°C for every 1,000 m increase in elevation (or 3.6°F per 1,000 feet),* a decrease equivalent to that over approximately a 13° change in latitude, or a

*This decrease in temperature, called *lapse rate,* is dependent on the humidity of the air and local winds. A more complete discussion of temperature change in mountains can be found in Chapter 2.

FIGURE 3.11 Mountain Biological Zones An elevational transect on the eastern slope of the southern Rocky Mountains passes through climatic conditions and biome-like assemblages similar to those found along a latitudinal gradient between Colorado and northern Canada. (Data from Marr 1967.)

Q Would you expect the same biological zonation on east-facing and west-facing slopes in a temperate mountain range near the west coast of a continent?

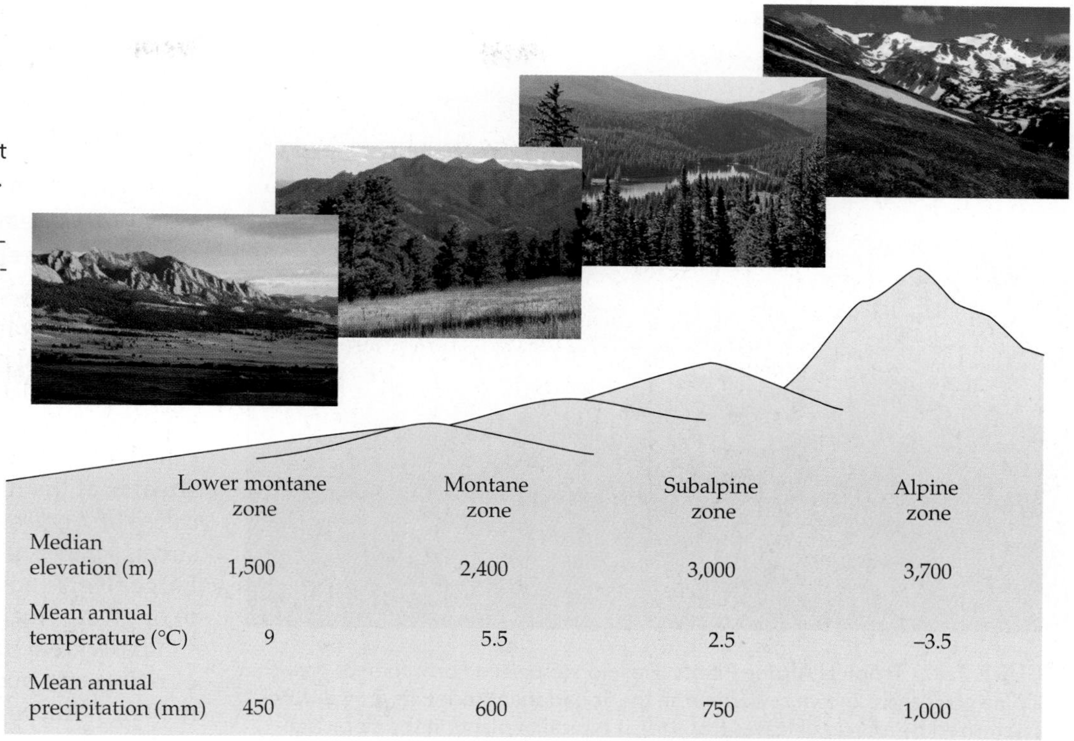

	Lower montane zone	Montane zone	Subalpine zone	Alpine zone
Median elevation (m)	1,500	2,400	3,000	3,700
Mean annual temperature (°C)	9	5.5	2.5	–3.5
Mean annual precipitation (mm)	450	600	750	1,000

distance of 1,400 km (870 miles). As we might expect from our consideration of biomes and their close association with climate, coarse biotic assemblages similar to biomes occur in elevational bands on mountains. Finer-scale biotic distinctions are found in association with slope aspect (e.g., north-facing versus south-facing), proximity to streams, and the orientation of slopes in relation to prevailing winds (see Concept 2.4).

The biological communities that occur from the base to the summit of a temperate-zone mountain range resemble what we would find along a latitudinal gradient to the north. An elevational transect on the eastern slope of the southern Rocky Mountains in Colorado, for example, includes grassland to alpine vegetation across a 2,200 m (7,200 feet) increase in elevation (**Figure 3.11**). The changes in climate and vegetation are similar to the transition from grassland to Arctic tundra that occurs with a 27° increase in latitude, from Colorado to the Northwest Territories of Canada. Grasslands occur at the base of the mountains, but give way to pine savannas on the initial slopes (the lower montane zone). Fire plays an important role in determining the vegetation structure of both montane grasslands and savannas. With increasing elevation, the pine savannas are replaced by denser stands of mixed pine–aspen forests (the montane zone), which resemble temperate evergreen and deciduous forest biomes. Spruce and fir trees make up the forests of the subalpine zone, which resemble the boreal forest biome. Mountain tree lines are similar to the transition from boreal forest to tundra, although topography can play an important role through its influence on snow distribution and avalanches. The alpine zone above the tree line includes diminutive plants such as sedges, grasses, and forbs, including some of the same species that occur in the Arctic tundra. Although the alpine zone resembles the Arctic tundra, its physical environment is different, with higher wind speeds, more intense solar radiation, and lower atmospheric partial pressures of CO_2.*

Mountains are found on all continents and at all latitudes. As indicated in the example above, the changes in climate associated with changes in elevation alter the composition of the local vegetation. However, not all of the vegetation assemblages that occur in mountains resemble major terrestrial biomes. Some mountain-influenced biological communities have no biome analogs. For example, daily temperature changes at high-elevation sites in the tropics (e.g., Mount Kilimanjaro and the tropical Andes) are greater than seasonal temperature changes. Subfreezing temperatures occur on most nights in the tropical alpine zone. As a result of these unique climatic conditions,

*The *partial pressure* of a gas is defined as the pressure exerted by a particular component of a mixture of gases. The concentrations of CO_2 and O_2 are the same at high elevations as they are at sea level, but their partial pressures are lower because total atmospheric pressure is lower. The exchange of a gas between an organism and the atmosphere is determined by its partial pressure rather than its concentration.

FIGURE 3.12 Tropical Alpine Plants Frailejón (*Espeletia hartwegiana*) grows in alpine grasslands, known as *páramo*, in the Equadorian Andes. Its growth form, characterized by a circle of leaves (rosette), is typical of plants in the tropical alpine zones of South America and Africa. The adult leaves help to protect the developing leaves and stems at the apex of the plant from nightly frosts. Such giant rosettes are found exclusively in the tropical alpine zone and do not have analogs in the Arctic or Antarctic.

tropical alpine vegetation does not resemble that of the temperate alpine zone or the Arctic tundra (**Figure 3.12**).

CONCEPT 3.2 Biological zones in freshwater ecosystems are associated with the velocity, depth, temperature, clarity, and chemistry of the water.

Freshwater Biological Zones

Although they occupy a small portion of the terrestrial surface, freshwater streams, rivers, and lakes are a key component in the connection between terrestrial and marine ecosystems. Rivers and lakes process inputs of chemical elements from terrestrial ecosystems and transport them to the oceans. The biota of these freshwater ecosystems reflect the physical characteristics of the water, including its velocity (flowing streams and rivers versus lakes and ponds), its temperature (including seasonal changes), how far light can penetrate it (clarity), and its chemistry (salinity, oxygen concentrations, nutrient status, and pH).

In this section we will explore the biota and associated physical conditions found in freshwater ecosystems in the next two sections. In contrast to terrestrial biomes, for which only plants are used as indicators, the biological assemblages of freshwater ecosystems are characterized by both plants and animals, reflecting the greater proportional abundance of animals in aquatic ecosystems.

Biological communities in streams and rivers vary with stream size and location within the stream channel

Water flows downhill over the land surface in response to the force of gravity. The land surface is partly shaped by the erosional power of water, which cuts valleys as it heads toward a lake or ocean. The descending water converges into progressively larger streams and rivers, called **lotic** (flowing water) ecosystems. The smallest streams at the highest elevations in a landscape are called *first-order streams* (**Figure 3.13**). Two first-order streams may converge to form a second-order stream. Large rivers such as the Nile or Mississippi are equal to or greater than sixth-order streams.

Individual streams tend to form repeated patterns of **riffles** and **pools** along their paths. Riffles are fast-moving portions of the stream flowing over coarse particles on the stream bed, which increase oxygen input into the water. Pools are deeper and water flows more slowly over a bed of fine sediments. Biological communities in lotic ecosystems are associated with different physical locations within the stream and their related environments (**Figure 3.14**). Organisms that live in the

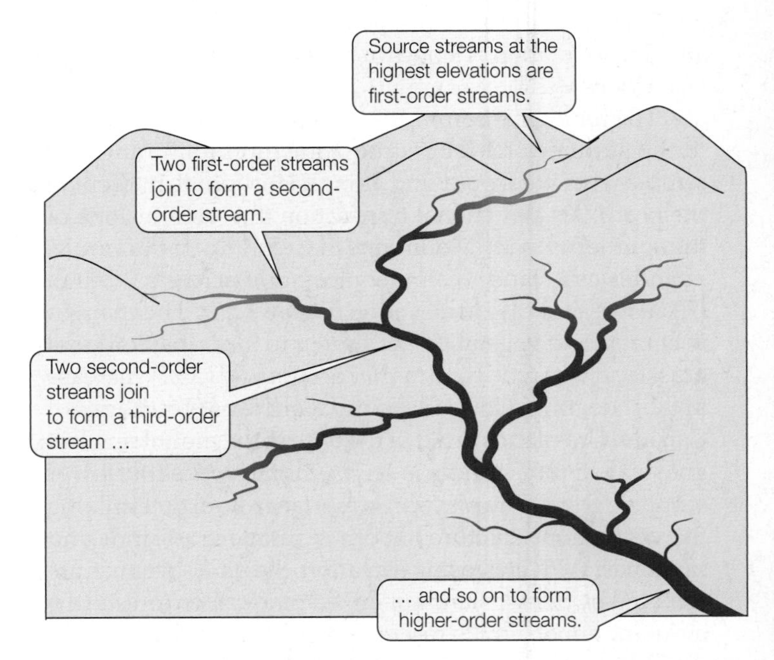

FIGURE 3.13 Stream Orders Stream order affects environmental conditions, community composition, and the energy and nutrient relationships of communities within the stream.

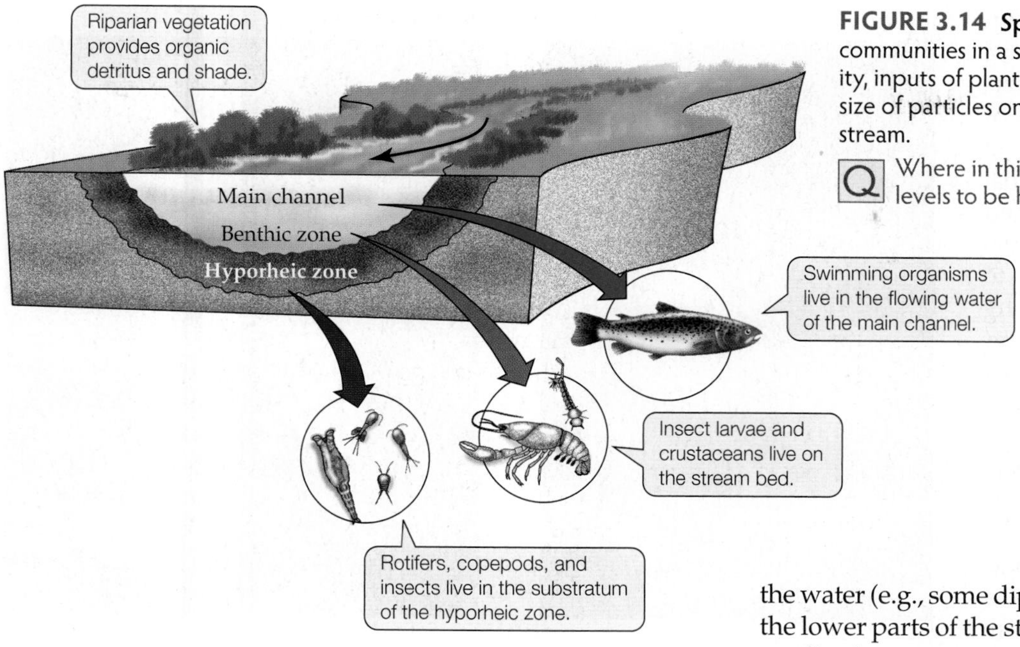

FIGURE 3.14 Spatial Zonation of a Stream Biological communities in a stream vary according to water velocity, inputs of plant material from riparian vegetation, the size of particles on the stream bed, and the depth of the stream.

Q Where in this stream would you expect oxygen levels to be highest and lowest?

flowing water of the main channel are generally swimmers, such as fish. The bottom of the stream, called the **benthic zone**, is home to invertebrates that consume **detritus** (dead organic matter), such as some mayfly and fly (dipteran) larvae, or hunt other organisms, such as some caddisflies and crustaceans. Some organisms, such as rotifers, copepods, and insects, are found in the zone below and adjacent to the stream, where water movement still occurs, either from the stream or from groundwater moving into the stream. This area is known as the **hyporheic zone**.

The composition of biological communities in streams and rivers changes with stream order (see Figure 3.13) and channel size. The *river continuum concept* was developed to describe these changes in both the physical and biological characteristics of a stream (Vannote et al. 1980). This conceptual model holds that as a stream flows downslope and increases in size, the input of detritus from the vegetation adjacent to the stream (known as *riparian vegetation*) decreases relative to the volume of water, and the particle size in the stream bed decreases, from boulders and coarse rock to fine sand. As a result, the importance of terrestrial vegetation as a food source for stream organisms decreases in the downstream direction. Coarse terrestrial detritus is most important near the stream source, while the importance of fine organic matter, algae, and rooted and floating aquatic vascular plants (known as **macrophytes**, from *macro*, "large"; *phyte*, "plant") increases downstream. The general feeding styles of organisms change accordingly as the river flows downstream. *Shredders*, organisms adapted to tear up and chew leaves (e.g., some species of caddisfly larvae), are most abundant in the higher parts of the stream, while *collectors*, organisms that collect fine particles from the water (e.g., some dipteran larvae), are most abundant in the lower parts of the stream. The river continuum concept applies best to temperate river systems, but not as well in boreal/Arctic and tropical rivers or in rivers with high concentrations of dissolved organic substances (including tannic and humic acids) derived from wetlands. Nonetheless, the model provides a basis for studying biological organization in stream and river systems.

Human effects on lotic ecosystems have been extensive. Most fourth- and higher-order rivers have been altered by the results of human activities, including pollution, increases in inputs of sediments, and introductions of non-native species. Streams and rivers have been used as conduits for the disposal of sewage and industrial wastes in most parts of the world inhabited by humans. These pollutants often reach levels that are toxic to many aquatic organisms. Excessive application of fertilizers to croplands results in runoff and leaching of nutrients into groundwater, which eventually reaches rivers. Inputs of nitrogen and phosphorus from fertilizers alter the composition of aquatic communities. Deforestation increases sediment inputs into nearby streams, which can reduce water clarity, alter benthic habitat, and inhibit gill function in many aquatic organisms. Introductions of non-native species, such as sport fishes (e.g., bass and trout), have lowered the diversity of native species in both stream and lake ecosystems. The construction of dams on streams and rivers tremendously alters their physical and biological properties, converting them into still waters—the topic of the next section.

Biological communities in lakes vary with depth and light penetration

Lakes and other still waters, called **lentic** ecosystems, occur where natural depressions have filled with water or where humans have dammed rivers to form reservoirs.

Lakes and ponds may be formed when glaciers gouge out depressions and leave behind natural dams of rock debris (moraines), or when large chunks of glacial ice break off, become surrounded by glacial debris, and then melt. Most temperate and polar lakes are formed by glacial processes. Lakes may also form when meandering rivers cease to flow through a former channel, leaving a section stranded, called an *oxbow lake*. Geologic phenomena, such as extinct volcanic calderas and sinkholes, form natural depressions that may fill with water. Lakes and ponds of biological origin, in addition to reservoirs, include beaver dams and animal wallows.

Lakes vary tremendously in size, from small, ephemeral ponds to the massive Lake Baikal in Siberia, which is 1,600 m (5,200 feet) deep and covers 31,000 km^2 (12,000 square miles). The size of a lake has important consequences for its nutrient and energy status, and therefore for the composition of its biological communities. Deep lakes with a small surface area tend to be nutrient-poor compared with shallow lakes with a large surface area (see Chapter 21).

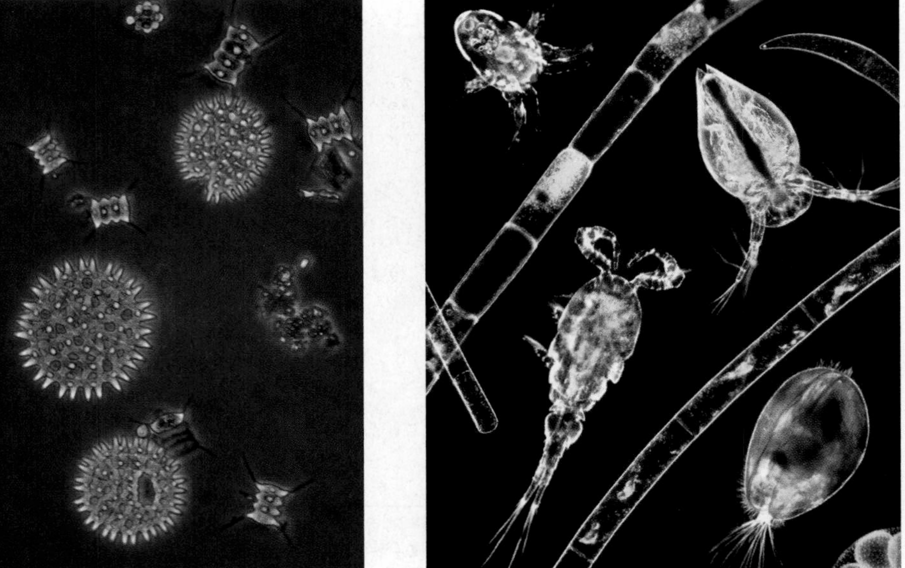

FIGURE 3.15 Examples of Lake Plankton (A) Green algae from a lake water sample, including spherical *Pediastrum* sp. and disklike *Scenedesmus* sp. Note the projections at the corners of the *Scenedesmus* and at the surface of *Pediastrum*, which help to suspend the cells in the water. (B) Zooplankton include a water flea (suborder Cladocera, upper left), a seed shrimp (*Cyclocypris* sp., upper right), a copepod (*Cyclops* sp., lower right) and the nauplius larva of a copepod (lower left). Strands of algae also stretch through the photo.

Lake biotic assemblages are associated with depth and degree of light penetration. The open water, or **pelagic zone**, is inhabited by **plankton**: small, often microscopic organisms that are suspended in the water (**Figure 3.15**). Photosynthetic plankton (called **phytoplankton**) are limited to the surface layer of water where there is enough light for photosynthesis, called the **photic zone**. **Zooplankton**—tiny animals and nonphotosynthetic protists—occur throughout the pelagic zone, as do other consumers such as bacteria and fungi, feeding on detritus as it falls through the water. Fish patrol the pelagic zone, scouting for food and predators that might eat them.

The nearshore zone where the photic zone reaches to the lake bottom is called the **littoral zone**. Here, macrophytes join with floating and benthic phytoplankton to produce energy by photosynthesis. Fish and zooplankton also occur in the littoral zone.

In the benthic zone, detritus derived from the littoral and pelagic zones serves as an energy source for animals, fungi, and bacteria. The benthic zone is usually the coldest part of the lake, and its oxygen concentrations are often low.

Let's move from fresh waters to the biological zones of the oceans. You will see that some of those zones have names and characteristics similar to those in freshwater lakes, but with much greater spatial cover. As in freshwater communities, physical characteristics are used to differentiate marine biological zones.

CONCEPT 3.3 Marine biological zones are determined by ocean depth, light availability, and the stability of the bottom substrate.

Marine Biological Zones

Oceans cover 71% of Earth's surface and contain a rich diversity of life. The vast area and volume of the oceans and their environmental uniformity make them considerably different from terrestrial ecosystems in terms of biological organization. Marine organisms are more widely dispersed, and marine communities are not as easily organized into broad biological units as terrestrial biomes are. Instead, marine biological zones are coarsely categorized by their physical location relative to shorelines and the ocean bottom (**Figure 3.16**). The distributions of the organisms that inhabit these zones reflect differences in temperature, as we saw for terrestrial biomes, as well as other important factors, including light availability, water depth, stability of the bottom substrate, and interactions with other organisms.

In this section, we will take a tour of the biological zones of the oceans, from the margins of the land to the deep, dark, cold ocean bottom. We will examine the physi-

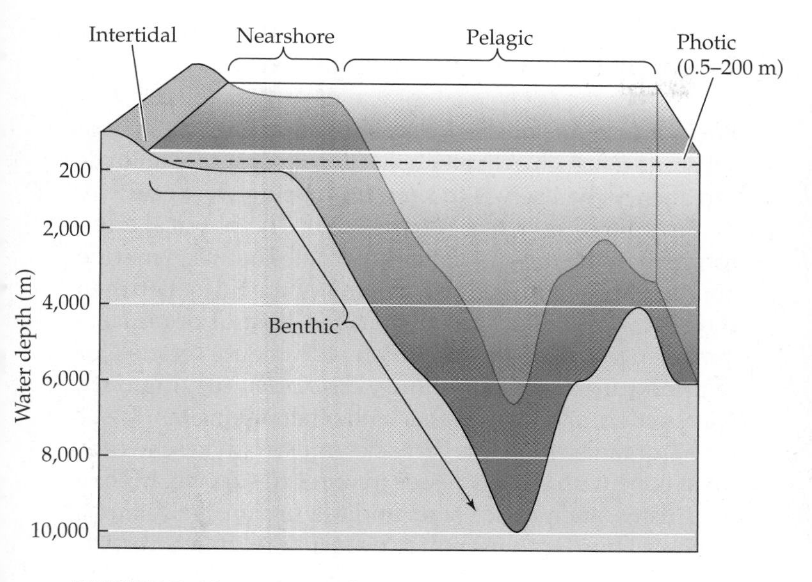

FIGURE 3.16 Marine Biological Zones Biological zones in the ocean are categorized by water depth and by their physical locations relative to shorelines and the ocean bottom.

FIGURE 3.17 Estuaries Are Junctions between Rivers and Oceans The mixing of fresh and salt water gives estuaries a unique environment with varying salinity. Rivers bring in energy and nutrients from terrestrial ecosystems.

cal and biological factors that characterize the different zones and the major organisms found in them.

Nearshore zones reflect the influence of tides and substrate stability

Marine biological zones adjacent to the continents are influenced by local climate, by the rise and fall of ocean waters associated with tides, and by wave action and the influx of fresh water and terrestrial sediments from rivers. **Tides** are generated by the gravitational attraction between Earth and the moon and sun. Ocean water rises and falls in most nearshore zones twice daily. The magnitude of the tidal range varies greatly among different locations because it is related to shoreline morphology and ocean bottom structure. Tides produce unique transition zones between terrestrial and marine environments and influence salinity and nutrient availability in these nearshore habitats.

Estuaries The junction of a river with the ocean is called an *estuary* (**Figure 3.17**). Estuaries are characterized by variations in salinity associated with the flow of fresh river water into the ocean and the influx of salt water flowing inland from the ocean as tides rise. Rivers also bring terrestrial sediments containing nutrients and energy (food) to the ocean, and the interaction of tidal and river flows acts to trap these sediments in estuaries, enhancing their productivity. The varying salinity of estuaries is an important determinant of the organisms that occur there. Many commercially valuable fish species spend their juvenile stages in estuaries, away from fish predators that are not as tolerant of the changes in salinity. Other inhabitants of estuaries include shellfish (e.g., clams and oysters), crabs, marine worms, and seagrasses. Estuaries are increasingly threatened by water pollution carried by rivers. Nutrients from upstream agricultural sources can cause local dead zones (see Chapter 24) and losses of biological diversity.

Salt marshes Terrestrial sediments carried to shorelines by rivers form shallow marsh zones (**Figure 3.18**) that are dominated by emergent vascular plants (*emergent* in this case refers to plants that rise out of the water), includ-

FIGURE 3.18 Salt Marshes Are Characterized by Salt-Tolerant Vascular Plants Emergent vascular plants form salt marshes in shallow nearshore zones.

ing grasses, rushes, and broad-leaved herbs. In these salt marshes, as in the estuaries that they often border, the input of nutrients from rivers enhances productivity. Periodic flooding of the marsh at high tide results in a gradient of salinity: the highest portions of the marsh are the most saline because infrequent flooding and evaporation of water from the soil leads to a progressive buildup of salts. Salt marsh plants grow in distinct zones that reflect this salinity gradient, with the most salt-tolerant species in the highest portions of the marsh. Salt marshes provide food and protection from predators for a wide variety of animals, including fish, crabs, birds, and mammals. Organic matter trapped in salt marsh sediments may serve as a nutrient and energy source for nearby marine ecosystems.

Mangrove forests Shallow coastal estuaries and nearby mudflats in tropical and subtropical regions are inhabited by salt-tolerant evergreen trees and shrubs (**Figure 3.19**). These woody plants are collectively referred to as "mangroves," but they do not make up a single taxonomic group; mangroves include species from 16 different plant families. Mangrove roots trap mud and sediments carried by the water, which build up and modify the shoreline. Like salt marshes, mangrove forests provide nutrients to other marine ecosystems and habitat for numerous animals, both marine and terrestrial. Among the unique animals associated with mangroves are manatees, crab-eating monkeys, fishing cats, and monitor lizards. Mangrove forests are threatened by human development of coastal areas—particularly the development of shrimp farms—as well as water pollution, diversion of inland freshwater sources, and cutting of the forests for wood.

FIGURE 3.19 Salt-Tolerant Evergreen Trees and Shrubs Form Mangrove Forests The mangrove roots trap mud and sediments and provide habitat for other marine organisms.

Rocky intertidal zones Rocky shorelines provide a stable substrate to which a diverse collection of algae and animals can anchor themselves to keep from being washed away by the pounding waves (**Figure 3.20**). The physical environment of the intertidal zone alternates between marine and terrestrial with the rise and fall of the tides. Between the high-tide and low-tide marks, a host of organisms are arranged in zones associated with their tolerance for temperature changes, salinity, *desiccation* (drying out), wave action, and interactions with other organisms. *Sessile* (fixed) organisms such as barnacles, mussels, and seaweeds must cope with these stresses in order to survive. Mobile organisms, such as sea stars and sea urchins, may move to pools in order to minimize exposure to these stresses.

FIGURE 3.20 The Rocky Intertidal Zone: Stable Substrate, Changing Conditions Rocky shorelines provide a stable substrate to which organisms can anchor themselves, but those organisms must cope with the shift from terrestrial to marine conditions that occurs with each tide, as well as wave action. Sessile organisms must be resistant to temperature changes and desiccation. Mobile organisms often take refuge in tide pools to avoid exposure to the terrestrial environment.

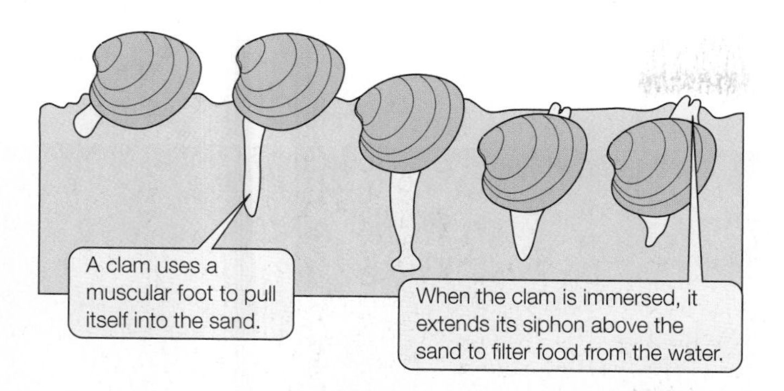

FIGURE 3.21 Burrowing Clams Clams, like most animals of sandy shorelines, live in the sandy substrate.

Sandy shores Except for a few scurrying crabs and shorebirds and the occasional bit of seaweed washed ashore, sandy beaches appear devoid of life. Unlike the rocky shore, the sandy substrate provides no stable anchoring surface, and the lack of attached seaweeds limits the supply of potential food for herbivorous animals. Tidal fluctuations and wave action further limit the potential for the development of biological communities. Beneath the sand, however, invertebrates such as clams, sea worms, and mole crabs find suitable habitat (**Figure 3.21**). Smaller organisms, such as polychaete worms, hydroids (small animals related to jellyfishes), and copepods (tiny crustaceans) live on or among the grains of sand. These organisms are protected from temperature changes and desiccation at low tide and from the turbulent water at high tide. When the sand is immersed in seawater, some of these organisms emerge to feed on detritus or other organisms, while others remain buried and filter detritus and plankton from the water.

Shallow ocean zones are diverse and productive

Near the coastline, enough light may reach the ocean bottom to permit the establishment of sessile photosynthetic organisms. Like terrestrial plants, these photosynthetic organisms provide energy that supports communities of animals and microorganisms as well as a physical structure that creates habitat for those organisms, including surfaces to which they can anchor and places where they can find refuge from predators. The diversity and complexity of the habitats provided by the photosynthesizers support considerable biological diversity in these shallow ocean environments.

Coral reefs In warm, shallow ocean waters, corals (animals related to jellyfishes), living in a close association with algal partners (symbiotic mutualism; see Chapter 14), form large colonies. The corals obtain most of their energy from algae that live within their bodies, while the algae receive protection from grazers and some nutrients from the corals. Many corals build a skeleton-like structure by extracting calcium carbonate from seawater. Over time, these coral skeletons pile up into massive formations called *reefs* (**Figure 3.22**). The formation of reefs is aided by other organisms that extract other minerals from seawater, such as sponges that precipitate silica. The unique association of these reef-building organisms gives rise to a structurally complex habitat that supports a rich marine community.

FIGURE 3.22 A Coral Reef Corals create habitat for a diverse assemblage of marine organisms.

Coral reefs grow at rates of only a few millimeters per year, but they have shaped the face of Earth (Birkeland 1997). Over millions of years, corals have constructed thousands of kilometers of coastline and numerous islands (**Figure 3.23**). The rate of production of living biomass in coral reefs is among the highest on Earth. The accretions

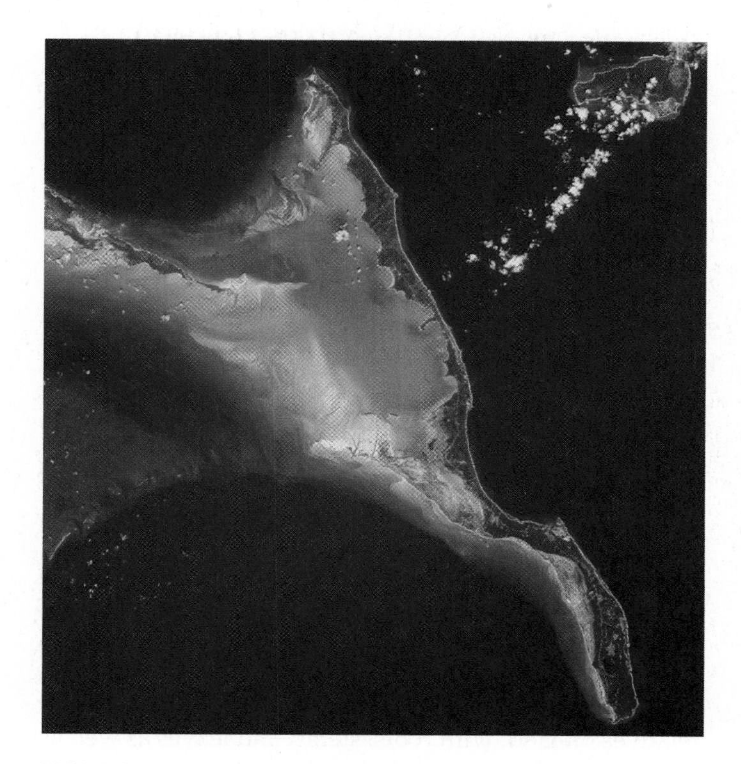

FIGURE 3.23 Coral Reefs Can Be Seen from Outer Space Long Island, in the Bahamas, was formed by coral reefs, which can be seen on the fringes of the island.

of coral skeletons are as much as 1,300 m thick in some places, and they currently cover a surface area of 600,000 km^2, approximately 0.2% of the ocean surface.

As many as a million species are found in coral reefs worldwide, including more than 4,000 fishes. Many economically important fish species rely on coral reefs for habitat, and reef fishes provide a source of food for fishes of the open ocean, such as jacks and tuna. The taxonomic and morphological diversity of animals in coral reefs is greater than in any other ecosystem on Earth (Paulay 1997). The full diversity of coral reefs has yet to be explored and described, however. The potential for development of medicines from coral reef organisms is great enough that the U.S. National Institutes of Health established a laboratory in Micronesia.

Human activities threaten the health of coral reefs in a number of ways. Sediments carried by rivers can cover and kill the corals, and excess nutrients increase the growth of algae on the surfaces of the corals, increasing coral mortality. Changes in ocean temperatures associated with climate change can result in the loss of the corals' algal partners, a condition called *bleaching*. Increased atmospheric CO_2 has increased ocean acidification (discussed in more detail in Chapter 24), which inhibits the ability of corals to form skeletons (Orr et al. 2005). Another threat is an increased incidence of fungal infections, possibly related to increased environmental stress.

Kelp beds In clear, shallow (< 15 m) temperate ocean waters, large stands of seaweed, known as kelp beds or kelp "forests" (**Figure 3.24**), support a rich and dynamic community of marine life. Kelp are large brown algae of several different genera. They have specialized tissues resembling leaves (fronds), stems (stipes), and roots (holdfasts). Kelp are found where a solid substrate is available for anchoring. Residents of kelp beds include sea urchins, lobsters, mussels, abalones, numerous other seaweeds, and sea otters. Interactions among these organisms, both direct and indirect, influence the abundance of the kelp (see the Case Study in Chapter 8 on p. 177). In the absence of grazing, kelp beds can become so dense that light reaching the bottom of the canopy is not sufficient to support photosynthesis.

FIGURE 3.24 A Kelp Bed Kelp are brown algae (order Laminariales) that attach themselves to the solid bottom in shallow ocean waters, providing food and habitat for many other marine organisms.

Seagrass beds While we typically associate flowering plants with terrestrial environments, some flowering plants are important components of shallow (< 5 m) subtidal communities. These submerged flowering plants are called seagrasses, although they are not closely related to plants in the grass family. Morphologically, they are similar to their relatives on land, with roots, stems, and leaves as well as flowers, which are pollinated under water. Seagrass beds are found in subtidal marine sediments composed of mud or fine sand. The plants reproduce primarily by vegetative growth, although they produce seeds as well. Marine algae and animals grow on the surfaces of the plants, and the larval stages of some organisms, such as mussels, are dependent on them for habitat. Inputs of nutrients from upstream agricultural activities can harm seagrass beds by increasing the density of algae in the water and on the surfaces of the seagrasses. Seagrasses are also susceptible to periodic outbreaks of fungal diseases.

Open ocean and deep benthic zones are determined by light availability and proximity to the bottom

The vastness and depth of the open ocean beyond the continental shelves, known as the pelagic zone, makes it difficult to differentiate distinct biological communities there. Light availability determines where photosynthetic organisms can occur, which in turn determines the availability of food for animals and microorganisms. Thus, the surface waters with enough light to support photosynthesis (the photic zone) contain the highest densities of organisms (see Figure 3.16). The photic zone extends about 200 m downward from the ocean surface, depending on water clarity. Below the photic zone, the supply of energy, mainly in the form of detritus falling from the photic zone, is much lower, and life is far less abundant.

The diversity of life in the pelagic zone varies considerably. Its **nekton** (swimming organisms capable of overcoming ocean currents) include cephalopods such as squids and octopuses, fishes, sea turtles, and mammals such as whales and porpoises. Most of the photosynthe-

FIGURE 3.25 Plankton of the Pelagic Zone (A) This sample of marine phytoplankton includes diatoms (*Odontella sinensis*, the rectangular cells with the concave ends) and dinoflagellates (*Biddulphia regia*). (B) Marine zooplankton include larval stages of various organisms, such as those shown here.

sis of the pelagic zone is carried out by phytoplankton, which include green algae, diatoms, dinoflagellates, and cyanobacteria (**Figure 3.25A**). Zooplankton include protists such as ciliates, crustaceans such as copepods and krill, and jellyfishes (**Figure 3.25B**). Many species of pelagic seabirds, including albatross, petrels, fulmars, and boobies, spend the majority of their lives flying over open ocean waters, feeding on marine prey (fish and zooplankton) and detritus found on the ocean surface.

Organisms that live in the pelagic zone must overcome the effects of gravity and water currents that could force them to progressively greater depths. Photosynthetic organisms, and those directly dependent on them as a food source, must stay in the photic zone where sunlight is sufficient to maintain photosynthesis, growth, and reproduction. Swimming is an obvious solution to this problem, used by organisms such as fishes and squids. Seaweeds such as *Sargassum* have gas-filled bladders that keep them buoyant. Large mats of *Sargassum* sometimes form "floating islands" that host rich and diverse biological communities. Some plankton retard their sinking by decreasing their density relative to seawater (e.g., through alteration of their chemical composition) or by adapting a shape that lowers their downward velocity (e.g., having a cell wall with projections).

Beneath the photic zone, the availability of energy decreases, and the physical environment becomes more demanding as temperatures drop and water pressure rises. As a result, organisms are few and far between. Crustaceans such as copepods graze on the rain of falling detritus from the photic zone. Crustaceans, cephalopods, and fishes are the predators of the deep sea. Some fishes take on frightening forms, appearing to be mostly mouth

FIGURE 3.26 A Denizen of the Deep Pelagic Zone Anglerfishes are named for their unique strategy for capturing prey. The bioluminescent organ dangling from the forehead of this female humpback anglerfish (*Melanocetus johnsoni*) is used to attract prey to a position where they are easily engulfed by her huge, toothy mouth. Some male anglerfishes attach themselves to a female once they reach maturity and spend the rest of their lives as parasites, sucking the blood of their mate.

(**Figure 3.26**). The scientific names given to some of the sea creatures at this depth, such as "vampire squid from hell" (*Vampyroteuthis infernalis*), "stalked toad with many filaments" (*Caulophryne polynema*), and "Prince Axel's wonder fish" (*Thaumatichthys axeli*), testify to the unusual forms found there. Most deep-sea fishes have weak bone structure to reduce their weight and lack the gas bladder found in most fishes, since the high pressures would collapse it.

The ocean bottom (the benthic zone) is also very sparsely populated. Temperatures are near freezing, and pressures are great enough to crush any terrestrial organism. Conversely, if deep-sea creatures adapted to these high pressures are brought to the surface, their bodies may expand and burst. The sediments of the benthic zone, which are rich in organic matter, are inhabited by bacteria and protists as well as sea worms. Sea stars and sea cucumbers graze the ocean floor, consuming organic matter or organisms in the sediments or filtering food from the water. Benthic predators, like those of the deep pelagic zone, use bioluminescence to lure prey. Because of the logistic difficulties involved, the deep-sea benthic zone is one of the least explored and poorly understood marine biological zones.

Marine biological zones have been impacted by human activities

Our discussion of marine biological zones has alluded to several services provided to humans. These services include food production (e.g., fisheries in the nearshore and open ocean zones), protection of coastal areas from erosion (e.g., mangrove forests), uptake and stabilization of pollutants and nutrients (estuaries and marshes), and recreational benefits (Barbier et al., 2011, in press). These services, along with ocean biodiversity, are increasingly threatened by human activities.

Despite the vastness of the ocean, human activities have affected it to varying degrees over the majority of its area (**Figure 3.27**). These include land-based activities such as inputs of nutrients and pollutants into rivers, ocean-based activities such as commercial fishing, and emissions of greenhouse gases. The impacts of these activities include changes in water temperature and ocean acidification due to increases in greenhouse gases, increases in UV radiation due to the loss of protective stratospheric ozone, inputs of pollutants, and overharvesting of sea creatures, particularly fish and whales (Halpern et al. 2008; see Chapter 24 for more discussion of ozone loss and the greenhouse effect). These impacts have the potential to influence the services on which humans depend, as well as the composition and abundance of the biota that inhabit different marine biological zones. These potential impacts have not been well characterized, however, primarily due to a lack of information. The greatest estimated impacts tend to be in nearshore marine ecosystems (estuaries, rocky intertidal zones, and sandy shores) near terrestrial regions producing high inputs of pollutants and nutrients, such as the regions adjacent to northern Europe and eastern Asia. Despite the widespread nature of human impacts, large areas of the ocean remain only moderately affected, and greater recognition of these impacts could lead to increased conservation and more sustainable use of ocean resources.

A CASE STUDY REVISITED

The American Serengeti—Twelve Centuries of Change in the Great Plains

Humans have been implicated in several major biological changes in the grasslands of the world. One of the earliest was the disappearance of large mammals from North America during the late Pleistocene. Paul Martin, an early

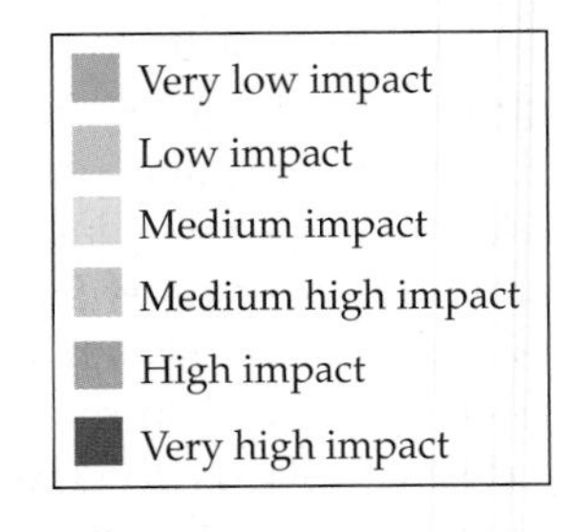

FIGURE 3.27 Human Impacts on the Oceans Activities such as greenhouse gas emissions, pollutant inputs, and overfishing have had varying impacts on different regions of the oceans. The colors represent the degree of impact, which was quantified using expert judgments of 17 different environmental impact factors. (From Halpern et al. 2008.)

proponent of this hypothesis, noted the strong correspondence between extinction events on several continents and the arrival of humans on those continents, principally Europe, North and South America, and Australia (Martin 1984, 2005). Martin suggested that the rapidity of the extinctions and the greater proportion of large animals that disappeared reflected the hunting efficiency of those early humans. Larger animals have lower reproductive rates than smaller animals, so they cannot recover from increases in predation as quickly. Martin's suggestion therefore took on the unfortunate label of "the overkill hypothesis."

Since it was first proposed, the overkill hypothesis has received increasing support. Archeological research has uncovered numerous butchering sites containing remains of extinct animals. Spearheads have been found among the bones, and some of them bear scrape marks made by tools found at the sites. Other strong evidence indicates that human arrival on small, isolated oceanic islands led to large numbers of extinctions due to predation by humans and by other animals they introduced (e.g., rats and snakes). While most scientists now accept that hunting of megafauna by humans had a role in some of the continental extinctions in the late Pleistocene, other mechanisms have been proposed as well. These mechanisms include the spread of diseases carried by humans and possibly the domesticated dogs that accompanied them (MacPhee and Marx 1997). Another hypothesis suggests that the loss of some animals on which other species depended, such as mastodons, led to more widespread extinctions (Owen-Smith 1987). No one hypothesis explains the extinctions of all of the megafauna on all of the continents, however. A combination of climate change and the arrival of humans probably contributed to their demise (Barnosky et al. 2004).

FIGURE 3.28 Buffalo Hunting The arrival of large numbers of Euro-Americans in the Great Plains in the nineteenth century led to a mass slaughter of bison, which was facilitated by the construction of railroad lines and the use of high-powered rifles.

Although the diversity of large mammals on the Great Plains was greatly diminished following the Pleistocene, large mammals were still abundant. Bison may have numbered as many as 30 million, and numerous elk (wapiti), pronghorn, and deer roamed the plains. These animals continued to be hunted by humans, who also began to use fire on the eastern edge of the Great Plains as a tool for managing the habitat of their prey, as well as for small-scale agriculture (Delcourt et al. 1998). The writings of travelers to the Great Plains in the early 1800s indicate that the western edge of the eastern deciduous forest was farther east than it is today, probably due to the influence of human-set fires.

Between 1700 and 1900, ecological changes occurred in the Great Plains that profoundly transformed both the plants and the animals. The reintroduction of horses into North America by Spanish explorers facilitated the development of a Native American culture centered on the hunting of bison. The arrival of Euro-Americans, and their subsequent conflicts with Native Americans, led to the near-extinction of bison and other large Plains animals by the late 1800s (**Figure 3.28**). With the arrival of cattle and mechanized agriculture after 1850, the Great Plains became a domesticated landscape. The moister, eastern tallgrass prairie was converted into monocultures of corn, wheat, soybeans, and other crops; today, only 1% of that grassland remains. A larger proportion of the mixed-grass and short-grass prairies to the west remained intact, but overgrazing and unsustainable agricultural practices led to serious degradation of some of these areas during the "Dust Bowl" of the 1930s, when drought and massive windstorms resulted in substantial losses of fertile topsoil (see Case Study in Chapter 24 on p. 525).

CONNECTIONS IN NATURE

Long-Term Ecological Research

Most terrestrial biomes and marine biological zones across the globe are experiencing changes due to human activities (see Figures 3.5 and 3.27). Even remote, seemingly pristine areas are subject to the effects of climate change and air pollution. Recognizing the effects of human activities on these systems, as well as our incomplete understanding of those effects, the U.S. National Science Foundation initiated a network of long-term ecological research (LTER) sites in 1980. Initially consisting of 5 sites, the network has grown to 26 sites representing a diversity of terrestrial biomes,

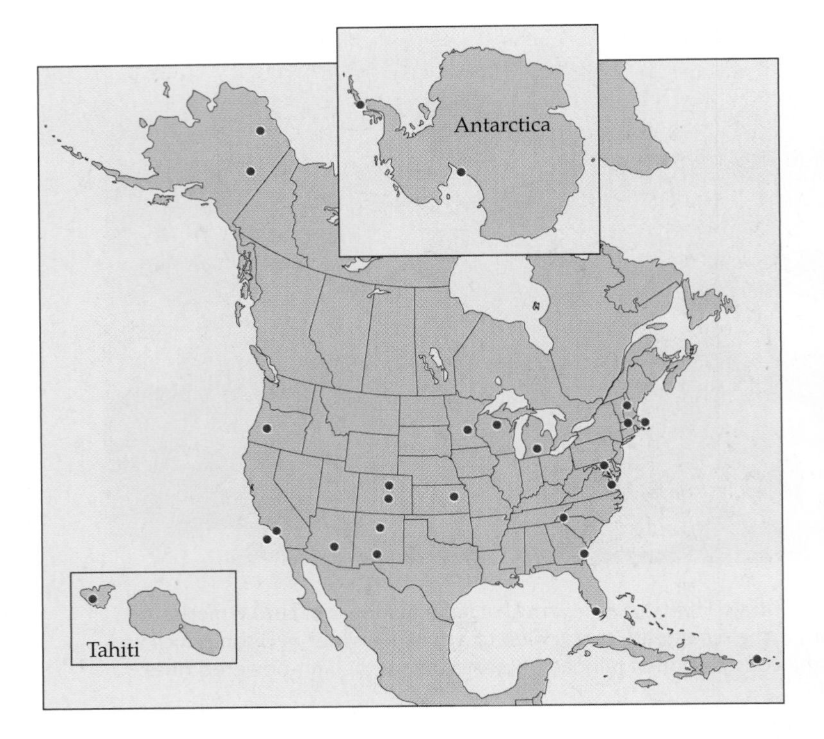

FIGURE 3.29 Long-Term Ecological Research Sites Twenty-six research sites constitute the U.S. Long-Term Ecological Research network. These sites encompass deserts, grasslands, forests, mountains, lakes, estuaries, agricultural systems, and cities. Researchers measure long-term changes in ecosystems and perform experiments at these sites to better understand ecological dynamics over decades to centuries.

from tropical to polar, as well as marine biological zones, croplands, and urban centers (**Figure 3.29**). The formation of the U.S. LTER program has spurred the formation of an international network of LTER sites, facilitating international collaborative research to better understand Earth's ecological systems.

Long-term ecological research has advanced our understanding of ecological changes that occur at decadal and longer time scales. For example, research at LTER sites in the western United States has led to an understanding of the influence of the El Niño Southern Oscillation and Pacific Decadal Oscillation, climate cycles discussed under Concept 2.5 on the grassland biome. The legacy of climate since the last glacial maximum, discussed in the Case Study at the opening of this chapter, is also better understood as a result of this research. Finally, research at LTER sites is providing a view of how environmental change, including climate change, may influence terrestrial biomes and marine ecosystems in the future.

In this chapter we've learned that grasslands are the most heavily impacted biome due to agricultural development. The Konza Prairie LTER site, located in the Flint Hills of northeastern Kansas, is a remnant tallgrass prairie—a

(A)

(B)

(C)

FIGURE 3.30 Research at the Konza Prairie LTER Site Long-term research and experiments are investigating the effects of the frequencies of (A) grazing, (B) fire, and (C) precipitation on the diversity and function of the tallgrass prairie ecosystem.

very heavily impacted grassland type with less than 4% of its original cover remaining. Research at the Konza Prairie site has focused on conserving this endangered biome in the face of rapid climate change and land use by examining the interactive roles of fire, grazing, and climate in the tallgrass prairie ecosystem. This research has included experiments varying the frequencies of fire and grazing in large landscape units to investigate their importance in maintaining the dominance of the grasses that characterize the grassland biome (**Figure 3.30**). Researchers have also examined the potential effects of changes in precipitation by varying the amount, intensity, and timing of watering. Results from this research have provided important insights into how climate change might affect the grassland biome, indicating that extremes in rainfall are important controls on its diversity and function (Knapp et al. 2002). Research at this and other LTER sites will enhance our ability to conserve native biodiversity in the face of accelerating environmental change.

SUMMARY

CONCEPT 3.1 Terrestrial biomes are characterized by the growth forms of the dominant vegetation.

- Terrestrial biomes are characterized by plant growth forms. These biomes reflect global patterns of precipitation and temperature.
- The potential and actual distributions of terrestrial biomes differ due to human activities, particularly conversion of land for agriculture, forestry, and grazing.
- There are nine major terrestrial biomes: tropical rainforests, tropical seasonal forests and savannas, and hot deserts in tropical and subtropical zones; grasslands, shrublands and woodlands, deciduous forests, and evergreen forests in the temperate zone, and boreal forests and tundra in polar regions.
- Biological communities in mountains occur in elevational bands associated with climatic gradients.

CONCEPT 3.2 Biological zones in freshwater ecosystems are associated with the velocity, depth, temperature, clarity, and chemistry of the water.

- Biological communities in streams and rivers vary with stream order and location within the stream channel.
- Biological communities in lakes vary with depth and light penetration.

CONCEPT 3.3 Marine biological zones are determined by ocean depth, light availability, and the stability of the bottom substrate.

- Estuaries, salt marshes, and mangrove forests occur in shallow zones at the margins between terrestrial and marine ecosystems. They are influenced by inputs of fresh water and sediments from nearby rivers.
- Biological communities at the shoreline reflect the influence of tides and the stability of the substrate (sandy versus rocky).
- Coral reefs and kelp and seagrass beds are productive communities with high diversity associated with the habitat complexity provided by their photosynthesizers.
- Biological communities of the open ocean and deep benthic zones contain sparse populations of organisms, whose distributions are determined by light availability and proximity to the bottom.

REVIEW QUESTIONS

1. Why are terrestrial biomes characterized using the growth forms of the dominant plants that occupy them?
2. Describe the close association between the distribution of biomes and the major climatic zones described in Chapter 2. In particular, consider how seasonality of both temperature and precipitation influence biome distribution.
3. As streams flow from their source to the oceans, what physical changes occur that affect the distribution of their biological communities?
4. Why do ocean depth and the stability of the substrate play important roles in determining the composition of marine biological communities?

ON THE COMPANION WEBSITE
sites.sinauer.com/ecology2e

The website includes Chapter Outlines, Online Quizzes, Flashcards & Key Terms, Suggested Readings, a complete Glossary, and the Web Stats Review. In addition, the following resources are available for this chapter:

HANDS-ON PROBLEM SOLVING

Changing Climate and Changing Tree Lines

This Web exercise explores connections between elevation of tree lines and climate patterns. You will read a paper that discusses factors determining upper tree lines and which types of tree lines are likely to advance with changes in temperature. You will then plot recent temperature changes in high elevation areas and discuss the probability of tree line advance there.

WEB EXTENSION

3.1 Climate Diagrams

CHAPTER

4

Coping with Environmental Variation: Temperature and Water

KEY CONCEPTS

- **CONCEPT 4.1** Each species has a range of environmental tolerances that determines its potential geographic distribution.
- **CONCEPT 4.2** The temperature of an organism is determined by exchanges of energy with the external environment.
- **CONCEPT 4.3** The water balance of an organism is determined by exchanges of water and solutes with the external environment.

Frozen Frogs: A Case Study

In the movie *2001: A Space Odyssey*, astronauts traveling to Jupiter are placed in a state of suspended animation while being transported across the vast distances of space to prevent them from aging. The idea of suspended animation—life being put on hold temporarily—has captured the imagination and hopes of people waiting for medical science to develop ways to cure untreatable diseases or reverse the ravages of aging. *Cryonics* is the preservation of the bodies of deceased people at subfreezing temperatures with the goal of eventually bringing them back to life and restoring them to good health. Proponents of cryonics exist throughout the world, some more visible than others. In Nederland, Colorado, each year, there is a "Frozen Dead Guy Days" festival, considered to be the "Mardi Gras of cryonics." This festival commemorates the efforts of a former resident who had his grandfather frozen immediately after his death from heart failure, hoping that one day his grandfather could be brought back to life and given a heart transplant (as also documented in the movie *Grandpa's Still in the Tuff Shed*).

To some, cryonics seems far-fetched, a thing of science fiction. Bringing life to a halt and then restarting it after a long period of quiescence doesn't seem plausible. Yet strange tales from nature provide examples of life apparently springing out of death. While seeking the existence of the Northwest Passage in the boreal and Arctic zones of Canada in 1769–1772, the English explorer Samuel Hearne found frogs under shallow layers of leaves and moss in winter "frozen as hard as ice, in which state the legs are as easily broken off as a pipe-stem" (Hearne 1911) (**Figure 4.1**). Hearne wrapped the frogs in animal skins and placed them next to his campfire. Within hours, the rock-hard amphibians came to life and began hopping around. The American naturalist John Burroughs found frozen frogs under a shallow cover of dead leaves in a New York forest in winter. Return visits to the same locations over a period of months indicated that the frogs hadn't moved, yet by spring they had disappeared. Could a complex organism like a frog, with a sophisticated circulatory and nervous system, have achieved cryonic preservation as an evolutionary response to a harsh winter climate?

Organisms of temperate and polar zones face tremendous challenges imposed by a seasonal climate that includes subfreezing temperatures in winter. Amphibians are unlikely candidates to have solved this challenge by allowing their bodies to partially freeze. Aside from their aforementioned complex organ–tissue

FIGURE 4.1 A Frozen Frog Wood frogs (*Rana sylvatica*) spend winter in a partially frozen state without breathing and with no circulation or heartbeat.

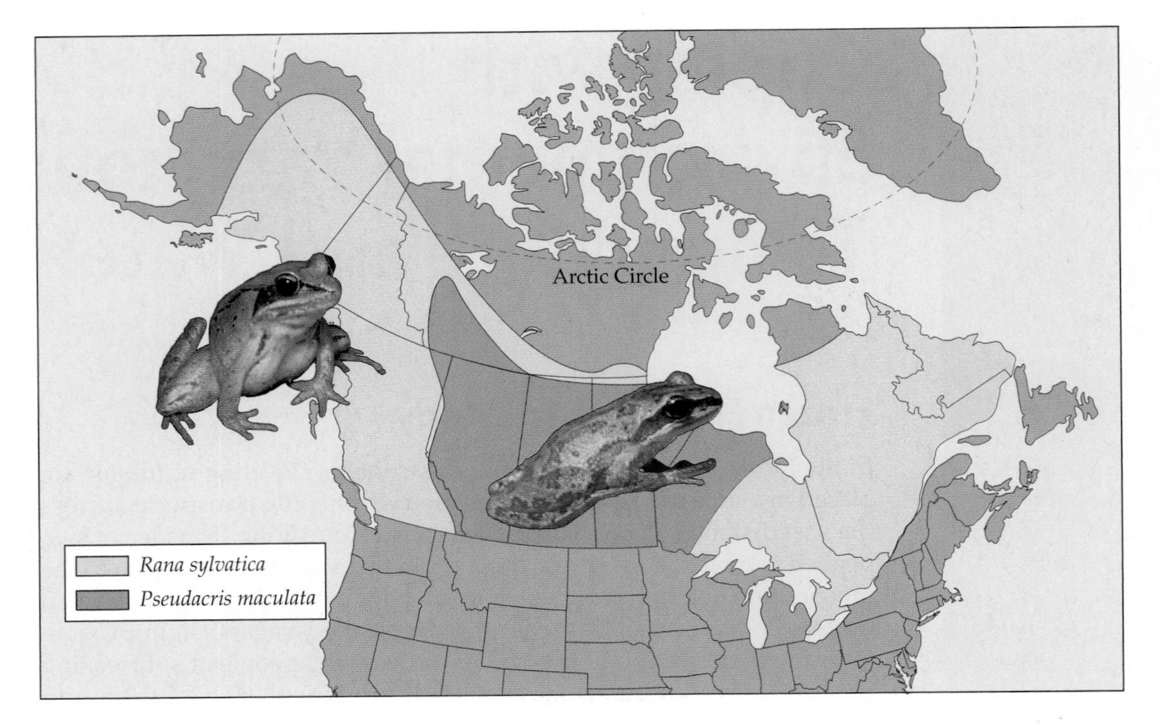

FIGURE 4.2 Northern Exposure Wood frogs (*Rana sylvatica*) and boreal chorus frogs (*Pseudacris maculata*) have geographic ranges that extend into the boreal forest and tundra biomes. (After Pinder et al. 1992.)

systems, amphibians are "cold-blooded" (generate little heat internally) and, as a group, first evolved in tropical and subtropical biomes. Yet two frog species, wood frogs (*Rana sylvatica*) and boreal chorus frogs (*Pseudacris maculata*), live in the Arctic tundra biome (**Figure 4.2**) (Pinder et al. 1992). These frogs survive extended periods of subfreezing air temperatures in shallow burrows in a semi-frozen state, with no heartbeat, no blood circulation, and no breathing. Among the vertebrates, only a few species of amphibians (four frogs and one salamander) and one turtle species can survive a long winter in a semi-frozen state. Freezing in most organisms results in substantial damage to tissues as ice crystals perforate cell membranes and organelles. How do these vertebrates survive being frozen without turning to mush once they thaw out, then reinitiate their blood circulation and breathing once spring arrives?

Introduction

Spruce trees (*Picea obovata*) of the boreal forest of Siberia experience the extreme range of seasonal temperatures characteristic of a continental climate. In winter, air temperatures regularly drop below –50°C (–58°F), and in summer, they reach 30°C (86°F). As an immobile tree, the spruce lacks the option to move to Florida for the winter or head to the coast to cool off in summer. The spruce must *tolerate* these temperature extremes, surviving the 80°C (144°F) seasonal change in its tissue temperatures, rather than *avoiding* them through some behavior or physiological change. These two options for coping with environmental change, **tolerance** and **avoidance**, provide a useful framework for thinking about how organisms cope with the environmental extremes they face.

The range of physical environmental conditions described in Chapter 2 establishes the variation in biomes and marine biological zones described in Chapter 3. In this chapter and the next, we will examine more thoroughly the interactions between organisms and the physical environment that influence their survival and persistence, and therefore their geographic range. The study of these interactions is known as *physiological ecology*.

CONCEPT 4.1 Each species has a range of environmental tolerances that determines its potential geographic distribution.

Response to Environmental Variation

A fundamental principle in ecology and biogeography is that the geographic ranges of species are related to constraints imposed by the physical and biological environments. In this section, we will discuss the general principles of organismal responses to the physical environment.

Species distributions reflect environmental influences on energy acquisition and physiological tolerances

The potential geographic range of an organism is ultimately determined by the physical environment, which influences an organism's *ecological success* (its survival and reproduction) in two important ways. First, the physical environment affects an organism's ability to obtain the energy and resources required to maintain its metabolic functions, and therefore to grow and reproduce. Rates of photosynthesis and abundances of prey, for example, are controlled by environmental conditions. An organism's ability to maintain a viable population is constrained at the limits of its geographic range. Second, an organism's survival can be affected by extreme environmental

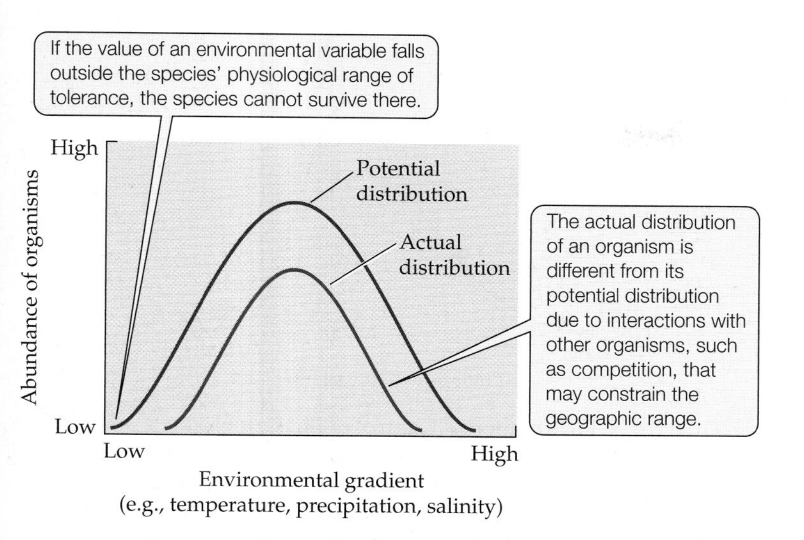

FIGURE 4.3 Abundance Varies across Environmental Gradients The abundance of an organism reaches a theoretical maximum at some optimal value across an environmental gradient and drops off at either end at values that constrain the potential geographic distribution of the organism. The actual abundance curve is likely to differ from the potential abundance curve due to biological interactions.

conditions. If temperature, water supply, chemical concentrations, or other physical conditions exceed what an organism can tolerate, the organism will die. These two influences—the availability of energy and resources and physical tolerance limits—are not mutually exclusive, as energy supply influences an organism's ability to tolerate environmental extremes. It is important to keep in mind that the *actual* geographic distribution of a species is also related to other factors, such as dispersal (Chapter 17), disturbance (e.g., fire, Chapter 16) and interactions with other organisms such as competition (Unit 3) (**Figure 4.3**).

As indicated in the introduction to this chapter and in the discussion of biomes in Chapter 3, the immobility of plants makes them good indicators of the physical environment. Farmers are acutely aware of the effects of extreme events on the survival of crop plants, which are often grown outside the geographic ranges where they evolved. Freezing events or extreme droughts can result in catastrophic crop losses. Aspen (*Populus tremuloides*) provides a good example of a native species with a geographic range related to climatic tolerance (**Figure 4.4**). Aspen occurs in boreal forests and mountain zones throughout North America. Its geographic

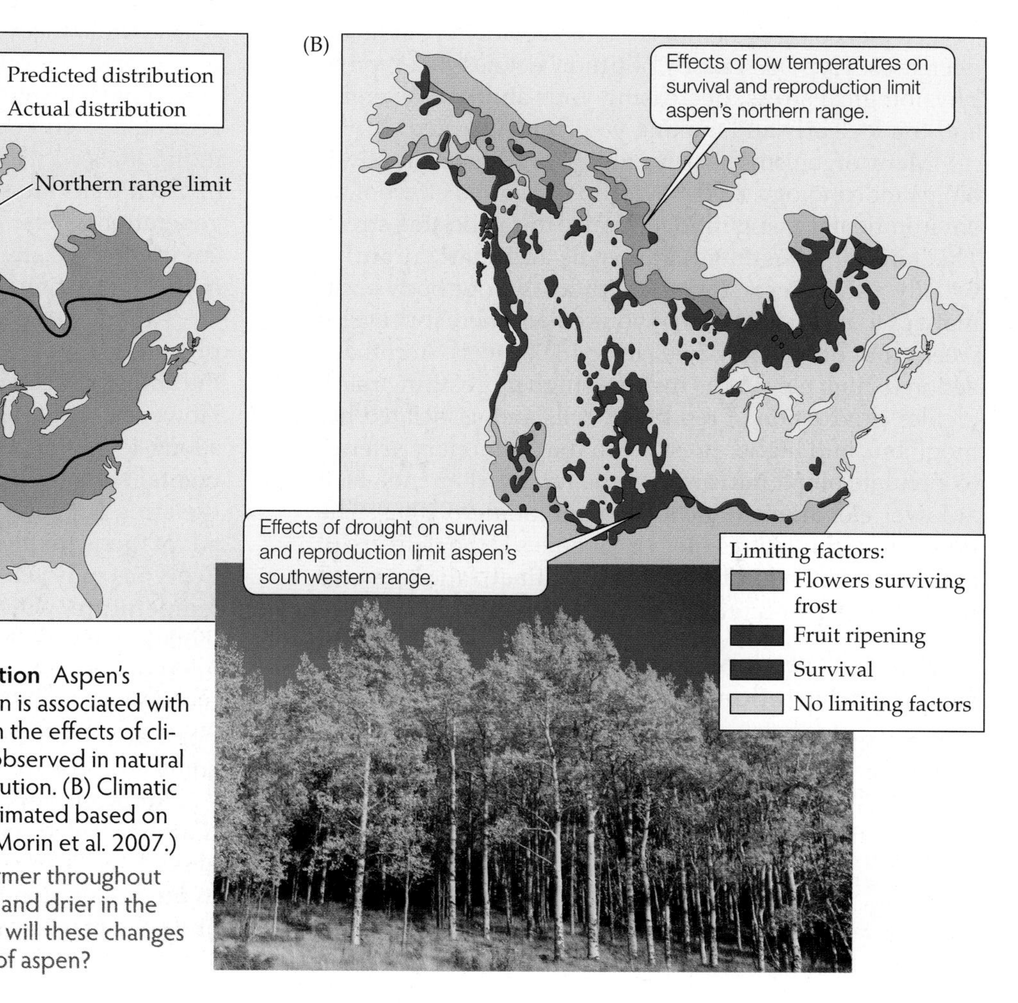

FIGURE 4.4 Climate and Aspen Distribution Aspen's (*Populus tremuloides*) geographic distribution is associated with climate. (A) Predicted distribution, based on the effects of climatic factors on survival and reproduction observed in natural populations, mapped with the actual distribution. (B) Climatic factors limiting the distribution of aspen, estimated based on observations of natural populations. (After Morin et al. 2007.)

The future climate is predicted be warmer throughout the interior of western North America and drier in the central portions of the continent. How will these changes influence the geographic distribution of aspen?

distribution can be predicted fairly accurately from the observed effects of climate on its survival and reproduction (Morin et al. 2007) (see Figure 4.4A). The climatic factors that limit its distribution include the effects of low temperatures on its reproductive success and the effects of drought and low temperatures on its survival (**Figure 4.4B**). The range of climatic conditions under which a species occurs—its *climate envelope*—provides a useful tool for predicting its response to climate change (see Chapter 24).

Individuals respond to environmental variation through acclimatization

Physiological processes such as growth or photosynthesis have a set of optimal environmental conditions most conducive to their functioning. Deviations from those optimal conditions cause a decrease in the rate of the process (**Figure 4.5**). **Stress** is the condition in which an environmental change results in a decrease in the rate of an important physiological process, thereby lowering the potential for an organism's survival, growth, or reproduction. For example, when you travel to high elevations, typically above 2,400 m (8,000 feet), the lower partial pressure of oxygen in the atmosphere (see p. 67) results in the delivery of less oxygen to your tissues by your circulatory system. This condition, known as *hypoxia*, results when the amount of oxygen picked up by hemoglobin molecules in your blood decreases. Hypoxia causes "altitude sickness," a type of physiological stress, decreasing your ability to exercise and think clearly and making you feel nauseated.

Many organisms have the ability to adjust their physiology, morphology, or behavior to lessen the effect of an environmental change and minimize the associated stress. This kind of adjustment, known as **acclimatization**,* is usually a short-term, reversible process. Your body acclimatizes to high elevations if you remain there for several weeks (but only below 5,500 m or 18,000 feet). Acclimatization to high elevations involves higher breathing rates, greater production of red blood cells and associated hemoglobin, and higher pressure in the pulmonary arteries to circulate blood into areas of the lung that are not used at lower elevations (Hochochka and Somero 2002). The outcome of these physiological changes is the delivery of more oxygen to your tissues. The acclimatization process reverses when you return to lower elevations.

Populations respond to environmental variation through adaptation

Within a species' geographic range, individual populations often occur in unique environments (e.g., cool climates, saline

*Animal physiologists use the term "acclimatization" to refer to the short-term response of an animal to changes in the physical environment under field conditions and "acclimation" to refer to a short-term response under controlled laboratory conditions.

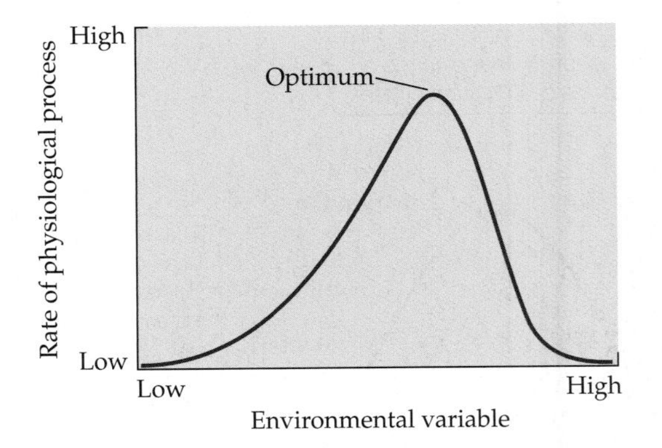

FIGURE 4.5 Environmental Control of Physiological Processes The rates of physiological processes are greatest under a set of optimal environmental conditions (e.g., optimal temperature, optimal water availability). Deviations from the optimum cause a decrease in the rates of physiological processes.

soils) that may have initially been stressful to the organisms when they first occupied them. Genetic variation among the individuals within such populations in physiological, morphological, or behavioral traits that influenced their survival and functioning in the new environment would have led to natural selection favoring those individuals whose traits made them best able to cope with the new conditions. The underlying genetic basis for these traits would have resulted in a change over generations in the genetic makeup of the population as the abundance of individuals with the favored traits increased (see Chapter 6). Such traits are known as **adaptations**. Through many generations, these unique, genetically based solutions to environmental stress would have become more frequent in the population.

Adaptation is similar to acclimatization in that both processes involve a change that minimizes stress, and the ability to acclimatize represents a type of adaptation. However, adaptation differs from acclimatization in being a long-term, genetic response of a population to environmental stress that increases its ecological success under the stressful conditions (**Figure 4.6**). Populations with adaptations to unique environments are called **ecotypes**. Ecotypes may represent responses to both abiotic (e.g., temperature, water availability, soil type, salinity) and biotic environmental factors (e.g., competition, predation). Ecotypes can eventually become separate species as the physiology and morphology of individuals in different populations diverge and the populations eventually become reproductively isolated.

Returning to our previous example, human populations have lived continuously in the Andean highlands for at least 10,000 years. When Spanish explorers first settled in the Andes alongside the native people in the sixteenth and seventeenth centuries, their birth rates were low for

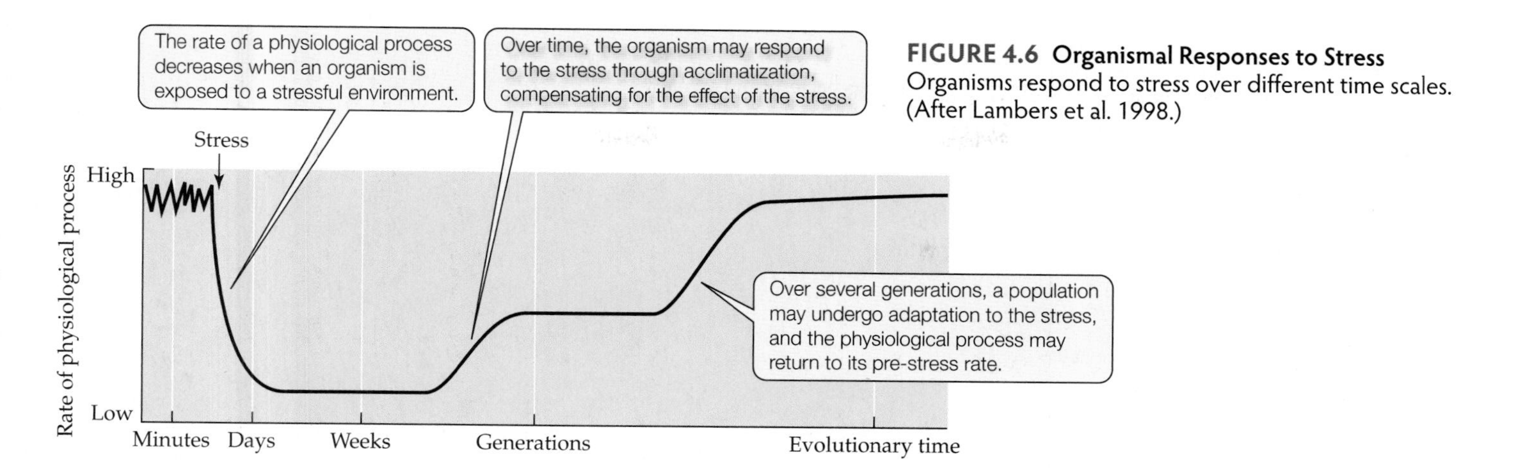

FIGURE 4.6 Organismal Responses to Stress Organisms respond to stress over different time scales. (After Lambers et al. 1998.)

2–3 generations, probably due to poor oxygen supply to developing fetuses (Ward et al. 1995). The same held true for the domesticated animals they brought with them. This observation provides anecdotal evidence for the adaptive response of the native Andean populations to the low-oxygen conditions at high elevations. Research in the twentieth century showed that their adaptations to high elevations include higher red blood cell production and greater lung capacity (Ward et al. 1995).

Adaptations to environmental stress can vary among populations. In other words, the solution to the environmental problem may not be the same for each population. This is demonstrated by a comparison of native human populations from the Andean and Tibetan highlands. The adaptations to high elevations in Andean populations (high red blood cell concentration and large lung capacity) are not the same as those found in Tibetan populations (Beall 2007). Tibetan populations have red blood cell concentrations similar to and blood oxygen concentrations lower than populations at sea level, but they have a higher breathing rate, which enhances the exchange of oxygen with the blood system, and higher blood flows, which enhance delivery of oxygen to vital organs such as the brain. Thus, there are at least two different ways that human populations have adapted to hypoxic stress imposed by living at high elevations.

Acclimatization and adaptation are not "free"; they require an investment of energy and resources by the organism (as in the maxim "there is no free lunch" presented in Chapter 1). They represent possible *trade-offs* with other functions of the organism that may also affect its survival and reproduction. Acclimatization and adaptation must therefore increase the survival and reproductive success of the organism under the specific environmental conditions in order to be favored over other patterns of energy and resource investment. (Trade-offs in energy and resource allocation are discussed in Chapter 7.)

In the remaining two sections, we will examine the factors that determine organisms' temperatures, water content, and water uptake, and we will consider examples of acclimatization and adaptation that allow organisms to function in the face of varying temperatures and water availability.

CONCEPT 4.2 The temperature of an organism is determined by exchanges of energy with the external environment.

Variation in Temperature

Environmental temperatures vary greatly throughout the biosphere, as we saw in Chapter 2. The boreal forest described earlier in this chapter represents one extreme of seasonal variation, with as much as an 80°C (144°F) swing from summer to winter. Tropical forests, on the other hand, experience far less seasonal variation in temperature, about 15°C (22°F). Soil environments, which are home to many species of microorganisms, plant roots, and animals, are buffered from aboveground environmental temperature extremes, although soil surface temperatures may change as much as or more than air temperatures. Aquatic environments also experience temperature changes over seasonal and daily time scales. Pelagic marine (open ocean) environments tend to have very little temporal variation in temperature due to the ocean's massive volume and heat capacity. In contrast, tide pools experience large variations in water temperature as the tides rise and fall, with as much as a 20°C (36°F) change over a 5-hour period.

The survival and functioning of organisms is strongly tied to their internal temperatures. The extreme upper limit for metabolically active multicellular plants and animals is about 50°C (122°F) (**Figure 4.7**). Some archaea and bacteria that live in hot springs can function at 90°C (194°F) (Willmer et al. 2005). The extreme lower limit for organismal function is tied to the temperature at which water in cells freezes, typically between –2°C and –5°C (28°F–23°F). Some organisms can survive periods of extreme heat or cold by entering a state of **dormancy**, in which little or no metabolic activity occurs.

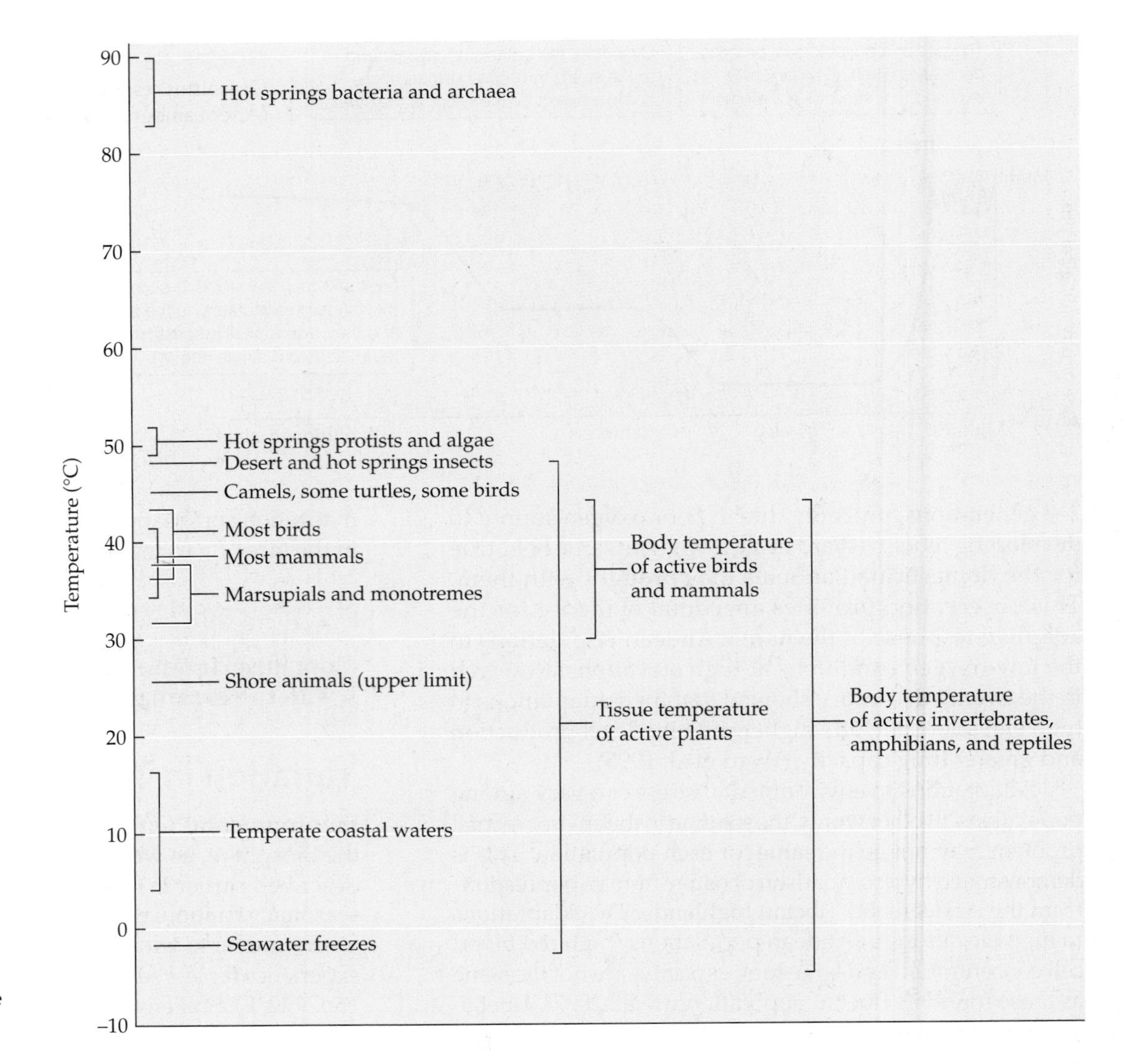

FIGURE 4.7 Temperature Ranges for Life on Earth

The internal temperature of an organism is determined by the balance between the energy it gains from and the energy it loses to the external environment. Thus, organisms must either tolerate changes in their internal temperature as the temperature of the external environment changes or modify their internal temperature by some physiological, morphological, or behavioral means. Environmental temperatures—particularly their extremes—are therefore important determinants of the distributions of organisms, as demonstrated by the relationship between biomes and global climatic patterns discussed in the Chapters 2 and 3.

Temperature controls physiological activity

The metabolic generation of chemical energy and the synthesis of organic compounds used for growth and reproduction depend on biochemical reactions that are temperature-sensitive. Each reaction has an optimal temperature that is related to the activity of *enzymes*, protein-based molecules that catalyze biochemical reactions. Enzymes are structurally stable under a limited range of temperatures. At high temperatures, the constituent proteins lose their structural integrity, or become *denatured*, as their bonds break. Most enzymes become denatured at temperatures ranging between 40°C and 70°C (104°F–158°F), but enzymes in bacteria inhabiting hot springs can remain stable at temperatures up to 100°C (212°F). The upper lethal temperature for most organisms is lower than the temperature at which their enzymes become denatured because metabolic coordination among biochemical pathways is lost at these temperatures. The extreme lower limit for enzyme activity is about –5°C (23°F) (Willmer et al. 2005). Antarctic fish and crustaceans may reach internal temperatures of –2°C because the salt concentration of the seawater in which they live lowers its freezing point. Some soil microbes are active at temperatures as low as –5°C.

Some species can produce different forms of enzymes (called *isozymes*) with different temperature optima as a means of acclimatization to changes in environmental temperature. For example, some fish species (e.g., trout, carp, goldfish) and trees (e.g., loblolly pine) can produce isozymes in response to seasonal changes in temperature. However, acclimatization to temperature changes using isozymes does not appear to be a common response in animals (Willmer et al. 2005).

Temperature also determines the rates of physiological processes by influencing the properties of membranes, particularly at low temperatures. Cell and organelle membranes are composed of two layers of lipid molecules. At low temperatures, these membranes can solidify, and proteins and enzymes embedded in them can lose their function, affecting processes such as mitochondrial respiration and photosynthesis. Membranes also lose their function as filters when they solidify, leaking cellular metabolites. Tropical plants may suffer loss of function associated with membrane disruption at temperatures as high as 10°C (50°F), while alpine plants can function at temperatures close to freezing. The sensitivity of membrane function to low temperatures is related to the chemical composition of the membrane lipid molecules. Plants of cooler climates have a higher proportion of unsaturated membrane lipids (with greater numbers of double bonds between carbon molecules) than plants of warmer climates.

Finally, temperature influences physiological processes in terrestrial organisms by affecting water availability. As we saw in Chapter 2, the warmer the air, the more water vapor it can hold. As a result, the rate at which terrestrial organisms lose water from their bodies is related to air temperature. We will return to this point later when we discuss how organisms cope with variations in water availability.

Organisms influence their temperature by modifying energy balance

On a hot day, jumping into a swimming pool and then sitting in the shade with a light wind brings relief from the oppressive heat. Elephants follow a similar routine, swimming and spraying water onto their backs with their trunks. This kind of behavior facilitates heat loss in several ways. The direct contact of your warm skin with the cool water causes heat energy to be lost from your body through the process of *conduction*, the transfer of energy from warmer, more rapidly moving molecules to cooler, more slowly moving molecules. When water and air move across the surface of your body, heat energy is carried away via *convection*. The change in the state of water from liquid to vapor as it evaporates on the surface of your skin absorbs heat (*latent heat transfer*). Finally, moving into the shade lowers the amount of energy you receive from solar radiation.

The balance between inputs and outputs of energy determines whether the temperature of any object, living or not, will increase or decrease. Archaea, bacteria, fungi, protists, and algae cannot avoid changes in their body temperature when the environmental temperature changes. They must tolerate variations in temperature through biochemical modifications. When temperatures outside of their range of tolerance occur, microorganisms often survive as dormant resistant spores. Plants and animals also exhibit varying degrees of tolerance of changes in body temperature. These organisms can also influence their temperature, and therefore their physiological processes, by adjusting their exchange of energy with the environment. Avoidance of stressful temperatures through behavioral and morphological modifications of energy balance is a strategy used by both plants and animals.

Let's examine some examples of modification of energy balance in plants and animals to better understand the adaptations that help them cope with temperature variation.

Plant modification of energy balance Among plants, temperature stress is experienced mainly in terrestrial environments. Marine and aquatic plants usually experience temperatures within the range that is conducive to their physiological functioning, although those in nearshore habitats can experience potentially lethal temperatures. The factors involved in the energy balance of terrestrial plants are shown in **Figure 4.8**. Energy inputs that warm the plant include sunlight and infrared radiation from surrounding objects. If the ground or air is warmer than the plant, energy inputs also include conduction and convection. Losses of energy include the emission of infrared

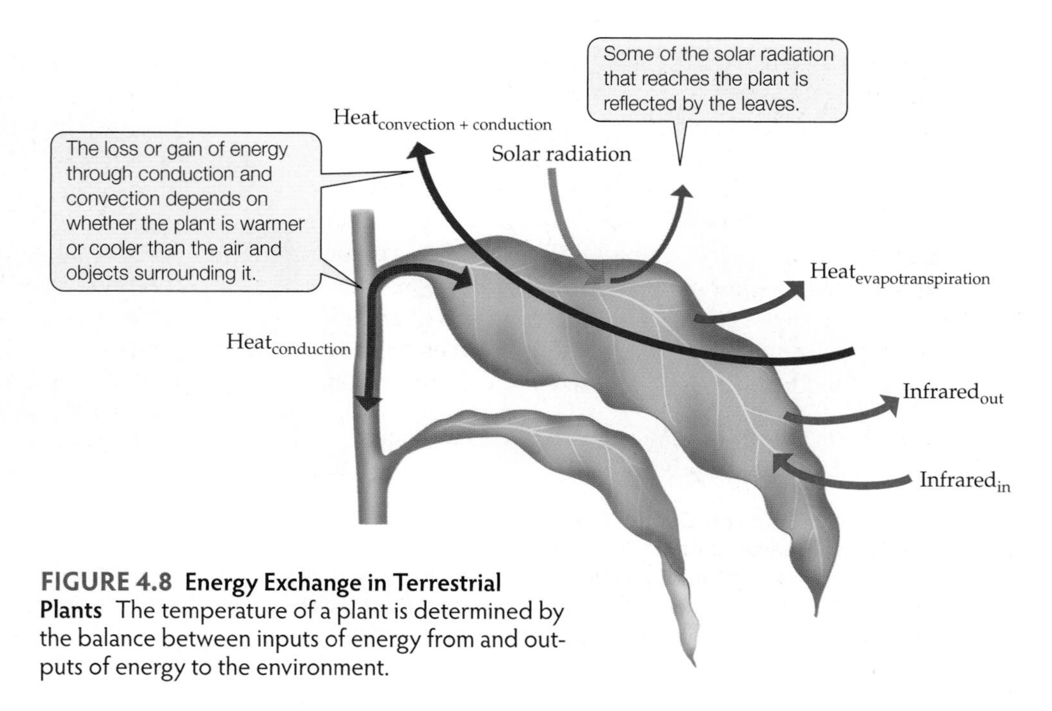

FIGURE 4.8 Energy Exchange in Terrestrial Plants The temperature of a plant is determined by the balance between inputs of energy from and outputs of energy to the environment.

radiation to the surrounding environment and, if the ground or air is cooler than the plant, conduction and convection. Heat loss also occurs through transpiration (evaporation of water from inside the plant) and evaporation, collectively referred to as *evapotranspiration*.

We can put these inputs and outputs together to determine whether the temperature of the plant is changing:

$$\Delta H_{plant} = SR + IR_{in} - IR_{out} \pm H_{conv} \pm H_{cond} - H_{et} \quad (4.1)$$

where ΔH_{plant} is the heat energy change of the plant (the Greek delta usually represents "change in"), SR is solar radiation, IR_{in} is the input of infrared radiation, IR_{out} is the output of infrared radiation, H_{conv} is convective heat transfer, H_{cond} is conductive heat transfer, and H_{et} is heat transfer through evapotranspiration. A negligible loss of energy occurs through the plant's use of solar radiation for photosynthesis. If the plant is warmer than the surrounding air, then H_{conv} and H_{cond} are negative. If the sum of the energy inputs exceeds the sum of the outputs, then ΔH_{plant} is positive, and the plant's temperature is increasing. Conversely, if more heat is being lost than gained, then ΔH_{plant} is negative, and the plant's temperature is decreasing.

Plants can modify their energy balance to control their temperature by adjusting these energy inputs and outputs. Leaves are most often associated with these adjustments because they are the primary photosynthetic organs of the plant and typically are the most temperature-sensitive tissue. The most important and common adjustments include changes in the rate of transpirational water loss. In addition, changes in leaf surface reflective properties or in leaf orientation toward the sun can alter the amount of solar radiation absorbed by the plant. Finally, changes in convective heat transfer can be accomplished by changing surface roughness.

Transpiration is an important evaporative cooling mechanism for leaves. Its effectiveness is especially evident in the canopies of tropical forests, which are subjected to warm air temperatures and high levels of solar radiation. Without transpirational cooling, the leaves of tropical canopy plants could reach lethal temperatures over 45°C (> 113°F). The rate of transpiration is controlled by specialized *guard cells* surrounding pores, called **stomates**, leading to the interior of the leaf. Stomates are the gateway for both transpirational water loss and the uptake of carbon dioxide from the atmosphere (we will return to the latter function in Chapter 5). Variation in the degree of stomatal opening, as well as in the number of stomates, controls the rate of transpiration and therefore exerts an important control on leaf temperature (**Figure 4.9**).

Transpiration requires a steady supply of water. Where the supply of water in the soil is limited—as it is over a substantial part of Earth's land surface—transpiration is not a reliable cooling mechanism. As we saw in Chapter 3, some plants shed their leaves during dry seasons, thereby avoiding both temperature stress and water stress. However, the high resource requirement for producing leaves may favor protecting existing leaves rather than shedding them. Plants that maintain their leaves during long dry periods require mechanisms other than transpiration to dissipate heat energy. One option is to alter the reflective properties of leaves via **pubescence**, which is the presence of hairs on the leaf surface. The light color associated with white hairs

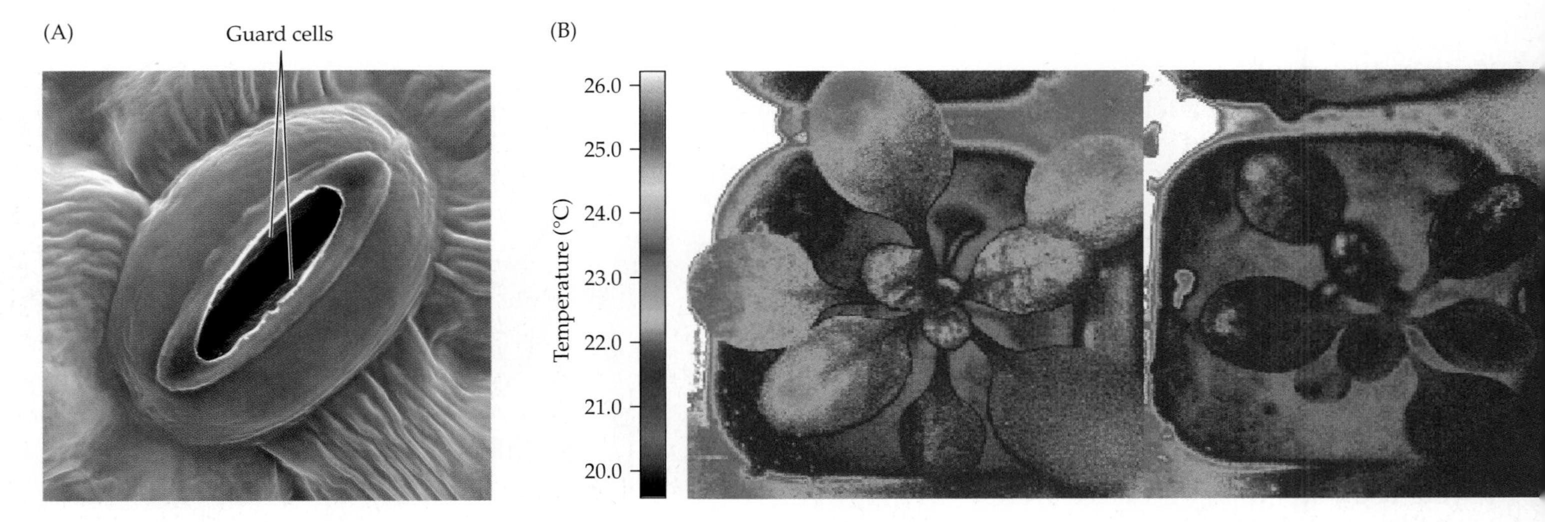

FIGURE 4.9 Stomates Control Leaf Temperature by Controlling Transpiration (A) Specialized guard cells surrounding the stomates control their degree of opening. Open stomates allow CO_2 to diffuse in for photosynthesis and allow water to transpire out, cooling the leaves. (B) Leaf temperatures vary according to the degree of stomatal opening. The plant on the right has open stomates and is transpiring freely, while the plant on the left, kept under identical conditions, has closed stomates, a lower transpiration rate, and a temperature 1°C–2°C (2°F–4°F) higher, as indicated by thermal infrared imaging.

Q Cooling of leaves using transpiration may be particularly important in what biomes?

lowers the amount of solar radiation absorbed by the leaf surface. Note, however, that pubescence can also lower the effectiveness of convective heat loss and thus represents a trade-off between two opposing heat exchange mechanisms.

One of the best studies addressing the adaptive significance of leaf pubescence for temperature regulation has focused on shrubs of the genus *Encelia* (members of the composite or daisy family). Jim Ehleringer and his colleagues described the role of pubescence in leaf temperature regulation among species of *Encelia* that occupy different geographic ranges. *Encelia farinosa*, a native of the Sonoran and Mojave deserts, maintains a high amount of leaf pubescence relative to *Encelia* shrubs from moister, cooler environments. Ehleringer and Craig Cook evaluated the relative roles of leaf pubescence and transpiration in the cooling of leaves of *E. farinosa* and two other species whose leaves lack pubescence: *E. frutescens*, which occurs in desert washes (which have more moisture than the rest of the desert), and *E. californica*, native to the cooler, moister coastal sage community of California and Baja California (Ehleringer and Cook 1990). To control for environmental variation that could influence the morphology and physiology of the plants, they grew plants of each species from seed together in *common gardens* in the Sonoran Desert and on the California coast. Half of their experimental plants were watered, and the other half were left under natural conditions. They measured leaf temperatures, the degree of stomatal opening, and the amount of sunlight absorbed.

The three *Encelia* species showed few differences in leaf temperature and stomatal opening when grown in the cooler, moister California coastal garden. In the desert garden, however, *E. californica* and *E. frutescens* shed their leaves during the hot summer months under natural conditions, while *E. farinosa* did not. *Encelia frutescens* did not shed its leaves when the shrubs were watered, and its leaves maintained sublethal temperatures using transpirational cooling. *Encelia farinosa* leaves reflected about twice as much solar radiation as the other two species (**Figure 4.10A**), which facilitated the shrub's ability to maintain leaf temperatures lower than the air temperature.

Ehleringer and Cook's common garden experiment provides correlative evidence of the adaptive value of leaf pubescence to *E. farinosa* under hot desert conditions. Additional work by Darren Sandquist and Ehleringer has supported its adaptive value, indicating that natural selection has acted on variation in pubescence among ecotypes of *E. farinosa*. Populations in drier environments have more leaf pubescence, and absorb less solar radiation, than populations from moister environments (Sandquist and Ehleringer 2003).

In addition to varying among species and populations, leaf pubescence can also vary seasonally in *E. farinosa*, exemplifying acclimatization to environmental conditions. *Encelia*

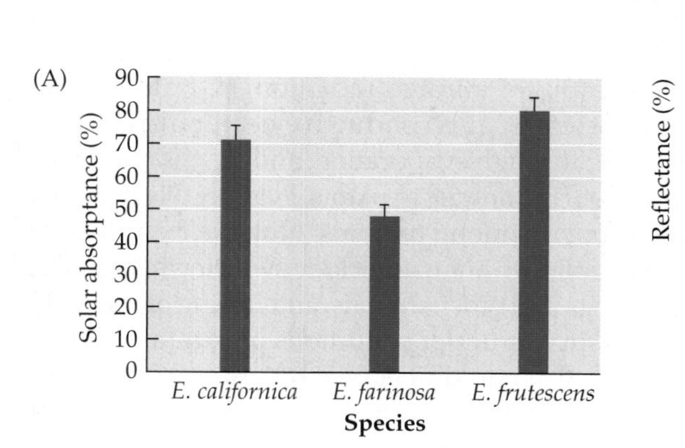

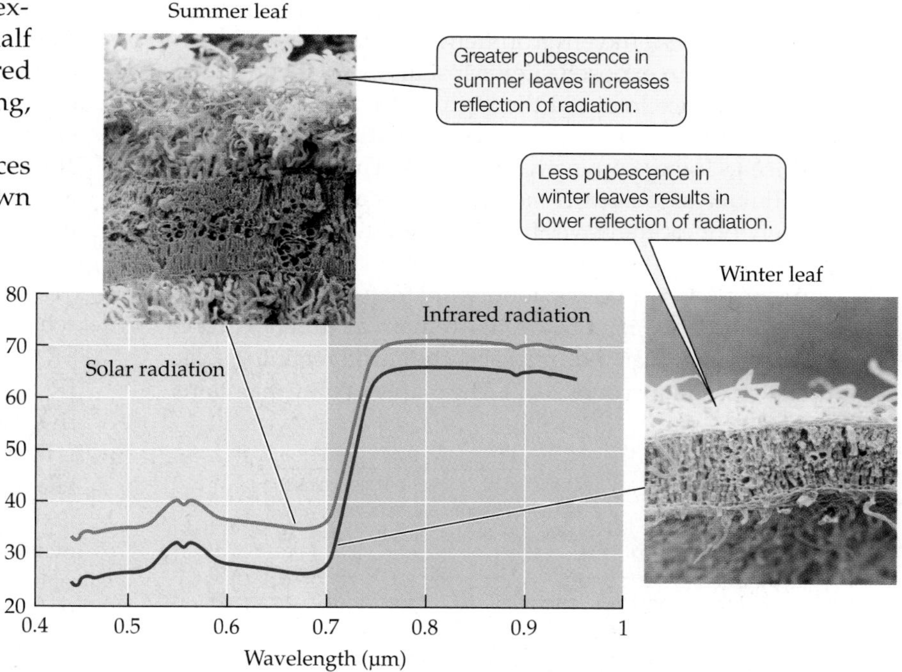

FIGURE 4.10 Sunlight, Seasonal Changes, and Leaf Pubescence (A) Solar heating of leaves varies according to the amount of pubescence on those leaves. The pubescent leaves of the desert shrub *Encelia farinosa* reflect a greater percentage of the incoming solar radiation than the leaves of two nonpubescent species: *E. californica*, native to the coastal sage community of California, and *E. frutescens*, an inhabitant of moister desert wash communities. *Encelia farinosa* is therefore less dependent on transpiration for leaf cooling than the other two species. Error bars show one SE of the mean. (B) *Encelia farinosa* produces greater amounts of pubescence on its leaves in summer than in winter, representing acclimatization to hot summer temperatures. The photos are scanning electron micrographs of leaf cross sections. (A after Ehleringer and Cook 1990.)

Why might temperature regulation associated with greater reflection of solar radiation via pubescence be more important in deserts than in a warm, moist climate such as the tropics?

farinosa shrubs produce smaller, more pubescent leaves in summer and larger, less pubescent leaves in winter (**Figure 4.10B**). There are costs to being pubescent, associated with the construction of the hairs and the loss of solar radiation that can be used for photosynthesis. Thus, when temperatures are cooler or when reliable soil water is present, *E. farinosa* plants construct leaves with fewer hairs.

Heat can be lost from a leaf by convection when the air temperature is lower than the temperature of the leaf. The effectiveness of convective heat loss is related to the speed of the air moving across a surface. As the moving air experiences more friction closer to the surface of an object, the flow becomes more turbulent, forming eddies (**Figure 4.11**). This zone of turbulent flow, called the **boundary layer**, lowers convective heat loss. The thickness of the boundary layer around a leaf is related to its size and its surface roughness. Small, smooth leaves have thin boundary layers and lose heat more effectively than large or rough leaves. This relationship between the boundary layer and convective heat loss is one reason for the rarity of large leaves in desert ecosystems.

Excessive heat loss by convection can be a problem for plants (and animals) in cold, windy environments such as the alpine zone of a mountain range. Convection is the most important source of heat loss from the land surface in temperate alpine environments, and high winds can shred leaves in exposed sites. Most alpine plants hug the ground surface to avoid the high wind velocities. Some alpine plants produce a layer of insulating hair on their surface to lower convective heat loss. The snow lotus of the Himalayas (*Saussurea medusa*) produces a series of very densely pubescent leaves that surround the flowers of the plant (**Figure 4.12**). Although they project above the ground surface and are exposed to more wind than ground-hugging plants, the flowers of *S. medusa* remain as much as 20°C (36°F) warmer than the air by absorbing and retaining solar radiation (Tsukaya et al. 2002). The plant not only maintains

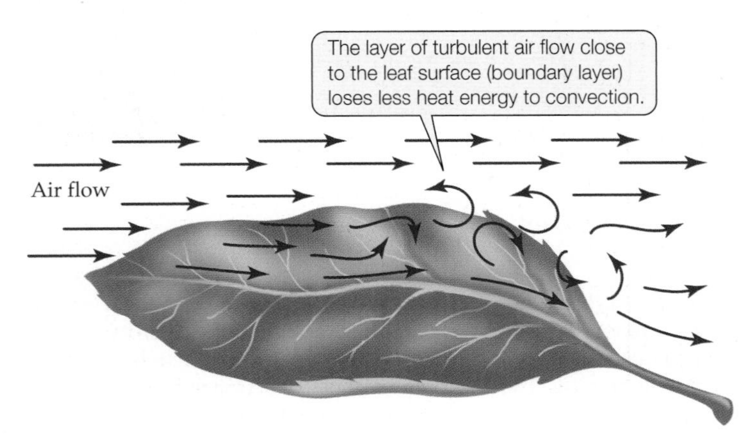

FIGURE 4.11 A Leaf Boundary Layer Air flowing close to the surface of a leaf is subject to friction, which causes the flow to become turbulent and lowers convective heat loss from the leaf to the surrounding air.

FIGURE 4.12 A Woolly Plant of the Himalayas The snow lotus (*Saussurea medusa*) has dense pubescence surrounding its emergent flowering stems, providing them with thermal insulation.

warmer photosynthetic tissues, but also provides a warm environment for potential pollinators, which are in short supply in cold, windy alpine environments.

Animal modification of energy balance Animals are subject to the same energy inputs and outputs described for plants in Equation 4.1, with one key difference: some animals—in particular, birds and mammals—have the ability to generate heat internally. As a result, another term is needed in the energy balance equation to represent this metabolic heat generation:

$$\Delta H_{animal} = SR + IR_{in} - IR_{out} \pm H_{conv} \pm H_{cond} - H_{evap} + H_{met} \quad (4.2)$$

where ΔH_{animal} is the heat energy change of the animal, SR is solar radiation, IR_{in} is the input of infrared radiation, IR_{out} is the output of infrared radiation, H_{conv} is convective heat transfer, H_{cond} is conductive heat transfer, H_{evap} is heat transfer through evaporation, and H_{met} is metabolic heat generation. In contrast to plants, evaporative heat loss is not widespread among animals. Notable examples of evaporative cooling in animals include sweating in humans, panting by dogs and other animals, and licking of the body by some marsupials under conditions of extreme heat.

The internal generation of heat by some animals represents a major ecological advance. Animals capable of metabolic heat generation can maintain relatively constant internal temperatures near the optimum for physiological functioning under a wide range of external temperatures, and as a result, can expand their geographic ranges. Varying degrees of reliance on internal heat generation exist throughout the animal kingdom. Animals that regulate their body temperature primarily through energy exchange with the external environment, which includes the majority of animal species, are called **ectotherms**. Animals

FIGURE 4.13 Internal Heat Generation as a Defense Bees can generate heat by contracting their flight muscles. Japanese honeybees (*Apis cerana*) use internal heat generation as a defense against hornets (*Vespa mandarina*) that attack bee colonies. (A) When a hornet enters a nest, the honeybees swarm the larger invader. (B) The defensive ball of bees surrounding an invading hornet generates enough heat that temperatures in the center exceed the upper lethal temperature for the hornet (about 47°C), resulting in the invader's death.

that rely primarily on internal heat generation, which are called **endotherms**, include, but are not limited to, birds and mammals. Internal heat generation is also found in some fish (e.g., tuna), insects (e.g., bees, which generate heat for metabolic function as well as for defense; **Figure 4.13**), and even a few plant species (e.g., skunk cabbage, *Symplocarpus foetidus*, which warms its flowers using metabolically generated heat during the spring).

Temperature regulation and tolerance in ectotherms

Generally, ectotherms have a greater tolerance for variation in their body temperature than endotherms, possibly because they are less able to adjust their body temperatures than endotherms. The exchange of heat between an animal and the environment, whether for cooling or heating, depends on the amount of surface area relative to the volume of the animal. A larger surface area relative to volume allows greater heat exchange, but makes it harder to maintain a constant internal temperature in the face of variable external temperatures. A smaller surface area relative to volume decreases the animal's ability to gain or lose heat. This relationship between surface area and volume imposes a constraint on the body size and shape of ectothermic animals. Generally speaking, the surface area-to-volume ratio decreases as body size increases, and the animal's ability to exchange heat with the environment decreases as well. As a result, large ectothermic animals are considered improbable. This conclusion has led to speculation that large dinosaurs may have had some degree of endothermy.

Small aquatic ectotherms (e.g., most invertebrates and fish) generally remain at the same temperature as the surrounding water. Some larger aquatic animals, however, can maintain a body temperature warmer than that of the surrounding water (**Figure 4.14**). For example, skipjack tuna (*Katsuwonis pelamis*) use muscle activity, in conjunction with heat exchange between blood vessels, to maintain

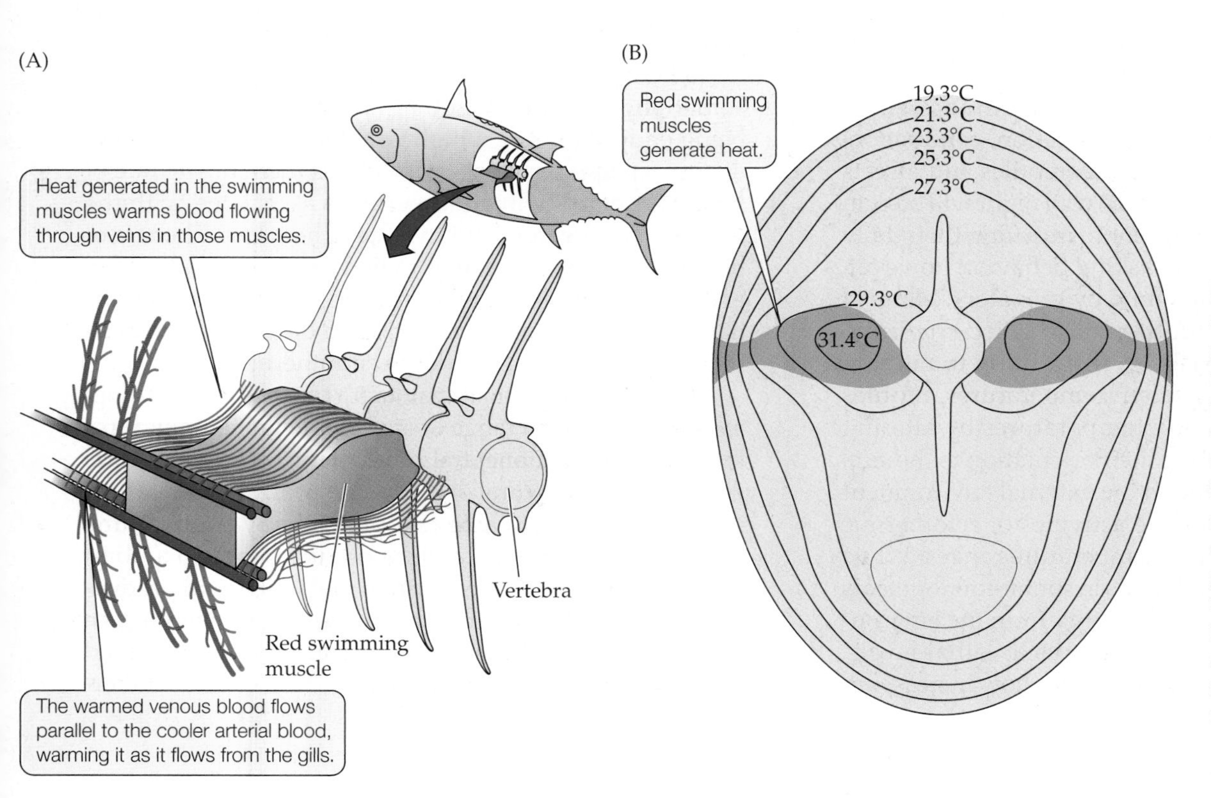

FIGURE 4.14 Internal Heat Generation by Tuna (A) Heat generated in the red swimming muscles of the skipjack tuna, used for cruising through the water, warms blood flowing through them, which is carried toward the body surface in veins. Those veins run parallel to arteries carrying cool oxygenated blood from the gills, warming that blood before it reaches the swimming muscles. (B) A cross section of the tuna shows that its core temperatures remain warmer than that of the surrounding water.

FIGURE 4.15 Mobile Animals Can Use Behavior to Adjust Their Body Temperature This ornate tree lizard (*Urosaurus ornatus*) has moved into a sunny location to raise its body temperature to a range suitable for undertaking its daily activities. Basking behavior can expose animals to the risk of predation, but camouflage (crypsis) helps minimize this risk.

Q What components of energy exchange are affected by this lizard's behavior?

a body temperature as much as 14°C (25°F) warmer than the surrounding seawater. Other large oceanic fishes use similar circulatory heat exchange mechanisms to keep their muscles warm. Such mechanisms are particularly important for predatory species that depend on rapid acceleration to capture prey, which is aided by having warmer muscles.

The mobility of many terrestrial ectotherms allows them to adjust their body temperature by moving to places that are warmer or cooler than they are. Basking in the sun or moving into the shade allows these animals to adjust their energy gains and losses via solar radiation, conduction, and infrared radiation. For example, reptiles and insects emerging from hiding places after a cool night will bask in the sun to warm their bodies prior to initiating their daily activities (**Figure 4.15**). This basking behavior, however, increases their risk of being found by predators. Many of these animals rely on camouflage (also called *crypsis*) to escape detection while basking. In addition to moving between locations with different temperatures, reptiles may also regulate their body temperatures by altering their coloration and changing their orientation to the sun.

Because they are reliant on the external environment for temperature regulation, the activities of ectothermic animals are limited to certain temperature ranges. When temperatures are warm, ectotherms in sunny environments (e.g., deserts) may gain enough energy from the environment to push their body temperatures to lethal levels. **Climate Change Connection 4.1** describes how increases in temperature associated with climate change over the past two decades appear to have limited the foraging periods of several species of Mexican lizards, whose abundances have decreased significantly during this period (discussed also on p. 537 in Chapter 24).

In temperate and polar regions, temperatures drop below freezing for extended periods. Ectotherms inhabiting these regions must either avoid or tolerate exposure to subfreezing temperatures. Avoidance may take the form of seasonal migration (e.g., moving to a lower latitude) or movement to local microhabitats where temperatures stay at or above freezing (e.g., burrowing into the soil). Tolerance of subfreezing temperatures involves minimizing the damage associated with ice formation in cells and tissues. If ice forms as crystals, it will puncture cell membranes and disrupt metabolic functioning. Some insects inhabiting cold climates contain high concentrations of glycerol, a chemical compound that minimizes the formation of ice crystals and lowers the freezing point of body fluids. These insects spend winter in a semi-frozen state, emerging in spring when temperatures are more conducive to physiological activity. Vertebrate ectotherms generally do not tolerate freezing to the degree that invertebrate ectotherms do because of their larger size and greater physiological complexity. A very few amphibians, however, can survive being partially frozen, as described in the Case Study at the opening of this chapter.

Temperature regulation and tolerance in endotherms As noted earlier, endotherms tolerate a narrower range of body temperatures (30°C–45°C, 86°F–113°F) than ectotherms. However, the ability of endotherms to generate heat internally has allowed them to greatly expand their geographic ranges and the time of year they can be active. Endotherms can remain active at subfreezing temperatures, something that most ectotherms cannot do. The cost of being endothermic is a high demand for food to supply energy to support metabolic heat production. The rate of metabolic activity in endotherms is associated with the external temperature and the rate of heat loss. The rate of heat loss, in turn, is related to body size due to its influence on surface area-to-volume ratio. Small endotherms have higher metabolic rates, require more energy, and have higher feeding rates than large endotherms.

Endothermic animals maintain a constant *basal* (resting) *metabolic rate* over a range of environmental temperatures known as the **thermoneutral zone**. Within the thermoneutral zone, minor behavioral or morphological adjustments are sufficient for maintaining an optimal body temperature. When the environmental temperature drops to a point at which heat loss is greater than metabolic heat production, the body temperature begins to drop, triggering an increase in metabolic heat generation. This point is called the **lower critical temperature** (**Figure 4.16A**). The thermoneutral zone and the lower critical temperature differ among mammal species (**Figure 4.16B**). As one would expect, mammals from the Arctic have critical temperatures below those of animals from tropical regions. Note also that the rate of metabolic activity increases more rapidly below the lower critical temperature in tropical than in Arctic mammals.

What causes the differences in lower critical temperatures and in metabolic rate increases between animals of different biomes? For endothermy to work efficiently, animals must be able to retain their metabolically generated heat. Thus, the evolution of endothermy in birds and mammals required insulation: feathers, fur, or fat. These insulating layers provide a barrier limiting conductive (and in some cases, convective) heat loss. Fur and feathers insulate primarily by providing a layer of still air, similar to a boundary layer, adjacent to the skin. Differences in insulation help explain the differences among the endotherms in Figure 4.16B. Arctic mammals generally maintain thick fur. In warmer climates, however, the ability to cool off through conduction and convection is inhibited by insulation, and thick fur can be an impediment to maintaining an optimal body temperature. Some endotherms acclimatize to seasonal temperature changes by growing thicker fur in winter and shedding fur when temperatures get warmer (a fact that most pet owners know well). Our human ancestors lost much of their hairy insulating layer about 2 million years ago, when they inhabited the hot tropical regions of Africa (Jablonski 2006).

Cold climates are tough on small endotherms. Small mammals, by necessity, have thin fur, since thick fur would inhibit their mobility. The high demand for metabolic energy below the lower critical temperature, the low insulation values of their fur, and their low capacity to store energy make small mammals improbable residents of polar, alpine, and temperate habitats. However, the faunas of many of these cold climates contain many small endotherms, sometimes in high abundances. What explains this apparent discrepancy? Small endotherms, such as rodents and hummingbirds, are able to alter their lower critical temperature during cold periods by entering a state of dormancy known as **torpor**. The body temperatures of animals in torpor may drop as much as 20°C (36°F) below their normal temperatures. The metabolic rate of an animal in torpor is 50%–90% lower than its basal metabolic rate, providing substantial energy savings (Schmidt-Nielsen 1997). However, energy is still needed to arouse the animal from torpor and

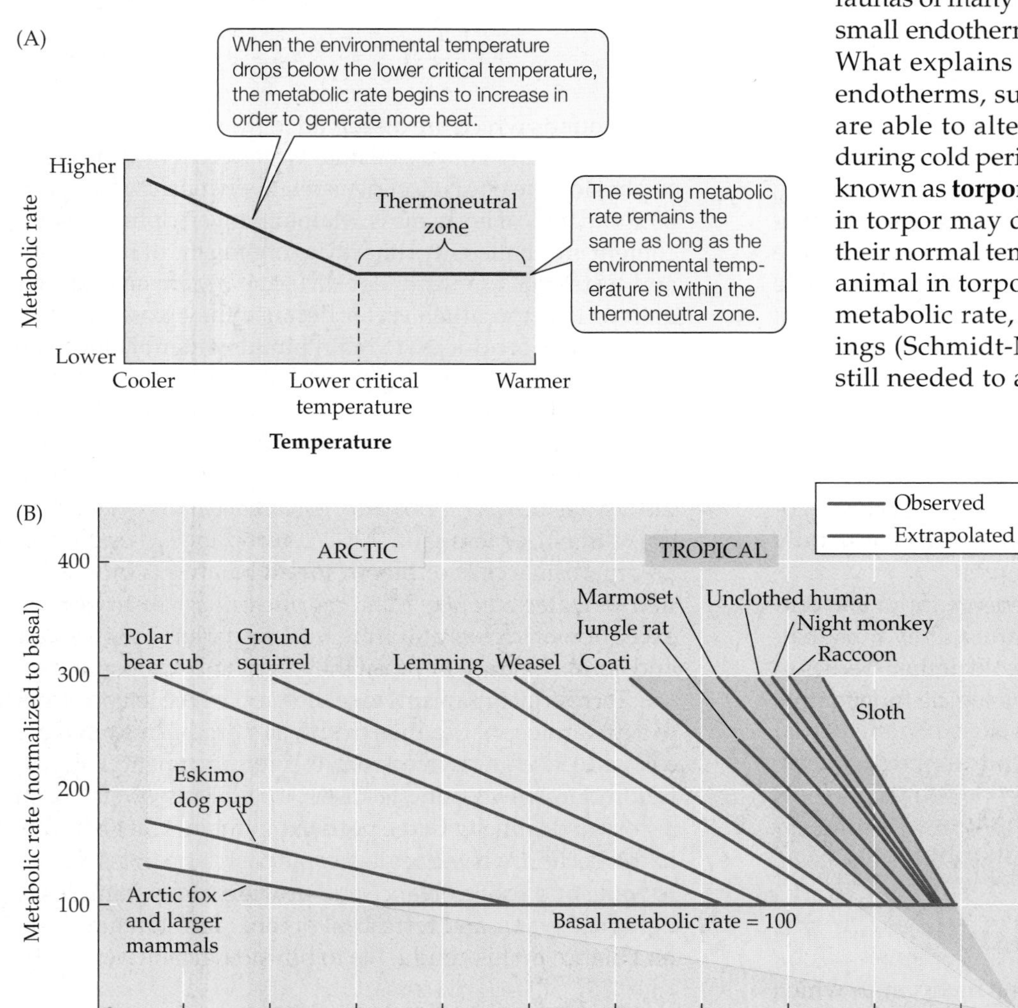

FIGURE 4.16 Metabolic Rates in Endotherms Vary with Environmental Temperatures (A) An endotherm's basal metabolic rate stays constant throughout a range of environmental temperatures known as the thermoneutral zone. When environmental temperatures reach a lower limit, known as the lower critical temperature, the endotherm's metabolic rate increases to generate additional heat. (B) The thermoneutral zones and lower critical temperatures of endotherms vary with their habitats. The lower critical temperatures of Arctic endotherms are lower than those of tropical endotherms, and their metabolic rates increase more slowly below those lower critical temperatures, as shown by the shallower slopes of the curves. (After Scholander et al. 1950.)

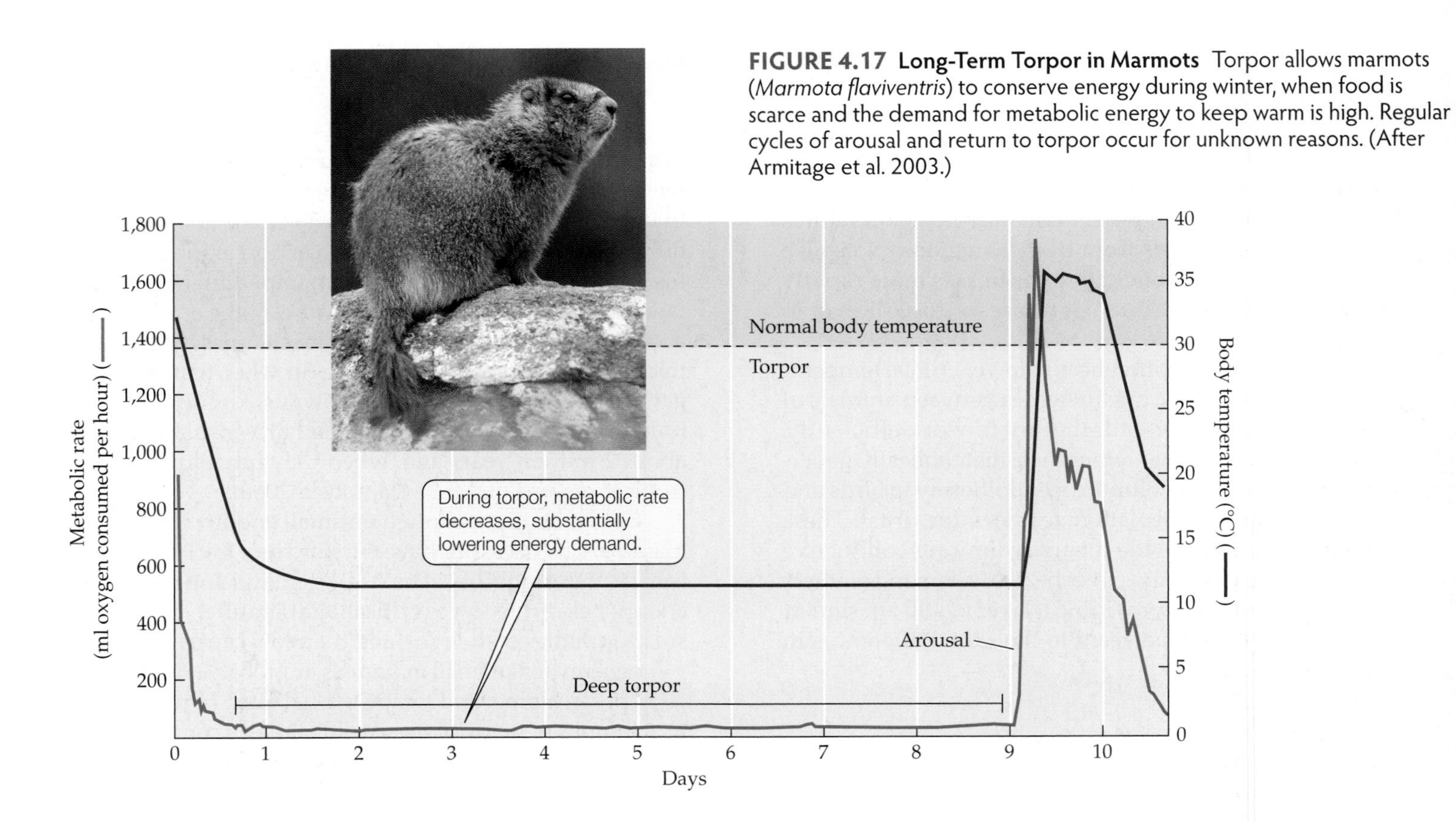

FIGURE 4.17 Long-Term Torpor in Marmots Torpor allows marmots (*Marmota flaviventris*) to conserve energy during winter, when food is scarce and the demand for metabolic energy to keep warm is high. Regular cycles of arousal and return to torpor occur for unknown reasons. (After Armitage et al. 2003.)

bring its body temperature back up to its usual set point. Thus, the length of time an animal can remain in torpor is limited by its reserves of energy. Small endotherms may undergo daily torpor to minimize the energy needed during cold nights. Torpor lasting several weeks during the winter, sometimes referred to as **hibernation**, is possible only for animals that have access to enough food and can store enough energy reserves, such as marmots (**Figure 4.17**). Hibernation is somewhat rare in polar climates because few animals have access to enough food to provide enough energy storage (in the form of fat) to get through winter without eating. Some large animals, such as bears, enter into a long-term winter sleep (sometimes called denning), during which the body temperature decreases only slightly, rather than going into torpor.

Just as organisms must balance energy input and output to maintain an optimal temperature, they must balance the movement of water into and out of their bodies to maintain optimal conditions for physiological functioning.

CONCEPT 4.3 The water balance of an organism is determined by exchanges of water and solutes with the external environment.

Variation in Water Availability

Water is essential for life. Water is the medium in which all biochemical reactions necessary for physiological functioning occur. Water has unique properties that make it a universal solvent for biologically important solutes (compounds which are dissolved in the solvent, including salts). The range of organismal water content conducive to physiological functioning is relatively narrow, between 60% and 90% of body mass. Maintaining an optimal water content is a challenge primarily to organisms of freshwater and terrestrial environments. Marine organisms seldom gain or lose too much water because they exist in a medium that is conducive to maintaining water balance: the oceans in which life first evolved.

In addition to a suitable water balance, organisms must also balance the uptake and loss of solutes, primarily salts. Aquatic environments may be more saline (*hyperosmotic* to the organism; *hyper*, "greater"), of similar salinity (*isoosmotic*; *iso*, "same"), or less saline (*hypoosmotic*; *hypo*, "less") than an organism's cells or blood, so salt balance is intimately tied to water balance. Most marine organisms (with the exception of vertebrates) rarely face problems with water and solute balance because they tend to be isoosmotic.

Terrestrial organisms face the problem of losing water to a dry atmosphere, while freshwater organisms may lose solutes to, and gain water from, their environment. The evolution of freshwater and terrestrial organisms is very much a story of dealing with the need to maintain water balance. In this section, we will review some basic principles related to water and solute balance and provide some examples of how freshwater and terrestrial organisms maintain a water balance that is conducive to physiological functioning.

Water flows along energy gradients

Water flows along gradients of energy, from high energy to low energy conditions. What is an energy gradient in the context of water? Gravity represents one example that is intuitively obvious—liquid water flows downhill, following a gradient of potential energy. Another type of energy influencing water movement is pressure. When elephants spray water out of their trunks, the water is flowing from a condition of higher energy, inside the trunk with muscles exerting pressure on it, to a condition of lower energy, outside of the trunk where that pressure is not acting on it.

Other, less obvious factors that influence the flow of water are important to organismal water balance. When solutes are dissolved in water, the solution loses energy. Thus, if the water in a cell contains more solutes than the water surrounding it, water will flow into the cell to equilibrate the energy difference. Alternatively, solutes may be lost to the surrounding medium, but most biological membranes selectively block the flow of many solutes. In biological systems, the energy associated with dissolved solutes is called **osmotic potential**. The energy associated with gravity is called **gravitational potential**, but in a biological context it is important in water movement only in very tall trees. The energy associated with the exertion of pressure is called **pressure** (or *turgor*) **potential**. Finally, the energy associated with attractive forces on the surfaces of large molecules inside cells or on the surfaces of soil particles is called **matric potential**.

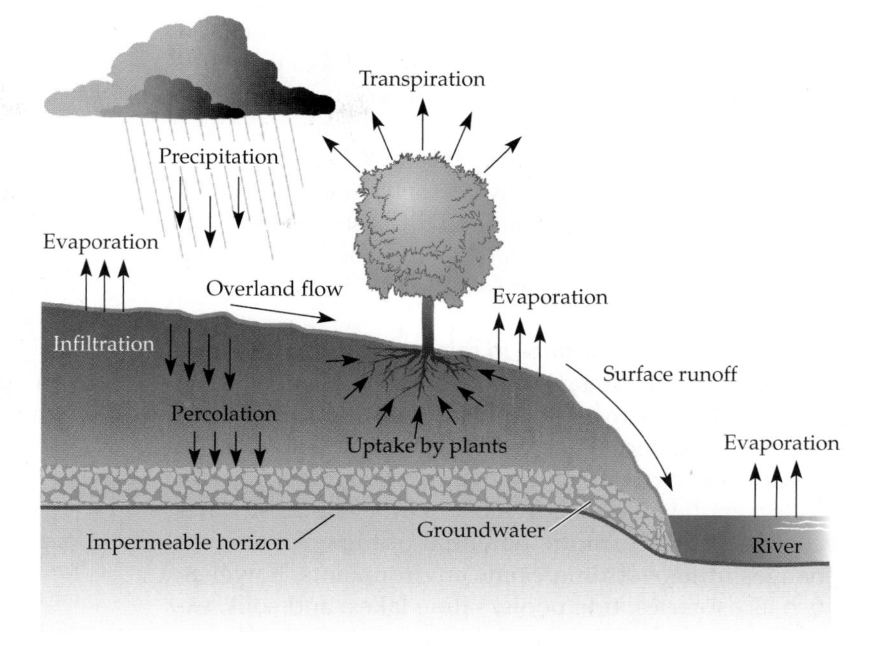

FIGURE 4.18 What Determines the Water Content of Soil? The water content of soil is determined by the balance between water inputs (infiltration of precipitation and overland flow of water) and outputs (percolation to deeper layers, evapotranspiration) and by the capacity of the soil to hold water. Soil water storage capacity and the rate of percolation are dependent on soil texture.

The sum of these energy components within an aqueous system determines its overall water energy status, or **water potential**. The water potential of a system can be defined mathematically as

$$\Psi = \Psi_o + \Psi_p + \Psi_m \qquad (4.3)$$

where Ψ is the total water potential of the system (in units of pressure; usually megapascals, MPa), Ψ_o is the osmotic potential (a negative value, because it lowers the energy status of the water), Ψ_p is the pressure potential (a positive value if pressure is exerted on the system; a negative value if the system is under tension), and Ψ_m is the matric potential (a negative value). Water will always move from a system of higher Ψ to a system of lower Ψ, following the energy gradient. This terminology is most often used in plant, microbial, and soil systems, but it works in animal systems as well.

The atmosphere has a water potential that is related to humidity. From a biological perspective, air with a relative humidity of less than 98% of saturation has a very low water potential, so the gradient in water potential between most terrestrial organisms and the atmosphere is very high. Without some barrier to water movement, terrestrial organisms would lose water rapidly to the atmosphere. Any force that impedes the movement of water (or other substances, such as CO_2) along an energy gradient is called **resistance**.* Barriers that increase organisms' resistance to water loss include the waxy cuticles of plants and insects and the skin of amphibians, reptiles, birds, and mammals.

Water losses and solute gains and losses must be compensated

Terrestrial plants and microorganisms rely on water uptake from soils to replace the water they lose to the atmosphere. Soils are important reservoirs of water that support a multitude of ecological functions, as we will see throughout this book. The water potential of most soils is dominated by the matric potential, which is related to the attractive forces of the soil particles. The osmotic potential of some soils can be important, particularly where dissolved salts are found, as in soils near marine environments or where salinization (see p. 43) has occurred. The amount of water that can be stored in soil is related to the balance between water inputs and outputs, soil texture, and topography (**Figure 4.18**). Inputs include precipitation that infiltrates the soil and overland flow of water. Outputs include percolation to deeper layers below the plant rooting zone and evapotranspiration. Sandy

*Many physiologists prefer using "conductance" rather than "resistance" to express the influence of a barrier on the movement of water or gases between an organism and its environment. Mathematically, conductance is the reciprocal of resistance.

soils store less water than fine-textured soils, but fine soil particles also have a higher matric potential, and thus hold onto water more tightly. Soils with mixed coarse and fine particles are generally most effective in storing water and supplying it to plants and soil organisms. When the volume of water in the soil drops below a certain point (25% of total soil mass in fine-textured soils and 5% in sandy soils), the matric forces are strong enough that most of the remaining water is unavailable to organisms.

Controls on water balance in microorganisms Single-celled microorganisms, including archaea, bacteria, algae, and protists, are active primarily in aqueous environments. Their water balance is dependent on the water potential of the surrounding environment, which is determined mainly by its osmotic potential. In most marine and freshwater ecosystems, the osmotic potential of the environment changes little over time. Some environments, however, such as estuaries, tide pools, saline lakes, and soils, experience frequent changes in osmotic potential due to evaporation or variable influxes of fresh and salt water. Microorganisms in these environments must respond to these changes by altering their cellular osmotic potential if they are to maintain a water balance suitable for physiological functioning. They accomplish this through **osmotic adjustment**, an acclimatization response that involves changing their solute concentration, and thus their osmotic potential. Some microorganisms synthesize organic solutes to adjust their osmotic potential, which also help to stabilize enzymes. Others use inorganic salts from the surrounding medium for osmotic adjustment. The ability to adjust osmotic potential in response to changes in external water potential varies substantially among microorganisms: some completely lack this ability, while others (such as *Halobacterium* spp.) can adjust to even the extremely saline conditions in landlocked saline lakes.

As noted above, terrestrial environments are too dry for any organism that does not have a barrier to prevent cellular water loss to the atmosphere. Many microorganisms avoid exposure to dry conditions by forming dormant resistant spores, encasing themselves in a protective coating that prevents water loss to the environment. Some microorganisms with filamentous forms, such as fungi and yeasts, are very tolerant of low water potentials and can grow in dry environments. Most terrestrial microorganisms, however, are found in soils, which have a higher water content and humidity than the air above them.

Controls on water balance in plants One of the distinguishing characteristics of plants is a rigid cell wall composed of cellulose. Bacteria and fungi also have cell walls, composed of materials such as chitin (in fungi) or peptidoglycans and lipopolysaccharides (in bacteria). Cell walls are important to water balance because they

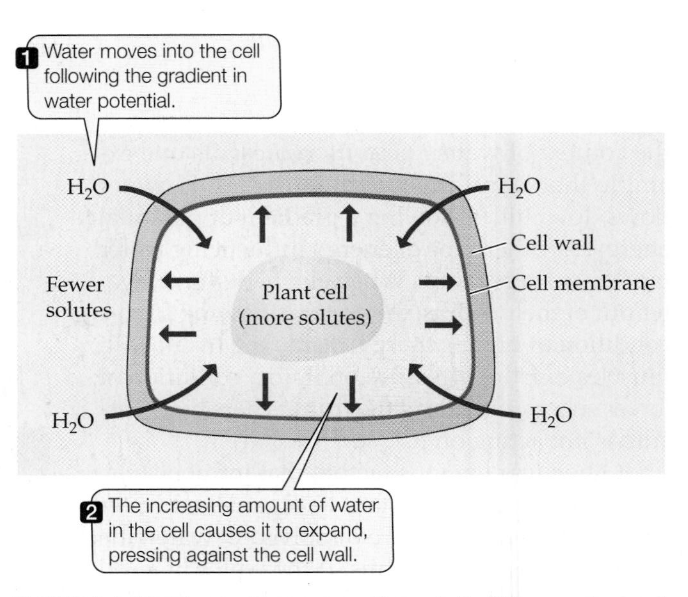

FIGURE 4.19 Turgor Pressure in Plant Cells When a plant cell is surrounded by water with a solute concentration lower than its own, water moves into the cell, while solutes in the cell are prevented from moving out by the cell membrane. The increasing amount of water in the cell causes the cell to expand, pressing against the cell wall.

result in the development of positive **turgor pressure**. When water follows a gradient of water potential into a plant cell, it causes the cell to expand and press against the cell wall, which resists the pressure due to its rigidity (**Figure 4.19**). Turgor pressure is an important structural component of plants, and it is also an important force for growth, promoting cell division. When nonwoody plants lose turgor pressure due to dehydration, they wilt. Wilting is generally a sign that a plant is experiencing water stress.

Plants take up water from sources with a water potential higher than their own. For aquatic plants, the source is the surrounding aqueous medium. In freshwater environments, the presence of solutes in the plant's cells creates a water potential gradient from the surrounding water to the plant. In marine environments, plants must lower their water potential below that of seawater to take up water. Marine plants, as well as terrestrial plants of salt marshes and saline soils, adjust their osmotic potential in a manner similar to microorganisms by synthesizing solutes and taking up inorganic salts from their environment. Inorganic salts must be taken up selectively, however, because some (e.g., Na^+, Cl^-) can be toxic at high concentrations. The cell membranes of plants act as a solute filter, determining the amounts and types of solutes that move into and out of the plant.

Terrestrial plants acquire water from the soil through their roots, as well as from mutualistic fungi that grow into their roots from the soil, called *mycorrhizae* (see Chapter 14). The earliest land plants, which had not yet evolved roots, used mycorrhizae to take up water and nutrients from the soil. The majority of terrestrial plants today use a combination of roots and mycorrhizae to take up water. Only the finest roots can take up water from the soil because older,

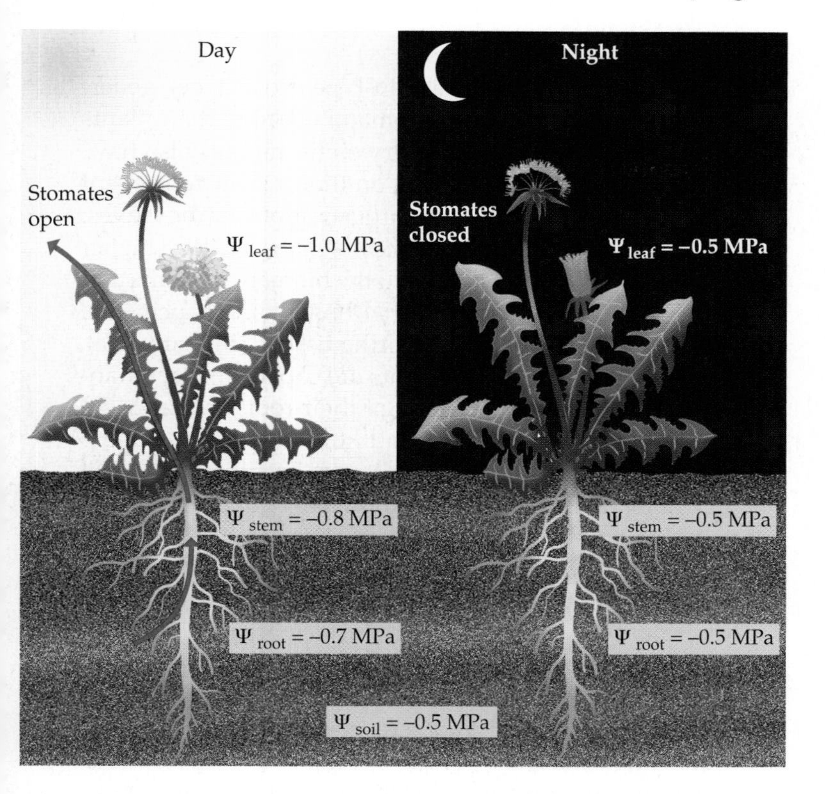

FIGURE 4.20 The Daily Cycle of Plant Dehydration and Rehydration During the day, when the stomates are open, transpiration results in a gradient of water potential from leaf to stem, stem to roots, and roots to soil. At night, when the stomates are closed, water potential equilibrates as the plant rehydrates.

thicker roots develop a water-resistant waxy coating that limits their ability to absorb water as well as to lose water to the soil. Mycorrhizae provide a greater surface area for absorption of water and nutrients for the plant and allow greater exploration of the soil for these resources. In turn, the mycorrhizal fungi obtain energy from the plant.

Plants lose water by transpiration when their stomates open to allow CO_2 from the atmosphere to diffuse into their leaves. Water moves out through the stomates, following the water potential gradient from the inside of the leaf (100% relative humidity) to the air. As we saw in the previous section, transpiration is an important cooling mechanism for leaves. The plant must replace the water lost by transpiration, however, if it is to avoid water stress. As a leaf loses water, the water potential of its cells decreases, creating a water potential gradient between the leaf and the xylem in the stem to which it is attached, so water moves through the xylem into the leaf. In this way, when the plant is transpiring, it creates a gradient of decreasing water potential from the soil through the roots and stems to the leaves (**Figure 4.20**). Water therefore flows from the soil, which has the highest water potential, into the roots, the xylem, and eventually the leaves, from which it is lost to the atmosphere via transpiration. Because there is greater resistance to the movement of water into the roots and through the xylem than out through the stomates, the water supply from the soil cannot keep up with water loss by transpiration. As a result, the water content of the plant decreases during the day. Extremely dry conditions can cause loss of xylem function (**Web Extension 4.1**). At night, however, the stomates close, and the water supply from the soil rehydrates the plant until it reaches near-equilibrium with the soil water potential. This daily cycle of daytime dehydration and nighttime rehydration can go on indefinitely if the supply of water in the soil is adequate.

The availability of water decreases when precipitation is not sufficient to replace the water lost from the soil through transpiration and evaporation. The water content of a plant will then decrease as its water potential increases, and its turgor pressure will decrease as its cells become dehydrated (**Figure 4.21**). To avoid reaching a detrimen-

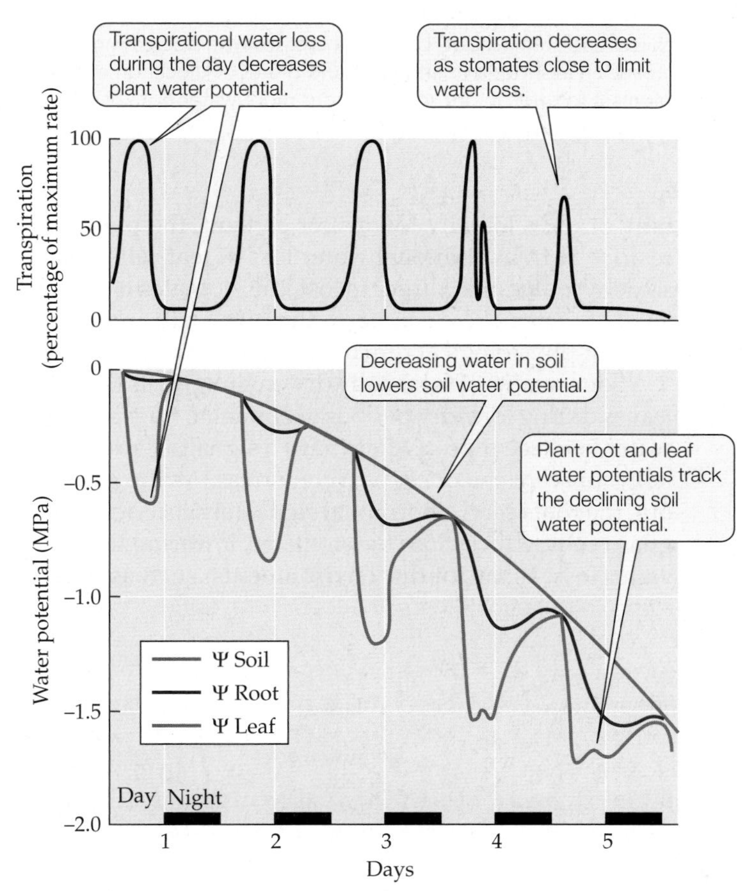

FIGURE 4.21 How Plants Cope with Depletion of Soil Water If soil water is not recharged, transpiration will deplete it, leading to progressive drying of the soil and a decrease in soil water potential.

 As the soil dries, midday closing of the stomates may occur, with reopening later in the afternoon, as seen on day 4 in the graph. Assuming the air temperature is cooler later in the day, what influence would this have on plant water loss?

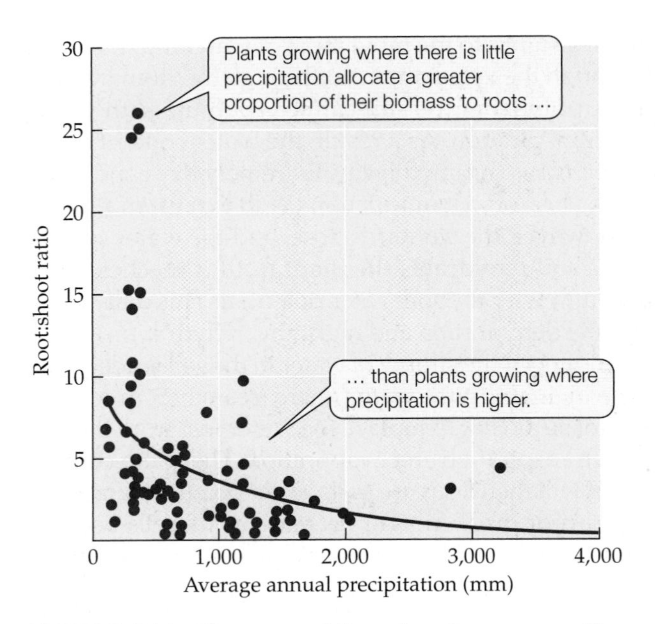

FIGURE 4.22 Allocation of Growth to Roots versus Shoots Is Associated with Precipitation Levels The ratio of root biomass to leaf and stem (shoot) biomass increases with decreasing precipitation in shrubland and grassland biomes. Allocation of more biomass to roots in dry soils provides more water uptake capacity to support leaf function. (After Mokany et al. 2006.)

tally or even lethally low water content, the plant must restrict its transpirational water loss. If leaf cells become so dehydrated that turgor is lost, the stomates close. This level of water stress can harm the plant, causing impairment of physiological functions such as photosynthesis.

Some plants of seasonally dry environments shed their leaves during long dry periods to eliminate transpirational water loss (see p. 56). Others have a signaling system that helps prevent the onset of water stress. As the soil dries out, the roots send a hormonal signal (abscisic acid) to the guard cells, which close the stomates, lowering the rate of water loss. Plants of dry environments, such as deserts, grasslands, and Mediterranean-type ecosystems, generally have better control of stomatal opening than plants of wetter climates. Plants of dry environments also have a thick waxy coating (cuticle) on their leaves to prevent water loss thorugh the nonporous regions of the leaves. Additionally, plants of dry environments maintain a higher ratio of root biomass relative to the biomass of stems and leaves than plants of moister environments, enhancing the rate of water supply to transpiring tissues (Mokany et al. 2006) (**Figure 4.22**). Some plants are capable of acclimatization by altering the growth of their roots to match the availability of soil moisture and nutrients.

Can plants have too much water? Technically, no, but saturation of soils inhibits the diffusion of oxygen and can cause hypoxic conditions. Thus, waterlogged soils inhibit aerobic respiration in roots. Wet soils also enhance the growth of harmful fungal species that can damage roots. Ironically, the combination of these factors can lead to root death, cutting off the supply of water to plants, and eventually to wilting. Adaptations to low oxygen concentrations in wet soils include root tissue containing air channels (called *aerenchyma*) as well as specialized roots in plants such as mangroves (see p. 72) that extend vertically above the water or waterlogged soil the plant is growing in.

Controls on water balance in animals Multicellular animals face the same challenges plants and microorganisms do in maintaining water balance. Water losses and gains in animals, however, are governed by a more diverse set of exchanges than in plants and microorganisms (**Figure 4.23**). Many animals have the added complexity of specialized organs for gas exchange, ingestion and digestion, excretion, and circulation, which create areas of localized water and solute exchange as well as gradients of water and solutes within the animal's body. Most animals are mobile and can seek out environments conducive to maintaining a favorable water and solute balance, an option not available to most microorganisms and plants.

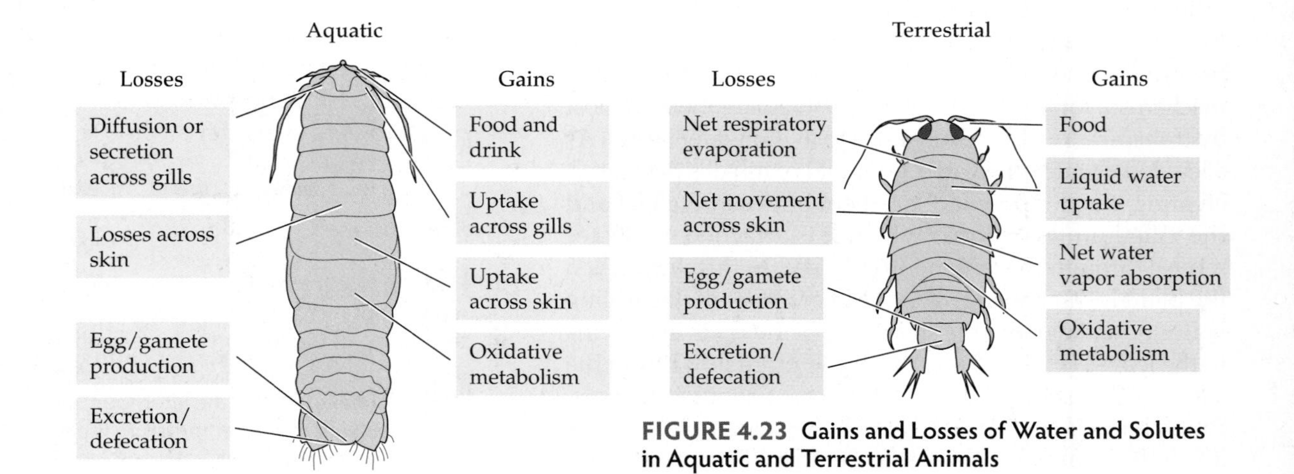

FIGURE 4.23 Gains and Losses of Water and Solutes in Aquatic and Terrestrial Animals

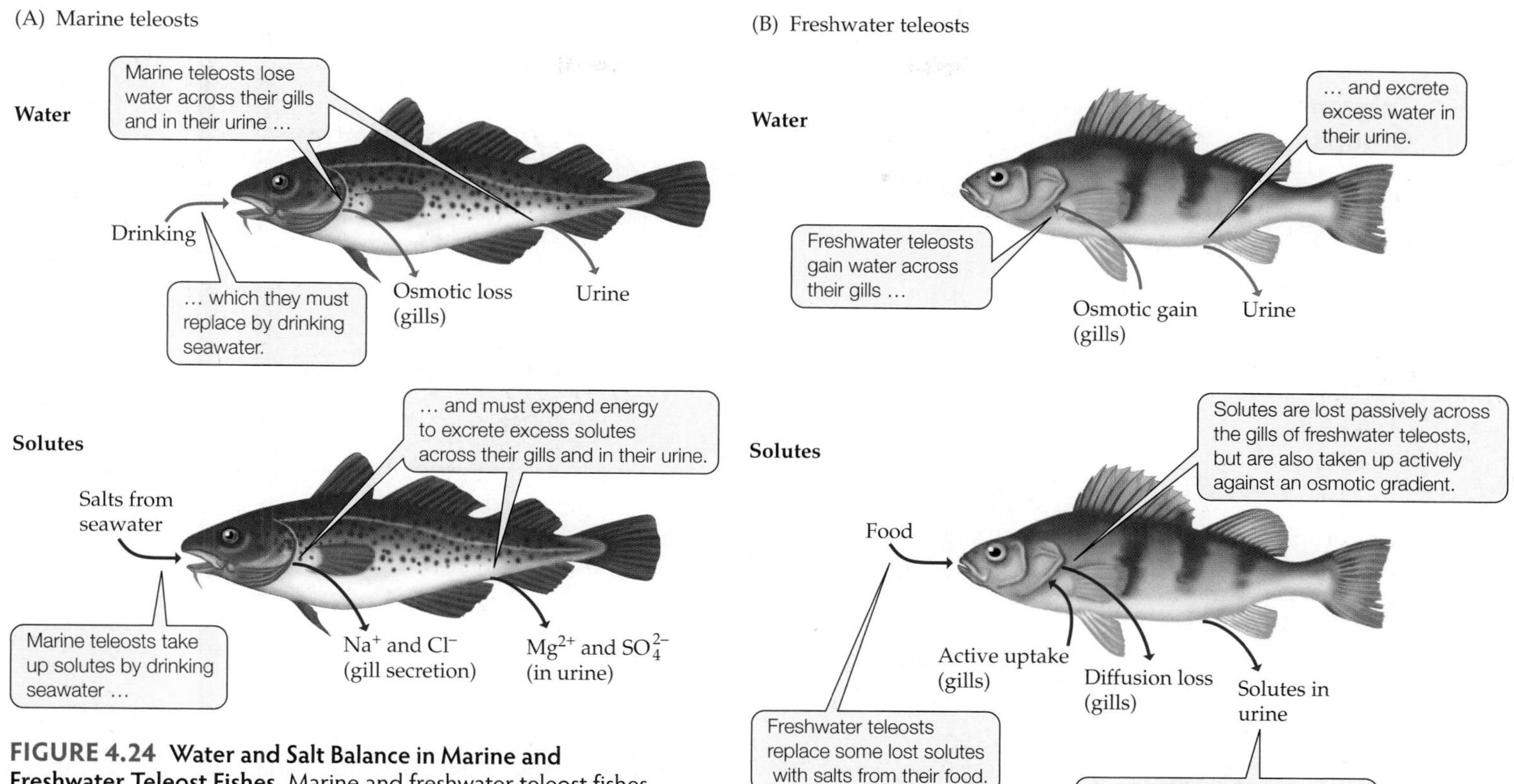

FIGURE 4.24 Water and Salt Balance in Marine and Freshwater Teleost Fishes Marine and freshwater teleost fishes face opposite challenges in maintaining water and solute balance. (A) Marine teleosts are hypoosmotic to their environment, so they tend to lose water and gain solutes. (B) Freshwater teleosts are hyperosmotic to their environment, so they tend to gain water and lose solutes.

Under conditions of varying salinity, animals must be able to adjust their gains and losses of water and solutes to maintain favorable water and solute balances. A marine animal that lacks this ability will die if transferred to brackish or fresh water. Although most marine invertebrates are isoosmotic to seawater, the specific types of solutes in their bodies can vary. Many invertebrates that are capable of adjusting to changes in the solute concentration of their environment do so by exchanging solutes with the surrounding medium. Like plants, these animals must selectively control the exchange of specific solutes with their environment because some external solutes are toxic at the concentrations at which they are found in seawater, and because some internal solutes are needed for biochemical reactions. Jellyfish, squid, and crabs, for example, have sodium (Na^+) and chloride (Cl^-) concentrations similar to those of seawater, but their sulfate (SO_4^{2-}) concentrations may be one-half to one-fourth of those found in seawater.

Marine vertebrates include animals that are isoosmotic and hypoosmotic to seawater. The cartilaginous fishes, including the sharks and rays, have blood solute concentrations similar to those of seawater, although, as in invertebrates, their concentrations of specific solutes differ from those in seawater. In contrast, marine teleost (bony) fishes and mammals evolved in fresh water and later moved into marine environments. Their blood is hypoosmotic to seawater. Fish exchange water and salts with their environment across the gills, which are also the organs of O_2 and CO_2 exchange, and through drinking and eating (**Figure 4.24A**). Salts that diffuse into or are ingested by marine teleost fishes must be continuously excreted in urine and through the gills against an osmotic gradient, which requires an expenditure of energy. Water lost across the gills must be replaced by drinking. Marine mammals, such as whales and porpoises, produce urine that is hyperosmotic to seawater and avoid drinking seawater to minimize salt uptake.

Freshwater animals are hyperosmotic to their environment; therefore, they tend to gain water and lose salts. Most salt exchange occurs at the gas exchange surfaces, including the skin of some invertebrates (e.g., freshwater worms) and the gills of many vertebrates and invertebrates. These animals must compensate for salt losses by taking up solutes in their food, and some groups, such as teleost fishes, must take up solutes actively through the gills against an osmotic gradient (**Figure 4.24B**). Excess

TABLE 4.1

Tolerances for Water Loss in Selected Animal Groups

Group	Weight loss (%)
INVERTEBRATES	
Mollusks	35–80
Crabs	15–18
Insects	25–75
VERTEBRATES	
Frogs	28–48
Small birds	4–8
Rodents	12–15
Human	10–12
Camel	30

Source: Willmer et al. 2005.

Note: Values are maximum percentages of body weight lost as water that can be tolerated, based on observations of a range of exemplary species in each group.

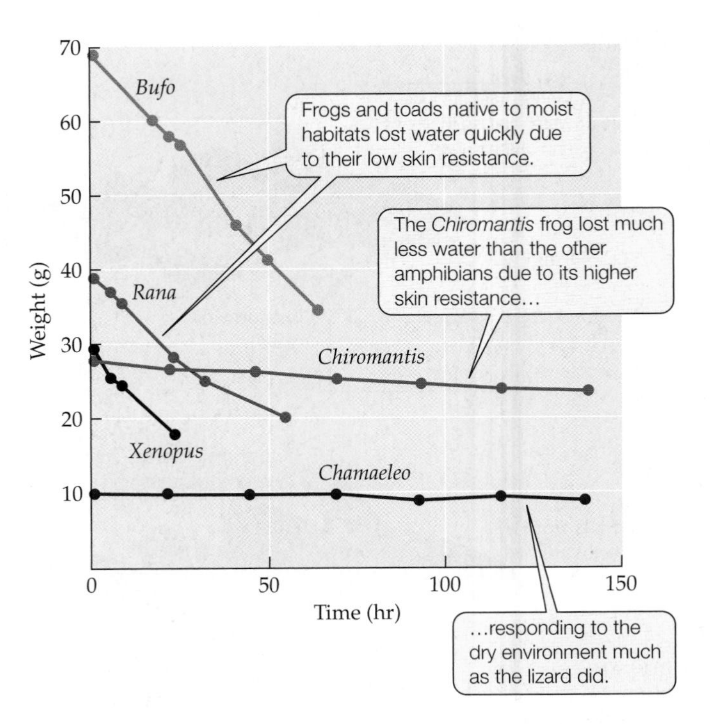

FIGURE 4.25 Resistance to Water Loss Varies among Frogs and Toads Amphibians were kept under uniform dry environmental conditions (25°C, 20%–30% relative humidity) to examine their rates of water loss, measured as loss of body weight. A lizard (*Chamaeleo*) was also tested for comparative purposes.

Q How could you estimate the resistance to water loss of these animals quantitatively using this graph?

water is excreted as dilute urine, from which the excretory system actively removes solutes to minimize their loss.

Terrestrial animals face the challenge of exchanging gases (O_2 and CO_2) in a dry environment with a very low water potential. These animals lower their evaporative water loss and exposure to water stress by having high skin resistance or by living in environments where they can compensate for high water losses with high water intake. Both approaches involve risks and trade-offs, however. A high resistance to water loss may compromise the ability to exchange gases with the atmosphere. Reliance on a steady water supply puts the animal at risk if the source of water fails (e.g., during a severe drought). Tolerance for water loss varies substantially among groups of terrestrial animals. Generally, invertebrates have a higher tolerance for water loss than vertebrates. Within the vertebrates, amphibians have a higher tolerance for, but lower resistance to, water loss than mammals and birds (**Table 4.1**).

Amphibians, including frogs, toads, and salamanders, rely primarily on stable water supplies to maintain their water balance. They can be found in a wide variety of biomes, from tropical rainforests to deserts, as long as there is a reliable source of water, such as regular rains or ponds. Amphibians depend on gas exchange through the skin to a greater degree than other terrestrial vertebrates. Therefore, amphibian skin is often thin, with a low resistance to water loss. However, some adult amphibian species have adapted to dry environments by developing specialized skin with higher resistance to water loss. For example, the southern foam-nest frog (*Chiromantis xerampelina*), which occurs throughout Africa, has skin that resists water loss in a manner similar to that of lizards (**Figure 4.25**). To compensate for reduced gas exchange through the skin, it has a higher breathing rate (Stinner and Shoemaker 1987). As a group, tree frogs have higher skin resistance to water loss than ground frogs, reflecting their drier habitat. Some ground frogs of seasonally dry environments, such as the northern snapping frog (*Cyclorana australis*) of Australia, lower their rates of water loss by forming a "cocoon" of mucous secretions consisting of proteins and fats that increases resistance to water loss.

Reptiles have been extremely successful at inhabiting dry environments. The thick skin of desert snakes and lizards provides protection for the internal organs as well as an effective barrier to water loss. The outer skin has multiple layers of dead cells with a fatty coating and is overlain by plates or scales. These layers give reptilian skin a very high resistance to water loss. Mammals and birds have skin anatomy similar to that of reptiles, but have hair or feathers covering the skin rather than scales. The presence of sweat glands in mammals represents a trade-off between resistance to water loss and evaporative cooling. The highest resistances to water loss among

TABLE 4.2

Resistance of External Coverings (Skin, Cuticle) of Animals to Water Loss

Group	Resistance (s/cm)
Crabs (marine)	6–14
Fish	2–35
Frogs	3–100
Earthworms	9
Birds	50–158
Desert tortoises	120
Desert lizards	1,400
Desert scorpions, spiders	1,300–4,000

Source: Willmer et al. 2005.

terrestrial animals are found in the arthropods (e.g., insects and spiders), which are characterized by an outer exoskeleton made of hard chitin and coated with waxy hydrocarbons that make it impervious to water movement (**Table 4.2**).

An instructive example of how animals use a variety of integrated adaptations to cope with arid environments involves the kangaroo rats (*Dipodomys* spp.) found throughout the deserts of North America. A combination of efficient water use and low rates of water loss greatly diminishes these rodents' water requirements (Schmidt-Nielsen and Schmidt-Nielsen 1951) (**Figure 4.26**). Kangaroo rats rarely drink water. A large proportion of their water requirement is met by eating dry seeds and by oxidative metabolism; that is, by metabolically converting carbohydrates and fats into water and carbon dioxide (Schmidt-Nielsen 1964). The animals also consume more water-rich foods, such as insects or succulent vegetation, if they are available. Kangaroo rats minimize water loss through several physiological and behavioral adaptations. During the hottest periods of the year, they are active only at night, when air temperatures are lowest and humidities highest. During the day, they stay in their underground burrows, which are cooler and more humid than the desert surface. In some parts of their geographic range, however, kangaroo rats experience high enough temperatures even in their burrows to expose them to significant evaporative water losses (Tracy and Walsberg 2002). To increase their resistance to these losses, kangaroo rats have thicker, oilier skin, with fewer sweat glands, than related rodents of moister environments. They minimize water losses in their urine and feces through effective removal of water by their kidneys and intestines. Kangaroo rats produce some of the most concentrated urine of any animal. The combination of these characteristics allows kangaroo rats to inhabit very arid environments without exposure to water stress, even without access to water for drinking.

A CASE STUDY REVISITED

Frozen Frogs

The existence of amphibians above the Arctic Circle seems improbable, given their reliance on a steady supply of liquid water to maintain their water balance and the high potential for damage associated with freezing. Several problems must be overcome in order for complex organisms to survive freezing. First, when water freezes, it forms needle-like crystals that can penetrate and damage or destroy cell membranes and organelles. Second, the supply of oxygen to tissues is severely restricted due to the lack of circulation and breathing. Finally, as ice forms, pure water is pulled from cells, resulting in shrinkage and an increase in solute concentration. Any one of these factors, or all of them working in combination, will kill tissues and organisms in subfreezing temperatures. Yet the frogs described in the Case Study, as well as many species of invertebrates, can tolerate the freezing of a substantial amount of their body water.

Wood frogs and other freeze-tolerant amphibians spend winter in shallow depressions under leaves, moss,

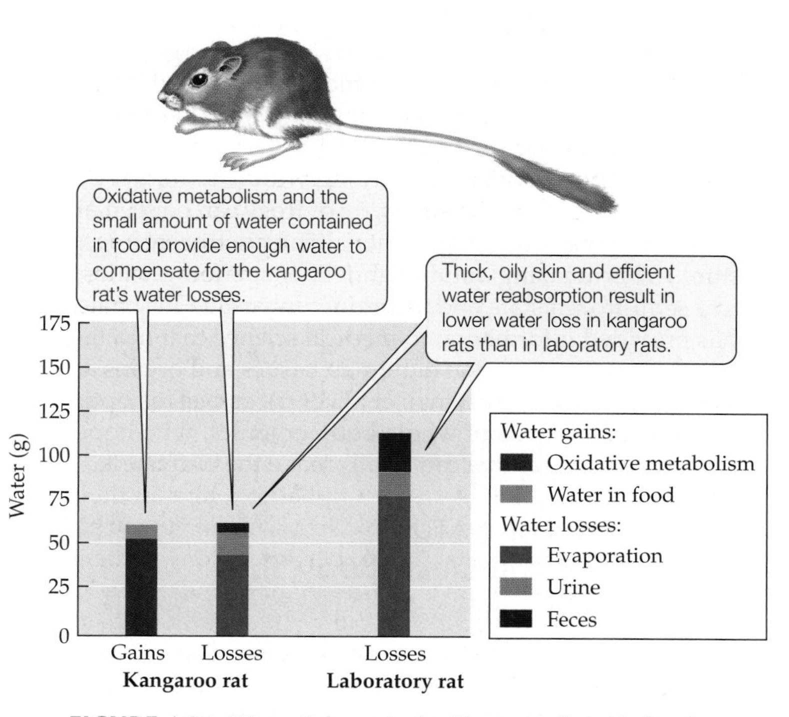

FIGURE 4.26 Water Balance in the Kangaroo Rat Under dry laboratory conditions (25°C, 25% relative humidity), kangaroo rats, native to deserts of western North America, do not require liquid water to survive. (After Tracy and Walsberg 2002.)

or logs, which do not protect them from subfreezing temperatures. Several adaptations facilitate the survival of these amphibians through the winter and allow them to emerge from their frozen state in spring unharmed. Freezing of water in these animals is limited to the spaces outside the cells. A substantial proportion of their body water, from 35% to 65% in "fully frozen" frogs, freezes (Pinder et al. 1992). If more than 65% of their body water is frozen, most individuals will die due to excessive cell shrinkage. The formation of ice outside the cells is enhanced by the existence of ice-nucleating proteins that serve as the site of slow, controlled ice formation (Storey 1990). Solute concentrations in the unfrozen cells increase as the cells lose water to extracellular ice formation. In addition, freeze-tolerant amphibians synthesize additional solutes, including glucose and glycerol derived from the breakdown of liver glycogen. The resulting increase in solute concentrations lowers the freezing point inside the cells, allowing the intracellular solution to remain liquid at subfreezing temperatures. The concentrated solutes also stabilize the cell volume and the structures of organelles, proteins, and enzymes. As freezing proceeds, the frog's heart stops, and its lungs cease to pump air. Once it reaches this semi-stable state of partial freezing, the frog can remain frozen for several weeks, as long as the temperature does not drop below about –5°C (23°F). Although their winter "quarters" are not far below the surface of the ground, the insulating cover of leaves and snow keeps the frogs above that temperature.

The freezing process is initiated in wood frogs within minutes of ice formation within the animal, although the full process occurs over several days to weeks (Layne and Lee 1995). Thawing, on the other hand, may be rapid, with normal body functioning returning within 10 hours. (A video of the freezing and thawing of frogs can be seen at http://www.pbs.org/wgbh/nova/nature/frozen-frogs.html.) This amazing amphibian feat of spending winter in a semi-frozen state and emerging unharmed in spring has provided information to medical science that has facilitated the preservation of human tissues and organs at low temperatures (Costanzo et al. 1995), as well as optimism to proponents of whole-body cryonics, who hope that someday Grandpa can finally leave the Tuff Shed.

CONNECTIONS IN NATURE

Desiccation Tolerance, Body Size, and Rarity

As we saw in Chapter 3 there is a close association between organismal adaptations to climatic conditions and their distribution among terrestrial biomes. While subfreezing temperatures are an important constraint on the distribution and functioning of organisms in high-latitude and high-elevation biomes, low water availability is a more widespread challenge. Arid conditions can potentially occur in most terrestrial biomes (see the climate diagrams under Concept 3.1), and they regularly occur over more than 60% of the land surface. As we have seen, the majority of terrestrial organisms, particularly animals, avoid exposure to dry conditions and rely on minimizing water losses to the environment. However, some organisms can tolerate arid conditions, much in the same way the frozen frogs tolerate subfreezing winter conditions, by entering a state of suspended animation while allowing themselves to dry out. This adaptive approach is common in microorganisms, including bacteria, fungi, and protists, but is also found in some multicellular animals and some plants, including mosses, liverworts, and a few flowering plants (Alpert 2006).

Desiccation-tolerant organisms can survive extreme dehydration, losing 80%–90% of their water as they equilibrate with the humidity of the air, then regain metabolic function shortly after they are rehydrated (**Figure 4.27**). As the cells dry out, the organisms synthesize sugars, which are the key to protecting cell and organelle structure (Alpert 2006). Once dehydration proceeds beyond a certain threshold, metabolism ceases, and the sugars and the small amount of remaining water form a glassy coating over the cellular constituents. As with recovery from freezing, recovery from dehydration is rapid, occurring in hours to days.

The prevalence of dry conditions in terrestrial environments suggests that desiccation tolerance should be more common than it is. Why haven't more plants and animals evolved to be more desiccation-tolerant (see Chapter 6)? A clue to this puzzle may be the small size of the organisms that are desiccation-tolerant (Alpert 2006). Small organisms (less than 5 mm in animals) do not require structural reinforcement, such as a skeletal system, which would restrict the necessary shrinking of the organism as it dehydrates. In addition, water loss during dehydration must be slow enough to allow the adaptive response of sugar synthesis to occur, but not so slow that the organism spends a long time with a low water content while metabolism is still occurring, which can cause physiological stress. Small organisms have surface area-to-volume ratios and thicknesses favorable for the water loss rates required.

These arguments explain why small size is favored in desiccation-tolerant organisms, but not why they are rare (see Chapter 22). The two characteristics—small size and rarity—are intimately linked. As we will see in Chapter 11, small size is often associated with slow growth rates and poor competitive ability under conditions of low resource availability. Thus, natural selection for desiccation tolerance may involve trade-offs with other ecological characteristics, such as competitive ability, that might prevent these organisms from being successful in competitive environments.

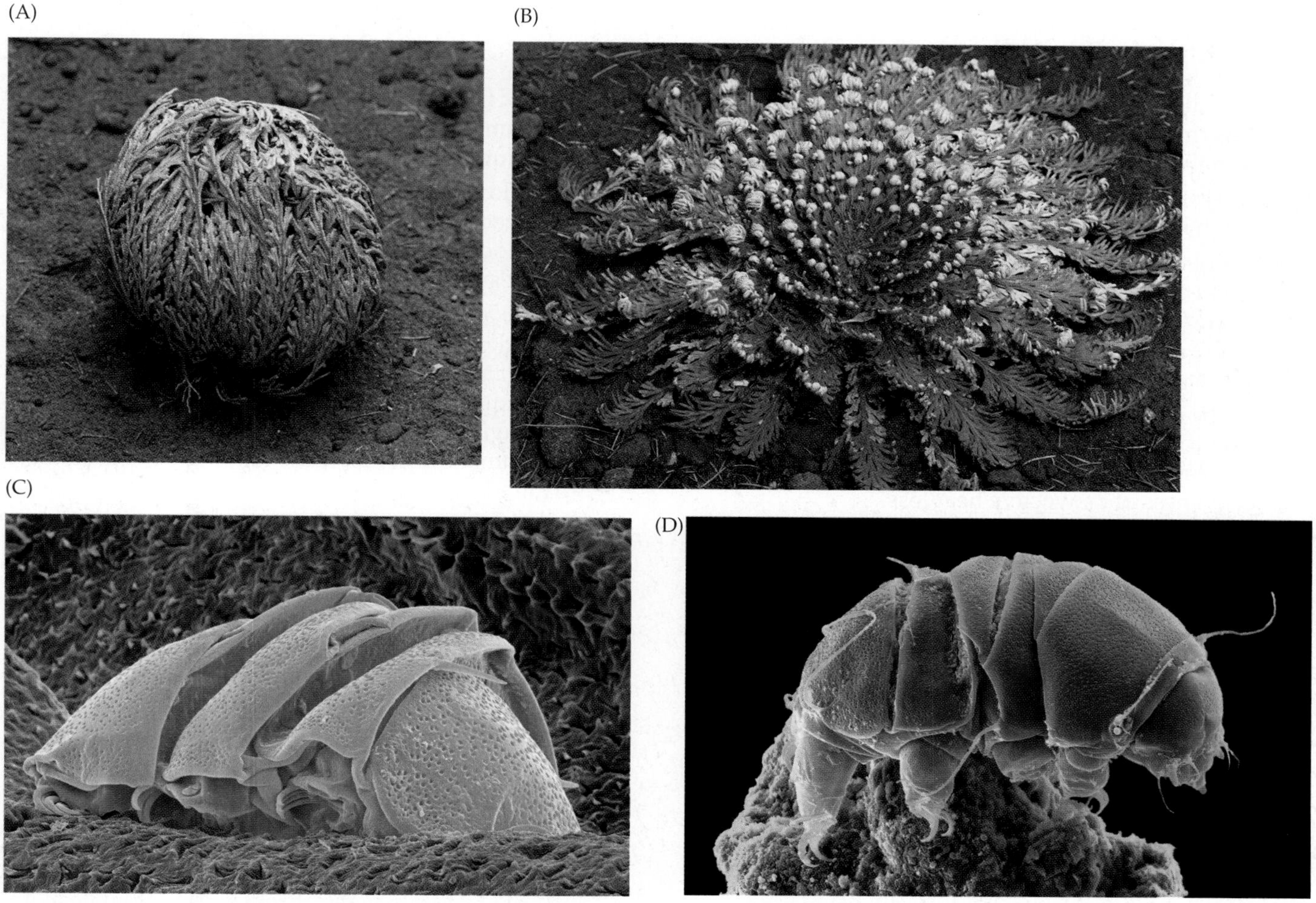

(A) (B) (C) (D)

FIGURE 4.27 Desiccation-Tolerant Organisms (A) The leaves of this club moss (*Selaginella lepidophylla*) reach a very low moisture content during prolonged periods without rain. (B) Within 6 hours of receiving water, the leaves are functional and carrying out photosynthesis. (C, D) Water bears (tardigrades) are small invertebrates (less than 1 mm in length) found in aqueous environments, including oceans, lakes and ponds, soil water, and the water films on vegetation. Water bears contract and cease metabolism when they and their environment dry up (C), but rehydrate when moisture returns (D).

SUMMARY

CONCEPT 4.1 Each species has a range of environmental tolerances that determines its potential geographic distribution.

- The physical environment affects an organism's ability to obtain energy and resources, thereby determining its growth and reproduction, and more immediately, its ability to survive the extremes of that environment. The physical environment is therefore the ultimate constraint on a species' geographic distribution.
- Individual organisms can respond to environmental change through acclimatization, a short-term adjustment of the organism's physiology, morphology, or behavior that lessens the effect of the change and minimizes the associated stress.
- A population may respond to unique environmental conditions through natural selection for physiological, morphological, and behavioral traits, known as adaptations, that enhance individuals' survival, growth, and reproduction under those conditions.

SUMMARY (continued)

CONCEPT 4.2 The temperature of an organism is determined by exchanges of energy with the external environment.

- Temperature controls physiological processes through its effects on enzymes and membranes.
- Gains of energy from and losses of energy to the external environment determine an organism's temperature. Modifying this exchange of energy with the environment allows an organism to control its temperature.
- Terrestrial plants may modify their energy balance by controlling transpiration, increasing or decreasing absorption of solar radiation, or adjusting the effectiveness of convective heat loss.
- Animals modify their energy balance mainly through behavior, adjusting the effectiveness of heat loss, and in the case of endothermic animals, generation of internal heat and insulation to lower heat loss.

CONCEPT 4.3 The water balance of an organism is determined by exchanges of water and solutes with the external environment.

- Water flows along energy gradients determined by solute concentrations (osmotic potential), pressure or tension (pressure potential), and the attractive force of surfaces (matric potential).
- Aquatic animals that are hypoosmotic to the surrounding water must expend energy to excrete salts against an osmotic gradient. On the other hand, aquatic animals that are hyperosmotic to their environment must take up solutes from the environment to compensate for solute losses to the surrounding water.
- Plants and microorganisms can influence water potential by adjusting the solute concentration in their cells (osmotic adjustment).
- Terrestrial organisms can alter their gains or losses of water by adjusting their resistance to water movement, as by the opening or closing of stomates in plants or variations in the thickness of skin in animals.

REVIEW QUESTIONS

1. Organisms subjected to stressful conditions exhibit different degrees of tolerance for environmental stresses. How does tolerance for variation in tissue temperature vary among plants, ectothermic animals, and endothermic animals? What factors influence the differences in tolerance among these groups? Can plants exhibit avoidance of temperature extremes?
2. Arctic foxes exhibit large seasonal changes in the thickness of their fur, while African bat-eared foxes lack this response. Use this example to describe the concepts of acclimatization and adaptation. Assume that the bat-eared fox evolved earlier than the arctic fox.
3. Organismal adaptations to environmental conditions often affect multiple ecological functions, leading to associated trade-offs. The following are two different trade-offs to consider.
   a. Plants transpire water through their stomates. What effects does transpiration have on temperature regulation in leaves? What is the trade-off with transpirational temperature regulation in terms of leaf physiological function?
   b. Animals can more effectively warm their bodies by absorbing solar radiation if they are a dark color. Many animals, however, are not dark-colored, but instead have a coloration close to that of their habitat (camouflage, as in the case of the basking lizard in Figure 4.15). What is the trade-off between animal coloration and heat exchange?
4. List several ways in which plants and animals in terrestrial environments influence their resistance to water loss to the atmosphere.

ON THE COMPANION WEBSITE
sites.sinauer.com/ecology2e

The website includes Chapter Outlines, Online Quizzes, Flashcards & Key Terms, Suggested Readings, a complete Glossary, and the Web Stats Review. In addition, the following resources are available for this chapter:

HANDS-ON PROBLEM SOLVING

Thermal Adaptations in Urban Ants

This Web exercise demonstrates thermal adaptations in ants that live in cities. You will investigate whether adaptation to heat in urban leaf-cutter ants reduces tolerance for cold temperatures.

WEB EXTENSION

4.1 Cavitation and the Loss of Xylem Function

CLIMATE CHANGE CONNECTIONS

4.1 Climate Change and Thermal Constraints on Foraging in Reptiles

CHAPTER

5

Coping with Environmental Variation: Energy

KEY CONCEPTS

CONCEPT 5.1 Organisms obtain energy from sunlight, from inorganic chemical compounds, or through the consumption of organic compounds.

CONCEPT 5.2 Radiant and chemical energy captured by autotrophs is converted into stored energy in carbon–carbon bonds.

CONCEPT 5.3 Environmental constraints have resulted in the evolution of biochemical pathways that improve the efficiency of photosynthesis.

CONCEPT 5.4 Heterotrophs have adaptations for acquiring and assimilating energy efficiently from a variety of organic sources.

Toolmaking Crows: A Case Study

Humans employ a multitude of tools to enhance our ability to gather food to meet our energy needs. We use a highly mechanized system of planting, fertilizing, and harvesting crops to feed ourselves or the livestock that we consume. For thousands of years, we have used specialized tools for hunting prey, including spears, bows and arrows, and rifles. We view our toolmaking capacity as something that differentiates us from other animals.

However, humans are not alone in using tools to enhance their food acquisition ability. In the 1920s, Wolfgang Köhler, a psychologist studying the behavior of chimpanzees, observed that chimps in captivity made tools to retrieve bananas stashed in areas that were difficult to reach (Köhler 1927). Jane Goodall, a prominent primatologist, reported observing chimpanzees in the wild using grass blades and plant stems to "fish" for termites in holes in the ground and in decaying wood (**Figure 5.1**). Although these reports challenged the commonly held belief that modern humans were the only makers of tools to enhance food acquisition, it was perhaps comforting to those clinging to this notion that the observations were associated with one of our closest extant relatives. No one would ever have suspected similar behavior in birds, touted as one of the least intelligent vertebrates, as evidenced by the dubious insult "birdbrain" exchanged between humans.

Among the families of birds, the corvids, which includes crows, ravens, magpies, jays, and jackdaws, have entered our cultural heritage with a reputation for being clever. However, the discovery that crows use food-collecting tools manufactured from plants was still quite unexpected. Gavin Hunt reported in 1996 that the crows (*Corvus moneduloides*) of New Caledonia, an island in the South Pacific, used tools to snag insect larvae, spiders, and other arthropods and pull them from the wood of living and decomposing trees (Hunt 1996) (**Figure 5.2A**). Hunt found that two types of tools were being used by the crows. The first was a hooked twig, fashioned from a shoot stripped of its leaves and bark (**Figure 5.2B**). The second was a "stepped-cut" serrated edge clipped from a *Pandanus* tree. The tools were therefore manufactured, rather than just collected from materials lying on the ground.

Hunt described a unique hunting style used by the New Caledonian crows. Birds probed tree cavities or areas of dense foliage using their tools as extensions of their bills. The birds used the tools repeatedly, carrying them from tree to tree while they foraged for food. The presence of hooks on both types of tools suggested an innovative element that

FIGURE 5.1 Nonhuman Tool Use Chimpanzees use a plant stem as a tool to forage for termites. Chimpanzees were the first nonhuman animals observed using tools to forage for food.

(A)

(B)

FIGURE 5.2 Tools Manufactured by New Caledonian Crows (A) Crows use the tools they make to probe for food in the cavities and crevices of trees. (B) Hooked twig tools, made from shoots of trees. The crows use their bills to form the hook while holding the stick with their feet. (B after Hunt 1996.)

might increase the birds' efficiency in extracting prey from their refuges in the trees. The tools also appeared to be uniform in their construction; Hunt examined 55 tools manufactured by different birds and found that they differed little. When New Caledonian crows were captured and brought into the laboratory, they made hooked tools from wire, and experiments showed that the tools increased their food retrieval efficiency (Weir et al. 2002).

Toolmaking at a skill level equivalent to that shown by the crows appeared in humans only in the late Stone Age, approximately 450,000 years ago (Mellars 1989). How could birds have achieved a similar level of sophistication in their tool construction? The high numbers of New Caledonian crows using tools, and the consistency in the construction of the tools, indicate a cultural phenomenon; that is, a skill learned socially within a population of animals—a phenomenon never before observed in birds. How much of an energetic benefit do the crows gain by using tools rather than just their bills? Hunt's discovery posed intriguing questions bridging psychology, anthropology, and behavioral ecology.

Introduction

Energy is the most basic requirement for all organisms. Physiological maintenance, growth, and reproduction all depend on energy acquisition. Organisms are complex systems, so if energy input stops, so does biological functioning. Enzyme systems fail if replacement proteins are not made. Cell membranes degrade and organelles cease to operate without energy to maintain and repair them. In this chapter, we will review the different ways in which organisms acquire energy to meet the demands of cellular maintenance, growth, reproduction, and survival. We'll focus on the major mechanisms that allow organisms to obtain energy from their environment, including the capture of sunlight and chemical energy and the acquisition and use of organic compounds synthesized by other organisms.

CONCEPT 5.1 Organisms obtain energy from sunlight, from inorganic chemical compounds, or through the consumption of organic compounds.

Sources of Energy

We sense energy in our environment in a variety of forms. Light from the sun, a form of *radiant energy*, illuminates our world and warms our bodies. Objects that are cold or warm to our touch have different amounts of *kinetic energy*, which is associated with the motion of the molecules that make up the objects. A grasshopper eating a leaf and a coyote eating a meadow vole both represent the transfer of *chemical energy*, which is stored in the food that is being consumed. Radiant and chemical energy are the forms organisms use to meet the demands of growth and maintenance, while kinetic energy, through its influence on the rate of chemical reactions and temperature, is important for controlling the rate of activity and metabolic energy demand of organisms. A cold endotherm needs to warm its body to the optimal temperature for physiological functioning. It does this by "burning" chemical energy from its food during cellular respiration. Ultimately, this food was derived from the radiant energy of sunlight, converted into chemical energy by plants. Even the energy used to support industrial development, fuel our cars, and heat our homes originated ultimately with photosynthesis, which produced the organisms that became the oil we pump out of the ground.

Autotrophs are organisms that assimilate energy from sunlight (*photosynthetic* organisms) or from inorganic chemical compounds.* in their environment (*chemosynthetic* archaea and bacteria). Autotrophs convert the energy of sunlight or inorganic compounds into chemical energy stored in the carbon–carbon bonds of organic compounds, typically carbohydrates. **Heterotrophs** are organisms that obtain their energy by consuming energy-rich organic compounds made by other organisms—energy that ultimately originated with organic compounds synthesized by autotrophs. Heterotrophs include organisms that consume nonliving organic matter (*detritovores*), such as earthworms and fungi that feed on soil detritus derived mainly from dead plants or bacteria in a lake that consume dissolved organic compounds. Heterotrophs also include organisms that consume living organisms, but do not necessarily kill them (*parasites* and *herbivores*), as well as consumers (*predators*) that capture and kill their food source (*prey*).

On the surface, the distinction between autotrophs and heterotrophs would seem to be clear-cut: all plants are autotrophs, all animals and fungi are heterotrophs, and archaea and bacteria can be both autotrophs and heterotrophs. Things are not always so simple, however. Some plants have lost their photosynthetic function and obtain their energy by parasitizing other plants. Such plants, known as *holoparasites* (*holo*, "entire, whole"), have no photosynthetic pigments, and are heterotrophs. Dodder (genus *Cuscuta*, with approximately 150 different species), for example, is a common plant parasite found throughout the world (**Figure 5.3A,B**) and is considered a major pest of agricultural species. Dodder attaches to its host plant by growing in spirals around the stem and penetrates the phloem of the host, using modified roots called haustoria, to take up carbohydrates. Other plants, known as *hemiparasites*, are photosynthetic, but obtain some of their energy, as well as nutrients and water, from host plants (**Figure 5.3C**).

Conversely, animals can act as autotrophs, although this phenomenon is relatively rare. Their photosynthetic capacity is acquired by consuming photosynthetic organisms or by living with them in a close relationship known as a *symbiosis* (see Chapter 14). Some sea slugs, for example, have fully functional chloroplasts that supply them with carbohydrates through photosynthesis. These animals, in the order Ascoglossa, take intact chloroplasts from the algae they feed on into their digestive cells (**Figure 5.4**). The chloroplasts are maintained intact for up to several months, providing energy as well as camouflage to the sea slug.

In the next two sections, we'll take a more detailed look at the mechanisms autotrophs use to capture energy and at some of the adaptations that make that process more efficient. We'll do the same more generally for heterotrophs in the final section of this chapter, but later chapters pro-

* *Organic* chemical compounds have carbon–hydrogen bonds and are usually biologically synthesized. All other compounds are considered *inorganic* compounds.

(A)

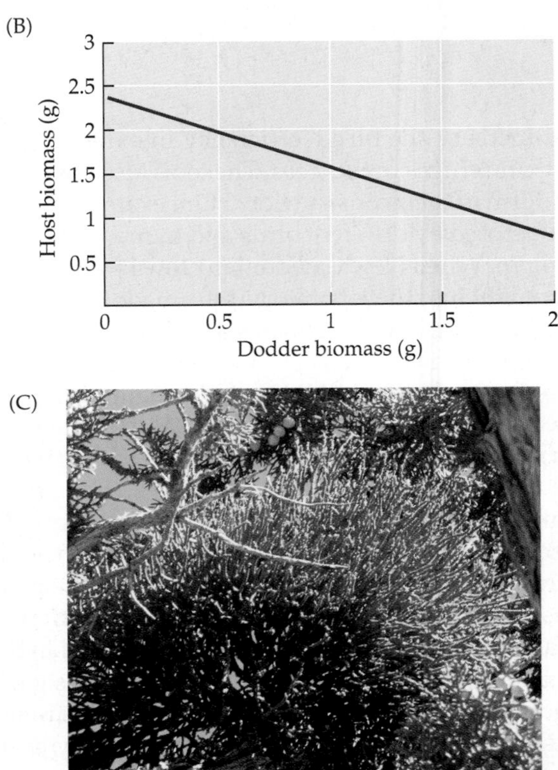

FIGURE 5.3 Plant Parasites (A) Dodder (*Cuscuta* sp.), a holoparasite that lacks chlorophyll, is shown here wrapped around the stem of a jewelweed plant. (B) Increasing amounts of dodder (*Cuscuta europaea*) biomass result in decreasing growth of its host plant, stinging nettle (*Urtica dioica*). (C) Mistletoe is a hemiparasite. This juniper mistletoe (*Phoradendron juniperinum*), growing on the branches of a Utah juniper (*Juniperus osteosperma*), is using its host to supply it with water and nutrients as well as some of its energy, despite having green photosynthetic tissues. (B after Koskela et al. 2002.)

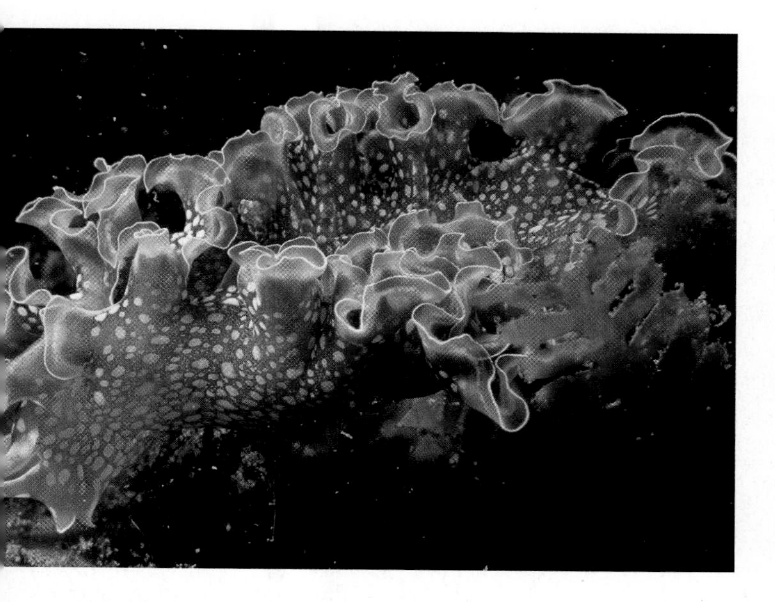

FIGURE 5.4 Green Sea Slug The green color of this lettuce sea slug (*Elysia crispata*) is associated with the chloroplasts in its digestive system. The chloroplasts can supply enough energy to the sea slug to maintain it for several months without food.

vide more detailed considerations of energy capture by heterotrophs (Chapters 12 and 13).

CONCEPT 5.2 Radiant and chemical energy captured by autotrophs is converted into stored energy in carbon–carbon bonds.

Autotrophy

The vast majority of the autotrophic production of chemical energy on Earth occurs through **photosynthesis**, a process that uses sunlight to provide the energy needed to take up CO_2 and synthesize organic compounds, principally carbohydrates. Although its contribution to the global energy picture is smaller, **chemosynthesis** (also known as *chemolithotrophy*), a process that uses energy from inorganic compounds to produce carbohydrates, is important to some key bacteria involved in nutrient cycling (see Chapter 21) and in some unique ecosystems, such as hydrothermal vent communities (see Chapter 19). Because the energy derived from photosynthesis and chemosynthesis is stored in the carbon–carbon bonds of the organic compounds produced by these processes, ecologists often use carbon as a measure of energy.

Chemosynthesis harvests energy from inorganic compounds

The earliest autotrophs on Earth were probably chemosynthetic bacteria or archaea that evolved when the composition of the atmosphere was markedly different than it is today: low in oxygen, but rich in hydrogen, with significant amounts of carbon dioxide and methane (CH_4). A diverse group of archaea and bacteria still use energy from inorganic elements and compounds in their environment to take up CO_2 and synthesize carbohydrates. Chemosynthetic bacteria are often named according to the inorganic substrate they use for energy (**Table 5.1**).

TABLE 5.1

Inorganic Substrates Used by Chemosynthetic Bacteria as Electron Donors for Carbon Fixation

Substrate (chemical formula)	Type of bacteria
Ammonium (NH_4^+)	Nitrifying bacteria
Nitrite (NO_2^-)	Nitrifying bacteria
Hydrogen sulfide (H_2S/HS^-)	Sulfur bacteria (purple and green)
Sulfur (S)	Sulfur bacteria (purple and green)
Ferrous iron (Fe^{2+})	Iron bacteria
Hydrogen (H_2)	Hydrogen bacteria
Phosphite (HPO_3^{2-})	Phosphite bacteria

Source: Madigan and Martinko 2005.

During chemosynthesis, organisms obtain electrons from the inorganic substrate, or in other words, they *oxidize** the inorganic substrate. They use the electrons to generate two high-energy compounds: adenosine triphosphate (ATP) and nicotinamide adenine dinucleotide phosphate (NADPH). They then use energy from ATP and NADPH for the uptake of carbon from gaseous CO_2 (known as **fixation** of CO_2). They use that carbon to synthesize carbohydrates or other organic molecules, which are then used for energy storage or biosynthesis (manufacture of chemical compounds, membranes, organelles, and tissues). Alternatively, some bacteria can use electrons from the inorganic substrate directly to fix carbon. The biochemical pathway most commonly used to fix carbon is the **Calvin cycle**, named for Melvin Calvin, the biochemist who first described it. The Calvin cycle is catalyzed by several enzymes, and it occurs in both chemosynthetic and photosynthetic organisms.

One of the most widespread and ecologically important groups of chemosynthetic organisms is the nitrifying bacteria (e.g., *Nitrosomonas*, *Nitrobacter*), which are found in both aquatic and terrestrial ecosystems. In a two-step process, these bacteria convert ammonium (NH_4^+) into nitrite (NO_2^-), then oxidize it to nitrate (NO_3^-). These chemical conversions of nitrogen compounds are an important

* *Oxidation–reduction reactions* involve the exchange of electrons between chemical compounds. The compound that gives up, or donates, electrons is *oxidized*, while the compound that accepts electrons is *reduced*.

FIGURE 5.5 Sulfur Deposits from Chemosynthetic Bacteria Sulfur bacteria thrive in sulfur hot springs with water temperatures as high as 110°C. They use hydrogen sulfide from these waters to generate chemical energy, leaving behind elemental sulfur.

component of nitrogen cycling and plant nutrition, and we will discuss them in more detail in Chapter 21. Another important chemosynthetic group is the sulfur bacteria, associated with volcanic deposits, sulfur hot springs, and acidic mine wastes. Sulfur bacteria initially use the higher-energy forms of sulfur, H_2S and HS^- (hydrogen sulfide), producing elemental sulfur, which is insoluble and highly visible in the environment (**Figure 5.5**). Once the H_2S and HS^- are exhausted, the bacteria use elemental S as an electron donor, producing SO_4^{2-} (sulfate).

Photosynthesis is the powerhouse for life on Earth

Prior to 1650, most people believed that plants obtained the raw material needed for their growth from the soil. Jan Baptist van Helmont (1579–1644), a Flemish scientist, tested this theory in a well-known experiment. Using a pot into which a carefully measured mass (200 pounds/91 kg) of dry soil was placed, he planted a willow sapling weighing 5 pounds (2.3 kg). Van Helmont watered the sapling using only rainwater for 5 years as it grew into a small tree. At the end of that time, the tree had gained 164 pounds (74 kg), and the soil had lost only 2 ounces (0.06 kg). Although he incorrectly concluded that the tree had gained its mass from the water, van Helmont's experiment established the basis for the later discovery that it was photosynthetic uptake of CO_2 from the air—not material from the soil—that was the source of the tree's weight gain.

The vast majority of biologically available energy on Earth is derived from the conversion of sunlight into energy-rich carbon compounds by photosynthesis. Photosynthetic organisms include some archaea, bacteria, and protists and most algae and plants. Leaves are the principal photosynthetic tissue in plants, but photosynthesis may also occur in stem and reproductive tissues. Like chemosynthesis, photosynthesis involves the conversion of CO_2 into carbohydrates used for energy storage and biosynthesis. Photosynthesis is also responsible for the largest movements of CO_2 between Earth and the atmosphere, and is critically important to the global climate system (Chapters 2 and 24). Here, we will briefly review the major steps of plant photosynthesis and consider some ecologically relevant constraints on photosynthetic rates. In the next subsection, we will examine some variations in plant photosynthetic pathways.

Light and dark reactions Photosynthesis has two major steps. The first is the harvesting of energy from sunlight, which is used to split water to provide electrons for generating ATP and NADPH. This step is often referred to as the "light reaction" of photosynthesis. The second step is the fixation of carbon and the synthesis of carbohydrates. This step is often referred to as the "dark reaction" of photosynthesis, despite its daytime occurrence in most plants.

Sunlight harvesting is accomplished by several pigments, principally chlorophyll. Chlorophyll gives photosynthetic organisms their green appearance because it absorbs red and blue light and reflects green wavelengths (**Figure 5.6**). Plants and photosynthetic bacteria have similar chlorophyll pigments, but they absorb light at slightly different wavelengths. Additional pigments associated with photosynthesis, called accessory pigments, include

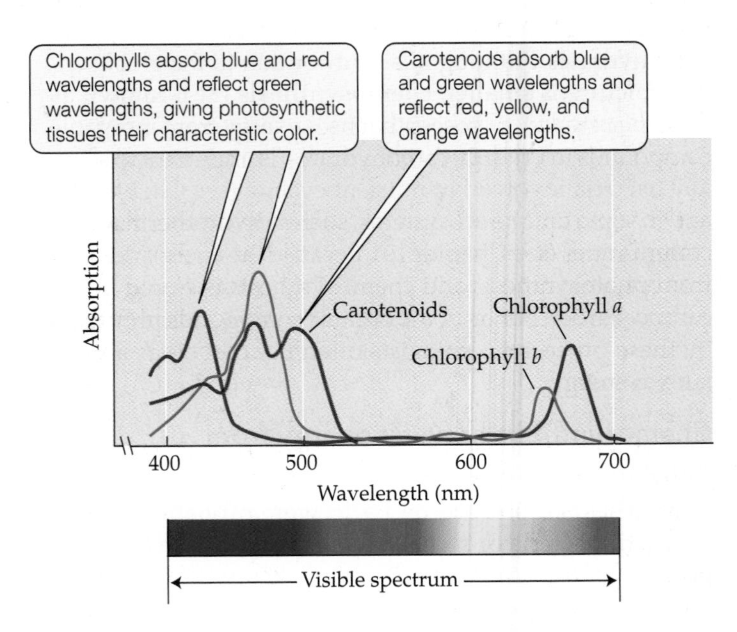

FIGURE 5.6 Absorption Spectra of Plant Photosynthetic Pigments Plants typically contain several light-absorbing pigments, each of which absorbs light of different wavelengths.

the carotenoids, which are characteristically red, yellow, or orange in appearance. These photosynthetic pigments are embedded in a membrane, along with other molecules involved in the light reaction. In plants, this membrane lies within specialized organelles called chloroplasts, while in photosynthetic bacteria the pigments are embedded in the cell membrane. The pigment molecules are arrayed like antennae, with each array containing between 50 and 300 molecules. The pigments absorb energy from discrete units of light, called *photons*. That energy is used to split water and provide electrons. The electrons are passed on to molecular complexes on the membranes, where they are used to synthesize ATP and NADPH.

The splitting of water to provide electrons for the light reaction generates oxygen (O_2). The evolution of photosynthesis, and the accompanying release of O_2 into the atmosphere, was a critical step in the development of the chemistry of the modern atmosphere and lithosphere as well as the evolution of life on Earth. Atmospheric oxygen led to the creation of a layer of ozone (O_3) high in the atmosphere that shields organisms from high-energy ultraviolet radiation. The evolution of aerobic respiration, in which O_2 is used as an electron acceptor, facilitated great evolutionary advances for life on Earth.

Energy from the high-energy compounds ATP and NADPH is used in the Calvin cycle to fix carbon. Carbon dioxide is taken up from the atmosphere through the stomates of vascular plants, or diffuses across the cell membranes in nonvascular plants, algae, and photosynthetic bacteria and archaea. A key enzyme associated with the Calvin cycle is ribulose 1,5 bisphosphate carboxylase/oxygenase, thankfully usually referred to by its abbreviation, *rubisco*. Rubisco, the most abundant enzyme on Earth, catalyzes the uptake of CO_2 and the synthesis of a three-carbon compound: phosphoglyceraldehyde, or PGA. PGA is eventually converted into a six-carbon sugar [glucose ($C_6H_{12}O_6$) in most plants]. The net reaction of photosynthesis is

$$6\ CO_2 + 6\ H_2O \rightarrow C_6H_{12}O_6 + 6\ O_2 \quad \textbf{(5.1)}$$

Environmental constraints and solutions The rate of photosynthesis determines the supply of energy and substrates for biosynthesis available to photosynthetic organisms, which in turn influences their growth and reproduction, often equated with their ecological success (abundance and geographic range). Thus, environmental controls on the rate of photosynthesis are a key topic in physiological ecology. It should be noted, however, that net energy (carbon) gain is also influenced by CO_2 losses associated with cellular respiration.

Light is clearly an important determinant of rates of photosynthesis in terrestrial and aquatic habitats. The relationship between the light level and a plant's photosynthetic rate can be portrayed by a *light response curve* (**Figure 5.7A**). When there is enough light that the plant's photosynthetic CO_2 uptake is balanced by its CO_2 loss

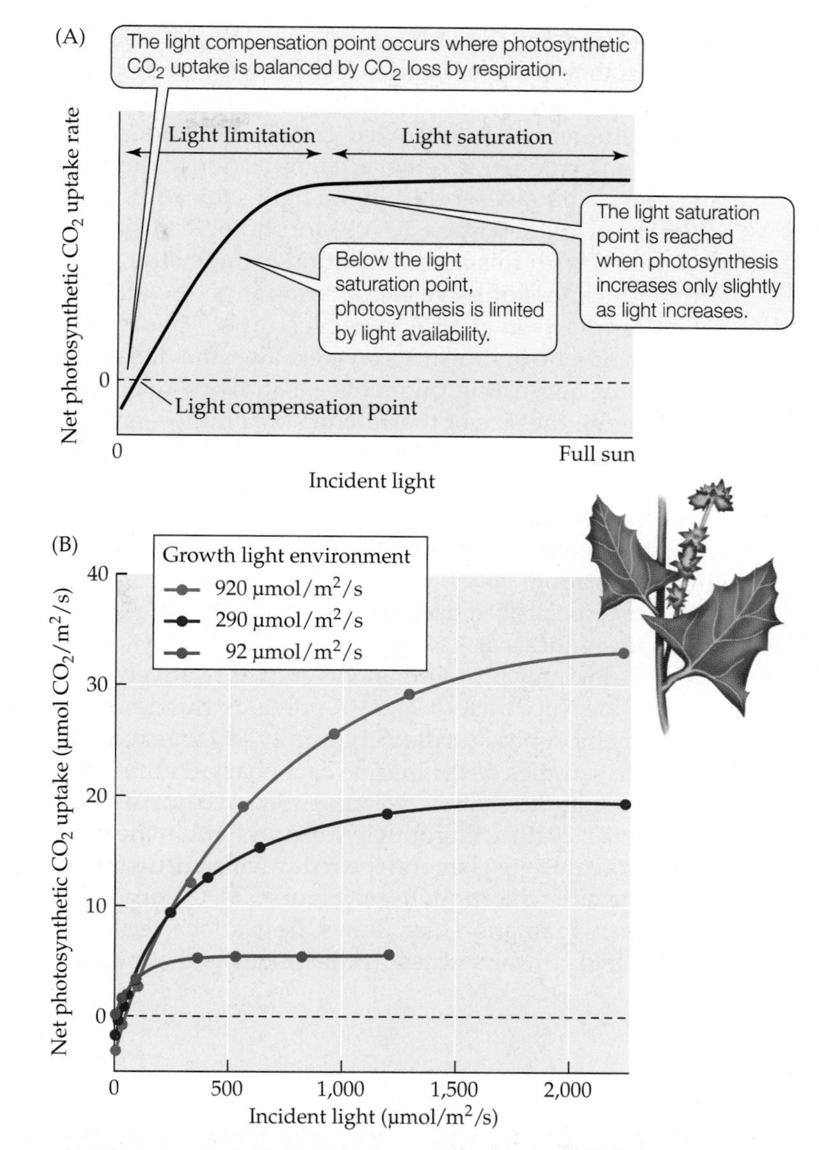

FIGURE 5.7 Plant Responses to Variations in Light Levels (A) Photosynthetic light response curve. (B) Spearscale (*Atriplex triangularis*) plants grown in different light environments in growth chambers acclimatized to those light levels. Their light response curves indicate that adjustments in the light saturation point occurred. Small, but ecologically significant, changes in the light compensation point occur in many other species, facilitating CO_2 uptake at low light levels. (B after Bjorkman 1981.)

Q Why might the light saturation point of a plant be below the maximum light level the plant is likely to be exposed to?

by respiration, the plant is said to have reached the *light compensation point*. As the light level increases above the light compensation point, the photosynthetic rate also increases; in other words, photosynthesis is *limited* by the availability of light. The photosynthetic rate levels off at a *light saturation point*, which is typically reached at a level below full sunlight.

How do plants cope with light variation in their environment? How would an understory forest plant, for example, respond to shading by canopy trees? Could that plant acclimatize to more light if the canopy tree fell and allowed full sunlight to reach the ground? In a series of classic studies using controlled growth conditions, Olle Bjorkman demonstrated that acclimatization to different light levels involves a shift in the light saturation point (Bjorkman 1981) (**Figure 5.7B**). Morphological changes associated with this acclimatization include alterations in the thickness of leaves and variation in the number of chloroplasts available to harvest light (**Figure 5.8**). Photosynthetic organisms may also alter the density of their light-harvesting pigments—a strategy analogous to changing the size of the antenna on a radio—and the amounts of photosynthetic enzymes available for the dark reaction. Typically, the average light level a plant experiences, integrated over the course of the day, is near the transition point between light limitation and light saturation.

Some specialized bacteria are especially well adapted to photosynthesis at low light levels, which allows them to thrive in dimly lit environments such as relatively deep ocean water (up to about 20 m). A previously undescribed form of chlorophyll (called chlorophyll *f*) was recently found in samples of the marine cyanobacteria that form sediments in the shallow waters of Shark Bay, Australia (Chen et al. 2010). Chlorophyll *f* absorbs light in the near-infrared region, just beyond the red wavelengths used by other forms of chlorophyll (see Figure 5.6). Chlorophyll *f* may be an adaptation that allows the cyanobacteria possessing it to grow underneath other photosynthetic organisms that use light in the blue and red wavelengths, as it lets them harvest energy at wavelengths that pass through those other photosynthetic organisms. The discovery of a pigment that can harvest near-infrared energy has implications for increasing the efficiency of photovoltaic panels used to generate electricity, which may help lower emissions of CO_2 (**Climate Change Connection 5.1**).

Water availability is an important control on the supply of CO_2 for photosynthesis in terrestrial plants. As we saw in Chapter 4, low water availability results in closure of the stomates, restricting the entry of CO_2 into leaves. Stomatal control represents an important trade-off for the plant: water conservation versus energy gain through photosynthesis as well as cooling of the leaf. Keeping stomates open while tissues lose water can permanently impair physiological processes in the leaf. Closing stomates, however, not only limits photosynthetic CO_2 uptake, but also increases the chances of light damage to the leaf. When the Calvin cycle is not operating, energy continues to accumulate in the light-harvesting arrays, and if enough energy builds up, it can damage the photosynthetic membranes. Plants have evolved a number of ways of dissipating this energy safely, including the use of carotenoids to release it as heat, as described in **Web Extension 5.1**.

Temperature influences photosynthesis in two main ways: through its effects on the rates of chemical reactions and by influencing the structural integrity of membranes and enzymes. Acclimatization and adaptation to temperature variation are associated with the properties of the Calvin cycle enzymes and the chemistry of the photosynthetic membranes. Different photosynthetic organisms have different forms of the same photosynthetic enzymes that operate best under the environmental temperatures where they occur. These differences result in markedly different temperature ranges for photosynthesis in organisms from different climates (**Figure 5.9A**). Lichens and plants of Arctic and alpine environments can photosynthesize at temperatures close to freezing, while desert plants may have their highest photosynthetic rates at temperatures that are hot enough to denature most other plants' enzymes (40°C–50°C, 104°F–122°F). Plants that acclimatize to changes in temperature synthesize different forms of photosynthetic enzymes with different temperature optima (**Figure 5.9B**). Temperature also influences the fluidity of the cell and organelle membranes (see Chapter 4). Cold sensitivity in plants of tropical and subtropical biomes is associated with loss of membrane fluidity, which inhibits the functioning of the light-harvesting molecules embedded in the chloroplast membranes. As we have seen, high temperatures, particularly in combination with intense sunlight, can damage photosynthetic membranes.

Nutrient concentrations in leaves reflect their photosynthetic potential because most of the nitrogen in plants

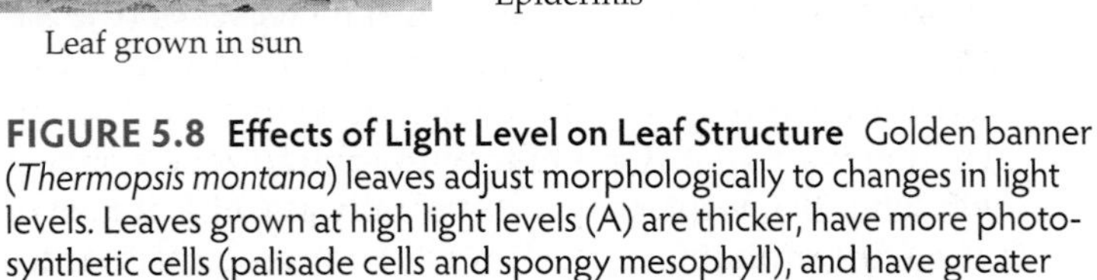

FIGURE 5.8 Effects of Light Level on Leaf Structure Golden banner (*Thermopsis montana*) leaves adjust morphologically to changes in light levels. Leaves grown at high light levels (A) are thicker, have more photosynthetic cells (palisade cells and spongy mesophyll), and have greater numbers of chloroplasts, than leaves grown at low light levels (B).

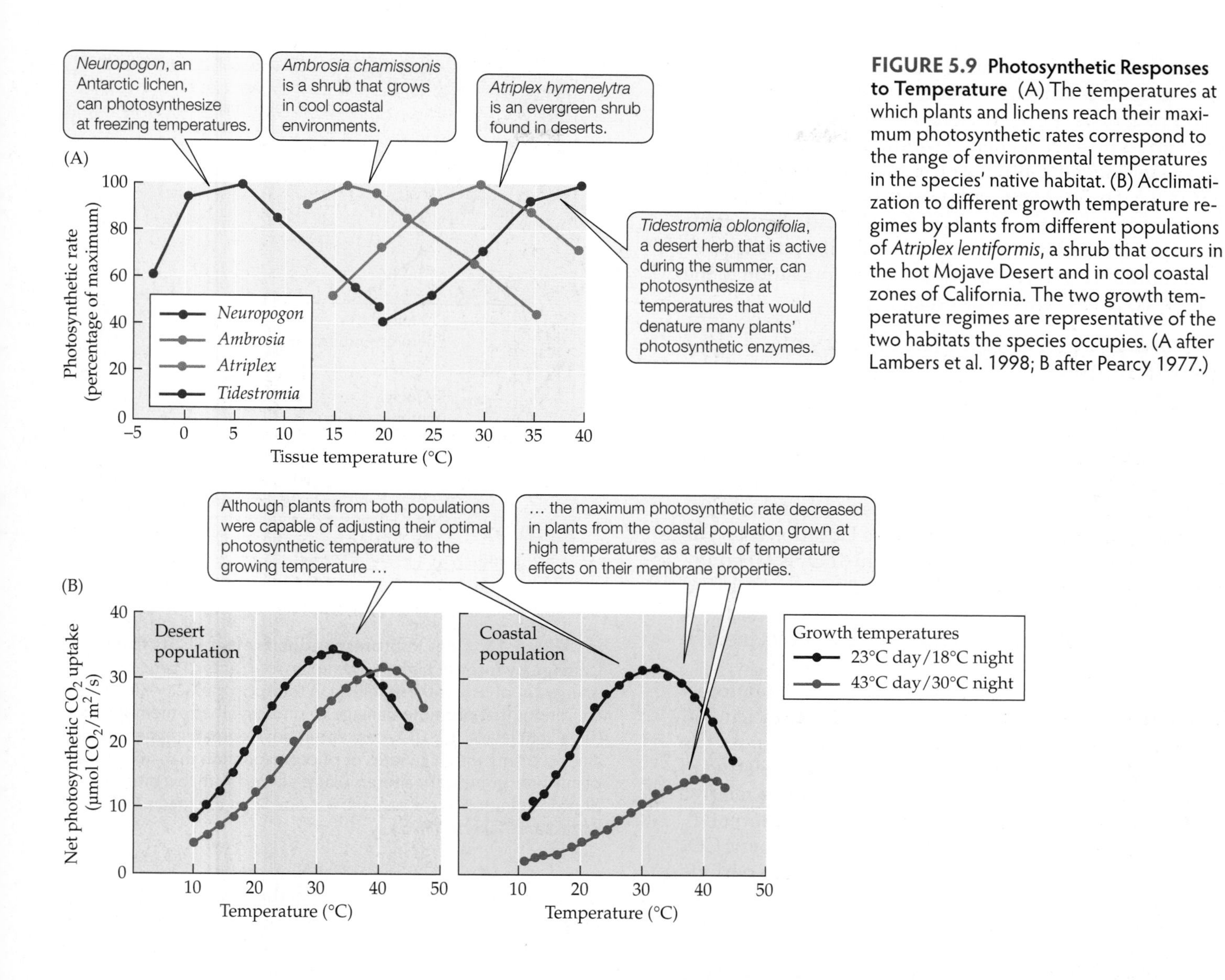

FIGURE 5.9 Photosynthetic Responses to Temperature (A) The temperatures at which plants and lichens reach their maximum photosynthetic rates correspond to the range of environmental temperatures in the species' native habitat. (B) Acclimatization to different growth temperature regimes by plants from different populations of *Atriplex lentiformis*, a shrub that occurs in the hot Mojave Desert and in cool coastal zones of California. The two growth temperature regimes are representative of the two habitats the species occupies. (A after Lambers et al. 1998; B after Pearcy 1977.)

is associated with rubisco and other photosynthetic enzymes. Thus, higher amounts of nitrogen in leaves are correlated with higher photosynthetic rates. Why, then, don't all plants allocate more nitrogen to their leaves to increase their photosynthetic capacity? There are two main reasons. First, the supply of nitrogen is low relative to the demand, and nitrogen is needed for growth and other metabolic functions in addition to photosynthesis (see Chapter 21). Second, increasing the nitrogen concentration of a leaf increases the risk that herbivores will consume the leaf, as plant-eating animals are also nitrogen-starved (see Chapter 12). Plants must balance the competing demands of photosynthesis, growth, and protection from herbivores.

Over evolutionary time, some plants have dealt with environmental constraints on photosynthesis with adaptations in their photosynthetic pathways, as we will see next.

CONCEPT 5.3 Environmental constraints have resulted in the evolution of biochemical pathways that improve the efficiency of photosynthesis.

Photosynthetic Pathways

Anything that influences energy gain by photosynthesis has the potential to affect the survival, growth, and reproduction of the organism. As we have just seen, rates of photosynthesis are influenced by environmental conditions, particularly temperature and water availability. In addition, an apparent biochemical inefficiency in the initial step of the Calvin cycle limits energy gain by photosynthetic organisms. In this section we will examine some evolutionary responses to these environmental constraints on photosynthesis. We will describe two specialized photosynthetic pathways, the C_4 pathway and crassulacean acid metabolism (CAM), that make photosynthesis more efficient under particular potentially stressful environmental conditions. Plants that lack these specialized pathways use the **C_3 photosynthetic pathway**. The C_3 and C_4 photosynthetic pathways take their names from the number of carbon atoms in their first stable chemical products. First, we'll examine photorespiration, a process that operates in opposition to the Calvin cycle and lowers its efficiency.

Photorespiration lowers the efficiency of photosynthesis

Earlier, we described a key enzyme in the Calvin cycle, rubisco, and noted that the "o" in the abbreviation stands for "oxygenase." Rubisco can catalyze two competing reactions. One is a carboxylase reaction, in which CO_2 is taken up, leading to the synthesis of sugars and the release of O_2 (i.e., photosynthesis; see Equation 5.1). The other is an oxygenase reaction, in which O_2 is taken up, leading to the breakdown of carbon compounds and the release of CO_2. This oxygenase reaction is part of a process called **photorespiration**, which results in a net loss of energy.

The balance between photosynthesis and photorespiration is related to two main factors: (1) the ratio of O_2 to CO_2 in the atmosphere and (2) temperature. As the atmospheric concentration of CO_2 decreases relative to that of O_2, the rate of photorespiration increases relative to the rate of photosynthesis. Since the evolution of C_3 photosynthesis over 3 billion years ago, atmospheric CO_2 concentrations have changed repeatedly over periods of hundreds of thousands of years in response to major global geologic and climatic events (see Chapter 24). These shifts in atmospheric CO_2 concentrations would have influenced the balance between photosynthesis and photorespiration. Furthermore, as temperatures increase, the rate of O_2 uptake catalyzed by rubisco increases relative to the rate of CO_2 uptake, and the solubility of CO_2 in the cytoplasm decreases more than that of O_2. As a result of these two processes, photorespiration increases more rapidly at high temperatures than photosynthesis does. Thus, energy loss due to photorespiration is particularly acute at high temperatures and low atmospheric CO_2 concentrations.

If photorespiration is detrimental to the functioning of photosynthetic organisms, why hasn't a new form of rubisco evolved that minimizes uptake of O_2? Is it possible that photorespiration provides some benefit to the plant? A possible clue comes from experiments with *Arabidopsis thaliana*, a small plant commonly used as a model organism in plant studies: plants with a genetic mutation that knocks out photorespiration die under normal light and CO_2 conditions (Ogren 1984). One hypothesis for a potential benefit of photorespiration is that it protects the plant from damage to the photosynthetic machinery at high light levels. This hypothesis is supported by the results of a study by Akiko Kozaki and Go Takeba, who used tobacco plants that had been genetically altered to elevate or lower their rates of photorespiration (Kozaki and Takeba 1996). They subjected these experimental plants to high-intensity light and recorded the damage to their photosynthetic machinery. Plants with higher rates of photorespiration showed less damage than control plants with normal rates of photorespiration (**Figure 5.10**) or plants with depressed rates of photorespiration.

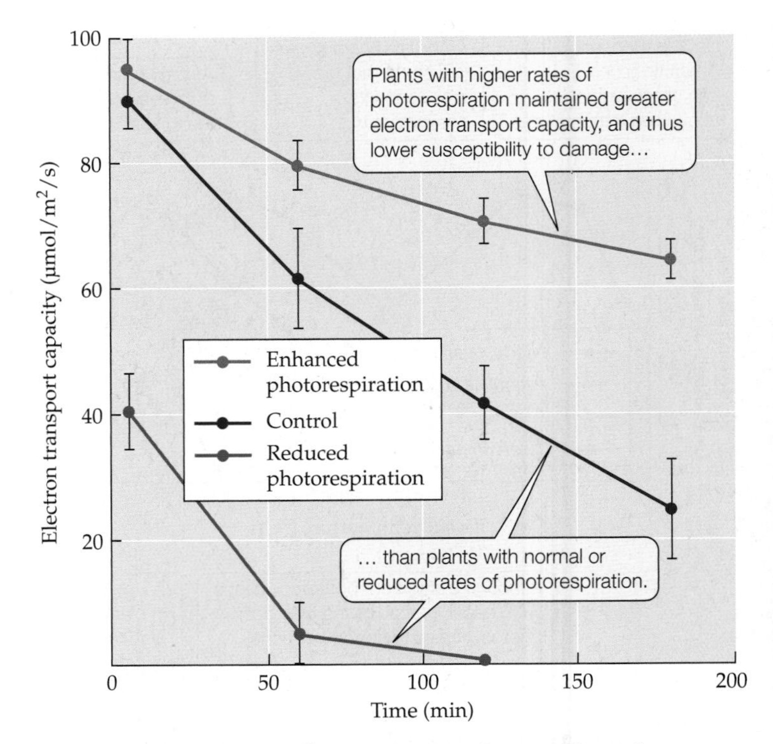

FIGURE 5.10 Does Photorespiration Protect Plants from Damage by Intense Light? The ability of plants to process light energy for photosynthesis (electron transport capacity) under conditions that promote damage to photosynthetic membranes (high light levels, low CO_2 concentrations) is greater in genetically altered plants with high rates of photorespiration than in control plants or in genetically altered plants altered with low rates of photorespiration. Error bars show ± one SE of the mean. (After Kozaki and Takeba 1996.)

Despite this possibility that photorespiration plays a role in protecting plants from damage at high light levels, there are conditions in which the decrease in photosynthetic CO_2 uptake it causes could be a serious problem for the plant. If atmospheric CO_2 concentrations are low and temperatures high, photosynthetic energy gain might not keep pace with photorespiratory energy loss. Such conditions existed 7 million years ago, at about the time when plants with a unique biochemical pathway, C_4 photosynthesis, first appeared (Cerling et al. 1997).

(A) Sugarcane (*Saccharum* sp.)

(B) *Amaranthus tricolor*

FIGURE 5.11 Examples of Plants with the C_4 Photosynthetic Pathway The C_4 photosynthetic pathway has evolved multiple times, and is found in plants in 18 different families encompassing a variety of growth forms.

C_4 photosynthesis lowers photorespiratory energy loss

The **C_4 photosynthetic pathway**, sometimes called the Hatch–Slack–Kortschack pathway to commemorate the plant physiologists who discovered it, reduces photorespiration. C_4 photosynthesis evolved independently several times in different plant species. It is found in 18 different plant families (**Figure 5.11**), but is most closely associated with the grass family. Well-known examples of crop plants with the C_4 pathway include corn, sugarcane, and sorghum.

C_4 photosynthesis involves both biochemical and morphological specialization (**Figure 5.12**). The biochemical specialization can be thought of as a pump that provides high concentrations of CO_2 to the Calvin cycle. This greater supply of CO_2 lowers the rate of O_2 uptake by rubisco, substantially reducing photorespiration. The morphological specialization involves spatial separation of the regions in the leaf where CO_2 is taken up (mesophyll) and where the

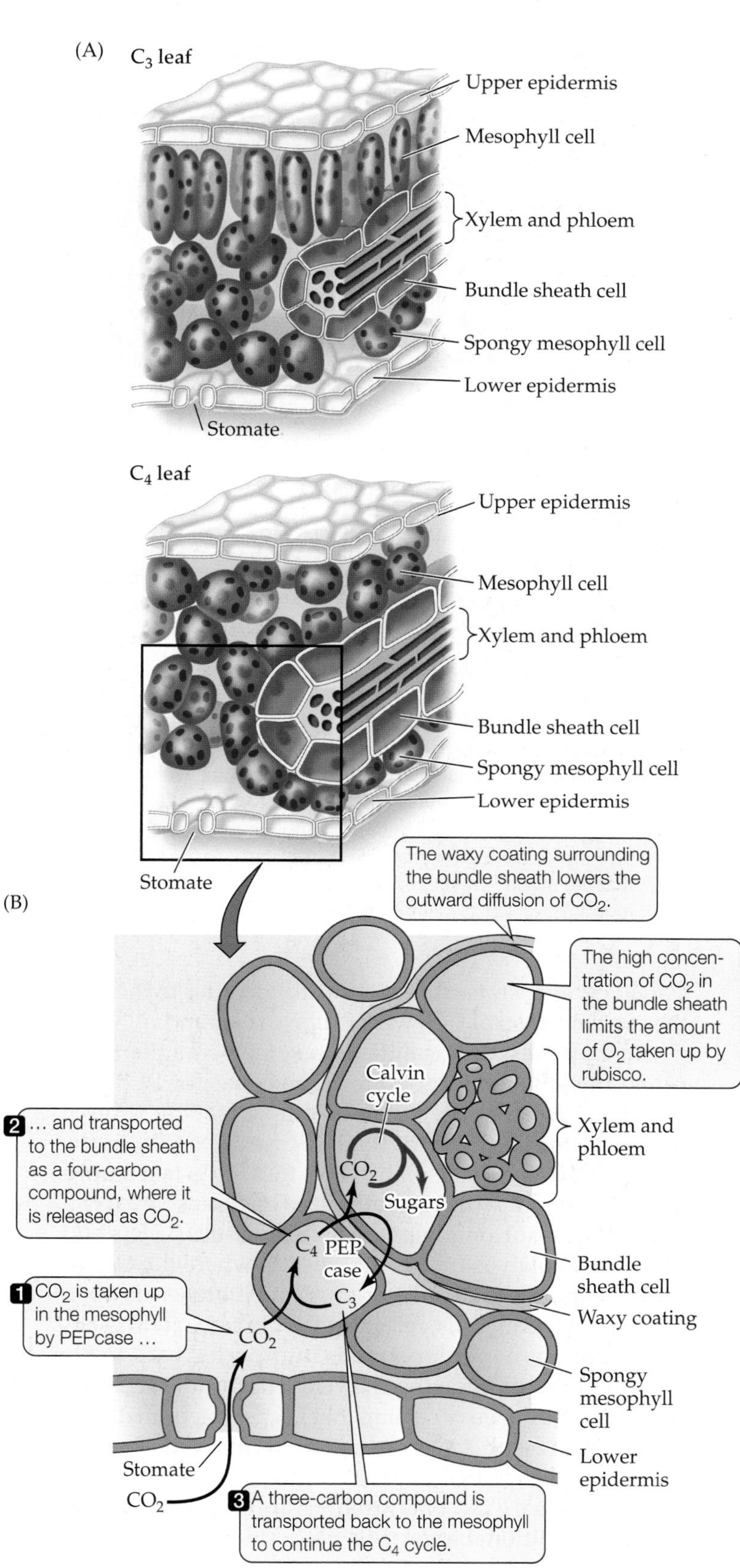

FIGURE 5.12 Morphological Specialization in C_4 Plants (A) Unlike C_3 plants, C_4 plants have well-differentiated mesophyll and bundle sheath tissues in their leaves. (B) The spatial separation of CO_2 uptake and the Calvin cycle effectively minimizes photorespiration.

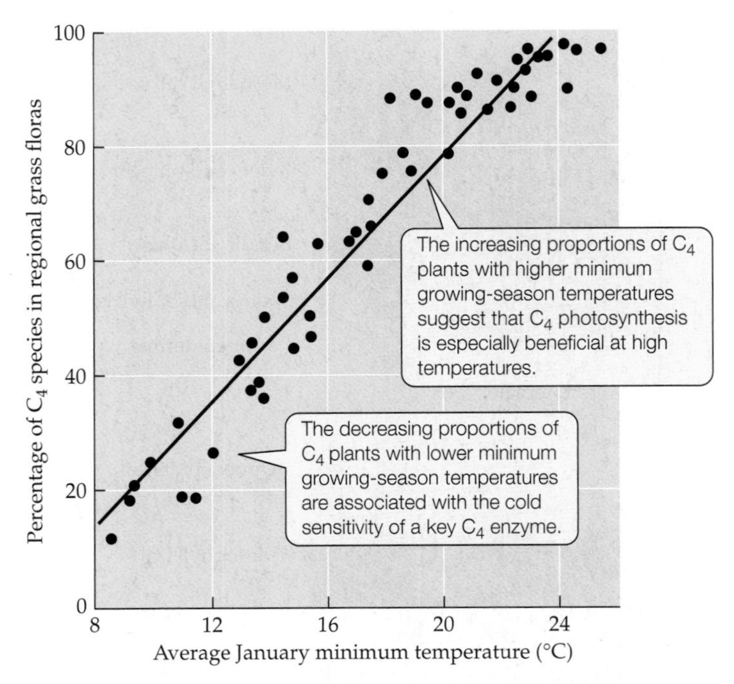

FIGURE 5.13 C_4 Plant Abundance and Growing-Season Temperature The proportions of C_4 plants in Australian grass- and sedge-dominated communities is correlated with the average minimum growing-season temperature. (After Henderson et al. 1995.)

Q Using the data shown in this graph and the seasonal temperature trends from the climate diagrams from Chapter 3 (assume that the monthly minimum temperature is 5°C cooler than the monthly average), what biome(s) should lack C_4 species?

Calvin cycle operates (bundle sheath) in order to increase the concentration of CO_2 where rubisco is found.

In C_4 plants, CO_2 is initially taken up by an enzyme called phosphoenol pyruvate carboxylase, or PEPcase, that has a greater capacity to take up CO_2 than rubisco and lacks oxygenase activity. PEPcase fixes CO_2 in the mesophyll tissue of the plant. Once the CO_2 is taken up, a four-carbon compound is synthesized and transported to a group of cells surrounding the vascular tissues (xylem and phloem), known as the bundle sheath, where the Calvin cycle occurs. The four-carbon compound is broken down in the bundle sheath cells, releasing CO_2 to the Calvin cycle, and a three-carbon compound is transported back to the mesophyll to continue the C_4 cycle. The bundle sheath is surrounded by a waxy coating that keeps CO_2 from diffusing out. As a result, CO_2 concentrations inside the bundle sheath may reach a high of 5,000 parts per million, even though external CO_2 concentrations are only 390 parts per million. Additional energy in the form of ATP must be expended to operate the C_4 photosynthetic pathway, but the increased efficiency of carbon fixation compensates for the higher energy requirement.

As is apparent from the discussion above, plants with the C_4 photosynthetic pathway can photosynthesize at higher rates than C_3 plants under environmental conditions that elevate rates of photorespiration, such as high temperatures. In addition, most C_4 plants have lower rates of transpiration at a given photosynthetic rate (known as *water use efficiency*) than C_3 plants. This difference is due to the ability of PEPcase to take up CO_2 under the lower CO_2 concentrations that exist when stomates are not fully open.

If we assumed that photosynthetic rates determine ecological success, we could use climatic patterns to predict where C_4 plants should predominate over C_3 plants. Such an analysis would be overly simplistic, however, because multiple factors other than temperature influence the biogeography of C_3 and C_4 plants, including abiotic factors such as light levels and biotic factors such as competitive ability and the pool of species available to colonize an area. However, analyses of similar communities across latitudinal and elevational gradients provide support for the benefit of C_4 photosynthesis at high temperatures and for its role in C_4 plant distribution (Ehleringer et al. 1997). In particular, studies of grass- and sedge-dominated communities in Australia suggest a close correlation between growing-season temperature and the proportion of C_3 and C_4 species in the community (**Figure 5.13**). As atmospheric CO_2 concentrations continue to increase, photorespiration rates are likely to decrease, and the advantages of C_4 over C_3 photosynthesis may be diminished, leading to changes in the proportions of C_3 and C_4 plants.

CAM photosynthesis enhances water conservation

When plants first colonized the terrestrial environment, they evolved adaptations to restrict water losses to a dry atmosphere. Among these adaptations is a unique photosynthetic pathway called **crassulacean acid metabolism (CAM)**, which occurs in over 10,000 plant species belonging to 33 families. While C_4 photosynthesis separates CO_2 uptake and the Calvin cycle spatially, CAM separates these two steps temporally (**Figure 5.14**). CAM plants open their stomates at night, when C_3 and C_4 plants have their stomates closed. Because air temperatures are cooler at night, humidity is higher. The higher humidity at night results in a lower water potential gradient between the leaf and the air (see Chapter 4), so the plant loses less water by transpiration than it would during the day. CAM plants close their stomates during the day, when the potential for water loss is highest.

During the night, when the stomates are open, CAM plants take up CO_2 using PEPcase and incorporate it into a four-carbon organic acid, which is stored in vacuoles (**Figure 5.15**). The resulting increase in acidity in the plants' tissues during the night is characteristic of CAM plants and is used to estimate their photosynthetic rates. During the day, when the stomates are closed, the organic acid is broken down, releasing CO_2 to the Calvin cycle. CO_2 con-

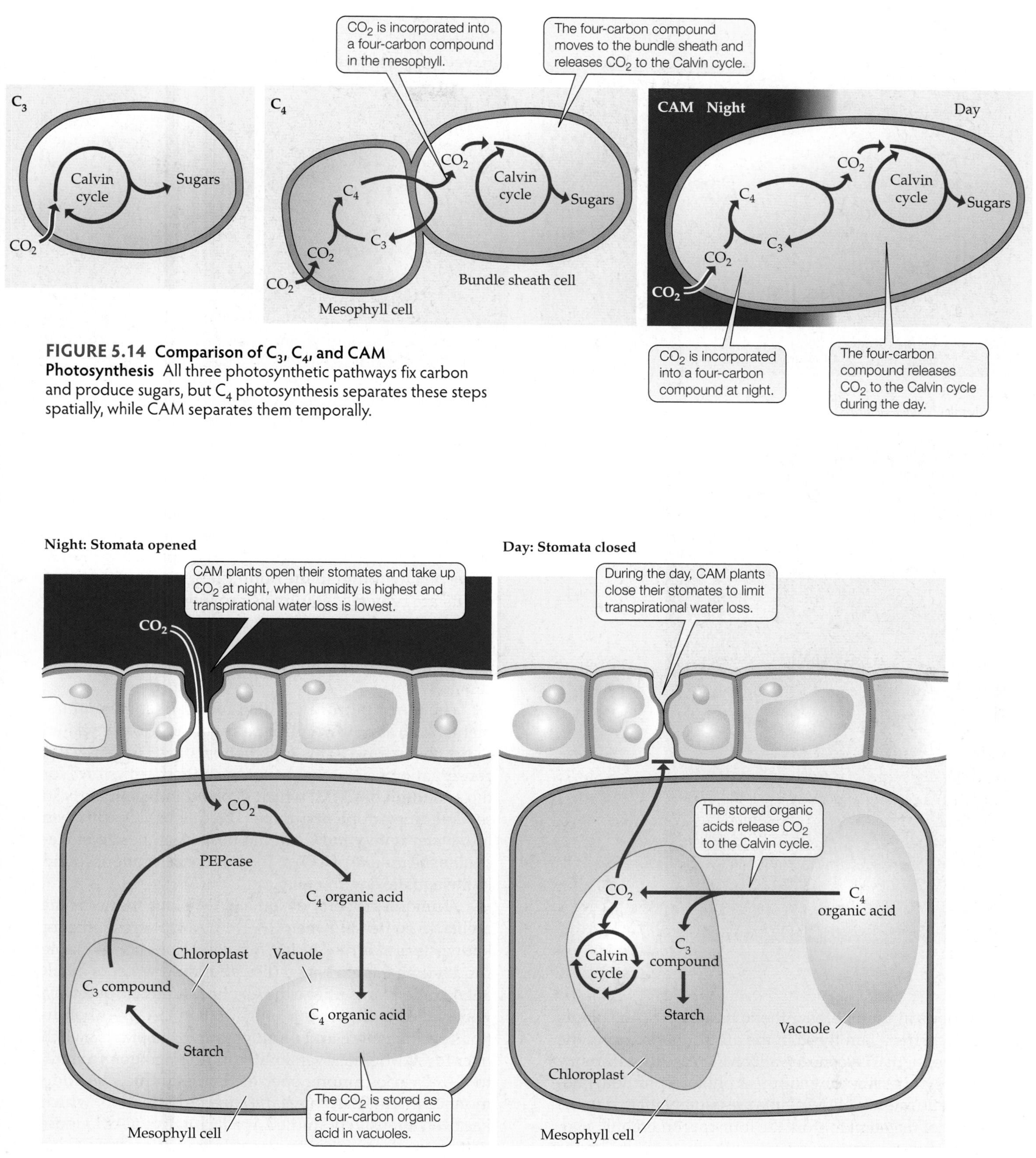

FIGURE 5.14 Comparison of C_3, C_4, and CAM Photosynthesis All three photosynthetic pathways fix carbon and produce sugars, but C_4 photosynthesis separates these steps spatially, while CAM separates them temporally.

FIGURE 5.15 CAM Photosynthesis CAM plants take up CO_2 at night and run the Calvin cycle during the day.

Hens and chicks (*Sempervivum* sp., Crassulaceae)

Ferocactus sp.

Pineapple (*Ananas comosus*)

FIGURE 5.16 Examples of Plants with the CAM Photosynthetic Pathway Most CAM plants are found in arid and saline regions or in other habitats where water availability is periodically low.

centrations in the photosynthetic tissues of CAM plants are thus higher than those in the atmosphere during the day. These high CO_2 concentrations increase the efficiency of photosynthesis as they suppress photorespiration. Photosynthetic rates in CAM plants are usually related to the capacity of the plant to store the four-carbon organic acid. CAM plants are often *succulent*, with thick, fleshy leaves or stems. The thickness of the tissues enhances their nighttime acid storage capacity.

CAM plants are typically associated with arid and saline environments, such as deserts and Mediterranean-type ecosystems (**Figure 5.16**). Some CAM plants, however, are found in the humid tropics. Tropical CAM plants are typically epiphytes growing on the branches of trees, without access to the abundant water stored in the soil. These epiphytes rely on rainfall for their water supply and may be subject to long periods without access to water.

The CAM pathway is also found in some aquatic plants, such as quillwort (*Isoetes*), which is closely related to the club mosses. This observation suggests that water conservation was probably not the only driving force for the evolution of CAM, which evolved independently in several different plant groups. The rate of CO_2 diffusion into water is low, and CAM has been hypothesized to facilitate the uptake of CO_2 at the low concentrations found in the aquatic environment.

A unique property of some CAM plant species is the ability to switch between C_3 and CAM photosynthesis, known as *facultative CAM*. When conditions are favorable for daytime gas exchange (i.e., abundant water is available), these plants utilize the C_3 photosynthetic pathway, which allows greater carbon gain than CAM. As conditions become more arid or more saline, the plants switch over to CAM. The reversibility of the transition from C_3 to CAM varies among species. For example, the common ice plant (*Mesembryanthemum crystallinum*), which has been intensively studied as a facultative CAM model system, undergoes an irreversible transition from C_3 to CAM photosynthesis when salinity increases or the soil dries out (Osmond et al. 1982). In contrast, some species in the genus *Clusia* can switch relatively rapidly between C_3 and CAM (Borland et al. 1992). These plants start out as epiphytes in canopy trees, but grow toward the base of their host tree, eventually strangling it and taking on a tree growth form. The capacity to switch between C_3 and CAM facilitates the change from epiphyte to tree form, and it supports continued photosynthesis during the transition from wet season to dry season characteristic of some tropical locations.

How can we tell what photosynthetic pathway a plant is using? The morphology of the plant gives us a clue: succulent plants suggest CAM photosynthesis, and plants with a well-developed bundle sheath suggest C_4 photosynthesis. These clues provide a starting point, but they are far from foolproof. We can measure the presence and activity of specific enzymes, but this approach requires substantial sample preparation and laboratory time. A simpler approach is to measure the proportions of stable carbon isotopes ($^{13}C/^{12}C$) in plant tissues. Although the isotopic technique uses sophisticated equipment, sample preparation is simple, and there are numerous laboratories that can routinely analyze plant tissue samples (**Ecological Toolkit 5.1**).

Now that we have reviewed the ways in which autotrophs acquire energy, let's turn our attention to how that energy is acquired by heterotrophs.

CONCEPT 5.4 Heterotrophs have adaptations for acquiring and assimilating energy efficiently from a variety of organic sources.

Heterotrophy

The first organisms on Earth were probably heterotrophs that consumed amino acids and sugars that formed spontaneously in the early atmosphere and rained down on the surface or formed in the oceans near hydrothermal vents. The diversity of heterotrophic strategies for obtaining energy has expanded tremendously since that time. The organic matter that provides energy for heterotrophs includes living and freshly killed organisms as well as organic material derived from dead organisms in various stages of decomposition (see Chapter 19). In this section we will examine the ways in which heterotrophs obtain energy. There is a wide range of variation in the complexity of heterotrophic energy acquisition and assimilation processes associated with heterotroph body size and physiology. In Chapters 12 and 13, we will take a more in-depth look at the various types of consumers (predators, herbivores, and parasites), and we will see how the food they consume affects their growth and reproduction as well as the distributions and abundances of the consumers themselves and their food resources (prey and hosts).

Food sources differ in their chemistry and availability

Heterotrophs consume energy-rich organic compounds (food) from their environment and convert them into usable chemical energy—primarily ATP—during catabolic processes such as *glycolysis*, which breaks down carbohydrates. The heterotroph's energy gain from food depends on the chemistry of the food, which determines its digestibility and the amount of energy per unit of mass it contains. The effort invested in finding and obtaining the food also influences how much benefit the heterotroph gets from consuming it. For example, microorganisms that consume detritus in the soil invest little energy in obtaining food. However, the energy content of this decomposing plant matter is low compared with the energy content of live organisms. Living food items (prey) are rarer than detritus, and they may have defensive mechanisms that consumers (predators) must expend energy to overcome. Thus, a cheetah hunting a gazelle invests substantial energy in finding, chasing, capturing, and killing its prey, but it obtains a substantial, energy-rich meal if the hunt is successful.

The chemistry of food is associated with the cell types of the organisms from which it is derived. Animal cells are generally more energy-rich than plant, fungal, or bacterial cells, which tend to have higher concentrations of structural components, such as cell walls, that are not easily digested. Most food consumed by heterotrophs consists of complex chemical compounds that must be chemically transformed into simpler compounds before they can be used as energy sources. Digestion breaks down proteins, carbohydrates, and fats into their component amino acids, simple sugars, and fatty acids. Each of these compounds has a different yield of energy. Fats are richer in energy than carbohydrates per unit of mass, and carbohydrates provide more energy than amino acids. However, amino acids also provide nitrogen, a nutrient often in high demand. In addition, some metabolic functions require specific energy-containing compounds. Insect flight, for example, has a high energy demand, and some insects must maintain fat storage bodies to supply the lipids required for initiation of flight. Humans require carbohydrates to fuel brain activity, which explains why low blood sugar can lead to poor cognitive ability.

Heterotrophs obtain food using diverse strategies

Heterotrophs vary in size from archaea and bacteria (as small as 0.5 μm) to blue whales (up to 25 m long). The ratio of their body size to the food they ingest varies widely, but generally increases as body size increases. Bacteria may be bathed in their food, while food for larger heterotrophs is usually more diffuse and smaller relative to the consumer. Feeding methods and the complexity of food absorption are accordingly very diverse among heterotrophs.

ECOLOGICAL TOOLKIT

5.1 Stable Isotopes

Many biologically important elements, including carbon, hydrogen, oxygen, nitrogen, and sulfur, have an abundant "light" isotopic form and one or more "heavy" non-radioactive isotopic forms, which contain additional neutrons. Because isotopes of these elements do not decay over time as radioactive isotopes do, they are referred to as **stable isotopes**. An example of a stable isotope is carbon-13 (^{13}C), which is heavier than the more abundant form, carbon-12 (^{12}C), because it has one more neutron. Groups of stable isotopes include hydrogen (H) and deuterium (D or ^{2}H); nitrogen-14 and -15 (^{14}N and ^{15}N); and oxygen-16, -17, and -18 (^{16}O, ^{17}O, and ^{18}O). The lighter isotopes of these elements are much more abundant than the heavier forms. For example, ^{12}C constitutes 98.9%, and ^{13}C only 1.1%, of the C on Earth. Similarly, ^{14}N constitutes 99.6%, and ^{15}N 0.4%, of the N on Earth.

The isotopic composition of a material is usually expressed as delta (δ), the difference between the ratio of the isotopic forms in a sample (R_{sample}) and that in a standard material ($R_{standard}$), divided by the ratio in the standard, multiplied by 1,000 (to give parts per thousand difference):

$$\delta = \frac{R_{sample} - R_{standard}}{R_{standard}} \times 1{,}000$$

Examples of the standards chosen for stable isotopes include a limestone rock from South Carolina for C, atmospheric N_2 for N, and ocean water for O and H.

These naturally occurring stable isotopes have become an important tool in ecological research (Fry 2007). Stable isotopes have been used to determine photosynthetic pathways in plants, identify food sources for animals, and track the movements of elements and rates of nutrient cycling in ecosystems. Because of the differences in their mass, the isotopes are affected differently by biological and physical processes. Generally, the heavier isotope is discriminated against, and the lighter isotope enriched. For example, when rubisco catalyzes the uptake of CO_2, it favors $^{12}CO_2$ over $^{13}CO_2$. As a result, plants are enriched in ^{12}C, and depleted in ^{13}C, relative to the C in atmospheric CO_2: atmospheric CO_2 has a $\delta^{13}C$ value of –7 parts per thousand (or 7 parts per thousand more depleted in ^{13}C than the standard), and C_3 plants have a $\delta^{13}C$ around –27 parts per thousand. C_4 and CAM plants, however, have less ^{12}C and more ^{13}C than C_3 plants. That is because initial CO_2 uptake in these plants is catalyzed by PEPcase, which discriminates against $^{13}CO_2$ less than rubisco does, and rubisco in C_4 and CAM plants takes up CO_2 in a semi-closed system (in the bundle sheath or with stomates closed), which inhibits enzymatic discrimination. As a result, measurement of the C isotope ratio in plant tissues can be used to determine the photosynthetic pathway used by a plant species, as shown in the figure.

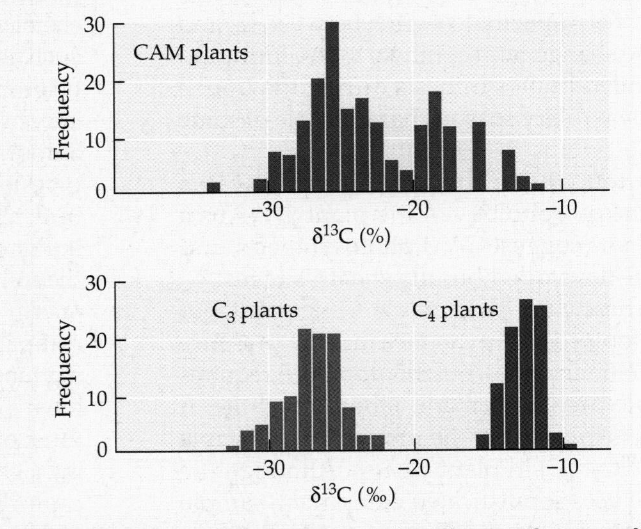

Carbon Isotopic Composition of Plants with Different Photosynthetic Pathways Plants with the C_3 photosynthetic pathway show the greatest discrimination against ^{13}C (and thus the most negative $\delta^{13}C$), while C_4 and CAM plants are more enriched in ^{13}C (have a less negative $\delta^{13}C$). (After Maslin and Thomas 2003.)

Q Why is the range of $\delta^{13}C$ values for CAM plants larger, bridging the values for C_3 and C_4 plants?

Stable isotopes have also been used to determine food sources for animals. The isotopic ratios of C, N, and S in potential food sources may differ significantly, and measurement of one or more of these isotopes in both potential food sources and consumer tissues can determine what is being eaten. For example, in this chapter's Case Study Revisited, we will see how isotopic ratios were used to determine the diet of New Caledonian crows. In Chapter 19, we will describe how measurement of N isotopes in ants was used to determine where ants in the rainforest canopy were getting their food.

Stable isotopes can also be added into the environment to help trace the movements of elements. This approach is often used to trace the fate of nutrients in ecosystems (as we'll see in Chapter 21).

Isotopic analysis of biological samples is relatively straightforward. For C and N, the samples are dried, ground, and burned in a closed furnace. The gases liberated by the combustion are then analyzed for isotopic composition using an instrument called a mass spectrometer. Many commercial laboratories specialize in the isotopic analysis of biological material, owing in part to the demand for such analyses from ecologists and other environmental scientists.

Prokaryotic heterotrophs typically absorb food directly through their cell membranes. Archaea, bacteria, and fungi excrete enzymes into the environment to break down organic matter, acting in effect to digest their food outside their bodies. Heterotrophic bacteria have adapted to a wide variety of organic energy sources and produce a large number of enzymes capable of breaking down organic compounds. This capacity of microorganisms to use diverse energy sources has been exploited in environmental waste management as an approach to cleaning up toxic chemical wastes, a process known as *bioremediation*. Spills of fuels, pesticides, sewage, and other toxins have been effectively contained by using microorganisms to break down these harmful compounds. Consumption of oil by marine bacteria is thought to have been critical in cleaning up the oil spill in the Gulf of Mexico that resulted when the *Deepwater Horizon* oil drilling rig exploded in 2010, releasing about 4.9 million barrels (780×10^3 m^3) of oil over a 4-month period.

The evolution of feeding in heterotrophic protists and animals is a story of increasing complexity in the ingestion and digestion of food. Small protozoans, such as amoebas and ciliates, ingest food particles into their cells, where the food is digested internally. With the advent of multicellular animals, specialized tissues for absorption, digestion, transport, and excretion evolved, and the efficiency of energy assimilation increased. Animals display tremendous diversity in their specialized morphological and physiological feeding adaptations, which reflect the diversity of the foods they consume. Several examples serve to demonstrate this diversification.

Insects display tremendous diversity in facial appearance, which reflects the diversity of their food sources, including dead organic matter, plants, and other animals. They may eat animal prey whole or suck their bodily fluids. All insects have the same basic set of mouthparts, consisting of several paired appendages that are used to seize, handle, and consume their food (**Figure 5.17**).

FIGURE 5.17 Variations on a Theme: Insect Mouthparts Differences in the morphology of insect mouth parts reflect different strategies for effectively capturing and consuming the food types they prefer.

FIGURE 5.18 Variations on a Theme: Bird Bills Bird bill morphology is associated with the feeding behavior of the species and enhances the capture of its preferred food resources.

Morphological variation in these mouthparts reflects the feeding specializations that have evolved within different insect groups. Common houseflies have "sponging" mouthparts that release saliva onto their food, then soak up and ingest the partially digested solution. Female mosquitoes and aphids have piercing and sucking mouthparts for extracting fluids from different living targets—blood from animals and sap from plants, respectively. Biting flies have razor-sharp appendages for cutting through skin to draw blood for drinking, similar to the cutting mouthparts of insects that consume leaves.

Birds display similar, although somewhat less complex, variation in their mouthparts (bills). Birds use their bills in a multitude of ways to capture, manipulate, and consume their food (**Figure 5.18**). The morphology of a bird's bill is closely associated with the taxonomic group to which the bird belongs. In other words, the flat bills of ducks and the hooked bills of raptors vary little within those taxonomic groups. However, subtle differences among the bills of closely related species reflect slight differences in food acquisition and handling. This variation in bill morphology reflects adaptations that help to optimize food acquisition and minimize competition among species (see Concept 11.4).

Craig Benkman studied the relationship between differences in bill morphology among crossbills and differences in the conifer seeds they use as food (Benkman 1993, 2003). As their name indicates, crossbills have unique asymmetrical bills with crossing tips (**Figure 5.19A**). Crossbills are adept at using their bills to open the cones of coniferous trees and pull out seeds for consumption. Across their geographic range, crossbills have multiple conifer species available as potential food sources. However, the dominant conifer species varies across the geographic range of crossbills. Benkman wondered if there were differences in the bill morphologies of the crossbills that were associated with the morphologies of the cones of their preferred conifer species.

Benkman tested this hypothesis experimentally using captive and wild birds from five incipient species (subspe-

(A)

(B)

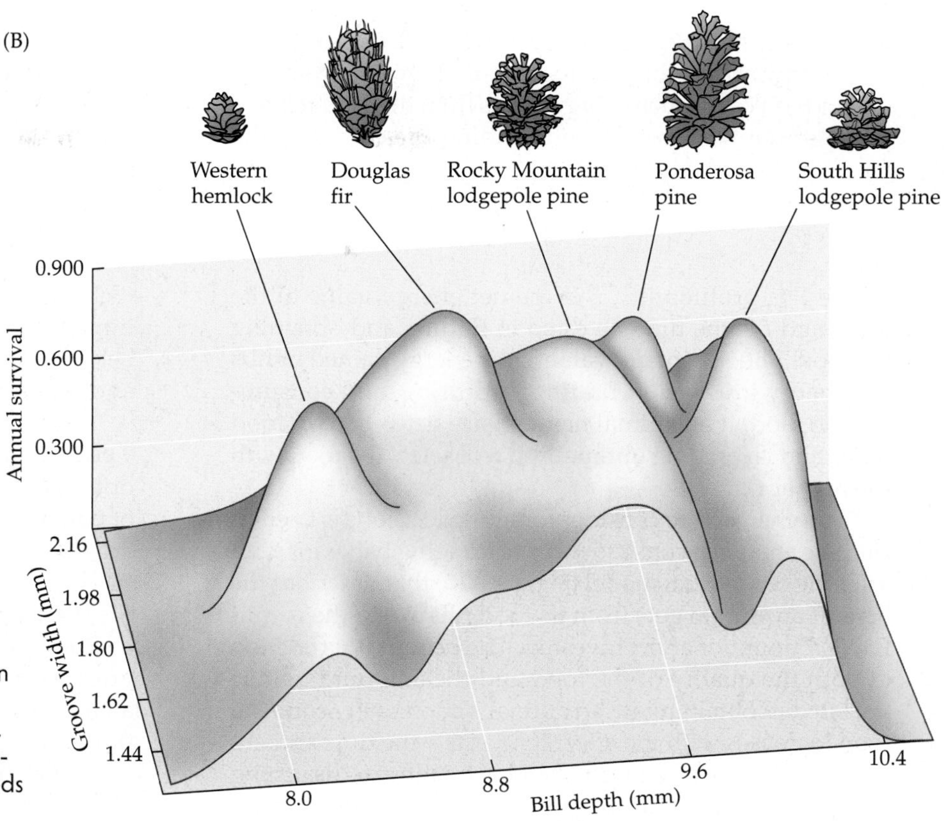

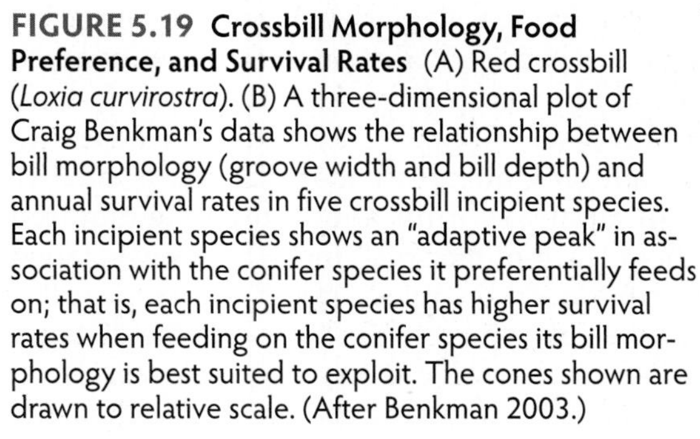

FIGURE 5.19 Crossbill Morphology, Food Preference, and Survival Rates (A) Red crossbill (*Loxia curvirostra*). (B) A three-dimensional plot of Craig Benkman's data shows the relationship between bill morphology (groove width and bill depth) and annual survival rates in five crossbill incipient species. Each incipient species shows an "adaptive peak" in association with the conifer species it preferentially feeds on; that is, each incipient species has higher survival rates when feeding on the conifer species its bill morphology is best suited to exploit. The cones shown are drawn to relative scale. (After Benkman 2003.)

cies that are in the process of becoming species) of the red crossbill species complex (*Loxia curvirostra*). He showed in a series of studies that a bird's speed of seed extraction from a given conifer's cone was associated with its bill depth. In addition, Benkman demonstrated that the speed of seed husking (removing the outer cover) was associated with the width of the groove in the bill where the seed is held (Benkman 1993, 2003). Each crossbill incipient species extracted and husked one conifer species' seeds more efficiently than other conifer species' seeds. The study showed an association between an incipient species' bill depth and the depth at which the seeds are held in the cones of its preferred conifer species. Benkman found that the annual survival rate for each crossbill incipient species was related to its feeding efficiency, which varied according to the conifer species it was feeding on. When he put these results together, Benkman found a series of five "adaptive peaks," showing that each incipient species' bill morphology was associated with the conifer that it fed most efficiently on, providing the highest survival rate (**Figure 5.19B**). Benkman (2003) concluded that red crossbills are currently undergoing evolutionary divergence (speciation) as a result of selection associated with differences in available food resources and the effects of those differences on bill morphology.

Optimal foraging theory addresses behavioral choices that enhance the rate of energy gain

The availability of food varies greatly over space and time. More food may be available in some locations than in others. For example, some areas of a landscape have a higher density of prey or host individuals than others due to differences in water or nutrient availability associated with differences in microclimate. In addition, some food items may be easier to obtain than others due to factors other than abundance, including how easy they are to detect, capture, or subdue. If energy is in short supply, then animals moving through a heterogeneous landscape should invest the majority of their time in acquiring the highest-quality food resources where they are most abundant and which are the shortest distance away. Such behavior should maximize the amount of energy obtained and minimize the risks involved, such as becoming food for another animal. These ideas are the essence of the theory of **optimal foraging**, which proposes that animals will maximize the amount of energy acquired per unit of time. Optimal foraging theory relies on the assumption that natural selection acts on the foraging behavior of animals to maximize their energy gain.

According to one formulation of optimal foraging theory, the profitability of a food item to a foraging animal depends on how much energy it gets from the food relative to the amount of time it spends finding and obtaining the food, or, in mathematical terms,

$$P = \frac{E}{t}$$

where P is profitability, E is the net energy value of the food, and t is the time invested in finding and obtaining the food. The net energy value is the energy gained minus the energy invested in finding, capturing, and consuming the food. If an animal has a choice between two food items and is foraging optimally, it will select the food with the higher P.

Another way to consider foraging decisions is to envision the energetic consequences of foraging behavior with a simple conceptual model (**Figure 5.20**) that describes the benefit an animal gets from its food relative to the cumulative amount of effort invested. The benefit is a function of both the quality of the food and the amount of effort invested in obtaining it. An animal's success in acquiring food increases with the effort it invests—the time and energy spent searching for food and extracting or disarming prey. At some point, however, a further increase in effort results in no incremental benefit, and the net energy gain begins to decrease. Several factors may be responsible for this decrease, including a limitation on how much food the animal can carry or ingest.

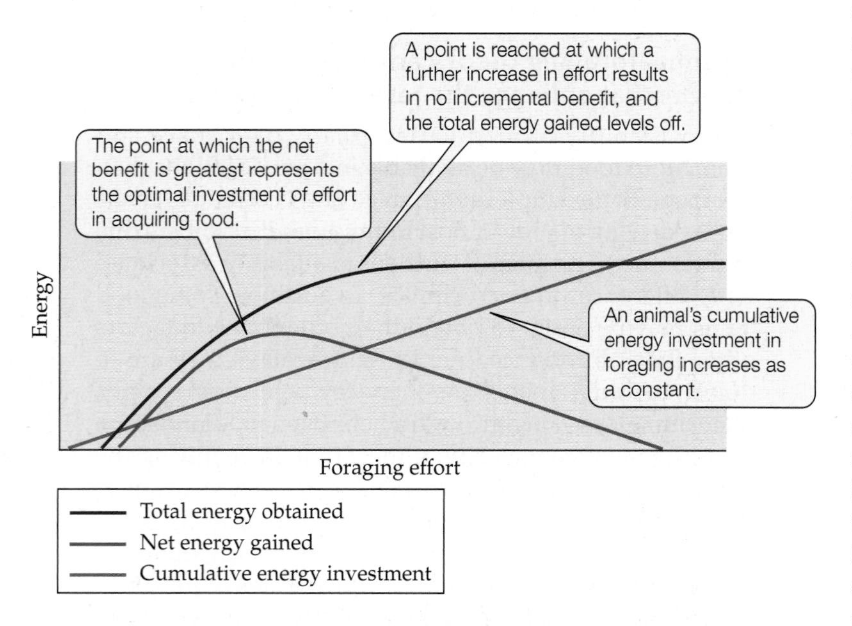

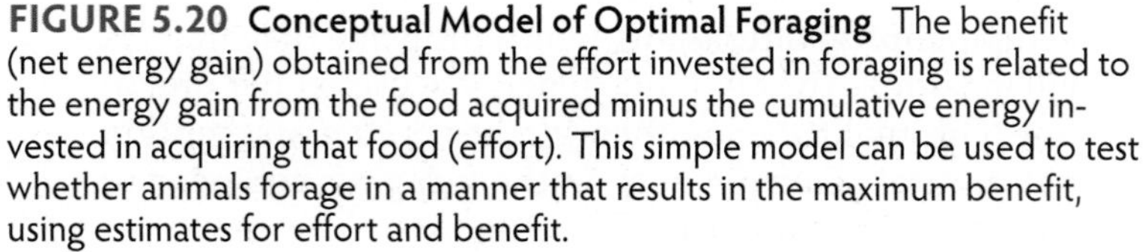

FIGURE 5.20 Conceptual Model of Optimal Foraging The benefit (net energy gain) obtained from the effort invested in foraging is related to the energy gain from the food acquired minus the cumulative energy invested in acquiring that food (effort). This simple model can be used to test whether animals forage in a manner that results in the maximum benefit, using estimates for effort and benefit.

While these models are rather simple, they provide a basis for making quantitative predictions about animal foraging behavior. More sophisticated models have been used to derive hypotheses that can be tested under field or laboratory conditions. An important component of these models is the currency that is used to determine the benefit, which might incorporate, for example, net energy gained, time spent feeding, and risk of predation (Schoener 1971). If foraging behavior is an adaptation to limited food supplies, then we must be able to relate the benefit (dependent variable in the model) to the survival and reproduction of the animal.

Tests of optimal foraging theory Research addressing optimal foraging has focused on diet selection, selection of patches to feed in, time spent in food patches, and prey movements (Pyke et al. 1977). John Krebs and colleagues (1977) devised a unique way to evaluate whether great tits (*Parus major*, a common bird found throughout much of Eurasia and northern Africa) selected prey types of greatest profitability. They placed captive birds next to a conveyer belt on which prey that differed in quality (large and small mealworms) and in the time required to obtain them (each of the small mealworms was taped to the surface) passed by. By changing the proportions of the prey types presented to the birds and the distance between adjacent prey on the conveyer belt (*encounter rates*), the researchers varied the profitability of the large mealworms. Using a model of optimal foraging and measurements of the times it took individual birds to subdue and consume the prey (*handling time*), they predicted how frequently the birds should select the large mealworms as encounter rates with the two prey types were varied. The model correctly predicted the frequency of consumption of the large mealworms as the profitability of those larger prey varied (**Figure 5.21**).

An example of a field study focusing on diet selection used the Eurasian oystercatcher (*Haematopus ostralegus*), a shorebird that eats bivalves (e.g., clams and mussels). A study by Meire and Ervynck (1986) demonstrated that oystercatchers select prey of the size that provides the most energy gain for the effort, despite the relatively low abundance of this prey size class (**Figure 5.22**). Oystercatchers must find a bivalve buried in the sand, lift it out, and open it before they can eat it. For bivalves below a certain size, the net energy gain from this effort is marginal, setting a lower limit on the bivalve size selected by the oystercatchers. Bivalves above a certain size have thicker shells and require more effort to open, setting an upper limit on the bivalve size selected by the birds.

Crows also exhibit feeding behavior that appears to support optimal foraging theory. As we saw in this chapter's Case Study, crows are clever birds, and they have developed a trick to circumvent the problem oystercatch-

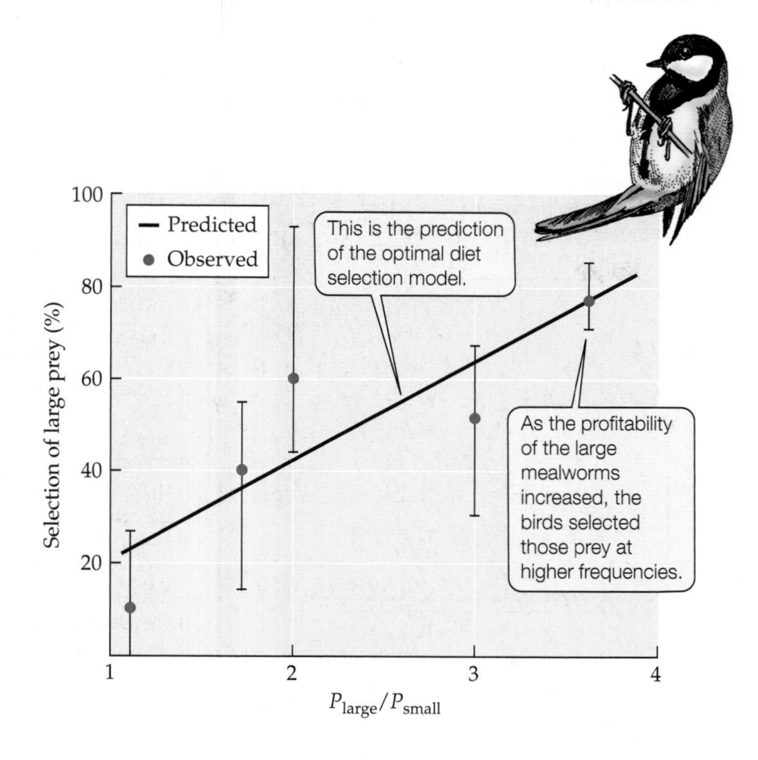

FIGURE 5.21 Effect of Profitability on Food Selection Krebs and colleagues used an optimal diet selection model, along with measurements of prey handling time for individual birds, to predict the rate at which great tits (*Parus major*) would select large over small mealworms as their encounter rate with the two prey types was varied (expressed as the calculated ratio of profitabilities of the prey types, P_{large}/P_{small}). Error bars show ± one SE of the mean. (After Krebs et al. 1977.)

ers had with opening large bivalves. Northwestern crows (*Corvus caurinus*) that forage for shellfish in intertidal and sandy shore zones use gravity as a means to open their shells. Crows pick up whelks or bivalves in their bills, fly into the air, and drop them on rocky surfaces to crack them open. They consume the shellfish once the protective shell has been shattered.

What determines which shellfish the crows will select to drop from on high? Bivalves of different sizes crack open equally well, but the net energy gain from larger bivalves is greater, although the time spent handling them is also greater (Richardson and Verbeek 1986). Howard Richardson and Nicolaas Verbeek estimated the energy costs associated with finding Japanese littleneck clams (*Venerupis japonica*), digging them up, carrying them into the air and dropping them, and extracting the body from the shell. They also estimated the amount of energy gained from consumption of clams of different sizes and the abundances of clams of different sizes in the crows' environment. These measures allowed them to estimate the clam size that provided the maximum net rate of energy gain for the crows (**Figure 5.23A**). Their estimate of the optimal clam size was very close to the size of the average clam selected, dropped, and eaten by crows in the wild (**Figure 5.23B**). Selecting clams smaller than the optimal size would result in too much energy investment for marginal energy gain, while larger clams were too rare to rely on and took more effort to find. Richardson and Verbeek found that the crows sometimes dug up clams smaller than the optimal size, but rejected them without investing the energy in dropping them.

Additional research demonstrated that shellfish size selection by crows overrides the influences of food quality. When provided access to different size classes of both clams (which have a higher energy content) and whelks (which have a lower energy content), crows consistently selected the larger shellfish, regardless of their identity, despite the greater potential energy gain of selecting small clams over large whelks (O'Brien et al. 2005).

Food density and the marginal value theorem Another aspect of optimal foraging theory considers the habitat in which an animal forages as a heterogeneous landscape made up of patches containing different amounts of food. To optimize its energy gain, an animal should forage in the most profitable patches—that is, those in which it can

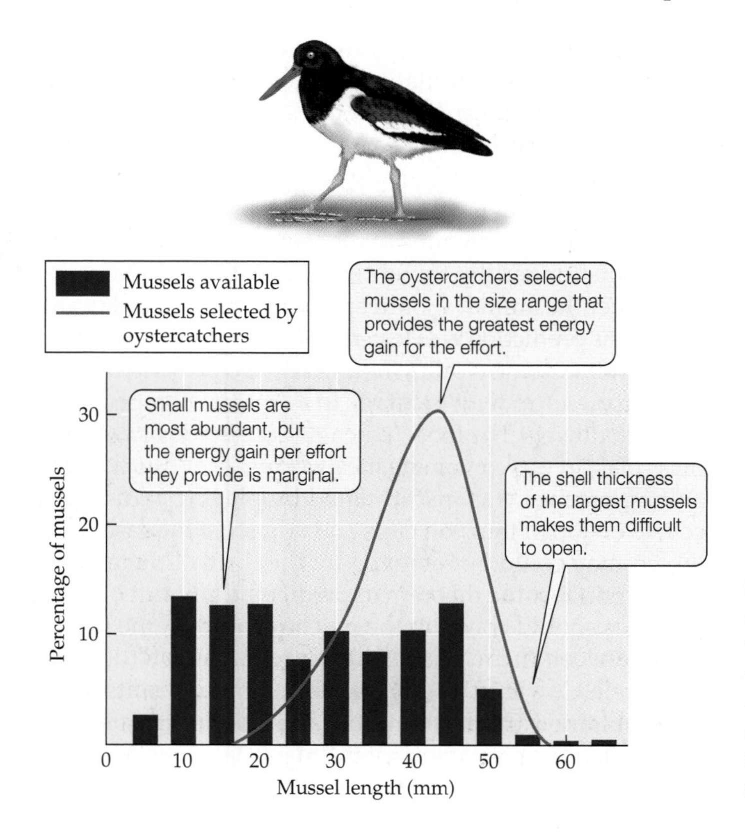

FIGURE 5.22 Food Size Selection by Oystercatchers Researchers determined the optimal food size range (25–60 mm) for Eurasian oystercatchers (*Haemotopus astralegus*) feeding on mussels (*Mytilus edulis*) using estimates of foraging effort (handling time, effort required to open the shell, and success rate) and benefit (energy gain) for mussels of each size class. They found that oystercatchers in the field selected mussels in the optimal size range, despite the greater abundance of smaller mussels. (After Meire and Ervynck 1986.)

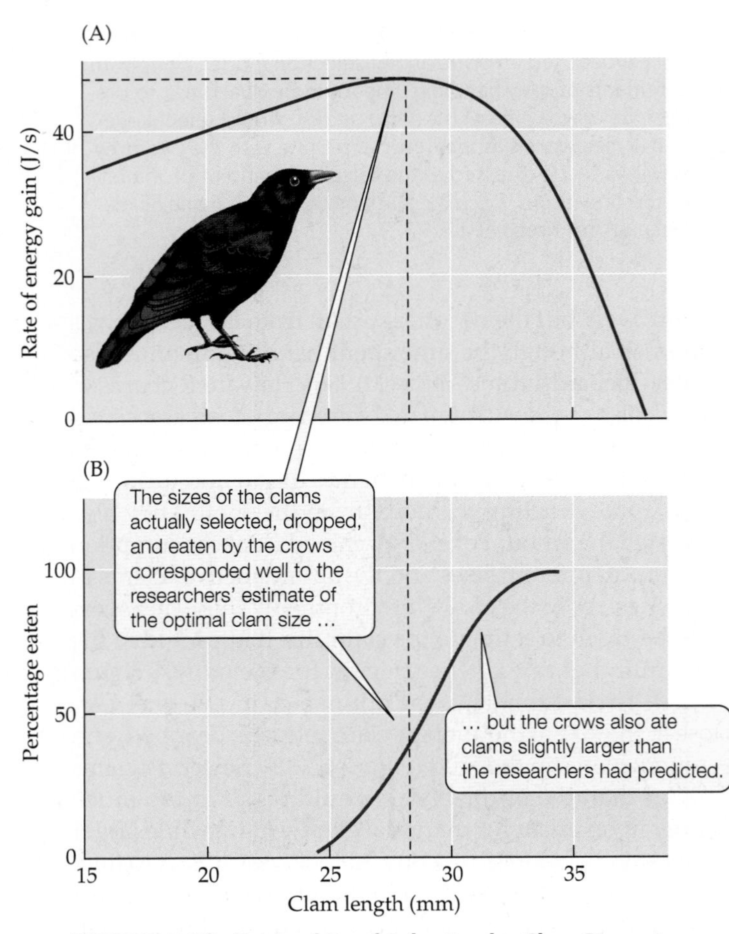

FIGURE 5.23 Optimal Food Selection by Clam-Dropping Crows (A) Richardson and Verbeek estimated net energy gains for Northwestern crows (*Corvus caurinus*) foraging on Japanese littleneck clams (*Venerupis japonica*) of different sizes as the energy gained from a clam minus the energy expended to find a clam, dig it up, carry it into the air and drop it to crack it open, and extract it from its shell. (B) The percentage of clams found by the crows that they dropped and ate varied with clam size. (After Richardson and Verbeek 1986.)

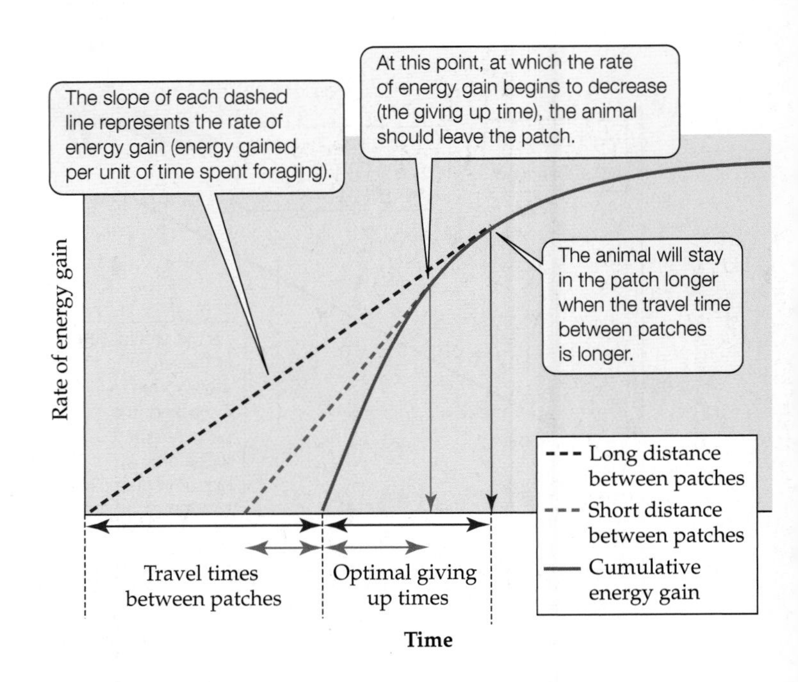

FIGURE 5.24 The Marginal Value Theorem The marginal value theorem assumes that a foraging animal will encounter patches containing varying amounts of food. The animal's rate of energy gain in a patch is initially high, but decreases as the animal depletes the food supply. The time the animal spends in a patch should optimize its rate of energy gain per unit of time spent foraging.

Q If prey density or prey quality is low, and the cumulative energy gain levels off at a lower level, how will this influence the giving up time?

achieve the highest energy gain per unit of time—until those patches offer the same energy gain per unit of time as the area surrounding them. We can also consider the benefit obtained by a foraging animal from the perspective of time spent in a patch (**Figure 5.24**). Once it finds a profitable patch, its rate of energy gain is initially high, but decreases and eventually becomes marginal as the forager depletes the food supply. A foraging animal should stay in a patch until the time when the rate of energy gain in that patch has declined to the average rate for the habitat (known as the *giving up time*), then depart for another patch. The giving up time should also be influenced by the distance to other patches, since effort must be invested in traveling to another patch. Therefore, the animal may accept a lower rate of energy if the distance between patches is greater. This conceptual model, called the **marginal value theorem**, was initially developed by Eric Charnov (Charnov 1976). It can be used to evaluate the influences of distance between patches, the quality of the food in a patch, and the animal's energy extraction efficiency on the giving up time.

One of the predictions of the marginal value theorem is that the longer the travel time between food patches, the longer the animal should spend in a patch (see Figure 5.24). This prediction was tested by Richard Cowie in 1977, soon after Charnov published his model. Cowie used a laboratory setup with great tits in a "forest" composed of wooden dowels. The food "patches" consisted of sawdust-filled plastic cups containing mealworms. The "travel time" among patches was manipulated by placing cardboard covers on top of the food cups and adjusting the ease with which they could be removed by the birds. Cowie used the marginal value theorem to predict the amount of time the birds should spend in the patches based on the travel time between them. His results matched his predictions very well (Cowie 1977) (**Figure 5.25**). Similar results have been obtained from other laboratory experiments that have tested the predictions of the marginal value theorem using bird and insect predators.

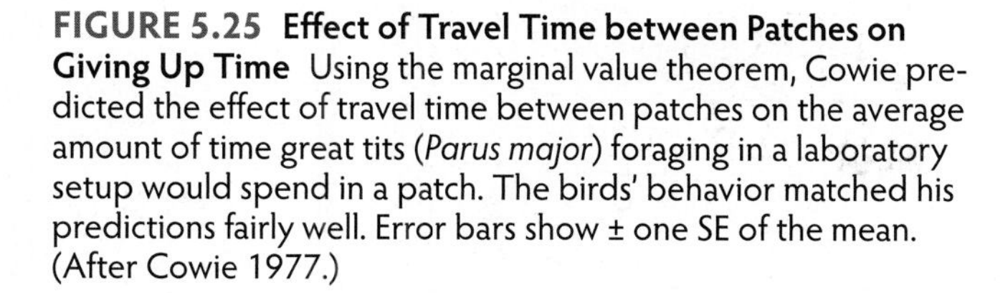

FIGURE 5.25 Effect of Travel Time between Patches on Giving Up Time Using the marginal value theorem, Cowie predicted the effect of travel time between patches on the average amount of time great tits (*Parus major*) foraging in a laboratory setup would spend in a patch. The birds' behavior matched his predictions fairly well. Error bars show ± one SE of the mean. (After Cowie 1977.)

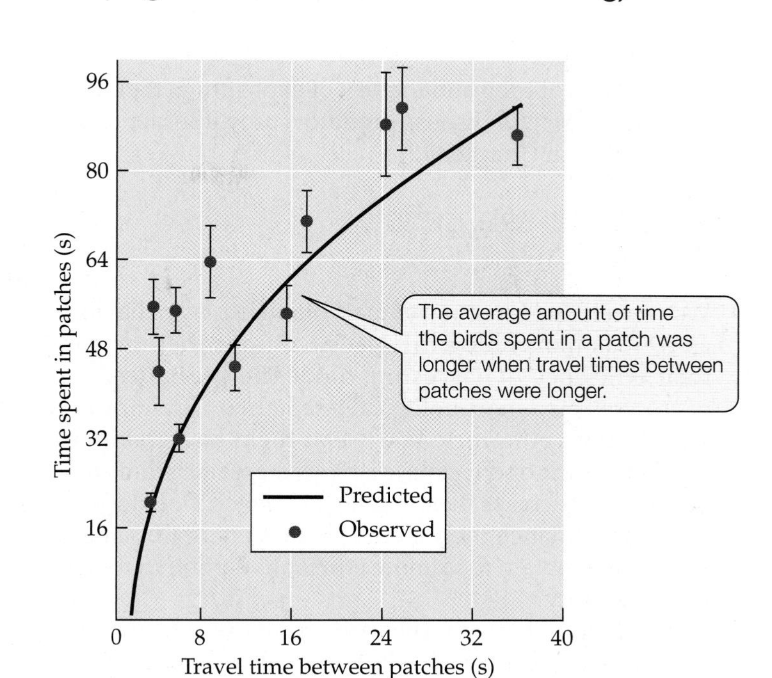

Despite their success in testing the predictions of optimal foraging theory and the marginal value theorem, laboratory experiments have been criticized for lacking any connection to "real-world" foraging animals. A test of the marginal value theorem in a natural setting was performed by James Munger, who studied the behavior of horned lizards (*Phrynosoma* spp.) foraging for ants in the Chihuahuan Desert (Munger 1984). Ant densities vary substantially across the desert landscape, depending on proximity to ant nests and the foraging trails used by worker ants. Munger observed the rate of ant consumption by horned lizards in various patches and compared that overall rate with the ant consumption rate at the time the lizards abandoned the patches. Using the marginal value theorem, he predicted that the consumption rate at the giving up time should be equal to the average consumption rate for the habitat, and that it should be higher at times when the overall ant density in the habitat was higher. Munger's predictions were largely borne out (**Figure 5.26**).

While ample evidence supports some aspects of optimal foraging theory, significant criticisms of the theory have been expressed. Optimal foraging best describes the foraging behavior of animals that feed on immobile prey and applies less well to animals feeding on mobile prey (Sih and Christensen 2001). In addition, the assumptions that energy is always in short supply, and that a shortage of energy dictates foraging behavior, may not always be correct. Carnivores, in particular, may not lack for food resources to the degree assumed in optimal foraging models (Jeschke 2007). Furthermore, resources other than energy may be involved in the selection of food items, particularly nutrients such as nitrogen and sodium. Additional considerations for foragers include

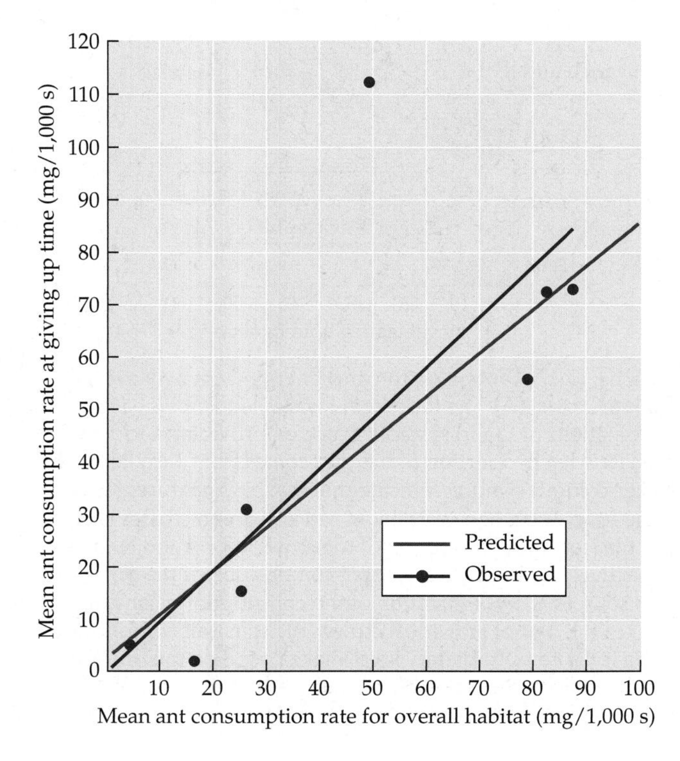

FIGURE 5.26 Giving Up Time in a Natural Setting Horned lizards (*Phrynosoma* spp.) feeding on patches of ants will remain in those patches until their rate of ant consumption approaches that in the surrounding habitat, as predicted by the marginal value theorem. The data in this graph have been corrected by removing observations of patch abandonment by lizards for reasons other than food depletion (e.g., temperature regulation). (After Munger 1984.)

the defenses of prey and the risk of exposure to their own predators. We will discuss predator–prey interactions in more detail in Chapter 12.

A CASE STUDY REVISITED

Toolmaking Crows

We've seen that foraging animals often display behavioral as well as morphological specializations that increase their efficiency in harvesting food. The specialized bill of crossbills is a morphological adaptation that improves their feeding efficiency. The use of flight and gravity by crows to crack open shells displays a greater reliance on behavior to increase their feeding efficiency. Does tool use by crows enhance their ability to gain energy by allowing them to obtain food more efficiently or obtain food of higher quality?

New Caledonian crows are omnivores with a wide variety of food sources to select from, including vertebrate and invertebrate prey, plants, and dead animals (carrion). As we discussed earlier, the benefit a foraging animal gets from its food is determined by the effort it invests in finding and obtaining the food as well as the quality of the food. There is a cost to tool use: collecting materials and fashioning the tools can be time consuming, and young crows may not initially be adept at using them. Evaluating the benefit of tool use to the crows requires knowledge of their energy requirements, the energetic benefits of their potential food sources, and the crow's actual diet.

The crows' shy nature and their tropical forest habitat make observational foraging studies difficult. To evaluate the energetic benefit of toolmaking and tool use, Christian Rutz and colleagues (2010) used stable isotope measurements (see Ecological Toolkit 5.1) to evaluate what the birds were eating, and used measurements of the lipid content of their potential food sources to estimate their energetic benefits. They also estimated the energy demands of the crows. Initial observations suggested that the crows relied on two high-quality food items (both with a lipid content of about 40%): nuts from candlenut trees, which they crack open by dropping them onto rocks, and beetle larvae obtained using their tools. Stable isotope measurements of N and C in the crows' blood and feathers and in their potential food sources indicated that the crows used a variety of food resources (**Figure 5.27A**), but that over 80% of their lipid intake was coming from the nuts and larvae (**Figure 5.27B**). This result indicates that a large proportion of the crows' energetic demands are met using two behaviors: tool use and nut cracking.

To address whether beetle larva extraction using tools could exclusively meet the energetic demand of the crows, Rutz and his colleagues determined the minimum number of beetle larvae needed on a daily basis to sustain a crow of average weight. They found that only

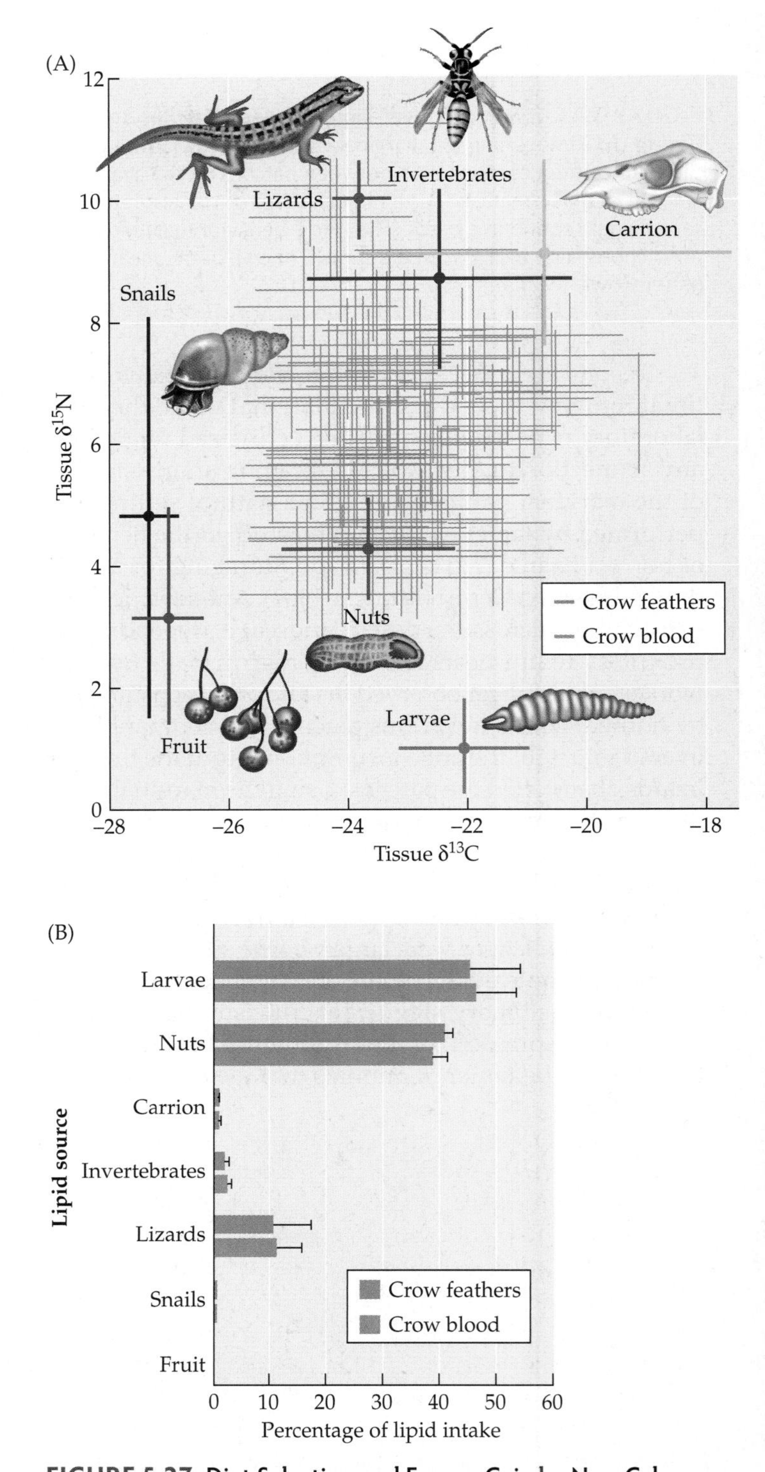

FIGURE 5.27 Diet Selection and Energy Gain by New Caledonian Crows (A) The food items available to crows have unique combinations of C and N stable isotopes, providing a tool to estimate what proportion of the diet comes from each item. The thicker colored symbols indicate the isotopic signatures for each of the food items, and the thinner red and green crosses indicate the isotopic composition of the crows' blood and feathers, respectively. Knowing the isotopic composition of the potential food sources allows estimation of the contribution of any one them to the diet of individual crows. (B) Estimated contributions of the food items to dietary lipid intake based on the isotopic composition of the crows' blood and feathers. Error bars show one SE of the mean. (After Rutz et al. 2010.)

three larvae per day were needed due to their high lipid content. Observations indicated that most adult crows can easily obtain three larvae per day; one competent adult crow was able to extract 15 larvae in 80 minutes. Tool use clearly provides a substantial benefit to the New Caledonian crows, giving them access to a high-quality food item that would otherwise not be available to them, or would at least require a very high investment of energy to obtain.

FIGURE 5.28 Untutored Tool Use in Captive Crows A captive New Caledonian crow (*Corvus moneduloides*) uses a stick tool to retrieve food from artificial crevices, despite never having been exposed to tool use, either by humans or by other birds. (From Kenward et al. 2005.)

CONNECTIONS IN NATURE

Tool Use: Adaptation or Learned Behavior?

How widespread is tool use among birds and other nonprimate animals? Many anecdotes of toolmaking and other innovative foraging techniques have been reported, but few have been examined thoroughly. The orange-winged sitella of Australia uses sticks to forage for insect larvae, much like the New Caledonian crows. Egyptian vultures crack open ostrich eggs using rocks. There are additional reports of tool use by insects, mammals, and other bird species (Beck 1980). The multitude of reports involving a wide range of animal species thoroughly dispels the notion of human monopoly on tool use. But how do these tool-using skills develop? Are these behaviors learned from other animals or are they innate (that is, determined genetically)? Several studies indicate that both learning and inherited behaviors can influence the development of tool use in animals.

As we learned above, tool use has a clear energetic benefit for the New Caledonian crows, but does that benefit exert strong enough selection pressure to have resulted in a behavioral adaptation—are the birds inheriting the ability to use tools? To address this question, Ben Kenward and colleagues reared New Caledonian crows in captivity, without exposure to adult birds. Some of the birds received "tutoring" in toolmaking and tool use by human foster parents, while another control group did not (Kenward et al. 2005). To evaluate the birds' toolmaking abilities, the researchers placed supplemental food in tight crevices in the birds' aviaries, where it was not accessible to the birds without the assistance of tools. Twigs and leaves were also left in their aviaries. The captive crows developed the ability to make and use tools to retrieve the food in the crevices, whether they had been tutored or not (**Figure 5.28**). Kenward and colleagues concluded that the ability of the New Caledonian crows to manufacture tools is at least partly inherited, rather than an acquired skill learned from adult birds in the wild. Very similar results were reported for experiments with captive woodpecker finches, birds endemic to the Galápagos archipelago that use twigs and cactus spines to forage for arthropods (Tebbich et al. 2001).

An additional twist to the crow toolmaking story is the apparent variation in tool styles among different crow populations on New Caledonia. In other words, there appears to be the potential for technological evolution in the styles of tools manufactured by crows. Gavin Hunt and Russell Gray conducted a survey of 21 sites on New Caledonia and examined 5,550 different stepped-cut tools constructed by crows from *Pandanus* tree leaves (Hunt and Gray 2003). They found three distinct widths of tools: wide, narrow, and stepped. Most of the tools found at a given site were very similar, and the geographic ranges of the tool types showed little overlap. There were no apparent correlations between where a tool type was found and local ecological factors such as forest structure or climate. Hunt and Gray suggested that the three tool designs were derived from a single original tool (of the wide type) subjected to additional modifications, including additional stripping of leaf material. Their study suggests ongoing innovation in toolmaking by the New Caledonian crows. This crow engineering challenges our traditional view of technological advancement in nonhuman animals.

Learned behavior is also important for toolmaking in some species. A notable example comes from studies of bottlenose dolphins in Shark Bay, Australia. Researchers observed that some dolphins swim with sponges plucked from the ocean floor on their noses (technically,

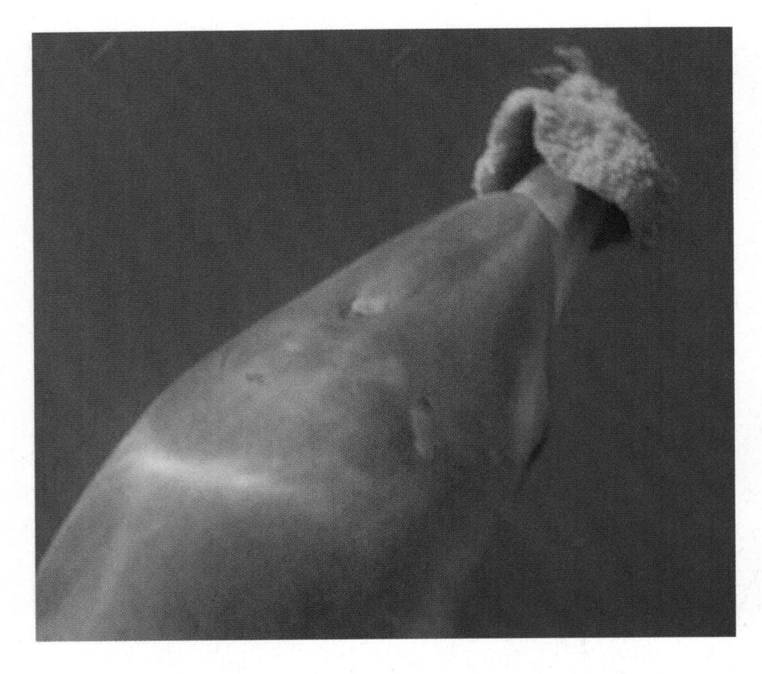

FIGURE 5.29 Dolphin Nose Gear in Shark Bay, Australia A bottlenose dolphin wears a sponge on its rostrum to protect it while foraging on the seafloor.

rostra) (**Figure 5.29**). The sponges appear to protect the dolphins' sensitive rostra from stinging animals, such as stonefish, and sharp objects as they probe the seafloor for fish to eat. The group of dolphins displaying this innovation is part of a larger group under study. The researchers' knowledge of the genetics and family structure of these dolphins allowed them to address the question of whether this unique behavior is learned or inherited. Michael Krützen and colleagues found that the majority of "sponging" dolphins were female. They reasoned that a single sex-linked gene (the kind of genetic basis one might expect for a trait occurring in only one sex) was a highly unlikely cause for a complex trait such as sponging. A comparison of the genetic fingerprints of individuals that sponged with those of nonsponging dolphins indicated that most of the sponging occurred within a single family line (Krützen et al. 2005). The combination of these results led Krützen and colleagues to conclude that sponging was a learned behavior passed from mother to daughter. This finding supports the idea of a cultural phenomenon in animals that influences the efficiency of their feeding behavior and challenges the notion that cultural learning is unique to humans.

SUMMARY

CONCEPT 5.1 Organisms obtain energy from sunlight, from inorganic chemical compounds, or through the consumption of organic compounds.

- Autotrophs convert energy from sunlight (by photosynthesis) or inorganic chemicals (by chemosynthesis) into energy stored in the carbon–carbon bonds of carbohydrates.
- Heterotrophs acquire energy by consuming organic compounds from other organisms, living or dead.

CONCEPT 5.2 Radiant and chemical energy captured by autotrophs is converted into stored energy in carbon–carbon bonds.

- During chemosynthesis, bacteria and archaea oxidize inorganic substrates to obtain energy, which they use to fix carbon and synthesize organic compounds.
- Photosynthesis has two main steps, the absorption of sunlight by pigments to produce energy in the form of ATP and NADPH (the light reaction), and the use of that energy in the Calvin cycle to fix CO_2 and synthesize carbohydrates (the dark reaction).
- Photosynthetic responses to variation in light levels, water availability, and nutrient availability include both short-term acclimatization and long-term adaptation.

CONCEPT 5.3 Environmental constraints have resulted in the evolution of biochemical pathways that improve the efficiency of photosynthesis.

- Photorespiration operates in opposition to photosynthesis, lowering the rate of energy gain, particularly at high temperatures and low atmospheric CO_2 concentrations.
- The C_4 photosynthetic pathway concentrates CO_2 at the site of the Calvin cycle, minimizing photorespiration.
- CAM plants reduce transpirational water loss by opening their stomates at night to take up CO_2 and releasing it to the Calvin cycle during the day, when the stomates are closed.

SUMMARY (continued)

CONCEPT 5.4 Heterotrophs have adaptations for acquiring and assimilating energy efficiently from a variety of organic sources.

- Variations in the chemistry and availability of food determine how much energy heterotrophs gain from different food sources.
- Heterotrophs display tremendous diversity in behavioral, morphological, and physiological adaptations that enhance their efficiency of energy acquisition and assimilation.
- Optimal foraging theory predicts that foraging animals will obtain the maximum amount of energy per unit of time and per unit of energy invested in seeking, capturing, and extracting food resources.
- The marginal value theorem suggests that an animal foraging in a heterogeneous environment should remain in a food patch until the rate of energy gained decreases to the level found in the habitat as a whole.

REVIEW QUESTIONS

1. Define autotrophy and heterotrophy and provide a few examples of each that illustrate the diversity of the ways in which organisms obtain energy.
2. Describe the biochemical and morphological specializations of the C_4 photosynthetic pathway. What are the ecological advantages of C_4 photosynthesis over C_3 photosynthesis?
3. How does the CAM photosynthetic pathway influence water loss from plants?
4. What are the trade-offs associated with heterotrophic consumption of live animals versus dead plant material?
5. Two bird species forage for insects that live in shrubs. The shrubs have a clumped, patchy distribution throughout their habitat. The two bird species have the same ability to locate, capture, and consume the insects. However, one species (species A) uses less energy to fly from patch to patch than the other species (species B). According to the marginal value theorem, which bird species should spend more time in each patch, and why?

ON THE COMPANION WEBSITE
sites.sinauer.com/ecology2e

The website includes Chapter Outlines, Online Quizzes, Flashcards & Key Terms, Suggested Readings, a complete Glossary, and the Web Stats Review. In addition, the following resources are available for this chapter:

HANDS-ON PROBLEM SOLVING

Optimal Foraging

This Web exercise illustrates patterns of movement predicted by optimal foraging. Foraging decisions are based on relative costs and benefits. You will manipulate the foraging decision rules of a predator to explore how distance to and size of prey influence foraging strategies and benefits.

WEB EXTENSION

5.1 How Do Plants Cope with Too Much Light?

CLIMATE CHANGE CONNECTIONS

5.1 Recently Discovered Pigment May Lead to More Efficient Solar Energy Systems

CHAPTER

6 Evolution and Ecology

KEY CONCEPTS

- **CONCEPT 6.1** Evolution can be viewed as genetic change over time or as a process of descent with modification.
- **CONCEPT 6.2** Natural selection, genetic drift, and gene flow can cause allele frequencies in a population to change over time.
- **CONCEPT 6.3** Natural selection is the only evolutionary mechanism that consistently causes adaptive evolution.
- **CONCEPT 6.4** Long-term patterns of evolution are shaped by large-scale processes such as speciation, mass extinction, and adaptive radiation.
- **CONCEPT 6.5** Ecological interactions and evolution exert a profound influence on one another.

Trophy Hunting and Inadvertent Evolution: A Case Study

Bighorn sheep (*Ovis canadensis*) are magnificent animals, beautifully suited for life in the rugged mountains in which they are found. Despite their substantial size (males can weigh up to 127 kg, or 280 pounds), these sheep can balance on narrow ledges and can leap 6 meters (20 feet) from one ledge to another. Bighorn sheep are also noted for the male's large curl of horns, which are used in combat over females (**Figure 6.1**). Rams run at speeds of up to 20 mph and crash their heads into each other, battling over the right to mate with a female.

Ram horns have been collected as trophies for centuries without drastically affecting sheep populations. Over the last 200 years, however, human actions such as habitat encroachment, hunting, and the introduction of domesticated cattle have reduced populations of bighorn sheep by 90%. As a result, the hunting of bighorn sheep has been restricted throughout North America. These restrictions make a world-class trophy ram (one with a large, full curl of horns) extremely valuable: permits to shoot one of these rams, which are sold at auction, can cost over $100,000.

Although funds raised by the auction of hunting permits are used to preserve bighorn sheep habitat, scientists have expressed concern that trophy hunting is having negative effects on today's small populations of bighorn sheep. Trophy hunting removes the largest and strongest males—the very males that would tend to sire large numbers of healthy offspring, thus helping sheep populations to recover. For example, in a population from which about 10% of the males were removed by hunting each year, both the average size of males and the average size of their horns decreased over a 30-year period (**Figure 6.2**).

Hunting, fishing, and other forms of harvest appear to be having similar effects on a wide range of other species including, fishes, invertebrates, and plants (Darimont et al. 2009). For example, by targeting older and larger fish, commercial fishing for cod has led to a reduction in the age and size at which these fish mature. This happens because cod that mature at a younger age and smaller size are more likely to reproduce before they are caught than are fish that mature when they are older and larger. Similarly, poaching for ivory appears to have caused the proportion of African elephants in a South African park that have tusks to decline from 90% to 62% over a 20-year period.

FIGURE 6.1 Fighting over the Right to Mate Large males with a full curl of horns—the same males prized by trophy hunters—are the most successful in contests over the right to mate with females.

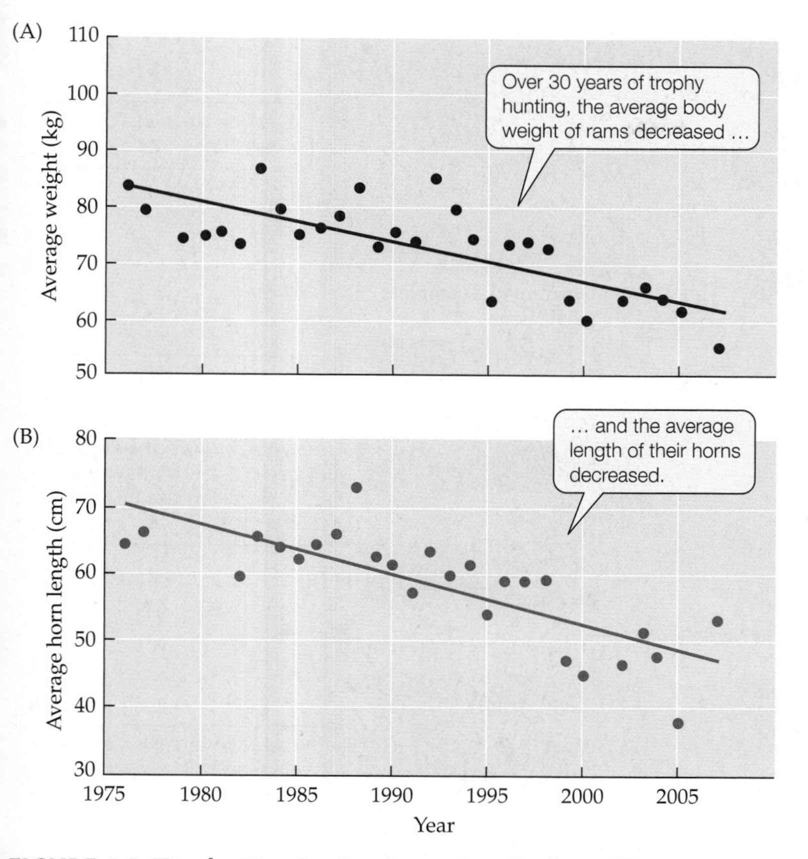

FIGURE 6.2 Trophy Hunting Decreases Ram Body and Horn Size Coltman and colleagues tracked the body weights (A) and horn lengths (B) of rams in a bighorn sheep population on Ram Mountain (Alberta, Canada) that was subjected to trophy hunting over a 30-year period. The changes shown here occurred across multiple generations of sheep, and thus indicate a change in the average characteristics of the sheep born from one generation to the next. (After Coltman et al. 2003.)

Harvesting has even affected sex determination in some species. Consider the cold-water rock shrimp of the northern Atlantic Ocean. Individuals of this species are born as males; later in life, when they are large enough to carry a clutch of eggs effectively, these males change into females. Beginning in the late 1950s, cold-water rock shrimp were harvested intensively. As in most shrimp fisheries, individuals with long tails were preferred. Thus, because females are larger than males, most of the shrimp harvested initially were females. As a result, females started to become rare. The shortage of females provided an advantage to those few individuals already present in the population that changed their sex from male to female at a smaller size. The resulting small females (and the genes that encoded the "early" switch) spread in the population, causing the abundance of females to bounce back but decreasing the number of offspring per female, since small females carry eggs less efficiently than larger females.

The unintended effects of human harvesting on bighorn sheep, cod, elephants, and shrimp illustrate how populations can change, or *evolve*, over time. What biological mechanisms cause these evolutionary changes? Do human actions other than harvesting produce evolutionary change?

Introduction

As news reports often emphasize, humans have a large effect on the environment. We change the global climate, pollute the water and air, convert large tracts of natural habitat into farmland and urban areas, drain wetlands, and reduce the population sizes of species we hunt for food (e.g., fishes) or use as resources (e.g., trees). Although we have taken steps to limit some of the damage we cause to biological communities, human actions have a pervasive consequence that we have barely begun to recognize, much less address: we cause evolutionary change.

In this chapter, we'll examine what evolution is, and we'll see how it affects ecological interactions and is affected by them. At the close of the chapter, we'll focus specifically on how humans cause evolutionary change. Our goal in this chapter is not to provide a comprehensive survey of evolutionary biology—for that, see the textbooks on evolution listed in the Suggested Readings on the book's website. Instead, our aim is to show that ecology and evolution are interconnected—a theme to which we will return in later chapters of this book. We'll begin by considering two ways in which evolution can be defined.

CONCEPT 6.1 Evolution can be viewed as genetic change over time or as a process of descent with modification.

What Is Evolution?

In the most general sense, biological evolution is change in organisms over time. Evolution includes the relatively small fluctuations that occur continually within populations, as when the genetic makeup of a population changes from one year to the next. But evolution can also refer to the larger changes that occur as species gradually become increasingly different from their ancestors. Let's explore these two ways of looking at evolution in more detail, focusing first on genetic fluctuations (*allele frequency change*) and then on how organisms accumulate differences from their ancestors (*descent with modification*).

Evolution is allele frequency change

Figure 6.2 shows that the average horn size of male bighorn sheep has decreased over time, but it does not reveal the cause of that decline. A clue to the cause comes from an additional observation (Coltman et al. 2003): horn size is

(A)

UNIQUE FOSSILS
These 10 million-year-old fossils of the stickleback fish *Gasterosteus doryssus* were collected from a lake bed in Nevada (USA). The fossils could be dated to the nearest 250 years because the rocks in which they were found showed exceptionally clear annual layers of sediments.

DESCENT
Fossil evidence suggests that *G. doryssus* colonized open waters of the lake about 10 million years ago. Over the next 16,000 years, many of the bones in these fish did not change in size, shape, or position. The resulting similarities in their overall bone structure illustrate common descent—the fish descended from the original colonists and hence shared many characteristics with them.

MODIFICATION
The fossil sticklebacks also show how organisms become modified from their ancestors over time. For example, in less than 5,000 years, the pelvic bone—originally the largest single bone in the body of these fish—became greatly reduced in size. This reduction has also occurred in modern lakes and probably resulted from natural selection.

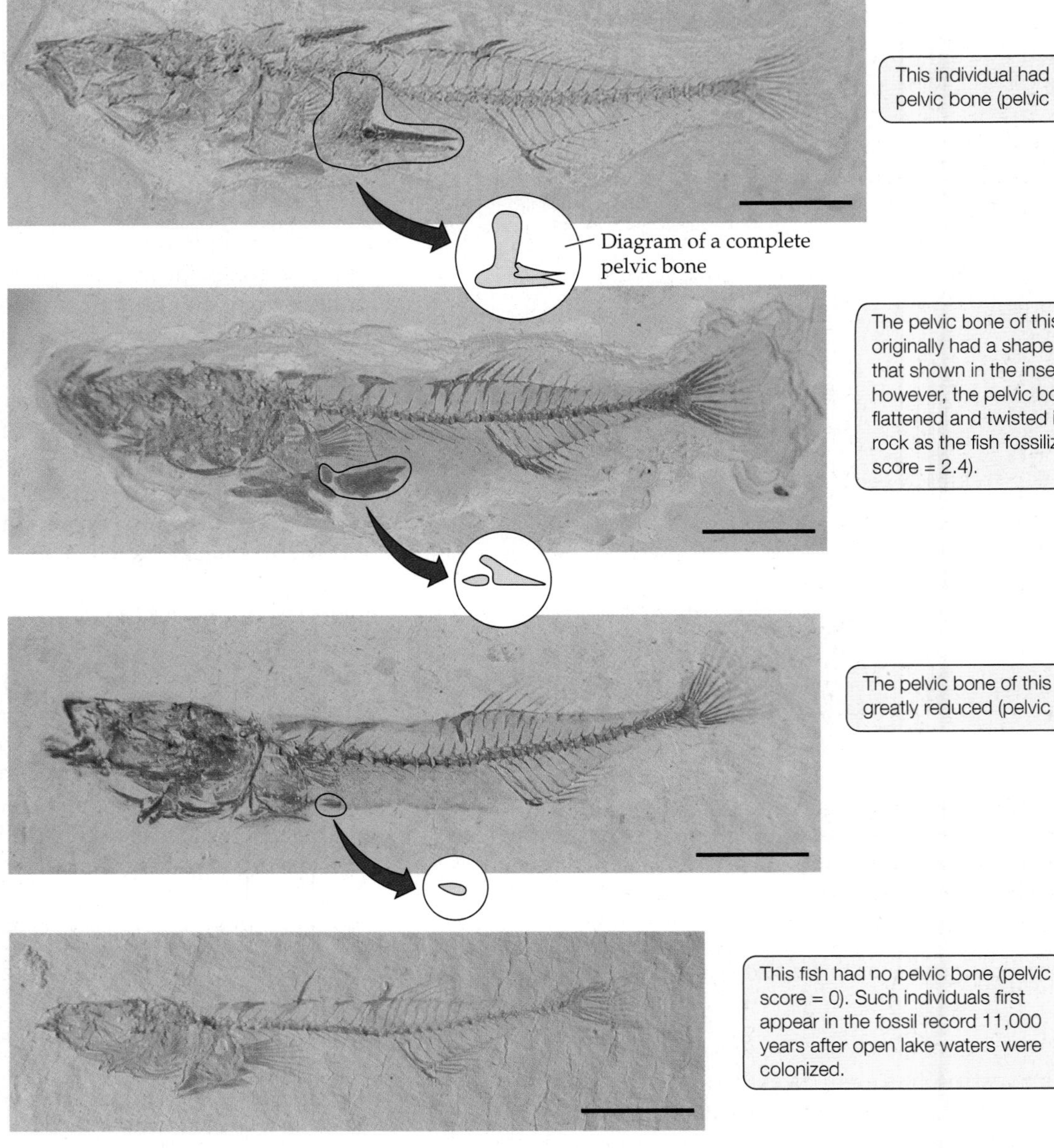

(B)

After 3,000 years, the average pelvic score began to drop substantially.

After 15,000 years, most members of the *G. doryssus* population had a highly reduced pelvis, causing the average pelvic score to hover near 1.

Pelvic score: 3, 2.5, 2, 1.5, 1, 0.5

Time (years since the open lake waters were colonized): 0, 5,000, 10,000, 15,000

FIGURE 6.3 Descent with Modification Michael Bell and colleagues have analyzed thousands of 10 million-year-old fossils of the stickleback fish *Gasterosteus doryssus*. Their specimens are unique in that the lake bed in which they were found is so finely layered that the ages of the fossils can be determined to the nearest 250-year interval. (A) Representative *G. doryssus* fossils showing how the pelvic bone became reduced over time; the scale bar for each fossil is 1 cm. (B) The average pelvic score at different times. Fossil pelvic bones were scored by size according to a scale that ranged from 3 (complete bone) to 0 (no bone). (After Bell et al. 2006.)

a heritable trait. This means that rams with large horns tend to have offspring that have large horns, and that rams with small horns tend to have offspring that have small horns. Because trophy hunting selectively eliminates rams with large horns, it favors rams whose genetic characteristics lead to the production of small horns. Hence, it seems likely that trophy hunting is causing the genetic characteristics of the bighorn sheep population to change, or evolve, over time.

As suggested by the trophy hunting example, biologists often define evolution in terms of genetic change. To make such a definition more precise (and to introduce terms that will be used throughout this chapter), let's review some principles from your introductory biology class:

- Genes are composed of DNA, and they specify how to build (encode) proteins.
- A given gene can have two or more forms (known as **alleles**) that result in the production of different versions of the protein that the gene encodes.
- We can designate the **genotype** (genetic makeup) of an individual with letters that represent the individual's two copies of each gene (one inherited from its mother, the other from its father). If a gene has two alleles, designated *A* and *a*, the individual could be of genotype *AA*, *Aa*, or *aa*.

With these principles as background, we can define **evolution** as change over time in the *frequencies* (proportions) of different alleles in a population. To illustrate how this definition is applied, consider a gene with two alleles (*A* and *a*) and a population of 1,000 individuals, in which there are 360 individuals of genotype *AA*, 480 of genotype *Aa*, and 160 of genotype *aa*. The frequency of the *a* allele in this population is 0.4, or 40%*; hence, since there are only two alleles in the population (*A* and *a*), the frequency of the *A* allele must be 0.6, or 60%. If the frequency of the *a* allele were to change over time—say, from 40% to 71%—then the population would have evolved at that gene. (In scientific studies, researchers often use an approach based on the *Hardy–Weinberg equation* to test whether a population is evolving at one or more genes; we describe this approach in **Web Extension 6.1**.)

Evolution is descent with modification

In many parts of this book, when we refer to evolution, we will be referring to allele frequency change over time. But evolution can also be defined more broadly as *descent with modification*. At the heart of this definition is the observation that populations accumulate differences over time, and hence, when a new species forms, it differs from its ancestors. The new species, however, typically differs from its ancestors in a relatively small number of ways. As a result, a new species not only differs from its ancestors, but also resembles its ancestors because it descended from them and continues to share many characteristics with them. Hence, when evolution occurs, both *descent* (shared ancestry, resulting in shared characteristics) and *modification* (the accumulation of differences) can be observed, as seen in the fossil fishes in **Figure 6.3**.

Charles Darwin (1859) used the phrase "descent with modification" to summarize the evolutionary process in his book *The Origin of Species*. Darwin proposed that populations accumulate differences over time primarily by **natural selection**, the process by which individuals with certain heritable characteristics survive and reproduce more successfully than other individuals because of those characteristics. We've already seen several examples of selection at work in the Case Study on pp. 132–133. In bighorn sheep populations, trophy hunting has selected for rams with small horns, while in the cod fishery, harvesting practices have selected for individuals that mature at a younger age and a smaller size.

How can natural selection explain the accumulation of differences between populations? Darwin argued that if two populations experience different environmental conditions, individuals with one set of characteristics may be favored by natural selection in one population, while individuals with a different set of characteristics may be favored in the other population (**Figure 6.4**). By

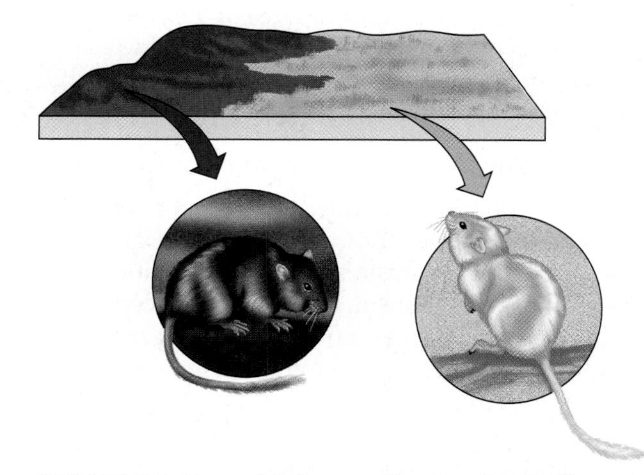

FIGURE 6.4 Natural Selection Can Result in Differences between Populations Populations of rock pocket mice (*Chaetodipus intermedius*) that live on dark lava formations in Arizona and New Mexico have dark coats, while nearby populations that live on light-colored rocks have light coats. In each population, natural selection has favored individuals whose coat colors match their surroundings, making them less visible to predators.

*There are 1,000 individuals in the population, and each individual carries two alleles. Thus, there are a total of 2,000 alleles in the population. Each of the 360 individuals of genotype *AA* has zero *a* alleles, each of the 480 individuals of genotype *Aa* has one *a* allele, and each of the 160 individuals of genotype *aa* has two *a* alleles. Thus, the frequency of the *a* allele is $(0 \times 360 + 1 \times 480 + 2 \times 160)/2{,}000 = 0.4$.

favoring individuals with different heritable characteristics in different populations, natural selection can cause populations to diverge genetically from one another over time; that is, each population will accumulate more and more genetic differences. Thus, natural selection can be responsible for the "modification" part of "descent with modification."

Populations evolve, individuals do not

Natural selection acts as a sorting process, favoring individuals with some heritable traits (e.g., those with small horns) over others (e.g., those with large horns). Individuals with the favored traits tend to leave more offspring than individuals with other traits. As a result, from one generation to the next, a greater proportion of the individuals in the population will have the traits favored by natural selection. This process ensures that the allele frequencies of the population will change over time, and hence, that the population will evolve. But the *individuals* in the population do not evolve—they either have the traits favored by selection or they don't.

FIGURE 6.5 Individuals in Populations Differ in Their Phenotypes These purple sea urchins (*Paracentrotus lividus*), native to the eastern Atlantic Ocean and Mediterranean Sea, show considerable variation in the color of their spines. Individuals in a population of these sea urchins would also be likely to differ in other morphological traits as well as in many biochemical, behavioral, and physiological traits.

CONCEPT 6.2 Natural selection, genetic drift, and gene flow can cause allele frequencies in a population to change over time.

Mechanisms of Evolution

Although natural selection is often the cause of evolutionary change, it is not the only one. In this section, we'll examine four key processes that influence evolution: mutation, natural selection, genetic drift, and gene flow. In broad overview, mutation is the source of the new alleles on which all of evolution depends, while natural selection, genetic drift, and gene flow are the main mechanisms that cause allele frequencies to change over time.

Mutation generates the raw material for evolution

Individuals in populations differ from one another in their observable characteristics, or **phenotype** (**Figure 6.5**). Many aspects of an organism's phenotype, including its physical features, metabolism, growth rate, susceptibility to disease, and behavior, are influenced by its genes. As a result, individuals differ from one another in part because they have different alleles of genes that influence their phenotype. These different alleles arise by **mutation**, a change in the DNA of a gene. Mutations result from events such as copying errors during cell division, mechanical damage when molecules and cell structures collide with DNA, exposure to certain chemicals (called *mutagens*), and exposure to high-energy forms of radiation such as ultraviolet light and X rays. As we'll see in Chapter 7, the environment can also affect an organism's phenotype. For example, a plant growing in nutrient-rich soil may grow larger than another individual of the same species growing in nutrient-poor soil, even if both have the same alleles of genes that influence size. In this chapter, however, we will focus on phenotypic differences that result from genetic, not environmental, factors.

The formation of new alleles by mutation is critical to evolution. In a hypothetical species in which there was no mutation, each gene would have only one allele, and all members of a population would be genetically identical. If this were the case, evolution could not occur: allele frequencies cannot possibly change over time unless the individuals in a population differ genetically. You may recall from your introductory biology class that the individuals in a population can differ genetically not only because of mutation, but also because of **recombination**, the production of offspring that have combinations of alleles that differ from those in either of their parents. We can think of mutation as providing the raw material (new alleles) on which evolution is based, and recombination as rearranging that raw material into unique new combinations. Together, these processes provide the genetic variation on which evolution depends.

Despite its importance to evolution, mutation occurs too rarely to be the direct cause of significant allele frequency

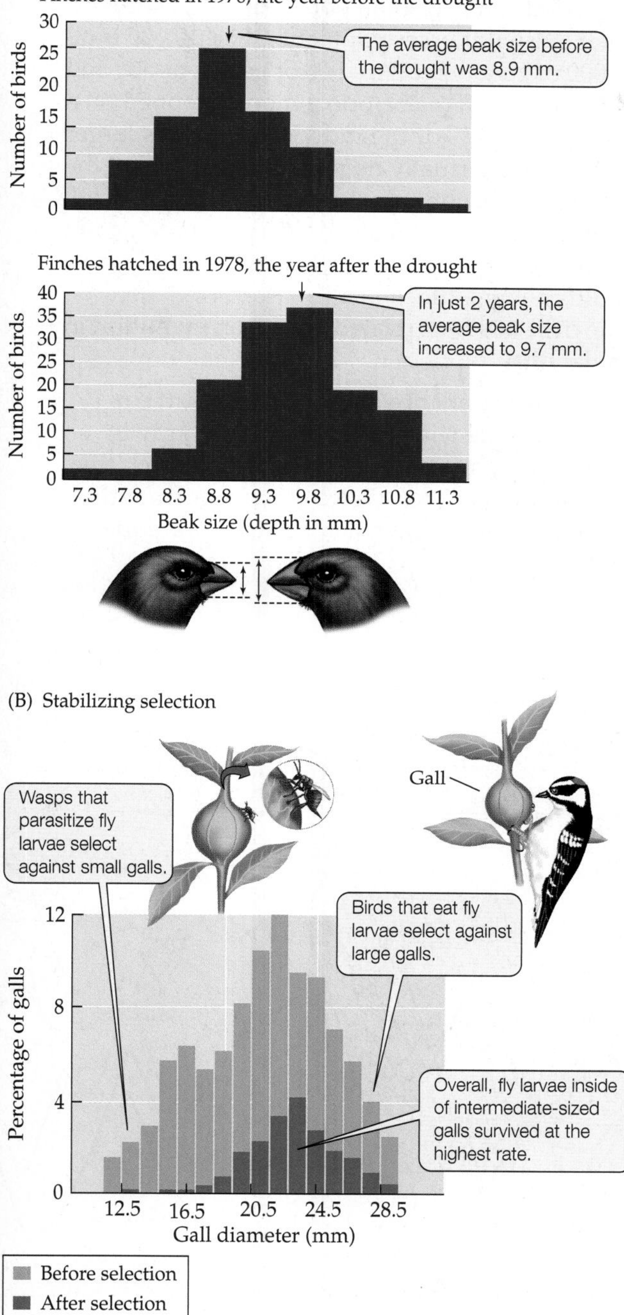

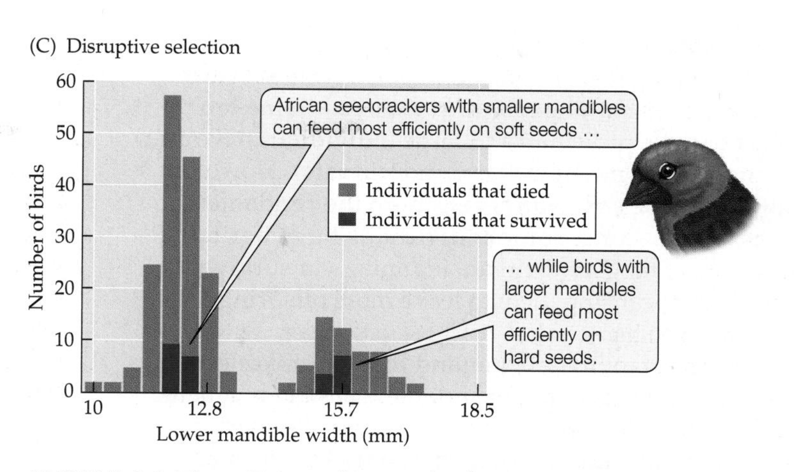

FIGURE 6.6 Three Types of Natural Selection (A) Directional selection favors individuals at one phenotypic extreme. A prolonged drought in the Galápagos archipelago resulted in directional selection on the beak size of the seed-eating medium ground finch (*Geospiza fortis*). As a result of the drought, most of the available seeds were large and hard to crack, so birds with large beaks, which could more easily crack those seeds, had an advantage over birds with smaller beaks. (B) Stabilizing selection favors individuals with an intermediate phenotype. *Eurosta* flies parasitize goldenrod plants, causing the plant to produce a gall in which the fly larva matures as it feeds on the plant. The preferences of *Eurosta*'s own predators and parasites result in stabilizing selection on gall size. Field observations showed that wasps that parasitize and kill the fly larvae prefer small galls, while birds that eat the fly larvae prefer large galls. As a result, larvae in galls of intermediate size have an advantage. (C) Disruptive selection favors individuals at both extremes. African seedcrackers (*Pyrenestes ostrinus*) depend on two major food plants in their environment. Birds with smaller mandible sizes can feed on one plant's soft seeds most efficiently, while birds with larger mandibles can feed on the other plant's hard seeds most efficiently. Thus, individuals with mandible sizes that are either relatively small or relatively large have an advantage. (A after Grant and Grant 2003; B after Weis and Abrahamson 1986; C after Smith 1993.)

Q In (B), do birds or wasps appear to provide stronger selection pressure on gall size? Explain.

change over short periods of time. Mutations typically occur at rates of 10^{-4} to 10^{-6} new mutations per gene per generation (Hartl and Clark 2007). In other words, in each generation, we can expect one mutation to occur in every 10,000 to 1,000,000 copies of a gene. At these rates, in one generation, mutation acting alone causes virtually no change in the allele frequencies of a population. Eventually, mutation can cause appreciable allele frequency change, but it takes thousands of generations for it to do so. Overall, in terms of its direct effects, mutation is a weak agent of allele frequency change. But because it provides new alleles on which natural selection and other mechanisms of evolution can act, mutation is central to the evolutionary process.

Natural selection increases the frequencies of advantageous alleles

Natural selection occurs when individuals with particular heritable traits consistently leave more offspring than do individuals with other heritable traits. But some traits may give organisms an advantage only under certain environmental conditions; indeed, as we'll see later in this chapter, traits that are advantageous in one environment can be disadvantageous in another.

Depending on what traits are favored, we can categorize natural selection into three types (**Figure 6.6**). **Directional selection** occurs when individuals with one extreme of a heritable phenotypic trait (for example, large size) are favored over other individuals (small and medium-sized individuals). In **stabilizing selection**, individuals with an intermediate phenotype (for example, medium-sized individuals) are favored, while in **disruptive selection**, individuals with a phenotype at either extreme are favored (for example, small and large individuals have an advantage over medium-sized individuals). However, in all three types of natural selection, the fundamental process is the same: some individuals have heritable phenotypes that give them an advantage in survival or reproduction, causing them to leave more offspring than other individuals.

It is important to bear in mind that natural selection operates on aspects of the phenotype that have a genetic basis. Selection favors individuals whose alleles encode phenotypic traits that make them more likely to survive and reproduce than are individuals with other alleles. In some cases, the end result of this process is that most or all of the individuals in a population have an allele favored by natural selection. A well-studied example is the Andean goose (*Chloephaga melanoptera*), which lives high in the Andes. These birds have evolved a version of the oxygen transport protein hemoglobin that has an unusually high affinity for oxygen and hence provides an advantage in their low-oxygen, high-altitude environment (Weber 2007; McCracken et al. 2009). The allele that encodes this version of hemoglobin occurs at a frequency of 100% in Andean goose populations.

At the population level, natural selection can cause the frequency of an allele that confers an advantage to increase over time, as has occurred in the Andean goose. We'll consider the consequences of such increases in the frequencies of advantageous alleles later in this chapter. But first, we'll look at two other mechanisms that can cause allele frequencies to change: *genetic drift* and *gene flow*.

Genetic drift results from chance events

Allele frequencies in populations can be influenced by chance events. Imagine a population of ten wildflowers in which three individuals have genotype *AA*, four have genotype *Aa*, and three have genotype *aa*. Thus, the initial frequency of the *A* allele is 50%, as is the frequency of the *a* allele. Assume that the *A* and *a* alleles encode two different versions of a protein that function equally well. Although neither allele is more advantageous than the other (and hence natural selection does not affect this gene), chance events could alter their frequencies. For example, suppose that a moose walking through the woods happened to step on four of the wildflowers—two of genotype *AA* and two of genotype *Aa*—killing them, but not harming any of the three wildflowers of genotype *aa*. As a result, the frequency of the *a* allele in the population would increase from 50% to 67% *by chance alone*.

When chance events determine which alleles are passed from one generation to the next, **genetic drift** is said to occur. Although chance events occur in populations of all sizes, genetic drift alters allele frequencies significantly over short periods only in small populations. To see why, imagine that our wildflower population had 10,000 individuals, 3,000 of genotype *AA*, 4,000 of genotype *Aa*, and 3,000 of genotype *aa*. If (as before) a moose stepped on a random sample of 40% of the individuals in this larger population, there is virtually no possibility that all of the 3,000 individuals of genotype *aa* would be spared. Instead, it is likely that many individuals of each genotype would be killed, and hence that the frequencies of the *A* and *a* alleles would change little, if at all.

Genetic drift has four related effects on evolution in small populations:

1. Because it acts by chance alone, genetic drift can cause allele frequencies to fluctuate randomly in small populations over time (**Figure 6.7**). When this occurs, eventually some alleles disappear from the population, while others reach **fixation** (a frequency of 100%).

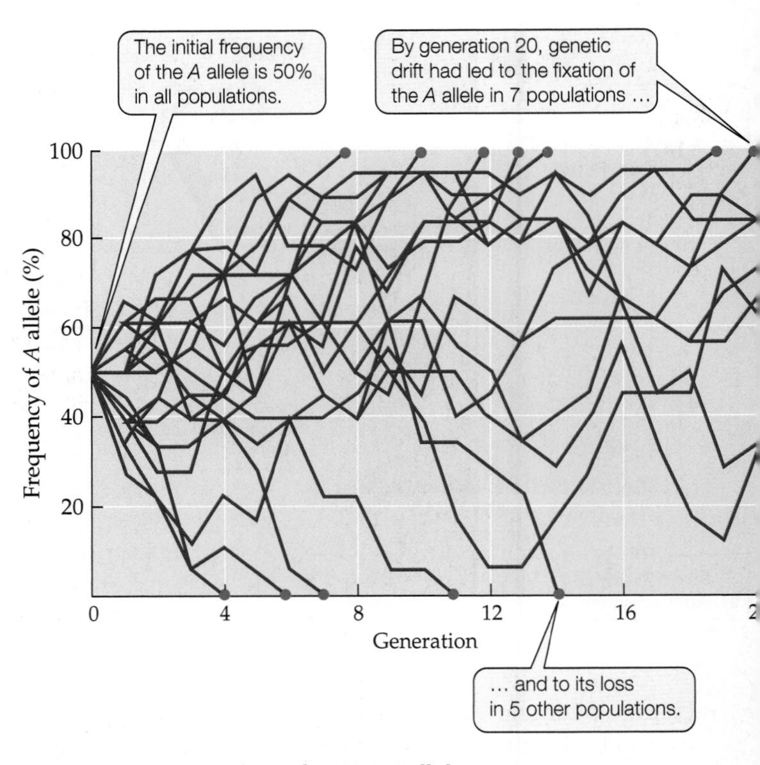

FIGURE 6.7 Genetic Drift Causes Allele Frequencies to Fluctuate at Random Results of a computer simulation of genetic drift in 20 populations for a gene with two alleles, *A* and *a*. Each population has 9 diploid individuals (18 alleles) each generation. In small populations such as these, genetic drift has rapid effects.

Q At the start of the simulation, how many *A* alleles and how many *a* alleles did each population have? At generation 20, how many populations still had both alleles? Predict what would eventually happen to the frequency of the *A* allele in those populations.

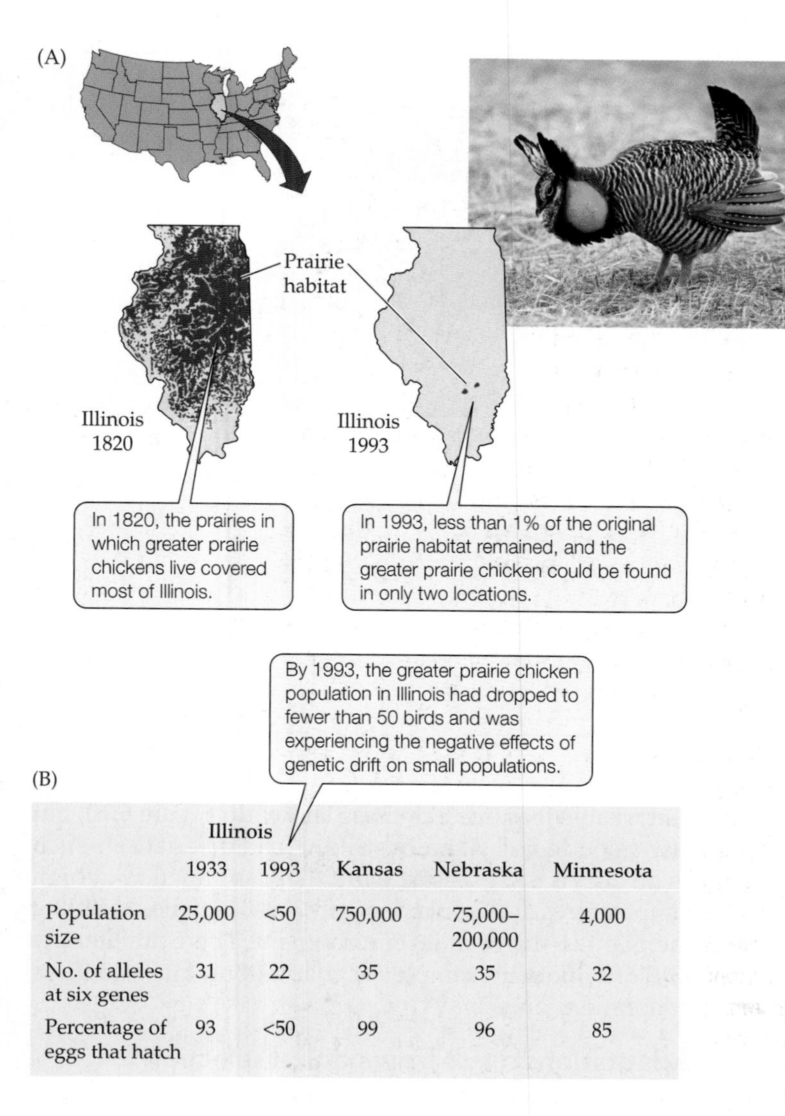

	Illinois				
	1933	1993	Kansas	Nebraska	Minnesota
Population size	25,000	<50	750,000	75,000–200,000	4,000
No. of alleles at six genes	31	22	35	35	32
Percentage of eggs that hatch	93	<50	99	96	85

FIGURE 6.8 Harmful Effects of Genetic Drift (A) The Illinois population of greater prairie chickens dropped from millions of birds in the 1800s to 25,000 in 1933, and, finally, to fewer than 50 birds in 1993. As the Illinois population shrank in size, genetic drift led to a loss of alleles at six genes and to a rise in the frequencies of harmful alleles, ultimately reducing egg hatching rates to less than 50%. (B) The table compares the 1993 Illinois populations with historical populations in Illinois and with populations in Kansas, Nebraska, and Minnesota, none of which experienced as severe a drop in population size. (B after Bouzat et al. 1998.)

2. By causing alleles to be lost from a population, genetic drift reduces the genetic variation of the population, making the individuals within the population more similar genetically to one another.
3. Genetic drift can increase the frequency of a harmful allele. This may seem counterintuitive, because in general, genetic drift acts on alleles that neither harm nor benefit the organism, and we would expect natural selection to reduce the frequency of a harmful allele. However, if the population size is very small and the allele has only slightly deleterious effects, genetic drift can "overrule" the effects of natural selection, causing the harmful allele to increase or decrease in frequency by chance alone.
4. Genetic drift can increase genetic differences between populations because chance events may cause an allele to reach fixation in one population, yet be lost from another population (see Figure 6.7).

The second and third of these effects can have dire consequences for small populations. A loss of genetic variation can reduce the capacity of a population to evolve in response to changing environmental conditions, potentially placing it at risk of extinction. Likewise, an increase in the frequency of harmful alleles in a population can hinder the ability of its members to survive or reproduce, again increasing the risk of extinction. This effect presents an ongoing problem for small populations. Although mutation is unlikely to produce harmful alleles of any particular gene from one generation to the next (because mutations are rare), it is highly likely to produce new deleterious alleles in *some* of an organism's many genes—and genetic drift can cause those alleles to increase in frequency.

Such negative effects of genetic drift are thought to have contributed to the near-extinction of the Illinois populations of the greater prairie chicken (*Tympanuchus cupido*). In the early 1800s, there were millions of these birds in Illinois. Over time, their numbers plummeted as more than 99% of the prairie habitat on which they depend was converted to farmland and other uses. By 1993, fewer than 50 greater prairie chickens remained in Illinois. By comparing the DNA of birds in the 1993 Illinois population with that of birds that lived in Illinois in the 1930s (obtained from museum specimens), Juan Bouzat and colleagues (1998) showed that the drop in population size had reduced the genetic variation of the population (**Figure 6.8**). In addition, more than 50% of the eggs laid by birds in the 1993 Illinois population failed to hatch, suggesting that genetic drift had led to the fixation of harmful alleles. This interpretation was strengthened by the results of experiments begun in 1992: when greater prairie chickens from other populations were brought to Illinois, new alleles entered the Illinois population, and egg hatching rates increased from less than 50% to more than 90% in just 5 years (Westemeier et al. 1998). (The discussion under Concept 10.3 covers the increased risk of extinction borne by small populations in greater detail.)

Gene flow is the transfer of alleles between populations

Gene flow occurs when alleles are transferred from one population to another via the movement of individuals or gametes (e.g., plant pollen). Gene flow has two important effects. First, by transferring alleles between populations, it tends to make populations more similar to one another genetically. This homogenizing effect of gene flow is one reason why individuals in different populations of the same species resemble one another: alleles are exchanged often

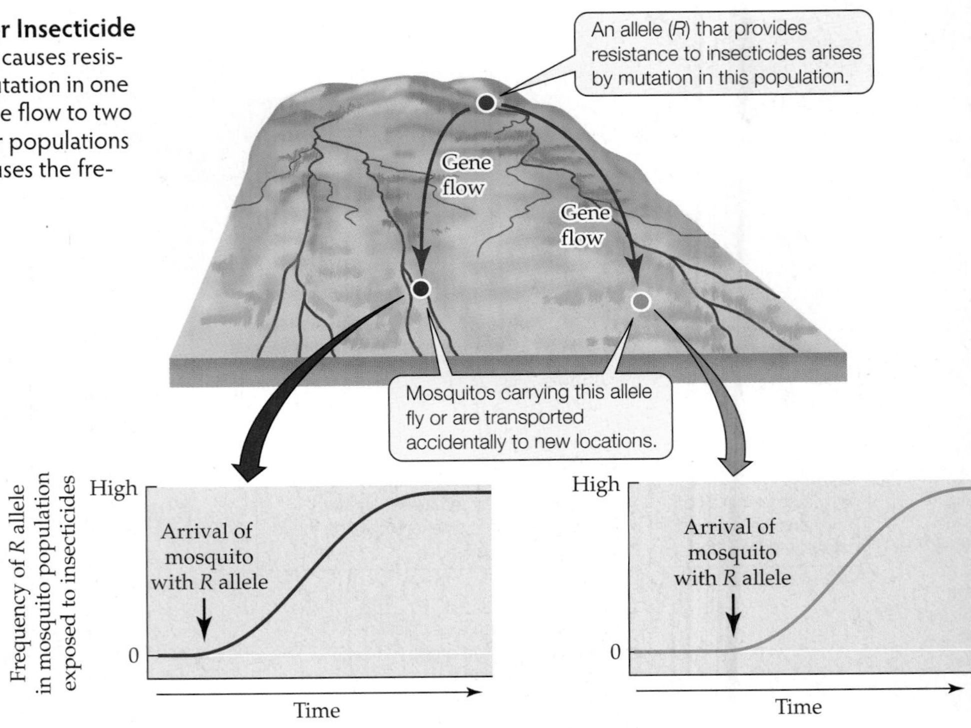

FIGURE 6.9 Gene Flow: Introducing Alleles for Insecticide Resistance In this idealized scenario, an allele that causes resistance to organophosphate insecticides arises by mutation in one population of mosquitoes and then spreads by gene flow to two other populations. If mosquitoes in those two other populations are exposed to the insecticide, natural selection causes the frequency of the resistance allele to increase rapidly.

enough that relatively few differences accumulate between the populations.

Second, gene flow can introduce new alleles into a population. When this occurs, gene flow acts in a manner similar to mutation (although mutation remains the original source of new alleles). This effect of gene flow can have considerable effects on human health. For example, before the 1960s, the mosquito *Culex pipiens* was not resistant to organophosphate insecticides. This mosquito transmits West Nile virus and other diseases, so insecticides were often used to destroy its populations. In the late 1960s, however, new alleles that provided resistance to organophosphate insecticides were produced by mutation in a few *C. pipiens* populations, probably in Africa or Asia (Raymond et al. 1998). Mosquitoes carrying these alleles were blown by storms or transported accidentally by humans to new locations, where they bred with mosquitoes from the local populations. In populations of mosquitoes exposed to insecticides, the frequency of these introduced alleles then increased rapidly because insecticide resistance was favored by natural selection (**Figure 6.9**). The global spread of these alleles by gene flow has allowed billions of mosquitoes to survive the application of insecticides that otherwise would have killed them.

Evolutionary change that results in a closer match between the traits of organisms and the conditions of their environment, such as the increase in the frequency of insecticide resistance in a *C. pipiens* population exposed to insecticides, is an example of adaptive evolution, the topic we'll consider next.

CONCEPT 6.3 Natural selection is the only evolutionary mechanism that consistently causes adaptive evolution.

Adaptive Evolution

The natural world is filled with striking examples of organisms that are well suited for life in their environments (**Figure 6.10**). This match between organisms and their environments highlights their *adaptations*, which are features of organisms that improve their ability to survive and reproduce in their environment (see the discussion under Concept 4.1). Examples of adaptations include remarkable features like those shown in Figure 6.10, but also include less visually striking characteristics—such as an enzyme in a desert plant that can function at high temperatures that would denature most enzymes, enabling the plant to thrive in its environment. There are literally millions of other examples of adaptations. How do these adaptations arise?

Adaptations result from natural selection

Unlike genetic drift, natural selection is not a random process. Instead, when natural selection operates, individuals with certain alleles consistently leave more offspring than do individuals with other alleles. By consistently favoring individuals with some alleles over individuals with other alleles, natural selection causes **adaptive evolution**, a process of change in which traits that confer survival or reproductive advantages tend to increase in frequency over time. This process tends to increase the effectiveness of an adaptation that selection acts on, causing the match between organisms and their environments to improve over time. Although gene flow and genetic drift *can* improve the effectiveness of an adaptation (by increasing the frequency of an advantageous allele), they can also do the reverse (by increasing the frequency of a disadvantageous allele). Thus, natural selection is the only evolutionary mechanism that consistently results in adaptive evolution.

An example of adaptive evolution is provided by changes in populations of the soapberry bug (*Jadera haematoloma*) (Carroll and Boyd 1992; Carroll et al. 1997). This insect uses its needle-like beak to feed on seeds located within the fruits of several different plant species. Soapberry bug populations in southern Florida feed on the seeds of the insect's native host, the balloon vine (*Cardiospermum corin-*

(A)

(B)

(C)

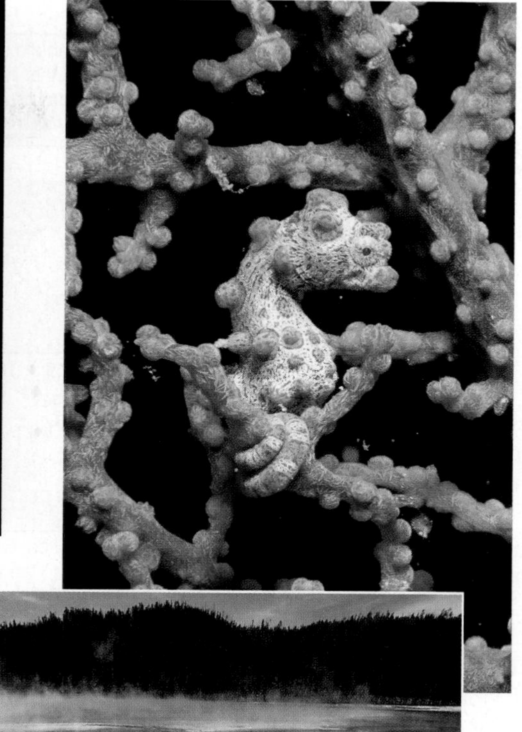

(D)

FIGURE 6.10 A Gallery of Adaptations (A) This archerfish (*Toxotes chatareus*) catches a spider by shooting a jet of water into the air. Field observations show that these fish will squirt repeatedly at potential prey, and that they can reliably hit targets at heights of up to eight times their body length. (B) This round-leaved sundew (*Drosera rotundifolia*) has captured a lacewing in a sticky substance the plant secretes from the hair-like structures on its leaves. Sundews, which live in nutrient-poor habitats such as bogs, feed on insects that are attracted to their bright red color and to a sugary solution that coats their leaves. (C) This small fish (2 cm), the pygmy seahorse (*Hippocampus bargibanti*), is only known to occur on corals of the genus *Muricella* in the western Pacific. Due to its effective camouflage, this species was discovered only after the coral on which the original specimen was living was collected and placed in an aquarium. (D) Temperatures of 64°C–87°C (147°F–188°F) are common at Grand Prismatic Spring, Yellowstone National Park, USA. Some species of algae and prokaryotes (bacteria and archaea), collectively known as "hyperthermophiles," have evolved remarkable structural and biochemical adaptations that enable life in water this hot. Extensive orange and brown microbial mats radiate from the hot spring.

dum). Balloon vines, however, are rare in central Florida. Thus, in that region, soapberry bugs do not feed on balloon vines, but instead feed on the seeds of a species introduced from eastern Asia, the goldenrain tree (*Koelreuteria elegans*). A few specimens of the goldenrain tree were brought to Florida in 1926, but it was not commonly planted until the 1950s. The oldest goldenrain trees in the central Florida populations studied by Carroll and colleagues were 35 years old, suggesting that the soapberry bugs there have fed on this species for 35 years or less.

Soapberry bugs feed most efficiently when the length of a bug's beak matches the depth to which it must pierce a fruit to reach the seeds. Since goldenrain tree fruits are smaller than balloon vine fruits, the introduction of the goldenrain tree 35 years ago can be viewed as a natural experiment on the effect of selection on the insect's beak length. Carroll and Boyd predicted that as a result of natural selection, beak lengths would evolve to be *shorter* in soapberry bug populations that fed on goldenrain tree fruits than in populations that fed on the native host, balloon vines. Carroll and Boyd also studied soapberry bugs in Oklahoma and Louisiana, where the insect had begun to feed on several other new host plants that had been introduced within the past 100 years. However, in Oklahoma and Louisiana, the fruits of the introduced hosts were larger than those of the native hosts, leading to the prediction that in those two states, the beak lengths of insects that ate the introduced species would be *longer* than those of insects that ate the native species.

In all three locations, Carroll and Boyd found that soapberry bug beak lengths evolved in the direction predicted

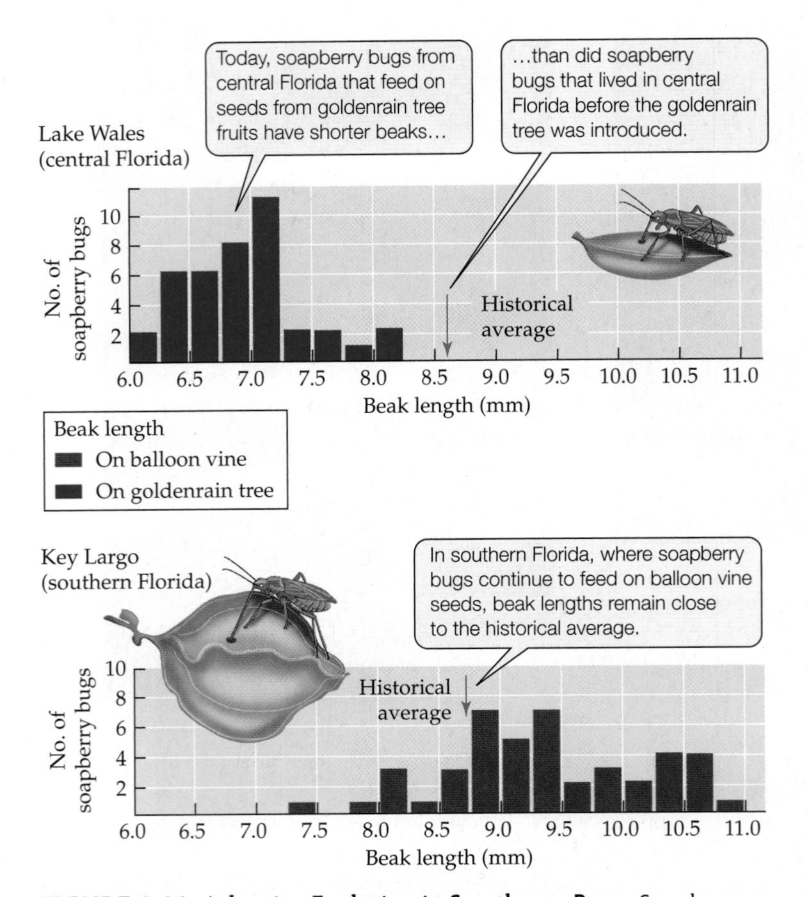

FIGURE 6.11 Adaptive Evolution in Soapberry Bugs Soapberry bug populations in southern Florida feed on the seeds of their native host, the balloon vine, while soapberry bug populations in central Florida feed on the seeds of an introduced plant, the goldenrain tree. The beak lengths of insects feeding on the goldenrain tree decreased by 26% in 35 years, providing a better match to the smaller fruits of this introduced plant. Red arrows indicate beak length historical averages (obtained from museum specimens collected before the introduction of goldenrain trees). (After Carroll and Boyd 1992.)

by fruit size, decreasing in central Florida (**Figure 6.11**) and increasing in both Oklahoma and Louisiana. The changes in beak length were substantial: compared with historical values, average beak lengths dropped by 26% in central Florida and increased by 8% (on one introduced host species) and 17% (on another introduced host species) in Oklahoma and Louisiana. In addition, Carroll et al. (1997) showed that beak length is a heritable characteristic, so the observed changes in beak length must have been due at least in part to changes in the frequencies of alleles that affect beak length. Thus, we can conclude that in a relatively short time (35–100 years), natural selection in soapberry bug populations caused adaptive evolution in which a characteristic of the organism (beak length) evolved to match an aspect of its environment (fruit size) more closely.

Adaptive evolution can occur rapidly

Soapberry bugs are not unique: studies on populations of a wide range of other organisms show that natural selection can lead to rapid increases in the frequency of advantageous traits. Examples include the evolution of increased antibiotic resistance in bacteria (in days to months); increased insecticide resistance in insects (in months to years); drabber coloration in guppies, which makes them harder for visually hunting predators to find (several years); and increased beak size in medium ground finches (several years; see Figure 6.6A). These and many other examples of apparently rapid evolution are described by Endler (1986), Thompson (1998), and Kinnison and Hendry (2001); collectively, these studies suggest that what we think of as "rapid" evolution may in fact be the norm, not the exception.

Rapid, apparently adaptive evolution has also been documented on a continental scale. A number of studies have focused on **clines**: patterns of change in a characteristic of an organism over a geographic region. For example, in the fruit fly *Drosophila melanogaster*, the alcohol dehydrogenase (*Adh*) gene exhibits a cline in which the Adh^S allele decreases in frequency as latitude increases (**Figure 6.12A**). This pat-

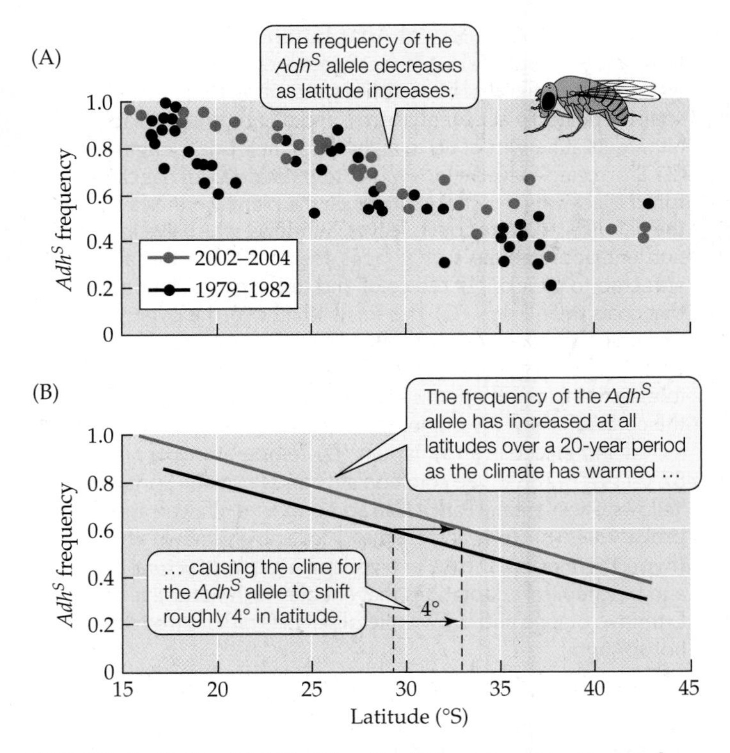

FIGURE 6.12 Rapid Adaptive Evolution on a Continental Scale The *Adh* gene encodes a metabolically important enzyme, alcohol dehydrogenase, used to detoxify alcohol. Previous field and laboratory studies indicate that the Adh^S allele of this gene is selected against in cooler environments, such as those found at high latitudes. (A) Frequencies of the Adh^S allele in coastal Australian *Drosophila melanogaster* populations in 1979–1982 and in 2002–2004. (B) Regression lines calculated from the data in part A show that between 1979–1982 and 2002–2004, the cline of the Adh^S allele shifted 4° toward the South Pole as the region's average temperatures increased by 0.5°C. (After Umina et al. 2005.)

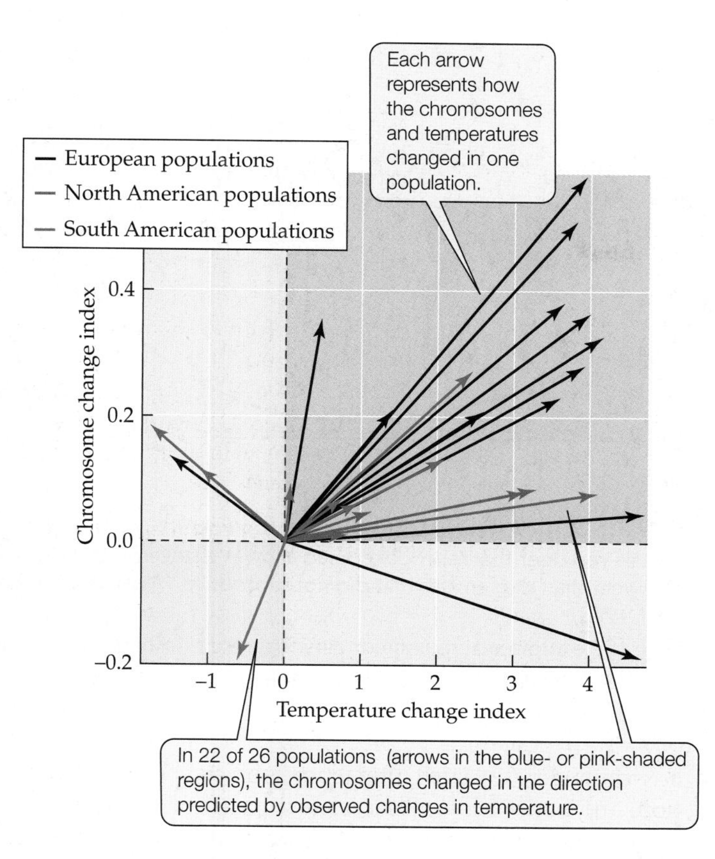

FIGURE 6.13 Global Evolutionary Responses Are Tracking Global Warming In 26 populations of the fruit fly *Drosophila subobscura*, historical chromosome genotypes were compared with current chromosome genotypes after periods of 13 to 46 years. During these periods, temperatures had increased in the environment of 22 of the 26 populations. Chromosome genotypes characteristic of warm climates had increased in frequency in the vast majority (21) of the 22 populations that experienced warming—an example of evolutionary responses across the globe that are tracking global warming. Positive values of the "Temperature change index" indicate populations that experienced warming, while negative values indicate populations that experienced cooling; positive values of the "Chromosome change index" indicate populations with an increase in genotypes characteristic of warm climates, while negative values indicate populations with an increase in genotypes characteristic of cool climates. (After Balanyá et al. 2006.)

tern has been found in both the Northern and Southern hemispheres. Previous studies indicated that this cline results from natural selection on the Adh^S allele, which is less effective in the colder temperatures at higher latitudes and hence is less common there.

Over the past 20 years in coastal Australia, the *Adh* cline has shifted about 4° in latitude toward the South Pole (Umina et al. 2005), a movement of roughly 400 km (**Figure 6.12B**). During the same period, mean temperatures in the region have increased by 0.5°C. Since the Adh^S allele is favored at higher temperatures, the 4° shift in latitude appears to be a rapid, adaptive increase in the frequency of this allele in response to climate change. Similarly, rapid evolutionary changes that are correlated with global warming have been observed in worldwide populations of another fruit fly species, *Drosophila subobscura* (**Figure 6.13**).

Evolutionary responses to climate change over short periods have also been documented in pitcher-plant mosquitoes (Bradshaw and Holzapfel 2001), red squirrels (Réale et al. 2003), and the mustard plant, *Brassica rapa* (Franks et al. 2007). In addition, hundreds of other species have altered the timing of key events in their lives in ways that may be a response to global warming, such as delaying the onset of winter dormancy or reproducing earlier in the spring (Parmesan 2006). In most of these cases, however, it is not yet known whether the observed changes are due to *phenotypic plasticity* (in which a single genotype produces different phenotypes in different environments; see the discussion under Concept 7.1) or to an evolutionary response (in which the genetic constitution of the population changes over time). (See **Climate Change Connection 6.1** for more information on evolutionary responses to climate change.)

Gene flow can limit local adaptation

Although many populations are strikingly well matched to their environments, others are not. Gene flow is one of the factors that can limit the extent to which a population is adapted to its local environment. For example, some plant species have tolerant genotypes that can grow on soils at former mine sites containing high concentrations of heavy metals; such soils are toxic to intolerant genotypes. On normal soils, the tolerant genotypes grow poorly compared with intolerant genotypes. Thus, we would expect the frequencies of tolerant genotypes to approach 100% on mine soils (where they are advantageous) and 0% on normal soils (where they are disadvantageous). Researchers found that a population of the bentgrass *Agrostis tenuis* growing on mine soils was dominated by tolerant genotypes, as expected. However, a population growing on normal soils downwind from the mine site contained more tolerant genotypes than expected (McNeilly 1968). Bentgrass is wind-pollinated, and each year, pollen from the plants growing on mine soils carried alleles for heavy metal tolerance into the population growing on normal soils, preventing that population from becoming fully adapted to its local conditions. The population growing on mine soils also received pollen from plants growing on normal soils. In this population, however, gene flow had relatively little effect on allele frequencies because selection against intolerant genotypes was so strong (they survived poorly on mine soils). In general, whenever alleles are transferred between populations that live in different environments, the extent to which adaptive evolution occurs in each population depends on whether natural selection is strong enough to overcome the effects of ongoing gene flow.

Adaptations are not perfect

As we have just seen, gene flow can limit the extent to which a population is adapted to its local environment. But even when gene flow does not have this effect, natural selection does not result in a perfect match between organisms and their environments. In part, this occurs because an organism's environment is not static—it is a moving target, because the abiotic and biotic components of the environment change continually over time. In addition, organisms face a number of constraints on adaptive evolution:

- *Lack of genetic variation.* If none of the individuals in a population has a beneficial allele of a particular gene that influences survival and reproduction, adaptive evolution cannot occur at that gene. For example, the mosquito *Culex pipiens* initially lacked alleles that provided resistance to organophosphate insecticides (see p. 140). For decades, this lack of genetic variation prevented adaptive evolution in response to insecticides, allowing humans to destroy mosquito populations at will—at least up until the time when insecticide resistance alleles arose by mutation and spread by gene flow. Note that in this and in all other cases, advantageous alleles arise by chance; they are not produced as needed or "on demand."
- *Evolutionary history.* Natural selection does not craft the adaptations of an organism from scratch. Instead, if the necessary genetic variation is present, it works by modifying the traits already present in an organism. Organisms have certain traits and lack others because of their ancestry. It would be advantageous, for example, for an aquatic mammal such as a dolphin to be able to breathe under water. Dolphins lack this capacity, however, in part because of constraints imposed by their evolutionary history: they evolved from terrestrial vertebrates that had lungs and breathed air. Natural selection can bring about great changes, as seen in the mode of life and streamlined body form of the dolphin, but it does so by modifying traits that are already present in the organism, not by creating advantageous traits de novo.
- *Ecological trade-offs.* To survive and reproduce, organisms must perform many essential functions, such as acquiring food, escaping predators, warding off disease, and finding mates. Energy and resources are required for each of these essential functions. Hence, as suggested by the maxim "There is no free lunch" (see Table 1.1), organisms face **trade-offs** in which the ability to perform one function reduces the ability to perform another (**Figure 6.14**). Trade-offs occur in all organisms, and they ensure that adaptations will never be perfect. Instead, adaptations represent compromises in the abilities of organisms to perform many different and sometimes conflicting functions.

Despite these pervasive constraints, adaptive evolution is a key component of the evolutionary process. What does the importance of adaptive evolution tell us about the link between ecology and evolution? As we saw in the case of soapberry bug populations (see Figure 6.11), natural selection, and the adaptive evolution that results, is driven by the interactions of organisms with one another and with their

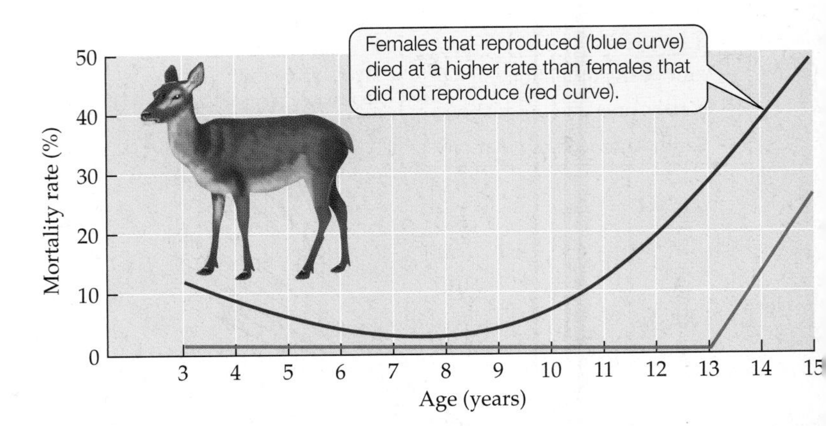

FIGURE 6.14 A Trade-Off between Reproduction and Survival Female red deer that reproduced had a lower chance of surviving to the next year than did females that did not reproduce. (After Clutton-Brock et al. 1983.)

Q Is the additional risk of mortality that results from reproduction the same for females of all ages? Explain.

environment. Any such interaction is an ecological interaction, and hence ecology serves as a basis for understanding natural selection. Next, we'll consider how ecological interactions influence broader evolutionary changes, such as the formation of new species and the great changes that have occurred during the history of life on Earth.

CONCEPT 6.4 Long-term patterns of evolution are shaped by large-scale processes such as speciation, mass extinction, and adaptive radiation.

The Evolutionary History of Life

Earth is home to roughly 2 million species that have been named by taxonomists and to millions more that have yet to be discovered or named. This tremendous diversity of species serves as a foundation for all of ecology, which, as we saw in the discussion under Concept 1.2, is the study of how species interact with one another and with their environment. But the causation runs both ways: while it is true that ecological interactions are influenced by the diversity of species, it is also true that ecological interactions have influenced the number of species alive today. To see why, let's examine the origin of species and some of the other processes that have affected the history of life on Earth.

The diversity of life results from speciation

Each of the millions of species alive today originated by **speciation**, the process by which one species splits into two or more species. A *species* can be defined as a group of organisms whose members have similar characteristics and can interbreed. Speciation most commonly occurs when a barrier prevents gene flow between two or more

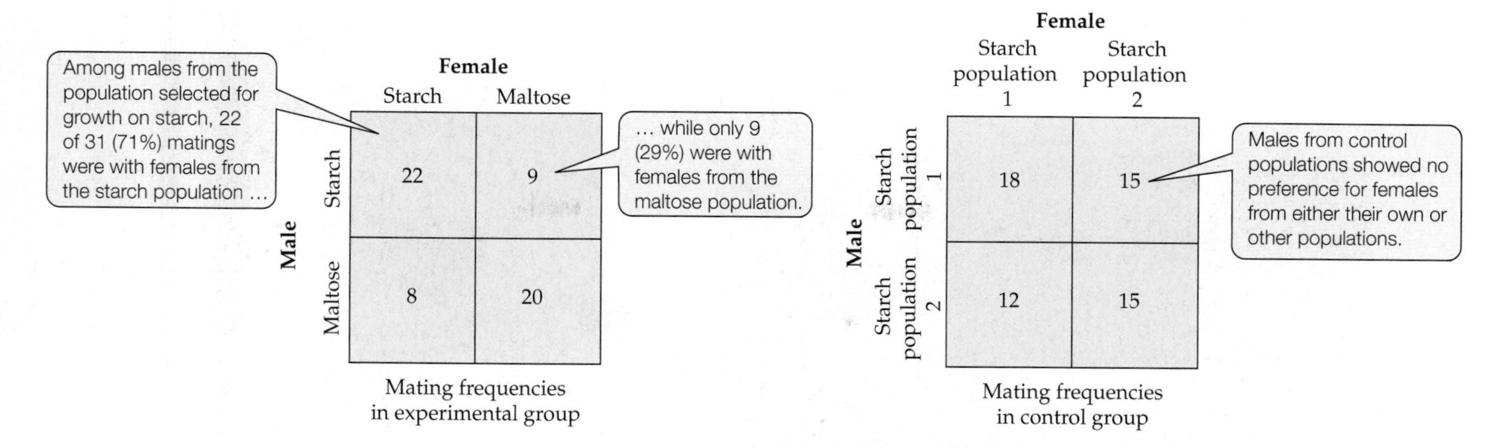

FIGURE 6.16 Reproductive Barriers Can Be a By-Product of Selection After 1 year (about 40 generations) in which experimental populations of *Drosophila pseudoobscura* fruit flies were selected for growth on different sources of food, most matings occurred between flies selected to feed on the same food source. No such mating preference was observed in control populations that were not subjected to selection, regardless of whether the control populations were reared on starch (shown here) or maltose (not shown). To reduce the chance that the food fly larvae ate would produce a body odor in adult flies that influenced the results, all flies used in the mating preference tests were reared for one generation on a standard cornmeal medium. (After Dodd 1989.)

populations of a species. The barrier may be geographic, as when a new population becomes established far from the parental population, or it may be ecological, as when some members of an insect population begin to feed on a new host plant. When a barrier to gene flow is established between populations, they diverge genetically over time (**Figure 6.15**). New species can also form in several other ways, such as when members of two different species produce fertile hybrid offspring (see p. 149 for an example in sunflowers). Whether it is produced by genetic divergence, hybridization, or other means, the key step in the formation of a new species is the evolution of barriers that prevent its members from breeding freely with members of the parental species. Such reproductive barriers arise when a population accumulates so many genetic differences from the parental species that its members rarely produce viable, fertile offspring if they mate with members of the parental species.

The accumulation of genetic differences that lead to the formation of a new species can be an incidental by-product of selection. For example, an experiment with fruit flies demonstrated the beginnings of reproductive barriers between populations selected for growth on different sources of food, but no such barriers were observed between control populations that had not been subjected to selection (**Figure 6.16**). Natural selection has produced similar changes, both in plant populations growing on soils with differing concentrations of heavy metals (Macnair and Christie 1983) and in frog populations living in environments with different temperatures (Moore 1957). In some cases, the trait favored by selection is the same trait that drives speciation. For example, the mosquitofish (*Gambusia hubbsi*) lives in small pools in the Bahamas. Some of these pools contain fish species that eat *Gambusia*, while others lack such predators. In pools with predatory fishes, the mosquitofish have evolved a body shape that allows high-speed escape swimming (Langerhans et al. 2007). Female mosquitofish prefer to mate with males whose body shape is similar to their own. Thus, natural selection favors different body shapes in pools with and without predators, and those different body shapes drive the early stages of speciation through their effects on mate choice.

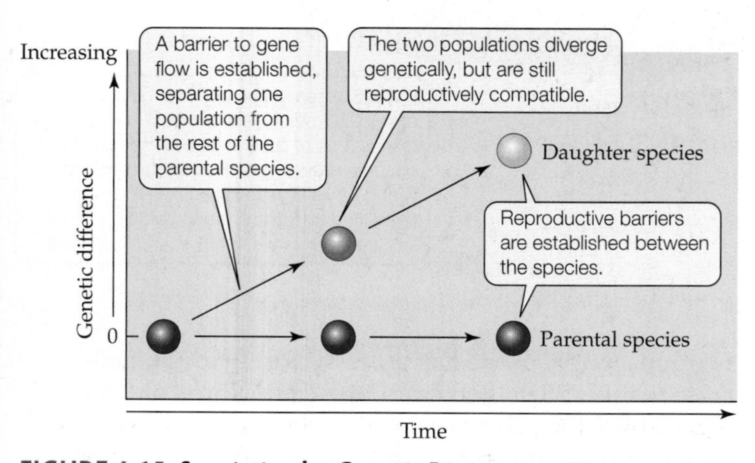

FIGURE 6.15 Speciation by Genetic Divergence Once genetic divergence begins, the time required for speciation varies tremendously, from a single generation (perhaps a single year), to a few thousand years, to millions of years in most cases.

In each of these cases, reproductive barriers arose as a by-product of selection in response to a feature of the environment, such as food source, heavy metal concentration, temperature, or presence of predators. As we've seen, genetic drift can also promote the accumulation of genetic differences between populations (see pp. 138–139). Thus, like natural selection, genetic drift can ultimately lead to the evolution of reproductive barriers, and hence to the formation of new species. In contrast to selection and drift, gene flow typically acts to slow down or prevent speciation because populations that exchange many

(A)

Fossil evidence indicates that the first organisms to live on Earth were prokaryotes (bacteria and archaea), such as those that formed this 3.5 billion-year-old fossil stromatolite from western Australia. Stromatolites, which are still being formed today in a few locations, are layered fossils that form when certain prokaryotes bind thin films of sediment to one another.

(B)

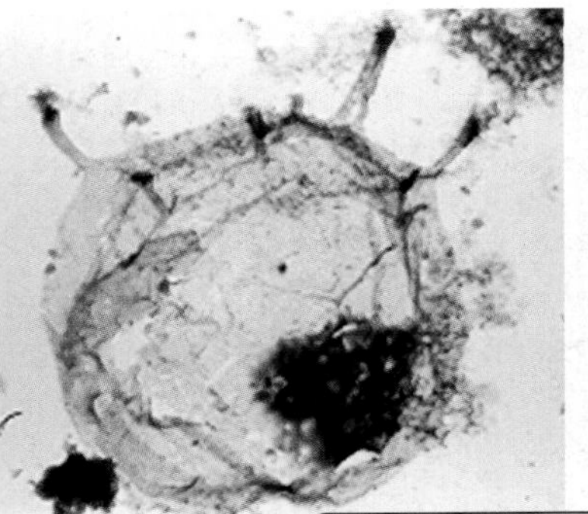

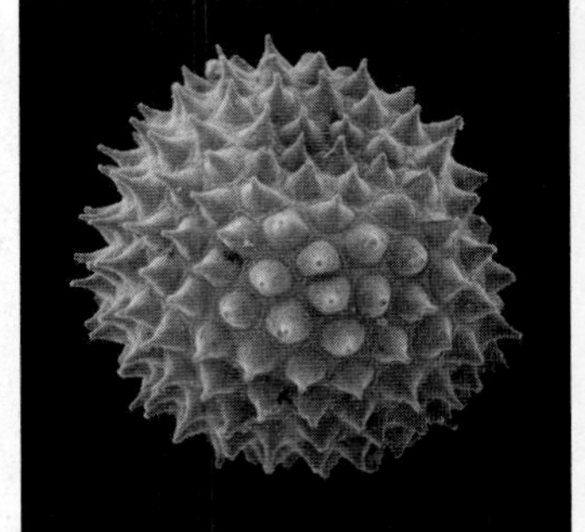

The oldest fossils of eukaryotes are 2.1 billion years old. Other early evidence of eukaryotes include this 1.5-billion-year-old fossil of the unicellular alga *Tappania* (left). Over time, some microscopic eukaryotes evolved anti-predator defenses, as seen above in this 625-million-year-old fossil from the Doushantuo Formation in South China.

(C)

This 530 million-year old fossil arthropod (*Waptia fieldensis*) from the Burgess Shale, British Columbia, is one of the many complex animals that originated during a 10 million-year burst of evolutionary activity known as the Cambrian Explosion.

(D)

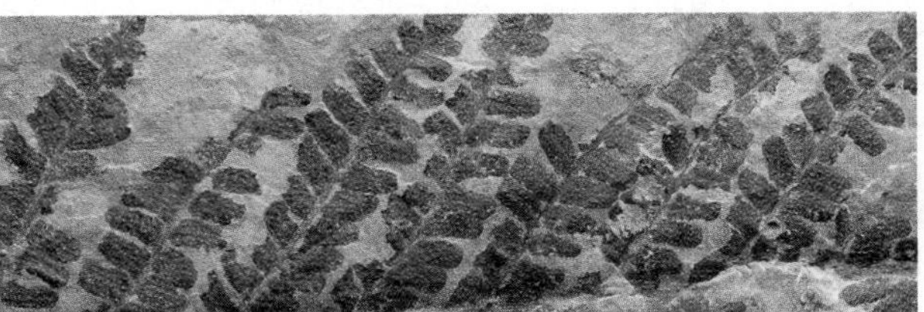

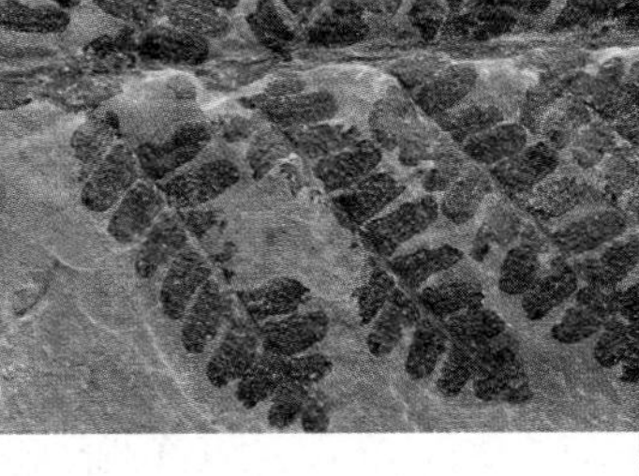

Once mistaken as a fern, this 350 million-year-old fossil is actually of an early seed plant, *Neuropteris*, a member of an extinct genus of "seed ferns." Although *Neuropteris* and other similar plants had leaves that resembled those of a fern, they also produced seeds—a trait found today only in angiosperms (flowering plants) and gymnosperms.

(E)

In 2006, researchers reported the discovery of this 380 million-year-old fossil of *Tiktaalik roseae*, one of dozens of fossil species that document the origin of tetrapods from lobe-finned fishes. The tetrapods are four-legged terrestrial vertebrates whose living members include amphibians, reptiles, and mammals.

FIGURE 6.17 Life Has Changed Greatly over Time

alleles tend to remain genetically similar to one another, making it less likely that reproductive barriers will evolve.

Mass extinctions and adaptive radiations have shaped long-term patterns of evolution

Thus far in this chapter, much of our focus has been on the *process* of evolution—the mechanisms by which evolutionary change occurs. But evolution can also be defined as an observed *pattern* of change. Evolutionary patterns are revealed by observations of the natural world, such as data on the changing allele frequencies of a population over time. Patterns of evolutionary change are also documented in the fossil record, which shows that life on Earth has changed greatly over long periods (**Figure 6.17**).

The earliest known fossils are those of 3.5 billion-year-old bacteria, while the most ancient fossils of multicellular organisms are 1.2 billion years old. Animals first appear in the fossil record about 600 million years ago, and complex animals with bilateral symmetry (in which the body has two equal but opposite halves, as in most living animals) arose roughly 25 million years later (Fedonkin et al. 2007; Chen et al. 2009). These and many other great changes in the history of life resulted from descent with modification as new species arose that differed from their ancestors. Over millions of years, these differences gradually accumulated, leading eventually to the formation of major new groups of organisms, such as terrestrial plants, amphibians, and reptiles.

For example, a rich variety of fossils have been discovered that illustrate steps in the origin of *tetrapods* (vertebrates with four limbs, a group whose living members include amphibians, reptiles, and mammals) from fishes. Similarly, the fossil record contains dozens of fossil species that show how mammals arose over a 120 million-year period (300 to 180 million years ago) from an earlier group of tetrapods, the synapsids (Allin and Hopson 1992; Sidor 2003). The fossil record also documents cases in which the rise to prominence of one group of organisms was associated with the decline of another group. For example, 265 million years ago, reptiles replaced amphibians as the ecologically dominant group of tetrapods, and then, 65 million years ago, the reptiles were replaced in turn by the mammals.

The rise and fall of different groups of organisms over time has been heavily influenced by mass extinctions and adaptive radiations. The fossil record documents five **mass extinction** events in which large proportions of Earth's species were driven to extinction worldwide in a relatively short time—a few million years or less, sometimes much less (**Figure 6.18**). The most recent mass extinction occurred 65 million years ago and may have been caused by a large asteroid that struck Earth, setting in motion cataclysmic environmental changes that led to the demise of dinosaurs and many other organisms.

Each of the five mass extinctions was followed by great increases in the diversity of some of the surviving groups of organisms; mammals, for example, increased greatly in diversity after the extinction of dinosaurs. Mass extinctions can promote such increases in diversity by removing competitor groups, thus allowing the survivors to expand into new habitats or new ways of life. Great increases in diversity can also occur when a group of organisms evolves major new adaptations, such as the stems, waxy cuticles, and stomates on leaves that provided early land plants with support against gravity and protection from desiccation (see Chapter 4). Whether stimulated by a mass extinction, new adaptations, or other factors (such as colonization of a new region that lacks competitors), an event in which a group of organisms gives rise to many new species that expand into new habitats or new ecological roles in a relatively short time is referred to as an **adaptive radiation**.

What can we learn about ecology and evolution from mass extinctions, adaptive radiations, and other great changes in the history of life? First, biological communities are devastated by mass extinction events (**Figure 6.19**).

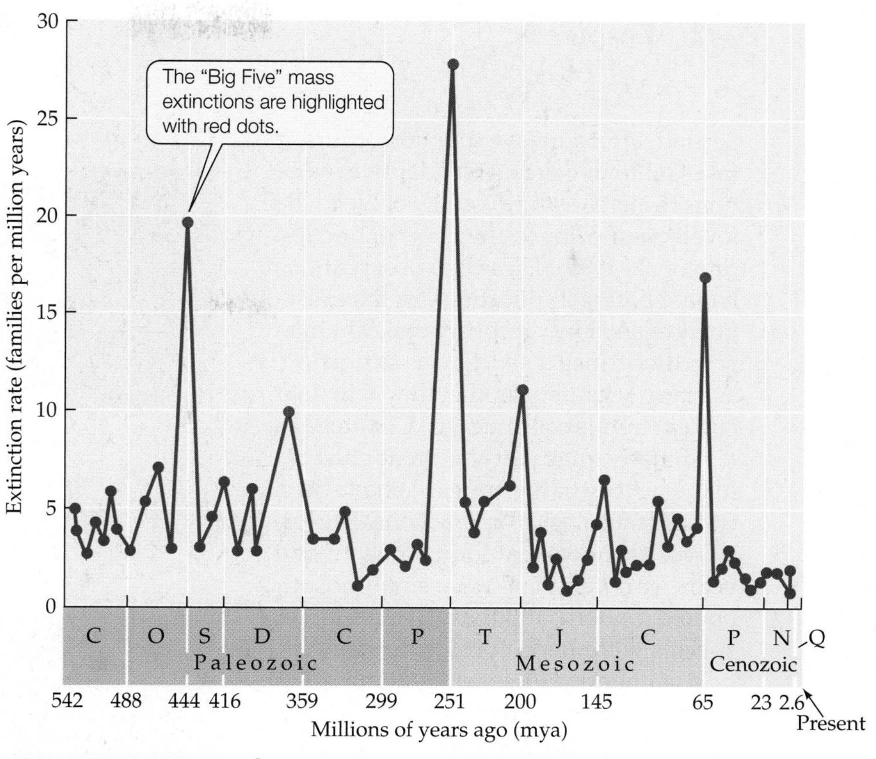

FIGURE 6.18 The "Big Five" Mass Extinctions Five peaks in extinction rates are revealed by a graph of extinction rates over time in families of marine invertebrates.

(A)

(B)

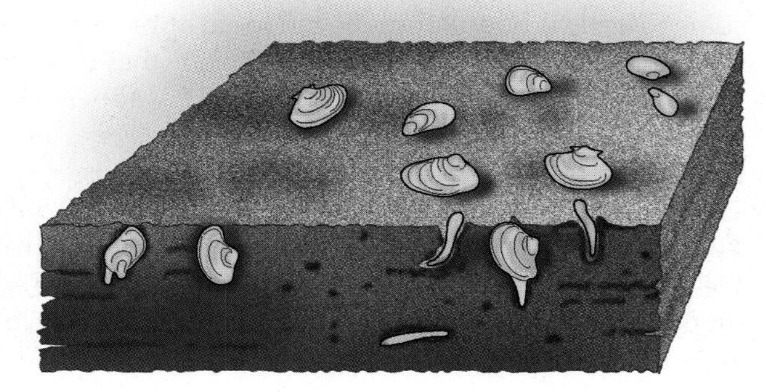

FIGURE 6.19 Devastating Effects of a Mass Extinction An ancient seabed (A) before and (B) after the end-Permian mass extinction 251 million years ago. Over 90% of marine species were driven to extinction, radically altering this biological community and many others worldwide.

Second, after a mass extinction occurs, it takes millions of years for adaptive radiations to increase the diversity of life to the levels seen prior to the mass extinction (Jablonski 1995). These two observations have sobering implications for the consequences and likely recovery time if human activities cause a sixth mass extinction, as many scientists predict they will do if current trends continue (see Chapter 22).

Finally, many of the great changes in the history of life appear to have been caused by ecological interactions. The fossil record shows that for over 60 million years, early animals were small or soft-bodied, or both, and that all of the larger species were herbivores, filter feeders, or scavengers. However, beginning 535 million years ago, this safe, soft-bodied world disappeared forever with the appearance of large, well-armed, mobile predators and large, well-defended prey. This major step in the history of life appears to have resulted from an "arms race" between predators and prey. Early predators equipped with claws and other adaptations for capturing large prey provided powerful selection pressure that favored heavily armored prey species. That armor, in turn, promoted further increases in the effectiveness of the predators, and so on. Such reciprocal evolutionary change in interacting species, known as *coevolution*, is discussed in more detail in Chapter 13.

Since that time, ecological interactions have continued to shape the history of life. For example, the origin of new species in one group of organisms can lead to increases in the diversity of other groups, especially those that can escape from, eat, or compete effectively with the new species (Farrell 1998; Benton and Emerson 2007). An example of this process can be seen in parasitic wasps that feed on the apple maggot fly (*Rhagoletis pomonella*), a species that eats fruits (**Figure 6.20**). Following the introduction of apple trees to North America 200 years ago, some *Rhagoletis* populations began to eat apples. As these populations adapted to their new food plant, they diverged from the parent species genetically and now appear to be well on the way to forming a new fly species (Feder 1998). In addition, populations of the wasp have emerged that specialize on the incipient fly species (Forbes et al. 2009). These wasps have become reproductively isolated from the parent wasp species, thereby providing evidence of a sequence of speciation events that is in progress today and appears to be driven by ecological interactions.

FIGURE 6.20 A Chain of Speciation Events Driven by Ecological Interactions? In the last 200 years, populations of the fly *Rhagoletis pomonella* that feed on apples have diverged genetically from their parent species, forming an incipient fly species. This change also appears to be leading to the formation of a new wasp species (*Diachasma alloeum*) that parasitizes members of apple-feeding *Rhagoletis* populations.

We turn next to a more detailed look at an idea that we have already encountered in this chapter: while ecological interactions influence evolution, evolution also influences ecological interactions.

CONCEPT 6.5 Ecological interactions and evolution exert a profound influence on one another.

Joint Effects of Ecology and Evolution

Ecological and evolutionary interactions can be so closely related as to be entangled. Consider the sunflower species *Helianthus anomalus*. This species originated from a speciation event in which two other sunflowers, *H. annuus* and *H. petiolaris*, produced hybrid offspring. As Loren Rieseberg and colleagues have shown in a series of experiments and genetic analyses (Rieseberg et al. 2003), the new gene combinations generated by hybridization appear to have facilitated a major ecological shift in *H. anomalus*. This hybrid species grows in a different and more extreme environment than either of its two parental species do (**Figure 6.21**)—an ecological shift that illustrates how evolution influences ecology. Simultaneously, however, life under different ecological conditions provided the selection pressures that molded the hybrid offspring of *H. annuus* and *H. petiolaris* into a new species, *H. anomalus* (see Rieseberg et al. 2003), showing how ecology influences evolution. Such joint ecological and evolutionary effects are common—as we should expect, given that both evolution and ecology depend on how organisms interact with one another and with their physical environment.

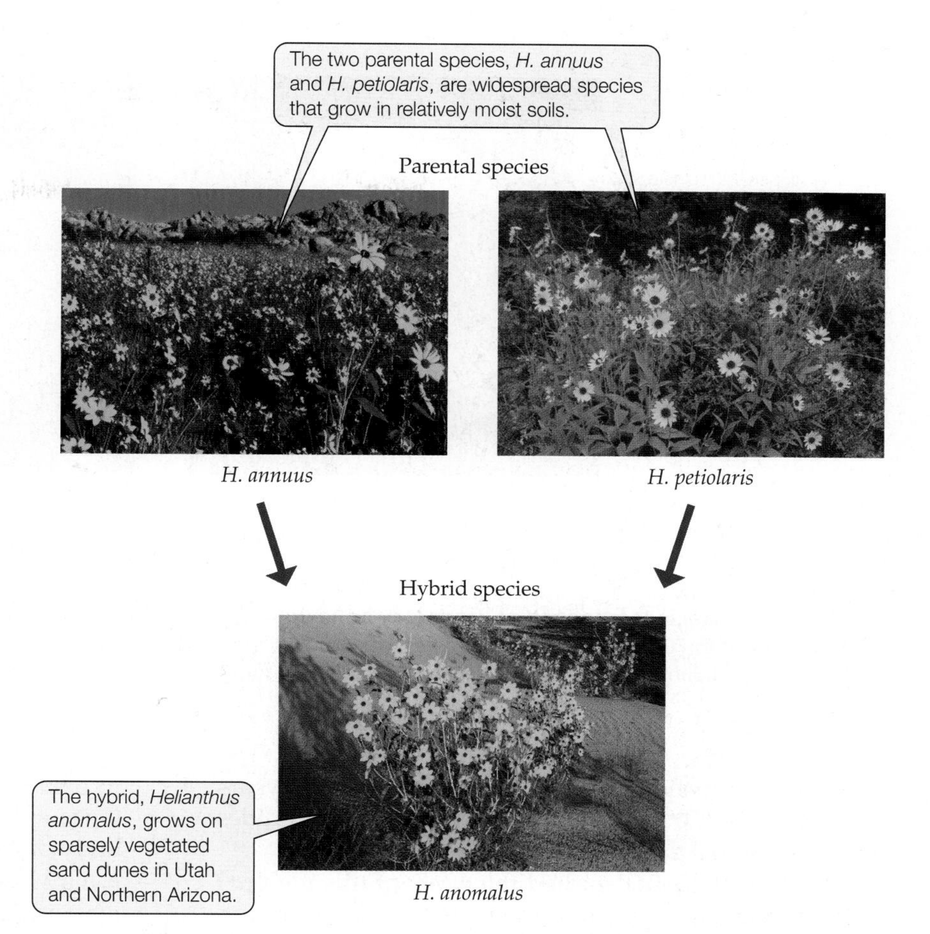

FIGURE 6.21 A Hybrid That Lives in a New Environment Two sunflower species, *Helianthus annuus* and *H. petiolaris*, gave rise to a new hybrid species, *H. anomalus*. This species grows in a drier environment than either of the two parental species.

Ecological interactions can cause evolutionary change

Much of the drama of the natural world stems from the efforts of organisms to do three things: to eat, to avoid being eaten, and to reproduce. As organisms interact with one another in this drama, a web of ecological interactions results. These interactions can drive evolutionary change. We've already seen (on the previous page) how predator–prey interactions (via natural selection) caused long-term, large-scale, reciprocal evolution in which predators became more efficient at capturing prey and prey became more adept at escaping their predators. Evolution can result from a broad range of other ecological interactions as well, including competition, herbivory, parasitism, and mutualism (see Unit 3).

Studies of speciation have led to a similar conclusion: it is common for speciation to be caused by ecological factors (Schluter 1998; Funk et al. 2006). The effect of ecology on evolution is also clear from studies of relatively small evolutionary changes in populations. Examples discussed earlier in this chapter include directional selection on soapberry bugs caused by interactions with their food plants (see Figure 6.11) and genetic drift in greater prairie chickens caused by habitat loss (see Figure 6.8).

Evolution can alter ecological interactions

Whenever a group of organisms evolves a new, highly effective adaptation, the outcome of ecological interactions may change, and that change may have a ripple effect that alters the entire community. For example, if a predator evolves a new way of capturing prey, some prey species may be driven to extinction, while others may decrease in abundance, migrate to other areas, or evolve new ways to cope with the more efficient predator. Similar changes can occur among species that compete for resources; we will discuss one such example in Chapter 11 (pp. 255–256), in which evolutionary changes in experimental populations of one fly species reversed the outcome of its competitive interactions with another fly species.

Processes that drive the patterns of evolution over long time scales can also have a profound effect on ecological interactions. For example, the adaptive radiation of terrestrial plants altered all aspects of life on land, from the stability of the soil to the sources of food available to other organisms to the cycling of nutrients within ecosystems. Likewise, mass extinctions are major events in which ecological communities are changed fundamentally (see Figure 6.19). Here, we'll take a closer look at the effects of

FIGURE 6.22 Ancient Evidence of Predatory Behavior Predatory snails can use either of two strategies when drilling into clam shells: (A) drilling through the shell wall or (B) drilling through the edge of the shell. Each strategy has certain advantages and disadvantages. Predatory snails have utilized these strategies for millions of years, as seen in these fossils.

a regional extinction event that occurred 2 million years ago, removing 70% of species from marine communities in the western Atlantic Ocean.

Dietl et al. (2004) studied the effect of this regional extinction event on predatory snails that drill through the shells of clams and other bivalves. Experiments with modern relatives of these snails that survived the extinction showed that it takes these snails roughly a week to drill through the wall of a clam shell (**Figure 6.22A**). Once it has drilled a hole, the snail eats the clam's soft body parts. While it is drilling, the snail faces two risks: it may lose its food to other snails, or it may lose its life by being eaten by a fish or a crab. Perhaps in response to these risks, predatory snails sometimes drill through the edge of a shell, rather than the wall (**Figure 6.22B**). Drilling through the edge of a shell decreases the time it takes to penetrate the shell by a factor of three. But edge drilling is not risk-free: as a snail drills through the edge of a shell, the bivalve may close its shell on the snail, amputating its feeding organ (proboscis).

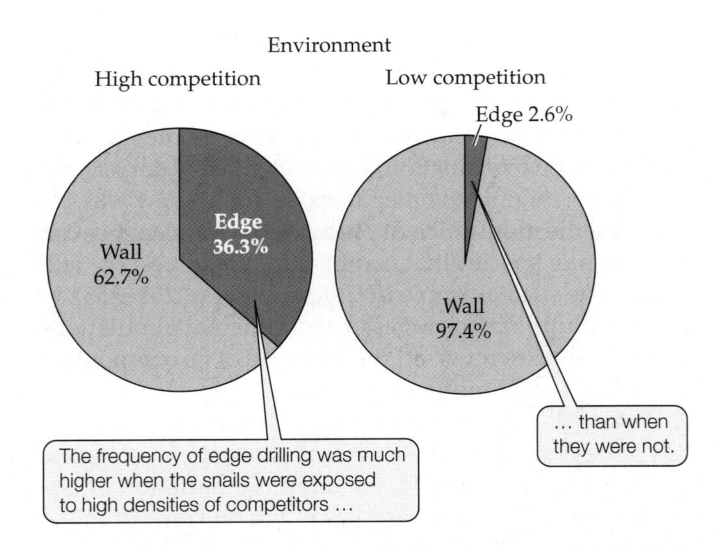

FIGURE 6.23 Edge Drilling Increases in High-Competition Environments In laboratory experiments, with modern predatory snails, researchers observed the responses of the snails' drilling behavior to environments with high and low densities of competitors. (After Dietl et al. 2004.)

In a species-rich community, edge drilling can be advantageous because it decreases the time during which the snail is vulnerable to its competitors and predators. Reasoning that edge drilling would provide less of an advantage after an extinction event (because the snails would face fewer competitor and predator species), Dietl and colleagues predicted that edge drilling would be uncommon after the regional extinction in the western Atlantic 2 million years ago. This prediction was supported in a survey of fossils from before and after the extinction: edge drilling was found in the species-rich communities that lived 5 million to 2 million years ago, but it stopped altogether after the extinction event. Thus, the extinction appears to have altered ecological interactions: it reduced the competition and predation experienced by shell-drilling snails, leading to a reduction in the frequency of edge drilling. This interpretation was strengthened by the results of experiments with modern snails shown in **Figure 6.23**.

A CASE STUDY REVISITED

Trophy Hunting and Inadvertent Evolution

Trophy hunters of bighorn sheep prefer to kill large males that carry a full curl of horns. The majority of these males are killed when they are between 4 and 6 years old, often before they have sired many offspring. As a result, hunting decreases the chance that alleles carried by males with a full curl of horns will be passed on to the next generation. Instead, it is males with relatively small horns who father most of the offspring, transmitting their alleles to the next generation. This change has caused the frequency of alleles encoding small horns to increase, thus leading to the observed 30-year decrease in average horn size (see Figure 6.2). Overall, trophy hunting has inadvertently caused directional selection in bighorn sheep, favoring small males with small horns and changing allele frequencies in the population over time.

Humans have caused unintended evolutionary changes in a wide variety of other populations. An early example was provided by the decline in the frequency of red foxes (*Vulpes fulva*) with coats that have a silver tint, a color preferred by hunters (**Figure 6.24**). In a medical setting, consider our use of antibiotics against bacteria that cause diseases and lethal infections. When antibiotics were first discovered (ca. 1940), they were hailed as miracle drugs. At first, bacterial populations were highly vulnerable to antibiotics, and their use saved many human lives. But the use of antibiotics provided a strong source of directional selection, leading to the evolution of antibiotic resistance in bacterial populations (see Figure 1.10). Today, as a result, antibiotic treatments sometimes fail, even when very high doses are administered. Antibiotic resistance also has enormous financial costs. The total magnitude of these costs is not known, but in the United States alone, efforts to cure patients infected with antibiotic-resistant strains of just one bacterial species (*Staphylococcus aureus*) result in an estimated $24–$31 billion in medical expenses *each year* (Palumbi 2001).

Although antibiotics still save many lives, for each antibiotic now in use, at least one bacterial species is resistant to it. This makes it vital that we develop new antibiotics and slow the evolution of resistance to those we already have. One way to achieve the latter goal would be to use antibiotics less frequently, thus reducing the extent to which bacterial populations are exposed to antibiotics and hence selected for resistance. We could, for example, limit the prescribing of antibiotics for people with colds and other respiratory infections, since these conditions are usually caused by viruses, and viruses are not affected by antibiotics.

We have seen throughout this chapter that human actions such as trophy hunting and antibiotic use act as selection pressures and hence may cause evolutionary change. But does our influence on evolution extend beyond cases in which we selectively kill other organisms?

CONNECTIONS IN NATURE

The Human Impact on Evolution

Many human actions alter the environment and hence have the potential to change the course of evolution. As we've seen in this chapter, actions such as trophy hunting and antibiotic use are themselves a powerful source of selection. Other human actions, such as emissions of pollutants or introductions of invasive species, change aspects of the abiotic or biotic environment. By changing the environment, we alter the selection pressures that organisms face, and hence may cause evolutionary change. Still other human actions, such as *habitat fragmentation* (in which portions of a species' habitat are destroyed, leaving spatially isolated fragments of the original habitat), can have a series of evolutionary effects (**Figure 6.25**).

Human actions that affect the environment can alter each of the three main mechanisms of evolution: natural selection, genetic drift, and gene flow. Because we know with certainty that our actions are causing great changes to environments worldwide, we can infer that they are also causing evolutionary changes in populations worldwide. Although we are just beginning to document their extent,

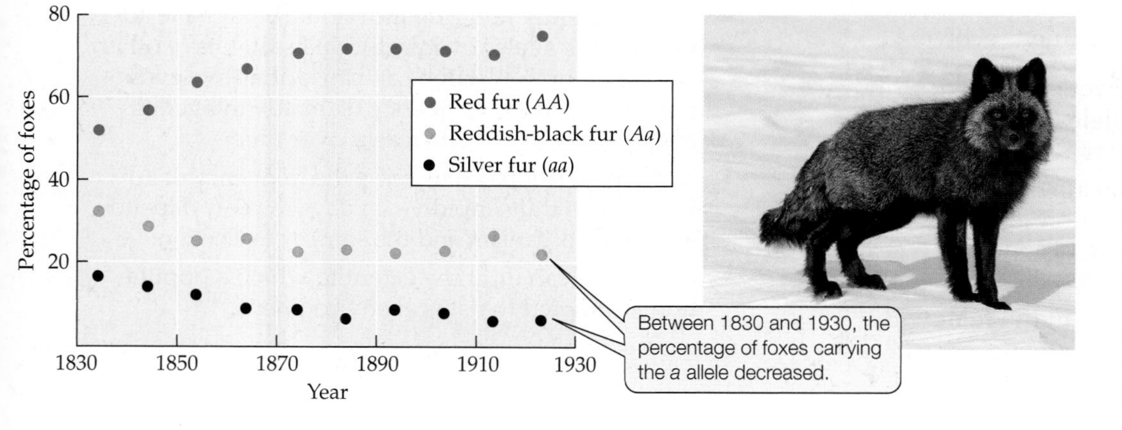

FIGURE 6.24 Hunting Resulted in the Decline of Silver Foxes Individual red foxes (*Vulpes fulva*) of genotype *AA* have red fur, and individuals of genotype *Aa* have reddish-black fur. Individuals of genotype *aa* are known as "silver foxes" because the tips of their hairs have a silver tint. Hunters preferentially killed silver foxes because their furs yielded 2.5–4 times the price of other red fox furs. (Data from Elton 1942.)

(A) Large populations

Habitat fragmentation

(B) Small populations

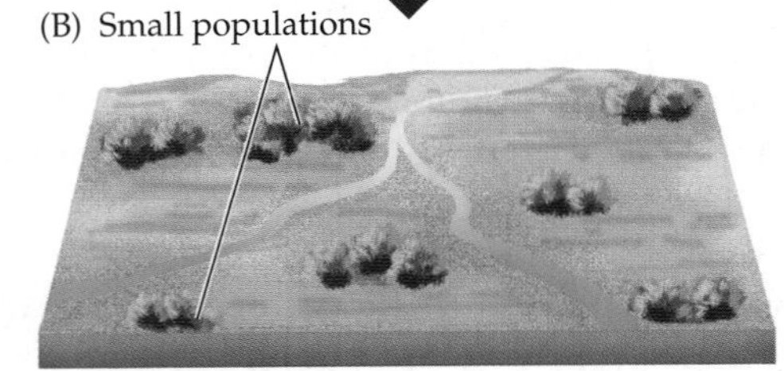

	Unfragmented habitat	Fragmented habitat
Population size	Large	Small
Distance between populations	Short	Long
Genetic drift	Low impact	High impact
Genetic variation within populations	High	Low
Gene flow	High	Low

FIGURE 6.25 Evolutionary Effects of Habitat Fragmentation on a Hypothetical Species (A) Prior to habitat fragmentation, there are many individuals in the species' populations, and the distances between populations are short. (B) When human activities remove large portions of the species' habitat, its population sizes shrink, and the distances between populations increase, causing evolutionary changes that decrease its potential for adaptive evolution and increase its risk of extinction.

such changes are likely to have unintended consequences for the species and ecosystems on which human societies depend. We also know that human actions have the potential to alter patterns of evolution over long time scales. As we will see in Chapters 22 and 23, human actions such as habitat destruction, overharvesting, and introductions of invasive species are among the main reasons why Earth is undergoing a biodiversity crisis (Chapter 18). The extinction rate of species today is 100 to 1,000 times higher than the usual or "background" extinction rate seen in the fossil record. Extinction is forever, so when human actions drive a species to extinction, the future course of evolution is altered in a way that cannot be reversed. If human activities cause a sixth mass extinction in the next few centuries or millennia, our actions will greatly and irreversibly change the evolutionary history of life on Earth.

SUMMARY

CONCEPT 6.1 Evolution can be viewed as genetic change over time or as a process of descent with modification.

- Biologists often define evolution in a relatively narrow sense as change over time in the frequencies of alleles in a population.
- Evolution can also be viewed as descent with modification, a process in which populations accumulate differences over time and hence differ from their ancestors.
- Natural selection modifies populations by favoring individuals with some heritable traits over others.
- Although natural selection acts on individuals, an individual does not evolve—it either has a favored trait or it does not. Only populations evolve.

CONCEPT 6.2 Natural selection, genetic drift, and gene flow can cause allele frequencies in a population to change over time.

- Mutation and recombination are the sources of new alleles and new combinations of alleles, thereby providing the genetic variation on which evolution depends.
- Natural selection occurs when individuals with certain heritable phenotypic traits survive and reproduce more successfully than individuals with other traits.
- Genetic drift, which occurs when chance events determine which alleles are passed from one generation to the next, can have negative effects on small populations.
- Gene flow, the transfer of alleles between populations, makes populations more similar to one another genetically and can introduce new alleles into populations.

CONCEPT 6.3 Natural selection is the only evolutionary mechanism that consistently causes adaptive evolution.

- By consistently favoring individuals that have advantageous alleles over individuals that have other alleles, natural selection can cause adaptive evolution, in which the frequency of an advantageous trait in a population increases over time.
- Natural selection can increase the frequency of advantageous traits rapidly—in days to years depending on the organism and the source of selection.
- Gene flow can limit the extent to which a population is adapted to its local environment.

SUMMARY (continued)

- Constraints on adaptive evolution result from factors such as lack of genetic variation, evolutionary history, and ecological trade-offs.

CONCEPT 6.4 Long-term patterns of evolution are shaped by large-scale processes such as speciation, mass extinction, and adaptive radiation.

- The diversity of life has resulted from speciation, the process by which one species splits into two or more species. Speciation requires the evolution of reproductive barriers between populations.
- Biological communities can lose much of their diversity in mass extinctions, global events in which large proportions of Earth's species are driven to extinction in a relatively short time.
- An adaptive radiation occurs when a group of organisms gives rise to many new species that expand into new habitat or fill new ecological roles.
- Adaptive radiations can be promoted by factors such as the removal of competitor groups by a mass extinction or by the evolution of a major new adaptation.

CONCEPT 6.5 Ecological interactions and evolution exert a profound influence on one another.

- Ecological interactions among organisms and between organisms and their environment can cause evolutionary changes, ranging from allele frequency changes in populations to the formation of new species.
- Similarly, evolutionary change can alter the outcome of ecological interactions, thus having a large influence on biological communities.

REVIEW QUESTIONS

1. Natural selection acts on individuals, yet one of the points made in this chapter is that *populations evolve, but individuals do not*. Explain how natural selection works and why the italicized statement is true.
2. What causes adaptive evolution? Explain in your answer why each of the three primary mechanisms of allele frequency change in populations causes or does not cause adaptive evolution.
3. What large-scale processes determine patterns of evolution observed over long time scales? Explain how each process that you describe has this effect.
4. Explain why ecological interactions and evolutionary change have joint effects, each affecting the other.

ON THE COMPANION WEBSITE
sites.sinauer.com/ecology2e

The website includes Chapter Outlines, Online Quizzes, Flashcards & Key Terms, Suggested Readings, a complete Glossary, and the Web Stats Review. In addition, the following resources are available for this chapter:

HANDS-ON PROBLEM SOLVING

Effects of Natural Selection and Genetic Drift

This Web exercise demonstrates how natural selection and genetic drift can alter the frequencies of alleles in populations. You will investigate the effects of manipulating population sizes (and thus the strength of genetic drift) and the force of selection.

WEB EXTENSION

6.1 Hardy–Weinberg Equation

CLIMATE CHANGE CONNECTIONS

6.1 Evolutionary Responses to Climate Change

UNIT 2
Populations

CHAPTER

7

Life History

KEY CONCEPTS

- **CONCEPT 7.1** Life history patterns vary within and among species.
- **CONCEPT 7.2** Reproductive patterns can be classified along several continua.
- **CONCEPT 7.3** There are trade-offs between life history traits.
- **CONCEPT 7.4** Organisms face different selection pressures at different life cycle stages.

Nemo Grows Up: A Case Study

Birds do it, bees do it, even educated fleas do it—they all produce offspring that perpetuate their species. But beyond that basic fact of life, the offspring produced by different organisms vary tremendously. A grass plant produces seeds a few millimeters long that can wait buried in the soil for years until conditions are right for germination. A sea star spews hundreds of thousands of microscopic eggs that develop adrift in the ocean. A rhinoceros produces one calf that develops in her womb for 16–18 months and can walk well several days after birth, but requires more than a year of care before it becomes fully independent (**Figure 7.1**).

Even this broad range of possibilities barely begins to describe the different ways in which organisms reproduce. In popular media, we humans often depict other animals as having family lives similar to ours. For example, in Pixar's animated film *Finding Nemo*, clownfish live in families with a mother, a father, and several young offspring. When Nemo the clownfish loses his mother to a predator, his father takes over the duties of raising him. But in a more realistic version of this story, after losing his mate Nemo's father would have done something less predictable: he would have changed sex and become a female.

Actually, the correspondence between the movie and biology breaks down long before Nemo loses his mother. Clownfish spend their entire adult lives within a single sea anemone (**Figure 7.2**). Anemones can be thought of as modified upside-down jellyfish with a central mouth ringed by stinging tentacles. In what appears to be a mutually beneficial relationship, the anemone protects the clownfish by stinging their predators, but the clownfish themselves are not stung. The clownfish, in turn, may help the anemone by eating its parasites or driving away its predators.

Two to six clownfish typically inhabit a single anemone, but they are far from a traditional human family—in fact, they are usually not related to one another. The clownfish that live in an anemone interact according to a strict pecking order that is based on size. The largest fish in the anemone is a female. The next fish in the hierarchy, the second largest, is the breeding male.

FIGURE 7.1 Offspring Vary Greatly in Size and Number Organisms produce a large range of offspring numbers and sizes. A rhinoceros produces a single calf that weighs 40–65 kg (90–140 pounds). On the other end of the spectrum, many plants produce hundreds to thousands of seeds that are less than a millimeter long and weigh as little as 0.8 μg (roughly one-50 billionth the weight of a rhinoceros calf).

FIGURE 7.2 Life in a Sea Anemone Clownfish (*Amphiprion percula*) form hierarchical groups of unrelated individuals that live and reproduce among the tentacles of their anemone host (*Heteractis magnifica*).

Q Predict the gender of each of these clownfish (assuming that they live together as a group of four fish in an anemone host). Explain.

The remaining fish are sexually immature *nonbreeders*. If the female dies, as in Nemo's story, the breeding male undergoes a growth spurt and changes sex to become a female, and the largest nonbreeder increases in size and becomes the new breeding male.

The breeding male clownfish mates with the female and cares for the fertilized eggs until they hatch. The hatchling fish leave the anemone to live in the open ocean, away from the predator-infested reef. The young fish eventually return to the reef and develop into juveniles. Then they must find an anemone to inhabit. When a juvenile fish enters an anemone, the resident fish allow it to stay there only if there is room. If there is no room, the young fish is expelled and returns to the dangers of an exposed existence on the reef.

This life cycle, with its expulsions, hierarchies, and sex changes, is certainly as colorful as the fish that live it. But why do clownfish engage in these complicated machinations just to produce more clownfish? Organisms have arrived at a vast array of solutions to the basic problem of reproduction. As we will see, these solutions are often well suited for meeting the challenges and constraints of the environment where a species lives.

Introduction

Human history is a record of past events. Your personal history might consist of a series of details about the course of your life: your birth weight, when you started walking and talking, your adult height, and other relevant information about your development. Similarly, an individual organism's **life history** is a record of major events related to its growth, development, reproduction, and survival.

In this chapter, we'll discuss traits that characterize the life history of a organism, including age and size at sexual maturity, amount and timing of reproduction, and survival and mortality rates. As we'll see, the timing and nature of life history traits, and therefore the life history itself, are products of adaptation to the environment in which the organism lives. We'll also consider how biologists analyze life history patterns in order to understand the trade-offs, constraints, and selection pressures imposed on different stages of an organism's life cycle.

CONCEPT 7.1 Life history patterns vary within and among species.

Life History Diversity

The study of life histories is concerned with categorizing variation in life history traits and analyzing the causes of that variation. In order to understand such analyses, we must first examine some of the broad life history patterns found within and among species.

Individuals within species differ in their life histories

Individual differences in life history traits are ubiquitous. Think about your own life experiences and those of your family and friends. Some members of your social group reached developmental milestones such as puberty earlier or later than others. Different women may have different numbers of children with different age gaps between them. Despite this variation, it is possible to make some generalizations about life histories in *Homo sapiens*: for example, women typically have one baby at a time, reproduction usually occurs between the ages of 15 and 45, and so on. Similar generalizations can be made for other species. The **life history strategy** of a species is the overall pattern in

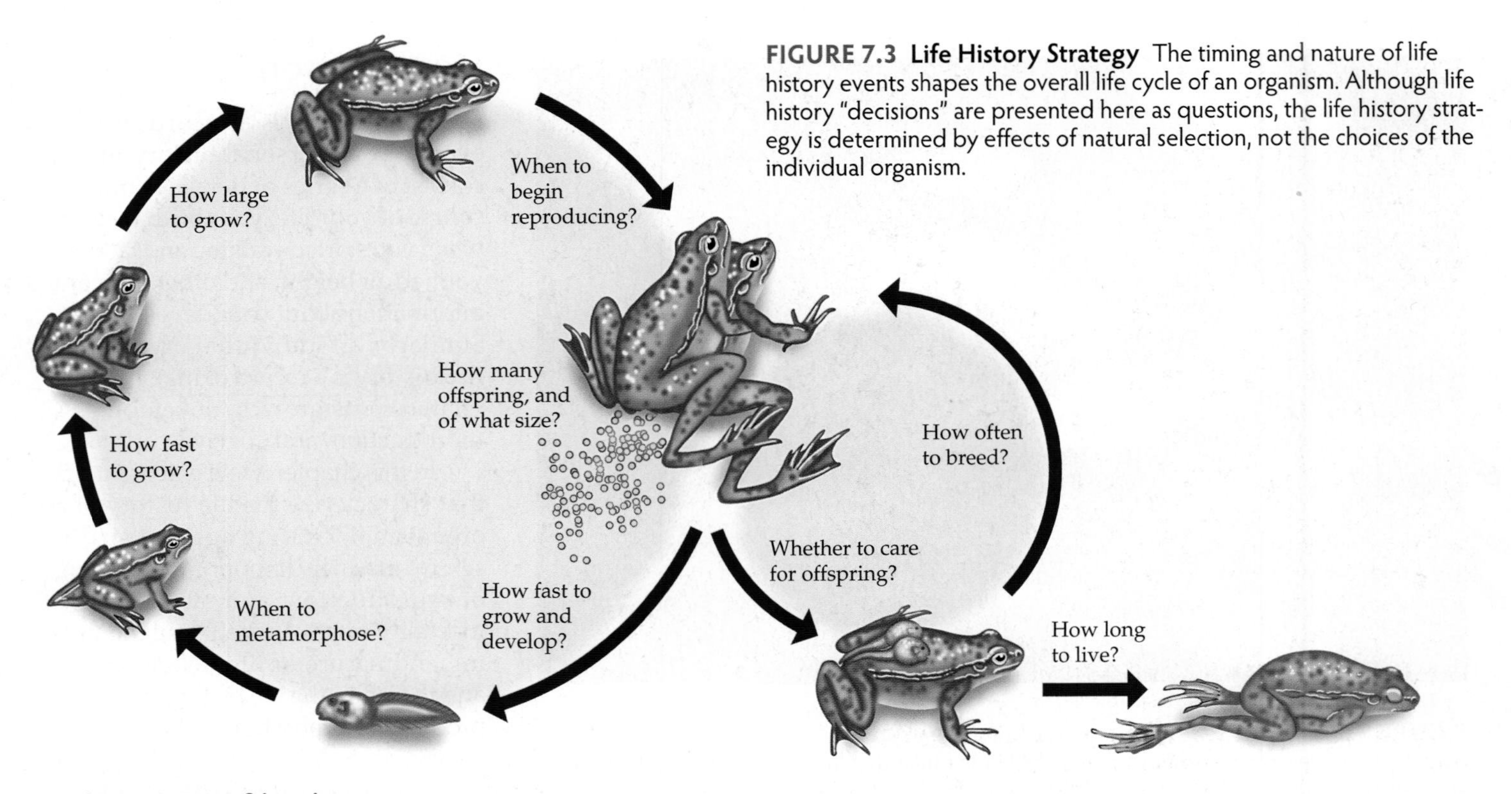

FIGURE 7.3 Life History Strategy The timing and nature of life history events shapes the overall life cycle of an organism. Although life history "decisions" are presented here as questions, the life history strategy is determined by effects of natural selection, not the choices of the individual organism.

Life History Strategy

the timing and nature of life history events averaged across all the individuals in the species (**Figure 7.3**).

The life history strategy is shaped by the way the organism divides its energy and resources between growth, reproduction, and survival. Within a species, individuals often differ in how they divide their energy and resources among these activities. Such differences may result from genetic variation, from differences in environmental conditions, or from a combination of both.

Genetic differences Some life history variation within species is determined genetically. Genetically influenced traits can often be recognized as those that are more similar within families than between them. Again, these kinds of traits are familiar in humans: for example, siblings often are similar in appearance and reach similar adult heights and weights. The same is true in other organisms. For example, in bluegrass (*Poa annua*), life history traits such as age at first reproduction, growth rate, and number of flowers produced are similar among sibling plants (Law et al. 1977). As with any other trait, heritable variation in life history traits is the raw material on which natural selection acts. Selection favors individuals whose life history traits result in their having a better chance of surviving and reproducing than do individuals with other life history traits.

Much of life history analysis is concerned with explaining how and why life history patterns have evolved to their present states. Ecologists sometimes describe life histories as *optimal* in that they are adapted to maximize **fitness** (the genetic contribution of an organism's descendants to future generations). However, no organism has a perfect life history—that is, one that results in the unlimited production of descendants. Instead, all organisms face *constraints* that prevent the evolution of a perfect life history. As we'll see in the discussion under Concept 7.3, these constraints often involve ecological trade-offs in which an increase in the performance of one function (such as reproduction) can reduce the performance of another (such as growth or survival). Thus, although life histories often serve organisms well in the environments in which they have evolved, they are optimal only in the sense of maximizing fitness subject to constraints.

Environmental differences A single genotype may produce different phenotypes under different environmental conditions, a phenomenon known as **phenotypic plasticity**. Almost every trait shows some degree of plasticity, and life history traits are no exception. For example, most plants and animals grow at different rates depending on temperature. They do so because development typically speeds up as the temperature warms, then slows down again due to heat stress as the temperature approaches the organism's upper lethal temperature (see the discussion under Concept 4.2 for more discussion of the effects of temperature).

Changes in life history traits often translate into changes in adult morphology. Slower growth under cooler

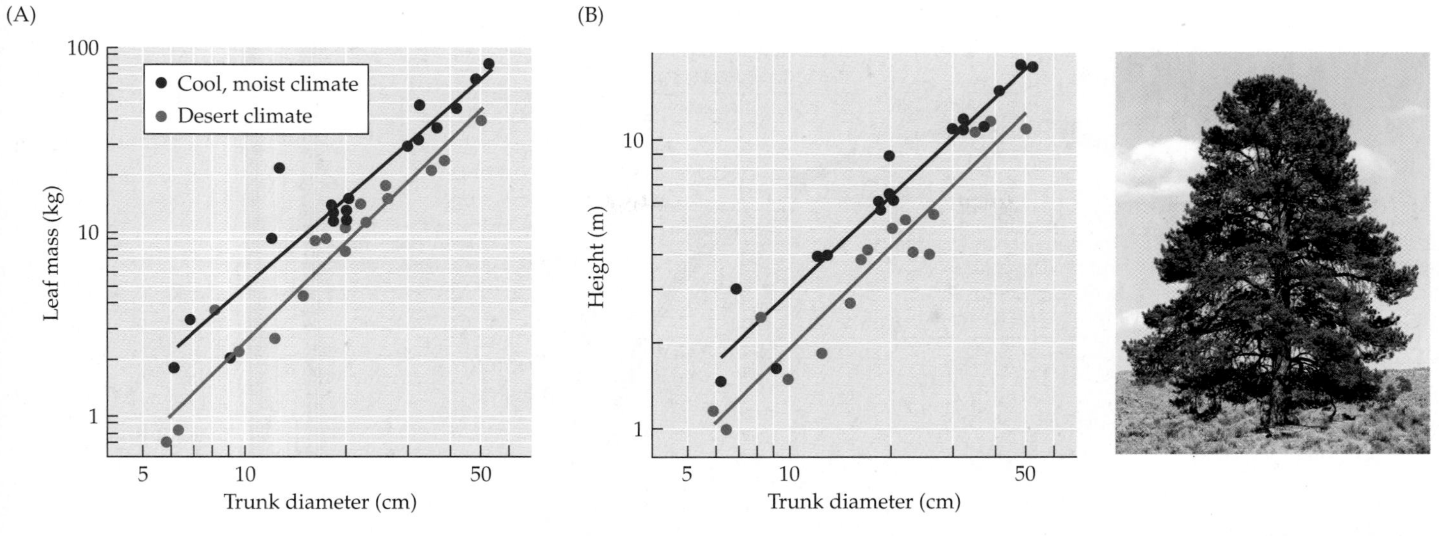

FIGURE 7.4 Plasticity of Growth Form in Ponderosa Pines (A) Ponderosa pine trees in cool, moist climates allocate more resources to leaf production than do trees in desert climates. (B) Desert trees are shorter than those grown in cooler climates, but for a given height, they have thicker trunks. (After Callaway et al. 1994.)

Q Use the solid (regression) lines in (B) to estimate the trunk diameter of a tree that is 5 m tall and grows in a cool, moist climate versus the trunk diameter of a tree of the same height that grows in a desert climate.

conditions, for example, may lead to a smaller adult size or to differences in adult shape. Callaway and colleagues (1994) showed that ponderosa pine (*Pinus ponderosa*) trees grown in cool, moist climates allocate more biomass to leaf growth relative to sapwood production than do those in warmer desert climates ("sapwood" refers to newly formed layers of wood that function in water transport). *Allocation* describes the relative amounts of energy or resources that an organism devotes to different functions. The result of allocation differences in ponderosa pines is that trees grown in different environments differ in their adult shape and size. Desert trees are shorter and squatter, with fewer branches. They also have lower photosynthetic rates and consume less CO_2 because they have fewer leaves (**Figure 7.4**).

Phenotypic plasticity that responds to temperature variation often produces a continuous range of sizes. In other types of phenotypic plasticity, a single genotype produces discrete types, or **morphs**, with few or no intermediate forms. Populations of spadefoot toad (*Spea multiplicata*) tadpoles in Arizona ponds contain two morphs: omnivore morphs, which feed on detritus and algae, and larger carnivore morphs, which feed on fairy shrimp and on other tadpoles (**Figure 7.5**). The differing body shapes of omnivores and carnivores result from differences in the relative growth rates of different body parts: carnivores have bigger mouths and stronger jaw muscles because of accelerated growth in those areas. Pfennig (1992) showed that omnivore tadpoles can turn into carnivores when fed on shrimp and tadpoles, and field studies show that the proportion of omnivore and carnivore morphs is affected by food supply. Carnivore tadpoles grow faster and are more likely to metamorphose before the ponds where they live dry up; thus, the rapidly growing carnivores

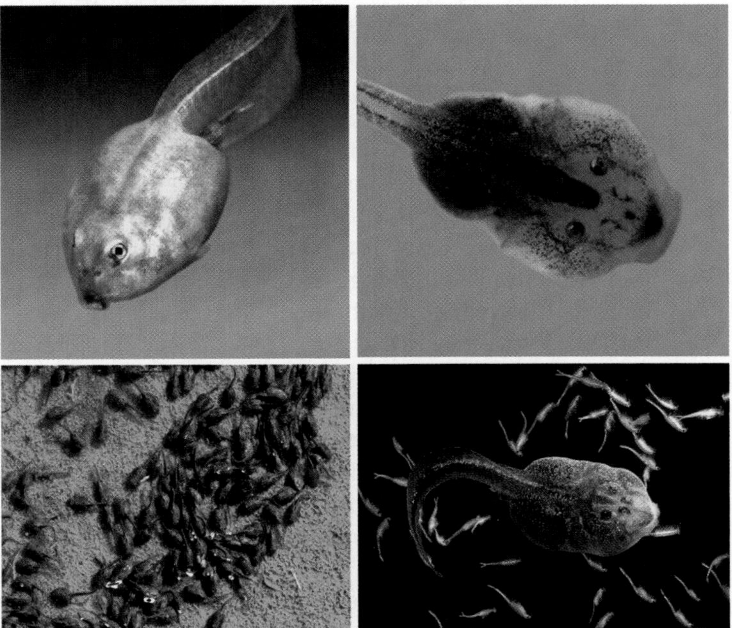

FIGURE 7.5 Phenotypic Plasticity in Spadefoot Toad Tadpoles Spadefoot toad (*Spea multiplicata*) tadpoles can develop into small-headed omnivores (A) or large-headed carnivores (B), depending on the food they consume early in development. Later in development, omnivores and carnivores feed on different food sources that are located in different portions of their habitat.

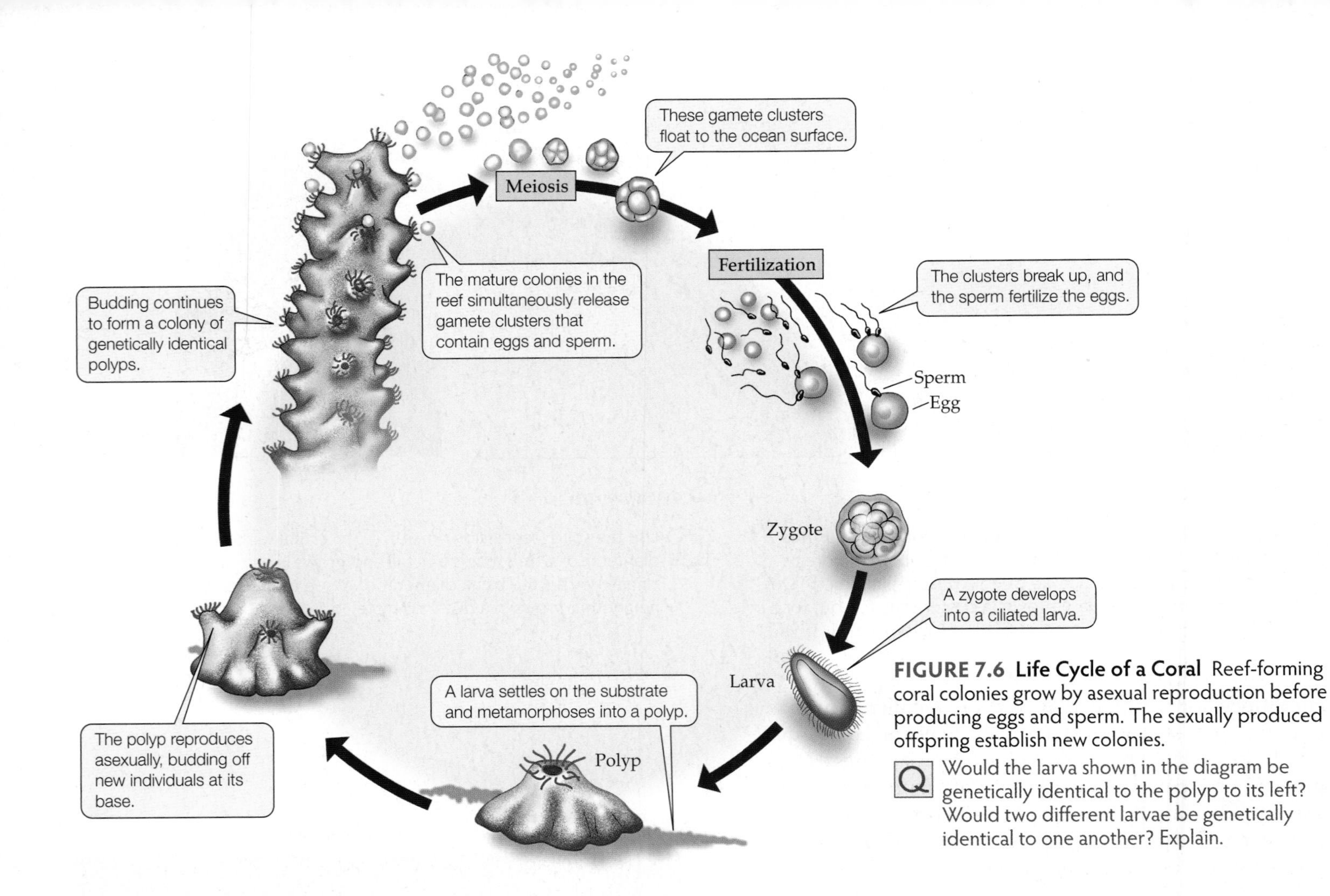

FIGURE 7.6 Life Cycle of a Coral Reef-forming coral colonies grow by asexual reproduction before producing eggs and sperm. The sexually produced offspring establish new colonies.

Q Would the larva shown in the diagram be genetically identical to the polyp to its left? Would two different larvae be genetically identical to one another? Explain.

are favored in ephemeral ponds. The more slowly growing omnivores are favored in ponds that persist longer because they metamorphose in better condition and thus have better chances of survival as juvenile toads.

The differences in morphology among ponderosa pines and between separate morphs of spadefoot toads both result from how changes in the environment affect the relative growth rates of different body parts. In the pines, the relative growth rates of leaves and sapwood determine body shape, while in the toads, the relative growth rates of the jaw and the rest of the body determine whether the tadpole is a carnivore or an omnivore. These patterns are examples of **allometry**, or differential growth of body parts that results in a change in shape or proportion with size. Allometry is a very common mechanism of variation within and among species.

When thinking about examples such as the omnivore and carnivore morphs of the spadefoot toad, it is tempting to assume that phenotypic plasticity is adaptive—that the ability to produce different phenotypes in response to changing environmental conditions increases the fitness of individuals. While that is often the case, adaptation must be demonstrated rather than assumed. For example, it may be adaptive for ponderosa pines to be stockier and have fewer leaves in hot, dry climates because these features could help reduce water loss. However, adaptation would have to be documented by measuring and comparing the survival and reproductive rates of stockier and taller trees in the desert environment. In some instances, phenotypic plasticity may be a simple physiological response, not an adaptive response shaped by natural selection. For example, as mentioned above, changes in growth rate due to temperature variation may occur because chemical reactions are slower at lower temperatures, and thus metabolism and growth are necessarily slower.

Mode of reproduction is a basic life history trait

At the most basic level, evolutionary success is determined by successful reproduction. Despite this universal reality, organisms have evolved vastly different mechanisms for reproducing—from simple asexual splitting to complex mating rituals and intricate pollination systems.

Asexual reproduction The first organisms to evolve on Earth reproduced asexually by *binary fission* ("dividing in half"). The sexual reproductive processes of meiosis, recombination, and fertilization arose later. Today, all prokaryotes and many protists reproduce asexually. While sexual reproduction is the norm in multicellular organisms, many can also reproduce asexually. For example, after they are initiated by a (sexually produced) founding polyp, coral colonies grow by asexual reproduction (**Figure 7.6**). Each individual polyp

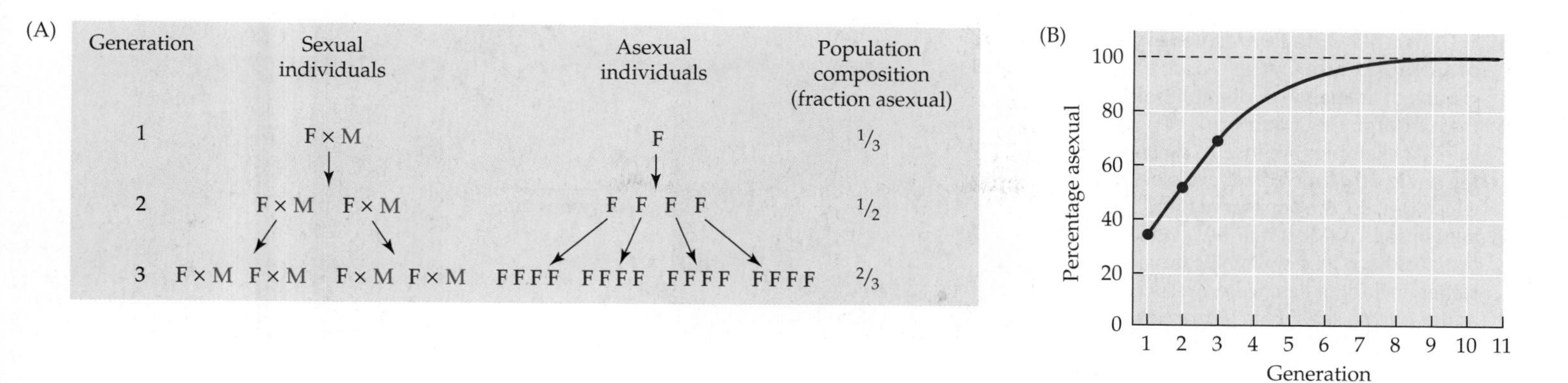

FIGURE 7.7 The Cost of Sex One cost of sex is referred to as the *cost of males*. Imagine a population in which there are sexual and asexual individuals. Assume that each sexual or asexual female can produce four offspring per generation, but half of the offspring produced by the sexual females are male and must pair with a female to produce offspring. Under these conditions, the asexual individuals will (A) increase in number more rapidly and (B) constitute nearly 100% of the population in less than ten generations.

Q In Generation 2 there are 4 sexual and 4 asexual individuals. How many sexual and asexual individuals are there in Generation 3? How many of each will there be in Generation 4? Explain your results in terms of the cost of males.

in a colony is produced when a multicellular bud splits off from a parent polyp to form a new polyp; as a result, each polyp is a genetically identical copy, or *clone*, of the founding polyp. Once the colony has grown to a certain size and conditions are right, the polyps reproduce sexually, producing offspring that develop into polyps that start their own new colonies of clones.

Sexual reproduction and anisogamy Sex has some clear benefits, including recombination, which promotes genetic variation and hence may increase the capacity of populations to evolve in response to environmental challenges such as drought or disease. Sex also has some disadvantages. Because meiosis produces haploid gametes that contain half the genetic content of the parent, a sexually reproducing organism can transmit only half of its genetic material to each offspring, whereas asexual reproduction allows transmission of the entire genome. In addition, the growth rate of sexually reproducing populations is only half that of asexually reproducing ones, all else being equal (**Figure 7.7**).

Sexual reproduction originated in single-celled protists. Protists such as the green alga *Chlamydomonas reinhardtii* (**Figure 7.8A**) have two different *mating types*, analogous to males and females, but they both produce gametes that are the same size. The production of equal-sized gametes is called **isogamy**. In most multicellular organisms, the two types of gametes are different sizes—a condition called **anisogamy**. Typically, the eggs are much larger than the sperm and contain more cellular and nutritional provisions for the developing embryo. The sperm are small and may be motile (**Figure 7.8B**). As we will see in this chapter's Case Study Revisited on p. 173, differences between the sexes in gamete size can influence other reproductive characteristics, such as the timing of sex changes.

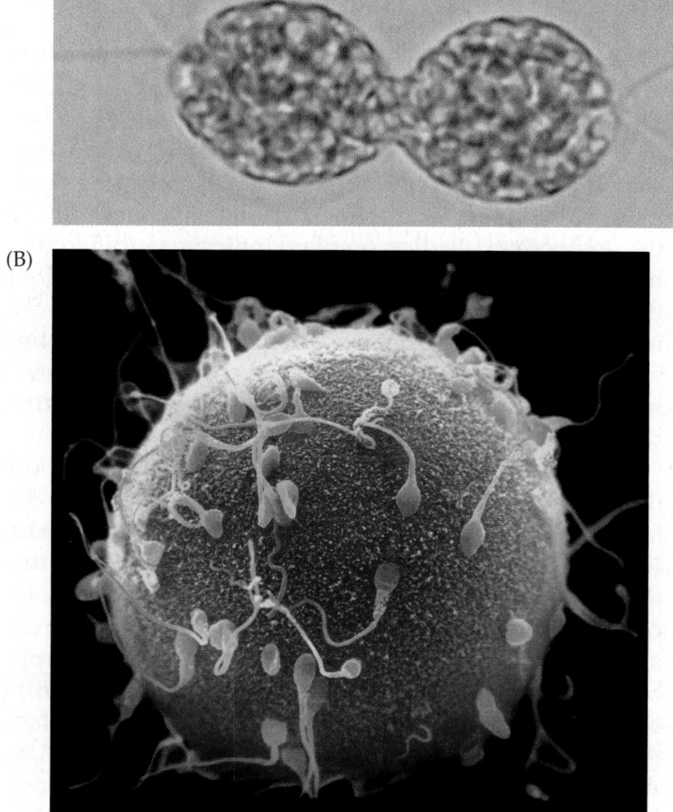

FIGURE 7.8 Isogamy and Anisogamy (A) An isogamous species: Two gametes of the single-celled alga *Chlamydomonas* fusing. (B) An anisogamous species: Fertilization of a human egg, showing the difference in size between egg and sperm.

FIGURE 7.9 The Pervasiveness of Complex Life Cycles Most groups of animals include members that undergo metamorphosis. (A) Familiar examples are insects such as the antlion (*Myrmeleon libelluloides*), which develops from a larva that lives in soil. (B) Most marine invertebrates have free-swimming larval stages, including echinoderms such as the purple sea urchin (*Paracentrotus lividus*).

Life cycles are often complex

The small, early stages of many animal life cycles look and behave completely differently from adult stages. They frequently eat different foods and prefer different habitats. For example, coral reef fishes such as the damselfish *Chromis atripectoralis* start life as hatchlings only a few millimeters long. The hatchlings live and grow in the open ocean, feeding on planktonic algae. When they have grown to about a centimeter in length, they return to the reef and begin to eat larger food items. This life cycle may have evolved in response to high levels of predation on young fish that stay on the reef; young fish that spend more time growing in the open ocean may have better chances of survival.

As corals (see Figure 7.6) and coral reef fishes both demonstrate, life cycles can involve stages that have different body forms or live in different habitats. A **complex life cycle** is a life cycle in which there are at least two distinct stages that differ in their habitat, physiology, or morphology. In many cases, the transitions between stages in complex life cycles are abrupt. For example, many organisms undergo **metamorphosis**, an abrupt transition in form from the larval to the juvenile stage that is sometimes accompanied by a change in habitat. As we will see in the discussion under Concept 7.4, complex life cycles and metamorphosis often result when offspring and parents are subjected to very different selection pressures.

Because most vertebrates have simple life cycles that lack an abrupt transition between habitats or forms, we humans tend to think of metamorphosis as an exotic and strange process. However, complex life cycles and metamorphosis can be found even among vertebrates, including some fishes and most amphibians. Most marine invertebrates produce microscopic larvae that swim in the open ocean before settling to the bottom at metamorphosis. Many insects also undergo metamorphosis—from caterpillars to moths, grubs to beetles, maggots to flies, and aquatic larvae to dragonflies and mayflies. In fact, Werner (1988) calculated that of the 33 phyla of animals recognized at that time, 25 contained at least some subgroups that have complex life cycles. He also noted that about 80% of all animal species undergo metamorphosis at some time in their life cycle (**Figure 7.9**).

Over the course of evolution, complex life cycles have been lost in some species that are members of groups in which such cycles are considered the ancestral condition. The resulting simple life cycles are sometimes referred to as **direct development** because development from fertilized egg to juvenile occurs within the egg prior to hatching and no free-living larval stage occurs. For example, most species in one group of salamanders, the plethodontids, lack the gilled aquatic larval stage that is typical of salamanders. Instead, they lay their eggs on land, where they hatch directly into small terrestrial juveniles.

Many parasites have evolved intricate and complex life cycles with one or more specialized stages for each host that they inhabit. For example, the worm parasite *Ribeiroia* has three specialized stages (see Figure 1.3). In *Ribeiroia* and other parasites, these stages are specialized

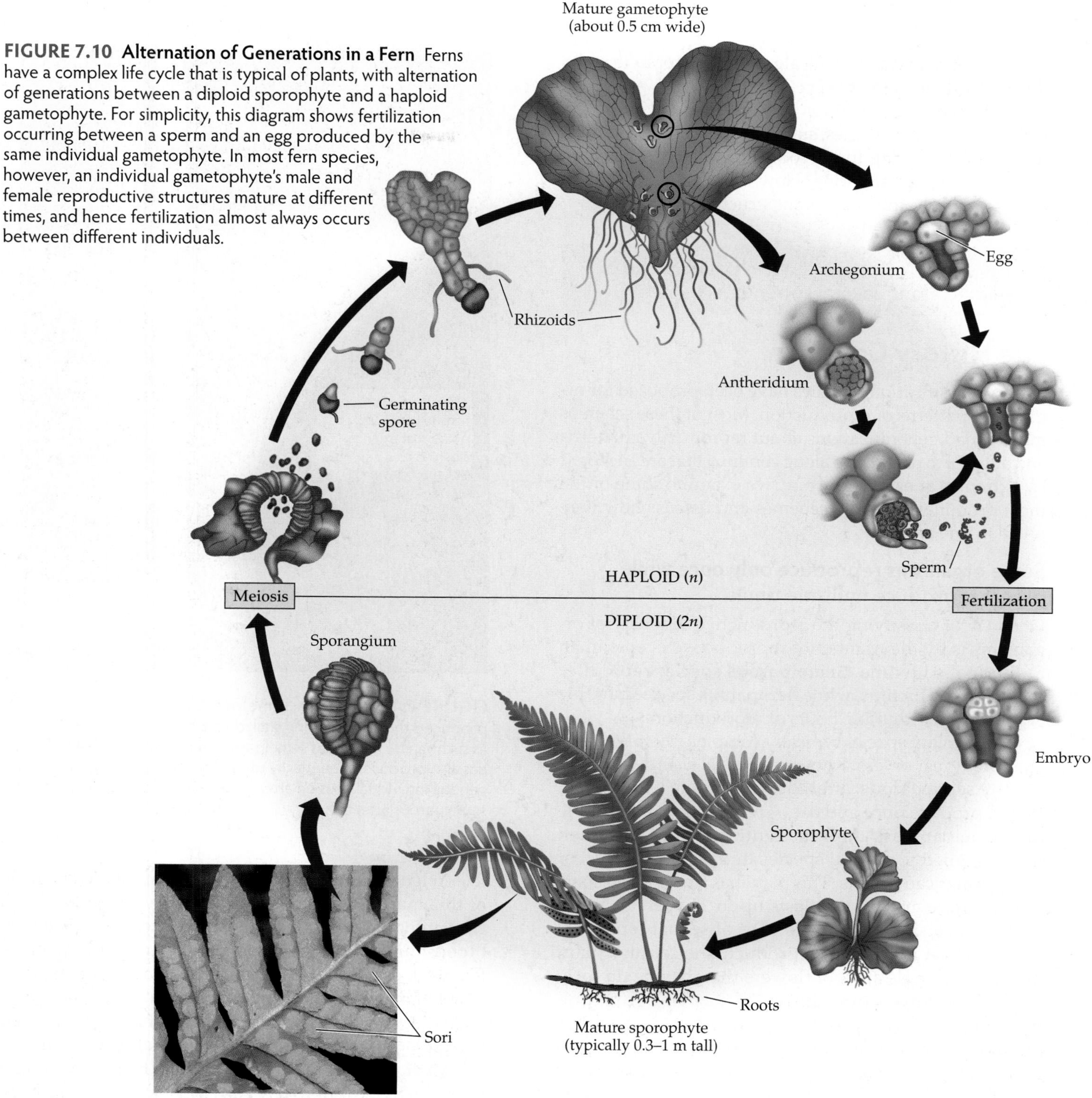

FIGURE 7.10 Alternation of Generations in a Fern Ferns have a complex life cycle that is typical of plants, with alternation of generations between a diploid sporophyte and a haploid gametophyte. For simplicity, this diagram shows fertilization occurring between a sperm and an egg produced by the same individual gametophyte. In most fern species, however, an individual gametophyte's male and female reproductive structures mature at different times, and hence fertilization almost always occurs between different individuals.

to perform essential functions such as asexual reproduction, sexual reproduction, and colonization of new hosts.

Complex life cycles also occur in many types of algae and plants, reaching some of their most elaborate forms in these groups. Some algae and all plants have complex life cycles in which a multicellular diploid *sporophyte* alternates with a multicellular haploid *gametophyte*. The sporophyte produces haploid spores that disperse and grow into gametophytes, and the gametophyte produces haploid gametes that combine in fertilization to form zygotes that grow into sporophytes (**Figure 7.10**). This type of life cycle, called **alternation of generations**, has been elaborated on in different plant and algal groups. In mosses and a few other plant groups, the gametophyte is larger, but in most plants and some algae, the sporophyte is the dominant stage of the life cycle.

As we've seen, organisms vary greatly in key aspects of their life history strategies, such as when they reproduce, how many offspring they produce, and how much care is allotted to each offspring. How can we organize these diverse patterns into a coherent structure?

CONCEPT 7.2 Reproductive patterns can be classified along several continua.

Life History Continua

Several classification schemes have been proposed for organizing patterns of reproduction. Most of these schemes make broad generalizations about reproductive patterns and attempt to place them along continua that are anchored by extremes at each end. Here, we examine some of the most prominent of these schemes and discuss how they relate to one another.

Some organisms reproduce only once while others reproduce multiple times

One way of classifying the reproductive diversity of organisms is by the number of reproductive events in an individual's lifetime. **Semelparous** species reproduce only once in a lifetime, while **iteroparous** species have the capacity to for multiple bouts of reproduction.

Many plant species typically complete their life cycle in a single year or less. Known as *annual plants*, such species are semelparous: after they germinate from a seed, they reproduce once and die. A more complex example of a semelparous plant is the century plant (a common name applied to several species in the genus *Agave*) of North American deserts. This plant has a prolonged stage of vegetative growth that lasts up to 25 years before it undergoes a single intensive bout of sexual reproduction. When it is ready to reproduce, the century plant produces a single stalk of flowers that may be up to 20 feet tall and towers over the rest of the plant. After pollination, the flowers produce clumps of seeds that drop off and take root around the parent plant. The portion of the plant that produced the tall stalk of flowers dies after this event; hence, it is semelparous. At the genetic level, however, a century plant individual does not die when it flowers because the plant also reproduces asexually, producing genetically identical clones that surround the original plant (**Figure 7.11**). In this sense, century plants are not semelparous after all—the clones survive after the flowering event and will eventually flower themselves.

FIGURE 7.11 Agave: A Semelparous Plant? The *Agave* that produced the tall flowering stalk dies shortly after it flowers, and as such can be viewed as semelparous. But the *Agave* that flowered has also produced genetically identical clonal offspring. Thus, the genetic individual lives on after it flowers and in that sense is not semelparous after all.

A striking example of a semelparous animal is the giant Pacific octopus (*Enteroctopus dofleini*), which can reach 25 feet in length and weigh nearly 400 pounds. The female of this marine invertebrate species lays a single clutch containing tens of thousands of fertilized eggs. She then broods the eggs for up to 6 months. During this time, the female does not feed at all, and she is a constant presence over her offspring, cleaning and ventilating them. The female dies shortly after the eggs hatch, having exhausted herself in this intense period of parental investment.

Most organisms do not invest so heavily in single reproductive events. Iteroparous organisms have multiple bouts of reproduction over the course of a lifetime. Examples of iteroparous plants are long-lived trees such as pines and spruces. Among animals, most large mammals are iteroparous. Of course, iteroparity can take a variety of forms, from plants that flower twice in a season and then die to trees that reproduce every year for centuries. We turn now to the first of several proposals that attempt to explain the wide variation in how organisms reproduce.

Live fast and die young or slow and steady wins the race?

One of the best-known schemes for classifying reproductive diversity was also one of the first proposed. In 1967, Robert MacArthur and Edward O. Wilson coined the terms *r*-selection and *K*-selection to describe two ends of a continuum of reproductive patterns. The *r* in the term *r*-selection refers to the *intrinsic rate of increase* of a population, a measure of how rapidly a population can grow. The term ***r*-selection** refers to selection for high population growth rates. This type of selection can occur in environments where population density is low—for example, in recently disturbed habitats that are being recolonized. In this type of habitat, genotypes that can grow and reproduce rapidly will be favored over those that cannot. In contrast, ***K*-selection** refers to selection for slower rates of increase, which occurs in populations that are at or approaching *K*, the carrying capacity or stable population size for the environment in which they live (see Chapter 9 for a more in-depth discussion of both *r* and *K*). *K*-selection occurs under crowded conditions, where genotypes that can efficiently convert food into offspring are favored. By definition, *K*-selected populations do not have high population growth rates because they are already near the carrying capacity for their environment and competition for resources can be intense.

One way to think of the *r*–*K* continuum is as a spectrum of population growth rates, from fast to slow. Organisms at the *r*-selected end of the continuum are often small and have short life spans, rapid development, early maturation, low parental investment, and high rates of reproduction. Examples of this "live fast, die young" end of the continuum include most insects, small short-lived vertebrates such as mice, and weedy plant species. In contrast, *K*-selected species tend to be long-lived, develop slowly, delay maturation, invest heavily in each offspring, and have low rates of reproduction. Examples of this "slow and steady" end of the continuum include large mammals such as elephants and whales, reptiles such as tortoises and crocodiles, and long-lived plant species such as oak and maple trees.

Like most classification schemes, the *r*–*K* continuum tends to emphasize the extremes. Most life histories are intermediate between these extremes, however, and hence the *r*–*K* approach is not informative in some situations. The distinction between *r*-selection and *K*-selection is perhaps most useful in comparing life histories in closely related species or species living in similar environments. For example, Braby (2002) compared three species of Australian butterflies in the genus *Mycalesis*. The species that occurs in the driest, least predictable habitats shows the most *r*-selected characteristics, including rapid development, early reproduction, production of many small eggs, and rapid population growth. In contrast, the two species found in more predictable, wet forest habitats have more *K*-selected characteristics.

Plant life histories can be classified based on habitat characteristics

In the late 1970s, Philip Grime (1977) developed a classification system specifically for plant life histories. The success of a plant species in a given habitat, he argued, is limited by two factors: stress and disturbance. Grime defined **stress** broadly as any external abiotic factor that limits growth. Under this definition, examples of stress include extreme temperatures, shading, low nutrient levels, water shortages, and any other characteristics of the abiotic environment that reduce vegetative growth. He also defined *disturbance* broadly as any process that destroys plant biomass; under Grime's definition, disturbance can result from biotic sources such as outbreaks of herbivorous insects and abiotic sources such as fire.

If we consider that in a given habitat, stress and disturbance may each be either high or low, then there are four possible habitat types: high stress–high disturbance, low stress–high disturbance, low stress–low disturbance, and high stress–low disturbance. If we further consider that most habitats with high stress *and* high disturbance will not be suitable for plant species, then there are three main habitat types to which plants may adapt. Grime developed a model for understanding the life history patterns that correspond to these three habitat types: competitive plants (low stress–low disturbance), ruderal plants (low stress–high disturbance, and stress-tolerant plants (high stress–low disturbance) (**Figure 7.12**).

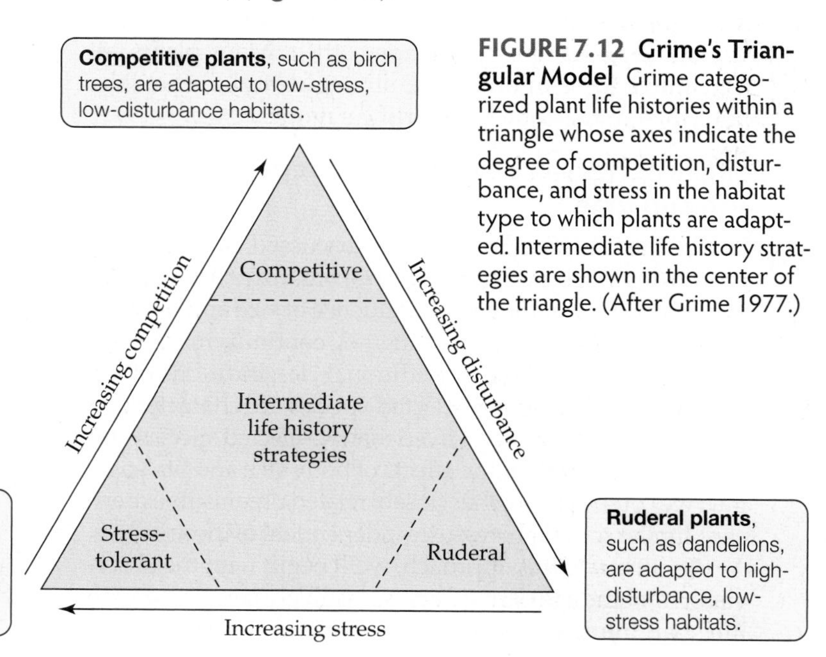

FIGURE 7.12 Grime's Triangular Model Grime categorized plant life histories within a triangle whose axes indicate the degree of competition, disturbance, and stress in the habitat type to which plants are adapted. Intermediate life history strategies are shown in the center of the triangle. (After Grime 1977.)

Grime defined competition between plants in a very specific manner as "the tendency of neighboring plants to utilize the same quantum of light, ion of a mineral nutrient, molecule of water, or volume of space." Under conditions of low stress and low disturbance, **competitive plants** that are superior in their ability to acquire light, minerals, water, and space should have a selective advantage.

Grime classified plants that are adapted to habitats with high levels of disturbance and low levels of stress as **ruderals**. The ruderal strategy generally includes short life spans, rapid growth rates, heavy investment in seed production, and seeds that can survive in the ground for long periods until conditions are right for rapid germination and growth. Ruderal species are often called "weedy" species and are adapted for brief periods of intense exploitation of favorable habitats after disturbance has removed competitors.

Finally, under conditions in which stress (in any form) is high and disturbance is low, stress-tolerant plants become ecologically dominant. Although stressful conditions may vary widely across habitats, Grime identified several features of **stress-tolerant plants**, including but not limited to slow growth rates, evergreen foliage, slow rates of water and nutrient use, low palatability to herbivores, and an ability to respond effectively to temporarily favorable conditions. Habitats favoring stress-tolerant plants might include places where water or nutrients are scarce or temperature conditions are extreme.

To summarize, the triangular model put forward by Grime posits that natural selection has resulted in three distinct yet very broad categories of life history strategies in plants. Although Grime focused on describing these three extreme strategies, he also recognized that intermediate strategies are commonly found. Indeed, various combinations of the three extreme strategies yield many possible intermediate strategies, such as competitive ruderals and stress-tolerant competitors, among others. Many plants fall into one of these intermediate strategies and may still be described in the context of Grime's model.

Life histories can be classified independent of size and time

Unlike the classification schemes discussed above, an approach described by Charnov (1993) organizes life histories in a manner that removes the influence of size and time. As we saw in our discussion of the *r*–*K* continuum, size and time play a critical role in traditional classifications of life histories. For example, *r*-selected species are characterized as smaller and more short-lived than *K*-selected species. But if we could control for the effects of body size and life span, then we could ask whether closely related organisms experience similar selection pressures independent of those factors.

To illustrate this approach, we'll begin with the observation that the age of maturity is positively correlated with life span in many species (Charnov and Berrigan 1990).

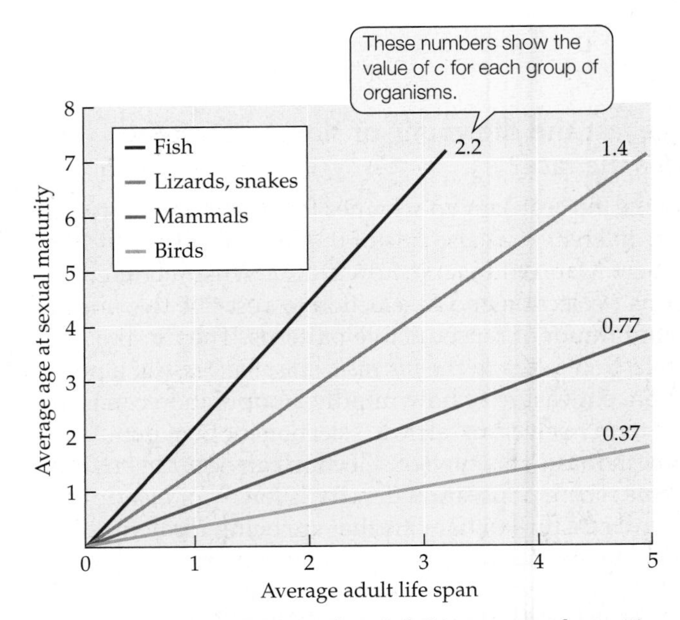

FIGURE 7.13 A Dimensionless Life History Analysis The average age at which females reach sexual maturity versus the average female life span is plotted for different groups of organisms. The slope of each line yields the dimensionless ratio *c*: the average age of maturity divided by the average life span. (After Charnov and Berrigan 1990.)

Q In groups of organisms for which $c > 1$, do most individuals live long enough to reproduce?

Such a correlation is not surprising: species with short life spans must mature in short periods, but the same is not true of species with long life spans; hence, a positive correlation can arise automatically. One way to remove this effect of life span is to divide a species' average age of maturity by its average life span. This division yields a *dimensionless ratio*; that is, a ratio in which the units in the numerator (e.g., age of maturity in *years*) are identical to and hence cancel the units in the denominator (e.g., life span, also in *years*).

By removing the effects of variables such as size or (in our case) time, a dimensionless ratio allows ecologists to compare the life histories of very different organisms. Charnov and Berrigan compiled data for a wide range of bird, mammal, lizard, and fish species. To remove effects of life span, they focused their analyses on the age of maturity:life span dimensionless ratio, which they denoted *c* (**Figure 7.13**). Their analysis revealed that *c* differed between ectothermic (fish, lizards, and snakes) and endothermic (mammals and birds) organisms. For example, if we compare organisms with a given life span, the values of *c* indicate that it takes fish 3–6 times longer to mature than mammals and birds, while it takes lizards and snakes 2–4 times longer. Such results can highlight major differences in the life histories of different groups of organisms, thus helping to make sense of life history variation.

It is probably inappropriate to think of this dimensionless approach as better or worse than classification schemes that incorporate time and size. Indeed, an emphasis on constant or "invariant" dimensionless life history parameters has been questioned by Nee et al. (2005), who argue that life history parameters can appear to be invariant simply as an artifact of the mathematical methods used to estimate them. In any case, there are many ways to organize the vast diversity of life history strategies. The classification scheme that is most useful in any given case will depend on the organisms and questions of interest. For example, the *r–K* continuum has a long history of use in relating life history characteristics to population growth characteristics, whereas Grime's scheme may be most appropriate for life history comparisons between groups of plants. Alternatively, dimensionless analyses may be most helpful when comparing life histories across broad ranges of taxonomy or size.

CONCEPT 7.3 There are trade-offs between life history traits.

Trade-Offs

No organism can invest unlimited amounts of energy in growth and reproduction. *Trade-offs* occur when organisms allocate their limited energy or other resources to one structure or function at the expense of another. As we'll see, trade-offs shape and constrain life history evolution.

There is a trade-off between number and size of offspring

Many organisms show a reproductive trade-off between their investment in each individual offspring and the number of offspring they produce. Investment in offspring includes energy, resources, time, and the loss of chances to engage in alternative activities such as foraging. Typically, organisms with a large investment in each offspring produce small numbers of large offspring.

We can think of the trade-off between the number and size of offspring in the context of a continuum like those presented in the previous section. At one end of the continuum are organisms that produce very large numbers of small offspring, and at the other end are those that produce very small numbers of large offspring. As before, these extremes represent only two ends of a continuous distribution of reproductive patterns.

Lack clutch size A classic example of the trade-off between investment per offspring and number of offspring was first described by David Lack in 1947. Lack asserted that clutch size in birds (the number of eggs per reproductive bout) is limited by the maximum number of young that the parents can raise at one time. If the parents rear fewer than this maximum number, then they will reduce their genetic representation in future generations. If they attempt to rear more than this maximum number, then they risk starvation and death for the entire clutch, in which case they will leave no representation in future generations.

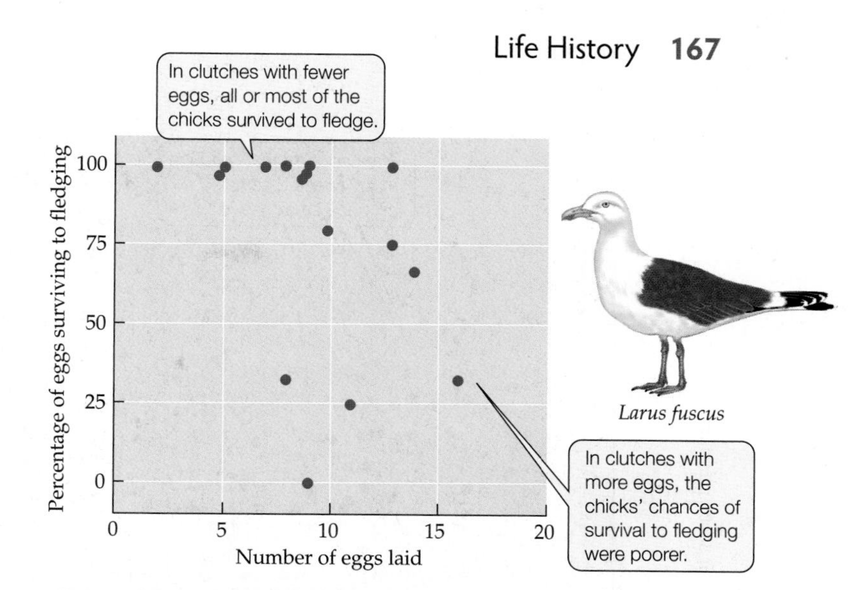

FIGURE 7.14 Clutch Size and Survival Lesser black-backed gulls typically lay three eggs in a clutch. However, when they are manipulated experimentally to produce larger clutches of eggs, their offspring have reduced chances of survival to fledging. (After Nager et al. 2000.)

Lack made careful observations of the breeding biology of bird species from the poles to the tropics. What struck him was that clutch size varied with latitude: at higher latitudes, birds could rear greater numbers of offspring. He hypothesized that the reason for larger clutches at higher latitudes was that such latitudes had longer periods of daylight during the breeding season. These longer days allowed parents more time for foraging, and they could therefore feed greater numbers of offspring.

The term "Lack clutch size" refers to the maximum number of offspring that a parent can successfully raise to maturity. Lack hypothesized that this most productive clutch size should be the clutch size that is most commonly observed in nature. Clutch size can be manipulated in birds by the addition and removal of eggs in order to examine whether there are costs to unusually large clutch sizes. For example, Nager and colleagues (2000) artificially increased the number of eggs in clutches laid by the lesser black-backed gull (*Larus fuscus*). They did this by removing eggs from nests, which stimulated the females to lay more eggs. Nager et al. found that the increased clutch size resulted in a drop in the nutritional quality of later-produced eggs (specifically, these eggs had a lower lipid content). To examine the consequences of this change, an experiment was performed in which eggs from clutches of different sizes were reared singly by foster parents. Chicks from larger clutches had reduced survivorship to fledging (the point at which wing feathers are developed enough for flight), indicating a trade-off between egg number and egg quality (**Figure 7.14**).

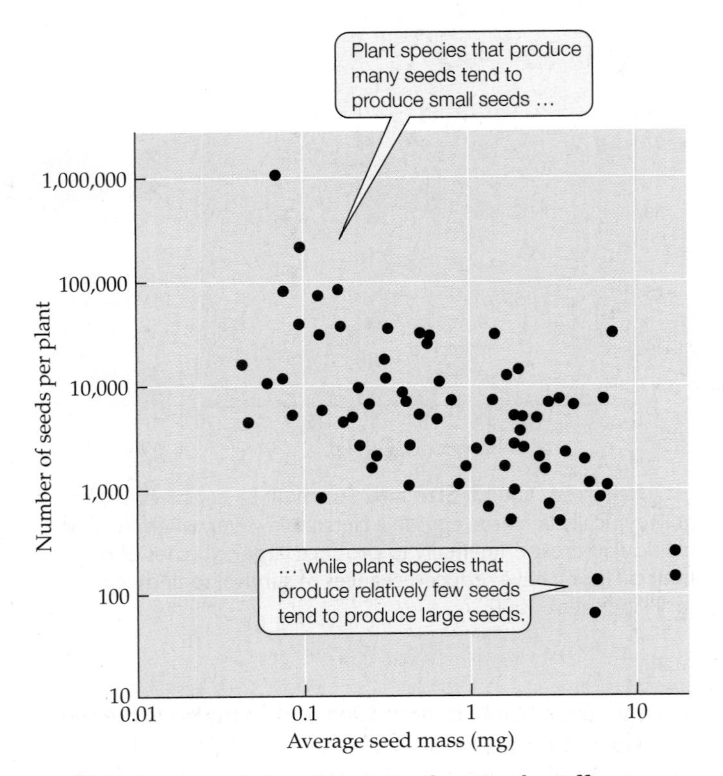

FIGURE 7.15 Seed Size–Seed Number Trade-Offs in Plants (After Stevens 1932.)

Size–number trade-offs in organisms without parental care Across the animal kingdom, parental care like that provided by birds and some other vertebrates is relatively rare. In organisms that do not provide parental care, resources invested in *propagules* (eggs or seeds) are the main measure of reproductive investment. In this case, the size of the propagule is the primary measure of parental investment, and propagule size is traded off against the number of propagules produced in a reproductive bout. In plants, the size of the seeds that a species produces is negatively correlated with the number of seeds it produces (**Figure 7.15**).

In some cases, the size–number trade-off also applies to variation within species. The western fence lizard (*Sceloporus occidentalis*), which is common throughout the coastal mountains of the western United States, does not provide parental care. Barry Sinervo (1990) found that lizard populations farther to the north laid more eggs per clutch (Washington: 12 eggs/clutch vs. California: 7 eggs/clutch), but laid smaller eggs (Washington: 0.40 g vs. California: 0.65 g) (**Figure 7.16**).

In order to determine the consequences of egg size for offspring performance, Sinervo raised fence lizard eggs in the laboratory. He artificially reduced the size of some of the eggs by using a syringe to remove some yolk from them. To control for any possible effects of this method on egg development, he inserted a syringe into some other eggs, but did not remove any yolk. These eggs that had been

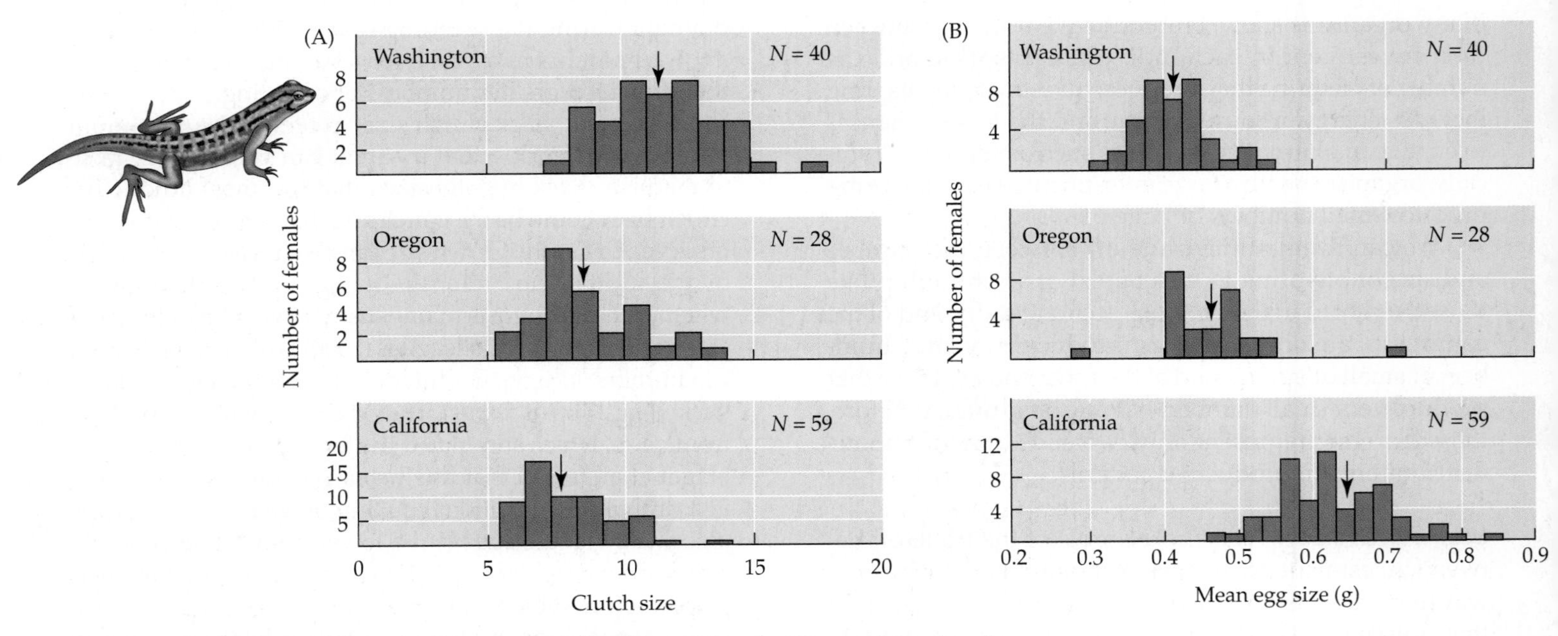

FIGURE 7.16 Egg Size–Egg Number Trade-Off in Fence Lizards Western fence lizards (*Sceloporus occidentalis*) in northern populations produced (A) larger clutches and (B) smaller eggs than those in southern populations. The arrow points to the average for each population. (After Sinervo 1990.)

poked, but not reduced, developed at the same rate as unmanipulated eggs, suggesting that insertion of the syringe was not the cause of differences between unmanipulated and reduced eggs. The reduced eggs developed faster than unmanipulated, larger eggs, but produced smaller hatchlings. These small hatchlings grew faster than their larger siblings, but were not able to sprint as fast to escape from predators. Many of the differences between the lizards hatched from the reduced eggs and from the unmanipulated eggs echoed differences observed between populations with naturally differing egg sizes. Sinervo speculated that the differences between populations in egg and hatchling size may be the result of selection favoring faster sprint speeds in the south, where there may be more predators, or of selection favoring earlier hatching and faster growth in the north, where the growing season is shorter.

There is a trade-off between current and future reproduction

Another important trade-off is that between an organism's current reproductive investment and the likelihood of future successful reproduction. This trade-off highlights the fact that maximizing current reproductive success does not always result in a maximization of lifetime reproductive success.

FIGURE 7.17 Evolving to Reproduce at an Earlier Age Fishing preferentially harvests large cod (because smaller individuals can escape through the mesh of fishing nets), providing an advantage to cod that reach sexual maturity at an earlier age (because such individuals are more likely than others to produce offspring before they are caught). As a result, the traditional measure of maturation (the age at 50% sexual maturity) has dropped in recent decades. (After Olsen et al. 2004.)

Age at sexual maturity The earlier an iteroparous organism reproduces, the more times it can reproduce over the course of its lifetime. However, not all reproductive events are equally successful, and often the number of offspring produced in a reproductive event increases with both the size and the age of the organism.

An example of an increase in reproductive output with age occurs in the Atlantic cod (*Gadus morhua*). Normally, female cod become sexually mature at a length of about 60 cm. By the time they reach 80 cm, their fecundity is approximately 2 million eggs per year. However, by the time they reach 120 cm, their annual reproductive output is about 15 million eggs! This pattern has had important consequences for Atlantic cod populations. Commercial fishing has not only dramatically reduced the size of these populations (see Figure 22.11), it has also selectively removed the older and larger fish. As a result, individuals that are genetically predisposed to reproduce at earlier ages and smaller sizes have been more likely to produce offspring than have other individuals. These selection pressures have produced evolutionary changes in the life history of cod (Olsen et al. 2004): both the age at which cod become sexually mature (**Figure 7.17**) and the size of these fish have declined in recent decades. Furthermore, by removing the larger and older fish, which also have the highest reproductive potential, commercial fishing has inadvertently reduced the quality and quantity of egg production. These evolutionary changes may delay or prevent the recovery of cod populations (Hutchings 2005).

In situations in which reproductive bouts become more productive with age, it may be advantageous for organisms to delay sexual maturation and focus their investment of energy on growth and survival instead. Imagine a fish with a 5-year life span whose reproductive output increases with age such that in year 1 it produces 10 offspring, in year 2 it produces 20 offspring, in year 3, 30 offspring, and so on. But now consider what would happen if the same fish could delay maturation by one year and devote the energy it would have invested in reproduction to growth. Now, by year 2, it may have grown to a large enough size that its reproductive output is 30 offspring. The size gain it made by delaying reproduction also carries over into subsequent years: in year 3 it produces 40 offspring, in year 4, 50 offspring, and in year 5, 60 offspring. In this scenario, if it reproduces in year 1, the fish produces a total of 150 offspring over its 5-year life span, whereas by delaying first reproduction for a year, it produces a total of 180 offspring, despite having one fewer reproductive bout. In this example, we assume that our hypothetical fish has a good chance of surviving for 5 years. If the fish has poor odds of surviving to reach larger sizes, however—because of overfishing or for any other reason—it is unlikely to realize the benefits of delaying reproduction. Under those circumstances, early reproduction would be favored.

Under what conditions should an organism allocate its energy to growth rather than to reproduction? A combination of a long life span, high adult survival rates, and increasing fecundity with increasing body size means that early investment in growth should be paid back by the success of future reproductive bouts. In contrast, if rates of adult survival from one time period to the next are low, then the benefits of future reproductive bouts may never be realized. In that case, investing in current reproduction rather than growth should lead to higher lifetime reproductive success. Of course, most organisms cannot make informed decisions about their likelihood of future reproduction. But natural selection can and does shape the life history of a species—as we've seen in the way fishing pressures have selected for cod that mature earlier and at smaller sizes and that grow more slowly.

Senescence A decline in the physiological function of an organism with age is called **senescence**. The onset of senescence often sets an upper age limit for reproductive events, but the reverse can also be true: reproduction can trigger senescence. For example, most semelparous organisms undergo very rapid senescence and death following a reproductive bout. In some iteroparous organisms, such as humans and elephants, senescence is a much longer stage. Many humans live well beyond their capacity for reproduction and then experience a gradual decline in physiological function with age. Because postreproductive individuals have already made their (genetic) contributions to the next generation, deleterious mutations that are not expressed until a late age cannot be removed from the population by natural selection. However, selection for delayed senescence might occur if postreproductive individuals contribute to the survival of their offspring and their offspring's offspring through parental and grandparental care or by contributing to the success of the social group in other ways. Some social mammals, including humans and elephants, do make such contributions—and those contributions may help to explain the gradual senescence found in those species.

While social interactions may contribute to delayed senescence in some species, many other aspects of an organism's ecology and life history can also be important. For example, David Reznick and colleagues (2004) have shown that wild populations of guppies that live in environments where mortality rates are low exhibit delayed senescence relative to populations from environments with high mortality rates. They have suggested that in populations in which mortality is high due to predation or starvation, guppies may be investing less energy in immune system development and maintenance, which results in higher rates of senescence due to disease.

CONCEPT 7.4 Organisms face different selection pressures at different life cycle stages.

Life Cycle Evolution

In the discussion under Concept 7.1, we saw that an organism's size may vary greatly over the course of its life cycle. This variation leads to differences among life cycle stages in habitat, food preferences, and vulnerability to predation. These differences mean that different morphologies and behaviors are adaptive at different life cycle stages. Differences in selection pressures over the course of the life cycle are responsible for some of the most distinctive patterns in the life histories of organisms.

Small size has benefits and drawbacks

Small, early life cycle stages are particularly vulnerable to predation because there are many predators that are big enough to consume them (although for some predators, small prey may be more difficult to detect). These small stages may also be poor competitors for food, and more susceptible to environmental perturbations that diminish food supply, because they have little storage capacity for nutrients to help them withstand starvation. These vulnerabilities are typically counterbalanced by behavioral, morphological, and physiological adaptations. Furthermore, in some organisms, small, mobile early stages can perform essential functions that are not possible for large adult stages. Here, we examine how organisms protect small life history stages and the important functions those stages can provide.

Parental investment In many organisms, the parents' main investment in their offspring is the provisioning of the eggs or embryos. Animals add yolk to their eggs, which helps their offspring survive and grow through the small, vulnerable stages of life. Female kiwis, for example, produce one very yolky egg at a time; the egg is so large that it makes up 15%–20% of the bird's body size (**Figure 7.18**). During the month that it takes her to make the egg, the female kiwi eats about three times as much as when she is not producing an egg. In many invertebrate groups, species with yolkier eggs develop more rapidly, and require less food during development, than those with less yolky eggs. Another pattern common among invertebrates is investment in energetically expensive egg capsules that protect the offspring during development. Plants provision the fertilized embryos in their seeds with **endosperm**, nutrient-rich material that sustains the developing embryo and often the young seedling. The starchy white part of corn kernels and the milk and meat of coconuts are examples of endosperm.

Another common mechanism for protecting small, vulnerable offspring is parental care. Birds and mammals

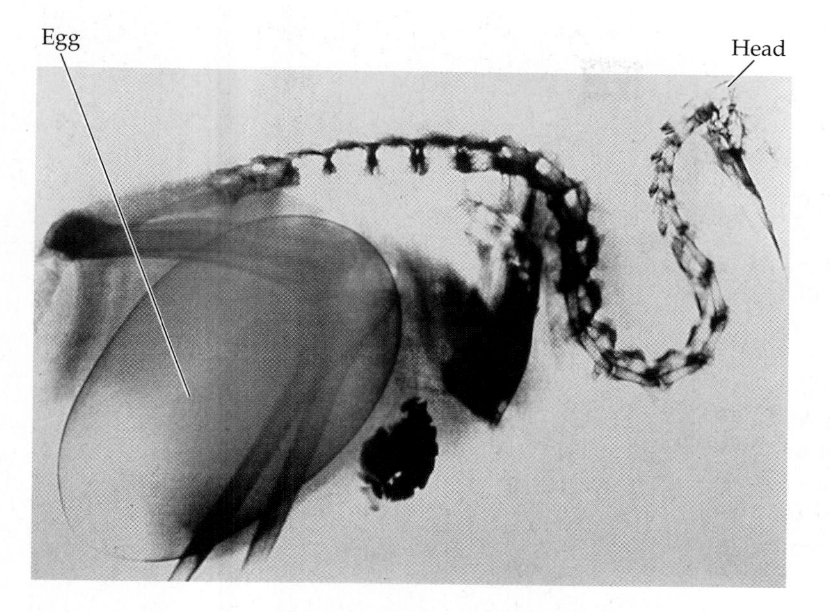

FIGURE 7.18 **Parental Investment in the Kiwi** This X-ray photograph shows the size of a kiwi egg in proportion to the female's body size.

are the most familiar examples of parental care because they invest large amounts of time and energy in protecting and feeding their relatively helpless offspring. Some fishes, reptiles, amphibians, and invertebrates also guard or brood their embryos and hatchlings, protecting them until they are big enough to be less vulnerable.

Dispersal and dormancy Although small offspring are vulnerable to many hazards, they are also well suited for several important functions, including dispersal and dormancy. **Dispersal**—the movement of organisms or propagules from their birthplace—is a key feature in the life history of all organisms. Dispersal provides a number of potential advantages: for example, it can reduce competition among close relatives, and it can allow organisms to reach new areas where they can grow and reproduce. In some circumstances, dispersal can increase the chance of escaping regions of high mortality, as when pathogens and other natural enemies are abundant at the location from which organisms disperse.

Plants, fungi, and many marine invertebrates that are sessile or move very little as adults typically disperse as they reproduce. The small pollen, seeds, spores, or larvae of these organisms can be carried long distances by water or wind or, in the case of pollen and seeds, by animals. In general, smaller propagules disperse more readily and can travel farther in a given amount of time.

The ability of an organism to disperse can have important evolutionary consequences. For example, Hansen (1978) compared the fossil records of extinct marine snails with typical swimming larvae to those of species that had lost their swimming larval stages and developed directly into crawling juveniles. He found that the species without swimming larvae tended to have smaller geographic distributions and were more prone to extinction (**Figure 7.19**). Hansen attributed these differences to differences in dispersal ability. Species with swimming larvae would have been able to move greater distances and hence had more broadly distributed populations that were less vulnerable to random events that could lead to extinction.

Small size also makes eggs and embryos well suited to *dormancy*—a state of suspended growth and development in which an organism can survive unfavorable conditions. Many seeds are capable of long periods of dormancy before germination. Many bacteria, protists, and animals can also undergo various forms of dormancy. The brine shrimp eggs that children purchase as "sea monkeys," for example, are in a dormant state that allows them to survive out of water, often for years. In general, small seeds, eggs, and embryos are better suited to dormancy than large multicellular organisms because they do not have to expend as much metabolic energy to stay alive. However, some animals do enter dormancy in mature stages in response to stressful environmental conditions (as described in the discussion under Concept 4.2).

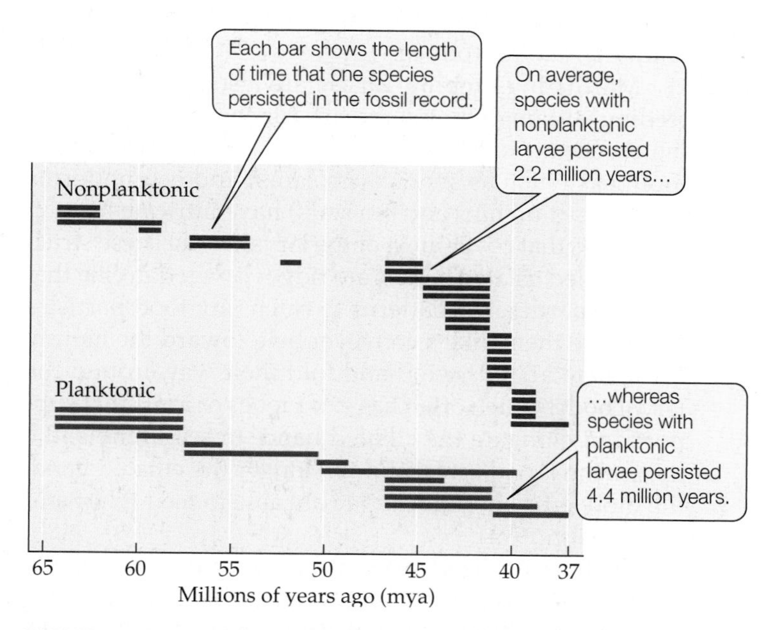

FIGURE 7.19 **Developmental Mode and Species Longevity** Species of marine snails that undergo direct development without a swimming larval stage (nonplanktonic) become extinct more rapidly than those with swimming larvae (planktonic). (After Hansen 1978.)

Complex life cycles may result from stage-specific selection pressures

Organisms with complex life cycles have multiple life stages, each adapted to its habitat and habits. This flexibility may be one of the reasons that complex life cycles are so common in so many groups of organisms. Because separate life history stages can evolve independently in response to size- and habitat-specific selection pressures, complex life cycles can minimize the drawbacks of small, vulnerable early stages.

Larval function and adaptation Functional specialization of particular life stages is a common feature of complex life cycles. Having multiple stages with largely independent morphological features can result in a pairing of particular functions with particular stages. Such a pairing can reduce some of the trade-offs that result from simultaneously optimizing multiple functions.

An example of this type of specialization occurs in many insects with complex life cycles. Such insects spend their entire larval stage in a very small area—sometimes on a single plant. Insect larvae such as caterpillars and grubs are specialized eating and growing machines. They spend almost all of their time taking in food and turning it into body mass, without forming many complex morphological structures other than jaws. Once they have accumulated sufficient mass, these larvae metamorphose into adult butterflies, moths, and beetles, whose main function is often to disperse, find a mate, and reproduce. In extreme cases, such as mayflies, the adults are incapable of feeding and live only the few hours or days it takes them to reproduce.

Marine invertebrate larvae are also specialized for feeding, although they serve this function while dispersing on ocean currents. For example, the larvae of many mollusks (such as snails and clams) and echinoderms (such as sea urchins and sea stars) have intricate feeding structures that cover most of the larval body. These structures, called ciliated bands, are ridges covered in cilia that beat in coordinated patterns to catch tiny food particles and move them, like a conveyer belt, toward the mouth. The ciliated bands wind and fold their way around the larval body, which often has extra lobes or arms that support and elongate the ciliated band. In sea urchins, the longer the larval arms, and the longer the ciliated band, the more efficiently the larvae are able to feed (Hart and Strathmann 1994).

Other specialized larval structures prevent small life cycle stages from becoming food for other organisms. For example, the skeletal structures of sea urchin larvae, which support the larval arms, the head spine of crab larvae (**Figure 7.20**), and the setae or bristles of polychaete worm larvae deter some predators by making the larva a large and uncomfortable mouthful.

FIGURE 7.20 Specialized Defensive Structures in Marine Invertebrate Larvae Crab larvae have defensive head spines that can make them difficult for fish to eat.

Timing of life cycle shifts Most organisms with complex life cycles use different habitats and food resources at different life stages; often such shifts occur abruptly, as in organisms that undergo metamorphosis. However, size-dependent differences in habitat and food preferences are not confined to species in which there are abrupt transitions between life stages. Even when growth and morphological change are gradual, different-sized and different-aged individuals of the same species may have very different ecological roles. We'll use the term *niche shift* to refer to such size- or age-specific changes in an organism's ecological function or habitat. (As we'll see in our discussion under Concept 8.5, an organism's *ecological niche* consists of the physical and biological conditions that the organism needs to grow, survive, and reproduce.)

In many organisms with complex life cycles, an abrupt metamorphosis occurs at the transition between stages, thus minimizing the time the organism spends in vulnerable stages that are intermediate between larva and adult. In theory, there should be an optimal time to undergo metamorphosis, or any niche shift, that maximizes survival over the course of the life cycle. Thus, we might expect a niche shift to occur when the organism reaches a size at which conditions are more favorable for its survival or growth in the adult habitat than in the larval habitat.

Dahlgren and Eggleston (2000) tested this idea for the Nassau grouper (*Epinephelus striatus*), an endangered coral reef fish that spends its juvenile stages in and around large clumps of algae. Smaller juveniles spend their time

FIGURE 7.21 Paedomorphosis in Salamanders The mole salamander (*Ambystoma talpoideum*) can produce both (A) paedomorphic aquatic adults and (B) terrestrial metamorphic adults.

hiding within the algae, whereas larger ones spend their time in rocky habitats near algal clumps. By tethering and enclosing juvenile fish of different sizes in the two habitats, Dahlgren and Eggleston were able to measure mortality and growth rates in each habitat. They found that smaller juveniles were very vulnerable to predation in the rocky habitats, while the larger juveniles were less vulnerable and were able to grow faster in the rocky habitats. Thus, the niche shift in this species appears to be timed to maximize growth and survival, as predicted.

In some cases, the larval habitat may be so favorable for growth and survival that metamorphosis is delayed—or even eliminated altogether. For example, most salamanders have aquatic larvae that metamorphose into terrestrial adults, but some salamanders, such as the mole salamander *Ambystoma talpoideum*, can become sexually mature while retaining gills and remaining in the aquatic habitat (**Figure 7.21**). These aquatic, gilled adults are referred to as **paedomorphic**, which means that they result from a delay of some developmental events (loss of gills, development of lungs) relative to sexual maturation. In the mole salamander, both aquatic paedomorphic adults and terrestrial metamorphic adults can exist in the same population. The frequency of paedomorphosis in these mixed populations seems to depend on factors such as predation, food availability, and competition—all of which influence survival and growth in the aquatic habitat.

A CASE STUDY REVISITED

Nemo Grows Up

Why does a male clownfish that has lost his mate become a female rather than simply finding a new partner? Consider the fact that, as in cod, the number of eggs a clownfish can produce is proportional to its body size. Thus, larger individuals can produce more eggs and presumably have a better chance of having some of their offspring survive. Smaller individuals are more easily able to make sperm cells, which are smaller and take fewer resources to produce. For these reasons, in clownfish and many other animals, females are larger than males.

Changes in sex during the course of the life cycle, called **sequential hermaphroditism**, are found in 18 fish families and in many invertebrate groups (**Figure 7.22**). Researchers have hypothesized that these sex changes should be timed to take advantage of the maximum reproductive potential of different sexes at different sizes, and in some cases they appear to do so. This hypothesis helps to explain sex changes in clownfish and the timing of those changes rela-

FIGURE 7.22 Sequential Hermaphroditism The slipper limpet (*Crepidula fornicata*) is a marine snail that exhibits sequential hermaphroditism. Slipper limpets live in stacks attached to rocks or shells. The largest individuals, at the bottom of the stack, are females, while the smaller upper individuals are males. As the males grow larger, they eventually reach a size (about 3 to 4 cm) at which they can produce more surviving offspring as females than as males; this is the size at which the sex change occurs.

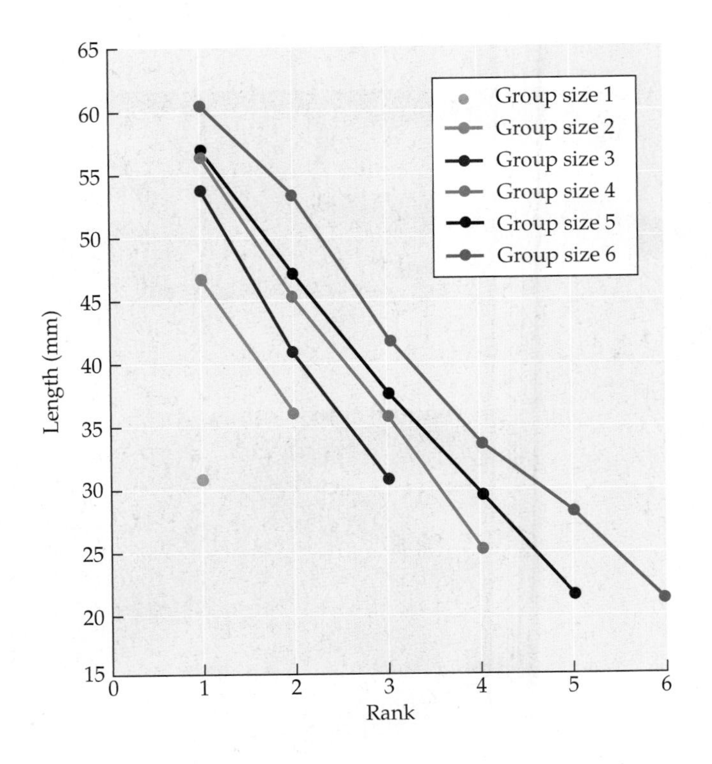

FIGURE 7.23 Clownfish Size Hierarchies Clownfish within an anemone regulate their growth to maintain a hierarchy in which each fish belongs to a distinct size class. Anemones may be home to between one and six fish, and the size of each fish is determined by that fish's rank and the size of the group in which it lives. (After Buston 2003a.)

tive to size, but it leaves unanswered the question of how and why a hierarchy of clownfish is maintained within each anemone.

As a graduate student at Cornell University, Peter Buston (2003a) conducted experiments on a clownfish species (*Amphiprion percula*) that lives on reefs in Papua New Guinea. He found that each clownfish strictly maintains the size hierarchy by remaining smaller than the fish ahead of it in line and bigger than the one behind it (**Figure 7.23**). If a fish grows to be too close in size to one of its anemone-mates, a fight results, which usually ends in the smaller fish being killed or expelled from the anemone. Buston suggested that the clownfish regulate their own growth to prevent such conflicts.

Buston also manipulated clownfish groups by removing the breeding males from anemones and measuring the growth of the remaining individuals. He found that the largest nonbreeder grew only enough to take the place of the breeding male; it avoided growing too big and threatening the female's dominance. Similarly, the next largest nonbreeder grew only enough to take the place of the fish that had become the breeding male, and so on. Thus, clownfish avoid conflict within their social groups by exerting remarkable control over their growth rates and reproductive status.

CONNECTIONS IN NATURE

Territoriality, Competition, and Life History

The physiology of clownfish growth regulation is not understood, but a more pressing ecological and evolutionary question is why the size hierarchy is maintained. What makes small clownfish bide their time as nonbreeders under the dominance of a single breeding female and male? The answer seems to lie in the clownfish's dependence on the protection of anemones for survival.

Clownfish are brightly colored, and they are poor swimmers. Outside the anemone's stinging tentacles, they are easy prey for larger fish on the reef. Thus, expulsion from the anemone is often a death sentence. So the stakes are very high in conflicts between fish within an anemone: the loser will probably die without reproducing. This situation exerts strong selection pressure on the fish to avoid conflicts by regulating their growth. In evolutionary terms, growth regulation mechanisms have evolved because individuals that avoid growing to a size that leads to conflict with other fish have higher survival and reproductive rates (we described this process of adaptive evolution in the discussion under Concept 6.2). Buston (2003b, 2004) demonstrated that remaining in an anemone as a nonbreeder is more advantageous than trying to leave the anemone and find a new one. Anemones are a limited resource for the clownfish, and those that bide their time once they find an anemone experience the highest lifetime fitness.

The scarcity of anemones also results in competition among clownfish at a key stage in their life history. As we have seen, hatchling clownfish disperse from their anemone and spend their early life stages in the plankton. When they return to the reef, their survival depends on their choice of an anemone. The number of fish in an anemone is generally correlated with the anemone's size. However, Buston found that at any given time some anemones are undersaturated, meaning that they have room for more fish. If a juvenile fish is lucky enough to enter such an anemone, it is allowed to stay, and it enters the line of succession toward becoming a breeder. If the juvenile enters a saturated anemone, however, it is expelled, and it often dies before it can find another anemone. Similar settlement lotteries play out in many organisms that live in crowded habitats and compete for space. For example, in environments such as tropical rainforests, where many long-lived tree species compete for limited space and sunlight, the success of any one seed or seedling can depend on chance events, such as the death of a nearby large tree that creates a gap in the canopy (Denslow 1987). As we'll see in Chapter 18, such settlement lotteries can play an important role in maintaining the diversity of species found in highly competitive environments.

SUMMARY

CONCEPT 7.1 Life history patterns vary within and among species.

- Life histories are diverse, varying between individuals of the same species as well as between species. The source of this variation may be genetic or environmental.
- Organisms may reproduce sexually or asexually. In many cases, the same organism can do both.
- Most organisms have complex life cycles with multiple stages that differ in size, morphology, and habitat.

CONCEPT 7.2 Reproductive patterns can be classified along several continua.

- Semelparous species reproduce only once in a lifetime, while iteroparous species reproduce multiple times.
- The terms *r*-selection and *K*-selection refer to two ends of a continuum of reproductive patterns based on population growth rates.
- Grime's triangular model categorizes plant life histories by the degree of competition, disturbance, and stress in the habitat type to which they are adapted.
- Charnov's dimensionless life history analysis attempts to remove the effects of size and time in order to compare life histories across a broad taxonomic range.

CONCEPT 7.3 There are trade-offs between life history traits.

- There is a trade-off between offspring size and number, such that organisms tend to produce larger numbers of relatively small offspring or smaller numbers of relatively large offspring.
- There is a trade-off between investment in current reproduction and investment in future reproductive bouts.
- As organisms age, they experience a decline in their physiological capabilities, termed senescence, the onset of which can set an upper age limit for reproductive events.

CONCEPT 7.4 Organisms face different selection pressures at different life cycle stages.

- The small sizes of early life cycle stages make them vulnerable to predation and food shortages.
- Small life cycle stages are well suited to some important functions, such as dispersal and dormancy.
- Complex life cycles allow life histories the flexibility to respond to differences in selection pressures on different life cycle stages.

REVIEW QUESTIONS

1. Many closely related animal species produce eggs of vastly different sizes. As discussed under Concept 7.3, one trade-off of producing larger eggs is that fewer eggs can be produced. Despite the apparent simplicity of this trade-off, it is still unclear why both strategies (many small eggs and few large eggs) are maintained in groups of closely related species. What are some other life history traits besides offspring number that might be correlated with egg size, and under what environmental conditions might those traits be advantageous? Can you think of any reasons why species that live in the same habitats continue to exhibit reproductive patterns that vary so widely?
2. Some animals exhibit both sexual and asexual reproduction, depending on the environmental conditions they experience. Rotifers are a classic example of this phenomenon. Females can produce diploid eggs by mitosis that hatch soon after release. In this manner, rotifer populations can double within hours. These same females, under other conditions, can produce haploid eggs that form males if unfertilized and form females if fertilized. What might be the reason for the maintenance in rotifers of both sexual and asexual reproduction?
3. The Nassau grouper is popular in the live-fish trade in Asia. Restaurant-goers in Hong Kong and other Asian cities can pick the grouper they want steamed for dinner from a selection of live fish swimming in

tanks. Adult groupers can grow to extremely large sizes (up to 3 feet long and 55 pounds), but those favored by restaurants are plate-sized juvenile and young adult fish. How would you expect the removal of these younger, smaller fish to affect the life history evolution of the remaining population? How might life history parameters such as age and size at reproduction and investment in growth versus reproduction evolve in response to fishing pressures?

ON THE COMPANION WEBSITE
sites.sinauer.com/ecology2e

The website includes Chapter Outlines, Online Quizzes, Flashcards & Key Terms, Suggested Readings, a complete Glossary, and the Web Stats Review. In addition, the following resources are available for this chapter:

HANDS-ON PROBLEM SOLVING

The Growth/Reproduction Trade-Off

This Web exercise explores the trade-off that species must make between growth and reproduction. You will investigate the effects of manipulating the set point at which fish start allocating resources into reproduction rather than further growth under different levels of predation.

CHAPTER

8 Population Distribution and Abundance

KEY CONCEPTS

CONCEPT 8.1 Populations are dynamic entities that vary in size over time and space.

CONCEPT 8.2 The distributions and abundances of organisms are limited by habitat suitability, historical factors, and dispersal.

CONCEPT 8.3 Many species have a patchy distribution of populations across their geographic range.

CONCEPT 8.4 The dispersion of individuals within a population depends on the location of essential resources, competition, dispersal, and behavioral interactions.

CONCEPT 8.5 Population abundances and distributions can be estimated with area-based counts, distance methods, mark–recapture studies, and niche modeling.

From Kelp Forest to Urchin Barren: A Case Study

Stretching over 1,600 kilometers across the Pacific Ocean to the west of Alaska, the mountainous Aleutian Islands are often shrouded in fog and battered by violent storms. The islands have few large trees, and except for the eastern islands that once were connected to the mainland, they have none of the terrestrial mammals that are found on the mainland, such as brown bears, caribou, or lemmings. There is abundant marine wildlife in the surrounding waters, however, including seabirds, sea otters, whales, and a variety of fish and invertebrates.

Although there are few trees on land, the nearshore waters of some Aleutian islands harbor fascinating marine communities known as kelp forests, made up of brown algae such as *Laminaria* and *Nereocystis*. Dense clusters of kelp rise from their holdfasts on the sea bottom toward the surface, producing what feels like an underwater forest (**Figure 8.1**). Other, nearby islands do not have kelp forests. Instead, the bottoms of their nearshore waters are carpeted with sea urchins and support few kelp or other large algae. Areas with large numbers of urchins are called "urchin barrens," since they contain far fewer species than do kelp forests. Why are some islands surrounded by kelp forests and others by urchin barrens?

One possibility is that islands with kelp forests differ from islands without kelp forests in terms of climate, ocean currents, tidal patterns, or physical features such as underwater rock surfaces. But no such differences have been found, leaving us to look for other reasons why some islands have kelp forests while others do not. Because urchins feed on algae and can eat vast quantities of it, investigators suspected that grazing by urchins might prevent the formation of kelp forests.

This hypothesis has been tested in two ways. First, studies in the Aleutian Islands and elsewhere along the Alaskan coast consistently showed that kelp forests were not found in regions where there were many large urchins. Although such correlations do not prove that urchins suppress kelp forests, the fact that a number of studies found the same result suggested that urchins might determine where kelp forests are located. Second, the effect of urchins was tested in an experiment that measured kelp densities in several 50 m^2 plots containing urchins and in similar, nearby 50 m^2 plots from which urchins were

FIGURE 8.1 Key Players in the Forests of the Deep The kelp *Nereocystis* is one of several species that make up the kelp forests found off the coasts of some Aleutian islands. Research shows that the presence or absence of kelp forests near these islands is influenced by the sea otter, *Enhydra lutris*.

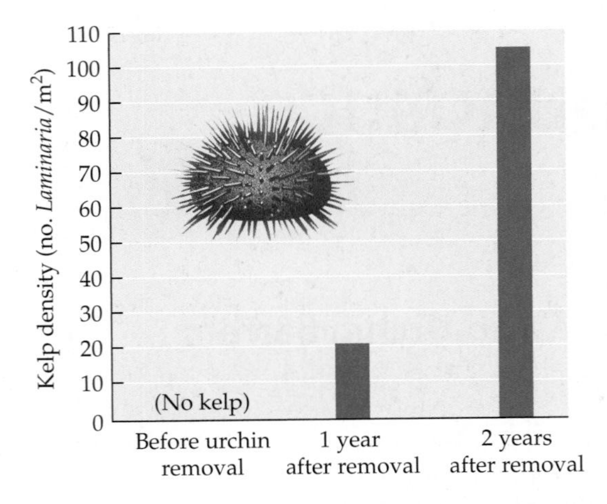

FIGURE 8.2 Do Sea Urchins Limit the Distribution of Kelp Forests? Mean densities of the kelp *Laminaria* increased dramatically after urchins were removed. (After Duggins 1980.)

removed (Duggins 1980). There were no kelp in any of the plots at the start of the experiment, and kelp densities remained at zero in the plots where urchins remained. In the plots from which urchins had been removed, however, the density of *Laminaria* rose to 21 individuals per square meter in the first year and reached 105 individuals in the second year (**Figure 8.2**). *Laminaria* is a dominant member of kelp forest communities, so these results suggested that kelp forests would grow in the absence of urchins.

These and other results indicated that the presence or absence of urchins is an answer to the question of why some Aleutian islands have kelp forests and others do not. But this answer just shifts the question from what determines the locations of kelp forests to what determines the locations of urchins. As we'll see, a more complete answer to our question about why kelp forests are found in some areas but not others turns out to depend on the voracious feeding habits of sea otters, which themselves may have become a meal of last resort for killer whales.

Introduction

In this chapter's Case Study, we discussed why kelp forests are found in some places and urchin barrens in others. That discussion focused on a fundamental ecological question: what determines a species' **distribution**, the geographic area where individuals of a species are present? In this chapter and throughout Unit 2, we will also be concerned with the related issue of what determines **abundance**, the number of individuals of a species that are found in a given area. These two issues are related: the distribution of a species can be viewed as a map of all areas where the abundance of the species is greater than zero.

Ecologists often seek to understand the factors that determine the distributions and abundances of organisms. This task can be challenging because populations are *dynamic*; that is, their distributions and abundances can change greatly over time and space. Such changes affect many aspects of biological communities, including the outcome of species interactions and ecosystem processes. Our ability to predict these changes serves as a "measuring stick" for how well we understand events in nature.

Knowledge of the factors that influence distributions and abundances also has practical importance for managing populations that we harvest (such as fish or trees) or seek to conserve (such as endangered species). As we have seen in earlier chapters, populations of many species are decreasing due to human actions (these declines will be addressed in detail in Chapter 22). We can best prevent the decline of populations if we have a clear understanding of what determines their distribution and abundance. We'll begin our exploration of the factors that influence distribution and abundance by describing populations in more detail.

CONCEPT 8.1 Populations are dynamic entities that vary in size over time and space.

Populations

A **population** is a group of individuals of the same species that live within a particular area and interact with one another. To explore this definition further, what exactly do we mean by "interact?" In species that reproduce sexually, a population might be defined as the group of individuals that interact by interbreeding. However, in species that reproduce asexually, such as dandelions or the fish *Poecilia formosa*, a population must be defined by other kinds of interactions, such as competition for common sources of food. Our definition of a population also focuses on the area over which members of a species interact. If that area is known, as in a population of lizards that live on and move throughout a small island, we can report population abundance either as **population size** (the number of individuals in the population) or as **population density** (the number of individuals per unit of area). For example, if there were 2,500 lizards on a 20-hectare island (roughly 50 acres), the population size would be 2,500 lizards, and the population density would be 125 lizards per hectare.

In some cases, the total area occupied by a population is not known. For example, when little is known about how far a sexually reproducing species or its gametes (e.g., plant pollen) can travel, it is difficult to estimate the area over which individuals interbreed frequently and hence represent a single population. For asexual species, similar problems are encountered when we try to estimate the area over which interactions other than interbreeding occur. When the area occupied by a population is not fully known, ecologists use the best available information about the biology of the species to delimit an area within which the size and density of the population can be estimated.

Abundances change over time and space

The number of individuals in a population changes over time. This is true whether abundances are measured on a small spatial scale, such as the number of plants found in a restricted area along a riverbank, or on a much larger spatial scale, such as the number of cod found in the North Atlantic Ocean. At any given time, abundances also differ from one place to another. Some populations differ little in abundance over time and space; others differ considerably.

For example, Richard Root and Naomi Cappuccino studied abundances of 23 species of herbivorous insects that fed on tall goldenrod (*Solidago altissima*). They studied these insects for at least 6 consecutive years at each of 22 sites located in the Finger Lakes region of New York State (**Figure 8.3**). These sites were no more than 75 km (47 miles) apart; hence, in any given year, all the sites experienced roughly the same climatic conditions. Nevertheless, insect abundances varied considerably from one site to another and from one year to the next. For some species, such as the ball gall fly (*Eurosta solidaginis*), abundances varied relatively little. The maximum abundance reached by

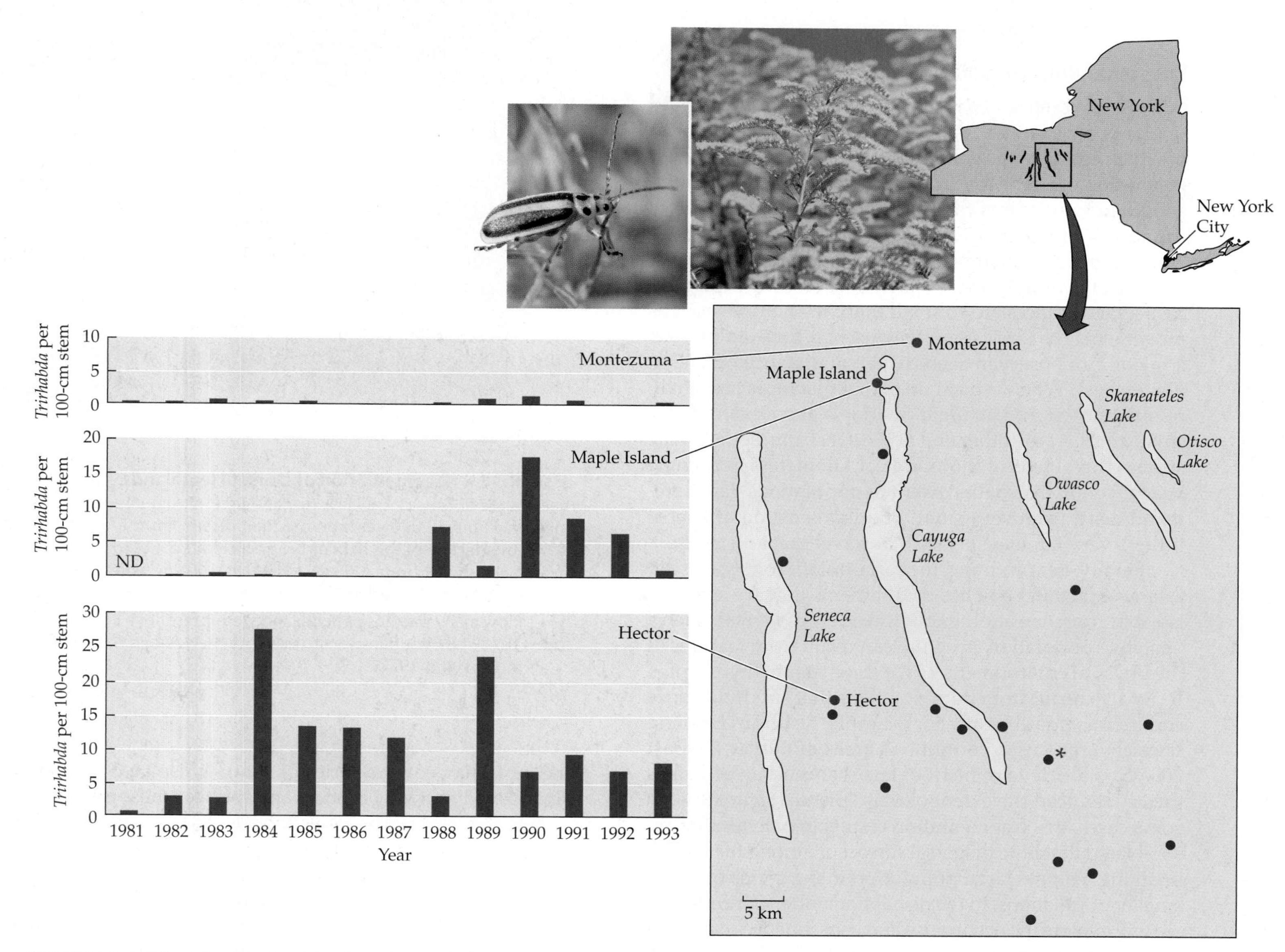

FIGURE 8.3 Abundances Are Dynamic Changes in abundances of the beetle *Trirhabda virgata* on tall goldenrod (*Solidago altissima*) plants over time at Montezuma, Maple Island, and Hector, three of the sites studied by Richard Root and Naomi Cappuccino. The 22 study sites are indicated by dots on the map; five study sites were located close to one another at the position marked with an asterisk. (After Root and Cappuccino 1992.)

Q In what year or years did *Trirhabda* abundance vary greatly over space? Explain.

Eurosta over a 6-year period varied sixfold across the 22 sites, from 0.05 insects per stem at the site with the fewest individuals to 0.3 insects per stem the site with the most individuals. Maximum abundances of other species, however, such as the beetle *Trirhabda virgata*, varied much more (by a factor of 336), ranging from 0.03 to 10.1 insects per stem. Overall, *Trirhabda* populations varied considerably in abundance, both from one site to another and over time (see Figure 8.3). We'll return to this subject in Chapters 9 and 10, where we'll delve into factors that cause populations to fluctuate in abundance.

Populations are dynamic in another sense as well: individuals move from one population to another, sometimes traveling great distances.

Dispersal links populations

Organisms differ greatly in their capacity for movement. In plants, for example, dispersal occurs when seeds, each of which contains the embryo of an individual of the next generation, move away from the parent plant. Although events such as storms can transport seeds long distances (hundreds of meters to many kilometers; see Cain et al. 2000), dispersal distances in plants are usually small (one to a few tens of meters). In some cases, typical seed dispersal distances are so small that they hardly count as movement. The forest plant *Viola odorata*, for example, has a maximum observed seed dispersal distance of 0.02 m (0.8 inches). When typical dispersal distances are small, populations of interacting individuals are often found in small areas. At the other end of the spectrum, some whale species travel tens of thousands of kilometers in a single year. Overall, the spatial extent of populations varies tremendously, from very small in organisms that disperse little to very large in species that travel great distances.

Finally, bear in mind that a population may exist in a series of habitat patches or fragments that are spatially isolated from one another but linked by dispersal. Such a "patchy" population structure can result from features of the physical environment, as we'll see later in this chapter. It can also result from human actions that subdivide once continuous populations. For example, heaths in England once covered large, continuous areas, but over the past 200 years the development of farms and urban areas has greatly reduced the extent of these plants (**Figure 8.4**). In some cases, this fragmentation results in patches that are so isolated that little dispersal can occur among them, thus breaking a single large population into a series of much smaller populations. In the discussion under Concept 10.4, we will explore the occurrence and consequences of patchy population structures (*metapopulations*) in more detail.

FIGURE 8.4 Fragmentation of Dorset Heathlands The heathlands of Dorset, England, reached their maximum extent in Roman times, 2,000 years before the present. From 1759 to 1978, the decline of this habitat type accelerated: the total area of heathlands shrank from 300 km^2 to less than 60 km^2 and the number of patches increased greatly. (After Webb and Haskins 1980.)

Q How many patches of heathland were present in 1759? In 1978? Use your answers to estimate the average patch size in 1759 and 1978.

What are individuals?

As we've seen, a population may cover a single area whose extent depends on the capacity of the species for dispersal, or it may cover a series of spatially isolated patches linked by dispersal. For the many species whose dispersal capabilities are poorly understood, it can be challenging to determine the spatial extent of a population. In addition, for many organisms, it can even be hard to determine what constitutes the individuals of the population.

How can there be confusion over what an individual is? Consider the aspen trees shown in **Figure 8.5**. Like many plant species, an aspen tree can produce genetically identical copies of itself, or **clones**. Aspens produce clones by forming new plants from root buds, while species such as clover and strawberries do so by forming new plants

FIGURE 8.5 Aspen Groves—One Tree or Many? These quaking aspen (*Populus tremuloides*) growing in Yellowstone National Park could represent over 20 different genetic individuals, each established from a seed. However, it is also possible that each of these aspens is actually part of one "tree," having been produced asexually from the root buds of a single genetic individual.

from buds located on horizontal stems, or "runners" (**Figure 8.6**). Among animals, many corals, sea anemones, and hydroids can form clones of genetically identical individuals, as can some frogs, fishes, lizards, and many insects. Some plant clones can grow to enormous sizes (e.g., 81 ha or 200 acres in aspen clones) or live for extremely long periods (e.g., 43,000 years in the king's holly, a Tasmanian shrub).

To cope with the complications that result from the formation of clones, biologists who study such organisms define individuals in several different ways. For example, an individual can be defined as the product of a single fertilization event. Under this definition, a grove of genetically identical aspen trees is actually a single genetic individual, or **genet**. However, members of a genet are often physiologically independent of one another, and they may in fact compete for resources. Such actually or potentially independent members of a genet are called **ramets**. In strawberries, for example, a rooted plant is considered a ramet because it can persist even if it is not connected to the rest of its genet (see Figure 8.6). Whether we view a patch of strawberries or a grove of aspen trees as one individual or many depends on what we are interested in. If we are interested in evolutionary change over time, the genet level may be most appropriate. In contrast, if we are interested in how independent physiological units compete, the ramet level may be most appropriate.

Now that we've defined populations and considered some of the issues that complicate their study, let's turn to the factors that influence where populations are found and how many individuals they contain.

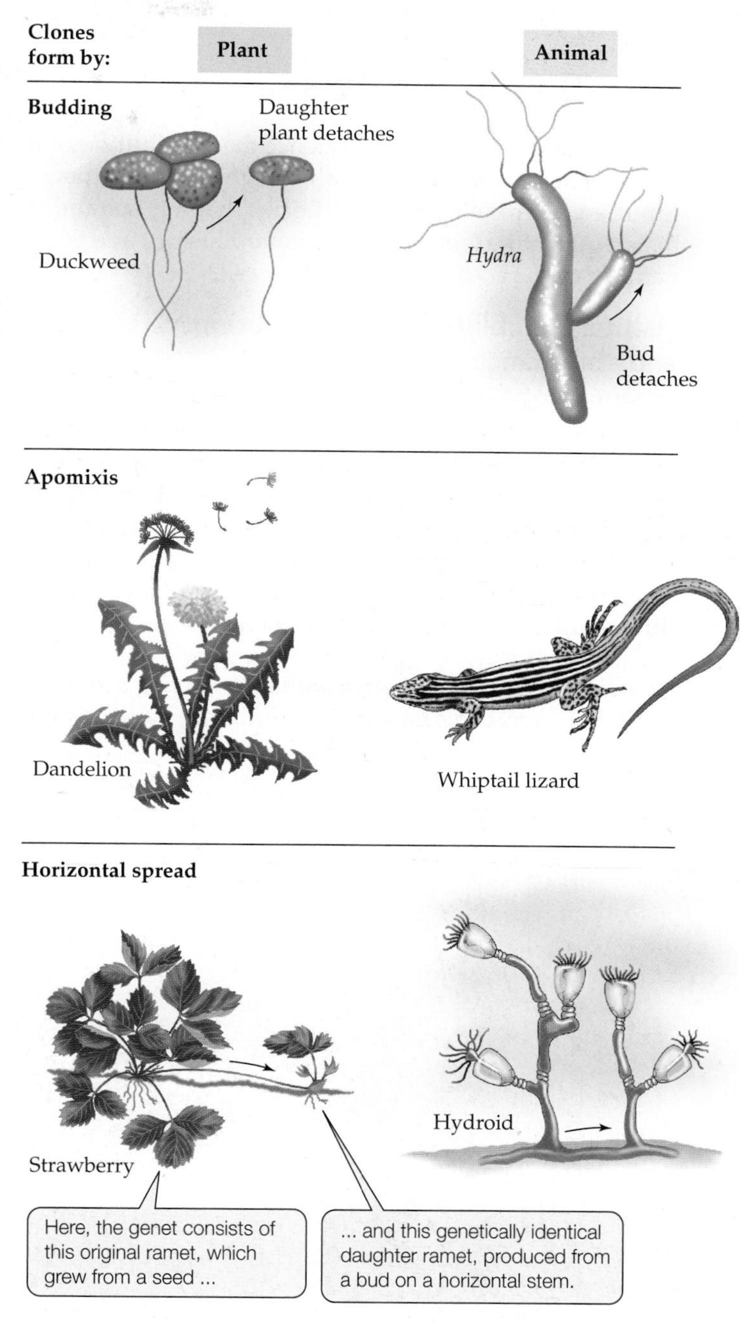

FIGURE 8.6 Plants and Animals That Form Clones Many plants and animals reproduce asexually, thereby forming clones of genetically identical individuals. Examples of asexual reproduction include *budding* (in which clonal offspring detach from the parent), *apomixis* (in which clonal offspring are produced from unfertilized eggs; also known as parthenogenesis), and *horizontal spread* (in which clonal offspring are produced as the organism grows).

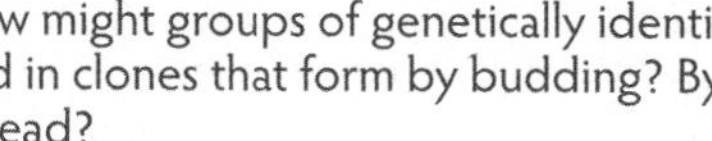

Q How might groups of genetically identical individuals be identified in clones that form by budding? By apomixis? By horizontal spread?

CONCEPT 8.2 The distributions and abundances of organisms are limited by habitat suitability, historical factors, and dispersal.

Distribution and Abundance

Many different factors can influence the distributions and abundances of organisms. We'll survey these factors by grouping them into three categories: habitat suitability, historical factors (such as evolutionary history and continental drift), and dispersal.

Habitat suitability limits distribution and abundance

Good and poor places to live exist for all species. A desert species is not likely to perform well in the Arctic, or vice versa. Even small differences in how well individuals survive or reproduce can cause a species' abundance to be high in certain environments and low in others. Thus, the distribution and abundance of a species should be influenced strongly by the presence of suitable habitat. But what factors make habitat suitable?

Abiotic features of the environment As we discussed in Unit 1, aspects of the abiotic (nonliving) environment, such as moisture, temperature, sunlight, soil pH, salt concentration, and available nutrients, set limits on whether a habitat will be suitable for a particular species. Some species can tolerate a broad range of physical conditions, while others have more narrow requirements.

Creosote bush (*Larrea tridentata*), for example, has a broad distribution in North American deserts, ranging across much of the southwestern United States and northwestern and central Mexico (**Figure 8.7**). Creosote bush is very tolerant of arid conditions: it uses water rapidly when it is available, then shuts down its metabolic processes during periods of extended drought. Creosote bush also tolerates cold well, so its populations thrive in high-elevation deserts where winter temperatures can remain below freezing for several days.

The saguaro cactus (*Carnegiea gigantea*), on the other hand, has a more limited distribution. Like creosote bush, saguaro flourishes under arid conditions, but it achieves its drought tolerance in different ways. Although saguaro does not have typical leaves, its spines are actually modified leaves whose low surface area reduces water loss. Furthermore, during wet periods, saguaro stores water in its massive trunk and arms, saving it for use during times of drought. Saguaro cannot tolerate cold, however: it is killed when temperatures remain below freezing for 36 hours or more. The importance of saguaro's sensitivity

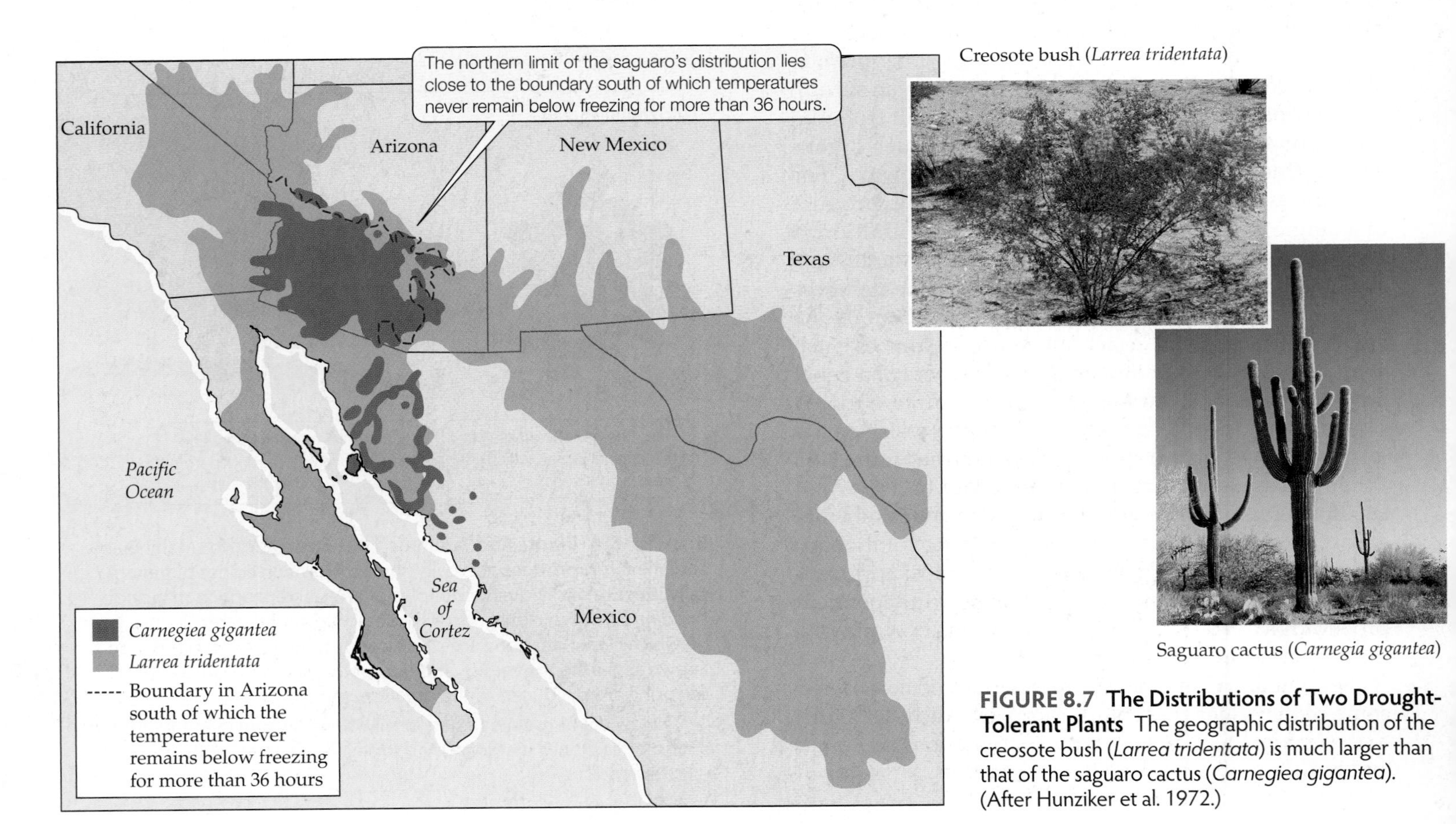

FIGURE 8.7 The Distributions of Two Drought-Tolerant Plants The geographic distribution of the creosote bush (*Larrea tridentata*) is much larger than that of the saguaro cactus (*Carnegiea gigantea*). (After Hunziker et al. 1972.)

(A)

(B)

FIGURE 8.8 Herbivores Can Limit Plant Distributions In Australia, the moth *Cactoblastis cactorum* was used to control populations of an introduced cactus, *Opuntia stricta*. (A) A dense thicket of *Opuntia* 2 months before the release of the moth. (B) The same stand 3 years later, after the moth had killed the cacti by feeding on their growing tips. (From Department of Natural Resources, Queensland, Australia.)

to cold is revealed by its distribution: the northern limit of its distribution corresponds closely to a boundary north of which temperatures occasionally remain below freezing for at least 36 hours (see Figure 8.7).

Biotic features of the environment The biotic environment also has important effects on distribution and abundance. Obviously, species that depend completely on one or a few other species for their growth, reproduction, or survival cannot live where the species on which they depend are absent. Organisms can also be excluded from an area by herbivores, predators, competitors, parasites, or pathogens, any of which can greatly reduce the survival or reproduction of members of a population.

A dramatic example of such exclusion is provided by the biological control of *Opuntia stricta*, an introduced cactus that spread rapidly to cover large areas in Queensland and New South Wales, Australia. The cactus was first imported from the southern United States in 1839 and planted as hedge. Within 40 years, *O. stricta* had become a pest species, and by 1925 it had spread to cover 243,000 km^2. The cactus can grow up to 2 m high, and in many areas it covered the ground with dense, spiny thickets, making the rangelands it occupied useless (**Figure 8.8A**). In the hope of controlling the cactus, an Argentinean moth (*Cactoblastis cactorum*) known to feed on *O. stricta* was released in 1926. By 1931, billions of cacti had been destroyed by the moth (**Figure 8.8B**). Since 1940, the cactus has persisted in small numbers, but its distribution and abundance have both been greatly reduced by the moth. Although the introduction of *Cactoblastis* as a means of biological pest control appears to have been a great success, in general such introductions must be undertaken cautiously because they can lead to unintended consequences, such as damage to native species (Louda et al. 1997).

Interactions between abiotic and biotic features In many cases, abiotic and biotic features of the environment act together to determine the distribution and abundance of a species (see Figure 4.3). For example, the barnacle *Semibalanus balanoides* cannot survive where summer air temperatures are above 25°C, and it cannot reproduce if winter air temperatures do not remain below 10°C for 20 days or more. On the Pacific coast of North America, temperatures are such that *S. balanoides* could be found 1,600 km farther south than it currently is. But this barnacle is absent from the region shown in purple in **Figure 8.9**, presumably be-

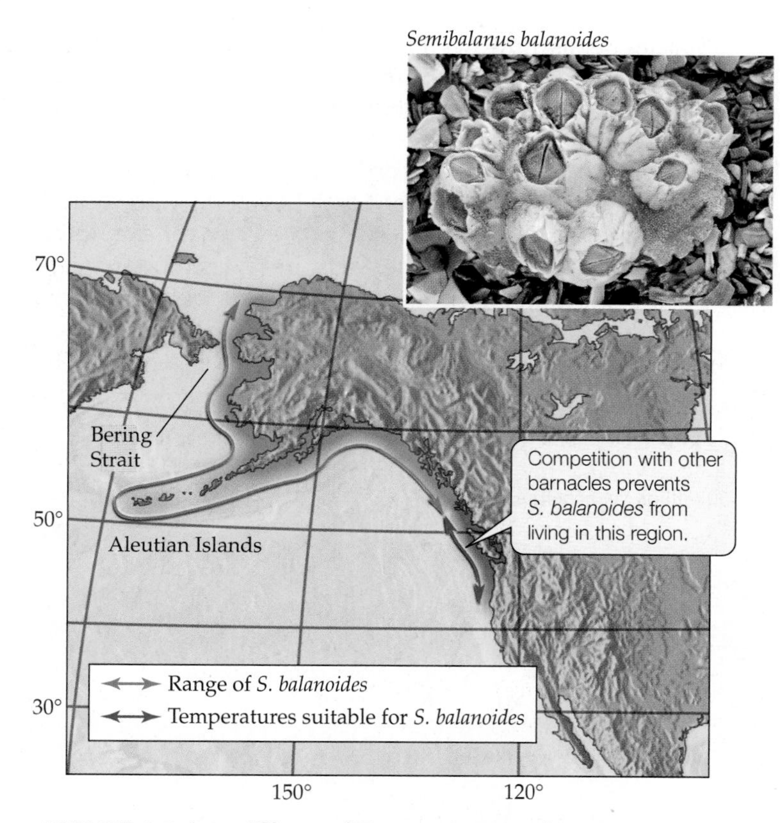

FIGURE 8.9 Joint Effects of Temperature and Competition on Barnacle Distribution Although temperatures are suitable for the barnacle *Semibalanus balanoides* as far south as the region shaded purple on the map, it is excluded from this region by its competitors. *S. balanoides* is found in the region shaded red, where temperatures are colder and it is a superior competitor.

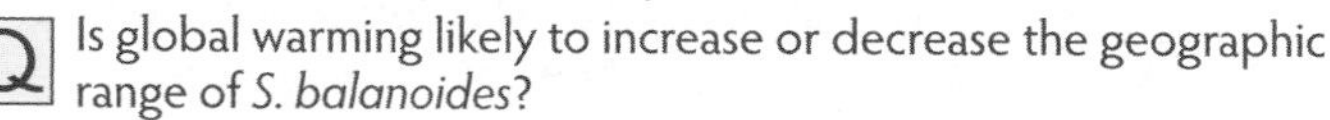

Q Is global warming likely to increase or decrease the geographic range of *S. balanoides*?

cause competition from other species of barnacles prevents it from living in what would otherwise be suitable habitat. To the north, as temperatures become increasingly colder, a point is reached where *S. balanoides* outcompetes the other barnacles and maintains healthy populations. Thus, the abiotic and biotic environments interact to determine where populations of this barnacle are found.

Disturbance The distributions of some organisms depend on regular forms of disturbance. A **disturbance** is an abiotic event that kills or damages some individuals and thereby creates opportunities for other individuals to grow and reproduce. Many plant species, for example, persist in an area only if there are periodic fires. If humans prevent fires, such species are replaced by species that are not as tolerant of fires but that are superior competitors in the absence of fires. Floods, windstorms, and droughts are other forms of disturbance that can exclude some species but give others an advantage. We'll discuss the role of disturbance in more detail in Chapter 16.

History and dispersal limit distribution and abundance

Species persist only in regions of suitable habitat, but they are not found in all such places. History and dispersal also play important roles in their distributions and abundances.

Evolution and continental drift Events in the evolutionary and geologic history of Earth have had a profound effect on where organisms live today. Why, for example, are polar bears found in the Arctic but not in Antarctica? Polar bears hunt on ice packs and eat seals, both of which abound in Antarctica. In part, the answer to our question lies in an accident of history. Fossil and genetic evidence indicate that polar bears evolved from brown bears in the Arctic (Lindqvist et al. 2010); hence they are in the Arctic because they originated there. As for their absence from Antarctica, polar bears can travel over 1,000 km in a year, but it appears they cannot or will not cross the tropical regions that separate the Arctic from Antarctica. Thus, the distribution of polar bear populations is influenced by evolutionary history and dispersal as well as by the presence of suitable habitat.

History also plays a key role in groups of organisms whose curious distributions had puzzled biologists for nearly a hundred years. Consider Alfred Russel Wallace's observation that the animals of a region can differ considerably across relatively short geographic distances (Wallace 1860). The mammal communities of the Philippines, for example, are more similar to those in Africa (88% overlap at the family level) than they are to those in New Guinea (64% overlap), despite the fact that Africa is 5,500 km away and New Guinea is only 750 km away. No explanation for this and other similar observations could be found until the discovery of *continental drift*, the gradual movement of continents over time (see Web Extension 17.1). This discovery led to the realization that the Philippines and New Guinea are on different tectonic plates and have been in close geographic contact for a relatively short time (**Figure 8.10**). We'll return to the roles of evolution and continental drift in Chapter 17, where we'll discuss biogeographic and historical approaches to community ecology.

Dispersal limitation As demonstrated by the polar bear's absence from Antarctica, a species' limited capacity for

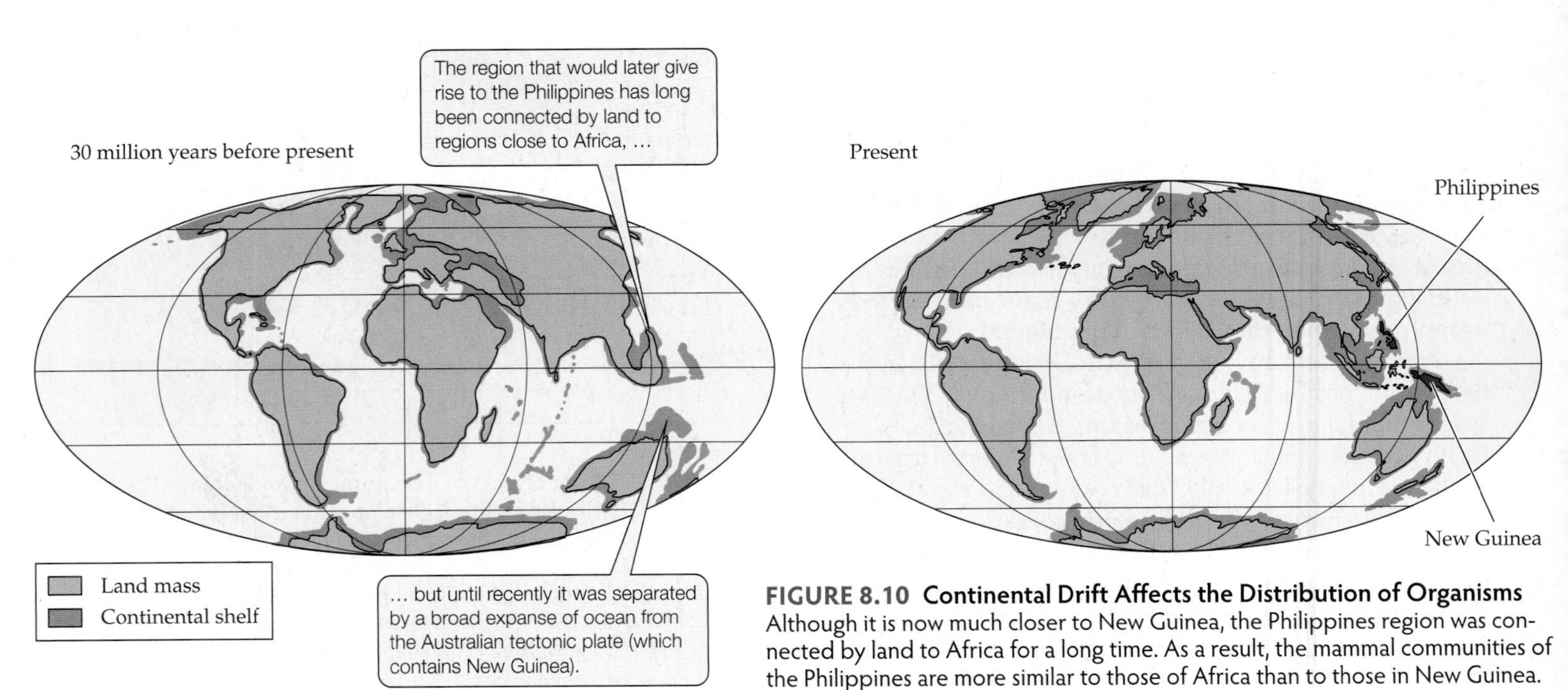

FIGURE 8.10 Continental Drift Affects the Distribution of Organisms Although it is now much closer to New Guinea, the Philippines region was connected by land to Africa for a long time. As a result, the mammal communities of the Philippines are more similar to those of Africa than to those in New Guinea.

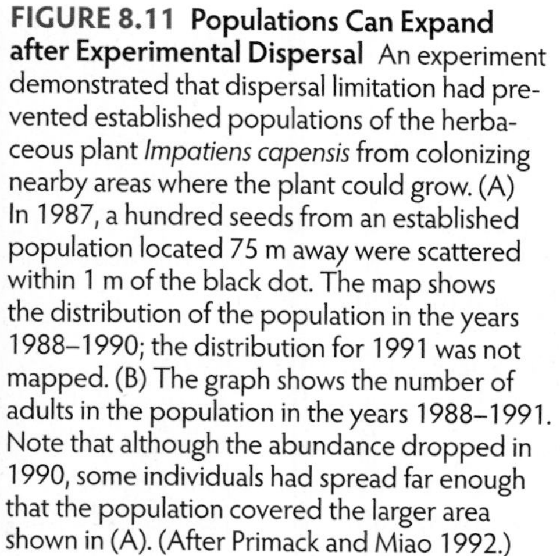

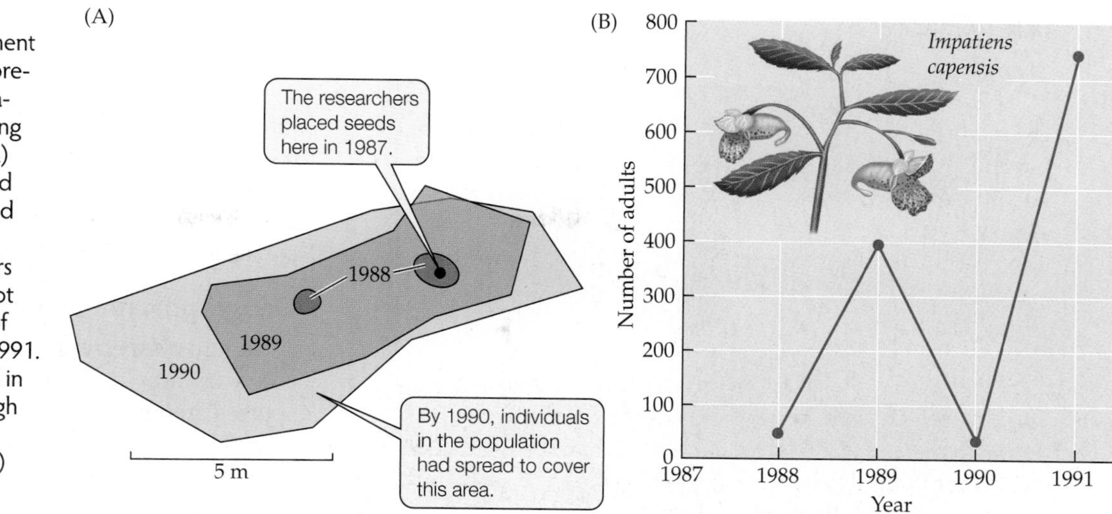

FIGURE 8.11 Populations Can Expand after Experimental Dispersal An experiment demonstrated that dispersal limitation had prevented established populations of the herbaceous plant *Impatiens capensis* from colonizing nearby areas where the plant could grow. (A) In 1987, a hundred seeds from an established population located 75 m away were scattered within 1 m of the black dot. The map shows the distribution of the population in the years 1988–1990; the distribution for 1991 was not mapped. (B) The graph shows the number of adults in the population in the years 1988–1991. Note that although the abundance dropped in 1990, some individuals had spread far enough that the population covered the larger area shown in (A). (After Primack and Miao 1992.)

dispersal can prevent it from reaching areas of suitable habitat—a phenomenon known as **dispersal limitation**. In another example, the Hawaiian Islands have only one native terrestrial mammal, the Hawaiian hoary bat (*Lasiurus cinereus*), which was able to fly to the islands. No other land mammals have been able to disperse to Hawaii on their own, although cats, pigs, wild dogs, rats, goats, mongooses, and other mammals now thrive in Hawaii following their introduction to the islands by people.

Dispersal limitation can also occur on smaller spatial scales, preventing populations from expanding to nearby areas of apparently suitable habitat. Primack and Miao (1992) documented such local dispersal limitation in the herbaceous plant *Impatiens capensis* (**Figure 8.11**). Similar results were obtained in a long-term study of the English bluebell, *Hyacinthoides non-scripta*. In 1960, 27 populations of 7–10 individuals each were established in apparently suitable forest habitat located near source populations (Van der Veken et al. 2007). Forty-five years later, 11 (41%) of these experimental populations persisted, and most contained hundreds or thousands of individuals. These results suggested that dispersal limitation had prevented the bluebells from reaching habitat where they could thrive. Sixteen (59%) of the experimental populations were extinct. The chance of extinction was not related to soil type or to natural changes in the vegetation, but was positively correlated with physical disturbances of the soil and with major changes in the overlying canopy resulting from human activities (including clear-cutting).

Dispersal and density Dispersal affects abundances as well as distributions. When individuals disperse from one population to another, the density of the population they leave decreases and the density of the population they join increases. Dispersal can also be affected *by* population density. For example, many species of aphids produce winged forms (which are capable of dispersing) in response to crowding (Harrison 1980). This point is illustrated by the bean aphid (*Aphis fabae*), in which the percentage of offspring that develop wings increases as the density of aphids increases (**Figure 8.12**).

FIGURE 8.12 Density Can Affect Dispersal The proportion of aphid offspring that develop wings in relation to the density at which they were reared, for mothers who were themselves reared at high (H) and low (L) densities. (After Shaw 1970.)

Q Do the results shown in the graph indicate that aphids are more likely to develop wings in response to high densities experienced by the offspring, by their mothers, or both? Explain.

FIGURE 8.13 Desert Pupfish Habitat Desert pupfish live in pools that are occasionally connected to one another by temporary streams.

Density and dispersal may play similar roles in populations of the desert pupfish (*Cyprinodon macularius*) (**Figure 8.13**). Following heavy rains, the pools in which these fish live are connected to one another by temporary streams. Dispersal is a risky enterprise for fish that live in desert pools, but under certain circumstances, it can be advantageous. As explored in **Web Extension 8.1**, results from experiments on desert pupfish suggest that dispersal may provide them with a greater chance for survival and reproduction than they would have if they remained in crowded pools with limited food (McMahon and Tash 1988). Many other organisms alter their rates of dispersal in response to conditions that affect survival and reproduction, such as increases in the abundance of predators or competitors or decreases in habitat quality or the availability of mates (Poethke et al. 2010).

CONCEPT 8.3 Many species have a patchy distribution of populations across their geographic range.

Geographic Range

As we saw in the previous section, a number of factors—including the presence of suitable habitat, geologic and evolutionary history, and dispersal—can limit the distribution of a species. Indeed, we expect that in many cases, several or perhaps all of these factors will operate at the same time to influence where a species is found. The net effect of these interacting factors limits some species to a small geographic region, while other species have much larger distributions. But whether a species inhabits a small or a large geographic area, no species can live everywhere, since much of Earth consists of unsuitable habitat for its populations.

Geographic ranges vary in size among species

Although no species is found everywhere, there is considerable variation in the sizes of species' geographic ranges. A species' **geographic range** is the entire geographic region over which that species is found. Examples of species with a small geographic range include the Devil's Hole pupfish (*Cyprinodon diabolis*), which lives in a single desert pool (7 × 3 m across and 15 m deep). Many tropical plants also have a small geographic range. This latter point was illustrated dramatically in 1978, when 90 new plant species were discovered on a single mountain ridge in Ecuador, each with a geographic range that was restricted to that ridge. Other species, such as coyotes, live over most of one continent (North America), while still others, such as gray wolves, live on small portions of several continents (North America and Eurasia). Relatively few terrestrial species are found on all or most of the world's continents. Notable exceptions include humans, Norway rats, and the bacteria *Escherichia coli*, which lives in the intestinal tracts of reptiles, birds, and mammals (including humans) and thus is found wherever its host organisms are found. Some marine species, including invertebrates with planktonic larvae (see Chapter 7) and whales, have large geographic ranges. But in the oceans, as on land, a similar pattern emerges: while range sizes vary greatly among species, most species have a relatively small geographic range (Gaston 2003).

The geographic range of a species includes the areas it occupies during all of its life stages. It is particularly important to keep this in mind for species that migrate and for species whose biology is poorly understood. For example, if we wish to protect monarch butterfly populations, we must ensure that conditions are favorable for them in both their summer breeding grounds and their overwintering sites (**Figure 8.14**). In some cases, we understand an organism's range poorly because it has life stages that are hard to find or study; this is true for many fungi, plants, and insects. We may know, for example, under what conditions the adult organism lives, yet have no idea where or how other life stages live. That was, in fact, long the case for the monarch butterfly. Biologists knew that these butterflies arrived each spring in eastern North America from the south, but it took almost 120 years of effort (from 1857 to 1975) before their overwintering sites were discovered in mountains west of Mexico City, Mexico.

Populations have patchy distributions

Even within a species' geographic range, much of the habitat is not suitable for the species. As a result, popula-

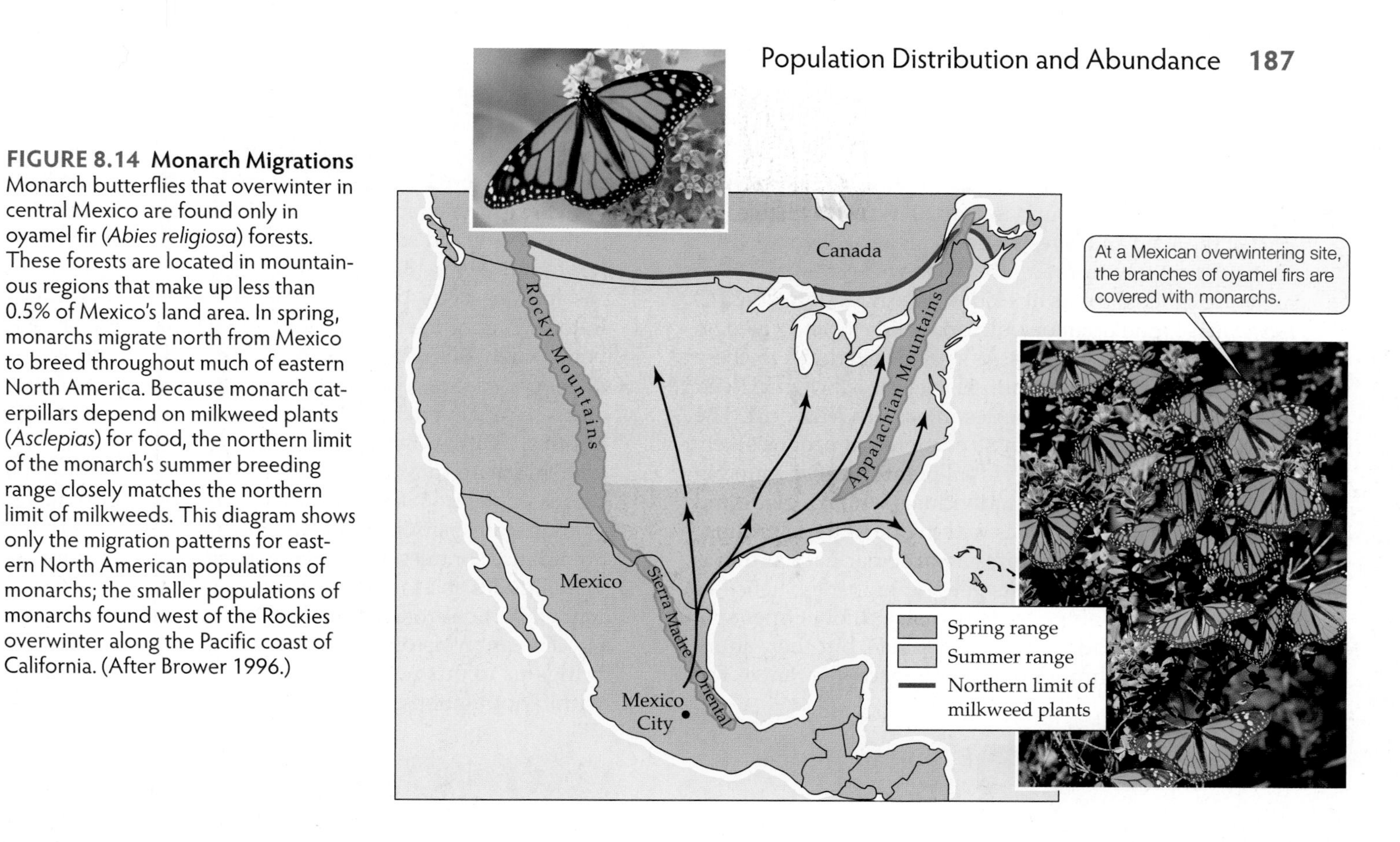

FIGURE 8.14 Monarch Migrations Monarch butterflies that overwinter in central Mexico are found only in oyamel fir (*Abies religiosa*) forests. These forests are located in mountainous regions that make up less than 0.5% of Mexico's land area. In spring, monarchs migrate north from Mexico to breed throughout much of eastern North America. Because monarch caterpillars depend on milkweed plants (*Asclepias*) for food, the northern limit of the monarch's summer breeding range closely matches the northern limit of milkweeds. This diagram shows only the migration patterns for eastern North American populations of monarchs; the smaller populations of monarchs found west of the Rockies overwinter along the Pacific coast of California. (After Brower 1996.)

tions tend to have a patchy distribution. This observation holds at both large and small spatial scales. On land, for example, at the largest spatial scales, climate constrains where populations of a species are located (see Chapter 3). At smaller spatial scales, factors such as topography, soil type, and the presence or absence of other species prevent populations from being spread evenly across the landscape.

A vivid example of patchiness at different spatial scales is provided by work Ralph Erickson performed on the herbaceous perennial, *Clematis fremontii* (**Figure 8.15**).

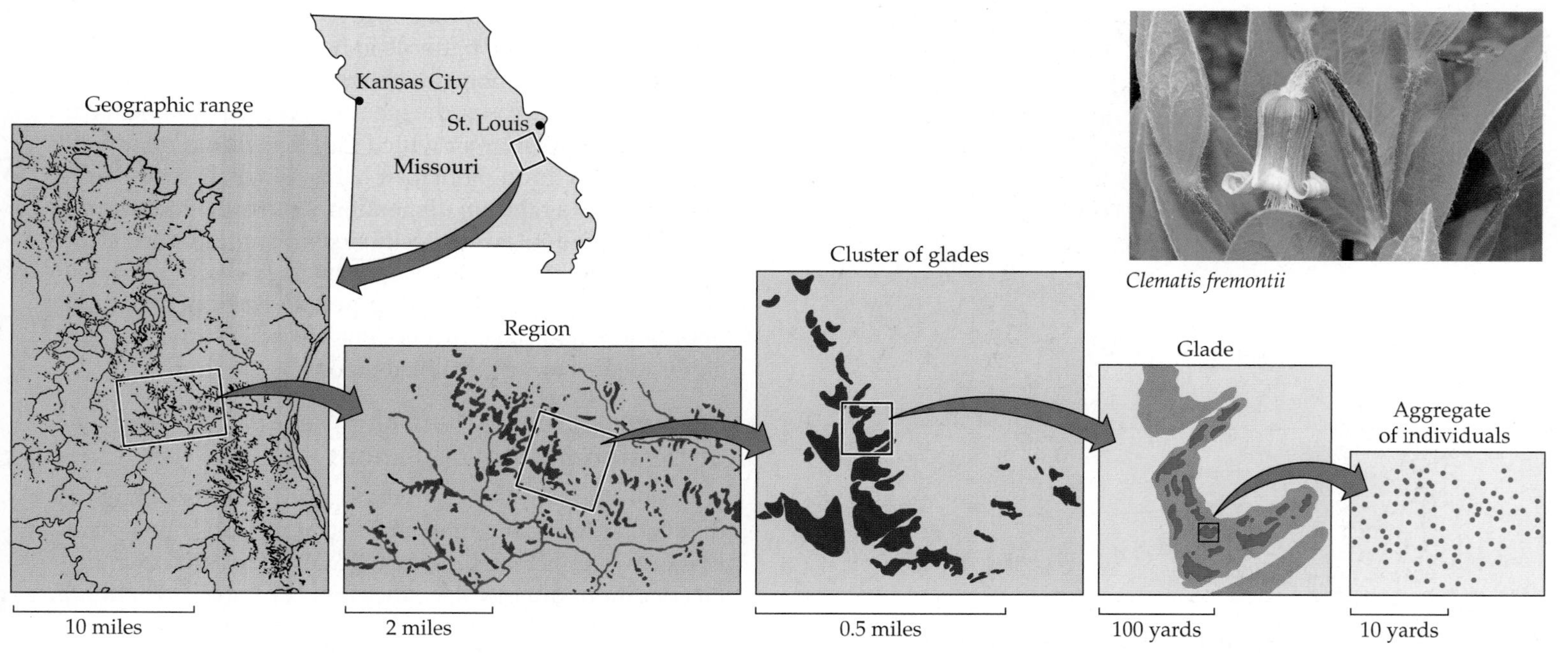

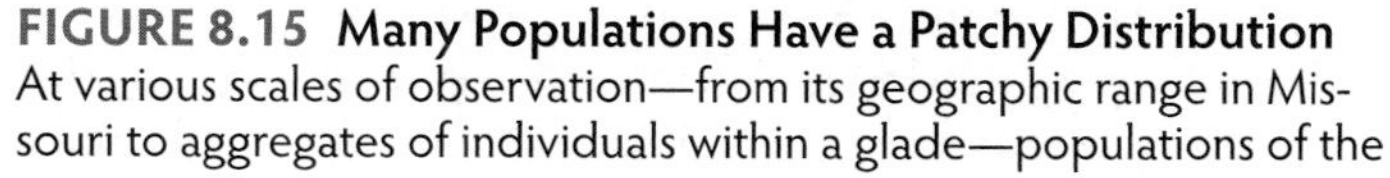

FIGURE 8.15 Many Populations Have a Patchy Distribution At various scales of observation—from its geographic range in Missouri to aggregates of individuals within a glade—populations of the herbaceous plant *Clematis fremontii* have a patchy distribution. The patchy distribution of this plant is controlled largely by the patchy distribution of suitable habitat. (After Erickson 1945.)

This species is found in portions of Kansas, Nebraska, and Missouri. Erickson studied the distribution of *C. fremontii* in Missouri, where it is found in a small region in the eastern part of the state and is restricted to areas of dry, rocky soil that support few trees in otherwise wooded areas; such areas are called barrens or glades. The glades on which *C. fremontii* is found occur on outcrops of limestone located on south- or west-facing slopes. As shown in Figure 8.15, these glades are clustered, or grouped in clumps, when viewed across the range of the species in eastern Missouri. The distribution of glades remains clustered at progressively smaller scales as well. Individual plants are also found in clusters, both as groups of individuals within a glade and as aggregates of individuals within one of these groups.

Clematis fremontii requires a particular habitat that is found only in portions of its geographic range; hence its populations have a very patchy distribution. Other species tolerate a broader range of habitats, but their abundances still vary throughout their geographic range. The red kangaroo (*Macropus rufus*) of Australia illustrates this point. The density of red kangaroos varies throughout their geographic range, which includes several regions of high density and several areas where red kangaroos are not found (**Figure 8.16**).

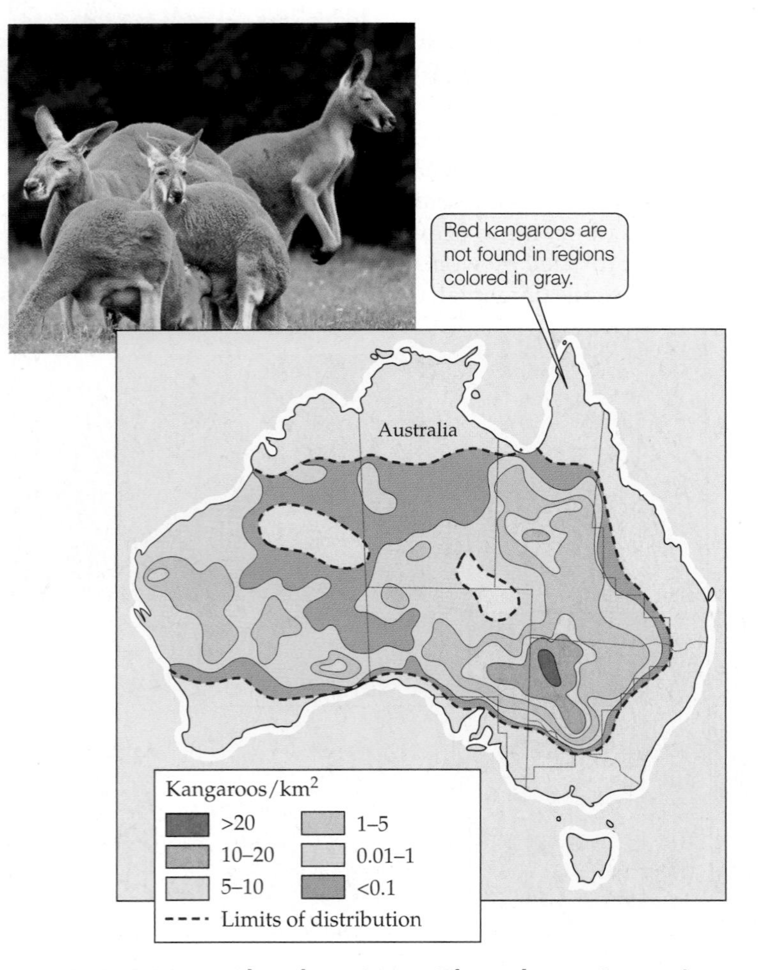

FIGURE 8.16 Abundance Varies Throughout a Species' Geographic Range The map shows abundances of the red kangaroo (*Macropus rufus*) throughout its range in Australia. These data were based on aerial surveys conducted from 1980 to 1982. (After Caughley et al. 1987.)

In some cases, population density tends to be greatest in the center of the range, declining gradually toward the boundaries. This pattern has been observed in many North American species, including plants, intertidal invertebrates, and terrestrial vertebrates. The mean density of the indigo bunting (*Passerina cyanea*), for example, declined smoothly when measured at various points along four lines (transects) that extended from the center of the bird's range. At smaller spatial scales, similar patterns have been found for the abundances of trees at different elevations along the sides of mountains and for mean densities of invertebrates found in the intertidal zone of the Gulf of California.

The analysis of abundances at small spatial scales brings us to our next topic, the location of individuals within populations.

CONCEPT 8.4 The dispersion of individuals within a population depends on the location of essential resources, competition, dispersal, and behavioral interactions.

Dispersion within Populations

The aggregates of *Clematis fremontii* individuals found in glades provide an example of the **dispersion**, or spatial arrangement, of individuals within a population. We can recognize three basic patterns in how the individuals of a population are positioned with respect to one another (**Figure 8.17**). In some cases, the members of a population have a **regular dispersion**, in which individuals are relatively evenly spaced throughout their habitat. In other cases, individuals show a **random dispersion**, similar to what would occur if individuals were positioned at locations selected at random. Finally, as in *Clematis fremontii*, individuals may be grouped together to form a **clumped dispersion**. In natural populations, clumped dispersions are more common than either regular or random dispersions.

Resources, competition, and dispersal affect dispersion within populations

A variety of processes can cause individuals to have a regular, random, or clumped dispersion. Consider a plant that grows poorly unless a particular set of environmental conditions are met (e.g., the right combination of light, temperature, and soil nutrients). In such a case, the spatial arrangement of individuals within a population would be likely to match the spatial arrangement of conditions suitable for growth. Since environmental conditions often vary at random or are clumped in space, our hypothetical plant

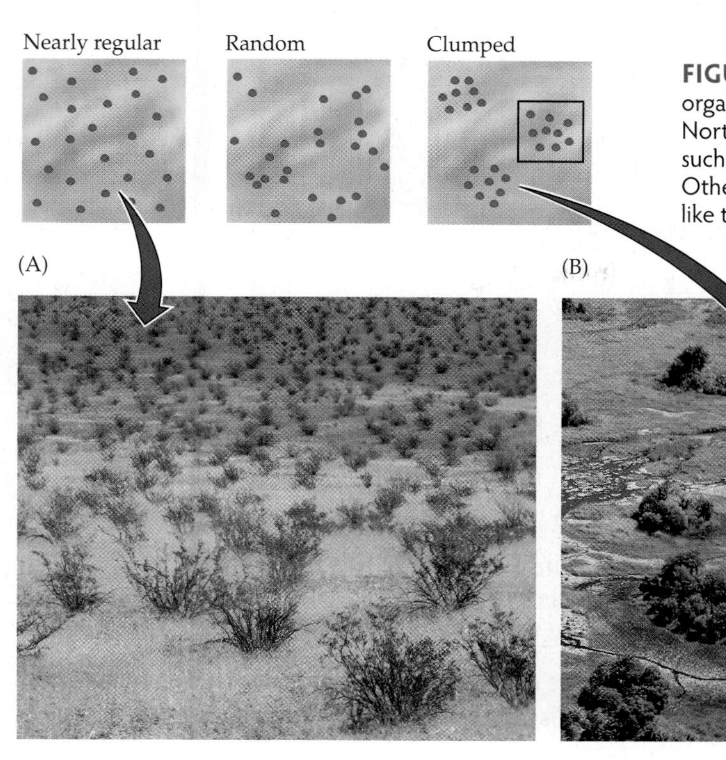

FIGURE 8.17 Dispersion of Individuals within Populations Some organisms, such as creosote bush (*Larrea tridentata*) in the Mojave Desert of North America (A), have a nearly regular dispersion. This plant probably has such a dispersion because individuals compete for limited water supplies. Other organisms have a random dispersion, and the majority of organisms, like these trees (B), have a clumped dispersion.

might be expected to have a random or clumped dispersion. Random or clumped dispersions can also occur as a result of dispersal (e.g., short dispersal distances can cause individuals to clump together). In some instances, competition for resources or space appears to have resulted in a nearly regular dispersion, as has been observed for the creosote bushes shown in Figure 8.17.

Individual behavior affects dispersion within populations

Interactions among individual organisms also influence dispersion patterns. Individuals may repel one another (to produce nearly regular dispersions) or attract one another (to produce clumped dispersions). Both of these tendencies can be seen in the Seychelles warbler (*Acrocephalus sechellensis*), an endangered songbird. In the 1950s, this bird nearly went extinct: its total world population was reduced to just 26 individuals located on Cousin Island in the Seychelles, a group of islands off the east coast of Africa. After the Seychelles warbler was legally protected in 1968, the Cousin Island population increased to about 300 birds, and the species was introduced successfully to two other islands.

Seychelles warblers are territorial: a breeding pair defends its territory against other birds of the species. This behavior causes the dispersion of individuals in the population to be somewhat regular (**Figure 8.18**). But not all territories are equal: some are of higher quality than others because they provide more food (e.g., insects). Birds that live in a high-quality territory live longer and produce more young. In addition, a breeding pair that lives in a high-quality territory often receives help from some of its offspring born in previous years. In a behavioral pattern known as **cooperative breeding**, these younger birds postpone breeding and instead help their parents raise offspring by performing such activities as nest building, feeding the young, defending the territory, and mobbing predators. Thus, high-quality sites serve to attract additional birds (offspring from previous years). Because the

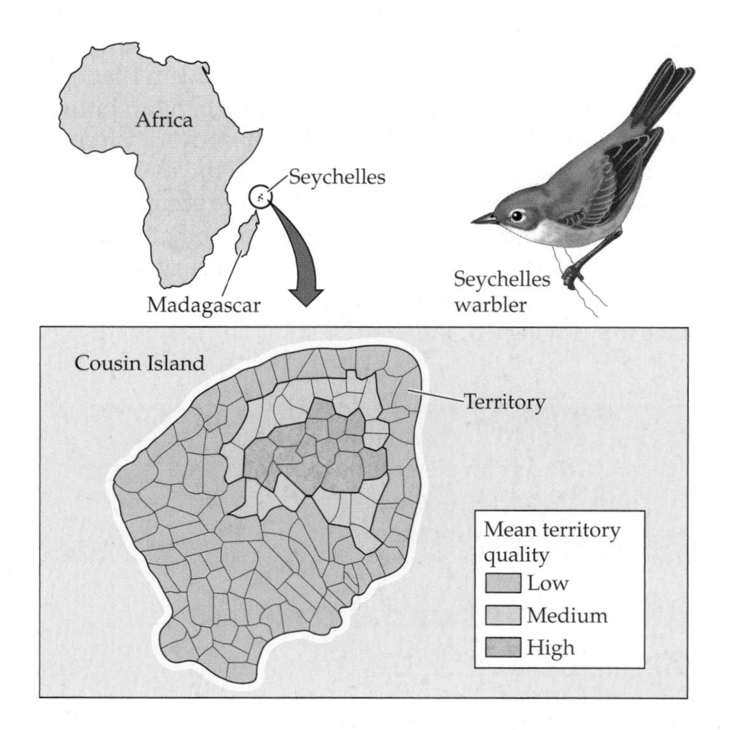

FIGURE 8.18 Territorial Behavior Affects Dispersion within Populations The mean quality across years of Seychelles warbler territories on Cousin Island from 1986 to 1990. Territory quality was calculated from data on territory size, vegetation cover, and insect abundance and grouped into three categories (high, medium, and low). High-quality territories were clustered inland; these territories had high vegetation cover, little wind, and abundant insects. Coastal areas had lower-quality territories because salt spray led to defoliation, which lowered insect abundance. (After Komdeur 1992.)

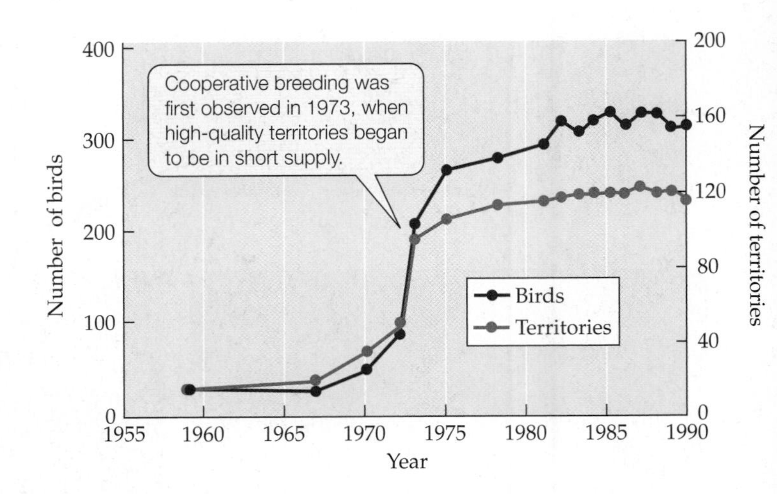

FIGURE 8.19 Cooperative Breeding Can Result from a Lack of Territories The number of Seychelles warblers increased over time, leading to a shortage of high-quality territories. (After Komdeur 1992.)

high-quality sites attract more birds and are aggregated toward one end of the island, differences in territory quality make the dispersion of individuals in the population more clumped than it otherwise would be.

Why do these young birds help their parents? The answer seems to hinge on the availability of high-quality territories. When high-quality territories are scarce, a young bird born in a high-quality territory produces more offspring over its lifetime if it remains with its parents and helps them for several years than if it leaves the nest right away and breeds in a territory of lower quality (Komdeur 1992). From this observation, we might predict that young birds should not help their parents if there is no shortage of high-quality breeding sites. Observations on Cousin Island are consistent with this prediction: as the population of Seychelles warblers increased from its low of 26, young birds did not start engaging in cooperative breeding until 1973, when high-quality territories began to be in short supply (**Figure 8.19**). Similar results were found in an experiment in which warblers were transferred from Cousin Island to a nearby island with apparently suitable habitat, but no Seychelles warblers. When the birds were first transferred, no cooperative breeding was observed. However, after 2 years, all of the high-quality territories were occupied, and, as predicted, cooperative breeding began at that time.

As we've seen, population sizes often vary greatly at a number of scales over time and space. Thus, a key challenge in studying a population is estimating its abundance.

CONCEPT 8.5 Population abundances and distributions can be estimated with area-based counts, distance methods, mark–recapture studies, and niche modeling.

Estimating Abundances and Distributions

The most direct way to determine how many individuals live in a population is to count all of them. This sounds simple enough, and it is possible in some cases, as for the Seychelles warblers on Cousin Island, the *Clematis fremontii* in a Missouri glade, and other organisms that are confined to a small area, are easy to see, or do not move. But complete counts of organisms are often difficult or impossible. Consider the chinch bug (*Blissus leucopterus*), an insect that attacks crops such as corn and wheat. This insect can cover large areas and reach densities that exceed 5,000 individuals per square meter, making it impractical to count all the individuals in a population. In such cases, a variety of methods can be used to estimate abundance.

Ecologists estimate abundance with area-based counts, distance methods, and mark–recapture studies

Many ecological studies require an estimate of a population's actual abundance, or **absolute population size**. For example, to quantify the extent to which the number of wolves affects and is affected by the number of their elk prey, we must estimate the absolute population size of both species. In other cases, it may be sufficient to estimate the **relative population size**, the number of individuals in one time period or place *relative to* the number in another. Estimates of relative population size are based on data that are presumed to be correlated with absolute population size but do not assess the actual number of individuals in the population. Examples of such data include the number of cougar tracks found in an area, the number of fish caught per unit of effort (e.g., per number of hooks trolled each day), or the number of birds observed while the observer walks a standard distance (or remains in one place for a standard time interval).

Relative population size estimates are usually easier and less expensive to obtain than are absolute estimates. While useful, estimates of relative population size must be interpreted carefully. The number of cougar tracks observed, for example, depends not only on cougar population density, but also on animal activity. Thus, if twice as many tracks were found in area A as in area B, we could not be confident that area A had twice as many cougars—there could be more or fewer than that, depending on whether cougars moved more frequently in one area than in another.

With the distinction between absolute and relative population size as background, we turn now to how ecologists estimate abundance. In **Ecological Toolkit 8.1**, we describe three common approaches: area-based counts, distance methods, and mark–recapture studies.

Area-based counts As described in Ecological Toolkit 8.1A, area-based counts are often used to estimate the population sizes of immobile organisms. In this approach, organisms are counted in a series of sample plots, or quadrats, and the

ECOLOGICAL TOOLKIT

8.1 Estimating Abundance

Methods for estimating abundance fall into three general categories: area-based counts, distance methods, and mark–recapture studies. Many variations on these approaches have been developed, and a wide range of statistical techniques are available for analyzing abundance estimates obtained using each of them (Krebs 1999; Williams et al. 2002).

A. Area-based counts In an *area-based count*, as its name suggests, the individuals in a given area or volume are counted. This method may make use of a **quadrat** (**Figure A**), which is a sampling area (or volume) of any size or shape, such as a 1 m^2 circular plot used to count small plants, a 0.1 ha plot used to count trees, or a soil core of a certain diameter and depth used to count soil organisms. The counts from several quadrats are then summed and averaged to estimate the size of the population.

Area-based counts are often used to estimate absolute population sizes of organisms that are sessile (e.g., plants) or can move only short distances during the time it takes to count the individuals in a quadrat (e.g., sea urchins). Area-based counts can also be used to estimate the abundances of more mobile organisms, as when large mammals are observed in aerial surveys. Area-based counts of highly mobile organisms can provide estimates of relative population sizes; further information (such as the probability that an organism will be present but not seen when surveyed by air) may be required before such counts can be used to estimate absolute population sizes.

B. Distance methods In *distance methods*, an observer measures the distances of individuals seen from a line or a point; these distances are then converted into estimates of the number of individuals per unit of area. For example, distance methods often use **line transects**, straight lines from which the distance to each individual is measured (**Figure B**). For organisms that move quickly or are hard to detect, the number of individuals observed along a line transect provides an estimate of relative population size. For both mobile and sessile organisms, distances recorded along a transect can also be used to estimate the absolute abundance; this conversion can be made if it is possible to determine a *detection function*, which accounts for how the chance of seeing an individual decreases with its distance from the transect. Other distance methods include *point sampling* techniques, in which the distance to the nearest (visible) individual is measured from a series of locations or "points"; as with line transect data, a detection function is used to convert these distances into estimates of the absolute population size (see Krebs 1999; Schwarz and Seber 1999).

C. Mark–recapture studies In *mark–recapture studies*, a subset of the individuals in a population is captured, marked (as with a tag or dot of paint) so that they can be recognized at a later time, and released (**Figure C**). After the marked individuals have been given enough time to recover and move throughout the population, individuals are captured a second time, and the proportion of marked individuals found in the second capture is used to estimate the total population size.

Figure C Release of Marked Salmon
To obtain mark–recapture estimates of salmon abundance, ecologists tag (see the two tags near the dorsal fin) and then release marked salmon.

Mark–recapture methods are used to estimate the absolute population size of mobile organisms; they are also used to obtain data on the survival or movement of individuals. The simplest mark–recapture method is summarized by Equation 8.3 (see p. 192); use of this equation assumes that (1) the population size does not change during the sampling period (no births, deaths, immigration, or emigration); (2) each individual has an equal chance of being caught; (3) marking does not harm individuals or alter their behavior (as by making them harder to recapture); and (4) marks are not lost over time. A wide range of other mark–recapture methods have been developed to address cases in which one or more of these assumptions are violated (Krebs 1999; Schwarz and Seber 1999; Williams et al. 2002).

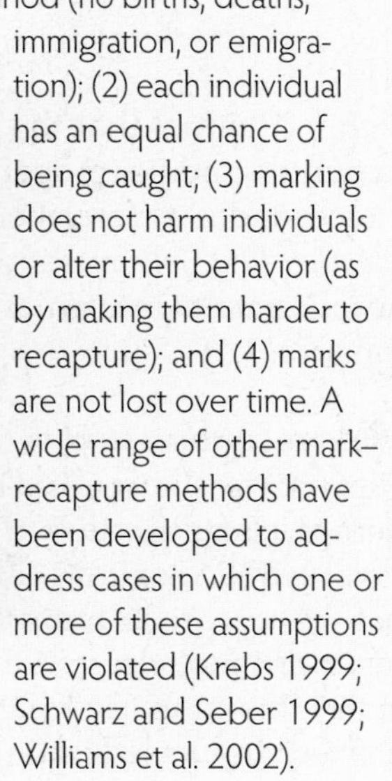

Figure A An Underwater Quadrat
A marine biologist counts the abundances of coral species found on a reef off the Caroline Islands, Micronesia.

Figure B Counting Acacia Trees from a Line Transect
The density of these camelthorn trees (*Acacia erioloba*) in Kgalagadi Transfrontier Park (South Africa) could be estimated using a line transect, as shown here.

resulting numbers are used to estimate the total population size. Suppose, for example, that a team of entomologists wants to estimate the population of chinch bugs in a 400 ha (ca. 1,000 acre) field of corn. If they counted chinch bugs in five 10 cm × 10 cm quadrats (i.e., five 0.01 m^2 quadrats), and their counts were 40, 10, 70, 80, and 50 chinch bugs, they would estimate that there were an average of

$$\frac{(40+10+70+80+50)/5}{0.01} = 5{,}000 \qquad \textbf{(8.1)}$$

chinch bugs per square meter. Thus, there would be an estimated 20 billion chinch bugs in the population.

Area-based methods work well if individuals can be counted accurately within the quadrats, and if the quadrats provide a good representation of the entire area covered by the population. To help ensure that the latter condition is met, ecologists use as many quadrats as is feasible, and they often place these quadrats at locations selected at random from the entire area covered by the population. Quadrats can also be placed in a variety of other ways, such as at evenly spaced locations along a rectangular grid.

Distance methods Estimates of abundance can also be based on various measurements of distance from a point or line. Such distance methods are "plotless" in that they are not made by counting individuals located within a given area or volume. For example, in the line transect approach, an observer travels by foot, horseback, or vehicle along a transect line, as shown here:

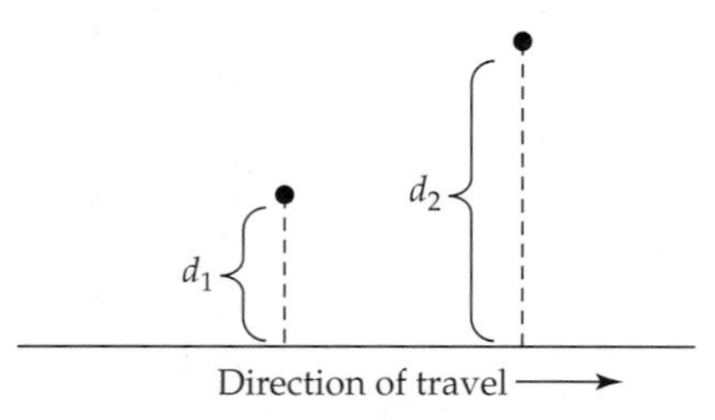

Each individual that the observer can see from the line is counted and its (perpendicular) distance from the line is recorded. As described in Ecological Toolkit 8.1B, a *detection function* must be used to convert such distance measurements into an estimate of the absolute population size.

Mark–recapture studies The mark–recapture approach relies on releasing marked individuals and then recapturing them at a later time to see what fraction of the population is marked (Ecological Toolkit 8.1C). Imagine, for example, that we capture 23 butterflies from a meadow, which we then mark and release. A day later, we sample the meadow again, this time catching 15 butterflies, of which 4 are marked. In our first sample, we caught and marked $M = 23$ butterflies from a total population of unknown size (N); thus, we initially caught a proportion M/N of the butterflies in the field. The second time butterflies were sampled, we caught $C = 15$ butterflies, of which 4 were marked and hence were recaptured ($R = 4$).

Assuming that no butterfly births, deaths, or movements into or out of the meadow have occurred since our first sample, the proportion of marked individuals captured in our second sample (R/C) should equal the original proportion we caught, M/N. Thus, we have the equation

$$M/N = R/C \qquad \textbf{(8.2)}$$

We can rearrange Equation 8.2 to estimate the total number of butterflies in the meadow as

$$N = (M \times C)/R \qquad \textbf{(8.3)}$$

which in this case would equal $(23 \times 15)/4 = 86$.

We'll close this section with two examples of how abundance and distribution data are collected and used, the first of which highlights how long-term ecological data sets can contribute to efforts to solve applied problems.

Ecologists used abundance data to track down a mysterious disease

In 1993, dozens of people in the Four Corners region of the southwestern United States became sick with flulike symptoms and shortness of breath, and 60% of them died within a few days of becoming ill. No one had seen this combination of symptoms before. An outbreak of a lethal, previously unknown disease appeared to be in progress, and there was no cure or successful treatment.

The U.S. Centers for Disease Control (CDC) quickly identified the disease agent as a new strain of hantavirus carried by the deer mouse (*Peromyscus maniculatus*). Seeking more information about the new disease, now known as hantavirus pulmonary syndrome, or HPS, the CDC contacted ecologists who had been studying mouse populations in the Southwest. Examination of deer mouse specimens collected between 1979 and 1992 revealed that the virus had been present in the area for more than 10 years prior to the outbreak. Why, then, did the outbreak of HPS occur in 1993 and not before?

To address this question, ecologists turned to data on the abundances of *Peromyscus* species collected at the nearby Sevilleta National Wildlife Refuge since 1989. These data showed that the densities of several *Peromyscus* species had increased 3–20-fold between 1992 and 1993. Next, a series of satellite images was used to develop an index of how much plant matter was available as food for *Peromyscus* at different times. When that index was compared with precipitation data, the results suggested that unusually high rainfall from September 1991 through May 1992 had led to enhanced plant growth in spring 1992 (**Figure 8.20**). In turn, the enhanced plant growth produced abundant food for rodents (seeds, berries, green plant matter, arthropods), which allowed mouse populations to increase in size by 1993—the year of the HPS outbreak.

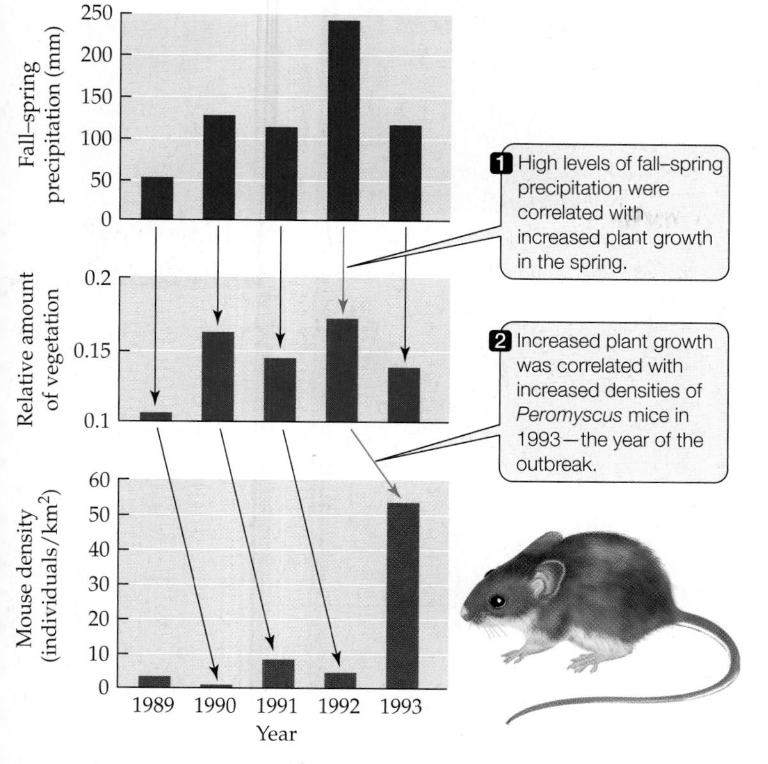

FIGURE 8.20 From Rain to Plants to Mice The outbreak of hantavirus pulmonary syndrome in the southwestern United States in 1993 may have been caused by a series of interconnected events. (After Yates et al. 2002.)

Rodents shed hantavirus in their urine, feces, and saliva; hence, high mouse numbers leading to increased mouse–human contact was thought to be the cause of the 1993 outbreak. For *Peromyscus* to spread HPS to people, however, the mice must be infected with hantavirus. Disease outbreaks are most likely to occur when both mouse densities and mouse infection rates are high, but the factors that cause that outcome remain only partially understood. Moreover, it is clear that the actual risk to people varies greatly with location and depends on such factors as habitat type (which can influence mouse movements), microclimate (e.g., in arid regions, nearby areas often experience very different amounts of rainfall), and local food abundance. Overall, we now know enough about this new disease to predict periods of heightened risk to human populations, but more remains to be learned before we can predict the specific locations most at risk.

How can we predict exactly where infected deer mice will be found? We turn next to a technique that has proved helpful in predicting the locations of other organisms: niche models.

Niche models can be used to predict where species can be found

To determine the geographic distribution of a species, scientists record all locations where the species is found. Most of our examples thus far in this chapter have involved species whose distributions are well understood. However, there are many species whose geographic ranges are not yet known. Furthermore, ecologists often want to predict *future* distributions of species—for example, whether and how a pest species will spread after it has been introduced to a new geographic region. Scientists and policymakers face similar challenges when they seek to predict how distributions of species will shift in response to global climate change.

How can we predict the current or future distributions of species? One way is to characterize a species' **ecological niche**, the physical and biological conditions that the species needs to grow, survive, and reproduce. For example, data collected from areas where a species is known to occur can be used to construct a **niche model**, a tool that predicts a species' geographic distribution based on the environmental conditions at locations the species is known to occupy.

Investigators from the United States and Mexico used such an approach to predict the distributions of chameleons in Madagascar (Raxworthy et al. 2003). The researchers obtained information about vegetation cover (from satellite images), temperature, precipitation, topography (elevation, slope, aspect), and hydrology (water flow, tendency to pool) from government and commercial sources. Values for these environmental variables were recorded for each of a series of 1×1 km^2 areas (referred to as "grid cells") that covered all of Madagascar. Next, for each of 11 chameleon species, rules were developed that described the environmental conditions where each species was most likely to be found; we'll refer to these rules as "habitat rules."

There are many different ways to develop such habitat rules. The chameleon study used a computer program that compared the environmental conditions of grid cells selected at random from a map of Madagascar with the environmental conditions of grid cells where a chameleon species was known to occur. The program then searched for accurate habitat rules using a flexible approach known as GARP (Genetic Algorithm for Rule-set Prediction). This search mechanism works by changing habitat rules in a way that mimics the occurrence of genetic mutations (random changes in the DNA sequence of an organism) and natural selection. For example, initially a habitat rule might state that a species should be found in regions where the temperature ranges from 15°C to 25°C and the elevation ranges from 300 to 550 m. This rule might "mutate" at random to have a temperature range of 15°C–30°C and an elevation range of 300–500 m. If the new rule improves the ability of the program to predict where the species is actually found, it is retained, and other, less successful rules are discarded.

For the Madagascar chameleons, the accuracy of the niche model developed using GARP was tested with chameleon location data that had not been entered into the program. The model performed well, correctly predicting where these chameleons lived 75%–85% of the time. Next, the model was used to predict the geographic ranges of each of the 11 chameleon species—information that will be useful in efforts to protect chameleon habitat. Finally, the researchers investigated an interesting "error" in the model: there were several overlapping areas in which the

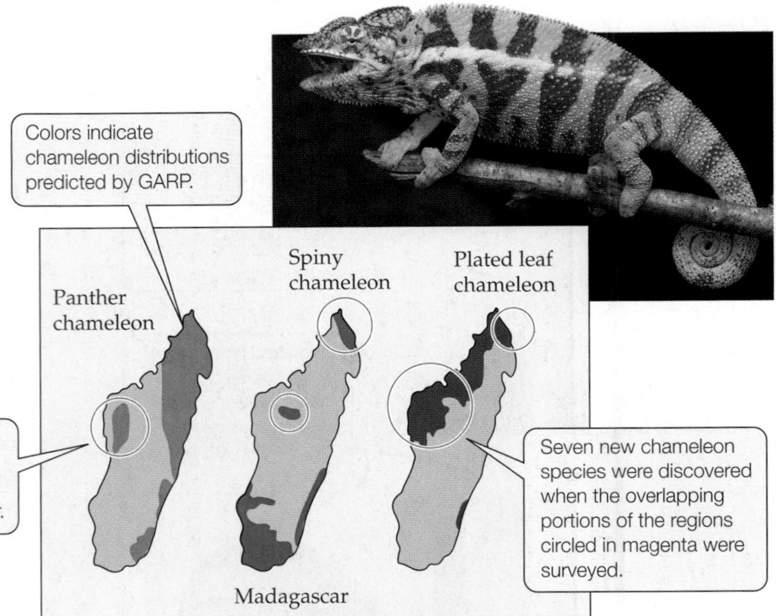

FIGURE 8.21 Predicted Distributions of Madagascar Chameleons The distributions predicted by GARP are shown for the panther chameleon (*Furcifer pardalis*), the spiny chameleon (*F. verrucosus*), and the plated leaf chameleon (*Brookesia stumpffi*), three of the eleven Madagascar chameleon species studied by Raxworthy et al. All eleven of the predicted distributions proved accurate. (After Raxworthy et al. 2003.)

model predicted that 2 or more of the 11 species would be found, but in which no chameleons were known to occur (**Figure 8.21**). When two of these overlapping areas were surveyed, seven previously unknown chameleon species were discovered. More intensive surveys conducted at the same time, but at sites outside these overlapping areas, found only two new species. Thus, the scientists were able to predict both the distributions of known chameleon species and the locations of habitats suitable for other chameleons, and the latter prediction led to the discovery of seven new chameleon species.

A CASE STUDY REVISITED

From Kelp Forest to Urchin Barren

When sea urchins graze kelp so heavily that kelp forests are replaced by urchin barrens, what happens next? We might expect that the urchins would starve because they have destroyed their food source. Field studies show that urchin barrens can persist for years on end, however, because urchins can use food sources other than kelp, including benthic diatoms, less preferred algae (including hard, encrusting forms that cover rock surfaces), and detritus. When food is extremely scarce, urchins can reduce their metabolic rate, reabsorb their sex organs (forgoing reproduction but increasing their chances of survival), and absorb dissolved nutrients directly from seawater.

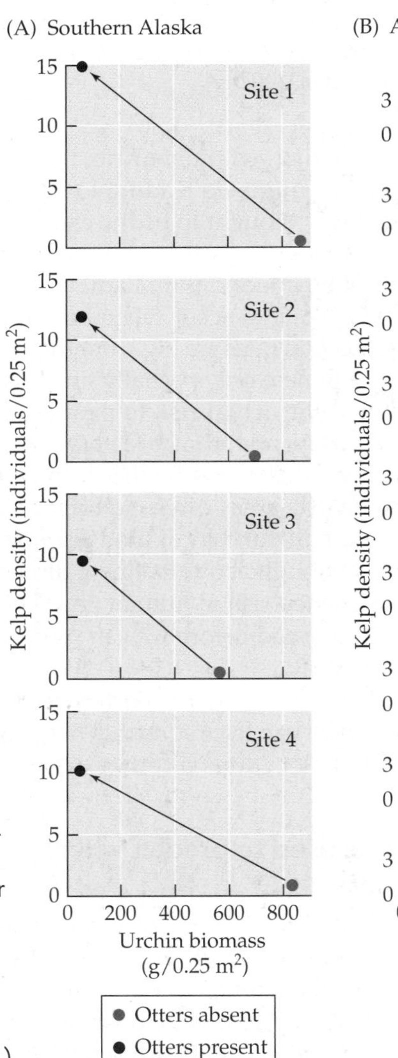

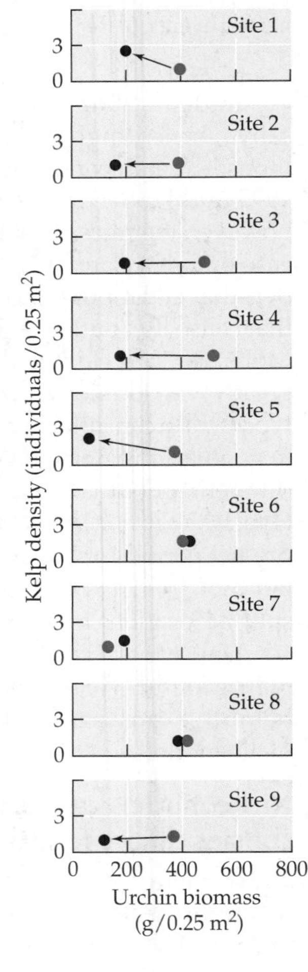

FIGURE 8.22 The Effect of Otters on Urchins and Kelp Plots of kelp density versus sea urchin biomass measured at sites in southern Alaska and in the Aleutian Islands before and 2 years after the return of otters. (A) Two years after otters colonized each of four sites in southern Alaska, urchin biomass had declined considerably, and kelp density had increased substantially. (B) Two years after otters colonized nine sites in the Aleutian Islands, sea urchin biomass had declined at six of the sites, but kelp showed clear signs of recovery only at sites 1 and 5. Arrows indicate a decline in urchin biomass and (only in some sites) an increase in kelp density in the presence of otters. (After Estes and Duggins 1995.)

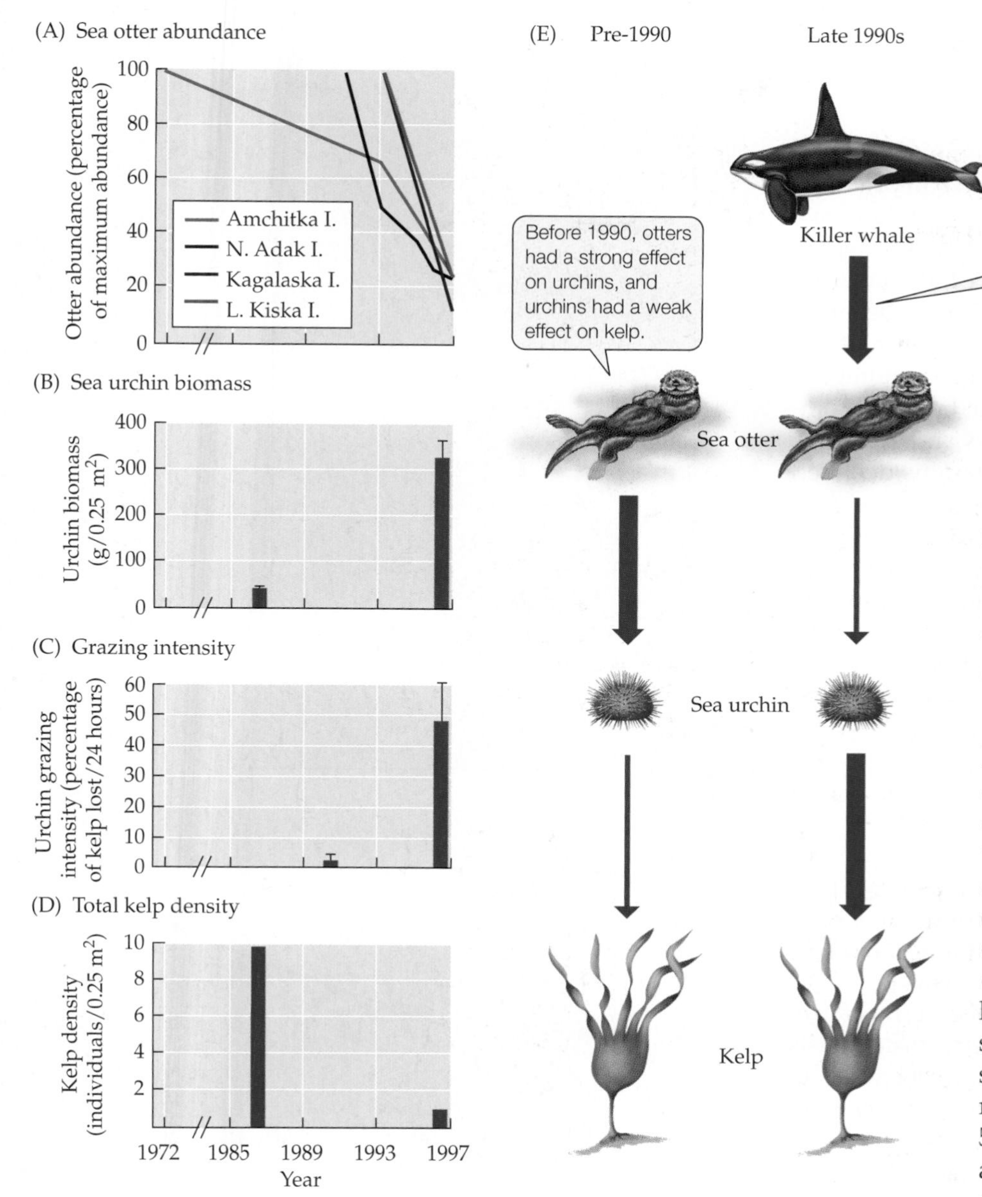

FIGURE 8.23 Killer Whale Predation on Otters May Have Led to Kelp Declines Declines in otter abundance over time (A) are associated with (B) a rise in urchin biomass, (C) an increase in the intensity of urchin grazing on kelp, and (D) a decrease in kelp density. (E) The proposed mechanisms for these changes. The strengths of the effects are indicated by the thickness of the arrows. Error bars in (B) and (C) show one SE of the mean. (After Estes et al. 1998.)

As tough and resilient as urchins are, they are vulnerable to predation by sea otters (*Enhydra lutris*), which function as impressive urchin-eating machines. Otters need to eat large quantities of food each day because they have a high metabolic rate and they store little energy as fat; in captivity, otters eat 20%–23% of their body weight in food each day. Urchins are a favorite food of otters, and since there are 20–30 otters per square kilometer around some Aleutian islands, the potential exists for otters to consume enormous quantities of urchins. These facts, coupled with the observation that urchins usually are common only where otters are absent, led investigators to suspect that otters might control the locations of urchins, and hence the locations of kelp forests.

To test this hypothesis, Estes and Duggins (1995) compared sites with and without otters, both in the Aleutian Islands and along the coast of southern Alaska. Confirming the results of previous studies, they found that sites where otters had been present for a long time usually had many kelp and few urchins, whereas sites without otters usually had many urchins and few kelp. Estes and Duggins also collected data from sites colonized by otters during the course of their study. At sites in southern Alaska, the arrival of otters had a rapid and dramatic effect: within 2 years, urchins virtually disappeared, and kelp densities increased dramatically (**Figure 8.22**). At Aleutian Islands sites, however, kelp recovered more slowly after the arrival of otters. At these sites, otters ate most of the large urchins, reducing urchin biomass by an average of 50%. However, in a twist that did not occur at the southern Alaska sites, the arrival of new urchin larvae (most likely via ocean currents) provided a steady supply of small urchins. These small urchins slowed the rate at which kelp forests replaced urchin barrens.

Historically, sea otters were abundant throughout the North Pacific, but by 1900 they had been hunted (for fur) to near-extinction. By 1911, when international treaties protected the sea otter, only about a thousand otters remained—less than 1% of their original numbers. Scattered colonies of otters survived around some Aleutian islands, causing the observed pattern of kelp forests around some islands, urchin barrens around others. By the 1970s, otter populations had recovered to between 110,000 and 160,000 individuals. In the 1990s, however, there was a sudden and unexpected decline in otter populations. Urchins made a comeback, and kelp densities were reduced (**Figure 8.23A–D**). The question now became, what caused the decline of sea otter populations?

James Estes and his colleagues have suggested that otters declined because of increased predation by the killer whale, *Orcinus orca* (**Figure 8.23E**). It is not known why killer whales began to eat more otters. Some researchers have argued that this change may have been part of a chain of events that began when commercial whaling drove populations of large whales to low numbers (Springer et al. 2003). According to this hypothesis, once their preferred prey (large whales) became rare, killer whales began to hunt a series of other species (first harbor seals, then fur seals, then sea lions), which then also declined in numbers. Other researchers dispute the connection between commercial whaling and the decline of seals and sea lions, suggesting that seal and sea lion populations declined for other reasons, such as a lack of food due to reduced fish populations in the open ocean (DeMaster et al. 2006). Whatever the cause, it was in the 1990s, when populations of harbor seals, fur seals, and sea lions had all declined to low levels, that killer whales were first seen attacking otters. Otters and killer whales had been observed in close proximity for decades, but within 10 years of the first known attack, otter populations crashed.

CONNECTIONS IN NATURE

From Urchins to Ecosystems

Urchins, otters, and perhaps killer whales and people play an important role in determining the distribution of kelp. But does the presence or absence of kelp matter? Do kelp have strong effects on nearshore ecosystems?

Indeed they do. Kelp forests are among the most productive ecosystems in the world, rivaling tropical forests in the amount of new biomass they produce each year (up to 2,000 grams of carbon per square meter per year). Kelp strands grow from their base, and their tips are constantly "eroded" by wave action and other physical forces. Thus, much of their biomass ends up as floating bits of detritus, which provides food for suspension feeders such as barnacles and mussels that filter food from the water. As a result, barnacles and mussels grow more rapidly and are more abundant in a kelp forest than in an urchin barren. Carbon-13 labeling studies (see Ecological Toolkit 5.1) have shown that the sugars kelp produce by photosynthesis provide a food source for a wide range of species (Duggins et al. 1989). Kelp forests also serve as nurseries for the young of many marine species and as havens from predators for the adults of still more species.

Overall, kelp has a large effect on nearshore ecosystems. Hence, the effects of otters on urchins matter. Otters (and perhaps killer whales and people) set into motion a chain of events that alters fundamental aspects of the marine ecosystem. Moreover, it appears that this phenomenon is not unique to Alaskan kelp forests. For example, a similar chain of events seems to have affected kelp ecosystems along the coast of Tasmania, Australia. However, the events in Tasmania may be driven by an additional factor: climate change.

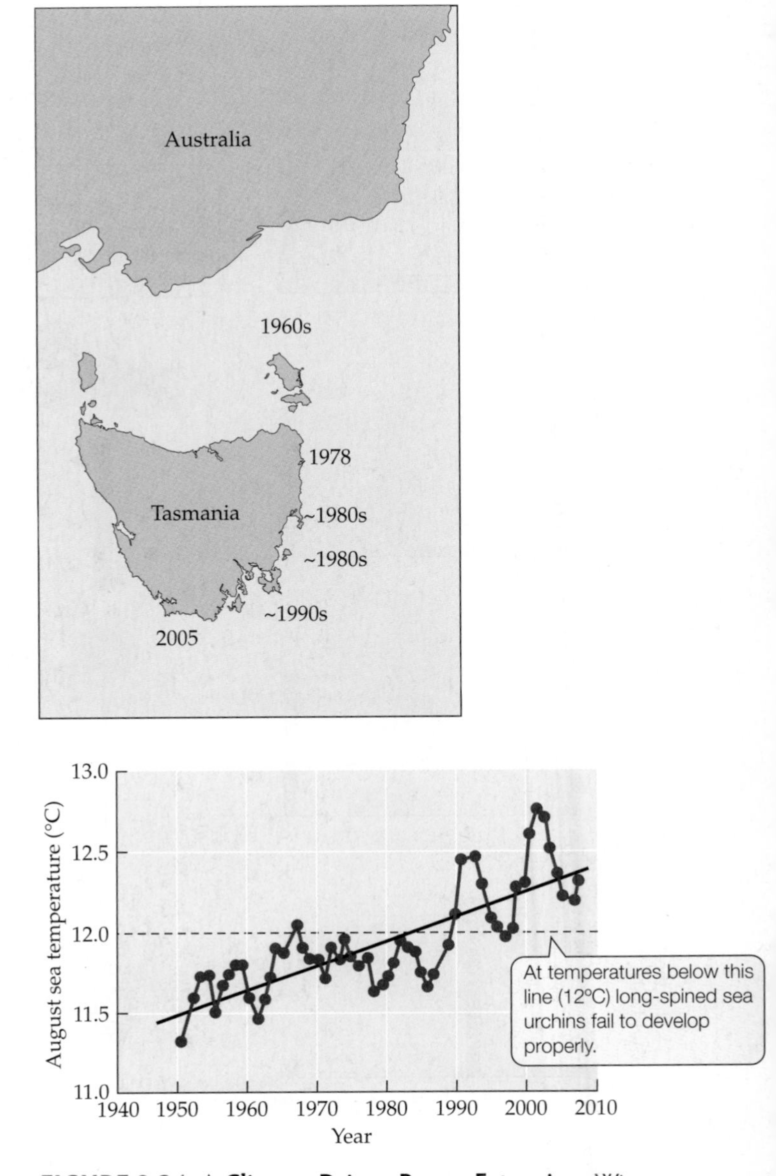

FIGURE 8.24 A Climate-Driven Range Extension Winter water temperatures along the east coast of Tasmania in August, the most important month for offspring production in long-spined sea urchins. The inset map shows the years in which long-spined sea urchins were first observed at points along the Tasmanian coast. (After Ling et al. 2009.)

The waters along the east coast of Tasmania have warmed considerably over the past 60 years (**Figure 8.24**). As has been observed in locations throughout the world (Parmesan and Yohe 2003), the warming in Tasmania has been associated with changes in the geographic distributions of species. In particular, as waters along the coast have warmed, the long-spined sea urchin (*Centrostephanus rodgersii*) has extended its range to the south (see the map in Figure 8.24). The changes in the distribution of this urchin are consistent with climate change being the underlying cause: the larvae of *C. rodgersii* fail to develop properly in waters colder than 12°C, and the urchin has moved into new regions as waters in those locations warmed to the point that they remained above that temperature. As *C. rodgersii* has expanded its range, it has established extensive urchin barrens in which all kelp have been removed by grazing (Ling 2008). Thus, through its effects on the long-spined sea urchin, ongoing climate change appears to be having a profound effect on kelp ecosystems along the Tasmanian coast. (For more information about how this example connects to other levels of the ecological hierarchy, see **Climate Change Connection 8.1**.)

SUMMARY

CONCEPT 8.1 Populations are dynamic entities that vary in size over time and space.

- Populations are groups of individuals of a species that live in the same area and interact with one another.
- The number of individuals in a population changes over time and from one place to another.
- Dispersal can link the populations of a species to one another.
- In species that can reproduce asexually, the members of a population can be defined in terms of genetic individuals (genets) or physiological individuals (ramets).

CONCEPT 8.2 The distributions and abundances of organisms are limited by habitat suitability, historical factors, and dispersal.

- The presence of suitable habitat limits the distribution and abundance of organisms.
- The suitability of habitat depends on abiotic and biotic features of the environment, the interaction between abiotic and biotic factors, and disturbance.
- Dispersal and events in the evolutionary and geologic history of Earth also influence the distribution and abundance of organisms.

CONCEPT 8.3 Many species have a patchy distribution of populations across their geographic range.

- No species is found everywhere because much of Earth is not suitable habitat for its populations.
- Geographic ranges vary considerably in size from one species to another.
- Many populations have a patchy structure at both small and large spatial scales.

CONCEPT 8.4 The dispersion of individuals within a population depends on the location of essential resources, competition, dispersal, and behavioral interactions.

- The dispersion of individuals within a population may be regular, random, or clumped. In the field, clumped dispersions are most common.
- A random or clumped dispersion may match the spatial arrangement of important resources. Clumping may also result from short dispersal distances, and competition can produce a nearly regular dispersion.
- Behavioral interactions in which individuals repel or attract one another can affect the dispersion of individuals within a population.

CONCEPT 8.5 Population abundances and distributions can be estimated with area-based counts, distance methods, mark–recapture studies, and niche modeling.

- The most direct way to determine the number of individuals in a population is to count all of them. When this is not possible or practical, area-based counts, distance methods, or mark–recapture studies can be used to estimate the number of individuals in a population.
- The geographic distribution of an organism can be analyzed in terms of its ecological niche, the physical and biological conditions of the environment that the organism needs to grow, survive, and reproduce.
- Niche models can be used to estimate the distribution of an organism when we have insufficient data on its geographic range or when we want to predict the future locations of its populations.

REVIEW QUESTIONS

1. Describe some of the complicating factors that can be encountered in studying a population.
2. No species is found everywhere on Earth. Why? Your answer should include an explanation of why organisms are not found in all places where you might expect them to thrive.
3. What are niche models? Describe how such a model could be used to predict the future distribution of an organism that is spreading into a new geographic region.
4. Sea otters can eat 20%–23% of their body weight in food each day. An average sea otter weighs 23 kg (roughly 50 pounds), and there are 20–30 otters per square kilometer where they are present. An average sea urchin weighs 0.55 kg. Assuming the otters eat only sea urchins, use these data to calculate a conservative estimate of the number of sea urchins per square kilometer that an otter population would be expected to eat each year.

ON THE COMPANION WEBSITE
sites.sinauer.com/ecology2e

The website includes Chapter Outlines, Online Quizzes, Flashcards & Key Terms, Suggested Readings, a complete Glossary, and the Web Stats Review. In addition, the following resources are available for this chapter:

HANDS-ON PROBLEM SOLVING

Effort and Accuracy of Population Estimates

This Web exercise illustrates the relationship between the effort required to obtain population size estimates and their accuracy. Species and population characteristics influence the ease of obtaining population estimates and the accuracy of those estimates. You will choose a method of population estimation and manipulate the amount of effort, to explore the effects on estimate accuracy.

WEB EXTENSION

8.1 An Experimental Study on Dispersal and Abundance in Desert Pupfish

CLIMATE CHANGE CONNECTIONS

8.1 Joint Effects of Climate Change and Overfishing

CHAPTER

9

Population Growth and Regulation

KEY CONCEPTS

- **CONCEPT 9.1** Life tables show how survival and reproductive rates vary with age, size, or life cycle stage.
- **CONCEPT 9.2** Life table data can be used to project the future age structure, size, and growth rate of a population.
- **CONCEPT 9.3** Populations can grow exponentially when conditions are favorable, but exponential growth cannot continue indefinitely.
- **CONCEPT 9.4** Population size can be determined by density-dependent and density-independent factors.
- **CONCEPT 9.5** The logistic equation incorporates limits to growth and shows how a population may stabilize at a maximum size, the carrying capacity.

Human Population Growth: A Case Study

Viewed from space, Earth appears as a beautiful ball of blue and white in a vast sea of black. If we use satellite images to explore the surface of this beautiful ball in more detail, we find clear signs of human impacts across the globe. These signs range from the clear-cutting of tropical forests, to rivers that once meandered but now flow straight in channels dug by people, to surrealistic patterns formed by agricultural fields (**Figure 9.1**).

People have a large effect on the global environment for two underlying reasons: our population has grown explosively, and so has our use of energy and resources. The human population crossed the 6.8 billion mark in 2010, more than double the 3 billion people alive in 1960 (**Figure 9.2**). Our use of energy and resources has grown even more rapidly. From 1860 to 1991, for example, the human population quadrupled in size, but our energy consumption increased 93-fold.

The addition of nearly 4 billion people since 1960 is remarkable. For thousands of years, the size of our population increased relatively slowly, reaching 1 billion for the first time in 1825 (Cohen 1995). The time we took to reach the 1 billion mark puts the current growth of our population in perspective: it took roughly 200,000 years (from the origin of our species to 1825) for the human population to reach its first billion, but now we are adding 1 billion people every 13 years. When did we switch from relatively slow to explosive increases in the size of our population?

No one knows for sure, since it is difficult to estimate population sizes from long ago. According to the best information we have, by 1550 there were roughly 500 million people alive, and the population was doubling every 275 years. By the time we reached our first billion in 1825, the human population was growing at a very rapid rate: it doubled from 1 to 2 billion by 1930, in just 105 years. Forty-five years later, it had doubled again, reaching 4 billion in 1975, at which time it was growing at an annual rate of nearly 2%. To appreciate what that means, a population with a 2% annual growth rate doubles in size every 35 years. If that rate of growth could be sustained, our population would double from 6.8 billion in 2010 to 13.6 billion in 2045, and double again to more than 27 billion in 2080. Should there be 27 billion people in 2080, a long-lived person born in 1975 and living to 2080 would have seen the human population increase nearly sevenfold in their lifetime.

What do you think the world would be like with 27 billion people? Already, with fewer than 7 billion people, we have transformed the planet. It is unlikely that

FIGURE 9.1 Transforming the Planet Irrigation systems that pivot around a central point create red circles of healthy vegetation in this satellite image of croplands near Garden City, Kansas.

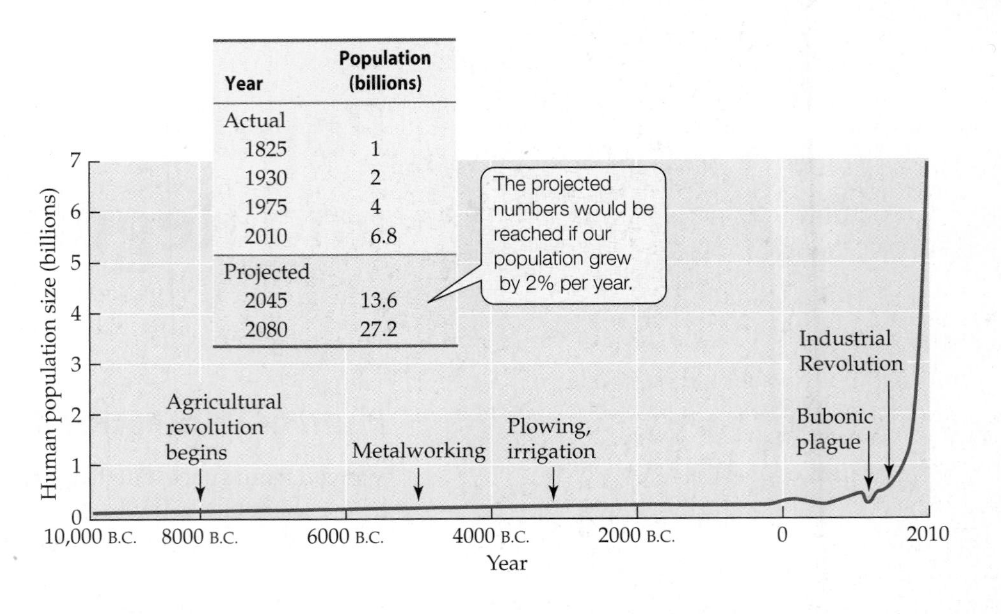

FIGURE 9.2 Explosive Growth of the Human Population The number of people in the human population increased relatively slowly until 1825, when the effects of the Industrial Revolution took hold, but since that time our population has increased in size by nearly 6 billion people.

there will be 27 billion people on Earth in 2080, however. Over the last 50 years, the annual rate of human population growth has slowed considerably, from a high of 2.2% per year in the early 1960s to the present rate of 1.18% annually. Currently, the human population increases by about 80 million people per year, more than 9,100 people each hour. Five countries account for almost half of this annual increase: India (21%), China (11%), Pakistan (5%), and Nigeria and the United States (4% each). If the current annual growth rate of 1.18% were maintained from now until 2080, there would be more than 15 billion people on Earth in 2080.

Could Earth support 15 billion people? Will there be that many people in 2080? Or will annual growth rates continue to fall? We'll return to these questions at the close of this chapter.

Introduction

One of the ecological maxims introduced in Table 1.1 reads, "No population can increase in size forever." Earth is a finite planet and hence cannot support ever-increasing numbers of any species. The limits imposed by a finite planet restrict what otherwise appears to be a universal feature of all species: a capacity for rapid population growth. As we saw in this chapter's Case Study, the human population is increasing rapidly. Other organisms, such as fungi known as "giant puffballs," have an even more impressive capacity to increase their numbers. One of these fungi produces so many offspring per individual (7 *trillion*) that if all of them reached adulthood, the descendants of two individuals would weigh more than the entire planet in just two generations. But Earth is not covered with giant puffballs, or even people. The challenge for ecologists—whether they are studying giant puffballs, people, or any other organism—is to understand what factors promote population growth and what factors limit population growth.

What we learn can surprise us. We may find, for example, that current methods of protecting an endangered species work poorly. Such was the case for loggerhead sea turtles, a rare species whose young often die as they crawl to the sea after hatching from nests dug in the sand (**Figure 9.3**). Efforts to protect loggerheads initially focused on protecting newborns. However, researchers found that even if newborn survival could be increased to 100%, loggerhead populations would continue to decline. Fortunately, the researchers were able to use methods described in this chapter to identify more effective ways to protect loggerheads (see p. 208).

How do scientists reach such conclusions? What data are needed? As we seek to understand population growth and its limits, some of the most powerful tools at our disposal are based on life tables, the topic we explore next.

FIGURE 9.3 Dash to the Sea In the nesting environment, the eggs and hatchlings of loggerhead sea turtles face threats from a variety of factors, including predators, beach development, and artificial lighting (which can disorient the hatchlings, preventing them from reaching the sea). Loggerheads also face threats in the marine environment from predators, commercial fisheries (turtles can be caught accidentally in nets and tarps), boat collisions, and chemical and solid-waste pollutants.

CONCEPT 9.1 Life tables show how survival and reproductive rates vary with age, size, or life cycle stage.

Life Tables

Information about patterns of births and deaths in a population is essential if we want to understand current population trends or predict future population sizes. To obtain such information for a plant, for example, we could mark (e.g., with a numbered tag) a large number of individual seedlings shortly after they germinated. We could then follow the fate of each seedling over the course of one or more growing seasons. By recording whether each seedling was alive or dead at various points in time, we could estimate how the chance of surviving from one time period to the next varied with plant age. Similarly, by recording how many seeds each plant produced at different times, we could estimate how reproduction varied with age.

Life tables can be based on age, size, or life cycle stage

Scientists have collected data on patterns of births and deaths for plants, people, sea turtles, and many other organisms, and they have used these data to construct life tables. A **life table** provides a summary of how survival and reproductive rates vary with the age of the organisms. For example, **Table 9.1** shows data collected for the grass *Poa annua*. These data were collected by marking 843 naturally germinating seedlings and then following their fates over time. The second column on the left, labeled N_x, shows the number of individuals alive at age x, where x was measured in 3-month periods. As individuals die over time, N_x decreases steadily from the original 843 individuals, reaching 0 at $x = 8$ (24 months).

The next two columns in Table 9.1, S_x and l_x, are calculated from the N_x data. S_x is the age-specific **survival rate**, which is the chance that an individual of age x will survive to be age $x + 1$. S_2, for example, equals 0.6, which indicates that, on average, an individual of age $x = 2$ (6 months) has a 60% chance of surviving to reach age $x = 3$ (9 months). The next column, l_x, represents **survivorship**, which is the proportion of individuals that survive from birth (age 0) to age x. For example, l_3 equals 0.375, indicating that 37.5% of newborns survive to reach age $x = 3$. The final column, F_x, represents **fecundity**, which is the average number of offspring produced by a female while she is of age x.

Table 9.1 is an example of a **cohort life table**, in which the fate of a group of individuals born during the same time period (a *cohort*) is followed from birth to death. Cohort life tables are often used for plants or other sessile organisms because individuals can be marked and followed over time relatively easily. For organisms that are highly mobile or have long life spans, however (e.g., trees that live much longer than people), it is hard to observe the fate of individuals from birth to death. In some of these cases, a **static life table** can be used, in which the survival and reproduction of individuals of different ages during a single time period are recorded. To construct a static life table, one must be able to estimate the ages of the organisms under observation. Estimating ages is difficult in some species, but for others, reliable indicators of age are known, including annual growth rings in fish scales and tree wood and tooth wear in deer. Once ages have been estimated, age-specific birth rates can be determined by counting how many offspring the individuals of different ages produce. Age-specific survival rates can also be determined from a static life table (see Review Question 1), but only if we assume that survival rates have remained constant during the entire time that the individuals in the population have been alive—an assumption that may not be correct.

In discussing life tables, we have emphasized the importance of age because in many species birth and death rates differ greatly among individuals of different ages. For other kinds of organisms, age is less important. In many plant species, for example, if conditions are favorable, a seedling may grow to full size relatively rapidly and reproduce at a young age. If conditions are not favorable, however, the plant may remain small for years and reproduce little or not at all; if conditions become favorable at a later time,

TABLE 9.1

Life Table for the Grass *Poa annua*

Age (in 3-month periods) x	Number alive N_x	Survival rate S_x	Survivorship l_x	Fecundity F_x
0	843	0.856	1.000	0
1	722	0.730	0.856	300
2	527	0.600	0.625	620
3	316	0.456	0.375	430
4	144	0.375	0.171	210
5	54	0.278	0.064	60
6	15	0.200	0.018	30
7	3	0.000	0.004	10
8	0		0.000	

Source: Data from Table 1.1 in Begon et al. 1996.

Note: Age (x) is measured in 3-month periods, so an individual of age $x = 5$, for example, is 15 months old. N_x = number of individuals alive at age x. S_x = proportion of individuals of age x that survive to age $x + 1$; $S_x = N_{x+1}/N_x$. l_x = proportion of individuals that survive from birth (age 0) to age x; $l_x = N_x/N_0$. F_x = average number of offspring born to a female while she is of age x.

the plant may then grow to full size and reproduce. For such species, whether an individual reproduces or not is more closely related to size than to age. When birth and death rates correlate poorly with age, or when age is difficult to measure, life tables based on the sizes or the life cycle stages (e.g., newborn, juvenile, adult) of individuals in the population can be constructed.

Extensive life table data exist for people

Many economic, sociological, and medical applications rely on human life table data. Life insurance companies, for example, use census data to construct static life tables that provide a snapshot of current survival rates; they use these data to determine the premiums they charge customers of different ages. Let's consider two examples of human life tables, one from the United States, the other from Gambia.

The U.S. Centers for Disease Control and Prevention periodically release reports that provide life table data for people in the United States. Reports released in 2009 and 2010 provide information on the survivorship (l_x), fecundity (F_x), and life expectancy (expected number of years of life remaining) of U.S. girls and women of different ages (**Table 9.2**). To make their interpretation easier, such data can be graphed, as in **Figure 9.4**, which plots l_x data for U.S. females. This curve shows that survival probabilities for U.S. females remain high for many years; in fact, as Table 9.2 reveals, these survival probabilities do not begin to drop sharply until around age 70.

The data from the United States are in stark contrast to data from Gambia, a country located on the west coast of Africa. Moore et al. (1997) analyzed birth and death records for 3,102 people born in three Gambian villages between 1949 and 1994. They found that the season of birth had a lasting impact on adult survivorship: children born during the annual "hungry season" (July–October, when food stored from the previous year is in low supply) had lower survivorship than children born at other times of the year (see Figure 9.4). Their data also reveal large differences between the survivorship of people in Gambia and in the United States. For example, only 47%–62% of Gambians (depending on season of birth) survived to reach age 45, whereas more than 96% of U.S. females survived to that age.

TABLE 9.2

Survivorship, Fecundity, and Life Expectancy by Age for U.S. Females in 2005

Age (in years) x	Survivorship l_x	Fecundity F_x	Life expectancy (at age x)
0	1.0	0.0	79.9
1	0.994	0.0	79.4
5	0.993	0.0	75.5
10	0.992	0.0	70.6
15	0.991	0.004	65.6
20	0.990	0.203	60.7
25	0.987	0.511	55.9
30	0.985	0.578	51.0
35	0.981	0.479	46.2
40	0.976	0.232	41.4
45	0.967	0.046	36.8
50	0.954	0.003	32.2
55	0.936	0.0	27.8
60	0.911	0.0	23.5
65	0.871	0.0	19.5
70	0.814	0.0	15.6
75	0.729	0.0	12.1
80	0.604	0.0	9.1
85	0.436	0.0	6.6
90	0.251	0.0	4.7
95	0.099	0.0	3.2
100	0.022	0.0	2.2

Source: Martin et al. (2009) and Arias et al. (2010).

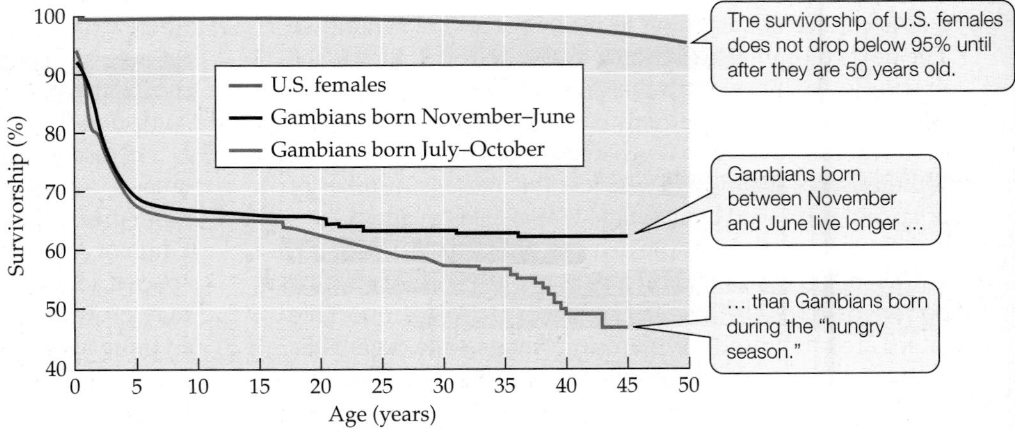

FIGURE 9.4 Survivorship Varies among Human Populations In the United States, survivorship (l_x) does not drop greatly until old age. In Gambia, many people die at much younger ages. (U.S. data from Arias 2010; Gambia data from Moore et al. 1997.)

The proportion of Gambians born in the hungry season that live to age 45 is roughly the same as the proportion of U.S. females that live to what age (see Table 9.2)?

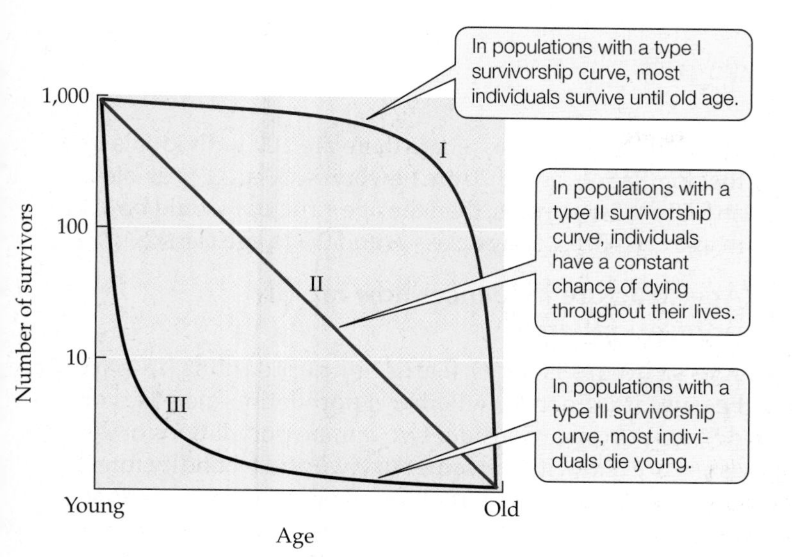

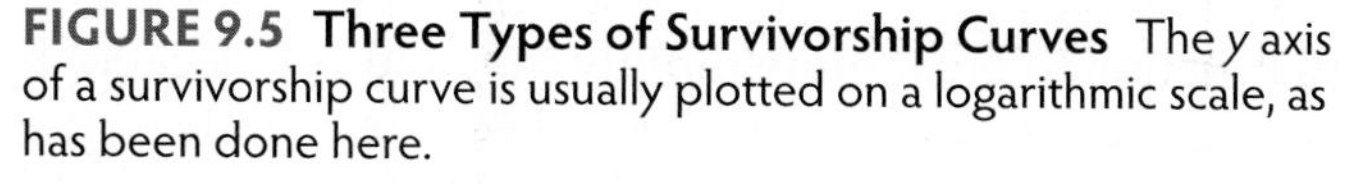

FIGURE 9.5 Three Types of Survivorship Curves The *y* axis of a survivorship curve is usually plotted on a logarithmic scale, as has been done here.

There are three types of survivorship curves

Survivorship data from life tables can be graphed as a **survivorship curve**. In such a curve, survivorship (l_x) data are used to plot the numbers of individuals from a hypothetical cohort (typically, of 1,000 individuals) that will survive to reach different ages. Results from studies on a variety of species suggest that survivorship curves can be classified into three general types, which indicate life stages at which high rates of mortality are most likely to occur (**Figure 9.5**).

In populations with a **type I survivorship curve**, newborns, juveniles, and young adults all have high survival rates; death rates do not begin to increase greatly until old age. Examples of populations with type I survivorship curves include U.S. females (see Figure 9.4) and Dall mountain sheep (**Figure 9.6A**). In populations with a **type II survivorship curve**, individuals have an approximately constant chance of surviving from one age to the next throughout their lives. Some bird species have a type II survivorship curve (**Figure 9.6B**), as do mud turtles (after their second year), some fish, and some plant species. Finally, in populations with a **type III survivorship curve**, individuals die at very high rates when they are young, but those that reach adulthood survive well later in life. Type III survivorship curves—the most common type observed in nature—are typical of species that produce large numbers of young. Examples include giant puffballs, oysters, marine corals, most known insects, and

FIGURE 9.6 Species with Type I, II, and III Survivorship Curves Survivorship curves in (A) the Dall mountain sheep (*Ovis dalli*), (B) the song thrush (*Turdus ericetorum*), and (C) the desert shrub *Cleome droserifolia*. (A, B after Deevey 1947; C after Hegazy 1990.)

Q What percentage of Dall mountain sheep survive to age 11?

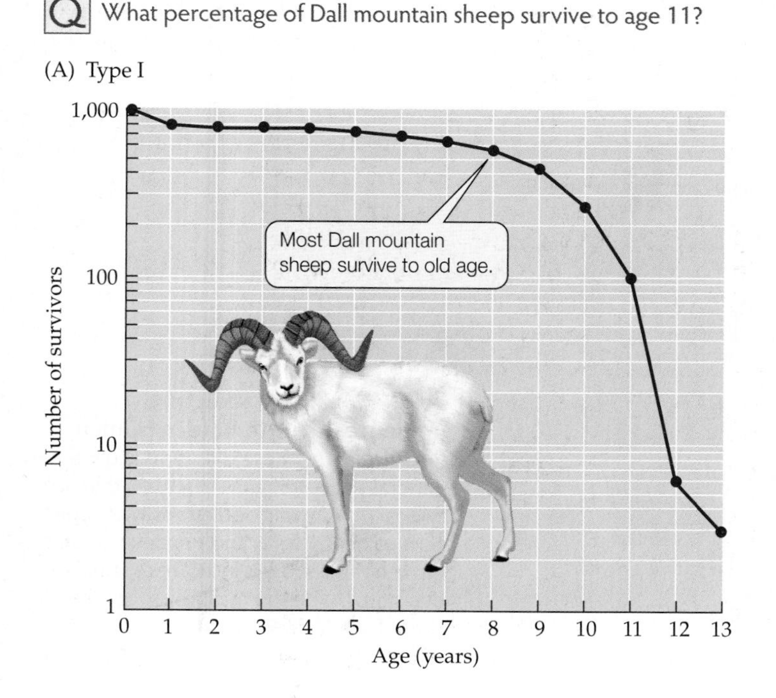

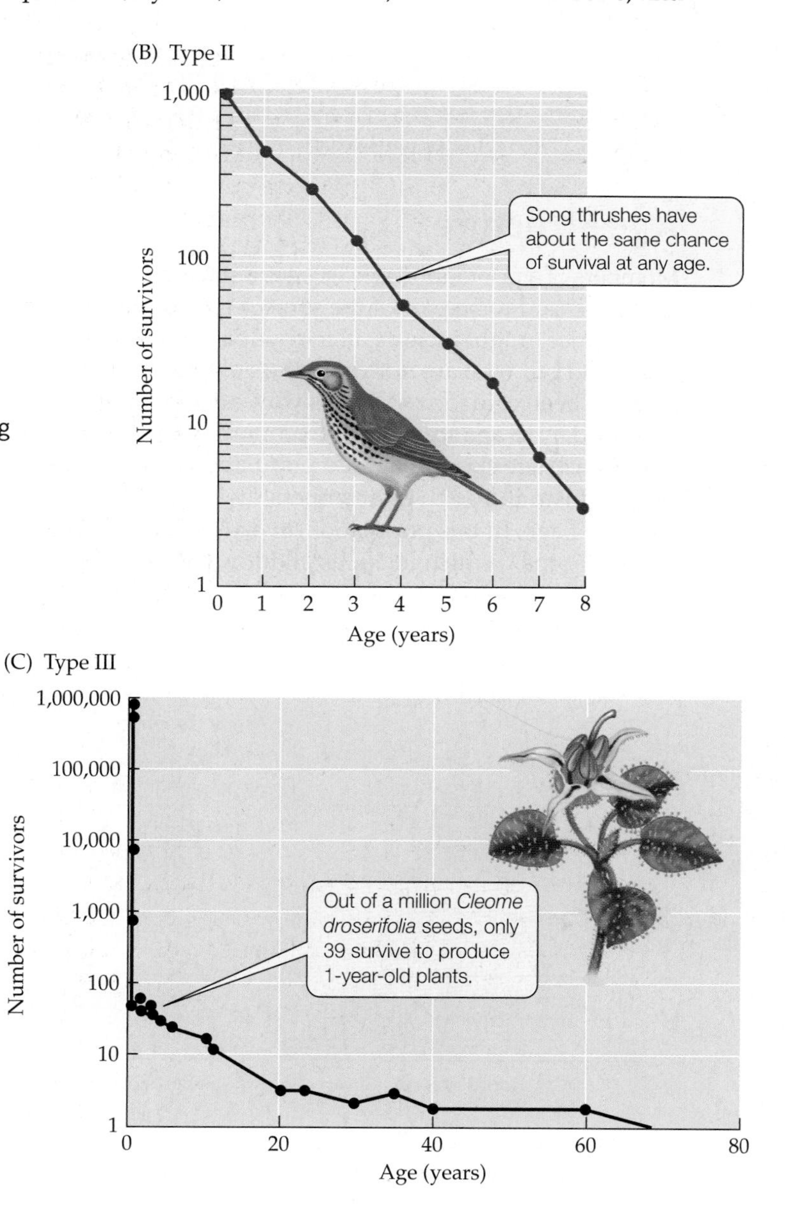

Fecundity - age specific birth rate

many plants, including the desert shrub *Cleome droserifolia* (**Figure 9.6C**). In this species, a population of 2,000 adults produces some 20 million seeds each year (roughly 10,000 seeds per adult), but only about 800 of those seeds survive to become juvenile plants.

We have discussed type I–III survivorship curves as if they were constant for each species, but that is not necessarily the case. Survivorship curves can vary among populations of a species, between males and females in a population, and among cohorts of a population that experience different environmental conditions (see Figure 9.4). In fact, by comparing birth and death rates in groups of individuals that experience different conditions, we can assess the effects of those conditions on populations. As we'll see in the next section, we can also use birth and death rates to predict how the size and composition of a population will change over time.

CONCEPT 9.2 Life table data can be used to project the future age structure, size, and growth rate of a population.

Age Structure

Members of a population whose ages fall within a specified range are said to be part of the same *age class*. Age class 1, for instance, might include all individuals who are at least 1 year old but who are not yet 2 years old. Once individuals have been categorized in this way, a population can be described by its **age structure**: the proportions of the population in each age class. Imagine a population of a hypothetical organism in which all members die before they reach 3 years of age. In this population, every individual will be 0 ("newborns," which includes all individuals less than 1 year old), 1, or 2 years old. If there are 100 individuals in the population, and if 20 are newborns, 30 are 1-year-olds, and 50 are 2-year-olds, then the age structure would be 0.2 in age class 0, 0.3 in age class 1, and 0.5 in age class 2.

Age structure influences how rapidly populations grow

Age structure is a key feature of populations, in part because it influences whether a population increases or decreases in size. Consider two human populations of the same size and with the same survival and fecundity rates, but with different age structures. If one of the populations had many people older than 55, while the other had many people between ages 15 and 30, we would expect the second population to grow more rapidly than the first because it contains more individuals of reproductive age. Indeed, human populations that are growing rapidly typically have a greater percentage of people in younger age classes than do populations that are growing slowly or are in decline (**Figure 9.7**). In general, age structure influences how rapidly any population grows—at least initially, as we'll see next by examining a hypothetical life table in some detail.

Age structure and population size can be predicted from life table data

Table 9.3 shows survival and fecundity rates for a population of a hypothetical organism that reproduces in the spring and dies before it reaches 3 years of age. How will the age structure and size of this population change over time?

To answer these questions, we'll represent the number of individuals in age class 0 (the newborns) by n_0, the number in age class 1 (the 1-year-olds) by n_1, and the number in age class 2 (the 2-year-olds) by n_2. Assume that our population begins with a total of 100 individuals, of which 20 are in age class 0 ($n_0 = 20$), 30 are in age class 1 ($n_1 = 30$), and 50

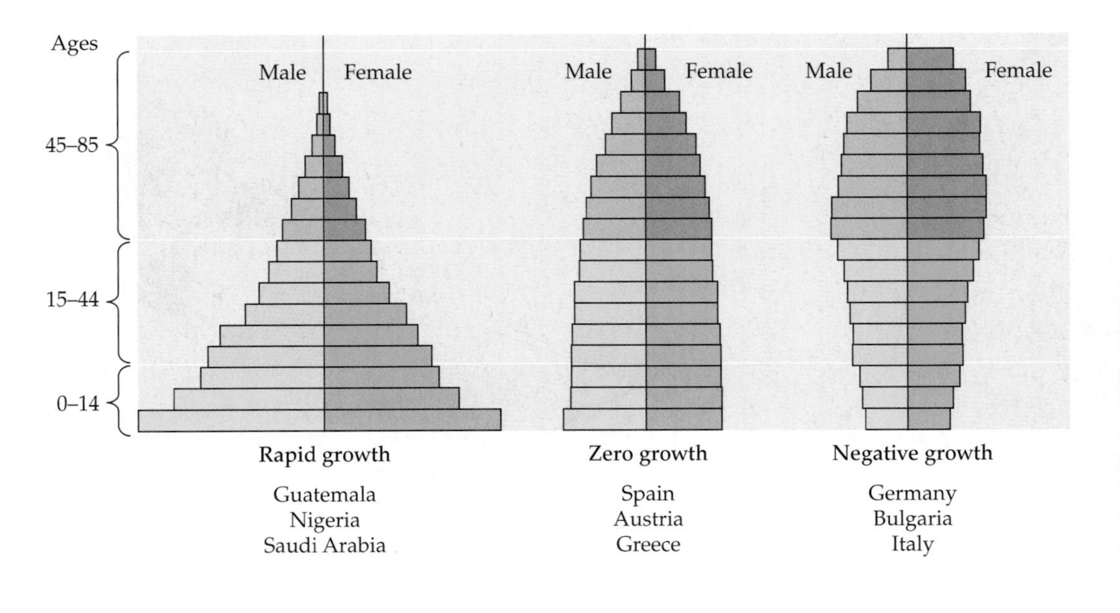

FIGURE 9.7 Age Structure Influences Growth Rate in Human Populations These diagrams show the general shapes of age structures that are typical of human populations with rapid, zero or nearly zero, and negative annual growth rates. Colors indicate reproductive status, with the main reproductive ages (15–44) shown in green. (After Miller 2007.)

TABLE 9.3

Life Table for a Hypothetical Organism

Age x	Survival rate S_x	Survivorship l_x	Fecundity F_x
0	0.3	1.00	0
1	0.8	0.30	2
2	0.0	0.24	4
3	—	0.0	—

Note: The organism reproduces in the spring and dies before it is 3 years old.

are in age class 2 ($n_2 = 50$). Assume further that all mortality occurs over the winter, before the next spring breeding season, and that individuals are counted immediately after the breeding season.* We can now use the information in Table 9.3 to predict how many individuals our population will have in the following year. To do this, we must calculate two things: (1) the number of individuals that will survive to the next time period (in this case, to the next year's breeding season), and (2) the number of newborns those survivors will produce in the next time period.

To calculate the number of individuals that will survive to the next time period, we multiply the number of individuals in each age class by the survival rate for that age class (**Table 9.4**). Thus, to determine the number of newborns that will survive to be 1-year-olds in the following year, we multiply $n_0 = 20$ by $S_0 = 0.3$ (see Table 9.3) to get 6 1-year-olds. Similarly, to determine how many of the currrent 1-year-olds will survive to be 2-year-olds in the following year, we multiply $n_1 = 30$ by $S_1 = 0.8$ to get 24 2-year-olds. Finally, to determine the number of newborns in the following year, we note that, on average, each 1-year-old has 2 offspring and each 2-year-old has 4 offspring (see Table 9.3). Since there will be 6 1-year-olds and 24 2-year-olds in the next year, the number of offspring produced at that time will be $(6 \times 2 + 24 \times 4) = 108$. Thus, in the following year, the total population size (N) will be $N = 138$ (108 newborns + 6 1-year-olds + 24 2-year-olds).

Now that we have predicted how the age structure and size of our hypothetical population will change in one year, we can extend those calculations to future years. Just as we used the data in Table 9.3 to predict that the population would increase from 100 individuals in the first year (at time $t = 0$) to 138 individuals in the next year (at time $t = 1$), we can perform calculations similar to those in Table 9.4 to determine how many individuals will be in each age class at time $t = 2$, then time $t = 3$, then time $t = 4$, and so on. We show the results of such calculations in Figure 9.8.

In **Figure 9.8A**, the numbers of individuals in each age class and in the population as a whole are plotted over time. Examining the curve for total population size (N), we see that with one exception ($t = 2$), the population rises steadily from its initial value of $N = 100$, reaching 1,361 individuals at time $t = 10$. By time $t = 4$, the numbers of individuals in each age class also rise steadily. In the first few years of population growth, however, the numbers of individuals in the different age classes vary considerably. For example, from its initial value of 50, n_2 drops to 24 individuals at time $t = 1$ and just 5 individuals at time $t = 2$; n_2 then rises substantially, reaching 136 individuals by time $t = 10$.

By time $t = 8$, the four curves in Figure 9.8A are roughly parallel to one another, indicating that all three age classes—

* Individuals could also be counted at other times, such as just before the breeding season. Although the calculations in the following paragraphs would be different, the final results would be the same; see Caswell 2001 for more information regarding census times.

TABLE 9.4

A Two-Step Method for Projecting the Size of the Hypothetical Population in Table 9.3

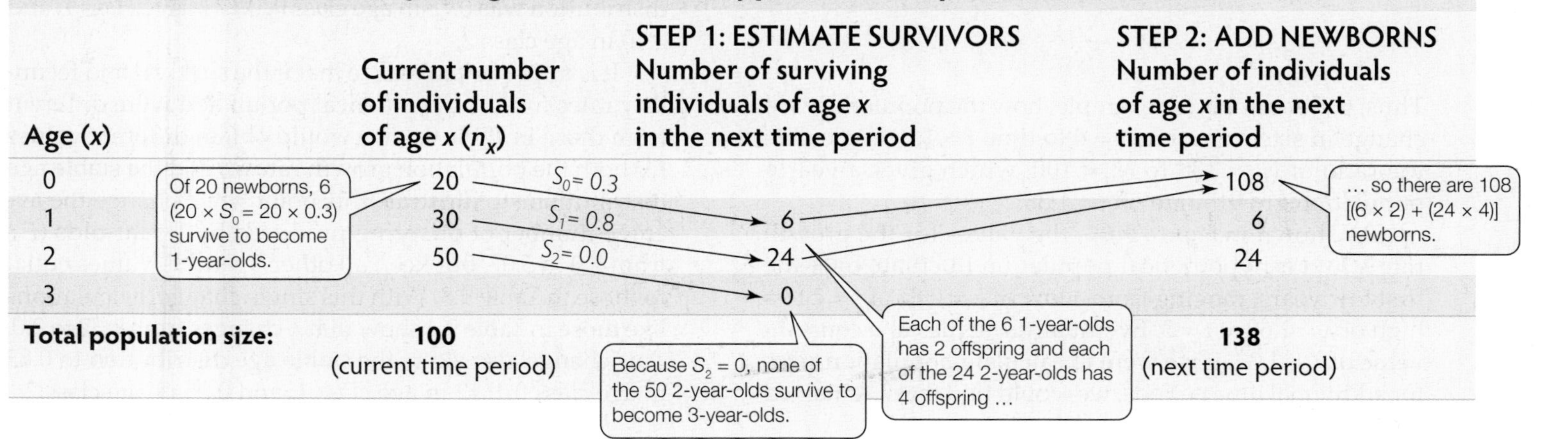

Age (x)	Current number of individuals of age x (n_x)	STEP 1: ESTIMATE SURVIVORS Number of surviving individuals of age x in the next time period	STEP 2: ADD NEWBORNS Number of individuals of age x in the next time period
0	20		108
1	30	6	6
2	50	24	24
3		0	
Total population size:	**100** (current time period)		**138** (next time period)

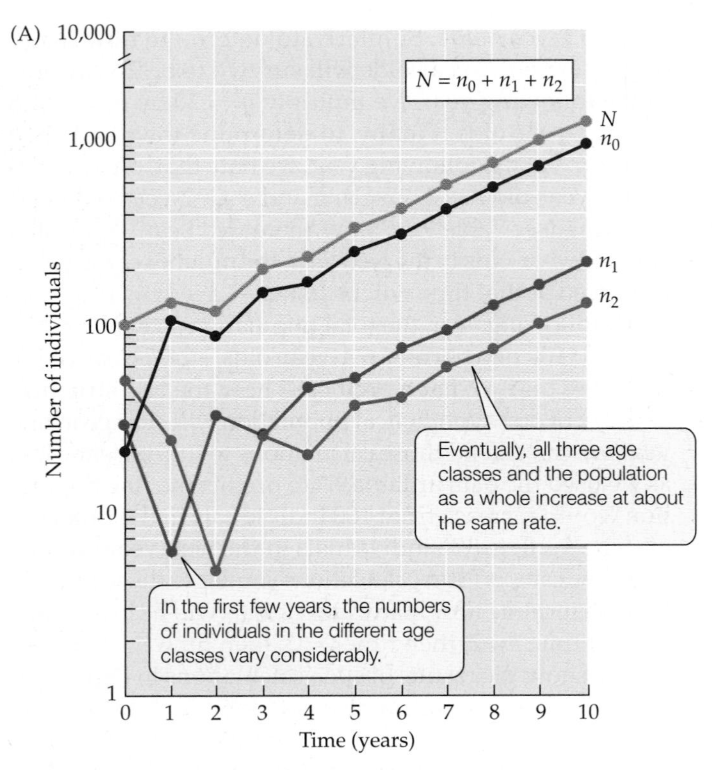

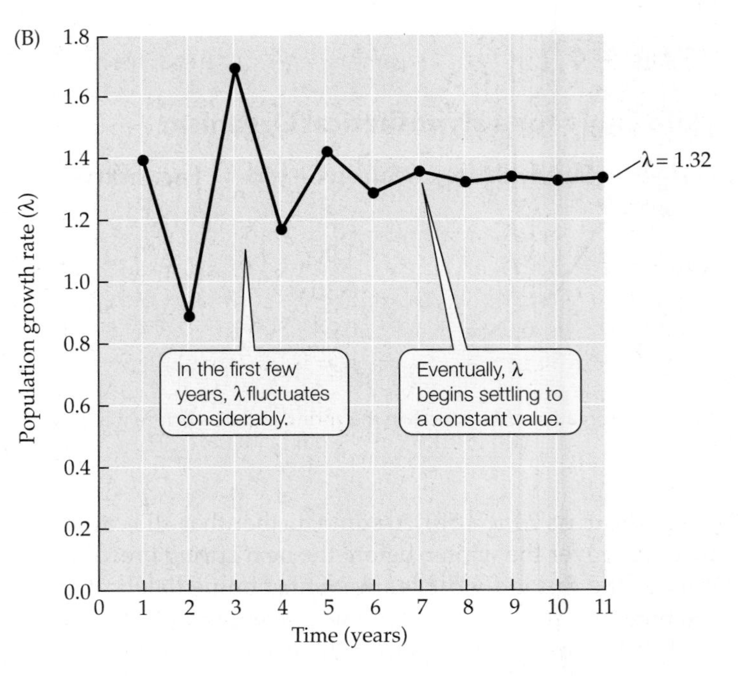

FIGURE 9.8 Growth of a Hypothetical Population The procedure described in Table 9.4 was used to calculate the growth of the hypothetical population whose life table data are shown in Table 9.3. These graphs plot (A) the number of individuals in each of the three age classes (n_0, n_1, n_2), as well as the total population size (N), at different times, and (B) the yearly rate of increase (λ) for the total population. Note that the number of individuals in (A) is plotted on a log scale.

Q Using the graph in (A), estimate λ from year 4 to year 5 for age class 2 (n_2).

and the total population size, N—are increasing at nearly the same rate from one year to the next. To examine this yearly rate of increase further, we can calculate the ratio of the population size in year $t + 1$, denoted N_{t+1}, to the population size in year t, denoted N_t. This ratio provides a measure of the year-to-year *population growth rate* and is designated by the Greek symbol λ (pronounced "lambda"):

$$\lambda = \frac{N_{t+1}}{N_t}$$

Thus, to determine, for example, how the population will change in size from time $t = 0$ to time $t = 1$, we calculate the ratio of $N_1 = 138$ to $N_0 = 100$, which gives a yearly population growth rate of $\lambda = 1.38$.

As shown in **Figure 9.8B**, the values for the growth rate λ in our hypothetical population fluctuate over the first few years, ranging from a low of $\lambda = 0.88$ at $t = 2$ to a high of $\lambda = 1.69$ at $t = 3$. Eventually, λ settles to a constant value of $\lambda = 1.32$; if we were to calculate population sizes for additional time periods, we would find that λ remained equal to 1.32 from time $t = 10$ forward. In addition, if we were to calculate λ for any of the age classes [e.g., calculate the ratio of the size of age class 1 (n_1) at time $t + 1$ to its size at time t], we would find that by time $t = 10$, λ would equal 1.32 for each age class.

Populations grow at fixed rates when age-specific birth and death rates are constant over time

If a population's age-specific survival and fecundity rates are constant over time, that population ultimately grows at a fixed rate from one year to the next. This was the case for our hypothetical population, which eventually grew at the fixed rate of $\lambda = 1.32$ (see Figure 9.8B), as did each of its age classes. Because the population and each of its age classes increase by a constant multiplier every year, the proportion of individuals in each age class remains constant as long as the multiplier (λ) remains constant. When the age structure of a population does not change from one year to the next, the population is said to have a **stable age distribution**. In our example, the stable age distribution was 0.73 in age class 0, 0.17 in age class 1, and 0.10 in age class 2.

It is important to realize that if the survival and fecundity rates for our hypothetical population were different from those in Table 9.3, we would obtain different values for both the population growth rate (λ) and the stable age distribution. To illustrate this point, we'll change the average number of offspring produced by 1-year-olds (F_1) from 2.0 to 5.07, but keep all other S_x and F_x values equal to those in Table 9.3. With this single change, calculations like those in Table 9.4 show that λ changes from 1.32 to 2.0. This change also alters the stable age distribution to 0.83 in age class 0, 0.12 in age class 1, and 0.05 in age class 2.

Birth and death rates—and hence population growth rates—can change when environmental conditions change

As the effects of our change to F_1 suggest, the growth rate and age structure of a population can change if the age-specific birth or death rates change. Ecologists and natural resource managers can apply this knowledge to modify an organism's biotic or abiotic environment in ways that are intended to change birth or death rates, with the ultimate goal of decreasing the size of a pest population or increasing the size of an endangered population. An efficient way to reach this goal is to identify the age-specific birth or death rates that most strongly influence the population growth rate. In one such example, life table data indicated that the most effective way to increase the growth rates of endangered turtle populations was to increase the survival rates of juvenile and mature turtles—a change from the common practice of protecting newborns (**Ecological Toolkit 9.1**, on the following page).

The birth or death rates of a population can be affected by a broad range of abiotic and biotic factors. Sudden changes in environmental factors can cause rapid and dramatic changes in birth or death rates, as when a catastrophic drought leads to massive die-offs in a population. But birth or death rates can also change more gradually over time. For example, over the past several decades, mortality rates have increased gradually in populations of coniferous forest trees across broad regions of the western United States (**Figure 9.9**). These increases occurred in stands of seemingly healthy forest that had not been cut for more than 200 years, leading researchers to ask, What was killing the trees?

A study by van Mantgem and colleagues (2009) ruled out several possible causes, including air pollution, forest fragmentation, changes in fire frequency, and within-stand increases in the intensity of competition. The researchers went on to note that during the time period covered by their study, regional temperatures in the western United States had increased at rates of 0.3°C–0.5°C per decade. These rapid temperature increases were associated with declines in the snowpack, earlier spring snowmelt, and a lengthening of the summer dry period. These changes caused an increase in the trees' *climatic water deficit* (the amount by which a plant's annual evaporative demand for water exceeds available water). Previous studies had shown that tree mortality rates tend to increase when climatic water deficit increases (Bigler et al. 2007). Overall, van Mantgem et al.'s study and others suggest that the rise in tree mortality rates may have been driven by regional warming and the ensuing drought stress. (We will continue our discussion of this example in **Climate Change Connection 9.1**, focusing on projected impacts on forest communities.)

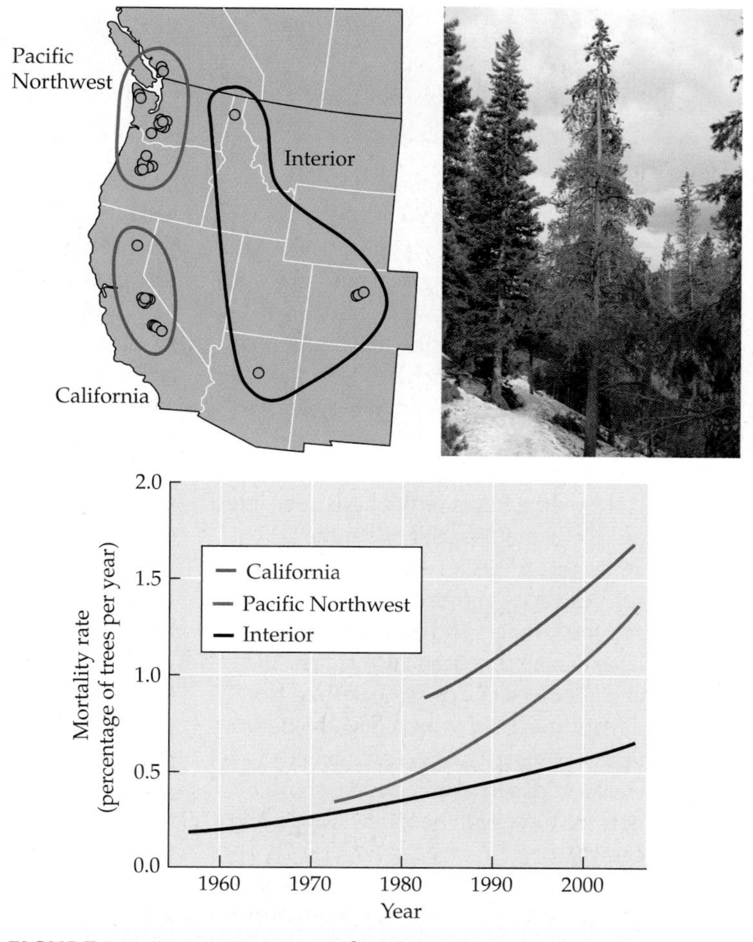

FIGURE 9.9 Rising Tree Mortality Rates Trends in tree mortality rates for 76 study plots located in the three regions shown on the map. (After van Mantgem et al. 2009.)

CONCEPT 9.3 Populations can grow exponentially when conditions are favorable, but exponential growth cannot continue indefinitely.

Exponential Growth

Many organisms produce large numbers of offspring, as illustrated by giant puffball fungi and the desert shrub *Cleome droserifolia*. In such cases, if even a small fraction of those offspring survive to reproduce, the population can increase in size very quickly. Even populations of people and other organisms that produce relatively few young can grow rapidly. In general, populations can grow rapidly whenever individuals leave an average of more than one offspring over substantial periods of time. In this section, we describe geometric growth and exponential growth, two related patterns of population growth that can lead to rapid increases in population size.

9.1 Estimating Population Growth Rates in a Threatened Species

Loggerhead sea turtles (*Caretta caretta*; **Figure A**) are large marine turtles that lay eggs in nests that adult females dig into sandy beaches. Newly hatched baby turtles weigh just 20 g (0.04 pounds) and have a shell length of 4.5 cm (1.8 inches). They reach adulthood after 20–30 years, at which point they can weigh up to 227 kg (500 pounds) and have a shell length of 122 cm (4 feet).

Loggerhead sea turtles have been listed as a threatened species under the U.S. Endangered Species Act since 1978. Many species eat loggerhead eggs or hatchlings, and the juveniles and adults are eaten by large marine predators such as tiger sharks and killer whales. Loggerheads also face threats from people, including the destruction of nesting sites by development as well as commercial fisheries (in whose nets sea turtles can become trapped and drown).

Early efforts to protect loggerhead sea turtles focused on the egg and hatchling stages, which suffer extensive mortality and are relatively easy to protect. To evaluate this approach, Crouse et al. (1987) and Crowder et al. (1994) used life table data to determine how population growth rates would change if new management practices improved the survival and fecundity rates of turtles of various ages. Their findings suggested that even if hatchling survival were increased to 100%, loggerhead populations would continue to decline. Instead, they found that the population growth rate was most responsive to decreasing the mortality of older juveniles and adults.

The results obtained by Crouse, Crowder and colleagues prompted the enactment of laws requiring Turtle Excluder Devices (TEDs) to be installed in shrimp nets (**Figure B**). TEDs function as a hatch through which juvenile and adult sea turtles can escape when caught in a net. Shrimp nets were singled out because the data suggested that shrimping accounted for more loggerhead deaths (from 5,000 to 50,000 deaths per year) than all other human activities combined.

Loggerheads are most easily counted when they nest, yet it takes 20–30 years for turtles to become sexually mature. As a result, it will be decades before we know whether TED regulations help turtle populations to increase in size. But early results are encouraging: the number of turtles killed in nets declined by about 44% after the TED regulations were implemented.

Figure B Turtle Excluder Device (TED)

Figure A Loggerhead Sea Turtle These endangered turtles face many threats in the marine environment, including predators, commercial fisheries (turtles can be caught accidentally in nets and tarps), collisions with boats, and chemical and solid-waste pollutants.

Populations grow geometrically when reproduction occurs at regular time intervals

Some species, such as cicadas and annual plants (and the hypothetical species described in Table 9.3), reproduce in synchrony at regular time intervals. These regular time intervals are called *discrete time periods*. If a population of such a species changes in size by a constant proportion from one discrete time period to the next, **geometric growth** is said to occur. The fact that the population grows by a constant proportion means that the number of individuals added to the population can become larger with each time period. As a result, the population can grow larger by ever-increasing amounts. When plotted on a graph, this growth pattern forms a J-shaped set of points (**Figure 9.10A**).

Mathematically, we can describe geometric growth as

$$N_{t+1} = \lambda N_t \qquad (9.1)$$

where N_t is the population size after t generations or, equivalently, after t discrete time periods (e.g., t years if there is one generation per year), and λ is any number

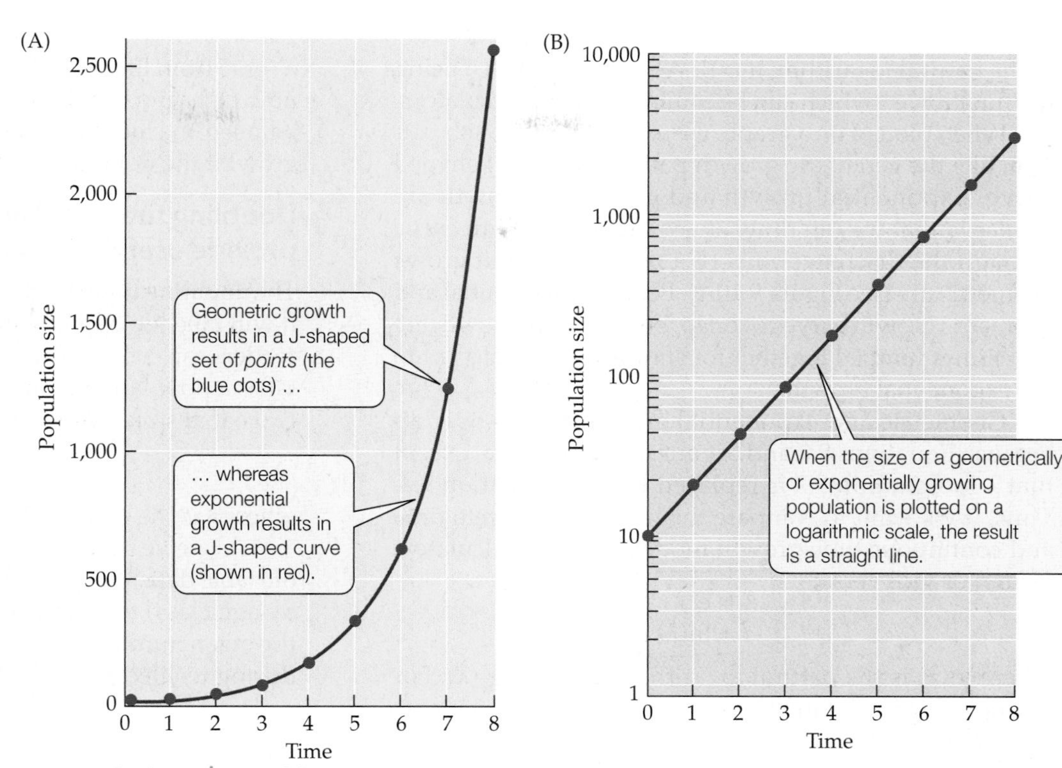

FIGURE 9.10 Geometric and Exponential Growth (A) The blue dots plot the size of a geometrically growing population that begins with 10 individuals and doubles in each discrete time period (i.e., $N_0 = 10$ and $\lambda = 2$). The red curve plots exponential growth in a comparable population that reproduces continuously, also beginning with 10 individuals and having a growth rate of $r = \ln(2) = 0.69$. (B) When the population sizes represented by the blue circles and the red curve in (A) are plotted on a logarithmic scale, the result is a straight line.

$0 < \lambda < 1$ = population decrease

greater than zero. In Equation 9.1, λ serves as a multiplier that allows us to predict the size of the population in the next time period. We'll refer to λ as the **geometric population growth rate**; λ is also known as the (per capita) **finite rate of increase**. We use this terminology by convention, but it can be confusing: we can see from Equation 9.1 that when the population "growth" rate λ is between 0 and 1, the population does not grow, but rather decreases in size over time.

Geometric growth can also be represented by a second equation,

$$N_t = \lambda^t N_0 \tag{9.2}$$

where N_0 is the initial population size (that is, the population size at time = 0).

The two equations for geometric growth (Equations 9.1 and 9.2) are equivalent in that each can be derived from the other (see **Web Extension 9.1**). Which one we use depends on what we are interested in. If we want to predict the population size in the next time period, and we know λ and the current population size, either equation can be used. If we know the population size in both the current and previous time periods, we can rearrange Equation 9.1 to get an estimate for λ ($\lambda = N_{t+1}/N_t$, as we saw on p. 206). Finally, we can use Equation 9.2 to predict the size of the population after any number of discrete time periods. If $\lambda = 2$, for example, then after 12 time periods, a population that begins with $N_0 = 10$ individuals will have $N_{12} = 2^{12}N_0$ individuals, which (as we can determine by using a calculator with a y^x function) equals 4,096 × 10, or 40,960.

Populations grow exponentially when reproduction occurs continuously

In contrast to the pattern described in the previous section, individuals in many species (including people) do not reproduce in synchrony at discrete time periods; instead, they reproduce at varying times. In such organisms, referred to as *continuously* reproducing species, generations can overlap. When a population of a species with continuous reproduction changes in size by a constant proportion at each instant in time, we refer to the growth that results as **exponential growth**. Mathematically, exponential growth can be described by the following two equations:

$$\frac{dN}{dt} = rN \tag{9.3}$$

and

$$N(t) = N(0)e^{rt} \tag{9.4}$$

where $N(t)$ is the population size at each instant in time, t.

In Equation 9.3, dN/dt represents the rate of change in population size at each instant in time; we see from the equation that dN/dt equals a constant rate (r) multiplied by the current population size, N. Thus, the multiplier r provides a measure of how rapidly a population can grow; r is called the **exponential population growth rate** or the (per capita) **intrinsic rate of increase**.

As we did for Equation 9.2, we can use Equation 9.4 to predict the size of an exponentially growing population at any time t, provided we have an estimate for r and know $N(0)$, the initial population size. The "e" in Equation 9.4 is a constant, approximately equal to 2.718 ["e" is the base of the natural logarithm, $\ln(x)$]. We can calculate e^{rt} using the function e^x, which can be found on many calculators.

When plotted on a graph, the exponential growth pattern, like the geometric growth pattern, forms a J-shaped curve. Exponential growth and geometric growth are similar in that we can draw an exponential growth curve through the discrete points of a population that grows geometrically (see Figure 9.10A). Because exponential and geometric growth curves overlap, both types of growth are sometimes lumped together for simplicity and referred to as "exponential growth."

Geometric and exponential growth curves overlap because Equations 9.2 and 9.4 are similar in form, except that λ in Equation 9.2 is replaced by e^r in Equation 9.4. Thus, if we want to compare the results of discrete time and continuous time growth models, we can calculate λ from r, or vice versa:

$$\lambda = e^r$$
$$r = \ln(\lambda)$$

where $\ln(\lambda)$ is the natural logarithm of λ, or $\log_e(\lambda)$. For example, if $\lambda = 2$ (as in Figure 9.10A), an equivalent value for r would be $r = \ln(2)$, which is approximately 0.69. **Figure 9.10B** illustrates a simple way to determine whether a population really is growing geometrically (or exponentially): plot the natural logarithm of population size versus time, and if the result is a straight line, the population is increasing by geometric or exponential growth.

Finally, look at Equations 9.1 and 9.3 for a moment. In Equation 9.1, which value of λ will ensure that the population does not change in size from one time period to the next? Similarly, in Equation 9.3, which value of r causes the population to remain fixed in size? The answers are $\lambda = 1$ (because then $N_{t+1} = N_t$) and $r = 0$ (because then the rate at which the population size changes is 0). When $\lambda < 1$ (or $r < 0$), the population will decline to extinction, whereas when $\lambda > 1$ (or $r > 0$), the population will increase exponentially (or geometrically) to form a J-shaped curve (**Figure 9.11**).

How can we estimate a population's growth rate (r or λ)? There are a variety of methods for doing so (see Caswell 2001), one of which we discussed on pp. 204–206: use life table data to predict future population sizes, graph those predicted population sizes versus time, and estimate the growth rate (λ) from the graph. Ecologists often estimate λ (or r) from life table data, since they can then determine how fast a population is growing. Life table data can also be used to calculate two other measures of population growth: the doubling time and the net reproductive rate.

Doubling times and net reproductive rates provide useful measures of population growth

The **doubling time** (t_d) of a population is the number of years it will take the population to double in size. As interested readers can confirm [by solving Equation 9.4 for the time it takes a population to increase from its initial size, $N(0)$, to twice that size, $2N(0)$], doubling times can be estimated as

$$t_d = \frac{\ln(2)}{r} \tag{9.5}$$

where r is the exponential growth rate.

As we've seen, r (and hence t_d) can be estimated from life table data. We can also use life table data (broken down by age class) to calculate the **net reproductive rate** (R_0): the mean number of offspring produced by an individual during its lifetime. R_0 is calculated as

$$R_0 = \sum_{x_{first}}^{x_{last}} l_x F_x \tag{9.6}$$

where x is age, x_{first} is the age of first reproduction, x_{last} is the age of last reproduction, and l_x and F_x are survivorship and fecundity, respectively, as defined in Table 9.1. Note that to estimate R_0, we multiply by l_x because the likelihood of surviving to each reproductive age is just as important as the number of offspring produced at that age (F_x). To check your understanding of Equation 9.6, use Table 9.1 to calculate an estimate of R_0 for the grass *Poa annua*; your calculations should yield $R_0 = 845.9$.

Whenever R_0 is greater than 1, measured from one generation to the next, λ will be greater than 1 (and $r > 0$). Under these conditions, populations have the potential to increase greatly in size, as we'll see in the next section.

Populations can grow rapidly because they increase by multiplication

Equations 9.1 and 9.3 show that populations increase by multiplication, not addition: at each point in time, the population changes in size according to the multiplier λ or r. As a result, populations have the potential to add large

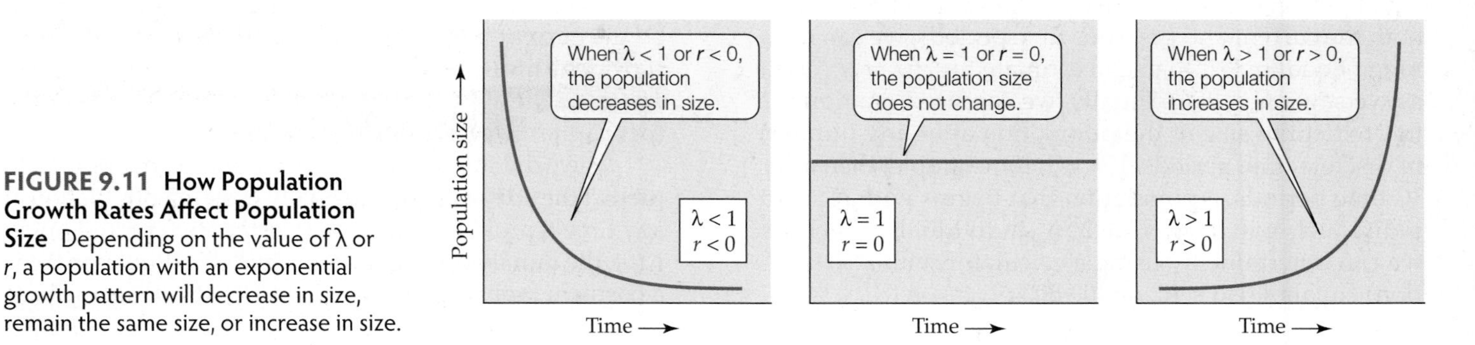

FIGURE 9.11 How Population Growth Rates Affect Population Size Depending on the value of λ or r, a population with an exponential growth pattern will decrease in size, remain the same size, or increase in size.

Year	Population growth rate (λ)
1990–1991	0.80
1991–1992	0.77
1992–1993	0.82
1993–1994	1.01
1994–1995	0.96

FIGURE 9.12 Some Populations Have Low Growth Rates The growth rates of a population of wild ginger (*Asarum canadense*) in a young forest vary from year to year. Growth rates in this forest are often less than 1.0, suggesting that the population will decline in size unless conditions improve. (Data from Damman and Cain 1998.)

numbers of individuals rapidly whenever $\lambda > 1$ or $r > 0$. The principle at work here is the same one that applies to interest on a savings account. Even when the interest rate is low, you can earn a lot of money each year if you have a large amount deposited in the bank, because savings, like populations, grow by multiplication. Similarly, the fact that populations grow by multiplication means that even a low growth rate can cause the size of a population to increase rapidly.

Consider our own population. On p. 200, we stated that the current annual growth rate of the human population was 1.18%. Such a growth rate implies that $\lambda = 1.0118$, and hence that $r = \ln(\lambda) = 0.0117$, a value that seems close to 0. If we set the year 2010 as time $t = 0$, we have $N(0) = 6.8$ billion, the size of the human population in 2010. Plugging these values of r and $N(0)$ into Equation 9.4, we calculate that the population size one year later should be $N(1) = 6.8 \times e^{0.0117}$, which rounds to 6.88 billion people. Thus, in 2010, the human population was increasing by about 80 million people per year (6.88 billion – 6.8 billion = 0.08 billion = 80 million). Since populations grow by multiplication, if r remained constant at 0.0117 for an extended period of time, the yearly increments to the human population would become astronomical. For example, after 215 years, there would be over 80 billion people, and our population would be increasing in size by a billion people *each year*.

Turning from humans to other species, what do field studies reveal about the growth rates of their populations? Some species, such as the woodland herb *Asarum canadense* (wild ginger), have maximum observed values of λ that are close to 1 ($\lambda = 1.01$ in young forests, $\lambda = 1.1$ in mature forests) (**Figure 9.12**). Similar values were observed for a population of 25 reindeer introduced to Saint Paul Island off the coast of Alaska in 1911. After 27 years, the population had increased from 25 to 2,046 individuals, which (when we solve for λ in Equation 9.2) yields $\lambda = 1.18$.

Considerably higher annual population growth rates have been observed in many species, including western grey kangaroos ($\lambda = 1.9$), field voles ($\lambda = 24$), and rice weevils ($\lambda = 10^{17}$), which are insect pests of rice and other grains. Some bacteria, such as the mammalian gut inhabitant *E. coli*, can double in number every 30 minutes, resulting in the unimaginably high annual population growth rate of $\lambda = 10^{5,274}$.

Recall that when $\lambda > 1$ (or $r > 0$) for an extended period of time, populations increase exponentially in size, forming a J-shaped curve like that in Figure 9.10A. In natural populations, $\lambda > 1$ (or $r > 0$) when key factors in the environment are favorable for growth, survival, and reproduction. But can such favorable conditions last for long?

There are limits to the growth of populations

An argument from basic principles suggests that the answer to the question we just posed is no. Physicists estimate that the known universe contains a total of 10^{80} atoms. Yet if favorable conditions persisted for long enough, allowing λ to remain greater than 1, even populations of relatively slowly growing species would eventually increase to more than 10^{80} individuals. For example, based on *Asarum*'s population growth rate of $\lambda = 1.01$ in young forests, a population that began with 2 plants would have more than 10^{82} plants after 19,000 years. For an extremely rapidly growing species such as *E. coli*, the numbers are even more absurd: it would take only 6 days for a population that began with a single bacterium to exceed 10^{80} individuals.

No population could ever come close to having 10^{80} individuals because there would be no atoms with which to construct their bodies. Thus, exponential growth cannot continue indefinitely. While these are extreme examples (because other difficulties would be encountered long before there was a shortage of atoms), it illustrates a fundamental point: there are limits to population growth, which cause it to slow and eventually stop. We'll look at some of those limits in the following section.

CONCEPT 9.4 Population size can be determined by density-dependent and density-independent factors.

Effects of Density

Although $\lambda > 1$ for all populations under favorable conditions (leading to exponential growth), conditions in nature are rarely favorable for long. For example, Damman and Cain (1998) calculated the geometric growth rate (λ) in each of 5 years for a population of the woodland herb

Asarum canadense located in a young forest. As mentioned above, the maximum growth rate was $\lambda = 1.01$. During the other 4 years, however, values for λ ranged from 0.77 to 0.96 (see Figure 9.12). Thus, far from threatening to overrun the planet with its offspring, we would expect this population to decline in the long run, unless conditions changed for the better.

What factors cause λ to fluctuate over time? We can explore answers to this question by asking whether the population growth rate changes independently of density or as a function of density.

Density-independent factors can determine population size

In many species, year-to-year variation in weather leads to dramatic changes in abundance and hence in population growth rates. For example, Davidson and Andrewartha (1948) studied how weather in Adelaide, Australia, affected populations of the insect *Thrips imaginis*, a pest of roses. By correlating weather conditions with thrips population sizes over a 14-year period, they showed that yearly fluctuations in population size could be predicted accurately by an equation that used temperature and rainfall data (**Figure 9.13**).

Factors such as temperature and precipitation, as well as catastrophic events such as floods or hurricanes, are often referred to as **density-independent** factors, meaning that their effects on birth and death rates are independent of the number of individuals in the population. Likewise, population growth rates (λ or r) are density-independent when they are not a function of population density (**Figure 9.14**).

As the *Thrips imaginis* data suggest, density-independent factors can have major effects on population size from one year to the next. In principle, such factors could account entirely for year-to-year fluctuations in the size of a population. But density-independent factors do not tend to increase the size of populations when they are small and decrease the size of populations when they are large. A factor that did consistently lead to such changes would cause the population growth rate to change as a function of density—that is, to be density-*dependent*, not density-independent.

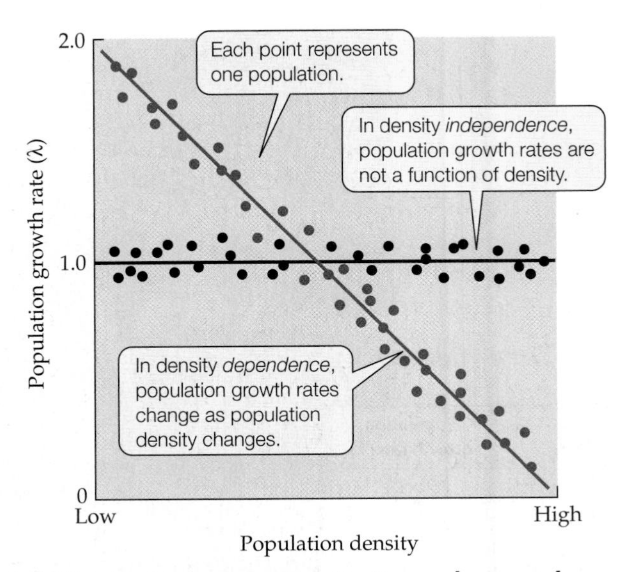

FIGURE 9.14 Comparing Density Dependence and Density Independence In the example of density dependence shown here, population growth rates decrease as population density increases. Depending on the species and the environmental conditions in its populations, λ may vary more or less than is shown in these two curves.

Density-dependent factors regulate population size

In many cases, the factors that influence population size are **density-dependent**, which means that they cause birth rates, death rates, and dispersal rates to change as the density of the population changes (see Figure 9.14). As densities increase, it is common for birth rates to decrease, death rates to increase, and dispersal from the population (emigration) to increase—all of which tend to decrease population size. When densities decrease, the opposite occurs: birth rates tend to increase and death and emigration rates decrease.

When one or more density-dependent factors cause population size to increase when numbers are low and decrease when numbers are high, **population regulation** is said to occur. Ultimately, when the density of any species becomes high enough, density-dependent factors decrease population size because food, space, or other essential resources are in short supply. Note that "regulation" has a particular meaning here, referring to the effects of factors that tend to increase λ or r when the population size is small and decrease λ or r when

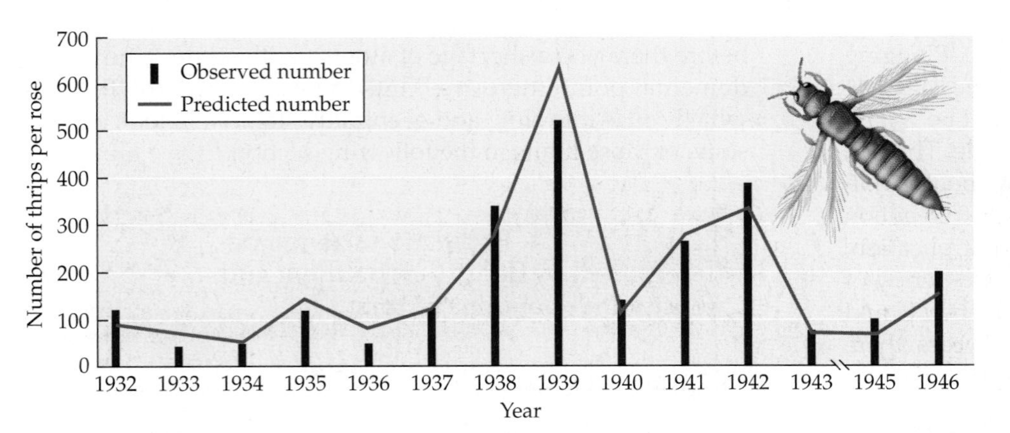

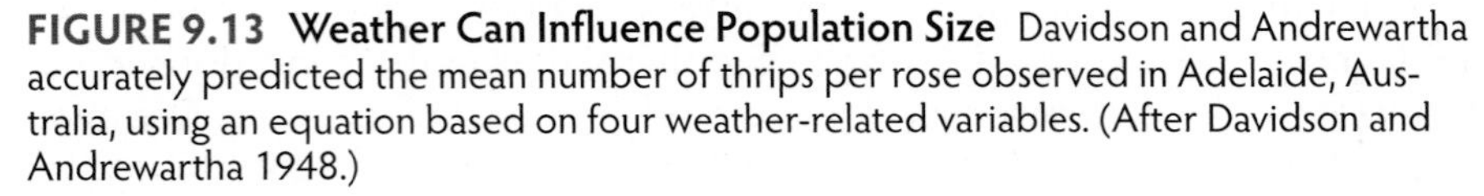
FIGURE 9.13 Weather Can Influence Population Size Davidson and Andrewartha accurately predicted the mean number of thrips per rose observed in Adelaide, Australia, using an equation based on four weather-related variables. (After Davidson and Andrewartha 1948.)

the population size is large. Density-independent factors can have large effects on population size, but they do not *regulate* population size because they do not consistently increase population size when it is small and decrease population size when it is large. Thus, by definition, only density-dependent factors can regulate population size.

Density dependence has been observed in many populations

Density dependence can often be detected in natural populations. For example, in a study that combined field observations with controlled experiments, Peter Arcese and James Smith (1988) examined the effect of population density on reproduction in the song sparrow (*Melospiza melodia*) on Mandarte Island, British Columbia. They found that the number of eggs laid per female decreased with density, as did the number of young that survived long enough to become independent of their parents (**Figure 9.15A**). Because Mandarte Island is small and the birds were likely to suffer food shortages at high densities, Arcese and Smith predicted that if they provided food to a subset of nesting pairs when densities were high, the birds that were fed should be able to rear more young to independence. That is exactly what happened: nesting pairs that were fed reared nearly four times as many young to independence as did control birds that were not fed (see Figure 9.15A).

In addition to the effect of density on reproduction, density-dependent mortality has been observed in many populations. For example, when Yoda et al. (1963) planted soybeans (*Glycine soja*) at various densities, they found that at the highest initial planting densities, many of the seedlings had died by 93 days of age (**Figure 9.15B**). Similarly, in an experiment in which eggs of the flour beetle *Tribolium confusum* were placed in glass tubes (each with 0.5 g of food), death rates increased as the density of eggs per tube increased—again revealing density dependence (**Figure 9.15C**). Density dependence has also been detected in populations whose abundance is strongly influenced by factors usually considered to

(A) *Melospiza melodia*

The number of young per female that survived to independence declined with population density.

In an experiment conducted at the high population densities observed in 1985, nesting pairs that were fed reared more young to independence than pairs that were not fed.

Independent young per female

Number of breeding females

'80 '81 '82 '76 '77 '78 '83 '84 '75 '86 '79 '85 (Fed) '85 (Control)

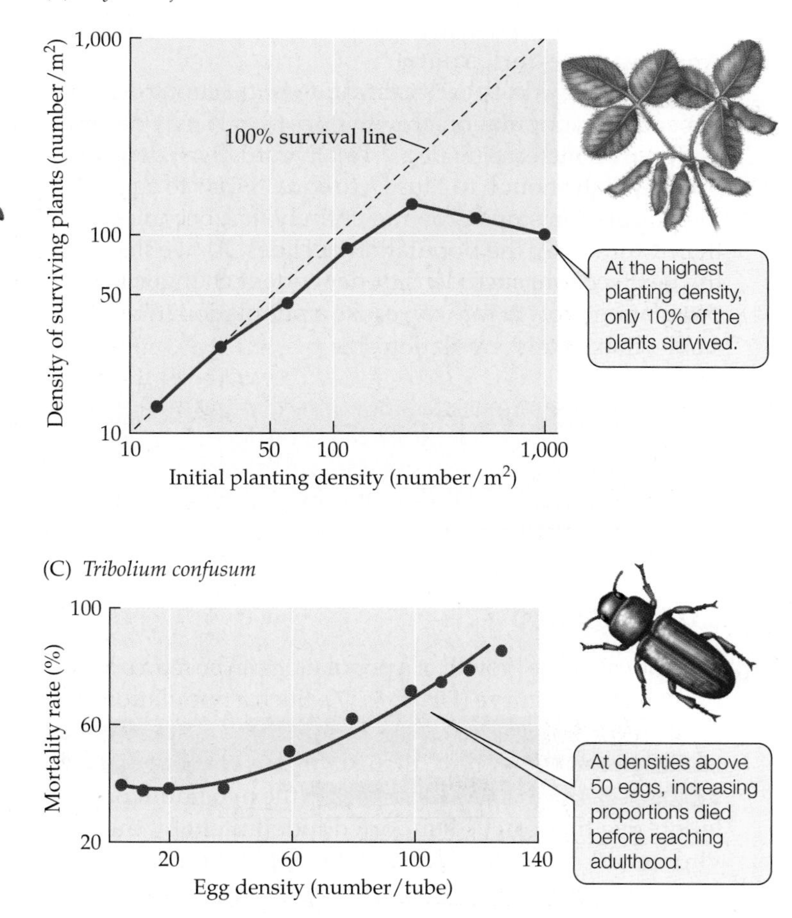

FIGURE 9.15 Examples of Density Dependence in Natural Populations (A) Numbers of young song sparrows reared to independence on Mandarte Island at different densities of breeding females. The number next to each point indicates the year of observation (1975–1986). (B) Density of surviving soybeans (log scale) 93 days after they were planted at densities ranging from 10 to 1,000 seeds per square meter. (C) Mortality rates in flour beetles at various egg densities. (A after Arcese and Smith 1988; B after Yoda et al. 1963; C after Bellows 1981.)

In (A), based on data from years other than 1975, how many young song sparrows per female would you have expected to have survived to independence in 1975? Explain your reasoning and describe factors that could have caused the observed results.

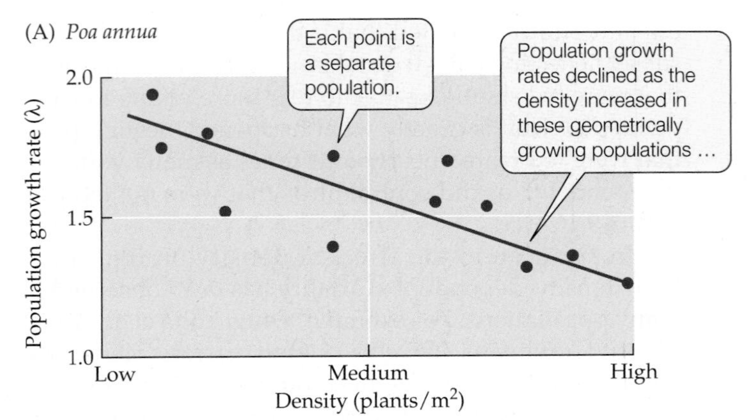

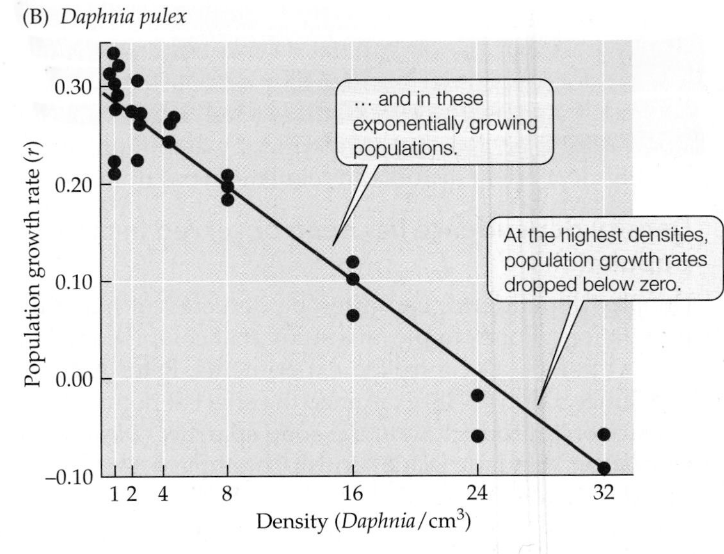

FIGURE 9.16 Population Growth Rates May Decline at High Densities (A) The geometric population growth rate (λ) of the grass *Poa annua* is density-dependent, as is (B) the exponential population growth rate (r) of the water flea *Daphnia pulex*. (A after Law 1975; B after Frank et al. 1957.)

act in a density-independent manner, such as temperature or precipitation; we describe one such example in **Web Extension 9.2**, in which Smith (1961) reanalyzed a classic example of density independence (Davidson and Andrewartha's thrips data).

When birth, death or dispersal rates show strong density dependence, population growth rates (λ or r) may decline as densities increase (**Figure 9.16**). Eventually, if densities become high enough to cause λ to equal 1 (or r to equal 0), the population stops growing entirely; if λ becomes less than 1 (or $r < 0$), the population declines. As we'll see in the next section, such density-dependent changes in the population growth rate can cause a population to reach a stable, maximum population size.

CONCEPT 9.5 The logistic equation incorporates limits to growth and shows how a population may stabilize at a maximum size, the carrying capacity.

Logistic Growth

In some cases, the growth of a population can be represented by an S-shaped curve (**Figure 9.17**). Such a population exhibits **logistic growth**, a pattern in which its abundance increases rapidly at first, then stabilizes at a population size known as the **carrying capacity** (the maximum population size that can be supported indefinitely by the environment). The growth rate of the population decreases as the population size nears the carrying capacity because resources such as food, water, or space begin to be in short supply. At the carrying capacity, the growth rate is zero, and hence the population size does not change.

The logistic equation models density-dependent population growth

To see how the idea of a carrying capacity can be represented in a mathematical model of population growth, let's reconsider Figure 9.16. The data in both graphs show that population growth rates (r or λ) decreased approximately as a straight line as population densities increased. But r is assumed to be constant in the exponential growth equation, $dN/dt = rN$. As we've seen, a constant value of $r > 0$ allows for unlimited growth in population size. Thus, to

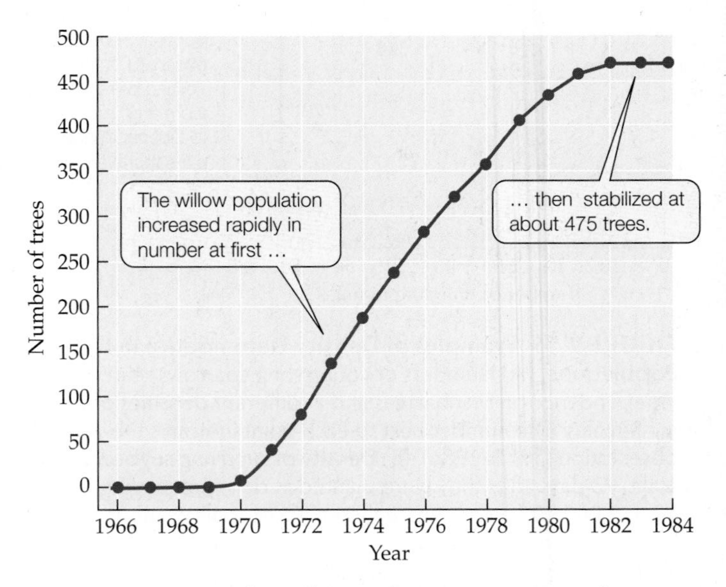

FIGURE 9.17 An S-Shaped Growth Curve in a Natural Population At a site in Australia, heavy grazing by rabbits had prevented willows from colonizing the area. The rabbits were removed in 1954. When willows colonized the area in 1966, ecologists tracked the growth of their population. (After Alliende and Harper 1989.)

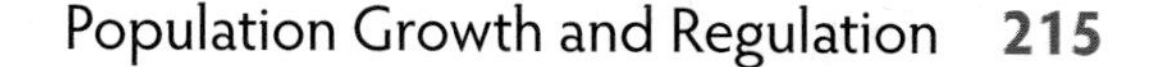

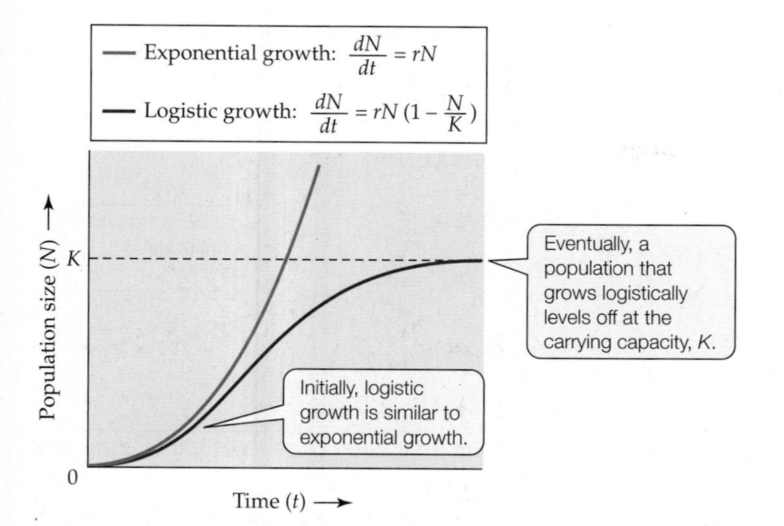

FIGURE 9.18 Logistic and Exponential Growth Compared Over time, logistic growth differs greatly from the unlimited growth of a population that increases exponentially.

Q In the logistic equation, as the population size (*N*) becomes increasingly close to the carrying capacity, *K*, how does that affect the term $(1 - N/K)$? Why does this cause *N* to stop increasing in size?

modify the exponential growth equation to make it more realistic, we replace the assumption that r is constant with the assumption that r declines in a straight line as density (N) increases. When we do this, as described in **Web Extension 9.3**, we obtain the *logistic equation*:

$$\frac{dN}{dt} = rN\left(1 - \frac{N}{K}\right) \quad \textbf{(9.7)}$$

where dN/dt is the rate of change in population size at time t, N is population density (also at time t), r is the population growth rate under ideal conditions, and K is the density at which the population stops increasing in size. K can be interpreted as the carrying capacity of the population, and the term $(1 - N/K)$ can be viewed as representing the net effect of factors that reduce the population growth rate from the constant rate (r) seen in exponential growth.

Logistic growth is similar to, but slightly slower than, exponential growth when densities are low (**Figure 9.18**). This occurs because when N is small, the term $(1 - N/K)$ is close to 1, and hence a population that grows logistically grows at a rate close to r. As the population density increases, however, logistic and exponential growth differ greatly. In logistic growth, the rate at which the population changes in size (dN/dt) approaches zero as the population size nears the carrying capacity, K. As a result, over time, the population size approaches K gradually, eventually remaining constant with K individuals in the population.

In Chapter 10 we'll discuss the extent to which the growth of natural populations can be described by the S-shaped curve that results from the logistic equation; here, we examine efforts to fit the logistic equation to U.S. census data.

Can logistic growth predict the carrying capacity of the U.S. population?

In a groundbreaking paper published in 1920, Pearl and Reed examined the fit of several different mathematical models to U.S. census data for the period 1790–1910. Several of the approaches they tested did a good job of matching the historical data, but none included limits to the eventual size of the U.S. population. To address this shortcoming, they derived the logistic equation, which, unknown to them, had been first described in 1838 by the Belgian mathematician P. F. Verhulst. Pearl and Reed argued that the logistic equation provided a sensible way to represent population growth because it included limits to growth. When they fit the census data to the logistic curve, they obtained an excellent match, from which they estimated that the U.S. population had a carrying capacity of $K = 197{,}274{,}000$ people (**Figure 9.19**).

The logistic curve estimated by Pearl and Reed provides a good fit to U.S. population data through 1950. After that time, however, the actual population size differed considerably from Pearl and Reed's projections. By 1967, the carrying capacity (197 million) they had predicted had been surpassed. Pearl and Reed intended their estimate of the carrying capacity to represent the number of people that could be supported in the United States in

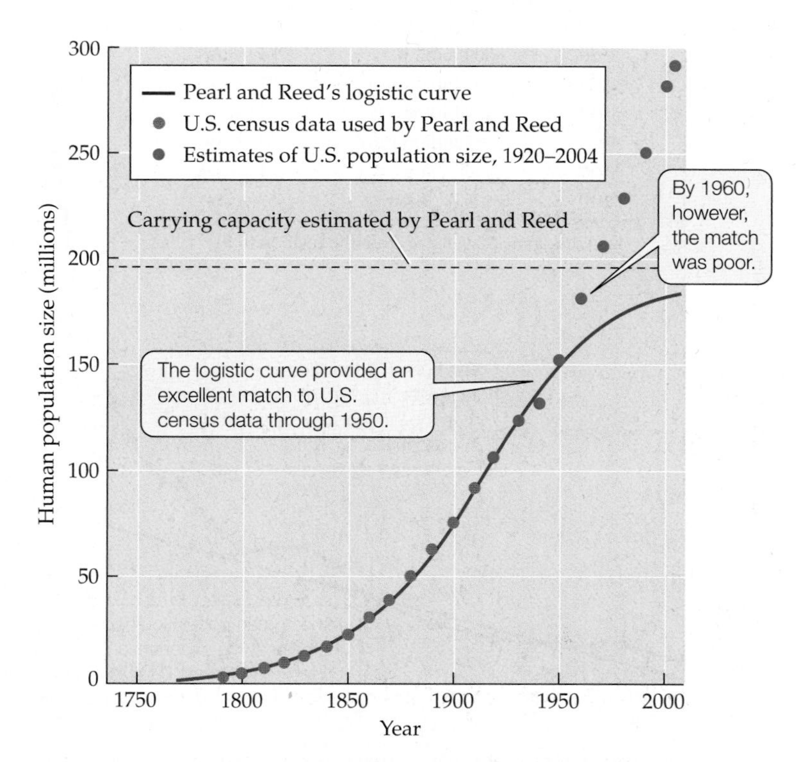

FIGURE 9.19 Fitting a Logistic Curve to the U.S. Population Size In 1920, Pearl and Reed fitted a logistic curve to U.S. census data for 1790–1910. The estimated carrying capacity (*K*) was 197 million people. (Data through 1910 from Pearl and Reed 1920; other data from Statistical Abstracts, U.S. Census Bureau.)

a self-sufficient manner. They recognized that if conditions changed—for example, if agricultural productivity increased or if more resources were imported from other countries—the population could increase beyond 197 million. These and other changes have occurred, leading some ecologists and demographers to shift their focus from the human carrying capacity to the area required to support a human population (the "ecological footprint," discussed on p. 217).

A CASE STUDY REVISITED

Human Population Growth

Media reports often state that the human population is growing exponentially. As we saw in Figure 9.10, a simple way to determine whether a population is growing exponentially is to plot the natural logarithm of population size versus time. If a straight line results, the population is growing exponentially. When we plot the natural logarithm of human population size versus time for the last 2,000 years, we see that our population sizes deviate considerably from the straight line expected in exponential growth (**Figure 9.20**). In fact, as fast as exponential growth is, historically the human population has increased even more rapidly than that.

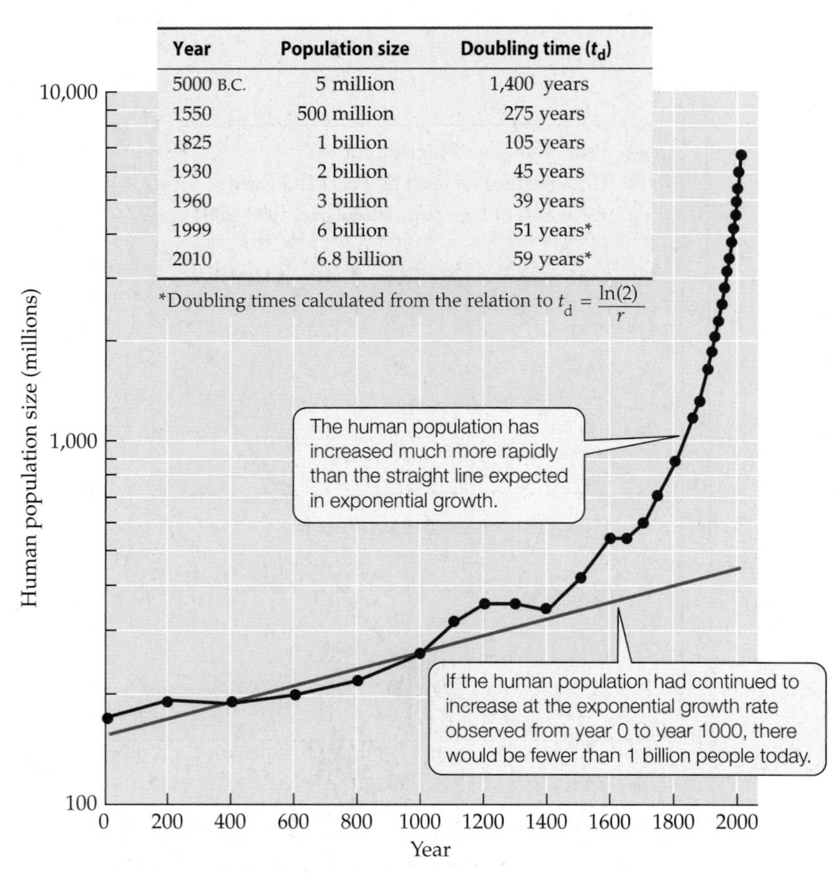

Year	Population size	Doubling time (t_d)
5000 B.C.	5 million	1,400 years
1550	500 million	275 years
1825	1 billion	105 years
1930	2 billion	45 years
1960	3 billion	39 years
1999	6 billion	51 years*
2010	6.8 billion	59 years*

*Doubling times calculated from the relation to $t_d = \frac{\ln(2)}{r}$

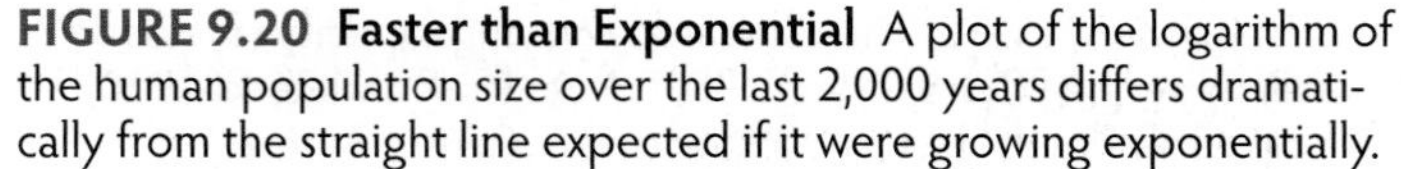

FIGURE 9.20 Faster than Exponential A plot of the logarithm of the human population size over the last 2,000 years differs dramatically from the straight line expected if it were growing exponentially.

The faster-than-exponential nature of human population growth is also evident from doubling times for the human population. Recall that in a population that grows exponentially, the doubling time remains constant. However, as shown in the inset of Figure 9.20, the doubling times observed for the human population dropped from roughly 1,400 years in 5000 B.C. to a mere 39 years in 1960—again indicating that our population has increased more rapidly than expected of exponential growth.

Projecting into the future, we can predict how long it will take our population to double in size at current rates of growth. To do this, the doubling time is estimated from the relation $t_d = \ln(2)/r$ (see Equation 9.5), where r is the current growth rate of the human population. Such estimates have shown that the human population was growing most rapidly in the early 1960s, with a doubling time of 32 years. Since then, the doubling time has increased (because r has decreased), reaching 59 years in 2010. So, returning to the question we asked on p. 200 (whether there would be 15 billion people in 2080), the answer is, probably not. U.S. Census Bureau projections indicate that population growth rates are likely to continue to fall over the next 40 years (**Figure 9.21**), leading to a predicted population size of 9.1 billion in 2050 (**Figure 9.22**). Extending that curve out to 2080 suggests that there will be roughly 10 billion people in that year. If these projections turn out to be correct, or nearly so, what will the future be like with that many people? Is 10 billion above the carrying capacity of the human population?

To answer these questions, we must determine the carrying capacity of the human population, but that is trickier than it may at first appear. Many researchers have

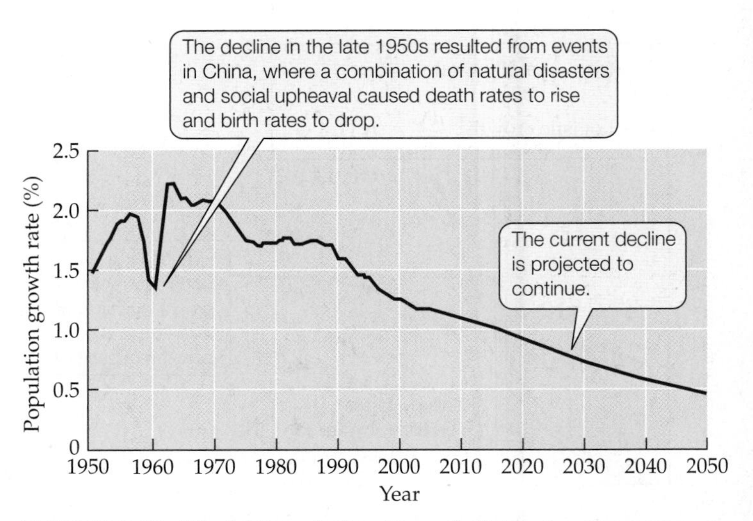

FIGURE 9.21 World Population Growth Rates Are Dropping Annual world population growth rates have declined since the early 1960s. After U. S. Census Bureau, International Data Base, December 2009 update.)

Q In 2050, will the human population still be increasing in size? Explain.

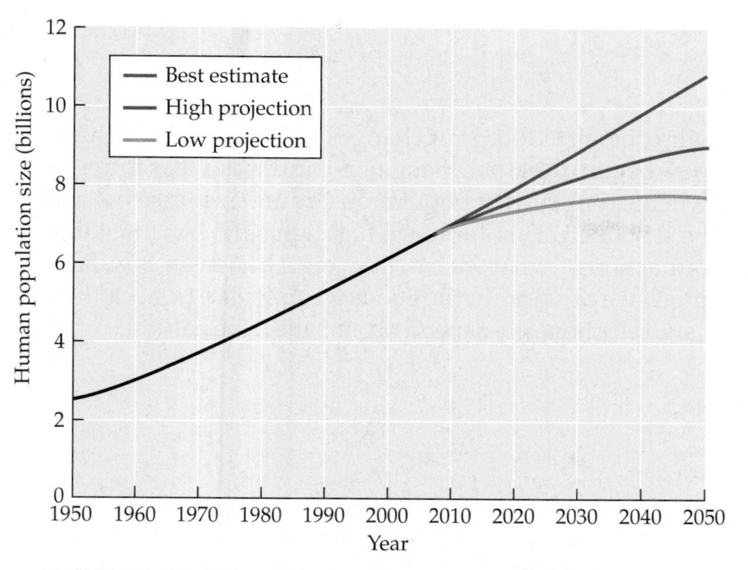

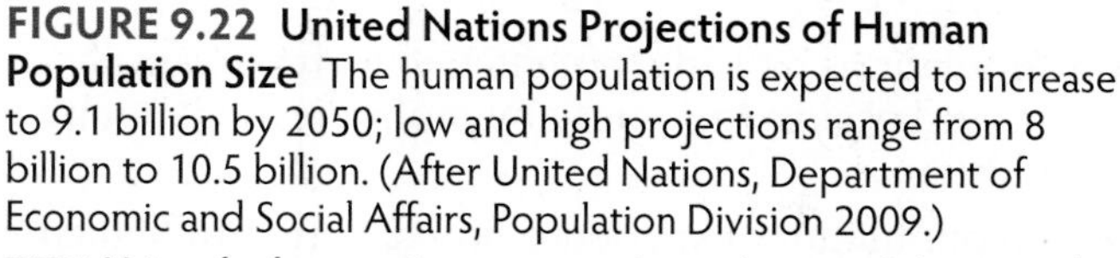

FIGURE 9.22 United Nations Projections of Human Population Size The human population is expected to increase to 9.1 billion by 2050; low and high projections range from 8 billion to 10.5 billion. (After United Nations, Department of Economic and Social Affairs, Population Division 2009.)

Q Using the best-estimate curve shown here and the annual growth rate estimated for the human population in 2050 (see Figure 9.21), approximately how large will our population be in 2051?

estimated the human carrying capacity, obtaining values that range from fewer than 1 billion to more than 1,000 billion (see Cohen 1995). This large variation is due in part to the fact that many different methods—from logistic models to calculations based on crop production and human energy requirements—have been used. In addition, different researchers have made different assumptions about how people would live and how technology would influence our future, assumptions that have a large effect on the estimated carrying capacity.

For example, using the "ecological footprint" approach described below, it has been estimated that Earth could support 1.3 billion people indefinitely if everyone used the amount of resources used by people in the United States in 2006 (Ewing et al. 2009). Toward the other end of the spectrum, if everyone used the amount of resources used by people in India in 2006, the world could support over 14 billion people. Thus, as we suggested in the introductory pages of this chapter, issues of human population size and resource use are linked inextricably: more people means that more resources will be used, but the degree to which our growing population affects the environment depends on the amount of resources used by each person.

CONNECTIONS IN NATURE

Your Ecological Footprint

When you turn on a light, purchase an appliance, drive a car, or eat fruit imported from another country, you may not think about the effects your actions have on the natural world. How, for example, does driving to the store to get groceries affect forests or coral reefs?

To answer this question, we must account for the resources required to support our actions. The grains we eat require farmland; the wood products we use require natural forests or plantations; the fish we eat require productive fishing grounds; the machines and appliances we purchase require raw materials and energy to build, as well as energy for their operation. Ultimately, every aspect of our economy depends on the land and waters of Earth (the services they provide for us will be discussed in Unit 6). Recognizing this, William Rees proposed that we measure the environmental impact of a population as its **ecological footprint**, which is the total area of productive ecosystems required to support that population (Rees 1992). The footprint approach turns the carrying capacity concept on its head: instead of asking how many people a given area can support, it asks how much area is required to support a given population.

Ecological footprints are calculated from national statistics on agricultural productivity, production of goods, and resource use. The area required to support these activities is then estimated. For example, the land required to support wheat consumption in 1993 by people in Italy was estimated by dividing the amount of wheat consumed (26,087,912 tons) by the amount of wheat produced per unit of land (2.744 tons/ha), resulting in 9,507,257 hectares, or 0.167 hectares per person (Wackernagel et al. 1999). To compare footprint calculations among nations and across different crops, such results are typically converted to *global hectares*, where a global hectare is defined as a hectare of world-average biological productivity (Kitzes and Wackernagel 2009).

Methods of calculating ecological footprints are still being refined, but results to date are sobering. In 2006, there were 11.9 billion global hectares of productive land available, and the ecological footprint of an average person was 2.6 global hectares (Ewing et al. 2009). This result suggests that Earth could have supported 4.6 billion people (11.9 billion ha/2.6 ha per person) for a long time. In fact, the human population in 2006 was 6.6 billion, more than a 40% overshoot of its carrying capacity. An overshoot of this magnitude indicates that in 2006, environmental resources were being used more rapidly than they could be regenerated, a pattern of use that cannot be sustained.

Similar calculations can be made for nations, cities, and even companies, schools, or individuals (see Review Question 5). In the United States, for example, the average ecological footprint was 9.0 global hectares per person in 2006. Since there were 1,330 million global hectares of productive land available in the nation, this calculation suggests that the carrying capacity of the United States in 2006 was 148 million people (1,330 million ha/9.0 ha per person); the actual population was 303 million, more than double the carrying capacity.

Human use of resources changes from year to year, depending on population size, per capita rates of resource use, and technology (i.e., the efficiency of production). In addition, the total area of productive ecosystems available to support our activities changes over time due to factors such as gain or loss of farmland, destruction of natural habitat, pollution, and extinctions of species. As a result, our ecological footprint changes over time. People have now begun to use our changing footprint as a way to assess whether our current population size and resource use can be sustained. This approach highlights the fact that all of our actions—what we eat, how big a house we buy, how much we drive or fly, the goods we purchase (e.g., clothes, cars, cell phones)—depend on and affect the natural world.

SUMMARY

CONCEPT 9.1 Life tables show how survival and reproductive rates vary with age, size, or life cycle stage.

- Cohort life tables can be constructed from data on the fates of individuals born during the same time period and used to calculate age-specific survival rate, survivorship, and fecundity.
- In highly mobile or long-lived organisms, a static life table may be constructed from data on the survival and fecundity of individuals of different ages during a single time period.
- In species for which age correlates poorly with survival and fecundity, life tables based on size or life cycle stage may be constructed.
- In populations with a type I survivorship curve, most individuals survive to old age. In populations with a type II survivorship curve, individuals experience a constant chance of surviving from one age to the next throughout their lives. In populations with a type III survivorship curve, death rates are very high for young individuals, but adults survive well later in life. Of the three types, type III is the most common.

CONCEPT 9.2 Life table data can be used to project the future age structure, size, and growth rate of a population.

- The age structure of a population influences the growth rate of that population over time.
- A population eventually grows at a fixed rate if age-specific survival rates and fecundities do not change over time.
- Any factor that changes age-specific survival rates or fecundities may alter a population's growth rate.

CONCEPT 9.3 Populations can grow exponentially when conditions are favorable, but exponential growth cannot continue indefinitely.

- Geometric growth occurs when a population of individuals that reproduce in synchrony at discrete time periods changes in size by a constant proportion from one discrete time period to the next.
- Exponential growth occurs when a population with continuous reproduction changes in size by a constant proportion at each instant in time.
- Populations have the potential to increase rapidly in size because they grow by multiplication, not by addition.
- All populations experience limits to growth, which ensure that exponential growth cannot continue indefinitely.

CONCEPT 9.4 Population size can be determined by density-dependent and density-independent factors.

- In many species, density-independent factors, such as temperature or precipitation, play a major role in determining year-to-year changes in population size.
- When the density of any species becomes high enough, a lack of food, space, or other resources causes birth rates to decrease, death rates to increase, or dispersal to increase.
- Population regulation occurs when one or more density-dependent factors tend to increase population size when densities are low and decrease population size when densities are high.

CONCEPT 9.5 The logistic equation incorporates limits to growth and shows how a population may stabilize at a maximum size, the carrying capacity.

- In some species, changes in population size over time can be described by an S-shaped curve in which the population increases rapidly at first, then stabilizes at a maximum level, the carrying capacity.
- The logistic equation can be used to represent density-dependent population growth.
- Logistic population growth provides a close fit to the size of the U.S. population up to 1950; after that time, the growth rate of the U.S. population was considerably greater than expected in logistic growth.

SUMMARY (continued)

CASE STUDY/CASE STUDY REVISITED Human Population Growth

- Over the past 2,000 years, the human population has increased in size even more rapidly than it would if it were growing exponentially.
- Estimates of the carrying capacity of the human population vary widely, from fewer than 1 billion people to more than 1,000 billion people.
- The carrying capacity concept applies poorly to human populations that import resources from outside the area in which the population is found.
- Ecological footprint analyses based on available productive land area and current patterns of resource use suggest that the global human population is 40% greater than the maximum number that could be sustained for a long time.

REVIEW QUESTIONS

1. For a field ecology project, you count the number of individuals of different ages found in a population during a single time period. There are 100 newborns, 40 1-year-olds, 15 2-year-olds, 5 3-year-olds, and 0 4-year-olds.

a. Use these data to fill in the N_x, S_x, and l_x columns of a static life table.

b. Explain the difference between a static life table and a cohort life table.

2.

Age(x)	S_x	l_x	F_x
0	0.33	1.0	0
1	0.50	0.333	3
2	0	0.167	2
3		0	0

a. A population that grows as shown in the life table given in this problem initially has 50 newborns, 50 1-year-olds, and 50 2-year-olds (i.e., at time 0, $N_0 = 50$, $N_1 = 50$, and $N_2 = 50$). Calculate how the age distribution changes from time $t = 0$ to time $t = 6$. What is your best estimate of the growth rate (λ) and stable age distribution of this population?

b. A second population also grows according to the life table given in this problem, but has a different initial age distribution: $N_0 = 80$, $N_1 = 50$, $N_2 = 20$. Calculate how the age distribution changes from $t = 0$ to $t = 6$, and estimate the growth rate (λ) and stable age distribution of this population.

c. Compare your answers to parts (a) and (b). Without performing further calculations, estimate the growth rate (λ) and stable age distribution of a third population that also grows according to the life table in this problem, but has an initial age distribution of $N_0 = 10$, $N_1 = 50$, $N_2 = 90$.

3. A population of insects triples every year. Initially, there were 40 insects.

a. How many insects would there be after 4 years?

b. How many insects would there be after 27 years? (Write your answer to this question as an equation.)

c. The habitat of the insect is degraded such that the population growth rate (λ) changes from 3.0 to 0.75. If there were 100 insects in the population when its habitat became degraded, how many insects would there be after 3 years?

4. What is the distinction between factors that regulate population size and factors that determine population size?

5. Calculate your ecological footprint at http://www.footprintnetwork.org/en/index.php/GFN/page/calculators

ON THE COMPANION WEBSITE
sites.sinauer.com/ecology2e

The website includes Chapter Outlines, Online Quizzes, Flashcards & Key Terms, Suggested Readings, a complete Glossary, and the Web Stats Review. In addition, the following resources are available for this chapter:

HANDS-ON PROBLEM SOLVING

Density-Dependent and Density-Independent Factors

This Web exercise explores the effects of density-dependent and density-independent factors on populations. You will analyze data from populations of arctic ground squirrels to determine the effects of these factors on weaning success, overwintering survival, and population growth rate.

WEB EXTENSIONS

9.1 Deriving the Geometric Growth Equations

9.2 Density Dependence in *Thrips imaginis*

9.3 Deriving the Logistic Equation

CLIMATE CHANGE CONNECTIONS

9.1 Consequences of Climate Change for Western Forest Communities

CHAPTER

10 Population Dynamics

KEY CONCEPTS

CONCEPT 10.1 Populations exhibit a wide range of growth patterns, including exponential growth, logistic growth, fluctuations, and regular cycles.

CONCEPT 10.2 Delayed density dependence can cause populations to fluctuate in size.

CONCEPT 10.3 The risk of extinction increases greatly in small populations.

CONCEPT 10.4 Many species have a metapopulation structure in which sets of spatially isolated populations are linked by dispersal.

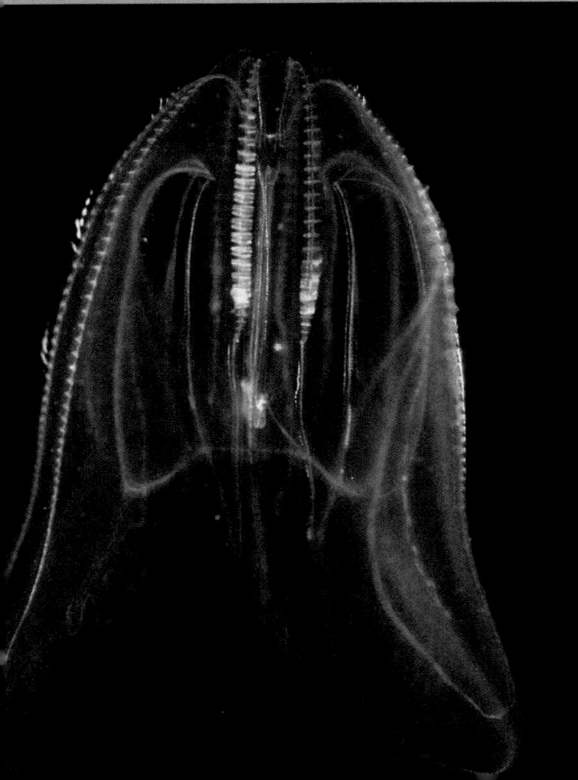

A Sea in Trouble: A Case Study

In the 1980s, the comb jelly *Mnemiopsis leidyi* (**Figure 10.1**) was introduced into the Black Sea, most likely by the discharge of ballast water from cargo ships. The timing of this invasion could hardly have been worse. At that time, the Black Sea ecosystem was already in decline due to increased inputs of nutrients such as nitrogen from sewage, fertilizers, and industrial wastes (as we'll discuss on pp. 237–238, overfishing also may have contributed to the ecosystem's decline). The increased supply of nutrients had devastating effects across the northern Black Sea, where the waters are shallow (less than 200 m deep) and prone to problems that stem from *eutrophication* (the addition of nutrients to water). As nutrient concentrations increased in these shallow waters, phytoplankton abundance increased, water clarity decreased, oxygen concentrations dropped, and fish populations experienced massive die-offs. Nutrient concentrations in deeper portions of the Black Sea also rose, causing increased phytoplankton abundance, but not fish die-offs.

Such was the situation when *Mnemiopsis* arrived. This marine invertebrate species is a voracious predator of zooplankton, fish eggs, and young fish. Furthermore, *Mnemiopsis* continues to feed even when it is completely full, which causes it to regurgitate large quantities of prey stuck in balls of mucus. Small prey encased in mucus survive poorly. As a result, the negative effect of *Mnemiopsis* on its prey outstrips even its considerable ability to digest food.

When well fed, an individual *Mnemiopsis* can produce up to 8,000 offspring just 13 days after its own birth. Following its arrival in the Black Sea in the early 1980s, *Mnemiopsis* gradually increased in numbers. Then, in 1989, *Mnemiopsis* populations exploded in size (**Figure 10.2**), reaching astonishing biomass levels (1.5–2.0 kg/m^2) throughout the sea. The total biomass of *Mnemiopsis* in the Black Sea was estimated at 800 million tons (live weight) in 1989—far greater than the world's entire annual commercial fish catch, which has never exceeded 95 million tons.

The enormous numbers of *Mnemiopsis* present in 1989, and again in 1990, compounded the effects of the Black Sea's ongoing problems. *Mnemiopsis* ate huge quantities of zooplankton, causing their populations to crash. Zooplankton eat phytoplankton, so *Mnemiopsis* indirectly caused phytoplankton populations to increase even more than they already had due to nutrient enrichment. Upon their deaths, the phytoplankton and *Mnemiopsis* provided food for bacterial decomposers. Bacteria use oxygen as they decompose dead organisms, so as bacterial activity increased, oxygen concentrations in the water decreased, harming some fish

FIGURE 10.1 A Potent Invader The comb jelly *Mnemiopsis leidyi* spread from ocean waters along the east coast of North America to the Black Sea, wreaking havoc in its new ecosystem upon its arrival.

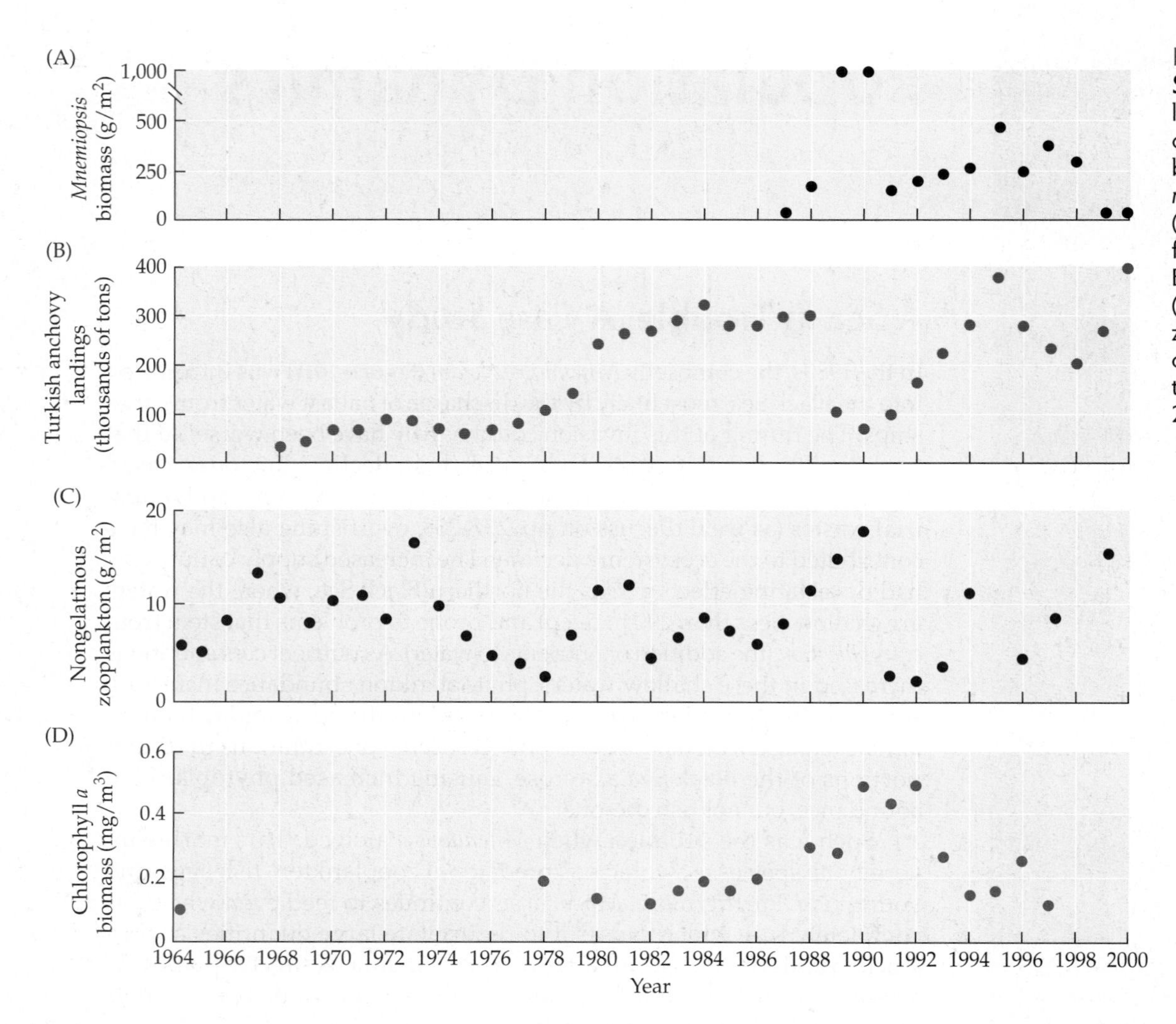

FIGURE 10.2 Changes in the Black Sea Ecosystem The graphs track long-term changes in four components of the Black Sea ecosystem: (A) mean biomass of the invasive species *Mnemiopsis leidyi* (first measured in 1987), (B) Turkish anchovy landings (Turkish fishermen have garnered most of the Black Sea anchovy catch since 1980), (C) mean biomass of nongelatinous zooplankton, and (D) mean biomass of chlorophyll *a* (an indicator of phytoplankton abundance). (After Kideys 2002.)

populations. In addition, by devouring the food supplies (zooplankton), eggs, and young of important commercial fishes such as anchovies, *Mnemiopsis* led to a rapid decline in fish catches (see Figure 10.2B), causing extensive losses in the Turkish fishing industry.

The combined negative effects of nutrient enrichment and invasion by *Mnemiopsis* posed a serious threat to the Black Sea ecosystem. Although it covers a large surface area (over 423,000 km²), the Black Sea is nearly landlocked and exchanges little of its water each year with other ocean waters. In addition, the Black Sea is unusual in that only the top 150–200 m of its waters (ca. 10% of its average depth) contain oxygen, effectively making the entire sea "shallow" for species that require oxygen. Its limited water exchange and anoxic deep waters make the Black Sea particularly vulnerable to the negative effects of nutrient enrichment.

Native predators and parasites had failed to regulate *Mnemiopsis* populations. Thus, in the early 1990s, the future of the Black Sea ecosystem looked bleak. Fortunately, there are signs of improvement today: *Mnemiopsis* and phytoplankton populations have fallen, paving the way for the recovery of the Black Sea. How did this happen?

Introduction

As we saw in the previous two chapters, populations can change in size as a result of four processes: birth, death, immigration, and emigration. We can summarize the effects of these four processes on population size with the following equation:

$$N_{t+1} = N_t + B + I - D - E$$

where N_t is the population size at time t, B is the number of births, I is the number of immigrants, D is the number of deaths, and E is the number of emigrants between time t and time $t + 1$. As implied by this equation, populations are open and dynamic entities. Individuals can move from one population to another, and the number of individuals in a population can change from one time period to the next.

Ecologists use the term *population dynamics* to refer to the ways in which populations change in abundance over time. In this chapter, we'll consider the dynamics of populations in more detail, placing special emphasis on two kinds of populations: small populations that face the threat of extinction, and sets of populations that are linked by dispersal (metapopulations). We'll begin our discussion of population dynamics by surveying patterns of population growth.

CONCEPT 10.1 Populations exhibit a wide range of growth patterns, including exponential growth, logistic growth, fluctuations, and regular cycles.

Patterns of Population Growth

Most observed patterns of population growth can be grouped into four major types: exponential growth, logistic growth, population fluctuations, and regular population cycles (a special type of fluctuation). Bear in mind, however, that a single population could experience each of these four types of growth at different times. Furthermore, these four patterns are not mutually exclusive. For example, as we will see shortly, a population may grow logistically yet fluctuate around the values expected in logistic growth.

Exponential growth can occur when conditions are favorable

In the first pattern, exponential growth, a population increases (or decreases) by a constant proportion at each point in time. As we saw in Chapter 9, exponential growth can be represented by either of two related equations, $N_{t+1} = \lambda N_t$ or $dN/dt = rN$, the choice of which depends on whether reproduction occurs at discrete time periods (the first equation) or is continuous (the second equation).

Exponential growth cannot continue indefinitely (see the discussion under Concept 9.3), but when conditions are favorable, a population can increase exponentially for a limited time. Such periods of exponential growth can occur within the established range of a species, as when good weather occurs for several years running. They can also occur when a species reaches a new geographic area, either by dispersing on its own or with human assistance. If conditions are favorable in the new environment, the population can increase rapidly until density-dependent factors (see the section under Concept 9.4) act to regulate its numbers.

An example of how dispersal can lead to exponential growth is provided by the cattle egret subspecies *Bubulcus ibis ibis*. These birds originally lived in the Mediterranean region and in parts of central and southern Africa. Since the late 1800s and early 1900s, however, they have colonized new regions on their own, including South Africa, South America, and North America. Typically, after the subspecies reached a new area, its population in that area increased exponentially as it became established in its new habitat. For example, after the cattle egret colonized Florida in the 1950s, its populations there grew exponentially for several decades, with an estimated intrinsic rate of increase of $r = 0.11$ for the period 1956–1971 (van den Bosch et al. 1992). By the 1980s, exponential growth had ceased and cattle egret numbers in Florida had stabilized (based on survey data reported by the USGS National Biological Service).

The spread of cattle egret populations illustrates some common aspects of dispersal to new regions (**Figure 10.3**). Species such as cattle egrets that successfully colonize new geographic regions do so by long-distance or **jump dispersal** events. Local populations in the new region then increase in size and expand (by relatively short-distance dispersal events) to occupy nearby areas of suitable habitat.

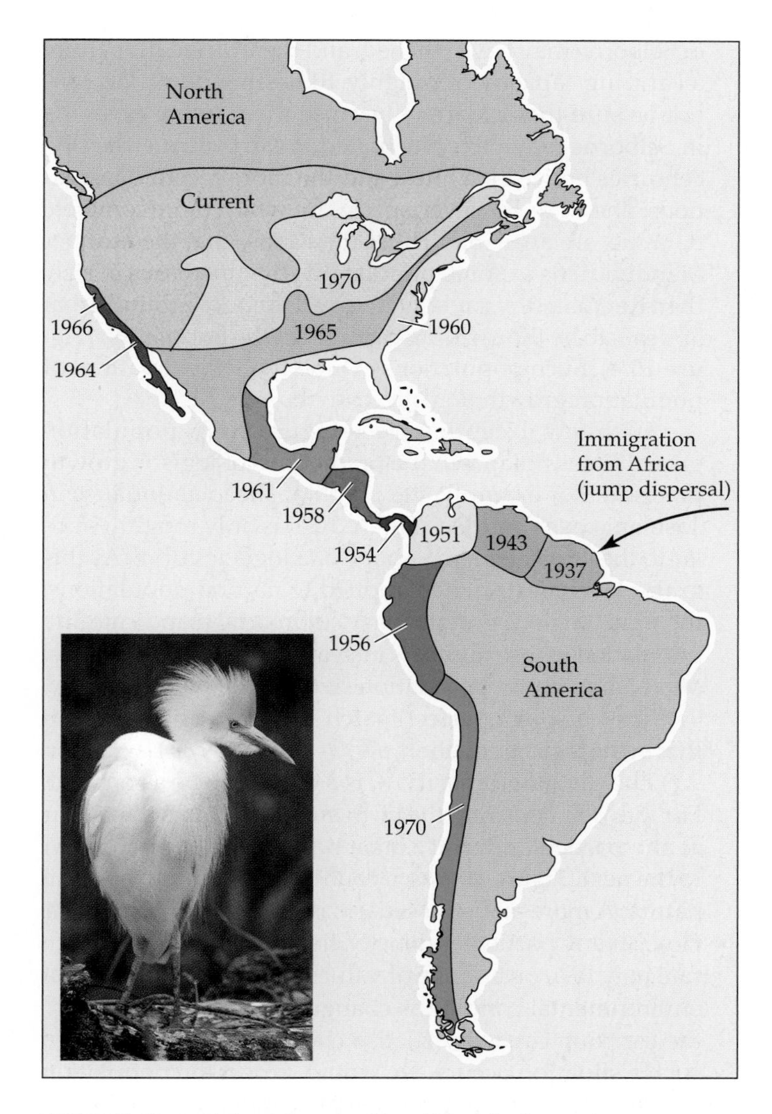

FIGURE 10.3 Colonizing the New World The cattle egret subspecies *Bubulcus ibis ibis* dispersed from Africa to South America in the late 1800s. Once it established colonies in the northeastern region of South America, it then spread rapidly to other parts of South and North America. The contour lines and dates show the edges of the cattle egret's range at different times. (After R. L. Smith 1974.)

In logistic growth, the population approaches an equilibrium

Some populations appear to reach a relatively stable population size, or *equilibrium*, that changes little over time. For example, in a letter written in 1778, the pastor and naturalist Gilbert White observed how many swifts (*Micropus apus*) were found each year in his village of Selborne (in southern England): "The number that I constantly find are eight pairs; about half of which reside in the church, and the rest in some of the lowest and meanest thatched cottages" (White 1789). In 1983, over 200 years later, several ecologists returned to Selborne and found 12 pairs of swifts, a number similar to that in White's time.

The 1778 and 1983 observations suggest that the swifts of Selborne may have reached, and then varied little from, a carrying capacity of roughly 10 nesting pairs. We cannot be sure that is correct because the number of swifts in Selborne may have fluctuated greatly during the two centuries between White's and the more recent observations. There are, however, species for which more complete data sets are available. Those data show that the number of individuals in some populations first increases in size, then fluctuates by a relatively small amount around what appears to be the carrying capacity of the population (**Figure 10.4**). Such populations exhibit the second pattern of population growth, logistic growth.

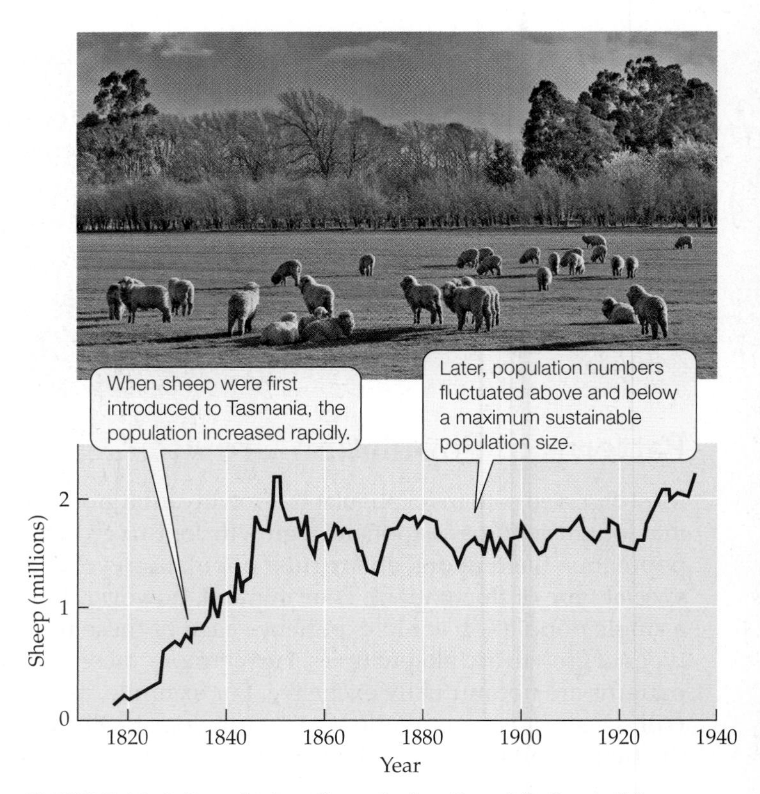

FIGURE 10.4 Population Growth Can Roughly Resemble a Logistic Curve Population growth in a few species matches a logistic curve closely (see Figure 9.17). More often, a species shows a pattern of growth (a rise in abundance, followed by a roughly stable population size) in which the match to a logistic curve is very rough, as seen here for sheep introduced to the island of Tasmania. (After Davidson 1938.)

With few exceptions (see Figure 9.17), population growth does not match the predictions of logistic growth precisely. For example, the graph of sheep abundance in Tasmania over time (see Figure 10.4) is only roughly similar to the characteristic S shape of a logistic curve. As this example suggests, when applied to natural populations, the term "logistic growth" is used broadly to indicate any population whose numbers increase initially but then level off at a maximum population size, the carrying capacity.

This lack of a perfect match is not surprising, given that in the logistic equation, $dN/dt = rN(1 - N/K)$ (Equation 9.7), the carrying capacity, K, is assumed to be a constant. For K to be a constant, the birth and death rates that occur at any particular density must not change from one year to the next (**Figure 10.5A**), a condition that rarely holds in nature. A more realistic scenario is shown in **Figure 10.5B**. Here, at any particular density, the birth rate or the death rate may take on a range of values, reflecting the fact that environmental conditions change over time. As a result, the carrying capacity (K) also changes over time. When such a situation occurs, we would expect the population size to fluctuate around an average value of K, as observed for sheep in Tasmania (see Figure 10.4).

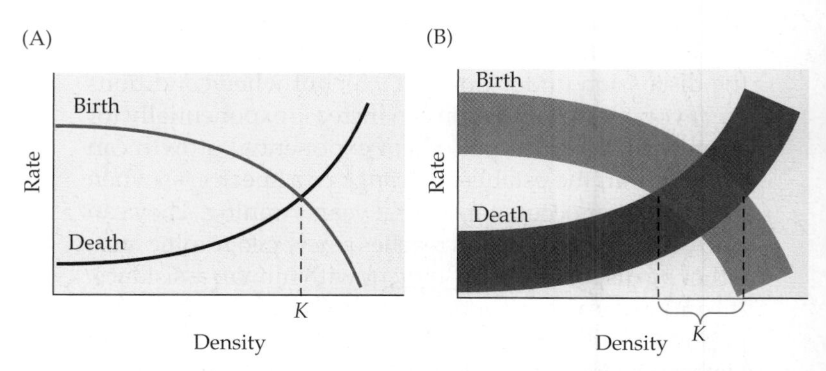

FIGURE 10.5 Why We Expect Carrying Capacity to Fluctuate The carrying capacity (K) of a population is the population size at which the birth rate equals the death rate. We can find K graphically by plotting the density at which the birth rate curve intersects the death rate curve. (A) Here we assume that at any given density, birth rates and death rates do not change over time. Hence, the two curves intersect at a single point, K, which is constant over time. (B) Here we assume that at any given density both the birth rate and death rate vary over time, as indicated by the broad bands. As a result, the birth and death rate curves can intersect at a broad range of values, causing K to take on a range of values (shown in red).

In (A), draw a second death rate curve that is similar in shape to the curve in purple but has a higher death rate at each density. Label the carrying capacity that results and compare it to the value of K shown in (A).

All populations fluctuate in size

Another characteristic of the sheep population in Tasmania is seen in all populations: their size rises and falls over time, illustrating the third and most common pattern of popu-

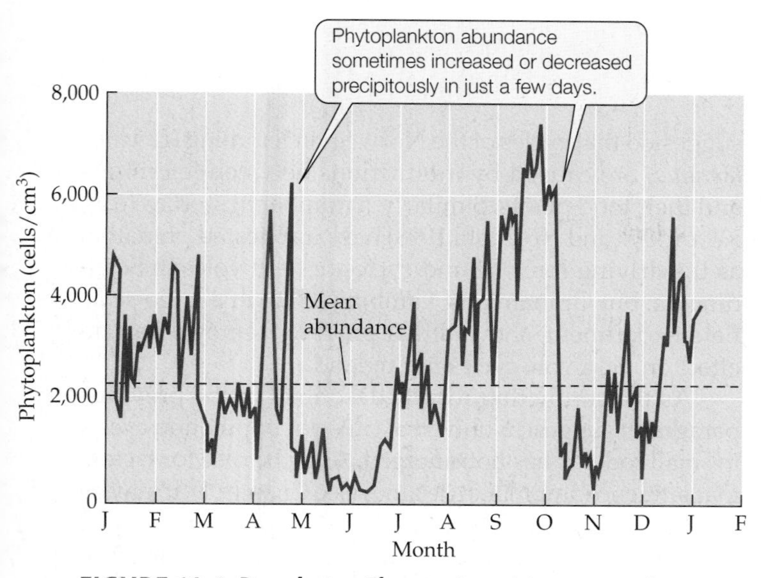

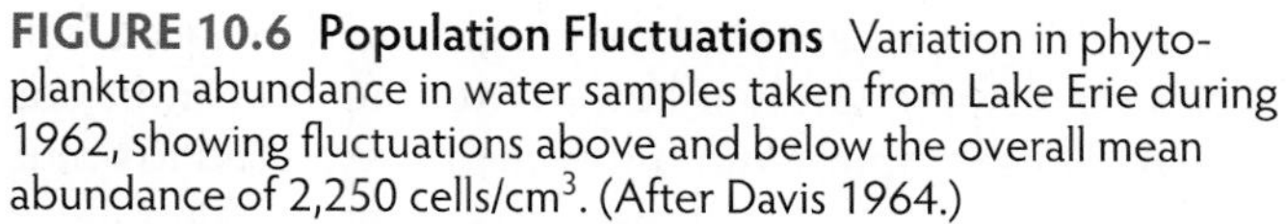
FIGURE 10.6 Population Fluctuations Variation in phytoplankton abundance in water samples taken from Lake Erie during 1962, showing fluctuations above and below the overall mean abundance of 2,250 cells/cm^3. (After Davis 1964.)

lation growth, **population fluctuations**. In some populations, fluctuations occur as erratic increases or decreases in abundance from an overall mean value (**Figure 10.6**). In other populations, fluctuations occur as deviations from a population growth pattern, such as exponential or logistic growth. If, for example, the growth of a population *exactly* matched a logistic curve, the population would not be said to fluctuate. But if population abundances rose above and fell below those expected in logistic growth (as in the Tasmanian sheep), the population would be said to fluctuate.

In some cases, population fluctuations are relatively small, as seen in Figure 10.4. In other cases, the number of individuals in a population can explode at certain times, causing a **population outbreak** (**Figure 10.7**). As we saw in Figure 10.2A, the biomass of the comb jelly *Mnemiopsis* increased a thousandfold during a 2-year outbreak in the Black Sea. Rapid variations in population sizes over time have also been observed in many terrestrial systems, especially in insects. Census data for the bordered white moth (*Bupalus piniarius*) collected from 1882 to 1940 in a German pine forest showed that the densities reached during outbreaks were up to 30,000 times as great as the lowest density observed. Such outbreaks can have wide-ranging ecological effects. For example, since 2000, an ongoing outbreak of the mountain pine beetle (*Dendroctonus ponderosae*) has killed hundreds of millions of trees across 16.3 million hectares (40 million acres) in British Columbia, Canada. The death of these trees has altered the composition of affected forests (**Figure 10.8**). Furthermore, as the dead trees decay, an estimated 17.6 megatons (Mt) of carbon dioxide is released into the atmosphere each year (Kurz et al. 2008)—an amount roughly equivalent to the yearly carbon emissions of all passenger cars in Great Britain.

Many different factors, both density-independent and density-dependent, can cause the size of a population to fluctuate. The increase in zooplankton populations in the Black Sea in the early 1980s probably occurred because their prey (phytoplankton) had increased in abundance (see Figure 10.2). Then, in 1991, zooplankton numbers plummeted, probably because of the spectacular increase

FIGURE 10.7 Populations Can Explode in Numbers As we saw in Chapter 9, all species have the potential for exponential growth. Hence, when conditions are favorable, a population outbreak can occur in which numbers increase very rapidly. The cockroaches covering the kitchen in this exhibit from the National Museum of Natural History represent the number that could have been produced by a single pregnant female in a few generations.

FIGURE 10.8 Consequences of an Insect Outbreak This aerial view shows red foliage of lodgepole pine (*Pinus contorta*) trees killed by an outbreak of mountain pine beetles in British Columbia, Canada.

in the abundance of their predator (*Mnemiopsis*) during the previous 2 years. The rapid changes in phytoplankton abundance in Lake Erie shown in Figure 10.6 could reflect changes in a wide range of environmental factors, including nutrient supplies, temperature, and predator abundance. In some cases, such factors produce a striking type of fluctuation: population cycles.

Some species exhibit population cycles

The fourth pattern of population growth is **population cycles**, in which alternating periods of high and low abundance occur after constant (or nearly constant) intervals of time. Such regular cycles have been observed in populations of small rodents such as lemmings and voles, whose abundances typically reach a peak every 3–5 years (**Figure 10.9**).

Population cycles are among the most intriguing patterns observed in nature. After all, what factors can cause numbers to fluctuate greatly over time, yet maintain a high degree of regularity? Possible answers to this question include both internal factors, such as hormonal or behavioral changes in response to crowding, and external factors, such as weather, food supplies, or predators. Gilg et al. (2003) used a combination of field observations and mathematical models to argue that the 4-year cycle of collared lemmings (*Dicrostonyx groenlandicus*) in Greenland is driven by predators, one of which, the stoat, specializes on lemmings (see Figure 10.9). Other investigators have suggested that cycles of the Norwegian lemming (*Lemmus lemmus*) are caused by interactions between lemmings and their food plants. Similarly, a number of studies (e.g., Korpimäki and Norrdahl 1998) have implicated predators as the driving force behind cycles of field voles in Scandinavia, but Graham and Lambin (2002), in a large-scale field experiment, showed that predator removal had no effect on field vole cycles in England.

As the conflicting results described in the previous paragraph suggest, a universal cause of population cycles in small rodents has not emerged. As is the case for factors that influence amphibian declines (see Chapter 1), it may be that the ecological mechanisms that drive population cycles differ from place to place and from one species to another.

In addition, recent evidence suggests that population cycles may stop entirely if key environmental conditions change. For example, lemming cycles have ceased to occur in some high-latitude locations (Kausrud et al. 2008; Gilg et al. 2009). Lemmings thrive when warmth from the ground melts a thin layer of the snow cover, leaving a small gap between the ground and the snow. In some regions over the past 10–15 years, warmer winter temperatures have caused the snow to melt and refreeze, preventing the formation of these gaps. A shortage of gaps has made it more difficult for lemmings to feed and has made lemmings easier for their predators to catch. As a result, lemming populations have ceased to rise in abundance every 3–4 years, halting the population cycles previously observed for this species (see Figure 10.9). In contrast, vole cycles in Finland have continued despite regional warming, indicating that the effect of climate change may depend on the species or on the particular mechanisms that drive the cycles (Brommer et al. 2010). We'll continue our exploration of factors that influence population cycles in Chapter 12, when we consider one of the most famous cycles of all, the hare–lynx cycle.

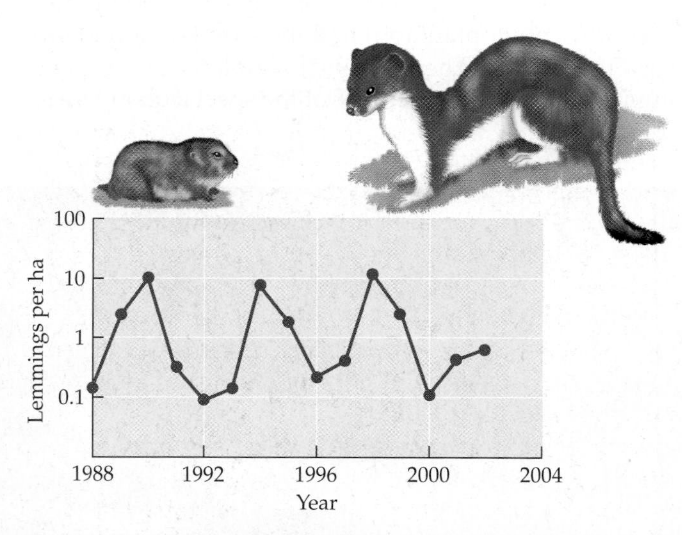

FIGURE 10.9 A Population Cycle In northern Greenland, collared lemming (*Dicrostonyx groenlandicus*, above left) abundance tends to rise and fall every 4 years. In this location, the population cycle appears to be driven by predators, the most important of which is the stoat (*Mustela erminea*, above right). In other regions, lemming population cycles may be driven by food supply. (After Gilg et al. 2003.)

Q Based on results from 1988 through 2000, how many lemmings per hectare would you have expected there to be in 2002? Explain your reasoning.

CONCEPT 10.2 Delayed density dependence can cause populations to fluctuate in size.

Delayed Density Dependence

Although relatively few populations exhibit regular population cycles, all populations fluctuate in size to some degree. As we've seen, such fluctuations can result from a variety of factors, including changes in food supply, temperature, or predator abundance. Population fluctuations can also be caused by delayed density dependence, the effects of which we examine here.

The effect of population density is often delayed in time

Delays, or time lags, are an important feature of interactions in nature. For example, when a predator or parasite feeds, it does not produce offspring immediately; thus, there is a built-in delay in the effect of food supply on birth rates. As a result, it is common for the number of individuals born in a given time period to be influenced by the population densities or other conditions that were present several time periods ago.

Beginning in the 1920s and 1930s, ecologists examined such time lags with mathematical models that incorporated **delayed density dependence** (delays in the effect that density has on population size). Results from these models indicated that delayed density dependence can contribute to population fluctuations.

Intuitively, this finding makes sense. Consider a population of predators that reproduce more slowly than their prey. If there are few predators initially, the prey population may increase rapidly in size. As a result, the predator population may also increase, reaching a point at which there are a large number of adult predators that survive well and produce many offspring. However, if the large population of predators eats so many prey that the prey population decreases sharply in size, there may be few prey available for the next generation of predators. In such a case, a mismatch in predator and prey numbers (high predator numbers, low prey numbers) occurs because there is a time lag in the response of predator numbers to prey numbers. When such a mismatch takes place, the predators may survive or reproduce poorly, and their numbers may drop. If prey numbers then increase (because there are now fewer predators), predator numbers may first rebound, then fall again due to the built-in time lag. Thus, in principle at least, it seems reasonable that a delay in the response of predators to prey density could cause predator numbers to fluctuate over time.

To examine further how delayed density dependence affects population fluctuations, consider the following version of the logistic equation, modified to include a time lag, τ:

$$\frac{dN}{dt} = rN\left[1 - \frac{N_{(t-\tau)}}{K}\right] \quad \textbf{(10.1)}$$

In this equation, dN/dt is the rate of change in population size at time t, r is the (per capita) population growth rate under ideal conditions, N is the population size at time t, and K is the carrying capacity. Equation 10.1 is identical to the logistic equation (Equation 9.7), except that within the term $(1 - N/K)$, N has been replaced by $N_{(t-\tau)}$, the population size at time $t-\tau$. Recall that the term $(1 - N/K)$ represents the net effect of factors that reduce the population growth rate from the constant rate (r) seen in exponential growth. Incorporating $N_{(t-\tau)}$ within this term indicates that the population growth rate is reduced by the size of the population at time $t - \tau$ in the past—not by the population's current size, N_t, as was assumed in Equation 9.7.

When we incorporate delayed density dependence into the logistic equation in this way, population fluctuations can result (**Figure 10.10**). Robert May (1976) studied the behavior of Equation 10.1 and showed that the occurrence of such fluctuations depends on the values of the parameters r and τ. Specifically, he found that when the product of those parameters is "small" ($0 < r\tau < 0.368$), the population increases smoothly to the carrying capacity, K (Figure 10.10A). At intermediate values of $r\tau$ ($0.368 < r\tau < 1.57$), delayed density dependence causes the population to fluctuate in size (Figure 10.10B). The result is a pattern of **damped oscillations**, in which the deviations from the carrying capacity gradually get smaller over time. Finally, when $r\tau$ is "large" ($r\tau > 1.57$), the population exhibits a regular cycle in which it fluctuates indefinitely about the carrying capacity, K (Figure 10.10C); such a pattern is called a **stable limit cycle**.

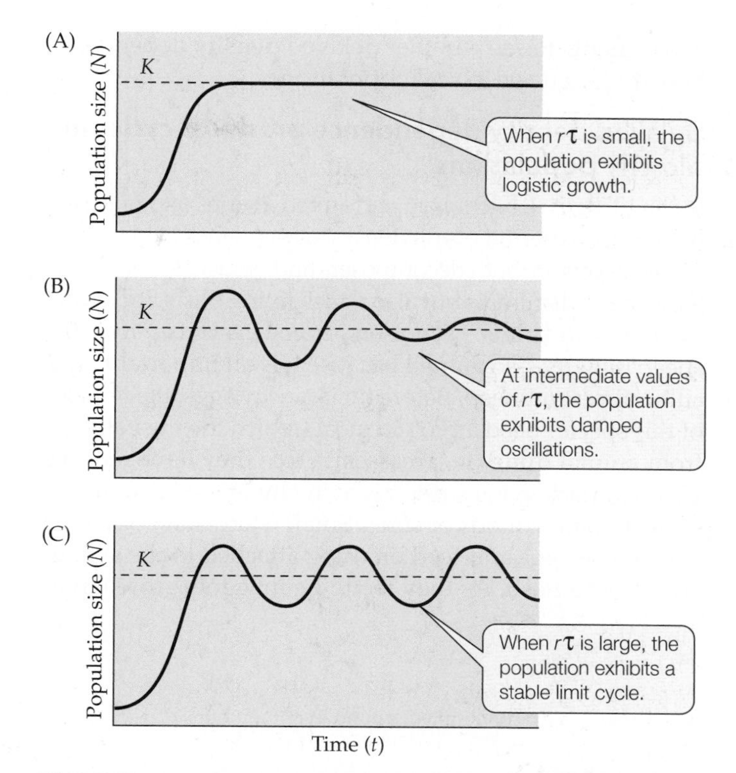

FIGURE 10.10 Logistic Curves with Delayed Density Dependence Depending on the values of the population growth rate (r) and the time lag (τ), adding delayed density dependence to the logistic equation can result in (A) an S-shaped logistic curve; (B) fluctuations about the carrying capacity, K, that become smaller over time; or (C) a regular cycle of ongoing fluctuations about the carrying capacity.

Overall, the results in Figure 10.10 indicate that population fluctuations become more pronounced as the product $r\tau$ increases. This observation makes intuitive sense: when a population grows very rapidly (large r), or when there is a very long time lag (large τ), the size of the population can become much larger than the carrying capacity before its numbers start to decline. We turn now to laboratory experiments to see whether delayed density dependence has similar effects in real populations.

Delayed density dependence produces cycles in blowfly populations

In the 1950s, A. J. Nicholson performed a series of pioneering laboratory experiments on density dependence in blowflies. These insects are both decomposers and parasites in that they feed on dead animals but also attack living hosts, including mammals and birds. Nicholson studied *Lucilia cuprina*—the sheep blowfly—so named because it is an important agricultural pest of sheep. Before they can lay eggs, the females of this species need a protein meal (which they usually get from animal dung or carcasses). Once they have fed, the females attack living sheep by laying their eggs near the tail or near open wounds or sores. Small white maggots hatch from those eggs and feed on dung attached to the skin or on exposed flesh. As they feed, the maggots grow larger and more voracious. At a certain point, the maggots burrow inside the sheep, where they feed on its internal tissues, causing severe lesions and sometimes death. Death can be caused directly by the maggots (as a result of their feeding activities) or by infections that spread through the lesions. The sheep blowfly's full life cycle (from egg to egg) can be completed in as little as 7 days.

In several of his laboratory experiments, Nicholson examined the effect of delayed density dependence on blowfly population dynamics. In the first of the two experiments that we will consider here, Nicholson provided adult blowflies with unlimited food (ground liver), but restricted maggots to 50 g of food per day. Because adults had abundant food, each female was able to lay many eggs. Thus, when there were many adults, enormous numbers of eggs were produced. When those eggs hatched, however, lack of food caused most or all of the maggots to die before they reached adulthood (**Figure 10.11A**). As a result, few adults were produced, and

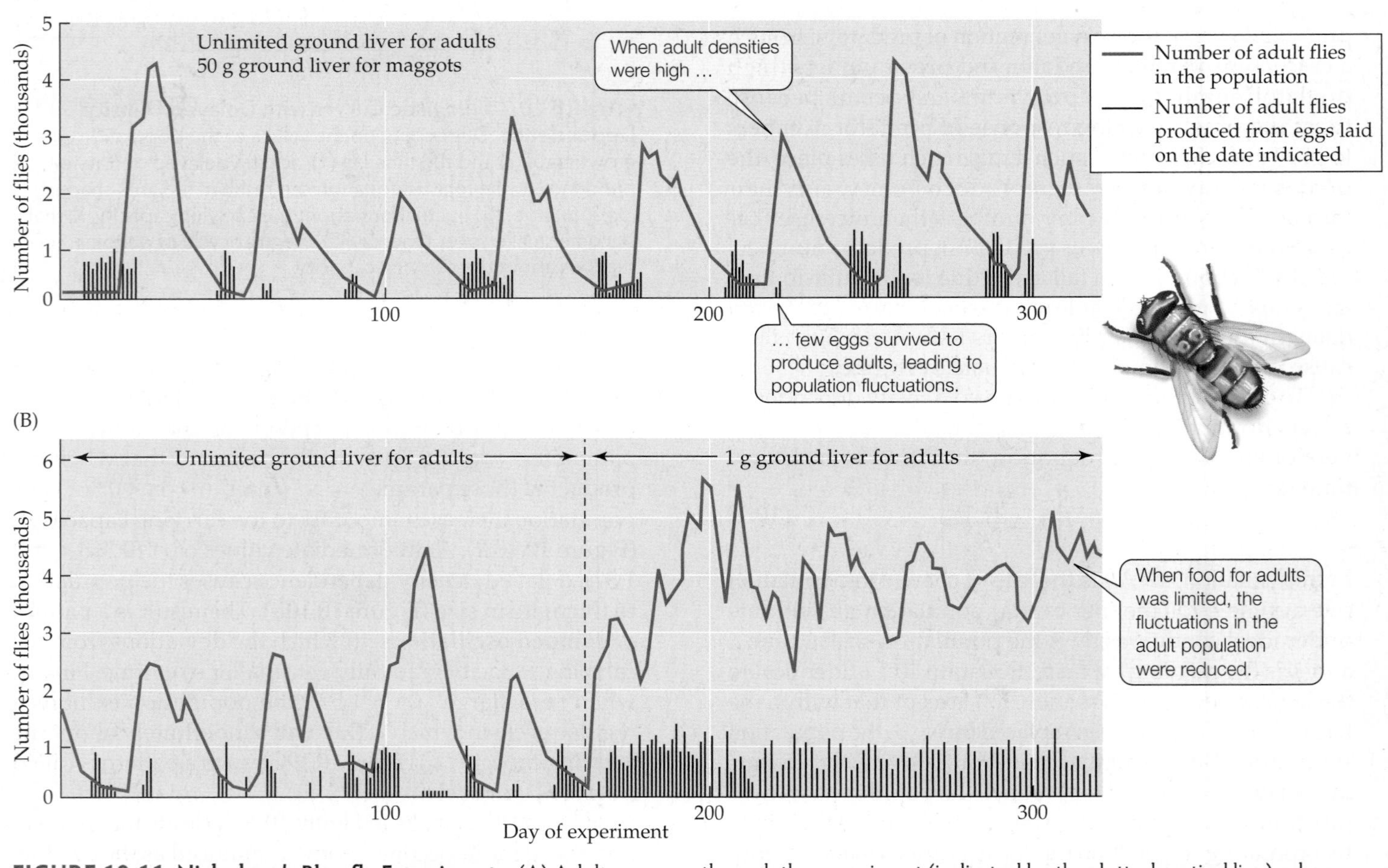

FIGURE 10.11 Nicholson's Blowfly Experiments (A) Adult blowflies were supplied with unlimited food, maggots with limited food. As a result, few or no adults were produced from the many eggs laid during periods of maximum adult abundance because the many maggots that hatched from those eggs had insufficient food to eat. (B) Experimental conditions were as in part (A) until roughly halfway through the experiment (indicated by the dotted vertical line), when the food supply for adults was also limited. (After Nicholson 1957.)

Which of the four population growth patterns discussed under Concept 10.1 best characterizes the results shown in (A)? In (B)? Explain.

the adult population invariably declined after reaching a peak. Eventually, the number of adults in the population reached such low levels that the few eggs they produced were able to give rise to a new generation of adults. Once this happened, the number of adults would begin to rise again, then crash, repeating the cycle just described.

Nicholson argued that delayed density dependence caused the number of adult blowflies to rise and fall repeatedly in this experiment. His reasoning was that because adults had unlimited food, the negative effects of high adult densities were not felt until a later time—that is, when the maggots hatched and began to feed. To test this idea, Nicholson performed a second experiment in which he removed some of the effects of delayed density dependence by providing both adults and maggots with a limited amount of food. When he did this, the adult population size no longer repeatedly rose and crashed. Instead, the number of adults increased and then fluctuated around an average of about 4,000 flies (**Figure 10.11B**). Taken together, the results shown in Figure 10.11 suggest that delayed density dependence can play a role in causing the pronounced fluctuations seen in some populations.

Delayed density dependence and other factors can cause a population to fluctuate in size because they can cause the growth, survival, or reproduction of individuals to vary over time, and that, in turn, can cause the population growth rate (λ) to vary significantly from one time period to the next. Next, we'll explore how such fluctuations in λ affect the risk that a population will become extinct.

CONCEPT 10.3 The risk of extinction increases greatly in small populations.

Population Extinction

Populations can be driven to extinction by many different factors, including predictable, or *deterministic*, changes in the environment. Consider a fish population that colonizes a temporary pond (one that forms during the rainy season but then dries out completely at other times of the year). The fish may thrive for a while, but as the water level drops, they are doomed. While such deterministic extinctions are both common and important, they are not our focus here. In this section, we'll look at how fluctuations in the population growth rate, population size, and chance events in the environment affect a population's risk of extinction.

To set the stage for this discussion, let's consider a version of the geometric growth equation (Equation 9.1: $N_{t+1} = \lambda N_t$) that includes random variation in the finite rate of increase, λ. From year to year, random changes in features of the environment (e.g., weather) may cause λ to fluctuate. There will be good years, in which λ is above its average value, and bad years, in which λ is below its average value. Imagine a population for which the average value of λ is greater than 1. If λ fluctuates little over time, then in most years λ will be greater than 1, and hence the population will usually increase in size. Under these circumstances, the population will deviate only slightly from a geometric growth pattern, and it will face little or no risk of extinction. However, if random variation in environmental conditions causes λ to change considerably from year to year, the population will fluctuate in size. What are the implications of such fluctuations?

Fluctuation in the population growth rate can increase the risk of extinction

To show what happens when λ fluctuates, computer simulations of geometric growth were performed for three populations in which λ was allowed to fluctuate at random. If we examine the results in **Figure 10.12**, we see that two of the populations recovered from low numbers, but one went extinct. These results support what common sense tells us: fluctuations increase the risk of extinction.

More specifically, what matters is the extent to which the population growth rate fluctuates over time. The growth

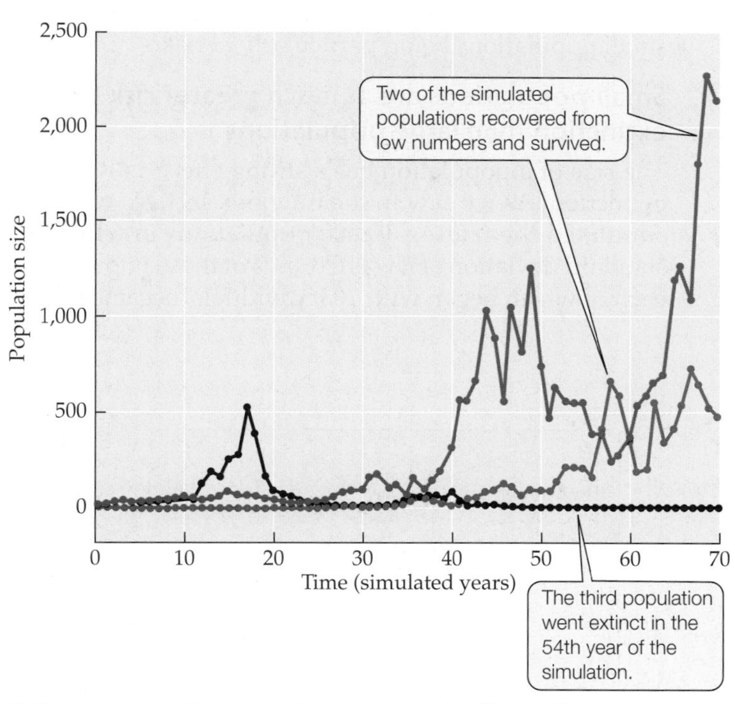

FIGURE 10.12 Fluctuations Can Drive Small Populations Extinct Simulated growth of three populations whose abundance changed according to the geometric growth equation ($N_{t+1} = \lambda N_t$) in which the value of λ varied at random from year to year. This variation in λ over time was intended to simulate random variation in environmental conditions. Each simulated population began with 10 individuals. For each of the three simulated populations shown here, the median value of λ was 1.05 and the standard deviation of λ was 0.4. Over time, roughly 70% of the simulated values of λ were between $\lambda = 0.70$ and $\lambda = 1.57$.

rate (λ) changed from year to year in each of the three populations in Figure 10.12, as indicated by the increases and decreases in their abundance over time. The extent of this variation in λ was determined by a variable controlled in the simulations, the standard deviation (σ) of the growth rate. In each of the three simulated populations shown in Figure 10.12, σ was set equal to 0.4. (See the **Web Stats Review** for more information about the standard deviation.)

To examine the effect of variation in λ more fully, we used the approach illustrated by Figure 10.12 to simulate 10,000 populations whose growth rate (λ) varied from year to year. Each of these 10,000 populations began with 10 individuals, and in each population, λ had a median value of 1.05 and a standard deviation of $\sigma = 0.2$. With this amount of variation in λ, only 0.3% of the populations went extinct in 70 years. When σ was increased to 0.4, however, 17% of the 10,000 populations went extinct in 70 years, and when it was increased still further to $\sigma = 0.8$, 53% of the populations went extinct.

The take-home message provided by these simulations is that when variable environmental conditions increase the extent to which a population's growth rate fluctuates over time, the risk of extinction also increases. This effect, however, is dependent on the size of the population, with small populations being particularly at risk.

Small populations are at much greater risk of extinction than large populations

The size of a population has a strong effect on its risk of extinction. As we saw in the previous section, when we simulated the fates of 10,000 populations in which the standard deviation of λ was 0.8, 53% of those populations, each of which began with 10 individuals, became extinct in 70 years. However, if we increase the initial population size to 100 and perform the simulation again, the chance of extinction drops from 53% to 29%. For initial population sizes of 1,000 or 10,000 individuals, the chance of extinction continues to fall (to 14% and 6%, respectively).

The simulation results we have just described suggest that small populations are much more prone to extinction than large populations. To see why, imagine that poor weather caused the growth rate (λ) of a population to be between 0.2 and 0.5 for 3 years in a row, thus causing the population to shrink considerably in size in each of those 3 years. In such a situation, a population with 10 individuals could easily be driven to extinction, whereas a population with 10,000 individuals would have survivors when conditions improved.

Similar patterns have been observed in real populations. For example, Jones and Diamond (1976) studied extinction in bird populations on the Channel Islands, located off the coast of California. By combining data from published articles (from 1868 on), museum records, unpublished field observations, and their own fieldwork, they showed that population size had a strong effect on the chance of extinction (**Figure 10.13**). They found that 39% of populations with fewer than 10 breeding pairs went extinct, whereas they observed no extinctions in populations with over 1,000 breeding pairs. Similar work by Pimm et al. (1988) showed that small populations can go extinct very rapidly: on islands off the coast of Britain, bird populations with 2 or fewer nesting pairs had a mean time to extinction of 1.6 years, while populations with 5–12 nesting pairs had a mean time to extinction of 7.5 years.

These findings for birds have been confirmed in other groups of organisms, including mammals, lizards, and

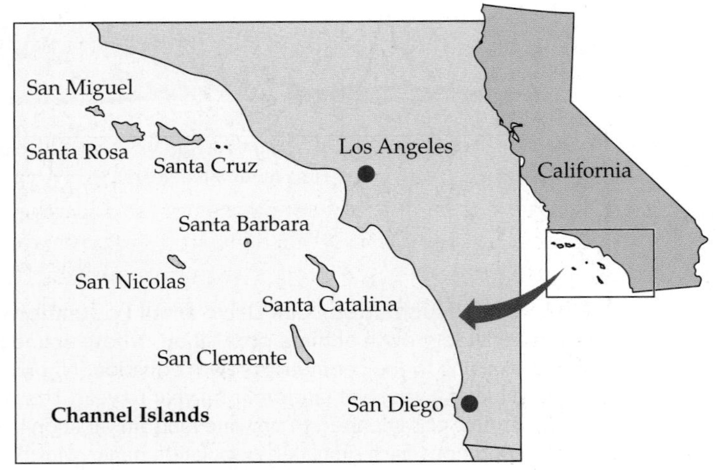

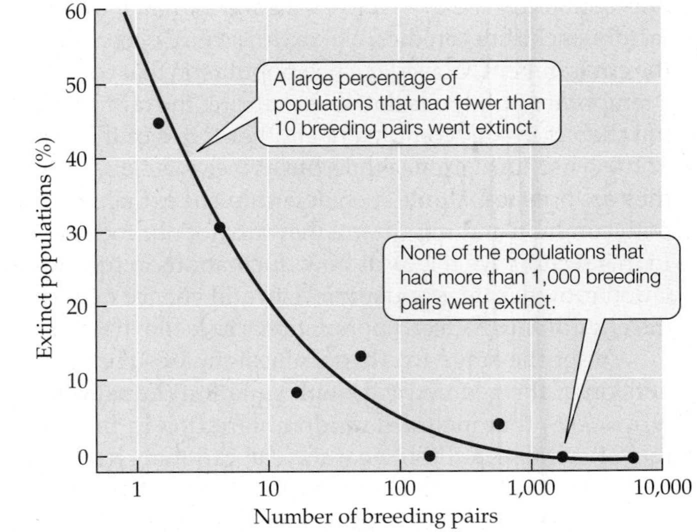

FIGURE 10.13 Extinction in Small Populations Among bird populations on the Channel Islands, the percentage of populations that went extinct declined rapidly as the number of breeding pairs in the population increased. (After Jones and Diamond 1976.)

insects. Overall, field data indicate that the risk of extinction increases greatly when population size is small. But what are the factors that place small populations at risk?

Chance events can drive small populations to extinction

If we could predict in advance how factors that influence the survival and reproduction of individuals would vary from one year to the next, we could also predict how population growth rates would change over time. But we cannot do this in real populations. There are too many factors that we cannot predict accurately, including variation in environmental conditions (e.g., temperature and rainfall) and variation in the fates of individuals (e.g., a pollinator such as a bee might land on plant A or plant B, thereby giving one individual, but not the other, the chance to reproduce). For simplicity, in the discussion that follows we'll refer to such unpredictable events as "chance events," even though they may ultimately have a deterministic cause. In particular, we'll consider the role of chance genetic, demographic, and environmental events in making small populations vulnerable to extinction.

Threats from genetic factors Small populations can encounter problems associated with genetic drift and inbreeding. Recall from the discussion under Concept 6.2 that *genetic drift* is the process by which chance events influence which alleles are passed on to the next generation. Genetic drift can occur in many ways, including chance events that determine whether individuals reproduce or die. Imagine, for example, that an elephant walks through a population of 10 small plants, 50% of which have white flowers (genotype *aa*) and 50% of which have red flowers (*AA*). If the elephant happens to crush more red-flowered than white-flowered plants, then by chance alone, there will be more copies of the *a* allele than of the *A* allele in the next generation. This scenario is just one of many possible examples of how genetic drift can cause allele frequencies to change at random from one generation to the next.

Genetic drift has little effect on large populations, but in small populations it can cause losses of genetic variation over time. For example, if genetic drift causes the frequency of two alleles (e.g., *A* and *a*) to change at random in each generation, one allele may eventually increase to a frequency of 100% (reach *fixation*), while the other is lost (see Figure 6.7). Drift can reduce the genetic variation of small populations rapidly: for example, after 10 generations, roughly 40% of the original genetic variation is lost in a population of ten individuals, while 95% is lost in a population of two individuals.

Small populations are vulnerable to extinction caused by genetic factors for three reasons. First, when genetic drift leads to a loss of genetic variation, the ability of a population to respond (via natural selection) to future environmental change is limited. Second, genetic drift can cause harmful alleles to occur at high frequencies, often causing individuals to suffer poor reproductive success (as in the case of the greater prairie chicken described on p. 139). Third, small populations show a high frequency of **inbreeding** (mating between related individuals). Inbreeding is common in small populations because after several generations at a small population size, most of the individuals in the population will be closely related to one another (to see why, answer Review Question 3). Inbreeding tends to increase the frequency of homozygotes, including those that have two copies of a harmful allele. Thus, like genetic drift, inbreeding can lead to reduced reproductive success, causing birth rates, and hence population growth rates, to drop.

The combined negative effects of genetic drift and inbreeding appear to have reduced the fertility of male lions that live on the floor of the Ngorongoro Crater, Tanzania (**Figure 10.14**). From 1957 to 1961, there were 60–75 lions living in the crater, but in 1962 an extraordinary outbreak of biting flies caused all but 9 females and 1 male to die. Seven males immigrated into the crater in 1964–1965, but no further immigration has occurred since that time. The population has increased in size since the 1962 crash. From 1975 to 1990, for example, the population fluctuated between 75 and 125 individuals. However, genetic analyses indicated that all of these individuals were descendants of just 15 lions (Packer et al. 1991). In a population of 15 individuals, genetic drift and inbreeding have powerful effects. Those effects appear to be the reason why the crater population has less genetic variation and more frequent sperm abnormalities than the large population of lions found nearby on the Serengeti Plain. In such a situation, all is not necessarily lost: in some cases, populations in decline because of drift and inbreeding have been "rescued" by introducing a small number of individuals from other, more genetically diverse populations (see the greater prairie chicken example on p. 139 and the florida panther example on p. 491).

Threats from demographic factors For an individual, survival and reproduction are all-or-nothing events: an individual either survives or it does not, and it either reproduces or it does not. At the population level, we can transform such all-or-nothing events into a probability that survival or reproduction will occur. For example, if 70 out of 100 individuals in a population survive from one year to the next, then (on average) each individual in the population has a 70% chance of survival.

In a small population, however, chance events related to the survival and reproduction of individuals (**demographic stochasticity**) can result in outcomes that differ from what such averages would lead us to expect. Consider a population of ten individuals for which previous data indicate

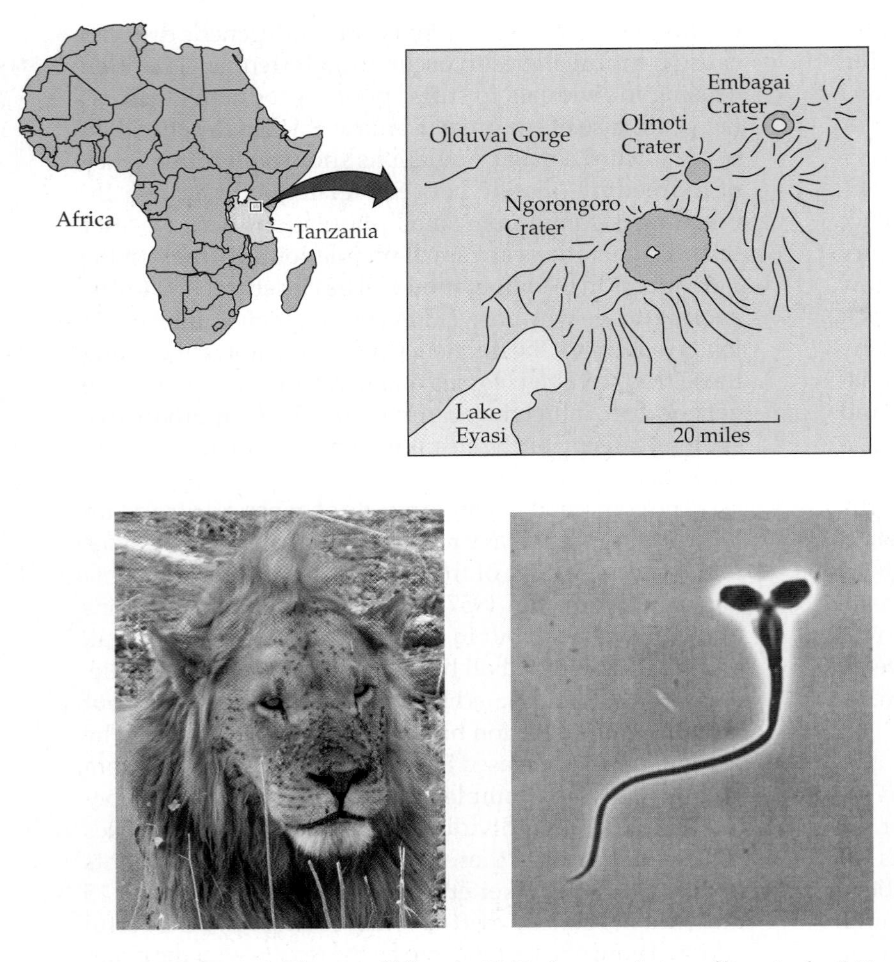

FIGURE 10.14 A Plague of Flies In 1962, the population of lions in the 260 km^2 (100-square-mile) Ngorongoro Crater of Tanzania was nearly driven to extinction by a catastrophic outbreak of biting flies similar to those on the face of this male. Lions became covered with infected sores and eventually could not hunt, causing many to die. In the population that descended from the few survivors, genetic drift and inbreeding have led to frequent sperm abnormalities, such as this "two-headed" sperm.

that, on average, each individual has a 70% probability of surviving from one year to the next. However, many chance events—such as whether an individual is struck by a falling tree—can cause the percentage of individuals that actually do survive to be higher or lower than 70%. For example, if six of the ten individuals experience (the ultimate) "bad luck" and die in chance mishaps, the observed survival rate (40%) would be much lower than the expected 70%. By affecting the survival and reproduction of individuals in this way, demographic stochasticity can cause the size of a small population to fluctuate over time. In one year the population may grow, while the next it may decrease in size, perhaps so drastically that extinction results.

In contrast, when the population size is large, there is little risk of extinction from demographic stochasticity. The fundamental reason for this has to do with laws of probability. You are, for example, much more likely to receive zero heads if you toss a fair coin 3 times than if you toss the same coin 300 times. Similarly, when we consider the demographic fates of individuals, we can see that chance events are much more likely to cause reproductive failure or poor survival in small populations than in large populations. If each individual in a population has a 1/3 chance of producing zero offspring, then if there are 2 individuals in the population, there is an 11% chance ($0.33 \times 0.33 = 0.33^2 = 0.11$) that no offspring will be produced—driving the population to extinction in one generation. Although demographic stochasticity could cause a population of 30 individuals to fluctuate in size (perhaps leading to eventual extinction), there is essentially no chance (0.33^{30}) that it could cause the population to go extinct in a single generation.

Demographic stochasticity is also one of several factors that can cause small populations to experience Allee effects. **Allee effects** occur when the population growth rate (r or λ) *decreases* as the population density decreases, perhaps because individuals have difficulty finding mates at low population densities (**Figure 10.15**). This phenomenon reverses the usual assumption that r and λ tend to *increase* as population density decreases (see Figure 9.16). Allee effects can be disastrous for small populations. If demographic stochasticity or any other factor decreases the population size, Allee effects can cause the population growth rate to drop, which causes the population size to decrease even further in a downward spiral toward extinction. We'll return to this issue in Chapter 22.

Threats from environmental variation As we have stressed repeatedly in this book, environmental conditions vary from year to year. Such variation can affect birth and death rates, leading to fluctuations in population size that can increase the risk of extinction. Our focus here will be on how two types of chance events, environmental stochasticity and natural catastrophes, can cause extinction in small populations.

Environmental stochasticity refers to erratic or unpredictable changes in the environment. In the simulations described above (see Figure 10.12), we've already seen (1) that variation in environmental conditions that causes fluctuations in population growth rates can lead to population size fluctuations and thus an increased risk of extinction, and (2) that such environmental variation is more likely to cause extinction when the population size is small. Many species face such risks from environmental stochasticity. For example, census data on female grizzly

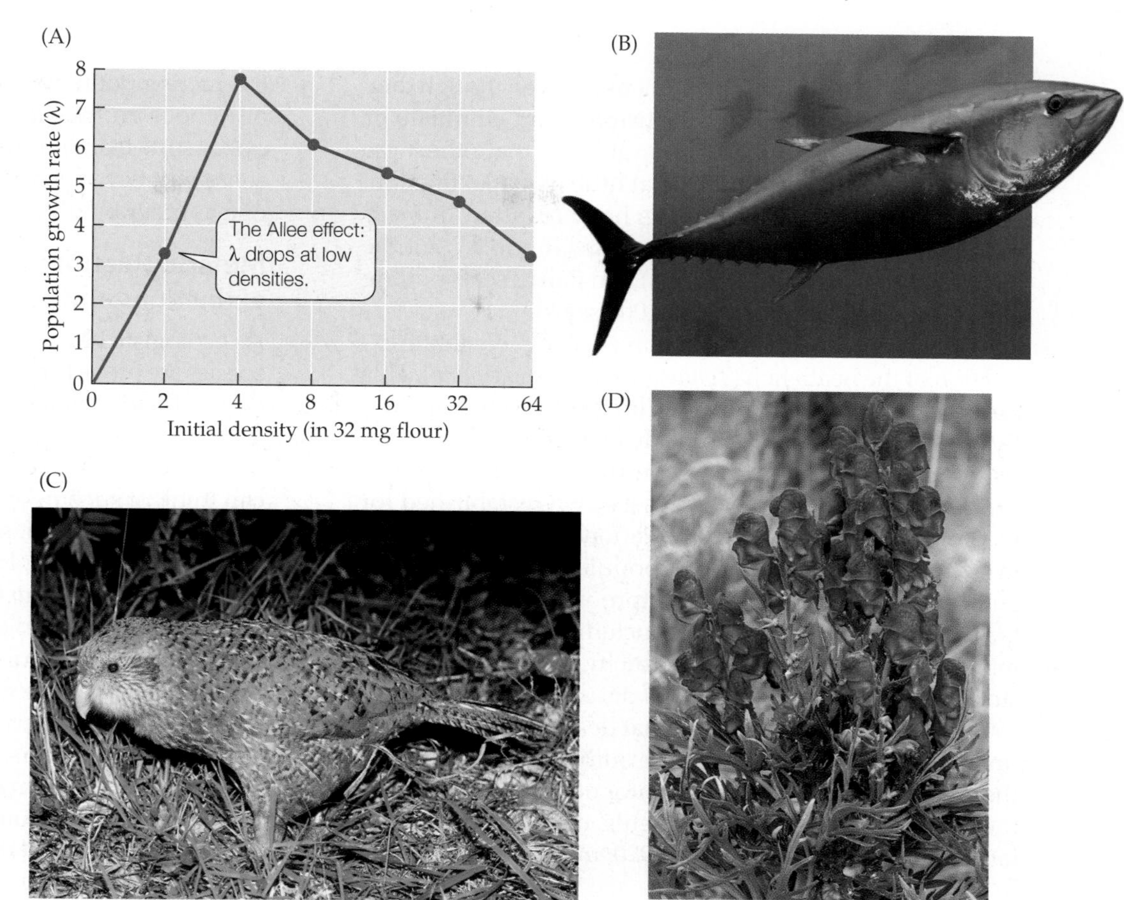

FIGURE 10.15 Allee Effects Can Threaten Small Populations Allee effects occur when the growth rate of a population decreases as population density decreases. (A) In laboratory experiments with the flour beetle *Tribolium*, population growth rates reached their lowest point at the lowest initial density. Allee effects can be important in animals such as (B) bluefin tuna (*Thunnus thynnus*) that form schools or herds whose protective or early warning systems function poorly at small population sizes, and in (C) kakapos (*Strigops habroptilus*), (D) monkshood aconite (*Aconitum napellus*), and many other species in which individuals have difficulty finding mates at low population densities. (A after Courchamp et al. 1999.)

bears (*Ursus arctos horribilis*) in Yellowstone National Park showed that the average population growth rate (r) was approximately 0.02, but that it varied from year to year. Despite the fact that the population tends to grow in size (because $r > 0$), researchers using a mathematical model found that random variation in environmental conditions could place the Yellowstone grizzly population at high risk of extinction, especially if the population size were to drop to 30 or 40 females from its 1997 level of 99 females (**Figure 10.16**).

Environmental stochasticity differs from demographic stochasticity in a fundamental way. Environmental stochasticity refers to changes in the average birth or death rate of a population that occur from one year to the next. These year-to-year changes reflect the fact that environmental conditions vary over time, affecting all the individuals in a population: sometimes there are good years and sometimes there are bad years. In demographic stochasticity, the average (population-level) birth and death rates may be constant across years, but the actual fates of individuals differ due to the random nature of whether each individual reproduces or not, and survives or not.

Populations also face risks from extreme environmental events such as floods, fires, severe windstorms, or

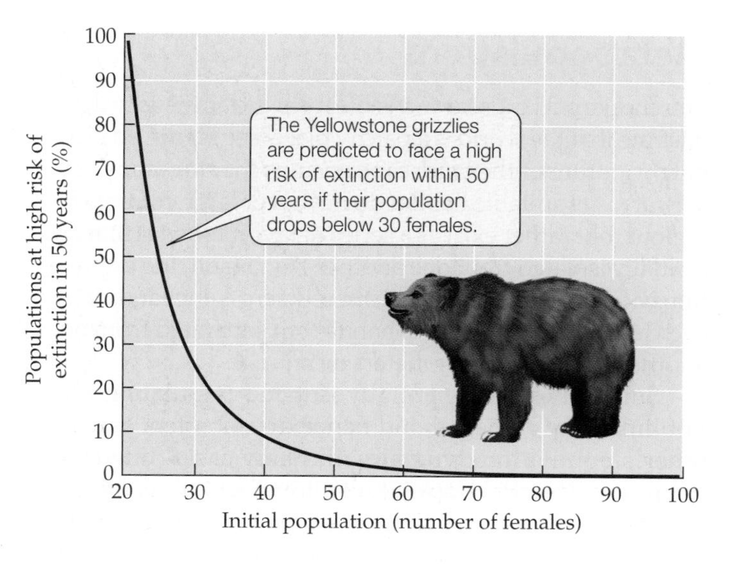

FIGURE 10.16 Environmental Stochasticity and Population Size This graph plots the risk that the Yellowstone grizzly bear population will be close to extinction in 50 years against the population size (number of females). By studying 39 consecutive years of census data, researchers found that the average population growth rate of Yellowstone grizzlies was $r = 0.02$—a rate that would lead to explosive growth if it remained constant from year to year. The risk of extinction was calculated from a mathematical model that examined the effect of environmental stochasticity by incorporating the variation in r observed over the 39 years of data. (After Morris and Doak 2002.)

outbreaks of disease or natural enemies. Even though they occur rarely, such **natural catastrophes** can eliminate or drastically reduce the size of populations that otherwise would seem large enough to be at little risk of extinction. For example, disease outbreaks have resulted in mass mortality in populations of sea urchins (up to 98% of the individuals in some populations) and Baikal seals (killing about 2,500 of a population of 3,000 seals).

Natural catastrophes also played a key role in the extinction of the heath hen (*Tympanuchus cupido cupido*). This bird was once abundant from Virginia to New England. By 1908, hunting and habitat destruction had reduced its population to 50 birds, all on the island of Martha's Vineyard, where a 1,600-acre reserve was established for its protection. Initially, the population thrived, increasing in size to 2,000 birds by 1915. A population of 2,000 seems large enough to be nearly "bulletproof" against problems that threaten small populations, including genetic drift and inbreeding, demographic stochasticity, and environmental stochasticity. However, a series of disasters struck between 1916 and 1920, including a fire that destroyed many nests, unusually cold weather, a disease outbreak, and a boom in the number of goshawks, a predator of heath hens. Due to the combined effects of these events, the heath hen population dropped to 50 birds by 1920 and never recovered. In 1932, the last heath hen died.

With the benefit of hindsight, we can see that heath hens were vulnerable in 1908 because they all lived in a single population. More typically, members of a species are found in multiple populations, which are often isolated from one another by regions of unsuitable habitat.

CONCEPT 10.4 Many species have a metapopulation structure in which sets of spatially isolated populations are linked by dispersal.

Metapopulations

The checkered landscapes visible from the air vividly demonstrate that the world is a patchy place (see Figure 9.1). The patchy nature of the landscape ensures that for many species, areas of suitable habitat do not cover large continuous regions, but rather exist as a series of favorable sites that are spatially isolated from one another. As a result, the populations of a species are often scattered across the landscape, each in an area of favorable habitat but separated from one another by hundreds of meters or more.

Sometimes these spatially isolated populations are not linked by dispersal and hence do not affect one another's population dynamics. In many cases, however, seemingly isolated populations do affect one another's dynamics because individuals (or gametes) occasionally disperse from one population to another. Such a group of interacting populations is called a **metapopulation**. Literally, the term "metapopulation" refers to a population of populations, but it is usually defined in a more particular sense as a set of spatially isolated populations linked to one another by dispersal (**Figure 10.17**).

Metapopulations are characterized by repeated extinctions and colonizations

As ecologists have long recognized, populations of some species are prone to extinction for two reasons: (1) the patchiness of their habitat makes dispersal between populations difficult, and (2) environmental conditions often change in a rapid and unpredictable manner. Metaphorically, we can think of such populations as a set of "blinking lights" that wink on and off, seemingly at random, as patches of suitable habitat are colonized and the populations in those patches then go extinct. Although the individual populations may be prone to extinction, the collection of populations—the metapopulation—persists because it includes populations that are going extinct and new populations established by colonization.

Building on this idea of random extinctions and colonizations, Richard Levins (1969, 1970) represented metapopulation dynamics in terms of the extinction and colonization of habitat patches:

$$\frac{dp}{dt} = cp(1-p) - ep \quad \textbf{(10.2)}$$

where p represents the proportion of habitat patches that are occupied at time t, while c and e are the patch colonization and patch extinction rates, respectively.

In deriving Equation 10.2, Levins made a number of assumptions: (1) that there is a very large (infinite) number of identical habitat patches, (2) that all patches have an equal chance of receiving colonists (hence the spatial arrangement of the patches does not matter), (3) that all patches have an equal chance of extinction, and (4) that once a patch is colonized, its population increases to its carrying capacity much more rapidly than the rates at which extinction and colonization occur (this assumption allows population dynamics within patches to be ignored).

Although some of the assumptions of Equation 10.2 are not realistic, it leads to a simple but fundamental insight: for a metapopulation to persist for a long time, the ratio e/c must be less than 1 (see **Web Extension 10.1** for a description of how this result was obtained). In words, this means that some patches will be occupied as long as the colonization rate is greater than the extinction rate; otherwise, the metapopulation will collapse and all populations in it will become extinct. Levins's groundbreaking approach focused attention on a number of key issues, such as how to estimate factors that influence patch colonization and extinction, the importance of the spatial arrangement of suitable patches, the extent to which the landscape between habitat patches affects dispersal, and the vexing problem of how to determine whether empty patches are

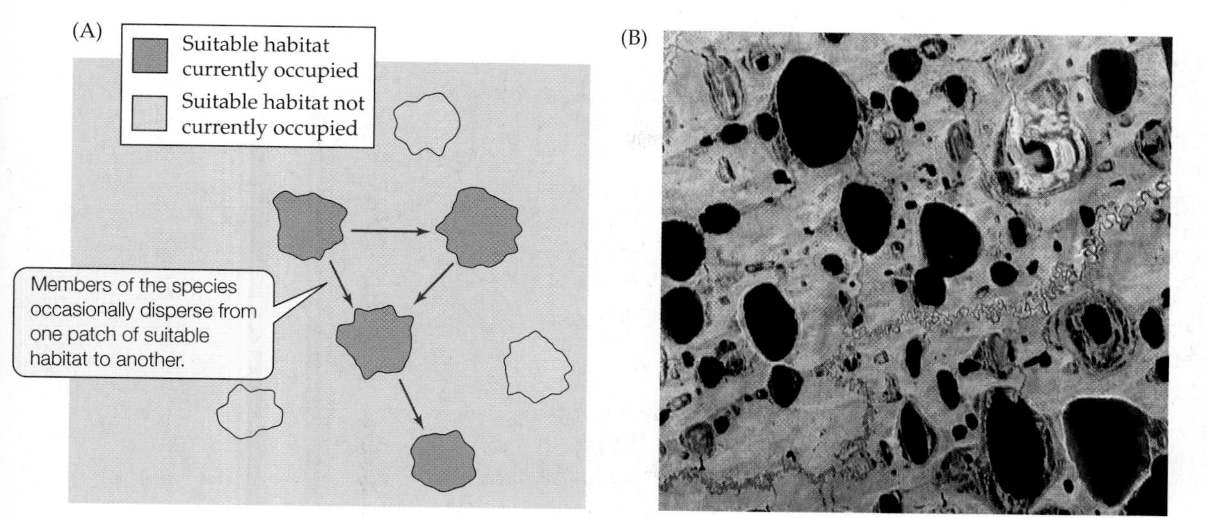

FIGURE 10.17 The Metapopulation Concept A metapopulation is a set of spatially isolated populations linked by dispersal. (A) Seven patches of suitable habitat for a species are diagrammed, four of which are currently occupied and three of which are not. The area outside of these seven patches represents unsuitable habitat. (B) A satellite image shows a group of lakes in northern Alaska that are sometimes connected to one another by temporary streams that form after the snow melts or after periods of heavy rainfall.

suitable habitat or not. Levins's rule for persistence also has applied importance, as we will see shortly.

A metapopulation can go extinct even when suitable habitat remains

Human actions (such as land development) often convert large tracts of habitat into a set of spatially isolated habitat fragments (see Figure 8.4). Such **habitat fragmentation** can cause a species to have a metapopulation structure where it did not have one before. If land development continues and the habitat becomes still more fragmented, the metapopulation's colonization rate (c) may decrease because patches become more isolated and hence harder to reach by dispersal. Further habitat fragmentation also causes the patches that remain to become smaller; as a result, the extinction rate (e) may increase because smaller patches have smaller populations, which have a higher risk of extinction. Both of these trends (an increase in e and a decrease in c) cause the ratio e/c to increase. Thus, if too much habitat is removed, the ratio e/c may shift suddenly from less than 1 to greater than 1, thereby dooming all populations—and the metapopulation—to eventual extinction, even though some habitat remains.

The idea that all populations in a metapopulation might go extinct while suitable habitat remains was developed further in studies on the northern spotted owl (**Figure 10.18**). The northern spotted owl (*Strix occidentalis caurina*) is found in the Pacific Northwest region of North America. It lives in old-growth forest, where nesting pairs establish large territories that range in size from 12 to 30 km^2 (territories are larger in poor-quality habitat). Lande (1988) modified Levins's model to include a description of how owls might search for vacant "patches," which were interpreted as sites suitable for individual territories. Lande estimated that the entire metapopulation would collapse if the logging of old-growth forest were to reduce the fraction of habitat patches that were suitable for the owl to less than 20%. This result had a powerful impact: it illustrated how a species might go extinct if its habitat dropped below a critical threshold (in this case, 20% suitable habitat), and it contributed to the 1990 listing of the northern spotted owl as a threatened species in the United States.

FIGURE 10.18 The Northern Spotted Owl The northern spotted owl (*Strix occidentalis caurina*) thrives in old-growth forests of the Pacific northwest; such forests include those that have never been cut, or have not been cut for 200 years or more.

Extinction and colonization rates often vary among patches

As the impact of Lande's work on the northern spotted owl suggests, the metapopulation approach has become increasingly important in applied ecology. But metapopulations in the field often violate the assumptions of Levins's model. For example, patches often differ considerably in population size and in the ease with which they can be reached by dispersal. As a result, extinction and colonization rates may vary greatly among patches. Therefore, most ecologists use more complex models (see Hanski 1999) when addressing practical questions in the field.

Consider the skipper butterfly *Hesperia comma*. In the early 1990s, this butterfly was found on grazed calcareous grasslands (i.e., grasslands growing in alkaline soils found on limestone or chalk outcrops) throughout a broad range of the United Kingdom. Starting in the 1950s, however, calcareous grasslands became overgrown because the numbers of cattle and other important grazers were reduced. As a result, *H. comma* populations began to decline. By the mid-1970s, the butterfly was found in only ten restricted regions, a very small fraction of its original range.

Things began to pick up for the butterfly in the early 1980s. By this time, habitat conditions had improved because livestock had been reintroduced. In 1982, Chris Thomas and Terésa Jones observed many habitat patches that appeared suitable for *H. comma* but did not support *H. comma* populations. To follow the fate of these patches over time, Thomas and Jones documented the locations of all patches containing *H. comma* populations and of all patches that appeared suitable for, but were not occupied by, *H. comma*. In 1991, they surveyed the patches again and noted which ones were occupied at that time. Their results highlight two important features of many metapopulations: isolation by distance and the effect of patch area.

Isolation by distance occurs when patches located far away from occupied patches are less likely to be colonized than are nearby patches. In *H. comma*, distance from occupied patches had a strong effect on whether patches vacant in 1982 were colonized by 1991: few patches separated by more than 2 km from an occupied patch were colonized during that period (**Figure 10.19**). Patch area also affected the chance of colonization: the majority of colonized patches were at least 0.1 ha in size. Patch area may have affected colonization rates directly because small patches may be harder for the butterflies to find than large patches. Alternatively, *H. comma* could have colonized small patches, but then suffered extinction in those patches by 1991 due to problems associated with small population size (see pp. 230–234); such patches would appear never to have been colonized because the sites were not sampled between 1982 and 1991.

Thomas and Jones also found that the chance of extinction was highest in small patches (most likely because small patches tend to have small population sizes) and in patches that were far from an occupied patch. Isolation by distance can affect the chance of extinction because a patch that is near an occupied patch may receive immigrants repeatedly, thereby increasing the patch population size and making extinction less likely. This tendency for high rates of immigration to protect a population from extinction (by reducing the problems associated with small population size) is known as the **rescue effect** (Brown and Kodric-Brown 1977).

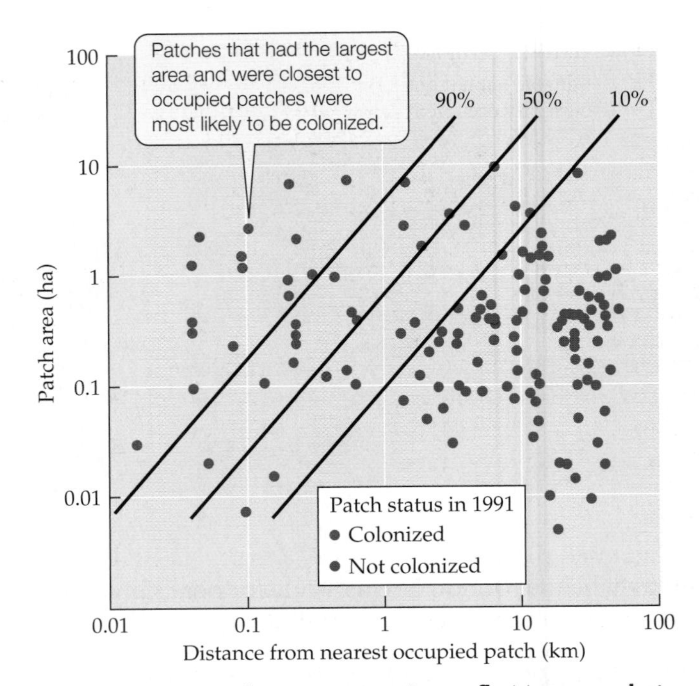

FIGURE 10.19 Colonization in a Butterfly Metapopulation Colonization of suitable habitat from 1982 to 1991 by the skipper butterfly *Hesperia comma* was influenced by patch area and patch isolation (distance to the nearest occupied patch). Each red or green circle represents a patch of suitable habitat that was not occupied by *H. comma* in 1982. The lines show the combinations of patch area and patch isolation for which there was a 90%, 50%, and 10% chance of colonization (as calculated from a statistical analysis of the data). (After Thomas and Jones 1993.)

Q Based on these results, estimate the chance of colonization for a 1 ha patch located 1 km away from the nearest occupied patch.

Finally, extinction and colonization can be influenced by nonrandom components of the environment. For example, primroses (*Primula vulgaris*) colonize patches on the forest floor where windstorms or other factors have killed trees, producing openings in the tree canopy above. While patch colonization can be viewed as a random event, patch extinction is not: as the forest regrows, the canopy closes, and the primroses die from lack of sunlight (Valverde and Silvertown 1997). Similar deterministic events affect a pool frog (*Rana lessonae*) metapopulation in ponds along the Baltic coast of Sweden. Over time, these ponds are filled in by silt and overtaken by vegetation, shrinking their size and leading to the extinction of frog populations (Sjögren-Gulve 1994). Pond colonization is also influenced by nonrandom features of the environment, such as water temperature. Certain ponds are consistently warmer than others, and relatively warm ponds are more likely to be colonized successfully because breeding success is greater in such ponds.

A CASE STUDY REVISITED
A Sea in Trouble

In the late 1980s and early 1990s, the Black Sea ecosystem was under severe duress from the combined effects of eutrophication and invasion by the comb jelly *Mnemiopsis leidyi*, as described in the Case Study on pp. 221–222. Although *Mnemiopsis* numbers declined sharply in 1991, they rose steadily again from 1992 to 1995, and then maintained high levels for several years—about 250 g/m^2, which translates to over 115 million tons of *Mnemiopsis* throughout the Black Sea. The situation did not look promising. But by 1999, matters were different: the Black Sea was showing signs of recovery.

The events that set the stage for the recovery of the Black Sea actually began prior to the first onslaught of *Mnemiopsis*. In the mid- to late 1980s, the amounts of nutrients added to the Black Sea began to level off. From 1991 to 1997, nutrient inputs actually declined, probably due to hard economic times in former Soviet Union countries coupled with national and international efforts to reduce nutrient inputs. The reduction had rapid effects: after 1992, phosphate concentrations in the Black Sea declined, phytoplankton biomass began to fall, water clarity increased, and zooplankton abundance increased. *Mnemiopsis* still posed a threat, however, as evidenced by its high biomass and by falling anchovy catches from 1995 to 1998 (see Figure 10.2). Scientists and government officials were gearing up to combat the threat from *Mnemiopsis* when the problem was inadvertently solved by the arrival of another comb jelly, the predator *Beroe* (**Figure 10.20**).

Beroe arrived in 1997. Like *Mnemiopsis*, *Beroe* was probably brought to the Black Sea in the ballast water of ships from the Atlantic. *Beroe* feeds almost exclusively on *Mnemiopsis*. It is such an effective predator that within 2 years of its arrival, *Mnemiopsis* numbers plummeted (see Figure 10.2A). Following the sharp decline in *Mnemiopsis*, the Black Sea population of *Beroe* also crashed, presumably because it depended on *Mnemiopsis* for food. The fall of *Mnemiopsis* led to a rebound in zooplankton abundance (which had dropped again from 1994 to 1996) and to increases in the population sizes of several native jellyfish species. In addition, after the *Mnemiopsis* population crashed, there was an increase in the anchovy catch and in field counts of anchovy egg densities. Overall, the decline of *Mnemiopsis* helped to improve the condition of the Black Sea ecosystem, including the fisheries on which people depend for food and income.

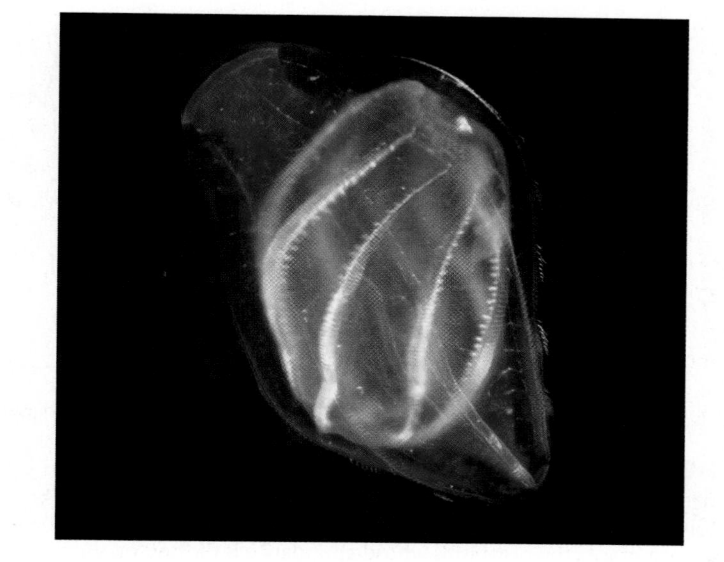

FIGURE 10.20 Invader versus Invader Another invasive comb jelly species, the predator *Beroe*, brought *Mnemiopsis* under control, thus contributing to the recovery of the Black Sea ecosystem.

CONNECTIONS IN NATURE
From Bottom to Top, and Back Again

The decrease in nutrient inputs by human activities and the control of *Mnemiopsis* by *Beroe* had rapid, beneficial effects on the entire Black Sea ecosystem. The speed and magnitude of the ecosystem's recovery provide a source of hope, suggesting that it may be possible to solve large problems in other aquatic communities. Note, however, that ecologists rarely attempt to solve such problems by deliberately introducing new predators, such as *Beroe*, because such introductions often have unanticipated negative effects.

The details of the fall and rise of the Black Sea ecosystem also illustrate two important types of causation in ecological communities: bottom-up and top-down controls. The fall of the Black Sea ecosystem began when increased nutrient inputs led to problems associated with eutrophication: increased phytoplankton abundance, increased bacterial abundance, decreased oxygen concentrations, and fish die-offs (see p. 221). The effect of adding nutrients to the Black Sea illustrates **bottom-up control**, which occurs when the abundance of a population is limited by nutrient supply or food availability. In this case, prior to nutrient enrichment, phytoplankton abundance—and thus the abundance of food for other organisms—was limited by the supply of nutrients.

Ecosystems are also affected by **top-down control**, which occurs when the abundance of a population is limited by predators. Recent evidence indicates that the decline of the Black Sea ecosystem may have been driven not only from the bottom up (by eutrophication), but also from the top down, by overfishing (Daskalov et al. 2007). Starting in the late 1950s, overfishing caused sharp drops in the

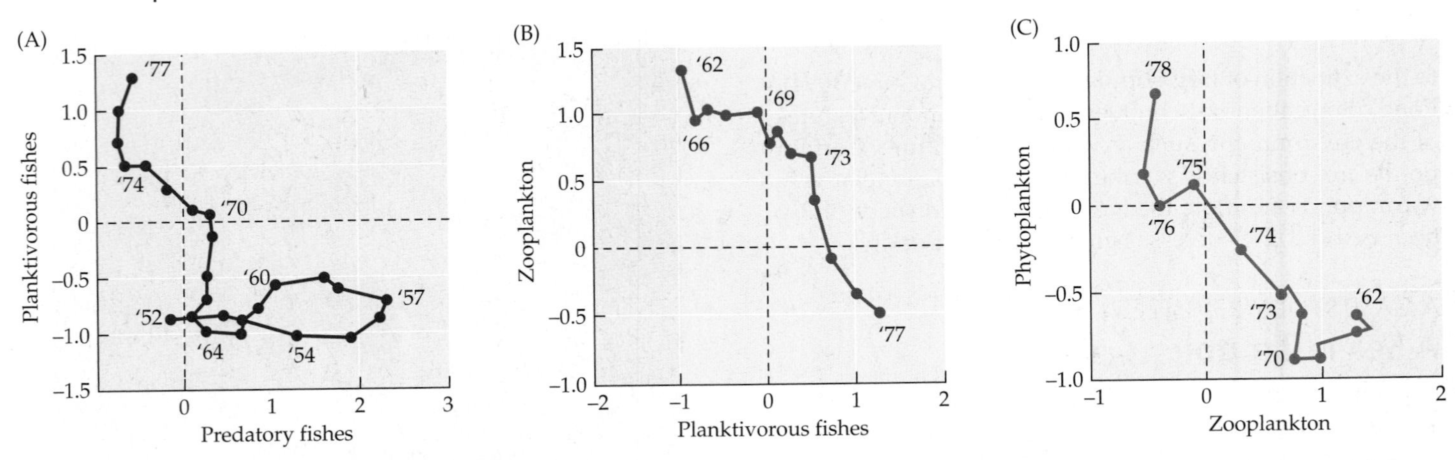

FIGURE 10.21 Ecosystem Changes in the Black Sea Abundance indices of (A) planktivorous and predatory fish, (B) zooplankton and planktivorous fish, and (C) phytoplankton and zooplankton. In each graph, the organisms whose abundance is plotted on the *y* axis are eaten by the organisms whose abundance is plotted on the *x* axis. (Planktivorous fish eat both zooplankton and phytoplankton, but they have a greater effect on zooplankton abundance than on phytoplankton abundance.) Numbers on the plots indicate years, beginning in 1952. In the abundance indices, data are standardized to have a mean of 0 and a variance of 1 (see **Web Stats Review** to learn how and why this is done). (After Daskalov et al. 2007.)

Q Referring to (A), describe predatory and planktivorous fish abundance from 1952 to 1957. Next, summarize how abundances of phytoplankton, zooplankton, planktivorous fish, and predatory fish changed in the 1970s. Finally, convert your summary into a chain of feeding relationships, where arrow thickness indicates the strength of each relationship; see Figure 8.22, in which similar chains are shown for Alaska. Is the chain you drew more similar to that in Alaska pre-1990 or that in the late 1990s? Explain.

abundances of predatory fish. As predatory fish populations declined, their prey, planktivorous (plankton-eating) fish, increased in number (**Figure 10.21A**). In turn, the increase in planktivorous fish was associated with declining numbers of zooplankton and increasing numbers of phytoplankton (**Figure 10.21B,C**), suggesting possible top-down control. Later, the arrival of the voracious predator *Mnemiopsis* also had a top-down effect, altering many key features of the ecosystem (e.g., zooplankton abundance, phytoplankton abundance, fish abundance). Top-down control also seems to have influenced ecosystem recovery: it took another predator, *Beroe*, to rein in *Mnemiopsis*. In many cases, as in the Black Sea, bottom-up and top-down controls interact to shape how ecosystems work. We'll return to bottom-up and top-down controls in Units 4 and 5, where we consider these important topics in more detail.

SUMMARY

CONCEPT 10.1 Populations exhibit a wide range of growth patterns, including exponential growth, logistic growth, fluctuations, and regular cycles.

- Most observed patterns of population growth can be grouped into four major types. These four patterns are not mutually exclusive, and a single population can experience each of them at different times.
- The first pattern, exponential growth, can occur for a limited time when conditions are favorable.
- The second pattern, logistic growth, is found in populations that increase initially and then level off at a maximum population size, the carrying capacity.
- The third pattern, population fluctuations, is found in all populations. Some populations fluctuate greatly over time; others fluctuate relatively little.
- The fourth pattern, regular population cycles, is a special type of fluctuation in which alternating periods of high and low abundance occur after nearly constant intervals of time.

CONCEPT 10.2 Delayed density dependence can cause populations to fluctuate in size.

- There is often a time lag between a change in population density and the effect that change has on future population densities.
- A version of the logistic equation that includes a time lag suggests that delayed density dependence can produce several types of population fluctuations, including damped oscillations and stable limit cycles.
- A series of pioneering experiments by A. J. Nicholson indicated that delayed density dependence was a cause of fluctuations in laboratory blowfly populations.

SUMMARY (continued)

CONCEPT 10.3 The risk of extinction increases greatly in small populations.

- The risk of extinction increases in populations whose growth rate (λ) varies considerably from one year to the next.
- Small populations are at much greater risk of extinction than large populations.
- Small populations can be driven to extinction by chance events associated with genetic drift and inbreeding, demographic stochasticity, environmental stochasticity, and natural catastrophes.

CONCEPT 10.4 Many species have a metapopulation structure in which sets of spatially isolated populations are linked by dispersal.

- Metapopulations are characterized by repeated extinctions and colonizations.
- A metapopulation can be doomed to extinction even when suitable habitat remains.
- Extinction and colonization rates often vary among a metapopulation's patches.

REVIEW QUESTIONS

1. Describe a factor that can cause time lags in the responses of natural populations to changes in population density. How do such time lags affect changes in abundance over time?
2. Summarize how chance events can threaten small populations.
3. A population consists of four unrelated individuals, two females (F1 and F2) and two males (M1 and M2). Individuals live only one year, and they mate only once, producing two offspring (one female, one male) from each mating. Individuals avoid mating with relatives if possible.
   a. Starting with individuals F1, F2, M1, and M2 as the parent generation, can the first two generations of offspring be born to parents that are not related to each other? You may find it helpful to construct a diagram to illustrate the two generations of parents and their offspring.
   b. If the second generation of offspring become parents, how many of the matings in this third generation of parents can occur between unrelated individuals? Generalizing from your results, is inbreeding likely to be common or uncommon in small populations?
4. a. Explain how a metapopulation can become extinct while suitable habitat remains.
   b. Imagine that human actions create a metapopulation from what was once continuous habitat. If many small and two large habitat patches remain, what arrangement of those patches would make it most likely that the metapopulation would *not* persist?

ON THE COMPANION WEBSITE
sites.sinauer.com/ecology2e

The website includes Chapter Outlines, Online Quizzes, Flashcards & Key Terms, Suggested Readings, a complete Glossary, and the Web Stats Review. In addition, the following resources are available for this chapter:

HANDS-ON PROBLEM SOLVING

Population Dynamics

This Web exercise explores how changes in population growth rates and the extent of delayed density dependence affect population dynamics. You will manipulate the values of these two factors, and run a simulation to determine their effects on a population.

WEB EXTENSION

10.1 Deriving Levins's Rule for Persistence

UNIT 3
Interactions among Organisms

CHAPTER

11 Competition

KEY CONCEPTS

CONCEPT 11.1 Competition occurs between individuals of two species that share the use of a resource that limits their growth, survival, or reproduction.

CONCEPT 11.2 Competition, whether direct or indirect, can limit the distributions and abundances of competing species.

CONCEPT 11.3 Competing species are more likely to coexist when they use resources in different ways.

CONCEPT 11.4 The outcome of competition can be altered by environmental conditions, species interactions, disturbance, and evolution.

Competition in Plants That Eat Animals: A Case Study

Despite repeated reports that plants could eat animals, early scientists were skeptical of those claims. Charles Darwin (1875) laid their doubts to rest by providing clear experimental evidence of carnivory by plants. Today, more than 600 species of plants that eat animals have been identified, including bladderworts, sundews, pitcher plants, and the well-known Venus flytrap.

Plants use a variety of mechanisms to eat animals. The Venus flytrap has modified leaves that look like fanged jaws, yet attract insects (and occasionally frogs) with a sweet-smelling nectar (**Figure 11.1**). The inner surface of the leaf has touch-sensitive hairs; if an insect trips those hairs, the leaf snaps shut in less than half a second. Once the insect is captured, the trap tightens further, forming an airtight seal around its victim, which is digested over the course of 5–12 days.

Other plants lack moving parts, yet still can eat animals. Consider pitcher plants, which can use nectar or visual cues to lure insects into a pitcher-shaped trap. The inside of the pitcher often has downward-facing hairs, which make it easy for the insect to crawl in, but hard to crawl out. What's more, in many pitcher plants, once it is about halfway down, the insect encounters a layer of flaky wax. An insect that steps onto this wax is doomed: the wax sticks to its feet, causing it to lose its grip and tumble into a vat that contains either water (in which it drowns) or deadly digestive juices.

Why do some plants eat animals? The answer may relate to the subject of this chapter: competition. Plants are rooted in the ground and cannot move in search of food. As a result, competition among plants can be intense where soil nutrients are scarce. Many carnivorous plants are found only in environments with nutrient-poor soils. Furthermore, evolutionary relationships among plants reveal that in nutrient-poor environments, the ability to eat animals has evolved multiple times, in a variety of independent plant lineages. Overall, these observations suggest that carnivory in plants may be an adaptation for life in nutrient-poor environments—perhaps providing a way to avoid competing with other plants for soil nutrients.

Turning to field evidence, does competition have a large effect on plants that eat animals? Typically, the root systems of carnivorous plants are less extensive than those of noncarnivorous plants that live in the same area. Hence, competition for soil resources could severely affect carnivorous plants, especially if animal prey were not available or were in short supply.

To test this idea, Stephen Brewer measured how the growth of the pitcher plant *Sarracenia alata* was affected when he cut

FIGURE 11.1 A Plant That Eats Animals Although the Venus flytrap typically captures insects, it can also feed on other animals—such as this hapless frog.

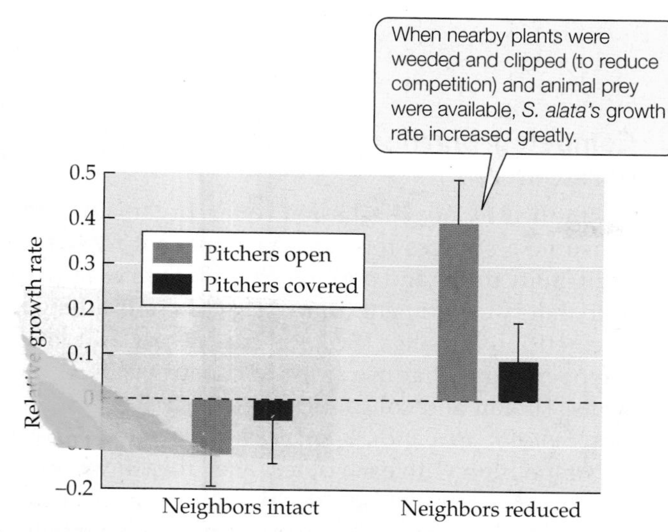

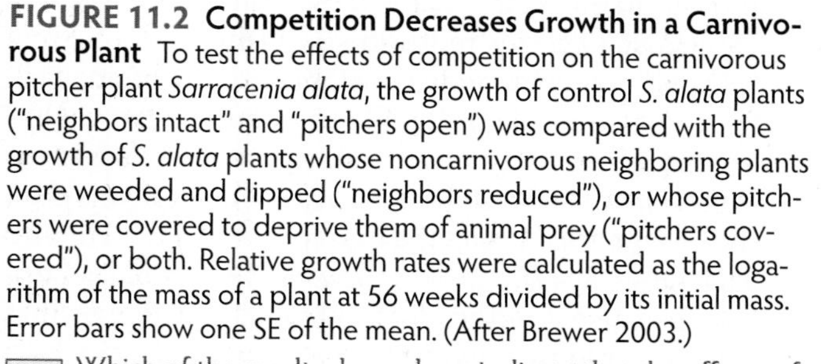

FIGURE 11.2 Competition Decreases Growth in a Carnivorous Plant To test the effects of competition on the carnivorous pitcher plant *Sarracenia alata*, the growth of control *S. alata* plants ("neighbors intact" and "pitchers open") was compared with the growth of *S. alata* plants whose noncarnivorous neighboring plants were weeded and clipped ("neighbors reduced"), or whose pitchers were covered to deprive them of animal prey ("pitchers covered"), or both. Relative growth rates were calculated as the logarithm of the mass of a plant at 56 weeks divided by its initial mass. Error bars show one SE of the mean. (After Brewer 2003.)

Which of the results shown here indicate that the effects of competition did not increase when pitchers were deprived of prey? Explain.

off its access to prey and when he reduced noncarnivorous competitor species by weeding and clipping. His results show that growth rates in *S. alata* increased dramatically when competitors were reduced (**Figure 11.2**), suggesting that competition had an important effect. But further examination of Figure 11.2 reveals that matters are not as simple as they may at first appear. Although the effect of competition for soil nutrients might be expected to increase when the plants were deprived of prey, that did not happen. Instead, it seems that pitcher plants can sidestep some of the negative effects of competition with noncarnivorous plants. How do they do it?

Introduction

In 1917, A. G. Tansley reported results from a series of experiments designed to explain the distribution in Britain of two species of bedstraw plants, *Galium hercynicum* and *G. pumilum* (then known as *G. sylvestre*). *G. hercynicum* was restricted to acidic soils, *G. pumilum* to calcareous soils (soils high in calcium, such as those derived from limestone). Even in places where the two species grew within inches of each other, each remained confined to its characteristic soil type. In his experiments, Tansley found that when grown alone, each species could survive on both acidic and calcareous soils. However, when he grew the species together on acidic soils, only *G. hercynicum* survived, while if he grew them together on calcareous soils, only *G. pumilum* survived. Tansley concluded that the two species competed with each other, and that when grown on its native soil type, each species drove the other to extinction. He also inferred that the restriction in nature of *G. hercynicum* and *G. pumilum* to a particular soil type resulted from competition between these species.

Tansley's work on bedstraws is one of the first experiments ever performed on **competition**, an interaction between individuals of two species in which each is harmed by their shared use of a resource that limits their ability to grow, survive, or reproduce (such a resource is called a *limiting resource*). Competition can also occur between individuals of a single species, in which case it is referred to as *intraspecific competition*. As a result of intraspecific competition, the resources available to members of a high-density population can be reduced to such an extent that growth, survival, or reproduction decreases or emigration increases. Thus, intraspecific competition can cause density-dependent reductions in population size, a topic we addressed in Chapter 9. This chapter focuses on competition between members of different species, or *interspecific competition*. We begin by describing how competition occurs.

CONCEPT 11.1 Competition occurs between individuals of two species that share the use of a resource that limits their growth, survival, or reproduction.

Competition for Resources

Organisms compete for **resources**, which are features of the environment that are required for growth, survival, or reproduction and which can be consumed or otherwise used to the point of depletion. What differentiates a resource from other components of the environment that organisms need to survive?

Organisms compete for resources such as food, water, light, and space

Food is an obvious example of a resource that organisms may compete for. Whether we consider the soil nutrients used by bacteria, the grass consumed by horses and rabbits, or the rodents eaten by foxes and owls, food is essential for life. When food is scarce, population growth rates plummet.

In terrestrial ecosystems—especially arid ones—water is also a resource. Food and water are intuitive examples of resources because organisms literally consume them: they are taken into the body and used to support metabolic activities. But an organism does not need to absorb, eat, or drink a substance for it to be a resource. Plants "consume" light in the sense that they use it to produce food and that

FIGURE 11.3 Space Can Be a Limiting Resource Competition for space can be intense when space is in short supply, as it is for the mosses and lichens on this boulder.

they can deplete (by shading) the supply available to other plants. Space can also be viewed as a resource. Plants, algae, and sessile animals (e.g., barnacles and corals) require space to grow, and competition for space can be intense (**Figure 11.3**). Mobile animals also compete for space as they seek access to good areas in which to hunt or attract mates, or to places that provide refuge from heat, cold, or predators. Although space is not consumed in the sense of being eaten, organisms can fill all the available space—thus depleting it—and when they do, population growth rates decrease. Thus, we can think of space as a resource that is consumed, analogous to food, water, or light.

Species are also strongly influenced by features of the environment that are not consumed, including temperature, salinity, and pH. Organisms may require particular temperatures for growth, survival, or reproduction, but they do not consume or deplete temperatures; hence, temperature is not a resource. The same is true for salinity and pH. Ecologists refer to features of the environment that affect population growth rates but are not consumed or depleted as *abiotic factors* or **physical factors**; their effects are covered in detail in Chapters 2–5.

The same substance can be a physical factor for some organisms and a resource for others. For example, terrestrial mammals and insects consume oxygen but usually do not deplete it; for them, oxygen is a physical factor, not a resource. In some aquatic and soil environments, however, organisms may consume oxygen more rapidly than it is replenished; for such organisms, oxygen is a resource, not a physical factor.

Competing organisms reduce the availability of resources

As implied by our discussion so far, organisms can consume resources to such an extent that resource availability drops and population growth rates fall. In a laboratory experiment, Tilman et al. (1981) examined competition for silica (SiO_2) in freshwater diatoms, a type of algae that use silica to construct their cell walls. Tilman and colleagues grew two diatom species, *Synedra ulna* and *Asterionella formosa*, alone and in competition with each other. They measured how the diatoms' population densities and silica concentrations in the water changed over time. When grown alone, each species reduced silica (the resource) to a low and approximately constant concentration; each species also reached a stable population size (**Figure 11.4**). *Synedra* had a lower stable population size than *Asterionella*, and it reduced silica to lower levels than did *Asterionella*. When the two species competed with each other, *Synedra* drove *Asterionella* to extinction, apparently because it reduced silica to such low levels that *Asterionella* could not survive.

Competition can increase in intensity when resources are scarce

Plants can compete for aboveground resources, such as light, as well as for belowground resources, such as soil nutrients. Researchers have suggested that the relative importance of aboveground and belowground competition in plants might change depending on whether aboveground or belowground resources are scarcer: belowground competition, for example, might be expected to increase in importance when the competing plants are growing in nutrient-poor soils. Scott Wilson and David Tilman tested this idea by performing transplant experiments with *Schizachyrium scoparium*, a perennial grass species native to their study site in Minnesota.

Wilson and Tilman selected a series of 5 × 5 m plots of natural vegetation growing in sandy, nitrogen-poor soils. For 3 years, they treated half of the plots with high-nitrogen fertilizer each year. This 3-year period gave the plant communities in the fertilized plots time to adjust to the experimentally imposed change in soil nitrogen levels. At the end of the 3-year period, they planted *S. scoparium* individuals in all the plots.

Once they were added to the high-nitrogen (fertilized) and low-nitrogen (unfertilized) plots, *S. scoparium* individuals were grown under three treatments: (1) with neighbors left intact, (2) with neighbor roots left intact but neighbor shoots tied back, or (3) with neighbor roots and shoots both removed. In treatment 1 there was both belowground and aboveground competition, while in treatment 3 there was no competition. In treatment 2, the tied-back neighbor shoots did not shade *S. scoparium*, but

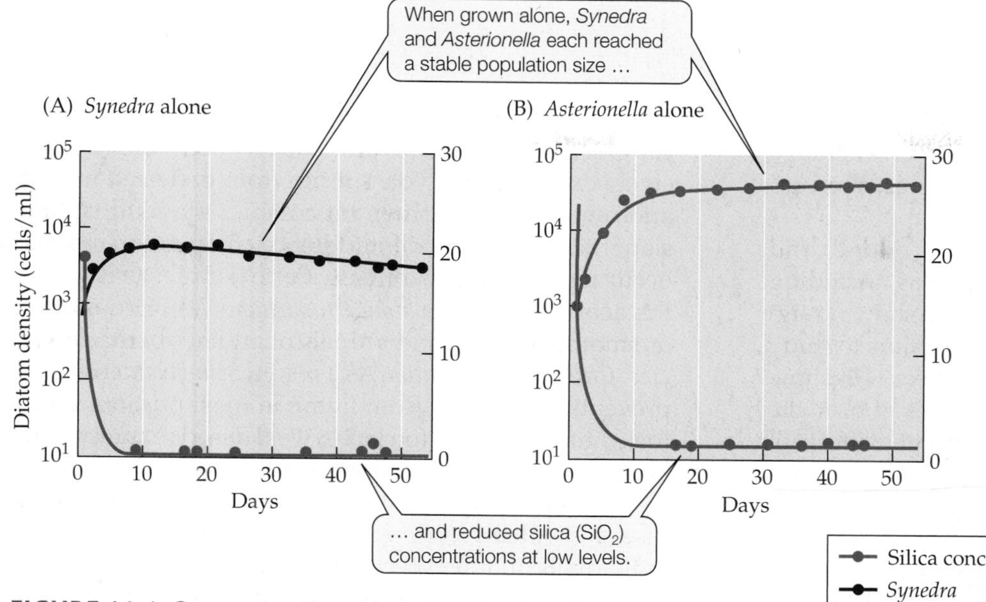

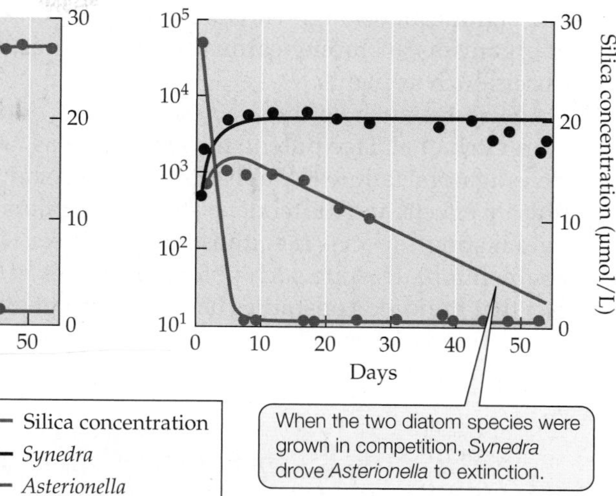

FIGURE 11.4 Competing Organisms Can Deplete Resources David Tilman and his colleagues demonstrated competition between two diatom species for silica by growing them alone and in competition with each other. *Synedra* (A) reduced silica concentrations to lower levels than did *Asterionella* (B). This result may explain why *Synedra* outcompeted *Asterionella* when the two species were grown together (C). (After Tilman et al. 1981.)

Q Suppose a third diatom species reduced the concentration of silica to 5 μmol/L when grown alone. Predict what would happen if this species was grown in competition with *Asterionella*.

the act of tying did not appear to affect neighbor roots, so that treatment was interpreted as belowground competition only. Aboveground competition was estimated as the amount of competition in treatment 1 minus the amount in treatment 2.

Wilson and Tilman found that while the total competition (the sum of belowground and aboveground competition) did not differ between the low- and high-nitrogen plots, belowground competition was most intense in the low-nitrogen plots (**Figure 11.5A**). They also found that aboveground competition for light increased when light levels were low (**Figure 11.5B**). Thus, their work demonstrates that the intensity of competition can increase when the resource being competed for is scarce.

Competition for resources is common in natural communities

How important is competition in natural communities? To answer this question, results from many field studies must be compiled and analyzed. The findings of three such analyses indicate that competition is common. For example, Schoener (1983) examined the results of 164 published studies on competition and found that of 390 species studied, 76% showed effects of competition under some circumstances, and 57% showed effects of competition under all circumstances tested. Connell (1983) examined the results of 72 studies and found that competition was important for 50% of the 215 species studied. Gurevitch et al. (1992) took a different approach: they did not report the percentage of species for which competition was important, but rather analyzed the magnitude of competitive effects found for 93 species in 46 studies published

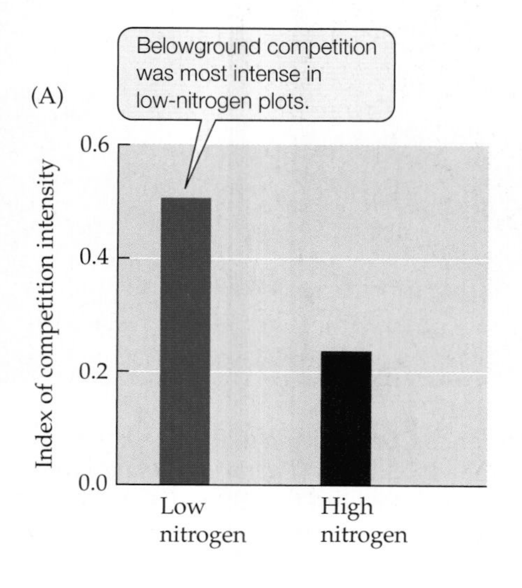

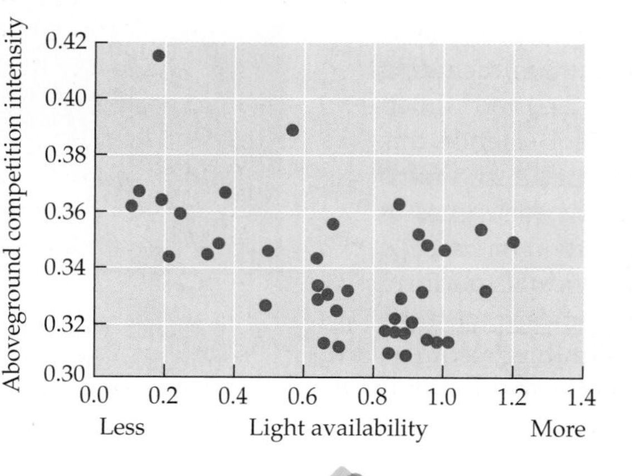

FIGURE 11.5 Resource Availability Affects the Intensity of Competition (A) In transplant experiments with the grass *Schizachyrium scoparium*, belowground competition between plant species for nutrients increased in intensity when soil nutrients were scarce. (B) Similarly, aboveground competition for light increased as light levels decreased. (After Wilson and Tilman 1993.)

between 1980 and 1989. They showed that competition had significant (though variable) effects on a wide range of organisms, including carnivores, herbivores, and producers such as plants.

Surveys such as those by Schoener, Connell, and Gurevitch et al. face potential sources of bias, including investigators' failure to publish studies that show no significant effects, and the tendency for investigators to study "interesting" species (i.e., those they suspect will show competition). Despite such potential sources of bias, the fact that hundreds of studies have documented effects of competition makes it clear that competition is common—though not ubiquitous—in nature.

CONCEPT 11.2 Competition, whether direct or indirect, can limit the distributions and abundances of competing species.

General Features of Competition

Since the beginning of ecology as a field of science, ecologists have thought that competition between species was important in natural communities. For example, although he often focused on competition within species, Charles Darwin (1859) also argued that competition between species could influence both evolutionary processes and the species found in different geographic regions. Tansley took Darwin's reasoning one step further by conducting experiments designed to demonstrate that competition could limit the distributions of species. In this section, we examine recent studies that have followed Tansley's lead in documenting how competition affects the distributions and abundances of organisms. But first we'll describe direct and indirect forms of competition, along with several other general features of competitive interactions.

Species may compete directly or indirectly

Species often compete indirectly through their mutual effects on the availability of a shared resource. Known as **exploitation competition**, this type of competition occurs simply because individuals reduce the supply of a resource as they use it. We have already discussed several examples of exploitation competition, including Brewer's work on pitcher plants (see p. 242) and Tilman et al.'s work on diatoms (see Figure 11.4). Hence, this section focuses on **interference competition**, which occurs when species compete directly for access to a resource that both require, such as food or space.

In interference competition, individuals perform antagonistic actions that directly interfere with the ability of their competitors to use a limiting resource. Such antagonistic actions are perhaps most familiar in mobile animals, as when a predator fights with a member of another species over prey that one of them has caught. Similarly, herbivores such as voles may aggressively exclude other vole species from preferred habitat, and members of warring ant colonies may kidnap and slaughter one another. Interference competition can also occur among sessile animals. For example, as it grows, the acorn barnacle *Semibalanus balanoides* often crushes or smothers nearby individuals of another barnacle species, *Chthamalus stellatus*. As a result, *Semibalanus* directly prevents *Chthamalus* from living in most portions of the rocky intertidal zone (on p. 248 we'll describe competition between these barnacles in more detail).

Interference competition also occurs in plants. In some cases, individuals of one species grow on or otherwise shade individuals of other species, reducing their access to light (**Figure 11.6**). There is also circumstantial evidence that interference competition can take the form of **allelopathy**, in which individuals of one species release toxins that harm individuals of other species. Although allelopathy appears to be important in some crop systems (Minorsky 2002; Belz 2007), there is little experimental evidence for allelopathy in natural communities. One reason for this lack of evidence is that in a species in which allelopathy is suspected, all members can usually produce the chemical that is thought to act as an allelopathic toxin—hence, it has not been possible to compare the performance of individuals that can produce the toxin with that of individuals that cannot. In a promising new line of research, genes that code for allelopathic toxins have been identified in some plant species. Using this information, researchers have developed genetic varieties in which these genes are disabled, or "silenced." In experiments using these genetic varieties, plants in which

FIGURE 11.6 Interference Competition in Plants A formidable competitor, the vine kudzu (*Pueraria montana*) has grown over and completely covered these South Carolina trees and shrubs, outcompeting them for light.

the production of allelopathic toxins has been silenced and plants able to produce these toxins are being grown with members of other species, thereby enabling a rigorous test of the effects of allelopathy in competitive environments.

Competition is often asymmetrical

When two species compete for a resource that is in short supply, each obtains less of the resource than it could have obtained if the competitor were not present. Because competition reduces the resources available for the growth, survival, and reproduction of both species, each species is harmed to some extent. In many cases, however, the effects of competition are unequal, or *asymmetrical*: one species is harmed more than the other. This asymmetry is especially clear in situations in which one competitor drives the other to extinction, as seen in Figure 11.4. Before it goes extinct, the inferior competitor uses some resources and hence reduces the resources available to the superior competitor. If we were to estimate how strongly each competitor affected the other, however, the effect of the superior competitor would be much greater than the effect of the inferior competitor.

Competition can occur between closely or distantly related species

We've seen that competition can occur between pairs of closely related species, such as the bedstraw species studied by Tansley. Brown and Davidson (1977) performed experiments with rodents and ants to test whether competition also occurs between groups of more distantly related species. They suspected that rodents and ants might compete because both eat the seeds of desert plants, and the sizes of the seeds they eat overlap considerably (**Figure 11.7**).

Brown and Davidson established experimental plots (each about 1,000 m^2 in area) in a desert region near Tucson, Arizona. Their experiment lasted 3 years and used four treatments: (1) plots in which a ¼-inch wire mesh fence excluded seed-eating rodents and from which rodents within the fence were removed by trapping; (2) plots in which seed-eating ants were excluded by applying insecticides; (3) plots in which both rodents and ants were excluded by fencing, trapping, and insecticides; and (4) plots in which both rodents and ants were left undisturbed (control plots).

The results indicated that rodents and ants compete for food. Relative to the control plots, the number of ant colonies increased by 71% in the plots from which rodents were excluded, and rodents increased by 18% in number and 24% in biomass in the plots from which ants were excluded. When both rodents and ants were excluded (treatment 3), the density of seeds increased by 450% compared with all other plots. Treatments 1 (no rodents), 2 (no ants) and 4 (the control plots, with both rodents and ants present) all resulted in similar densities of seeds. These results suggested that when either rodents or ants were removed, the group that remained ate roughly as many seeds as rodents and ants combined ate in the control plots. Thus, under natural conditions, each group would be expected to eat fewer seeds in the presence of the other group than it could eat when alone.

It is not surprising that species as different as ants and rodents compete. After all, people differ greatly from bacteria, fungi, and insects, yet we compete with these organisms for food in farm fields, in grain storage bins—even in our refrigerators. Overall, irrespective of whether they are closely or distantly related, organisms can compete if they share the use of a resource whose supply is limited.

Competition can determine the distributions of species

As we saw in the previous section, Schoener (1983), Connell (1983), and Gurevitch et al. (1992) analyzed the results of hundreds of field studies and found that competition had important effects on many, but not all, species. Typically, in species for which it was important, competition reduced the growth, survival, or reproduction of one or both competitors, thus limiting their abundances. In some cases, competition also affected where the species lived. As we'll see, competition may restrict species to certain portions of a particular habitat, or it may determine the broad geographic regions in which species are found (or both).

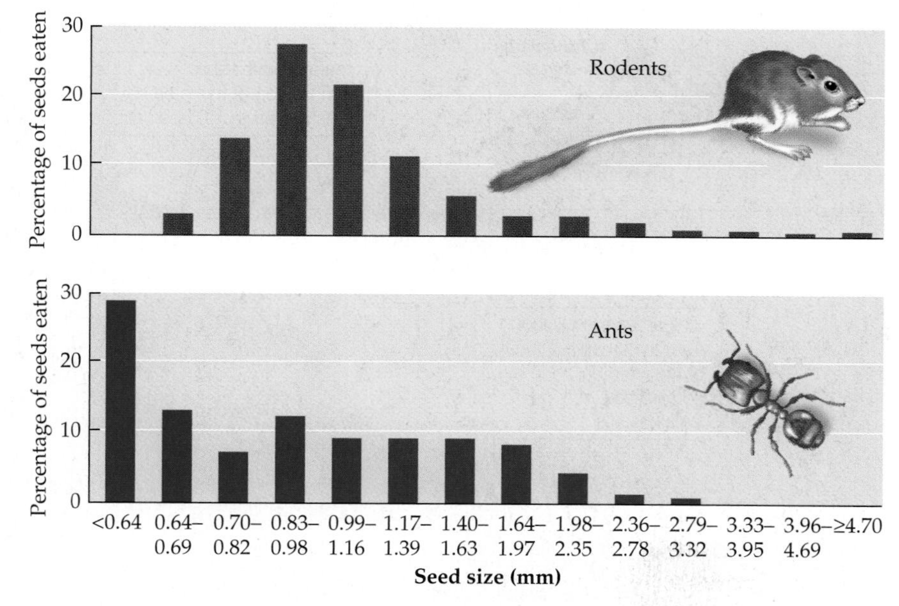

FIGURE 11.7 Ants and Rodents Compete for Seeds There is extensive overlap in the sizes of seeds eaten by ants and by rodents. Removal experiments showed that these two distantly related groups compete for this food source. (After Brown and Davidson 1977.)

Local effects In a series of classic experiments, Joseph Connell (1961a,b) examined factors that influenced the distribution, survival, and reproduction of two barnacle species, *Chthamalus stellatus* and *Semibalanus balanoides* (formerly *Balanus balanoides*). The larvae of these and other barnacles drift through ocean waters, then settle on rocks or other surfaces (such as boat hulls), where they metamorphose into adults, forming a hard outer shell.

At Connell's study site along the coast of Scotland, the distributions of *Chthamalus* and *Semibalanus* larvae overlapped considerably: the larvae of both species were found throughout the upper and middle intertidal zones. However, adult *Chthamalus* usually were found only near the top of the intertidal zone, whereas adult *Semibalanus* were not found there, but were found throughout the rest of the intertidal zone (**Figure 11.8**). What accounted for these differences in distribution?

To answer this question, Connell examined the effects of competition and of physical factors such as desiccation (drying out due to exposure to air, which is greatest in the upper intertidal zone). To test the importance of competition, he chose some individual young barnacles of each species that had settled in each zone and removed all nearby members of the other species. For other focal individuals, he left nearby members of the other species in place. He found that competition with *Semibalanus* excluded *Chthamalus* from all but the top of the intertidal zone. As they grew, *Semibalanus* smothered (by growing on top of), removed (by growing underneath, hence prying off the rocks), and crushed the *Chthamalus*. Averaging across all regions of the intertidal zone, only 14% of *Chthamalus* survived their first year when faced with competition from *Semibalanus*, whereas 72% survived where Connell had removed *Semibalanus*. *Chthamalus* that survived a year of competition with *Semibalanus* were small and reproduced poorly.

Semibalanus, in contrast, was not affected strongly by competition with *Chthamalus*. However, whether *Chthamalus* was removed or not, *Semibalanus* dried out and survived poorly near the top of the intertidal zone. Thus, *Semibalanus* appears to have been excluded from that zone not by competition, but by its sensitivity to desiccation.

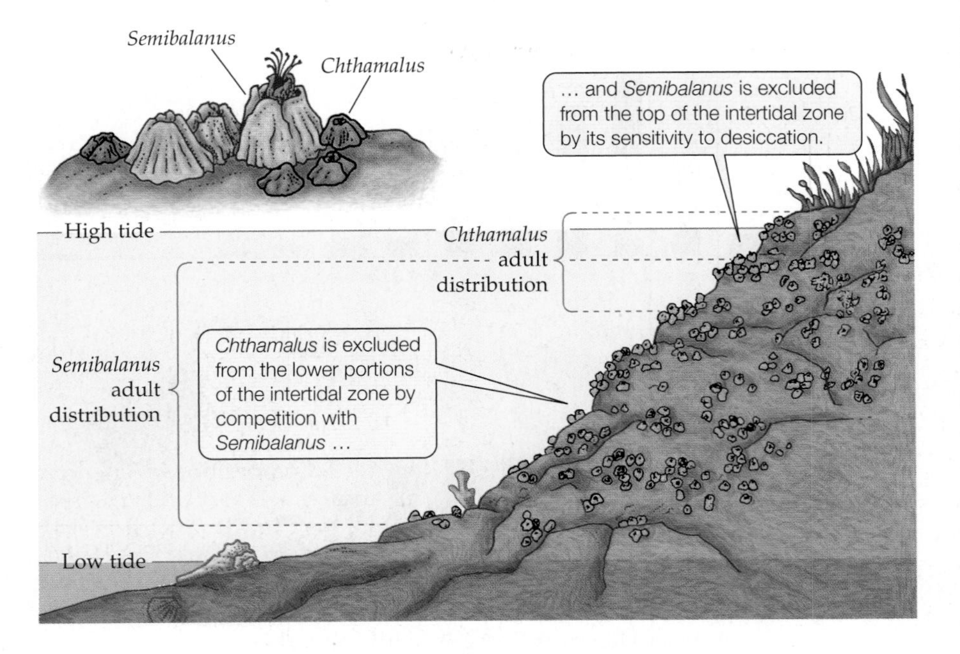

FIGURE 11.8 Squeezed Out by Competition Removal experiments at a field site in Scotland showed that competition determines the local distribution of the barnacle *Chthamalus stellatus*.

Regional effects As observed for Tansley's bedstraw plants and Connell's barnacles, competition can restrict the local distribution of a species to a particular set of environmental conditions—the bedstraws, for example, could be growing inches away from each other, but each species was restricted to a particular soil type. Competition has also been shown to prevent a wide range of species, including mammals, marine invertebrates, birds, and plants, from occupying geographic regions in which they would otherwise thrive.

In some cases, a "natural experiment"—a situation in nature that is similar in effect to a controlled removal experiment—provides evidence that competition affects geographic distributions. Consider a hypothetical pair of species that compete with each other, species A and species B. Assume that, simply by chance, some regions of suitable habitat are occupied only by species A, while other similar regions are occupied only by species B, and still other regions are occupied by both species. If the absence (i.e., "removal") of species A is consistently associated with an expansion of the area occupied by species B (or vice versa), that would suggest that competition affects the geographic distributions of these species.

Such a result was found for chipmunks in the genus *Tamias* (previously known as *Neotamias* or *Eutamias*). These chipmunks live in forests on mountains in the southwestern United States, where mountain ranges are separated from one another by desert flatlands. Patterson (1980, 1981) studied the distributions of *Tamias* chipmunks and found that when a species lived alone on a mountain range, it consistently occupied a broader range of habitats and elevations than when it lived with a competitor species (**Figure 11.9**). As in Connell's removal experiments, this result suggests that competition may have prevented some *Tamias* chipmunk species from living in areas of otherwise suitable habitat.

As illustrated by Patterson's chipmunks and Connell's barnacles, a species may limit the distribution of a competitor without driving the

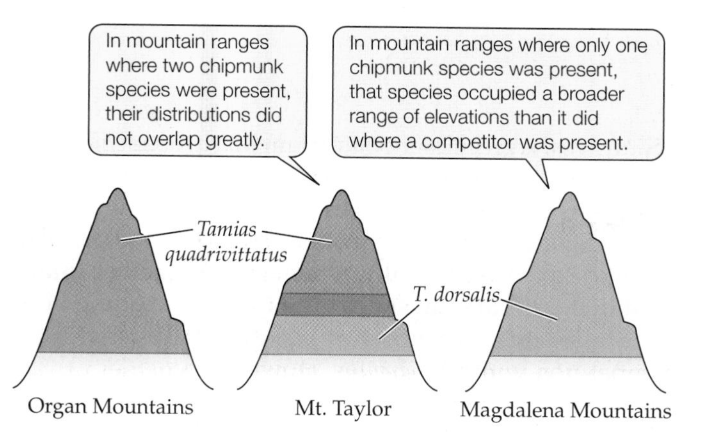

FIGURE 11.9 A Natural Experiment on Competition between Chipmunk Species Observations of the distributions of *Tamias* chipmunks in New Mexico suggest that competition may restrict the habitats in which they live. Similar results were obtained for *Tamias* species living in Nevada.

competitor to extinction. On the other hand, Tilman's experiments with diatoms suggest that competition can lead to the complete removal of one of the competing species. Why do competitors coexist in some situations, but not others?

CONCEPT 11.3 Competing species are more likely to coexist when they use resources in different ways.

Competitive Exclusion

Connell's work on barnacles provides a starting point for understanding the different outcomes that can occur when species compete for resources. In habitats where both species could thrive, *Semibalanus* typically drove *Chthamalus* to extinction, yet *Chthamalus* could persist because it could grow and reproduce in habitats that were too dry for *Semibalanus*. The picture that emerges from Connell's results is this: if the overall ecological requirements of a species—its *ecological niche* (see the definition under Concept 8.5)—are very similar to those of a superior competitor, that competitor may drive it to extinction. We turn now to experiments, field observations, and mathematical models that examine this idea. In particular, we will see how the similarity of the resources the competitors use affects the likelihood that one will drive the other to extinction.

The competitive exclusion principle states that complete competitors cannot coexist

In the 1930s, the Russian ecologist G. F. Gause performed laboratory experiments on competition using three species of the single-celled protist *Paramecium*. He constructed miniature aquatic ecosystems by growing *Paramecium* in tubes filled with a liquid medium that contained bacteria and yeast cells as a food supply. He found that populations of each of the three *Paramecium* species reached a stable carrying capacity when grown alone (**Figure 11.10A–C**). But when pairs of these species competed with each other, several different outcomes were obtained.

For example, when *P. aurelia* was grown in competition with *P. caudatum*, *P. aurelia* drove *P. caudatum* to extinction (**Figure 11.10D**). These two species may have been unable to coexist because they both fed primarily on bacteria floating in the medium, which led to considerable overlap in their

complete competitors

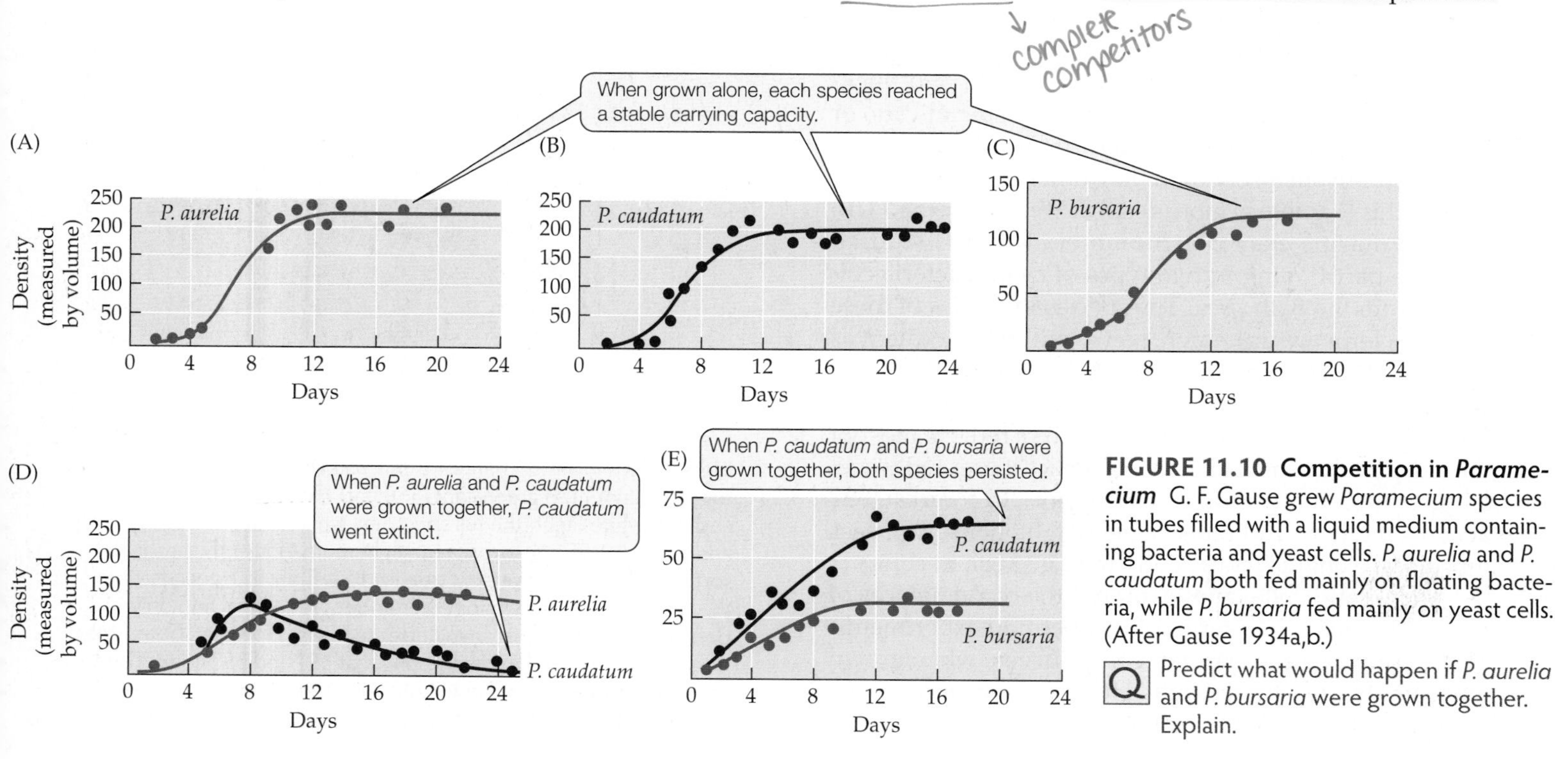

FIGURE 11.10 Competition in *Paramecium* G. F. Gause grew *Paramecium* species in tubes filled with a liquid medium containing bacteria and yeast cells. *P. aurelia* and *P. caudatum* both fed mainly on floating bacteria, while *P. bursaria* fed mainly on yeast cells. (After Gause 1934a,b.)

Q Predict what would happen if *P. aurelia* and *P. bursaria* were grown together. Explain.

food requirements. In contrast, when *P. caudatum* was grown with *P. bursaria*, neither species drove the other to extinction (**Figure 11.10E**). Although *P. caudatum* and *P. bursaria* coexisted, it was clear that they competed for one or more resources because the carrying capacity of each was lowered by the presence of the other. Gause suggested that *P. caudatum* and *P. bursaria* could coexist because *P. caudatum* usually ate bacteria floating in the medium, while *P. bursaria* usually fed on yeast cells that settled to the bottom of the tubes.

Experiments with a wide range of other species (e.g., algae, flour beetles, plants, and flies) have yielded similar results: one species drives the other to extinction unless the two species use the available resources in different ways. Such results led to the formulation of the **competitive exclusion principle**, which states that two species that use a limiting resource in the same way cannot coexist indefinitely. As we'll see next, field observations are consistent with this explanation of why competitive exclusion occurs in some situations, but not others.

Competitors may coexist when they use resources differently

NOT COMPLETE

In natural communities, many species use the same limiting resources, yet manage to coexist with one another. This observation does not violate the competitive exclusion principle because a key point of that principle is that species must use limiting resources *in the same way*. Field studies often reveal differences in how species use limiting resources. Such differences are referred to as **resource partitioning**.

Thomas Schoener studied resource partitioning in four lowland *Anolis* lizard species that live on the West Indian island of Jamaica. Although these species all live together in trees and shrubs and eat similar foods, Schoener (1974) found differences among them in the height and thickness of their perches and in the time they spent in sun or shade. As a result, members of the different *Anolis* species competed less intensely than they otherwise would. (We explore this example in more detail in **Web Extension 11.1**.)

In a marine example, Stomp et al. (2004) studied resource partitioning in two types of cyanobacteria collected from the Baltic Sea. The species identities of these cyanobacteria are unknown, so we will refer to them as BS1 and BS2 (standing for Baltic Sea 1 and Baltic Sea 2). BS1 absorbs green wavelengths of light efficiently, which it uses in photosynthesis. However, BS1 reflects most of the red light that strikes its surface; hence it uses red wavelengths inefficiently (and is red in color). In contrast, BS2 absorbs red light and reflects green light; hence BS2 uses green wavelengths inefficiently (and is green in color).

Stomp and colleagues explored the consequences of these differences in a series of competition experiments. They found that each species could survive when grown alone under green or red light. However, when they were grown together under green light, the red cyanobacterium BS1 drove the green cyanobacterium BS2 to extinction (**Figure 11.11A**)—as might be expected, since BS1 uses green light more efficiently than does BS2. Conversely, under red light, BS2 drove BS1 to extinction (**Figure 11.11B**), as also might be expected. Finally, when grown together under "white light" (the full spectrum of light, including both

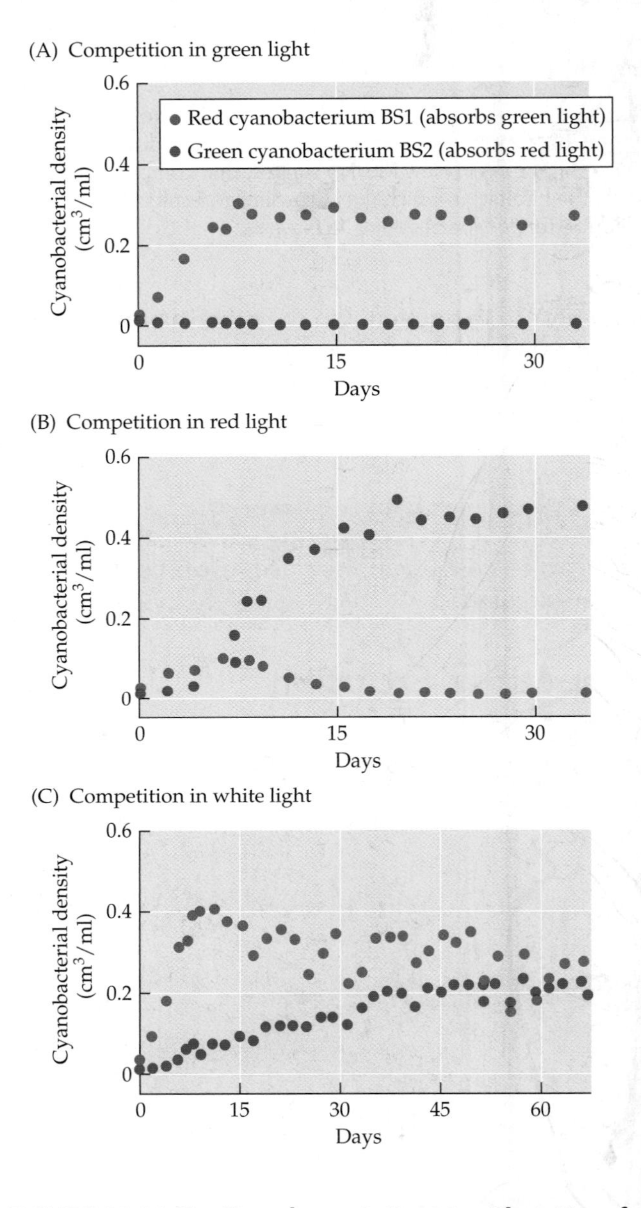

FIGURE 11.11 Do Cyanobacteria Partition Their Use of Light? Two types of cyanobacteria, BS1 and BS2, were grown together under (A) green light (550 nm), (B) red light (635 nm), and (C) "white" light (the full spectrum, which includes both green and red light). BS1 absorbs green light more efficiently than it absorbs red light; the reverse is true for BS2. Only BS1 persists when the two types are grown together under green light, and only BS2 persists when they are grown under red light. However, both types persist under white light, suggesting that BS1 and BS2 coexist by partitioning their use of light. Population densities are expressed in biovolumes. (After Stomp et al. 2004.)

green and red light), both BS1 and BS2 persisted (**Figure 11.11C**). Taken together, these results suggest that BS1 and BS2 coexist under white light because they use different wavelengths of light in photosynthesis.

Following up on their laboratory experiments, Stomp et al. (2007) analyzed the cyanobacteria present in 70 aquatic environments that ranged from clear ocean waters (where green light predominates) to highly turbid lakes (where red light predominates). As seen in **Figure 11.12**, only red cyanobacteria were found in clear waters, and only green cyanobacteria were found in highly turbid waters. However, as could be predicted from Figure 11.11C, both types were found in waters of intermediate turbidity, where both green and red light were available. Thus, the laboratory experiments and field surveys conducted by Stomp and colleagues suggest that red and green cyanobacteria coexist because they partition the use of a key limiting resource: the underwater light spectrum.

Evidence for resource partitioning has been found in many other species, including protists, birds, fishes, crustaceans, and plants. Overall, studies of resource partitioning suggest that species can coexist if they use resources in different ways—an inference that is also supported by results from mathematical models of competition.

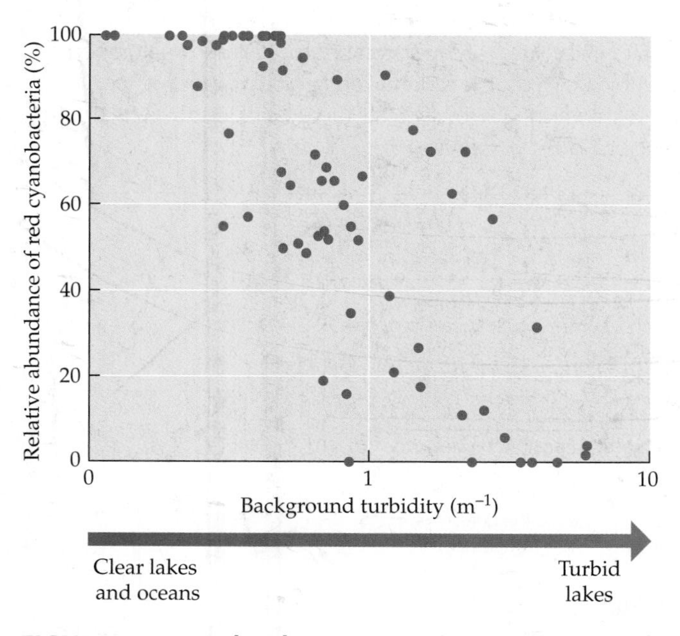

FIGURE 11.12 Red and Green Cyanobacteria Coexist in Nature The percentage of red cyanobacteria found in aquatic environments is highest in very clear waters (low turbidity) and decreases in more turbid waters. Clear waters are characterized by green light, which red cyanobacteria are proficient at using in photosynthesis; in contrast, highly turbid waters are characterized by red light, which red cyanobacteria use much less efficiently. As the percentage of red cyanobacteria drops from 100% in clear waters to 0% in highly turbid waters, the percentage of green cyanobacteria rises from 0% to 100% (not shown). (After Stomp et al. 2007.)

Competition can be modeled by modifying the logistic equation

Working independently of each other, A. J. Lotka (1932) and Vito Volterra (1926) both modeled competition by modifying the logistic equation that we discussed under Concept 9.5; we describe their approach in **Web Extension 11.2**. Their equation, now known as the **Lotka–Volterra competition model**, can be written as

$$\frac{dN_1}{dt} = r_1 N_1 \left[1 - \frac{(N_1 + \alpha N_2)}{K_1}\right]$$
$$\frac{dN_2}{dt} = r_2 N_2 \left[1 - \frac{(N_2 + \beta N_1)}{K_2}\right]. \quad \textbf{(11.1)}$$

In these equations, N_1 is the population density of species 1, r_1 is the intrinsic rate of increase of species 1, and K_1 is the carrying capacity of species 1; N_2, r_2, and K_2 are similarly defined for species 2. The **competition coefficients** (α and β) are constants that describe the effect of one species on the other: α is the effect of species 2 on species 1, and β is the effect of species 1 on species 2. We explore the meaning of these competition coefficients in more detail in **Web Extension 11.3**.

Analyses of Equation 11.1 support the idea that *competitive exclusion is likely when competing species require very similar resources*. In the remainder of this section, we'll describe the two-step process by which this support is obtained. First, we'll see how Equation 11.1 can be used to predict the outcome of competition; then we'll explore how the chance of competitive exclusion is affected by patterns of resource use.

Predicting the outcome of competition The outcome of competition could be predicted if we knew how the densities of species 1 and species 2 were likely to change over time. For example, if the abundance of species 2 was likely to increase while that of species 1 decreased to zero, then species 2 should drive its competitor to extinction, thus "winning" the competitive interaction. A computer could be programmed to solve Equation 11.1, thereby predicting the population densities of species 1 and 2 at different times. Here, however, we use a graphical approach to examine the conditions under which each species is expected to increase or decrease in abundance.

We begin by determining when the population of each species stops changing in size, using the approach described in **Ecological Toolkit 11.1**. Equations 1 and 2 in the toolkit are straight lines, written with N_2 as a function of N_1. These lines are called **zero population growth isoclines**, so named because the population does not increase or decrease in size for any combination of N_1 and N_2 that lies on these lines.

We can use these zero growth isoclines to determine the conditions under which each species will increase or

11.1 Isocline Analyses

Ecologists often want to predict the outcome of competition under different conditions. As we'll describe here, one way to do this is based on a graphical analysis of how the abundances of the interacting species are likely to change over time.

The first step in this approach is to determine the conditions under which the population size of each species stops changing over time. As we saw in the discussion under Concept 9.3, *dN*/*dt* represents the rate of change in population size at each instant in time. When *dN*/*dt* = 0, the population stops changing in size; as a result, it remains constant in size unless perturbed in some way (e.g., if many individuals die in a catastrophic fire or storm).

If we apply this approach to the Lotka–Volterra competition equations (Equation 11.1), the population density of species 1 does not change over time when $dN_1/dt = 0$,

$$\frac{dN_1}{dt} = r_1 N_1 \left[1 - \frac{(N_1 + \alpha N_2)}{K_1} \right] = 0$$

In this equation, $dN_1/dt = 0$ when $N_1 = 0$ or when

$$1 - \frac{(N_1 + \alpha N_2)}{K_1} = 0$$

The first case ($N_1 = 0$) just states the obvious (a population with no members does not grow), so we rearrange the equation for the second case to find

$$N_2 = \frac{K_1}{\alpha} - \frac{1}{\alpha} N_1 \qquad \textbf{(1)}$$

If we use a similar approach for species 2, we find that $dN_2/dt = 0$ when

$$N_2 = K_2 - \beta N_1 \qquad \textbf{(2)}$$

Equations 1 and 2 represent straight lines when values of N_2 are plotted on the *y* axis and values of N_1 are plotted on the *x* axis of a graph (as we do in Figure 11.13). The *y* intercept of Equation 1 is K_1/α, and the slope of Equation 1 is $-1/\alpha$, indicating that the line falls from left to right. For Equation 2, the *y* intercept and slope are K_2 and $-\beta$, respectively.

Equation 1 specifies all the combinations of the densities of species 1 and 2 for which the population density of species 1 does not change over time—this is true because we obtained Equation 1 by *assuming* that the population density of species 1 does not change (i.e., that $dN_1/dt = 0$). Similarly, Equation 2 specifies all the combinations of the densities of species 1 and 2 for which the population density of species 2 does not change. Curves such as Equations 1 and 2 that are obtained by setting *dN*/*dt* = 0 are known as *zero population growth isoclines*.

Isocline analyses have been used in studies on competition between two desert gerbils, *Gerbillus allenbyi* and *G. pyramidum* (Abramsky et al. 1991). As the densities of the competitors changed, the inferior competitor (*G. allenbyi*) foraged in less preferred habitats, causing its isocline to have the curved shape shown in the figure.

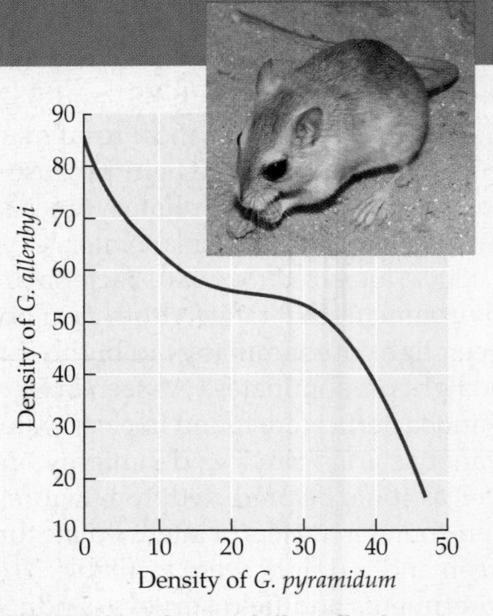

The Isocline for *Gerbillus allenbyi*

decrease in abundance. For example, in **Figure 11.13A**, at point A to the right of the N_1 isocline, there are $N_{1(A)}$ individuals of species 1. Because $N_{1(A)}$ is greater than the number of individuals that would produce zero population growth ($\hat{N}_1$), the density of species 1 will decrease at point A on the graph. This is true for the entire region shaded in blue: the density of species 1 decreases for all points to the right of the N_1 isocline. In contrast, when the density of species 1 is to the left of the N_1 isocline, the density of species 1 increases. Similar reasoning applies to species 2, only in this case—because N_2 is plotted on the *y* axis—the density of species 2 decreases in regions above the N_2 isocline and increases in regions below the N_2 isocline (**Figure 11.13B**).

The graphical approach we have just described can be used to predict the end result of competition between species. To do this, we plot the N_1 and N_2 isoclines together. Because there are four possible ways

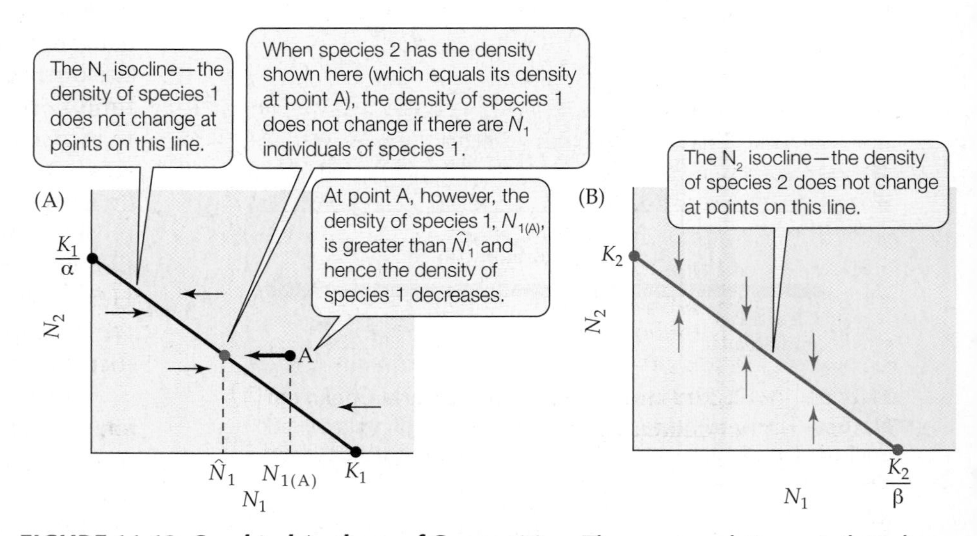

FIGURE 11.13 Graphical Analyses of Competition The zero population growth isoclines from the Lotka–Volterra competition model can be used to predict changes in the density of competing species. (A) The N_1 isocline. The density of species 1 increases at all points in the yellow region to the left of the N_1 isocline and decreases at all points in the blue region to the right of the isocline. (B) The N_2 isocline. The density of species 2 increases at all points in the yellow region and decreases at all points in the blue region.

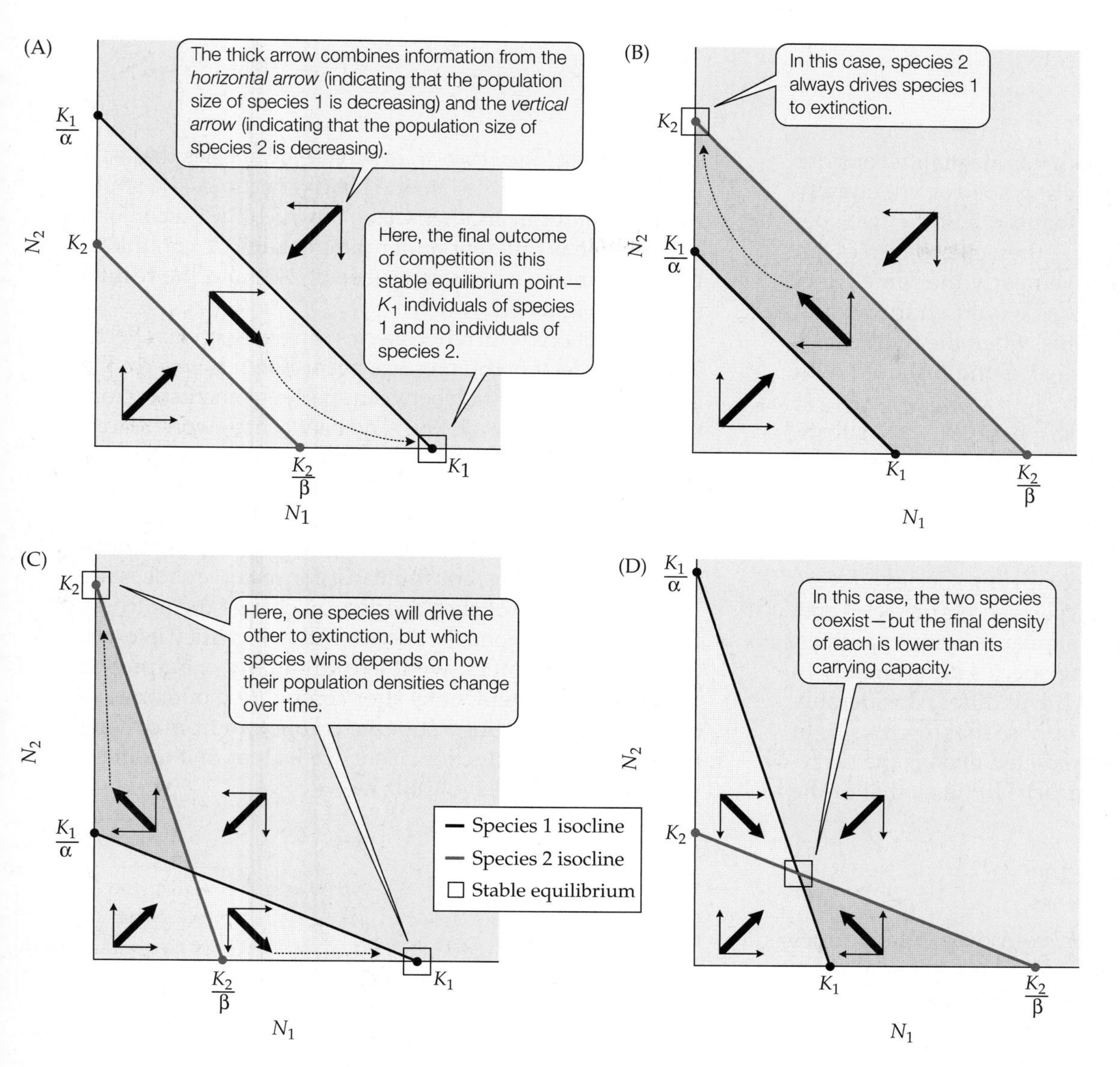

FIGURE 11.14 Outcome of Competition in the Lotka–Volterra Competition Model The outcome of competition depends on how the N_1 and N_2 isoclines are positioned relative to one another. (A) Competitive exclusion of species 2 by species 1; species 1 always wins. (B) Competitive exclusion of species 1 by species 2; species 2 always wins. (C) The two species cannot coexist; either species 1 or species 2 wins. (D) Species 1 and species 2 coexist. Each box indicates a *stable equilibrium point*—a combination of densities of the two species that once reached, does not change over time.

Q In (B), if K_2 = 1,000 and if species 1 went extinct when N_2 = 1,200, how would the density of species 2 change after the extinction of species 1?

that the N_1 and N_2 isoclines can be arranged relative to each other, we must make four different graphs. In two of these graphs, the isoclines do not cross, and competitive exclusion results: depending on which isocline is above the other, either species 1 (**Figure 11.14A**) or species 2 (**Figure 11.14B**) always drives the other to extinction. Note that in the regions shaded in blue, the densities of both species are greater than the densities on their isoclines, and hence both species *decrease* in number (as indicated by the thick black arrows). Similarly, in the regions shaded in yellow, the densities of both species are less than those on their isoclines, and hence both species *increase* in number. In the regions shaded in gray, one species increases in number (because its densities are less than those on its isocline) while the other decreases.

Competitive exclusion also occurs in the third graph (**Figure 11.14C**), but which species "wins" depends on whether the changing population densities of the two species first enter the region shown in dark gray (in which case species 2 drives species 1 to extinction) or the region shown in light gray (in which case species 1 drives species 2 to extinction). Finally, **Figure 11.14D** shows the only case in which the two species coexist, and hence competitive exclusion does not occur. Although in this case neither species drives the other to extinction, competition still has an effect: the final or equilibrium density of each species (indicated by the box in the figure) is lower than its carrying capacity, as in Gause's experiments with *Paramecium* (compare Figures 11.10B, C, and E).

How resource use affects competitive exclusion Now that we've seen the four possible outcomes predicted by the Lotka–Volterra competition model, let's revisit our goal of using the model to explain why competitive exclusion occurs in some situations, but not others. To do this, we'll focus on the single case in which competitive exclusion does *not* occur. As described in **Web Extension 11.4**, we can use Figure 11.14D to show that coexistence occurs when the values of α, β, K_1, and K_2 are such that the following inequality holds:

$$\alpha < \frac{K_1}{K_2} < \frac{1}{\beta} \quad (11.2)$$

To see what we can learn from this inequality, consider a situation in which the competing species are equally strong competitors, indicating that $\alpha = \beta$. If the two species are also very similar in how they use resources, an individual of species 1 will have nearly the same effect on the growth rate of species 2 as would an individual of species 2 (and vice versa). Thus, when the two species use resources in very similar ways, α and β should both be close to 1 (see Web Extension 11.3).

Suppose, for example, that $\alpha = \beta = 0.95$. If we substitute these values for α and β into Equation 11.2, we obtain

$$0.95 < \frac{K_1}{K_2} < 1.053$$

This result suggests that when competing species are very similar in how they use resources, coexistence is predicted only when the two species also have similar carrying capacities.

In contrast, if the competing species differ greatly in how they use resources, α and β will differ considerably from 1. To illustrate this case, suppose that $\alpha = \beta = 0.1$. In this situation, coexistence is predicted even if the carrying capacity of one species is nearly 10 times that of the other species, namely,

$$0.1 < \frac{K_1}{K_2} < 10$$

As you can demonstrate on your own, other values for the competition coefficients α and β yield similar results. Taken together, such analyses of the Lotka–Volterra competition model suggest the following refinement of the competitive exclusion principle: competing species are more likely to coexist (and hence competitive exclusion is *less* likely) when they use resources in very different ways.

A variety of factors can influence how species divide their use of resources, thereby preventing one competitor from driving the other to extinction. As we'll see in the next section, some of these factors can alter the outcome of competition entirely, turning the inferior competitor into the superior one.

CONCEPT 11.4 The outcome of competition can be altered by environmental conditions, species interactions, disturbance, and evolution.

Altering the Outcome of Competition

The outcome of competition between species can be changed by a broad suite of factors, including physical features of the environment, interactions with other species, disturbance, and evolution. For example, a difference in abiotic conditions—as might occur from one place to another—can cause a *competitive reversal*, in which the species that was the inferior competitor in one habitat becomes the superior competitor in another. Cases in which the outcome of competition is different under different abiotic conditions include Tansley's bedstraws (see p. 243) and the results shown in Figure 8.9.

Interactions with other species can have similar effects. The presence of herbivores has been shown to reverse the outcome of competition between species of encrusting marine algae (Steneck et al. 1991) and between ragwort (*Senecio jacobaea*) and other plant species (**Figure 11.15**). Herbivores can have this effect if they prefer to feed on the superior competitor, thereby reducing the growth, survival or reproduction of that species. What is true of herbivores is also true of predators, pathogens, and mutualists: an increase or decrease in the abundance of such species can change the outcome of competition among the species with which they interact.

In later chapters, we'll explore many examples in which species interactions alter competitive outcomes—sometimes preventing a superior competitor from driving other species to extinction. Here, we'll focus on the effects of disturbance and evolution.

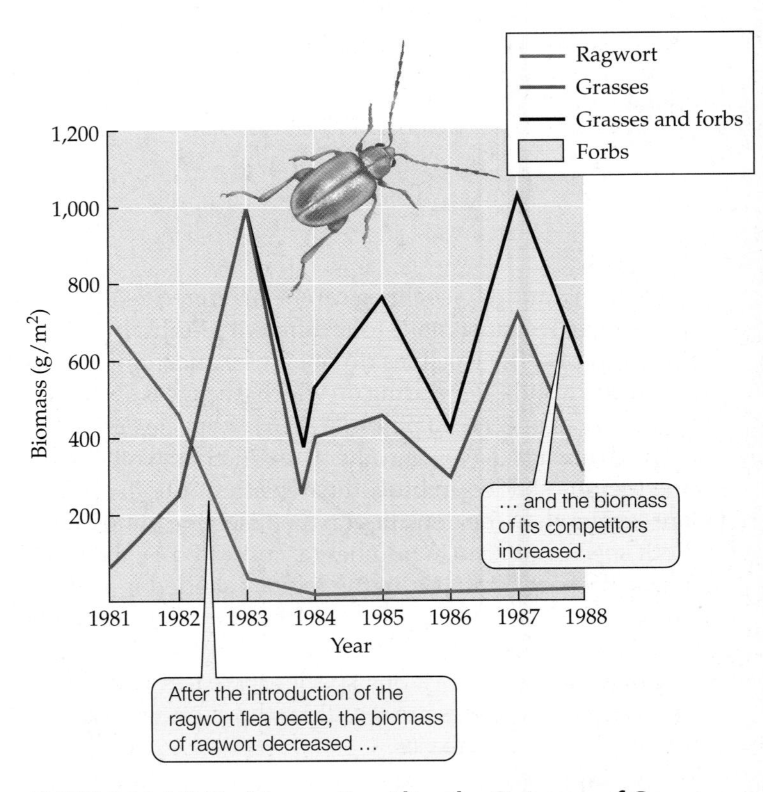

FIGURE 11.15 Herbivores Can Alter the Outcome of Competition Ragwort flea beetles are herbivores that feed on ragwort (*Senecio jacobaea*), an invasive plant species. In the absence of the flea beetle, ragwort is a superior competitor. The graph tracks the biomasses of ragwort, grasses, and forbs (broad-leaved herbaceous plants) at a site in western Oregon after the beetle was introduced there in 1980. (After McEvoy et al. 1991.)

Disturbance can prevent competition from running its course

As we saw in the discussion under Concept 8.2, a *disturbance* such as a fire or major storm may kill or damage some individuals while creating opportunities for others. Some species can persist in an area only if such disturbances occur regularly. Forests, for example, contain some herbaceous plant species, such as the primroses described on p. 233, that require abundant sunlight and are therefore found only in areas where wind or fire has created an opening in the tree canopy. Over time, a population of such plants is doomed: as trees recolonize the area, shade increases to a point at which the species cannot persist. Such species are called **fugitive species** because they must disperse from one place to another as environmental conditions change.

Robert Paine has described how periodic disturbance allows a fugitive algal species, the sea palm (*Postelsia palmaeformis*), to coexist with a competitively dominant species, the mussel *Mytilus californianus*. The sea palm is a brown alga that lives in the intertidal zone and must attach itself to rocks to grow. It competes for attachment space with mussels. Although a sea palm can outcompete an individual mussel (by growing on top of it), the sea palm is eventually displaced by other mussels that grow in from the side.

Competition with mussels causes sea palm populations to decline over time (**Figure 11.16**). Hence, if competition ran its course, mussels would drive sea palm populations to extinction. That is exactly what happens on low-disturbance shorelines (with a mean rate of 1.7 disturbances per year), where waves only occasionally tear patches of mussels from the rocks. However, sea palms can persist in shoreline areas where high-energy waves remove mussels more frequently (with a mean rate of 7.7 disturbances per year), thereby creating temporary openings for sea palm individuals. Similarly, in ecosystems such as grasslands or savannas, in which certain plants or animals depend on periodic fires to lower the abundance of a superior competitor, disturbance prevents competition from running its course and thus promotes species coexistence.

Evolution by natural selection can alter the outcome of competition

When two species compete for resources, natural selection may favor individuals whose phenotype either (1) allows them to be superior competitors or (2) allows them to partition the available resources, thus decreasing the intensity of competition. In principle, then, competition has the potential to cause evolutionary change, and evolution has the potential to alter the outcome of competition. The interplay between competition and evolution can be observed in both experimental and observational studies.

Competitive reversal Pimentel and colleagues studied the effect of natural selection on two competing fly species, the housefly *Musca domestica* and the green blowfly *Lucilia sericata* (referred to as *Phaenicia sericata* in Pimentel et al. 1965). The larvae of these species compete for foods such as decaying animal carcasses, garbage, and the feces of chickens, humans, and carnivores. Houseflies and blowflies have similar life cycles (taking about 14 days to

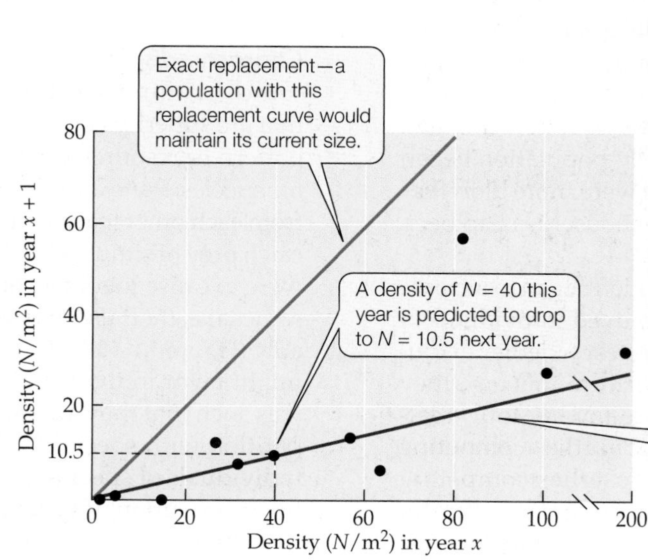

FIGURE 11.16 Population Decline in an Inferior Competitor In this graph, each point represents an observed change in density (N, the number of individuals per square meter) from one year (year x) to the next (year x + 1) at sites where sea palms are growing in competition with mussels. These points can be used to estimate a *replacement curve*, which shows the extent to which sea palm individuals replace themselves over time. The exact replacement curve shows the densities at which the population size would not change from one year to the next. (After Paine 1979.)

Q Based on the observed replacement curve (the blue line), how many years would it take for a sea palm population to decline from 100 individuals to fewer than 20 individuals?

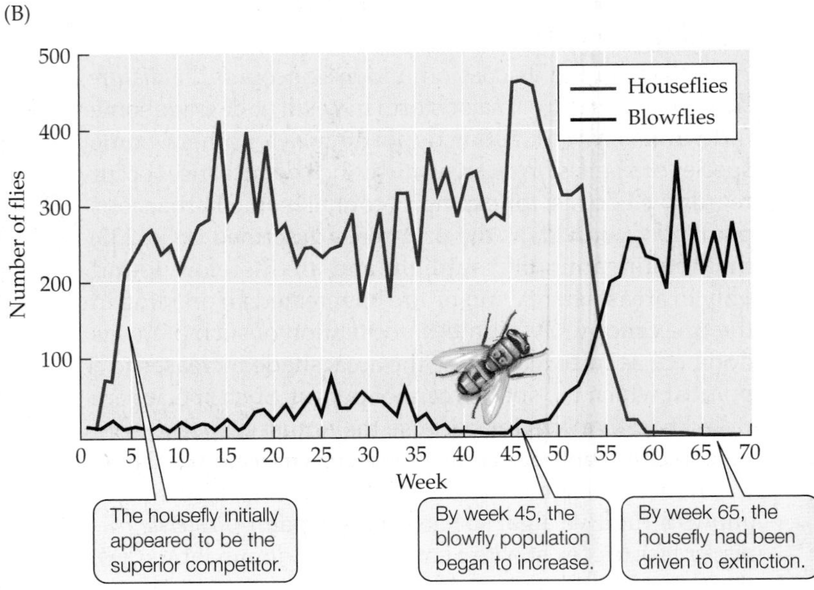

FIGURE 11.17 A Competitive Reversal (A) The 16-cell cage used by Pimentel and colleagues in their competition experiments on houseflies and blowflies. (B) After a period of initial decline, blowflies outcompeted houseflies in the 16-cell cage. Results from additional single-cage experiments suggest that the blowfly evolved to become a better competitor during the first 38 weeks of the 16-cage experiment. (After Pimentel et al. 1965.)

complete) and can be reared in laboratory cages, which makes them excellent study organisms.

Pimentel and colleagues reared houseflies and blowflies in laboratory cages to study competition between them. Larvae were fed a gel (consisting of nonfat dried milk, brewer's yeast, and agar) topped with a gram of liver; adults were fed lumps of sugar. The researchers released 100 male and 100 female houseflies into one end of an experimental arena that consisted of 16 cages linked by tubes through which the flies could crawl; the same number of blowflies were released into the opposite end of the arena (**Figure 11.17A**). The housefly initially appeared to be the superior competitor, increasing rapidly in density while the blowfly was restricted to a few cages along one side of the arena. Over time, however, the situation changed dramatically: by week 45, the blowfly population began to increase in size; by week 55, there were more blowflies than houseflies; and by week 65, the houseflies had been driven to extinction (**Figure 11.17B**).

Shortly before the blowfly population began to increase in size, Pimentel and colleagues removed individuals of both species from the 16-cell cage to test whether evolutionary change had occurred. To perform this test, they established 5 cages for each of the following four treatments: (1) "wild" (newly caught) houseflies competing against wild blowflies; (2) wild houseflies competing against "experimental" blowflies (drawn from the 16-cell cage); (3) experimental houseflies competing against wild blowflies; and (4) experimental houseflies competing against experimental blowflies.

In this experiment, treatments 3 and 4 used houseflies that had had a chance to evolve in competition with blowflies; similarly, treatments 2 and 4 used blowflies that had had a chance to evolve in competition with houseflies. The results showed that wild and experimental housefly populations usually outcompeted wild blowfly populations (7 out of 10 times). The situation was reversed, however, when experimental blowflies were used: the experimental blowfly populations always outcompeted housefly populations (10 out of 10 times). These results suggest that the blowflies raised in competition with houseflies had evolved to become better competitors, although the underlying mechanisms of competition and the associated genetic changes are not known.

Character displacement When species compete for resources, natural selection may favor not only individuals that are superior competitors, but also individuals whose pattern of resource use differs from that of the competitor species. For example, when two fish species live apart from each other (each in its own lake), the two species may catch prey of similar size. If some factor (such as dispersal) were to cause members of these two species to live in the same lake, their use of resources would overlap considerably (**Figure 11.18A**). In such a situation, natural selection might favor individuals of species 1 whose morphology was such that they ate smaller prey, hence reducing competition with species 2; similarly, selection might favor individuals of species 2 that ate larger prey, hence reducing competition with species 1. Over time, such selection pressures could cause species 1 and species 2 to evolve to

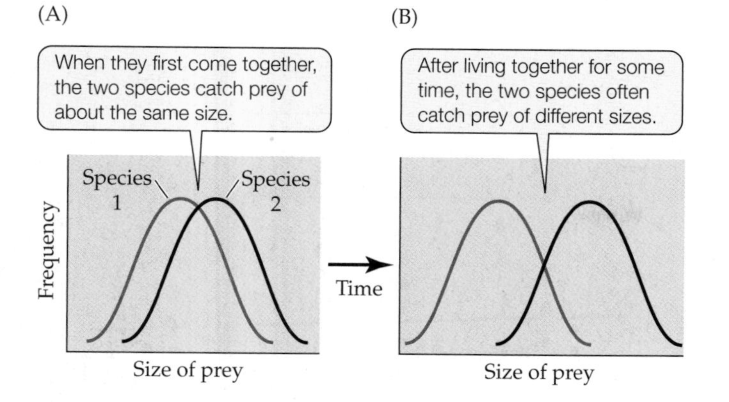

FIGURE 11.18 Character Displacement Competition for resources can cause competing species to become more different over time. Imagine that two fish species that once lived apart and tended to catch prey of about the same size are brought together in a single lake. (A) When the two species first come together, there is considerable overlap in the resources they use. (B) As the two species interact over time, their forms might evolve such that they tend to catch prey of different sizes.

become more different when they live together than when they live apart (**Figure 11.18B**). Such a process illustrates **character displacement**, which occurs when competition causes the phenotypes of competing species to evolve to become more different over time.

Character displacement appears to have occurred in two species of finches on the Galápagos archipelago. Specifically, the beak sizes of the two species, and hence the sizes of the seeds the birds eat, are more different on islands where both species live than on islands that have only one of the two species (**Figure 11.19**). Field observations suggest that these two finch species probably differ when they live together because of competition, not because of other factors such as differences in food supplies (Schluter et al. 1985; Grant and Grant 2006).

Data suggestive of character displacement have also been observed in plants, frogs, fishes, lizards, birds, and crabs: in each of these groups, there are pairs of species that consistently differ more where they live together than where they live apart. Additional evidence is needed, however, if we are to make a strong argument that such differences result from competition (as opposed to other factors). Strong support for the role of character displacement can come from experiments designed to test whether competition occurs and has a selective effect on morphology. Such experiments were conducted on sticklebacks of the genus *Gasterosteus*, a group of fish species whose morphology varies most when different species live in the same lake (Schluter 1994). The results indicated that individuals whose morphology differed the most from the morphology of their competitors had a selective advantage: they grew more rapidly than did individuals whose morphology was more similar to that of their competitors. Support for character displacement has also been found in field experiments with spadefoot toad tadpoles (Pfennig et al. 2007) and in laboratory experiments with the bacterium *Escherichia coli* (Tyerman et al. 2008). In each of these studies, experimental results suggest that competition caused the observed morphological differences—that is, that character displacement occurred.

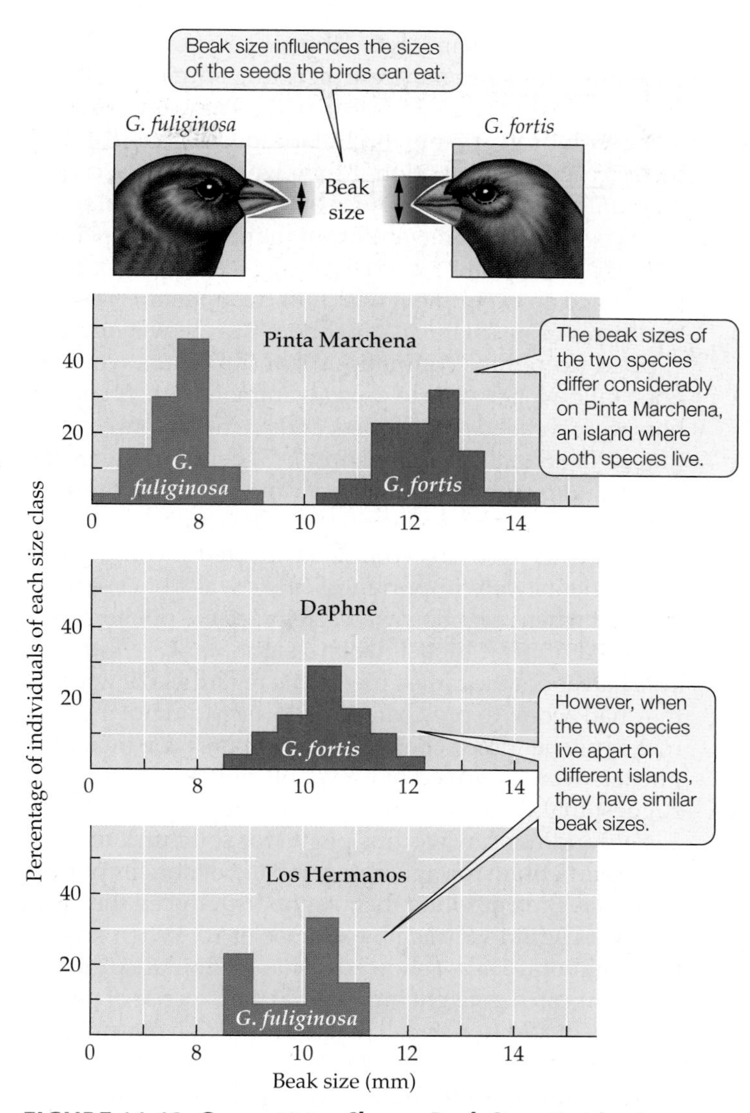

FIGURE 11.19 Competition Shapes Beak Size On islands harboring both species, competition between two species of Galápagos finches, *Geospiza fuliginosa* and *G. fortis*, may have had a selective effect on the sizes of their beaks.

A CASE STUDY REVISITED

Competition in Plants That Eat Animals

In plants, belowground competition can be especially intense in nutrient-poor soils (see Figure 11.5). Carnivorous plants live in such soils, and their root systems are usually less well developed than the root systems of their noncarnivorous neighbors. These observations suggest that carnivorous plants may be poor competitors for soil nutrients and hence may rely on eating animals to obtain the nutrients they need for growth. They also suggest that carnivorous plants might be especially hard-hit by belowground competition if they were denied access to their unique, alternative source of nutrients (animal prey).

When Stephen Brewer studied the effect of competition on the pitcher plant *Sarracenia alata* (see Figure 11.2), he performed an experiment with four treatments: (1) nearby noncarnivorous plants ("neighbors") were weeded and clipped (to reduce competition), and pitchers were left open (allowing *S. alata* to capture animal prey); (2) neighbors were weeded and clipped, and pitchers were covered (depriving *S. alata* of animal prey); (3) neighbors were left intact, and pitchers were covered; and (4) neighbors were left intact, and pitchers were open (the control).

Contrary to what would be expected if competition for nutrients were important, *S. alata* was not especially hard-hit when neighbors were left intact and pitchers were deprived of prey. In fact, when neighbors were intact, *S. alata* had about the same growth rate regardless of whether they had access to prey. Moreover, *S. alata* did not increase root or pitcher production when neighbors were intact and pitchers were deprived of prey. Collectively, these results suggest that there is relatively little competition between *S. alata* and noncarnivorous plants for soil nutrients.

Competition for light appeared to be more important to *S. alata*. Brewer found that neighbors reduced the availability of light to *S. alata* by a factor of 10. When shaded by neighbors, *S. alata* showed a shade-avoidance growth response: pitcher height increased at the expense of pitcher volume. In addition, recall that the reduction of neighbors greatly increased growth rates in *S. alata*, but only when the pitcher plants could capture animal prey (see Figure 11.2). Thus, prey deprivation reduced the ability of the pitcher plants to grow in response to increased light following competitor reduction. Finally, Brewer observed that when neighbors were left intact, natural differences in the light available to pitcher plants had no effect on their growth rates when pitchers were covered (**Figure 11.20**). In plants that had access to prey, however, growth rates increased as the amount of light increased. Hence, just as they did when neighbors were reduced, *S. alata* responded to higher light levels when neighbors were intact by growing more rapidly—but only if prey were available to supply the extra nutrients they needed for such growth.

Overall, it appears that *S. alata* competes with its neighbors for light, but avoids competition for soil nutrients by eating animals and by using changes in light levels as a cue for growth. When light levels are low—as would be the case when it is shaded by competitors—*S. alata* grows little, and hence requires few nutrients. In such a situation, prey deprivation has little effect because the plant does not need extra nutrients. When light levels are high, however—as would occur after a fire or whenever few competitors are present—*S. alata* is stimulated to grow. Under these circumstances, prey deprivation has a major effect because animal prey supply most of the nutrients that *S. alata* uses for growth.

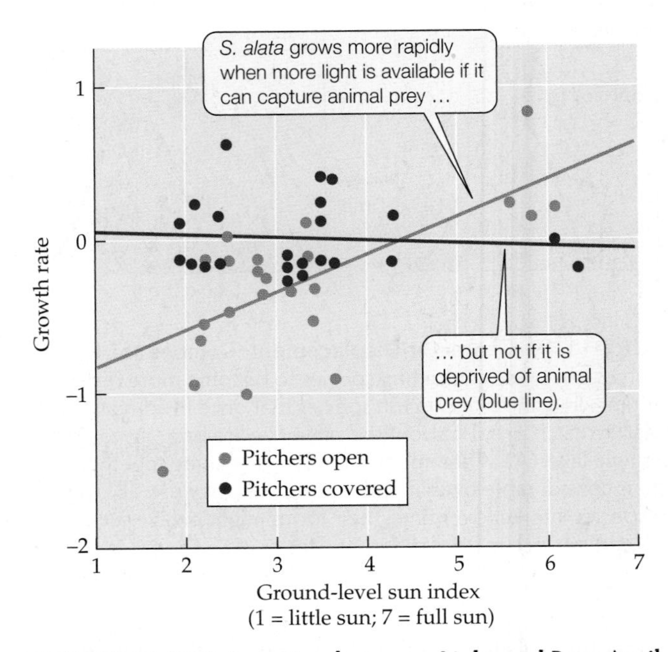

FIGURE 11.20 Interaction between Light and Prey Availability The pitcher plant *Sarracenia alata* grows more rapidly when more light is available, but only when pitchers are open, allowing it to eat animals and thereby obtain the nutrients it needs to support rapid growth. (After Brewer 2003.)

CONNECTIONS IN NATURE

The Paradox of Diversity

As we've seen, some field data show that superior competitors can drive inferior competitors extinct—which is exactly what the competitive exclusion principle states *should* happen whenever two or more species use the same set of limiting resources. Natural communities, however, contain many species that share the use of scarce resources without driving one another to extinction, as we'll see in the discussion under Concept 18.2. We saw an example in the case of the pitcher plants just discussed: pitcher plants were predicted to be inferior competitors for soil nutrients, yet they coexist with a diverse group of other species (**Figure 11.21**). In the context of Brewer's experiments on pitcher plants, let's reconsider why superior competitors do not always drive inferior competitors to extinction.

The concept of resource partitioning suggests that coexistence in nutrient-poor environments could occur if species avoided competition for scarce resources by acquiring nutrients in different ways. This idea helped

FIGURE 11.21 Coexistence in a Nutrient-Poor Environment The pitcher plant *Sarracenia alata* coexists with noncarnivorous plants that can outcompete it for both nutrients and light.

to motivate Brewer's study: he wanted to know whether differences in their means of nutrient acquisition could explain the coexistence of carnivorous and noncarnivorous plants. To find out, Brewer deprived carnivorous plants of their unique source of nutrients (animal prey), thus increasing the overlap between the ways in which carnivorous and noncarnivorous plants acquired nutrients. If competition for nutrients was important, pitcher plants that were deprived of prey should have experienced more severe competitive effects, or they should have compensated for reduced nutrient intake by increasing their production of roots or pitchers. Neither of these outcomes occurred, so Brewer sought other explanations of species coexistence.

As we'll see in the discussion under Concept 18.1, environmental variation provides a second mechanism for the coexistence of species in communities: if environmental conditions fluctuate over space or time (or both), species may coexist if different species are superior competitors under different environmental conditions. The bedstraw example (discussed on p. 243) illustrates how differences in soils can alter the outcome of competition, thus promoting coexistence in environments that vary over space. With respect to variation over time, an inferior competitor could persist whenever competition failed to run its course. Consider a species such as the sea palm (see p. 255), which competes poorly but tolerates disturbance well. Such a species can persist if a disturbance periodically "resets the clock" by decreasing the abundance of a superior competitor before that species drives the inferior competitor to extinction. Such a scenario may also apply to the pitcher plant *S. alata*. The habitat in which it lives is prone to fire, and *S. alata* tolerates fire well and uses changes in light levels as a cue for growth. As a result, *S. alata* grows primarily when its competitors are reduced by fire. This growth strategy may allow *S. alata* to escape competition for nutrients by reducing its demand for scarce nutrients when competition is potentially most intense (i.e., in years without fire) and increasing its demand for nutrients when competitors have been reduced (years with fire).

SUMMARY

CONCEPT 11.1 Competition occurs between individuals of two species that share the use of a resource that limits their growth, survival, or reproduction.

- Organisms compete for resources such as food, water, light, and space, all of which can be consumed or otherwise used to the point of depletion.
- When organisms compete, they reduce the availability of resources.
- If resource levels become sufficiently low, the intensity of competition can increase.
- Competition for resources is common—though not ubiquitous—in natural communities.

CONCEPT 11.2 Competition, whether direct or indirect, can limit the distributions and abundances of competing species.

- The most common form of competition is exploitation competition, which occurs when species compete indirectly as they share the use of a limiting resource. Another form of competition, called interference competition, occurs when species compete directly for access to resources.
- Competition is often asymmetrical, affecting one competitor more strongly than the other.
- Competition can occur between closely or distantly related species.
- Competition can determine the abundance of competing species.
- Competition can determine both where a species lives within a local habitat and the extent of its geographic distribution.

SUMMARY (continued)

CONCEPT 11.3 Competing species are more likely to coexist when they use resources in different ways.

- The competitive exclusion principle states that if competing species use the same limiting resource in the same way, they cannot coexist.
- Field studies have revealed many examples of resource partitioning, in which competing species use one or more shared resources in different ways.
- Lotka and Volterra modeled the effects of competition by modifying the logistic equation.
- Graphical analyses of the Lotka–Volterra competition model suggest a refinement of the competitive exclusion principle: the more similar two competing species are in their use of resources, the more likely it is that one will drive the other to extinction.

CONCEPT 11.4 The outcome of competition can be altered by environmental conditions, species interactions, disturbance, and evolution.

- The outcome of competition can be altered by changes in the physical conditions of the environment or by changes in the competitors' interactions with other species.
- Periodic disturbances that remove a superior competitor can allow an inferior competitor to persist.
- Evolutionary change within a population of an inferior competitor that causes it to become a better competitor can reverse the outcome of competition.
- In character displacement, competition causes the phenotypes of competing species to evolve to become more different from each other over time, thereby reducing the intensity of competition.

REVIEW QUESTIONS

1. a. Explain the difference between a resource and a physical factor.

   b. Plant species require nitrogen to grow and reproduce. Assume that a single application of high-nitrogen fertilizer is added to a low-nitrogen, sandy soil in which two plant species are found. Predict how the intensity of competition for soil nitrogen will change over time (and explain your prediction).

2. List four general features of competition described in the text under Concept 11.2 and provide an example of each.

3. a. As described in the text under Concept 11.3, laboratory experiments, field observations, and mathematical models have been used to explain why competing species coexist in some situations, but not others. Describe a result from each approach that helps to explain when competing species are likely to coexist. Taken together, do these three approaches to studying competition yield similar or different explanations for why coexistence occurs in some situations, but not others?

   b. If $\alpha = 0.8$, $\beta = 1.6$, $N_1 = 140$, and $N_2 = 230$, are individuals of species 1 or species 2 having a greater effect on the growth rate of species 2?

   c. Based on graphical analyses of the Lotka–Volterra competition model, evaluate the following statement: If $\alpha < \beta$, species 1 will always drive species 2 to extinction. Explain your answer.

4. Suppose that each of 20 meadows contains a population of plant species 1, a population of plant species 2, or a population of both plant species. Species 1 and 2 are known to compete with each other. Each meadow is separated from the others by areas in which neither species 1 or species 2 can grow or survive.

   a. List three possible reasons why the meadows contain different combinations of the two plant species.

   b. Describe an experiment that would help to evaluate one or more of the reasons you have listed.

ON THE COMPANION WEBSITE
sites.sinauer.com/ecology2e

The website includes Chapter Outlines, Online Quizzes, Flashcards & Key Terms, Suggested Readings, a complete Glossary, and the Web Stats Review. In addition, the following resources are available for this chapter:

HANDS-ON PROBLEM SOLVING

Character Displacement

This Web exercise explores character displacement and phenotypic plasticity for a spadefoot toad species. You will interpret data collected on different populations of toads and the results of common garden experiments to assess genetic effects.

WEB EXTENSIONS

11.1 Resource Partitioning in *Anolis* Lizards

11.2 How Should Competition Be Modeled?

11.3 What Do the Competition Coefficients α and β Represent?

11.4 Deriving the Conditions for Coexistence in the Lotka–Volterra Competition Model

CHAPTER

12 Predation and Herbivory

KEY CONCEPTS

- **CONCEPT 12.1** Most predators have broad diets, whereas a majority of herbivores have relatively narrow diets.
- **CONCEPT 12.2** Organisms have evolved a wide range of adaptations that help them obtain food and avoid being eaten.
- **CONCEPT 12.3** Predation and herbivory can affect ecological communities greatly, in some cases causing a shift from one community type to another.
- **CONCEPT 12.4** Population cycles can be caused by exploitative interactions.

Snowshoe Hare Cycles: A Case Study

In 1899, a fur trader in northern Ontario reported to the Hudson's Bay Company that "Indians are bringing poor hunts. They have been starving all spring. Rabbits being scarce" (Winterhalder 1980). The "hunts" referred to were pelts of beavers and other fur-bearing animals trapped by members of the Ojibwa tribe, and the "rabbits" were actually snowshoe hares (*Lepus americanus*) (**Figure 12.1**). Collectively, 200 years of such reports showed that hare populations increased and decreased regularly. When hares were abundant, the Ojibwa had enough food to spend time trapping for pelts, which they then traded to the Hudson's Bay Company. But when hares were scarce, tribal members concentrated on gathering food, rather than trapping animals that provided pelts, but little meat.

Beginning in the early 1900s, wildlife biologists used the Hudson's Bay Company's careful records to estimate abundances of snowshoe hares and their lynx predators (*Lynx canadensis*). Both species exhibited regular population cycles, with abundances peaking about every 10 years and then falling to low levels (**Figure 12.2A**). Snowshoe hares constitute a major portion of the lynx diet, so it was not surprising that numbers of lynx should rise and fall with numbers of hares. But what drove the cyclic fluctuations in the hare population? Adding to the mystery, hare population sizes rise and fall in synchrony across broad regions of the Canadian forest, so explanations of hare cycles should account for that synchrony as well.

One approach to finding the cause of hare population cycles is to document the changes in birth, death, and dispersal rates that are associated with increasing or declining numbers of hares. Dispersal plays a relatively small role: it may alter local population sizes, but hares do not move far enough to account for the simultaneous changes in their abundance seen across broad geographic regions. In contrast, consistent patterns of birth and death rates have been found across different regions of Canada. Snowshoe hares can raise up to three or four litters over the summer, with an average of five young per litter. Hare reproductive rates reach their highest levels (ca. 18 young/female) several years before hare density reaches a maximum. They then begin to fall, reaching their lowest levels 2–3 years after hare density peaks (**Figure 12.2B**). Hare survival rates show a similar pattern: they are highest several years before hare density peaks, then they fall and do not rise again until several years after hare density peaks.

Together, the changes over time in hare birth and survival rates drive the hare population cycle. But what causes these

FIGURE 12.1 Predator and Prey A snowshoe hare (*Lepus americanus*) flees from its specialist predator, the Canadian lynx (*Lynx canadensis*).

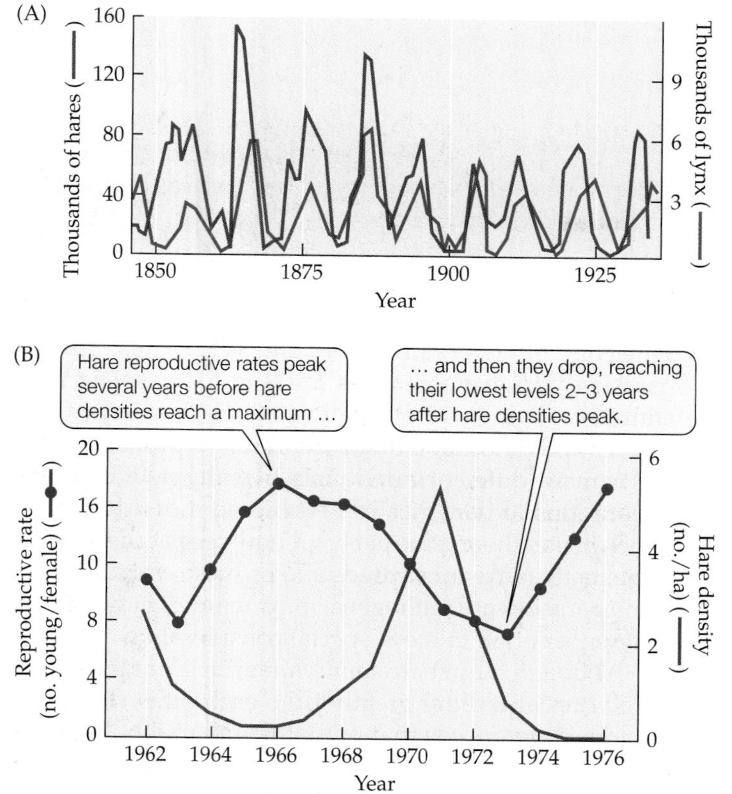

FIGURE 12.2 Hare Population Cycles and Reproductive Rates (A) Historical trapping data from the Hudson's Bay Company indicate that numbers of both hares and lynx fluctuate in a 10-year cycle. (B) The highest hare reproductive rates do not coincide with the highest hare densities. (B after Cary and Keith 1979.)

Q In (A), does the peak abundance of one species typically occur after the peak abundance of the other species? Describe the observed pattern and hypothesize why it might occur.

rates to change? Several hypotheses have been proposed, one of which focuses on food supplies. Large numbers of hares consume prodigious amounts of vegetation, and studies have shown that food can be limiting at peak hare densities (up to 2,300 hares/km^2). Two observations, however, indicate that food alone does not drive the hare cycle: first, some declining hare populations do not lack food, and second, the experimental addition of high-quality food does not prevent hare populations from declining.

A second hypothesis focuses on predation. Many hares (up to 95% of those that die) are killed by predators such as lynx, coyotes, and birds of prey. In addition, lynx and coyotes kill more hares per day during the peak and decline phases of the hare cycle than during the increase phase. But questions remain. The killing of hares by predators explains the drop in survival rates as hare numbers decline, but by itself does not explain (1) why hare birth rates drop during the decline phase of the cycle, (2) why hare numbers sometimes rebound slowly after predator numbers plummet, or (3) why the physical condition of hares worsens as hares decrease in numbers. What other factors are at work?

Introduction

Over half of the species on Earth sustain themselves by feeding on other organisms. Some kill other organisms, then eat them, while others "graze" on living organisms by eating their tissues or internal fluids. As we will see, those millions of species interact with the organisms they eat in a rich variety of ways. But all of these interactions share a common feature: they are all forms of **exploitation**, a +/– interaction in which one organism benefits by feeding on, and thus directly harming, another.

Over the course of this and the next chapter, we will consider three broad categories of organisms that harm the organisms they feed on: herbivores, predators, and parasites (**Figure 12.3**). An **herbivore** eats the tissues or internal fluids of living plants or algae. A **predator** kills and eats other organisms, referred to as its **prey**. A *parasite* typically lives in or on another organism (its *host*), feeding on parts of the host, such as its tissues or body fluids. Parasites harm, but

FIGURE 12.3 Three Ways to Eat Other Organisms (A) Herbivores such as these sawfly larvae (*Croesus latitarsus*) eat leaves or other plant parts. (B) This predator, a common caiman (*Caiman crocodilus*), kills and consumes fish and other animals. (C) This isopod parasite (*Anilocra* sp.) attaches to and feeds on the tissues of its host, the four-eye butterfly fish (*Chaetodon capistratus*).

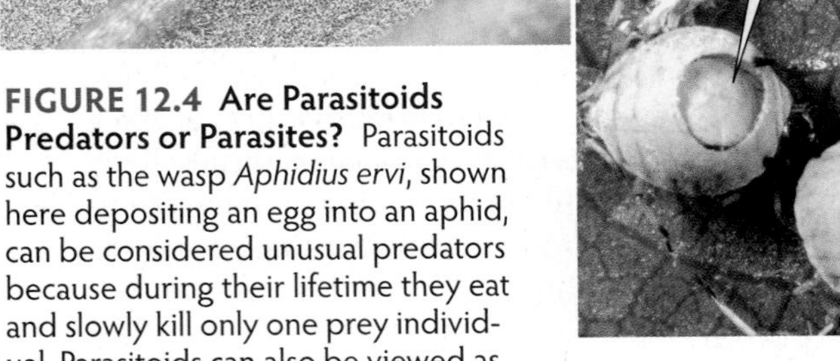

FIGURE 12.4 Are Parasitoids Predators or Parasites? Parasitoids such as the wasp *Aphidius ervi*, shown here depositing an egg into an aphid, can be considered unusual predators because during their lifetime they eat and slowly kill only one prey individual. Parasitoids can also be viewed as unusual parasites that eat all or most of their host, thereby killing it.

usually do not kill, the organisms they feed on; some parasites (called *pathogens*) cause disease. Individual parasites usually feed on only one or a few host individuals.

These definitions seem simple, and it is easy to think of examples: an insect that eats a plant leaf, a lion that kills and eats a zebra, a tapeworm that robs a dog of nutrients in its digested food. But the natural world defies such simple categorization. Is an insect that spends its entire life eating the leaves of one tree an herbivore, a parasite, or both? Or consider those prototypical herbivores, sheep: they get most of their food from plants, but they have also been known to eat the helpless young of ground-nesting birds. Conversely, predators can act like herbivores: wolves, for example, will eat berries, nuts, and leaves. And some organisms do not fit neatly into any category. **Parasitoids** are insects that typically lay one or a few eggs on or in another insect (the host) (**Figure 12.4**). After they hatch from their eggs, the parasitoid larvae remain with the host, which they eat and usually kill. Parasitoids can be considered unusual parasites (because they consume most or all of their host, almost always killing it) or unusual predators (because over the course of their lives they eat only one host individual, killing it slowly).

Despite these and other complications, we will approach the rich variety of exploitative interactions by organizing Chapters 12 and 13 around three basic types: predation, herbivory, and parasitism. This chapter is focused on predators and herbivores, the next on parasites and pathogens. We will begin by exploring some aspects of the natural history of predators and herbivores.

CONCEPT 12.1 Most predators have broad diets, whereas a majority of herbivores have relatively narrow diets.

Predators and Herbivores

This chapter covers predators and herbivores, but not parasites, for several reasons. Parasites often spend their entire lives living with and eating a single host individual, whereas predators and herbivores more typically eat several to many different individuals. In many cases, multiple generations of parasites live within an individual host, which makes them more likely to show special adaptations to hosts than are most predators or herbivores. Finally, some parasites are pathogens, the dynamics of which are different enough to merit special consideration.

Although they share some similarities, predators and herbivores also differ greatly from each other. The most obvious difference is that predators invariably kill their prey, while herbivores usually do not kill their food plants, at least not immediately. Here too there are exceptions, however. Herbivores that eat seeds can be considered predators: each seed contains a unique genetic individual that is killed when the seed is eaten. And, as we will see later in this chapter, some herbivores can strip entire regions bare of vegetation, causing many plants to die.

We begin with an overview of how predators obtain food and what they eat, and then follow with a similar discussion of herbivores.

Some predators move in search of prey, others sit and wait

Many predators forage throughout their habitat, moving about in search of prey. Examples of species that hunt in this way include wolves, sharks, and hawks. Other predators, called **sit-and-wait predators**, remain in one place and attack prey that move within striking distance (as do moray eels and sessile filter feeders, such as barnacles) or enter a trap (such as a spider's web or the modified leaf of a carnivorous plant).

In either case, predators tend to concentrate their efforts in areas that yield abundant prey. Birds that eat insects forsake patches with few prey to forage in patches where more food is available. Wolf packs follow the seasonal migrations of elk herds. Sit-and-wait predators, including spiders, relocate more often from areas where prey are scarce to areas where prey are abundant than vice versa. As a result, areas with more prey also have more spiderwebs.

Many predators have broad diets

Most predators eat prey in relation to their availability without showing a preference for any particular prey species. A predator can be said to show a preference for

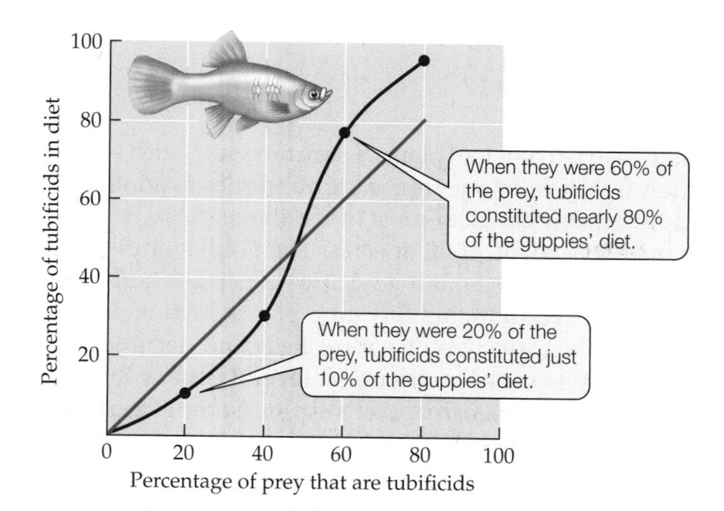

FIGURE 12.5 A Predator That Switches to the Most Abundant Prey Guppies focused their foraging efforts on whichever prey species was most common in their habitat: tubificids (aquatic worms) or fruit flies. The solid green line indicates the results that would have been expected if the guppies had captured tubificids according to their availability instead of switching to whichever prey species was more abundant. (After Murdoch et al. 1975.)

a particular prey species if it eats that species more often than would be expected based on that prey's availability. Some predators do show a strong preference for certain prey species, and in that sense they can be considered *specialist predators*. Lynx and coyotes, for example, eat more hares than would be expected based on their availability; even when hares constitute only 20% of the available food, they constitute 60%–80% of the diet of lynx and coyotes.

Some predators concentrate their foraging on whatever prey is most plentiful. When researchers provided guppies with two kinds of prey, fruit flies (floating on the water surface) and tubificids (aquatic worms found on the bottom), the guppies ate disproportionate amounts of whichever prey was most abundant (**Figure 12.5**). Predators like these guppies that focus on abundant prey tend to "switch" from one prey species to another. Such switching may occur because the predator forms a search image of the most common prey type and hence tends to orient toward that prey, or because learning enables it to become increasingly efficient at capturing the most common prey type. As we saw in the discussion under Concept 5.4, in some cases predators switch from one prey species to another in a manner consistent with the predictions of optimal foraging theory.

Many herbivores specialize on particular plant parts

Plants have a simple body plan, with three main organs (roots, stems, leaves) linked to one another by a series of vessels (xylem and phloem) that transport water and nutrients. Seed production occurs in modified leaves (such as the flowers of flowering plants). With this simple body plan as their target food source, herbivores can be grouped according to what part of the plant they eat. Some herbivores that are large relative to their food plants eat all aboveground parts. Most herbivores, however, specialize on particular plant parts, such as leaves, roots, stems, seeds, or internal fluids (e.g., nutrient-containing sap).

More herbivores eat leaves than any other plant part. Leaves are abundant and are available year-round in many places; leaves are also more nutritious than other plant parts (except for seeds) (**Figure 12.6**). Herbivores that eat leaves range from large browsers such as deer or giraffes, to grasshoppers and herbivorous fish, to tiny "leaf miners" such as fly larvae that enter a leaf and eat it from the inside. By removing photosynthetic tissues, leaf-eating herbivores can reduce the growth, survival, or reproduction of their food plants.

Belowground herbivory can also have major effects on plants, as illustrated by the 40% reduction in growth observed in bush lupine plants after 3 months of herbivory by caterpillars of the root-killing ghost moth, *Hepialus cali-*

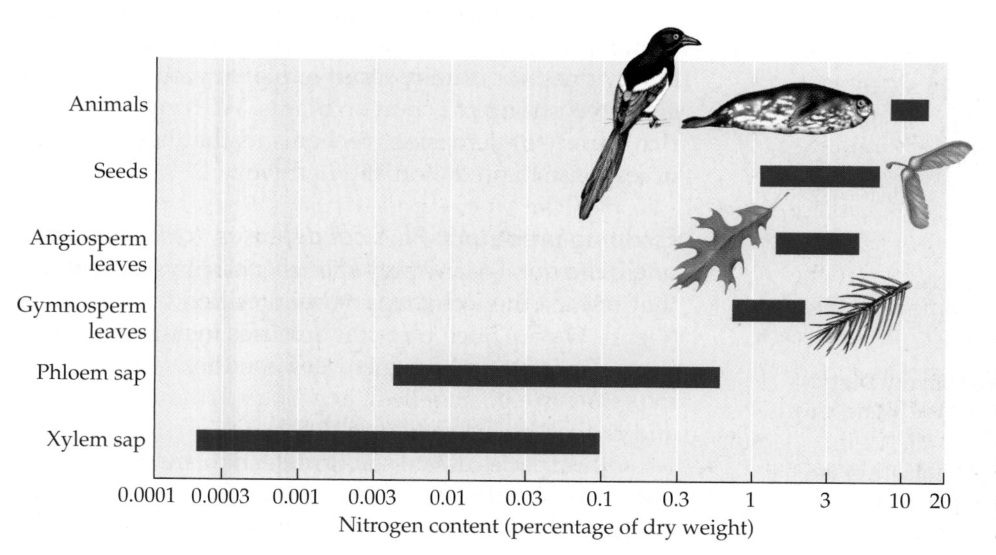

FIGURE 12.6 The Nitrogen Content of Plant Parts Varies Considerably Nitrogen is an essential component of any animal's diet. Leaves tend to have the highest nitrogen concentration of any plant parts other than seeds. Compared with those in animal bodies, however, nitrogen concentrations in plant parts are low. (After Mattson 1980.)

fornicus. Similarly, herbivores that eat seeds can have large effects on plant reproductive success, sometimes reducing it to zero. The effects of herbivores that feed on internal fluids are not always obvious (because visible plant parts are not removed), but they can be considerable. For example, Dixon (1971) showed that although the lime aphid *Eucallipterus tiliae* did not reduce aboveground growth in lime trees during the year of infestation, the roots of trees infested with aphids did not grow that year, and a year later, their leaf production dropped by 40%.

Most herbivores have relatively narrow diets

While most predators eat a broad range of prey species, the majority of herbivores feed on a comparatively restricted set of plant species. This statement is true largely because of the insects: there are an enormous number of herbivorous insect species, and most of them live on and eat only one (or a few) plant species. For example, most species of agromyzid flies, whose larvae are leaf miners, feed on only one or a few plant species (**Figure 12.7**). Similar results have been found for leaf-feeding beetles in the genus *Blepharida*: among 37 species of these beetles, 25 feed on a single plant species, 10 feed on 2–4 plant species, and 2 feed on a relatively broad suite of plants (12–14 species) (Becerra 2007).

There are numerous examples of herbivores that eat many plant species, however. Grasshoppers, for example, graze on a broad range of plant species, and even among the leaf miners shown in Figure 12.7, several species eat more than ten different plants. Large browsers, such as deer, often switch from one tree or shrub species to another; in addition, they eat all or most of the aboveground parts of many herbaceous plant species. The golden apple snail (*Pomacea canaliculata*) is a voracious generalist herbivore, capable of removing all the large plants from wetlands; the snail then survives by eating algae and detritus.

Now that we have discussed several general features of predation and herbivory, let's explore some of those features in more detail. We'll begin by focusing on adaptations found in predators, herbivores, and their food organisms.

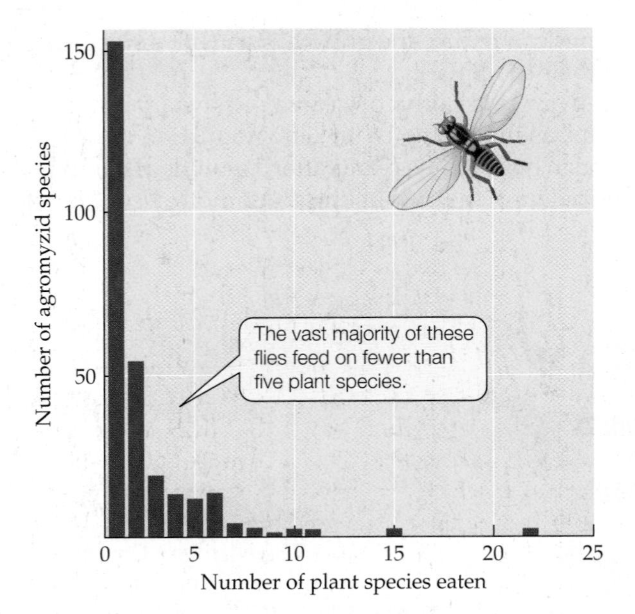

FIGURE 12.7 Most Agromyzid Flies Have Narrow Diets
The larvae of agromyzid flies are leaf miners that live inside leaves and feed on leaf tissue. (After Spencer 1972.)

Q Using the data in the graph, make a rough estimate of the percentage of agromyzid fly species that feed on fewer than five host plant species.

CONCEPT 12.2 Organisms have evolved a wide range of adaptations that help them obtain food and avoid being eaten.

Adaptations to Exploitative Interactions

Among the other challenges it faces, an organism must obtain food while striving not to become food for another organism. This ongoing drama has resulted in the evolution of a dazzling array of defensive mechanisms in prey animals and food plants—as well as novel ways of overcoming those defenses in predators and herbivores.

Predators and herbivores exert strong selection on their food organisms

Life on Earth changed radically with the appearance of the first macroscopic predators (organisms large enough to visible to the naked eye) roughly 530 million years ago. Before that time, the seas were dominated by soft-bodied organisms. Within a few million years after the evolution of the first large predators, however, many prey had evolved formidable defenses, such as body armor and spines. This explosive increase in prey defenses occurred because predators exert strong selection pressure on their prey: if prey are not well defended, they die. Although herbivory does not usually result in plant death, herbivores also exert strong selection on plants. We turn now to the rich variety of defensive mechanisms that have evolved in response to predation and herbivory.

Escaping predators: Physical defenses, toxins, mimicry, and behavior Many prey species have physical features that reduce their chances of being killed by predators (**Figure 12.8A**). Such physical defenses include large size (e.g., elephants), a body plan designed for rapid or agile movement (e.g., gazelles), and body armor (e.g., snails and the sea urchin in Figure 12.8A).

Other species use poisons to defend themselves against predators. Species that contain powerful toxins are often brightly colored (**Figure 12.8B**). Such **warning** (**aposematic**)

(A)

(B)

(D)

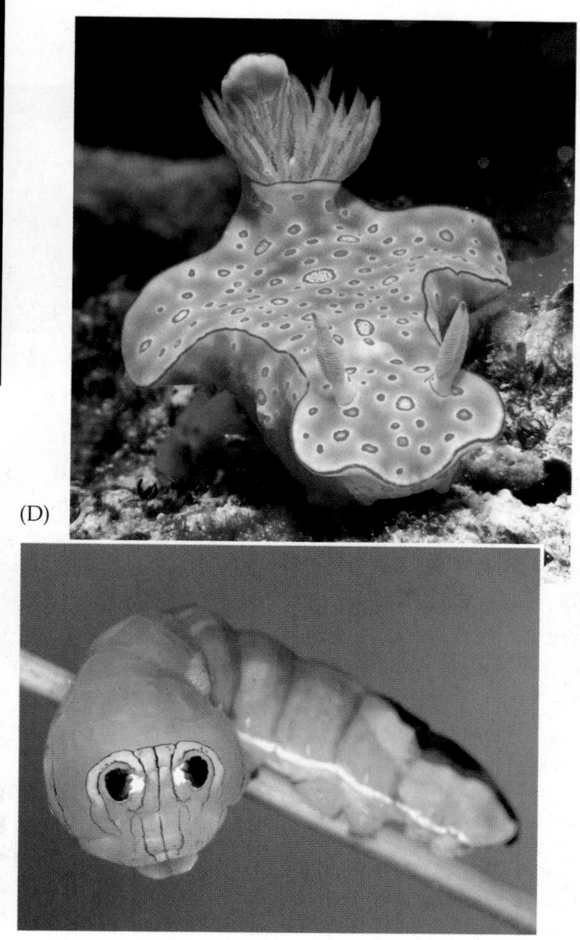

(C)

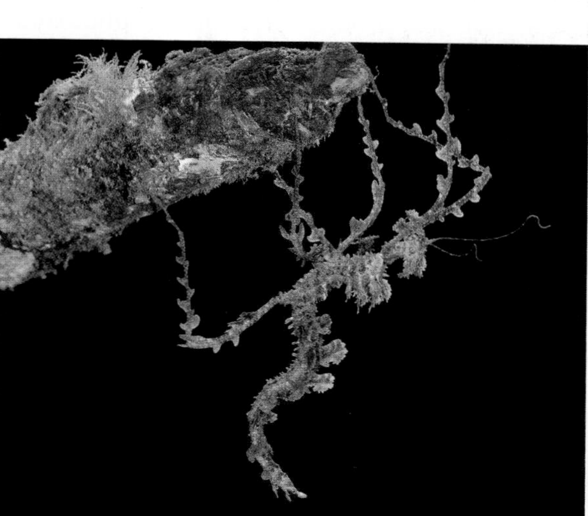

(E)

FIGURE 12.8 Adaptations to Escape Being Eaten Prey have evolved a wide range of mechanisms to escape from predators, including (A) physical features, such as the hard "spines" of sea urchins (*Echinometra* sp.); (B) toxins, advertised by bright warning colors such as those of this nudibranch (*Ceratosoma trilobatum*); (C) crypsis or camouflage, as in this stick insect (family Bacillidae), which blends in with moss and lichen in a forest; (D) mimicry, as in this harmless caterpillar that resembles the head of a snake; and (E) behavioral defenses, as in these muskoxen (*Ovibos moschatus*) forming a defensive circle.

coloration can itself provide protection from predators, which may instinctually avoid or learn from experience not to eat prey that are brightly colored.

Other prey species use **mimicry** as a defense: by resembling less palatable organisms or physical features of their environment, they cause potential predators to mistake them for something less desirable to eat. There are many forms of mimicry. Some species have a shape or coloration that provides camouflage, allowing them to avoid detection by predators; this form of mimicry is called **crypsis** (from *cryptic*, "hidden"). Examples include a stick insect that blends in with moss and lichens in a forest (**Figure 12.8C**), a flounder that is virtually indistinguishable from the ocean bottom on which it rests, and an octopus that changes its coloration rapidly to match its background. Other prey species use mimicry as a form of "false advertising": their shape and coloration mimic that of a species that is fierce or that contains a potent toxin. Examples include harmless flies that resemble yellow jacket wasps and caterpillars that resemble the head of a snake (**Figure 12.8D**).

Finally, some species change their behavior to avoid predation. When predators are abundant, snowshoe hares forage less in open areas (where they are most vulnerable to attack). When threatened, muskoxen form a defensive

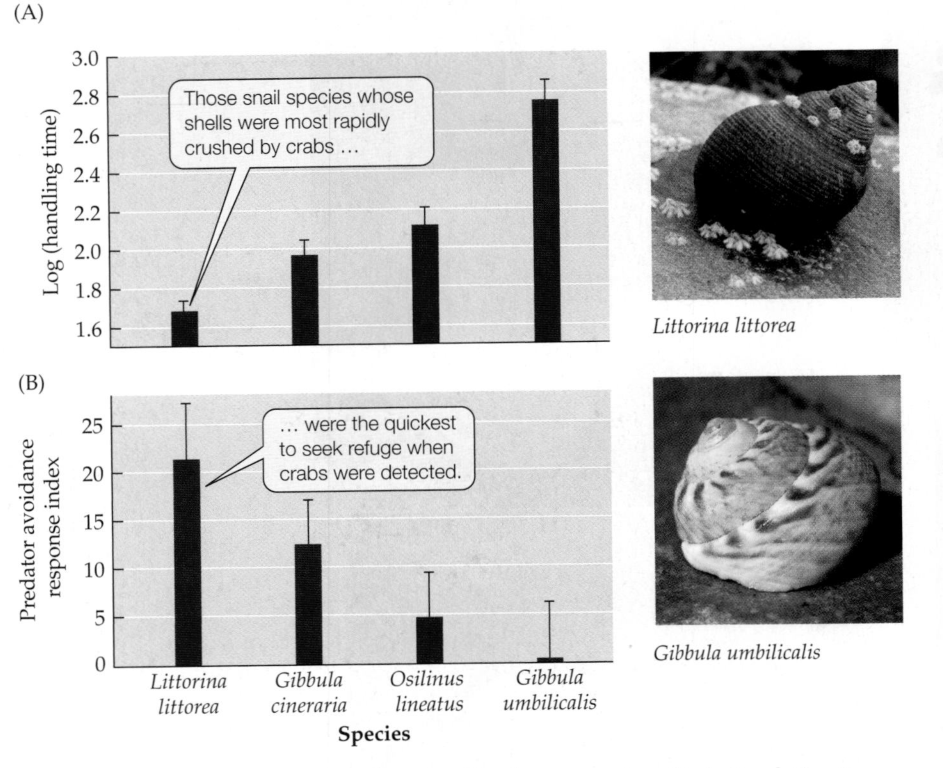

FIGURE 12.9 A Trade-Off in Snail Defenses against Crab Predation? (A) Time taken by shore crabs (*Carcinus maenas*) to manipulate and crush the shells of each of four snail species. (B) Index of the strength of the predator avoidance response of each of four snail species; larger values indicate a more rapid behavioral response to crabs. Error bars show one SE of the mean. (After Cotton et al. 2004.)

circle, making them a difficult target (**Figure 12.8E**). In some cases, there may be a trade-off between physical and behavioral defenses. For example, among four species of marine snails eaten by the shore crab (*Carcinus maenas*), the species whose shells could be crushed most rapidly by crabs were the quickest to take refuge when crabs were detected, and vice versa (**Figure 12.9**). The exact negative correlation between these two defensive traits suggests that there may be a trade-off between a snail's physical and behavioral defenses.

Reducing herbivory: Avoidance, tolerance, and defenses

Some plants avoid herbivory by producing great numbers of seeds in some years and few or no seeds in other years, a phenomenon known as *masting*. Up to a hundred years may pass between bouts of seed production, as in the mass flowering of bamboos in China. Masting allows plants to hide (in time) from seed-eating herbivores, then overwhelm them by sheer numbers. Plants can also avoid herbivores in other ways, such as by producing leaves at times of the year when herbivores are scarce.

Other plants have adaptive growth responses that allow them to compensate for, and hence tolerate, the effects of herbivory—at least up to a point. **Compensation** occurs when removal of plant tissues stimulates a plant to produce new tissues, allowing for relatively rapid replacement of the material eaten by herbivores; when *full compensation* occurs, herbivory causes no net loss of plant tissue. Compensation may occur, for example, when removal of leaf tissue decreases self-shading, resulting in increased plant growth, or when removal of apical buds (those at the end of a branch or shoot) allows lower buds to open and grow. Beech trees respond to simulated herbivory (clipping) by increasing both their leaf production and their photosynthetic rate. Similarly, moderate to high levels of herbivory may benefit field gentians (*Gentianella campestris*) under some circumstances (**Figure 12.10**). In this case, the timing of herbivory is critical: early in the growing season (up to July 20), the plant more than fully compensates for the lost tissue, but later in the season (July 28), it does not. If the amount of material removed from a field gentian—or any other plant—is large enough, however, or if insufficient resources are available for growth, the plant cannot fully compensate for the damage.

Finally, plants use an enormous array of structural and chemical defenses to ward off herbivores (Pellmyr et al. 2002; Agrawal and Fishbein 2006). A stroll through many plant communities makes this readily apparent: the leaves of many plants are tough, and many plant bodies are covered with spines, thorns, sawlike edges, or pernicious (nearly invisible) hairs that can pierce the skin like miniature porcupine quills. In some cases, such structures are an **induced defense** (stimulated by herbivore attack), as illustrated by individual cacti that increase their production of spines only after they have been grazed (Myers and Bazely 1991).

Plants also produce a wide variety of chemicals, called **secondary compounds**, that function to reduce herbivory. Some secondary compounds are toxic, protecting the plant from all but the relatively small number of herbivore species that can tolerate them. Others serve as chemical cues that attract predators or parasitoids to the plant, where they attack herbivores (Schnee et al. 2006). Some plant species produce secondary compounds constantly, regardless of whether herbivores have attacked the plant. In other species, the production of secondary compounds is an induced defense.

Consider the native tobacco *Nicotiana attenuata*, an annual plant species found in the western United States. When attacked by herbivores, *N. attenuata* produces two induced defenses: toxic secondary compounds that deter herbivores directly, and compounds that deter herbivores indirectly by attracting predators and parasitoids. Acting together, these defenses are very effective in reducing losses

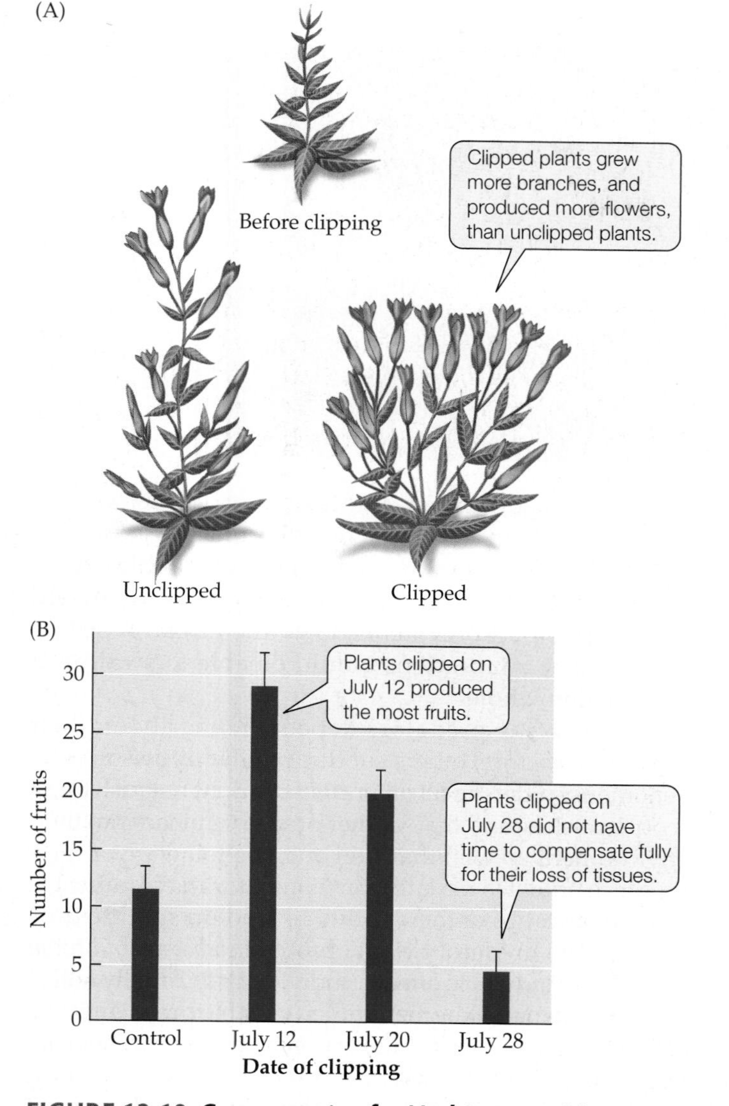

FIGURE 12.10 Compensating for Herbivory Field gentians (*Gentianella campestris*) were clipped at different times during the growing season to simulate herbivory. (A) The shape and production of flowers in unclipped (control) and clipped plants. (B) Numbers of fruits produced by control plants and plants clipped on different dates. Error bars show one SE of the mean. (After Lennartsson et al. 1998.)

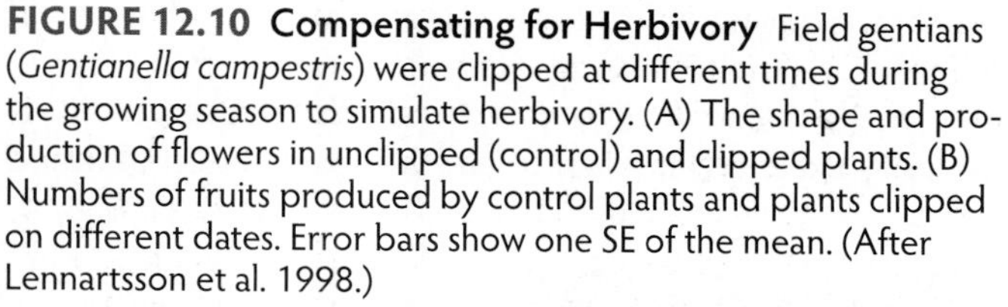

Q How many fruits would you expect to be produced by a plant that compensates fully for clipping? Explain your reasoning.

of tissue to herbivores. In one experiment, the application of compounds that are normally induced by herbivory to the stems of *N. attenuata* caused the numbers of a leaf-feeding herbivore on the plants to drop by more than 90% (Kessler and Baldwin 2001).

To evaluate the genetic basis for this phenomenon, André Kessler and colleagues (2004) used a gene silencing approach. They developed three *N. attenuata* varieties in each of which one of three genes (*LOX3*, *HPL*, or *AOS*) was disabled. Each of these genes is part of a chemical pathway thought to control the induction of both direct defenses (toxins) and indirect defenses (predator and parasitoid attractants) against herbivore attack. When grown in the field, the *not-LOX3* variety (in which the *LOX3* gene was silenced) suffered more damage from herbivores than control plants (in which no genes were silenced) or the other two experimental varieties (**Figure 12.11**). Kessler et al. also found that a greater number of herbivore species could feed on the *not-LOX3* plants than could feed on any of the other three varieties. These results showed that changes in a single gene can alter both the intensity of herbivory and the community of herbivores that attack a plant. More generally, this study revealed the power of combining molecular genetic techniques with ecological field experiments: by doing so, the effects of particular genes could be examined in a natural setting.

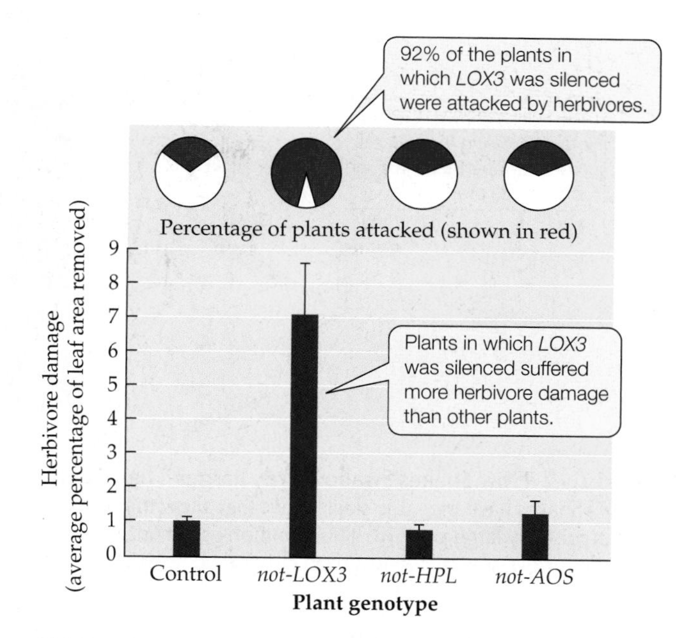

FIGURE 12.11 Herbivores Damage Plants Lacking an Induced-Defense Gene Herbivore damage was measured in wild *Nicotiana attenuata* (tobacco) plants (controls) and in three experimental varieties of *N. attenuata*. In each experimental variety, one of three genes suspected to be involved in the plant's induced defenses (*LOX3*, *HPL*, or *AOS*) was silenced; these genotypes are referred to as *not-LOX3*, *not-HPL*, and *not-AOS*. The bar graphs show herbivore damage (average percentage of leaf area removed); error bars show one SE of the mean. (After Kessler et al. 2004.)

Predators and herbivores have adaptations to overcome the defenses of their food organisms

As we have just seen, predators and herbivores exert strong selection pressure on their prey and food plants, which can lead to improvements in the defenses of those organisms. Those improvements, in turn, impose strong selection pressure on predators and herbivores—those organisms must eat, and a well-defended prey or plant species could provide an abundant food source for any species able to overcome its defenses. Evidence of increases in the ability of predators and herbivores to counter the defenses of their

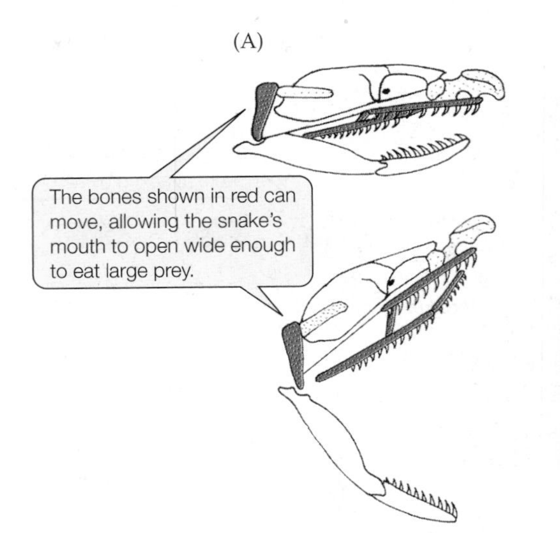

FIGURE 12.12 How Snakes Swallow Prey Larger Than Their Heads (A) Snakes have movable skull bones that allow them to swallow surprisingly large prey. (B) This common garter snake (*Thamnophis sirtalis*) is swallowing a toad larger than its head.

food organisms can be found throughout the history of life on Earth; here we consider just a few specific examples.

Prey-capture adaptations: Physical features, toxins, mimicry, and detoxification For any given defensive mechanism used by prey, there is usually a predator that has a countervailing offensive mechanism. Cryptic prey, for example, may be detected by a predator that uses its sense of smell or touch.

Many predators have unusual physical features that help them capture prey. The body form of the cheetah, for example, enables great bursts of speed that allow it to catch rapidly fleeing gazelles. In another example, most snakes can swallow prey that are considerably larger than their heads (**Figure 12.12**). Unlike those of other terrestrial vertebrates, the bones of a snake's skull are not rigidly attached to one another. This unique feature allows the snake to open its jaws to a seemingly impossible extent. Curved teeth mounted on bones that can move inward then help to pull prey items down the throat. A person with similar adaptations would be able to swallow a watermelon whole.

While some predators depend primarily on their physical structure, others subdue prey with poison (e.g., venomous spiders). Still others use mimicry: ambush bugs, scorpionfishes, and many other predators blend into their environment so well that prey may be unaware of their presence until it is too late. Some predators have inducible traits that improve their ability to feed on specific prey species. The predatory ciliate protist *Lembadion bullinum* has such an inducible offense: individuals gradually adjust their size to match the size of the available prey. Thus, if a ciliate is small but the available prey in its environment are large, the ciliate increases in size—and vice versa. Finally, some predators can detoxify or tolerate prey chemical defenses, as the following example shows.

The garter snake (*Thamnophis sirtalis*) (**Figure 12.13**) is the only predator known to eat the toxic rough-skinned

FIGURE 12.13 A Nonvenomous Snake and Its Lethal Prey The garter snake (*Thamnophis sirtalis*) is the only predator known that can eat the highly toxic, rough-skinned newt (*Taricha granulosa*).

newt (*Taricha granulosa*). In some of its populations, the skin of this newt contains large amounts of tetrodotoxin (TTX), an extremely potent neurotoxin. TTX binds to sodium channels in nerve and muscle tissue, thus preventing nerve signal transmission and causing paralysis and death. A single newt can contain enough TTX to kill 25,000 mice—far more than enough to kill a person, as was tragically demonstrated in 1979 when a 29-year-old man died after eating a rough-skinned newt on a dare.

The garter snakes in some populations, however, can eat rough-skinned newts because they can tolerate TTX. These snakes have TTX-resistant sodium channels (Geffeney et al. 2005). Although these garter snakes are protected from the lethal effects of TTX, those individuals that can tolerate the highest concentrations of TTX move more slowly than less resistant individuals—a trade-off between tolerance for the poison and speed of locomotion. In addition, once they swallow a poisonous newt, the snakes are immobilized for up to 7 hours. During that time, the snakes are vulnerable to predation themselves and may also suffer from heat stress.

Overcoming plant defenses: Structural, chemical, and behavioral adaptations The defenses used by plants prevent most herbivores from eating most plants. But for any given plant species, there are some herbivores that can cope with its defensive mechanisms. A plant covered with spines may be attacked by an herbivore that can avoid or tolerate those spines. Many herbivores have evolved digestive enzymes that enable them to disarm or tolerate plant chemical defenses. Such herbivores may gain a considerable advantage: they can eat plants that other herbivores cannot, and thereby have access to an abundant food resource.

Finally, some herbivores use behavioral responses to circumvent an otherwise effective plant defense. For example, some beetles use a behavioral response to cope with the defenses of tropical plants in the genus *Bursera*. These plants combine the production of toxic secondary compounds with a high-pressure delivery system: they store a toxic, sticky resin in a network of canals that runs through their leaves and stems (**Figure 12.14**). If an insect herbivore chews through one of these canals, the resin squirts from the plant under high pressure and may repel or even kill the insect (the resin hardens after it is exposed to air, so if an insect is drenched in resin, it can be entombed). Yet some tropical beetles in the genus *Blepharida* have evolved an effective counterdefense (Becerra 2003). Their larvae chew slowly through the leaf veins where the resin canals are located, releasing the pressure so gradually that the resin does not squirt from the plant. It often takes a beetle larva more than an hour to "disarm" a leaf in this manner; once that job is done, the larva eats the leaf in 10–20 minutes.

FIGURE 12.14 Plant Defense and Herbivore Counter-Defense Some plants in the genus *Bursera* store toxic resin under high pressure in leaf canals. (A) When herbivores eat the leaves, they chew through these canals, causing the resin to be squirted up to 2 m from the leaf. (B) The larvae of some beetles in the genus *Blepharida* can disable this defense by chewing slowly through the canals, releasing the pressure in a gradual and harmless way.

CONCEPT 12.3 Predation and herbivory can affect ecological communities greatly, in some cases causing a shift from one community type to another.

Effects of Exploitation on Communities

A general theme that runs through this book is that ecological interactions affect the distributions and abundances of the interacting species. In turn, such effects on populations can alter ecological communities. The community-level consequences of predation and herbivory can be profound, in some cases causing major shifts in the types of organisms found at a given location.

Exploitation can reduce the distributions and abundances of food organisms

All exploitative interactions have the potential to reduce the growth, survival, or reproduction of the organisms that are eaten. These effects can be dramatic, as demonstrated in the case of a leaf-feeding beetle (*Chrysolina quadrigemina*) that rapidly reduced the density of Klamath weed, an invasive plant that is poisonous to livestock (**Figure 12.15**). Predators and parasitoids can also have dramatic effects when they are introduced as biological pest controls. In six cases, introductions of wasps that preyed on crop-eating insects decreased the herbivores' densities by more than 95%, thus greatly reducing the economic damage caused by those pests.

The distributions and abundances of organisms can also be altered by the effects of predation or herbivory on other ecological interactions. As we've seen, predators and herbivores can change the outcome of competition (see Figure 11.15), thereby affecting the distributions or abundances of competitor species. In particular, inferior competitors may increase in abundance when they are in the presence of a predator or herbivore that decreases the abundance or performance of a dominant competitor. Paine (1974) found such a result: he showed that the removal of a sea star predator (*Pisaster*) led to the local extinction of all large invertebrates but one, a mussel. The mussel was a dominant competitor that, in the absence of the sea star, drove all the other large invertebrates to extinction.

We turn now to two other examples of how exploitation can affect the distributions and abundances of food organisms.

Effects of predators *Anolis* lizards are predators that eat a broad range of prey species, including spiders. Thomas Schoener and David Spiller studied the effects of lizard predators on their spider prey in the Bahamas. They selected twelve small islands and divided them into four groups of three islands each that were similar in size and vegetation. Initially, each group of three islands contained one island with lizards and two without. One of the latter two islands was then chosen at random to have two male and three female adult *Anolis sagrei* lizards introduced to it; the other island was left as a control where lizards were absent naturally.

The introduced lizards greatly reduced the distributions and abundances of their spider prey (Schoener and Spiller 1996). Before the experiment began, the numbers of spider species and the overall densities of spiders were similar among the eight islands that lacked lizards. By the end of the experiment, however, the introduction of lizards to four islands had reduced the numbers and densities of spider species to the levels found on the four islands where lizards were present naturally. The proportion of spider species that went extinct was nearly 13 times higher on islands where lizards were introduced than on islands without lizards (**Figure 12.16**). Similarly, the density of spiders was about 6 times higher on islands without lizards than on islands that had lizards (either naturally or experimentally). The introduction of lizards reduced the densities of both common and rare spider species, and most of the rare species went extinct. Similar experimental results have been obtained for beetles eaten by rodents and grasshoppers eaten by birds.

Effects of herbivores Lesser snow geese (*Chen caerulescens*) migrate from their overwintering grounds in the United States to breed in salt marshes that border Canada's Hudson

(A)

(B)

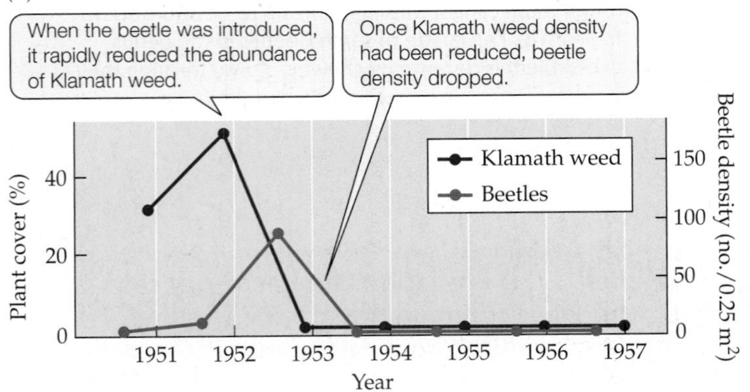

FIGURE 12.15 A Beetle Controls a Noxious Rangeland Weed Klamath weed (*Hypericum perforatum*), which poisons cattle, once covered about 4 million acres of rangeland in the western United States. (A) This photograph, taken in 1949, shows a field completely covered with Klamath weed at the flowering stage. (B) The leaf-feeding beetle *Chrysolina quadrigemina* was introduced in 1951 in the hope of controlling Klamath weed. This graph tracks densities of beetles and of Klamath weed (as a percentage of plant cover) in ¼ m^2 plots after the beetle's introduction. (After Huffaker and Kennett 1957.)

Q Explain how a plant community might change after *Chrysolina* reduced the density of Klamath weed.

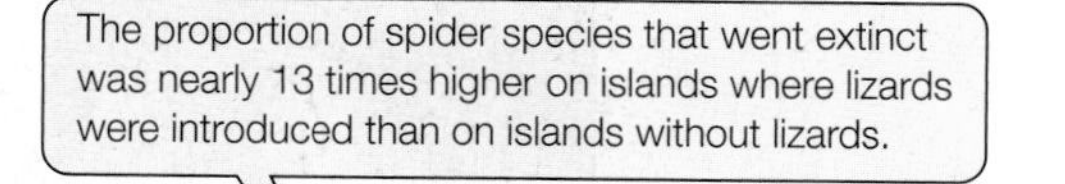

FIGURE 12.16 Lizard Predators Can Drive Their Spider Prey to Extinction The experimental introduction of lizards to small islands in the Bahamas greatly increased the rate at which their spider prey became extinct; error bars show one SE of the mean. The photograph shows Thomas Schoener on one of the study islands. (After Schoener and Spiller 1996.)

Bay. During the summer, the geese graze on marsh grasses and sedges. Historically, although the geese removed considerable plant matter, their presence benefited the marshes. Nitrogen, a limiting resource for plant growth, is scarce in salt marshes. As they eat, the geese defecate every few minutes, thereby adding nitrogen to the soil (nitrogen moves into the soil from goose feces more rapidly than it does from the decomposing leaves of marsh plants). The plants absorb the added nitrogen, which allows them to grow rapidly after being grazed. Overall, low to intermediate levels of grazing by geese lead to increased plant growth (Jefferies et al. 2003). For example, net primary productivity (NPP, measured as the annual amount of new aboveground plant growth) was higher in lightly grazed plots than in ungrazed plots (**Figure 12.17A**).

About 40 years ago, however, the situation described in the previous paragraph started to change. Beginning around 1970, lesser snow goose densities increased exponentially. This increase probably occurred because increased crop production near their overwintering sites provided the geese with a superabundant supply of food. The ensuing high densities of geese no longer benefited marsh plants. The geese completely removed the vegetation, drastically changing the distributions and abundances of marsh plant species (**Figure 12.17B**). Of an original 54,800 ha (135,400 acres) of intertidal marsh in the Hudson Bay region, geese are estimated to have destroyed 35% (19,200 hectares or 47,400 acres). An additional 30% (16,400 ha or 40,500 acres) of the original marsh has been badly damaged by the geese. Controlled hunts (from 1999 on) have slowed goose population growth; this strategy may eventually lead to marsh recovery.

Exploitation can alter ecological communities

Schoener and Spiller's work on the effects of lizard predators on spiders shows that the direct effects of a predator

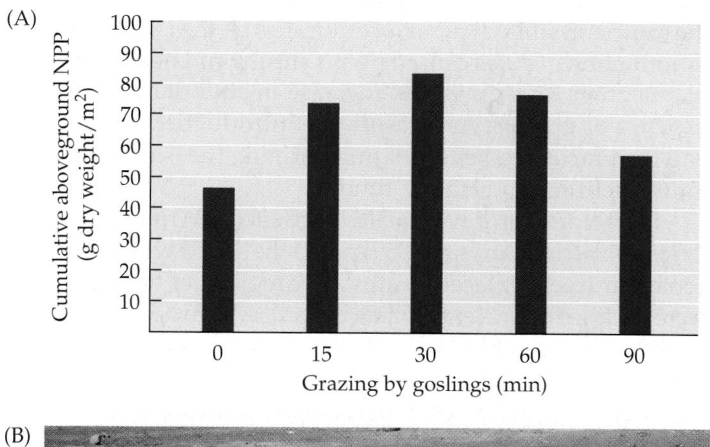

(B)

FIGURE 12.17 Snow Geese Can Benefit or Decimate Marshes (A) When lightly grazed (for a single 15–90-minute episode) by snow goose goslings, salt marsh plants increased their cumulative production of new biomass over a 60-day period. (B) Adult geese, however, can remove all plant matter from a square meter in an hour. Heavy grazing by high densities of snow geese can convert salt marshes to mudflats, as seen by comparing this small remnant of marsh (protected from geese) with the surrounding mudflat (a former marsh that was grazed heavily by geese). (After Hik and Jefferies 1990.)

can greatly reduce the diversity (i.e., the number) of prey species in a community. In other cases, a predator that suppresses a dominant competitor can (indirectly) cause the diversity of species in a community to increase (as in the sea star and mussel example). Indirect effects of predators can also alter ecological communities by affecting the transfer of nutrients from one ecosystem to another, as the following study on arctic foxes illustrates.

In the late nineteenth and early twentieth centuries, humans introduced arctic foxes (*Alopex lagopus*) to some of the Aleutian Islands off the coast of Alaska. Other islands remained fox-free, either because foxes were never introduced there or because the introductions failed. Taking advantage of this inadvertent large-scale experiment, Croll et al. (2005) determined that, on average, the introduction of foxes to an island reduced the density of breeding seabird populations by nearly a hundredfold. The decrease in seabird numbers, in turn, reduced the input of guano (bird feces) to an island from roughly 362 to 6 grams per square meter. Seabird guano, which is rich in phosphorus and nitrogen, transfers nutrients from the ocean (where seabirds feed) to the land. By reducing the amount of guano that fertilized the (nutrient-limited) plant communities on the islands, the introduction of foxes caused dwarf shrubs and herbaceous plants other than grasses to increase in abundance at the expense of grasses. As a result, the introduction of foxes had the unexpected effect of transforming the island communities from grassland to tundra.

Herbivores can have equally large effects. Writing in *The Origin of Species*, Darwin (1859) noted the speed with which Scotch fir trees replaced heaths after regions of heathland were enclosed to prevent grazing by cattle. When he observed heathlands grazed by cattle, "on looking closely between the stems of the heath, I found a multitude of seedlings and little trees, which had been perpetually browsed down by the cattle. In one square yard … I counted thirty-two little trees; and one of them, judging from the rings of growth, had during 26 years tried to raise its head above the stems of the heath, and had failed." Darwin concluded that seeds dispersed from trees located at the edge of the heath would germinate and overgrow the heath if not for grazing by cattle. Thus, the very existence of the heath community in that area depended on herbivory.

Herbivores can also have pronounced effects in aquatic environments. The golden apple snail was introduced into Taiwan from South America in 1980 for local consumption and export. The snail escaped from cultivation and spread rapidly through Southeast Asia (**Figure 12.18**). Its spread caught the attention of researchers and government officials because it proved to be a serious pest of rice. The snail has also been found in Hawaii, the southern United States, and Australia and is expected to reach Bangladesh and India (Carlsson et al. 2004).

Most freshwater snails eat algae, but the golden apple snail prefers to eat aquatic plants, including those that float on the water surface and those that attach themselves to the bottom. However, as mentioned on p. 266, golden apple snails are generalists, and if plants are not available, they can survive on algae and detritus. As a result, these snails are resilient and hard to get rid of.

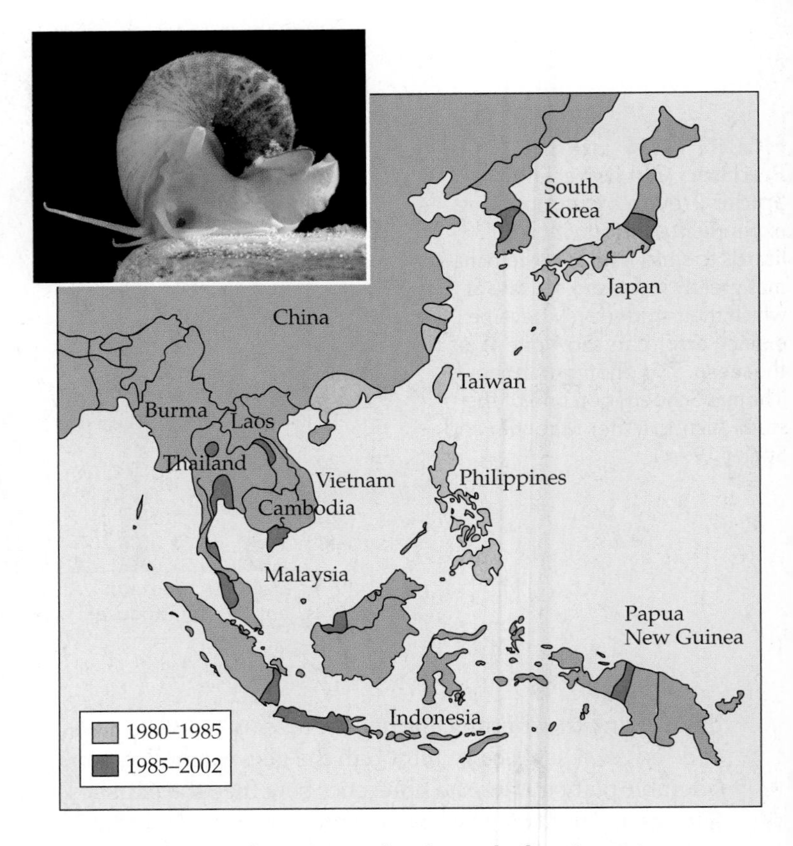

FIGURE 12.18 The Geographic Spread of an Aquatic Herbivore Since its introduction to Taiwan in 1980, the golden apple snail (*Pomacea canaliculata*) has spread rapidly across parts of Southeast Asia, threatening rice crops and native plant species. The map shows the regions the snail had occupied by 1985 and by 2002.

As a first step toward assessing how the snail had affected natural communities, Nils Carlsson and colleagues surveyed 14 wetlands in Thailand with varying densities of snails. They found that wetland communities with high densities of snails were characterized by few plants, high nutrient concentrations in the water, and a high biomass of algae and other phytoplankton (**Figure 12.19**).

To test whether the trends observed in their survey could have been caused by the snail, Carlsson et al. (2004) placed 24 1 × 1 × 1 m enclosures in a wetland in which snail densities were low. To each enclosure, they added about 420 grams of water hyacinth (*Eichhornia crassipes*), one of the most abundant plant species in many Southeast Asian wetlands. Next, they added 0, 2, 4, or 6 snails to the enclosures; there were six replicates of each of the four snail density treatments. Carlsson and colleagues then measured the effects of the snails on plant biomass and phytoplankton biomass. Water hyacinth biomass increased in the enclosures where no snails were present, but decreased in all the other enclosures. At the highest snail density tested (6 snails/m^2), phytoplankton biomass increased.

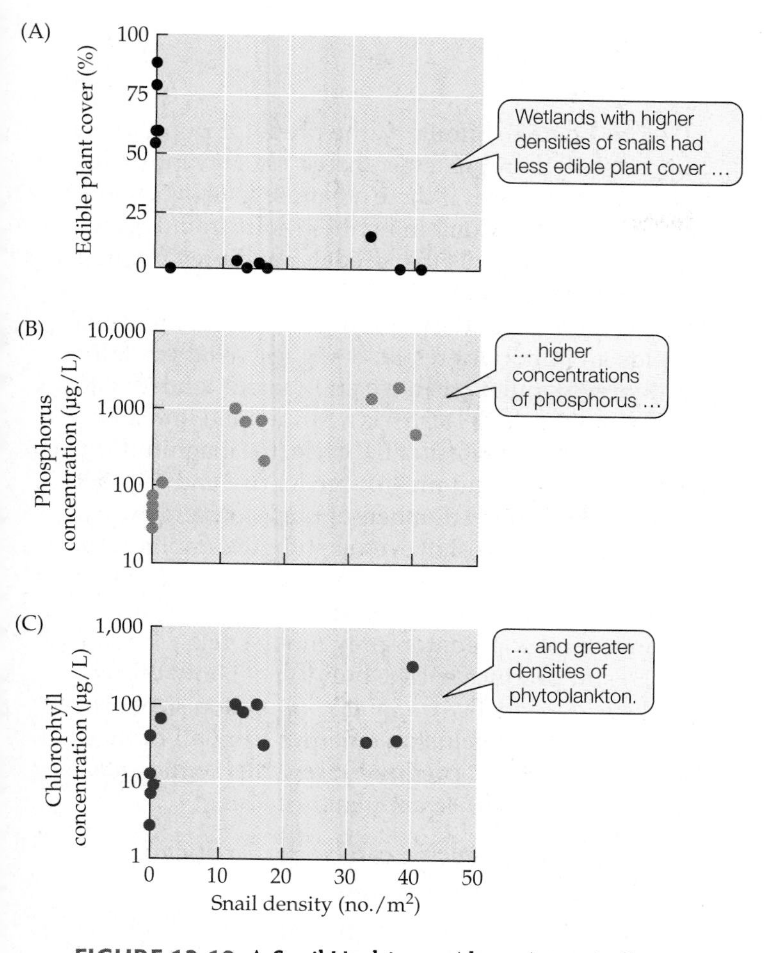

FIGURE 12.19 A Snail Herbivore Alters Aquatic Communities Nils Carlsson and colleagues measured several characteristics of 14 natural wetlands in Thailand that differed in their densities of golden apple snails (*Pomacea canaliculata*). (A) Percentage of the wetlands covered by edible plant species. (B) Concentrations of phosphorus in the water. (C) Chlorophyll concentrations (an indicator of phytoplankton biomass). Note the log scale for parts (B) and (C). Experiments conducted separately indicated that all of the trends shown here could have been caused by the snail. (After Carlsson et al. 2004.)

Q In (B), compare the average total phosphorus concentration in wetlands without snails with that in wetlands with snails.

The results of the survey and the experiment concur in suggesting that the golden apple snail can have an enormous effect on wetland communities, causing a complete shift from a wetland with clear water and many plants to a wetland with turbid water, few plants, high nutrient concentrations, and high phytoplankton biomass. It is likely that this shift occurs because the snails suppress plants directly (by eating them) and because they release the nutrients they obtain from the plants into the water, thus providing improved growth conditions for algae and other phytoplankton.

In this section, we have seen how exploitation can alter the distributions and abundances of organisms, in some cases resulting in major changes to ecological communities. We turn next to a more specific effect of exploitation: population cycles.

CONCEPT 12.4 Population cycles can be caused by exploitative interactions.

Exploitation and Population Cycles

We introduced population cycles in Chapter 10 (see Figure 10.9), and in the Case Study at the opening of this chapter (p. 262), we described the most famous one of all, the hare–lynx cycle. Cyclic fluctuations in abundance are one of the most intriguing patterns in nature. After all, what could cause populations to change so considerably in size over time, yet in such a regular manner? We will return to the mechanisms that underlie the hare–lynx cycle on p. 278, but first we'll describe some insights into the causes of population cycles that have come from models, experiments, and field observations of predator–prey interactions.

Predator–prey cycles can be modeled mathematically

One way to evaluate possible causes of population cycles is to investigate the issue mathematically. In the 1920s, Alfred Lotka and Vito Volterra independently represented the dynamics of predator–prey interactions with what is now called the Lotka–Volterra predator–prey model:

$$\frac{dN}{dt} = rN - aNP$$

$$\frac{dP}{dt} = baNP - mP \qquad \textbf{(12.1)}$$

In these equations, N represents the number of prey individuals and P represents the number of predator individuals. The equation for change in the prey population over time (dN/dt) assumes that when predators are absent ($P = 0$), the prey population grows exponentially (i.e., $dN/dt = rN$, where r is the population growth rate). When predators are present ($P \neq 0$), the rate at which they kill prey depends in part on how frequently predators and prey encounter one another. This frequency is expected to increase with the number of prey (N) and with the number of predators (P), so a multiplicative term (NP) is used in the equation for dN/dt. The rate at which predators kill prey also depends on the efficiency with which predators capture prey; this capture efficiency is represented by the constant a, so that the overall rate at which predators remove individuals from the prey population is aNP.

Predators starve when there are no prey. Thus, the equation for change in the predator population over time (dP/dt) assumes that in the absence of prey ($N = 0$), the number of predators decreases exponentially with a mortality rate of m (i.e., $dP/dt = -mP$). When prey are present ($N \neq 0$), individuals are added to the predator population according to the number of prey that are killed (aNP) and the efficiency

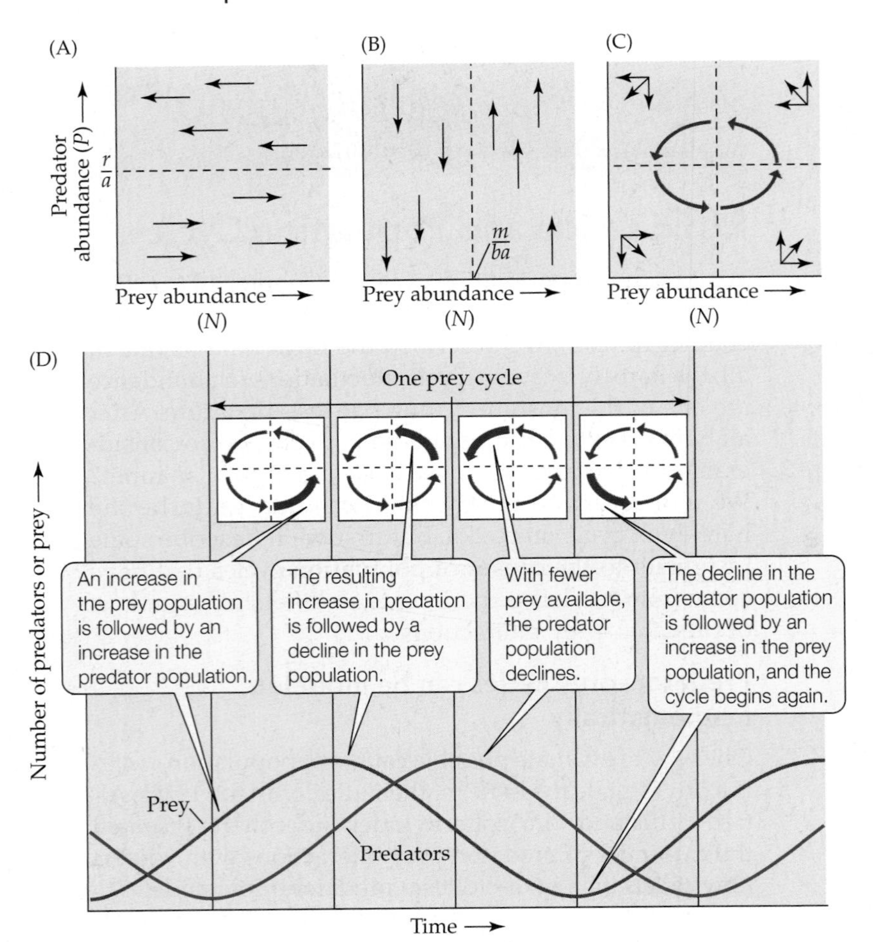

FIGURE 12.20 The Lotka–Volterra Predator–Prey Model Produces Population Cycles (A) Considering the prey population first, the abundance of prey does not change when $dN/dt = 0$, which occurs when $P = r/a$ (see Equation 12.1). (B) Similarly, considering the predator population, the abundance of predators does not change when $dP/dt = 0$, which occurs when $N = m/ba$. Combining the results in parts (A) and (B) suggests that predator and prey populations have an inherent tendency to cycle. These cycles are shown here in two ways: (C) by plotting the abundance of predators versus the abundance of prey, and (D) by plotting the abundance of both predators and prey versus time; the four inset diagrams in (D) plot the abundance of predators vs. the abundance of prey. In (D), note that the predator abundance curve is shifted ¼ of a cycle behind the prey abundance curve.

with which prey are converted into predator offspring (represented by the constant b). Thus, the rate at which individuals are added to the predator population is $baNP$.

What are the implications of the relationships represented by Equation 12.1? As with the Lotka–Volterra competition model (see Equation 11.1), we can answer this question by using zero population growth isoclines to determine what happens to predator and prey populations over long periods (to review how this is done, see Ecological Toolkit 11.1). We find that the prey population decreases if the number of predators is greater than r/a, whereas it increases if there are fewer than r/a predators (**Figure 12.20A**). Similarly, the predator population decreases if there are fewer than m/ba prey individuals, and it increases if there are more than m/ba prey (**Figure 12.20B**). Combining the results from Figures 12.20A and B reveals that predator and prey populations tend to cycle (**Figure 12.20C,D**).

The Lotka–Volterra predator–prey model thus yields an important result: it suggests that predator and prey populations have an inherent tendency to cycle. But the model also has a curious and unrealistic property: the *amplitude* of the cycle (the magnitude by which predator and prey numbers rise and fall) depends on the initial numbers of predators and prey. If the initial numbers shift even slightly, the amplitude of the cycle will change (see **Web Extension 12.1** to find out what features of the model cause this change). More complex predator–prey models (e.g., Turchin 2003) still produce cycles, but do not show this unrealistic dependence on initial population sizes. The same general conclusion emerges from all of these models, however: predator–prey interactions have the potential to cause population cycles.

Predator–prey cycles can be reproduced under laboratory conditions

Can the cycling behavior of predator–prey models be reproduced in the laboratory? Experiments show that such cycles can be difficult to achieve. When prey are easy for predators to find, predators typically drive prey to extinction, then go extinct themselves. Such was the case in C. B. Huffaker's experiments with the herbivorous six-spotted mite (*Eotetranychus sexmaculatus*) and a predatory mite (*Typhlodromus occidentalis*) that eats it (Huffaker 1958). In an initial set of experiments, Huffaker released 20 six-spotted mites on a tray with 40 positions, a few of which contained oranges, which these herbivorous mites could eat (**Figure 12.21A**). At first, the six-spotted mite population increased, in some cases reaching densities of 500 mites per orange. Eleven days after the start of the experiment, Huffaker released two predatory mites on the tray. Both prey and predator populations increased for a time, then declined to extinction (**Figure 12.21B**).

Huffaker observed that the prey persisted longer if the oranges were widely spaced—presumably because it took the predators more time to find their prey. He tested this idea in a follow-up experiment in which he increased the complexity of the habitat in the following way. First he added strips of Vaseline that partially blocked the predatory mites as they crawled from one orange to another. Then he placed small wooden posts in an upright position on some of the oranges; these posts allowed the six-spotted mites to take advantage of their ability to spin a silken thread and float on air currents over the Vaseline barriers. Thus, he altered the experimental environment to favor dispersal of

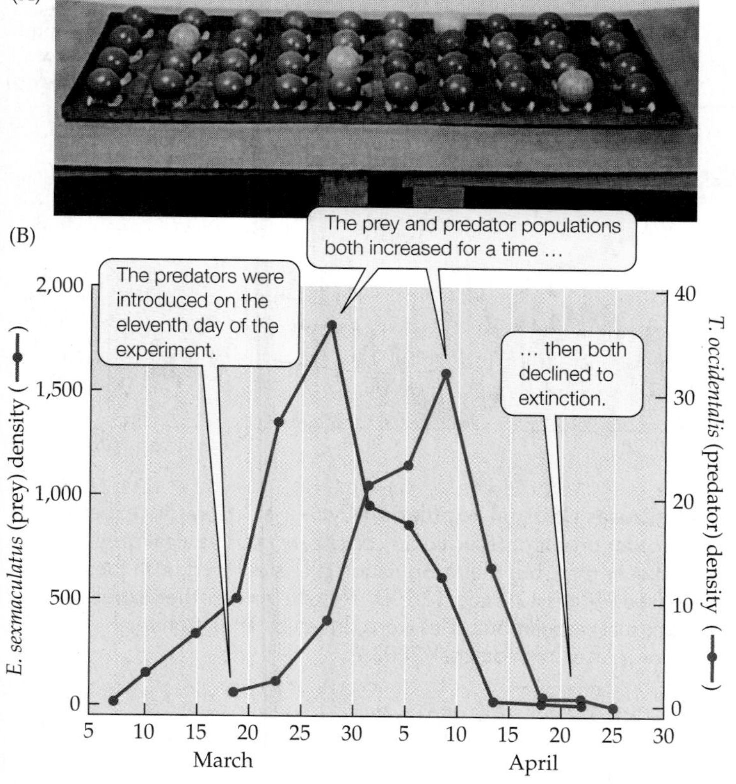

FIGURE 12.21 In a Simple Environment, Predators Drive Prey to Extinction (A) C. B. Huffaker constructed a simple laboratory environment to test for conditions under which predators and prey would coexist and produce population cycles. He placed oranges in a few positions in an experimental tray to provide food for the herbivorous six-spotted mite (*Eotetranychus sexmaculatus*); the remainder of the positions contained inedible rubber balls. (B) When a predatory mite (*Typhlodromus occidentalis*) was introduced into this simple environment, it drove the prey to extinction, causing its own population to go extinct as well. (After Huffaker 1958.)

the six-spotted mite and impede dispersal of the predatory mite. Under these conditions, the prey and the predators both persisted, illustrating a form of "hide-and-seek" dynamics that produced population cycles (**Figure 12.22**). The six-spotted mites dispersed to unoccupied oranges, where their numbers increased. Once the predators found an orange with six-spotted mites, they ate them all, causing both prey and predator numbers on that orange to plummet. In the meantime, however, some six-spotted mites dispersed to other portions of the experimental environment, where they increased in number until they too were discovered by the predators.

Predator–prey cycles can persist in the field

Natural populations of predators and prey can coexist and show dynamics similar to those of Huffaker's mites. Clumps of mussels off the coast of California, for example, can be driven to extinction locally by sea star predators. However, mussel larvae float in ocean currents and hence disperse

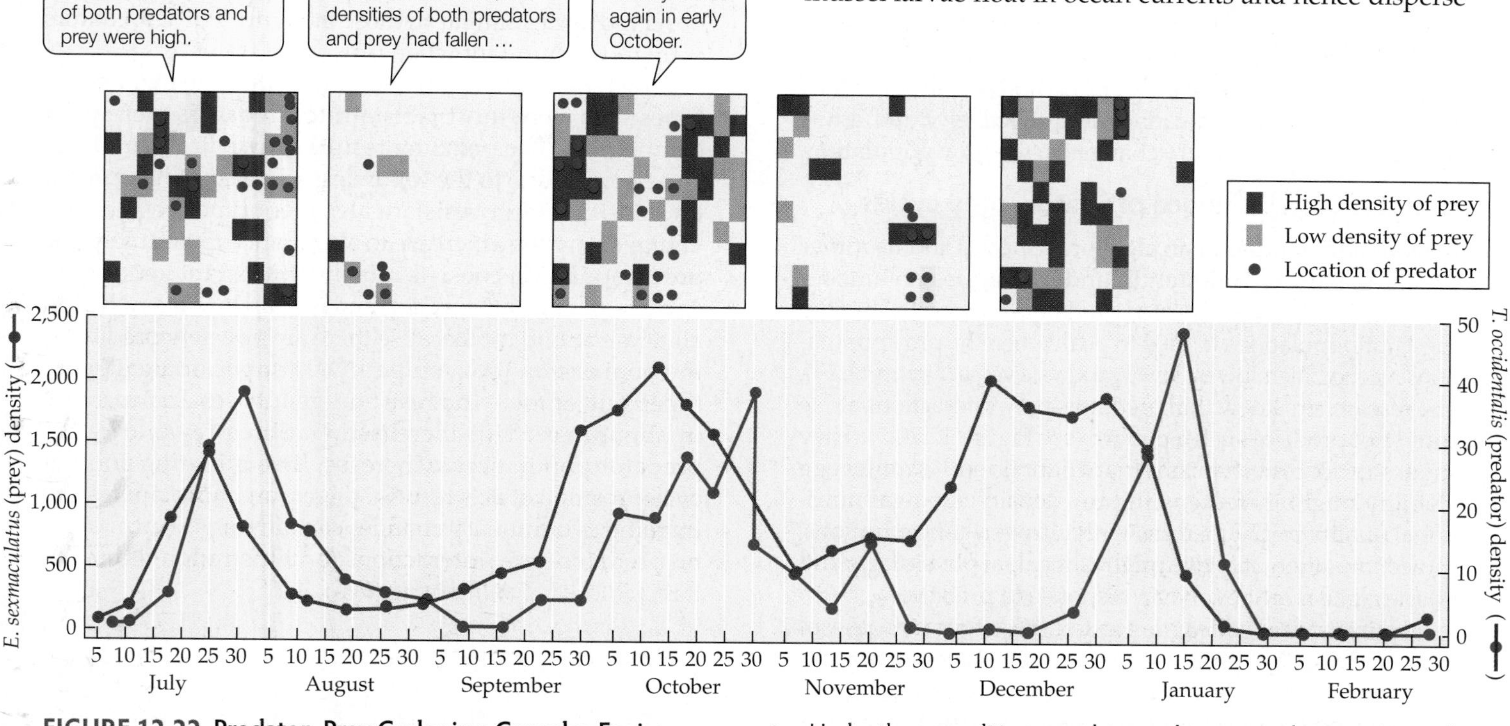

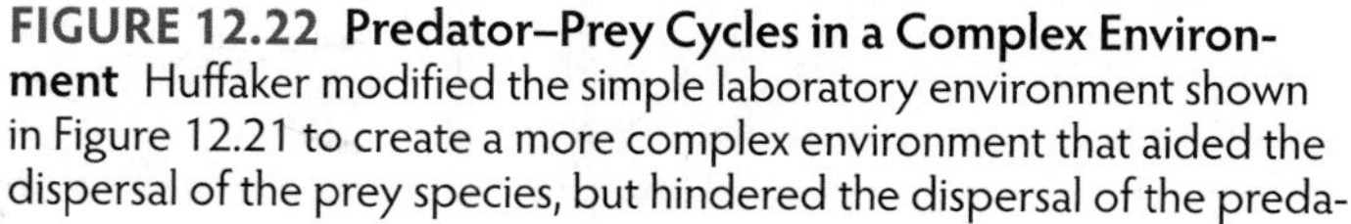

FIGURE 12.22 Predator–Prey Cycles in a Complex Environment Huffaker modified the simple laboratory environment shown in Figure 12.21 to create a more complex environment that aided the dispersal of the prey species, but hindered the dispersal of the predator. Under these conditions, predator and prey populations coexisted, and their abundances cycled over time. The top panels show the locations within the environment of prey (shaded regions) and predators (circles) at five different points in time. (After Huffaker 1958.)

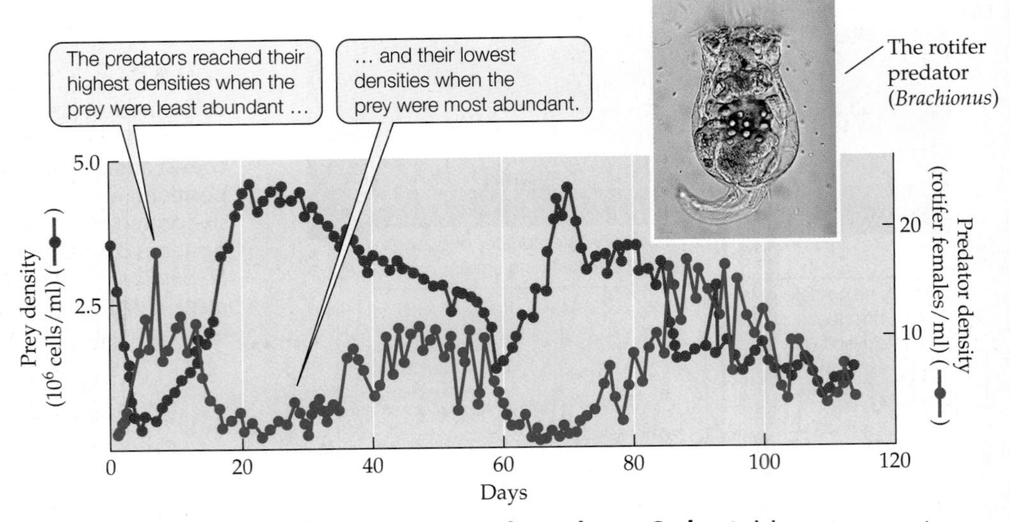

FIGURE 12.23 Evolution Causes Unusual Population Cycles In laboratory experiments, the abundances of a rotifer predator (*Brachionus calyciflorus*) and its algal prey (*Chlorella vulgaris*) fluctuated over time, but their population cycles differed from the typical predator–prey cycle (see Figures 12.2 and 12.20D). Results from further experiments indicated that these unusual population cycles were driven by evolutionary changes in the prey population. (After Yoshida et al. 2003.)

more rapidly than the sea stars. As a result, the mussels continually establish new clumps that flourish until they are discovered by sea stars. Thus, like the six-spotted mites in Huffaker's experiments, the mussels persist because portions of their population escape detection by predators for a time.

Field studies have also shown that predators influence population cycles in species such as southern pine beetles, voles, collared lemmings, snowshoe hares, and moose (Turchin 2003; Gilg et al. 2003). But predation is not the only factor that causes population cycles in these species. The supply of food plants for the herbivorous prey can also play an important role, and in some cases, social interactions are important as well. Thus, reality is not as simple as implied by the results of predator–prey models (in which cycles are maintained purely by predator–prey interactions). In the field, some population cycles may be caused by three-way feeding relationships—by the effects of predators and prey on each other, coupled with the effects of prey and their food plants on each other.

Whether their populations cycle or not, a variety of factors can prevent predators from driving prey to extinction. Such factors include habitat complexity and limited predator dispersal (as in Huffaker's mites), prey switching behavior in predators (see Figure 12.5), spatial refuges (i.e., areas in which predators cannot hunt effectively), and, as we will see next, evolutionary changes in the prey population.

Evolution can influence predator–prey cycles

In laboratory studies of an algal prey species and its rotifer predator, Nelson Hairston Jr. and colleagues obtained a puzzling result: they observed predator–prey cycles, but the predator populations tended to peak when the prey populations reached their lowest levels, and vice versa (**Figure 12.23**). The researchers knew that predator–prey interactions alone could not produce such a pattern (see Figure 12.20D). They suggested four mechanisms that might do so: (1) rotifer egg viability might increase with prey density; (2) algal nutritional quality might increase with nitrogen concentrations; (3) accumulation of toxins might alter algal physiology; and (4) the algae might evolve in response to predation.

Hairston and colleagues tested these four hypotheses in two ways. First, they compared their data with results from four mathematical models (one for each mechanism). Only the model that included evolution in the prey population provided a good match to their data. Second, they performed an experiment in which they manipulated the ability of the prey population to evolve; the idea was to see whether the puzzling results of Figure 12.23 would be duplicated only when the prey population could evolve freely. That is exactly what happened (Yoshida et al. 2003). In treatments in which prey evolution was restricted (because only a single algal genotype was used), they observed typical predator–prey cycles; that is, predator abundance peaked shortly after prey abundance peaked (as in Figure 12.20D). In contrast, when the prey population could evolve freely (because multiple genotypes were used), they observed cycles similar to those in Figure 12.23: predators were most abundant when prey were scarce.

Yoshida et al. (2003) also found that the algal genotypes that were most resistant to predators were poor competitors. The puzzling results shown in Figure 12.23 appear to be due to the following scenario: When predator density is high, resistant algal genotypes have an advantage, and they increase in abundance. Eventually, the prey population consists mostly of resistant genotypes, and predator numbers drop and remain low, even though algae are abundant. Because there are now few predators, the nonresistant but competitively superior algal genotypes outcompete the resistant genotypes and increase in abundance. This increase in edible prey allows the predator population to increase, thus initiating another cycle. Yoshida et al.'s results suggest an important lesson: ongoing evolutionary changes can have a powerful effect on predator–prey interactions (an illustration of maxim 5 in Table 1.1, *Evolution matters*).

A CASE STUDY REVISITED

Snowshoe Hare Cycles

As we saw on pp. 262–263, neither the food supply hypothesis nor the predation hypothesis alone can explain hare population cycles. However, much of the variation in hare densities can be explained when we combine these two hypotheses—and add even more realism with a few new twists.

Charles Krebs and colleagues performed an experiment designed to determine whether food, predation, or their interaction caused population cycles in hares. The sheer scope of the experiment was impressive: the experimental treatments were performed in seven 1 × 1 km blocks of forest located in an isolated region of Canadian wilderness. Three blocks were not manipulated and were used as controls. Food for hares was added to two blocks (the "+Food" treatment). In 1987, an electric fence 4 km in length was constructed to exclude predators from one block of forest (the "–Predators" treatment). In the following year, a second 4 km fence was built; in the block of forest enclosed by this fence, food was added and predators were excluded (the "+Food/–Predators" treatment). The two fences (with a total length of 8 km) had to be monitored daily during the winter, when temperatures could plummet to –45°C (–49°F); this monitoring required so much time that the researchers could not replicate either fenced treatment. The survival rates and densities of hares in each block of forest were observed for 8 years.

Compared with the control blocks, hare densities were considerably higher in the +Food, –Predators, and +Food/–Predators blocks (**Figure 12.24**). The most pronounced effects were seen in the +Food/–Predators block, where, on average, hare densities were 11 times those in the control blocks. The strong effect of jointly adding food and removing predators suggests that hare population cycles are influenced by both food and predation.

This conclusion was supported by results from a mathematical model that examined feeding relationships across three levels: vegetation (the hares' food), hares, and predators (King and Schaffer 2001). Field data were used to estimate the model parameters, and the model's predictions were compared with the actual results for Krebs et al.'s four treatments. Although the match was not exact, there was reasonably good agreement between the model and the results, again suggesting that both food and predators influence hare population cycles (**Figure 12.25**).

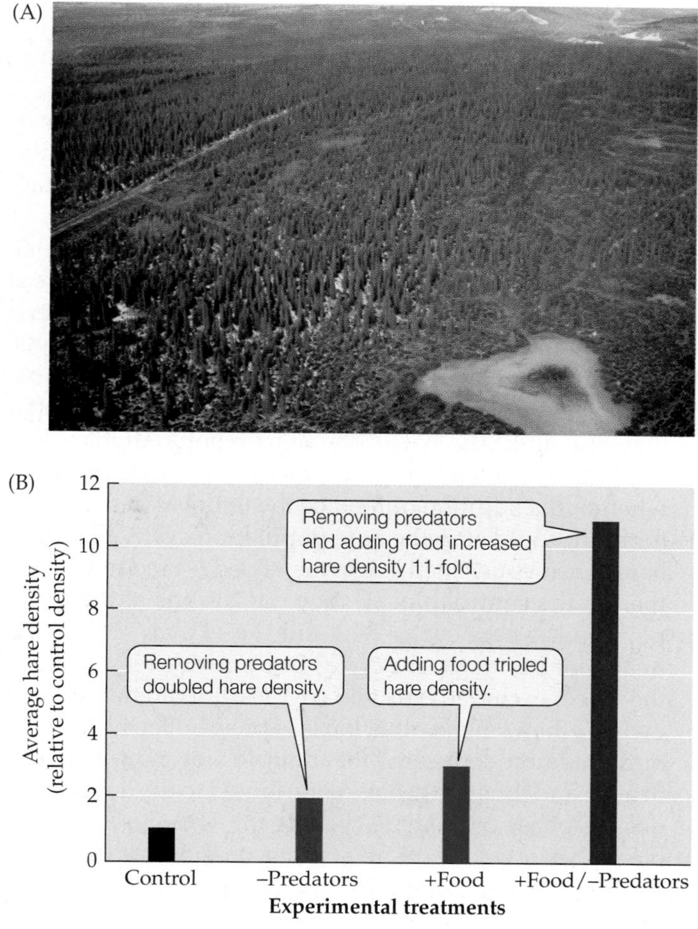

FIGURE 12.24 Both Predators and Food Influence Hare Density (A) This aerial photograph shows one of the 1 km^2 snowshoe hare study sites described in the text. (B) Average hare densities relative to their densities in control blocks of forest. (B after Krebs et al. 1995.)

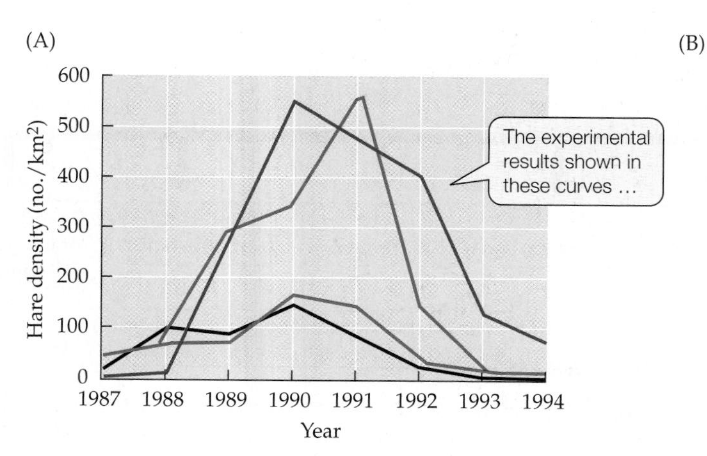

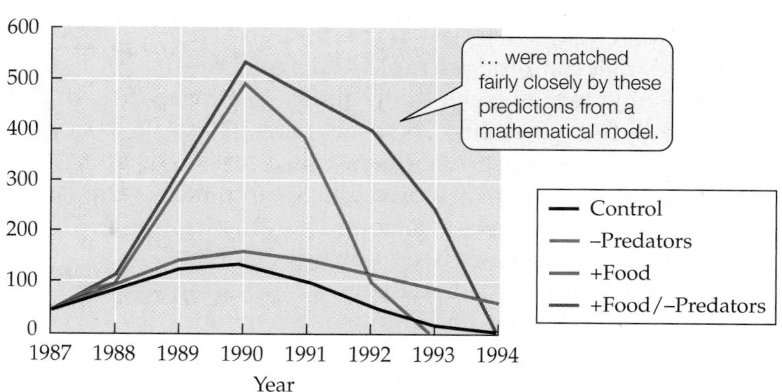

FIGURE 12.25 A Vegetation–Hare–Predator Model Predicts Hare Densities Accurately The model assumed that hare population densities are influenced by feeding relationships across three levels: vegetation, hares, and predators. Parameters for the model were estimated from field data. When the investigators compared the predictions of their model with the experimental results of Krebs et al. (1995), they found a reasonably good match between (A) the experimental results and (B) the model's predictions. (After King and Schaffer 2001.)

While much progress has been made in the study of snowshoe hare population cycles, some questions remain. We do not yet have a complete understanding of the factors that cause hare populations across broad regions of Canada to cycle in synchrony. Lynx can move from 500 to 1,100 km. If lynx move from areas with scarce prey to areas with abundant prey on a scale of hundreds of kilometers, their movements might be enough to cause geographic synchrony in hare cycles. In addition, large geographic regions in Canada experience a similar climate, and that may also affect the synchrony of hare population cycles.

Finally, the Krebs et al. experiment provided a test of whether the addition of food or the removal of predators (or both) could stop the hare population cycle. Although hare densities declined less in the +Food/–Predators block than in the control blocks, they did decline at the usual point in the hare cycle. Why did the +Food/–Predators treatment fail to stop the cycle? One possible reason is that the fences excluded lynx and coyotes, but did not exclude owls, goshawks, and other birds of prey. Collectively, these bird predators accounted for about 40% of snowshoe hare deaths, and thus could have contributed to the onset of the decline phase of the hare cycle in the +Food/–Predators block. Next, we'll explore another possible explanation: stress caused by the fear of predator attack.

CONNECTIONS IN NATURE

From Fear to Hormones to Population Dynamics

Predators not only affect their prey directly (by killing them), but also influence them indirectly—for example, by altering their behavior (see pp. 267–268). Boonstra et al. (1998) tested snowshoe hares for another possible indirect effect of predators: fear. Their results hint at a fascinating way in which predation might influence the decline phase of the hare cycle.

When humans are in a dangerous situation, we often engage a set of "fight-or-flight" responses that can produce rapid and sometimes astonishing results (such as the ability to move unusually heavy weights). Snowshoe hares have a similar stress response. A hormone called cortisol stimulates the release of stored glucose into the blood, where it becomes available to the muscles; cortisol also suppresses body functions that are not essential for immediate survival, including growth, reproduction, and immune system function (**Figure 12.26**).

The stress response works well for immediate, or *acute*, forms of stress, such as an attack by a predator. Energy is provided to the muscles rapidly to help the animal deal with the threatening situation. Shortly thereafter, the response is shut down by a negative feedback process. The stress response works less well for long-term, or *chronic*, stress, however. In such cases, the negative feedback signals are weak, and the stress response is maintained for a long time. A failure to "turn off" the stress response can have harmful effects, including decreased growth and reproduction and increased susceptibility to disease. Collectively, such effects can reduce a population's survival and reproductive rates.

When predators are abundant, they can cause up to 95% of snowshoe hare deaths. At such times, hares are at increased risk of encountering predators; hares would also be likely to see or hear predators killing other hares and to find the remains of hares that had been killed by predators. Reasoning that the fear provoked by such events could trigger chronic stress, Boonstra and colleagues measured the hormonal and immune responses of hares exposed in

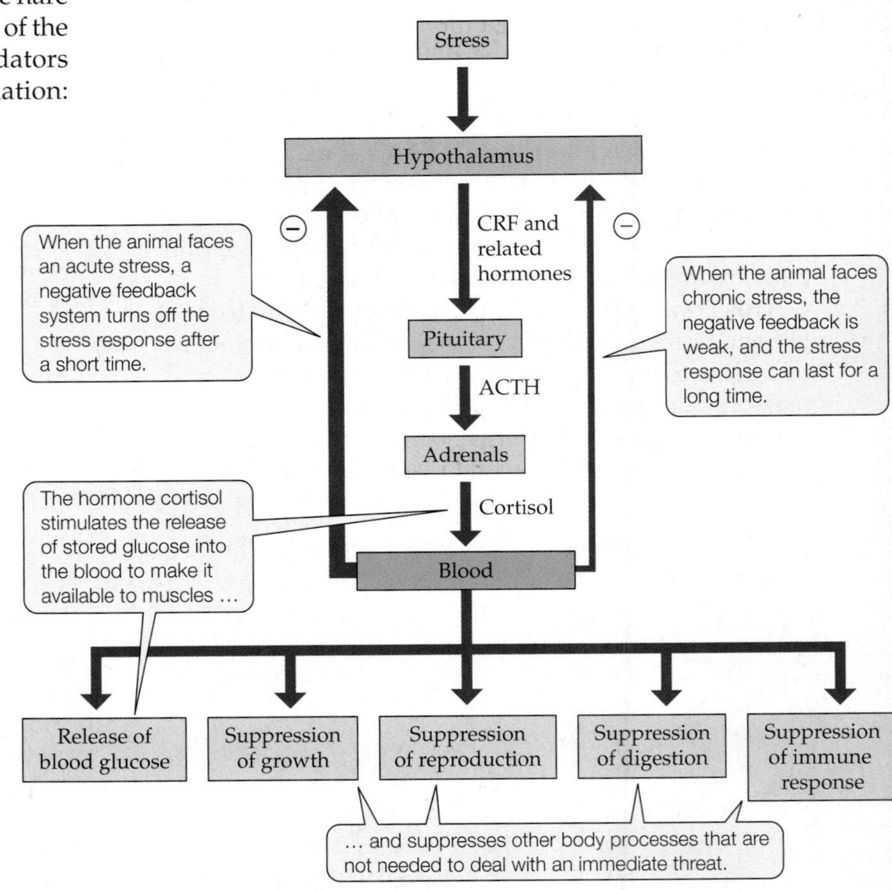

FIGURE 12.26 The Stress Response When an animal is stressed, the hypothalamus releases a hormone called CRF, which stimulates a cascade of reactions that affect a number of body processes. (After Boonstra et al. 1998.)

the field to high versus low numbers of predators. In the decline phase of the hare cycle (when hares are exposed to many predators), cortisol levels increased, blood glucose levels increased, reproductive hormone levels decreased, and overall body condition worsened—as expected for hares experiencing chronic stress (see Figure 12.26). Further experiments showed that a predator-induced increase in cortisol levels led to a drop in the number and size of offspring produced by female snowshoe hares (Sheriff et al. 2009).

Overall, chronic stress induced by predation may help to explain some of the puzzling observations mentioned on p. 263, including why birth rates drop during the decline phase of the hare cycle and why hare numbers sometimes rebound slowly after predator numbers plummet. If future studies confirm the results of Boonstra et al. (1998) and Sheriff et al. (2009), their work will provide a clear example of how predation risk can alter the physiology of individual prey, thereby changing prey population dynamics and influencing predator–prey cycles.

SUMMARY

CONCEPT 12.1 Most predators have broad diets, whereas a majority of herbivores have relatively narrow diets.

- Over half of the organisms on Earth are predators, herbivores, or parasites that sustain themselves by feeding on other organisms.
- Many predators move about their habitat in search of prey; others, known as sit-and-wait predators, remain in one place and attack or trap prey that move within striking distance.
- Most predators do not specialize on particular prey species; instead, they typically eat prey in relation to their availability across a broad range of prey species.
- Many herbivores specialize on particular plant parts, such as leaves (the most common food source), roots, stems, seeds, or internal fluids.
- Herbivorous insects, which constitute a majority of herbivore species, tend to have relatively narrow diets, feeding on only one or a few plant species.

CONCEPT 12.2 Organisms have evolved a wide range of adaptations that help them obtain food and avoid being eaten.

- In response to strong selection pressure exerted by predators and herbivores, prey and food plant species have evolved a rich variety of defensive mechanisms.
- Prey may rely on physical defenses, toxins, mimicry, or behavioral responses to escape predation.
- Plants cope with herbivory via masting and other forms of avoidance, compensation (a form of tolerance), and secondary chemicals that deter herbivores.
- Predators have a wide range of adaptations for overcoming the defenses of their prey, including physical features, toxins, mimicry, and tolerance or detoxification of prey toxins.
- Although a plant's defensive mechanisms prevent most herbivores from eating it, typically some herbivores can overcome the plant's defenses by structural, chemical, or behavioral means.

CONCEPT 12.3 Predation and herbivory can affect ecological communities greatly, in some cases causing a shift from one community type to another.

- Predators and herbivores may have direct effects on the distributions and abundances of their food organisms as well as indirect effects on other species in their communities.
- Predators can cause dramatic declines in the distributions and abundances of their prey; similarly, herbivores can decimate their food plants.
- Exploitation can also alter the composition of ecological communities, in some cases changing one community type to another.

CONCEPT 12.4 Population cycles can be caused by exploitative interactions.

- Results from mathematical predator–prey models, laboratory experiments, and field observations suggest that population cycles can be caused by exploitation.
- Whether predators and prey can coexist may depend on several factors, including habitat complexity and dispersal.
- Evolutionary change can affect predator–prey population dynamics, in some cases producing unusual population cycles.

REVIEW QUESTIONS

1. Compare and contrast the diet breadth of predators and herbivores.
2. Summarize the evolutionary effects that predators and herbivores have on their food organisms, as well as the effects that food organisms have on the predators and herbivores that eat them. Explain why these effects are pervasive and pronounced.
3. In this chapter, we claim that exploitative interactions such as predation and herbivory can have strong effects on ecological communities.
   a. Provide a logical argument to support this claim.
   b. Does the scientific evidence support or contradict this claim? Explain.
4. Based on the relationships shown in Equation 12.1, use the zero population growth isocline approach described in Ecological Toolkit 11.1 to show that the prey population is expected to *decrease* if there are more than r/a predators.

ON THE COMPANION WEBSITE
sites.sinauer.com/ecology2e

The website includes Chapter Outlines, Online Quizzes, Flashcards & Key Terms, Suggested Readings, a complete Glossary, and the Web Stats Review. In addition, the following resources are available for this chapter:

HANDS-ON PROBLEM SOLVING

Cascading Effects of Predators

This Web exercise explores the effect of predators through multiple trophic levels. You will read a recent paper on predator-driven cascades in marine systems. You will then use data from a system in which wolves are the top predator to test for a trophic cascade.

WEB EXTENSION

12.1 Modifying the Lotka–Volterra Predator–Prey Model

CHAPTER

13 Parasitism

KEY CONCEPTS

- **CONCEPT 13.1** Parasites typically feed on only one or a few host individuals.
- **CONCEPT 13.2** Hosts have adaptations for defending themselves against parasites, and parasites have adaptations for overcoming host defenses.
- **CONCEPT 13.3** Host and parasite populations can evolve together, each in response to selection pressure imposed by the other.
- **CONCEPT 13.4** Parasites can reduce the sizes of host populations and alter the outcomes of species interactions, thereby causing communities to change.
- **CONCEPT 13.5** Simple models of host–pathogen dynamics suggest ways to control the establishment and spread of diseases.

Enslaver Parasites: A Case Study

In science fiction books and movies, villains sometimes use mind control or physical devices to break the will and control the actions of their victims. In these stories, a person may be forced to perform strange or grotesque actions, or to harm themselves or others—all against their will.

Real life can be just as strange. Consider the hapless cricket shown in the video found in **Web Extension 13.1**. This cricket does something that a cricket ordinarily would never do: it walks to the edge of a body of water, jumps in, and drowns. Shortly afterward, a hairworm begins to emerge from the body of the cricket (**Figure 13.1**). For the worm, this is the final step in a journey that begins when a terrestrial arthropod—such as a cricket—drinks water in which a hairworm larva swims. The larva enters the cricket's body and feeds on its tissues, growing from microscopic size into an adult that fills all of the cricket's body cavity except its head and legs. When fully grown, adult hairworms must return to the water to mate. After the adults mate, the next generation of hairworm larvae are released to the water, where they will die unless they are ingested by a terrestrial arthropod host.

Has the hairworm "enslaved" its cricket host, forcing it to jump into the water—an act that kills the cricket, but is essential for the hairworm to complete its life cycle? The answer appears to be yes. Observations have shown that when crickets infected with hairworms are near water, they are much more likely to enter the water than are uninfected crickets (Thomas et al. 2002). Furthermore, in ten out of ten trials, when infected crickets were rescued from the water, they immediately jumped back in. Uninfected crickets do not do this.

Hairworms are not the only parasites that enslave their hosts. Maitland (1994) coined the term "enslaver parasites" for several fungal species that alter the perching behavior of their fly hosts in such a way that fungal spores can be dispersed more easily after the fly dies (**Figure 13.2**). The fungus *Ophiocordyceps unilateralis* also manipulates the final actions of its host, the ant *Camponotus leonardi*. First, an infected ant climbs down from its home in the upper branches of trees and selects a leaf in a protected environment about 25 cm above the soil (Andersen et al. 2009). Then, just before the fungus kills it, the ant bites into the selected leaf with a "death grip" that will hold its body in place after it is dead. The fungus grows well in such protected environments but cannot survive where the ant usually lives—at the tops of trees, where the temperature and humidity are more variable. Thus, while the ant's final actions do not benefit the ant, they do allow the fungus to complete its life cycle in a favorable environment.

FIGURE 13.1 Driven to Suicide The behavior of this wood cricket (*Nemobius sylvestris*) was manipulated by the hairworm (*Paragordius tricuspidatus*) emerging from its body. By causing the cricket to jump into water (where it drowns), the hairworm is able to continue its life cycle.

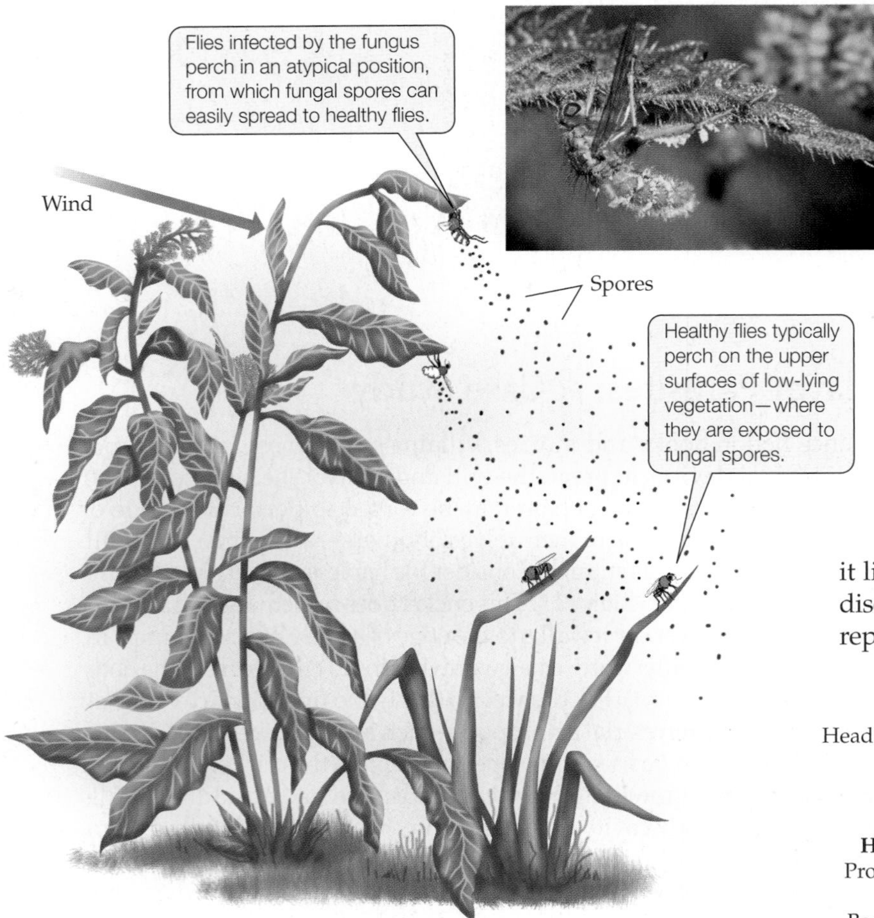

FIGURE 13.2 Enslaved by a Fungus Shortly before they die from the infection, yellow dungflies infected by the fungus *Entomophthora muscae* move to the downwind side of a relatively tall plant and perch on the underside of one of its leaves. This position increases the chance that fungal spores released by *Entomophthora* will land on healthy yellow dungflies. (After Maitland 1994.)

Even vertebrates can be enslaved by parasites. Rats typically engage in predator avoidance behaviors in areas that show signs of cats. However, rats infected with the protist parasite *Toxoplasma gondii* behave abnormally: they do not avoid cats, and in some cases they are actually attracted to cats. While such a behavioral change can be a fatal attraction for the rat, it benefits the parasite because it increases the chance that the parasite will be transmitted to the next host in its complex life cycle—a cat.

How do some parasites enslave their hosts? Can the hosts fight back? More generally, what can these remarkable interactions tell us about host–parasite relationships?

Introduction

More than half of the millions of species that live on Earth are **symbionts**, meaning that they live in or on other organisms. To begin to understand how many symbionts there are, we need look no further than our own bodies (**Figure 13.3**). Our faces are home to mites that feed on exudates from the pores of our skin and on secretions at the base of our eyelashes. There are bacteria and fungi that grow on our skin and under our toenails. Arthropods such as lice may live on our heads, pubic regions, and other parts of our bodies. Moving inward, our tissues, organs, and body cavities can be infested with a rich variety of organisms, from bacteria to worms to fungi to protists.

Although some symbionts are mutualists (as we will see in Chapter 14), the majority of them are parasites. A **parasite** consumes the tissues or body fluids of the organism on or within which it lives, its **host**; some parasites, called **pathogens**, cause diseases. Unlike predators, parasites usually have a higher reproductive rate than their hosts. Also unlike predators,

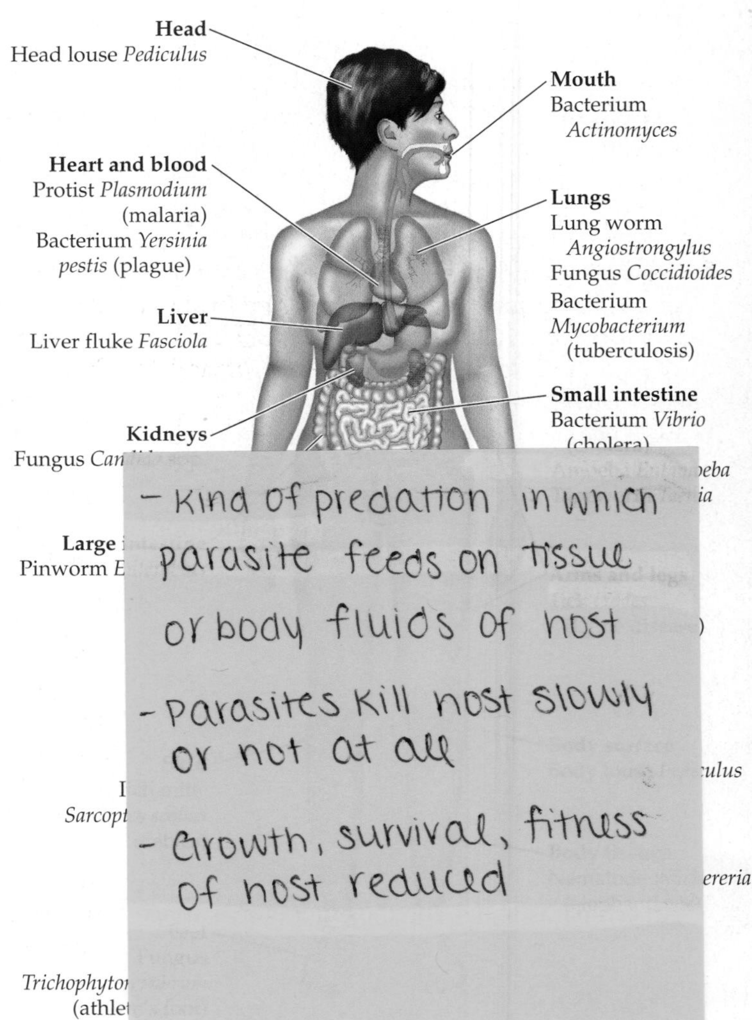

FIGURE 13.3 The Human Body as Habitat Different parts of our bodies provide suitable habitat for a wide range of symbionts, many of which are parasites; only a few examples are shown here. Some of these organisms are pathogens that cause disease.

parasites typically harm, but do not immediately kill, the organisms they eat. The negative effects of parasites on their hosts vary widely, from mild to lethal. We see this variation in our own species, for which some parasites, such as the fungus that causes athlete's foot, are little more than a nuisance. Others, such as the protist *Leishmania tropica*, can cause disfigurement, and still others, such as *Yersinia pestis*, the bacterium that causes the plague, can kill. There is similar variation in the degree of harm caused by parasites that infect other species. Parasites vary in many other ways, as we'll see next as we examine their basic biology.

CONCEPT 13.1 Parasites typically feed on only one or a few host individuals.

Parasite Natural History

Parasites vary in size from relatively large species (**macroparasites**), such as arthropods and worms, to species too small to be seen with the naked eye (**microparasites**), such as bacteria, protists, and unicellular fungi. But whether they are large or small, parasites typically feed on only one or a few host individuals over the course of their lives. Thus, defined broadly, parasites include herbivores, such as aphids or nematodes, that feed on only one or a few host plants, as well as *parasitoids* (see p. 264), insects whose larvae feed on a single host, almost always killing it.

Most species are attacked by more than one parasite (**Figure 13.4**), and even parasites have parasites. Because parasites spend their lives feeding on one or a few host individuals, they tend to have a close relationship to the organisms they eat. For example, many parasites are closely adapted to particular host species, and many attack only one or a few host species. This specialization at the species level helps to explain why there are so many species of parasites—many host species have at least one parasite that eats *only* them. Overall, although the total number of parasite species is not known, a rough estimate is that 50% of the species on Earth are parasites (Windsor 1998).

Parasites are also specialized for living on or eating certain parts of the host's body. We'll focus next on this aspect of parasite specialization by describing both ectoparasites and endoparasites.

Ectoparasites live on the surface of their host

An **ectoparasite** lives on the outer body surface of its host (**Figure 13.5**). Ectoparasites include plants such as dodder

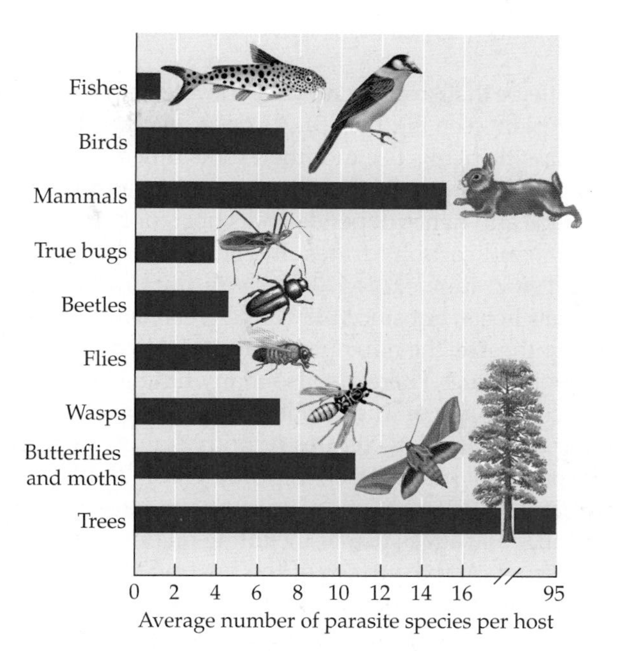

FIGURE 13.4 Many Species Are Host to More Than One Parasite Species In a study conducted in Britain, most host species were found to harbor more than one parasite species. The number of parasite species shown here for fishes, birds, and mammals includes only helminth worm parasites and hence is likely to underestimate the actual number of parasite species found in these vertebrates.

Q Averaging across the six groups of organisms other than vertebrates (which we exclude because the data underestimate the true number of parasites), what is the average number of parasite species per host? Suppose the number of parasite species were determined for a previously unstudied host from one of the six groups. Is it likely that the number of parasites in that host would be close to the average you calculated? Explain.

FIGURE 13.5 Ectoparasites A wide range of parasites live on the outer surfaces of their hosts, feeding on host tissues. Examples include (A) the orange rust fungus (*Gymnoconia peckiana*), which grows on the leaves of its blackberry host, and (B) the velvet mite (*Trombidium* sp.), which in its larval form feeds parasitically on the blood of insects, such as this sawfly larva.

and mistletoe that grow on, and obtain water and food from, another plant (see Figure 5.3). As described in Chapter 5, such parasitic plants use modified roots called *haustoria* to penetrate the tissues of their host. Dodder cannot photosynthesize and hence depends on its host for both mineral nutrients and carbohydrates. In contrast, mistletoes are *hemiparasitic*: they extract water and mineral nutrients from their hosts, but since they have green leaves and can photosynthesize, they do not rely exclusively on their host for carbohydrates.There are also many fungal and animal parasites that live on the surfaces of plants and animals, feeding on their hosts' tissues or body fluids. More than 5,000 species of fungi attack important crop and horticultural plants, causing billions of dollars of damage each year. Some fungi that attack plants, including mildews, rusts, and smuts, grow on the surface of the host plant and extend their hyphae (fungal filaments) within the plant to extract nutrients from its tissues. Plants are also attacked by numerous animal ectoparasites, including aphids, whiteflies, and scale insects, which are found on stems and leaves, and nematode worms, beetles, and (juvenile) cicadas, which are found on roots. Animals like these that eat plants and live on their outer surfaces can be thought of both as herbivores (because they eat plant tissues) and as parasites (especially if they remain on a single host plant for much of their lives).

A similar array of fungal and animal ectoparasites can be found on the surfaces of animals. Familiar examples include *Trichophyton rubrum*, the fungus that causes athlete's foot, and fleas, mites, lice, and ticks, which feed on the tissues or blood of their hosts (see Figure 13.5B). Some of these parasites also transmit diseases to their hosts, including fleas that spread the plague and ticks that spread Lyme disease.

Endoparasites live inside their host

If we ignore the details of their shape, we can think of people and most other animals as being constructed in a similar way: their bodies consist of tissues that surround an open tube called the *alimentary canal*. The alimentary canal runs through the middle of the body, from the mouth to the anus. Parasites that live inside their hosts, called **endoparasites**, include species that inhabit the alimentary canal as well as species that live within host cells or tissues (**Figure 13.6**).

The alimentary canal provides excellent habitat for parasites. The host brings in food at one end (the mouth) and excretes what it cannot digest at the other (the anus). Parasites that live within the alimentary canal often do not eat host tissues at all; instead, they rob the host of nutrients. A tapeworm, for example, has a *scolex*, a structure with suckers (and sometimes hooks) that it uses to attach itself to the inside of the host's intestine (see Figure 13.6A). Once it is attached, the tapeworm simply absorbs food that the host has already digested. Tapeworms that infect humans can grow up to 10–20 m (33–66 feet) long; large tapeworms such as these can block the intestines and cause nutritional deficiencies.

Many other endoparasites live within the cells or tissues of animal hosts, causing a wide range of symptoms

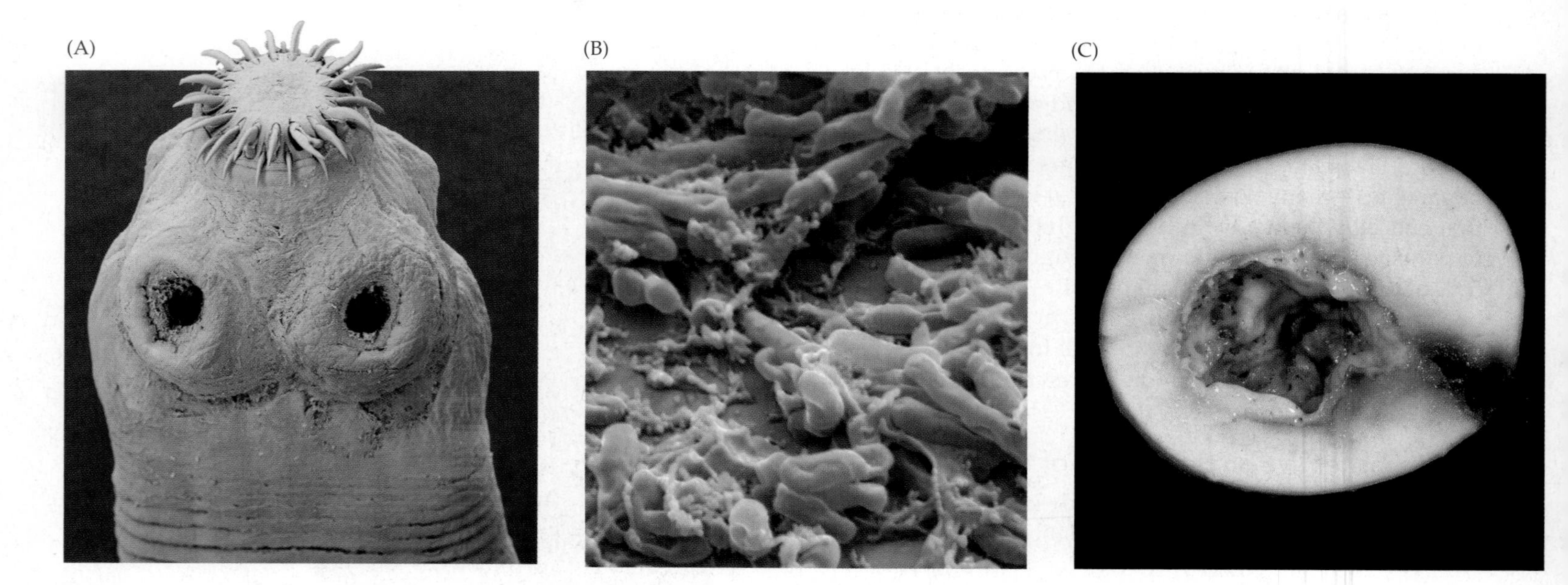

FIGURE 13.6 Endoparasites Many parasites live within the body of their host, feeding on the host's tissues or robbing it of nutrients. (A) The tapeworm *Taenia taeniformis* uses the suckers and hooks shown here to attach to the intestinal wall of its mammalian host, often a rodent, rabbit, or cat. Once attached, an adult can grow to over 5 m (16 feet) in length. (B) The bacterium *Mycobacterium tuberculosis* causes the lung disease tuberculosis, which kills 1 to 2 million people each year. (C) This section of a potato tuber shows the destruction wrought by *Erwinia carotovora*, a bacterium that causes soft rot. Affected areas become soft with decay and develop a distinctive foul odor.

as they reproduce or consume host tissues. Examples in humans include *Yersinia pestis*, the bacterium that causes the plague, and *Mycobacterium tuberculosis*, the bacterium that causes tuberculosis (TB; see Figure 13.6B). TB is a potentially fatal lung disease, aptly referred to as the "Captain of Death"; with the possible exception of malaria, it has killed more people than any other disease in human history. TB continues to kill 1 to 2 million adults each year (a number comparable to the roughly 2 million that currently die each year from AIDS).

Plants too are attacked by a wide variety of endoparasites, including bacterial pathogens that cause soft rot in various plant parts, such as fruits (tomatoes) or storage tissues (potatoes; see Figure 13.6C). Other plant pathogens include fungi that cause plant parts to rot from the inside out. Some bacteria invade plant vascular tissues, where they disrupt the flow of water and nutrients, causing wilting and often death. Plant pathogens can have large effects on natural communities, as illustrated by the funguslike protist *Phytophthora ramorum* that causes sudden oak death (SOD), a disease that has recently killed more than a million oaks and other trees in California and Oregon (see also the discussion under Concept 13.4).

Endoparasitism and ectoparasitism have advantages and disadvantages

There are advantages and disadvantages to living in or on a host (**Table 13.1**). Because ectoparasites live on the surface of their host, it is relatively easy for them or their offspring to disperse from one host individual to another. It is much more difficult for endoparasites to disperse to new hosts. Endoparasites solve this problem in a variety of ways. Some, like the enslaver parasites discussed in the Case Study at the opening of this chapter (see p. 283), alter the physiology or behavior of their host in ways that facilitate their dispersal. Other examples include the bacterium (*Vibrio cholerae*) that causes cholera and the amoeba (*Entamoeba histolytica*) that causes amoebic dysentery. People with cholera and dysentery have diarrhea, a symptom that increases the chance that the parasite will contaminate drinking water and thereby spread to new hosts. Other endoparasites have complex life cycles that include stages that are specialized for dispersing from one host species to another (see Figure 13.9 below).

Although dispersal is relatively easy for ectoparasites, there are costs to life on the surface of a host. Compared with endoparasites, ectoparasites are more exposed to natural enemies such as predators, parasitoids, and parasites. Aphids, for example, are attacked by ladybugs, birds, and many other predators, as well as by lethal parasitoids (see Figure 12.4) and by parasites such as mites that suck fluids from their bodies. Endoparasites, in contrast, are safe from all but the most specialized predators and parasites. Endoparasites are also relatively well protected from the external environment, and they have relatively easy access to food—unlike an ectoparasite, an endoparasite does not have to pierce the host's protective outer surfaces to feed. But living within the host does expose endoparasites to a different sort of danger: attack by the host's immune system. Some parasites have evolved ways to tolerate or overcome immune-system defenses, as we shall see in the following section.

TABLE 13.1

Advantages and Disadvantages of Living in or on a Host

	Ectoparasitism	Endoparasitism
Advantages	Ease of dispersal Safe from host's immune system	Ease of feeding Protected from external environment Safer from natural enemies
Disadvantages	Vulnerability to natural enemies Exposure to external environment Feeding more difficult	Vulnerability to host's immune system Dispersal more difficult

CONCEPT 13.2 Hosts have adaptations for defending themselves against parasites, and parasites have adaptations for overcoming host defenses.

Defenses and Counterdefenses

As we saw in Chapter 12, predators and herbivores exert strong selection pressure on their food organisms, and vice versa. The prey species and plants eaten by predators and herbivores have adaptations that help them avoid being eaten; similarly, predators and herbivores have adaptations that help them to overcome the defenses of their prey or food plants. The same is true of parasites and their hosts: hosts have evolved ways to protect themselves against parasites, and parasites have evolved countermeasures to circumvent host defenses.

Immune systems and biochemical defenses can protect hosts against parasites

Host organisms have a wide range of defensive mechanisms that can prevent or limit the severity of parasite attacks.

For example, a host may have a protective outer covering, such as the skin of a mammal or the hard exoskeleton of an insect, that can keep ectoparasites from piercing its body or make it difficult for endoparasites to enter. Endoparasites that do manage to enter the host's body are often killed or rendered less effective by the host's immune system or biochemical defenses.

Immune systems The vertebrate immune system includes specialized cells that allow the host to recognize microparasites to which it has been previously exposed; in many instances, the "memory cells" of the immune system are so effective that the host has lifelong immunity against future attack by the same microparasite. Other immune system cells engulf and destroy parasites or mark them with chemicals that target them for later destruction.

Plants can also mount highly effective responses to invasion by parasites. Some plants have resistance genes, the different alleles of which provide protection against microparasites that have particular genotypes; we will describe this defense system in more detail on pp. 292–293. Plants are not helpless, however, even when they lack alleles that provide resistance to a specific attacker. In such a case, the plant relies on a nonspecific immune system that produces antimicrobial compounds, including some that attack the cell walls of bacteria and others that are toxic to fungal parasites (**Figure 13.7**). The plant may also produce chemical signals that "warn" nearby cells of imminent attack, and still other chemicals that stimulate the deposition of lignin, a hard substance that provides a barricade against the invader's spread.

Biochemical defenses Hosts have ways of regulating their biochemistry to limit parasite growth. Bacterial and fungal endoparasites, for example, require iron to grow. Vertebrate hosts—including mammals, birds, amphibians, and fishes—have a protein called transferrin that removes iron from their blood serum (where parasites could use it) and stores it in intracellular compartments (where parasites cannot get to it). Transferrins are so efficient that the concentration of free iron in mammalian blood serum is only 10^{-26} *M*—so low that parasites cannot grow in vertebrate blood unless they can somehow outmaneuver the host. To do this, some parasites steal iron from the transferrin itself and use it to support their own growth.

Similar biochemical battles occur between plants and their parasites. As we saw in Chapter 12, plants use a rich variety of chemical weapons to kill or deter the organisms that eat them. Plant defensive secondary compounds are so effective that some animals eat specific plants in order to treat or prevent parasite infections. For example, when parasitic flies lay eggs on their bodies, woolly bear caterpillars switch from their usual food plant (lupines) to a diet of poisonous hemlock (Karban and English-Loeb 1997). The new diet does not kill the parasites, but it does increase the chance that the caterpillar will survive the attack and metamorphose into an adult tiger moth (*Platyprepia virginalis*). Chimpanzees infected with nematodes (*Oesophagostomum stephanostomum*) specifically seek out and eat a bitter plant that scientists have learned contains compounds that kill or paralyze the nematodes and can also deter many other parasites (Huffman 1997) (**Figure 13.8**). Humans do essentially the same thing: we spend billions of dollars each year on pharmaceuticals that are based on compounds originally obtained from plants.

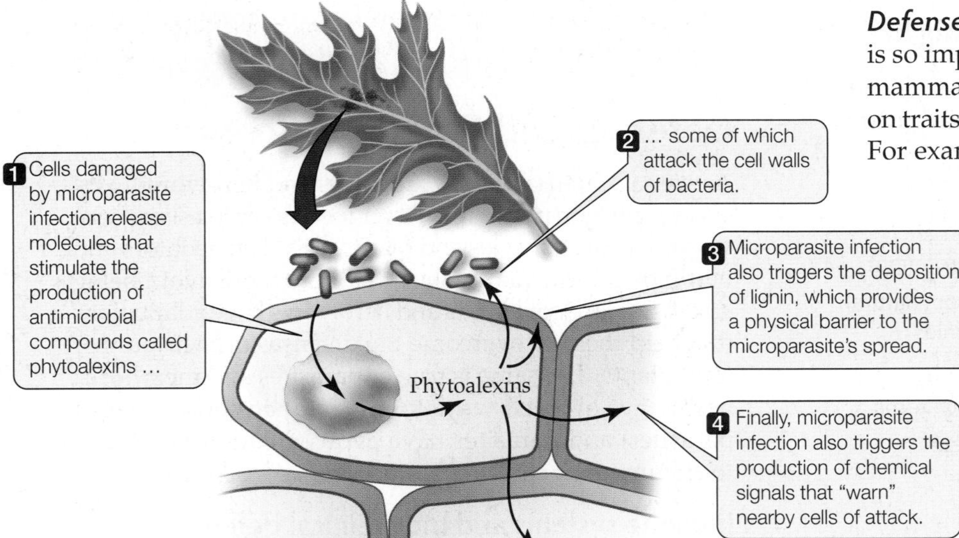

FIGURE 13.7 Nonspecific Plant Defenses Plants can mount a nonspecific defensive response that is effective against a broad range of fungal and bacterial microparasites.

Defenses and mate choice Defense against parasites is so important to individual fitness that in some bird, mammal, and fish species, females select mates based on traits that indicate that a male has effective defenses. For example, a group of proteins known as the major histocompatibility complex (MHC) is a key component of the vertebrate immune system. Typically, the larger number of different MHC proteins an individual has, the better protected it is against a wider range of parasites. Female stickleback fish prefer to mate with males that have many MHC proteins (Reusch et al. 2001). The females use their sense of smell to detect which males have many MHC proteins—and are thus likely to have few parasites.

Other species may use aspects of male performance to assess parasite loads. In *Copadichromis eucinostomus*, a cichlid fish from Lake Malawi, Africa, males court fe-

FIGURE 13.8 Using Plants to Fight Parasites A chimpanzee in Mahale National Park, Tanzania, chews on the bitter pith of *Vernonia amygdalina*, which produces a compound that is toxic to nematode parasites.

males by building a sand structure called a *bower*. Females prefer to mate with males that make large, smooth bowers. Researchers have found that such males have fewer tapeworm parasites than do males that make smaller bowers, perhaps because males with many tapeworms must spend more of their time eating to compensate for the nutrients they lose to the parasites, and hence cannot build large bowers. Similar results have been found in birds, in which traits such as beak color reflect parasite loads and are used by females as key criteria for mate choice.

Parasites have adaptations that circumvent host defenses

To survive and reproduce, a parasite must be able to tolerate or evade its host's defensive mechanisms. Aphids and other ectoparasites, for example, must be able to pierce the protective outer covering of the host, and they must be able to tolerate whatever chemical compounds are present in the host tissues or body fluids that they eat. Viewed broadly, the challenges faced by ectoparasites are similar to those faced by herbivores and predators as they attempt to cope with the toxins and physical structures that their food organisms use to defend themselves. We discussed such challenges under Concept 12.2, so here we focus on how endoparasites cope with defenses found inside the host.

Counterdefenses against encapsulation Endoparasites face formidable challenges from host immune systems and related aspects of host biochemistry. Host species typically have a number of ways to destroy parasite invaders. In addition to the strategies we have already described, some hosts can cover parasites or parasite eggs with capsules that kill them or render them harmless, a process called *encapsulation*.

Encapsulation is used by some insects to defend themselves against macroparasites. Insect blood cells can engulf small invaders, such as bacteria, but they cannot engulf large objects, such as nematodes or parasitoid eggs. However, some insects have *lamellocytes*, which are blood cells that can form multicellular sheaths (capsules) around large objects. When an insect mounts such an encapsulation defense, most or all of the attacking parasites may be destroyed. As a result, the parasites are under strong selection to develop a counterdefense.

For example, *Drosophila* fruit flies can mount an effective defense against wasp parasitoids by encapsulating (and hence killing) their eggs. Parasitoid wasps that attack fruit flies avoid encapsulation in several different ways. When wasps in the genus *Leptopilina* lay their eggs inside a fruit fly host, they also inject viruslike particles into the host. These particles infect the host's lamellocytes and cause them to self-destruct, thus weakening the host's resistance and increasing the percentage of wasp eggs that survive (Rizki and Rizki 1990). Other parasitoid wasps, such as *Asobara tabida*, lay eggs covered with filaments. These filaments cause the eggs to stick to and become embedded in fat cells and other host cells, where they are not detected by circulating lamellocytes.

Counterdefenses involving hundreds of genes Some endoparasites have a complex set of adaptations that allows them to thrive inside their host. One such endoparasite is *Plasmodium*, the protist that causes malaria, a disease that kills 1 to 2 million people each year. *Plasmodium*, like many endoparasites, has a complex life cycle with specialized stages that allow it to alternate between a mosquito and a human host (**Figure 13.9**). Infected mosquitoes contain one specialized *Plasmodium* stage, called a *sporozoite*, in their saliva. When an infected mosquito bites a human, sporozoites enter the victim's bloodstream and travel to the liver, where they divide to form another stage, called a *merozoite*. The merozoites penetrate red blood cells, where they multiply rapidly. After 48–72 hours, large numbers of merozoites break out of the red blood cells, causing the periodic chills and fever that are associated with malaria. Some of the offspring merozoites attack more red blood cells, while others transform into gamete-producing cells. If another mosquito bites the victim, it picks up some of the gamete-producing cells, which enter its digestive tract and form gametes. After fertilization occurs, the resulting zygotes produce thousands of sporozoites, which then migrate to the mosquito's salivary glands, where they await their transfer to another human host.

Plasmodium faces two potentially lethal challenges from its human host. First, red blood cells do not divide or grow, and hence they lack the cellular machinery needed to import nutrients necessary for growth. A *Plasmodium* merozoite inside a red blood cell would starve if it did not have a way to obtain essential nutrients. Second, after 24–48 hours, infection by *Plasmodium* causes red blood cells to have

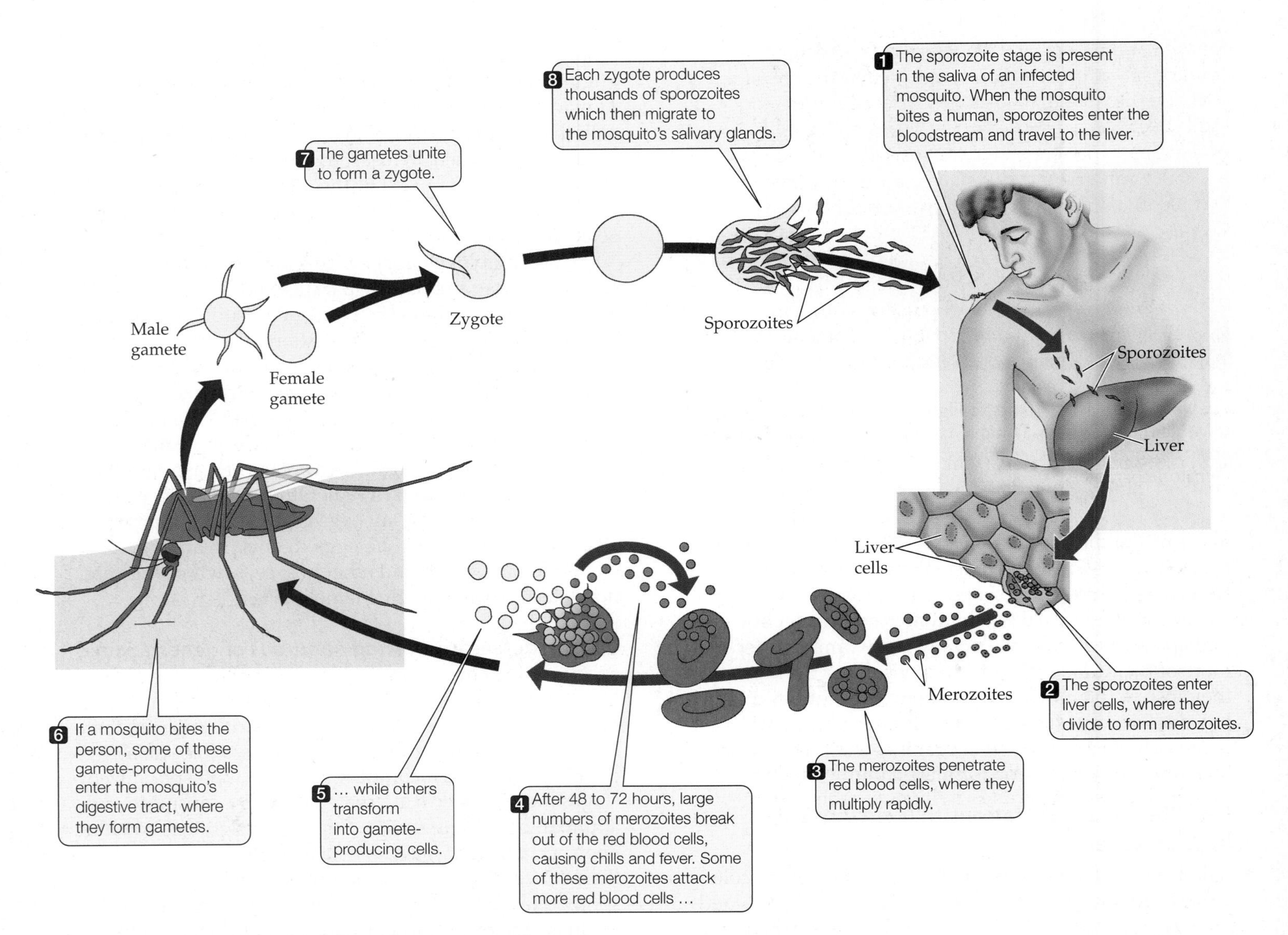

FIGURE 13.9 Life Cycle of the Malaria Parasite, *Plasmodium* The life cycle of the protist *Plasmodium* includes specialized stages that facilitate the dispersal of this endoparasite from one host to another. The sporozoite stage, for example, enables the parasite to disperse from an infected mosquito to a human host.

Q Which stage in the life cycle enables the parasite to disperse from a human host to a mosquito?

an abnormal shape. The human spleen recognizes and destroys such deformed cells, along with the parasites inside.

Plasmodium addresses these challenges by having hundreds of genes whose function is to modify the host red blood cell in ways that allow the parasite to obtain food and escape destruction by the spleen (Marti et al. 2004; Hiller et al. 2004). Some of these genes cause transport proteins to be placed on the surface of the red blood cell, thereby enabling the parasite to import essential nutrients into the host cell. Other genes guide the production of unique knobs that are added to the surface of the red blood cell. These knobs cause the infected red blood cell to stick to other human cells, thereby preventing it from traveling in the bloodstream to the spleen, where it would be recognized as infected and destroyed. The proteins on these knobs vary greatly from one parasite individual to another, making it difficult for the human immune system to recognize and destroy the infected cells.

CONCEPT 13.3 Host and parasite populations can evolve together, each in response to selection pressure imposed by the other.

Parasite–Host Coevolution

As we have just seen, *Plasmodium* has specific adaptations that enable it to live inside a red blood cell. When a parasite and its host each possess such specific adaptations, that observation suggests that the strong selection pressure that hosts and parasites impose on each other has caused both of their populations to evolve. Such changes have been directly observed in Australia, where the myxoma virus was introduced to control populations of the European rabbit (*Oryctolagus cuniculus*).

European rabbits were first introduced to Australia in 1859, when 24 wild rabbits were released at a ranch in Victoria. Within a decade, rabbit populations had grown so large, and were consuming so much plant material, that they posed a threat to cattle and sheep pasturelands and wool production. A series of control measures were enacted, including introductions of predators, shooting and poisoning of rabbits, and the building of fences to limit the spread of rabbits from one region to another (Fenner and Ratcliffe 1965). None of these methods worked. By the 1900s, hundreds of millions of rabbits had spread throughout much of the continent.

After years of investigation, Australian government officials settled on a new control measure: introduction of the myxoma virus. A rabbit infected with this virus may suffer from skin lesions and severe swellings, which can lead to blindness, difficulty with feeding and drinking, and death (usually within 2 weeks of infection). The virus is transmitted from rabbit to rabbit by mosquitoes. In 1950, when the virus was first used to control rabbit populations, 99.8% of infected rabbits died. In the ensuing decades, millions of rabbits were killed by the virus, and the sizes of rabbit populations dropped dramatically throughout the Australian continent. Over time, however, rabbit populations evolved resistance to the virus, and the virus evolved to become less lethal (**Figure 13.10**). The myxoma virus is still used to control rabbit populations, but doing so requires a constant search for new, lethal strains of the virus to which the rabbit has not evolved resistance.

The increased resistance of the rabbit and the reduced lethality of the virus illustrate **coevolution**, which occurs when populations of two interacting species evolve together, each in response to selection pressure imposed by the other. The outcome of coevolution can vary greatly depending on the biology of the interacting species. In the European rabbit, selection favored the evolution of increased resistance to viral attack, as you might expect. In addition, viral strains of intermediate lethality predominated, perhaps because such strains allowed rabbits to live long enough for one or more mosquitoes to bite them and transmit the virus to another host (mosquitoes do not bite dead rabbits). In other cases of host–parasite coevolution, the parasite evolves counterdefenses to overcome host resistance mechanisms, as the following examples illustrate.

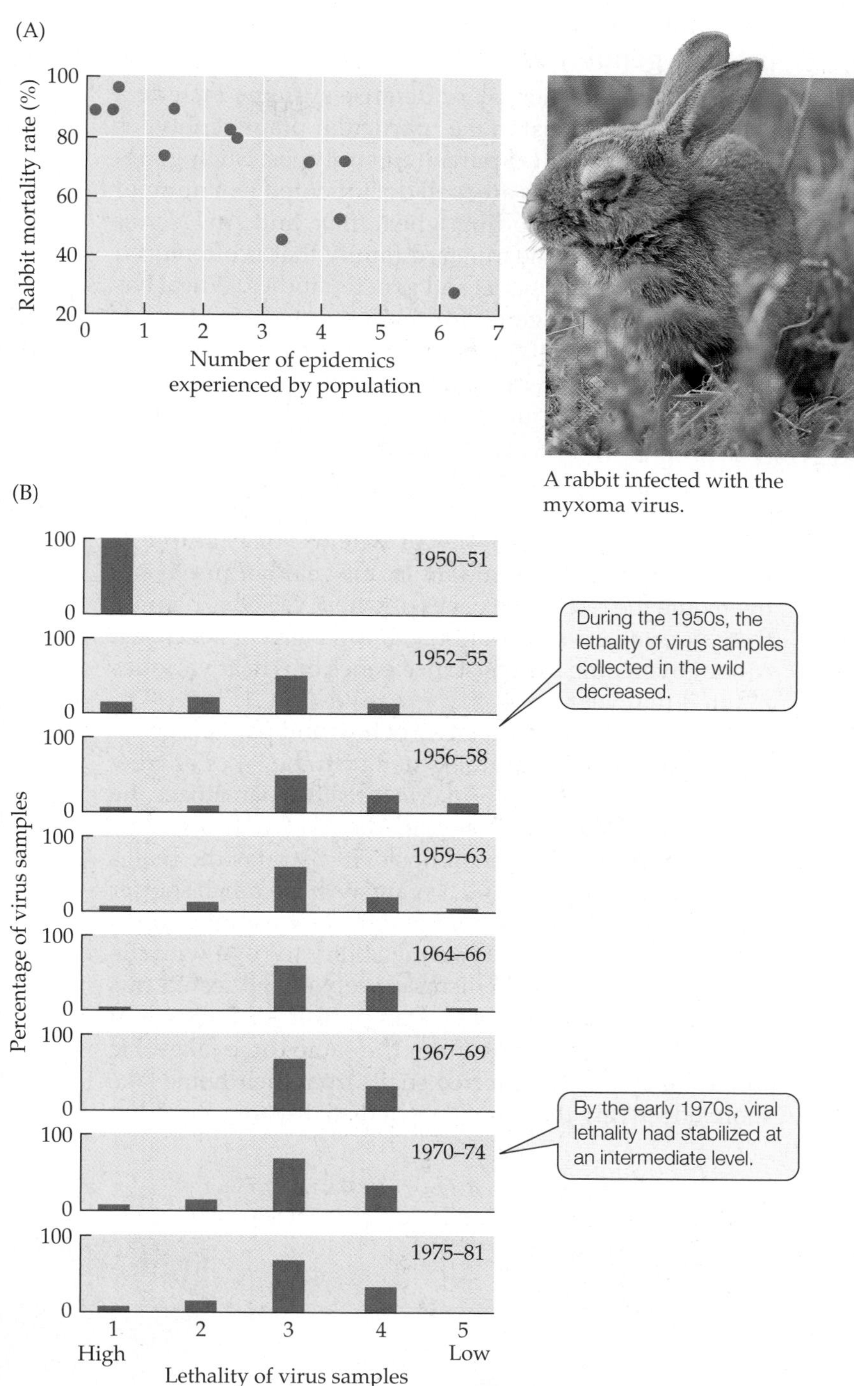

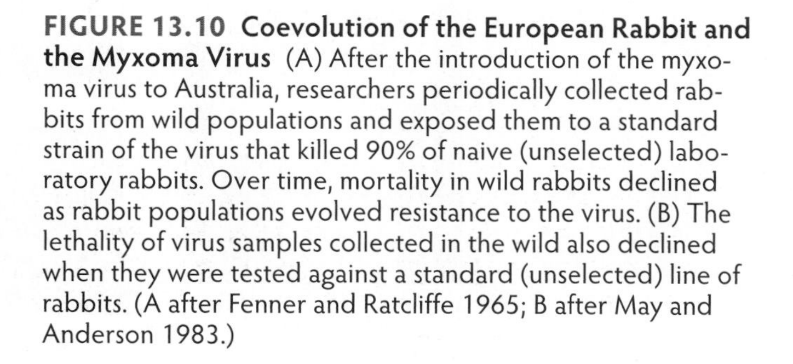
FIGURE 13.10 Coevolution of the European Rabbit and the Myxoma Virus (A) After the introduction of the myxoma virus to Australia, researchers periodically collected rabbits from wild populations and exposed them to a standard strain of the virus that killed 90% of naive (unselected) laboratory rabbits. Over time, mortality in wild rabbits declined as rabbit populations evolved resistance to the virus. (B) The lethality of virus samples collected in the wild also declined when they were tested against a standard (unselected) line of rabbits. (A after Fenner and Ratcliffe 1965; B after May and Anderson 1983.)

Selection can favor a diversity of host and parasite genotypes

As mentioned earlier, plant defense systems include a specific response that makes particular plant genotypes resistant to particular parasite genotypes. Such **gene-for-gene interactions** are well documented in a number of plant species, including wheat, flax, and *Arabidopsis thaliana*, a plant in the mustard family that is often used in laboratory experiments and genetic studies. Wheat has dozens of different genes for resistance to fungi such as wheat rusts (*Puccinia*). Different wheat rust genotypes can overcome different wheat resistance genes, however, and periodically, mutations occur in wheat rusts that produce new genotypes to which wheat is not resistant. Studies have shown that the frequencies of wheat rust genotypes vary considerably over time as farmers use different resistant varieties of wheat. For example, a rust variety may be abundant in one year because it can overcome the resistance genes of wheat varieties planted that year, yet less abundant the following year because it cannot overcome the resistance genes of wheat varieties planted that year.

Changes in the frequencies of host and parasite genotypes also occur in natural systems. In the lakes of New Zealand, a trematode worm, *Microphallus*, parasitizes the snail *Potamopyrgus antipodarum*. *Microphallus* has serious negative effects on its snail hosts: it castrates the males and sterilizes the females. The parasite has a much shorter generation time than its host, and hence we might expect that it would rapidly evolve the ability to cope with the snail's defensive mechanisms. Lively (1989) tested this idea in an experiment that pitted parasites from each of three lakes against snails from the same three lakes. He found that parasites infected snails from their home lake more effectively than they infected snails from the other two lakes (**Figure 13.11**). This observation suggests that the parasite genotypes in each lake had evolved rapidly enough to overcome the defenses of the snail genotypes found in that lake.

The snails also evolved in response to the parasites, albeit more slowly. Dybdahl and Lively (1998) documented the abundances of different snail genotypes over a 5-year period in another New Zealand lake. The snail genotype that was most abundant changed from one year to the next. Moreover, roughly a year after a genotype was the most abundant one in the population, snails of that genotype had a higher than typical number of parasites. Together with Lively's earlier study (1989), these results suggest that parasite populations evolve to exploit the snail genotypes found in their local environment. Refining this idea further, Dybdahl and Lively hypothesized that it should be easier for parasites to infect snails with a common genotype than it is for them to infect snails with a rare genotype. That is exactly what they found in a laboratory experiment (**Figure 13.12**). Hence, snail genotype frequencies may change from year to year because common genotypes are attacked by many parasites, placing them at a disadvantage and driving down their numbers in future years.

Host defenses and parasite counterdefenses both have costs

Parasites and hosts have such a powerful effect on each other that we might expect an ever-escalating "arms race" in which host resistance and parasite counterdefenses both get stronger and stronger over time. But such an outcome rarely occurs. In some cases—as in Dybdahl and Lively's snails and trematodes—host genotypes that are common decrease in frequency because they are attacked by many parasites, leading to an increase in the frequency of a previously rare genotype, so that the arms race continually

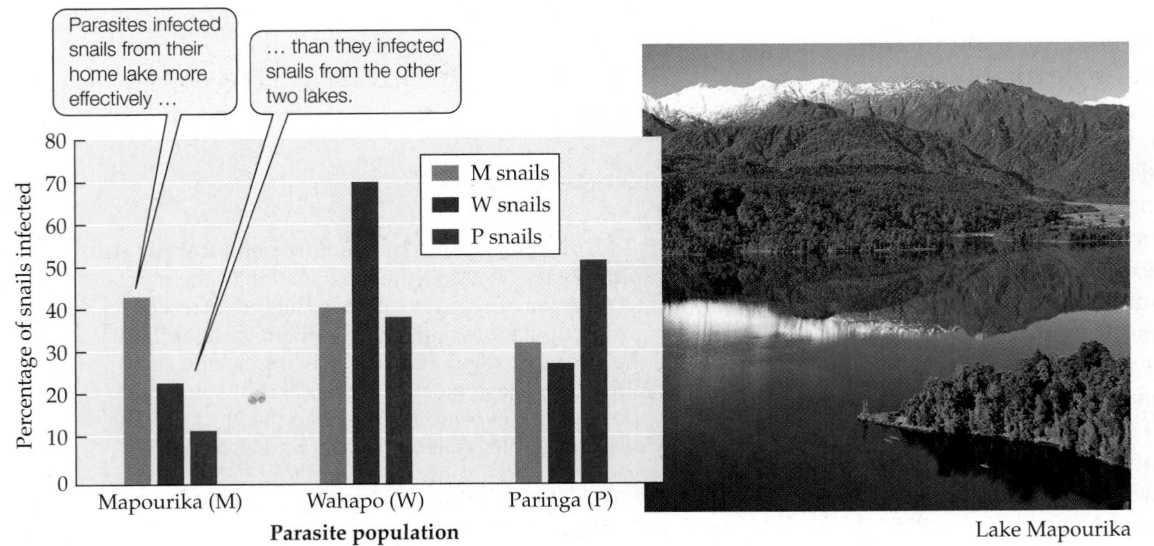

FIGURE 13.11 Adaptation by Parasites to Local Host Populations The graph shows the frequencies with which *Microphallus* parasites from three lakes in New Zealand (Lake Mapourika, Lake Wahapo, and Lake Paringa) were able to infect snails (*Potamopyrgus antipodarum*) from the same three lakes. (After Lively 1989.)

Q Are snails that are poorly defended against parasites from their own lake also poorly defended against parasites from other lakes? Explain.

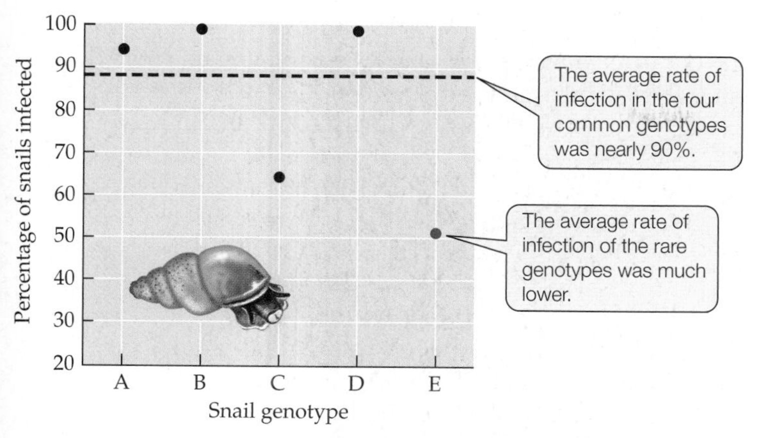

FIGURE 13.12 Parasites Infect Common Host Genotypes More Easily Than Rare Genotypes In a laboratory experiment, Dybdahl and Lively compared rates of *Microphallus* infection in four common snail genotypes (A–D, represented by blue dots) and a group of 40 rare snail genotypes (E, represented by a red dot). The parasites and snails in this experiment were all taken from the same lake. (After Dybdahl and Lively 1998.)

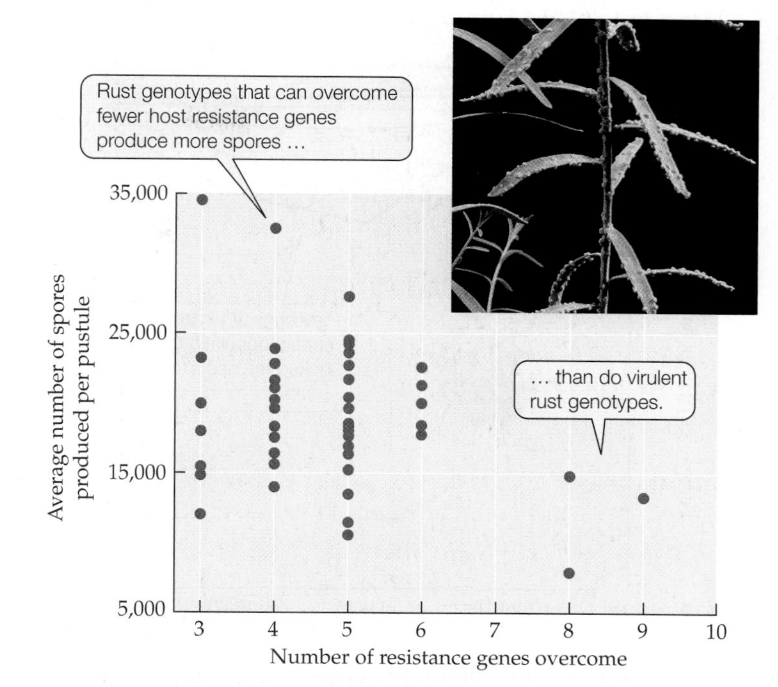

FIGURE 13.13 A Cost of Virulence Genotypes of a rust pathogen (*Melampsora lini*) that attacks wild flax (*Linum marginale*) show a trade-off between virulence (the ability to overcome many flax resistance genes) and spore production. Each red dot represents one rust genotype. The photo shows a wild flax plant covered with rust pustules. (After Thrall and Burdon 2003.)

begins anew. An arms race may also stop because of trade-offs: a trait that improves a host's defenses or a parasite's counterdefenses may have costs that reduce other aspects of the organism's growth, survival, or reproduction.

Such trade-offs have been documented in a number of host–parasite systems, including *Drosophila* fruit flies and the parasitoid wasps that attack them (described on p. 293). Alex Kraaijeveld and colleagues have shown that selection can increase both the frequency with which fruit fly hosts encapsulate wasp eggs (from 5% to 60% in five generations) and the ability of wasp eggs to avoid encapsulation (from 8% to 37% in ten generations). But they have also shown that there are costs to these defenses and counterdefenses. For example, fruit flies from lineages that can mount an encapsulation defense have lower larval survival rates when they compete for food with flies of the same species that cannot. Similarly, wasp eggs that avoid encapsulation by becoming embedded in host tissues take longer to hatch than do other eggs (Kraaijeveld et al. 2001).

Similar results were obtained by Peter Thrall and Jeremy Burdon, who studied populations of wild flax (*Linum marginale*) and its rust pathogen, *Melampsora lini*. Some rust genotypes are more virulent than others, meaning that they can overcome more plant resistance genes. Thrall and Burdon (2003) showed that virulent rust genotypes were common only in host populations dominated by plants with many resistance genes. Here too, a trade-off appears to be at work. Virulent rust genotypes produce fewer spores than do rust genotypes that can overcome only a few resistance genes (**Figure 13.13**). Spores are the means by which rusts disperse and reproduce. In flax populations dominated by plants with few or no resistance genes, the ability of virulent genotypes to overcome many resistance genes confers a disadvantage (reduced spore production), but no advantage (since there are few or no resistance genes to overcome). Apparently as a result of this trade-off, virulent rust genotypes are not common in low-resistance plant populations.

The evolutionary changes in host and parasite populations that we've discussed in this section reflect the profound effects these organisms have on each other. Next, we'll focus on some of the ecological consequences of host–parasite interactions.

CONCEPT 13.4 Parasites can reduce the sizes of host populations and alter the outcomes of species interactions, thereby causing communities to change.

Ecological Effects of Parasites

As we've seen, parasites can reduce the survival, growth, or reproduction of their hosts—an observation that is illustrated clearly by the large drop in reproductive success that a sexually transmitted mite inflicts on its beetle host

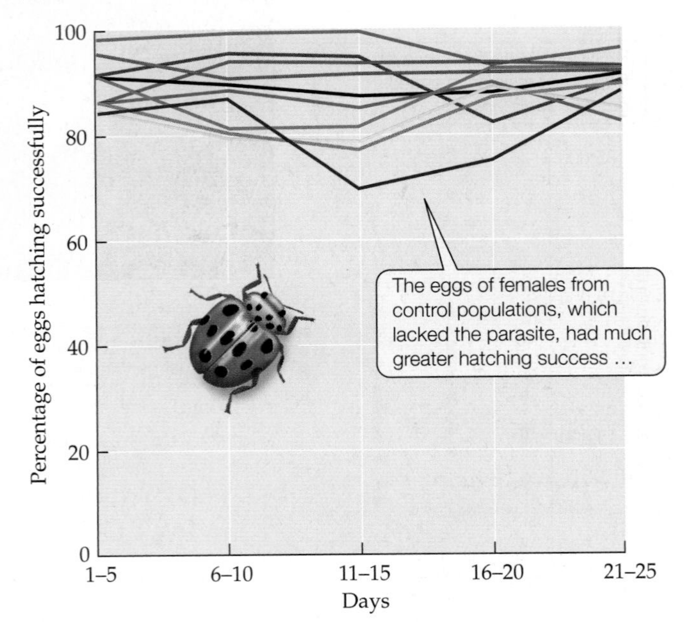

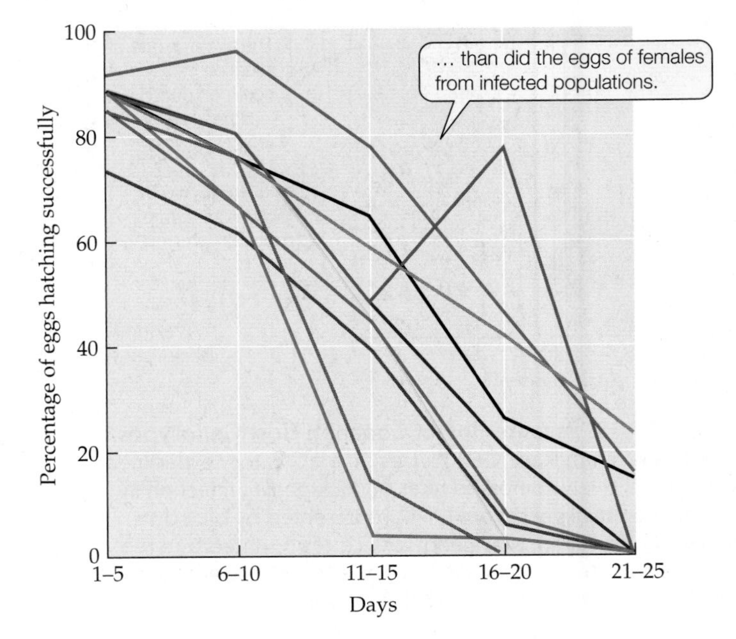

FIGURE 13.14 Parasites Can Reduce Host Reproduction Researchers infected experimental populations of the beetle *Adalia decempunctata* with a sexually transmitted mite parasite (*Coccipolipus hippodamiae*). Over the next 25 days, they monitored the proportions of the eggs laid by female beetles from infected and control populations that hatched. Each curve represents the eggs laid by a single female. (After Webberley et al. 2004.)

(**Figure 13.14**). At the population level, the harm that parasites cause host individuals translates into a reduction of the host population growth rate, λ (see Chapter 9). As we will see in this section, the reduction in λ can be drastic: parasites may drive local host populations extinct or even reduce the geographic range of the host species. In other, less extreme cases, parasites may reduce host abundances or otherwise alter host population dynamics, but do not cause the extinction of host populations.

Parasites can drive host populations to extinction

The amphipod *Corophium volutator* lives in North Atlantic tidal mudflats. *Corophium* is small (about 1 cm long) and often very abundant, reaching densities of up to 100,000 individuals per square meter. *Corophium* builds tubular burrows in the mud, from which it feeds on plankton suspended in the water and on microorganisms found in sediments near the burrow opening. *Corophium* is eaten by a wide range of organisms, including migratory birds and trematode parasites. The parasites can reduce the size of *Corophium* populations greatly, even to the point of local extinction. For example, in a 4-month period, attack by trematodes caused the extinction of a *Corophium* population that initially had 18,000 individuals per square meter (Mouritsen et al. 1998). We'll return to this example on p. 297.

Parasites can also drive host populations to extinction over a large geographic region. The American chestnut (*Castanea dentata*) once was a dominant member of deciduous forest communities in eastern North America (**Figure 13.15**), but the parasitic fungus *Cryphonectria parasitica* changed that completely. This fungal pathogen causes chestnut blight, a disease that kills chestnut trees. The fungus was introduced to New York City from Asia in 1904 (Keever 1953). By midcentury, the fungus had wiped out most chestnut populations, greatly reducing the geographic range of this once-dominant species.

Isolated chestnut trees still can be found in North American forests, and some of these trees show signs of resistance to the fungus. But it is likely that many of the standing trees simply have not yet been found by the fungus. Once the fungus reaches a tree, it enters the tree through a hole or wound in the bark, killing the aboveground portion of the tree in 2–10 years. Before they die, infected trees may produce seeds, which may germinate and give rise to offspring that live for 10–15 years before they are killed by the fungus in turn. Some infected trees also produce sprouts from their roots, but these are usually killed a few years after they appear aboveground. Efforts are under way to breed resistant chestnut varieties, but at present it is not known whether chestnut populations will ever recover from the onslaught of the chestnut blight fungus.

Parasites can influence host population cycles

Ecologists have long sought to determine the causes of population cycles. As we saw in the discussion under Con-

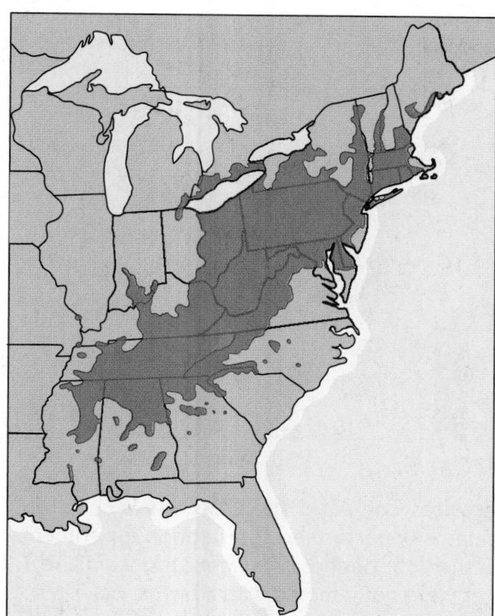

FIGURE 13.15 Parasites Can Reduce Their Host's Geographic Range (A) The original distribution of the American chestnut (*Castanea dentata*) is shown in red. Although a few chestnut trees remain standing, a fungal parasite drove this once-dominant species virtually extinct throughout its entire former range. (B) Chestnuts were once important timber trees (note the two loggers shown in the photograph).

cept 12.4, such cycles may be caused by three-way feeding relationships—by the effects that predators and herbivorous prey have on each other, coupled with the effects that those prey and their food plants have on each other.

Population cycles can also be influenced by parasites. Consider the work of Peter Hudson and colleagues, who manipulated the abundances of parasites in red grouse (*Lagopus lagopus*) populations on moors in northern England. In this region, red grouse populations tend to crash every 4 years. Previous studies had shown that a parasitic nematode, *Trichostrongylus tenuis*, decreased the survival and reproductive success of individual red grouse. Hudson et al. (1998) investigated whether this parasite might also cause grouse populations to cycle.

The researchers studied changes in red grouse numbers in six replicate populations over the course of two population cycles. Long-term data on grouse population cycles indicated that these populations were likely to crash in 1989 and again in 1993. In two of the six study populations, the researchers treated as many grouse as they could catch in 1989 and 1993 with a drug that killed the parasitic nematodes. In two of the other study populations, grouse were caught and treated for parasites in 1989 only. The remaining two populations served as unmanipulated controls. Because each replicate population covered a very large area (17–20 km^2), it was not possible to count red grouse directly. Instead, Hudson and colleagues used the number of red grouse shot by hunters as an index of the actual population size.

In the control populations, red grouse numbers crashed as predicted in 1989 and 1993 (**Figure 13.16**). Although parasite removal did not completely stop the red grouse

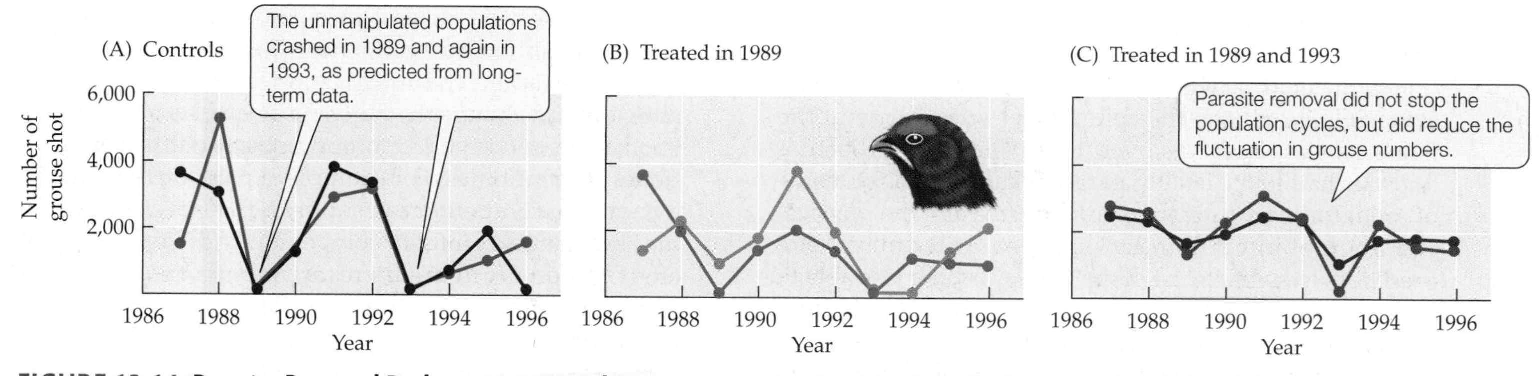

FIGURE 13.16 Parasite Removal Reduces Host Population Fluctuations Hudson et al. studied the effects of parasites on the cycling of six red grouse populations subjected to three treatments: (A) two control populations, (B) two populations treated for nematode parasites in 1989, and (C) two populations treated for parasites in 1989 and 1993. Each of the six replicate populations is designated by a different color. (After Hudson et al. 1998.)

Q If parasite removal completely stopped the population cycles, how might the results in (C) differ from those actually obtained?

population cycle, it did reduce the fluctuation in grouse numbers considerably; this was particularly true for the populations that were treated for parasites in both 1989 and 1993. Thus, the experiment provided strong evidence that parasites influence—and may be the primary cause of—red grouse population cycles.

Parasites can change ecological communities

The effects of parasites on their hosts can have ripple effects: by reducing the performance of host individuals and the growth rates of host populations, parasites can change the outcome of species interactions, the composition of ecological communities, and even the physical environment in which a community is found.

Changes in species interactions When two individual organisms interact with each other, the outcome of that interaction depends on many features of their biology. An individual predator that is young and healthy may be able to catch its prey—even though the prey organism literally "runs for its life"—whereas a predator that is old or sick may go hungry. Similarly, an individual that is in good condition may be able to compete effectively with others for resources, while an individual in poor condition may not.

Because they can affect host performance, parasites can affect the outcome of interactions between their hosts and other species. Thomas Park conducted a series of experiments on factors that influenced the outcome of competition between flour beetle species. In one of those experiments, Park (1948) examined how the protist parasite *Adelina tribolii* affected the outcome of competition between two species of flour beetles, *Tribolium castaneum* and *T. confusum*. In the absence of the parasite, *T. castaneum* usually outcompeted *T. confusum*, driving it to extinction in 12 of 18 cases (**Figure 13.17**). The reverse was true when the parasite was present: *T. confusum* outcompeted *T. castaneum* in 11 of 15 cases. The outcome of competition was reversed because the parasite had a large negative effect on *T. castaneum* individuals, but virtually no effect on *T. confusum*. Parasites can also affect the outcome of competition in the field, as when the malaria parasite *Plasmodium azurophilum* reduced the competitive superiority of the lizard *Anolis gingivinus* over its smaller counterpart, *A. wattsi* (Schall 1992). Finally, parasites can alter the outcome of predator–prey interactions: by decreasing the physical condition of infected individuals, parasites may make predators less able to catch their prey, or prey less able to escape predation.

In the examples described in the previous paragraph, parasites affected the outcome of species interactions by altering the physical condition of their host. Parasites can also alter the outcome of species interactions by changing host behavior. For example, when infected by a parasite, the host may behave in an unusual manner that makes it more vulnerable to predation. There are numerous examples of this phenomenon, including the protist parasites described in the Case Study that make rats less wary of cats (see p. 284). Some worm parasites cause amphipods to move from sheltered areas to areas of relatively bright light, where the amphipods are more likely to be seen and eaten by fish or bird predators. In both of these cases, the parasite induces a change in host behavior that makes the host more likely to be eaten by a species that the parasite requires to complete its life cycle.

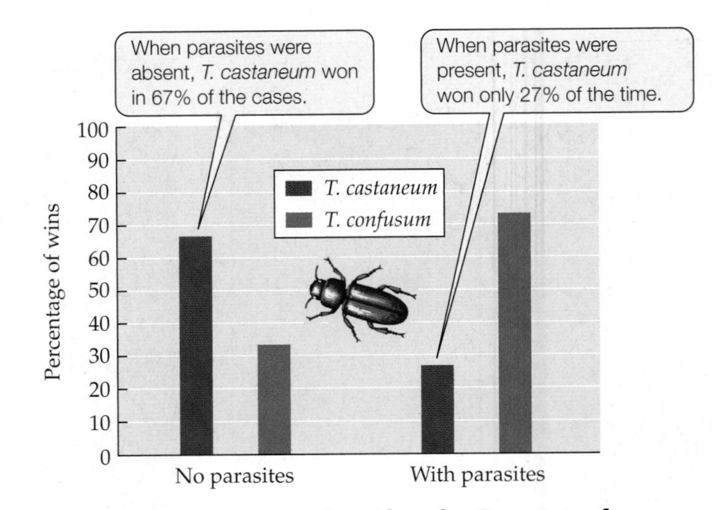

FIGURE 13.17 Parasites Can Alter the Outcome of Competition Thomas Park performed competition experiments using populations of the flour beetles *Tribolium castaneum* and *T. confusum* that were or were not infected with a protist parasite. (After Park 1948.)

Changes in community structure As we'll discuss in Chapter 15, communities can be characterized by the number and relative abundances of the species they contain as well as by physical features of the environment. Parasites can alter each of these aspects of ecological communities.

We have seen several cases in which a parasite reduced the abundance of its host (for example, chestnut blight fungus and chestnuts), and we have also seen that parasites can change the outcome of species interactions. Such changes can have profound effects on the composition of communities. For example, a parasite that attacks a dominant competitor can suppress that species, causing the abundances of inferior competitors to increase. Such an effect was observed in six stream communities studied by Kohler and Wiley (1997). Prior to recurrent outbreaks of a fungal pathogen, the caddisfly *Glossosoma nigrior* was the dominant herbivore in each of the six communities. The fungus devastated *Glossosoma* populations, reducing their densities nearly 25-fold, from an average of 4,600 individuals per square meter to an average of 190 individuals per square meter. This drastic reduction in *Glossosoma* density

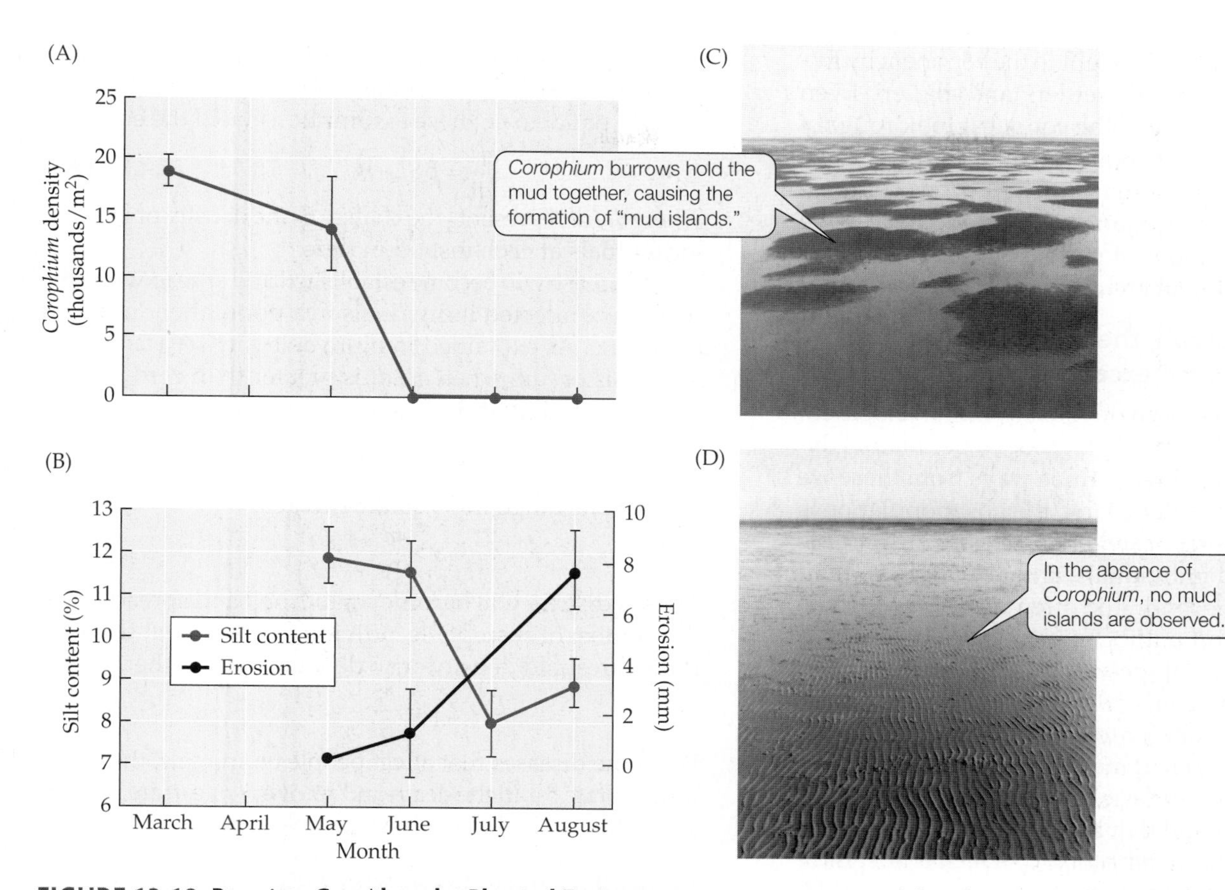

FIGURE 13.18 Parasites Can Alter the Physical Environment Infection of the amphipod *Corophium volutator* by a trematode parasite affects not only the host, but its entire tidal mudflat community. (A) The trematode can drive amphipod populations to local extinction. (B) In the absence of *Corophium*, the erosion rate increases and the silt content of the mudflats decreases. (C, D) The overall physical structure of the mudflats also changes (compare [C] with [D]). Error bars show ±one SE of the mean. (After Mouritsen and Poulin 2002.)

allowed increases in the abundances of dozens of other species, including algae, grazing insects that ate algae, and filter feeders such as blackfly larvae. In addition, several species that previously were extremely rare or absent from the communities were able to establish thriving populations, thus increasing the diversity of the communities.

Parasites can also cause changes in the physical environment. This can happen when a parasite attacks an organism that is an *ecosystem engineer*, a species whose actions change the physical character of its environment, as when a beaver builds a dam (see Chapter 15). The amphipod *Corophium* (see p. 294) sometimes functions as an ecosystem engineer in its tidal mudflat environment: in some circumstances, the burrows it builds hold the mud together, preventing the erosion of silt and causing the formation of "mud islands" that rise above the surface of the water at low tide. As described earlier, trematode parasites can drive local *Corophium* populations to extinction (**Figure 13.18A**). When this happens, erosion rates increase, the silt content of the mudflats decreases, and the mud islands disappear (**Figure 13.18B–D**). Along with these physical changes, the abundances of ten large species in the mudflat community change considerably, including one species (a ribbon worm) that was driven to local extinction (K. N. Mouritsen, personal communication).

CONCEPT 13.5 Simple models of host–pathogen dynamics suggest ways to control the establishment and spread of diseases.

Dynamics and Spread of Diseases

As we've seen, parasites that cause diseases (pathogens) can greatly affect the population dynamics of both wild and domesticated plant and animal species. Pathogens also have large effects on human populations—so much so that they are thought to have played a major role in the rise and fall of civilizations throughout the course of human history (McNeill 1976; Diamond 1997). Examples include the European conquest of North America, where up to 95% of the native population (19 million of the original 20 million) was killed by new diseases brought to the continent by European trappers, missionaries, settlers, and soldiers. Even with such massive mortality, the conquest took roughly 400 years; without it, the conquest would certainly have taken longer, and might have failed. Pathogens continue to be a major source of human mortality today. Despite medical advances, millions of people die each year from diseases such as AIDS, tuberculosis, and malaria.

For a disease to spread, the density of susceptible hosts must exceed a critical threshold

Considerable effort has been devoted to the development of mathematical models of host–pathogen population dynamics. These models differ in three ways from those we have seen in earlier chapters. First, the host population is subdivided into different categories, such as susceptible individuals, infected individuals, and recovered and immune individuals. Second, it is often necessary to keep track of both host and pathogen genotypes because, as we have seen, host genotypes may differ greatly in their resistance to the pathogen, and pathogen genotypes may differ greatly in their ability to cause disease. Third, depending on the pathogen, it may be necessary to account for other factors that influence its spread, such as (1) differences in the likelihood that hosts of different ages will become infected, (2) a *latent period* in which a host individual is infected but cannot spread the disease, and (3) *vertical transmission*, the spread of the disease from mother to newborn, as can occur in AIDS.

Models that include all of these factors can be very complicated. Here we'll consider a simple model that ignores most of these complicating factors, yet still yields an important insight: a disease will spread only if the density of susceptible hosts exceeds a critical **threshold density**.

Modeling host–pathogen population dynamics To develop a model that can be used to estimate the threshold density, we must determine how to represent the transmission of the disease from one host individual to the next. We'll denote the density of susceptible individuals by S and the density of infected individuals by I. For a disease to spread, infected individuals must encounter susceptible individuals. Such encounters are assumed to occur at a rate that is proportional to the densities of susceptible and infected individuals; here, we'll assume that this rate is proportional to the product of their densities, SI. Diseases do not spread with every such encounter, however, so we multiply the encounter rate (SI) by a transmission coefficient (β) that indicates how effectively the disease spreads from infected to susceptible individuals. Thus, the key term in the model—disease transmission—is represented by the term βSI.

The density of infected individuals increases when the disease is transmitted successfully (at the rate βSI) and decreases when infected individuals die or recover from the disease. If we set the combined death and recovery rate equal to m, these assumptions yield the equation

$$\frac{dI}{dt} = \beta SI - mI \tag{13.1}$$

where dI/dt represents the change in the density of infected individuals at each instant in time.

A disease will become established and spread when the density of infected individuals in a population increases over time. As explained in more detail in **Web Extension 13.2**, this occurs when dI/dt is greater than zero, which, according to Equation 13.1, occurs when

$$\beta SI - mI > 0$$

We can rearrange this equation to get

$$S > \frac{m}{\beta}$$

Thus, a disease will become established and spread when the number of susceptible individuals exceeds m/β, which is the threshold density (also denoted S_T). In other words,

$$S_T = \frac{m}{\beta}$$

For some diseases that affect people or animals, the transmission rate β and the death and recovery rate m are known, permitting estimation of the threshold density.

Controlling the spread of diseases As Equation 13.1 suggests, to prevent the spread of a disease, the density of susceptible individuals must be kept below the threshold density (S_T). There are several ways of achieving this goal. People sometimes slaughter large numbers of susceptible domesticated animals to reduce their density below S_T and hence prevent disease spread. This is typically done when the disease in question can spread to humans, as in highly virulent forms of bird flu. In human populations, if an effective and safe vaccine is available, the density of susceptible individuals can be reduced below S_T by a mass vaccination program. Such programs work, as illustrated by the dramatic results of a measles vaccination program in Romania (**Figure 13.19**).

In addition, public health measures can be taken to raise the threshold density, thereby making it more difficult for the disease to become established and spread. For example, the threshold density can be raised by taking actions that increase the rate at which infected individuals recover and become immune (thereby increasing m and hence increasing $S_T = m/\beta$). One way to increase the recovery rate is to improve the early detection and clinical treatment of the disease. The threshold density can also be raised if β, the disease transmission rate, is decreased. This can be achieved by quarantining infected individuals or by convincing people to engage in behaviors (such as hand washing or condom use) that make it more difficult for the disease to be transmitted from one person to the next.

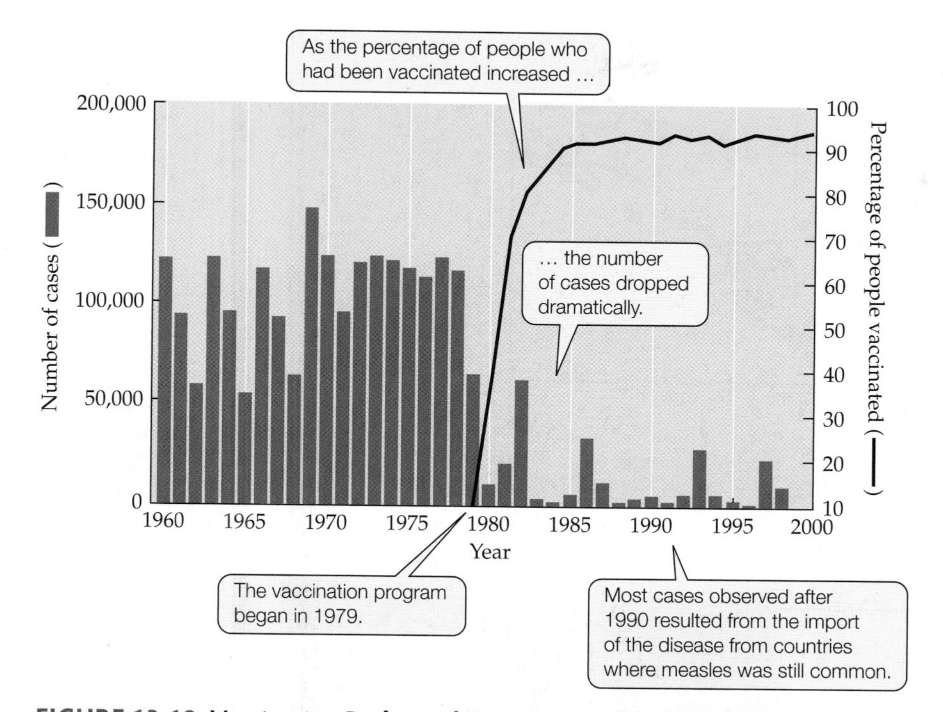

FIGURE 13.19 Vaccination Reduces the Incidence of Disease The results of a measles vaccination program in Romania show that lowering the density of susceptible individuals can control the spread of a disease. Measles often kills (especially in populations that are poorly nourished or that lack a history of exposure to the disease) and can cause severe complications in survivors, including blindness and pneumonia. (After Strebel and Cochi 2001.)

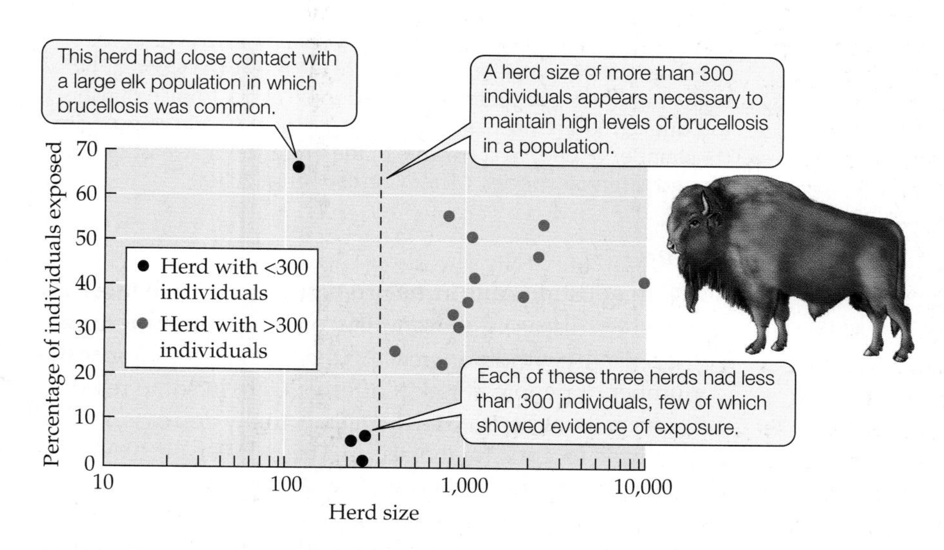

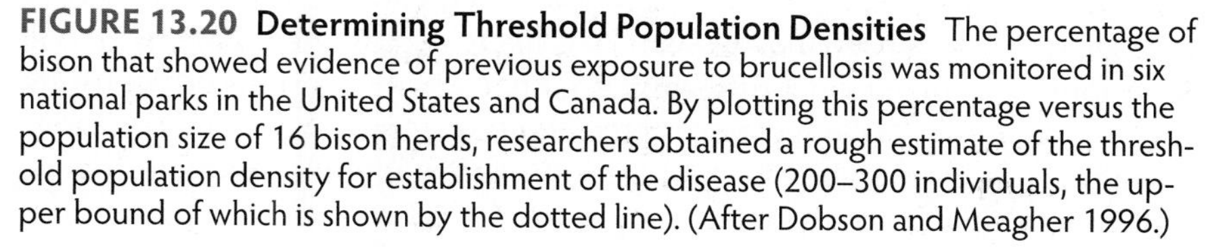
FIGURE 13.20 Determining Threshold Population Densities The percentage of bison that showed evidence of previous exposure to brucellosis was monitored in six national parks in the United States and Canada. By plotting this percentage versus the population size of 16 bison herds, researchers obtained a rough estimate of the threshold population density for establishment of the disease (200–300 individuals, the upper bound of which is shown by the dotted line). (After Dobson and Meagher 1996.)

The same principles can be applied to wild populations. Dobson and Meagher (1996) studied bison populations to determine how best to prevent the spread of the bacterial disease brucellosis. Using data from previous studies in which 16 bison herds in six national parks in Canada and the United States had been tested for exposure to the disease, they found that the threshold density (S_T) for disease establishment appeared to be a herd size of 200–300 bison (**Figure 13.20**). This field-based estimate of S_T was very similar to the estimated threshold density of 240 individuals calculated from a model similar to Equation 13.1. Many of the herds in the six national parks had 1,000–3,000 individuals, so reducing herd sizes below a threshold value of 200–300 individuals would require implementing a vaccination program or killing large numbers of bison. An effective vaccine was not available, and killing many bison was not acceptable, either politically or ecologically (since herds as small as 200 individuals would face an increased risk of extinction). Thus, Dobson and Meagher concluded that it would be difficult to prevent the establishment of brucellosis in wild bison populations.

Climate change is altering the distribution and incidence of some diseases

Climate affects the physiology of organisms, the distribution and abundance of populations, and the outcome of interactions between species (see Chapter 2). As a result, changes in climate are expected to have wide-ranging effects on ecological communities. In particular, because mosquitoes and other vectors that transmit pathogens are often more active or produce more offspring under warm conditions, scientists have predicted that ongoing climate change may cause the incidence of some diseases to rise in human and wildlife populations (Epstein 2000; Harvell et al. 2002). A growing body of evidence supports this prediction. In one such study, increases in ocean temperatures were strongly correlated with increases in coral diseases along Australia's Great Barrier Reef (Bruno et al. 2007). Similar results have been found in corals at other locations, as well as in a variety of amphibian and shellfish populations (Harvell et al. 2009).

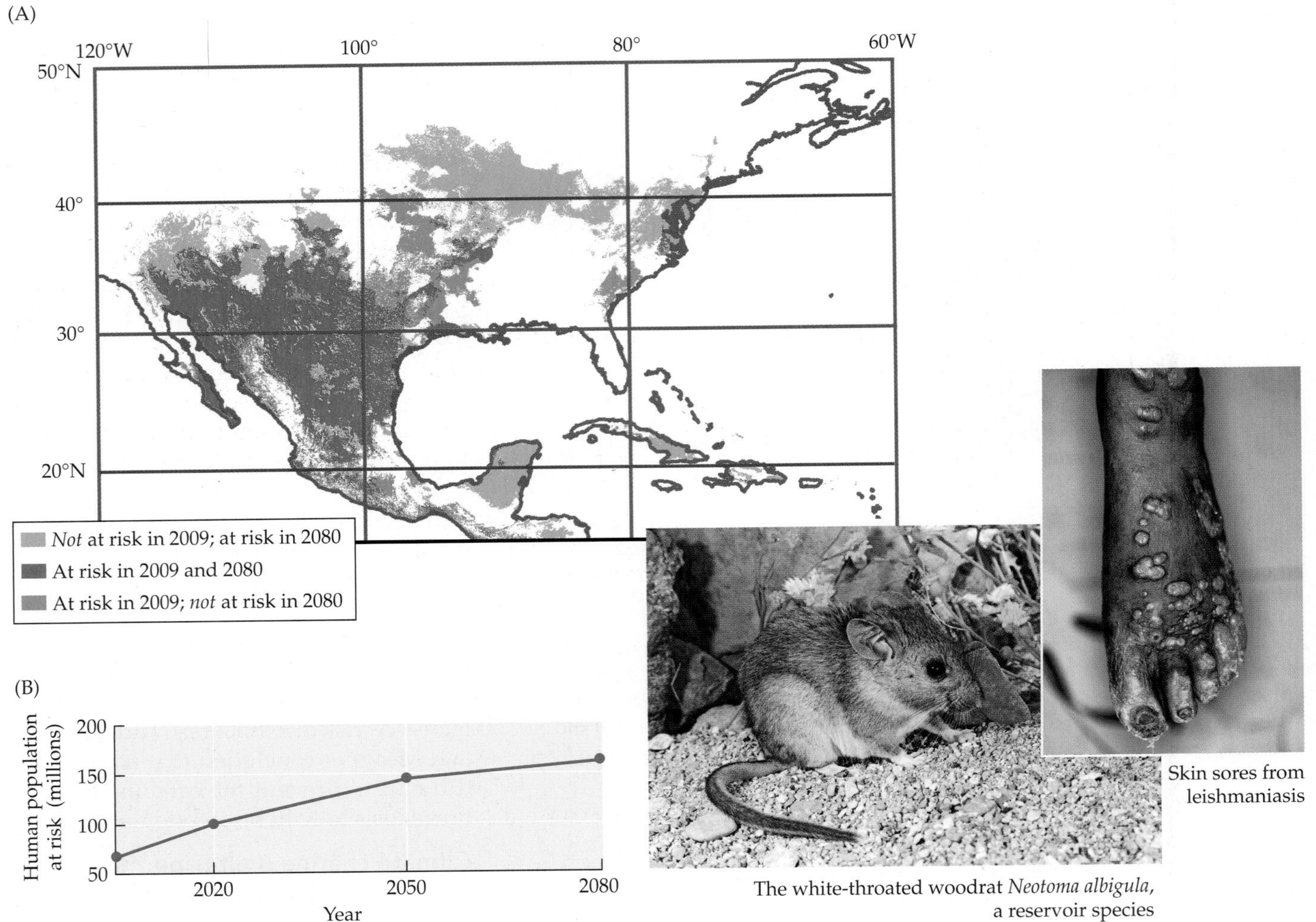

FIGURE 13.21 Climate Change May Increase the Risk of Leishmaniasis in North America Leishmaniasis can cause severe skin sores (see inset), difficulty breathing, immune system impairment, and other complications that can lead to death. There are currently 2 million new cases each year. Leishmaniasis is caused by protists in the genus *Leishmania* and spread by sand flies (bloodsucking insects in the genera *Lutzomya* and *Phlebotomus*); in addition to infecting humans, the pathogen can persist in several reservoir species (rodents in the genus *Neotoma*; see inset). (A) Geographic regions in which people are predicted to be at risk from leishmaniasis due to the presence of at least one vector and reservoir species. (B) Number of people at risk due to the presence of at least one vector and reservoir species. (After González et al. 2010.)

Climate change is also expected to change the distributions of some pathogens and their vectors by changing the locations where conditions are suitable for those organisms. For example, González et al. (2010) found that climate change is likely to increase the risk of leishmaniasis in North America by increasing the geographic ranges of its reservoir species (rodents in the genus *Neotoma* that can harbor the pathogen) and its sand fly vectors (**Figure 13.21**). Similarly, the number of people at risk from malaria, cholera, and the plague may increase as global temperatures continue to warm (see citations in Ostfeld 2009).

Although climate change has favored and will continue to favor the spread of some pathogens, a variety of complicating factors may influence how climate change affects any given disease. For example, a pathogen might shift geographically without increasing the area of its range or the number of people at risk (Lafferty 2009); in some regions, the number of people at risk might actually decrease, as was predicted recently for malaria in Africa (Peterson 2009). The effects of climate change on disease may also be altered by control efforts and by the effects of dispersal limitation and ecological interactions on both pathogens and their hosts. Nevertheless, while it is likely that different pathogens will respond differently to climate change, serious public health challenges are expected in many cases. (See **Climate Change Connection 13.1** for more information on climate change and disease.)

A CASE STUDY REVISITED

Enslaver Parasites

Returning to a question we posed in the opening pages of this chapter, how do enslaver parasites manipulate the behavior of their hosts? In some cases, we have hints of how this occurs. Consider the tropical parasitoid wasp *Hymenoepimecis argyraphaga* and its host, the orb-weaving spider *Plesiometa argyra*. The larval stage of this wasp attaches to the exterior of a spider's abdomen and sucks the spider's body fluids. When fully grown, the wasp larva induces the spider to make a special "cocoon web" (**Figure 13.22**). Once the spider has built the cocoon web, the larva kills and eats the spider. The larva then spins a cocoon and attaches it to the cocoon web. As the larva completes its development within the cocoon, the cocoon web serves as a strong support that protects the larva from being swept away by torrential rains.

A parasitized spider builds normal webs right up to the night when the wasp induces it to make a cocoon web. This sudden change in the spider's web-building behavior suggested that the wasp might inject the spider with a chemical that alters its behavior. To test this idea, William Eberhard (2001) removed wasp larvae from their hosts several hours before the time when a cocoon web would usually be made. Wasp removal sometimes resulted in the construction of a web that was very similar to a cocoon web, but more often resulted in the construction of a web that differed substantially from both normal and cocoon webs. In the days that followed the removal of the parasite, some spiders partially recovered the ability to make normal webs. These results are consistent with the idea that the parasite induces construction of a cocoon web by injecting a fast-acting chemical into the spider. The chemical appears to act in a dose-dependent manner; otherwise, we would expect spiders exposed to the chemical to build only cocoon webs, not webs that are intermediate in form. Spiders build cocoon webs by repeating the early steps of their normal web-building sequence a large number of times; thus, the chemical appears to act by interrupting the spiders' usual sequence of web-building behaviors.

Other enslaver parasites also appear to manipulate host body chemistry. On p. 283, we described hairworm parasites that cause crickets to commit suicide by jumping into water. Thomas and colleagues (2003) have shown that the hairworm causes biochemical and structural changes in the brain of its cricket host. The concentrations of three amino acids (taurine, valine, and tyrosine) in the brains of parasitized crickets differed from those in crickets that were not parasitized. Taurine, in particular, is an important neurotransmitter in insects, and it also regulates the brain's ability to sense a lack of water. Hence, it is possible that the parasite induces its host to commit suicide by causing biochemical changes in its brain that alter the host's perception of thirst.

The papers by Eberhard (2001) and Thomas et al. (2003) suggest that some parasites enslave their hosts by manipulating them chemically. In other cases, however, little is known about how enslaver parasites alter the

FIGURE 13.22 Parasites Can Alter Host Behavior The parasitic wasp *Hymenoepimecis argyraphaga* dramatically alters the web-building behavior of the orb-weaving spider *Plesiometa argyra*. (A) The web of an uninfected spider. (B) The "cocoon web" of a parasitized spider.

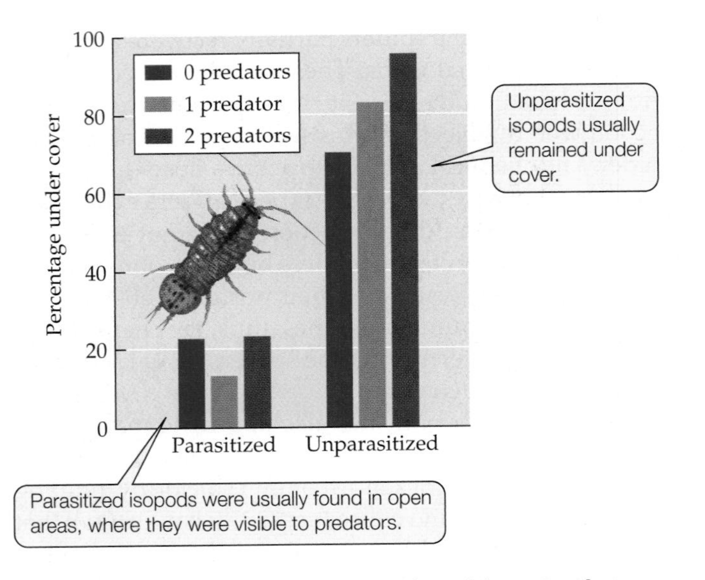

FIGURE 13.23 Making the Host Vulnerable to Predation Isopods that are not infected by worm parasites tend to remain under cover, where food is available and where they are not exposed to fish predators. Hechtel et al. exposed parasitized and unparasitized isopods to 0, 1, or 2 predator individuals and observed the influence of parasites on their behavior. (Data from Hechtel et al. 1993.)

Q Do either parasitized or unparasitized isopods alter their behavior if predators are present?

behavior of their hosts. For example, Laura Hechtel and colleagues studied isopods (*Caecidotea intermedius*) parasitized by the worm *Acanthocephalus dirus*. They observed that, compared with unparasitized isopods, those that were parasitized spent little time under cover, where food was available and where they could hide from a fish predator, the creek chub—the next host in the parasite's complex life cycle (**Figure 13.23**). Hechtel and colleagues also performed a choice experiment in which isopods could move to one side of an aquarium, where a chub was present (but separated from the isopod by a net), or to the other side, where no chubs were present. They found that unparasitized isopods avoided the chubs, but parasitized isopods were actually drawn toward the chubs—a behavior that in the wild would benefit the parasite but be disastrous for the isopod.

The mechanism by which the parasite induces changes in isopod behavior is not known. Even in Eberhard's work, which suggests that the wasp injects a chemical into its spider host that induces it to make a cocoon web, the chemical in question has not been found. If this chemical were known, it could be injected into unparasitized spiders; if those spiders constructed cocoon webs, we would have a clear understanding of how the parasite manipulates the spider. Such a definitive experiment has yet to be performed for any parasite that enslaves its host.

CONNECTIONS IN NATURE
From Chemicals to Evolution and Ecosystems

Enslaver parasites that manipulate their hosts exert strong selection pressure on host populations. As such, resistance to the manipulations of enslaver parasites might be expected to evolve in host populations—for example, selection would favor host individuals with the ability to recognize and destroy chemicals that a parasite uses to alter its behavior. Likewise, enslaver parasites might be expected to evolve the ability to overcome host resistance mechanisms. To date, we know of no such evidence of ongoing host–enslaver parasite coevolution.

However, interactions between enslaver parasites and their hosts do provide evidence of previous evolutionary change. Like any parasite, an enslaver parasite has adaptations that allow it to cope with host defenses (otherwise it would not survive). More specifically, an enslaver parasite that uses a chemical to manipulate a specific host behavior is beautifully adapted to take advantage of the body chemistry of its host. Such evolutionary links between enslaver parasites and their hosts illustrate a central feature of both ecology and evolution: ecological interactions affect evolution, and vice versa, at times making it difficult to distinguish one from the other (see the discussion under Concept 6.5). As we've seen in this chapter, the outcome of such ecological and evolutionary interactions can have profound effects on individuals, populations, communities, and ecosystems. As evolutionary change tips the balance back and forth, first in favor of the host, then in favor of the parasite, we can expect concomitant changes in the population dynamics of other species, such as those that compete with or eat the host or the parasite. Viewed in this way, communities and ecosystems are highly dynamic, always shifting in response to the ongoing ecological and evolutionary changes that are occurring within them.

SUMMARY

CONCEPT 13.1 Parasites typically feed on only one or a few host individuals.

- Parasites usually feed on only one or a few host individuals during the course of their lives. Many parasites are closely adapted to particular host species.
- Some parasites are ectoparasites that live on the surface of their host; others are endoparasites that live within the body of their host.
- Endoparasitism and ectoparasitism each have advantages and disadvantages. It is easier for ectoparasites or their offspring to disperse from one host individual to another; however, ectoparasites are at greater risk from natural enemies than are endoparasites.

CONCEPT 13.2 Hosts have adaptations for defending themselves against parasites, and parasites have adaptations for overcoming host defenses.

- Many host organisms have immune systems that allow them to recognize and defend against endoparasites. Biochemical conditions inside the host's body can also provide protection against parasites.
- In some birds, mammals, and fishes, females select their mates based on traits that indicate whether a male is well defended from parasites.
- Parasites have a broad suite of adaptations that allow them to circumvent host defenses, from relatively simple counterdefenses against encapsulation to more complex counterdefenses that involve hundreds of genes.

CONCEPT 13.3 Host and parasite populations can evolve together, each in response to selection pressure imposed by the other.

- Host–parasite interactions can result in coevolution, in which populations of the host and parasite evolve together, each in response to selection pressure imposed by the other.
- Selection can favor a diversity of host and parasite genotypes. A rare host genotype may increase in frequency because few parasites can overcome its defenses; as a result, parasite genotypes that can cope with the now-common host genotype's defenses may also increase in frequency.
- Host–parasite interactions can exhibit trade-offs in which a trait that improves host defenses or parasite counterdefenses has costs that reduce other aspects of the organism's growth, survival, or reproduction.

CONCEPT 13.4 Parasites can reduce the sizes of host populations and alter the outcomes of species interactions, thereby causing communities to change.

- Parasites can reduce the abundances of host populations, in some cases driving local host populations to extinction or changing the geographic distributions of host species.
- Evidence suggests that parasites can influence host population cycles.
- Parasites can affect the outcomes of interactions between their hosts and other species; for example, a species that is a dominant competitor may become an inferior competitor when infected by a parasite.
- The effects of parasites can alter the composition of ecological communities and change features of the physical environment.

CONCEPT 13.5 Simple models of host–pathogen dynamics suggest ways to control the establishment and spread of diseases.

- Some models of host–pathogen population dynamics subdivide the host population into susceptible individuals, infected individuals, and recovered and immune individuals; track different host and pathogen genotypes; and take into account factors such as host age, latent periods, and vertical transmission.
- A simple mathematical model of host–pathogen dynamics (Equation 13.1) yields an important insight: for a disease to become established and spread, the density of susceptible hosts must exceed a critical threshold density.
- To control the spread of a disease, efforts may be made to lower the density of susceptible hosts (by slaughtering domesticated animals or undertaking vaccination programs) or to raise the threshold density (by increasing the recovery rate or decreasing the transmission rate).
- Ongoing climate change may cause the incidence of some diseases to rise.

REVIEW QUESTIONS

1. Define endoparasites and ectoparasites, giving an example of each. Describe some advantages and disadvantages associated with each of these two types of parasitism.
2. Given the effects that parasites can have on host individuals and host populations, would you expect that parasites could also alter the outcomes of species interactions and the composition of ecological communities? Explain.
3. a. What is meant by the concept of a threshold density for the establishment and spread of a disease? Why is this concept important?
   b. Explain the logic and show the algebraic steps by which Equation 13.1 can be used to calculate that the threshold density (S_T) has the value $S_T = m/\beta$.
4. a. Summarize the mechanisms that host organisms use to kill parasites or reduce the severity of their attack.
   b. With your answer to part (a) as background material, do you think the following statement from a news report could be true?

   The parasite has a mild effect on a plant species in Australia, but after it was introduced for the first time to Europe, it had devastating effects on European populations of the same plant species.

   Explain your reasoning, and illustrate your argument with an example of how a plant defensive mechanism might work—or fail to work—in a situation such as this.

ON THE COMPANION WEBSITE
sites.sinauer.com/ecology2e

The website includes Chapter Outlines, Online Quizzes, Flashcards & Key Terms, Suggested Readings, a complete Glossary, and the Web Stats Review. In addition, the following resources are available for this chapter:

HANDS-ON PROBLEM SOLVING

Dynamics of Disease

This Web exercise explores the dynamics of host–pathogen systems. You will manipulate traits of the interacting species to simulate various strategies of hosts and pathogens. You will also explore the spread or decline of pathogens based on the population size of vulnerable hosts, and discuss the ecological and evolutionary implications of parasitic interactions.

WEB EXTENSIONS

13.1 Enslaved by a Hairworm Parasite

13.2 When Will a Disease Become Established and Spread?

CLIMATE CHANGE CONNECTIONS

13.1 Climate Change and Disease

CHAPTER

14 Mutualism and Commensalism

KEY CONCEPTS

CONCEPT 14.1 In positive interactions, neither species is harmed and the benefits of the interaction are greater than the costs for at least one species.

CONCEPT 14.2 Each partner in a mutualistic interaction acts in ways that serve its own ecological and evolutionary interests.

CONCEPT 14.3 Positive interactions affect the abundances and distributions of populations as well as the composition of ecological communities.

The First Farmers: A Case Study

Humans first began to farm about 10,000 years ago. Agriculture was a revolutionary development that led to great increases in the size of our population as well as to innovations in government, science, the arts, and many other aspects of human societies. But people were far from the first species to farm. That distinction goes to ants in the tribe Attini, a group of 210 species, most of which live in tropical forests of South America. These ants, known informally as the attines or fungus-growing ants, started cultivating fungi for food at least 50 million years before the first human farmers (**Figure 14.1**).

Like human farmers, the ant farmers nourish, protect, and feed on the species they grow, forming a relationship that benefits both the farmer and the crop. The attines cannot survive without the fungi they cultivate; many of the fungi depend absolutely on the ants as well. When a virgin queen ant leaves her mother's nest to mate and begin a new colony, she carries in her mouth some of the fungi from her birth colony. The fungi are cultivated by the colony in subterranean gardens (**Figure 14.2**). An ant colony may contain hundreds of gardens, each roughly the size of a football; these hundreds of gardens can provide enough food to support 2–8 million ants.

Some attines occasionally replace the fungi in their gardens with new, free-living fungi that they gather from surrounding soils. Other species, such as leaf-cutter ants in the genera *Atta* and *Acromyrmex*, do not cultivate fungi found in the environment. Instead, the fungi in their gardens come only from propagules passed from a parent ant colony to each of its descendant colonies.

As their name suggests, leaf-cutter ants cut portions of leaves from plants and feed them to the fungi in their gardens. Back at the nest, the ants chew the leaves to a pulp, fertilize them with their own droppings, and "weed" the fungal gardens to help control bacterial and fungal invaders. In turn, the cultivated fungi produce specialized structures, called *gongylidia*, on which the ants feed. The partnership between leaf-cutter ants and fungi has been called an "unholy alliance" because each partner helps the other to overcome the formidable defenses that protect plants from being eaten. The ants, for example, scrape a waxy covering from the leaves that the fungi have difficulty penetrating, while the fungi digest and render harmless the chemicals that plants use to kill or deter insect herbivores.

FIGURE 14.1 Collecting Food for Their Fungi Fungus-growing ants (*Atta cephalotes*) in Costa Rica carry leaf segments to their colony, where the leaves will be fed to the fungi the ants cultivate for food. The ants riding on the leaves are of the same species as the leaf carriers and are called *minims* (in reference to their small size). Minims guard the ants carrying leaves from parasitic flies.

(A)

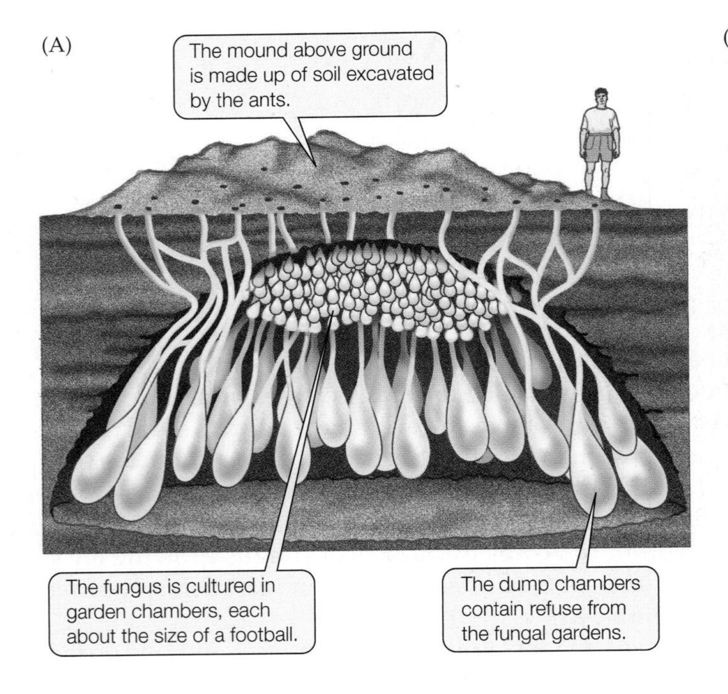

(B)

FIGURE 14.2 The Fungal Garden of a Leaf-Cutter Ant (A) A diagrammatic representation of a large *Atta* leaf-cutter ant colony. (B) This photo shows a cutaway view of a garden chamber in a central Paraguay colony of the leaf-cutter ant *Atta laevigata*. Several winged ants can be seen hiding from the disturbance created by excavating the garden: they have placed their heads into crevices of the garden, where they will remain relatively motionless for a short time.

But all is not perfect in the gardens. Nonresident fungi, which themselves would benefit from ant cultivation, periodically invade leaf-cutter ant colonies. Furthermore, pathogens and parasites that attack the cultivated fungi occasionally outstrip the ants' ability to weed them out. What prevents such unwanted guests from destroying the gardens?

Introduction

Chapters 11–13 emphasized interactions between species in which at least one member is harmed (competition, predation, herbivory, and parasitism). But life on Earth is also shaped by **positive interactions**, those in which one or both species benefit and neither is harmed. Most vascular plants, for example, form beneficial associations with fungi that improve the growth and survival of both species. In fact, fossil evidence indicates that the earliest vascular plants formed similar associations with fungi more than 400 million years ago (Selosse and Le Tacon 1998). These early vascular plants lacked true roots, so their interactions with fungi may have increased their access to soil resources and aided their colonization of land.

As this example suggests, positive interactions have influenced key events in the history of life as well as the growth and survival of organisms living today. As we'll see in this chapter, positive interactions can also influence the outcome of other types of interactions among organisms, thus shaping communities and influencing ecosystems. We will begin our study of positive interactions with definitions of some key terms and an overview of the scope of these interactions in ecological communities.

CONCEPT 14.1 In positive interactions, neither species is harmed and the benefits of the interaction are greater than the costs for at least one species.

Positive Interactions

There are two fundamental types of positive interactions: mutualism and commensalism. **Mutualism** is a mutually beneficial interaction between individuals of two species (a +/+ relationship). **Commensalism** is an interaction between individuals of two species in which individuals of one species benefit, while those of the other species do not benefit and are not harmed (a +/0 relationship). Many ecologists refer to mutualism and commensalism collectively as *facilitation*, a synonym for positive interactions.

In some cases, the species involved in a positive interaction form a **symbiosis**, a relationship in which individuals of the two species live in close physiological contact with each other. Examples include the relationships between corals and algae (described on the next page) and between humans and bacteria (we have a diverse set of bacteria living in our guts, many of which are beneficial). However, parasites also form symbiotic associations with their hosts (see Figure 13.3). Thus, symbiotic relationships can range from parasitism (+/–) to commensalism (+/0) to mutualism (+/+).

In mutualism and commensalism, the growth, survival, or reproduction of individuals of one or both spe-

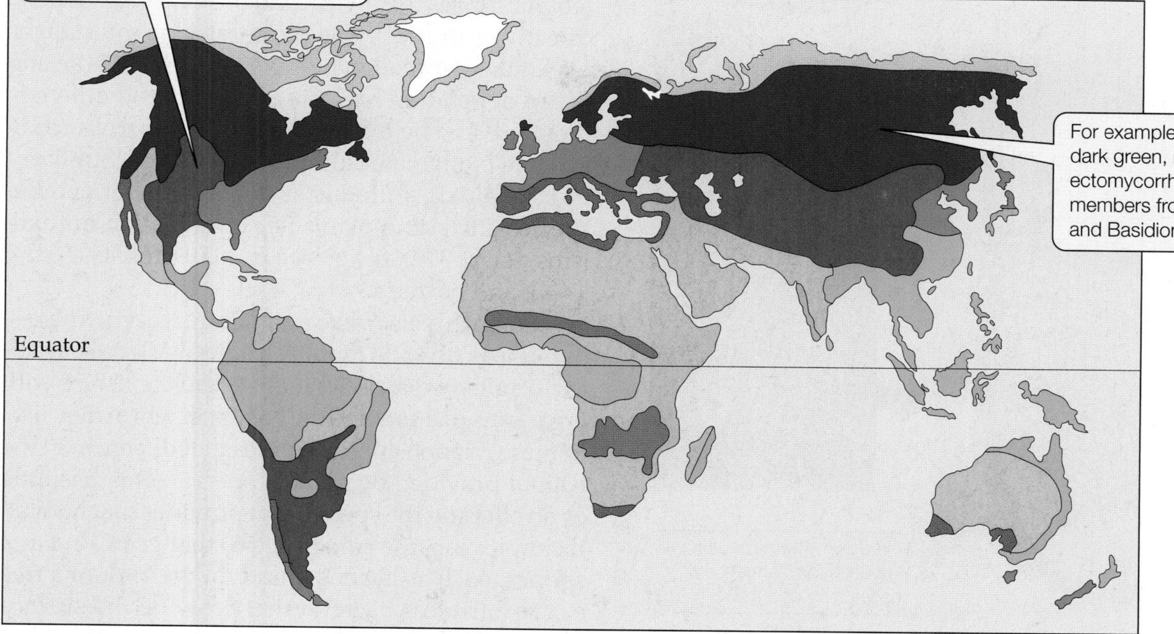

FIGURE 14.3 Mycorrhizal Associations Cover Earth's Land Surface Each color on the map shows the region in which one of eight major types of mycorrhizal associations is found (see Fitter 2005 to learn which fungi are involved in each of these eight mycorrhizal associations). Notice that the locations of the different types of mycorrhizal associations correspond fairly closely to the locations of major terrestrial biomes (see Figure 3.5). (After Fitter 2005.)

Q What types of plants are likely to be involved in the mycorrhizal association shown in light green (see Figure 3.5)?

cies is increased by their interaction with the other species. Such benefits can take a variety of forms. A species may provide its partner with food, shelter, or a substrate to grow on; it may transport its partner's pollen or seeds; it may reduce heat or water stress; or it may decrease the negative effects of competitors, herbivores, predators, or parasites. In mutualism, there can be costs to an organism that provides a benefit to its partner, as when supplying food to its partner reduces its own opportunity for growth. Nevertheless, the net effect of the interaction is positive because the benefits are greater than the costs for each of the partners in a mutualism.

In the remainder of this section, we will discuss some general observations that apply to both mutualism and commensalism; in the discussion under Concept 14.2, we'll examine some characteristics that are specific to mutualism.

Mutualism and commensalism are ubiquitous

Mutualistic associations literally cover the land surface of Earth (**Figure 14.3**). Most vascular plant species, including those that dominate terrestrial ecosystems, form **mycorrhizae**, symbiotic associations between plant roots and various types of fungi that are usually mutualistic. The fungi benefit the plants by increasing the surface area over which the plants can extract water and nutrients from the soil; in some cases, over 3 m of fungal filaments, known as *hyphae*, may extend from 1 cm of plant root. The fungi may also protect the plants from pathogens. Mycorrhizae provide clear benefits to the plants, improving their growth and survival in a wide range of habitats (Smith and Read 2008; Booth and Hoeksema 2010). The plants typically benefit the fungi by supplying them with carbohydrates.

There are two major types of mycorrhizae (**Figure 14.4**). In **ectomycorrhizae**, the fungal partner typically grows between root cells and forms a mantle around the exterior of the root; hyphae in the mantle often extend short distances into the soil. In **arbuscular mycorrhizae**, the fungal partner grows farther into the soil, and it grows between some root cells while penetrating the cell walls of others.* About 80% of angiosperms (flowering plants) and all gymnosperms (e.g., conifers, cycads, and the ginkgo) form mycorrhizal associations.

Mutualistic associations can be found in many other organisms and habitats. In the oceans, corals form a mutualism with symbiotic algae, as mentioned in Chapter 3. The coral provides the algae with a home, nutrients (nitrogen and phosphorus), and access to sunlight; the algae provide the coral with carbohydrates produced by photosynthesis. All of the numerous invertebrate and ver-

*Arbuscular mycorrhizae were previously known as "endomycorrhizae" (from the Greek *entos*, "in"), a term that referred to the fact that their hyphae can penetrate root cells. The term "endomycorrhizae" has been abandoned by many researchers, however, because the hyphae of some ectomycorrhizae can also penetrate root cells.

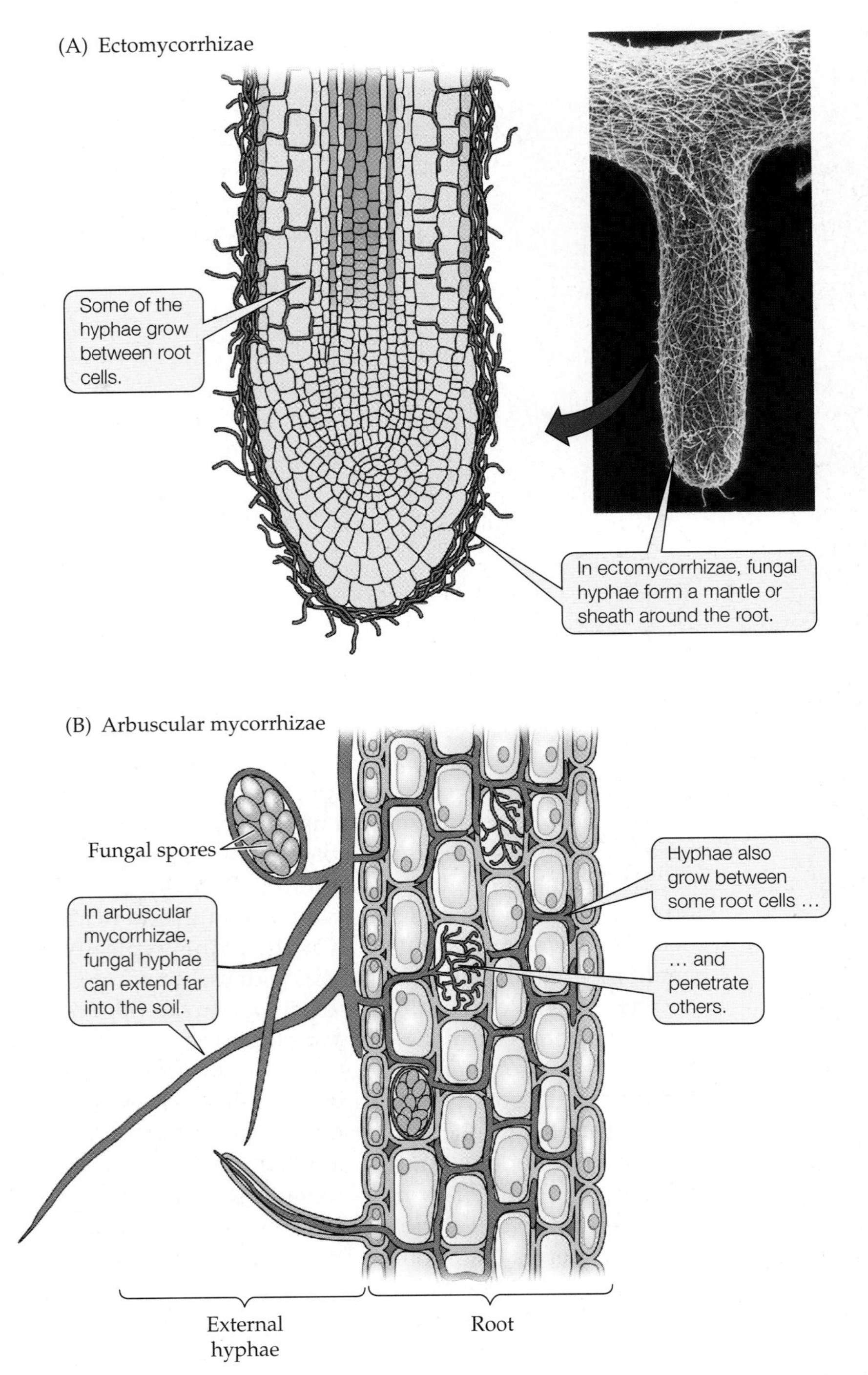

FIGURE 14.4 Two Major Types of Mycorrhizae Mycorrhizae can be classified as (A) ectomycorrhizae or (B) arbuscular mycorrhizae. In arbuscular mycorrhizae, hyphae that enter root cells penetrate the cell wall, but not the plasma membrane. (A after Rovira et al. 1983; B after Mauseth 1988.)

Q Suppose mycorrhizal hyphae extend 20 cm from a plant root into the soil. Is this likely an ectomycorrhizal association or an arbuscular mycorrhizal association? Explain.

tebrate species that live in and on coral reefs depend directly or indirectly on the coral–alga mutualism. On land, mammalian herbivores such as cattle and sheep depend on bacteria and protists that live in their guts and help them metabolize otherwise indigestible plant material, such as cellulose. Similarly, insects rely on mutualisms with a number of other species, including plants (e.g., pollination mutualisms, see p. 309), fungi (see p. 305), protists (**Figure 14.5**), and bacteria.

Commensalism, like mutualism, is everywhere—the ecological world is built on it. As we'll see in Chapter 15, millions of species form +/0 relationships with organisms that provide the habitat in which they live. In these relationships, a species that depends on the habitat provided by another species often has little or no effect on the species that provides that habitat. Examples include small species that grow on large species, such as lichens found on the bark of a tree or the harmless bacteria that grow on the surface of your skin. Many algae, invertebrates, and fishes found in marine kelp forests go locally extinct if the kelp are removed (see p. 177); such species depend on the kelp for a home, but most of them do not harm or benefit the kelp. Likewise, although the numbers are quite uncertain, there may be more than a million insect species and thousands of understory plant species that live in tropical forests and nowhere else. These insects and small plants depend on the forest for their habitat, yet many have little or no effect on the trees that tower above them.

Mutualism and commensalism can evolve in many ways

Many different types of ecological interactions can evolve into commensalism or mutualism. For example, a lichen that grows on the surface of a tree's leaves can harm the tree by reducing its access to light. Over time, however, this relationship may evolve toward a commensalism if the tree gains the ability to tolerate the lichen's presence. Some trees have done so: the Australian palm *Calamus australis*, for example, increases the concentration of chlorophyll in portions of its leaves that are covered with lichens (Anthony et al. 2002). This response improves the efficiency with which the palm absorbs the little light that passes through the lichen. As a result, the photosynthetic capacity of leaves covered by lichens is the same as that of leaves lacking lichens.

It is also possible for a mutualism to emerge from what begins as a host–parasite interaction. In 1966, Kwang Jeon observed the spontaneous infection of a strain of *Amoeba proteus* by a rod-shaped bacterium. Large numbers of bacteria (ca. 40,000) infected each individual amoeba. Initially, these bacteria had a negative effect on their hosts: they often killed

FIGURE 14.5 A Protist Gut Mutualist This wood-eating cockroach (like other wood-eating insects, such as termites) would starve if gut mutualists such as the protist shown here (a hypermastigote) did not help it to digest wood. The hypermastigote can break down cellulose, a major structural component of wood that the cockroach cannot digest on its own.

the host, and they caused infected hosts to be smaller, grow more slowly, and starve more easily than uninfected amoebas. As discussed under Concept 13.3, however, parasites and hosts often coevolve, each in response to selection pressure imposed by the other. Indeed, 5 years later, the bacterium had evolved to become harmless to the amoeba, and the amoeba had evolved such that its nucleus depended on the bacterium for normal metabolic function. As Jeon (1972) showed in a series of experiments, neither species could survive without the other. Thus, what began as a parasitic relationship evolved into a mutualism in which each species provided the other with a clear benefit (in this case, the ability to survive). Such a change is not unique to the species studied by Jeon: *Wolbachia* bacterial symbionts of the fruit fly *Drosophila simulans* also appear to have evolved rapidly from parasite to mutualist (Weeks et al. 2007).

Positive interactions can be obligate and coevolved or facultative and loosely structured

Mutualism and commensalism include a broad set of interactions, ranging from those that are species-specific, obligate (that is, not optional for either species), and coevolved to those that show none of these three characteristics. The leaf-cutter ant–fungus mutualism discussed in the Case Study at the opening of this chapter (p. 305) illustrates one end of this spectrum: the ants and the fungi they cultivate have a highly specific, obligate relationship in which neither partner can survive without the other, and their interaction has led each partner to evolve unique features that benefit the other species.

Similarly, many tropical fig trees are pollinated by one or a few species of fig wasps. These relationships are mutually beneficial and obligate for both species in that neither species can reproduce without the other. Fig–fig wasp interactions also show clear signs of coevolution (Bronstein 1992). Fig flowers are contained within structures of fleshy stem tissue known as *receptacles* (**Figure 14.6**). In monoecious figs (those in which each tree has separate male and female flowers), the male and female flowers are located in different parts of the receptacle, and the male flowers mature after the female flowers. The forms of female flowers range from those with short styles to those with long styles.

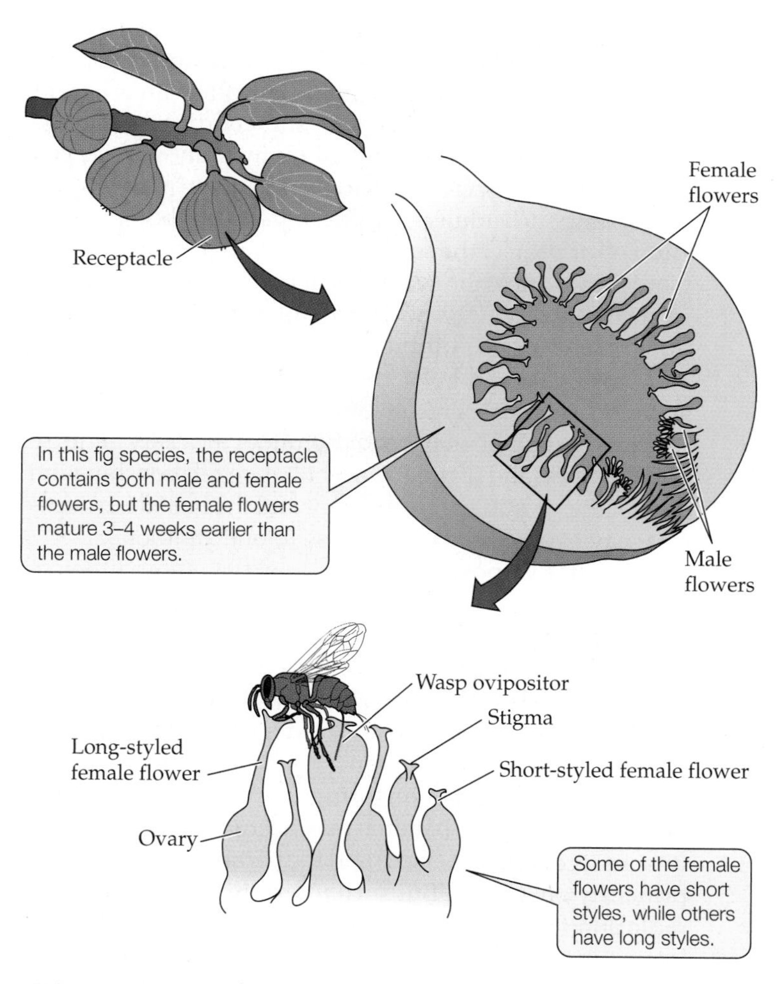

FIGURE 14.6 Fig Flowers and the Wasp That Pollinates Them The receptacle and flowers of a typical monoecious fig tree, *Ficus sycomorus*. (After Bronstein 1992.)

A female fig wasp enters the receptacle, carrying pollen she collected from male flowers in another receptacle. Once inside, the wasp inserts her ovipositor through the styles of the flowers to lay eggs in the ovaries (see Figure 14.6). She then deposits pollen on the stigmas of both long- and short-styled flowers, and hence both flower types develop seeds. Perhaps because wasp ovipositors are not long enough to reach the ovaries of long-styled flowers, wasp larvae typically develop within short-styled flowers and feed on some of their seeds.

When the young wasps complete their development, they mate, the males burrow through the receptacle, and the females exit through this passageway. Before the females leave the receptacle, however, they visit male flowers (which are now mature), collect pollen from them, and store it in a specialized sac for use when they lay their eggs in another receptacle. The wasp's reproductive behavior is a remarkable example of specialized behavior that provides a benefit to another species.

Unlike the ant–fungus and fig–fig wasp mutualisms, many mutualisms and commensalisms are facultative (not obligatory) and show few signs of coevolution. In desert environments, for example, the soil beneath an adult plant is often cooler and moister than the soil of an adjacent open area. These differences in soil conditions may be so pronounced that the seeds of many plant species can germinate and survive only in the shade provided by an adult plant; such adults are called *nurse plants* because they "nurse" or protect the seedlings. A single species of nurse plant may protect the seedlings of many different species. Desert ironwood (*Olneya tesota*), for example, serves as a nurse plant for 165 different species, most of which can also germinate and grow under other plant species. This situation is typical of facultative interactions: a species that requires "nursing" may be found under a variety of nurse plant species (and hence has a facultative relationship with each of them), and the nurse plant and the beneficiary species may evolve little in response to one another.

Facultative positive interactions that show little coevolution also occur in forest communities. For example, large herbivores such as deer or moose may inadvertently eat the seeds of small herbaceous plants whose leaves they feed on. The seeds may pass unharmed through the herbivore's digestive tract and be deposited with its feces, often far from the parent plant (**Figure 14.7**). As we saw in Chapter 7, dispersal of offspring away from parents may be advantageous, so benefits may accrue to both the plant (whose seeds are dispersed) and the herbivore (which feeds primarily on leaves). But such interactions are often sporadic and facultative, and there is little evidence that the interacting species have evolved in response to each other.

Overall, mutualisms have the potential to be obligate for both species and highly coevolved, but not all mutualisms show these characteristics. In a commensalism, the relationship is always facultative for the species that does not benefit; in addition, coevolution does not occur because natural selection has little or no effect on the species that does not benefit from the relationship.

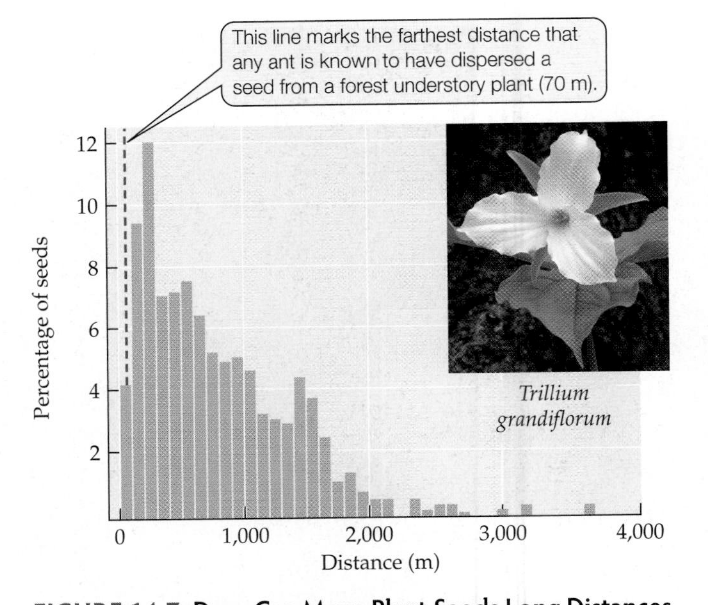

FIGURE 14.7 Deer Can Move Plant Seeds Long Distances These estimates of the distances that white-tailed deer disperse the seeds of the forest understory plant *Trillium grandiflorum* are based on observations of deer movements and of the length of time that deer retain plant seeds in their digestive tracts (from the time they eat the seeds until they defecate them). Although *T. grandiflorum* seeds are also dispersed by ants, deer move the seeds much farther. (After Vellend et al. 2003.)

Positive interactions can cease to be beneficial under some circumstances

Interactions between two species can be categorized by determining for each species whether the outcome of the interaction is positive (benefits > costs), negative (costs > benefits), or neutral (benefits = costs). However, the costs and benefits experienced by interacting species can vary from one place and time to another (Bronstein 1994). Thus, depending on the circumstances, an interaction between two species may have either positive or negative outcomes.

Soil temperature, for example, influences whether a pair of wetland plant species interact as commensals or competitors (Callaway and King 1996). Some wetland plants aerate hypoxic soils by passively transporting oxygen through air channels in their leaves, stems, and roots. Oxygen leaked into the soil from the roots of such plants can become available to other plant species, thereby reducing the negative effects of the hypoxic soil conditions. In a greenhouse experiment, Ragan Callaway and Leah King grew *Typha latifolia* (cattail), a species that has extensive air channels, together with *Myosotis laxa* (small-flowered forget-me-not), a species that lacks air channels. They grew these plants under two different temperature regimes (11°C–12°C and 18°C–20°C) in pots filled with a mix of natural pond soil and peat, with the soil in the pots submerged under

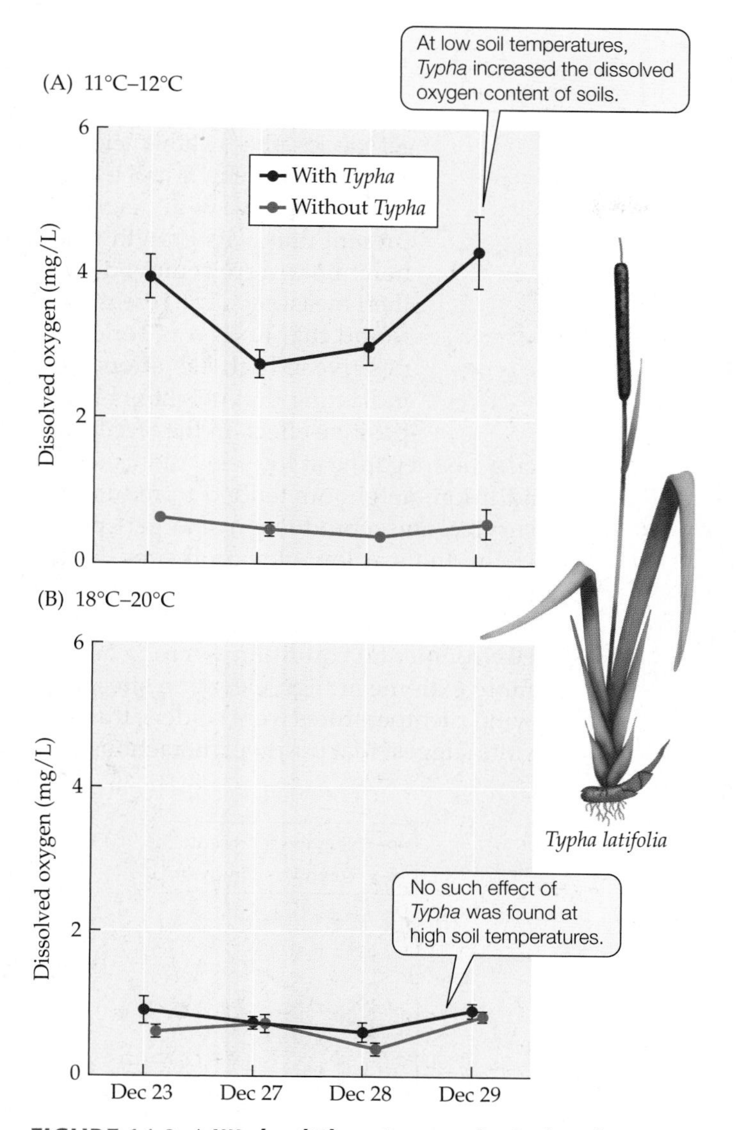

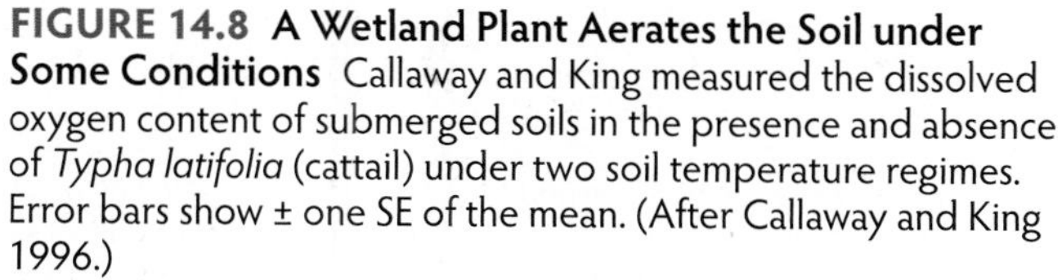

FIGURE 14.8 A Wetland Plant Aerates the Soil under Some Conditions Callaway and King measured the dissolved oxygen content of submerged soils in the presence and absence of *Typha latifolia* (cattail) under two soil temperature regimes. Error bars show ± one SE of the mean. (After Callaway and King 1996.)

Q Does the temperature of the soil, acting alone, have a strong effect on its dissolved oxygen content? Explain.

1–2 cm of water to make it hypoxic. They also grew some pots of *Myosotis* without *Typha* under the same conditions.

At the low soil temperatures, the dissolved oxygen content of the soil increased when *Typha* was present (**Figure 14.8A**), but that did not happen at the high soil temperatures (**Figure 14.8B**). How did these different oxygen levels affect the outcome of the *Myosotis–Typha* interaction? At the low soil temperatures, the growth of *Myosotis* roots and shoots increased when *Typha* was present (**Figure 14.9A**). At the high soil temperatures, however, *Myosotis* root mass decreased when *Typha* was present (**Figure 14.9B**). Overall, these results suggest that at the low soil temperatures, *Typha* provided benefits to *Myosotis* (perhaps by aerating the soil), while at the high temperatures, *Typha* had a negative effect on *Myosotis*—just one example of how a change in environmental conditions can alter the outcome of an ecological interaction (other examples are discussed under Concepts 15.3 and 16.3 and in Bronstein 1994).

Positive interactions may be more common in stressful environments

In recent decades, studies have shown that positive interactions are important in a number of ecological communities, such as oak woodlands, coastal salt marshes, and marine intertidal communities. Many of these studies have focused on how individuals of a target species are affected by nearby individuals of one or more other species. These effects can be assessed by comparing the performance of the target species when neighbors are present with its performance when neighbors are removed. Although results

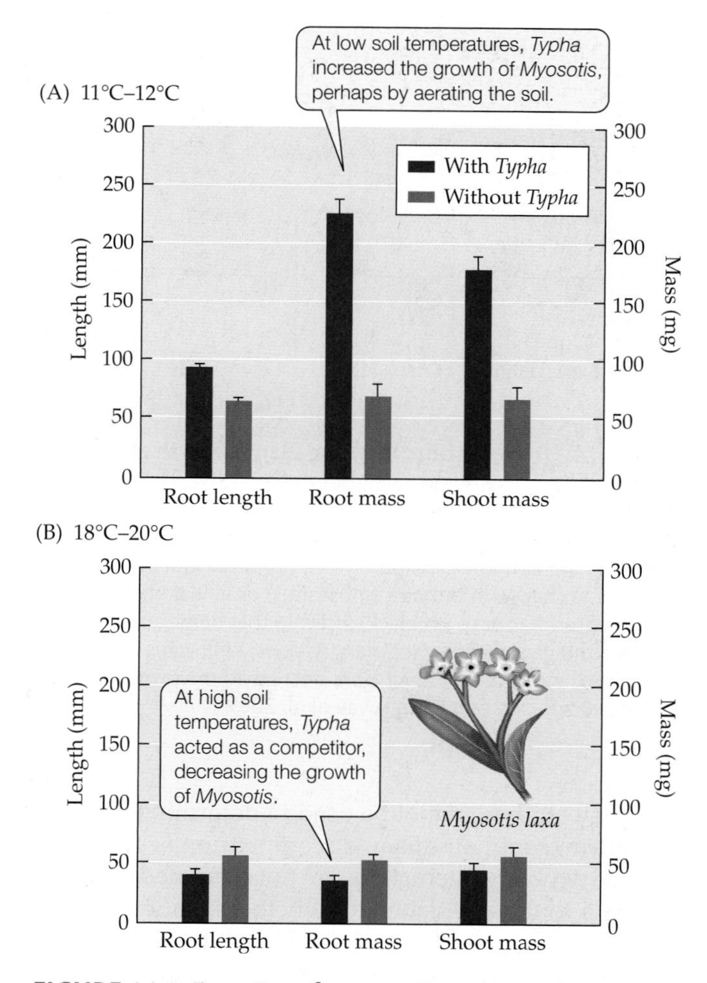

FIGURE 14.9 From Benefactor to Competitor The growth of *Myosotis laxa* under two temperature regimes in the presence and absence of *Typha latifolia* was measured by changes in three parameters: root length (left *y* axis), root mass (right *y* axis), and shoot mass (right *y* axis). Error bars show one SE of the mean. (After Callaway and King 1996.)

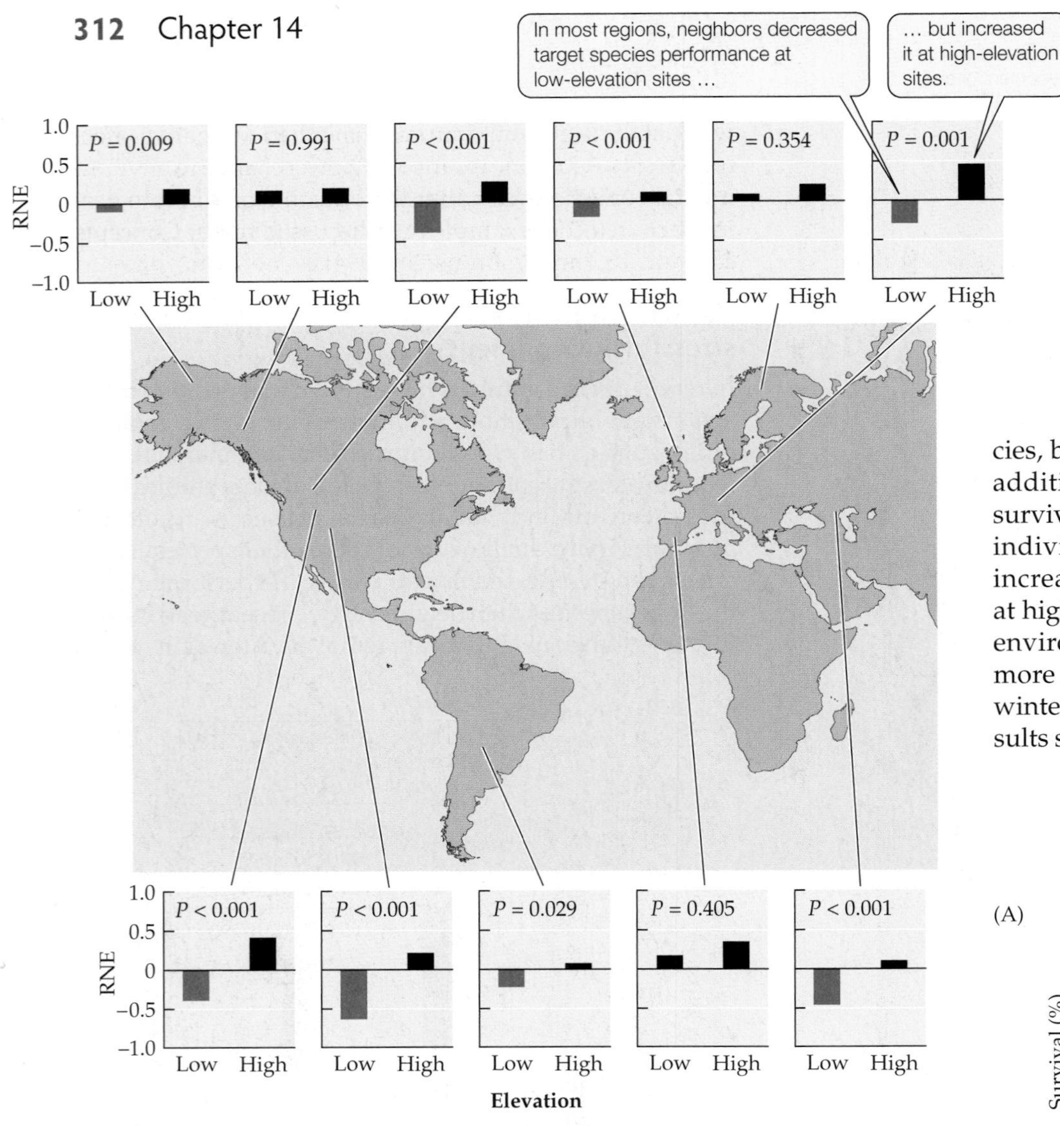

FIGURE 14.10 Neighbors Increase Plant Growth at High-Elevation Sites The relative neighbor effect (RNE, defined as the growth of the target plant species when neighbors are present minus its growth when neighbors are removed) was measured in plots at high and low elevations in 11 regions. Plant growth was measured as change in biomass (most sites) or in leaf number. RNE values greater than zero (in black) indicate that neighboring plants increased the growth of target plant species; RNE values less than zero (in red) indicate that neighbors decreased the growth of target plant species. (After Callaway et al. 2002.)

from such studies cannot be used to determine whether mutualism, commensalism, or competition is occurring (because two-way interactions are not examined), they do provide a rough assessment of whether positive interactions are common in ecological communities.

In one of the most comprehensive studies of this type, an international group of ecologists tested the effects that neighboring plants had on a total of 115 target plant species in 11 regions worldwide (**Figure 14.10**). In 8–12 replicate plots of each treatment for each target species, neighbors were either left in place or removed from the vicinity of the target species. The "relative neighbor effect" (RNE, defined as the target species' growth with neighbors present minus its growth when neighbors were removed) was then measured. The researchers found that RNE was generally positive at high-elevation sites, indicating that neighbors had a positive effect on the target species, but negative at low-elevation sites. In addition, neighbors tended to reduce the survival and reproduction of target-species individuals at low-elevation sites, but to increase their survival and reproduction at high-elevation sites (**Figure 14.11**). Since environmental conditions were generally more extreme at high-elevation sites (e.g., winter temperatures were colder), these results suggest that positive interactions may

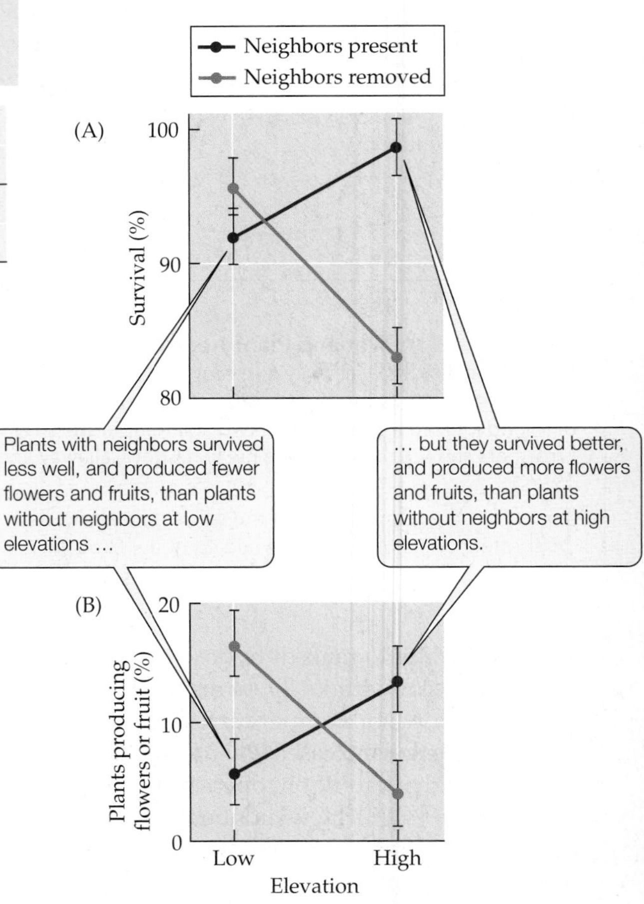

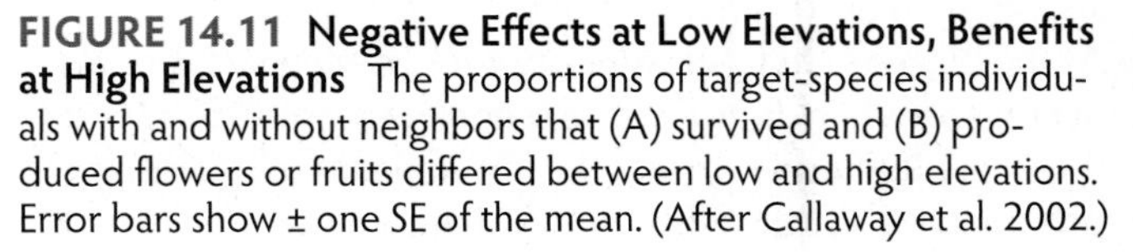

FIGURE 14.11 Negative Effects at Low Elevations, Benefits at High Elevations The proportions of target-species individuals with and without neighbors that (A) survived and (B) produced flowers or fruits differed between low and high elevations. Error bars show ± one SE of the mean. (After Callaway et al. 2002.)

be more common in stressful environments. Similar results have been found in intertidal communities (Bertness 1989; Bertness and Leonard 1997).

With this discussion of positive interactions as background, let's examine some of the characteristics that are unique to mutualism. Our discussion will place special emphasis on what can be learned from studies that document the costs and benefits of mutualistic interactions.

CONCEPT 14.2 Each partner in a mutualistic interaction acts in ways that serve its own ecological and evolutionary interests.

Characteristics of Mutualism

In the previous section we discussed some features that apply to both mutualism and commensalism: these two types of positive interactions are ubiquitous, they can evolve in many ways, and they can cease to be beneficial under some conditions. However, because mutualism is a reciprocal relationship in which both parties benefit, some of its characteristics differ from those of commensalism. A mutualism has costs as well as benefits, and if the costs exceed the benefits for one or both partners, their interaction will change. Before we describe the special characteristics of mutualism, however, we'll begin with a discussion of how mutualisms are classified.

Mutualisms can be categorized according to the benefits they provide

Mutualisms are often categorized by the types of benefits that the interacting species provide to each other, such as food, a place to live, or an ecological service. Although such categories can be helpful, they are not mutually exclusive. As we'll see, one partner in a mutualism may receive one type of benefit (such as food) while the other receives a different benefit (such as a place to live). In such cases, the mutualism could be classified in two different ways.

There are many **trophic mutualisms**, in which a mutualist receives energy or nutrients from its partner. In the leaf-cutter ant–fungus mutualism described in this chapter's Case Study (p. 305), each partner feeds the other. (Recall that the ant and the fungus also help each other to overcome plant defenses, so each also provides the other with an ecological service.) In other trophic mutualisms, one organism may receive an energy source while the other receives limiting nutrients. This is the case in mycorrhizae, in which the fungus receives energy in the form of carbohydrates and the plant may get help in taking up water or a limiting nutrient, such as phosphorus. An exchange of energy for limiting nutrients also occurs in the coral–alga symbiosis, in which the coral receives carbohydrates and the alga receives nitrogen.

In **habitat mutualisms**, one partner provides the other with shelter, a place to live, or favorable habitat. Alpheid (pistol) shrimps form a habitat mutualism with some gobies (fishes of genera *Cryptocentrus* or *Vanderhorstia*) in environments with abundant food but little protective cover. The shrimp digs a burrow in the sediments, which it shares with a goby, thus providing the fish with a safe haven from danger. For its part, the goby serves as a "seeing-eye fish" for the shrimp, which is nearly blind. Outside the burrow, the shrimp keeps an antenna on the fish (**Figure 14.12**); if a predator or some other form of disturbance causes the fish to move suddenly, the shrimp darts back into the burrow.

In other habitat mutualisms, a species may provide its partner with favorable habitat by altering local environmental conditions or by improving its partner's tolerance of existing conditions. The grass *Dichanthelium lanuginosum* grows next to hot springs in soils whose temperatures can be as high as 60°C (140°F). Regina Redman, Russell Rodriguez, and colleagues performed laboratory and field experiments in which this grass was grown with and without *Curvularia protuberata*, a symbiotic fungus that grows throughout the plant body (such fungi are called *endophytes*). In the laboratory, 100% of the grass plants

FIGURE 14.12 A Seeing-Eye Fish In environments with little protective cover, a habitat mutualism between an alpheid (pistol) shrimp and a goby benefits both partners.

that had the *Curvularia* endophyte survived intermittent soil temperatures of 60°C, while none of the plants without the endophyte survived (Redman et al. 2002). In field experiments in which soil temperatures reached up to 40°C (104°F), plants with endophytes had greater root and leaf mass than plants without endophytes. In soils above 40°C, the grass plants with endophytes continued to grow well, but all of the plants without endophytes died. Thus, *Curvularia* increased the ability of its grass host to tolerate high soil temperatures. *Curvularia* is not alone: many other fungal endophytes can increase the tolerance of their host plant for soils that are high in temperature or salinity (Rodriguez et al. 2009), as can some mycorrhizal fungi (Bunn et al. 2009).

Our final category, **service mutualisms**, includes interactions in which one partner performs an ecological service for the other. Mutualists perform many such services, including pollination, dispersal, and defense against herbivores, predators, or parasites. We have already discussed several examples of service mutualisms (e.g., the fig–fig wasp pollination mutualism) and will consider others later in this chapter (e.g., the cleaner–client mutualism among coral reef fish described on p. 318). Many service mutualisms can also be viewed as trophic mutualisms; for example, a species may perform a service such as pollination in exchange for food.

FIGURE 14.13 A Facultative Mutualism Ants often form facultative mutualisms with insects that secrete honeydew, a solution high in sugar on which the ants feed. In one such case, the ants (family Formicidae) shown here will protect this Peruvian treehopper from predators and parasites.

Mutualists are in it for themselves

Although both partners in a mutualism benefit, that does not mean that a mutualism has no costs for the partners. In the coral–alga mutualism, for example, the coral receives benefits in the form of energy, but incurs the costs of supplying the alga with nutrients and space. Likewise, the alga gains limiting nutrients, but provides the coral with energy that it could have used to support its own growth and metabolism. The costs of mutualism may be especially clear when one species provides the other with a "reward" such as food for a service such as pollination. For example, during flowering, milkweeds use up to 37% of the energy that they gain from photosynthesis to produce the nectar that attracts insect pollinators such as honeybees.

Neither partner in a mutualism is in it for altruistic reasons. For an ecological interaction to be a mutualism, the net benefits must exceed the net costs for both partners. Should environmental conditions change so as to reduce the benefits or increase the costs for one of the partners, the outcome of the interaction may change. This is especially true if the interaction is not obligate. Ants, for example, often form facultative relationships in which they protect other insects from competitors, predators, and parasites. In one such case, ants protect treehoppers from predators, and the treehoppers secrete "honeydew" (a solution high in sugars), which the ants feed on (**Figure 14.13**). Treehoppers always secrete honeydew, so the ants always have access to this food source. However, in years when predator abundances are low, the treehoppers may receive no benefit from the ants. In such years, the outcome of the interaction may shift from +/+ (a mutualism) to either +/0 (a commensalism) or +/– (parasitism), depending on whether the consumption of honeydew by ants reduces treehopper growth or reproduction.

Finally, under certain environmental conditions, a mutualist may withdraw the reward that it usually provides to its partner. In high-nutrient environments, some plants reduce the carbohydrate rewards that they usually provide to mycorrhizal fungi. In such environments, the plant can obtain ample nutrients on its own, and hence the fungus is of little benefit. Thus, when nutrients are plentiful, the plant may cease to reward the fungus because the costs of supporting fungal hyphae are greater than the benefits the fungus can provide.

Some mutualists have mechanisms to prevent overexploitation

As we've seen, there is an inherent conflict of interest between the partners in a mutualism: the benefit to each species comes at a cost to the other. In such a situation, natural selection may favor **cheaters**, individuals that

FIGURE 14.14 Yuccas and Yucca Moths *Yucca filamentosa* has an obligate relationship with its exclusive pollinator, the yucca moth *Tegeticula yuccasella*. (A) The female moth collects pollen from a yucca flower in specialized mouthparts. She may carry a load of up to 10,000 pollen grains, nearly 10% of her own weight. (B) The moth at the lower right of this photo is laying eggs in the ovary of a yucca flower; the moth at the top is placing pollen on the stigma.

increase their production of offspring by overexploiting their mutualistic partner. When one of the partners in a mutualism overexploits the other, it becomes less likely that the mutualism will persist. But mutualisms do persist, as the 50 million-year association between fungus-growing ants and the fungi they cultivate readily attests. What factors allow a mutualism to persist in spite of the conflict of interest between the partners?

One answer is provided by "penalties" imposed on individuals that overexploit a partner. If those penalties are high enough, they can reduce or remove any advantage gained by cheating. Olle Pellmyr and Chad Huth documented such a situation in an obligate, coevolved mutualism between the yucca plant *Yucca filamentosa* and its exclusive pollinator, the yucca moth *Tegeticula yuccasella* (Pellmyr and Huth 1994). Female yucca moths collect pollen from yucca plants with their unique mouthparts (**Figure 14.14A**). After collecting pollen, a female moth typically moves to another plant, lays eggs in the ovary of a flower, and then walks up to the top of the style. There, the moth deliberately places some of the pollen she carries on the stigma, thus pollinating the plant (**Figure 14.14B**). The larvae that hatch from the moth's eggs complete their development by eating some of the seeds that develop in the ovary of the flower.

The moth and the plant depend absolutely on each other for reproduction. However, the mutualism is vulnerable to overexploitation by moths that lay too many eggs and hence consume too many seeds. Yuccas have a mechanism to prevent such overexploitation: they selectively abort flowers in which female moths have laid too many eggs (**Figure 14.15**). On average, yuccas retain 62% of the flowers that contain 0–6 moth eggs, but only 11% of the flowers that contain 9 or more eggs. When the yucca aborts a flower, it does so before the moth larvae hatch from their eggs. Although the cue that determines flower abortion is not known, it is clear that it is a powerful mechanism for reducing overexploitation: all the moth larvae in an aborted flower die.

Few other clear cases of a penalty for cheating have been documented, so we do not yet know whether such penalties are common in nature. Be that as it may, the yucca–yucca moth interaction illustrates the theme that runs throughout this section: the partners in a mutualism are not altruistic.

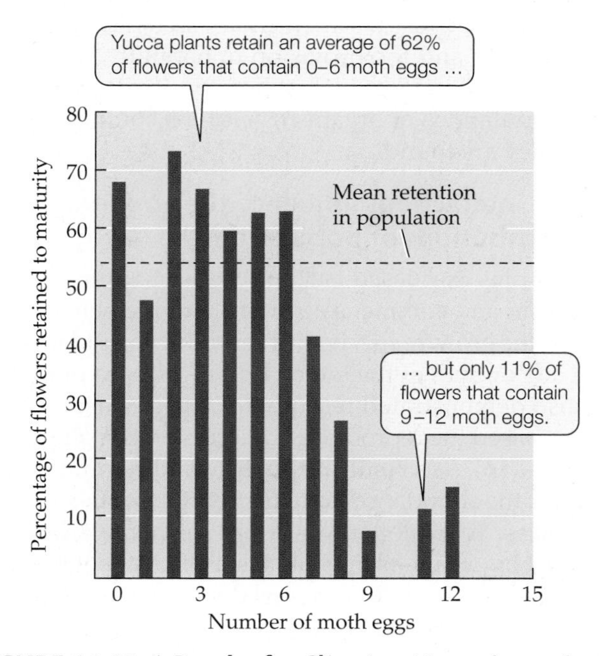

FIGURE 14.15 A Penalty for Cheating Yucca plants selectively abort flowers in which yucca moths have laid too many eggs. (After Pellmyr and Huth 1994.)

Instead, the yucca takes actions that promote its own interests, and the yucca moth does the same. In general, a mutualism evolves and is maintained because its net effect is advantageous to both parties. If the net effect of a mutualism were to impair the growth, survival, or reproduction of one of the interacting species, the ecological interests of that species would not be served, and the mutualism might break down, at least temporarily. Should such a situation continue, the longer-term or evolutionary interests of that species might also fail to be served, and the mutualism might break down on a more permanent basis.

Although it is possible for a mutualism to break down, we've also seen that mutualism and commensalism are very common, and that some of these interactions have been maintained for millions of years. Let's turn now to the ecological effects of these pervasive interactions.

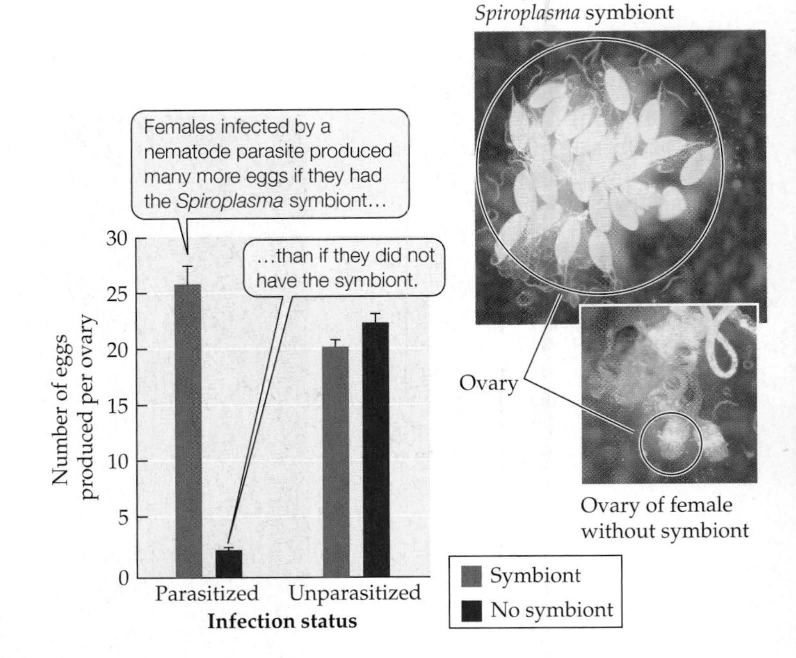

FIGURE 14.16 A Symbiont Increases the Fertility of its Host Bacteria in the genus *Spiroplasma* are symbionts that live within the cells of their host, the fruit fly *Drosophila neotestacea*. The graphs show the number of eggs produced by laboratory-reared female flies that either had *Spiroplasma* symbionts (red bars) or did not have *Spiroplasma* symbionts (blue bars), and that either were infected by a nematode parasite ("Parasitized") or were not infected by the nematode parasite ("Unparasitized"). The nematode parasite can sterilize female flies and reduce the mating success of male flies. Error bars show one SE of the mean. (After Jaenike et al. 2010.)

CONCEPT 14.3 Positive interactions affect the abundances and distributions of populations as well as the composition of ecological communities.

Ecological Consequences of Positive Interactions

So far in this chapter, we've discussed features that are common to commensalism and mutualism as well as characteristics that are unique to mutualism. At various points in these discussions, we've mentioned some ecological consequences of positive interactions, including increased survival rates and the provision of habitats. In this section, we'll take a closer look at how positive interactions affect populations of organisms and the communities in which they are found.

Positive interactions influence the abundances and distributions of populations

As examples discussed earlier in this chapter suggest, mutualism and commensalism can provide benefits that increase the growth, survival, or reproduction of individuals in one or both of the interacting species—a point that was also demonstrated recently for a bacterial symbiont that increased the reproductive success of its fruit fly host (**Figure 14.16**). As a result, mutualism and commensalism can affect the abundances and distributions of the interacting species. To explore these issues further, we will first examine how an ant–plant mutualism affects the abundance of its members. We will then consider how mutualism and commensalism influence the distributions of organisms.

Effects on abundances The effects of mutualism on abundance can be seen in the obligate relationship between the ant *Pseudomyrmex ferruginea* and the bullhorn acacia (*Acacia cornigera*). This plant has unusually large thorns, which provide a home for the ants (**Figure 14.17A**). The thorns have a tough, woody covering but a soft, pithy interior that is easy for the ants to excavate. A queen ant establishes a new colony by burrowing into a green thorn, removing some of its pithy interior, and laying eggs inside the thorn. As the colony grows, it eventually occupies all of the acacia's thorns.

The ants feed on nectar that the plant secretes from specialized nectaries and on modified leaflet tips called Beltian bodies, which are high in protein and fat (**Figure 14.17B**). The ants aggressively attack insect and even mammalian herbivores (such as deer) that attempt to eat the plant. The ants also use their mandibles to maul other plants that venture within 10–150 cm of their home acacia, thus providing the acacia with a competitor-free zone in which to grow (**Figure 14.17C**).

Do the services provided by the ants benefit the acacias? To find out, Dan Janzen removed ants from some acacia plants and compared the growth and survival of those plants with that of plants that had ant colonies. The results were striking. On average, bullhorn acacias with ant colonies weighed over 14 times as much as acacia plants

FIGURE 14.17 An Ant–Plant Mutualism (A) Acacia ants (*Pseudomyrmex spinicola*) tending to larvae and pupae inside an acacia thorn. (B) A nectary at the base of a leaf and Beltian bodies at the leaflet tips. (C) Ants have removed the plants that grew near this acacia, creating a competitor-free zone for the plant.

that lacked colonies; acacias with ants also had higher survival rates (72% vs. 43%) and were attacked by insect herbivores much less frequently (Janzen 1966).

If a bullhorn acacia lacks an ant colony, the repeated loss of its leaves and growing tips to herbivores often kills the plant in 6 to 12 months. The ants, in turn, depend on the acacias for food and a home, and they cannot survive without these plants. Thus, the ant–acacia mutualism appears to be obligate for both partners. Furthermore, the ant and the plant have each evolved unusual characteristics that benefit their partner. For example, *Pseudomyrmex* ants that depend on acacias are highly aggressive, remain active for 24 hours a day (patrolling the plant surface), and attack vegetation that grows near their home plant; *Pseudomyrmex* species that do not form mutualisms with acacias show none of these traits. Similarly, acacias that form mutualisms with ants have enlarged thorns, specialized nectaries, and Beltian bodies on their leaves; few nonmutualistic acacia species show these traits. Overall, both the ant and the acacia appear to have evolved in response to their partner, making this an example of an obligate and coevolved mutualism.

Effects on distributions There are literally millions of positive interactions in which one species provides another with favorable habitat and thus influences its distribution. Specific examples include corals that provide their algal symbionts with a home and fungal symbionts that enable plants to live in environments they otherwise could not tolerate (such as the *Curvularia* fungi that enable the grass *Dichanthelium* to live in high-temperature soils). Of course, obligate mutualisms, such as the fig–fig wasp mutualism discussed earlier, have a profound influence on the geographic distribution of the interacting species because neither can live where its partner is absent.

It is very common for a group of dominant species, such as the trees in a forest, to determine the distributions of other species by physically providing the habitat on which they depend. Many plant and animal species are found only in forests. Such "forest specialists" either cannot tolerate the physical conditions of more open areas (such as a nearby meadow), or they are prevented from living in those open areas by competition with other species. Similarly, at low tide in marine intertidal communities, many species (e.g., crabs, snails, sea stars, sea urchins, barnacles) can be found under the strands of seaweeds that are attached to the rocks. The seaweeds provide a moist and relatively cool environment that enables some species to live in higher regions of the intertidal zone than they otherwise could. Finally, many sandy and cobblestone beaches are stabilized by grasses such as *Ammophila breviligulata* and *Spartina alterniflora*. By holding the substrate together, these species enable the formation of entire communities of plants and animals.

Many forest specialists have little direct effect on the trees under which they live; hence, they have a commensalism with the trees of the forest. The same is true of many marine species that seek shelter under seaweeds and of many of the organisms that depend on substrate stabilization by grasses. In each of these cases, a positive interaction (often, a commensalism) allows one species to have a larger distribution than it otherwise would.

Positive interactions can alter communities and ecosystems

The effects that commensalism and mutualism have on the abundances and distributions of species can affect interactions among species, and those effects, in turn, can have a large influence on a community. For example, if a dominant competitor depends on a mutualist, loss of the mutualist may reduce the performance of that dominant species and increase the performance of other species—thus changing the mix of species in the community or their relative abundances. As we'll see, when the composition of a community changes, properties of the ecosystem may also change.

Community diversity Coral reefs are known for their astonishing beauty, and they are exceptional ecologically in that their fish communities are the most diverse vertebrate communities in the world. One of the most common interactions among these diverse coral reef fish is a service mutualism in which a small species (the "cleaner") removes parasites from a larger fish (the "client"). The cleaner often ventures into the mouth of the client (**Figure 14.18A**). What prevents the client from simply eating the cleaner?

The answer appears to be that the benefit a client receives from cleaning (parasite removal) is greater than the energy benefit it could gain by eating the cleaner. In the Great Barrier Reef of Australia, individuals of the cleaner species *Labroides dimidiatus* were visited by an average of 2,297 clients each day, from which the cleaner removed (and ate) an average of 1,218 parasites per day (0.53 parasites per client). To determine whether the activities of cleaners were translated into a reduction in the number of parasites found on clients, Alexandra Grutter experimentally removed *L. dimidiatus* from three of five small reefs. After 12 days, on the reefs from which the cleaners were removed, there were 3.8 times more parasites on *Hemigymnus melapterus* fish than on the control reefs. In follow-up studies, Grutter and colleagues (2003) examined the effect of *L. dimidiatus* on the number of species and the total abundance of fish found on coral reefs. The results were dramatic: removal and exclusion of *L. dimidiatus* for a period of 18 months caused large drops in both the number of fish species and the total abundance of fish found on the reefs (**Figure 14.18B,C**).

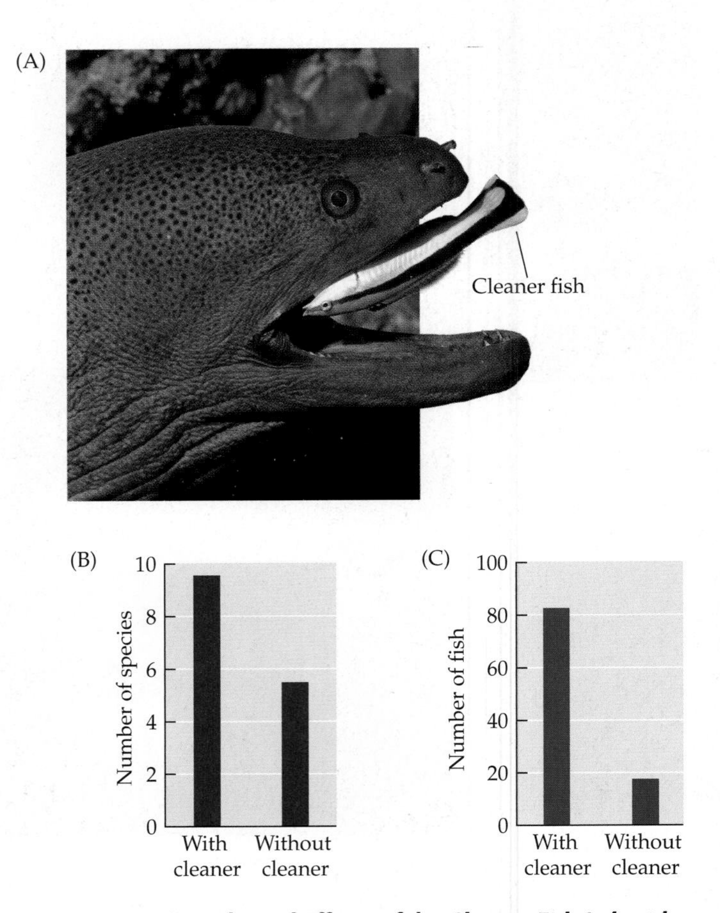

FIGURE 14.18 Ecological Effects of the Cleaner Fish *Labroides dimidiatus* (A) As it looks for parasites, a cleaner places its head within the mouth of a much larger client fish, such as this moray eel. The experimental removal of *L. dimidiatus* from small reefs within the Great Barrier Reef of Australia led to (B) a drop in the number of fish species found on the reefs and (C) a decrease in the total abundance of fish on the reefs. (B, C after Grutter et al. 2003.)

Grutter's work shows that a service mutualism can have a major effect on the diversity of species found in a community. Most of the species lost from the reefs without cleaners were species that typically move among reefs, including some large predators. Large predators can themselves affect the diversity and abundance of species, so the removal of cleaner fish could also result in further, but difficult to predict, long-term changes to the community.

Species interactions and ecosystem properties Barbara Hetrick and colleagues (1989) performed greenhouse experiments in which the presence of mycorrhizal fungi altered the outcome of competition between two prairie grasses, big bluestem (*Andropogon gerardii*) and junegrass (*Koeleria macrantha*). Big bluestem dominated when mycorrhizal fungi were present, and junegrass dominated when they were not. In a natural prairie community of which big bluestem was a dominant member, when David Hartnett and Gail Wilson (1999) suppressed mycorrhizal fungi with

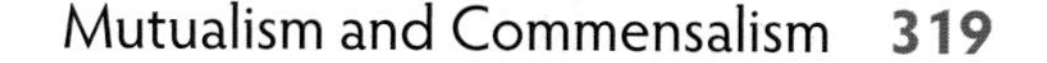

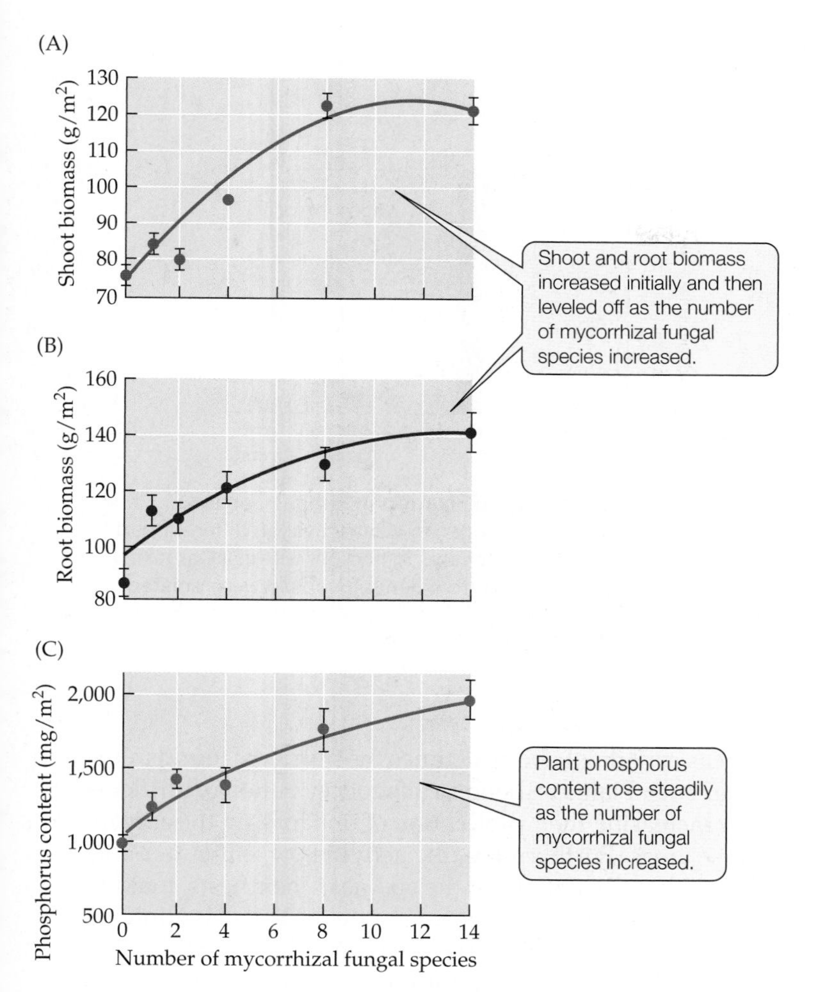

FIGURE 14.19 Mycorrhizal Fungi Affect Ecosystem Properties Researchers measured the effects of the number of mycorrhizal fungal species in the soil on (A) average shoot biomass, (B) average root biomass, and (C) phosphorus content in 15 species of plants grown from seed in a field experiment. Error bars show ± one SE of the mean. (After van der Heijden et al. 1998.)

a fungicide, the performance of big bluestem decreased. At the same time, the performance of a variety of other plant species, including both grasses and wildflowers, increased. Hartnett and Wilson suggested that big bluestem's dominance may have come from a competitive advantage conferred by its association with mycorrhizal fungi, and that removal of these fungi removed that advantage and released the inferior competitors from the negative effects of competition.

Mycorrhizal associations can also affect certain features of ecosystems, as shown in a 1998 study by Marcel van der Heijden, John Klironomos, and colleagues. In a large-scale field experiment, these scientists manipulated the number of species of mycorrhizal fungi found in soils in which identical mixtures of the seeds of 15 plant species had been sown. After one growing season, plant dry weights and phosphorus content were measured. Plant root and shoot biomass increased as the number of species of fungi increased (**Figure 14.19A,B**), as did the efficiency of phosphorus uptake by plants (**Figure 14.19C**). These results show that mycorrhizal fungi can influence key features of ecosystems such as net primary productivity (measured as the amount of new plant growth over one growing season) and the supply and cycling of nutrients such as phosphorus.

A CASE STUDY REVISITED
The First Farmers

The fungal gardens of leaf-cutter ants represent an enormous food resource for any species able to overcome the ants' defenses. As we saw in Chapter 13, roughly half of the world's species are parasites, and many of them have remarkable adaptations for evading host defenses. Are there any parasites that specialize in attacking fungal gardens?

Although you might expect that the answer would be yes, for more than a hundred years after the fungus-growing role of leaf-cutter ants was discovered (Belt 1874), no such parasites were known. That changed in the early 1990s, when Ignacio Chapela observed that leaf-cutter ant gardens were plagued by a virulent parasitic fungus of the genus *Escovopsis* (see also Currie et al. 1999a). This parasite can spread from one garden to the next, and it can rapidly destroy the gardens it invades, leading to the death of the ant colony.

Leaf-cutter ants respond to *Escovopsis* by increasing the rate at which they weed their gardens (**Figure 14.20**)

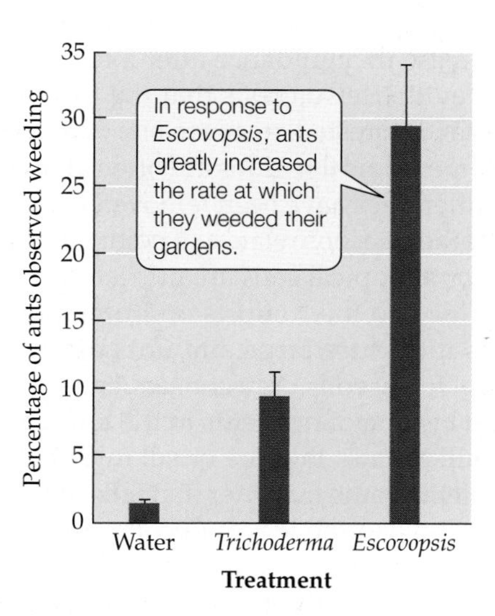

FIGURE 14.20 A Specialized Parasite Stimulates Weeding by Ants Currie and Stuart measured the frequency with which the leaf-cutter ant *Atta colombica* weeded its fungal gardens after colonies were exposed to water, *Trichoderma viride* (a generalist fungal parasite), and the specialized fungal parasite *Escovopsis*. Error bars show one SE of the mean. (After Currie and Stuart 2001.)

Suppose 2% of ants were observed weeding in colonies exposed to water, 20% in colonies exposed to *Trichoderma*, and 20% in colonies exposed to *Escovopsis*. Propose a hypothesis that might explain these results.

and, in some cases, by increasing how often they dose the garden with antimicrobial toxins that they produce in specialized glands (Fernández-Marín et al. 2009). The ants also enlist the help of other species (Currie et al. 1999b). On the underside of the ant's body is a bacterium that produces chemicals that inhibit *Escovopsis*. The queen carries this bacterium on her body when she begins a new colony. While the ant clearly benefits from the use of these fungicides, what of the bacterium? Recent work (Currie et al. 2006) indicates that the bacterium also benefits: the ant provides it with both a place to live (it is housed in specialized structures called *crypts* that are located on the ant's exoskeleton) and a source of food (glandular secretions). Thus, the bacterium appears to be a third mutualist that benefits from, and contributes to, these unique fungal gardens.

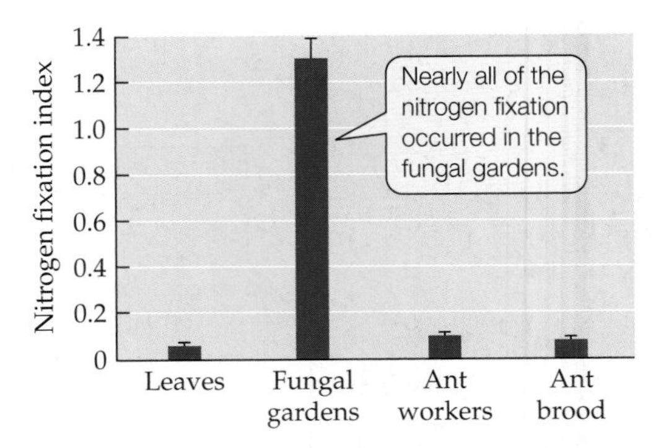

FIGURE 14.21 Nitrogen Fixation in Fungal Gardens Researchers measured nitrogen fixation activity in different parts of the colonies of leaf-cutter ants to find out where it was taking place. In addition, bacteria from genus *Klebsiella* were isolated from the fungal gardens and shown to fix nitrogen. Error bars show one SE of the mean. (After Pinto-Tomás et al. 2009.)

CONNECTIONS IN NATURE

From Mandibles to Nutrient Cycling

While you have been reading this chapter, billions of pairs of leaf-cutter ant mandibles have been removing leaves from the forests of the Americas. The workers of a single colony can harvest as much plant matter each day as it would take to feed a cow. People have long known that leaf-cutter ants are potent herbivores. Weber (1966) describes reports—the earliest from 1559—of leaf-cutter ants destroying the crops of Spanish colonists, and they still plague farmers today. In tropical regions, these ants tend to increase in abundance after a forest is cut down. Anecdotal evidence suggests that the thriving ant colonies found in deforested areas are one of the reasons why farms in some tropical regions are often abandoned just a few years after trees have been removed to make room for farmland (other reasons relate to a point made in Chapters 3 and 21: some tropical soils are nutrient-poor).

In addition to their effects on human farmers, leaf-cutter ants introduce large amounts of organic matter into tropical forest soils. As a consequence, they affect the supply and cycling of nutrients in the forest ecosystem (a topic we will discuss in more detail in Chapter 21). Normally, nutrients in the leaf litter that falls to the forest floor enter the soil when the leaves decompose. Bruce Haines (1978) compared the amounts (g/m^2) of 13 mineral nutrients contained in leaf litter with the amounts of the same nutrients found in above-ground areas where colonies of the leaf-cutter ant *Atta colombica* deposit refuse (other *Atta* species deposit refuse below-ground, as shown in Figure 14.2A). Averaged across the 13 nutrients, the ants' refuse areas contained about 48 times the nutrients found in the leaf litter. Plants respond to this concentration of nutrients by increasing their production of fine roots in the *Atta* refuse areas. Furthermore, the activities of leaf-cutter ants have the effect of tilling the soil near their nests, making it easier for plant roots to penetrate the soil (Moutinho et al. 2003). Moutinho and colleagues also found that the leaf material ants bring into their colonies fertilizes the soil, causing soils beneath ant colonies to be 3 to 4 times richer in calcium and 7 to 14 times richer in potassium than are soils that are 15 m away from the nest. Finally, recent evidence suggests that the fungal gardens tended by ants may also house nitrogen-fixing bacteria (**Figure 14.21**). These nitrogen-fixing bacteria may be part of yet another mutualism found in the gardens—a mutualism that may prove to be an important source of nitrogen in tropical ecosystems.

The overall effects of leaf-cutter ants on the ecosystems in which they live are complex. In forest ecosystems, net primary productivity (NPP) is usually measured as new aboveground plant growth (see Chapter 19); root growth is often ignored, since it is difficult to measure in trees. Although leaf-cutter ants reduce NPP by harvesting leaves, some of the ants' other activities (e.g., tillage, fertilization) may increase NPP. As a result, the net effect of the ants on the NPP of their ecosystem is difficult to estimate. While it may prove possible to disentangle such effects in future studies, there is no doubt that the ants and their partners have considerable effects on the ecosystems in which they are found.

SUMMARY

CONCEPT 14.1 In positive interactions, neither species is harmed and the benefits of the interaction are greater than the costs for at least one species.

- Mutualism and commensalism are ubiquitous interactions that are important in both terrestrial and aquatic communities.
- Mutualism and commensalism can evolve from other kinds of ecological interactions; for example, over time, a host–parasite interaction may evolve to become a mutualistic interaction.
- Mutualism may or may not be species-specific, obligate, and coevolved. Commensalism is always facultative for the species that does not benefit and is not coevolved.
- The costs and benefits of a positive interaction can vary from one place and time to another; as a result, a positive interaction may cease to be beneficial under some circumstances.
- Positive interactions may be more common in stressful environments.

CONCEPT 14.2 Each partner in a mutualistic interaction acts in ways that serve its own ecological and evolutionary interests.

- Mutualisms can be categorized by whether one partner provides the other with food (a trophic mutualism), a place to live (a habitat mutualism), or an ecological service (a service mutualism).
- The partners in a mutualism are in it for themselves—it is not an altruistic interaction.
- Some mutualists have mechanisms to prevent overexploitation by cheaters.

CONCEPT 14.3 Positive interactions affect the abundances and distributions of populations as well as the composition of ecological communities.

- Positive interactions can provide benefits that increase the growth, survival, or reproduction of one or both of the interacting species.
- As a result of such demographic effects, positive interactions can determine the abundances and distributions of populations of the interacting species.
- Positive interactions can also affect interactions among organisms and hence the composition of ecological communities and the properties of the ecosystems of which those communities are a part.

REVIEW QUESTIONS

1. Summarize the key features of positive interactions described under Concept 14.1.
2. Researchers who study mutualism do not think of it as an altruistic interaction. Explain why.
3. High water temperature is one of several stressors that can cause *coral bleaching*, a process in which a coral expels its algal mutualists and hence loses its color. If bleaching occurs repeatedly, coral death may result. Some corals are more sensitive than others to high water temperatures. If a reef containing a mixture of coral species was exposed to increasingly high water temperatures over a series of years, how might the community change over time?

ON THE COMPANION WEBSITE
sites.sinauer.com/ecology2e

The website includes Chapter Outlines, Online Quizzes, Flashcards & Key Terms, Suggested Readings, a complete Glossary, and the Web Stats Review. In addition, the following resources are available for this chapter:

HANDS-ON PROBLEM SOLVING

Population Dynamics of a Mutualism

This Web exercise explores the dynamics of a mutualism between cacti and moths. You will read a recent paper that presents a model of this mutualism and discusses the natural history of the two species. Using the model, you will explore the effects of starting size and proportion of the two species on population dynamics.

UNIT 4
Communities

CHAPTER

15 The Nature of Communities

KEY CONCEPTS

CONCEPT 15.1 Communities are groups of interacting species that occur together at the same place and time.

CONCEPT 15.2 Species diversity and species composition are important descriptors of community structure.

CONCEPT 15.3 Communities can be characterized by complex networks of direct and indirect interactions that vary in strength and direction.

"Killer Algae!": A Case Study

In 1988, a French marine biology student dove into the crystal clear water of the Mediterranean Sea and made an unusual discovery. On the seafloor, just below the cliffs on which stood the palatial Oceanographic Museum of Monaco, grew an unusual seaweed, *Caulerpa taxifolia* (**Figure 15.1**), a native of the warm tropical waters of the Caribbean. The student told Alexandre Meinesz, a leading expert on tropical algae and a professor at the University of Nice, about the unusual species. Over the following year, Meinesz confirmed its presence and determined that its feathery fluorescent green fronds, interconnected by creeping underground stems called rhizomes, carpeted an underwater area in front of the museum.

Meinesz was astonished by his observations because this species had never been seen in such cold waters, and it had certainly never reached the high densities he recorded. As it later turned out, earlier sightings from 1984 allowed Meinesz to calculate a spread of more than 1 hectare in 5 years! Over the next few months, he asked himself and his colleagues some important questions. First, how did the seaweed get to the Mediterranean in the first place, and how could it survive in temperatures as cold as 12°C–13°C (given that its normal temperature range is 18°C–20°C)? Second, did this species occur anywhere else in the Mediterranean, and was it spreading beyond the soft-sediment habitats located in front of the museum? Most importantly, at such high densities, how was it interacting with native algae and seagrasses, both of which serve as critical habitat and food for fish and invertebrate species?

A definitive answer to the second question came in July 1990, when the alga was found 5 km east of the museum, at a popular fishing location. Evidently, fragments had been caught on the gear and anchors of fishing vessels and transported to new sites of colonization. The find generated media coverage that included information on the toxicity of the seaweed, which produces a peppery secondary compound to deter the fish and invertebrate herbivores that abound in the tropics. The press sensationalized *Caulerpa*'s natural toxicity with headlines such as "Killer algae!" a misleading title that suggested the seaweed was toxic to humans (it is not). As the news spread, so did the sightings of the fluorescent green alga. By 1991, 50 sightings had been reported in France alone. The alga had indiscriminately colonized muddy, sandy, and rocky bottoms at 3 to 30 meters water depth. By 2000, the alga had moved from France to Italy, then to Croatia to the east and Spain to the west, eventually spreading as far as Tunisia (**Figure 15.2**).

FIGURE 15.1 Invading Seaweed *Caulerpa taxifolia* rapidly invaded and dominated marine communities in the Mediterranean Sea.

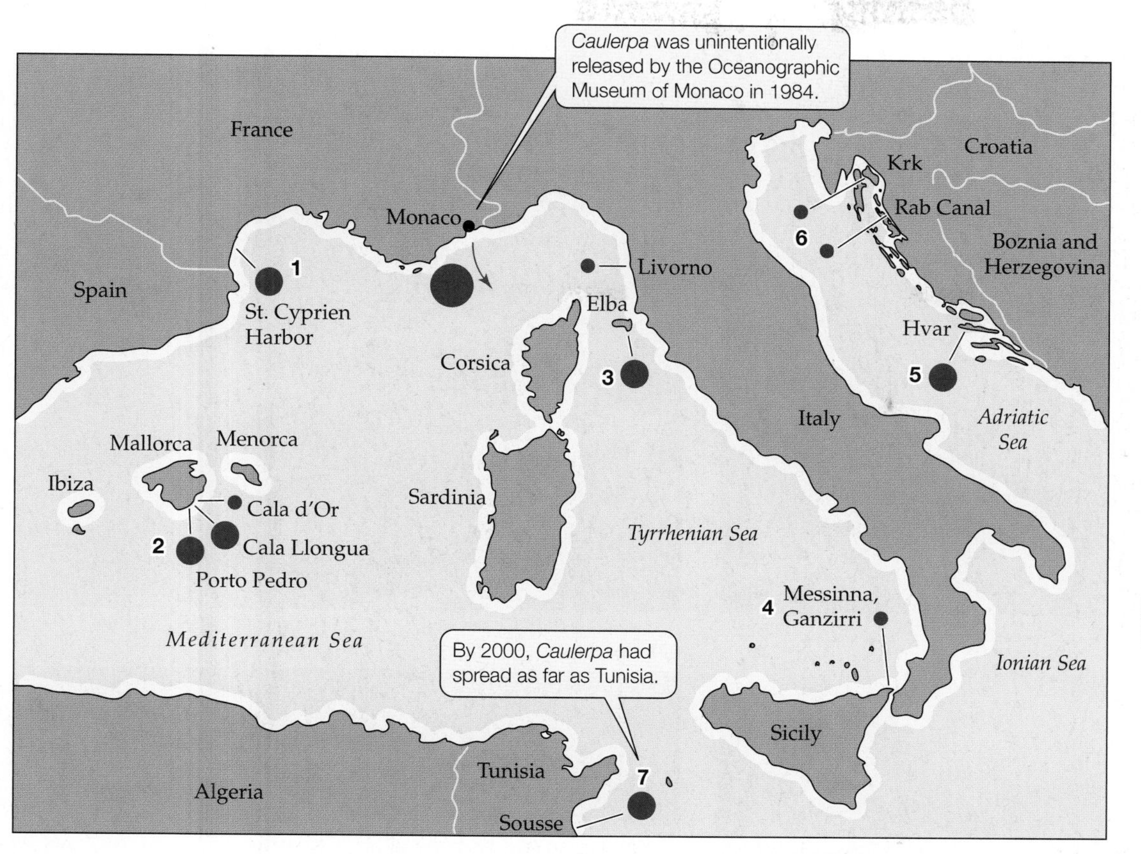

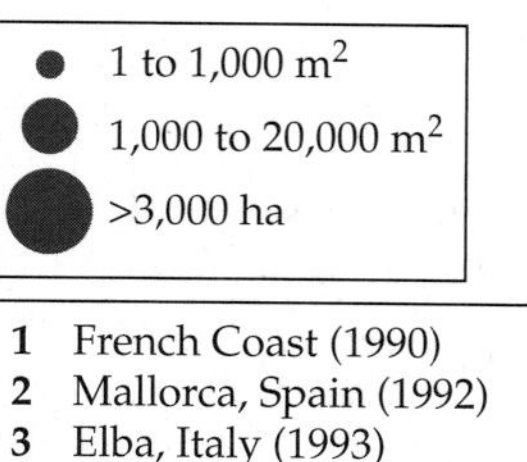

1 French Coast (1990)
2 Mallorca, Spain (1992)
3 Elba, Italy (1993)
4 Sicily, Italy (1993)
5 Island of Hvar, Croatia (1995)
6 Upper Adriatic Sea, Croatia (1996)
7 Sousse, Tunisia (2000)

FIGURE 15.2 Spread of *Caulerpa* in the Mediterranean Sea *Caulerpa* invaded Monaco and France first. It has reached Croatia and Tunisia more recently. (After Meinesz 2001.)

Q Describe the possible invasion pathways of *Caulerpa* within this region.

It had invaded roughly 3,000 hectares, despite frantic but futile efforts to remove it.

From the very beginning, Meinesz suspected that the answer to his first question lay with the museum. In 1980, a cold-resistant strain of *Caulerpa taxifolia* had been discovered and propagated in the tropical aquariums at the Wilhelma Zoo in Stuttgart, Germany. Cuttings were sent to other aquariums, including the one in Monaco, to be grown as aesthetically pleasing backdrops to tropical fish displays. The museum admitted to unintentionally releasing *Caulerpa* in the process of cleaning tanks, but believed the alga would die in the cold waters of the Mediterranean.

Given that *Caulerpa* did not die, but instead quickly invaded and overtook shallow areas of the Mediterranean, scientists and fisherman alike wanted to understand how this abundant and fast-spreading seaweed would affect marine habitats and the fisheries dependent on them. How do interactions with one very abundant species influence the hundreds of other species with which it shares a community?

Introduction

We have emphasized throughout this book that species are connected with one another and with their environment. Ecology is, at its very essence, a study of these interconnections. In Unit 3, we looked at interactions between species as two-way relationships, with one species eating, competing with, or facilitating another species. For ease of mathematical modeling, we considered these pairwise interactions in isolation, even though we have emphasized that, in reality, species experience multiple interactions. In this chapter, we will explore multiple species interactions and how they shape the nature of communities. We will consider the various ways in which ecologists have defined communities, the metrics used to measure community structure, and the types of species interactions that characterize communities.

CONCEPT 15.1 Communities are groups of interacting species that occur together at the same place and time.

What Are Communities?

Ecologists define **communities** as groups of interacting species that occur together at the same place and time. Interactions among multiple species give communities their character and function. These interactions are *synergistic*, which means that they make communities into something more than the sum of their parts. For example, we all know that the human body, made up of various limbs and organs, assumes true structure and function only when all the parts interact. These interactions can be negative, positive, direct, or indirect, but they all work in a synergistic fashion to produce a functioning human individual. This synergistic and interactive view applies to communities as well. Whether we are dealing with a desert, a kelp forest, or the gut of an ungulate, the existence of the community is dependent on the individual species

(A) Desert

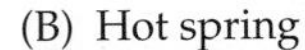

(B) Hot spring

(C) Tropical rainforest

(D) Coral reef

FIGURE 15.3 Defining Communities Ecologists often delineate communities based on their physical attributes or their biological attributes.

Q Of the four communities shown in this figure, which are defined by physical attributes and which are defined by biological attributes?

that are present and how they interact with one another and their physical surroundings.

Ecologists often delineate communities by their physical or biological characteristics

The technical definition of a community given above is more theoretical than operational. In practical terms, ecologists often delineate communities using physical or biological characteristics as a guide (**Figure 15.3**). A community may be defined by the physical characteristics of its environment; for example, a physically defined community might encompass all the species in a sand dune, a mountain stream, or a desert. The biomes and aquatic biological zones described in Chapter 3 are based largely on the physical characteristics thought to be important in defining communities. Similarly, a biologically defined community might include all the species associated with a kelp forest, a freshwater bog, or a coral reef. This way of thinking uses the presence and implied importance and interaction of abundant species, such as kelp, wetland plants, or corals, as the basis for community delineation.

In most cases, however, communities end up being defined somewhat arbitrarily by the ecologists who are studying them. For example, if an ecologists are interested in studying the marine invertebrates living in seagrasses, they are likely to restrict their definition of the community to that particular interaction. Unless they broaden their question, they are unlikely to consider the roles of birds that forage in mussel beds or other inherently important aspects of the rocky intertidal zone in which they are working. Thus, is it important to recognize that ecologists typically define communities based on the questions they are posing.

Regardless of their definition, ecologists interested in knowing which species are present in a community must

FIGURE 15.4 Subsets of Species in Communities Ecologists may use subsets of species to define communities. These examples show three ways in which such subsets could be designated. (A) All the bird species in a community could be grouped together by taxonomic affinity. (B) All the species that use pollen as a resource form a guild. (C) All the species that feed using stylet mouthparts, even though they eat different foods, could be placed together in a functional group.

contend with the difficult issue of accounting for them. Merely creating a species list for a community is a huge undertaking, and one that is essentially impossible to complete, especially if small or relatively unknown species are considered. As we will see in Chapter 22, taxonomists have officially described about 2 million species, but we know from sampling studies of tropical insects and microorganisms that this number greatly underestimates the actual number of species on Earth, which could be up to 15 million species or even more. For this reason, and because of the difficulty of studying many species at one time, ecologists usually consider a subset of species when they define and study communities.

Ecologists may use subsets of species to define communities

One common way of subdividing a community is based on taxonomic affinity (**Figure 15.4A**). For example, a study of a forest community might be limited to all the bird species within that community (in which case an ecologist might speak of "the forest bird community"). Another useful subset of a community is a **guild**, a group of species that use the same resources, even though they might be taxonomically distant (**Figure 15.4B**). For example, some birds, bees, and bats feed on flower pollen, thus forming a guild of pollen-eating animals. Finally, a **functional group** is a subset of a community that includes species that function in similar ways, but do not necessarily use the same resources (**Figure 15.4C**). For example, mosquitoes and aphids both have stylet mouthparts, although one feeds on mammalian blood and the other feeds on plant phloem. Especially in the case of plants, some functional groups do use similar resources; for example, nitrogen-fixing plants function similarly and also use the same set of resources (p. 457).

There are other subsets of communities, such as food webs (**Figure 15.5A**), that allow ecologists to organize species based on their *trophic*, or energetic, interactions. Food webs can be further organized into **trophic levels**, or groups of species that have similar ways of obtaining energy. The lowest trophic level contains *primary producers*, which are autotrophs such as plants. The primary producers are fed on by organisms at

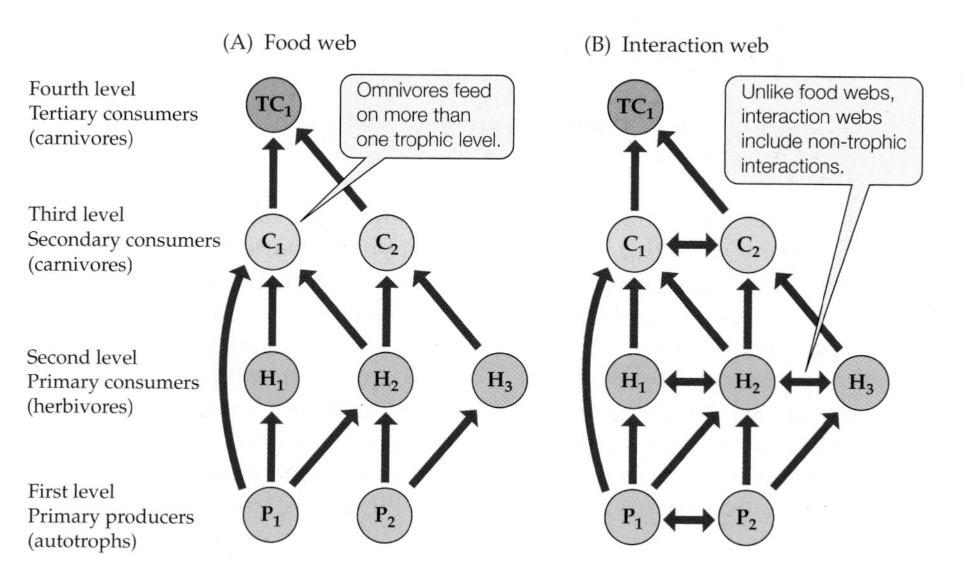

FIGURE 15.5 Food Webs and Interaction Webs (A) Food webs describe trophic interactions among species. (B) Interaction webs include both trophic interactions (vertical arrows) and non-trophic (horizontal) competitive and positive interactions.

the second level, the *primary consumers*, which are herbivores. The third level contains *secondary consumers*, which are carnivores, which are fed on in turn by *tertiary consumers*, also carnivores. Almost all food webs that have been studied in any detail appear to have two to five trophic levels.

Traditionally, food webs have been used as a descriptive or idealized method of understanding the trophic relationships among the species in a community. Food webs tell us little, however, about the strength of those interactions or their importance in the community. In addition, the use of trophic levels can create confusion for a number of reasons: for example, some species span two trophic levels (e.g., corals can be classified as both herbivores and carnivores because they have symbiotic algae), some species change their feeding status as they mature (e.g., amphibians can be herbivores as tadpoles and carnivores as adults), and some species are **omnivores**, feeding on more than one trophic level (e.g., some fish feed on both algae and invertebrates). Finally, food webs do not include non-trophic interactions (so-called **horizontal interactions**, such as competition and some positive interactions), which, as we have seen in Unit 3, can also influence community character. The concept of an **interaction web** has been introduced to more accurately describe both the trophic (vertical) and non-trophic (horizontal) interactions among the species in a traditional food web (**Figure 15.5B**). Despite these drawbacks, the food web concept remains a strong one, if only for its visual representation of important consumer relationships within a community.

We will learn much more about food webs in Chapter 20. Let's move from defining what a community is to considering the important properties of communities that allow us to characterize them and to distinguish one from another.

CONCEPT 15.2 Species diversity and species composition are important descriptors of community structure.

Community Structure

We have seen that communities vary greatly in the number of species they contain. A tropical rainforest, for example, has many more tree species than a temperate rainforest, and a midwestern prairie has many more insect species than a New England salt marsh. Ecologists have devoted substantial effort to measuring this variation at a number of spatial scales. Species diversity and species composition are important descriptors of **community structure**: the set of characteristics that shape a community. Community structure is descriptive in nature, but provides the necessary quantitative basis for generating hypotheses and experiments directed at understanding how communities work.

Species diversity is an important measure of community structure

Species diversity is the most commonly used measure of community structure. Even though the term is often used generally to describe the number of species within a community, it has a more precise definition. **Species diversity** is a measure that combines the number of species (species richness) and their relative abundances compared with one another (species evenness). **Species richness** is the easiest metric to determine: one simply counts all the species in the community one has delineated. **Species evenness**, which tells us about the commonness or rarity of species, requires knowing the abundance of each species relative to those of the other species within the community, a harder value to obtain. (See Ecological Toolkit 8.1 for methods of estimating abundances in terms of number, biomass, or percentage of cover.)

The contributions of species richness and species evenness to species diversity can be illustrated using a hypothetical example (**Figure 15.6**). Let's imagine two mushroom communities, each containing four species. Both communities have the same species richness, but their species evenness differs. In community A, one mushroom species constitutes 85% of mushroom abundance, while the others constitute only 5% each; thus, species evenness is low. In community B, mushroom abundances are evenly divided among the four species (25% each), so species evenness is high. In this case, even though each community has the same species richness (four species), community B has the higher species diversity because it has higher species evenness.

There are a number of species diversity indices that can be used to describe species diversity quantitatively. By far the most commonly used is the **Shannon index**,

$$H = \sum_{i=1}^{s} p_i \ln(p_i)$$

where

H = the Shannon index value

p_i = the proportion of individuals found in the ith species

ln = the natural logarithm

s = the number of species in the community

We can calculate the Shannon index for our two mushroom communities in Figure 15.6 as shown in **Table 15.1**. These calculations show that community A has the lower Shannon index value (H) of the two communities, confirming mathematically that this community has lower species diversity than community B. The lowest possible value of H is zero. The higher a community's H value, the greater its species diversity.

As we mentioned earlier, the term "species diversity" is often used imprecisely to describe the number of species in

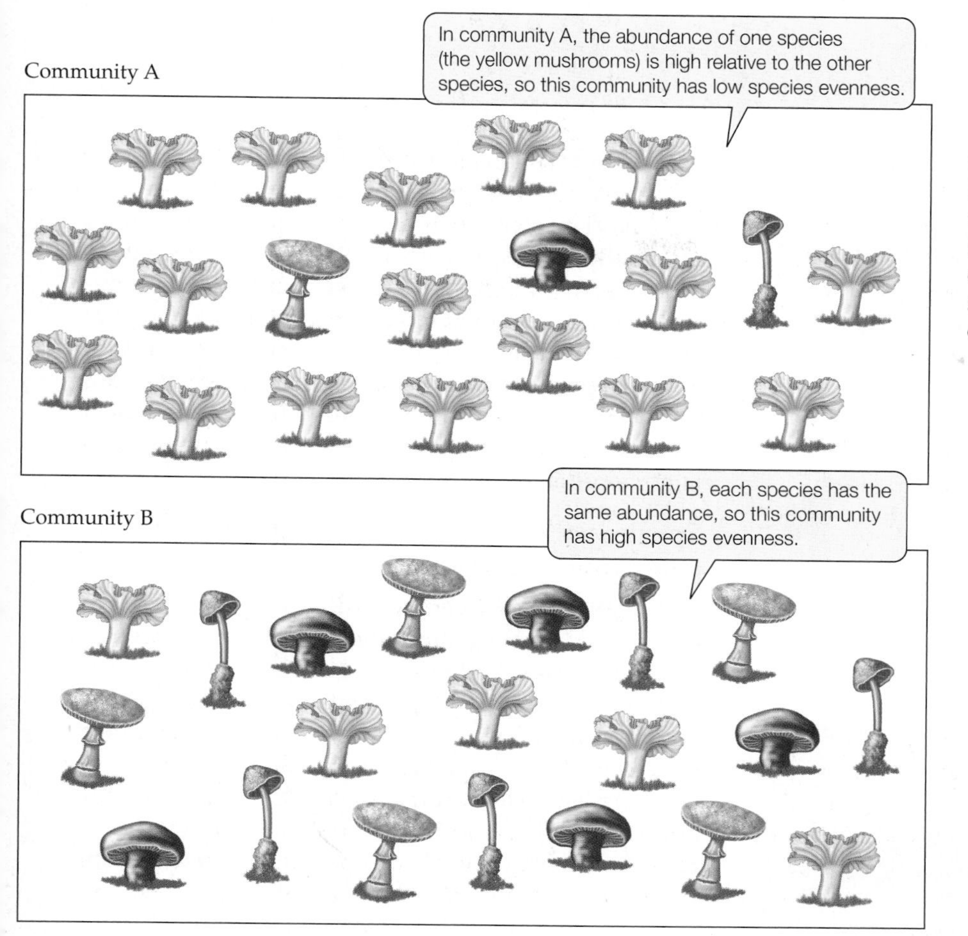

FIGURE 15.6 Species Richness and Species Evenness The two hypothetical mushroom communities shown here each have the same number of species (species richness), but different relative abundances (species evenness). Community A has lower species evenness than community B; thus, species diversity is lower in community A (see Table 15.1).

TABLE 15.1

Calculation of the Shannon Index for Communities A and B in Figure 15.6

COMMUNITY A

Species	Abundance	Proportion (p_i)	ln (p_i)	p_i ln (p_i)
Yellow	17	0.85	–0.163	–0.139
Orange	1	0.05	–2.996	–0.150
Purple	1	0.05	–2.996	–0.150
Brown	1	0.05	–2.996	–0.150
Total	**20**	**1.00**		**–0.589**

$$H = -\sum_{i=1}^{s} p_i \ln(p_i) = 0.589$$

COMMUNITY B

Species	Abundance	Proportion (p_i)	ln (p_i)	p_i ln (p_i)
Yellow	5	0.25	–1.386	–0.347
Orange	5	0.25	–1.386	–0.347
Purple	5	0.25	–1.386	–0.347
Brown	5	0.25	–1.386	–0.347
Total	**20**	**1.00**		**–1.388**

$$H = -\sum_{i=1}^{s} p_i \ln(p_i) = 1.388$$

To calculate the Shannon Index (H), the natural logarithm (ln) is applied to p_i for each species (i) ...

... and then this value is multiplied by p_i once again.

All the values are summed for all the species in the community ...

... and multiplied by –1 to get H.

Community B has higher species diversity than community A.

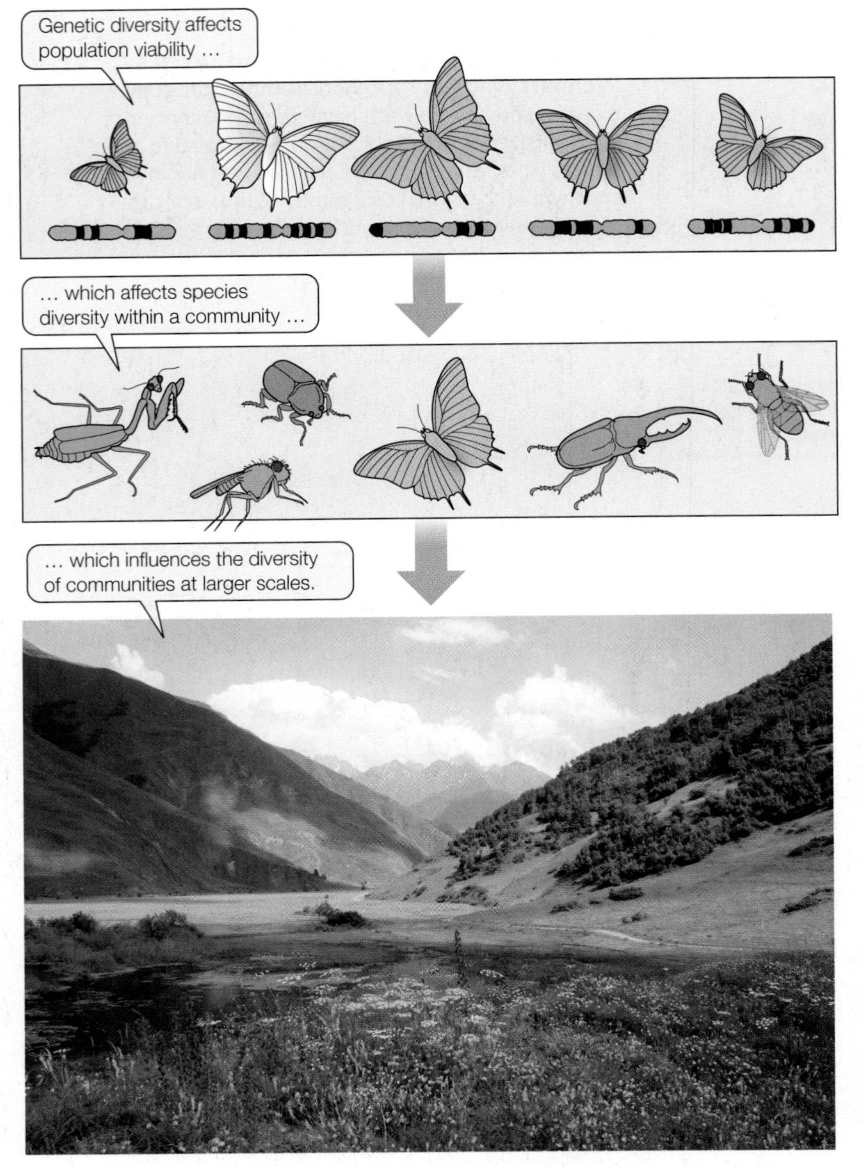

FIGURE 15.7 Biodiversity Considers Multiple Spatial Scales Diversity can be measured at different spatial scales that range from genes to species to communities.

a community without regard to the relative abundances of species or species diversity indices. For example, one commonly hears the assertion that "species diversity" is higher in tropical communities than in temperate communities, without any accompanying information about the actual relative abundances of species in the two community types. Another term that is often used interchangeably with "species diversity" is "biodiversity." Technically, **biodiversity** is a term used to describe the diversity of important ecological entities that span multiple spatial scales, from genes to species to communities (**Figure 15.7**). Implicit in the term is the interconnectedness of individuals, populations, species, and even community-level components of diversity. As we saw in Chapter 10, the genetic variation among individuals within a population influences that population's *viability* (its chance of persistence). Population viability, in turn, has important consequences for species persistence, and ultimately for species diversity within communities. Moreover, the number of different kinds of communities in an area is critical to diversity at larger regional and latitudinal scales. We will discuss the importance of spatial scale and biodiversity in chapters to come, but it is worth understanding some of the ways in which the term "diversity" is used as a starting point for those later discussions.

Species within communities differ in their commonness or rarity

Although species diversity indices allow ecologists to compare different communities, graphical representations of species diversity can give us a more explicit view of the commonness or rarity of the species in communities. Such graphs, called **rank abundance curves**, plot the proportional abundance of each species (p_i) relative to the others in rank order, from most abundant to least abundant (**Figure 15.8**). If we use rank abundance curves to compare our two mushroom communities from Figure 15.6 and Table 15.1, we can see that community A has one abundant mushroom species (i.e., the yellow species) and three rare species (i.e., the orange, purple, and brown species), whereas in community B, all the species have the same abundance.

These two patterns could suggest the types of species interactions that might occur in these two communities. For example, the dominance of the yellow mushroom species in community A might indicate that it has a strong negative effect on the three rare species in the community. In community B, where all the species have the same abundance, their interactions might be fairly equivalent, with no one species dominating the others. To test these hypotheses, we can design manipulative experiments to explore relationships between species abundances and the types of interactions that occur among the species in a community. As we will see in the next section, experiments of this kind typically involve adding or removing a species and measuring the responses of other species in the community to the manipulation.

For simplicity, we have considered a simple hypothetical example of species diversity patterns in mushroom communities. What do real communities reveal in this regard? Let's consider a study in which Allison McCaig and colleagues (1999) determined species diversity and rank abundance curves for two soil bacterial communities in pastures in Scotland. They began by sampling bacteria from the soils of an undisturbed pasture and a pasture that had been fertilized regularly. How does one go about identifying different species of bacteria or determining their abundances? It is impractical to do this visually, but we

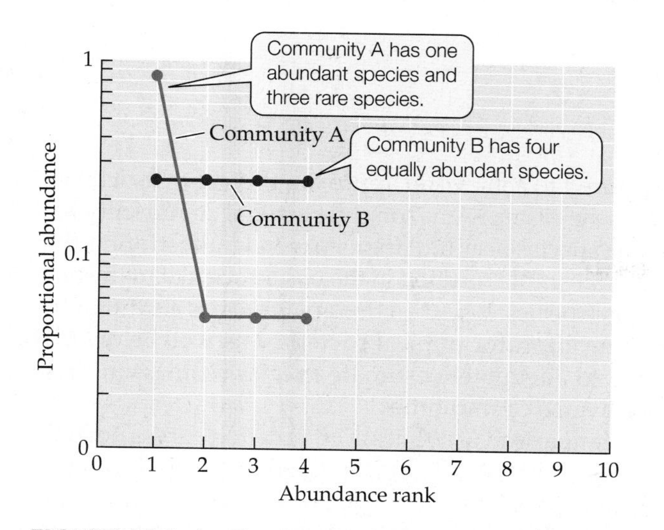

FIGURE 15.8 Are Species Common or Rare? Using rank abundance curves, we can see that the two hypothetical mushroom communities in Figure 15.6 differ in the commonness and rarity of the same four mushroom species.

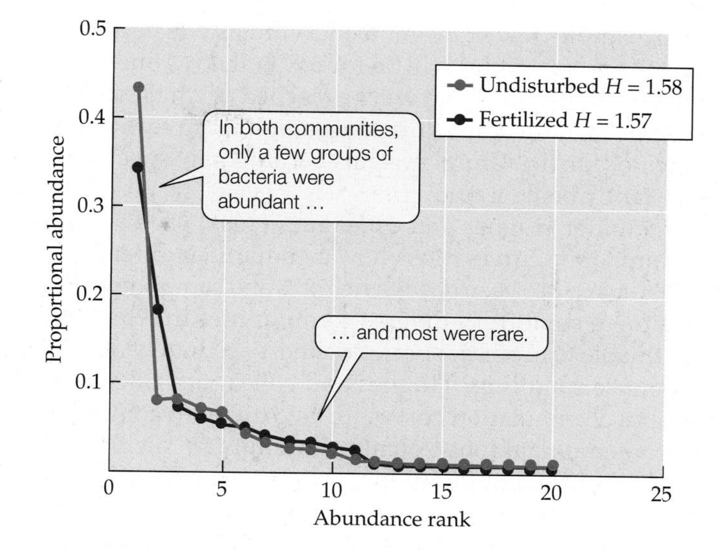

FIGURE 15.9 Bacterial Diversity in Pastures in Scotland When Allison McCaig and colleagues calculated rank abundance curves and the Shannon index (H) for communities of bacteria in undisturbed and in fertilized soils, they found that the two communities did not differ significantly by either measure. (After McCaig et al. 1999.)

can use molecular techniques that allow DNA sequences to be determined (in this study, a technique known as 16S ribosomal DNA sequence analysis was used). Unique DNA sequences can then be combined into taxonomic groups of bacteria using phylogenetic analysis. Phylogenetic analysis involves the use of statistical techniques to understand the evolutionary relationships and common ancestry of organisms. As is generally true for microbial communities, it is often difficult to identify bacteria at the species level, so microbiologists usually group them into higher taxonomic levels using phylogenetic analysis.

McCaig and colleagues' molecular work identified 275 unique DNA sequences, which they grouped into 20 taxonomic groups of bacteria. Each unique sequence can be thought of as an "individual" within its group, thus allowing abundance to be determined for each group. When they calculated rank abundance curves and diversity indices, they saw that there was little difference in the structure of the two bacterial communities (**Figure 15.9**). In addition, they found a pattern that is consistent with most communities in which larger organisms are the focus: a few species were abundant, and the majority of the rest were rare. Specifically, one group of bacteria, the -proteobacteria, was abundant (constituting 43% and 34% of the community in the undisturbed and fertilized pastures, respectively), and the other groups were rare. As mentioned earlier, whether this pattern of a few abundant species and many rare species tells us something about the importance of the species and their interactions is largely unknown, especially for microbial communities.

Species diversity estimates vary with sampling effort and scale

Let's imagine that you are sampling your backyard for insect species. It makes sense that the more samples you collect, the more species you are likely to find. However, eventually you reach a point in your sampling effort at which any additional sampling will reveal so few new species that you could stop sampling and still have a good notion of the species richness of your backyard. That point of "no significant return" for your effort can be determined using a **species accumulation curve** (**Figure 15.10**). These curves are calculated by plotting species richness as a function of the sampling effort. In other words, each data point on a species accumulation curve represents the total number of individuals and sampling effort up to that point. The

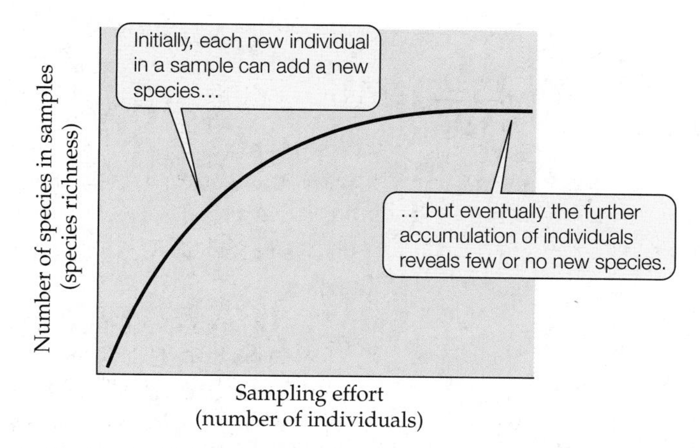

FIGURE 15.10 When Are All the Species Sampled? Species accumulation curves can help us determine when most or all of the species in a community have been observed. In this hypothetical example, the number of new species observed in each sample decreases after about half the individuals in the samples have accumulated.

more samples taken, the more individuals will be added, and the more species will be found. In theory, one could imagine that a threshold will be reached at which no new species are added by additional sampling. In reality, this never occurs in natural systems because new species are constantly being found.

Jennifer Hughes and colleagues (2001) used species accumulation curves to ask how communities differ in the relationship between species richness and sampling effort. Are there some very diverse communities in which we are unable to estimate species richness accurately despite intensive sampling? Hughes and colleagues calculated species accumulation curves for five different communities: a temperate forest plant community in Michigan, a tropical bird community in Costa Rica, a tropical moth community in Costa Rica, a bacterial community from the human mouth, and a bacterial community from tropical soils in the eastern Amazon (**Figure 15.11**). To compare the curves properly, given that the communities differed substantially in organismal abundance and species richness, the data sets were standardized by calculating for each data point the proportions of the total number of individuals and species that had been sampled up to that point. The results showed that the species richnesses of the Michigan forest plant and Costa Rican bird communities were adequately represented well before half the individuals were sampled. Human oral bacteria and Costa Rican moth communities had species accumulation curves that never completely leveled off, suggesting that their species richness was high and that additional sampling would be required to achieve an approximation of that richness. Finally, the eastern Amazon soil bacterial community had a linear species accumulation curve, demonstrating that each new sample resulted in the observation of many new bacterial species. Based on this analysis, it is clear that the sampling effort for tropical bacteria was well below that needed to adequately estimate species richness in these hyperdiverse communities.

A comparison of species accumulation curves not only provides valuable insight into the differences in species richness among communities, but also demonstrates the influence of the spatial scale at which sampling is carried out. For example, if we were to sample the richness of bacteria in tropical soils at the same scale at which we sampled Costa Rican moths, the bacterial richness would be immense in comparison. But such comparisons do suggest that our ability to sample all the bacteria in the human mouth is roughly equivalent to our ability to sample all the moth species in a few hundred square kilometers of tropical forest. The work of Hughes et al. also reminds us how little we know about the community structural characteristics of rarely sampled assemblages, such as microbial communities.

Species composition tells us who is in the community

A final element of a community's structure is its **species composition**: the identity of the species present in the community. Species composition is an obvious but important characteristic that is not revealed in species diversity

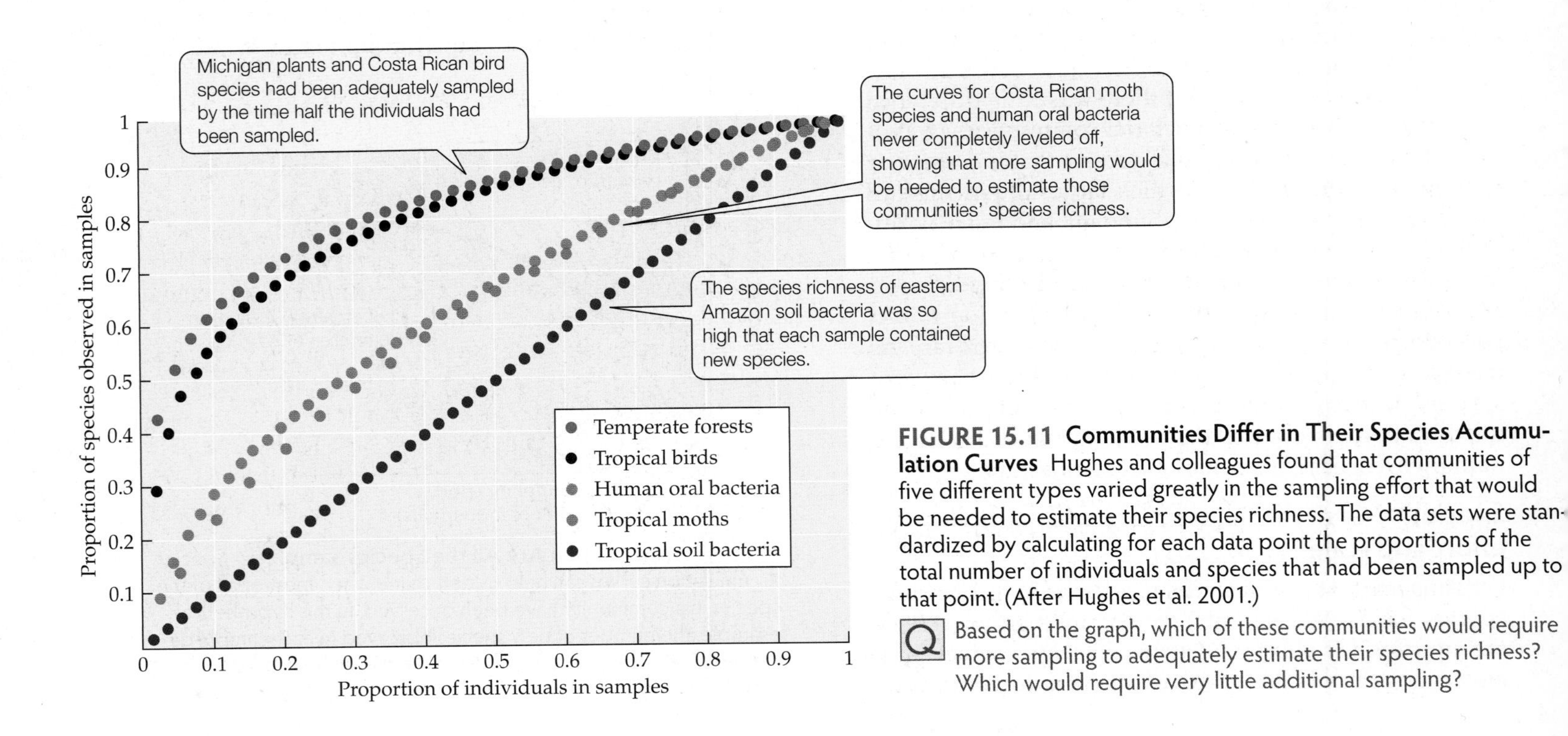

FIGURE 15.11 Communities Differ in Their Species Accumulation Curves Hughes and colleagues found that communities of five different types varied greatly in the sampling effort that would be needed to estimate their species richness. The data sets were standardized by calculating for each data point the proportions of the total number of individuals and species that had been sampled up to that point. (After Hughes et al. 2001.)

Q Based on the graph, which of these communities would require more sampling to adequately estimate their species richness? Which would require very little additional sampling?

indices. For example, two communities might have the same species diversity values, but have completely different members. In the case of the bacterial communities in Scottish pastures that we considered above (McCaig et al. 1999), although diversity indices for the two communities were nearly identical (see Figure 15.9), their composition differed. Five taxonomic groups out of the 20 the researchers found were present in one or the other pasture, but not in both.

As we will see in the next section and the next few chapters to come, knowing the identity and abundance of the species in a community makes all the difference in our understanding of the types of interactions they are having with one another and the significance of those interactions for the community as a whole. In many ways, community structure is the starting point for more interesting questions: How do species in the community interact with one another? Do some species play greater roles in the community than others? How is species diversity maintained? How does this information shape our view of communities in terms of conservation and the services they provide to humans? Let's move from the rather static view of communities as groups of species occurring together at the same place and time to a more active view of them as complex networks of species with connections and interactions that vary in strength and direction.

CONCEPT 15.3 Communities can be characterized by complex networks of direct and indirect interactions that vary in strength and direction.

Interactions of Multiple Species

The way we think about interactions between species changes dramatically when we consider that they are embedded in a community of multiple interactors. Instead of a particular species experiencing a single, direct interaction with another species, we are now dealing with multiple species interactions that generate a multitude of connections—some direct, but many indirect (**Figure 15.12**). **Direct interactions** occur between two species and include trophic and non-trophic interactions—the interactions we explored in Unit 3. **Indirect interactions** occur when the relationship between two species is mediated by a third (or more) species. The simple addition of a third species to a two-species interaction creates many more effects, both direct and indirect, which have the potential to change the outcome of the original interaction dramatically.

A social interaction analogy fits well here. Consider Figure 15.12B. Let's say you are person A and you have a good friend (person B) with whom you interact well. Now, suppose this friend meets another person (person

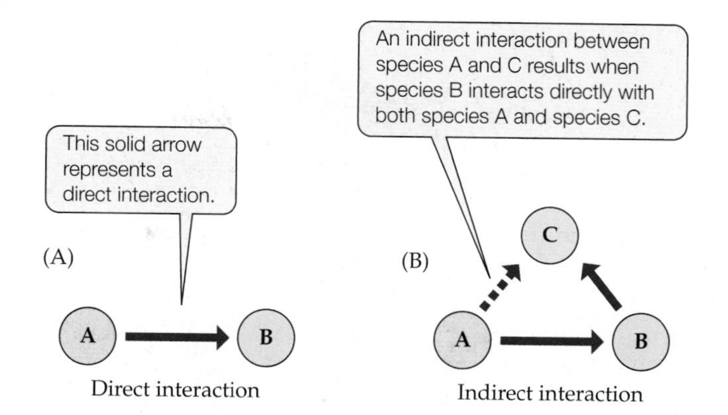

FIGURE 15.12 Direct and Indirect Species Interactions (A) A direct interaction occurs between two species. (B) An indirect interaction (dashed arrow) occurs when the direct interaction between two species is mediated by a third species.

C) who dominates your friend's time. They go to dinner, the movies, and bowling—all things you and your friend enjoyed together—without you. At some point, this new friend might begin to interfere with your friendship, possibly compromising it to the point at which it becomes antagonistic. Sadly, the indirect effect of person C changes your friendship irreparably. You might say that "the friend of my friend is my enemy." Likewise, one could imagine gaining a friend indirectly, if your foe had a foe (in this case, "the enemy of my enemy is my friend"). The point is that simply adding another person to the social circle can change the outcome of your relationship completely. The same is true of species interactions when we view them in the community context, rather than as isolated entities.

Indirect species interactions can have large effects

Charles Darwin was one of the first to convey the importance of indirect interactions in *The Origin of Species*, published in 1859. Darwin set the scene by describing the role of bees in flower pollination, and hence in seed production, among native plants in the region of England where he lived. In the book, he established the hypothesis that the number of bees is dependent on the number of field mice, which prey on the combs and nests of bees. Mice, in turn, are eaten by cats, leading Darwin to muse, "Hence it is quite credible that the presence of a feline animal in large numbers in a district might determine, through the intervention first of mice and then of bees, the frequency of certain flowers in that district!" (Darwin 1859, p. 59).

It is only recently that the sheer number and variety of effects of indirect interactions have been documented (Menge 1995). In many cases, indirect effects are discovered almost by accident when species are experimentally removed to study the strength of a direct negative inter-

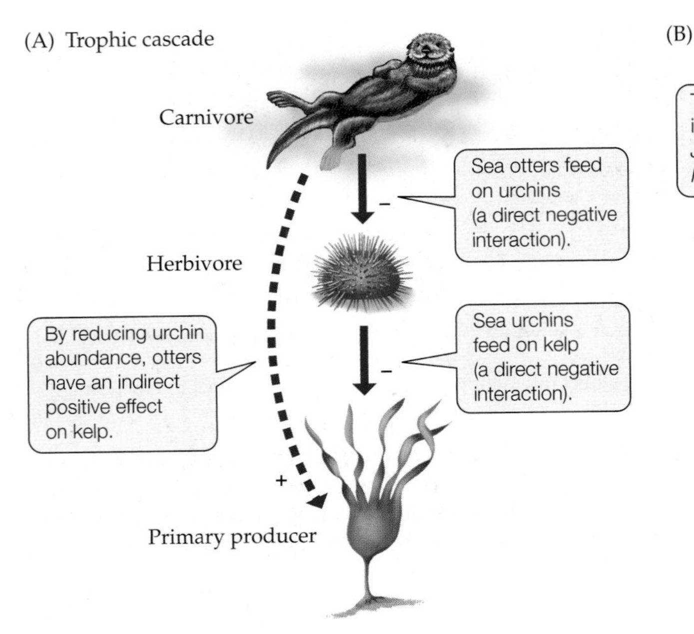

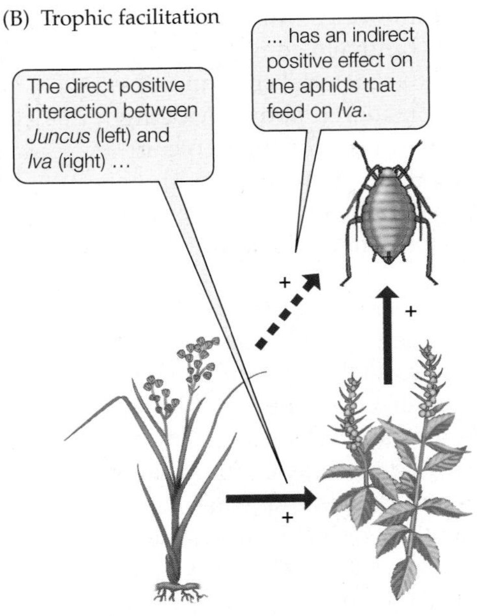

FIGURE 15.13 Indirect Effects in Interaction Webs (A) A trophic cascade occurs when a carnivore feeds on an herbivore and thus has an indirect positive effect on a primary producer that is eaten by that herbivore. (B) Trophic facilitation occurs when a consumer is indirectly facilitated by a positive interaction between its prey and another species.

action such as predation or competition. A good example of this type of indirect effect comes in the form of an interaction web called a trophic cascade (**Figure 15.13A**). A **trophic cascade** occurs when the rate of consumption at one trophic level results in a change in species abundance or composition at lower trophic levels. For example, when a carnivore eats an herbivore (having a direct negative effect on the herbivore) and decreases its abundance, there may be an indirect positive effect on a primary producer that was eaten by that herbivore. One of the best-known examples is the indirect regulation of kelp forests by the sea otter (*Enhydra lutris*) through its direct interaction with sea urchins (*Strongylocentrotus* sp.) along the west coast of North America (see the Case Study Revisited on p. 194 in Chapter 8) (Simenstad et al. 1978). Two direct trophic interactions, those of sea otters feeding on sea urchins and sea urchins feeding on kelp, generate indirect positive effects, including the effect of sea otters on kelp (via their reduction of urchin abundance) and kelp on sea otters (via the food they provide for the urchins). Furthermore, kelp can positively affect the abundances of other seaweeds, which serve as habitat and food for many marine invertebrates and fishes. The indirect effects generated in this simple food web are just as important as the direct effects in determining whether the ecosystem will be a kelp forest or an urchin barren (see Figure 8.23). We will explore the effects of indirect interactions on species diversity (see Chapter 18) and food webs (see Chapter 20) in more detail later in the book.

Indirect effects can also emerge from direct positive interactions called trophic facilitations. A **trophic facilitation** occurs when a consumer is indirectly facilitated by a positive interaction between its prey and another species (**Figure 15.13B**). An example of this type of indirect effect was demonstrated by Sally Hacker (Oregon State University) and Mark Bertness (Brown University), who studied salt marsh plant and insect interaction webs in New England. Their research showed that a commensal interaction between two salt marsh plants—a rush, *Juncus gerardii*, and a shrub, *Iva frutescens*—has important indirect effects on aphids feeding on *Iva* (Hacker and Bertness 1996).

To explore these findings in greater detail, let's first consider the commensal interaction between the two plant species. When *Juncus* was experimentally removed, the growth rate of *Iva* decreased (**Figure 15.14A**). In contrast, removing *Iva* had no effect on *Juncus*. In the absence of *Juncus*, soil salinity increased and oxygen content decreased considerably around *Iva*, suggesting that the presence of *Juncus* ameliorated harsh physical conditions for *Iva*. *Juncus*, by shading the soil surface and thus decreasing water evaporation from the surface of the marsh, decreases salt buildup. *Juncus* also has specialized tissue called aerenchyma, through which oxygen can move into the belowground parts of the plant, thus keeping it from drowning during daily high tides. Some of the oxygen "leaks" out of the plant and can be used by other neighboring plants, such as *Iva*.

To understand the importance of this direct positive interaction, Hacker and Bertness (1996) measured the population growth rate of aphids on *Iva* growing with

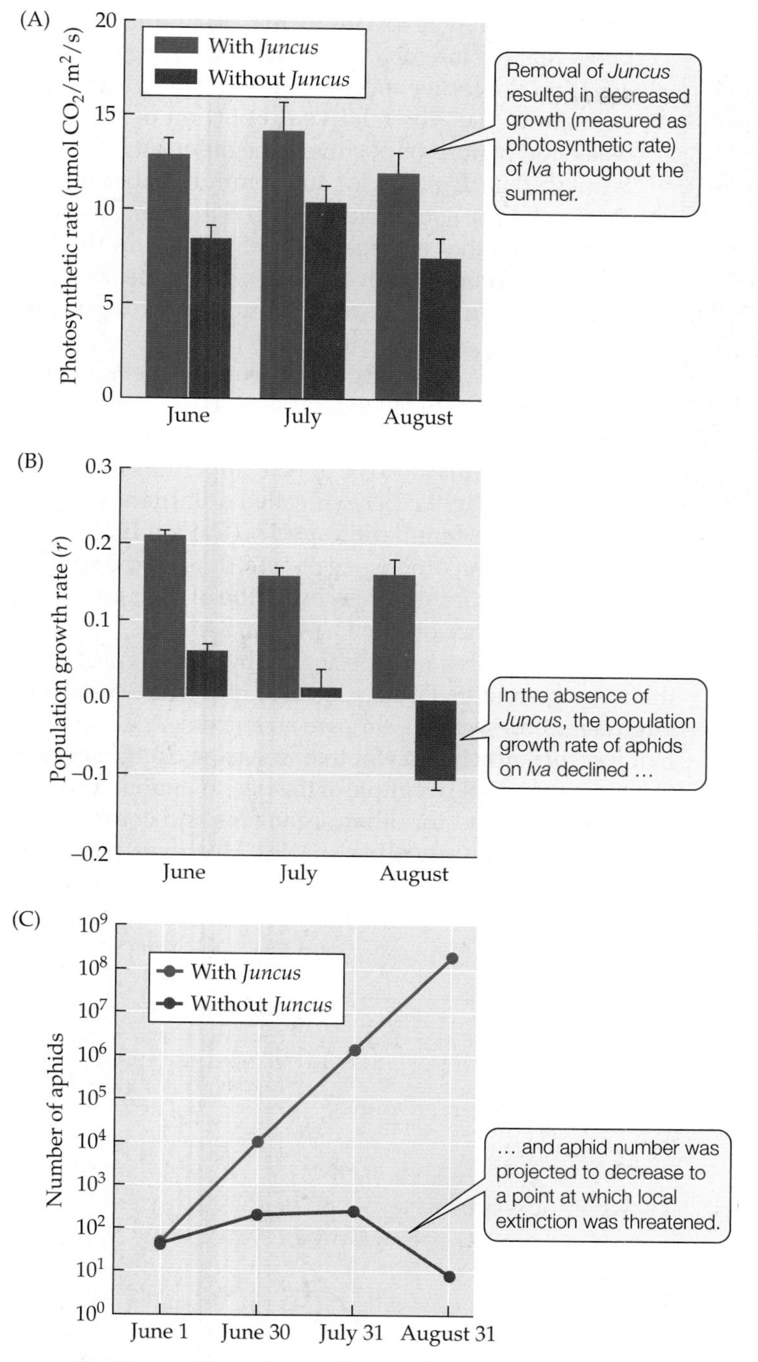

FIGURE 15.14 Results of Trophic Facilitation in a New England Salt Marsh Removal experiments demonstrated that aphids are indirectly facilitated by the rush *Juncus gerardii*, which has a direct positive effect on the shrub *Iva frutescens*, on which the aphids feed. (A) Photosynthetic rate of *Iva* with and without *Juncus*. (B) Growth rate of aphid populations with and without *Juncus*. (C) Projected numbers of aphids with and without *Juncus*. Error bars show one SE of the mean. (After Hacker and Bertness 1996.)

and without *Juncus*. They found that aphids had a much harder time finding shrubs in the presence of the rush, but that once they did, their population growth rates were significantly higher (**Figure 15.14B**). Using the exponential growth equation, they predicted that aphids would become locally extinct in the salt marsh without the indirect positive effects of *Juncus* (**Figure 15.14C**). It is clear from this example that interactions in trophic facilitation webs can have both positive effects (as when *Juncus* improves soil conditions for *Iva*) and negative effects (as when *Juncus* facilitates aphids that feed on *Iva*), but it is the sum total of these effects that determine whether the interaction is beneficial or not. Given that the ultimate fate of *Iva* without *Juncus* is death, the positive effects greatly outweigh the negative ones.

Finally, important indirect effects can arise from multiple species interactions at one trophic level (i.e., the horizontal interactions in Figure 15.5B). Buss and Jackson (1979), looking for an explanation for the coexistence of competitors, hypothesized that **competitive networks**—competitive interactions among multiple species in which every species negatively interacts with every other species—might be important in maintaining species richness in communities. A network, as opposed to a hierarchy, is an interaction web that is circular rather than linear (**Figure 15.15A**). The idea is that networks of interacting species indirectly buffer strong direct competition, thus making competitive interactions weaker and more diffuse. So, for example, species A may have the potential to outcompete species B, and species B may have the potential to outcompete species C, but because species C also has the potential to outcompete species A, no one species dominates the interaction. This is clearly an example of the "enemy of my enemy is my friend" effect described earlier. All else being equal, a hierarchical view of competition, with species A outcompeting B and B outcompeting C (**Figure 15.15B**), always results in species A dominating the interaction.

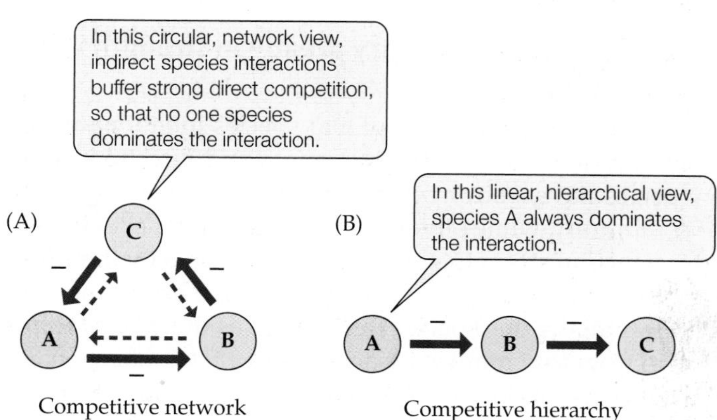

FIGURE 15.15 Competitive Networks versus Competitive Hierarchies

FIGURE 15.16 **Competitive Networks in Coral Reef Communities** Encrusting invertebrates and algae compete for space on coral reefs by overgrowing one another, but no one species consistently "wins" this competition.

Buss and Jackson (1979) tested this hypothesis using encrusting invertebrates and algae that live on the undersides of coral reefs in Jamaica (**Figure 15.16**). These species compete for space by growing over one another. The researchers collected samples at the margins between species, where one species overgrows another, for as many pairs of species as possible to determine the proportion of wins (species on top) to losses (species on the bottom) for each species. Their results showed that every species either overgrew or was overgrown by another species, and that no one species consistently won the competition. The species interacted in a circular network rather than a linear hierarchy. These observations demonstrate how competitive networks, by fostering diffuse and indirect interactions, can promote diversity in communities.

Species interactions vary greatly in strength and direction

You should realize by now that species interactions in a community vary greatly in strength and direction. Some species have a strong negative or positive effect on the community, while others probably have little or no effect. **Interaction strength**, the effect of one species on the abundance of another species, can be measured experimentally by removing one species (referred to as the *interactor species*) from the community and looking at the effect on the other species (the *target species*, as described in **Ecological Toolkit 15.1**). If the removal of the interactor species results in a large decrease in the target species, we know that the interaction is strong and positive. However, if the abundance of the target species increases significantly, we know that the interactor species has a strong negative effect on the target species. The interaction strength "dynamic" (i.e., the relative proportion of strong to weak interactions or positive to negative interactions) is not well understood for any community because of the numbers of species involved and the many indirect interactions that emerge. As you will see in the following chapters, however, we can get an idea of what species are "in charge" of communities through both observations and experiments.

There are some large or abundant species, such as trees, that are likely to have large community-wide effects by virtue of providing habitat or food for other species. They may also be good competitors for space, nutrients, or light. These so-called **dominant species**, also known as **foundation species** (Dayton 1971), have large effects on other species, and thus on the species diversity of communities, by virtue of their considerable abundance or biomass (**Figure 15.17**).

Some dominant species act by "bioengineering" their environment. These species, known as **ecosystem engineers** (Jones et al. 1994), are able to create, modify, or maintain physical habitat for themselves and other species. Consider the simple example of the trees mentioned above. Trees provide food for other organisms and compete for resources just like any other species. However, trees also engineer their environment in subtle but important ways

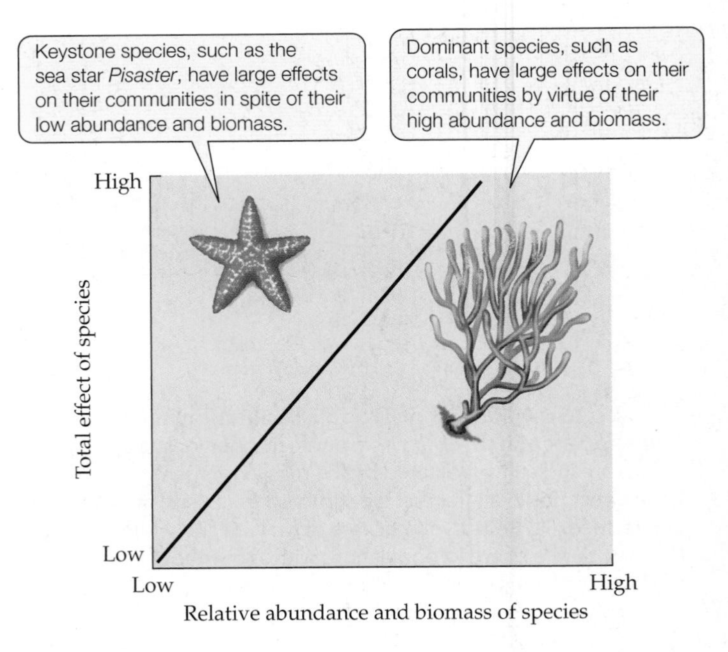

FIGURE 15.17 **Dominant versus Keystone Species** Species that have large effects on their communities may or may not do so by virtue of their high abundance or biomass. (After Power et al. 1996b.)

15.1 Measurements of Interaction Strength

We can measure interaction strength by experimentally manipulating species interactions. The usual procedure involves either the removal or the addition of one of the species involved in the interaction (the *interactor species*) and measurement of the response of the other species (the *target species*). There are several different ways to calculate interaction strength, but the most common is to determine the per capita interaction strength using the following equation:

$$\text{per capita interaction strength} = \frac{\ln\left(\frac{C}{E}\right)}{I}$$

where

C = the number of target individuals in the presence of the interactor

E = the number of target individuals in the absence of the interactor

I = the number of interactor individuals

ln = the natural logarithm

Interaction strength can vary depending on the environmental context in which the interaction is measured. For example, Menge et al. (1996) measured the interaction strength of sea star (*Pisaster ochraceus*) predation on mussels (*Mytilus trossulus*) in wave-exposed versus wave-protected areas of the shoreline at Strawberry Hill on the coast of Oregon (part i of the figure). Sea stars were excluded from some mussel beds using cages in both exposed and protected areas. At the end of the experiment, the number of mussels in the cages (E) was compared with the number of mussels in control plots (C) that had been exposed to sea star predation (parts ii and iii of the figure). The value of I was determined by counting all the sea stars near the plots in each type of area (exposed and protected).

An equation similar to the one above was used to calculate the per capita interaction strength of sea star predation on mussels. The results (part iv of the figure) showed that interaction strength was greater in wave-protected than in wave-exposed areas. Sea stars probably cannot feed as efficiently when subjected to the crashing waves characteristic of wave-exposed areas. Thus, this study demonstrates the importance of environmental context (in this case, wave exposure) to the strength of species interactions. It also shows how those interactions can change over relatively small spatial scales (e.g., between wave-exposed and wave-protected areas of the Strawberry Hill shoreline).

More recently, ecologists have been considering how ocean acidification caused by increasing atmospheric concentrations of CO_2 may affect species interactions such as the one between *Pisaster* and *Mytilus*. As ocean waters become increasingly acidic, the calcium carbonate skeletons of marine invertebrates are predicted to be weakened. Will this change allow *Pisaster* to more easily open the shells of mussels, or, because the sea star has an internal calcium carbonate skeleton, might the strength of its interaction with mussels decline? These are difficult questions to answer, but they are becoming increasingly important given the predicted acidification of the oceans (see Chapter 24 for more information on climate change and ocean acidification).

(i)

(ii)

(iii)

(iv)

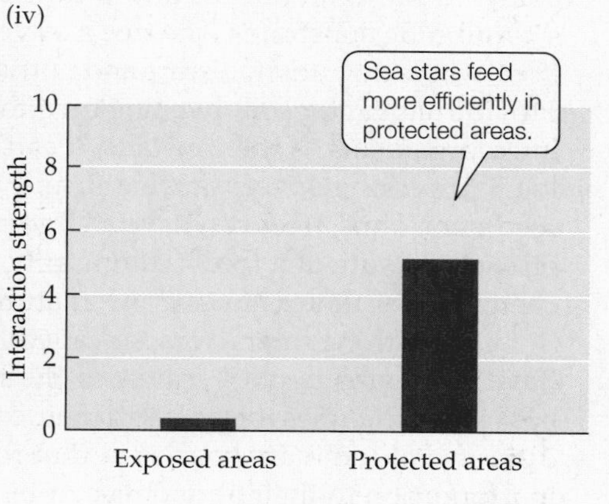

How Much Does Predation by Sea Stars Matter? It Depends
(i) The shoreline at Strawberry Hill, Oregon. (ii, iii) Plots (ii) with and (iii) without cages that excluded sea stars were set up in both wave-exposed and wave-protected areas along the shoreline. (iv) When mussels were counted and interaction strengths calculated, the results showed that interaction strength was greater in protected than in exposed areas. (iv after Menge et al. 1996.)

FIGURE 15.18 Trees Are Dominant Species and Ecosystem Engineers Trees not only provide food for and compete with other species, but also act as ecosystem engineers by creating, modifying, or maintaining physical habitat for themselves and other species. (After Jones et al. 1997.)

(**Figure 15.18**; Jones et al. 1997). The trunk, branches, and leaves of a tree provide habitat for a multitude of species, from birds to insects to lichens. The physical structure of the tree reduces sunlight, wind, and rainfall, influencing temperature and moisture levels in the forest. The roots of the tree can increase weathering and aeration of the soil, and they can anchor and bind rocks that stabilize surrounding substrates. The tree's leaves fall to the forest floor, where they add moisture and nutrients to the soil and provide habitat for soil-dwelling invertebrates, seeds, and microorganisms. If the tree falls, it can become a "nurse log," providing space, nutrients, and moisture for tree seedlings. Thus, trees can have a large physical influence on the structure of a forest community, which obviously changes over time as trees grow, mature, and die.

Other strong interactors, so-called **keystone species**, have large effects not because of their abundance, but because of the roles they play in their communities. They differ from dominant species in that their effect is large in proportion to their abundance or biomass (see Figure 15.17). Keystone species usually influence community structure indirectly, via trophic means, as we saw in the case of sea otters (see Figure 15.13A and Chapter 8). Sea otters are considered keystone species because, by preying on sea urchins, they indirectly control the presence of kelp, which provides important habitat for many other species. We will consider the role of keystone species in more detail in Chapter 20.

There are also keystone species that act as ecosystem engineers. A great example is the beaver, a species in which just a few individuals can have dramatic effects on the landscape. Beavers dam streams with cut trees and woody debris. Very quickly, flooding ensues and sediment accumulates as the increasing number of woody obstacles slows the water flow. Eventually, the once swiftly flowing stream is replaced by a wetland, containing plants that can deal with flood conditions; plants that cannot do so, such as trees, are lost from the community. At the landscape level, by creating a mosaic of wetlands within a larger forested community, beavers can increase regional species diversity significantly (**Figure 15.19**). Naiman et al. (1988) showed that there was a 13-fold increase in wetland area in one region of Minnesota (from roughly 200 ha to 2,600 ha) when beavers were allowed to recolonize areas where they had been hunted nearly to extinction some 60 years previously.

Environmental context can change the outcome of species interactions

As we have seen in this section, interactions among multiple species can vary in strength and direction, and their outcome is highly dependent on the influence of each of the species in the community. As we have seen in Chapter 14 and Ecological Toolkit 15.1, another important factor in the outcome of species interactions is the environmental context in which they occur. For example, under benign environmental conditions that are favorable for population growth, it makes sense that species will thrive and be limited by resources, thus engaging in negative interactions such as competition or predation. Under harsh environmental conditions, species will naturally be more limited by physical factors, thus interacting either weakly or positively with other species.

This view of species interactions as *context-dependent*, or changeable under different environmental conditions, is relatively new to ecology, but a number of important examples of context dependence exist. Most of these examples involve keystone or dominant species that play important roles in their communities in one context, but not in another. Mary Power, a professor at the University

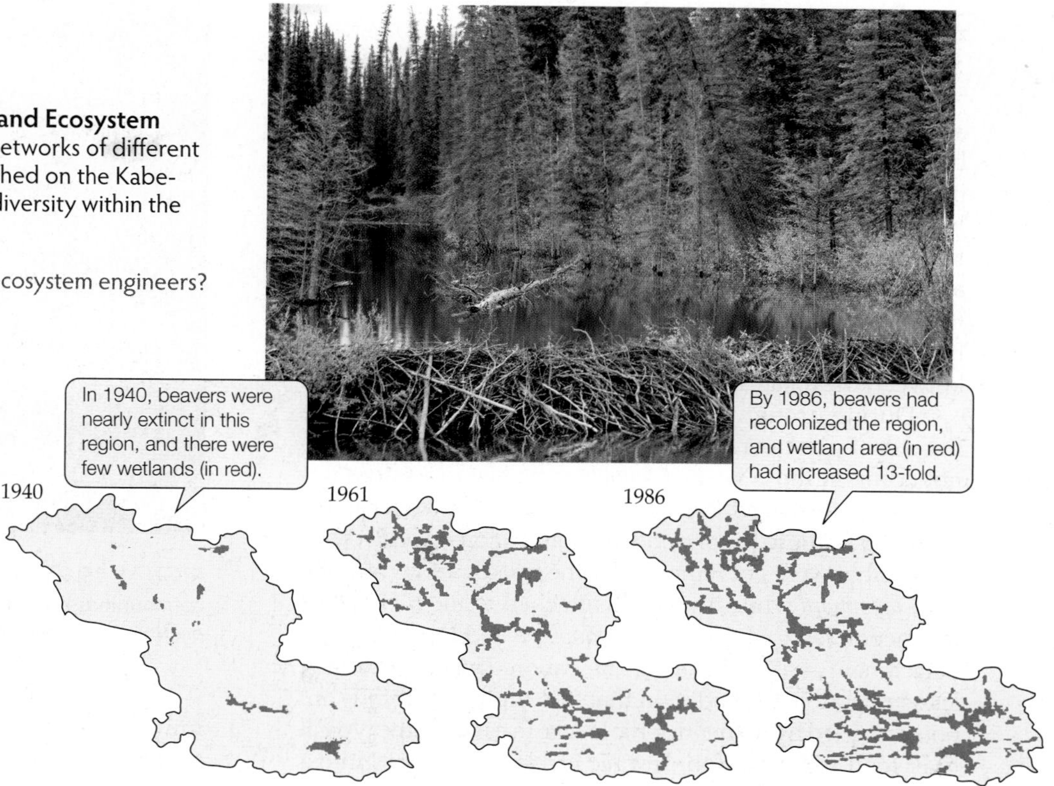

FIGURE 15.19 Beavers Are Keystone Species and Ecosystem Engineers By damming streams, beavers created networks of different types of wetlands (shown in red) in a 45 km^2 watershed on the Kabetogama Peninsula in Minnesota, thus increasing biodiversity within the region. (After Naiman et al. 1988.)

Q Why are beavers both keystone species and ecosystem engineers?

of California, Berkeley, who works on stream communities in northern California, has shown that the role of fish predators (roach, *Hesperoleucas symmetricus*, and steelhead, *Oncorhynchus mykiss*) changes from year to year. The role of these predators shifts from that of keystone species following winters of scouring floods to that of weak interactors during years with winter droughts and in places where flood control is operating (**Figure 15.20**; Power et al. 2008).

In the northern California rivers where Power works, there is a natural winter flood regime that produces dramatic population cycles of a green filamentous alga (*Cladophora glomerata*). In most years, scouring winter floods remove most of the inhabitants—particularly armored herbivorous insects—from the river bottom. In the following spring, there are large blooms of *Cladophora*. Increased light and nutrients and the lack of invertebrate herbivores allow *Cladophora* mats to grow profusely over the rocks, producing filaments up to 8 m (26 ft) long. By midsummer, these mats detach from the rocks and cover large portions of the river, at which time midge larvae, which feed on the floating alga and use it to weave

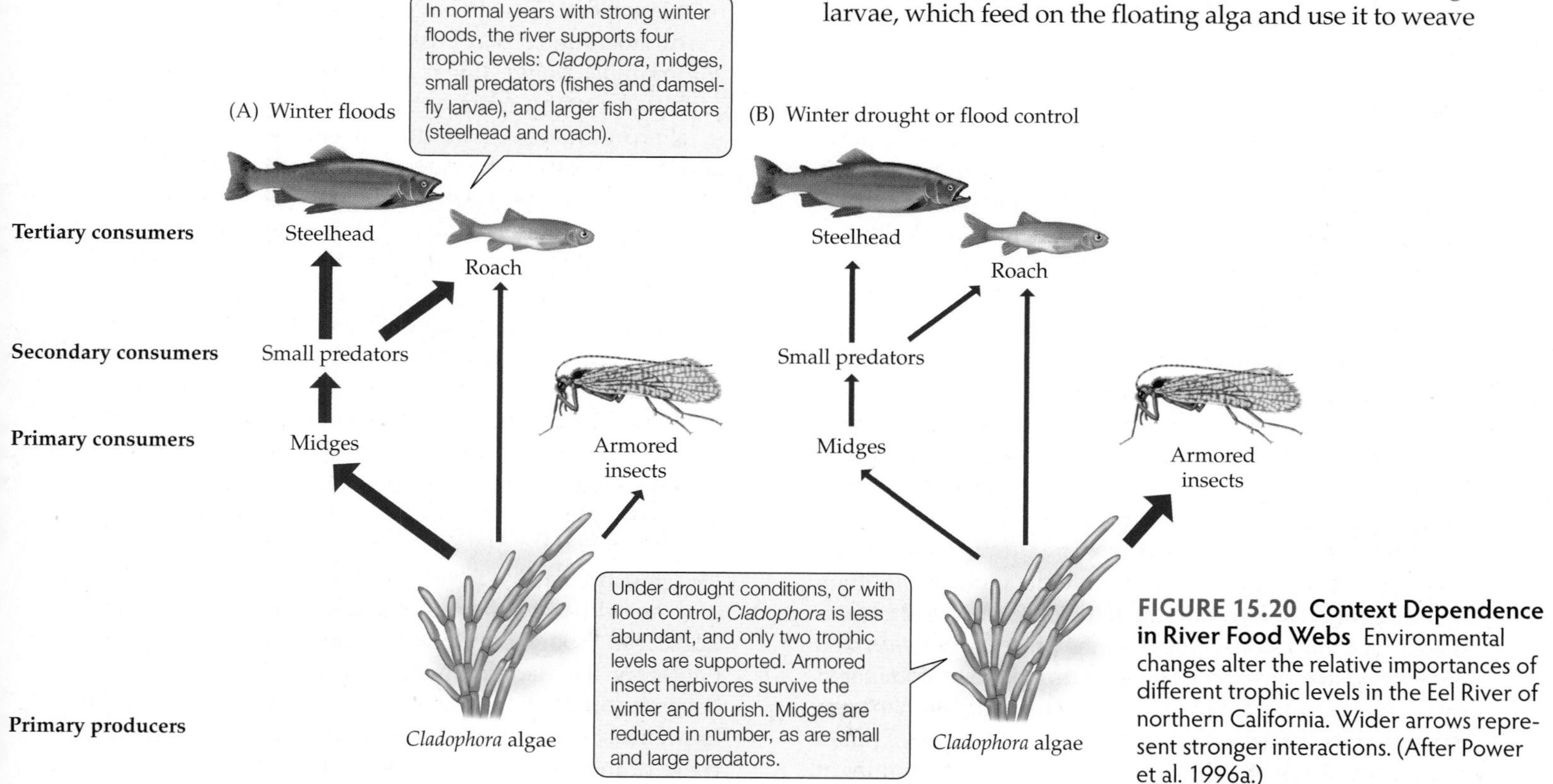

FIGURE 15.20 Context Dependence in River Food Webs Environmental changes alter the relative importances of different trophic levels in the Eel River of northern California. Wider arrows represent stronger interactions. (After Power et al. 1996a.)

small homes, increase in number. The midges are fed on by small fish and damselfly larvae, which in turn are eaten by steelhead and roach (a trophic cascade with four trophic levels; see Figure 15.20A). The steelhead and roach are able to decrease the size of the algal mats by eating small fish and damselfly larvae, which feed on midge larvae, which feed on the mats. The roach also fed on the algae directly, but only to a small degree, represented by the narrow arrow between the algae and roach (see Figure 15.20).

During drought years, however, and in rivers where flood control is operating, scouring of the river bottom does not occur. In those years, *Cladophora* persists, but does not form large, lush mats. Power and colleagues showed that this change was due to the presence of more armored herbivorous insects that were not removed by the floods, and which ate the *Cladophora* while it was still attached to the rocks. This interaction led to declines in *Cladophora* and the loss of the detachment phase of the alga. The armored insects are much less susceptible to predation than the midges and thus are not controlled by higher trophic levels. In essence, the typical river food web with four trophic levels is converted into a two-level food web during drought years, and the steelhead and roach, which once were keystone predators, are now minor players in the food web (Figure 15.20B).

In upcoming chapters, we will consider in much more detail the effects of disturbance, stress, and predation, as well as physical factors, on the outcome of species interactions and ultimately on the species diversity of communities.

A CASE STUDY REVISITED
"Killer Algae!"

The introduction of *Caulerpa taxifolia* into the Mediterranean Sea in the early 1980s set in motion a series of unfortunate events that resulted in large carpets of fluorescent green algae dominating formerly species-rich nearshore marine communities. *Caulerpa* thrived because humans facilitated its dispersal and its physiological tolerance. Even in the early stages of the invasion, Meinesz documented the seaweed in at least three types of communities, with different species compositions, on rocky, sandy, and muddy substrates. Together, these communities are home to several hundred species of algae and three marine flowering plants, as well as a number of animal species. Once *Caulerpa* arrived, native competitor and herbivore species were unable to keep it from spreading.

The invasion of *Caulerpa* has dramatically changed the ways in which the native species interact with one another, and thus the structure and function of the native communities. One obvious consequence of the presence of *Caulerpa* is the overgrowth and decline of seagrass meadows dominated by *Posidonia oceanica* (**Figure 15.21**). This seagrass has been likened to an "underwater tree" because of its long life span and slow growth (patches grow to 3 meters in diameter in 100 years). Just like forests, seagrass meadows support a multitude of species that use the vegetation as habitat. Research showed that *Posidonia* and *Caulerpa* have different growth cycles: *Posidonia* loses blades in the summer, when *Caulerpa* is most productive. Over time, this asynchronous growth pattern results in *Caulerpa* overgrowing the existing seagrasses and establishing itself as the dominant species. Additional research has shown that *Caulerpa* acts as an ecosystem engineer, accumulating sediments around its roots more readily than *Posidonia*, which dramatically changes the species composition of the small invertebrates that live on the seafloor. Surveys have revealed a significant drop in the numbers and sizes of fish that use the *Caulerpa*-invaded communities, suggesting that these habitats may no longer suitable for many commercially important species.

FIGURE 15.21 A Mediterranean Seagrass Meadow Native communities like this one, dominated by the seagrass *Posidonia oceanica*, are quickly overcome by invasive *Caulerpa*.

Future changes in Mediterranean seagrass meadows, and in the species dependent on them, will be difficult to predict, given the sheer number of species that are potentially affected by *Caulerpa*, the indirect effects that will be generated by changing interactions, and the relatively short time that has elapsed since the invasion began. A scientific approach, guided by a combination of theory and real-world observations, will be necessary if future predictions are to be made about the ultimate effect of *Caulerpa* on this vanishing underwater community.

CONNECTIONS IN NATURE
Stopping Invasions Requires Commitment

Invasive species pose threats to biodiversity in many regions of the world, as we'll see in the discussion under Concept 22.3. Even though it may be too late to stop the invasion of *Caulerpa* in the Mediterranean, the lessons learned there have been important in other regions of the world. In 2000, just as Meinesz was making progress in banning international trade of the alga, he received an e-mail from an environmental consultant in San Diego, California. While surveying eelgrass in a lagoon, she had noticed a large patch of what was later identified as *Caulerpa taxifolia*. Acting on Meinesz's recommendation, a team of scientists and managers from county, state, and federal agencies immediately assembled to design an eradication plan. This plan involved treating the alga with chlorine gas injected under tarps placed on top of algal patches. More than $1 million was initially budgeted for the project in 2000, but it eventually took 6 years and $7 million to eradicate the alga. The invasion was widely publicized, resulting in the discovery of another patch of *Caulerpa* in another lagoon near Los Angeles, which was also eradicated. The California experience is a rare success story only because immediate action was taken by scientists, managers, and politicians to deal with the invasion before eradication became an ecological and fiscal impossibility. It is estimated that control of invasive species in the United States costs $22 billion per year (Pimentel et al. 2000).

To determine the origin of the *Caulerpa* that invaded California, molecular evidence was needed. This shift in the team's focus from communities to genes illustrates a point made in Chapter 1: that ecologists must study interactions in nature across many levels of biological organization. The team sent specimens of the alga to geneticists at two universities, who analyzed the sequences of its ITS ribosomal DNA and quickly determined that they were identical to those of *Caulerpa* from the Mediterranean, the Wilhelma Zoo (where the strain was first cultivated), and many other public aquariums around the world (Jousson et al. 2000). Unfortunately, it is still unknown how the species was introduced into the two California lagoons, but hypotheses range from amateur aquarists cleaning their tanks in the lagoons to an accidental release from aquariums onboard a Saudi Arabian prince's yacht, which was being repainted in San Diego at about the time the alga probably arrived. Through the use of DNA analysis, it has been determined that the *Caulerpa* involved in subsequent invasions in Australia and Japan are genetically identical to the original German strain. The molecular evidence makes it clear that the trade of this alga in aquarium circles poses a global threat to nearshore temperate marine environments. Legislation is now in place to ban the "killer alga" from a number of other countries where it has a good chance of invading successfully.

SUMMARY

CONCEPT 15.1 Communities are groups of interacting species that occur together at the same place and time.

- Communities can be delineated by the characteristics of their physical environment or by biological characteristics, such as the presence of abundant species.
- Ecologists often use subsets of species to define and study communities because it is impractical to count or study all the species within a community, especially if they are small or undescribed.
- Subsets of species used to study communities include taxonomic groups, guilds, functional groups, and food and interaction webs.

CONCEPT 15.2 Species diversity and species composition are important descriptors of community structure.

- Species diversity, the most commonly used measure of community structure, is a combination of the number of species (species richness) and the abundances of those species relative to one another (species evenness).
- Communities differ in the commonness or rarity of their species, such that even in communities with the same numbers of species, one community might have a few very abundant species and many rare species, while another might have species of equal abundance.
- Species richness estimates for communities improve with increased sampling effort up to a certain point, at which additional samples reveal few or no new species.
- Species composition—the identity of the species present in a community—is an obvious but important characteristic of community structure that is not revealed in measures of species diversity.

CONCEPT 15.3 Communities can be characterized by complex networks of direct and indirect interactions that vary in strength and direction.

- Indirect species interactions, in which the relationship between two species is mediated by a third (or more) species, can have large effects on the outcomes of direct species interactions.
- Some species have a strong negative or positive effect on their communities, but others probably have little or no effect.
- Species that have large effects on their communities by virtue of their abundance or biomass are known as dominant species. Those that have large effects due to the roles they play in their communities are known as keystone species.
- Ecosystem engineers create, modify, or maintain physical habitat for themselves and other species.
- The environmental context of species interactions can modify them enough to change their outcome.

REVIEW QUESTIONS

1. What is the formal definition of a community? Why is incorporating species interactions into that definition important?
2. Species diversity measurements take into account both species richness and species evenness. Why would these measurements be preferred to species richness alone? What do rank abundance curves add to one's knowledge about community structure?
3. Species vary in the strength of their interactions with other species. Species that interact strongly with other species include dominant species, keystone species, and ecosystem engineers. Describe the differences between these three types of species. Can dominant and keystone species also be ecosystem engineers?
4. The interaction web at right is common in the rocky intertidal zone of Washington and Oregon. The arrows and their signs (+ or –) represent interactions that occur between species in the rocky intertidal zone community.
   a. Suppose seagulls were removed from this community. Determine the two direct effects this manipulation would have on limpets and gooseneck barnacles. Would it increase, decrease, or not change the abundances of those species, and through what type of interaction (i.e., by predation, competition, or facilitation)?
   b. Determine four indirect effects of removing seagulls from the community. How would each species respond (i.e., would it increase, decrease, or not change in abundance), and through what type of interaction (i.e., predation, competition, or facilitation)?

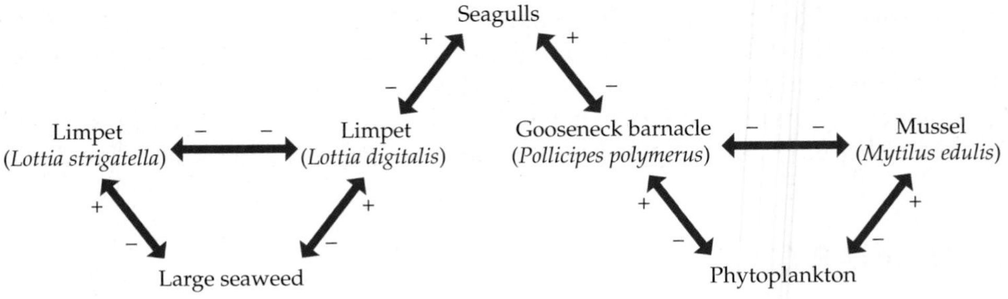

ON THE COMPANION WEBSITE
sites.sinauer.com/ecology2e

The website includes Chapter Outlines, Online Quizzes, Flashcards & Key Terms, Suggested Readings, a complete Glossary, and the Web Stats Review. In addition, the following resources are available for this chapter:

HANDS-ON PROBLEM SOLVING
Measuring Marine Species Diversity

This Web exercise explores various methods of measuring species diversity. You will interpret data from a recent paper that measured marine species diversity, with emphasis on the benthic fauna of the continental shelf of Norway, using the Shannon index as well as other indices of diversity.

CHAPTER

16 Change in Communities

KEY CONCEPTS

- **CONCEPT 16.1** Agents of change act on communities across all temporal and spatial scales.
- **CONCEPT 16.2** Succession is the process of change in species composition over time as a result of abiotic and biotic agents of change.
- **CONCEPT 16.3** Experimental work on succession shows its mechanisms to be diverse and context-dependent.
- **CONCEPT 16.4** Communities can follow different successional paths and display alternative states.

A Natural Experiment of Mountainous Proportions: A Case Study

The eruption of Mount St. Helens was a defining moment for ecologists interested in natural catastrophes. Mount St. Helens, located in Washington State, is part of the geologically active Cascade Range, located in the Pacific Northwest region of North America (**Figure 16.1**). The once frosty-topped mountain had a rich diversity of ecological communities. If you had visited Mount St. Helens in the summer, you could have seen alpine meadows filled with colorful wildflowers and grazing elk. At lower elevations, you could have hiked across the cool fern- and moss-covered forest floor under massive old-growth trees. You could have swum in the blue, clear water of Spirit Lake, or fished along its shores. But a few minutes after 8:30 A.M. on May 18, 1980, all that was living on Mount St. Helens would be gone. On the north side of the mountain, a huge magma-filled bulge had been forming for months, and it gave way that morning in an explosive eruption and the largest avalanche in recorded history.

Photos of the eruption show that mud and rock flowed down the face of Mount St. Helens and was deposited tens of meters deep in some areas (**Figure 16.2**). The wave of debris that passed over Spirit Lake was 260 m (858 ft) deep and decreased the lake's water depth by 60 m (200 ft). The bulk of the avalanche traveled 23 km (14 mi) in about 10 minutes to the North Fork Toutle River, where it scoured the entire valley, from floor to rim, with material from the volcano and left a truly massive pile of tangled vegetation at its tail end. In addition to the avalanche, the blast produced a cloud of hot air that burned forests to ash near the mountain, blew down trees over a large area, and left dead but standing trees stretching for miles away from the mountain. Ash from the explosion blanketed forests, grasslands, and deserts located hundreds of kilometers away.

The destruction that ensued on that day created whole new habitats on Mount St. Helens, some of which were completely devoid of any living organisms. At one extreme, there was the Pumice Plain, a large, gently sloping moonscape of a place below the volcano that was pelted with hot, sterilizing pumice (see Figure 16.2). This harsh and geologically monotonic environment lacked life, and even organic matter, of any form. All life in Spirit Lake was extinguished, and huge amounts of woody debris were deposited there, some of which still floats on top of the lake today. But not surprisingly, given the large forests that had surrounded the

FIGURE 16.1 Once a Peaceful Mountain Before the eruption shown here on May 18, 1980, Mount St. Helens, in southwestern Washington State, had a diversity of communities, including alpine meadows, old-growth forests, and lakes and streams.

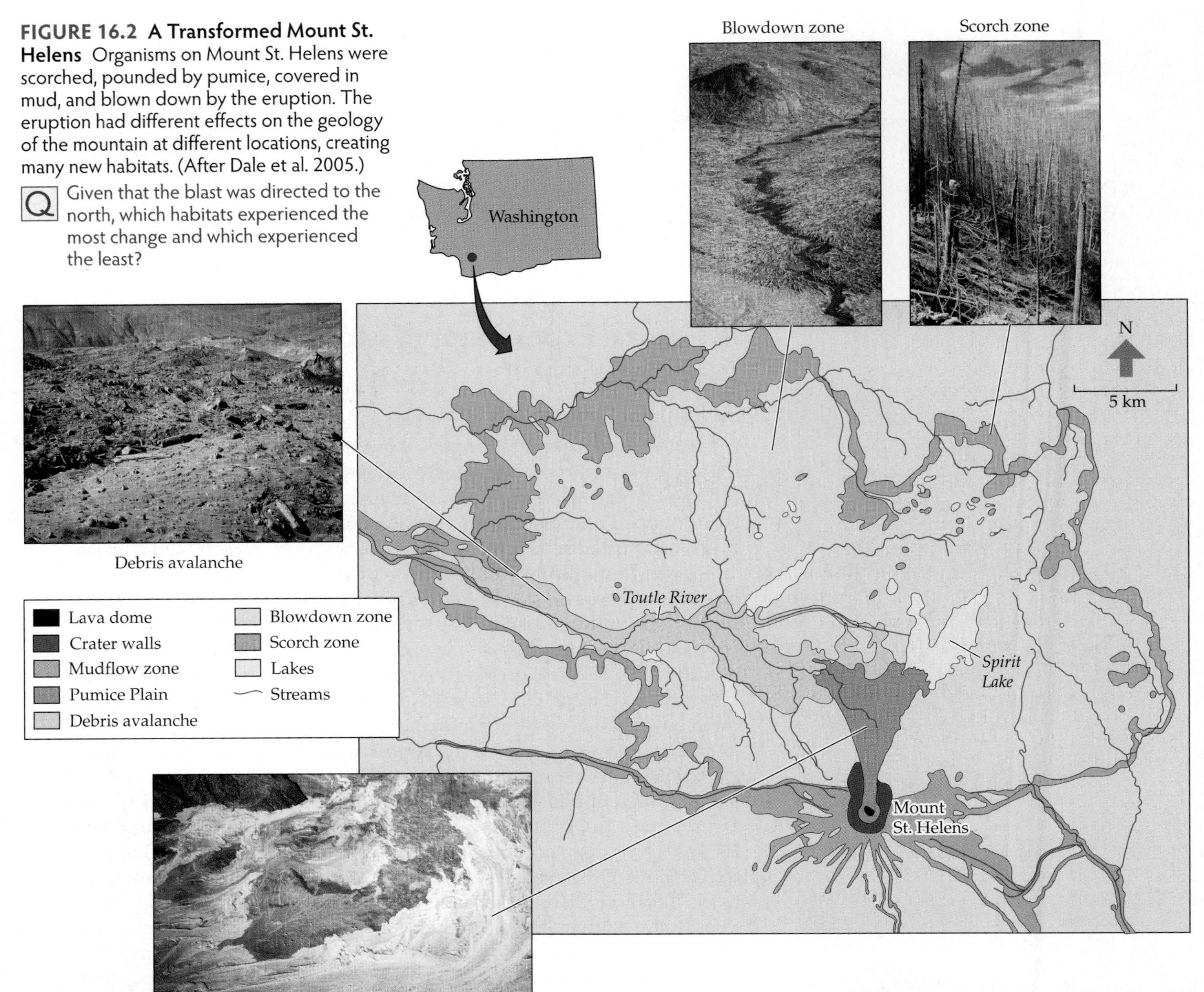

FIGURE 16.2 A Transformed Mount St. Helens Organisms on Mount St. Helens were scorched, pounded by pumice, covered in mud, and blown down by the eruption. The eruption had different effects on the geology of the mountain at different locations, creating many new habitats. (After Dale et al. 2005.)

Q Given that the blast was directed to the north, which habitats experienced the most change and which experienced the least?

mountain, the majority of the landscape now consisted of downed or denuded trees, covered with rock, gravel, and mud tens of meters deep in some places (see Figure 16.2). Compared with the Pumice Plain, this blowdown zone had some hope of a biological legacy buried under the piles of trees and ash.

Shortly after the eruption, helicopters delivered the first scientists to the mountain to begin studying what was essentially a natural experiment of epic proportions. A few lucky ecologists recorded the first observations of the sequence of biological changes that began soon after the eruption. Field excursions in the summers of 1980 and 1981 were organized, and valuable baseline data were collected. Now, more than 30 years later, hundreds of ecologists have studied the reemergence of life on Mount St. Helens. For many, the experience has been life-changing, and their careers have been consumed by research on this fascinating study system. Much of what has been learned has been unexpected and has changed the way we view the recovery of communities and the persistence of life on Earth.

Introduction

One constant that all ecologists can agree on is that communities are always changing. Some communities show more dynamism than others. For example, it is hard to imagine how deserts, with their large, stoic cacti, have changed much over time. This is especially true if you compare deserts with, for example, high mountain streams or rocky intertidal zones, where species are coming and going on a regular basis. But community change is relative, and there is no question that even deserts change, though at a much slower pace than we might realize on the basis of one visit, or even one ecological study. In the words of one of the ecological maxims presented in Table 1.1, time matters.

Unfortunately, we humans cannot deny that our actions are becoming one of the strongest forces of change in communities, and that we are taking those actions with an imperfect understanding of their consequences. In this chapter, we will consider the agents of change in communities, from subtle to catastrophic, and their effects on community structure over time.

CONCEPT 16.1 Agents of change act on communities across all temporal and spatial scales.

Agents of Change

Let's imagine for a moment that you have the ability to look back in time and follow the change in a typical coral reef community in the Indian Ocean (**Figure 16.3**). Over the last few decades, you might have seen considerable change, both subtle and catastrophic. Subtle changes might include the slow rise to dominance of certain coral species, and the slow decline of others, due to the effects of competition, predation, and disease. More catastrophic changes might include the massive deaths of corals in the last decade due to coral bleaching (loss of symbiotic algae, as described on p. 525) and the great tsunami of 2004, resulting in the replacement of some coral species with other species, or no replacement at all. Taken together, these changes make the community what it is today: a reef that has many fewer coral species than it did a few decades ago, due to a combination of natural and human-caused agents of change.

Succession is change in the species composition of communities over time. Succession is the result of a variety of abiotic (physical and chemical) and biotic agents of change. In the following sections, we will consider the theory behind succession and examples that illustrate how it works in a variety of systems. But first, in this section, we will identify and define the agents of change that are most responsible for driving succession.

Agents of change can be abiotic or biotic

Communities, and the species contained within them, change in response to a number of abiotic and biotic factors (**Table 16.1**). We have considered many of these factors in previous chapters. In Unit 1, we learned that abiotic factors, in the form of climate, soils, nutrients, and water, vary over daily, seasonal, decadal, and even 100,000-year time scales. This variation has important implications for community change. For example, in our Indian Ocean coral reef community (see Figure 16.3), unusually high water temperatures driven by large-scale climate change have been implicated in recent losses of symbiotic algae from corals, resulting in coral bleaching. If the symbiotic algae do not return, the corals will eventually die, thus creating the conditions for species replacement. Likewise, increases in sea level can decrease the amount of light that reaches the corals. If light availability falls below the physiological limits of some coral species, they could slowly be replaced by more tolerant species, or even by macroalgae

FIGURE 16.3 Change Happens Coral reef communities in the Indian Ocean have experienced significant changes over the last few decades. The agents of change have been both subtle and catastrophic, natural and human-caused.

TABLE 16.1

Examples of Abiotic and Biotic Agents of Change and Their Effects on Organisms

Agent of change	Effects	Examples
ABIOTIC FACTORS		
Waves, currents	Organisms are detached, injured, or killed	Storms, hurricanes, floods, tsunamis, ocean upwelling
Wind	Organisms are detached, injured, or killed	Storms, hurricanes, wind-driven sediment scouring
Water supply	Organisms grow slowly, are injured, or are killed	Droughts, floods, mudslides
Chemical composition	Organisms grow slowly, are injured, or are killed	Pollution, acid rain, high or low salinity, high or low nutrient supply
Temperature	Organisms grow slowly, are injured, or are killed	Freezing, snow and ice, avalanches, excessive heat, fire, sea level rise or fall
Volcanic activity	Organisms are injured or killed	Lava, hot gases, mudslides, flying rocks and debris, floods
BIOTIC FACTORS		
Negative interactions	Organisms grow slowly, are injured, or are killed	Competition, predation, herbivory, disease, parasitism, trampling, digging, boring

Source: Adapted, with additions, from Sousa 2001.

(seaweeds). Finally, increasing ocean acidification can dissolve the skeletons of corals, hindering their growth (see Chapter 24 for more information on climate change and ocean acidification). Because these abiotic conditions are constantly changing, communities are doing the same, at a pace consistent with their environment.

Abiotic agents of change can be placed into two categories: disturbance and stress, which differ in the effects they have on species. A **disturbance** is an abiotic event that physically injures or kills some individuals and creates opportunities for other individuals to grow or reproduce. Some ecologists also consider biotic events such as digging by animals to be disturbances. In our coral reef example, the 2004 tsunami can be viewed as a disturbance because the force of water passing over the reef injured and killed many coral individuals. Likewise, a biotic event such as coral boring by snails could be considered a disturbance because it injures and weakens coral skeletons. *Stress*, on the other hand, occurs when some abiotic factor reduces the growth or reproduction of individuals and creates opportunities for other individuals. A stress in our coral reef might be the effect of water temperature or sea level rise on the growth and reproduction of corals. Examples of other stresses and disturbances are included in Table 16.1. Both disturbance and stress are believed to play critical roles in driving succession.

How do biotic factors influence community change? In Unit 3, we saw that species interactions, both negative and positive, can result in the replacement of one species with another (see Table 16.1). In our coral reef (see Figure 16.3), change might be driven by competition between, for example, platelike corals and branched corals, with the platelike forms eventually dominating over time. Coral diseases are another example of a species interaction that can initiate change in communities by causing particular coral species to grow more slowly or eventually die. Equally common agents of change are the actions of ecosystem engineers and keystone species (see Figures 15.18 and 15.19). Both types of species have large effects on other species that result in community change.

Finally, it is important to realize that abiotic and biotic factors often interact to produce change in communities. We can see this interaction in the case of ecosystem engineers such as beavers, which cause changes in abiotic conditions that in turn cause species replacement (see Figure 15.19). Similarly, abiotic factors such as wind, waves, or temperature can act by modifying species interactions, either positively or negatively, thus creating opportunities for other species. We have seen examples of this kind of effect on sea palms in the rocky intertidal zone (see p. 255), plants in alpine regions (see Figure 14.10), and stream insects in northern California (see Figure 15.20).

Agents of change vary in their intensity, frequency, and extent

As you might guess, how often, and at what magnitude and areal extent, agents of change act largely determines the tempo of succession. For example, when the avalanche

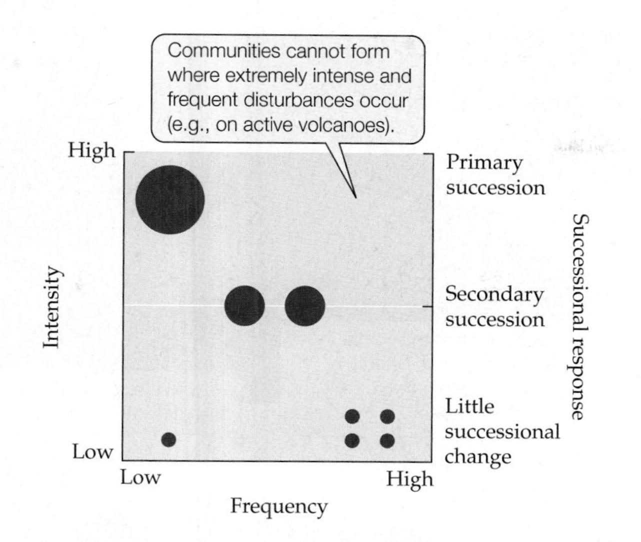

FIGURE 16.4 The Spectrum of Disturbance How much biomass is removed (the intensity of disturbance) and how often it is removed (the frequency of disturbance) can influence the amount of disturbance (represented by the red circles) that occurs and the type of succession that is possible afterward.

Q Describe how the type of organism being studied might influence whether we speak of a disturbance as being intense or frequent.

produced by Mount St. Helens ripped through the alpine community back in 1980, it produced a disturbance that was larger and more severe than any others that had occurred that year, that decade, or that century. The *intensity* of that disturbance—the amount of damage and death it caused—is huge, both because of the massive physical force involved and because of the area covered. In contrast, the *frequency* of that disturbance is low because such eruptive episodes are so rare (occurring once every few centuries). Extremely intense and infrequent events, such as the eruption of Mount St. Helens, are at the far end of the spectrum of disturbances species experience in communities (**Figure 16.4**). In this case, the entire community is affected, and recovery involves the complete reassembly of the community over time. We will consider the mechanisms of that reassembly in the following sections.

At the other end of the spectrum are weak and frequent disturbances that may have more subtle effects or affect a smaller area (see Figure 16.4). Prior to the eruption of Mount St. Helens, such disturbances might have included wind blowing down old trees living in the Douglas fir forests surrounding the mountain. These more frequent disturbances form patches of available resources that can be used by individuals of the same or different species. A mosaic of disturbed patches can promote species diversity in communities over time, but may not lead to much successional change. We will learn more about these smaller disturbances and their effects on species diversity in Chapter 18.

Let's turn our attention now from the agents of change to their consequences for community succession.

CONCEPT 16.2 Succession is the process of change in species composition over time as a result of abiotic and biotic agents of change.

Basics of Succession

At the most basic level, the term "succession" refers to the process by which the species composition of a community changes over time. Mechanistically, succession involves the colonization and extinction of species in a community due to abiotic and biotic agents of change. Even though studies of succession often focus on vegetative change, the roles of animals, fungi, bacteria, and other microbes are equally important.

Theoretically, succession progresses through various stages that include a *climax* stage (**Figure 16.5**). The climax is thought to be a stable end point that experiences little change until a particularly intense disturbance sends the community back to an earlier stage. As we will see in the following sections, there is some argument about whether succession can ever lead to a stable end point.

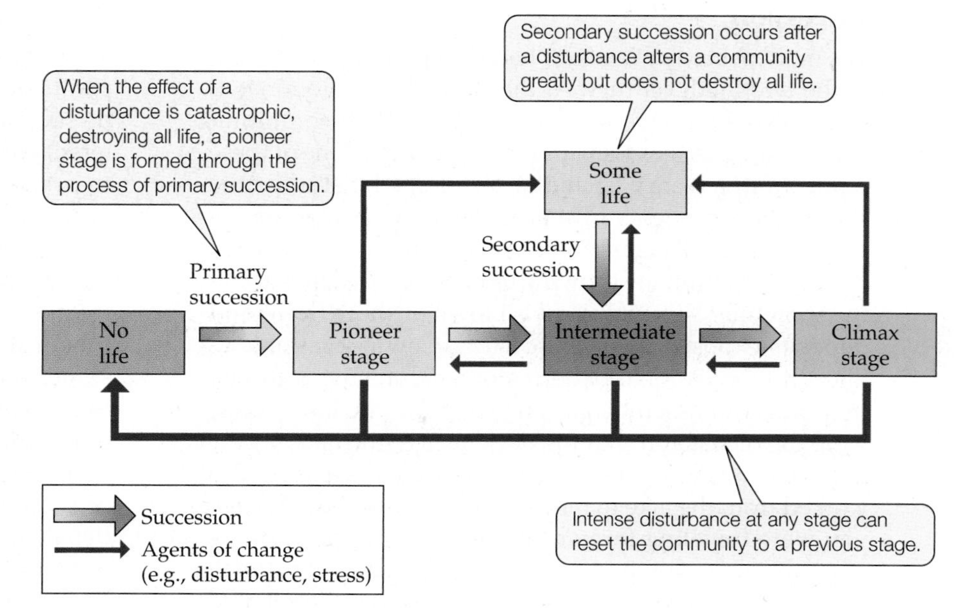

FIGURE 16.5 The Trajectory of Succession A simple model of succession involves transitions between stages driven by species replacements over time. Theoretically, these changes ultimately result in a climax stage that experiences little change.

Primary succession and secondary succession differ in their initial stages

Ecologists recognize two types of succession that differ in their initial stages. The first type, **primary succession**, involves the colonization of habitats that are devoid of life (see Figure 16.5), either as a result of catastrophic disturbance, as we see on the Pumice Plain at Mount St. Helens, or because they are newly created habitats, such as volcanic rock. As you can imagine, primary succession can be very slow because the first arrivals (known as *pioneer* or *early successional* species) typically face extremely inhospitable conditions. Even the most basic resources needed to fuel life, such as soil, nutrients, and water, may be lacking. The first colonizers, then, tend to be species that are capable of withstanding great physiological stress and transforming the habitat in ways that benefit their further growth and expansion (and that of other species, as we will see).

The other type of succession, known as **secondary succession**, involves the reestablishment of a community in which most, but not all, of the organisms or organic constituents have been destroyed (see Figure 16.5). Agents of change that can create such conditions include fire, hurricanes, logging, and herbivory. Despite the catastrophic effect of the eruption on Mount St. Helens, there were many areas, such as the blowdown zone, where some organisms survived and secondary succession took place. As you might expect, the legacy of the preexisting species and their interactions with colonizing species play larger roles in the trajectory of secondary succession than in primary succession.

The early history of ecology is a study of succession

The modern study of ecology had its beginnings at the turn of the twentieth century. At that time, it was dominated by scientists who were fascinated with plant communities and the changes they undergo over time. One of these pioneers was Henry Chandler Cowles, who studied the successional sequence of vegetation in sand dunes on the shore of Lake Michigan (**Figure 16.6**). In this ecosystem, the dunes are continually growing as new sand is deposited at the shoreline. This new sand is blown onshore from sand deposits exposed during droughts. Cowles was able to infer the successional pattern along a dune by assuming that the plant assemblages farthest from the lake's edge were the oldest and that the ones nearest the lake, where new sand was being deposited, were the youngest. As you walked from the lake to the back of the dune, he believed, you were traveling forward in time and able to imagine what the areas you had just passed through would look like in centuries to come. The first stages were dominated by a hardy ecosystem engineer, American beach grass (*Ammophila breviligulata*). *Ammophila* (whose name literally means "sand lover") is excellent at trapping sand and

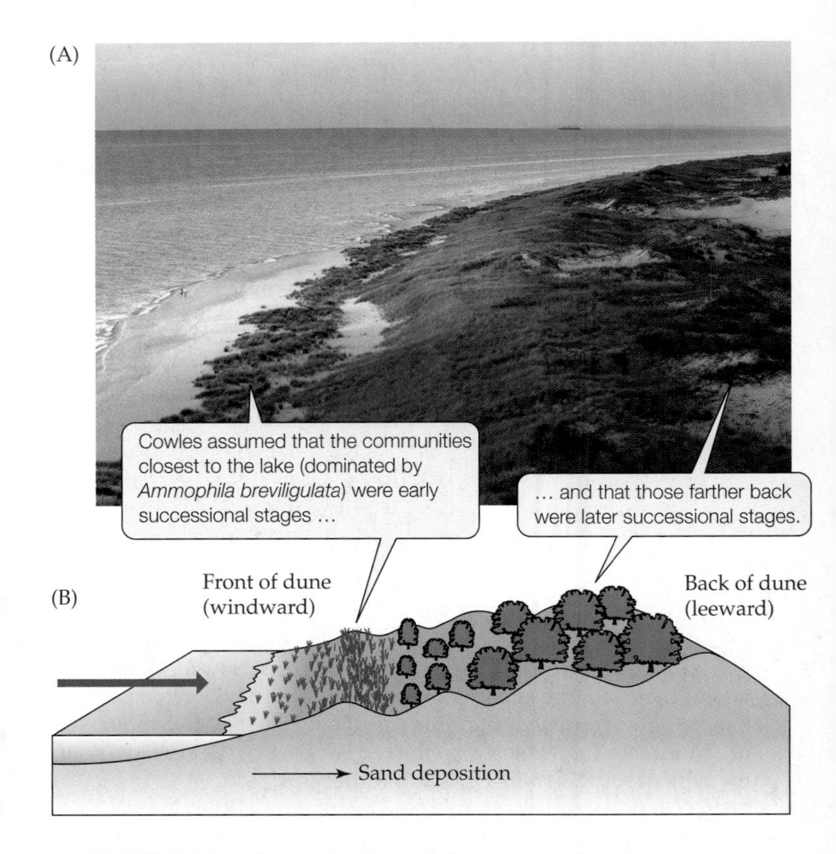

FIGURE 16.6 Space for Time Substitution (A) The portion of a dune nearest the shoreline on Lake Michigan is covered with *Ammophila*. (B) When Henry Chandler Cowles studied succession on these dunes, he assumed that the earliest successional stages occurred on the newly deposited sand at the front of the dune, and that later successional stages occurred at the back of the dune.

creating hills that provide refuge on their leeward side for plants less tolerant of the constant burial and sand scouring experienced on the beachfront.

Cowles (1899) made the assumption that the different plant assemblages—or "societies," as he called them—that he saw in different positions on a dune represented different successional stages. That assumption allowed him to predict how a community would change over time without actually waiting for the pattern to unfold, which would have taken decades to centuries. This idea, known as the "space for time substitution" (Pickett 1989), is used frequently as a practical way to study communities over time scales that exceed the life span of an ecologist. It assumes that time is the main factor causing communities to change and that unique conditions in particular locations are inconsequential. These are big assumptions, and they have fueled a current debate about the predictability of community dynamics over time. We will discuss this debate in more detail under Concept 16.4, when we deal with alternative stable state theory.

Henry Cowles was not alone in his interest in plant succession. His peers included Frederick Clements and Henry Gleason, two men who had completely different and contentious views on the mechanisms driving succession

(Kingsland 1991). Clements, one of the first to write a formal book in 1907 on the new science of ecology, believed that plant communities were like "superorganisms," groups of species that worked together in a mutual effort toward some deterministic end. Succession was similar to the development of an organism, complete with a beginning (embryonic stage), middle (adult stage), and end (death). Clements (1916) thought that each community had its own predictable life history and, if left undisturbed, ultimately reached a stable end point. This "climax community" was composed of species that dominated and persisted over many years and provided the type of stability that could potentially be maintained indefinitely.

Gleason (1917) thought that viewing a community as an organism, with various interacting parts, ignored the responses of individual species to prevailing conditions. In his view, communities were not the predictable and repeatable result of coordinated interactions among species, but rather the random product of fluctuating environmental conditions acting on individual species. Each community was the product of a particular place and time, and was thus unique in its own right.

Looking back, it is clear that Gleason and Clements had extreme views of succession. As we will see in the next section, we can find elements of both theories in the results of studies that have accumulated over the last century. First, however, it is important to mention one last ecologist, Charles Elton (**Figure 16.7A**), whose perspective on succession was shaped not only by those of the botanists who came before him, but also by his interest in animals. He wrote his first book, *Animal Ecology* (1927), in 3 months' time at the age of 26. The book addresses many important ideas in ecology, including succession. Elton believed that organisms and the environment interact to shape the direction succession will take. He presented an example from pine forests in England that were being subjected to deforestation. After the felling of the pines, the trajectory of succession varied depending on the moisture content of the environment (**Figure 16.7B**). Wetter areas developed into sphagnum bogs, while slightly drier areas developed into wetlands containing rushes and grasses. Eventually, these communities all became birch scrub, but then ultimately diverged into two types of forest. Through these observations, Elton demonstrated that the only way to predict the trajectory of succession was to understand the biological and environmental context in which it occurred.

Elton's greatest contribution to the understanding of succession was his acknowledgment of the role of animals. Up to that point, most ecologists believed that plants drove succession, while animals were passive followers. Elton provided many examples showing how animals could create successional patterns by eating, dispersing, trampling, and destroying vegetation in ways that greatly affected the sequence and timing of succession. We will review some examples of animal-driven succession in the next section, but it is clear that the observations and conclusions Elton made 80 years ago still hold today.

Multiple models of succession were stimulated by lack of scientific consensus

Fascination with the mechanisms responsible for succession, and attempts to integrate the controversial theories

(A)

(B)

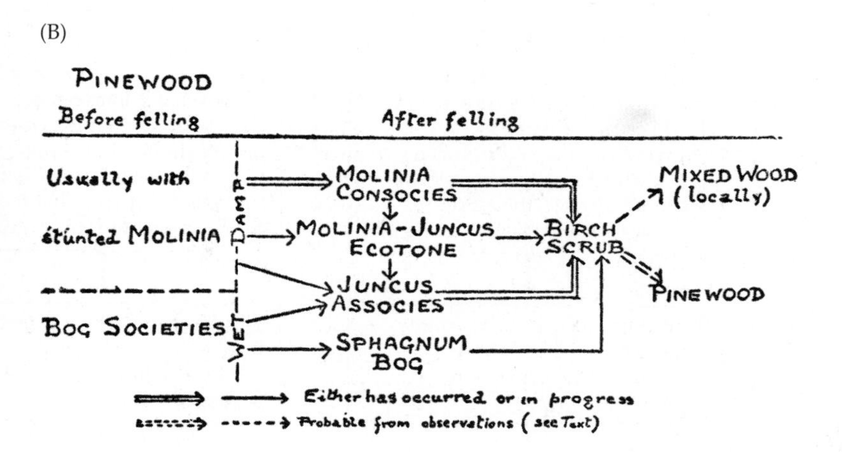

FIGURE 16.7 Elton's Trajectory of Pine Forest Succession (A) Charles Elton at the age of 25, a year before the publication of his first book, *Animal Ecology* (1927). (B) Elton's book contained this diagram of succession in pine forests after logging. The successional trajectory differs depending on the moisture content of a particular area: wetter areas become sphagnum bogs, while slightly drier areas become wetlands containing rushes (*Juncus*) and grasses (*Molinia*). Eventually, these communities all became birch scrub, but then ultimately diverged into pine woods or mixed woods, again depending on moisture. (B from Elton 1927.)

of Clements, Cowles, and Elton, led ecologists to use more scientifically rigorous methods to explore succession, including comprehensive reviews of the literature and manipulative experiments. Joseph Connell and his collaborator Ralph Slatyer (1977) surveyed the literature and proposed three models of succession that they believed to be important (**Figure 16.8**).

The *facilitation model*, inspired by Clements, describes situations in which the earliest species modify the environment in ways that ultimately benefit later species, but hinder their own continued dominance. These early species are usually stress-tolerant and are good at engineering the habitat and ameliorating the harsh physical conditions often characteristic of early successional stages. Eventually, however, a sequence of species facilitations leads to a climax community composed of species that no longer facilitate other species and are displaced only by disturbances.

The *tolerance model* also assumes that the earliest species modify the environment, but in neutral ways that neither benefit nor inhibit later species. These early successional species have life history strategies that allow them to grow and reproduce quickly. Later species persist merely because they have life history strategies that allow them to tolerate environmental or biological stresses that would hinder early successional species.

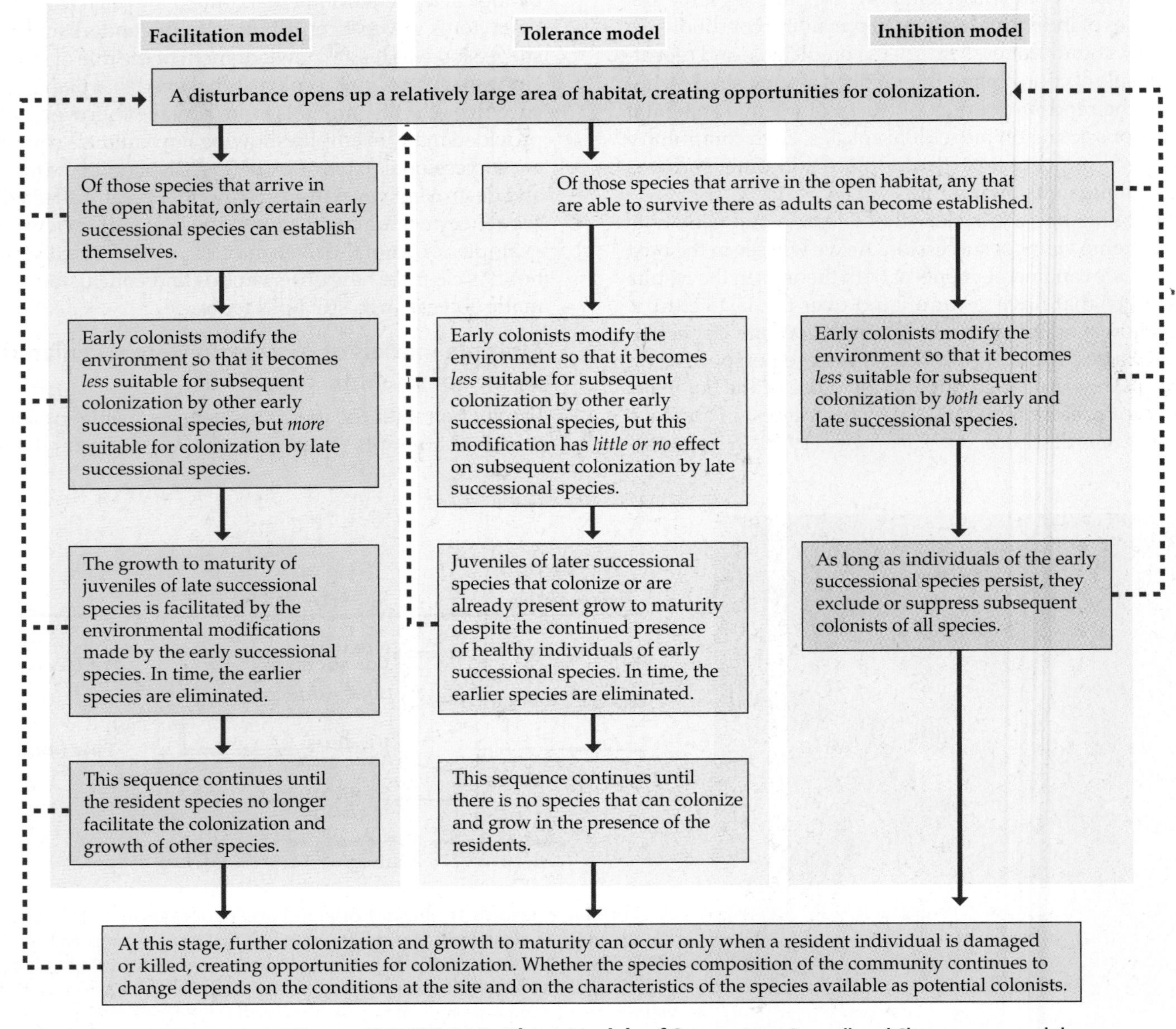

FIGURE 16.8 Three Models of Succession Connell and Slatyer proposed three conceptual models—the facilitation, tolerance, and inhibition model—to describe succession. (After Connell and Slatyer 1977.)

The *inhibition model* assumes that early successional species modify the environment in ways that hinder later successional species. This suppression of the next stage of succession is broken only when stress or disturbance decreases the abundance of the inhibitory species. As in the tolerance model, later species persist merely because they have life history strategies that allow them to tolerate environmental or biological stresses that would hinder early successional species.

Although Connell and Slatyer focused on plant communities, it is important to consider the role of animals as drivers of succession. Animals not only act as ecosystem engineers, but also change assemblages of plants in the ways Elton described, leading to transitions from one successional stage to another. In the next section, we will see examples of the role of animals in succession.

CONCEPT 16.3 Experimental work on succession shows its mechanisms to be diverse and context-dependent.

Mechanisms of Succession

More than thirty years have gone by since Connell and Slatyer wrote their influential theoretical paper on succession. Since that time, there have been a number of experimental tests of their three models. As we will see in this section, these studies show that the mechanisms driving succession rarely conform to any one model, but instead are dependent on the community and the context in which experiments are conducted.

Studies of succession show that no one model fits any one community

To illustrate the types of successional mechanisms that have been observed using experiments, we will focus on three examples: communities that form (1) after glacial retreat in Alaska, (2) after vegetation disturbance in salt marshes in New England, and (3) after wave disturbance in the rocky intertidal of the Pacific coast of the U.S.

Primary succession in Glacier Bay, Alaska One of the best-studied examples of primary succession occurs in Glacier Bay, Alaska, where the melting of glaciers has led to a sequence of community change that reflects succession over many centuries (**Figure 16.9**). Captain George Vancouver

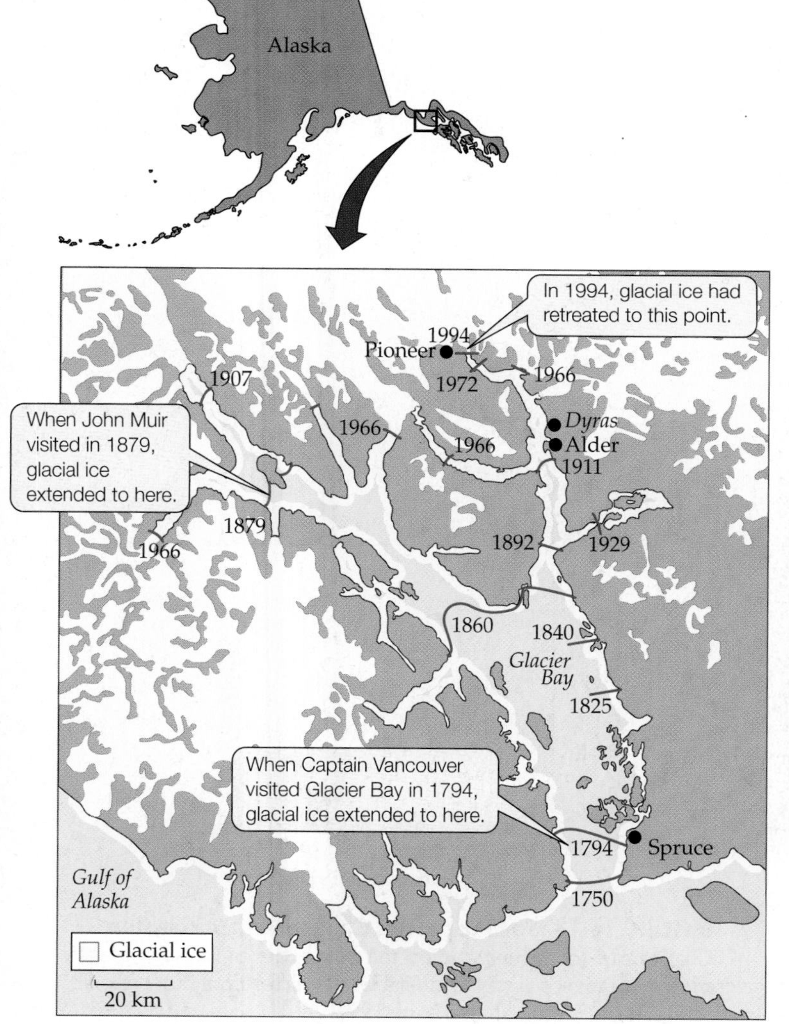

FIGURE 16.9 Glacial Retreat in Glacier Bay, Alaska Over more than 200 years, the melting of glaciers has exposed bare rock to colonization and succession. (After Chapin et al. 1994.)

Q Based on the locations of the glaciers over time, describe where the oldest and youngest communities are located.

Youngest near the receeding front

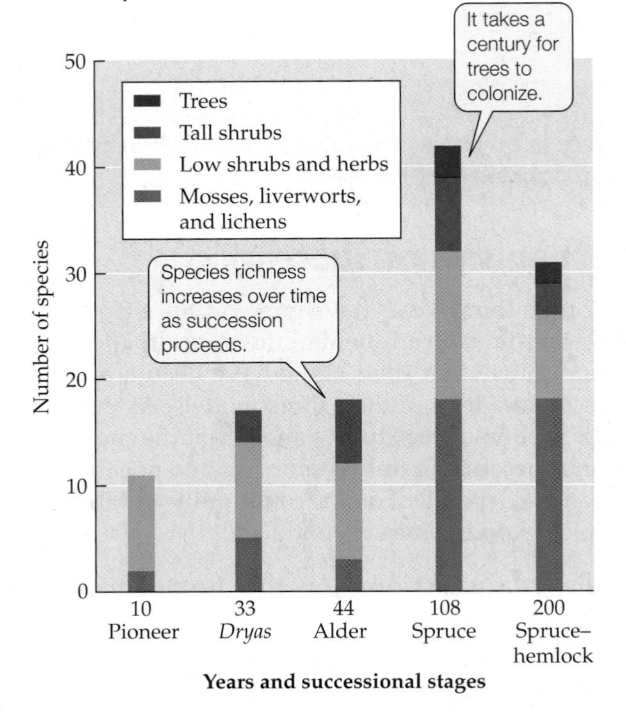

FIGURE 16.10 Successional Communities at Glacier Bay, Alaska Plant species richness has generally increased over the 200 years following glacial retreat. (Data from Reiners et al. 1971.)

first recorded the location of glacial ice there in 1794, while exploring the west coast of North America. Over the last 200 years, those glaciers have retreated up the bay, leaving behind bare, broken rock (known as *glacial till*). John Muir, in his book *Travels in Alaska* (1915), first noted how much the glaciers had melted since Vancouver's time. When he visited Glacier Bay in 1879, he camped among ancient tree stumps that had once been covered in ice and saw forests that had grown up in previously glaciated areas. He was impressed with the dynamic nature of the landscape and how the plant community responded to the changes.

Muir's book spiked the interest of William S. Cooper (1923a), who began his studies of Glacier Bay in 1915. A former student of Henry Chandler Cowles, Cooper saw Glacier Bay as an example of the "space for time substitution" so well documented by his advisor in the Lake Michigan dunes. He established permanent plots (Cooper 1923b) that have allowed researchers to observe the pattern of community change along the bay from Vancouver's time to today. This pattern is generally characterized by an increase in plant species richness and a change in plant species composition with time and distance from the melting ice front (**Figure 16.10**). In the first years after new habitat is exposed, a primary or **pioneer stage** develops, dominated by a few species that include lichens, mosses, horsetails, willows, and cottonwoods. Roughly 30 years after exposure, a second community develops, named the *Dryas* stage after the small shrub (*Dryas drummondii*) that dominates this community. In this stage, species richness increases, with willows, cottonwoods, alders (*Alnus sinuata*), and Sitka spruce (*Picea sitchensis*) sparsely distributed among the carpets of *Dryas*. After about 50 years (or some 20 km from the ice front), alders dominate, forming the third community, referred to as the alder stage. Finally, a century after glacial retreat, a mature Sitka spruce forest (the spruce stage) is in place, which fosters a diverse array of lichens, low shrubs, and herbs. Two hundred years after exposure, Reiners et al. (1971) documented that species richness decreases somewhat as Sitka spruce forests are transformed into forests of longer-living western hemlocks.

The mechanisms underlying succession in this system have been studied extensively by F. Stuart Chapin and colleagues (1994). They wondered, given the harsh physical conditions experienced by most species in the pioneer stage, whether the facilitation model could explain the pattern of succession observed by Cooper and Reiners et al. First, they analyzed the soils of the different successional stages.

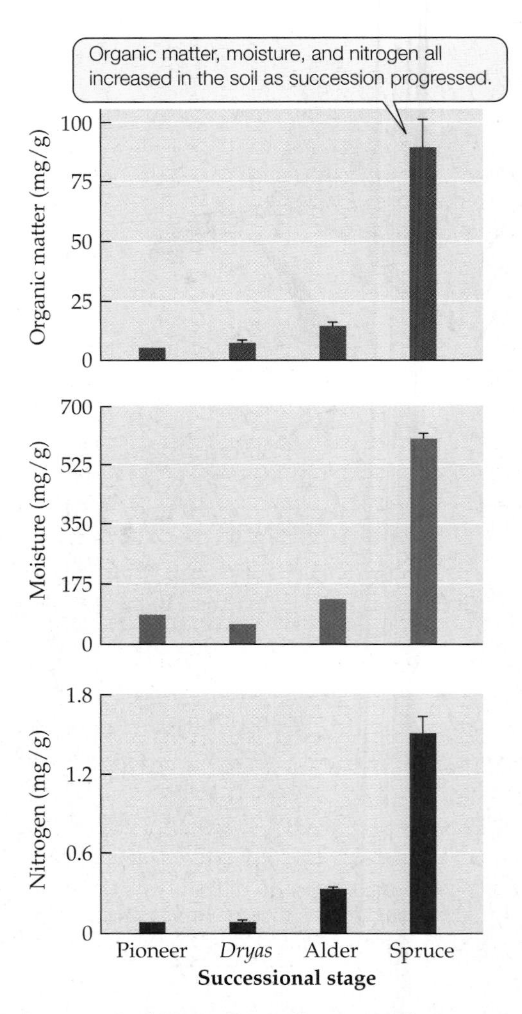

FIGURE 16.11 Soil Properties Change with Succession Chapin and colleagues studied the properties of the soils in each of four successional stages at Glacier Bay. Error bars show one SE of the mean. (After Chapin et al. 1994.)

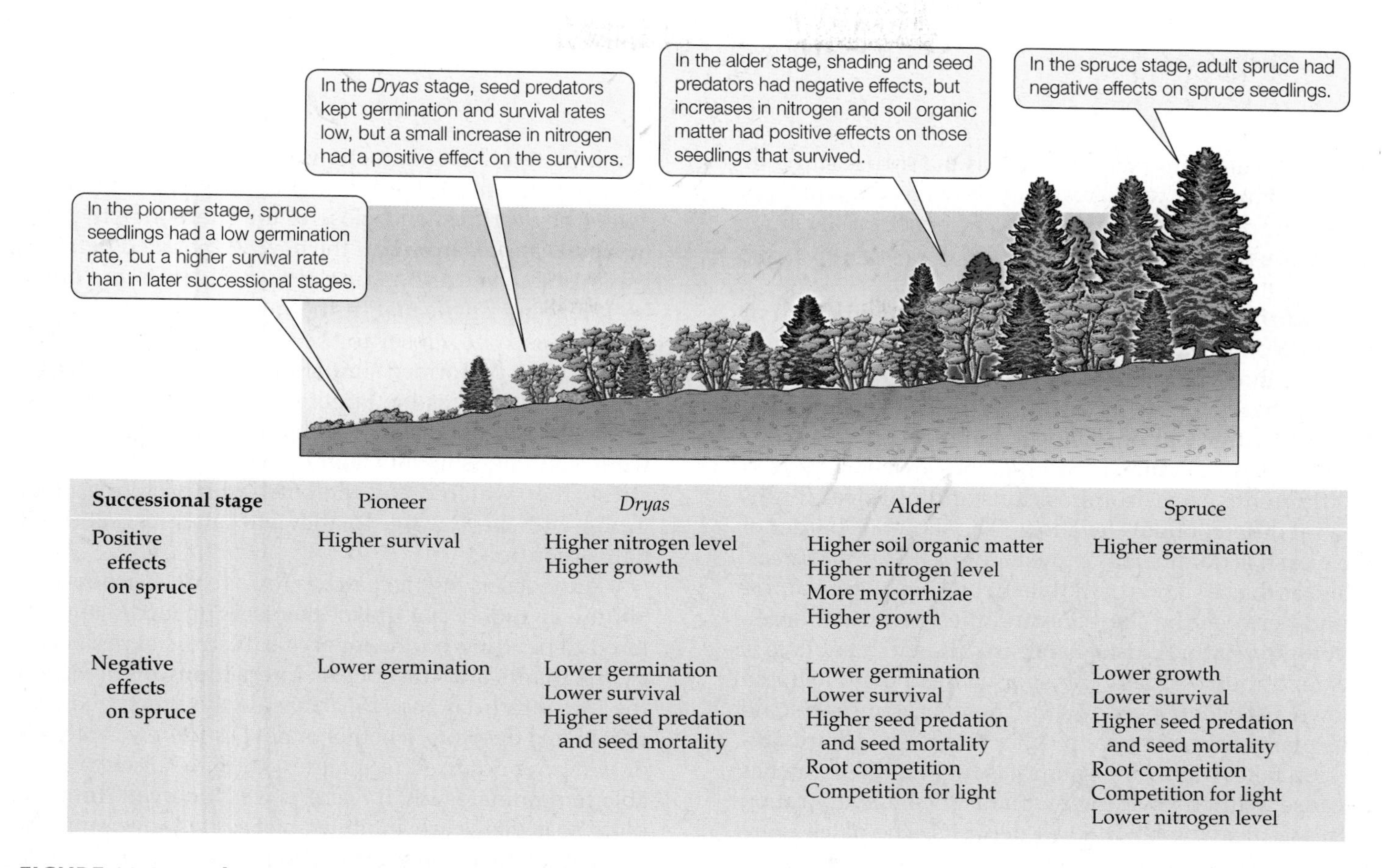

Successional stage	Pioneer	*Dryas*	Alder	Spruce
Positive effects on spruce	Higher survival	Higher nitrogen level Higher growth	Higher soil organic matter Higher nitrogen level More mycorrhizae Higher growth	Higher germination
Negative effects on spruce	Lower germination	Lower germination Lower survival Higher seed predation and seed mortality	Lower germination Lower survival Higher seed predation and seed mortality Root competition Competition for light	Lower growth Lower survival Higher seed predation and seed mortality Root competition Competition for light Lower nitrogen level

FIGURE 16.12 Both Positive and Negative Effects Influence Succession The relative contributions of positive and negative effects of other species on spruce seedling establishment changed across successional stages in Glacier Bay, Alaska. Positive effects equaled or outweighed negative effects in the first three stages, but the opposite was seen in the spruce stage. (After Chapin et al. 1994.)

They found significant changes in soil properties that were coincident with the increases in plant species richness (**Figure 16.11**). Not only were there increases in soil organic matter and soil moisture in later stages of succession, but nitrogen increased more than fivefold from the alder stage to the spruce stage. (This increase resulted from the action of nitrogen-fixing bacteria associated with plant roots, which we'll describe in more detail in this chapter's Connections in Nature on p. 361.) Chapin hypothesized that the assemblage of species at each stage of succession was having effects on the physical environment that largely shaped the pattern of community formation. The question remained, however, whether those effects were facilitative or inhibitory, and how they varied across the different successional stages.

To test their facilitation hypothesis, Chapin et al. (1994) conducted manipulative experiments. They added spruce seeds to each of the successional stages and observed their germination, growth, and survival over time. These experiments, along with observations of unmanipulated plots, showed that neighboring plants had both facilitative and inhibitory effects on the spruce seedlings, but that the directions and strengths of those effects varied with the stage of succession (**Figure 16.12**). For example, in the pioneer stage, spruce seedlings had a low germination rate, but a higher survival rate than in later successional stages. In the *Dryas* stage, spruce seedlings had low germination and survival rates due to increases in seed predators, but those individuals that did survive grew better due to the presence of nitrogen fixed by symbiotic bacteria associated with *Dryas*. In the alder stage, a further increase in nitrogen (alders also host nitrogen-fixing bacteria) and an increase in soil organic matter had positive effects on spruce seedlings, but shading and seed predators led to overall low germination and survival rates. In this stage, alders had a net positive effect on spruce seedlings that germinated before alders were able to dominate. Finally, in the spruce stage, the effects of large spruce on spruce seedlings were mostly negative and long-lasting. Growth and survival rates were low due to competition with adult spruce for light, space, and nitrogen. Interestingly, seed production by adults was enhanced, which led to relatively high seedling numbers merely as a consequence of the many more seeds available for germination.

Thus, in Glacier Bay, the mechanisms outlined in Connell and Slatyer's models were operating in at least some stages of succession. Early on, aspects of the facilitation model were seen as plants modified the habitat in positive ways for other plants and animals. Species such as alders had negative effects on later successional species unless they were able to colonize early, supporting the inhibition model. Finally, some stages—such as the spruce stage, in which dominance was a result of slow growth and long life—were driven by life history characteristics, a signature of the tolerance model.

Secondary succession in a New England salt marsh

What do other studies show with regard to Connell and Slatyer's three models? Mark Bertness and Scott Shumway studied the relative importance of facilitative versus inhibitory interactions in controlling secondary succession in a New England salt marsh. Salt marshes are characterized by different species compositions and physical conditions at different tidal elevations. The shoreline border of the marsh is dominated by the cordgrass *Spartina patens*, whereas dense stands of the black rush *Juncus gerardii* are found between the shoreline and the terrestrial border. A common natural disturbance in salt marsh habitats is the deposition of tidally transported dead plant material known as *wrack* (**Figure 16.13**). The wrack smothers and kills plants, creating bare patches where secondary succession takes place. Soil salinity is high in these patches because, without shading by plants, water evaporation increases, leaving behind salt deposits. The patches are initially colonized by the spike grass *Distichlis spicata*, an early successional species that is eventually outcompeted by *Spartina* and *Juncus* in their respective zones.

Bertness and Shumway (1993) hypothesized that *Distichlis* could either facilitate or inhibit later colonization by *Spartina* or *Juncus* depending on the salt stress experienced by the interacting plants. To test this idea, they created bare patches in two zones of a marsh and manipulated plant interactions shortly after the patches had been colonized (**Figure 16.14**). In the low intertidal zone (the *Spartina* zone, close to the shoreline), they removed *Distichlis* from half the newly colonized patches, leaving *Spartina*, and removed *Spartina* from the other half, leaving *Distichlis*. In the middle intertidal zone (the *Juncus* zone, closer to the terrestrial border of the marsh), they performed similar manipulations, with *Juncus* and *Distichlis* as the target species. Control patches, in which the colonization process was not manipulated, were maintained in both zones. Then, to manipulate salt stress, they watered half the patches in each treatment group with fresh water to alleviate salt stress, and left half as controls.

After observing the patches for 2 years, Bertness and Shumway found that the mechanisms of succession differed depending on the level of salt stress experienced by the plants and the species interactions involved. In the low intertidal zone, *Spartina* always colonized and dominated the plots, whether or not *Distichlis* was present or watering occurred (see Figure 16.14A). *Distichlis* was able to dominate only if *Spartina* was removed from the plots, so it was clearly inhibited by *Spartina*, the dominant competitor. In the middle intertidal zone, *Juncus* was able to colonize only if *Distichlis* was present or watering occurred (see Figure 16.14B). Measurements of soil salinity confirmed that the presence of *Distichlis* helped to shade the soil surface, thus decreasing salt accumulation and reducing stress for *Juncus*. *Distichlis*, however, was able to colonize plots with *Juncus* only when salt stress was high—that is, under the control conditions. If plots were watered, *Distichlis* was easily outcompeted by *Juncus*.

These experimental manipulations confirmed that the mechanisms important to succession are context-dependent. No single model is sufficient to explain the underlying causes of succession. In the middle intertidal zone, *Distichlis* was a strong facilitator of colonization by *Juncus*. Once this facilitation occurred, the balance was tipped in favor of *Juncus*, which outcompeted *Distichlis* (see Figure 16.14B). In the low intertidal zone, *Distichlis* and *Spartina* were equally able to colonize and grow in salty patches. If *Spartina* arrived first, it inhibited *Distichlis* colonization. If *Distichlis* arrived first, it persisted only if *Spartina* did not arrive and displace it (see Figure 16.14A).

FIGURE 16.13 Wrack Creates Bare Patches in Salt Marshes A tidal deposit of wrack at Rumstick Cove, Rhode Island, where Bertness and Shumway conducted their research on secondary succession. This dead plant material smothers living plants, creating bare patches with high soil salinity.

Primary succession in rocky intertidal communities

Our final examples come from an environment where succession has been studied extensively: the rocky intertidal zone. Here, disturbances are created mainly by waves, which can tear organisms from the rocks during storms or propel objects such as logs or boulders into them. In addition, stresses caused by low tides that expose organisms to high or low air temperatures can easily kill them or cause them to lose their attachment

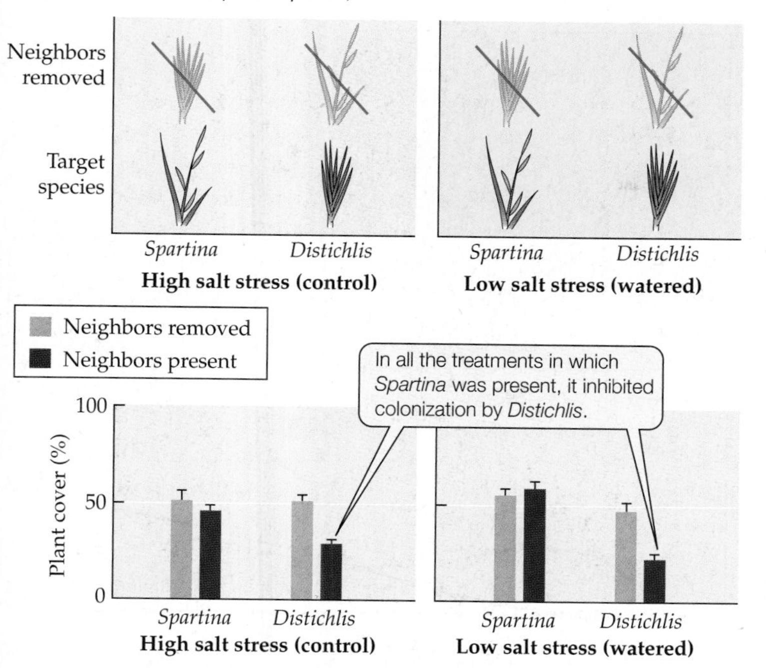

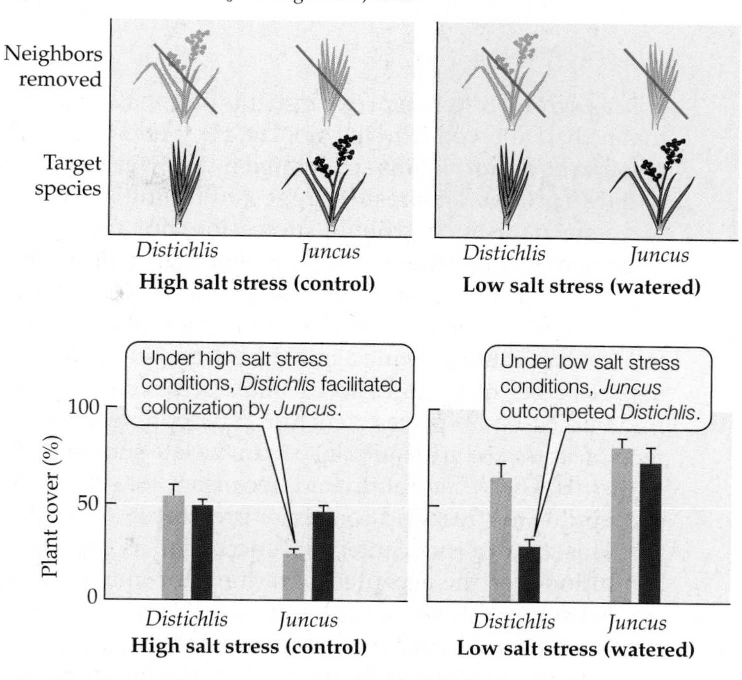

FIGURE 16.14 New England Salt Marsh Succession Is Context-Dependent The trajectory of succession in salt marshes depends on soil salinity and the physiological tolerances of plant species. The kinds of interactions observed differed between the low intertidal zone (A) and the middle intertidal zone (B). Error bars show one SE of the mean. (Data from Bertness and Shumway 1993.)

to the rocks. The resulting bare rock patches become active areas of colonization and succession.

Some of the first experimental work on succession in the rocky intertidal zone was done on boulder fields in southern California by Wayne Sousa, a graduate student at the time. Sousa (1979b) noticed that the algae-dominated communities on these boulders experienced disturbance every time the boulders were overturned by waves. When he cleared some patches on the boulders and observed succession in those patches over time, he found that the first species to colonize and dominate a patch was always the bright green alga *Ulva lactuca* (**Figure 16.15A**). It was followed by the red alga *Gigartina canaliculata*. To understand the mechanisms controlling this successional sequence,

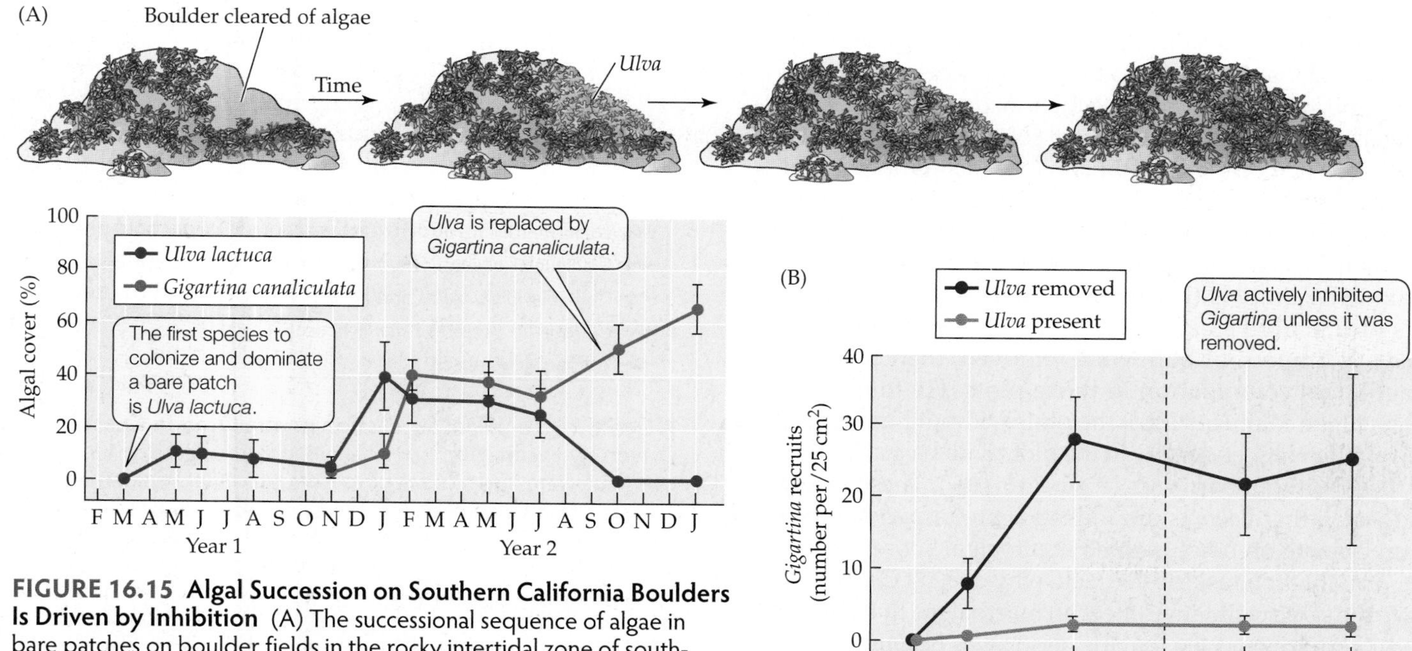

FIGURE 16.15 Algal Succession on Southern California Boulders Is Driven by Inhibition (A) The successional sequence of algae in bare patches on boulder fields in the rocky intertidal zone of southern California. (B) Sousa performed removal experiments on concrete blocks to understand the mechanisms of succession in this ecosystem. Error bars show ± one SE of the mean. (Data from Sousa 1979b.)

Sousa performed removal experiments on concrete blocks that he had allowed *Ulva* to colonize. He found that colonization by *Gigartina* was accelerated if *Ulva* was removed (**Figure 16.15B**). This result suggested inhibition as the main mechanism controlling succession, but a question remained: if *Ulva* is able to inhibit other seaweed species, why doesn't it always dominate? Through a series of further experiments, Sousa found that grazing crabs preferentially fed on *Ulva*, thus initiating a transition from the early *Ulva* stage to other mid-successional algal species. In turn, the mid-successional species were more susceptible to the effects of stress and parasitic algae than the late successional *Gigartina*. *Gigartina* dominated because it was the least susceptible to stress and consumer pressures.

This view of rocky intertidal succession as driven by inhibition was the accepted paradigm for many years. Facilitation and tolerance were thought to be much less important in a system where competition for space was strong. More recent work by Terence Farrell and others (e.g., Berlow 1997) demonstrated that the relative importance of inhibition is probably much more context-dependent than previously thought. In the more productive rocky intertidal zone of the Oregon coast, the communities include many more sessile invertebrates, such as barnacles and mussels, than Sousa's communities of the southern California coast, where seaweeds dominate. In the high intertidal zone of Oregon, Farrell (1991) found that the first colonizer of bare patches was a barnacle, *Chthamalus dalli*. *Chthamalus* was replaced by another, larger barnacle species, *Balanus glandula*, which was then replaced by three species of macroalgae, *Pelvetiopsis limitata*, *Fucus gardneri*, and *Endocladia muricata*. A series of removal experiments showed that *Chthamalus* did not inhibit colonization by *Balanus*, but that *Balanus* was able to outcompete *Chthamalus* over time, thus supporting the tolerance model. Likewise, *Balanus* did not hinder macroalgal colonization, but in fact facilitated it, lending credibility to the facilitation model.

But why and how would *Balanus* facilitate macroalgal colonization? Farrell suspected that *Balanus* protected the algae in some way, possibly from desiccation stress or grazing by limpets (herbivorous marine snails). To test this idea, Farrell created experimental plots from which *Balanus*, limpets, or both were removed, then observed macroalgal colonization in those plots. He found that macroalgae colonized all of the plots without limpets, but had a much higher density in the plots with barnacles than in those without barnacles (**Figure 16.16A**). These results suggested that *Balanus* did indeed act to impede limpets from grazing on newly settled macroalgal sporelings.

You might be asking yourself, why doesn't *Chthamalus* have the same facilitative effect on macroalgae that *Balanus* does? Ferrell suspected that the reason was *Balanus*'s larger size (it is nearly three times wider than *Chthamalus*). By using plaster casts to mimic barnacles that were slightly

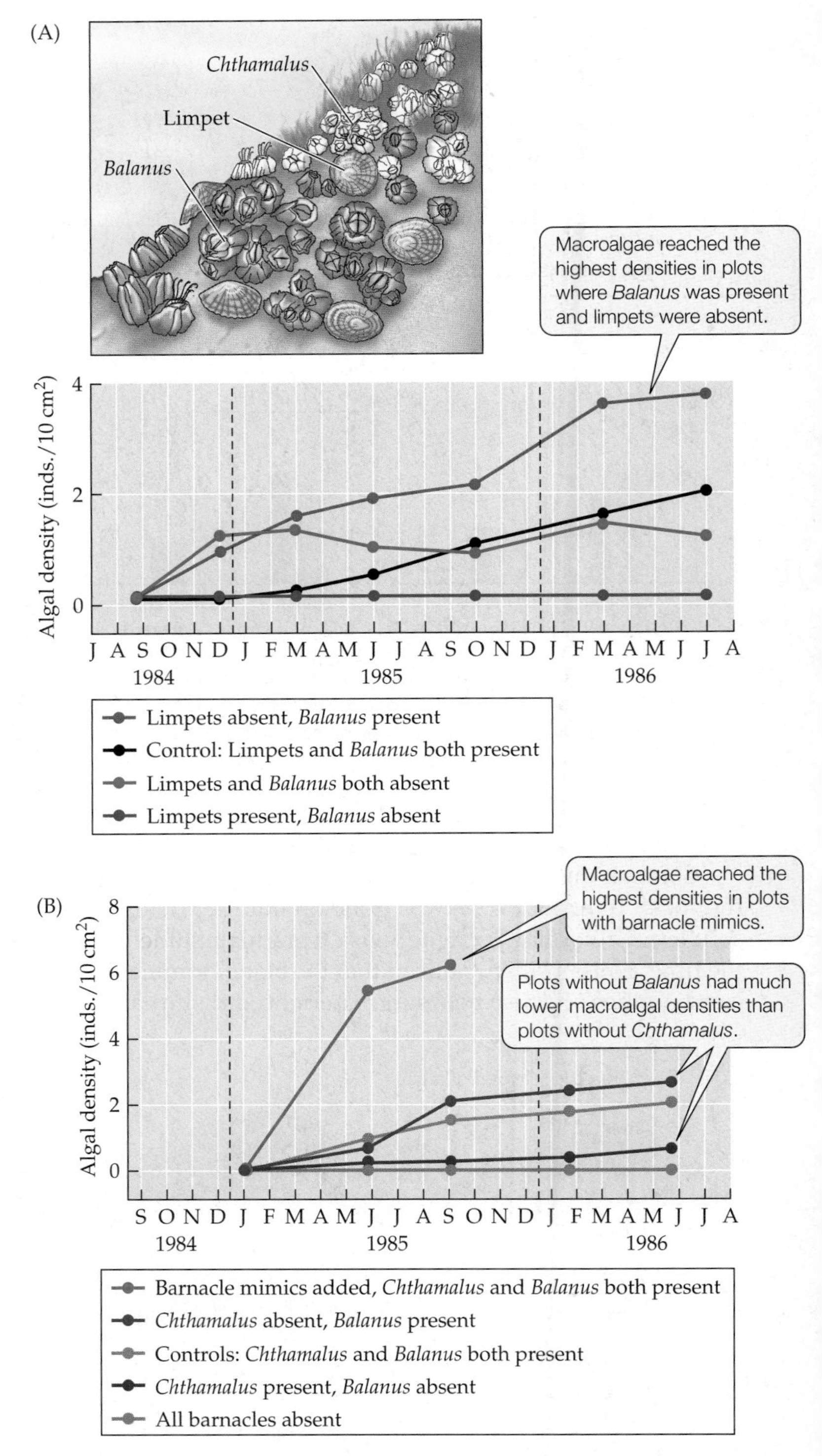

FIGURE 16.16 Algal Succession on the Oregon Coast Is Driven by Facilitation (A) Changes in macroalgal densities over time were measured in plots from which *Balanus* barnacles, limpets, or both had been removed. The results suggested that *Balanus* facilitates macroalgae by reducing limpet grazing. (B) To understand the mechanisms of facilitation, Farrell added large barnacle mimics to some plots and compared them with plots from which the real barnacle species—*Balanus*, *Chthamalus*, or both—had been removed. The results suggested that the larger the barnacle species, the better it protects macroalgae against limpet grazing and desiccation. (Data from Farrell 1991.)

larger than *Balanus*, Ferrell found that these barnacle mimics had an even more positive effect on macroalgal colonization than did smaller-sized live barnacles of either species (**Figure 16.16B**). It seems likely that the smaller and smoother *Chthamalus* does not retain as much moisture, or block as many limpets, as the larger and more sculpted *Balanus*—or the mimics, for that matter.

Experiments show facilitation to be important in early stages

A number of experimental studies like the ones we have just described, initially stimulated by Connell and Slatyer's thought-provoking paper (1977), suggest that succession in any community is driven by a complex array of mechanisms. No one model fits any one community; instead, each community is characterized by elements of all three of Connell and Slatyer's models. In most successional sequences, especially those in which a pioneer stage is exposed to physically stressful conditions, facilitative interactions are important drivers of early succession. Organisms that can tolerate and modify these physically challenging environments will thrive and facilitate other organisms that lack those capabilities. As succession progresses, slow-growing and long-lived species begin to dominate. Those species tend to be larger and more competitively dominant than early successional species. For this reason, one might expect competition to play a more dominant role than facilitation later in succession.

As succession proceeds, species richness typically increases (see Figure 16.10); thus, we must recognize that vast arrays of both positive and negative interactions are operating in mid- to late successional stages. We will learn more about the mechanisms responsible for controlling species diversity in Chapter 18, but let's turn our attention next to the question of whether succession always takes one predictable path, as Clements believed, or whether other paths are possible.

CONCEPT 16.4 Communities can follow different successional paths and display alternative states.

Alternative Stable States

Up to this point, we have assumed that the trajectory of succession is repeatable and predictable. But what if, for example, a boulder in the rocky intertidal zone of southern California turns over and, instead of a seaweed community forming, as Sousa (1979b) observed, a sessile invertebrate community forms instead? Or what if *Dryas* never colonizes the till left behind by a glacier at Glacier Bay, but is replaced by a grass that competes with, rather than facilitates, later successional species such as Sitka spruce? Might spruce forests never develop? Possibly. There are cases in which different communities develop in the same area under similar environmental conditions. Ecologists refer to such alternative scenarios as **alternative stable states**. Richard Lewontin (1969) was one of the first to formally define and model alternative stable states in natural communities.

A community is said to be **stable** when it returns to its original state after some perturbation. How stable are natural communities? This question has perplexed ecologists for some time, partly because the notion of stability depends on spatial and temporal scale. At a small spatial scale, such as a 1 m^2 plot in a midwestern prairie, there might be considerable change or instability over time. If all the plants were removed from the plot, it is unlikely that all the same species would recolonize that particular plot, and certainly not in the exact same locations. However, if a larger area is considered (e.g., a 100 m^2 plot), the chance of finding those same species increases. Similarly, if one followed the plot for a short time, the chance that its species composition would change would be low. But the more time that elapsed, the more likely it would be that the community would change and thus appear unstable. With these caveats in mind, let's take a closer look at examples of communities that, once disturbed, do not revert to previous states, but instead show alternative stable states.

Alternative states are controlled by strong interactors

John Sutherland (1974) studied alternative states in marine fouling communities: the sponges, hydroids, tunicates, and other invertebrates that encrust ships, docks, and other hard surfaces in bays and estuaries. He suspended ceramic tiles in early spring from the dock at Duke Marine Laboratory in Beaufort, North Carolina, and allowed them to be colonized by planktonic invertebrate larvae (**Figure 16.17A**). At the end of 2 years, even though a handful of species had colonized the tiles, most of them were dominated by *Styela*, a solitary tunicate species. The dominance of *Styela* was not universal, however. *Styela* actually declined on Sutherland's tiles during the first winter and was replaced by the hydroid species *Tubularia*. This effect was due to the annual nature of *Styela*, which dies off in winter, and it quickly regained dominance the next spring, when larvae started to settle.

By placing new tiles out periodically, Sutherland also concluded that *Styela* was able to persist despite the existence of other potential colonizers. These colonizers fouled the new tiles, but were unable to colonize those dominated by *Styela*. As such, Sutherland viewed this fouling community as stable. Within a few months, he also identified what he believed to be another stable fouling community, this one dominated by *Schizoporella*, an en-

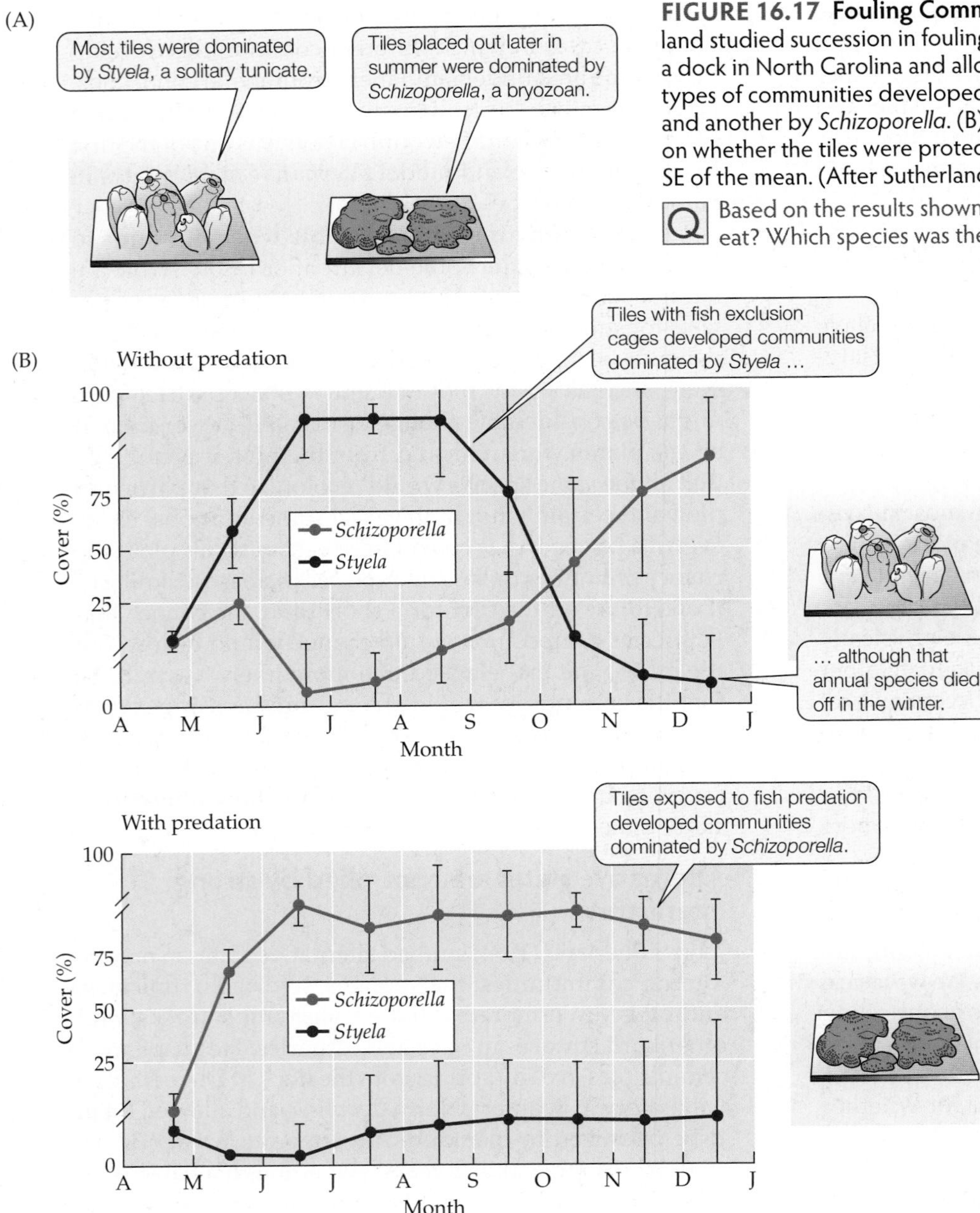

FIGURE 16.17 Fouling Communities Show Alternative States John Sutherland studied succession in fouling communities by suspending ceramic tiles from a dock in North Carolina and allowing invertebrates to colonize them. (A) Two types of communities developed on the tiles over time, one dominated by *Styela* and another by *Schizoporella*. (B) Different communities developed depending on whether the tiles were protected from fish predation. Error bars show ± one SE of the mean. (After Sutherland 1974.)

Q Based on the results shown in (B), which fouling species did fish prefer to eat? Which species was the competitive dominant?

crusting bryozoan (see Figure 16.17A). This community developed on new tiles suspended from the dock in late summer and was also impervious to colonization by other species, including *Styela*.

To understand what might be controlling these two alternative outcomes of succession, Sutherland submerged new tiles at the same spot on the dock, but excluded fish predators from half of the tiles by surrounding them with cages (**Figure 16.17B**). After a year, Sutherland found that the tiles protected from fish predation had formed communities dominated by *Styela*, while those exposed to fish predation had formed communities dominated by *Schizoporella*. He also noticed that the abundances of both species on the tiles protected from predators were reversed when *Styela* began to die off in the winter. These results suggested that *Styela* is competitively dominant if left undisturbed, but is outcompeted by *Schizoporella* when disturbed. Sutherland explained his original observations of *Styela* dominance by suggesting that fish predation was spotty and that the tunicates themselves, once they reached a certain large size, might have acted as a natural "cage" or predator exclusion mechanism.

Lewontin (1969) and Sutherland (1974) both believed that multiple stable states existed in communities and could be driven by the addition or exclusion of particularly strong interactors. If those species were missing or ineffective, communities could follow alternative successional trajectories that might never lead back to the original community type, but might instead form a new community type. We can visualize the theory behind alternative stable states by imagining a landscape in which different valleys represent the range of possible community states and a ball represents a community occupying a particular state at a moment in time (**Figure 16.18A**). The ball can move from one valley to another, depending on the presence or absence of strong interactors and how they shape the community (**Figure 16.18B**). For example, it might take only a slight change in the abundance of one or more dominant species to force the ball into a new valley, or it might require complete removal of a species to cause this change. If we use Sutherland's work as an example, we can think of the *Styela* and *Schizoporella* communities as two different valleys. Whether the ball resides in the *Schizoporella* valley or the *Styela* valley depends on the presence of fish predators. Interestingly, in this system, the ball may not simply move back to the *Schizoporella* valley if access by

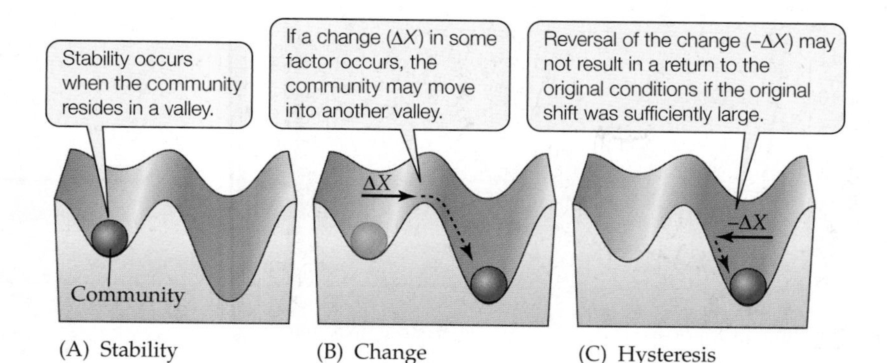

FIGURE 16.18 A Model of Alternative Stable States (A) A community is represented by a ball that moves along a landscape of community states (valleys). (B) Notice that the valleys can be shallower or deeper, suggesting the magnitude of change (ΔX) needed to shift the community from one state to another. (C) Hysteresis occurs when reversal of the change ($-\Delta X$) does not return the community to its original state. (After Beisner et al. 2003.)

fish predators is restored (**Figure 16.18C**). As Sutherland noted, *Styela* is able to escape predation once it reaches a certain size. Thus, this system might show **hysteresis**, or an inability to shift back to the original community type, even when the original conditions are restored.

Connell and Sousa (1983) were skeptical that Sutherland had demonstrated the existence of alternative stable states for several reasons. First, they thought the tile communities did not persist long enough, or have a spatial scale large enough, to be considered stable. If the tiles could be followed over multiple years, they asked, would they not all end up being dominated by one or the other species? In addition, they wondered whether the fouling communities could have been sustained outside of an experimental setting in which predators were removed. Their final argument, although it was not a criticism of Sutherland's study per se, was that alternative stable states could be driven only by species interactions and not by physical changes in the community. They argued that many of the examples Sutherland had used to bolster the importance of alternative stable states fell into the latter category. Their requirement that the physical environment not change is problematic because it excludes as drivers of succession all species that interact with other species by modifying their physical environment—that is, all ecosystem engineers. We know that ecosystem engineers can have strong effects on communities, so excluding them is unrealistic to most ecologists.

Human actions have caused communities to shift to alternative states

The stringent requirements suggested by Connell and Sousa had the effect of delaying alternative stable state research for two decades. Recently, however, there has been renewed interest in alternative stable states, spurred by the increasing evidence that human activities, such as habitat destruction, species introductions, and overharvesting of wild species, are shifting communities to alternative states. We have already seen examples of such changes in several of the Case Studies in this book, including the change from kelp forests to urchin barrens due to the decline of sea otters (see Chapter 8, p. 177), the crash of the anchovy fishery in the Black Sea due to the introduction of *Mnemiopsis* (see Chapter 10, p. 221), and the invasion of the aquarium strain of *Caulerpa taxifolia* in the Mediterranean, Australia, Japan, and North America (see Chapter 15, p. 324). These so-called *regime shifts* are caused by the removal or addition of strong interactors that maintain one community type over others. Once communities have been "manipulated" by human activities and a new regime is in place, ecologists are uncertain whether the results can be reversed. Will recolonization by sea otters rejuvenate kelp forests? Will the cessation of nutrient enrichment in the Black Sea revitalize the anchovy fishery? And will the removal of *Caulerpa* restore seagrass meadows? These are all questions whose answers may be found in a better understanding of the factors that drive alternative stable states and of the role restoration of the original conditions can play in reversing the effects of those factors.

A CASE STUDY REVISITED

A Natural Experiment of Mountainous Proportions

On the twentieth anniversary of the eruption of Mount St. Helens, in 2000, a group of ecologists gathered on the once smoking and ash-covered volcano to participate in a week-long field camp. They gathered their gear, including tape measures, quadrat frames, and maps, and visited the same sites they had explored two decades earlier. This visit, termed a "pulse," was an opportunity to establish a 20-year benchmark of data comparable to those first collected in 1980 and 1981. Many of the participants had spent the past 20 years—for some, their entire careers—studying recolonization and succession patterns in those once-devastated landscapes. When they departed, they agreed to write a book, the chapters of which would contain all that was known about the extraordinary ecology of this ecosystem, with the hope that young ecologists would be motivated to continue the research and carry on their legacy. The book, *Ecological Responses to the 1980 Eruption of Mount St. Helens* (Dale et al. 2005) was published 5 years later.

What does the book tell us about succession on Mount St. Helens? First, the eruption created disturbances that varied in their effects depending on distance from the volcano and habitat type (e.g., aquatic versus terrestrial)

(**Table 16.2**). Although areas close to the summit, such as the Pumice Plain, were literally sterilized by the heat of the eruption, ecologists were surprised to discover how many species actually survived on the mountain. Because the eruption occurred in spring, many species were still dormant under the winter snows. Survivors included plants with underground buds or rhizomes, animals such as rodents and insects with burrows, and fish and other aquatic species in ice-covered lakes. In the blowdown zone, large trees and animals perished while smaller organisms survived in the protection of their larger neighbors. The opposite was true in areas outside the blowdown zone, where falling rocks and ash smothered smaller plants and animals, but not larger organisms.

A second important research discovery from Mount St. Helens is the role survivors have played in controlling the pace and pattern of succession. In many cases, these species were thrust into novel physical environments and species assemblages without time to adapt over evolutionary time scales. Some species thrived, while others fared poorly, but their adaptability and unpredictability were surprising. Unlikely alliances were formed that hastened succession in particular habitats. For example, newly formed and isolated ponds and lakes were colonized by amphibians much faster than had been thought possible (**Figure 16.19**). Scientists discovered that frogs and salamanders were using tunnels created by northern pocket gophers (*Thomomys talpoides*) to make their way from one pond to another across the arid landscape (Crisafulli et al. 2005). The gophers were particularly successful on Mount St. Helens, both because they survived the eruption in their tunnels and because

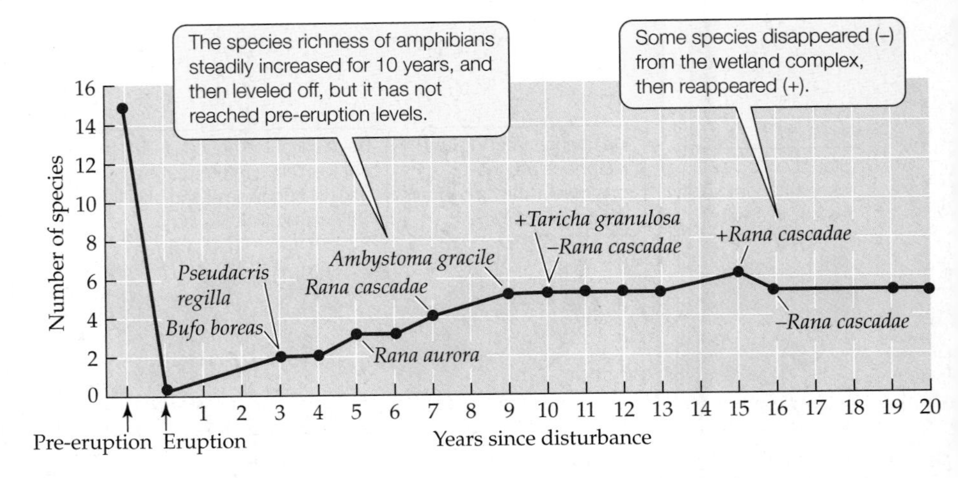

FIGURE 16.19 Rapid Amphibian Colonization Frog and salamander species rapidly colonized a wetland complex in the Pumice Plain on Mount St. Helens. (Data from Crisafulli et al. 2005.)

grassy meadows—their preferred habitat—expanded greatly after the eruption. Interestingly, the gophers were also responsible for facilitating plant succession: their burrowing activity brought to the soil surface organic matter, seeds, and fungal spores buried deep under the volcanic rock and ash (Crisafulli et al. 2005; **Figure 16.20**).

A third important discovery was the realization that multiple mechanisms were responsible for primary succession on Mount St. Helens. Facilitation on the Pumice Plain was exemplified by the dwarf lupine (*Lupinus lepidus*), the first plant to arrive there. Dwarf lupines trapped seeds and detritus and increased the nitrogen content of the soil though their symbiotic association with nitrogen-fixing bacteria (del Moral et al. 2005). The lupines, in turn, were inhibited by multiple insect herbivores, which essentially controlled the pace of primary succession (Bishop et al. 2005). Tolerance was evident in some primary successional habitats, where Douglas fir lived in concert with annual

TABLE 16.2

Surviving Organisms Found on Mount St. Helens within a Few Years after the Eruption

Disturbance zone	Mean vegetation cover (%)	Average number of plant species/m^2	Animal species: Small mammals	Large mammals	Birds	Lake fish	Amphibians	Reptiles
Pumice Plain	0.0	0.0	0	0	0	0	0	0
Mudflow zone (central flow path)	0.0	0.0	0	0	0	—	0	0
Blowdown zone			8	0	0	4	11	1
Pre-eruption clear-cut	3.8	0.0050						
Forest without snow	0.06	0.0021						
Forest with snow	3.3	0.0064						
Scorch zone	0.4	0.0039	—	0	0	2	12	1

Source: Adapted from Crisafulli et al. 2005.

FIGURE 16.20 Pocket Gophers to the Rescue The burrowing activity of northern pocket gophers, some of which survived the eruption underground, brought organic matter, seeds, and fungal spores to the soil surface, creating microhabitats, like this one in the Pumice Plain, where plants could grow.

herbs. The diversity of strategies species used, and the resulting communities, never ceased to amaze ecologists, who were guided, up to that point, mostly by the models of Connell and Slatyer (1977).

Despite decades of data and a trove of novel discoveries, research on Mount St. Helens has only just begun. A similar "pulse" visit occurred in 2010 to establish a 30-year benchmark since the eruption. Will the communities there follow paths of succession that lead to predictable and repeatable outcomes? Or will they form alternative states that are highly dependent on their historical legacies? Geologic work suggests that Mount St. Helens erupts roughly every 300 years. The life span of its community succession thus greatly exceeds our own life span by hundreds of years. As such, we must be content with the limited knowledge we have gained from studying what is arguably the most interesting phase of succession on Mount St. Helens and with the hope that ecologists will continue the tradition of research there for years to come.

CONNECTIONS IN NATURE

Primary Succession and Mutualism

We saw in Chapter 14 that positive relationships can alter communities, and that they may be particularly important in stressful environments. Primary succession in terrestrial environments illustrates both of these effects: some of the examples presented in this chapter involve plants that interact in a mutualistic way with symbiotic nitrogen-fixing bacteria. These bacteria form nodules in the roots of their plant hosts, where they convert nitrogen gas from the atmosphere (N_2) into a form that is usable by plants (ammonia; NH_4^+). The plants provide the bacteria with sugars produced by photosynthesis. This interaction appears to be extremely important for plants and animals colonizing completely sterile environments. We have seen that *Dryas* and alders, both species that form tight mutualisms with nitrogen-fixing bacteria, were some of the first species to colonize the till left behind by glaciers at Glacier Bay, Alaska. Similarly, *Lupinus lepidus* was able to use the nitrogen produced by its bacterial symbionts to colonize the sterile Pumice Plain of Mount St. Helens after the eruption. Lupines were the major source of nitrogen for subsequent plants and herbivorous insects for many years. Thus, lupines and their symbiotic bacteria play a large role in controlling the rate of primary succession on Mount St. Helens.

The nitrogen-fixing bacteria involved in symbioses are extremely diverse. Only a few groups of bacteria live in root nodules; all the rest are associated either with the surfaces of roots or the guts of ruminants. The nodule-forming bacteria include the rhizobia, a taxonomic group associated with legumes (such as lupines), and actinomycetes of the genus *Frankia*, which are associated with woody plants such as alders and *Dryas*. Nodule formation involves a complex series of chemical and cellular interactions between the root and the bacteria (**Figure 16.21**). Free-living bacteria are attracted to root exudates, which cause them to attach to the roots and multiply. Sets of genes are activated in both bacterial and root cells that allow the bacteria to enter the root, the root cells to divide, and the nodule to be formed.

The enzyme involved in nitrogen fixing (known as nitrogenase) is highly sensitive to oxygen and requires anaerobic conditions. Thus, wherever nitrogen-fixing symbioses occur, there are some structural components to the interaction, such as membranes within the root nodules, that produce anaerobic conditions. The bacteria, however, need oxygen to metabolize, so a hemoglobin protein known as leghemoglobin, which has a high affinity for oxygen, is produced in the nodules to deliver oxygen to the bacteria in an essentially anaerobic environment. The nodules often have an eerie pink color that is associated with the leghemoglobin. In addition, a specialized vascular system develops that supplies sugars to the bacteria and carries fixed nitrogen to the plant.

Maintaining a symbiosis with nitrogen-fixing bacteria is costly to plants. Estimates suggest that creating and maintaining the nodules alone costs a plant 12%–25% of its total photosynthetic output. Plants may be able to shoulder this cost, especially if it allows them to live in environments free of competitors and herbivores. But as they increase the nitrogen content of the soils in which they live, plants with symbionts make conditions better for other plant species as well—some of which are likely to be competitors. Thus, plants face a trade-off between improving the environment for themselves and competing with other species, which makes their role in early successional environments important, if somewhat ironic.

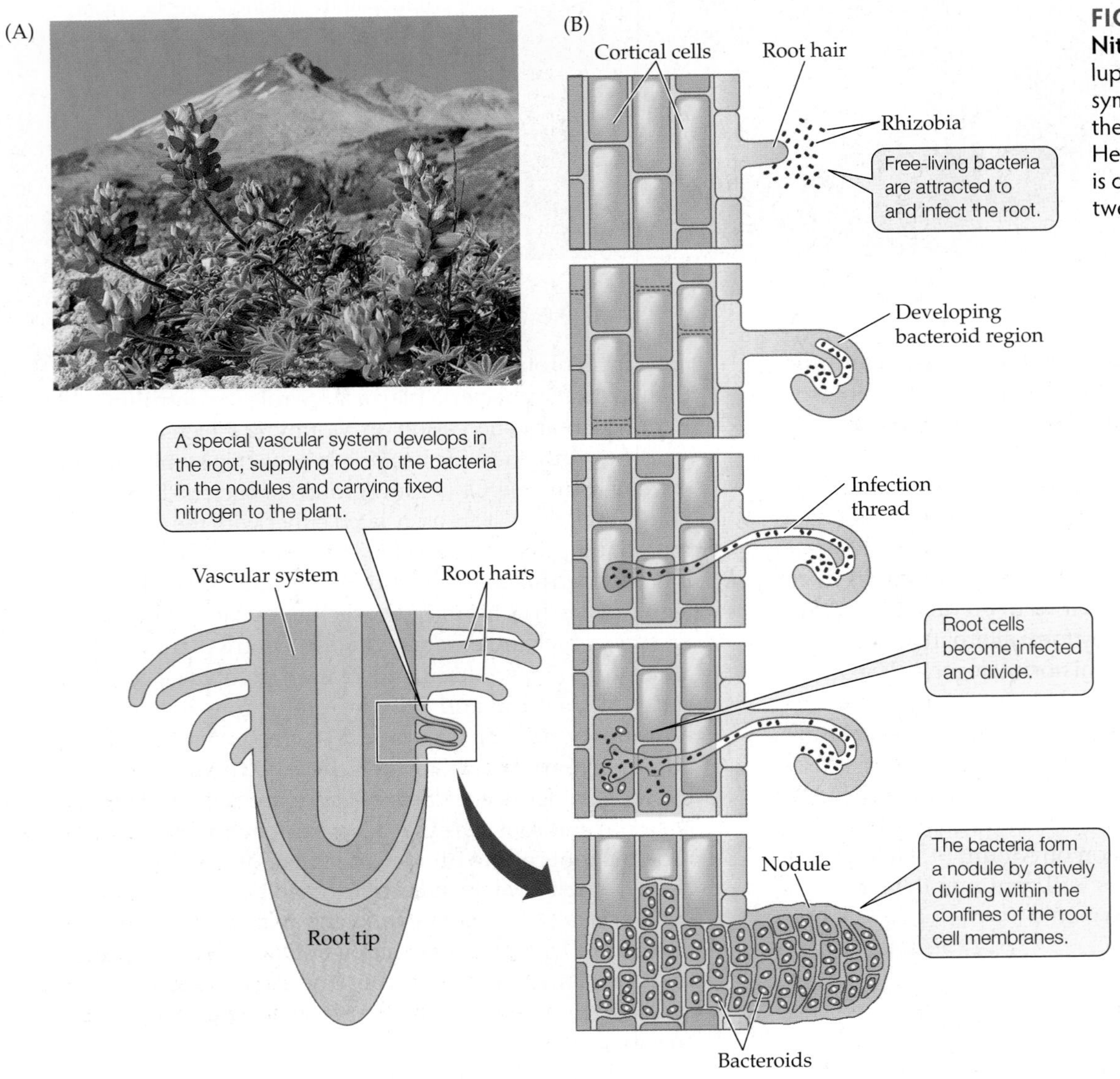

FIGURE 16.21 Dwarf Lupines and Nitrogen-Fixing Bacteria (A) Dwarf lupine (*Lupinus lepidus*), a legume with symbiotic nitrogen-fixing bacteria, was the first plant to colonize Mount St. Helens. (B) Root nodule development is controlled by a strong interaction between the plant and the bacteria.

SUMMARY

CONCEPT 16.1 Agents of change act on communities across all temporal and spatial scales.

- Agents of change include both abiotic and biotic factors.
- Abiotic agents of change can act as disturbances (injuring or killing organisms) or as stresses (reducing the growth or reproduction of organisms).
- Biotic agents of change include negative species interactions such as, for example, competition, predation, and trampling. Ecosystem engineers and keystone species are common agents of change.
- Agents of change vary in their intensity, frequency, and areal extent.

CONCEPT 16.2 Succession is the process of change in species composition over time as a result of abiotic and biotic agents of change.

- Theoretically, succession involves a series of stages that include a stable end point, or climax stage.
- Primary succession involves the colonization of habitats that are devoid of life.
- Secondary succession involves the reestablishment of a community in which most, but not all, of the organisms or organic constituents have been destroyed.
- Early ecologists were fascinated with succession, but disagreed about whether it proceeded in deterministic or random ways.

SUMMARY (continued)

- Connell and Slatyer proposed three models of succession in 1977, known as the facilitation model, tolerance model, and inhibition model.

CONCEPT 16.3 Experimental work on succession shows its mechanisms to be diverse and context-dependent.

- Multiple studies of succession have shown that no one model fits any one community. Aspects of the facilitation, tolerance, and inhibition models can be seen in almost all systems studied.
- Generally, experiments show that facilitation tends to be important in early stages of succession and competition in later stages of succession.

CONCEPT 16.4 Communities can follow different successional paths and display alternative states.

- Alternative stable states occur when different communities develop in the same area under similar environmental conditions.
- In communities that experience alternative states, succession is typically controlled by strong interactors.
- Human activities have caused regime shifts in communities that may or may not be reversible.

REVIEW QUESTIONS

1. List some abiotic and biotic agents of change in communities. Describe the intensities and frequencies with which they are likely to act.
2. Describe the differences between primary and secondary succession and what those differences mean for colonizing species.
3. Connell and Slatyer proposed three separate models of succession: the facilitation model, the tolerance model, and the inhibition model. Choose a hypothetical community and describe the different circumstances that would be required to support each of the models.
4. Why is it hard to determine whether a community is stable? Do you think Sutherland was able to demonstrate alternative stable states on his ceramic tiles? Why or why not?

ON THE COMPANION WEBSITE
sites.sinauer.com/ecology2e

The website includes Chapter Outlines, Online Quizzes, Flashcards & Key Terms, Suggested Readings, a complete Glossary, and the Web Stats Review. In addition, the following resources are available for this chapter:

HANDS-ON PROBLEM SOLVING

Soil Invertebrates and Succession

This Web exercise explores how invertebrates that live in the soil could affect patterns of succession. You will interpret data from a recent paper that shows the effects of soil invertebrates on the growth of early- and mid-succession plants.

CHAPTER

17

Biogeography

KEY CONCEPTS

- **CONCEPT 17.1** Patterns of species diversity and distribution vary at global, regional, and local spatial scales.
- **CONCEPT 17.2** Global patterns of species diversity and composition are influenced by geographic area and isolation, evolutionary history, and global climate.
- **CONCEPT 17.3** Regional differences in species diversity are influenced by area and distance, which determine the balance between immigration and extinction rates.

The Largest Ecological Experiment on Earth: A Case Study

There is probably only one place on Earth where a person could hear the calls of a hundred species of birds or smell the fragrances of a thousand species of flowering plants or see the leaf patterns of three hundred species of trees, all in 1 hectare (2.5 acres) of land. That place is the Amazon, where half the world's remaining tropical rainforests and species reside. One hectare of rainforest in the Amazon contains more plant species than all of Europe! Of course, not all of the species diversity of the Amazon is confined to the rainforest itself. The Amazon Basin contains the largest watershed in the world; one-fifth of all the fresh water on Earth falls on its slopes, collects in over a thousand forested tributaries, and eventually flows into the Amazon River and out to sea. A trip to a fish market in Manaus, Brazil, would reveal the amazing aquatic diversity in these rivers (**Figure 17.1**). The number of fish species in the Amazon River exceeds the total number found in the entire Atlantic Ocean.

Ironically, with this incredible species diversity can come devastating species losses when these ecosystems are disturbed. The main destructive force has been deforestation, which began in earnest with the building of roads in the 1960s (Bierregaard et al. 2001). Before then, most of the Amazon had no roads and was relatively isolated from the rest of society. In 50 years' time, 15% of the rainforest has been converted to pastureland, towns, roads, and mines. Although this percentage might seem modest, it is deceivingly so, both because of the sheer number of species involved and because of the pattern of deforestation. Logging practices have caused extreme habitat fragmentation, sometimes resulting in a "fishbone" pattern in which thin linear fragments of rainforest are surrounded by strips of nonforested land (see Figure 3.6). As we will see, habitat fragmentation can have serious consequences for the maintenance of species diversity.

It was this fragmentation of the Amazon rainforest that motivated Thomas Lovejoy and his colleagues to initiate one of the largest and longest-running ecological experiments ever conducted. When the Biological Dynamics of Forest Fragments Project (BDFFP) began in 1979, Lovejoy seized a unique opportunity to find out what was happening to the species diversity of the Amazon as more and more forests were being logged. He was guided by an elegant model in Robert MacArthur and Edward O. Wilson's 1967 book *The Theory of Island Biogeography*, which presents an explanation for the observation that more species are found on large islands than on small islands. By taking advantage of a Brazilian law requiring landowners to leave half of their land as forest, Lovejoy was able to arrange for the designation of different-sized forest plots that would be surrounded by either forested (controls) or deforested land (fragments) (**Figure 17.2**). The control plots and the fragments were of four

FIGURE 17.1 Diversity Abounds in the Amazon Freshwater fish caught in the Amazon River on display in a market in Manaus, Brazil.

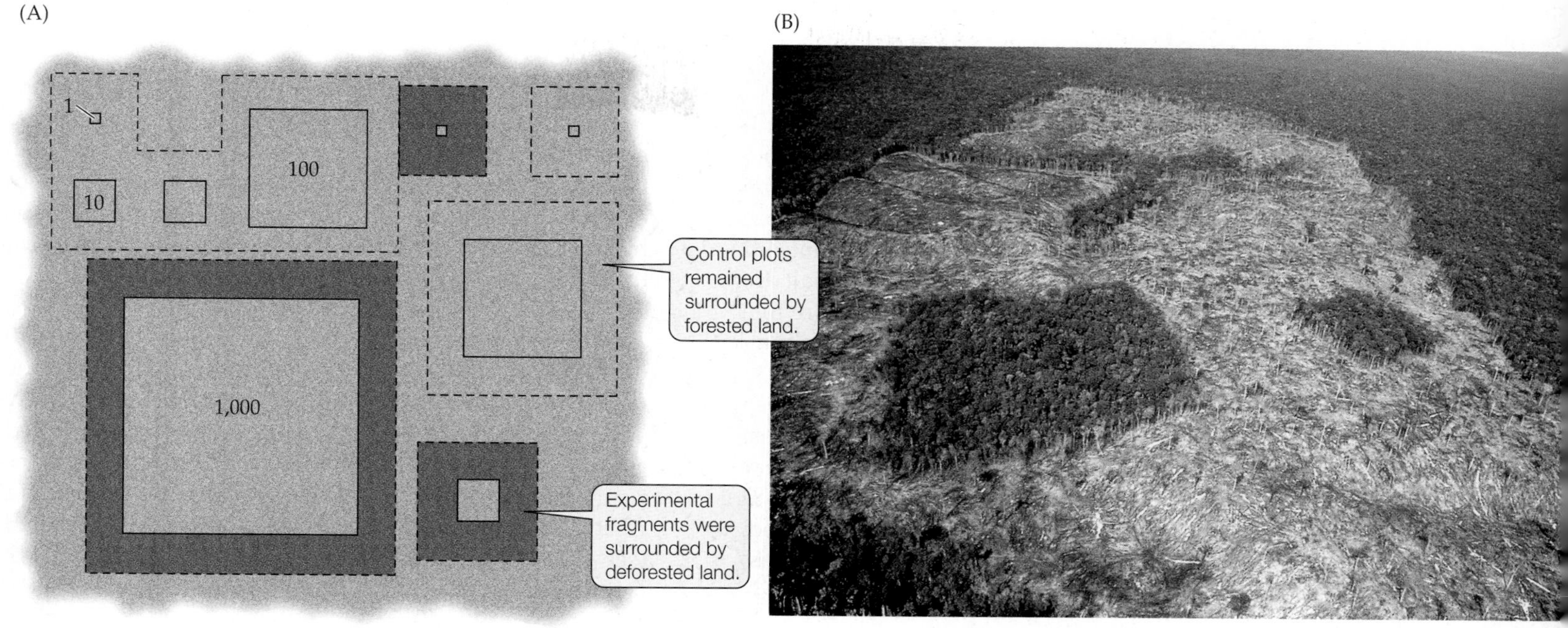

FIGURE 17.2 Studying Habitat Fragmentation in Tropical Rainforests The Biological Dynamics of Forest Fragments Project (BDFFP) near Manaus, Brazil, was designed to study the effects of habitat fragment size on species diversity. (A) Plots of four sizes (1, 10, 100, and 1,000 ha) were designated before logging took place, then either isolated by logging or left surrounded by forest as controls. (B) Aerial photo of a 10 ha and 1 ha fragment isolated in 1983. (A after Bierregaard et al. 2001.)

Q Why didn't the experimental manipulation involve removing forest from the fragments?

sizes: 1, 10, 100, or 1,000 hectares. Baseline data collected at the start of the experiment (after logging) showed little difference in species diversity between control plots and fragments. By the mid-1980s, the ecologists had a fully replicated experiment at a scale unimaginable in the past.

Over the last 25 years, the BDFFP has evolved from a study that asks the simple question, "What is the minimum area of rainforest needed to maintain species diversity?" to one that asks, "What roles do the shape, configuration, and connectivity of forest fragments play in maintaining species diversity? How does the surrounding habitat influence that diversity? And what is the prognosis for the Amazon rainforest, one of the most deforested but species-rich terrestrial biomes on Earth?"

Introduction

Looking out over a community such as a rocky intertidal zone on the northern California coast, it is obvious that the locations of species on the shoreline are influenced not only by physical factors, such as tide height and wave action, but also by a variety of biological interactions. Sea stars eat sessile mussels in the low intertidal zone, thus limiting them to the higher intertidal zones. In those zones, the crevices between mussels provide habitat for many species that otherwise would be absent. Local conditions such as these are important regulators of species distributions. However, as important as these conditions appear to us, we must always be cognizant of the influence of processes operating at larger geographic scales. Oceanographic processes, such as currents and upwelling, regulate the delivery of invertebrate larvae to rocky shorelines. At a global scale, oceanic circulation patterns control current direction. By limiting dispersal, those patterns can isolate species over ecological and evolutionary time. As a result, the local assemblage of species on the northern California coast is ultimately based on a foundation of global and regional processes. In this chapter, we will consider the effects of these large-scale geographic processes on one of the most recognizable ecological patterns known: that of the distribution and diversity of species on Earth.

CONCEPT 17.1 Patterns of species diversity and distribution vary at global, regional, and local spatial scales.

Biogeography and Spatial Scale

One of the most obvious ecological patterns on Earth is the variation in species composition and diversity among geographic locations. The study of this variation is known as **biogeography**. Pretend for a moment that you have a lifelong desire to see all the forest biomes on Earth. In this imaginary scenario, you have the ability to move from one geographic region on Earth to another. Think Google Earth,* but with the ability to fly down into a community and see species up close. You start in the tropics at 4° S

*See http://earth.google.com.

latitude and 60° W longitude and fly into the Amazon rainforest, the most species-rich forest on Earth (**Table 17.1**). At 20 m altitude, you fly through the middle of the humid forest, and as you travel over each hectare, you see new tree species (**Figure 17.3A**). You may have encountered half of them in the previous hectare, but at least half are completely new. The more area you cover, the more tree species you see. The richness is almost mind-numbing, and the heat and humidity are stifling, so you decide to head north to drier climes.

You arrive at 35° N, 125° W. This is the southern coast of California, where the forests are oak woodland—a dry biome, as we learned in Chapter 3. Most of the trees and shrubs are evergreen, but they are not conifers. Instead, they are flowering plants with small, tough (*sclerophyllous*) leaves. The woodlands are interspersed with grasslands (**Figure 17.3B**). Flying down through the vegetation, you notice the many kinds of trees and shrubs, all with small leaves and thick bark. The woodland is aromatic because of the volatile oils contained in the shrubs and herbaceous plants. Plant species richness is high, but just a fraction of that in the Amazon (see Table 17.1).

It's still warm, so you decide to head north to 45° N, 123° W, where the forest is cool and very wet. You are in the Pacific Northwest region of North America, where the forests are dominated by large conifers. As you fly through, you notice the lushness of the forest, with its lichen-filled canopy and fern-covered floor (**Figure 17.3C**). Tree species richness in these lowland temperate evergreen forests is a fraction of that in the two previous forests you've visited (see Table 17.1). There are only a handful of tree species: Douglas fir, western hemlock, western red cedar, red alder,

(A)

(B)

(C)

(D)

FIGURE 17.3 Forests around the World Forest biomes vary greatly in their species composition and species richness. (A) A tropical rainforest in Brazil. (B) Oak woodland in southern California. (C) Lowland temperate evergreen forest in the Pacific Northwest. (D) Boreal spruce forest in northern Canada.

TABLE 17.1

Tree Species Richness in Different Forests around the World

Forest location/ forest type	Latitude, longitude	Approximate tree species richness	Source
Amazon, Brazil	4°S, 60°W	1,300	Laurance 2001
Southern California, USA	35°N, 125°W	57	Allen et al. 2007
Pacific Northwest, USA			Franklin and Dyrness 1988
Douglas fir forest	45°N, 123°W	7	
Garry oak forest	45°N, 123°W	4	
Boreal forest, Canada	64°N, 125°W	2	Kricher 1998
New Zealand			Dawson and Lucas 2000
Beech forest	45°S, 170°E	20	
Flowering tree forest	35°S, 170°E	100	

and big leaf maple. What these forests lack in species richness, however, is made up in their huge biomass.

You want to see the extremes in species richness, so your next stop is 60° N, 125° W, in the boreal forests of Canada. Flying over the cold landscape, you notice rows and rows of identical spruce trees, broken once in a while by large wetlands (**Figure 17.3D**). Dipping down into the canopy, you are struck by the dense and monotonous nature of the forest. It's dark down under those spruce boughs, but low-lying berry bushes are a reminder that light does penetrate the canopy, especially in the summer months. You continue to fly north, and the forests thin until the landscape is one long expanse of treeless tundra.

Your trip could end here, but you have always wanted to visit New Zealand, so you take the time to fly back to the Southern Hemisphere. New Zealand was separated from the ancient continent of Gondwana roughly 80 million years ago, and since that time, evolution has produced unique forests there (**Figure 17.4**). Roughly 80% of the species in New Zealand are **endemic**,

FIGURE 17.4 Forests of the North and South Islands, New Zealand The two islands of New Zealand span a large latitudinal gradient (35°–47° S) and thus have different forest types. (A) The forests of the South Island are dominated by beeches. (B) The forests of the warmer North Island have greater tree species diversity and a different species composition than those on the South Island (see Table 17.1).

meaning that they occur nowhere else on Earth. Dialed into 45° S, 170° E, you fly through the Southern Hemisphere equivalent of the Pacific Northwest. Instead of conifers, the forests are dominated by four species of southern beech trees with billowy layers of twisted branches (see Figure 17.4A). Below the canopy are "divaricating" shrubs, with multiple-angled branches that give them a zigzag appearance. Plants with this growth form are found in highest abundance in New Zealand. Although temperate evergreen forests in the Northern and Southern hemispheres are similar in some ways (for example, each have low tree species richness compared with forests in the tropics), they are made up of completely different species assemblages with very different evolutionary histories.

Even within New Zealand, over a distance that extends from 35° S to 47° S (a latitudinal distance identical to that from southern California to British Columbia in the Northern Hemisphere), there are big differences in tree species richness and composition. The North Island is warmer (closer to the equator) than the South Island and has more diverse forests, consisting of many flowering tree species with a few tall emergent conifers (see Figure 17.4B). These forests have a tropical feel to them because of all the flowering trees and the multitude of vines and epiphytes (plants and lichens that live on larger plants). The tree ferns growing here are similar to those that were dominant 100 million years ago, during the age of the dinosaurs. One of the most extraordinary trees is the kauri (*Agathis australis*), which is among the largest tree species on Earth (interestingly, the largest is the Giant Sequoia *Sequoiadendron giganteum*, which occurs at roughly the same latitude in the Northern Hemisphere). Some kauri trees are 60 m (200 feet) high and 7 m (23 feet) in diameter. Unfortunately, like redwoods, kauris have been extensively logged, and they exist in a forest community in only two small reserves, 100 km^2 in total size. Given that old-growth stands of kauris take 1,000–2,000 years to generate, these forests are irreplaceable. If we contrast the tree species richness of the forests characteristic of the North Island with those on the South Island, we find more than 100 tree species in the warmer northern forests, compared with the 10–20 species in the less diverse beech forests characteristic of the temperate south (see Table 17.1).

With our world forest tour at its end, what can we conclude about biogeographic patterns on Earth, assuming that forest communities are good global representatives?

- First, species richness and composition tend to vary with latitude: the lower tropical latitudes have many more, and different, species than the higher temperate and polar latitudes.
- Second, species richness and composition vary from continent to continent, even where longitude or latitude is roughly similar.
- Third, the same community type or biome can vary in species richness and composition depending on its location on Earth.

As we will see in the rest of this chapter, these are reliable patterns that have been demonstrated over and over again for many regions of the world and many community types. What has puzzled naturalists for centuries is just what processes control these biogeographic patterns. Why are more species found in some areas than in other areas? Why do some regions harbor species assemblages that are not found anywhere else on Earth?

A number of hypotheses have been proposed to explain biogeographic variation in species composition and diversity. As we'll see, these hypotheses are highly dependent on the spatial scale at which they are applied. In other words, as one of the ecological maxims presented in Table 1.1 states, space matters. Let's deal with the issue of spatial scale before we go any further.

Patterns of species diversity at different spatial scales are interconnected

On our world forest tour, we saw that patterns of species diversity and composition varied at global, regional, and local spatial scales. We can think of these spatial scales as interconnected in a hierarchical way, with patterns of species diversity and composition at one spatial scale setting the conditions for patterns at smaller spatial scales (Whittaker et al. 2001) (**Figure 17.5**). Let's start here with the largest spatial scale and work downward.

The *global scale*, as the term suggests, includes the entire world, a huge geographic area over which there are major variations with latitude and longitude (see Figure 17.5A). Species have been isolated from one another, often on different continents or in different oceans, by long distances and over long periods. As such, differences in the rates of three processes—speciation, extinction, and dispersal—help determine differences in species diversity and composition at the global scale. We will consider these processes in more detail in the following section.

The **regional scale** encompasses smaller geographic areas in which the climate is roughly uniform and to which species are restricted by dispersal limitation (see the discussion of dispersal limitation under Concept 8.2) (Figure 17.5B). All the species contained within a region are known as the **regional species pool** (sometimes called the **gamma diversity** of the region). Earth's regions differ in species diversity and composition due to differences in the rates of speciation, extinction, and dispersal at the global scale, as mentioned above. The Amazon, for example, has many more species, and thus a larger species pool, than the Canadian boreal forest.

The physical geography of a region, such as the number, area, and distance from one another of mountains, valleys, deserts, islands, and lakes—referred to collectively

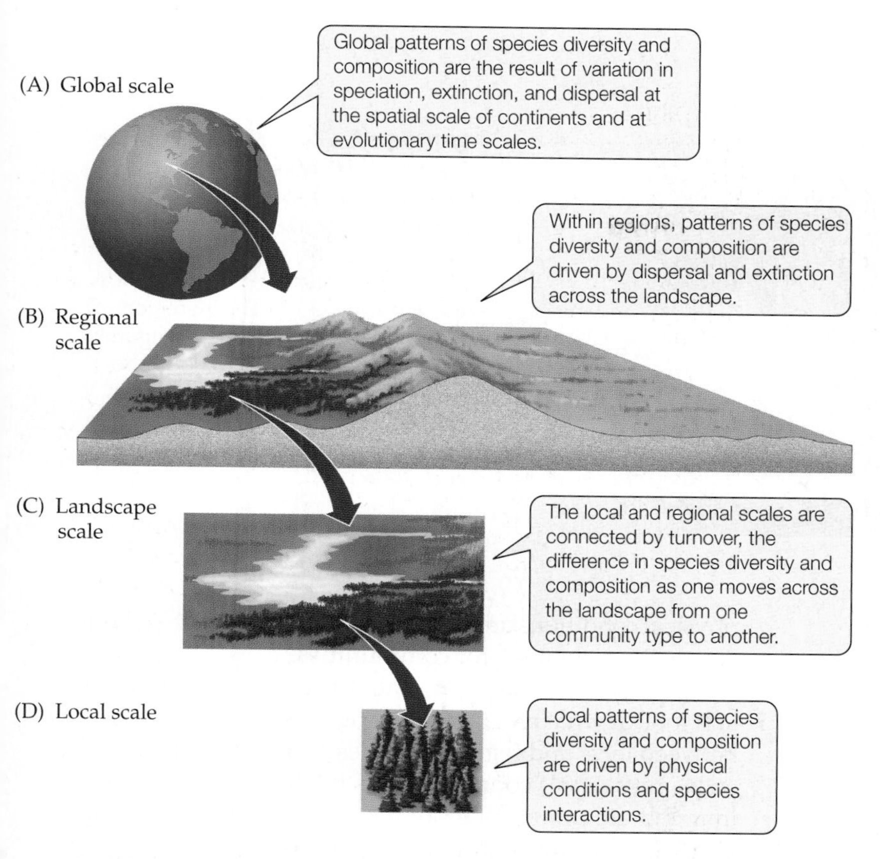

FIGURE 17.5 Interconnected Spatial Scales of Species Diversity The arrows represent the relationships between, and processes important to, species diversity and composition at (A) global, (B) regional, (C) landscape, and (D) local scales.

as the *landscape*—is critical to within-region biogeography. Species composition and diversity vary within a region depending on how the landscape shapes the rate of extinction in, and the rates of immigration to and emigration from, local habitats. Ecologists consider within-region biogeography in two related ways:

- The **local scale**, which is essentially equivalent to a community, reflects the suitability of the abiotic and biotic characteristics of habitats for species from the regional species pool once they reach those habitats through dispersal (Figure 17.5D). Species physiology and interactions with other species both influence species diversity at the local scale (sometimes called **alpha diversity**).
- The connection between local and regional scales of species diversity is expressed by a measurement known as **beta diversity**. Beta diversity tells us the change in species composition, or **turnover** of species, across the landscape as one moves from one local community to another (Figure 17.5C).

Knowing how spatial scales are related to one another in a hierarchical way is important, but are there actual area values one could apply to local and regional spatial scales? For example, how much area does a region or locality encompass? The answer is highly dependent on the species and communities of interest. For example, Shmida and Wilson (1985) suggest that terrestrial plants might have a local scale of 10^2–10^4 m^{-2} and a regional scale of 10^6–10^8 m^{-2}. But for bacteria, the local scale might be something more like 10^2 cm^{-2}. As we will see, the actual area we use to define species diversity measurements can be critical to our interpretation of the processes controlling biogeographic patterns.

Local and regional processes interact to determine local species diversity

Figure 17.5 illustrates the concept that patterns of species diversity, and the processes that control them, are interconnected across spatial scales. Given these interconnections, ecologists are interested in knowing just how much variation in species diversity at the local scale is dependent on larger spatial scales. The regional species pool provides the raw material for local species assemblages and sets the theoretical upper limit on species richness for communities in the region. But is local species richness also determined by local conditions, including species interactions and the physical environment?

One way we can consider this question quantitatively is by plotting local species richness for a community against the regional species richness for that community (**Figure 17.6**). Three basic types of relationships can be seen in such plots. First, if local species richness and regional species richness are equal (slope = 1), then all the species within a region will be found

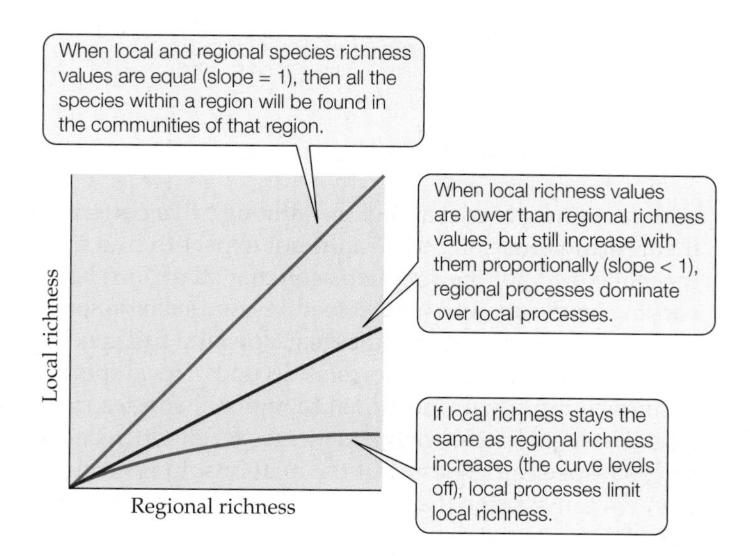

FIGURE 17.6 What Determines Local Species Richness? The relative influences of local and regional processes in a community can be determined by plotting local species richness against regional species richness.

Q Would you ever have a local to regional species richness relationship that had a slope of more than 1? Why or why not?

(A) Study sites

1	Gulf of Maine
2	Iceland
3	Northeastern Pacific
4	Galápagos archipelago
5	Chilean Patagonia
6	Antarctic Peninsula
7	Eastern Caribbean
8	Southwestern Africa
9	Southwestern New Zealand
10	Seychelles
11	Norfolk Island
12	Palau

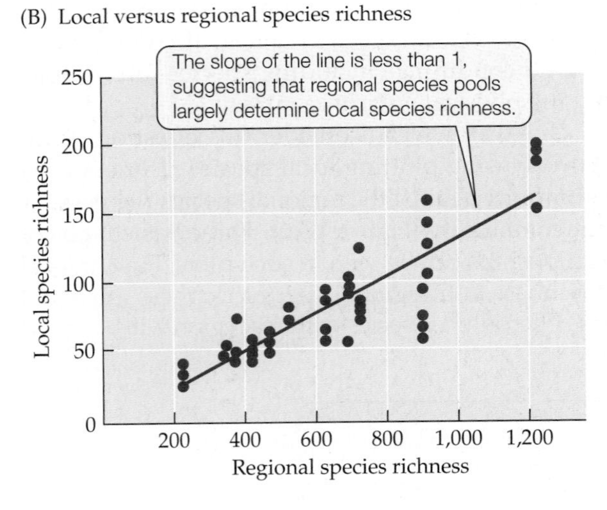

FIGURE 17.7 Marine Invertebrate Communities May Be Limited by Regional Processes Among shallow subtidal marine invertebrate communities, regional species richness explains approximately 75% of the local species richness. (A) The 12 regions of the world where the 49 sampling sites were located. (B) A plot of local species richness against regional species richness. Each dot represents one of the 49 sampling sites. (After Witman et al. 2004.)

in the community of that region. Although this pattern is theoretically possible, we would not expect to find it in the real world, for the simple reason that all regions have varying landscape and habitat features that exclude some species from some communities (e.g., lowland tree species will not be found in alpine forests). Second, if local species richness is simply proportional to regional species richness (i.e., local species richness increases with increasing regional species richness, but the relationship is not 1:1), then we can assume that local species richness is largely determined by the regional species pool, with local processes such as species interactions and physical conditions playing some role. Finally, if local species richness levels off despite an increasing regional species pool, then local processes can be assumed to limit local species richness. The degree to which local richness levels off can tell us something about how important species interactions and physical conditions are in setting a *saturation point*—a limit on species richness—for communities.

Let's move away from these theoretical constructs and look at what real data show us about the relationship between local and regional species richness. Witman and colleagues (2004) considered this relationship for marine invertebrate communities living on subtidal rock walls at a variety of locations throughout the world (**Figure 17.7A**). At 49 local sites in 12 regions, they surveyed species richness in 0.25 m^2 plots on rock walls at a 10–15 m (33–50 feet) water depth. They then compared the local species richness values they found at the sites with regional species richness values from published lists of invertebrate species capable of living on hard substrates at similar depths. A plot of local versus regional species richness at all the sites (**Figure 17.7B**) showed that local species richness was always proportionally lower than regional species richness. Furthermore, local species richness never leveled off—that is, the communities never became saturated—at high regional richness values. Instead, regional species richness explained approximately 75% of the variation in local species richness. The results of this study suggest that regional species pools largely determine the number of species present in these marine invertebrate communities.

Does the lack of saturation detected in this study and others indicate that local processes are unimportant in determining local species richness? The answer is no, for at least two reasons. First, there was still considerable unexplained variation among local communities within regions, which could be attributable to the effects of local processes such as species interactions, abiotic conditions, or dispersal limitation (see Figure 18.4). The effects of species interactions, in particular, are likely to be highly sensitive to the local spatial scale chosen. Although the small spatial scale of Witman and colleagues' study is probably appropriate for species interacting on subtidal rock walls, other studies have used inappropriate (usually too large) spatial scales that were unlikely to detect local effects. Nevertheless, the strong influence of regional-scale processes on local species richness suggests that both marine and terrestrial communities are likely to be much more susceptible to changes such as species invasions from outside their regions than previously thought.

In the remainder of this chapter, we will explore the factors controlling variation in species diversity at global and regional biogeographic scales. Chapter 18 will delve in more detail into the causes and consequences of species diversity differences at the local scale.

CONCEPT 17.2 Global patterns of species diversity and composition are influenced by geographic area and isolation, evolutionary history, and global climate.

Global Biogeography

It must have been incredible to be a scientific explorer 250 years ago. You would have left the safety of your home in Europe to travel by ship to a destination largely unknown. You would have had to endure seasickness, disease, accidents of all kinds, and years away from your family, friends, and colleagues. You might have had many years of financial debt to repay unless you were independently wealthy or could sell your collections. But you would have been the first scientist to document and collect animal and plant species of beauty, novelty, and rarity. It was under these circumstances that the science of biogeography was born and many important discoveries were made. Up to that point, European scientists had very little information about the natural history and ecology of other parts of the world; most was second-hand or anecdotal. What these early naturalists were able to bring back were specimens, and most of all, theories to help make sense of their observations.

Although not the first in his field, Alfred Russel Wallace (1823–1913) rightly earned his place as the father of biogeography (**Figure 17.8**). Inspired by naturalists such as Alexander von Humboldt, Charles Darwin, and Joseph Hooker, Wallace came on the scene with considerably less wealth or education, but his intellect and motivation more than made up for what he lacked in financial resources and training. Wallace is best known, along with Charles Darwin, as the codiscoverer of the principles of natural selection, although he has always stood in the shadow of Darwin in that regard. But his main contribution was the study of species distributions across large spatial scales.

Wallace left England for Brazil in 1848 and explored the Amazon rainforest for 4 years. On his way back, the ship he was traveling on burned in the middle of the Sargasso Sea, destroying all his specimens and most of his notes and illustrations. After 10 days in a lifeboat, he was rescued and made his way back to England, where he published an impressive six papers on his observations.

(A)

(B)

FIGURE 17.8 Alfred Russel Wallace and His Collections (A) A photograph of Wallace taken in Singapore in 1862, during his expedition to the Malay Archipelago. (B) Part of Wallace's rare beetle collection from the Malay Archipelago, found in an attic by his grandson in 2005.

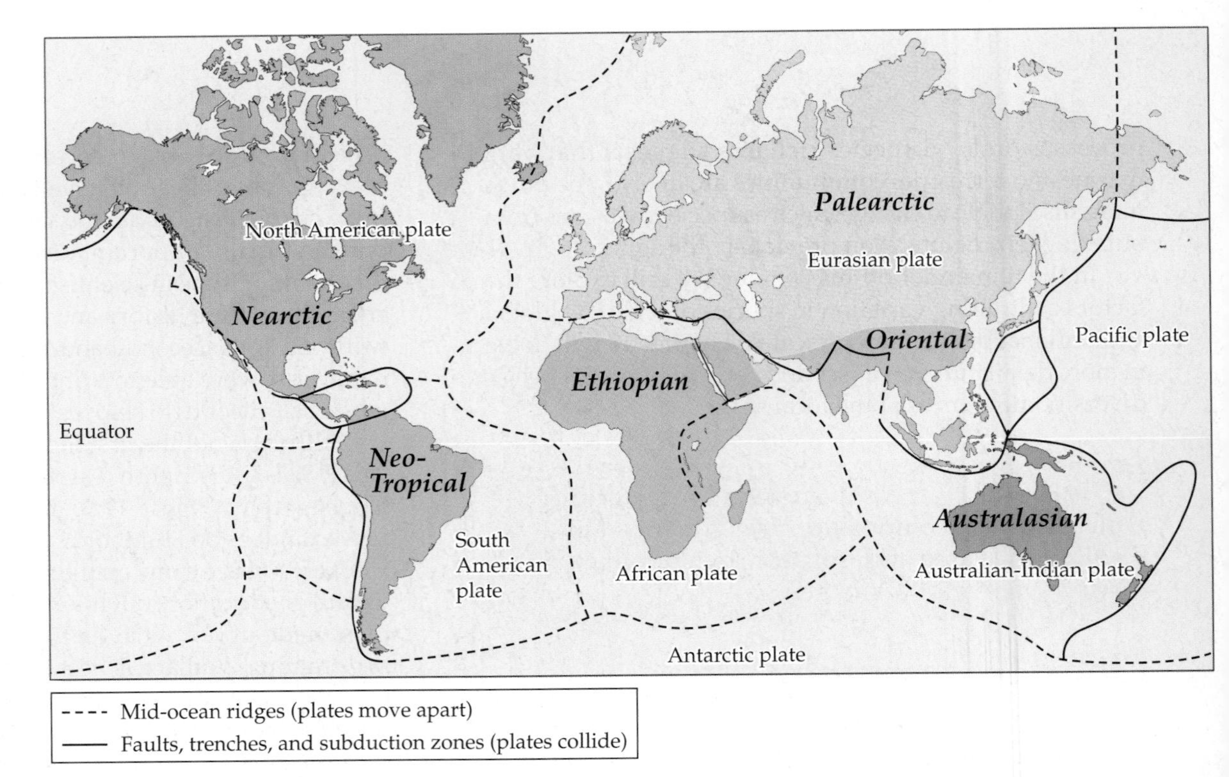

FIGURE 17.9 Six Biogeographic Regions Wallace identified six biogeographic regions using the distributions of terrestrial animals. These six regions roughly correspond to Earth's major tectonic plates.

Even though he had vowed never to travel again, Wallace left, in 1852, for the Malay Archipelago (present-day Indonesia, the Philippines, Singapore, Brunei, East Malaysia, East Timor). It was here that he made the puzzling observation described in his book *The Malay Archipelago* (Wallace 1869): that the mammals of the Philippines were more similar to those in Africa (5,500 km away) than they were to those in New Guinea (750 km away) (see Figure 8.10). Wallace was the first to notice the clear demarcation between these two faunas, which came to be known as Wallace's line. It turns out, as we'll see shortly, that these separate groups of mammals evolved on two different continents that have come into close proximity only within the last 15 million years.

Wallace's biogeographic research culminated in the publication of a two-volume work called *The Geographical Distribution of Animals*, published in 1876. In this book, Wallace overlays species distributions on top of geographic regions and reveals two important global patterns:

- Earth's land masses can be divided into six recognizable **biogeographic regions** containing distinct biotas that differ markedly in species composition and diversity.
- There is a gradient of species diversity with latitude: species diversity is greatest in the tropics and decreases toward the poles.

These two patterns are necessarily interrelated: the latitudinal gradient is superimposed over the biogeographic regions. For ease of explanation, we'll begin by describing the biogeographic regions described by Wallace and the underlying forces that created them. We will then consider some of the processes likely to be responsible for the latitudinal gradient in species diversity.

The biotas of biogeographic regions reflect evolutionary isolation

Earth's six biogeographic regions are the Nearctic (North America), Neotropical (Central and South America), Palearctic (Europe and parts of Asia and Africa), Ethiopian (most of Africa), Oriental (India, China, and Southeast

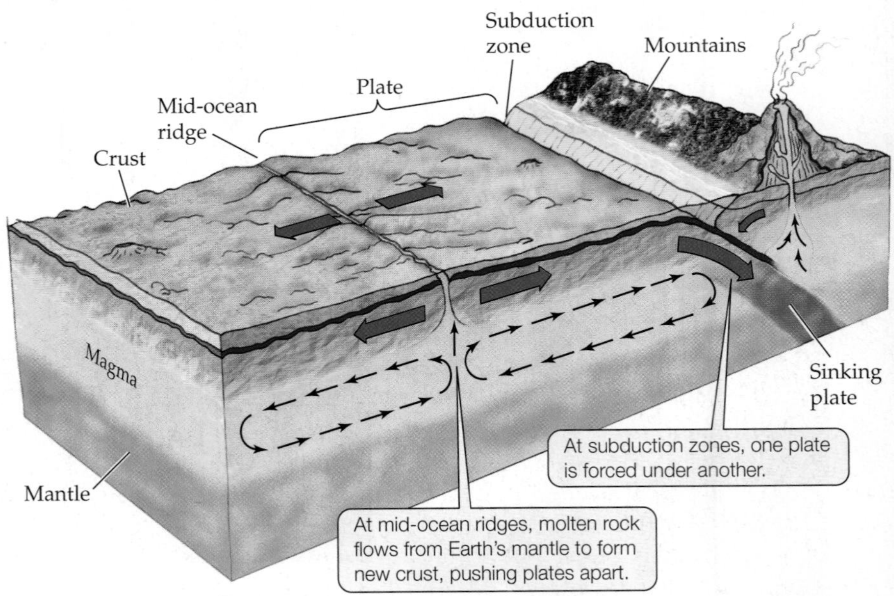

FIGURE 17.10 Mechanisms of Continental Drift Over geologic time, currents generated deep within Earth's molten rock mantle move sections of Earth's crust across its surface.

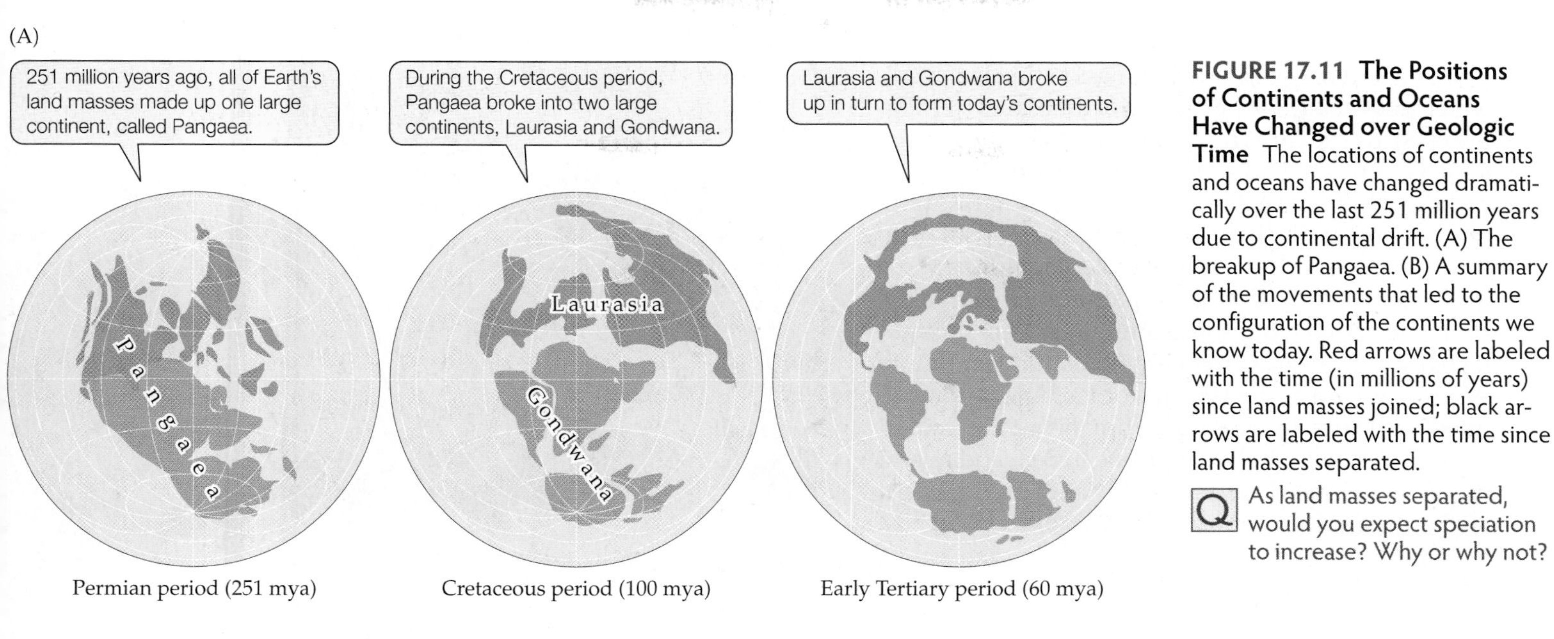

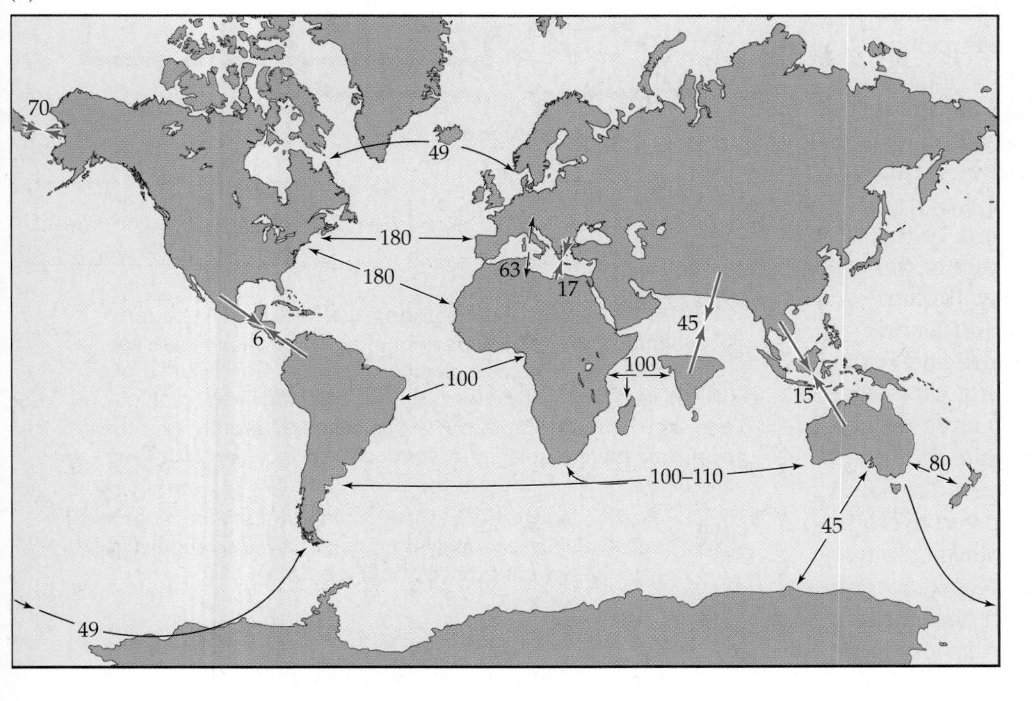

FIGURE 17.11 The Positions of Continents and Oceans Have Changed over Geologic Time The locations of continents and oceans have changed dramatically over the last 251 million years due to continental drift. (A) The breakup of Pangaea. (B) A summary of the movements that led to the configuration of the continents we know today. Red arrows are labeled with the time (in millions of years) since land masses joined; black arrows are labeled with the time since land masses separated.

Q As land masses separated, would you expect speciation to increase? Why or why not?

Asia), and Australasian (Australia, the Indo-Pacific, and New Zealand) (**Figure 17.9**). It is no coincidence that these regions correspond roughly to Earth's six major tectonic plates. These plates are sections of Earth's crust that move across Earth's surface through the action of currents generated deep within its molten rock mantle (**Figure 17.10**). Before scientists understood the processes driving the movement of these plates, they hypothesized that the continents drifted over Earth's surface, thus the name **continental drift** was given to the early theory describing these movements. There are three major types of boundaries between tectonic plates. In areas known as *mid-ocean ridges*, molten rock flows out of the seams between plates and cools, creating new crust and forcing the plates apart in a process called *seafloor spreading*. In some areas where two plates meet, known as *subduction zones*, one plate is forced downward under another plate. These areas are associated with strong earthquakes, volcanic activity, and mountain range formation. In other areas where two plates meet, the plates slide sideways past each other, forming a *fault*.

As a result of processes such as seafloor spreading and subduction, the positions of the plates, and of the continents that sit on them, have changed dramatically over geologic time (for a video, see **Web Extension 17.1**). For our purposes here, let's consider the movements of the major tectonic plates since the end of the Permian period (251 million years ago), when all of Earth's land masses made up one large continent, called Pangaea (**Figure 17.11A**). At this time, there was a massive extinction event (see Figure 6.18), eventually leading to the rise of the first archosaurs (precursors to dinosaurs) and the cynodonts (precursors to mammals). About 144 million years ago, during the Cretaceous period, Pangaea began to split into two land masses, Laurasia to the north and Gondwana to the south. During this time, dinosaurs were in their heyday and mammals were small and a relatively minor component of the fauna. The end of the Cretaceous period was marked by another mass extinction event, which caused the disappearance of dinosaurs (see Figure 6.18). By the early Tertiary period (60 million years ago), Gondwana had separated into the present-day continents of South America, Africa, India, Antarctica, and Australia. Laurasia eventually split apart to form North America, Europe, and Asia. Most of these movements resulted in the separation of continents from one another, but some continents were brought together (**Figure 17.11B**). For example, North and South America joined at the Isthmus of Panama, India collided with Asia to create the Himalayas, Africa and Europe united at the Mediterranean Sea, and a land bridge was formed between North America and Asia at the Bering Strait (for a video, see **Web Extension 17.2**).

The movement of Earth's tectonic plates thus separated the terrestrial biota of Pangaea, united by geography and phylogeny, into six biogeographically distinct groups of species by isolating them on different continents. The sequence and tempo of the continental movements has resulted in some biogeographic regions having very different flora and fauna than others. For example, the Neotropical, Ethiopian, and Australian regions, all once part of Gondwana, have been isolated for quite some time and have very distinctive forms of life. In other cases, however, distinct groups of species have been united. For example, the biota of the Nearctic region differs substantially from that of the Neotropical region despite their modern-day proximity. Because North America was part of Laurasia and South America was part of Gondwana, North and South America had no contact until about 6 million years ago. Within that time, however, many species have moved from one continent to the other (for example, mountain lions, wolves, and the precursors of llamas spread to South America, while armadillos and opossums spread to North America), somewhat homogenizing the biotas of the two regions. Interestingly, there is also evidence that several families of terrestrial mammals went extinct once the two continents merged, suggesting that ecological coexistence was not possible for some species (Flessa 1975). Finally, the Nearctic and Palearctic, both part of ancient Laurasia, have similarities in biota across what is now Greenland as well as across the Bering Strait, where a land bridge has intermittently allowed exchanges of species over the last 100 million years.

The legacy of continental movements can be found in a number of existing taxonomic groups as well as in the fossil record. The evolutionary separation of species due to barriers such as those formed by continental drift is known as **vicariance**. A good example of the effects of vicariance can be seen among the large flightless birds known as ratites, all of which probably had a common ancestor from Gondwana (**Figure 17.12**). Over the millennia, as Gondwana broke up, the rheas of South America, the ostriches of Africa, the cassowaries and emus of Australia, and the moas of New Zealand became isolated from one another. They all evolved unique characteristics in isolation, but still retained their large size and inability to fly. Interestingly, the kiwis of New Zealand are more closely related to ostriches, cassowaries, and emus than they are to moas, despite their co-occurrence with moas on New Zealand. This observation suggests that kiwis evolved elsewhere and dispersed to New Zealand sometime after the breakup of Gondwana.

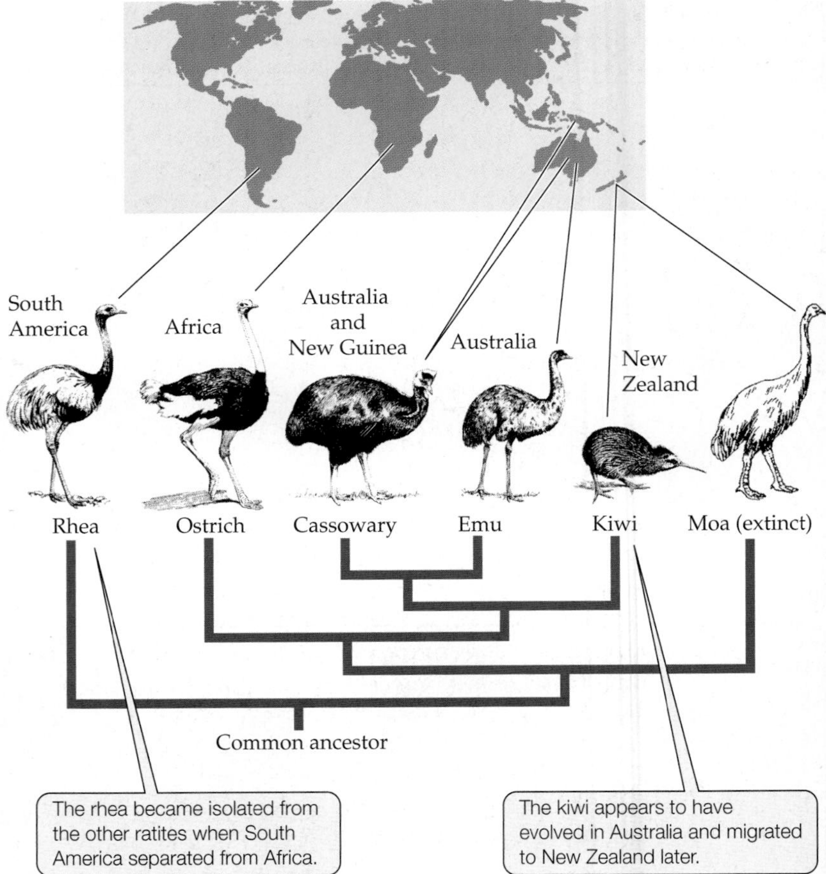

FIGURE 17.12 Vicariance among the Ratites The pattern of evolutionary relationships among the ratites shown here corresponds to the pattern of continental drift as Gondwana broke up. These large flightless birds share a common ancestor that once lived on Gondwana, but they evolved differently after their populations were isolated by continental drift. (After Van Tyne and Berger 1959; Haddrath and Baker 2001.)

Q Referring to Figure 17.11B, estimate how long the moas had been evolving separately from the cassowaries and emus before they went extinct (about A.D. 1400).

Tracing the threads of vicariance over large geographic areas and long periods provided important evidence for early theories of evolution. For example, as Wallace began to amass knowledge of the distributions of more and more species and make geographic connections between them, his ideas about the origin of species started to solidify. In an 1855 paper titled "On the law which has regulated the introduction of new species," he wrote, "Every species has come into existence coincident both in space and time with a pre-existing closely allied species." Despite the biogeographic evidence of evolutionary connections among species, it took a few more years for the mechanism of evolution (i.e., natural selection) and its role in the origin of new species to be formally proposed by both Wallace (1858) and Darwin (1859).

Before we move on, it is important to consider whether the biogeographic regionalism first identified by Wallace is present in the oceans as well as on land. After all, the oceans make up 71% of Earth's surface area and, just as we have seen for continents, they are dynamic, in the sense that they are created, merged, or destroyed by the movements of Earth's tectonic plates (see Figure 17.11). The main question, then, is whether there are barriers to dispersal between oceans as there are between continents. Despite their appearance of connectivity, oceans do have significant impediments to the exchange of biotas in the form of continents and currents; thermal, salinity, and oxygen gradients; and differences in water depth. Oceanographic discontinuities have isolated species from one another, allowed for evolutionary change, and created unique oceanic biogeographic regions (Briggs 2006). Unfortunately, delineation of marine biogeographic regions has been hindered by the extra complicating factor of water depth and by our basic lack of natural history and taxonomic knowledge of the deep oceans. One recent model by Adey and Steneck (2001) identifies 24 recognizable biogeographic regions for intertidal benthic marine macroalgae. Although it is hard to compare these macroalgal regions with terrestrial biogeographic regions, the analysis does suggest that the marine realm has much more biogeographic variation than we realized.

Species diversity varies with latitude

If you recall our Google Earth–style tour of the globe in the previous section, it was clear that plant species diversity and community composition changed dramatically with latitude: species diversity was highest at tropical latitudes and decreased toward the poles. Wallace and other nineteenth-century European scientific explorers became keenly aware of this pattern as they collected thousands of species in the tropics and compared them with their more meager European collections. As more data have accumulated over the last 200 years, the latitudinal gradient in species diversity has been more firmly established (**Figure 17.13**). Willig and colleagues (2003) tallied the results of 162 studies on a variety of taxonomic groups extending over broad spatial scales (20° latitude or more) that considered whether diversity and latitude showed a negative relationship (with diversity decreasing toward the poles), a positive relationship (increasing toward the poles), a unimodal relationship (increasing toward mid-latitudes and then declining toward the poles), or no relationship. Negative relationships were by far the most often seen, but unimodal relationships were also evident.

In addition to this undeniably strong latitudinal gradient, biogeographers have observed an important pattern of longitudinal variation. Gaston et al. (1995) measured the numbers of families along multiple transects running north to south and separated by 10° longitude. Families of seed plants, amphibians, reptiles, and mammals all increased in number toward the equator and declined at higher latitudes in both the Northern and Southern hemispheres. These researchers determined, however, that the number of families also depended on the longitude chosen. Their observations point to the importance of so-called *hot spots*, or areas of high species richness, that occur at particular longitudes, sometimes secondary to latitude.

Of course, as Figure 17.13 indicates, not all groups of organisms show decreases in species richness at higher latitudes. Some groups display the opposite pattern. Seabirds, for example, have their highest diversity at temperate and

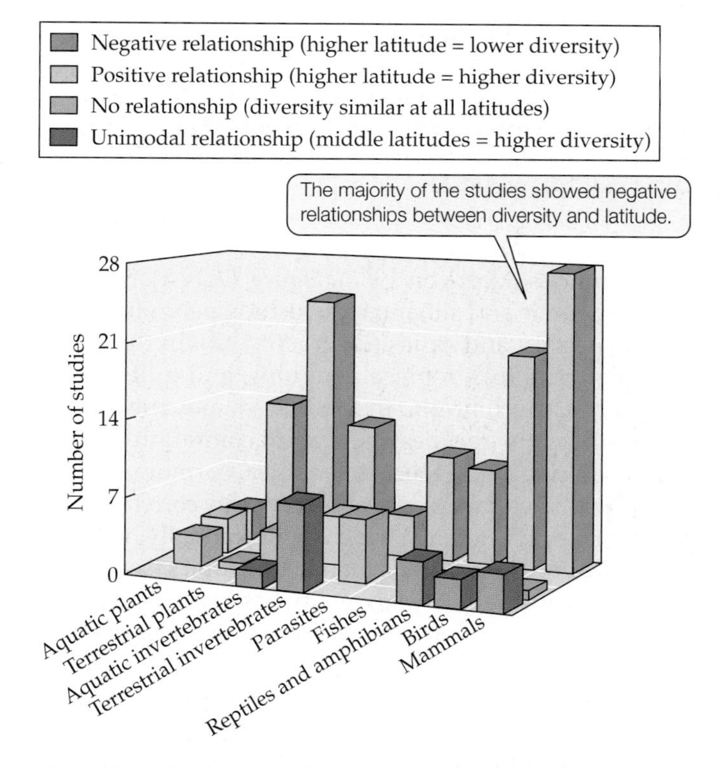

FIGURE 17.13 Studies of Latitude and Species Diversity Confirm Conventional Wisdom Willig and colleagues tallied the results of a number of studies of relationships between species diversity and latitude. The studies included a variety of taxonomic groups, but all were done at broad latitudinal scales (> 20°). (After Willig et al. 2003.)

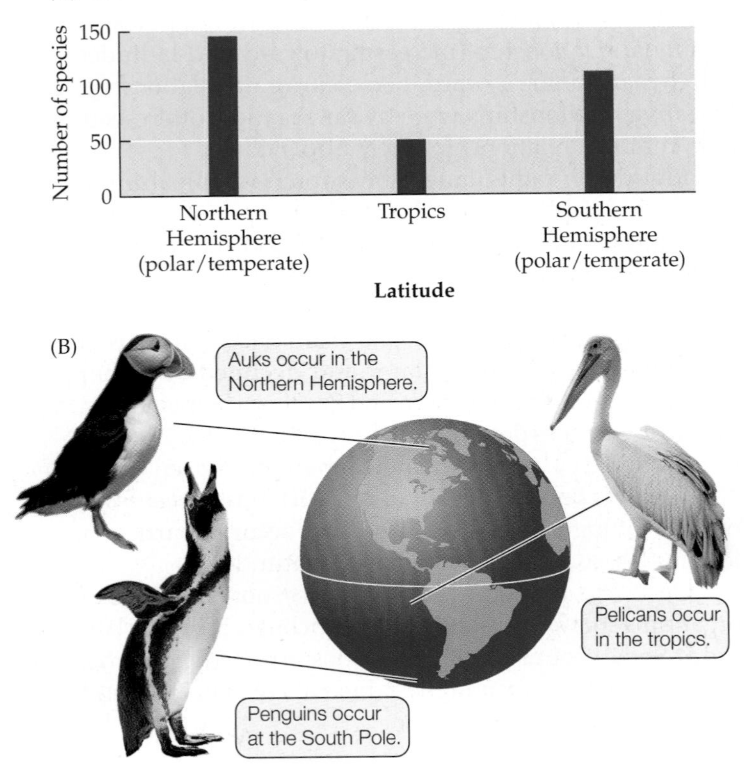

FIGURE 17.14 Seabirds Defy Conventional Wisdom Global seabird species richness shows a latitudinal pattern opposite to that of most faunas. (A) Species richness among seabirds is high in temperate and polar regions and much lower in the tropics. (B) Species composition also shows strong latitudinal differences. (Data from Harrison 1987.)

polar latitudes (Harrison 1987) (**Figure 17.14A**). Seabirds of the Antarctic and subantarctic include penguins, albatrosses, petrels, and skuas (**Figure 17.14B**). In the Arctic and subarctic, auks replace penguins, and gulls, terns, and grebes are common. In the tropics and subtropics, seabird diversity declines: the seabird community here is composed mostly of pelicans, boobies, cormorants, and frigatebirds. This pattern of seabird diversity correlates well with marine productivity, which is substantially higher in temperate and polar oceans than in the tropics (see Figure 19.9). The same pattern of diversity has been observed in marine benthic communities, which also experience much higher productivity at higher latitudes.

As we will see, productivity differences are one possible cause of latitudinal gradients in species diversity. Let's turn now to some other possible explanations.

Latitudinal gradients have multiple, interrelated causes

As we have seen, global patterns of species richness are ultimately controlled by the rates of three processes: speciation, extinction, and dispersal. Let's assume here for simplicity's sake that the dispersal rate is roughly the same worldwide. We can then predict that the number of species at any particular location will reflect a balance, or equilibrium, between the rates of two fundamental processes: speciation and extinction. Subtracting the extinction rate from the speciation rate gives us the rate of *species diversification*: the net increase or decrease of species over time. What ultimately controls this rate? Do species diversification rates vary with geographic location? If so, what causes these differences? Dozens of hypotheses have been proposed to explain them, but there is very little agreement among biogeographers and ecologists. Part of the reason lies in the fact that there are multiple and confounding latitudinal gradients in area, evolutionary age, and climate that are correlated with species diversity gradients. In addition, because speciation and extinction occur at a global spatial scale and over evolutionary time scales, it is impossible to conduct manipulative experiments to isolate various factors and separate correlation from causation.

Gary Mittelbach and colleagues (2007), in an effort to summarize the most convincing ideas, suggested that hypotheses proposed to explain latitudinal gradients in species richness fall into three broad categories (**Figure 17.15**). The first category of hypotheses is based on the assumption that the rate of species diversification in the tropics is greater than that in temperate regions (see Figure 17.15A). The second category of hypotheses suggest that the rates of diversification in the tropics and at higher latitudes are similar, but that the evolutionary time available for diversification has been much greater in the tropics (see Figure 17.15B). The third category of hypotheses suggest that resources are more plentiful in the tropics due to higher productivity, and thus that species there have a higher carrying capacity and a greater ability to coexist (see Figure 17.15C). Let's take a look at each category of hypotheses in more detail.

Species diversification rate There are a number of hypotheses that seek to explain why species diversification might be higher in the tropics. One hypothesis relates diversification to geographic area and temperature. John Terborgh (1973) and Michael Rosenzweig (1992) proposed that terrestrial species diversity is highest in the tropics because the tropics have the largest land area (**Figure 17.16A**). Rosenzweig calculated that the region between 26° N and S has 2.5 times more land area than any other latitude range on Earth. This makes intuitive sense, given that this latitude range is at the middle, and thus at the widest part, of the planet. Equally interesting are data showing that this very large area is also the most thermally homogeneous region on Earth (**Figure 17.16B**). A plot of average annual temperature against latitude by Terborgh showed that land temperatures are remarkably uniform

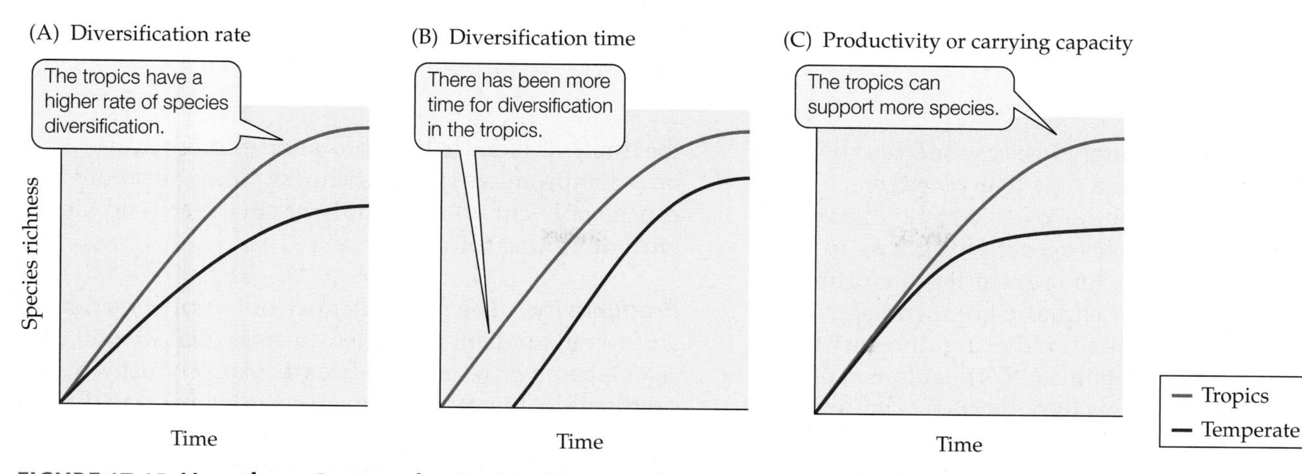

FIGURE 17.15 Hypotheses Proposed to Explain the Latitudinal Gradient in Species Richness (A) The tropics have a higher diversification rate (speciation rate – extinction rate) than temperate areas do, so they have accumulated species faster. (B) The tropics have had more time for diversification than the temperate areas have; thus they have accumulated more species. (C) Because their productivity is higher, the tropics have a higher carrying capacity than temperate areas; thus, more species can coexist there. (After Mittelbach et al. 2007.)

over a wide area between 25° N and S, but then drop off rapidly at higher latitudes.

Why would a larger land area and more constant temperatures foster greater species diversity? Rosenzweig suggested that these two factors combine to decrease extinction rates and increase speciation rates in tropical regions. He argued that a larger and more thermally stable area should decrease extinction rates in two ways: first, by increasing the sizes of species' populations (assuming that their densities are the same worldwide), and thus decreasing their risk of extinction due to chance events, and second, by increasing the geographic ranges of species, and thus decreasing their chances of extinction by spreading the risk over a larger geographic area (see the discussion of extinction under Concept 10.3). He further suggested that speciation should increase in larger areas because species should have larger geographic ranges, and thus should have a greater chance of reproductive isolation of populations and speciation (see the discussion under Concept 6.4). Rosenzweig's theory is controversial for a number of reasons, however. **Web Extension 17.3** describes several alternative hypotheses proposed to explain species diversification rates in the tropics.

Species diversification time The second type of hypothesis, which proposes that latitudinal gradients in species diversity are influenced by evolutionary history, was first championed by Wallace (1878). He suggested that tropical regions, because they are thought to have been more climatically stable over time (see Figure 17.16B), could have considerably longer evolutionary histories than temperate or polar regions, where severe climatic conditions (such as ice ages) might have disrupted species diversification. Thus, even if rates of speciation and extinction were the same worldwide, the tropics should have accumulated

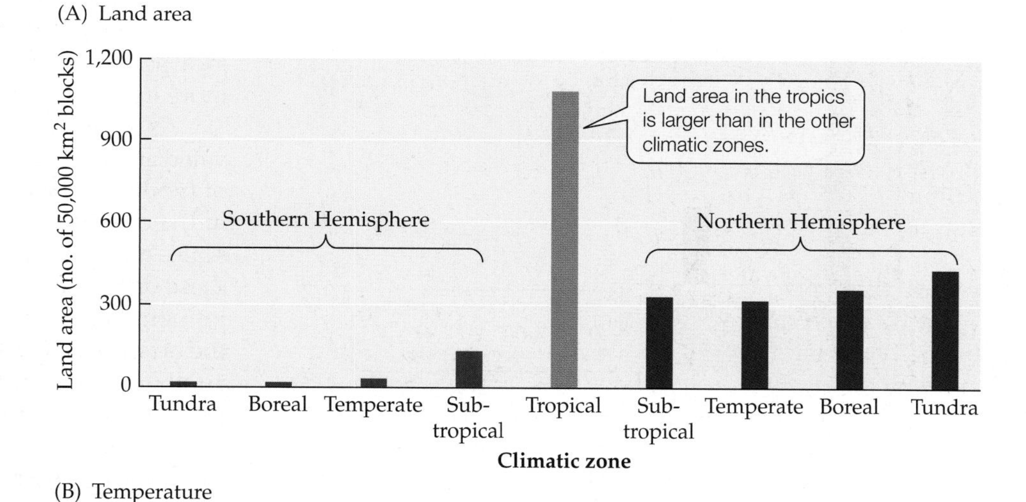

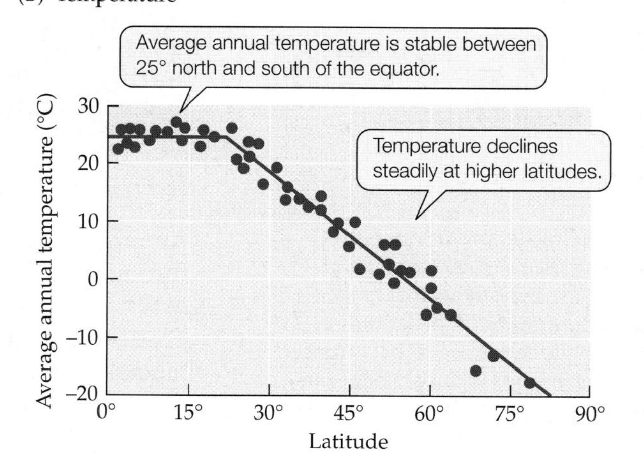

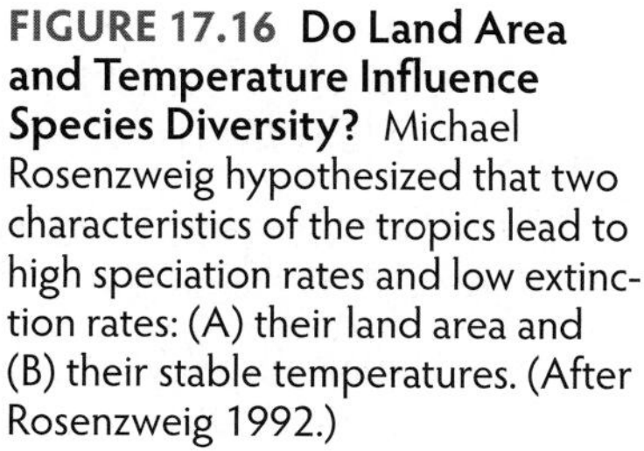

FIGURE 17.16 Do Land Area and Temperature Influence Species Diversity? Michael Rosenzweig hypothesized that two characteristics of the tropics lead to high speciation rates and low extinction rates: (A) their land area and (B) their stable temperatures. (After Rosenzweig 1992.)

more species over time merely because species should have had more uninterrupted time to evolve there.

With these ideas in mind, we can consider another possibility: that most species actually originate in the tropics and then move to temperate regions during warmer periods of greater climatic homogeneity. The idea that the tropics serve as a "cradle" for diversity was originally proposed by Stebbins (1974). Jablonski et al. (2006) recently examined this hypothesis by comparing modern-day marine bivalve faunas with marine bivalve fossils from as far back as 11 million years ago. They found that the majority of extant marine bivalve taxa originated in the tropics (**Figure 17.17A**) and spread toward the poles (**Figure 17.17B**), but without losing their tropical presence. Thus, in this particular case, we can think of the tropics as a cradle of species diversity because the majority of extant taxa originated there. But, as Jablonski and colleagues also pointed out, the tropics can serve as a "museum" as well as a "cradle." If extinction rates in the tropics are low, then species that diversify there will tend to stay there "on display," if you will. Jablonski and colleagues suggested that the current loss of biodiversity in the tropics is likely to have profound effects because it not only compromises species richness today, but could also conceivably cut off the supply of new species to higher latitudes in the future.

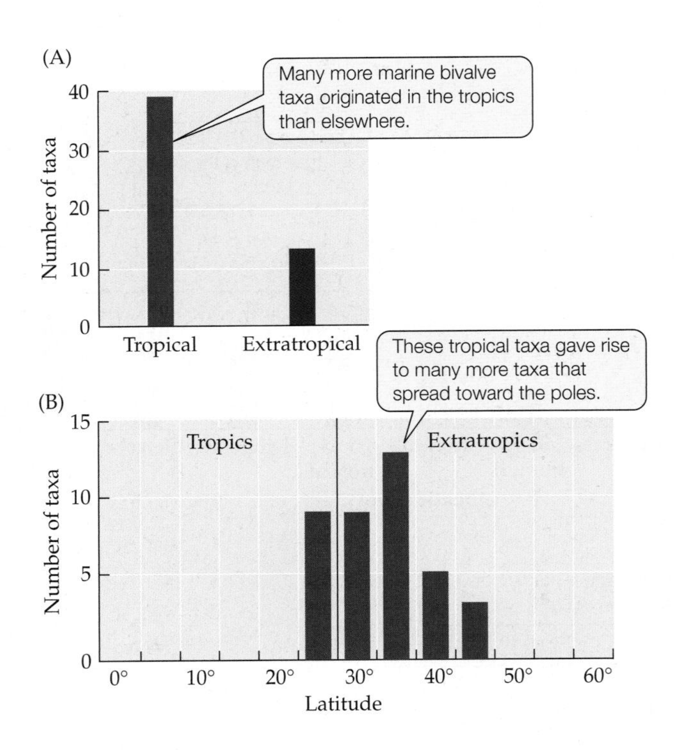

FIGURE 17.17 The Tropics Are a Cradle and a Museum for Speciation Jablonski and colleagues examined extant and fossil marine bivalve taxa to evaluate the hypothesis that longer evolutionary histories in the tropics contribute to the latitudinal gradient in species diversity. (A) Climatic zones of first occurrence of marine bivalve taxa (based on families of fossils). (B) Range limits of modern-day marine bivalve taxa with tropical origins. (After Jablonski et al. 2006.)

Productivity The final category of hypotheses for the latitudinal gradient in species diversity that we will consider is based on resources—in particular, productivity. The productivity hypothesis, proposed as long ago as 1959 by G. E. Hutchinson, posits that species diversity is higher in the tropics because that is where productivity is highest, at least for terrestrial systems (see Figure 19.7). The thought is that higher productivity promotes larger population sizes because species will have higher carrying capacities. This higher productivity will lead to lower extinction rates, greater species coexistence, and overall higher species richness. The productivity hypothesis might also explain why we see a reversal in the latitudinal gradient for some marine organisms, such as seabirds (see Figure 17.14). Productivity is generally higher in temperate coastal marine habitats than in tropical regions (see Figure 19.9). But we also know that some of the most productive habitats on Earth, such as estuaries, typically have very low species diversity. Suffice it to say, the productivity hypothesis is complex and unsatisfactory in many cases. In Chapter 18, we will consider how productivity influences diversity at local scales, where manipulative experiments can give us more insight into its effects.

As we have seen, biogeographic patterns have motivated and inspired some of the best and brightest scientists of modern times. Their fascination with the differences in the numbers and kinds of species at large geographic scales and their overwhelming drive to understand why these differences exist have resulted in some of the most influential scientific theories of all time, including that of the origin of species. In the next section, we will consider another important theory that strives to understand species diversity at smaller spatial scales.

CONCEPT 17.3 Regional differences in species diversity are influenced by area and distance, which determine the balance between immigration and extinction rates.

Regional Biogeography

An important thread that runs through this chapter, and through biogeography generally, is the relationship between species richness and geographic area. We saw in the Case Study at the opening of this chapter that larger fragments of Amazon rainforest had greater species richness than smaller fragments. In our global tour of the world's forests, we saw that species diversity was greatest in the tropics (see Table 17.1), the climatic zone whose geographic area is largest (see Figure 17.16A). We further learned that this pattern has held over evolutionary time (see Figure 17.17). This so-called **species–area relationship**, in which species richness increases with the area sampled, has been documented at a variety of spatial scales, from small ponds to whole continents. Most studies of species–area relationships have been targeted at regional spatial scales, where they tend to be good predictors of differences in species richness.

Species richness increases with area and decreases with distance

In 1859, H. C. Watson plotted the first curve showing a quantitative species–area relationship—in this case, for plants within Great Britain (**Figure 17.18**; Williams 1943). The curve starts with a small "bit" of Surrey County and expands to ever-increasing areas that eventually encompass all of Surrey County, southern England, and finally Great Britain. With each increase in area, species richness increases until it reaches a maximum number bounded by the largest area considered. (**Ecological Toolkit 17.1** describes how species–area curves are plotted and interpreted.)

Most species–area relationships have been documented for islands (**Figure 17.19**). Islands, in this case, include all kinds of isolated areas surrounded by a "sea" of dissimilar habitat (referred to as *matrix* habitat). So "islands" can include real islands surrounded by ocean, lake "islands" surrounded by land, or mountain "islands" surrounded by valleys. They can also include habitat fragments, like those produced by the deforestation of the Amazon (see Figure 17.2). Nonetheless, all of these islands and island-like habitats display the same basic pattern: large islands have more species than small islands.

In addition, because of the isolated nature of islands, species diversity on islands shows a strong negative relationship to distance from the main source of species. For example, Lomolino et al. (1989) found that mammal species richness on mountaintops in the American Southwest decreases as a function of the distance from the main source of species—in this case, two large mountain ranges in the region. This and other examples generally show that islands more distant from source populations, such as those in mainland areas or unfragmented habitats, have

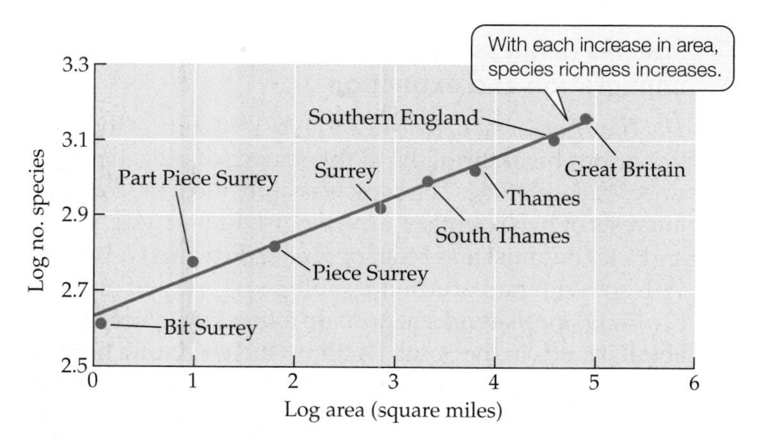

FIGURE 17.18 The Species–Area Relationship The first species–area curve, for British plants, was constructed by H. C. Watson in 1859. (After Williams 1943.)

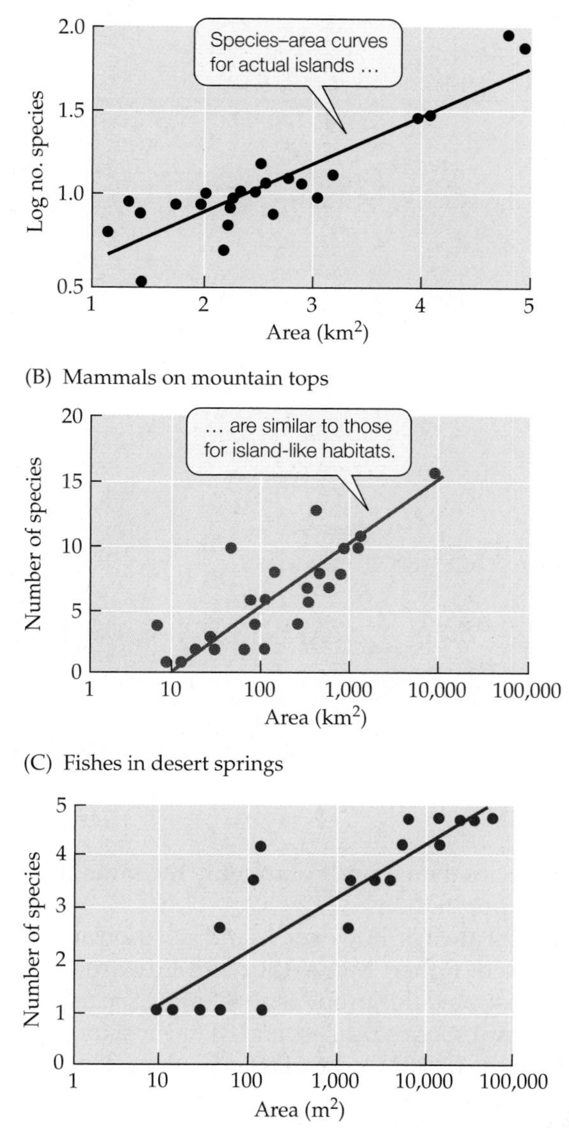

FIGURE 17.19 Species–Area Curves for Islands and Island-Like Habitats Species–area curves plotted for (A) reptiles on Caribbean islands, (B) mammals on mountaintops in the American Southwest, and (C) fishes living in desert springs in Australia all show a positive relationship between area and species richness. (A after Wright 1981; B after Lomolino et al. 1989; C after Kodric-Brown and Brown 1993.)

ECOLOGICAL TOOLKIT

17.1 Species–Area Curves

Species–area curves are the result of plotting the species richness (S) of a particular sample against the area (A) of that sample. A linear regression equation estimates the relationship between S and A in the following manner:

$$S = zA + c$$

where z is the slope of the line and c is the y intercept of the line.

Because species–area data are typically nonlinear, ecologists transform S and A into logarithmic values so that the data fall along a straight line and conform to a linear regression model.

The figure shows species–area curves for plants on the Channel Islands (off the coast of France) and on the French mainland (Williams 1964). Log transformations were conducted on both the island and mainland data, the two data sets were plotted separately, and a linear model was used to estimate the best curve fit to each of the data sets.

An important characteristic of species–area curves for islands versus mainlands is evident in this figure. The steeper the slope of the line (that is, the greater the z value), the greater the difference in species richness among the sampling areas. The Channel Islands have a much steeper slope than the French mainland areas, for the reasons outlined on p. 383.

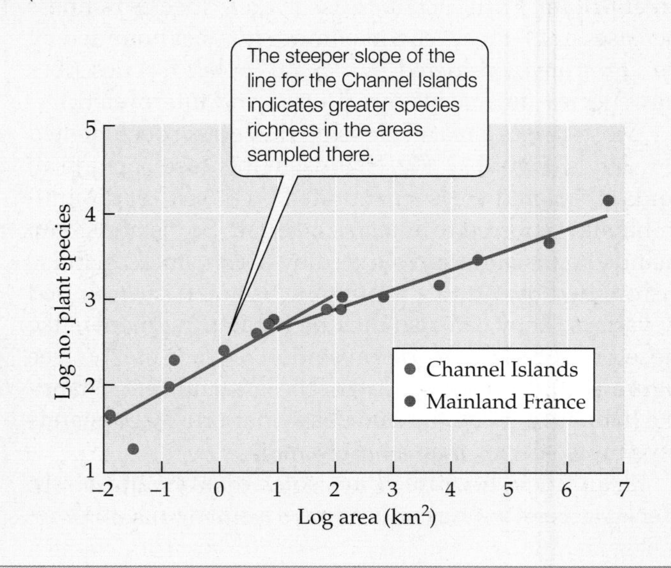

Species–Area Relationships of Island versus Mainland Areas Species–area curves plotted for plant species on the Channel Islands and in mainland France show that the slope of a linear regression equation (z) is greater for the islands than for the mainland areas. (After Williams 1964.)

fewer species than islands of roughly the same size closer to source populations.

Almost always, however, island isolation and size are confounded. Robert MacArthur and Edward O. Wilson (1963) illustrated this problem by plotting the relationship between bird species richness and island area for a group of islands in the Pacific Ocean off New Guinea (**Figure 17.20**). Here, the islands varied both in size and in degree of isolation from the mainland, but some patterns were evident. For example, if we compare islands of equivalent size, the island farthest from the source of species (New Guinea) has fewer bird species than the island nearest the source.

Let's turn now to the question of how island area and isolation could together act to produce these commonly observed species diversity patterns.

Species richness is a balance between immigration and extinction

The Theory of Island Biogeography (1967) was one of the most important breakthroughs in the science of biogeography since Wallace's time. The book was born out of the common interests of two scientists: an ecologist, Robert MacArthur, and a taxonomist and biogeographer, Edward O. Wilson. Wilson, who had studied the biogeography of ants for his Ph.D. thesis work, had made a few key observations about islands in the South Pacific, which he found himself discussing with MacArthur when they met at a scientific meeting (Wilson 1994). The first observation was that for every tenfold increase in island area, there was a rough doubling of ant species number. The second was that as ant species spread from mainland areas to islands, the new

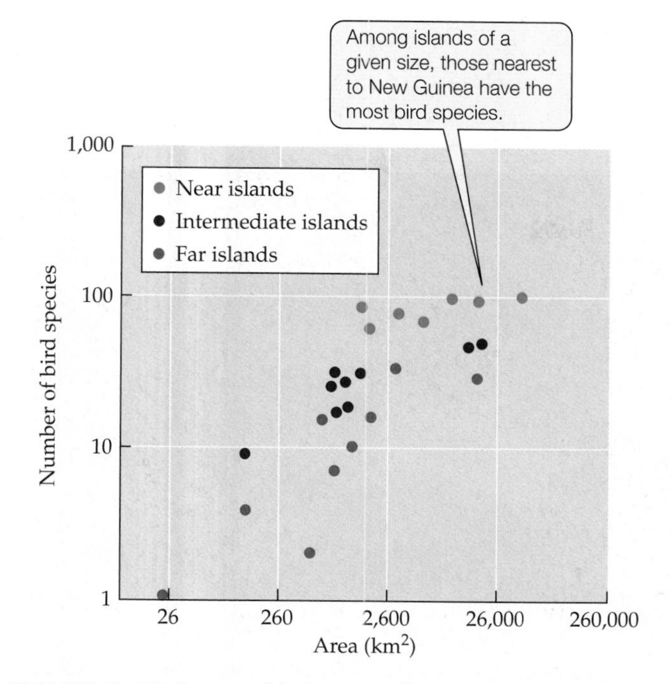

FIGURE 17.20 Area and Isolation Influence Species Richness on Islands MacArthur and Wilson plotted species–area relationships for birds on islands of different sizes and at different distances from New Guinea. (Data from MacArthur and Wilson 1963.)

species tended to replace the existing species, but there was no net gain in species richness. There appeared to be an equilibrium number of species on the islands, which was dependent on their size and distance from the mainland, but species composition on the islands could, and did, change over time.

MacArthur, a gifted mathematical ecologist, was just 31 years old when he and Wilson developed these observations into the beginnings of a simple but elegant theoretical regional biogeographic model. The model, published in their book 5 years later, became more commonly known as the **equilibrium theory of island biogeography**. The theory is based on the idea that the number of species on an island, or in an island-like habitat, depends on a balance between immigration or dispersal rates and extinction rates. The theory works something like this: Imagine an empty island open for colonization by species from mainland, or source, populations. As new species arrive on the island, by whatever means necessary, the island starts to fill up. The rate of immigration (the number of new species arriving) decreases over time as more and more species are added, eventually reaching zero when the entire pool of new species that could reach the island and be supported there is exhausted. But as the number of species on the island increases, there should also be an increase in the rate of extinction. This assumption makes sense according to the simple principle of balance mentioned above: with more species, there are more species extinctions. Additionally, as the number of species increases, the population size of each species should get smaller. Conceivably, this could occur for two reasons. First, competition may increase, thus decreasing the population sizes of species as they vie for the same space and resources. Second, predation may increase as more consumer species are added to the island. The result of either interaction is smaller population sizes and thus a greater risk of species extinction. If we plot the immigration rate against the extinction rate, the actual number of species on the island should fall where the two curves intersect, or where species immigration and extinction are in balance (**Figure 17.21**). This equilibrium number is the number of species that should theoretically "fit" on the island, irrespective of the turnover, or replacement of one species with another, that occurs on the island over time.

To understand the influence of island size and isolation on island species richness, MacArthur and Wilson simply adjusted their curves up or down to reflect their effects (see Figure 17.21). They assumed that island size mainly controls the extinction rate. They reasoned that small islands should have higher extinction rates than large islands, for the same two reasons described above, resulting in an extinction curve for small islands that is higher than that for large islands. Likewise, they reasoned that the distance of an island from the mainland mainly

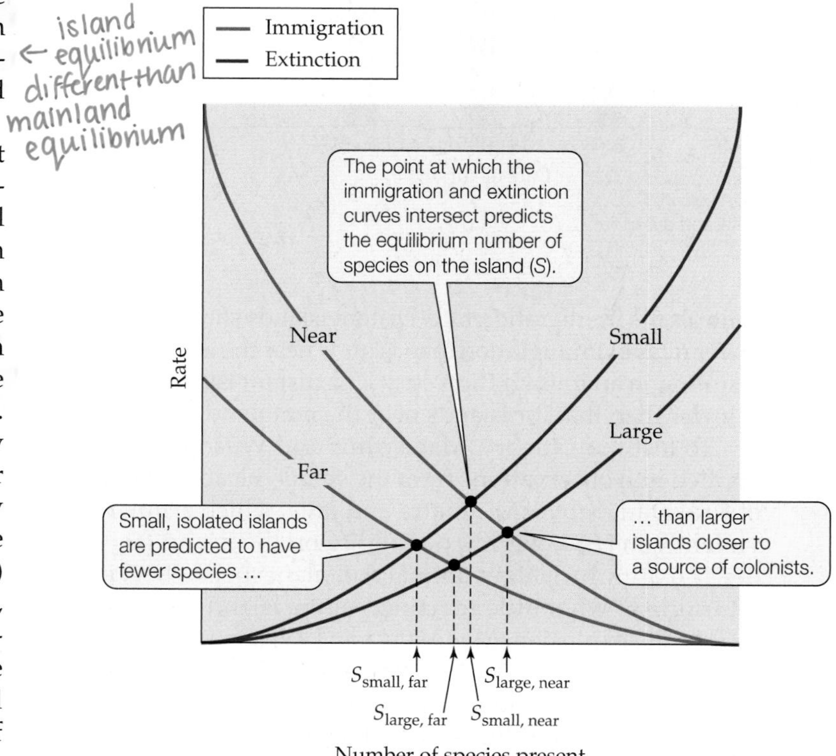

FIGURE 17.21 The Equilibrium Theory of Island Biogeography MacArthur and Wilson's theory emphasized the balance between species immigration rates and species extinction rates for islands of different sizes and at different distances from a source of colonizing species. (After MacArthur and Wilson 1967.)

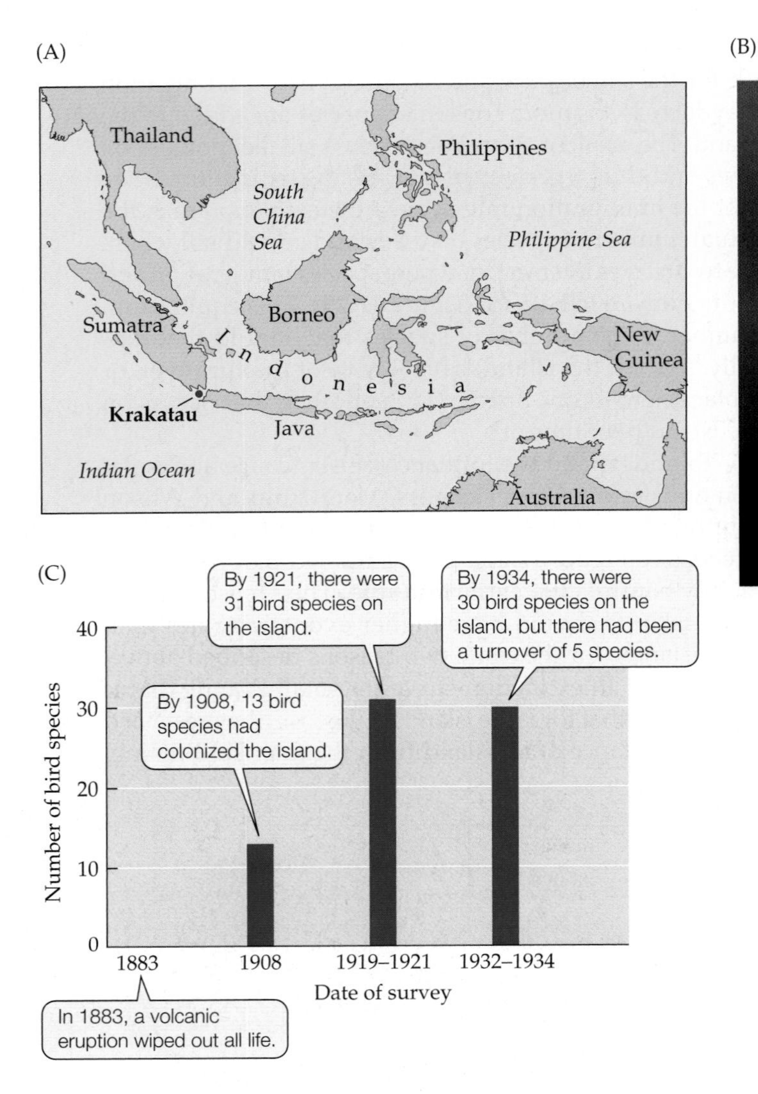

FIGURE 17.22 The Krakatau Test (A) The eruption of the small volcanic island of Krakatau, near Sumatra and Java, in 1883 provided a natural test of the equilibrium theory of island biogeography. (B) Krakatau is still an active volcano, as this recent photo shows. (C) By 1921, the number of bird species had reached 31, and in 1934, it was at 30—the equilibrium number predicted by MacArthur and Wilson's theory. Turnover, however, was five times higher than the theory had predicted. (Data from MacArthur and Wilson 1967.)

controls the immigration rate. Distant islands should have lower rates of immigration than islands near the mainland, resulting in an immigration curve for distant islands that is lower than that for islands near the mainland.

To test their theory, MacArthur and Wilson (1967) applied it to observations from the small volcanic island of Krakatau, between Sumatra and Java, which erupted violently in 1883, wiping out all life on the island (**Figure 17.22**). Surprisingly, animal and plant species began returning to what little remained of the island within a year of the explosion. MacArthur and Wilson used data from three surveys at various times since the eruption to calculate the immigration and extinction rates of birds on the island. Based on these rates, they predicted that the island should sustain roughly 30 bird species at equilibrium, with a turnover of 1 species. The data showed that bird species richness on the island had indeed reached 30 species within 40 years after the eruption and had remained close to that number thereafter. However, they also found that turnover was much higher, at 5 species. Whether this difference was due to a sampling error or a problem with the model is unknown, but this example motivated Wilson and others (for example, the BDFFP researchers whose work is described in this chapter's Case Study) to start testing the model using manipulative experiments.

One of the best-known experiments to test the equilibrium theory of island biogeography was conducted by Daniel Simberloff and his advisor, Edward O. Wilson, on small mangrove islands and their arthropod inhabitants in the Florida Keys (Simberloff and Wilson 1969; Wilson and Simberloff 1969). These islands were scattered at various distances from large "mainland" mangrove stands (**Figure 17.23A**). After surveying species richness on the islands, Simberloff and Wilson manipulated a handful of them by fumigating them with an insecticide to remove all of their insects and spiders (**Figure 17.23B**). They then surveyed the defaunated islands over a year-long period (**Figure 17.23C**). By the end of the year, species numbers on the islands were similar to those before the defaunation; furthermore, the island closest to a source of colonists had the most species, and the farthest island had the least (**Figure 17.23D**). Interestingly, the farthest island had not quite regained its original species richness even after 2 years. All the islands showed considerable turnover of species, as might be expected for small islands where extinction rates are predicted to be high (see Figure 17.21).

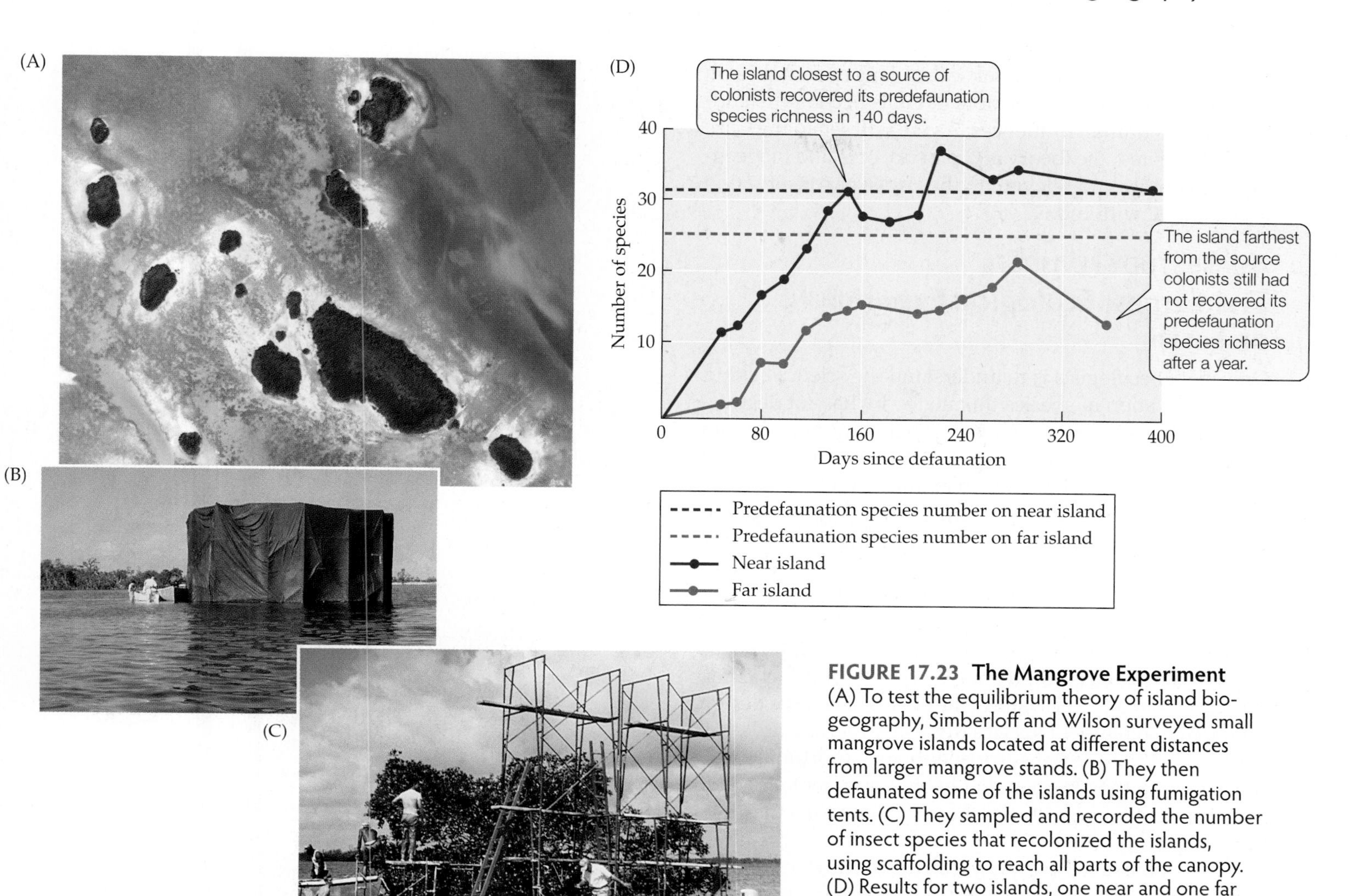

FIGURE 17.23 The Mangrove Experiment (A) To test the equilibrium theory of island biogeography, Simberloff and Wilson surveyed small mangrove islands located at different distances from larger mangrove stands. (B) They then defaunated some of the islands using fumigation tents. (C) They sampled and recorded the number of insect species that recolonized the islands, using scaffolding to reach all parts of the canopy. (D) Results for two islands, one near and one far from a source of colonists. (After Simberloff and Wilson 1969.)

The equilibrium theory of island biogeography holds true for mainland areas

Do the effects of area and isolation influence differences in species richness in mainland areas as well as on islands? As we saw in Watson's graph of plant species richness in Great Britain (Figure 17.18), the species–area relationships observed on islands also hold for mainland areas. How, then, does the biogeography of mainland areas differ from that of islands and island-like areas?

Let's consider a plot of plant species richness on the Channel Islands versus that in mainland areas of France (see Ecological Toolkit 17.1). Williams (1964) showed that plant species richness increases with area in both locations, but that the slope of the line representing the increase is steeper for the Channel Islands than for the French mainland (i.e., the z value was greater on the islands). How can we interpret this difference? In mainland areas, just as on islands, species richness is controlled by rates of immigration and extinction. In mainland areas, however, these rates are different from those on islands. Immigration rates are greater in mainland areas because the barriers to dispersal are lower. Species can move from one area to the next, presumably through continuous, non-island habitat. In addition, extinction rates are much lower in mainland areas because of the continual immigration of new individuals from the larger mainland population. Species will always have a good chance of being "rescued" from local extinction by other population members. The end result of these higher immigration and lower extinction rates in mainland areas is a lower rate of increase in species richness with increasing area, and thus a gentler slope, than in island areas.

slow rate of increase

We have seen over and over again in this chapter that geographic area has a large influence on species diversity at global and regional spatial scales. This effect takes on heightened significance as more habitats become "island-like" due to human influences. As we will see in the Case Study Revisited, the theory and practice of island biogeography is timely and relevant to the issues of conservation that we deal with today.

A CASE STUDY REVISITED
The Largest Ecological Experiment on Earth

One goal of ecologists is to understand the science behind the conservation of species threatened by habitat destruction and fragmentation. As we set aside more and more reserves to protect species diversity, the areas around those reserves continue to be changed by human activities, leaving many of them islands in a matrix of degraded habitat that is unsuitable for the species they contain. Thus, it is critical that we understand reserve design if we are to maximize our conservation goals. When Lovejoy and colleagues embarked, nearly 30 years ago, on the Biological Dynamics of Forest Fragments Project in the Amazon, as described in the Case Study at the opening of this chapter, one of their goals was to study the effects of reserve design on the maintenance of species diversity (Bierregaard et al. 2001). What they found surprised them. Unfortunately, as it turned out, they learned that habitat fragmentation had even more negative and complicated effects than they had originally anticipated.

One of the first things they learned was that forest fragments needed to be quite large and close together to effectively maintain their original species diversity. For example, in a study of forest understory birds, Ferraz et al. (2003) found that even the largest fragments they surveyed (100 ha) lost 50% of their species within 15 years. Given the regeneration time for these tropical rainforests, which ranges from several decades to a century, they projected that even 100 ha fragments would be ineffective at maintaining bird species richness until forest regeneration could "rescue" species within the fragments. The ecologists calculated that over 1,000 ha would be needed to maintain bird species richness until the forests could be regenerated, an area far greater than the average Amazon rainforest fragment in existence today (Gascon et al. 2000). If forest regeneration did not occur—as is likely when the land around a forest fragment is developed or used for agriculture—the fragment would have to contain 10,000 ha or more to maintain most of its bird species over more than 100 years of isolation (although even a fragment of that size could not sustain them all).

The researchers of the BDFFP were also quite surprised at the short distances between fragments that resulted in almost complete isolation of species. Clearings even 80 m (265 feet) wide hindered the recolonization of fragments by birds, insects, and arboreal (tree-dwelling) mammals (Laurance et al. 2002). It seemed that animals avoided entering the clearings for a number of interrelated reasons, the most obvious of which is that they have no innate reason to do

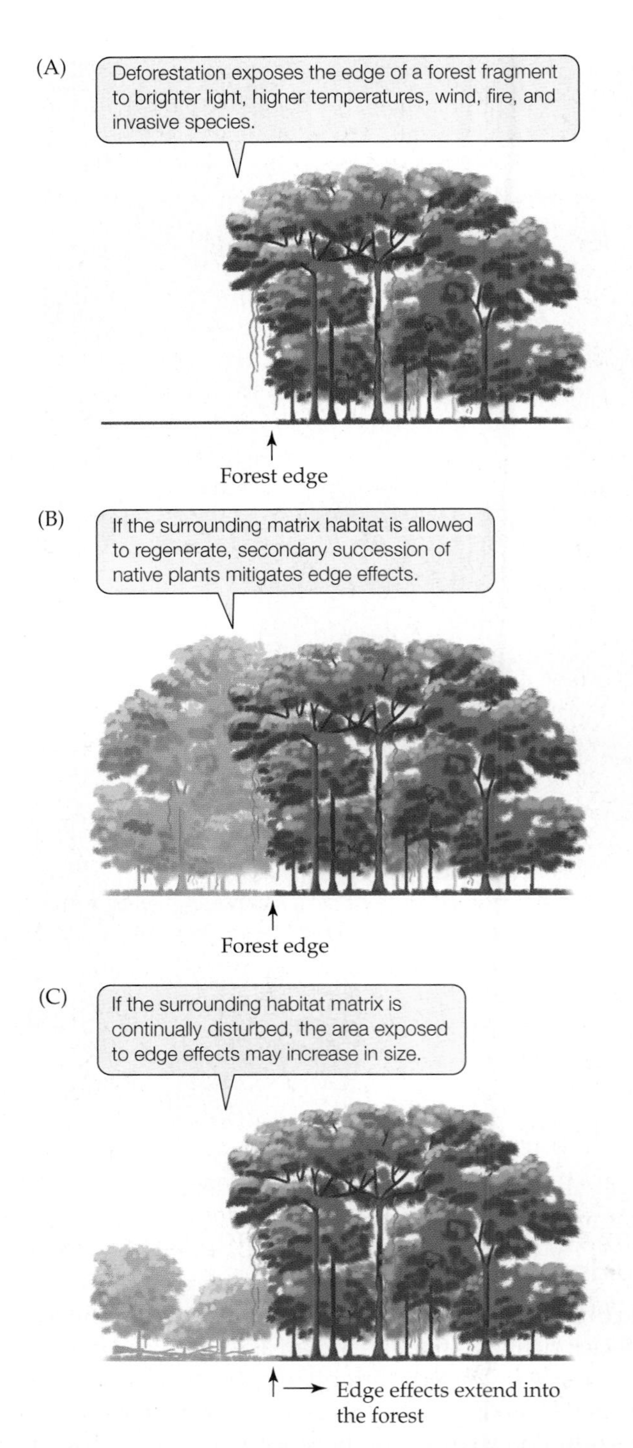

FIGURE 17.24 Tropical Rainforests on the Edge The BDFFP's research showed that deforestation subjects the forest fragments that remain to negative edge effects. (After Gascon et al. 2000.)

so, having evolved within large, continuous, and climatically stable habitats that lacked the fragmentation imposed on them by deforestation. Moreover, even if some animals were inclined to venture into the clearings, specific requirements for their movement, such as trees for arboreal mammals, would not be present to facilitate their travel to other forest patches.

A second major finding of the BDFFP was that habitat fragmentation exposes the species within a fragment to a wide variety of potential hazards, including harsh environmental conditions, fires, hunting, predators, diseases, and invasive species. These *edge effects*, which occur at the transition between forest and nonforested matrix habitat, can act together to increase local species extinctions. Trees, for example, can be killed or damaged by their sudden exposure to brighter light, higher temperatures, wind, fire, and diseases (**Figure 17.24**). Over time, depending on the surrounding matrix habitat, the ultimate influences of edge effects are revealed. If the matrix habitat is left undisturbed, secondary succession occurs, as described in Chapter 16, reducing edge effects. If the matrix habitat continues to be disturbed, however, then the area subjected to edge effects may increase in size. For example, Gascon et al. (2000) describes forest fragments in the southern Amazon that are embedded in huge non-native sugarcane and *Eucalyptus* plantations where burning is used regularly for crop rotation. The burning keeps the forest edges in a constant state of disturbance. Fire-tolerant plant species, many of them non-native, become more common at the edges and act as conduits for more fires. This positive feedback loop ends up decreasing the effective size of the forest fragments and continually increasing the area subjected to edge effects. Some edge effects can extend a kilometer or more into a fragment, essentially influencing the entire area of a 1,000 ha fragment.

The results of the BDFFP have made an immense and sobering contribution to our understanding of forest fragmentation. As Laurance et al. (2002) point out, the BDFFP is a controlled experiment that probably provides a conservative estimate of species losses. The BDFFP has shown us that most of the forest fragments human activities are creating are too small to maintain all their original species; thus, habitat fragmentation is likely to result in the loss of many species. We'll see how the BDFFP's findings are being applied to reserve design and other conservation efforts when we discuss habitat fragmentation and edge effects in more detail under Concept 23.2.

CONNECTIONS IN NATURE

Tropical Rainforest Diversity Benefits Humans

Why do we care when species go extinct in a rainforest far away? As we will see in the discussion under Concept 22.1, such extinctions raise ethical and aesthetic concerns, similar to those that arise when great works of art or antiquities are lost to society. But, in addition, there are economic concerns about the loss of important *ecosystem services* produced by natural systems, that help sustain human health and well-being (read more about this under Concept 22.1). For example, tropical deforestation raises concerns about losses of important foods and medicines, which have their origins in rainforests. At least 80% of the developed world's diet originated in tropical rainforests, including corn, rice, squash, yams, oranges, coconuts, lemons, tomatoes, and nuts and spices of all kinds. Twenty-five percent of all commercial pharmaceuticals are derived from tropical rainforest plants, but less than 1% of tropical rainforest plants have been tested for their potential medical uses.

These statistics beg the question, "How does the economic value of tropical rainforest plants used for non-timber purposes compare to the value of deforestation?" It turns out that there have been very few economic analyses of this type. A few studies come from the Millennium Ecosystem Assessment (2005), a synthesis of studies on the use of the environment and its relationship to human needs created by leading scientists from around the world. An example comes from Cambodia, where the total economic value of traditional forest products (e.g., fuelwood, rattan and bamboo, malva nuts, and medicines) was compared with that of unsustainable forest harvesting. The value of traditional forest products is four to five times greater ($700–$3,900 per ha) than that of unsustainable forest harvesting ($150–$1,100 per ha).

Recognition of the economic benefits of changing our resource management practices has only just begun. Why is this? Part of the answer lies in our not formally recognizing the economic value of the services provided to humans by species or whole communities. Tropical rainforests provide food, medicine, fuel, and a destination for tourists, all of which can be obtained without complete deforestation. Rainforests also regulate water flow, climate, and atmospheric CO_2 concentrations. Assigning a value to any of these important services is difficult compared with setting the market price of timber or agricultural products. For that reason, it is easier to justify the use of rainforest timber and land (and even of some sustainable forest products) for private profit than the conservation of rainforests for the ecological services they provide to society in general. If private landowners are not given incentives to value the larger social benefits of ecological services, then maximization of personal gain will often drive their decisions. Given the importance of ecological services to our planet, we cannot afford to ignore these economic trade-offs any longer.

SUMMARY

CONCEPT 17.1 Patterns of species diversity and distribution vary at global, regional, and local spatial scales.

- Biogeography is the study of variation in species composition and diversity among geographic locations.
- Patterns of species composition and diversity at different spatial scales are connected to one another in a hierarchical way.
- The global spatial scale includes the entire world, a huge geographic area over which there are major differences in climate and in species diversity and composition.
- The regional spatial scale encompasses a smaller geographic area in which the climate is roughly uniform and the species contained therein are bound by dispersal limitation to that region.
- The local spatial scale encompasses the smallest geographic area and is essentially equivalent to a community.
- Beta diversity is the change in species number and composition, or turnover of species, across the landscape from one local community to another.
- Studies show that regional species pools largely determine the numbers of species present in local communities, but that local conditions are also important and cannot be discounted.

CONCEPT 17.2 Global patterns of species diversity and composition are influenced by geographic area and isolation, evolutionary history, and global climate.

- Earth's land mass can be divided into six biogeographic regions that vary markedly in species diversity and composition.
- The biotas of the six biogeographic regions reflect a history of isolation due to continental drift caused by the movements of Earth's tectonic plates.
- Tracing the threads of vicariance over large geographic areas and long time periods provided important evidence for early theories of evolution.
- Species diversity is greatest in the tropics and declines at higher latitudes.
- A number of hypotheses, involving species diversification rate, species diversification time, and productivity, have been proposed to explain the latitudinal gradient in species diversity.

CONCEPT 17.3 Regional differences in species diversity are influenced by area and distance, which determine the balance between immigration and extinction rates.

- Species richness tends to increase with the area sampled and tends to decrease with distance from a source of species.
- Most species–area relationships have been documented for islands, which include all kinds of isolated areas surrounded by dissimilar habitat.
- The equilibrium theory of island biogeography predicts that a balance between immigration and extinction rates controls species diversity on islands or in island-like areas.
- According to the theory, larger islands closer to a source of species have more species than smaller islands that are more distant from a source of species because they have higher immigration rates and lower extinction rates.
- The same species–area relationship observed on islands also holds for mainland areas, but the rate of increase in species richness with increasing area is lower than on islands and in island-like areas.

REVIEW QUESTIONS

1. Spatial scale is important to the biogeographic patterns of species diversity and composition that we see on Earth. Define the various spatial scales that are important to biogeography, describe how they are related to or interconnected with one another.
2. Describe the factors believed to have created biogeographic regions on land and in the oceans.
3. Latitudinal gradients in species diversity and composition are strong global features of biogeography. Describe three hypotheses proposed to explain why species diversity is higher in the tropics and decreases toward the poles for the majority of taxonomic groups.
4. Imagine that you work for your government's park service. You are presented with three proposed designs for a new park that is intended to preserve an important type of rare habitat. All of the designs consist of fragments of rare habitat surrounded by a matrix of less desirable habitat. You must choose the park design that will best preserve the species richness of the rare habitat. Using MacArthur and Wilson's equilibrium theory of island biogeography, justify which of the three proposed park designs you would most recommend to the government.

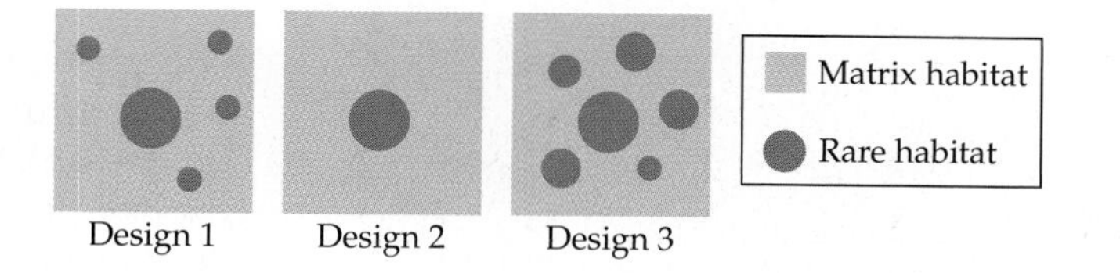

ON THE COMPANION WEBSITE
sites.sinauer.com/ecology2e

The website includes Chapter Outlines, Online Quizzes, Flashcards & Key Terms, Suggested Readings, a complete Glossary, and the Web Stats Review. In addition, the following resources are available for this chapter:

HANDS-ON PROBLEM SOLVING

Island Biogeography

This Web exercise explores the factors that determine the number of species that can live on different islands according to the theory of island biogeography. In a series of simulations, you will manipulate the size of an island and the distance from the island to the mainland to demonstrate how these factors affect the equilibrium number of species on that island.

WEB EXTENSIONS

17.1 Animation of Continental Drift

17.2 Animation of the Bering Land Bridge after Glaciers Retreated

17.3 Alternative Hypotheses Posed for the Species Diversification Rate in the Tropics

CHAPTER

18 Species Diversity in Communities

KEY CONCEPTS

- **CONCEPT 18.1** Species diversity differs among communities due to variation in regional species pools, abiotic conditions, and species interactions.
- **CONCEPT 18.2** Resource partitioning among the species in a community reduces competition and increases species diversity.
- **CONCEPT 18.3** Processes such as disturbance, stress, predation, and positive interactions can mediate resource availability, thus promoting species coexistence and species diversity.
- **CONCEPT 18.4** Many experiments show that species diversity is positively related to community function.

Powered by Prairies? Biodiversity and Biofuels: A Case Study

A hundred and twenty years ago, as the last covered wagons crossed the vast and beautiful prairies of the North American Great Plains on their way out west (**Figure 18.1**), two German inventors toiled in a laboratory to develop the first gasoline-powered car. In 1889, Gottlieb Daimler, along with his design partner, Wilhelm Maybach, built the first automobile from the ground up; it had a four-speed transmission and could reach speeds of 10 mph. By the turn of the twentieth century, automobiles were being produced by the thousands, and a century later, by the millions.

Although automobiles dominate our lives today, in large part because of their convenience and affordability, their negative environmental impacts cannot be denied. Internal combustion engines create nearly 80% of the human emissions of CO_2 into the atmosphere that are contributing to global warming (IPCC 2007). Furthermore, the world's supplies of fossil fuels are not endless, and their costs will no doubt continue to rise. These circumstances have led to the development of alternative sources of energy, one of which is biofuels.

Biofuels are liquid or gas fuels made from plant material (biomass). Two basic kinds of biofuels are being made in the United States today: ethanol and biodiesel, produced mostly from corn and soybeans, respectively. Biofuels have two potential advantages over fossil fuels. First, they aim to be **carbon neutral**, meaning that the amount of CO_2 produced by burning them matches the amount taken up by the plants from which they are made. Second, their supply is limitless as long as crops can be grown to fulfill production needs. Unfortunately, their downsides are also numerous. The growing of corn and soybeans for biofuels competes for land and water that could be used for food production. Furthermore, fossil fuels are needed to produce the fertilizers that are required to grow these crops. Large amounts of fossil fuels are also required to produce and transport the biofuels themselves. For these reasons, biofuels, as they are currently produced, have their own negative environmental impacts and reduce fossil fuel use only modestly (for more information

FIGURE 18.1 Powered by Prairies? A native prairie on the North American Great Plains. Could these plants be used to produce biofuels more sustainably?

on climate change and the carbon cycle see the discussion under Concept 24.2).

One promising advance in biofuel technology involves the use of alternative forms of biomass that are more sustainable than food crops. These materials include by-products of food or lumber production, such as straw, corn stalks, or waste wood. This is also where those vast prairies, once an inspiration to the settlers of western North America, come into play. Most of the land on the Great Plains that was once prairie has been converted to agriculture. Years of cultivation have degraded some of this land, making it unsuitable for food crops. This situation inspired David Tilman, a community ecologist at the University of Minnesota, to consider how prairie communities, once symbols of the limitless potential of the land and its resources, might be used to make biofuels in a more sustainable manner.

Tilman had been studying prairie communities for some time and was interested in how plant species diversity influenced the productivity of these communities. Working with a set of experimental plots on abandoned agricultural land at Cedar Creek, Minnesota (**Figure 18.2**), he and his colleagues showed that plots with more plant species (up to 16 species in some plots) produced greater biomass for a given amount of water or nutrients than did plots with fewer species (Tilman and Downing 1994; Tilman et al. 1996). Tilman reasoned that if diverse assemblages of prairie plants could be grown on degraded land, they might be a good source of biomass for biofuel production, given the low energy inputs required by the prairie ecosystem.

Can prairie plants be used in biofuel production, and are they a viable substitute for traditional crops such as corn and soybeans? As we will see, prairie plant species diversity makes all the difference in the answers to these questions.

Introduction

Communities vary tremendously in the numbers and kinds of species they contain. In Chapter 17, our worldwide tour of forest communities demonstrated the wide variation in species diversity that occurs both globally and regionally. We saw that communities in the tropics (such as the Amazon rainforest) had many more tree species than those at higher latitudes (such as the forests of the Pacific Northwest or New Zealand). Moreover, we found that regional species pools had an important, but not an exclusive, influence over the number of species within a community. In New Zealand, for example, regional tree diversity is generally high, but there are many more species in the northern flowering tree communities than in the southern beech communities (see Table 17.1) because local abiotic and biotic conditions have an important role in mediating tree diversity.

In this chapter, we will focus on species diversity at the local scale. We will ask two important questions: First, what are the factors that control species diversity within communities? Second, what effects does species diversity have on the functioning of communities?

FIGURE 18.2 Plant Diversity Matters Experimental plots used to investigate the relationship between plant species richness and productivity at Cedar Creek, Minnesota.

CONCEPT 18.1 Species diversity differs among communities due to variation in regional species pools, abiotic conditions, and species interactions.

Community Membership

If you looked across a landscape from the top of a mountain, you would see a patchwork of different communities that might consist of, say, forests, meadows, lakes, streams, and marshes (**Figure 18.3**). You could be sure that each of those communities would have a different species composition and a different species richness. The meadow would probably be dominated by a variety of grasses, herbs, and terrestrial insects. The lake might be filled with different species of fish, plankton, and aquatic insects, and might possibly harbor as many species as the meadow. Even though some species might be able to move from one community to another (such as amphibians), the two communities would still be highly distinct.

How do collections of species end up coming together and forming communities with different species compositions and richnesses? One way to answer this question is to consider the factors that control species membership in communities. If you think about the sheer number of species that coexist within any community, it is clear that no one process is responsible for all the species we find

FIGURE 18.3 A View from Above Looking at these mountains in Glacier National Park, Montana, USA, it is easy to see that the landscape is made up of a patchwork of communities of different types.

there. As we saw in Chapters 8 and 17, the distributions and abundances of organisms within communities are dependent on three interacting factors: (1) regional species pools and dispersal ability (species supply), (2) abiotic conditions, and (3) species interactions. We can think of these three factors as "filters" that act to exclude species from (or include them in) particular communities (**Figure 18.4**). Let's briefly consider each of them in more detail.

Species supply is the "first cut" to community membership

In our discussion under Concept 17.1, we saw that the regional species pool provides an absolute upper limit on the numbers and types of species that can be present within communities (see Figure 17.6). Not surprisingly, we saw that regions of high species richness tend to have communities of high species richness (see Figure 17.7). This relationship is due to the role of the regional species pool and, more specifically, of dispersal in "supplying" species to communities (see Figure 18.4A). Nowhere is the controlling effect of dispersal on community membership more evident than in the invasion of communities by non-native species.

As ecologists are beginning to learn, humans have greatly expanded the regional species pools of communities by serving as vectors of dispersal. For example, we know that many aquatic species travel to distant parts of the world that they could not otherwise reach in the ballast water carried by ships (**Figure 18.5A**). Seawater is pumped into and out of ballast tanks, which serve to balance and stabilize cargo-carrying ships, all over the world. Most of the time, the water—along with the organisms it contains (from bacteria to planktonic larvae to fish)—is taken up and released close to ports, where some of the organisms have the opportunity to colonize nearshore communities. As many as 5,000 freshwater and marine species are transported in the ballast water of oceangoing vessels each day. Ballast water introductions have increased substantially over the past few decades because ships are larger and faster,

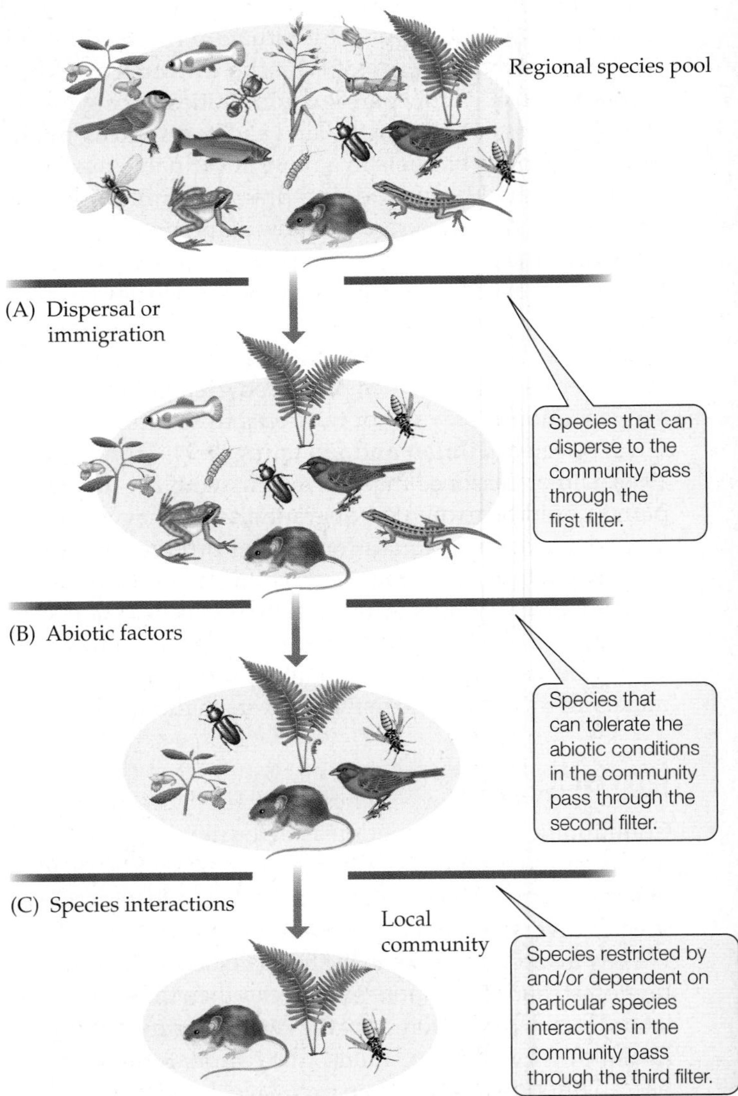

FIGURE 18.4 Community Membership: A Series of Filters Species end up in a local community by passing through a series of "filters" that determine community membership. Species are lost at each filter, so local communities contain only a fraction of the species in the regional pool. In practice, all the filters work at the same time, rather than in series as the figure suggests. (After Lawton 2000.)

Would it make sense for the fish and frog species in the regional pool to be present in the local community given in the figure? Why or why not?

(A)

(B)

FIGURE 18.5 Humans Are Vectors for Invasive Species (A) Large and fast oceangoing ships are carrying marine species to all parts of the world in their ballast water. (B) The zebra mussel (*Dreissena polymorpha*), a destructive invader of the inland waterways of the United States, was carried there from Europe in ballast water.

so that more species can be taken up and more survive the trip. In 1993, Carlton and Geller listed 46 known examples of ballast-mediated invasions in the previous 20 years. One species, the zebra mussel (*Dreissena polymorpha*), arrived in North America in the late 1980s in ballast water discharged into the Great Lakes (**Figure 18.5B**). It has had community-changing effects on inland waterways and native bivalves. Another example of a ballast water introduction with negative ecological consequences, which we learned about in the Case Study in Chapter 10 (p. 221), is the release of the comb jelly *Mnemiopsis leidyi* into the Black Sea.

Let's turn our attention next to the role of local conditions, particularly the abiotic and biotic characteristics of communities that help determine their structure.

Abiotic conditions play a strong role in limiting community membership

A species may be able to get to a community, but may fail to become a member of the community because it is physiologically unable to tolerate the abiotic conditions there (see Figure 18.4B). Such physiological constraints can be quite obvious. For example, if we return to our thought experiment of viewing a landscape from the top of a mountain, it is reasonable to assume that the abiotic attributes of the lakes we see make them good places for fishes, plankton, and aquatic insects, but not for terrestrial plants. Similarly, lakes might be good habitat for certain species of fish, plankton, and aquatic insects, but not for all of them. Some of these species depend on fast-flowing water and are thus restricted to streams. These differences among abiotic environments are obvious constraints (or requirements, depending on how you look at it) that largely determine where particular species can and cannot occur within a region. There are many examples throughout this book that demonstrate how physiological constraints can control the distributions and abundances of species [see, for example, the discussions of aspen (p. 83), creosote bush and saguaro cactus (p. 182), and barnacle *Semibalanus balanoides* (p. 183)].

In our earlier discussion of species introductions by ballast water, it was clear that humans transport many more species than can actually survive in the new locations to which they are carried. For example, the majority of organisms released with ballast water find themselves in coastal waters that may not have the temperature, salinity, or light regimes they need to survive or grow. Luckily, many of these individuals die before they can become a threat to the native community. But ecologists know, based on examples such as the *Caulerpa* invasion in the Mediterranean (see the Case Study in Chapter 15 on p. 324), that it is not wise to rely on physiological constraints to exclude potential invaders from a community. It may be that, with multiple introductions, particular individuals with slightly different physiological capabilities can survive and reproduce in an environment once thought uninhabitable by their species. Moreover, there is growing evidence that climate change—and in particular, rising temperatures—may facilitate the invasions of species that would be unable to survive under cooler conditions. For example, Jay Stachowicz and colleagues (2002) found that the recruitment and growth of invasive marine invertebrates known as ascidians (sea squirts) in New England was dependent on warm winter and summer water temperatures. These warmer temperatures gave the non-native organisms an earlier start in spring and increased the magnitude of their growth relative to the native sea squirts, thus allowing them to gain a foothold and compete against the native species. It seems likely that as climate changes, more invasions will be possible (see **Climate Change Connection 18.1** for more information on climate change and species invasions).

As implied by the sea squirt example, species may be hindered by particular interactions with other species that may compete with them for limited resources, or they may even require interactions with other species to help ameliorate unfavorable conditions. We turn next to species interactions, the final factor determining community membership.

Who interacts with whom makes all the difference in community membership

Even if species can disperse to a community and cope with its potentially restrictive abiotic conditions, the final cut to community membership is coexistence with other species (see Figure 18.4C). Clearly, if a species depends on other species for its growth, reproduction, and survival, those other species must be present if it is to gain membership in a community. Equally importantly, some species may be excluded from a community by competition, predation, parasitism, or disease. For example, returning to our thought experiment, we might assume that lakes are suitable habitats for many fish species, but could those species all live together in one lake, given that resources are limiting? A simple view suggests that the best competitors or predators should dominate the lake, thus excluding weaker competitors and resulting in a low-diversity ecosystem. But we know that most communities are full of species that are actively interacting and coexisting. So what allows this coexistence? There are a number of important mechanisms that allow species to coexist, and we will spend the next two sections considering them. But first, let's ask how species might be excluded from communities by biological interactions—a question that is a bit different, but equally relevant.

The best examples of species exclusion from communities by other species again come from the invasive species literature. The failure of some non-native species to become incorporated into communities has been attributed to interactions with native species that exclude or slow the population growth of the non-native species—a phenomenon that ecologists call **biotic resistance**. Multiple studies in a variety of communities have shown that native herbivores have the ability to reduce the spread of non-native plants in substantial ways. Maron and Vila (2001) found that mortality of non-native plants due to native herbivores can be quite high (about 60%), especially at the seedling stage (up to 90% in some studies). But while native herbivores can kill individual non-native plants, it is still unknown how important native species are in completely excluding non-native species from a community. For example, Faithfull (1997) found that in Australia, adults and larvae of the native Lucerne seed web moth (*Etiella behrii*) breed and feed on the seed pods of the invasive gorse shrub (*Ulex europaeus*), but the plant still continues to spread (**Figure 18.6**). This lack of knowledge about biotic resistance may be an artifact of ecologists being more likely to study why a particular non-native species does or does not spread once it becomes a provisional member of the community than to study all the cases in which it is unable to gain a foothold because of interactions with native species. It may also be true that most failed introductions of non-native species go completely undetected.

Studying invasions gives us valuable insights into the question of how species are excluded from communities by other species. But the flip side of that question, which asks how species are included in communities, or better yet, how they coexist, can be more complicated. In the next two sections, we will consider theories that seek to explain how species coexist with one another and form communities with varying species diversity. We will start

FIGURE 18.6 Stopping Gorse Invasion? Herbivory by adults and larvae of the native Lucerne seed web moth (*Etiella behrii*) has slowed, but not stopped, an invasion of the non-native gorse shrub (*Ulex europaeus*; the plants with yellow flowers in the photo) in Australia.

by revisiting the concept of resource partitioning (introduced in our discussion of Concept 11.3), which relies on ecological and evolutionary "compromises" that result in divergence in resource use as a mechanism for coexistence. We will then explore alternative theories and studies that consider the importance of disturbance, stress, predation, and even positive interactions, to the coexistence of species and, ultimately, the species richness of communities. Let's consider resource partitioning first.

CONCEPT 18.2 Resource partitioning among the species in a community reduces competition and increases species diversity.

Resource Partitioning

Differences in how species use limiting resources

In our discussion of competition under Concept 11.3, we used the Lotka–Volterra competition model to show that competing species are more likely to coexist if they use resources in different ways. Let's consider how this concept, known as **resource partitioning**, has been used both in theory and in practice to explain species diversity within communities.

Resource partitioning allows more species to coexist along a resource spectrum

A simple model of resource partitioning envisions each type of resource available in a community as varying along a "resource spectrum" (**Figure 18.7**). The resource spectrum could represent, for example, different nutrients, prey sizes, or habitat types (note that the spectrum represents the variability of an available resource, but not the amount). We can assume that each species' resource use falls somewhere along this spectrum and overlaps with the resource uses of other species to varying degrees (see Figure 18.7A). The more overlap, the more competition between species, with the extreme being complete overlap and competitive exclusion. The less overlap, the more partitioning of resources has occurred, and the less strongly the species will compete with one another.

Using this guiding principle, we can consider some of the possible ways in which resource partitioning might result in higher species richness in some communities than in others. First, species richness could be high in some communities because species show a high degree of partitioning along the resource spectrum (see Figure 18.7B). More species can be "packed" into a community if the overlap in resource use among the species is low, leading to less competition and ultimately higher species richness. This lower overlap could be due to the evolution of specialization or character displacement (see Figure 11.18), which reduces competition over time. Second, species richness could be high in some communities because the resource spectrum is broad (see Figure 18.7C). A broader resource spectrum would make a diversity of resources available to be used by a wider variety of species, resulting in higher species richness.

At this point, let's turn our attention away from models and take a look at some real communities to see how resource partitioning might work in practice.

Specialization is the main mechanism of coexistence

As we learned earlier from the two-species studies of Gause (1934a) on *Paramecium* (p. 249) and Connell (1961b) on barnacles (p. 248), species that compete with each other can coexist by using slightly different resources. Robert MacArthur, whose work on the equilibrium theory of island biogeography we described under Concept 17.3, played a pioneering role in understanding how this principle might be applied to whole communities, where multiple species interactions are occurring all at once.

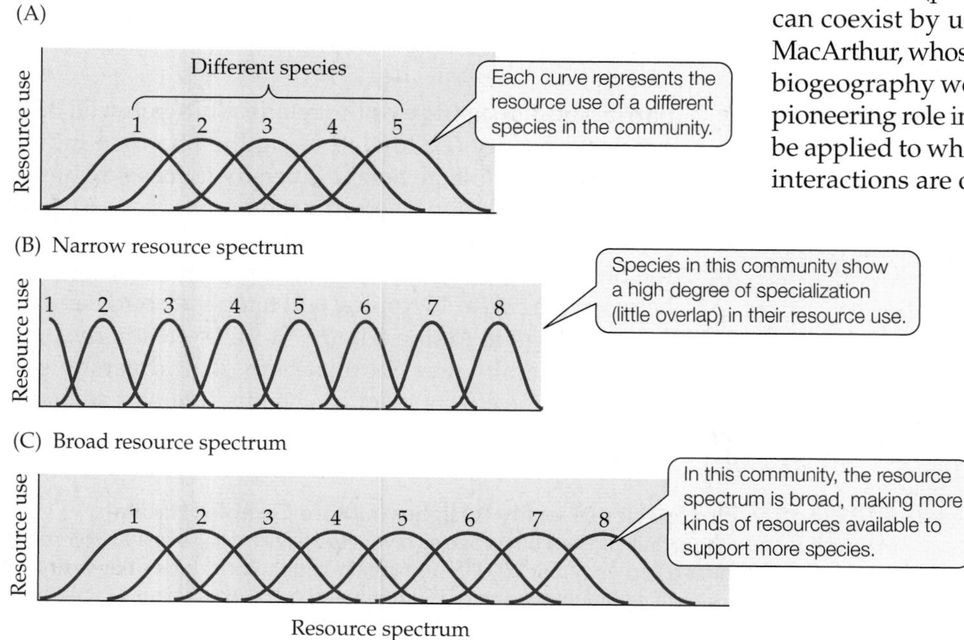

FIGURE 18.7 Resource Partitioning Species coexistence within communities may depend on how the species divide resources. (A) The principle of resource partitioning along a resource spectrum. (B, C) Two characteristics of communities that can result in higher species richness.

Q Which plot shows the most resource partitioning? Which shows the least?

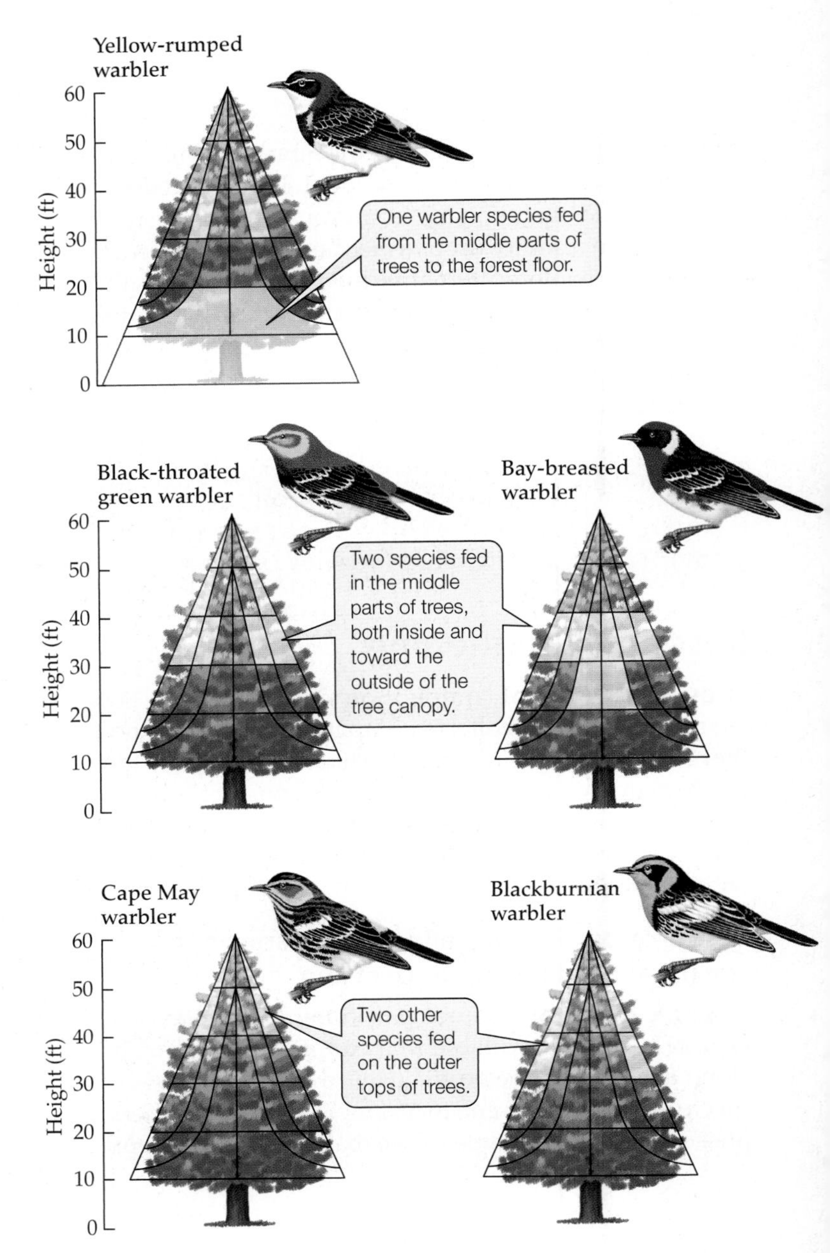

FIGURE 18.8 Resource Partitioning by Warblers Robert MacArthur studied the habitat and food choices of five species of warblers in New England forests. He found that the warblers partition resources by feeding in different parts of the same trees. The shaded areas in each tree diagram represent the parts of trees where each warbler species fed most often. (After MacArthur 1958.)

MacArthur studied warblers, small and brightly colored birds that co-occur in the forests of northern North America. The idyllic New England forests that MacArthur studied are home to an array of warbler species (*Dendroica* spp.) that migrate from the tropics each spring to breed and feed on insects. Through a series of detailed natural history observations in the summers of 1956 and 1957 in Maine and Vermont, MacArthur (1958) recorded the feeding habits, nesting locations, and breeding territories of five species of warblers to find out how they might coexist in the face of very similar resource needs.

MacArthur began mapping the locations of warbler activity in tree canopies and found that the warblers were using different parts of the habitat in different ways (**Figure 18.8**). For example, yellow-rumped (*D. coronata*) warblers fed from the middle parts of trees to the forest floor, while bay-breasted (*D. castanea*) and black-throated green (*D. virens*) warblers fed more in the middle of a tree, both inside and toward the outside of the tree canopy. Blackburnian (*D. fusca*) and Cape May (*D. tigrina*) warblers both fed on the outside tops of trees, often catching their prey in midflight. MacArthur found that the nesting heights of the five warbler species also varied, as did their use of breeding territories. Taken together, these observations supported his hypothesis that the warblers, although using the same habitat and food resources, were able to coexist by partitioning those resources in slightly different ways. His work, which was part of his Ph.D. thesis, earned him the Mercer Award, a prestigious annual award for the best paper in ecology.

MacArthur, along with John MacArthur (MacArthur and MacArthur 1961), extended these ideas about resource partitioning in a study of the relationship between bird species diversity (calculated using the Shannon index; see p. 328) and foliage height diversity (a measure of the number of vegetation layers in a community that serves as an indication of habitat complexity, also calculated using the Shannon index). They found a positive relationship between the two in 13 tropical and temperate bird habitats from Panama to Maine (**Figure 18.9**). Interestingly, bird species diversity was not related to plant diversity per se, beyond the effects of foliage height diversity, suggest-

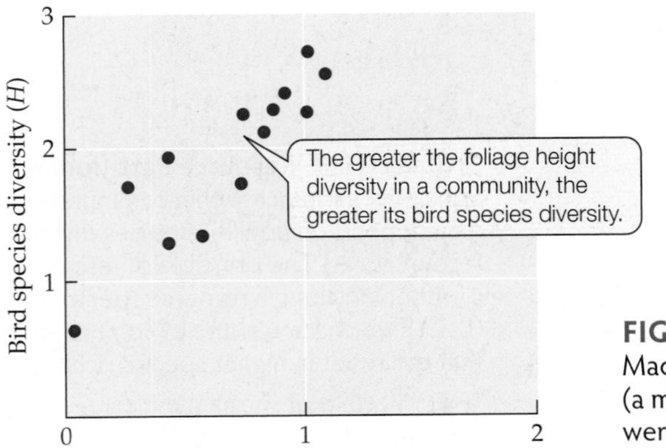

FIGURE 18.9 Bird Species Diversity Is Higher in More Complex Habitats MacArthur and MacArthur plotted bird species diversity against foliage height diversity (a measure of habitat complexity) for 13 different communities. Both kinds of diversity were calculated for each community using the Shannon index (*H*). (Data from MacArthur and MacArthur 1961.)

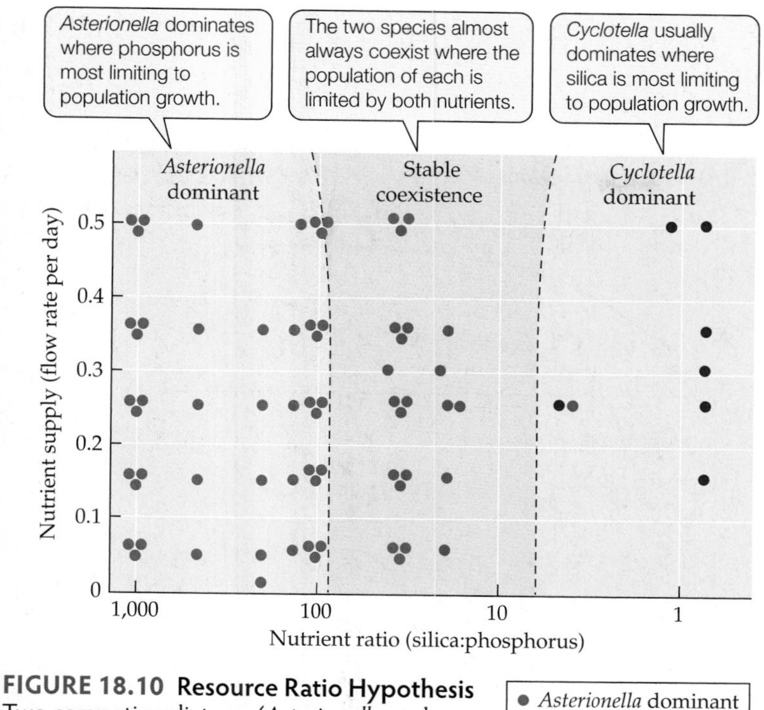

FIGURE 18.10 Resource Ratio Hypothesis Two competing diatoms (*Asterionella* and *Cyclotella*) can coexist by using slightly different ratios of phosphorus and silica. (Data from Tilman 1977.)

ing that tree species identity was less important than the structural complexity of the habitat.

Another important resource partitioning study comes from phytoplankton communities. In our discussion of Concept 11.1, we learned about David Tilman and colleagues' (1981) study of two species of diatoms that competed for silica (which diatoms use to build their cell walls). When the two species were grown together in a laboratory environment with limited supplies of silica, one outcompeted and excluded the other (see Figure 11.4). How, then, do diatom species coexist in nature? Tilman (1977) proposed what has become known as the **resource ratio hypothesis**, which posits that species coexist by using resources in different ratios or proportions. He predicted that diatoms would be able to coexist, despite using the same set of limiting nutrients, by acquiring those nutrients in different ratios (**Figure 18.10**). By growing two diatom species, *Cyclotella* and *Asterionella*, in laboratory environments that differed in their ratios of silica (SiO_2) to phosphorus (PO_4), Tilman found that *Cyclotella* was able to dominate only when the ratio of silica to phosphorus was low (approximately 1:1). When the ratio of silica to phosphorus was high (more like 1,000:1), *Asterionella* outcompeted *Cyclotella*. Only when the ratios of silica and phosphorus were limiting to both species (in the range of 100:1 to 10:1, green area on Figure 18.10) could they coexist with one another. Even though both species needed the same set of nutrients, it was the way in which they partitioned those resources that allowed them to coexist.

Outside of a laboratory setting, this type of partitioning would work best if resources naturally varied within the environment. What is the support for this possibility in the field? In a detailed survey, Robertson and colleagues (1988) mapped resource distribution in an abandoned agricultural field in Michigan that had been colonized by grassland plants. They found considerable variation in soil nitrogen and moisture at spatial scales of a meter or less (**Figure 18.11**). These patches of nitrogen and water

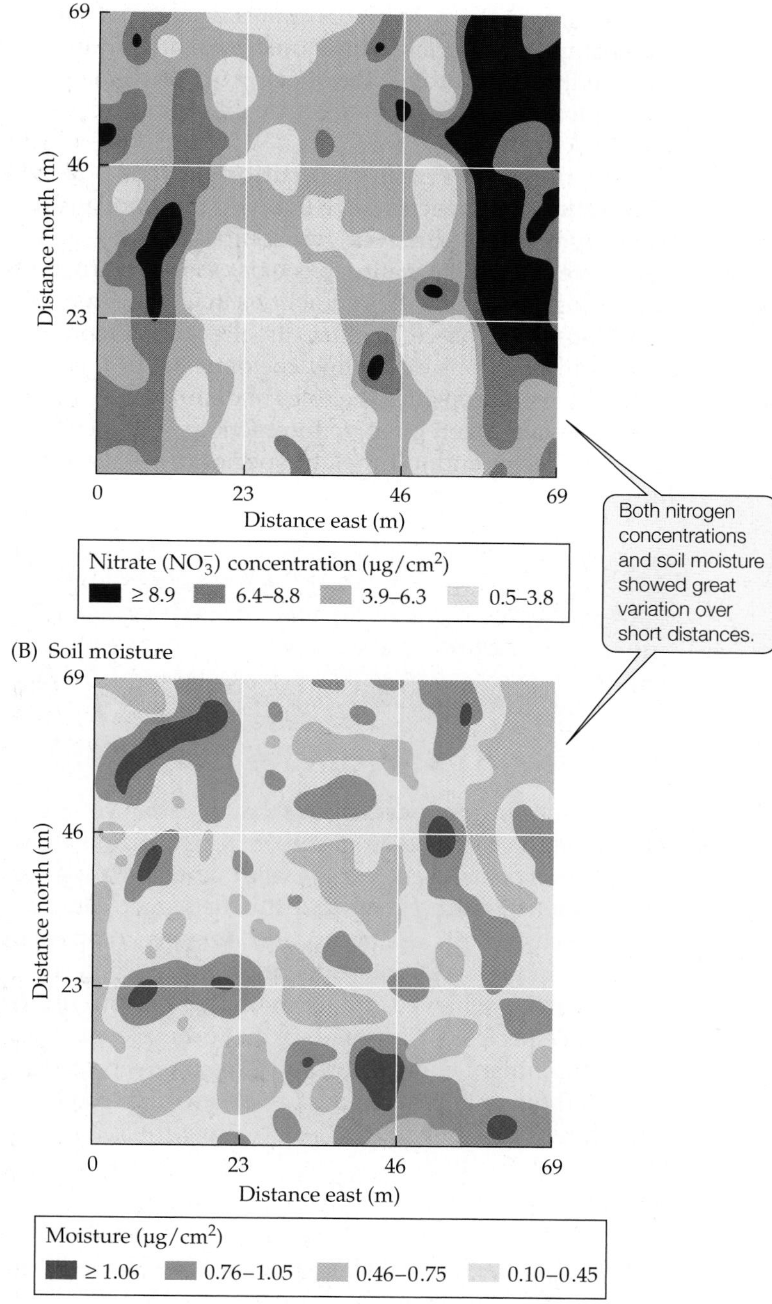

FIGURE 18.11 Resource Distribution Maps Mapping of (A) nitrogen concentrations and (B) soil moisture in an abandoned agricultural field revealed considerable small-scale variation. (From Robertson et al. 1988.)

resources did not necessarily correspond to topographic differences, and they were not correlated with each other. If we were to overlay the nitrogen map on the water map, we would find even smaller patches corresponding to different proportions of these two resources. Whether this spatial distribution of resources affected or reflected the structure of the plant community was not investigated, but it does suggest that resource partitioning could occur in plants (see Figure 21.9). Some of the best evidence of resource partitioning in plants comes from experiments that manipulate species richness and measure productivity, as in the Case Study at the opening of this chapter. We will explore this evidence in more detail in our discussion of Concept 18.4.

The theory of resource partitioning relies on the assumption that species have evolved mechanisms for using resources in different, but complementary, ways, thus increasing their ability to coexist. As we learned in our discussion of species interactions in Unit 3, there are numerous other processes that can alter the outcome of species interactions and allow coexistence. In the next section, we will consider the roles of disturbance, stress, predation, and even positive interactions in the coexistence of species and, ultimately, species diversity at the local scale.

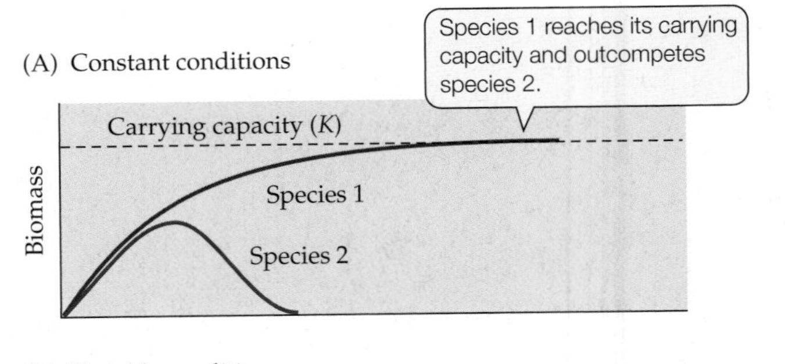

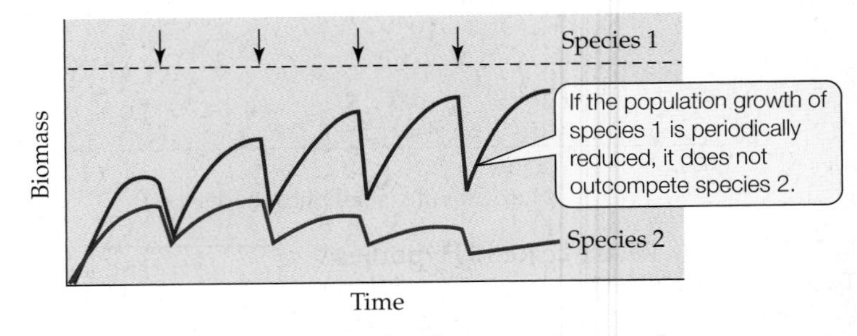

FIGURE 18.12 The Outcome of Competition under Constant and Variable Conditions (A) Under constant conditions, species 1 (the dominant competitor) outcompetes species 2 when it reaches its own carrying capacity (K). (B) If disruptive processes such as disturbance, stress, or predation (represented by the arrows) reduce the population growth of species 1, it will not reach its carrying capacity and will not outcompete species 2, thus allowing coexistence. (After Huston 1979.)

CONCEPT 18.3 Processes such as disturbance, stress, predation, and positive interactions can mediate resource availability, thus promoting species coexistence and species diversity.

Processes That Promote Coexistence

We have seen in previous chapters that disturbance, stress, and predation can modify species interactions and allow for species coexistence. We saw that when two species are competing with each other for the same resource, as in the case of the sea palms and mussels competing for space in the rocky intertidal zone (see p. 255), coexistence can be achieved if the population growth of the dominant species is disrupted. In that example, mussels are the dominant competitors, and sea palms can coexist with them only where the mussels are disturbed frequently enough by wave action to allow the sea palms to acquire space. In this and many other examples in this book, as long as the dominant competitor is unable to reach its own carrying capacity because of reductions in its abundance due to disturbance, stress, or predation, competitive exclusion cannot occur, and coexistence will be maintained (**Figure 18.12**).

We have also explored the effect of positive interactions between species in ameliorating extreme conditions and allowing coexistence. For example, we saw in the cases of salt marsh plants (Figure 16.14) and plants at high elevations (Figure 14.11) that species that might normally be unable to tolerate stressful conditions can maintain viable populations under these conditions because of the facilitative effects of other species.

Let's expand these ideas about modification of species interactions to whole communities and ask how disturbance, stress, predation, and positive interactions influence species diversity.

Processes that free up resources allow species to coexist

There is an old adage among ecologists that goes something like this: "If you think it's a new idea, check out Darwin. He probably discovered it first." In fact, when it comes to theories that explain coexistence, Darwin was the first to formally recognize disturbance as a mechanism for the maintenance of species diversity. In *The Origin of Species* (1859, p. 55), he noted the following results after an impromptu experiment in which he left a meadow on his property undisturbed by mowing: "Out of twenty species growing on a little plot of mown turf (three feet by four) nine species perished, from the other species being allowed to grow freely." Without mowing, the dominant

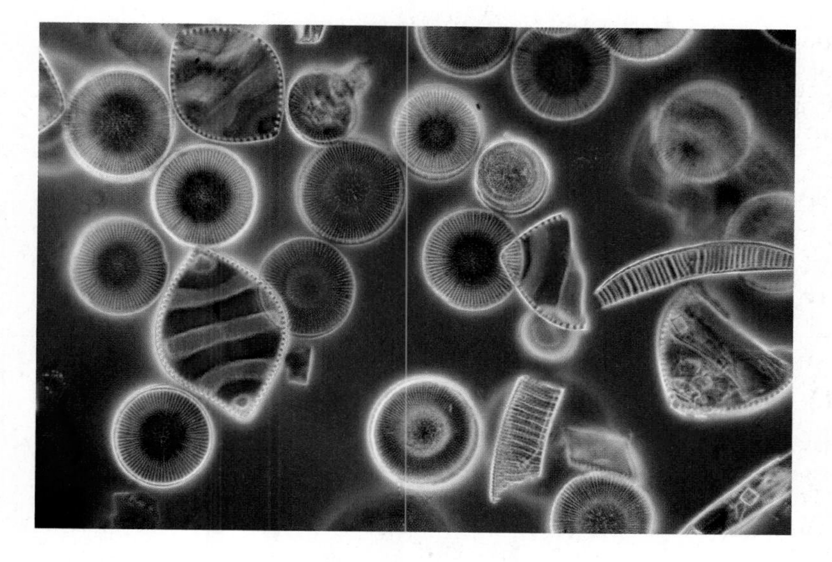

FIGURE 18.13 Paradox of the Plankton Phytoplankton from a freshwater lake. How can so many species coexist using the same set of basic resources? G. E. Hutchinson suggested that the answer is the influence of environmental variation over time.

competitors in the meadow community competitively excluded weedy plants and cut species richness nearly in half. Darwin used this example, along with a multitude of others, to support the argument that nature applies limits to the tendency of species to increase in abundance and outcompete other species. His hypothesis was that species struggle for existence, a necessary first piece to his theory of natural selection.

In 1961, G. E. Hutchinson revived this idea in a paper titled "The Paradox of the Plankton." Hutchinson, an influential community ecologist from Yale University (and major professor to Robert MacArthur), provided one of the first mechanistic descriptions of how coexistence could be maintained under fluctuating environmental conditions. He focused on phytoplankton communities in temperate freshwater lakes (**Figure 18.13**). The simple idea behind Hutchinson's model was the seeming paradox of the presence of 30–40 species of phytoplankton given the relatively limited resources at their disposal. He reasoned that all of the phytoplankton compete for the same array of resources, including carbon dioxide, nitrogen, phosphorus, sulfur, and trace elements, which are likely to be evenly distributed in lakes. How could so many species manage to coexist with so few resources and in such a structurally simple environment as a lake? Hutchinson hypothesized that the conditions in the lake changed seasonally and over longer periods, and that those changes kept any one species from outcompeting the others. As long as conditions in the lake changed before the competitively superior species eliminated others, coexistence would be possible.

Hutchinson's model has two components that interact to control coexistence among species. One is the time required for one species to competitively exclude another species (t_c), which depends on the population growth rates of the two competing species. The second is the time it takes for environmental variation to act on the population growth of the two competing species (t_e). Hutchinson predicted that when competitive exclusion occurs more rapidly than environmental conditions can change ($t_c << t_e$), coexistence cannot be achieved. One could imagine this occurring in communities where there is little environmental change or where the dominant competitor has very rapid growth rates. Conversely, in a fluctuating environment to which the competitors are adapted (where $t_c >> t_e$), environmental variation does not affect the competitive interactions, and competitive exclusion occurs. One could imagine this pattern in environments with frequent, low-intensity environmental fluctuations and long-lived species. Hutchinson argued that it is only when the time it takes for competitive exclusion to occur is roughly equal to the time it takes for environmental variation to interrupt the competitive interaction (when $t_c = t_e$) that competitive exclusion is thwarted and coexistence occurs. Hutchinson argued that this condition is likely to be met often in lake phytoplankton communities; otherwise, very few species, rather than tens of species, would coexist.

Hutchinson proposed the idea that competitive exclusion is rare in nature, but did not test it. It was Robert Paine's work in the rocky intertidal zone of the west coast of North America in the late 1960s (mentioned in our discussion of Concept 12.3) that provided some of the most rigorous and convincing evidence that coexistence could be maintained by disruptive processes such as predation or disturbance. Paine (1966) manipulated population densities of *Pisaster*, a predatory sea star that feeds preferentially on the mussel *Mytilus californianus*. In plots from which *Pisaster* was removed, species richness decreased as mussels outcompeted barnacles and other competitively inferior species. In plots where *Pisaster* was present, species richness was enhanced. There are a number of important aspects to Paine's work, including the keystone species concept and the effects of indirect interactions, but we will consider those aspects in more detail in Chapter 20, when we discuss food webs. For now, let's concentrate on an idea that arose from the work of Darwin, Hutchinson, and Paine: the intermediate disturbance hypothesis.

The intermediate disturbance hypothesis explains species diversity under variable conditions

The **intermediate disturbance hypothesis** was proposed to explain how gradients in disturbance (although we can

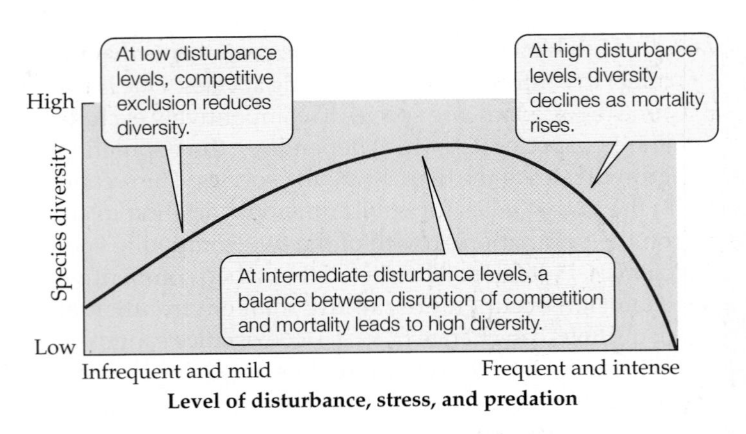

FIGURE 18.14 The Intermediate Disturbance Hypothesis Species diversity is expected to be greatest at intermediate levels of disturbance, stress, or predation. (After Connell 1978.)

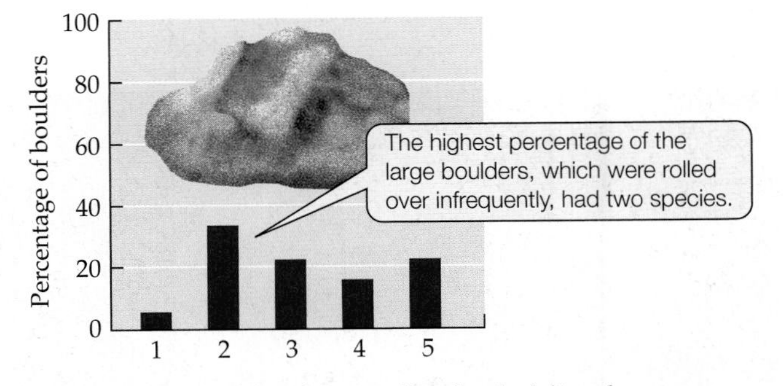

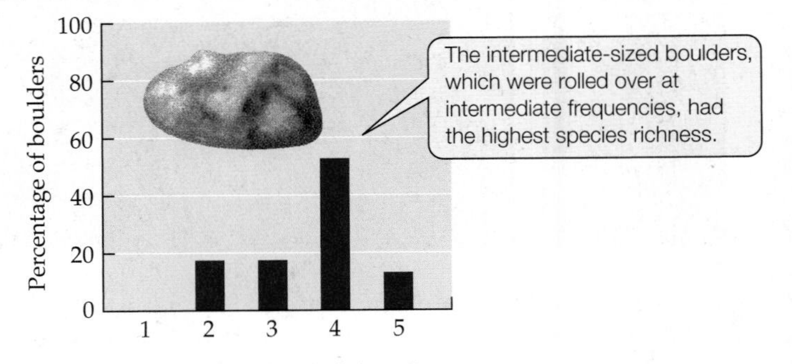

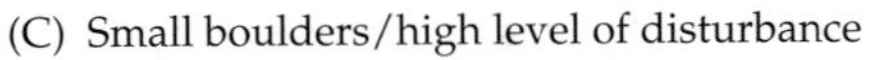

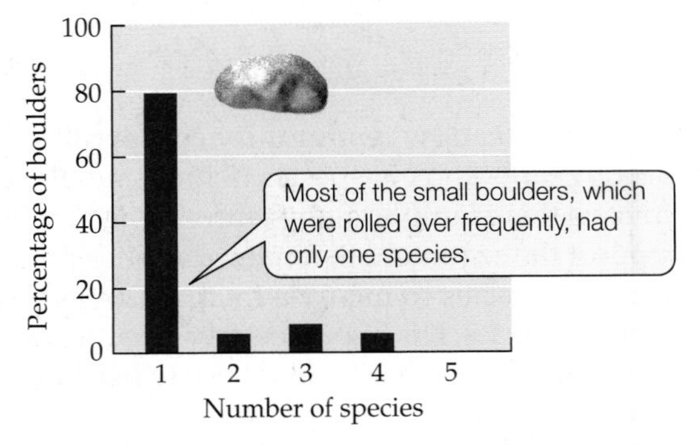

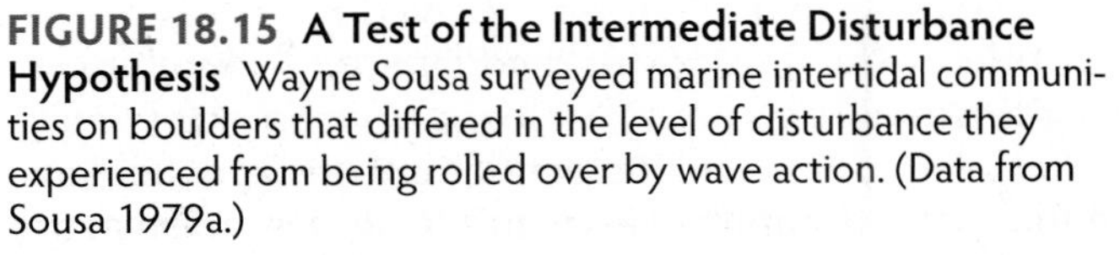
FIGURE 18.15 A Test of the Intermediate Disturbance Hypothesis Wayne Sousa surveyed marine intertidal communities on boulders that differed in the level of disturbance they experienced from being rolled over by wave action. (Data from Sousa 1979a.)

Q Assuming that the maximum number of species that could colonize the boulders is 5, which size of boulder had the lowest species richness, and why?

easily include stress and predation in this model) affect species diversity in communities (**Figure 18.14**). This hypothesis was first formally proposed by Joseph Connell, a contemporary of Paine's and author of the classic work on barnacle competition (described in the discussion of Concept 11.2). Connell (1978) recognized that the level of disturbance (its frequency and intensity; see Figure 16.4) experienced by a particular community could have dramatic effects on its species diversity. He hypothesized that species diversity would be greatest at intermediate levels of disturbance and lowest at high and low levels of disturbance. Why would this be the case? At low levels of disturbance, competition would regulate species diversity because dominant species would be free to exclude competitively inferior species. At high levels of disturbance, on the other hand, species diversity would decline because many individuals would die and some species would become locally extinct as a result. At intermediate levels of disturbance, species diversity would be maximized simply due to the balance between disruption of competition and mortality due to disturbance.

The intermediate disturbance hypothesis is highly amenable to testing. One such test was carried out by Wayne Sousa (1979a), who studied succession in intertidal boulder fields in southern California (see Figure 16.15). In a different but related study, Sousa measured the rate of disturbance of communities growing on the boulders and documented their species richness (**Figure 18.15**). Small boulders were rolled over frequently by waves, and thus constituted highly disturbed environments for the marine algae and invertebrate species that lived on them. The opposite was true for large boulders, which rarely experienced wave forces large enough to dislodge them. Intermediate-sized boulders, of course, were rolled over at intermediate frequencies. After 2 years, Sousa found that most of the small boulders had only one species (early successional species: the macroalga *Ulva* or the barnacle *Chthamalus*), while the greatest percentage of the large boulders had two species (late successional species: the macroalga *Gigartina canaliculata* and others). The greatest percentage of the intermediate-sized boulders had four species, but some had up to seven species (a mixture of early, mid-, and late successional species). Sousa's study is just one of many that have demonstrated the highest diversity at intermediate disturbance levels.

There have been several elaborations on the intermediate disturbance hypothesis

The intermediate disturbance hypothesis is a simple model that relies on variation in disturbance levels to explain species diversity in communities. A handful of ecologists have used it as a foundation for adding more complexity and realism to their theories. One of the first to elaborate on the model was Michael Huston (1979), who acknowledged the effect of disturbance on competition, but reasoned that a second process, competitive displacement, could be an important mediating factor. *Competitive displacement* can be thought of as the growth rate of the strongest competitors in a community and is dependent on the productivity of the environment. Huston's **dynamic equilibrium model** considers how the frequency or intensity of disturbance and the rate of competitive displacement combine to determine species diversity (**Figure 18.16**). Like Hutchinson's model, Huston's model predicts maximum species diversity when the level of disturbance and the rate of competitive displacement are roughly equivalent (thus the term "equilibrium" in the model name). Species diversity will be highest when the frequency or intensity of disturbance and the rate of competitive displacement are both at low to intermediate levels (Figure 18.16, point AC). Moreover, species diversity will be lowest either when disturbance is high and competitive displacement is low (point AD), or competitive displacement is high and disturbance is low (point BC). When both processes are high and roughly similar (point BD), we expect species diversity to be relatively low because both high mortality and competitive displacement will be acting to reduce species diversity. Perhaps because of its added complexity, there have been few observational or experimental studies of the dynamic equilibrium model. One example, which comes from an observational study of riparian wetlands in Alaska by Pollock et al. (1998), can be found in **Web Extension 18.1**.

Another elaboration of the intermediate disturbance hypothesis comes from Hacker and Gaines (1997), who incorporated positive interactions into their model. If we think back to Chapters 14, 15, and 16, we learned that species interactions are highly context-dependent, varying in direction and strength depending on certain physical and biological factors. Theory and experiments both suggest that positive interactions should be more common under relatively high levels of disturbance, stress, or predation—all circumstances in which associations among species could increase their growth and survival. Hacker and Gaines reasoned that positive interactions might be particularly important in promoting species diversity at intermediate to high levels of disturbance (or stress or predation) for two reasons (**Figure 18.17**). First, at high levels of disturbance, positive interactions should increase the survival of individuals of the interacting species through both the amelioration of harsh conditions and associational defenses. Second, even at intermediate levels of disturbance, species involved in positive interactions should be released from competition, an effect that should further increase species diversity.

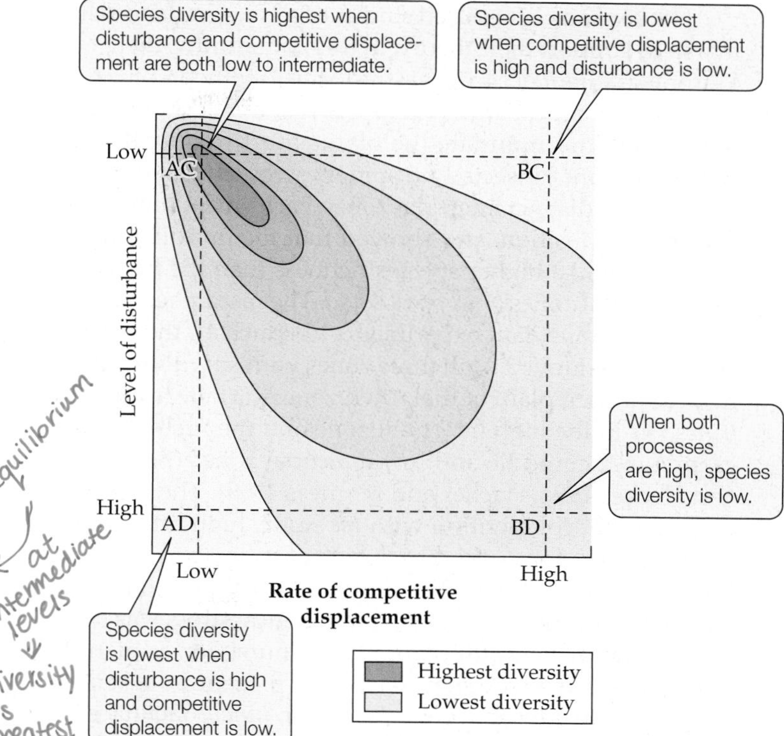

FIGURE 18.16 The Dynamic Equilibrium Model The dynamic equilibrium model predicts that species diversity will be highest when the frequency and intensity of disturbance and the rate of competitive displacement are low to intermediate. (After Huston 1979.)

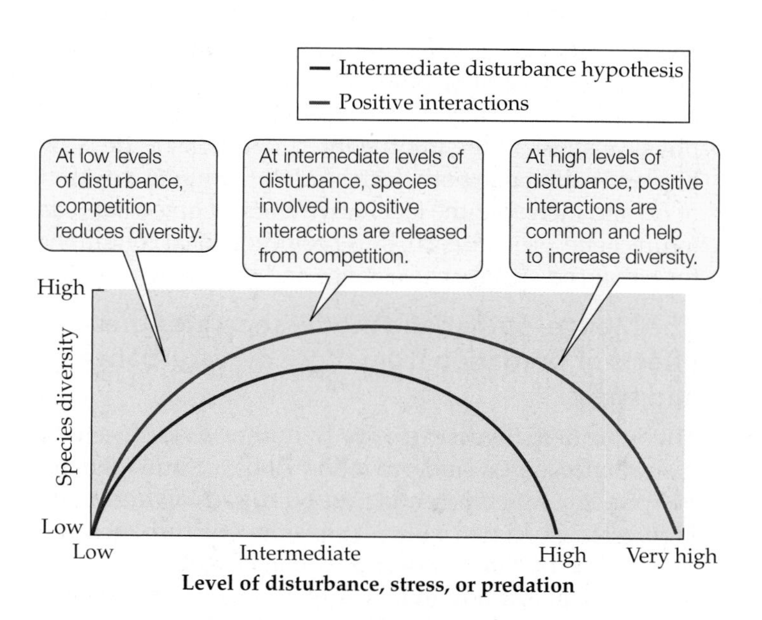

FIGURE 18.17 Positive Interactions and Species Diversity The intermediate disturbance hypothesis has been elaborated to include positive interactions. (After Hacker and Gaines 1997.)

Hacker and Gaines used studies of a New England salt marsh to support their theory. In this community, there is a strong gradient of physical stress due to saltwater inundation. The highest stress occurs closest to the shoreline, where the tides inundate the plants most frequently. A survey of plants, insects, and spiders across the marsh revealed three distinct intertidal zones, each with a different species composition, and showed that the middle intertidal zone had a higher species richness than the high or low intertidal zones (**Figure 18.18A**). The researchers then conducted transplant experiments in which all the plant species were moved to all three zones with or without the most abundant plant of their own zone: the tall shrub *Iva frutescens* in the high intertidal zone and the rush *Juncus gerardii* in the middle and low intertidal zones (Bertness and Hacker 1994; Hacker and Bertness 1999). The results revealed that competition with *Iva* in the high intertidal zone led to the competitive exclusion of most plant species transplanted there, whether or not *Juncus* was also present. In the low intertidal zone, physiological stress was the main factor in controlling population numbers, as many individuals died whether *Juncus* was present or absent. In the middle intertidal zone, however, *Juncus* facilitated other plant species. Without *Juncus*, mortality was 100% for most species by the end of the summer. The mechanism of facilitation, described under Concept 15.3, was amelioration of both hypoxia and salt stress by *Juncus* (see p. 334). Additionally, as we saw in that discussion, *Juncus* indirectly facilitates an aphid herbivore that depends on *Iva* for survival. It turns out that such indirect interactions affect a number of insect herbivores that feed on a variety of other plants facilitated by *Juncus* in the marsh. Hacker and Gaines (1997) concluded, based on these studies, that positive interactions are critically important in maintaining species diversity, especially at intermediate levels of physical stress (**Figure 18.18B**). They recognized that physical stress in the middle intertidal zone of the New England salt marsh both decreased the competitive effect of *Iva* and increased the facilitative effect of *Juncus* (and its indirect effects on insects), thus providing ideal conditions for enhanced species coexistence and diversity.

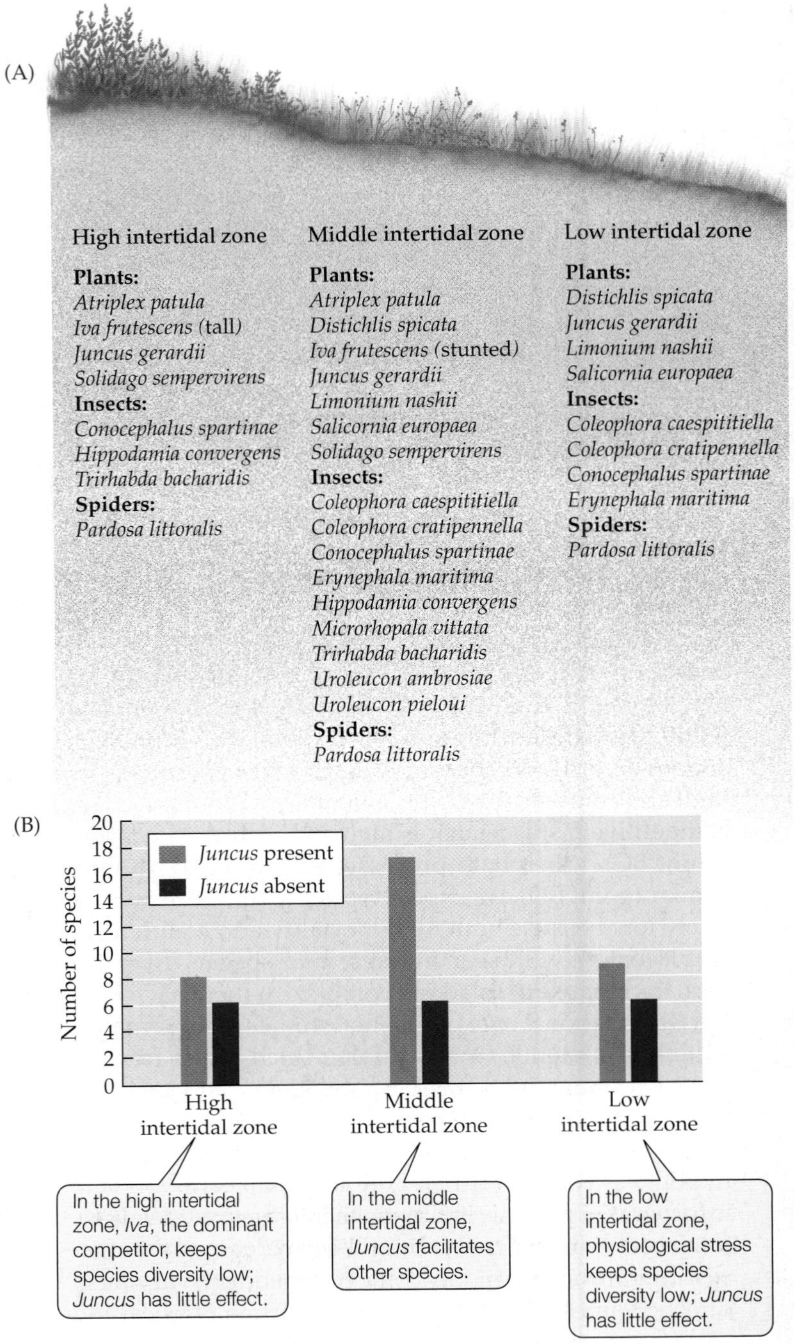

(A)

High intertidal zone
Plants: *Atriplex patula*, *Iva frutescens* (tall), *Juncus gerardii*, *Solidago sempervirens*
Insects: *Conocephalus spartinae*, *Hippodamia convergens*, *Trirhabda bacharidis*
Spiders: *Pardosa littoralis*

Middle intertidal zone
Plants: *Atriplex patula*, *Distichlis spicata*, *Iva frutescens* (stunted), *Juncus gerardii*, *Limonium nashii*, *Salicornia europaea*, *Solidago sempervirens*
Insects: *Coleophora caespititiella*, *Coleophora cratipennella*, *Conocephalus spartinae*, *Erynephala maritima*, *Hippodamia convergens*, *Microrhopala vittata*, *Trirhabda bacharidis*, *Uroleucon ambrosiae*, *Uroleucon pieloui*
Spiders: *Pardosa littoralis*

Low intertidal zone
Plants: *Distichlis spicata*, *Juncus gerardii*, *Limonium nashii*, *Salicornia europaea*
Insects: *Coleophora caespititiella*, *Coleophora cratipennella*, *Conocephalus spartinae*, *Erynephala maritima*
Spiders: *Pardosa littoralis*

(B)

Number of species (0–20); *Juncus* present / *Juncus* absent
High intertidal zone — Middle intertidal zone — Low intertidal zone

In the high intertidal zone, *Iva*, the dominant competitor, keeps species diversity low; *Juncus* has little effect.

In the middle intertidal zone, *Juncus* facilitates other species.

In the low intertidal zone, physiological stress keeps species diversity low; *Juncus* has little effect.

FIGURE 18.18 Positive Interactions: Key to Diversity in Salt Marsh Communities? (A) Surveys of plant and insect species diversity in a New England salt marsh show diversity to be greatest in the middle intertidal zone. (B) Experiments suggest that this high diversity is controlled by the direct and indirect effects of the facilitating rush species *Juncus gerardii* as well as by a decrease in the effect of the dominant competitor, *Iva frutescens*, due to physical stress. (After Hacker and Gaines 1997.)

The Menge–Sutherland model separates the effects of predation from those of disturbance and stress

The intermediate disturbance hypothesis assumes that disturbance, stress, and predation all have similar effects on species competition, and thus on species diversity (see Figure 18.17). In particular, it considers disturbance and predation to be similar processes—that is, processes that act to kill or damage dominant competitors and thereby create opportunities for subordinate species. This equating of disturbance and predation ignores an important difference between them: disturbance is a physical process, whereas predation is a biological one. Menge and Sutherland (1987) have argued that because predation is a biological interaction, it is independently affected by physical disturbance and stress and thus should be considered separately (**Figure 18.19**). The Menge–Sutherland model predicts that predation should be relatively important in maintaining species

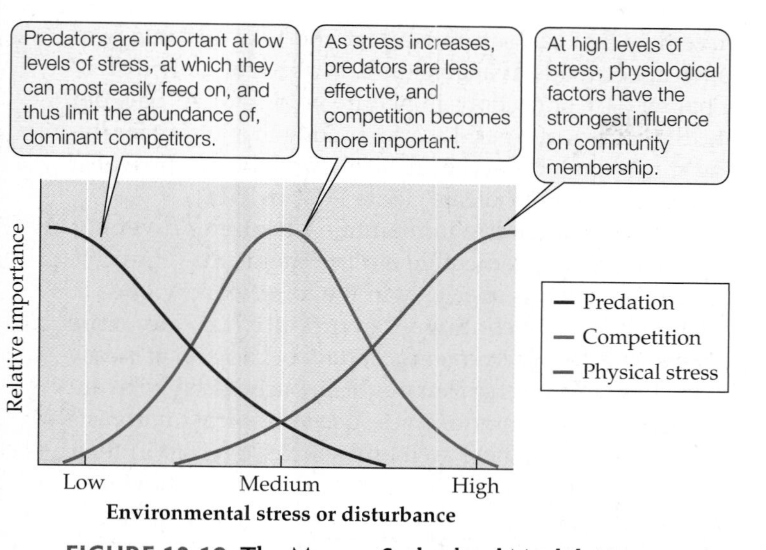

FIGURE 18.19 The Menge–Sutherland Model Menge and Sutherland's model of influences on community diversity is similar to the intermediate disturbance hypothesis (see Figure 18.14), but it accounts for the effect of predation separately from that of physical stress or disturbance. Menge and Sutherland also considered the effect of recruitment, which reduces the importance of competition when it is low. (After Menge and Sutherland 1987.)

richness at low levels of environmental stress (or disturbance), at which predators can most easily feed on, and thus limit the abundance of, competitively dominant species. As environmental stress increases, the effect of predation decreases as predators become less able to inflict damage at lower trophic levels. These lower trophic levels, which are predicted by the model to be more tolerant of physical stress, are released to compete for resources, causing the influence of competition on species diversity to increase. Finally, as environmental stress increases to high levels, both predation and competition become unimportant as more and more species are excluded from the community by their physiological limitations. As with the intermediate disturbance hypothesis, the influence of positive interactions, especially important at either extreme of predation or physical stress, have since been incorporated into the Menge–Sutherland model (Bruno et al. 2003), leading to conclusions similar to those of Hacker and Gaines (1997) (see Figure 18.17).

Another important factor that Menge and Sutherland considered in their model was the influence of *recruitment*: the influx of young individuals into a population. They predicted that if recruitment was low, competition might not be particularly important in determining species diversity because resources would not be limiting. Instead, the interplay between predation under benign environmental conditions and physical stress under extreme conditions would be the most influential factors regulating community membership. If recruitment increased, however, the role of competition would also increase, ultimately resulting in predictions similar to those in Figure 18.19. Thus, Menge and Sutherland suggest that recruitment can be another important influence on community diversity, as mentioned in the discussion of Concept 18.1.

The intermediate disturbance hypothesis and the Menge–Sutherland model assume that there is an underlying competitive hierarchy among species—that is, that some species are much stronger competitors than others and thus dominate communities if they are not kept in check by disruptive processes. What happens if we assume that there is no competitive hierarchy among species? If species have equivalent interaction strengths, then the ability of any one species to live in a community will depend more on chance than on "conflict resolution." Let's spend a moment discussing this unconventional theory of species diversity.

The lottery model relies on equality and chance

A final group of models that help explain coexistence are so-called **lottery models** (Sale 1977; Chesson and Warner 1981). As the name suggests, these models emphasize the role of chance in the maintenance of species diversity. Lottery models assume that resources in a community made available by the effects of disturbance, stress, or predation are captured at random by recruits from a larger pool of potential colonists. For this mechanism to work, species must have fairly similar interaction strengths and population growth rates, and they must have the ability to respond quickly, by dispersing, to disturbances that free up resources. If there is a large disparity in competitive abilities among species, the dominant competitor will have a greater chance of obtaining resources and eventually monopolizing them. In the lottery model, it is the equal chance of all species to obtain resources that allows species coexistence.

Lottery models have most often been applied to highly diverse communities. Peter Sale (1977, 1979) conducted one of the earliest and best-known tests of the lottery model on fishes of the Great Barrier Reef of Australia. Fish diversity on this reef ranges from 1,500 species in the north to 900 species in the south. On any one small patch reef (3 m; 10 ft in diameter), up to 75 species might be recorded. In the reef ecosystem, there is strong habitat fidelity and severe space limitation, such that many individual fish spend their entire lives in roughly the same spot on the reef. Given these conditions, Sale asked the obvious question: How could so many species coexist in such a small space for so long?

Sale reasoned that only a portion of the coexistence among these fishes could be explained by resource partitioning because the species tended to have very similar diets. He noted that vacant sites or territories were highly desirable and were made available rather unpredictably by the deaths of individual occupants (due, for example, to predation, disturbance, starvation, or disease). To look at this system in more detail, Sale observed losses of occupants and recruitment to newly vacated sites among three species of territorial pomacentrid fishes (*Eupomacentrus apicalis*, *Plectroglyphidodon lacrymatus*, and *Pomacentrus wardi*). He found the pattern of occupation to be random (**Figure 18.20**): the species that had previously occupied a site had no bearing on which species was recruited to that site when it became vacant. One species, *P. wardi*, both lost and occupied sites at a greater rate than the other two species, but this had no effect on its overall ability to coexist with the other two species. Sale noted that one important component of this lottery system is that fishes produce many, highly mobile juveniles that can saturate a reef and take advantage of open space made available (as described for clownfish in Chapter 7's Connections in Nature on p. 156). As Sale put it, "the species of a guild are competing in a lottery for living space in which larvae are tickets and the first arrival at a vacant site wins that site" (Sale 1977, p. 351).

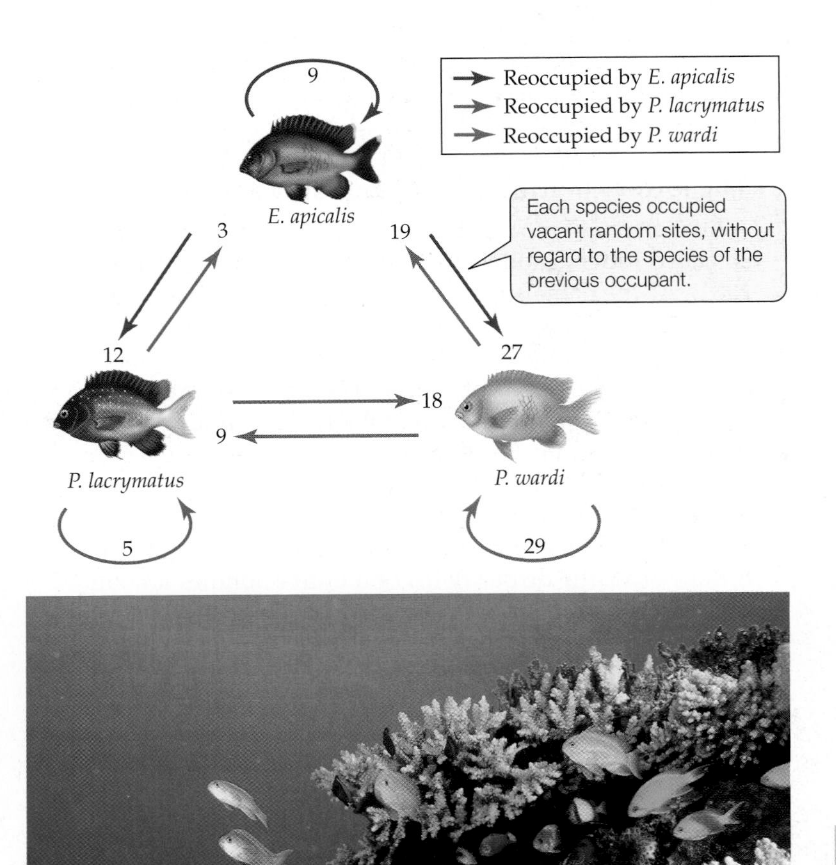

FIGURE 18.20 A Test of the Lottery Model Peter Sale tested the lottery model using coral reef fishes living on the Great Barrier Reef of Australia. By counting the individuals of three fish species (*Eupomacentrus apicalis*, *Plectroglyphidodon lacrymatus*, and *Pomacentrus wardi*) that occupied vacated sites, he found that the species of the new occupant was random and unrelated to the species that had previously occupied the site. (Data from Sale 1979.)

The role of chance in maintaining species diversity, especially in unpredictable environments, has intuitive appeal. As long as species win the lottery every once in a while, they will continue to reproduce (i.e., buy more tickets) and be able to enter the lottery once again. It is easy to see how this mechanism might be particularly relevant in highly diverse communities such as tropical rainforests and grasslands, where so many species overlap in their resource requirements. Its relevance decreases, however, in communities where species have large disparities in interaction strength. In those communities, it appears that the "great equalizers" are processes that decrease competitive exclusion, such as disturbance, stress, or predation, or increase inclusion, such as positive interactions.

Ecologists are a long way from agreeing on any one theory to explain why certain species end up coexisting in space and time. Instead, they continue to strive for generalities while recognizing that the relative importance of different mechanisms of species diversity may depend on the characteristics of the community in question.

Up to this point in the chapter, we have focused on the causes of species diversity at the community level. We have asked, "Why and how does species diversity differ among communities?" In the next section, we will shift gears and instead ask what might be considered the flip side of that question. We want to know, given the variation in species diversity among communities (and the current losses of species diversity due to human activities), whether species diversity matters. In other words, what do species do in communities? Does species diversity have functional significance?

CONCEPT 18.4 Many experiments show that species diversity is positively related to community function.

The Consequences of Diversity

In the Case Study at the opening of this chapter, we described a case in which prairie communities with higher species richness produced more biomass than those with lower species richness. These results support the notion that species diversity can control certain ecological functions of a community. Those functions are numerous, but include plant productivity, soil fertility, water quality and availability, atmospheric gas exchange, and even **resistance** to disturbance (and the speed of recovery afterward, also known as *resilience*). Many of these functions of communities provide valuable services to humans, such as food and fuel production, water purification, O_2 and CO_2 exchange, and protection from catastrophic events such as floods or tsunamis. The Millennium Ecosystem Assessment (2005), a synthesis of studies produced under the auspices of the United Nations, details the importance of these ecosystem services to humans. The assessment predicts that if the current losses of species diversity continue, the world's human populations will be severely affected by the loss of the services those species, and the communities in which they live, provide. What evidence underlies these dire predictions? Recent research has attempted to look at the connections between species diversity and community function, not only to seek basic insights into community ecology, but also because of concerns over these species losses.

Relationships between species diversity and community function are positive

David Tilman and colleagues performed the first experimental test of a long-standing theory, proposed by both Robert MacArthur (1955) and Charles Elton (1958), that species richness is positively related to community stability. **Stability**, in this case, is defined as the tendency of a community to remain the same in structure and function.

The diversity–stability theory remained "conventional wisdom" until the mid-1970s, when it was tested mathematically using food web models that varied in species richness and complexity. We will consider these models in more detail in Chapter 20. But when Tilman and Downing (1994) noticed that some of their experimental plots at Cedar Creek (see Figure 18.2) seemed to be responding to a drought differently from others, they were reminded of these models and of MacArthur and Elton's conventional wisdom. A survey of their plots showed that plots with higher species richness were better able to withstand drought than plots with lower species richness (but the same density of plants) (**Figure 18.21A**). Total plant biomass decreased less due to the drought in species-rich plots than in species-poor ones, resulting in a positive, curvilinear relationship between species richness and drought resistance (measured as the difference between biomass before and after the drought). Tilman and Downing reasoned that a curvilinear relationship would be expected if additional species beyond some threshold (the point at which the curve levels off; roughly 10–12 species in this study) had little additional effect on drought resistance. These species could be considered redundant in the sense that they had essentially the same effects on drought resistance as other species. Tilman and Downing suggested, however, that once the number of species in a plot declined below that threshold, each additional species lost from the plot would result in a progressively greater negative effect of drought on the community.

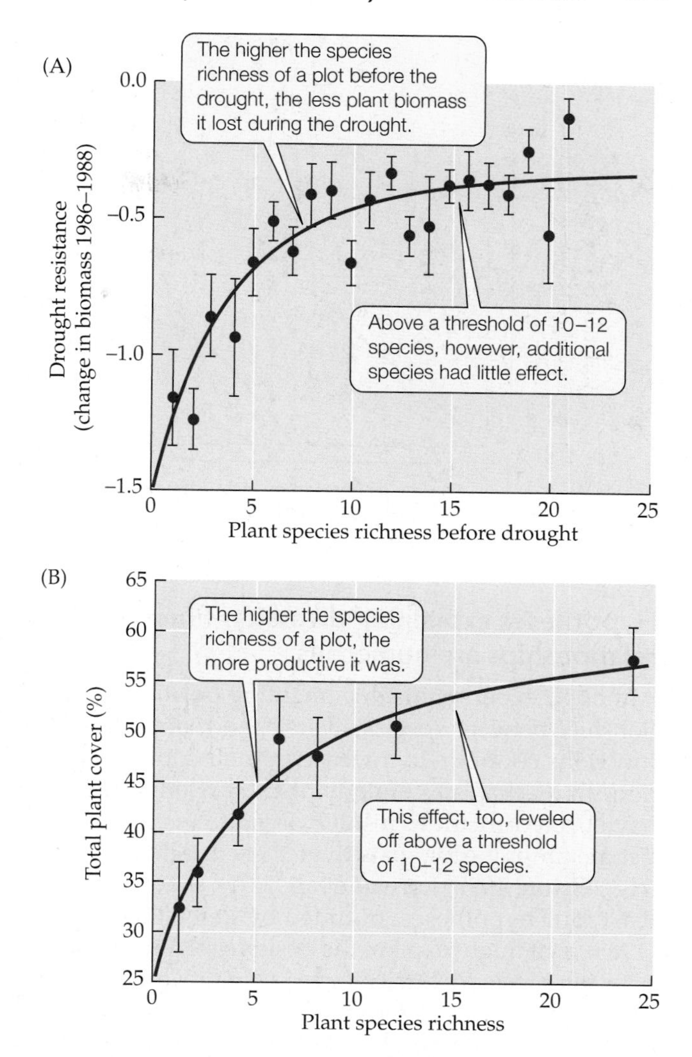

FIGURE 18.21 Species Diversity and Community Function Tilman and colleagues used their prairie plots at the Cedar Creek site in Minnesota (see Figure 18.2) to test the effects of species richness on community function. (A) First, they measured the effects of a drought on plant biomass in plots that varied in species richness. (B) They then created plots that varied in species richness, but all had the same density of individual plants, and measured biomass in those plots after 2 years of growth. Error bars show ± one SE of the mean. (A after Tilman and Downing 1994; B after Tilman et al. 1996.)

To test this idea more rigorously, Tilman et al. (1996) conducted a well-replicated experiment in which species diversity was directly manipulated in the same prairie ecosystem. A series of plots that differed in plant species richness, but not in the number of individual plants, was created by randomly selecting sets of species from a pool of 24 species. Each plot was provided with the same amounts of water and nutrients, and biomass in the plots was measured after 2 years of growth. The results confirmed the curvilinear effect of species richness on biomass (**Figure 18.21B**) and additionally showed that nitrogen was more efficiently used as species richness increased.

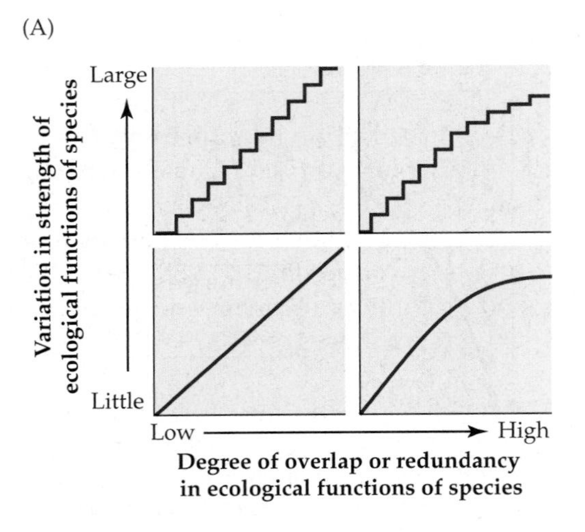

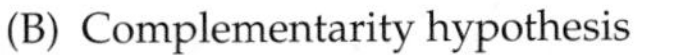

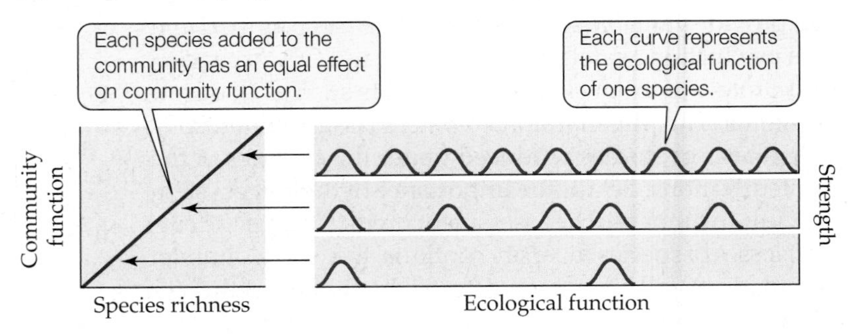

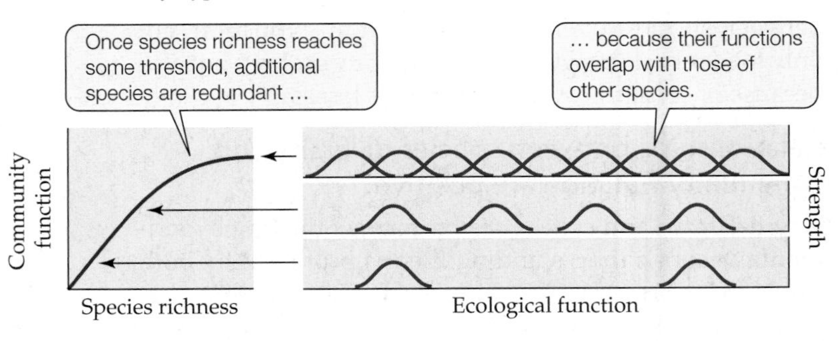

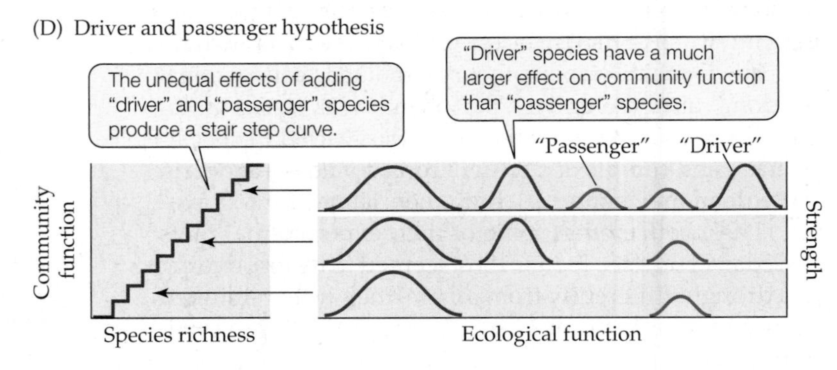

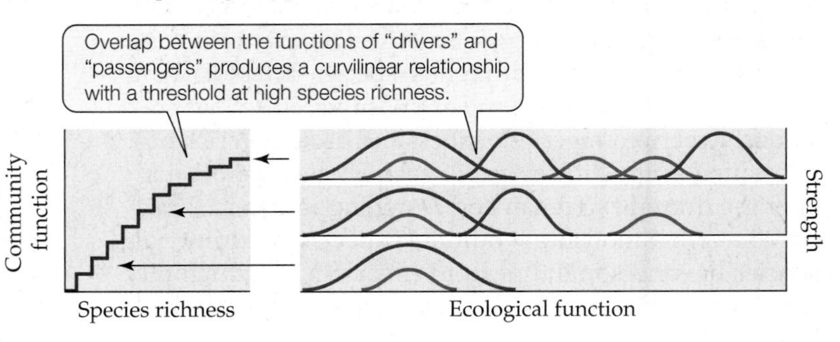

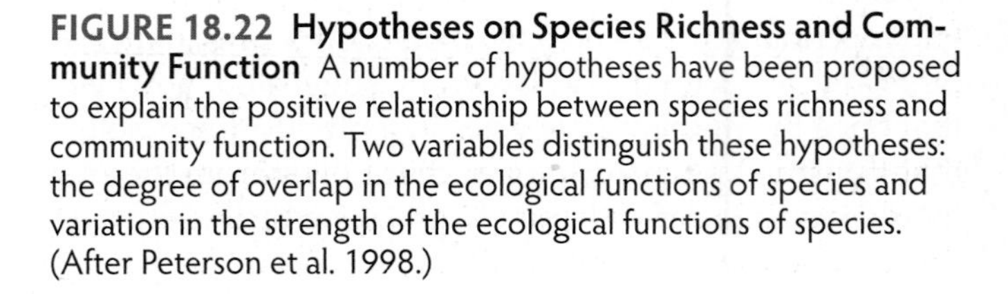

FIGURE 18.22 Hypotheses on Species Richness and Community Function A number of hypotheses have been proposed to explain the positive relationship between species richness and community function. Two variables distinguish these hypotheses: the degree of overlap in the ecological functions of species and variation in the strength of the ecological functions of species. (After Peterson et al. 1998.)

Hypotheses explaining diversity–function relationships are numerous

Although experiments documenting positive relationships between species diversity and community functions continue to increase in their sophistication, the mechanisms underlying these relationships have been difficult to resolve. Despite the intense debate among ecologists over these mechanisms (e.g., Grime 1997; Loreau et al. 2001), there are at least four hypotheses, outlined by Peterson et al. (1998), that might explain the positive relationships between species diversity and community function. These hypotheses differ in their assumptions concerning two variables: (1) the degree of overlap, or redundancy, in the ecological functions of species, and (2) variation in the strength in the ecological functions of species (**Figure 18.22A**). These different assumptions result in diversity–function curves with differing shapes.

The first hypothesis, known as the **complementarity hypothesis**, proposes that as species richness increases, there will be a linear increase in community function (**Figure 18.22B**). In this case, each species added to the community will have a unique and equally incremental effect on community function. We might expect this type of pattern if we assume that species are equally partitioning their functions within a community. For example, as more and more species are added to the community, each of their unique individual functions will contribute to an ever-increasing community function value.

The second model, known as the **redundancy hypothesis**, relies on assumptions similar to those of the complementarity hypothesis, but places an upper limit on the effect of species richness on community function (**Figure 18.22C**). This model best fits the results of Tilman and colleagues described above (see Figure 18.21), in which the functional contribution of additional species reaches a threshold. This threshold is reached because as more species are added to the community, there is overlap in their function—essentially, there is redundancy among species. In this model, species can be thought of as belonging to certain functional groups (see Figure 15.4C). As long as all the important functional groups are represented, the actual species composition of the community is of little importance to its overall function.

The third model, known as the **driver and passenger hypothesis**, proposes that the strengths of the effects of species' ecological functions vary dramatically among species (**Figure 18.22D**). "Driver" species have a large effect on community function, while "passenger" species have a minimal effect. The addition of driver and passenger species to a community will therefore have unequal effects on community function, producing a "stair step" curve, as shown in Figure 18.22D. The shape of the curve—the rise and run of the stair steps—will depend on the numbers of driver and passenger species added and on how much their effects differ. If communities are assembled in such a way that there are only a few driver species (e.g., keystone or dominant species; see Figure 15.17) but many passenger species, then one would expect community function values to vary dramatically with species richness; that is, the stair steps will vary in both rise and run, depending on whether driver species are present or not. As species richness increases, however, the chance that the driver species will be present becomes very high, and the variation in community function values—in the rise and run of the stair steps—should decrease.

The fourth model is a variation on the driver and passenger hypothesis. It assumes that there could be overlap in the function of drivers and passengers, and thus predicts a curvilinear relationship, with a threshold at high species richness (**Figure 18.22E**).

Although these models provide a theoretical foundation to understand how species contribute to community function, testing them is logically challenging because of the number of species involved and the variety of community functions that could be considered. In many ways, these experiments and models are at the heart of modern community ecology, not only because they tell us something about how communities work, but also because they may be able to tell us what the future holds for communities that are both losing (by extinction) and gaining (by invasions) species through human influence.

A CASE STUDY REVISITED

Powered by Prairies? Biodiversity and Biofuels

The potential value of understanding how species diversity controls community function is limitless when we consider the services communities provide to humans. As we have seen, these services are numerous and diverse. One potential ecosystem service that has been overlooked until recently is biofuel production. As we saw in the Case Study at the opening of this chapter, Tilman and colleagues' prairie studies showed that biomass production is greater in high-diversity plots than in single-species plots (*monocultures*). In a more recent study (2006), they established that their high-diversity plots produced nearly 238% more biomass per input of energy than did single-species plots. They used these experimental results to compare the amounts of biofuels that could be produced from traditional crops and from high-diversity crops of prairie plants. In particular, they compared three biomass types—soybeans, corn, and low-input, high-diversity (LIHD) biomass from their prairie plots—and three types of biofuel products—biodiesel, ethanol, and synfuel (a type of synthetic gasoline)—produced from each of those crops (**Figure 18.23**). When they compared the biofuels made from LIHD prairie biomass with those made from traditional crops, they found that synfuel production from prairie biomass netted the most energy (measured as "net energy balance," or NEB, which is the amount of biofuel produced minus the amount of fossil fuel used to produce it). A comparison of NEB among the biomass types showed that corn ethanol and soybean biodiesel netted 34% and 48%, respectively, less energy than LIHD synfuel, and equivalent or even lower than LIHD ethanol (see Figure 18.23).

The higher energy yields of LIHD biomass resulted from two main differences between it and the two traditional biofuel crops. First, energy inputs were lower for the prairie plants because they are perennials (so they don't have to be replanted each year) and they are grasses that require little water, fertilizer, or pesticides—all of which

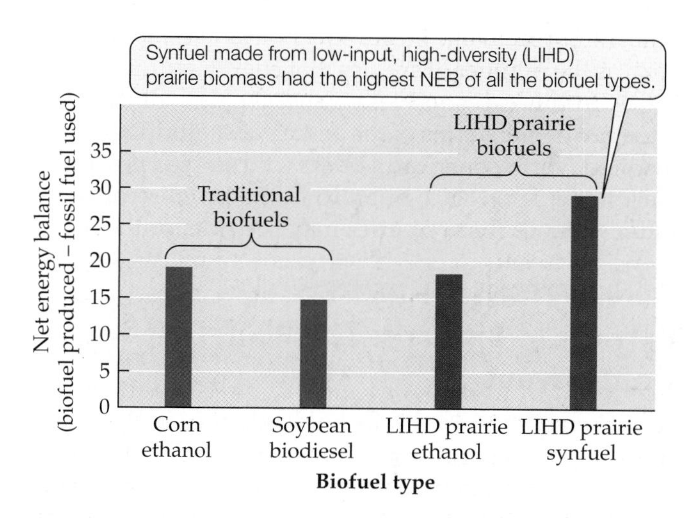

FIGURE 18.23 Biofuel Comparisons Net energy balance (NEB), a measure of the biofuel produced minus the fossil fuels needed to produce them, for corn ethanol, soybean biodiesel, and two biofuel types made from LIHD prairie biomass. (After Tilman et al. 2006.)

require energy to produce or deliver (**Figure 18.24A,B**). Second, the prairie plants had a very high yield due to diversity effects on biomass and the fact that all of the aboveground plant material (rather than just the seeds, as in the case of corn and soybeans) can be used.

Prairie plants are also superior to corn and soybeans in their ability to take up and store CO_2. This difference has important implications for the reduction of greenhouse gas emissions. Tilman et al. (2006) found that LIHD prairie plots sequestered 160% more CO_2 in plant roots and soil than did single-species prairie plots. When they calculated the amounts of CO_2 and other greenhouse gases produced and consumed in the process of producing and burning the different biofuels, they found that the traditional biofuels are net carbon sources, while LIHD biofuels are carbon neutral. Greenhouse gas emission reductions relative to the burning of fossil fuels were 6 to 16 times greater for LIHD fuels than for corn ethanol or soybean biodiesel (**Figure 18.24C**). In addition, because LIHD prairie biomass can be grown on degraded former agricultural land, it need not take up land that could be used for growing food.

The results of this study are promising, but can LIHD biomass provide enough fuel to make a difference? Tilman et al. (2006) estimated that about 5×10^8 hectares of abandoned agricultural land could produce enough fuel to substitute for about 13% of global petroleum consumption and 19% of global electricity consumption. This strategy could also eliminate 15% of global CO_2 emissions. Even if these are optimistic values, it is clear that this new twist on an emerging technology deserves more attention.

By applying basic principles of ecology to the biofuels debate, Tilman and colleagues showed elegantly that we cannot merely compare the attributes of natural ecosystems with those of agricultural ecosystems. We must consider what might seem like inconsequential and esoteric details, such as the number of species that coexist within communities. In this case, species richness makes all the difference, not only in terms of the beauty and productivity that inspired North American settlers who crossed the prairies back in the late 1880s, but also in the promise of energy independence those prairies may provide for the future.

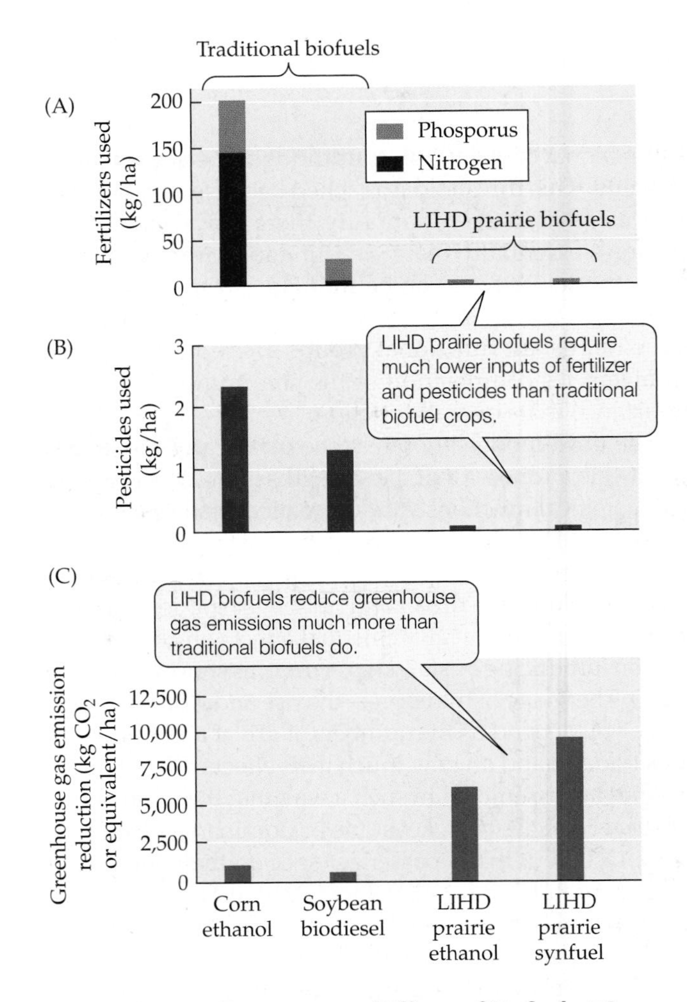

FIGURE 18.24 Environmental Effects of Biofuels Tilman and colleagues compared the environmental effects of biofuels made from corn, soybeans, and LIHD prairie biomass. (A) Fertilizer requirements. (B) Pesticide requirements. (C) Reduction of greenhouse gas emissions relative to the combustion of an equal amount of fossil fuel, taking into account the uptake and release of CO_2 and other greenhouse gases through the entire life cycle of the biofuel, from crop production to combustion. (After Tilman et al. 2006.)

CONNECTIONS IN NATURE

Barriers to Biofuels: The Plant Cell Wall Conundrum

Biofuels vary considerably in the biomass needed to produce them and the energy required to refine them. Ethanol is most commonly made from corn grains that are fermented and distilled. The energy costs associated with growing the grain and producing the ethanol are high, so there is only a small net energy gain in ethanol production (see Figure 18.23). Additionally, ethanol production from corn competes with the food needs of humans. For example, an acre of corn produces about 440 gallons of ethanol, which represents 4–5 months of driving for the average individual in the United States. The same amount of corn could feed one person for 20–27 years.

Biodiesel is easily produced from oil seed crops such as soybeans (most common in the United States) and palms (most common in Malaysia and Indonesia). However, even though soybeans are legumes, and thus have symbiotic nitrogen-fixing bacteria in their roots (see p. 361), they can promote soil erosion and require lots of water for irrigation. Biodiesel production has the potential to compete with food production as well.

For these reasons, and for the reasons given in the Case Study Revisited, scientists are turning to nonfood biomass, such as crop residues, logging wastes, and prairie plants, to produce what is called *cellulosic ethanol*. Unlike corn ethanol, which is made of sugars broken down from starch, cellulosic ethanol is made of sugars liberated from cellulose. As we will learn under Concept 21.2, breaking down and transforming some organic matter such as cellulose—the major component of plant cell walls—is extremely difficult and requires special enzymes produced by microbes. In fact, no multicellular organisms have the biochemical capability of extracting energy from cellulose. Plants have evolved many protective mechanisms to resist digestion by both bacteria and herbivorous animals, including epidermal waxes, thick and reinforced cell walls, enzyme-insoluble surfaces, and inhibitors of fermentation.

Himmel et al. (2007) have outlined some of the techniques that molecular biologists are developing to overcome this recalcitrance, most of which involve genetically engineered enzymes that work on the plant both externally and internally. The thought is that, one day, plant cells might even be genetically programmed to self-destruct before harvest. It is clear that for biofuel production to be a viable alternative to fossil fuels, ecologists and molecular biologists will have to work together to break down the current barriers to cellulosic biofuel production.

SUMMARY

CONCEPT 18.1 Species diversity differs among communities due to variation in regional species pools, abiotic conditions, and species interactions.

- The regional species pool and the dispersal abilities of species play important roles in supplying species to communities.
- Humans have greatly expanded the regional species pools of communities by serving as vectors for the dispersal of non-native species.
- Local abiotic conditions act as a strong "filter" for community membership.
- When species depend on other species for their growth, reproduction, and survival, those other species must be present if they are to gain membership in a community.
- Species may be excluded from communities by competition, predation, parasitism, or disease.

CONCEPT 18.2 Resource partitioning among the species in a community reduces competition and increases species diversity.

- Resource partitioning predicts that species must use resources slightly differently if they are to avoid competitive exclusion.
- One model of resource partitioning states that the less overlap between the species in a community in their use of resources along a resource spectrum, the more species can coexist in the community.
- The resource ratio hypothesis posits that species that use the same set of resources are able to partition them by using them in different proportions.

CONCEPT 18.3 Processes such as disturbance, stress, predation, and positive interactions can mediate resource availability, thus promoting species coexistence and species diversity.

- Disturbance, stress, or predation can cause species' populations to fluctuate, making resources available and increasing species coexistence.
- The intermediate disturbance hypothesis states that intermediate levels of disturbance, stress, or predation promote species diversity by reducing competitive exclusion. Species diversity is low at low levels of disturbance due to competitive exclusion and at high levels of disturbance due to high mortality.
- The dynamic equilibrium model predicts that species diversity will be highest when the level of disturbance and the rate of competitive displacement are roughly equivalent.
- Positive interactions can promote species diversity, particularly at intermediate to high levels of disturbance, stress, or predation.
- The Menge–Sutherland model is similar to the intermediate disturbance hypothesis except that it separates the effect of predation from that of physical disturbance.
- Lottery models assume that resources made available by disturbance, stress, or predation are captured at random by recruits from a larger pool of colonists, all of which have an equal chance to do so.

CONCEPT 18.4 Many experiments show that species diversity is positively related to community function.

- Species diversity can control numerous functions of communities, including productivity, soil fertility, water quality and availability, atmospheric gas exchange, and responses to disturbance.
- Many manipulative experiments in different communities have shown that as species diversity increases, so does community function.
- Hypotheses proposed to explain the positive relationship between species diversity and community function fall into four general categories that include different assumptions about the degree of overlap in the ecological functions of species and about variation in the strength of ecological functions of species.

REVIEW QUESTIONS

1. Suppose you are an ecologist studying prairie grassland communities in Minnesota. As you are doing your fieldwork, grass seeds with hooked spines attach themselves to your shoes. You then travel to New Zealand to study the grasslands on the South Island. When you enter the Customs area in the Auckland airport, the officers in charge ask if you have visited a natural area or farm recently. You say "yes," and they tell you to take off your shoes and wait while they disinfect them with bleach. Given what you know about the mechanisms important to community membership, is it worth the time and money required to clean all that footwear before allowing it into New Zealand?
2. We know that species diversity varies greatly among communities. Describe how some of the models proposed to explain this variation differ in their explanations of the mechanisms controlling species diversity within communities.
3. Suppose you are studying a tropical rainforest community in Panama. You obtain a 50-year data set for the forest that records both the mortality of adult trees and the emergence of new tree seedlings. As you analyze the data, you try to determine whether there is a pattern of species replacement, with individuals of one species generally replacing one another in the same sites in preference to individuals of other species. After much work, you are convinced that no pattern of replacement exists in this forest—instead, sites are colonized in an entirely random fashion, with no one species having an advantage. What model of species diversity best describes your observations, and why?
4. Recent experimental work in communities has shown positive relationships between species diversity and community function. We learned that there is considerable debate about the mechanisms controlling these relationships and that at least four hypotheses have been developed to explain them. Below are four graphs (A, B, C, and D) of species diversity–community function relationships that vary in the shapes of their curves. Describe which hypothesis best fits each curve, and why.

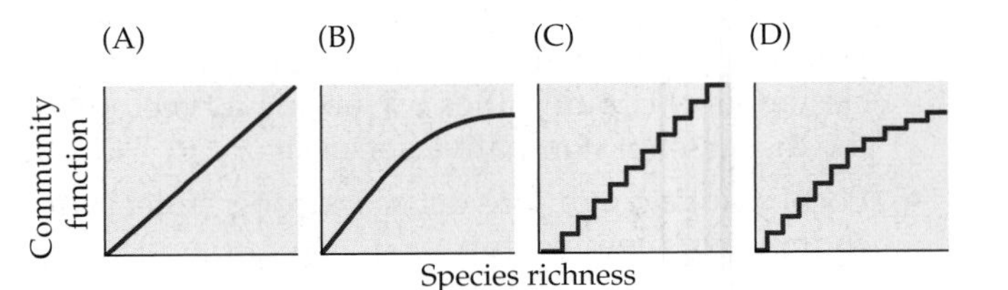

ON THE COMPANION WEBSITE
sites.sinauer.com/ecology2e

The website includes Chapter Outlines, Online Quizzes, Flashcards & Key Terms, Suggested Readings, a complete Glossary, and the Web Stats Review. In addition, the following resources are available for this chapter:

HANDS-ON PROBLEM SOLVING

Periodic Disturbance and Its Effect on Species

This Web exercise demonstrates how periodic disturbance can maintain species in a community that otherwise could not coexist. You will manipulate the frequency and intensity of disturbances to investigate this effect.

WEB EXTENSION

18.1 Testing the Dynamic Equilibrium Model

CLIMATE CHANGE CONNECTIONS

18.1 Species Invasion and Climate Change

UNIT 5
Ecosystems

CHAPTER

19 Production

KEY CONCEPTS

- **CONCEPT 19.1** Energy in ecosystems originates with primary production by autotrophs.
- **CONCEPT 19.2** Net primary production is constrained by both physical and biotic environmental factors.
- **CONCEPT 19.3** Global patterns of net primary production reflect climatic constraints and biome types.
- **CONCEPT 19.4** Secondary production is generated through the consumption of organic matter by heterotrophs.

Life in the Deep Blue Sea, How Can It Be?: A Case Study

Ecologists once considered the deep sea to be the marine equivalent of a desert. The physical environment at depths between 1,500 and 4,000 m did not seem conducive to life as we knew it. It is completely dark, so photosynthesis is not possible. The water pressure reaches values 300 times greater than those at the surface of the ocean, similar to the pressure used to crush cars at a junkyard. Organisms living on the bottom of the deep sea were thought to obtain energy exclusively from the sparse rain of dead algae and zooplankton falling from the photic zone in the upper layers of the ocean, where phytoplankton carry out photosynthesis. Most of the known deep-sea organisms were detritus feeders such as echinoderms (e.g., sea stars), mollusks, crustaceans, and polychaete worms.

Our view of deep-sea life was changed dramatically in 1977 when an expedition using the submersible craft *Alvin*, led by Robert Ballard of the Woods Hole Oceanographic Institution, visited a mid-ocean ridge near the Galápagos archipelago (**Figure 19.1**). The *Alvin* was in search of deep-sea hot springs thought to occur along mid-ocean ridges. These ridges lie at the junctions of tectonic plates, where the seafloor spreads as the plates are pushed apart by molten rock rising from Earth's mantle (see Figure 17.10). Because the mid-ocean ridges are volcanically active, geologists and oceanographers had hypothesized that seawater seeping into cracks in the ocean floor near the ridges would be superheated by pockets of magma, chemically transformed, and then ejected as hot springs. These hot springs were considered potential sources of chemicals for the ocean system as well as sources of heat. Despite their hypothesized existence, no such hot springs had ever been located.

Ballard's group did indeed find hot springs, known as *hydrothermal vents*, but from an ecological perspective, this geochemical finding paled in comparison to their biological discovery: the areas around the hydrothermal vents were teeming with life. Dense assemblages of tube worms (e.g., genus *Riftia*), giant clams (e.g., genus *Calyptogena*), shrimps, crabs, and polychaete worms were found in the areas surrounding the vents (**Figure 19.2**). The density of organisms was unprecedented on the deep, dark seafloor.

The discovery of these diverse and productive hydrothermal vent communities posed an immediate question: how did the organisms obtain the energy needed to sustain themselves in such abundance? The rate at which dead organisms from the upper zones of the ocean accumulate on the bottom is very low (0.5–0.01 mm/year). The newly formed areas of seafloor where the vents are located are only decades old, and thus the amount of food that

FIGURE 19.1 ***Alvin* in Action** The deep-sea submersible craft *Alvin* was instrumental in locating and exploring the first known hydrothermal vent site in 1977. The *Alvin* can carry two scientists and is equipped with video cameras and robotic arms for collecting specimens from the seafloor.

FIGURE 19.2 Life around a Hydrothermal Vent Tube worms over 2 m in length surround a "black smoker" hydrothermal vent. The vent is spewing superheated water as hot as 400°C, which contains high concentrations of dissolved metals and chemicals, particularly hydrogen sulfide.

would have accumulated should not be enough to sustain these high densities of organisms. Photosynthesis in the surface waters therefore did not appear to be the energy source supporting these hydrothermal vent communities.

The water being emitted from the hydrothermal vents also posed a problem: its chemical composition would be toxic to most organisms. The water emitted by the vents is rich in sulfides as well as heavy metals such as lead, cobalt, zinc, copper, and silver, which inhibit metabolic activity in most organisms.

Hydrothermal vent communities thus pose two mysteries: first, what is the source of energy that sustains them, and second, how do the organisms tolerate the high concentrations of potentially toxic sulfides in the water? As we shall see, the answers to these two questions are intimately related.

Introduction

In 1942, a controversial paper authored by Raymond Lindeman, describing energy flow in a Minnesota lake, was published in the journal *Ecology*. Lindeman had studied the energy relationships among the organisms and nonliving components in a lake ecosystem. Rather than grouping its component plants, animals, and bacteria according to their taxonomic categories, Lindeman grouped them into categories based primarily on how they obtained their energy (**Figure 19.3**). His views on the importance of the energy base of the system ("ooze"—particulate and dissolved dead organic matter) and on the efficiency of energy transfer among the biological components of the system were groundbreaking. Lindeman's treatment of energy flow in the lake was considered too theoretical at the time, and his paper was initially rejected, although the publishers later reconsidered after Lindeman's mentor, G. E. Hutchinson, advocated its acceptance.

Lindeman's pioneering paper was among the first in the area of ecosystem science, and it is now considered a fundamental paper in the discipline. The term **ecosystem** was first coined by A. G. Tansley, a plant ecologist, to refer to all of the components of an ecological system, biotic and abiotic, that influence the flow of energy and elements (Tansley 1935). The "elements" considered in ecosystem studies are primarily nutrients, but also include pollutants; the movements of those elements through ecosystems are the topic of Chapter 21. The ecosystem concept is now well

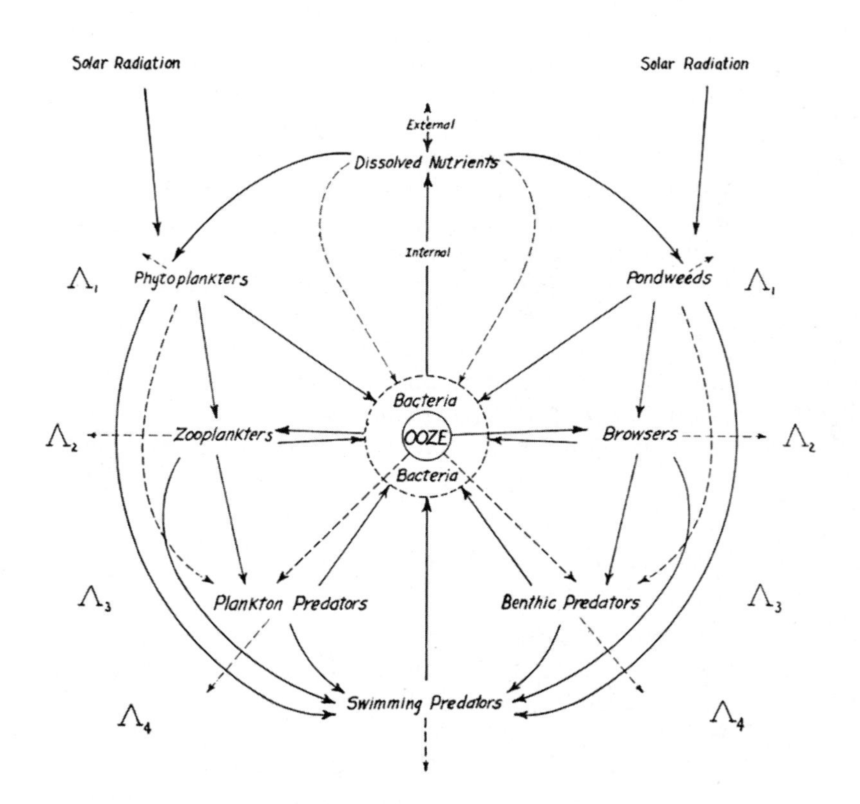

FIGURE 19.3 Energy Flow in a Lake Raymond Lindeman's diagram describes the movement of energy among groups of organisms at Cedar Bog Lake, Minnesota. Note the general functional categories of organisms Lindeman used, as well as the central position of "ooze" (organic matter) in the diagram. The subscripts next to the uppercase Greek lambdas represent trophic levels. (From Lindeman 1942.)

established and has become a powerful tool for integrating ecology with other disciplines such as geochemistry, hydrology, and atmospheric science.

In Chapter 5, we described the physiological basis for the capture of energy through photosynthesis and chemosynthesis by autotrophs, and we explained how heterotrophs obtain that energy by consuming autotrophs. In this chapter, we return to the topic of energy as we review how energy enters ecosystems, how it is measured, and what controls rates of energy flow through ecosystems. In Chapter 20, we will describe the movements of energy through the biotic components of ecosystems and the importance of those movements to the dynamics of communities and ecosystems.

CONCEPT 19.1 Energy in ecosystems originates with primary production by autotrophs.

Primary Production

The chemical energy generated by autotrophs, known as **primary production**, is derived from the fixation of carbon during photosynthesis and chemosynthesis (see Chapter 5). (Chemosynthesis accounts for a very small amount of this energy globally, although it can be important locally, as we will see.) Primary production represents an important energy transition: the conversion of light energy from the sun into chemical energy that can be used by autotrophs and consumed by heterotrophs. Primary production is the source of energy for all organisms, from bacteria to humans (nearly all of the energy we use is derived from sunlight, including fossil fuels and hydroelectricity). The energy assimilated by autotrophs is stored as carbon compounds in plant tissues; therefore, carbon (C) is the currency used for the measurement of primary production. The rate of primary production is sometimes referred to as *primary productivity*.

Gross primary production is total ecosystem photosynthesis

The amount of carbon fixed by the autotrophs in an ecosystem is called **gross primary production** (**GPP**). The GPP in most terrestrial ecosystems is equivalent to the total of all plant photosynthesis.

The GPP of an ecosystem is controlled by the rate of photosynthesis (which is strongly influenced by climate, as we saw under Concept 5.2) and by the leaf area of the plants per unit of ground area, known as the **leaf area index** (**LAI**, a metric that lacks units, since it is an area divided by an area). The leaf area index varies among biomes, from less than 0.1 in Arctic tundra (i.e., less than 10% of the ground surface has leaf cover) to 12 in boreal and tropical forests (i.e., on average, there are 12 layers of leaves between the canopy and the ground). Shading of the leaves below the uppermost layer increases with the addition of each new leaf layer, so the incremental gain in photosynthesis for each added leaf layer decreases (**Figure 19.4**). Eventually, the respiratory costs associated with adding leaf layers outweigh the photosynthetic benefits. Plants generally optimize their leaf area index in relation to the climatic conditions and the supply of resources, particularly water and nutrients, available to them.

A plant uses approximately half of the carbon it fixes by photosynthesis in cellular respiration to support biosynthesis and cellular maintenance. All living plant tissues lose carbon via respiration, but not all of them acquire carbon via photosynthesis. Thus, plants that have a large proportion of nonphotosynthetic stem tissue, such as trees and shrubs, tend to have higher overall respiratory carbon losses than herbaceous plants. Plant respiration rates increase with increasing temperatures, and as a result, respiratory carbon losses are higher in tropical forests than in temperate and boreal forests.

Net primary production is the energy remaining after respiratory losses

The balance between GPP and autotroph respiration is called **net primary production** (**NPP**):

$$NPP = GPP - respiration$$

NPP is the amount of energy captured by autotrophs that results in an increase in living plant matter, or **biomass**. In other words, NPP is the energy left over for plant growth, reproduction, and consumption by herbivores and detritivores. It also represents the total input of carbon into ecosystems.

Carbon not used in respiration can be allocated to growth and reproduction, storage, and defense against herbivory. Furthermore, plants can respond to varying environmental conditions by allocating carbon to the growth of different tissues. The allocation of carbon within a plant varies considerably according to the species, the availability of resources, and the climate. Allocation of carbon to photosynthetic tissues is an investment in potential future NPP, but the demands of the plant for other resources, particularly water and nutrients, as well as biological interactions such as herbivory influence whether that investment pays off.

A plant's allocation of NPP to the growth of leaves, stems, and roots is generally balanced so as to maintain supplies of water, nutrients, and carbon to match the plant's requirements. For example, plants growing in desert, grassland, and tundra ecosystems are regularly exposed to shortages of water or nutrients. Plants in these ecosystems may allocate a greater proportion of NPP to root growth, relative to the growth of shoots (leaves and stems), than plants growing in ecosystems with higher soil water and

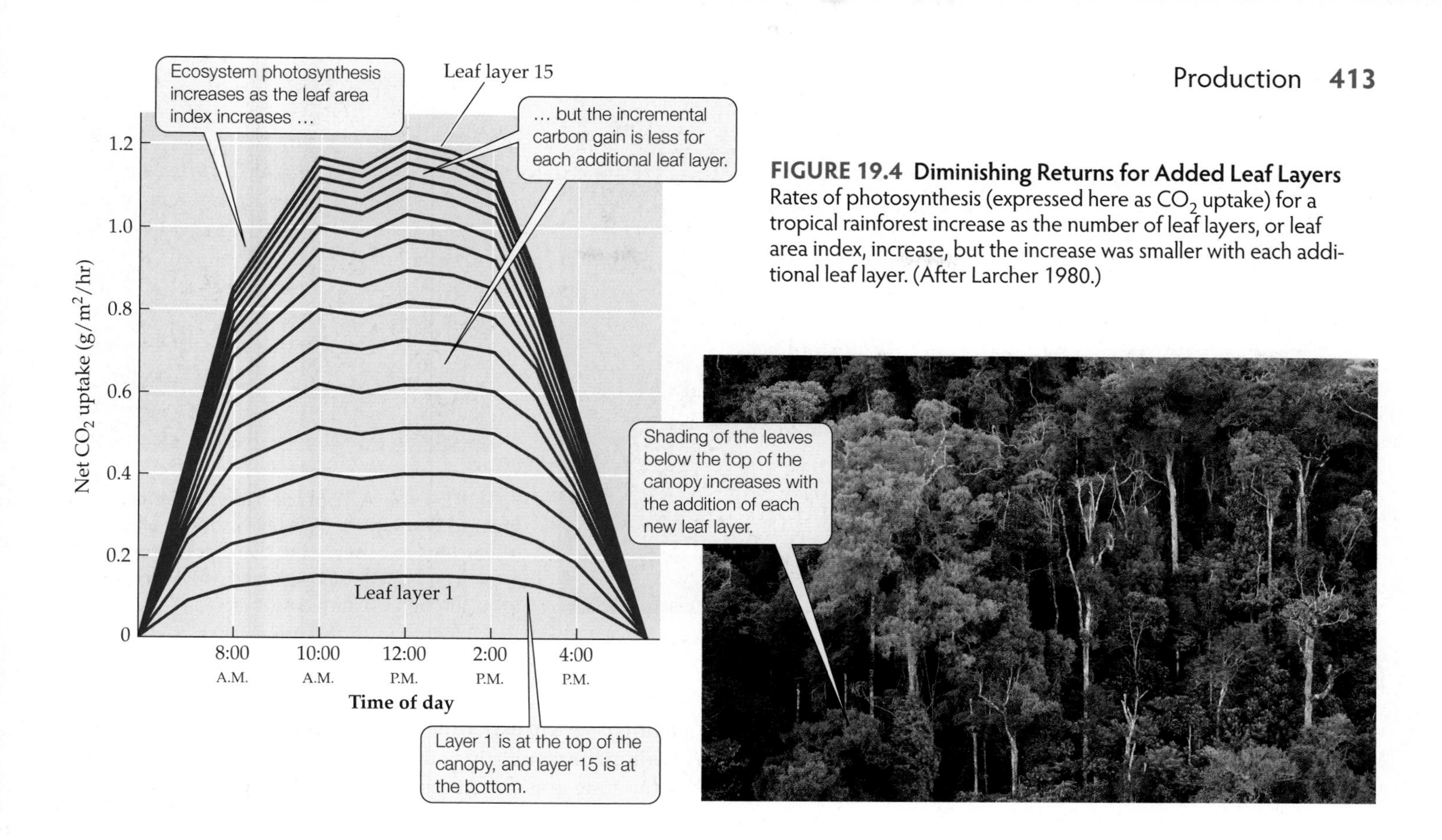

FIGURE 19.4 Diminishing Returns for Added Leaf Layers Rates of photosynthesis (expressed here as CO_2 uptake) for a tropical rainforest increase as the number of leaf layers, or leaf area index, increase, but the increase was smaller with each additional leaf layer. (After Larcher 1980.)

nutrient availability (**Figure 19.5**). This greater allocation to roots facilitates their acquisition of the resources that are in short supply. In contrast, plants growing in dense communities with neighbors that may shade them may allocate NPP preferentially to stems and leaves in order to capture more sunlight for photosynthesis. In other words, plants tend to allocate the most NPP to those tissues that acquire the resources that limit their growth.

Allocation of NPP to storage compounds such as starch and carbohydrates provides insurance against losses of tissues to herbivores, disturbances such as fire, and climatic events such as frost. These compounds are usually stored in stems of woody plants or in belowground stems and roots of herbaceous plants. Where levels of herbivory are high, plants may allocate a substantial amount of NPP (up to 20%) to defensive secondary compounds such as tannins or terpenes that inhibit grazing.

NPP changes during ecosystem development

As ecosystems develop during primary or secondary succession (see Chapter 16), NPP changes as the leaf area index, the ratio of photosynthetic to nonphotosynthetic tissue, and plant species composition all change. Most ecosystems have their highest NPP at mid-successional stages. Several factors contribute to this pattern, including the tendency for the proportion of photosynthetic tissues, plant diversity, and nutrient supply to be highest at mid-successional stages. In forest ecosystems, the leaf area index and the photosynthetic rates of leaves decrease in old-growth

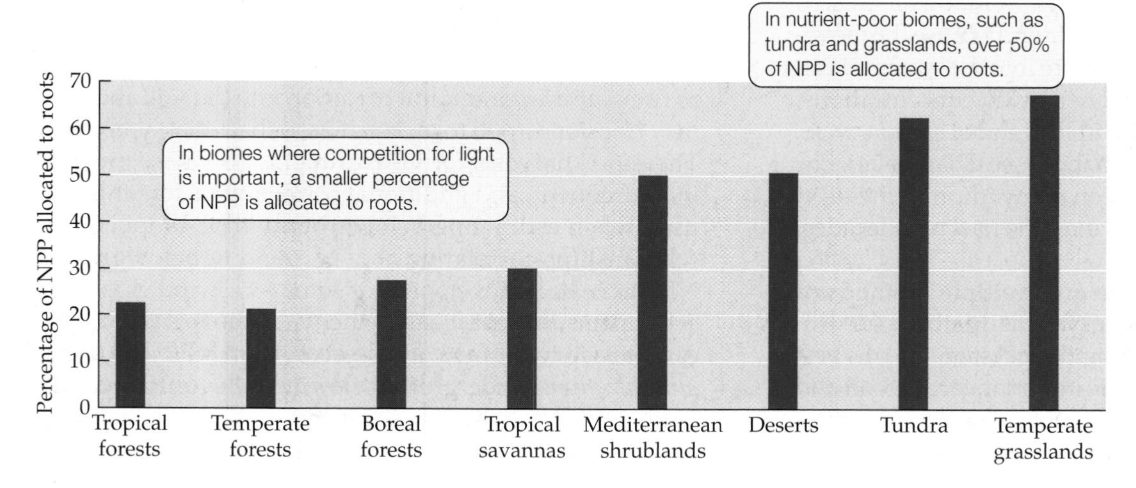

FIGURE 19.5 Allocation of NPP to Roots The proportion of NPP that plants allocate to roots varies with the resources available to them. (After Saugier et al. 2001.)

Q In addition to low supplies of resources in the soil, what other factors might favor greater allocation of NPP to tissues below the soil surface?

FIGURE 19.6 A Tool for Viewing Belowground Dynamics (A) Minirhizotrons allow researchers to observe the dynamics of root growth and death belowground. (B) A view of roots through a minirhizotron.

stands, lowering GPP and thus NPP. In some grasslands, such as the tallgrass prairies of the central United States, the accumulation of dead leaves near the ground surface and the development of a closed upper canopy of leaves decrease light availability to short plants, lowering the photosynthetic carbon gain of the ecosystem. However, the decrease in NPP over time is far less pronounced in grasslands than in forest ecosystems. Although NPP may decrease in late successional stages, these old-growth ecosystems have large pools of stored carbon and nutrients and provide habitat for late successional animal species.

NPP can be estimated by a number of methods

There are a number of reasons why it is important to be able to measure NPP in an ecosystem. As we have seen, NPP is the ultimate source of energy for all organisms in an ecosystem and thus determines the amount of energy available to support that ecosystem. It varies tremendously over space and time. Year-to-year variation in NPP provides a metric for examining ecosystem health because changes in primary productivity can be symptomatic of stresses such as drought or acid rain. In addition, because the movement of carbon from the atmosphere into ecosystems is an important control over atmospheric CO_2 concentrations, NPP is intimately associated with the global carbon cycle, and thus with global climate change (see Chapter 24). For all these reasons, there have been many improvements in techniques for estimating NPP over the past two decades.

Terrestrial ecosystems There are multiple methods of estimating NPP in terrestrial ecosystems. Methods for estimating NPP in forest and grassland ecosystems are the best developed due to their economic importance for wood and forage production. Traditional techniques measure the increase in plant biomass during the growing season by harvesting plant tissues in experimental plots, measuring biomass, and scaling the results up to the ecosystem level. For example, in temperate grassland ecosystems, the aboveground biomass can be harvested from plots at the start of the growing season and again when the amount of plant biomass reaches its maximum. The difference in plant biomass between the two harvests is used as an estimate of NPP. In forests, the radial growth of wood must be included in estimates of NPP. In the tropics, plants may continue to grow throughout the year, and tissues that die decompose rapidly, making the use of harvest techniques problematic. Despite these shortcomings, harvest techniques still provide reasonable estimates of aboveground NPP, particularly if corrections are made for tissue loss to herbivory and mortality.

Measuring the allocation of NPP to growth belowground is more difficult because root growth is more dynamic than the growth of leaves and stems. As noted earlier, the proportion of NPP in roots exceeds that in aboveground tissues in some ecosystems. In grassland ecosystems, root growth may be twice that of the aboveground leaves, stems, and flowers combined. The finest roots *turn over* more quickly than shoots; that is, more roots are "born" and die during the growing season than stems and leaves. In addition, roots may exude a large amount of carbon into the soil, and they may transfer carbon to mycorrhizal or bacterial symbionts. Therefore, harvests for measuring root biomass must be more frequent, and additional correction factors must be used when estimating belowground NPP. Proportional relationships correlating aboveground to belowground NPP have been developed for some forest and grassland ecosystems, so that measurements of aboveground NPP can be used to estimate whole-ecosystem NPP. The use of *minirhizotrons*, underground viewing tubes outfitted with video cameras, has led to advances in the understanding of belowground production processes (**Figure 19.6**).

The labor-intensive and destructive nature of harvest techniques makes them impractical for estimating NPP over large areas or in biologically diverse ecosystems. Several nondestructive techniques have been developed that allow more frequent estimation of NPP over much larger spatial scales, although with lower precision than harvest techniques. Some of these techniques, which include remote sensing and atmospheric CO_2 measurements, provide a quantitative index rather than an absolute measure of NPP. Some techniques require a combination of data collection and modeling of plant physiological and climatic processes to accurately estimate the actual fluxes of carbon associated with NPP.

The concentration of the photosynthetic pigment chlorophyll in a plant canopy provides a proxy for photosynthetic biomass that can be used to estimate GPP and NPP. Chlorophyll concentrations can be estimated using remote sensing techniques that rely on the reflection of solar radiation (**Ecological Toolkit 19.1**). Remote sensing allows NPP to be measured frequently, at spatial scales up to the entire globe, using satellite-based sensors (**Figure 19.7**). Indicators of NPP that are based on chlorophyll concentrations can overestimate NPP if the vegetation is not physiologically active, as in boreal forests in winter, but remote sensing generally provides the best estimate for NPP at regional to global scales.

NPP can also be estimated from direct measurements of its components: GPP and plant respiration. This approach typically involves measuring the change in CO_2 concentration in a closed system, which can be created by placing a chamber around stems and leaves, whole plants, or whole stands of plants. For example, Howard Odum estimated NPP for a tropical forest in Puerto Rico by enclosing a stand of trees inside a 200 $m^2 \times 20$ m tall clear plastic "tent" (Odum and Jordan 1970). The sources of CO_2 added to the atmosphere in such a closed system are respiration by the plants and heterotrophs, including microorganisms in the soil and animals in the forest. Uptake of CO_2 from the atmosphere results from photosynthesis. Thus, the net change in CO_2 inside the system results from the balance between GPP and total respiratory release by the plants and the heterotrophs. This net exchange of CO_2 is called **net ecosystem production** or **net ecosystem exchange (NEE)**. Heterotrophic respiration must be subtracted from NEE to obtain NPP; as a result, NEE provides a more refined estimate of ecosystem carbon storage than NPP. Carbon movement into and out of ecosystems, such as carbon lost through leaching from the soil or through disturbances (e.g., fire) can influence estimates of NPP.

Another noninvasive approach to estimating NEE uses frequent measurements of CO_2 and microclimate at various heights throughout a plant canopy and into the open air above the canopy. This technique, known as *eddy covariance* or *eddy correlation*, takes advantage of the gradi-

FIGURE 19.7 Remote Sensing of Terrestrial NPP Global terrestrial NPP for the period 2000–2005, estimated using a satellite-based sensor [Moderate resolution Imaging Spectroradiometer (MODIS)]. Note the latitudinal patterns in NPP corresponding to climatic zones. (From Zhao et al. 2006.)

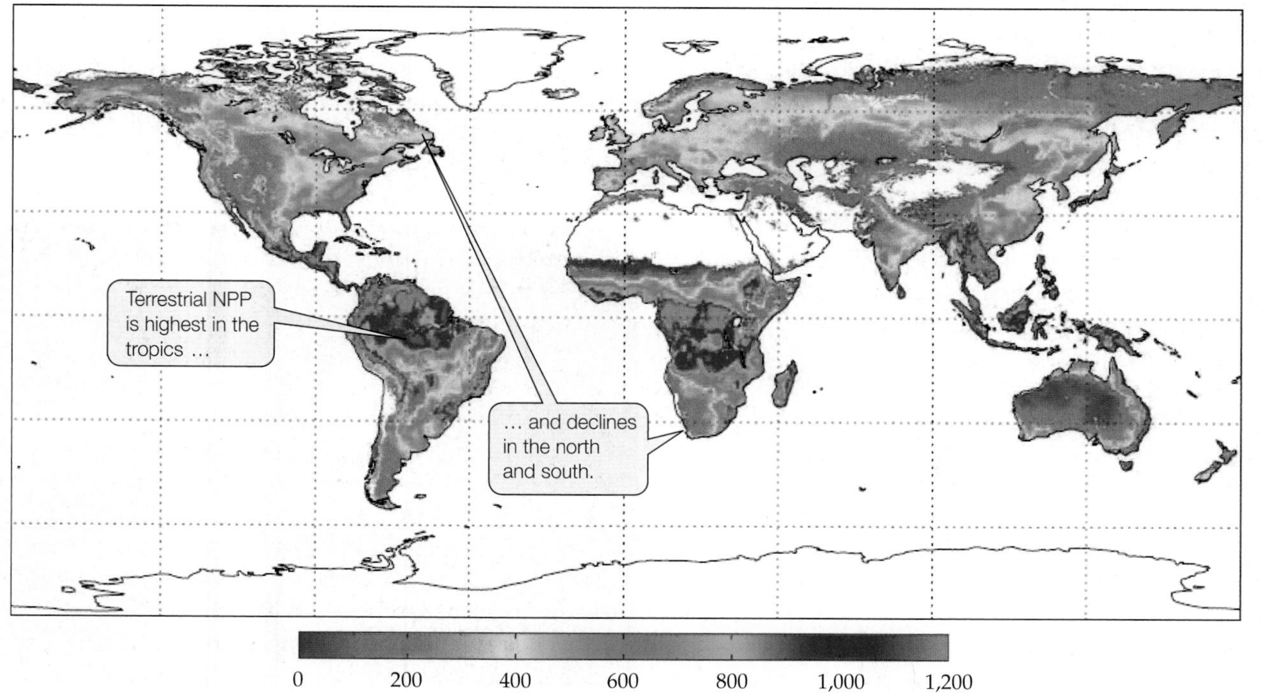

ECOLOGICAL TOOLKIT

19.1 Remote Sensing

When sunlight strikes an object, it is absorbed or scattered in such a way that the amount and quality of the light that reflects off of the object is changed. For example, when sunlight strikes a clear lake, about 5% of the visible light is reflected, while a light-colored sandy soil, such as might be found in a desert, reflects back as much as 40%. The amount of light reflected depends on the wavelengths of the light: different kinds of objects absorb or reflect some wavelengths more than others. The atmosphere scatters more blue wavelengths than red or green, and therefore the sky appears blue to our eyes. The lake, however, appears blue because most of the red and green light is absorbed by the water before it can be scattered back to our eyes. Lakes with high concentrations of phytoplankton appear green because much of the blue light is absorbed by the phytoplankton, leaving only the green light to be scattered back to our eyes.

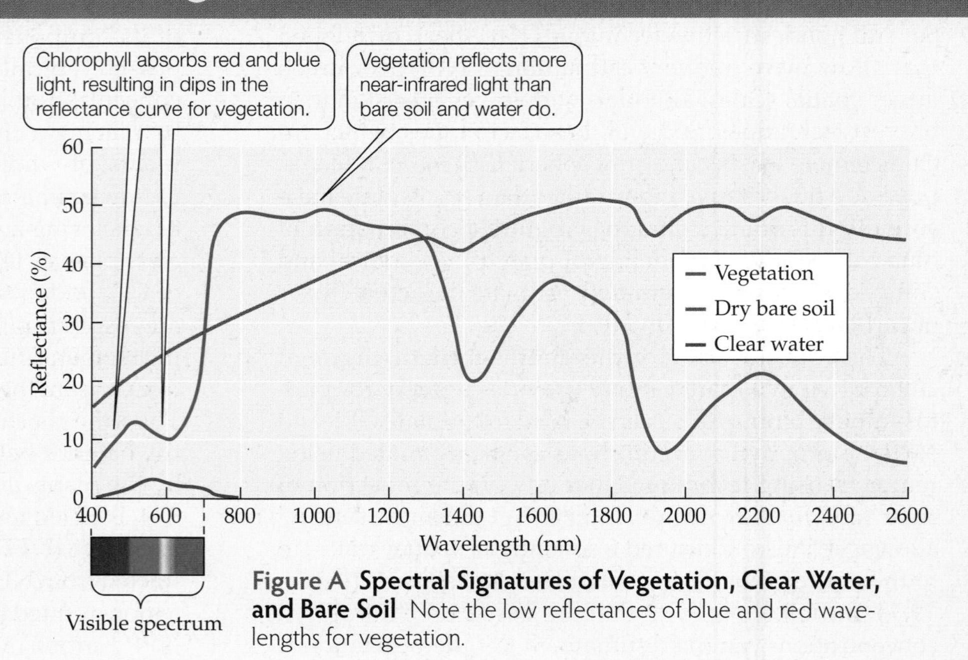

Figure A Spectral Signatures of Vegetation, Clear Water, and Bare Soil Note the low reflectances of blue and red wavelengths for vegetation.

Remote sensing is a technique that takes advantage of light reflection and absorption to estimate the density and composition of objects on Earth's surface, in waters, and in the atmosphere. Ecologists use remote sensing to estimate NPP by taking advantage of the unique reflectance pattern of chlorophyll-containing plants, algae, and bacteria (**Figure A**). Because chlorophyll absorbs visible solar radiation in blue and red wavelengths, it has a characteristic *spectral signature* with greater reflection of green wavelengths. In addition, vegetation has higher absorption of red wavelengths than do bare soil or water.

Ecologists can measure the reflection of specific wavelengths from a land or water surface and estimate NPP using several indices that have been developed. One of the most commonly used indices is the *normalized difference vegetation index*, or NDVI, which uses differences between visible-light and near-infrared reflectance to estimate the density of chlorophyll:

$$\text{NDVI} = \frac{(\text{NIR} - \text{red})}{(\text{NIR} + \text{red})}$$

where "NIR" is the near-infrared wavelength band (700–1,000 nm) and red is the red wavelength band (600–700 nm). Note that the spectral signature of vegetation in Figure A shows a large difference between reflectance of red and of near-infrared relative to the spectral signatures of water and soil, which gives vegetation a high NDVI value and water and soil low NDVI values. NDVI is coupled with estimates of the efficiency of light absorption to estimate photosynthetic CO_2 uptake.

Remote sensing of light reflectance from Earth's surface and atmosphere can be done at large spatial scales using satellites (**Figure B**), which transmit their measurements to receiving stations. Depending on the spatial resolution of the surface measurement and the number of wavelengths measured, satellite remote sensing can generate massive amounts of data that need to be processed. Advances in computing power have enhanced the spatial and temporal capabilities of remote sensing, making it a powerful tool for measuring NPP as well as deforestation, desertification, atmospheric pollution, and many other phenomena of interest.

Figure B Remote Sensing by Satellite Remote sensing instruments mounted on satellites can measure the reflectance of solar radiation from Earth to provide ecologists with large-scale measurements of NPP and other phenomena.

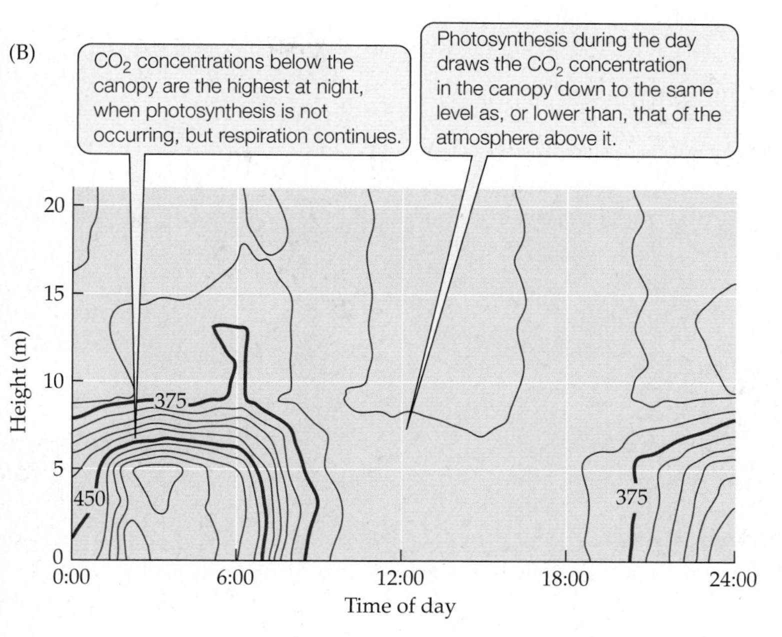

FIGURE 19.8 Eddy Covariance Estimates of NPP (A) A tower projecting above a subalpine forest on Niwot Ridge, Colorado. Attached to the tower are instruments that measure microclimate (temperature, wind speed, radiation) and atmospheric CO_2 concentrations at frequent intervals. These measurements are used for estimating net ecosystem exchange of CO_2. (B) Concentration of CO_2 (parts per million) from the ground surface to above the canopy in a boreal forest in Siberia, measured over the course of a 24-hour period in the summer. Average canopy height was 16 m. (B after Hollinger et al. 1998.)

Q What would the daily pattern of CO_2 concentrations look like during the summer in a community made up primarily of cacti?

ent in CO_2 concentration between the plant canopy and the atmosphere that develops because of photosynthesis and respiration. During the day, when plants are photosynthetically active, the concentration of CO_2 is lower in the plant canopy than in the air above the plant canopy. At night, when photosynthesis shuts down but respiration continues, the CO_2 concentration in the canopy is higher than that in the atmosphere. Instrument-bearing towers established in forest, shrubland, and grassland canopies have been used to measure the net ecosystem exchange of CO_2 over long periods (**Figure 19.8**). Depending on the tower height, eddy covariance can provide an integrated NEE for up to several square kilometers of the surrounding area. A network of eddy covariance sites in the Americas (Ameriflux: http://public.ornl.gov/ameriflux/) has been established to help researchers better understand the uptake and fate of carbon in terrestrial ecosystems and how carbon uptake is influenced by climate.

Aquatic ecosystems The dominant autotrophs in both freshwater and marine ecosystems are phytoplankton, including algae and bacteria. These organisms have much shorter life spans than terrestrial plants, so the biomass present at any given time is very low compared with NPP; therefore, harvest techniques are not used to estimate NPP for phytoplankton, although they can be used for seagrasses and macroalgae. Instead, rates of photosynthesis and respiration are measured in water samples collected in bottles and incubated at the collection site with light (for photosynthesis) and without light (for respiration). The difference between the two rates is equal to NPP. Although there are errors associated with the artificial environment of the bottles, as well as the inclusion of respiration by heterotrophic bacteria and zooplankton in the bottles, this technique is used widely in freshwater and marine ecosystems.

Remote sensing of chlorophyll concentrations in the oceans using satellite-based instruments provides good estimates of marine NPP (**Figure 19.9**). As described for terrestrial remote sensing, indices based on absorption and reflection of light of different wavelengths are used to indicate how much light is being absorbed by chlorophyll, which is then related to NPP by using a light utilization coefficient, a term that incorporates the efficiency of light absorption into photosynthetic CO_2 uptake.

As Figure 19.7 shows, there can be as much as a 50-fold difference in NPP between Arctic and tropical ecosystems. In the following section we will investigate the role of abiotic and biotic factors that influence differences in NPP among ecosystems.

FIGURE 19.9 Remote Sensing of Marine NPP Primary production in the oceans, estimated using a satellite-based sensor [Sea-viewing Wide Field-of-view Sensor (SeaWiFS)].

Q In addition to zones of upwelling, what other kinds of coastal areas with high rates of NPP might be visible in this image?

CONCEPT 19.2 Net primary production is constrained by both physical and biotic environmental factors.

Environmental Controls on NPP

As we have seen, NPP varies substantially over space and time. Much of this variation is associated with differences in climate, such as latitudinal gradients in temperature and precipitation. In this section, we explore the factors that constrain rates of NPP.

NPP in terrestrial ecosystems is controlled by climatic factors

Variation in terrestrial NPP at the continental to global scales correlates well with variation in temperature and precipitation. NPP increases as average annual precipitation increases up to a maximum of about 2,400 mm per year, after which it decreases in some ecosystems (e.g., highland tropical forests), but not in others (e.g., lowland tropical forests) (**Figure 19.10A**). At very high precipitation levels, NPP may decrease due to several factors. Cloud cover over long periods lowers available sunlight. High amounts of precipitation leach nutrients from soils, and high soil water content results in hypoxic conditions that cause stress for both plants and decomposers. NPP increases with average annual temperature (**Figure 19.10B**). This does not mean, however, that ecosystem carbon storage (NEE, discussed earlier) would increase with increasing temperature. The loss of carbon from ecosystems due to respiration of heterotrophic organisms also increases at warmer temperatures, so NEE may potentially decrease. Several lines of evidence suggest that climate change over the past decades had changed NEE in some ecosystems. For example, Arctic tundra sites that were once carbon sinks (with greater GPP than carbon loss due to respiration) are now carbon sources (with greater respiratory carbon loss than GPP). These changes are increasing CO_2 losses to the atmosphere, as **Climate Change Connection 19.1** explains.

These correlations of NPP with climate suggest that NPP is directly linked to water availability and temperature. Such links makes sense when we consider the direct influence of water availability on photosynthesis via the opening and closing of stomates and the influence of temperature on the enzymes that facilitate photosynthesis (see Chapter 5). In deserts and in some grassland ecosystems, water availability has a clear, direct influence on NPP. In other ecosystems where water limitation is not as severe, the causal connection between precipitation and NPP is less clear.

The links between climate and NPP may also be indirect, mediated by factors such as nutrient availability or the particular plant species found within an ecosystem. How can we detect whether the influence of climate on NPP is direct or indirect? Several approaches, both observational and ex-

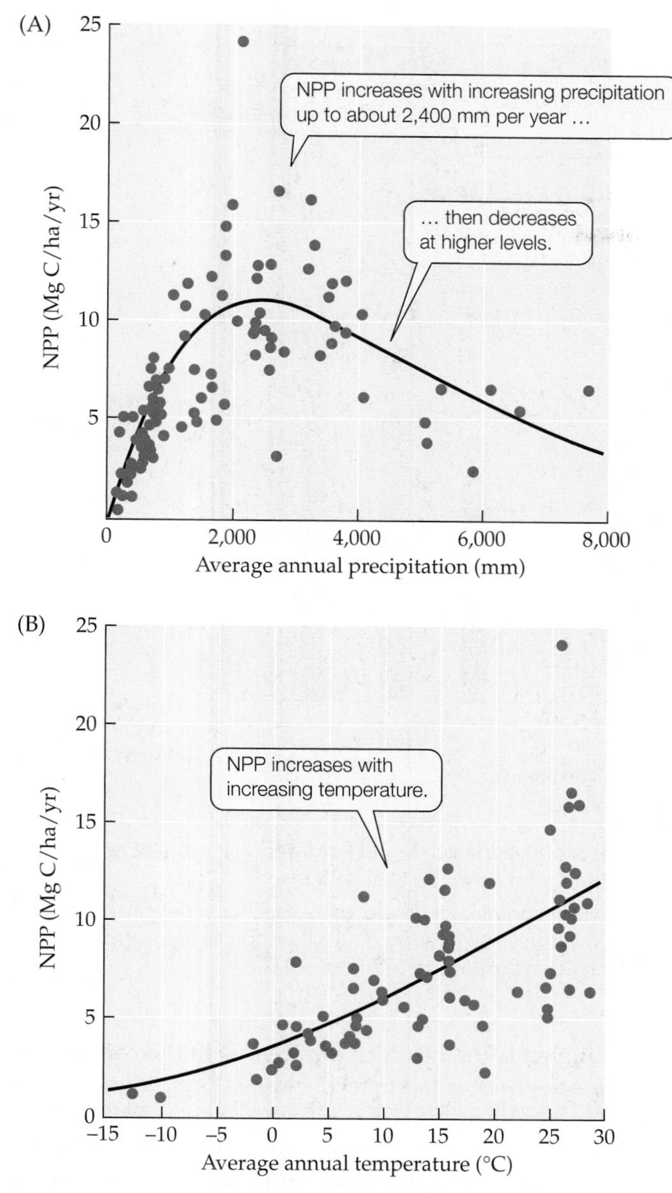

FIGURE 19.10 Global Patterns of Terrestrial NPP Are Correlated with Climate These graphs show the relationships between NPP and (A) precipitation and (B) temperature in terrestrial ecosystems worldwide. (Mg = 10^6 g). (After Schuur 2003.)

perimental, have been used. William Lauenroth and Osvaldo Sala examined how NPP in a short-grass steppe ecosystem responded to year-to-year variation in precipitation (Lauenroth and Sala 1992). They also examined the average annual NPP and precipitation across several grassland ecosystems at different locations in the central United States. When they compared the correlations between NPP and precipitation in their two analyses, they found that NPP increased more as precipitation increased for the site-to-site comparison than for the comparison among years in the short-grass steppe (**Figure 19.11**). They attributed the difference in the response of NPP to precipitation to variation in the species that made up the various U.S. grasslands. Some grass species have a greater capacity than others to grow with enhanced water availability, associated with greater ability to produce new shoots and flowers. Lauenroth and Sala also suggested that there was a time lag in the response to increased precipitation in the short-grass steppe ecosystem; that is, the increase in NPP in response to an increase in precipitation did not occur in the same year, but was delayed one to several years. Within the grassland biome, differences in species' abilities to respond to climatic variation can contribute to site-to-site variation in NPP, influencing the correlation between climate and NPP among sites.

Experimental manipulations of water, nutrients, CO_2, and plant species composition have been used to examine the direct influence of those factors on NPP. The results of numerous experiments indicate that nutrients, particularly nitrogen, control NPP in terrestrial ecosystems. For example, William Bowman, Terry Theodose, and their colleagues used a fertilization experiment in alpine communities of the southern Rocky Mountains to determine whether the supply of nutrients limits NPP (Bowman et al. 1993). They knew that spatial differences in NPP among alpine communities were correlated with differences in soil water availability, as in the grassland ecosystems described above. Bowman and colleagues' fertilization experiment was performed in two communities, a dry meadow and a wet meadow, which differed in soil moisture and nutrient availability. They sought to determine whether these soil resources and the species composition of the communities influenced their responses to added nutrients. They added nitrogen, phosphorus, and both nitrogen and phosphorus to different plots in both communities, and they maintained plots with no nutrient additions as controls. Their results indicated that the supply of nitrogen limited NPP in the dry meadow (which had lower soil moisture

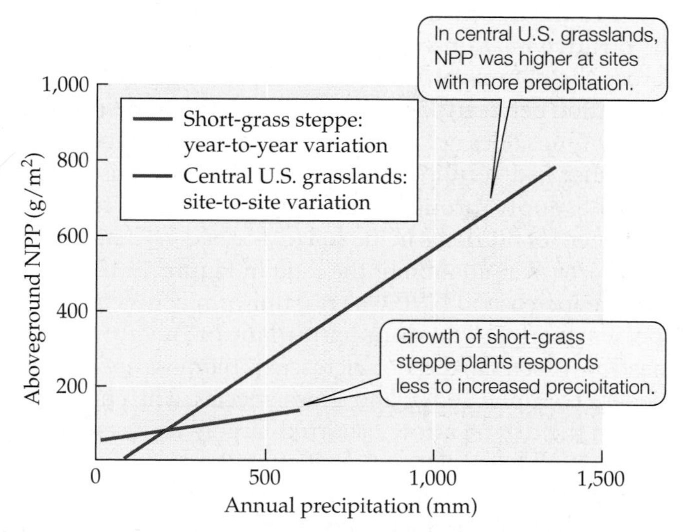

FIGURE 19.11 The Sensitivity of NPP to Changes in Precipitation Varies among Grassland Ecosystems The relationship between aboveground NPP and precipitation is shown for a short-grass steppe ecosystem and for several grassland ecosystems of different types at different sites in the central United States. (After Lauenroth and Sala 1992.)

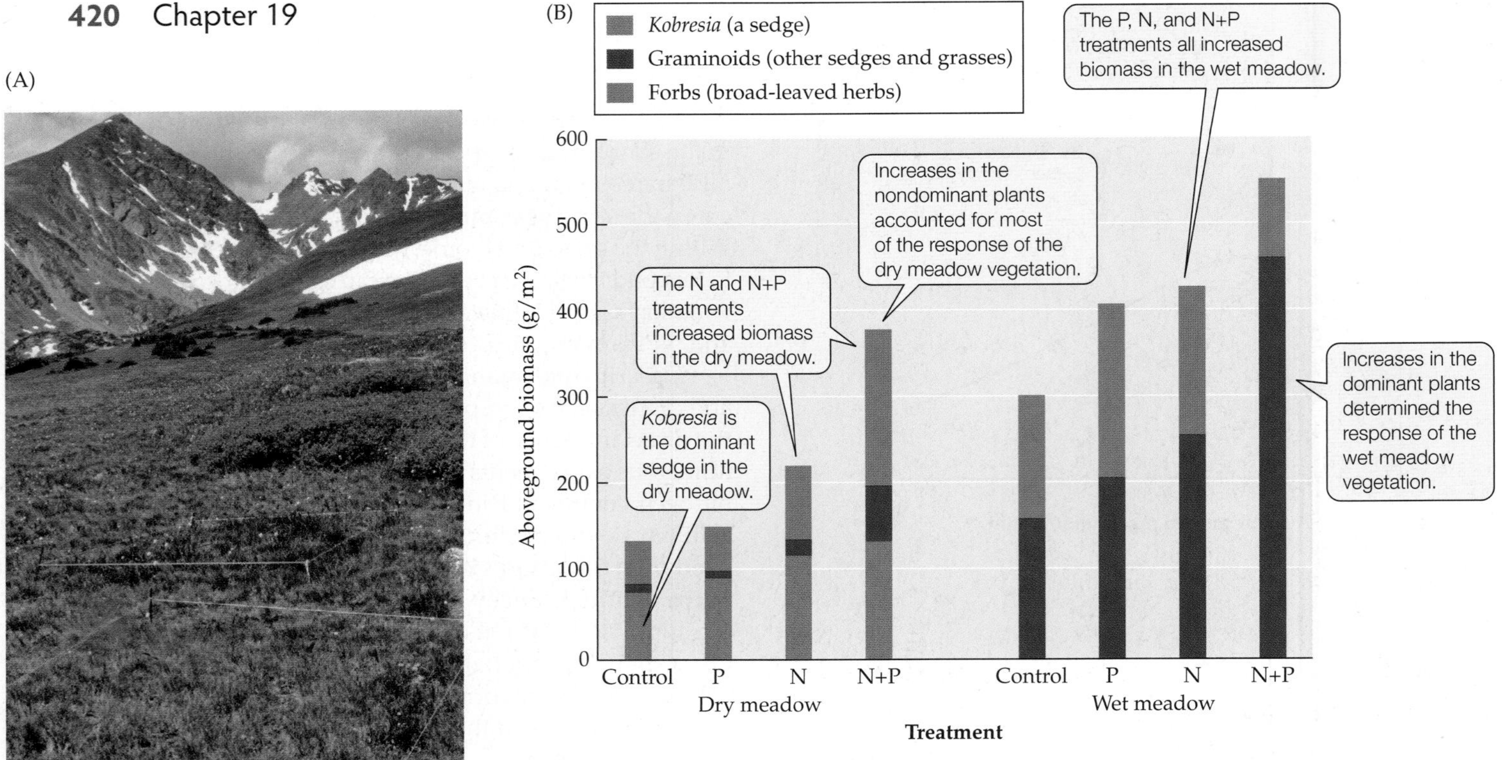

FIGURE 19.12 Nutrient Availability Influences NPP in Alpine Communities (A) Fertilized plots in an alpine dry meadow community in the Colorado Rocky Mountains, dominated by sedges, forbs, and grasses (see Figure 3.11). (B) Fertilization of plots in a resource-poor dry meadow and a resource-rich wet meadow with nitrogen (N), phosphorous (P), and both N and P showed that nutrient availability limits NPP. (B after Bowman et al. 1993.)

Q In which community would you expect a higher proportion of belowground NPP? Would the allocation to belowground NPP change in response to fertilization?

and nutrient availability), while nitrogen and phosphorus both limited NPP in the wet meadow (which had higher soil moisture and nutrient availability) (**Figure 19.12**). An additional experiment indicated that the addition of water to the dry meadow did not increase NPP, despite the positive relationship between NPP and soil moisture across the communities. These results suggest that the correlation between soil moisture and NPP in these alpine communities does not indicate a direct causal relationship, but rather is determined by the effect of soil moisture on nutrient supply through its effects on decomposition and movement of nutrients in the soil (described in Chapter 21).

Closer examination of the data in Figure 19.12 shows that the increase in NPP was not uniform across all plant species groups. The dominant plant of the alpine dry meadow (*Kobresia*) did not increase its biomass as much as the less common sedge and grass species. The change in NPP in the dry meadow occurred largely as a result of a change in plant species composition within the experimental plots. This was not the case in the wet meadows, where the dominant sedges increased their growth more than the subdominant forb species. These results are consistent with the general trend of results from many fertilization experiments, which indicate that plant species from resource-poor communities have lower growth responses to fertilization than species from resource-rich communities. This apparent contradiction is the result of differences in the capacity of plant species to respond to fertilization. Plants of resource-poor communities tend to have low intrinsic growth rates, a characteristic that lowers their resource requirements. Plants of resource-rich communities tend to have higher growth rates, which make them better able to compete for resources, particularly light. Although NPP increases in nutrient-poor communities when they are fertilized, the change in plant species composition that occurs in many such experiments indicates that plant species composition can determine the intrinsic capacity of an ecosystem to increase its NPP when resources are increased (**Figure 19.13**).

NPP is often limited by nutrients in nondesert terrestrial ecosystems. Some general differences among terrestrial ecosystem types have emerged from resource manipulation experiments and measurements of plant and soil chemistry. In lowland tropical rainforests, NPP is often limited by the supply of phosphorus, since the relatively old, leached tropical soils in which they grow are low in available phosphorus relative to other nutrients. Other nutrients, such as calcium and potassium, can also limit production in lowland tropical ecosystems. Montane tropical ecosystems, and most temperate and Arctic ecosystems, are limited by the supply of nitrogen, and occasionally by phosphorus. Even in some desert ecosystems, NPP is co-limited by water and nitrogen.

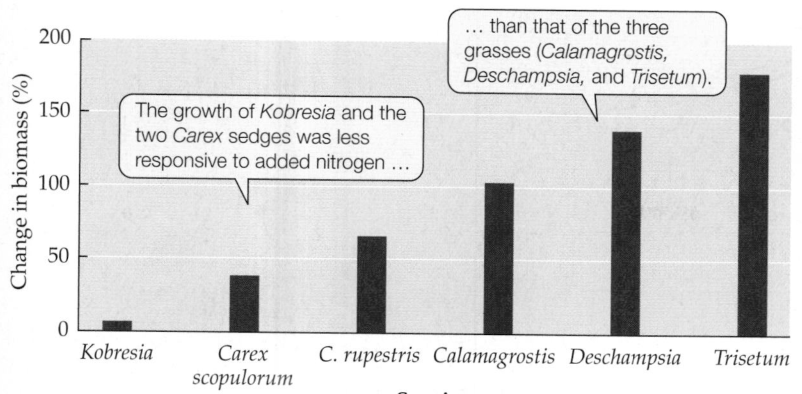

FIGURE 19.13 Growth Responses of Alpine Plants to Added Nitrogen The percent change in growth from low to high nitrogen levels (with all other nutrients maintained at optimal concentrations) indicated that alpine plant species vary substantially in their ability to increase growth in response to variation in nitrogen availability. (After Bowman and Bilbrough 2001.)

FIGURE 19.14 Limnocorral Fertilization Studies Student assistants add nutrients to an experimental enclosure in Redfish Lake, Idaho. The experiment tested whether nutrients stimulate NPP in the lake in the hope of assisting the recovery of the endangered Snake River sockeye salmon (*Oncorhynchus nerka*).

NPP in aquatic ecosystems is controlled by nutrient availability

The primary producers in lake ecosystems are phytoplankton and rooted macrophytes. NPP in lake ecosystems is often limited by the supply of both phosphorus and nitrogen, as we know not only from the results of experimental manipulations, but also from unintentional "experiments" set in motion by wastewater discharges into lakes (see p. 469). A common approach to determining the response of NPP in lakes to changes in nutrient supply is to incubate translucent or open-top containers, sometimes referred to as "limnocorrals," of lake water, amended with one or more nutrients, in the lake (**Figure 19.14**). The NPP response is measured by changes in chlorophyll concentrations or numbers of phytoplankton cells.

One of the most convincing studies of the effect of nutrients on NPP in lakes was a series of whole-lake fertilization experiments by David Schindler (Schindler 1974). The experiments were initiated in 1969 in the Experimental Lakes Area in Ontario, a series of 58 small lakes set aside for experimental manipulations. Concern over declining water quality in the lakes of North America and Europe motivated Schindler and his colleagues to establish several experiments to determine whether inputs of nutrients in wastewater were involved in the dramatic increases in the growth of phytoplankton that had been observed. They added nitrogen, carbon, and phosphorus to all or half of several individual lakes. The results of these experiments provided strong evidence for phosphorus limitation of NPP (**Figure 19.15**). Massive increases in the abundances of cyanobacteria were responsible for the increase in NPP in response to phosphorus addition. Evidence for nitrogen limitation of NPP in high-elevation lakes also exists based on small-scale fertilization experiments and measurements of the ratio of nitrogen to phosphorus in the water (Elser et al. 2007).

FIGURE 19.15 Response of a Lake to Phosphorus Fertilization Experimental Lake 226 was divided into two sections as part of David Schindler's experiments on the effects of nutrient availability on NPP.

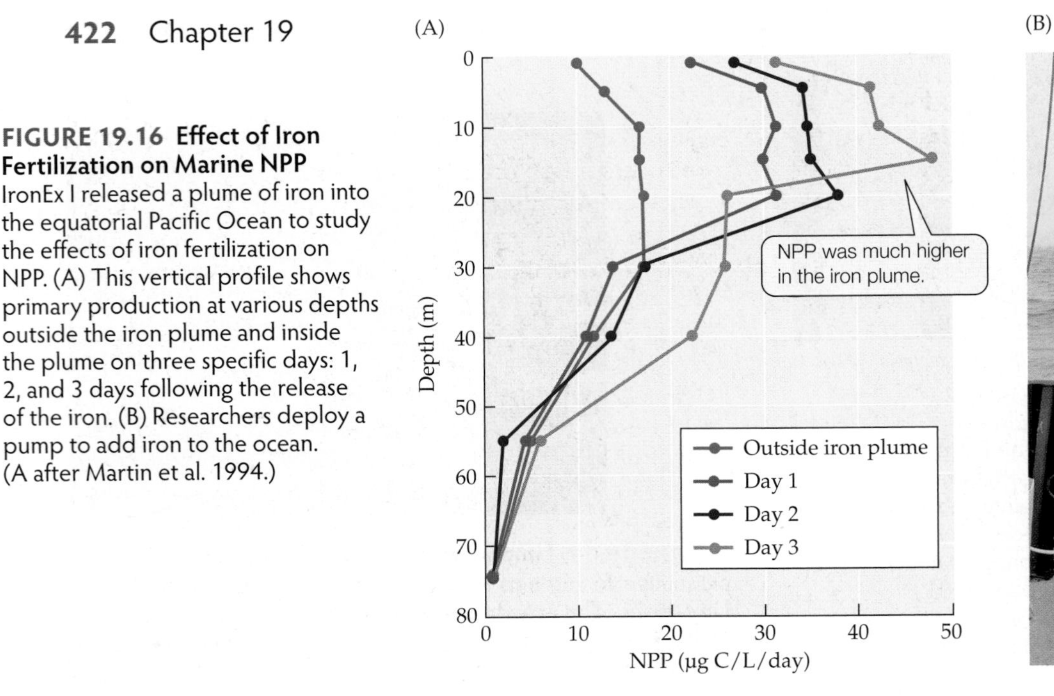

FIGURE 19.16 Effect of Iron Fertilization on Marine NPP IronEx I released a plume of iron into the equatorial Pacific Ocean to study the effects of iron fertilization on NPP. (A) This vertical profile shows primary production at various depths outside the iron plume and inside the plume on three specific days: 1, 2, and 3 days following the release of the iron. (B) Researchers deploy a pump to add iron to the ocean. (A after Martin et al. 1994.)

NPP in streams and rivers is often low, and the majority of the energy in those ecosystems is derived from terrestrial organic matter (see Chapter 20). Water movement limits the abundance of phytoplankton, except where the water velocity is relatively low. In our discussion of Concept 3.2, we introduced the *river continuum concept*, which describes the increasing importance of in-stream NPP as the river flows downstream. Most of the NPP in streams and rivers comes from photosynthesis by macrophytes and algae attached to the bottom in shallow areas where there is enough light for photosynthesis. Suspended sediment in rivers can limit light penetration; thus, turbidity often controls NPP. Nutrients, particularly nitrogen and phosphorus, can also limit NPP in streams and rivers.

Marine NPP is usually limited by nutrient supply, but the specific limiting nutrients vary among marine ecosystem types. Estuaries, the zones where rivers empty into the ocean (described under Concept 3.3), are rich in nutrients relative to other marine ecosystems. Variation in NPP among estuaries is correlated with variation in nitrogen inputs from rivers. Agricultural and industrial activities have increased riverine inputs of nitrogen into estuaries, which have caused periodic "blooms" of algae. These blooms have been implicated in the development of "dead zones"—areas of high fish and zooplankton mortality—in over 400 nearshore ecosystems worldwide, as we'll see in Chapter 24.

NPP in the open ocean is derived primarily from phytoplankton, including a group referred to as the *picoplankton*, consisting of cells smaller than 1 µm. Picoplankton contribute as much as 50% of the total marine NPP. Smaller contributions come from floating mats of seaweeds such as *Sargassum*. Near the coast, kelp forests may have leaf area indices and rates of NPP as high as those of tropical forests. "Meadows" of seagrasses such as eelgrass (genus *Zostera*) are also important contributors to NPP in shallow nearshore zones.

In much of the open ocean, NPP is limited by nitrogen. In the equatorial Pacific Ocean, however, detectable concentrations of nitrogen can be found in the water even when peak NPP occurs, suggesting that some other factor limits NPP. John Martin and colleagues measured the concentrations of nutrients in the open waters of the Pacific and performed bottle incubation experiments with iron added. They found that adding iron to the bottles increased NPP (Martin et al. 1994). Based on this evidence that iron limits NPP in some ocean regions, Martin suggested that aeolian (windblown) dust from Asia, a source of iron for the open ocean, could play an important role in the global climate system through its influence on marine NPP, and thus on atmospheric CO_2 concentrations. During glacial periods, large areas of the continents lacking vegetative cover could have contributed aeolian dust that would have fertilized the ocean. As NPP in marine ecosystems increased, those ecosystems might have taken up more CO_2 from the atmosphere, reducing its atmospheric concentration and serving as a positive feedback to cool the climate further. Martin suggested that these findings might be applied to address global warming, saying at the time, "Give me half a tankerload of iron, and I'll give you an Ice Age." He recommended the use of large-scale experiments to investigate the influence of iron on ocean NPP. Unfortunately, Martin died in 1993, before his ambitious experiments could be carried out.

Martin's colleagues subsequently performed the first of several experiments in 1993, adding iron sulfate to surface waters of the equatorial Pacific west of the Galápagos archipelago. This experiment was alternatively referred to as IronEx I or the "Geritol solution"* to global climate

*Geritol is a dietary supplement once widely believed to help cure "iron-poor, tired blood."

change. During IronEx I, a 64 km^2 area was fertilized with 445 kg of iron, which resulted in a doubling of phytoplankton biomass and a fourfold increase in NPP (**Figure 19.16**). Three other iron fertilization experiments were subsequently performed, one in 1995 (IronEx II), which produced a tenfold increase in phytoplankton biomass; a second in 1999 in the Southern Ocean; and the last in 2002, also in the Southern Ocean. While the iron limitation hypothesis has been strongly supported by these experiments, fertilizing large areas of the ocean is unlikely to provide a solution to increasing atmospheric CO_2 concentrations and global climate change. Some of the CO_2 taken up by phytoplankton is eventually reemitted to the atmosphere via respiration by zooplankton and bacteria that consume the phytoplankton. In addition, the iron is lost relatively quickly from the surface photic zone, sinking to deeper layers where it is unavailable to support phytoplankton photosynthesis and growth. Iron fertilization on a large scale could also have detrimental effects on ocean biodiversity and could create large "dead zones" similar to those generated by nitrogen inputs into estuaries.

The development of remote sensing and eddy covariance techniques has improved our ability to discern global patterns of NPP. We'll examine those patterns in the next section.

CONCEPT 19.3 Global patterns of net primary production reflect climatic constraints and biome types.

Global Patterns of NPP

Initial estimates of global NPP were based on compilations of plot-level measurements from different biomes, scaled up using estimates of the spatial distributions of those biomes. These estimates were subject to error associated with the uncertainty of the actual area covered by each biome type, as well as with the potential for overestimating NPP if undisturbed, old-growth study plots were selected to represent a biome. Remote sensing data now provide direct measurements of NPP independent of estimates of biome type or landscape condition.

Total terrestrial and oceanic NPP are nearly equal

Chris Field and colleagues estimated total planetary NPP to be 105 petagrams (1 Pg = 10^{15} g) of carbon per year, based on remote sensing data collected over multiple years (Field et al. 1998). They determined that 54% of this carbon is taken up by terrestrial ecosystems, while the remaining 46% is taken up by primary producers in the oceans. Their estimate of oceanic NPP (which comes to 48 Pg C/year) was considerably higher than previous estimates. Despite the similar contributions of land and oceans to total global NPP, the average rate of NPP on the land surface (426 g C/m^2/year) is higher than that in the oceans (140 g C/m^2/year). The lower rate in the oceans is compensated for by the greater percentage (70%) of Earth's surface it covers.

Most of the surface of both oceans and land is dominated by areas with relatively low NPP (see Figures 19.7 and 19.9). The highest rates of NPP on land are found in the tropics (**Figure 19.17**). This pattern results from latitudinal variation in climate and in the length of the growing season. Higher latitudes have shorter growing seasons, and low temperatures constrain nutrient supply by lowering decomposition rates, which in turn limits NPP. Tropical zones have long growing seasons and high rates of precipitation, promoting high rates of NPP. NPP declines to the north and south of the tropics at about 25°, reflecting the increasing aridity at those latitudes, associated with the high-pressure zones generated by the descending air of the Hadley cells (described under Concept 2.2). Another peak in terrestrial NPP occurs at the northern mid-latitudes, where the temperate forest biome is found. NPP in the mid- to high latitudes shows strong seasonal trends, with peaks in summer and declines in winter. In contrast, seasonal trends in the tropics are often slight and are associated with wet–dry cycles.

Oceanic NPP peaks at the mid-latitudes between 40° and 60° (see Figure 19.17). These peaks are associated with zones of upwelling, areas where ocean currents bring nutrient-rich deep water to the surface (as described under

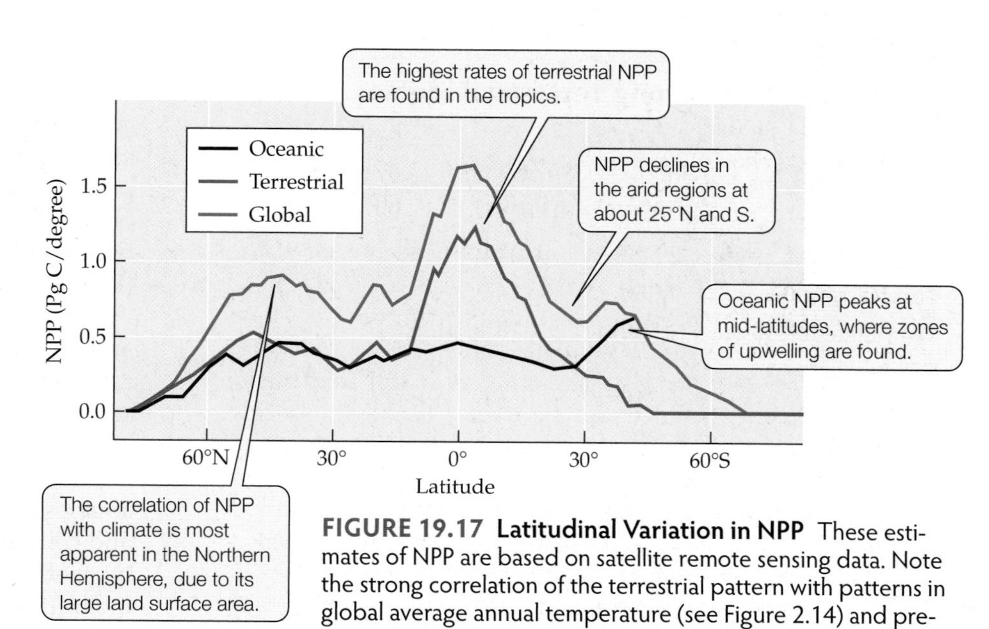

FIGURE 19.17 Latitudinal Variation in NPP These estimates of NPP are based on satellite remote sensing data. Note the strong correlation of the terrestrial pattern with patterns in global average annual temperature (see Figure 2.14) and precipitation (see Figure 2.16). (After Field et al. 1998.)

Concept 2.2). High NPP is also associated with estuaries at these latitudes. Seasonal trends in NPP occur in the oceans, but their magnitude is less than on the land surface.

Differences among biomes in NPP reflect climatic and biotic variation

It is not surprising that NPP varies among biomes, since biomes are associated with latitudinal climatic variation. For example, the high NPP in the tropics is associated with tropical ecosystems, including forests, grasslands, and savannas. The low NPP at high latitudes is associated with boreal forests and tundra. Tropical forests and savannas contribute approximately 60% of terrestrial NPP (**Table 19.1**), or about 30% of total global NPP. In the oceans, zones of upwelling have high rates of NPP, but cover less than 5% of the ocean surface. Coastal zones, including estuaries, account for approximately 20% of oceanic NPP, or about 10% of total global NPP. Despite its low rate of NPP, the vast area of the open ocean accounts for the majority of oceanic NPP and approximately 40% of total global NPP.

As noted under Concept 19.1, much of the variation in NPP among terrestrial biomes is associated with differences in leaf area index. In addition, the length of the growing season varies markedly among biomes, from year-round in some tropical ecosystems to 100 days or less in Arctic tundra. Variation associated with different plant growth forms (e.g., C_3 versus C_4 plants; grasses versus shrubs versus trees) is also important, but contributes less to variation among biomes than do growing season and leaf area index. Variation in NPP among aquatic ecosystems, as we also saw under Concept 19.1, is primarily related to variation in inputs of nutrients.

TABLE 19.1

Variation in NPP among Terrestrial Biomes

Biome	NPP (g/m²/yr)	Total NPP (Pg/yr)	Percentage of terrestrial NPP
Tropical forest	2,500	21.9	35
Tropical savanna	1,080	14.9	24
Temperate forest	1,550	8.1	13
Temperate grassland	750	5.6	9
Boreal forest	390	2.6	4
Shrubland	500	1.4	2
Tundra	180	0.5	1
Desert	250	3.5	5
Crops	610	4.1	6

Source: Saugier et al. 2001.

What happens to all of this NPP? In the next section, we will introduce some of the concepts associated with secondary production. We will cover energy flow among organisms and its consequences for population growth, community dynamics, and ecosystem function in Chapter 20.

CONCEPT 19.4 Secondary production is generated through the consumption of organic matter by heterotrophs.

Secondary Production

Energy that is derived from the consumption of organic compounds produced by other organisms is known as **secondary production**. Organisms that obtain their energy in this manner are known as *heterotrophs*, and they include animals, archaea, bacteria, fungi, and even a few plants (see the Case Study in Chapter 11, p. 242).

Heterotrophs are classified according to the type of food they consume. The most general categories, introduced in Chapter 5, are *herbivores*, which consume plants and algae; *carnivores*, which consume live animals; and *detritivores*, which consume dead organic matter (detritus). Organisms that consume live organic matter from both plants and animals are called *omnivores*. Further refinement of feeding preferences is sometimes incorporated into the terminology used to describe heterotrophs; insect eaters, for example, are referred to as *insectivores*.

Heterotroph diets can be determined from the isotopic composition of food sources

Determining what organisms eat may be as simple as watching them feed. Such observations, however, may be a time-consuming and imprecise exercise. An alternative method of determining an organism's diet involves measuring stable isotopes (see Ecological Toolkit 5.1). The concentrations of naturally occurring stable isotopes of carbon (^{13}C), nitrogen (^{15}N), and sulfur (^{34}S) differ among potential food items. Measuring the isotopic composition of the feeding organism and its potential food sources can help us identify the food sources that make up its diet (Peterson and Fry 1987).

Isotopic measurement was used by Diane Davidson and colleagues to address a feeding mystery in tropical rainforests. Ants are abundant in the tree canopy, representing 80%–90% of the arthropod biomass (Davidson et al. 2002). Thus, the abundance of suitable arthropod prey is low compared with the ants' abundance. What are the food sources that allow the ants to be so abundant? Davidson and colleagues suspected that the ants must be obtaining most of their food directly or indirectly from plant sources, including plant tissues as well as phloem sap exuded by aphids and other sap-feeding insects. To evaluate this possibility, they examined the correspondence between the nitrogen isotopic ratios ($^{15}N/^{14}N$, see Ecological Toolkit 5.1) of different groups of ants and their potential food sources. These food sources included plants, sap-feeding insects (both the exuded sap and the insects themselves), herbivores, and predatory arthropods (**Figure 19.18**). The nitrogen isotopic ratios of the ants' bodies indicated that the majority of their nitrogen, and thus their diet, came from sap exuded by sap-feeding insects. Many groups of ants form mutualistic relationships with sap-feeding insects, protecting them from predators and parasites and feeding on the sap they produce. Some of the ant groups that tended sap-feeding insects had N isotopic ratios that suggested they also fed on insects. There was also some evidence that nitrogen-fixing bacterial symbionts in the guts of some ant groups provided them with nitrogen. Davidson and colleagues' hypothesis was supported: most of the diet of the ants was derived from plant sources.

Net secondary production is equal to heterotroph growth

Not all of the organic matter consumed by heterotrophs is incorporated into heterotroph biomass. Some is used in respiration, and some is egested (lost in urine and feces).

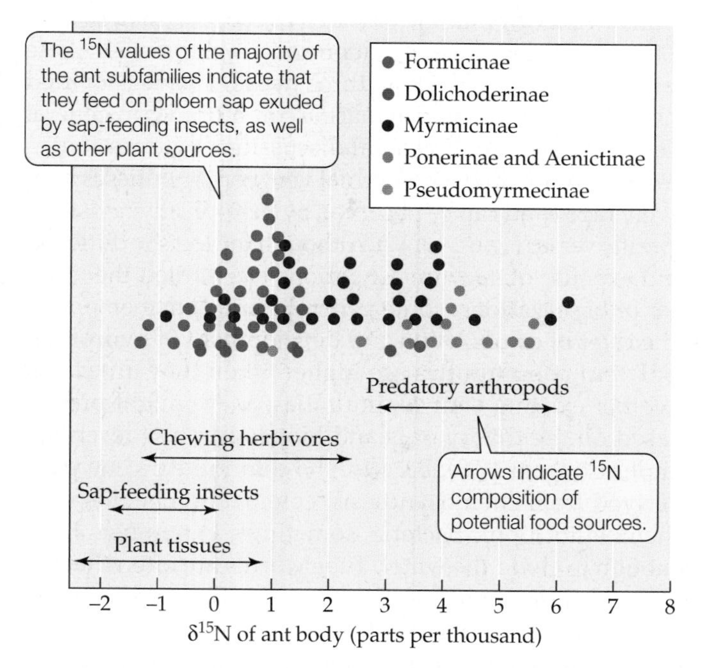

FIGURE 19.18 Nitrogen Isotopic Ratios of Ants and Their Diets Each dot represents a different ant subfamily collected from tropical forest canopies in Brunei. Its position on the *x* axis indicates the ^{15}N values found in the ants' bodies. Arrows indicate the ^{15}N values associated with various potential ant foods. Similar data were obtained from ant communities in tropical forest canopies of Peru. (After Davidson et al. 2002.)

Net secondary production is therefore the balance between ingestion, respiratory loss, and egestion:

$$\text{net secondary production} = \text{ingestion} - \text{respiration} - \text{egestion}$$

Net secondary production by a heterotroph depends on the quality of its food, in terms of digestibility and nutrient content. In addition, the physiology of the heterotroph influences how effectively its food intake is channeled into growth. Animals with high respiration rates (e.g., endotherms) have less energy left over to allocate to growth.

Net secondary production in most terrestrial ecosystems is a small fraction of NPP due to predation on herbivores and plant defenses, as we'll see in Chapter 20. Net secondary production in aquatic ecosystems represents a greater fraction of NPP than it does in terrestrial ecosystems. The majority of net secondary production in most ecosystems is associated with detritivores, primarily bacteria and fungi.

A CASE STUDY REVISITED

Life in the Deep Blue Sea, How Can It Be?

In this chapter, we have emphasized the importance of photosynthetic autotrophs as the source of energy for ecosystems, since the vast majority of the energy that enters ecosystems is derived from visible solar radiation. Here and in Chapter 5, however, we have alluded to another source of energy for ecosystems: chemosynthesis. Some bacteria can use chemicals such as hydrogen sulfide (H_2S and related chemical forms, HS^- and S^{2-}) as electron donors to take up carbon dioxide and convert it into carbohydrates:

$$S^{2-} + CO_2 + O_2 + H_2O \rightarrow SO_4^{2-} + (CH_2O)_n$$

Bacteria that provide energy for ecosystems via chemosynthesis are known as *chemoautotrophs*. The existence of chemoautotrophic bacteria was known for at least a century before the discovery of hydrothermal vents, which we described in this chapter's Case Study on p. 410, but their role in providing energy for the vent communities was uncertain.

Initially, hypotheses suggested that the high velocity of water flow around the hydrothermal vents helped direct organic matter from the photic zone toward the filter-feeding invertebrates. However, several lines of evidence suggested that chemoautotrophs were the major source of energy for these ecosystems. First, the carbon isotopic ratios ($^{13}C/^{12}C$) in the bodies of the vent invertebrates are different from those of phytoplankton in the photic zone (see Ecological Toolkit 5.1). Second, the tube worms collected from the vents (genus *Riftia*) lack mouths and digestive systems. These gutless tube worms also have structures called trophosomes, made up of highly vascularized tissues with specialized cells containing large

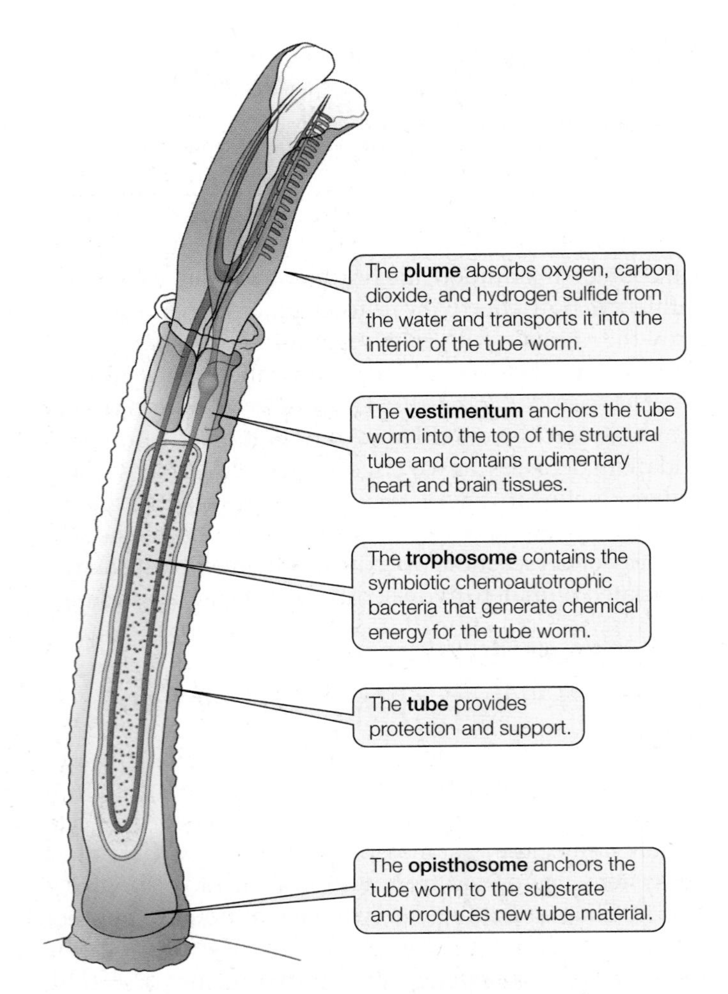

FIGURE 19.19 *Riftia* Anatomy *Riftia* tube worms have a number of specialized structures that make them well adapted to their hydrothermal vent environment.

amounts of bacteria (**Figure 19.19**). Elemental sulfur was found in the trophosomes, suggesting that sulfides were being chemically transformed in the tube worms' bodies. Finally, enzymes associated with the Calvin cycle, the biochemical pathway used by autotrophs to synthesize carbohydrates (described under Concept 5.2), as well as enzymes involved in sulfur metabolism were found in the trophosomes. Furthermore, the clams and other mollusks collected from the vent communities lacked some of the critical tissues for filter feeding, and they also had large amounts of bacteria in specialized tissues, as well as enzymes associated with the Calvin cycle.

All of this evidence pointed to the conclusion that deep-sea hydrothermal vent communities derive their energy from chemoautotrophic bacteria. These bacteria also aid in detoxifying the sulfides in the water, which would normally inhibit aerobic respiration. Furthermore, many of the abundant organisms have symbiotic relationships with the bacteria—that is, they house the chemoautotrophs in their bodies, often in specialized structures. Is this interaction a mutualistic symbiosis of the kind described in Chapter 14? The tube worms and clams housing the bacteria benefit by obtaining carbohydrates to fuel their metabolic processes, growth, and reproduction, as well as from detoxification of the sulfides. Do the bacteria derive any benefit from the invertebrates? The answer is yes: the invertebrates provide them with a chemical environment unlike that found in the surrounding water, supplying them with more carbon dioxide, oxygen, and sulfides than they could obtain if they were free-living in the water or the sediments surrounding the vent. The symbiosis between the bacteria and the invertebrates therefore results in higher productivity than if the organisms lived separately.

CONNECTIONS IN NATURE

Energy-Driven Succession and Evolution in Hydrothermal Vent Communities

Hydrothermal vent environments are dynamic, born with the eruption of new hot springs, which eventually cease to emit sulfide-laden water as the subsurface water channels are altered and the underlying magma cools (Van Dover 2000). When the hot springs no longer emit water, and the sulfide in the seawater has been consumed, the communities surrounding the vents collapse as their energy source disappears and the physical substrate falls apart. The life span of vent communities varies from approximately 20 to 200 years. Studies of colonization and development in these communities over the past three decades have provided insights into succession in marine communities in general (see Chapter 16 for a general discussion of succession).

Succession in hydrothermal vent communities is relatively rapid and can be observed by periodically revisiting specific vents (**Figure 19.20**). Although the logistic difficulty and expense of such investigations has limited the number of observations, some general trends have emerged. The rates of colonization and development of hydrothermal vent communities are higher when they are closer to other existing vent communities, as we might predict based on the theory of island biogeography (described under Concept 17.3). Because the community's energy is derived from chemosynthesis, colonization begins with chemoautotrophic bacteria, sometimes in numbers large enough to cloud the water. Tube worms are often the first invertebrates to arrive. Clams and other mollusks are thought to be stronger competitors for sites with optimal temperatures and water chemistry, and over time they increase in abundance at the expense of the tube worms. A few scavengers and carnivores, such as crabs and lobsters, are found in the developing community, although at low abundances. As the tube worm and bivalve populations decline with the drop in sulfide input when water flow from the vent decreases, the abundance of scavenger organisms increases until the energy available in the form of detritus is gone.

The pattern of succession in hydrothermal vent communities is subject to the same random factors that influence succession in other habitats: the order of arrival of organisms at a site can influence the long-term dynamics of the community (see the discussion under Concept 16.4). Neighboring vent communities found in the same area of a mid-ocean ridge may show different stages of succession, associated with the stages of hot spring development, as well as different trajectories of succession due to differences in the organisms present. Thus, collections of hydrothermal vents within the same general area are a mosaic of communities at different successional stages, similar to those in terrestrial forest patches, albeit separated by greater distances than the patches within a forest.

The unique nature of the energy supply in hydrothermal vent communities would suggest strong evolutionary divergence between the organisms that inhabit the vents and their nearest non-vent relatives (described under Concept 6.4). Where phylogenetic relationships between the vent organisms and their non-vent relatives have been worked out, the divergence is indeed deep, usually at the level of genus, family, or order. Since the discovery of hydrothermal vents, approximately 500 new vent species have been described; of these species, about 90% are endemic to hydrothermal vents. However, large areas of mid-ocean ridges potentially containing hydrothermal vents have yet to be explored.

The close association between the chemoautotrophic bacteria and their invertebrate hosts suggests the potential for a coevolutionary relationship of the type described under Concept 14.1. Have the invertebrates and their chemosynthetic bacterial symbionts evolved in concert? To address this question, Andrew Peek and colleagues compared the evolutionary relationships (phylogenetic trees) of vent-dwelling clams in the family Vesicomyidae with those of their symbiotic bacteria (Peek et al. 1998). Clams in this family transfer bacteria to their offspring in the cytoplasm of their eggs. Peek and colleagues collected eight species of clams in three genera from hydrothermal vent communities at latitudes ranging from 18° N to 47° N and at depths ranging from 500 to 6,370 m. Ribosomal DNA taken from the clams and the bacteria was used to construct the phylogenetic trees. The two trees showed remarkable congruence (**Figure 19.21**), providing strong evidence that speciation in the clams and in their bacterial symbionts has occurred synchronously. Other vent groups lack this apparent coevolutionary relationship, however. For example, three different species of tube worms found in different geographic locations have been found to contain the same species of sulfur-oxidizing bacteria.

Recently, it has been suggested that hydrothermal vents are a potential site for the origin of life on Earth. The

The site has been colonized by tube worms in the genus *Tevnia* (bottom right) (March 1992).

FIGURE 19.20 Succession in Hydrothermal Vent Communities The species composition and abundances in a hydrothermal vent community change over time following the eruption of a hot spring.

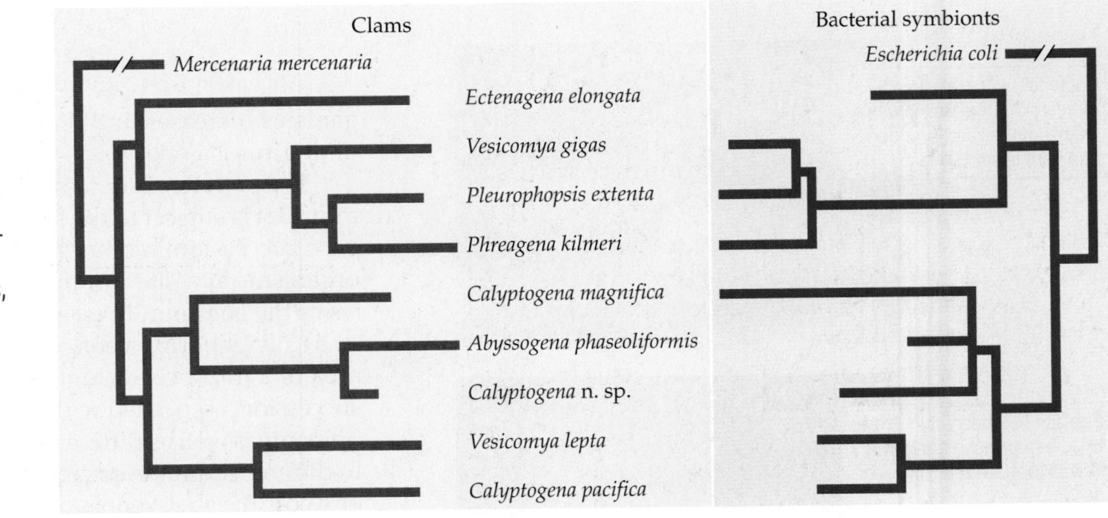

FIGURE 19.21 Coevolution of Vent Clams and Their Symbiotic Bacteria The phylogenetic trees of vesicomyid clams collected from hydrothermal vents and their accompanying chemoautotrophic bacterial symbionts show remarkable parallels, suggesting that these species have coevolved. (After Peek et al. 1998.)

reducing (i.e., electron-donating) geochemical environment of hydrothermal vents is conducive to the abiotic synthesis of amino acids, which would have been required for the development of living systems. Although amino acids are not stable in ocean water under the high pressures and temperatures found at some deep-sea hydrothermal vents, there are vents with lower temperatures at shallower depths where amino acid genesis could (and does) occur. As Cyndy Lee Van Dover (2000) so eloquently stated, "Vent water may be the ultimate soup in the sorcerer's kettle."

SUMMARY

CONCEPT 19.1 Energy in ecosystems originates with primary production by autotrophs.

- Gross primary production (GPP) is the total amount of carbon fixed by the autotrophs in an ecosystem.
- The GPP of an ecosystem is determined by the rate of photosynthesis and the leaf area index.
- Net primary production (NPP) is equal to GPP minus autotroph respiration.
- NPP changes during succession due to changes in leaf area index and in the balance between photosynthetic and nonphotosynthetic tissues.
- Researchers have developed diverse approaches to measuring NPP at different spatial and temporal scales.

CONCEPT 19.2 Net primary production is constrained by both physical and biotic environmental factors.

- Variation in terrestrial NPP is associated with variation in temperature and precipitation, both of which affect resource availability and the types and abundances of plants.
- The intrinsic growth rates of different plant species influence spatial variation in NPP and its response to variation in resource availability.
- NPP in aquatic ecosystems is controlled by the supply of nutrients, particularly phosphorus and nitrogen.

CONCEPT 19.3 Global patterns of net primary production reflect climatic constraints and biome types.

- Terrestrial and oceanic NPP contribute nearly equal proportions of global NPP.
- The majority of terrestrial NPP occurs in the tropics.
- Differences in NPP among terrestrial biomes reflect differences in leaf area index and in the length of the growing season.
- Although zones of upwelling and coastal zones have the highest rates of NPP, the open ocean accounts for the majority of oceanic NPP due to its larger area.

CONCEPT 19.4 Secondary production is generated through the consumption of organic matter by heterotrophs.

- Heterotrophs derive energy from the consumption of live or dead organic matter.
- Heterotroph diets can be determined by measuring and comparing the ratios of stable isotopes in the tissues of the feeding organism and in those of its potential food sources.
- Net secondary production is the energy ingested by heterotrophs minus the energy used in respiration and egested in feces and urine.

REVIEW QUESTIONS

1. Why is it important to know how much primary production occurs in ecosystems?
2. Plants allocate the energy they acquire through photosynthesis to different functions, including growth and metabolism. The allocation to growth can go preferentially to particular organs, such as leaves, stems, roots, or flowers. How would you expect the allocation of energy among plant organs to change as the amount of terrestrial ecosystem NPP increased? Explain why you would expect the allocation pattern you describe.
3. Ecologists interested in the underlying factors that control variation in NPP in Arctic tundra measured the growth of plants at several locations over multiple years. They also measured air and soil temperatures, wind speed, solar radiation, and soil moisture. When they analyzed all of their data, they found that the best correlation between NPP and any of the physical environmental factors they measured was with soil temperature. The researchers concluded that NPP in Arctic tundra is controlled by the effect of soil temperature on root growth. Is this conclusion correct?
4. What are some of the benefits and drawbacks associated with measuring NPP using (a) harvest techniques and (b) remote sensing?
5. Stable isotope ratios are often used as indicators of diet preference in animals. Would you expect the isotopic ratio of ^{15}N in a population of an omnivore species to vary more or less among individuals than that in a population of an herbivore species?

ON THE COMPANION WEBSITE
sites.sinauer.com/ecology2e

The website includes Chapter Outlines, Online Quizzes, Flashcards & Key Terms, Suggested Readings, a complete Glossary, and the Web Stats Review. In addition, the following resources are available for this chapter:

HANDS-ON PROBLEM SOLVING

Drought Reduces Productivity across Europe

This Web exercise examines the effect of the 2003 European drought on primary productivity. You will interpret data from a recent paper that demonstrated that the drought substantially reduced primary productivity at various sites in Europe.

CLIMATE CHANGE CONNECTIONS

19.1 The Transformation of Arctic Ecosystems from Carbon Sinks to Carbon Sources

CHAPTER

20 Energy Flow and Food Webs

KEY CONCEPTS

CONCEPT 20.1 Trophic levels describe the feeding positions of groups of organisms in ecosystems.

CONCEPT 20.2 The amount of energy transferred from one trophic level to the next depends on food quality and consumer abundance and physiology.

CONCEPT 20.3 Changes in the abundances of organisms at one trophic level can influence energy flow at multiple trophic levels.

CONCEPT 20.4 Food webs are conceptual models of the trophic interactions of organisms in an ecosystem.

Toxins in Remote Places: A Case Study

The Arctic is considered one of most pristine and remote regions on Earth. Human effects on its environment are thought to be slight relative to those in the temperate and tropical zones where the vast majority of humans live. Thus, the Arctic is one of the last places one would expect to find high levels of pollutants in living organisms.

In the mid-1980s, Eric Dewailly was studying concentrations of polychlorinated biphenyls (PCBs) in the breast milk of mothers in southern Quebec. PCBs belong to a group of chemical compounds called persistent organic pollutants (POPs) because they remain in the environment for a long time. POPs originate from industrial and agricultural activities and from the burning of industrial, medical, or municipal wastes. Exposure to PCBs has been linked to increased incidence of cancer, impairment of the ability to fight infections, decreased learning ability in children, and lower birth weights in newborns.

Dewailly was seeking a human population from a pristine area that could be used as a control in his study. He enlisted the help of some Inuit mothers from northern Canada. The Inuit are primarily subsistence hunters of the Arctic, and they have no developed industry or agriculture that would expose them to POPs (**Figure 20.1**). Dewailly therefore assumed that Inuit mothers would have few or no PCBs in their breast milk, providing a benchmark against which to compare populations in more industrialized areas.

What Dewailly found was startling: the Inuit women had concentrations of PCBs in their breast milk that were seven times higher than those in women of southern Quebec (**Figure 20.2**) (Dewailly et al. 1993). These alarming findings were reinforced by the work of Harriet Kuhnlein, who at the same time found that approximately two-thirds of the children from an Inuit community in northeastern Canada had PCB levels in their blood that exceeded Canadian health guidelines (Kuhnlein et al. 1995). More extensive surveys found that POPs were widespread in Inuit populations. As many as 95% of the people in Inuit communities of Greenland had blood levels of PCBs that exceeded health standards (Pearce 1997).

How were these toxins finding their way into the Arctic environments where the Inuit live? The POPs that were found in the tissues of Inuit populations occur in gaseous form at most environmental temperatures. Produced in lower-latitude industrial areas,

FIGURE 20.1 Subsistence Hunting Inuit hunters peel layers of skin and fat from a slaughtered bowhead whale in a remote, very sparsely populated Arctic region.

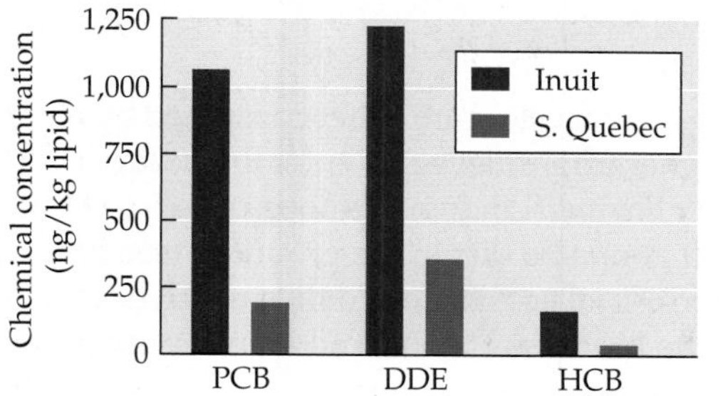

FIGURE 20.2 Persistent Organic Pollutants in Canadian Women The breast milk of Inuit mothers from northern Canada was found to contain substantially higher concentrations of polychlorinated biphenyls (PCBs) and two other POPs—dichlorodiphenyldichloroethylene (DDE, a pesticide similar to DDT), and hexachlorobenzene (HCB, an agricultural fungicide)—than that of mothers from southern Quebec. (After Dewailly et al. 1993.)

these compounds enter the atmosphere under warmer conditions, but when carried by atmospheric circulation patterns into the colder atmosphere of the Arctic, they condense into liquid forms and fall from the atmosphere, sometimes in snowflakes. The manufacture and use of most POPs has been banned in North America since the 1970s. Some developing countries continue to produce POPs, however, and they are important sources of the compounds found in Arctic regions. Although emissions of POPs have decreased, these compounds may remain in Arctic snow and ice for many decades, being released slowly during snowmelt every spring and summer.

While the source of the POPs was certain, the high concentrations of these compounds in the Inuit were a mystery. The concentrations of POPs in their drinking water were not high enough to explain this phenomenon. One hint came from the correlation between the levels of the toxins in people and in their preferred diets. Communities that had traditionally relied on marine mammals for their food tended to have the highest levels of POPs, while communities that consumed herbivorous caribou tended to have lower levels. We will discover the ecological basis for this difference as we trace the flow of energy and materials through ecosystems in this chapter.

Introduction

To begin our discussion of energy flow in ecosystems, we move from the Arctic to a much warmer place: a North American desert. Despite their aridity, deserts contain diverse assemblages of plants, animals, and microorganisms. This diversity is reflected in the variation in size, shape, and physiology of the animals making up the desert fauna, from nematodes in the soil to grasshoppers in the plant canopy to hawks in the sky. What links these animals together in the context of ecological functioning isn't necessarily their physical appearances or their taxonomic relationships. Rather, their ecological roles are determined by what they eat and by what eats them—that is, by their feeding, or *trophic*, interactions. In other words, the influence an organism has on the movement of energy and nutrients through an ecosystem is determined by the type of food it consumes, as well as by what consumes it. For example, grasshoppers and scorpions are both arthropods, with similar morphology and physiology, yet their ecological effects on energy flow through the desert ecosystem are quite different. In the context of energy flow, grasshoppers are more similar to mule deer than to scorpions. Grasshoppers and mule deer are both generalist herbivores that consume a variety of desert plant species. The scorpion, by contrast, is a carnivorous arthropod, feeding primarily on other insects, and has an ecological role more similar to that of a kestrel than to that of a grasshopper.

In this chapter, we will continue our discussion of energy, describing its flow through ecosystems and the factors that control its movement through different trophic levels. We will also look at the feeding relationships in an ecosystem as an intricate web of interactions between species, a view that has important implications for energy flow and ecosystem function as well as species interactions and community dynamics (the topics of Chapters 15 and 18).

CONCEPT 20.1 Trophic levels describe the feeding positions of groups of organisms in ecosystems.

Feeding Relationships

In Chapter 19, we introduced Ray Lindeman's simplified approach to categorizing groups of organisms in an ecosystem according to how they obtain energy (see Figure 19.3). Rather than grouping them by their taxonomic identity, he grouped them into categories based on their roles in moving energy through the ecosystem. In this section, we'll take a closer look at these feeding categories.

Organisms can be grouped into trophic levels

Each feeding category, or **trophic level**, is based on the number of feeding steps by which it is separated from autotrophs (**Figure 20.3**). The first trophic level consists of the autotrophs, the primary producers that generate chemical energy from sunlight or inorganic chemical compounds. The first trophic level also generates most of the dead organic matter in an ecosystem. In our desert ecosystem, the first trophic level includes all of the plants, which we lump together to form a single component, regardless of their taxonomic identity. In Lindeman's lake ecosystem (see Figure 19.3), the first trophic level was associated primarily with dead organic matter, which Lindeman poetically referred to as "ooze," as well as with autotrophs such as phytoplankton and pondweeds. The second trophic level is composed of the herbivores that consume autotroph biomass—which in our desert ecosystem would include grasshoppers and mule deer—as well as the detritivores that consume dead organic matter. The remaining trophic levels (third and up) contain the carnivores that consume animals at the trophic level below them. The primary carnivores constituting the third trophic level in our desert ecosystem would include small birds and scorpions, while examples of the secondary carnivores making up the fourth trophic level would include foxes and birds of prey. Most ecosystems have four or fewer trophic levels.

Some organisms do not fit conveniently into the trophic levels we have defined here. Coyotes, for example, operate as opportunistic feeders, consuming vegetation, mice, other carnivores, and old leather boots. Such omnivores* defy our attempt to group organisms into simple feeding categories. However, their diets can be partitioned to reflect how much energy they consume within each trophic level (Pimm 2002). This partitioning is facilitated by the use of stable isotopes to trace food sources (Post 2002a; see Ecological Toolkit 5.1). Omnivory is common in many ecosystems. These organisms occupy intermediate trophic levels as determined by the proportions of the food they consume.

All organisms are either consumed or end up as detritus

All organisms in an ecosystem are either consumed by other organisms at higher trophic levels or enter the pool of dead organic matter, or *detritus* (Lindeman's ooze, or as Tom Waits put it, "we're all gonna be dirt in the ground") (see Figure 20.3). In most terrestrial ecosystems, only a relatively small proportion of the biomass is consumed, and most of the energy flow passes through detritus (**Figure 20.4**). Because most of this energy flow occurs in the soil, we are not always aware of its magnitude and importance. Dead plant, microbial, and animal matter, as well as feces, are consumed by a multitude of organisms, known as detritivores (primarily bacteria and fungi), in a process known as *decomposition*. We will describe decomposition in more detail in Chapter 21 in the context of nutrient cycling. Detritus is part of the first trophic level, and detritivores are placed with herbivores in the second trophic level. Although autotroph-based and detritus-based trophic levels are sometimes considered separately, they are tightly linked through primary production, nutrient cycling, and the many organisms that acquire energy from both plants and detritus.

Energy flow through detritus is important in both terrestrial and aquatic ecosystems. Detritus in terrestrial ecosystems comes primarily from plants within the ecosystem. On the other hand, a large proportion of the input of detritus into stream, lake, and estuarine ecosystems is derived from terrestrial organic matter, which is considered external to the aquatic ecosystem. External energy inputs are re-

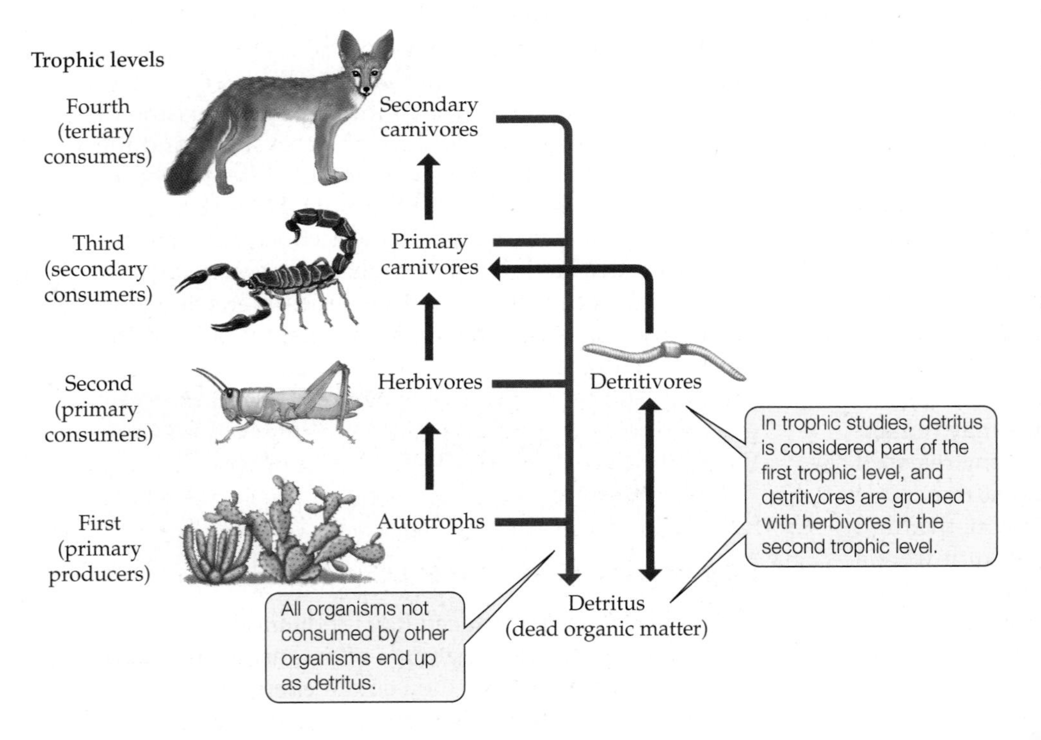

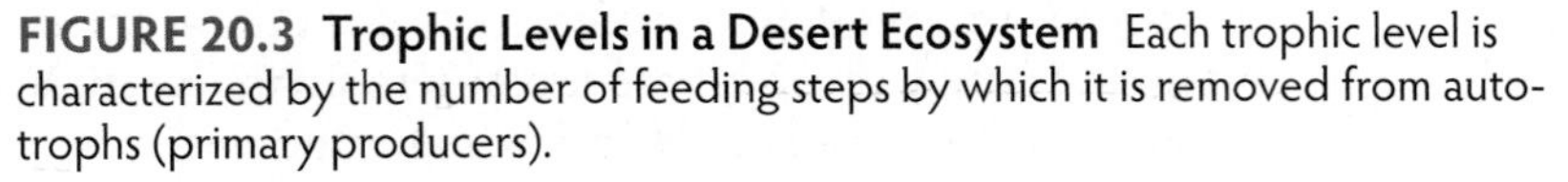
FIGURE 20.3 Trophic Levels in a Desert Ecosystem Each trophic level is characterized by the number of feeding steps by which it is removed from autotrophs (primary producers).

*In trophic studies, the term "omnivory" may be used to refer to feeding at multiple trophic levels, in contrast to our earlier definition of omnivores in Chapter 19 as heterotrophs consuming both plants and animals.

(A)

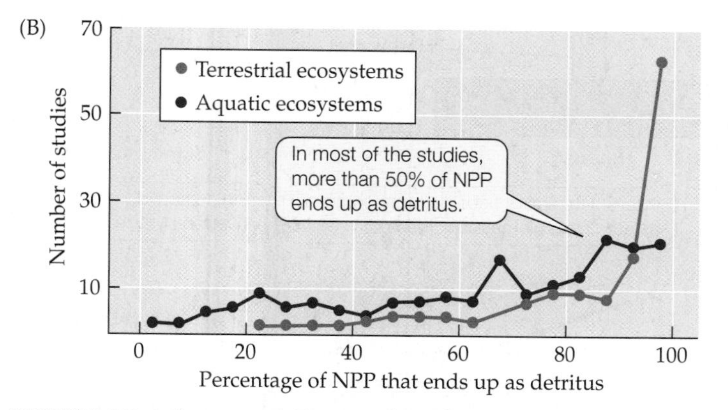

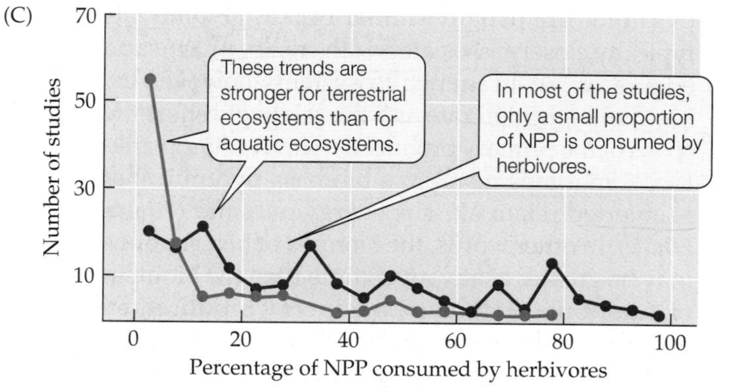

FIGURE 20.4 Ecosystem Energy Flow through Detritus (A) Detritus is consumed by a multitude of organisms, including fungi (turkey tail fungus, *Trametes versicolor*) and arthropods (such as this unidentified millipede). (B) Numerous trophic studies of both terrestrial and aquatic ecosystems have found that in most ecosystems, most of the NPP ends up as detritus. (C) Similarly, relatively little of the NPP in an ecosystem is consumed by herbivores. (B and C after Cebrian and Lartigue 2004.)

ferred to as **allochthonous** inputs, while energy produced by autotrophs within the system is known as **autochthonous** energy. Allochthonous inputs into aquatic ecosystems include plant leaves, stems, wood, and dissolved organic matter that fall in from adjacent terrestrial ecosystems or flow in via ground water. Allochthonous inputs tend to be more important in stream and river ecosystems than in lake and marine ecosystems. For example, Bear Brook, a headwater stream in New Hampshire, receives 99.8% of its energy as allochthonous inputs; the rest is NPP derived from benthic algae and mosses in the stream (Fisher and Likens 1973). In contrast, autochthonous energy accounts for almost 80% of the energy in nearby Mirror Lake (Jordan and Likens 1975). Allochthonous energy is often of lower quality, however, due to the chemical composition of the carbon compounds that enter the system. As a result, the fraction of allochthonous energy that is actually used is lower than the inputs indicate (Pace et al. 2004). The importance of autochthonous energy inputs usually increases from the headwaters toward the middle reaches of a river, in concert with decreases in water velocity and increases in nutrient concentrations, as suggested by the river continuum concept (described under Concept 3.2).

As this aquatic example shows, grouping organisms into trophic levels makes it easier to trace the flow of energy through an ecosystem. That flow is the topic to which we'll turn next.

CONCEPT 20.2 The amount of energy transferred from one trophic level to the next depends on food quality and consumer abundance and physiology.

Energy Flow among Trophic Levels

The second law of thermodynamics states that during any transfer of energy, some energy is dispersed as unusable energy due to the tendency toward an increase in disorder (entropy). Thus, we can expect that available energy will decrease with each trophic level as we move from the first trophic level upward. We know from our discussion of primary production in Chapters 5 and 19 that autotrophs lose chemical energy through cellular respiration, lowering the amount of energy available to the second trophic level. In this section, we will examine more closely the factors influencing energy movement between trophic levels.

Energy flow between trophic levels can be depicted using energy or biomass pyramids

A common approach to conceptualizing trophic relationships in an ecosystem is to construct a stack of rectangles, each of which represents the amount of energy or biomass within one trophic level. When assembled from lower to higher trophic levels, these rectangles form a **trophic pyramid**. By portraying the relative amounts of energy or biomass at each trophic level, these pyramids show us how energy flows through the ecosystem.

As we have noted, a proportion of the biomass at each trophic level is not consumed, and a proportion of the energy at each trophic level is lost in the transfer to the next trophic level. Therefore, the sizes of the energy rectangles always decrease as we move from one trophic level to the one above it. In terrestrial ecosystems, energy and biomass pyramids are usually similar because biomass is typically closely associated with energy production (**Figure 20.5A**). In aquatic ecosystems, however, the high consumption rate and the relatively short life spans of the primary producers, mainly phytoplankton, sometimes result in a biomass pyramid that is inverted relative to the energy pyramid (**Figure 20.5B**). In other words, the biomass of heterotrophs may be greater at any given time than the biomass of autotrophs. However, the energy produced by the autotrophs is still greater than that produced by the heterotrophs. This tendency toward inverted biomass pyramids is greatest where productivity is lowest, such as in nutrient-poor regions of the open ocean (**Figure 20.5C**). The higher proportion of primary consumer biomass relative to producer biomass in these nutrient-poor regions results from a more rapid turnover of phytoplankton, which have higher growth rates and shorter life spans than phytoplankton of more nutrient-rich waters. Phytoplankton in nutrient-poor regions thus provide a greater energy supply per unit of time (Gasol et al. 1997). In addition, detritus makes a higher proportional contribution to energy flow in these nutrient-poor waters than in nutrient-rich waters.

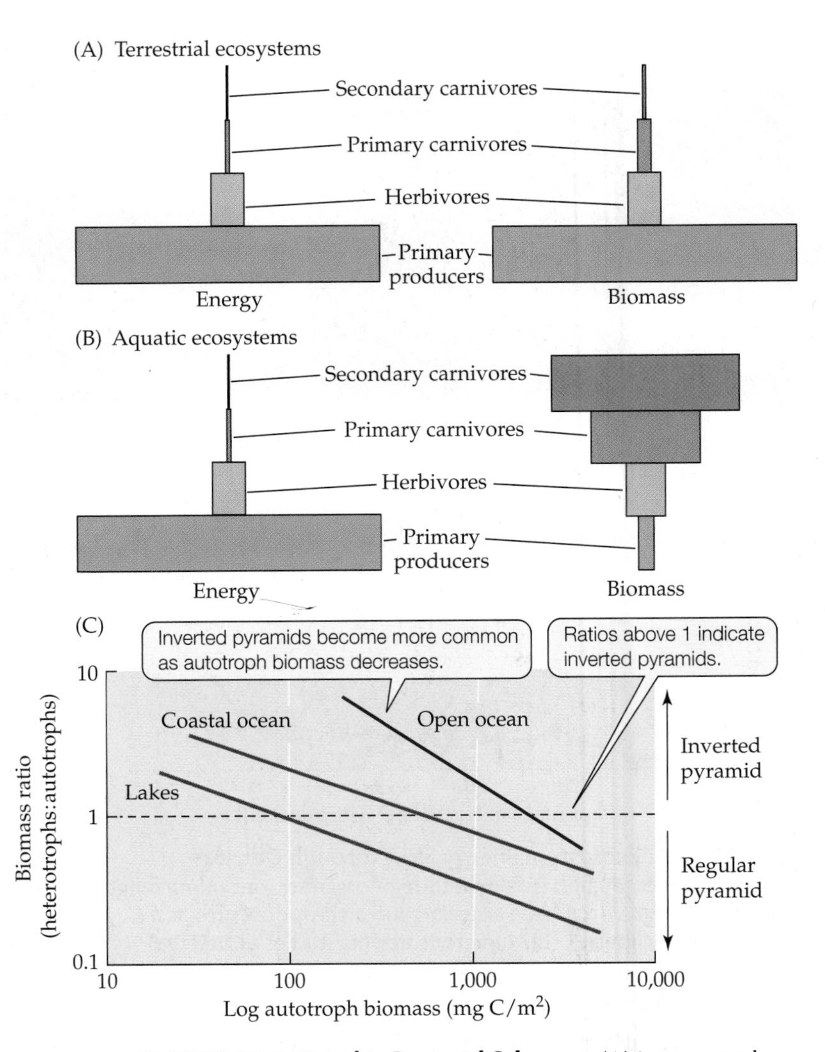

FIGURE 20.5 Trophic Pyramid Schemes (A) In terrestrial ecosystems, energy and biomass pyramids are usually similar. (B) In many aquatic ecosystems, the biomass pyramid is inverted relative to the energy pyramid. (C) Inverted biomass pyramids in aquatic ecosystems are more common in nutrient-poor waters with low autotroph biomass. (C after Gasol et al. 1997.)

Energy flow between trophic levels differs among ecosystem types

What factors determine the amount of energy that flows from one trophic level to the next? In our discussion of Concept 19.2, we evaluated the factors that influence NPP in terrestrial and aquatic ecosystems, emphasizing abiotic factors such as climate and nutrient availability as well as differences in the inherent ability of autotroph species to produce biomass. It would be reasonable to assume that the flow of energy to higher trophic levels is associated with the amount of NPP at the base of the food web. As we will see, however, the situation is not quite so simple. The proportion of each trophic level consumed by the one above it, the nutritional content of autotrophs, detritus, and prey, and the efficiency of energy transfers also play roles in determining the flow of energy between trophic levels.

A comparison of the proportions of autotroph biomass consumed in terrestrial and in aquatic ecosystems provides some insight into the factors that influence energy flow between trophic levels. When viewed from space, some parts of Earth's terrestrial surface appear green, while the ocean appears blue. Why is the land surface green and the ocean blue? Furthermore, in Chapter 19, we saw that very productive lakes (e.g., those that are experimentally fertilized; see Figure 19.15) can appear green. What these green areas have in common is primary productivity that far exceeds rates of herbivory. Herbivores on land consume a much lower proportion of autotroph biomass than do herbivores in most aquatic ecosystems. On average, about 13% of terrestrial NPP is consumed (range 0.1%–75%), while in aquatic ecosystems, an average of 35% of NPP is consumed (range 0.3%–100%) (Cebrian and Lartigue 2004).

There is a positive relationship between NPP and the *amount* of biomass consumed by herbivores (**Figure 20.6**). This relationship, which holds across most ecosystem types, would seem to suggest that herbivore production is limited by the amount of food available. Why, then, is the *proportion* of autotroph biomass consumed in terres-

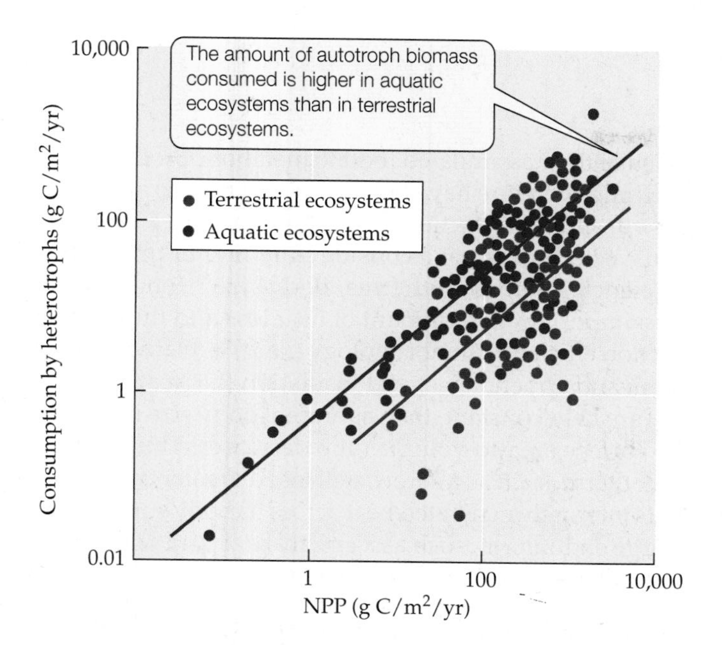

FIGURE 20.6 Consumption of Autotroph Biomass Is Correlated with NPP The amount of autotroph biomass consumed increases with increasing NPP in both terrestrial and aquatic ecosystems. (After Cebrian and Lartigue 2004.)

trial ecosystems relatively low? If herbivore production is limited by the supply of energy and nutrients from plants, why don't terrestrial herbivores consume a greater proportion of the biomass that is available?

Several hypotheses have been proposed to explain the lower proportion of autotroph biomass consumed in terrestrial ecosystems. First, Nelson Hairston and others (1960) have argued that the population growth of herbivores is constrained by predation, which keeps population sizes below their potential carrying capacities. Predator removal experiments such as those described in Chapter 12 and later in this chapter have provided support for this hypothesis in some ecosystems by demonstrating that removal of predators leads to increases in populations of their prey. Second, defenses against herbivory, such as the secondary compounds and structural defenses described under Concept 12.2, lower the amount of autotroph biomass that is consumed. Plants of resource-poor environments, such as desert and tundra, tend to be more strongly defended against herbivory than plants from resource-rich environments. This greater allocation to defense may explain why the proportion of plant biomass consumed is lower in resource-poor terrestrial environments. Third, the chemical composition of phytoplankton makes them more nutritious for herbivores than terrestrial plants. Terrestrial plants contain nutrient-poor structural materials such as stems and wood, which are typically absent in aquatic autotrophs. Herbivores typically require large amounts of nutrients such as nitrogen and phosphorus to meet their demands for structural growth, metabolism, and protein synthesis. The ratio of carbon to nutrients in a food (with carbon representing energy) is thus an important measure of its quality. Carbon:nutrient ratios differ markedly between autotrophs in terrestrial and in freshwater ecosystems. Freshwater phytoplankton have carbon:nutrient ratios closer to those of herbivores than to those of terrestrial plants (Elser et al. 2000), better meeting the nutritional needs of the herbivores. Each of these factors—predation, plant defenses, and food quality—contributes to differences in the proportion of NPP consumed among ecosystems, in particular the greater consumption of autotroph biomass in aquatic ecosystems.

The efficiency of energy transfer varies among consumers

Not all of the energy consumed by a heterotroph gets incorporated into heterotroph biomass. We can use the concept of *energy efficiency*, defined as the output of energy per unit of energy input, to characterize the transfer of energy between trophic levels. In studies of energy transfer in trophic systems, the concept of **trophic efficiency** is used, defined as the amount of energy at one trophic level divided by the amount of energy at the trophic level immediately below it. Trophic efficiency incorporates the proportion of available energy that is consumed (consumption efficiency), the proportion of ingested food that is assimilated by the consumer (assimilation efficiency), and the proportion of assimilated food that goes into producing new consumer biomass (production efficiency) (**Figure 20.7**).

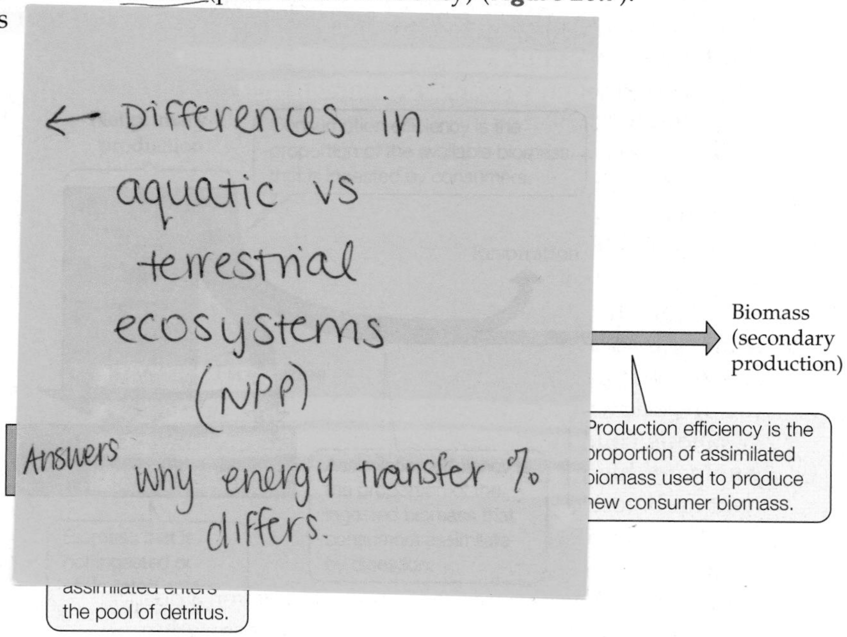

FIGURE 20.7 Energy Flow and Trophic Efficiency The proportion of energy transferred between trophic levels depends on efficiencies of consumption, assimilation, and production.

Q How do the trends in consumption efficiency vary in Figures 20.4 and 20.6? What does this variation suggest about differences in consumption efficiency in aquatic versus terrestrial ecosystems?

As we saw above, not all of the biomass available at one trophic level is consumed by the next trophic level. The proportion of the available biomass that is ingested is the **consumption efficiency**. We have seen that consumption efficiency is typically higher in aquatic ecosystems than in terrestrial ecosystems. Consumption efficiencies also tend to be higher for carnivores than for herbivores, although a systematic survey comparing the two groups has not been done.

Once biomass is ingested by the consumer, it must be assimilated by the digestive system before the energy it contains can be used to produce new biomass. The proportion of the ingested food that is assimilated is the **assimilation efficiency**. Food that is ingested but not assimilated is lost as feces to the environment, entering the pool of detritus. Assimilation efficiency is determined by the quality of the food (its chemical composition) and the physiology of the consumer.

The quality of the food available to herbivores and detritivores is generally lower than that of the food available to carnivores. Plants and detritus are composed of relatively complex carbon compounds, such as cellulose, lignins, and humic acids, that are not easily digested. In addition, plants and detritus have low concentrations of nutrients. Animal bodies, on the other hand, have a carbon:nutrient ratio that is usually very similar to that of the animal consuming them and so are assimilated more readily. Assimilation efficiencies of herbivores and detritivores vary between 20% and 50%, while those of carnivores are about 80%.

The digestive capacity of consumers is associated with their thermal physiology. Endotherms (animals that generate heat internally, as described under Concept 4.2) tend to digest food more completely than ectotherms (animals that rely on heat exchange with the environment for thermoregulation) and therefore have higher assimilation efficiencies. Additionally, some herbivores have mutualistic symbionts that help them digest cellulose. For example, ruminants such as cattle, deer, and camels have a modified foregut that contains bacteria and protists that increase the breakdown of cellulose-rich foods. This mutualistic symbiosis, coupled with a longer period of digestion, gives ruminants higher assimilation efficiencies than nonruminant herbivores.

Assimilated food can be used to produce new biomass in the form of consumer growth and production of new consumer individuals (reproduction). However, a portion of the assimilated food must be used for respiration associated maintenance of existing molecules and tissues as well as with construction of new biomass (see Chapter 5). The proportion of the assimilated food that is used to produce new consumer biomass is the **production efficiency**.

Production efficiency is strongly related to the thermal physiology and size of the consumer. Endotherms allocate much of their assimilated food to metabolic production of heat, and therefore have less energy left over to allocate to growth and reproduction than ectotherms do (**Table 20.1**). Thus, ectotherms have considerably higher production efficiencies than endotherms. Body size in endotherms is an important determinant of heat loss and thus of production efficiency. If morphology (i.e., the relative sizes of trunk and appendages) and insulation (fat, feathers, and fur) are held constant, then as animal body size increases, the surface area-to-volume ratio decreases. Thus, a small endotherm, such as a shrew, will lose a greater proportion of its internally generated heat across its body surface than a large endotherm, such as a grizzly bear, and will tend to have a lower production efficiency.

Trophic efficiencies can influence population dynamics

Changes in food quantity and quality, and the resulting changes in trophic efficiency, can determine the consumer population sizes that can be sustained and the health of the individuals in consumer populations. Here we'll examine the potential contribution of changes in food quality to the decline in numbers of Steller sea lions (*Eumetopias jubatus*) in Alaska.

From the late 1970s into the 1990s, the total population of Steller sea lions in the Gulf of Alaska and the Aleutian Islands decreased by about 80%, from approximately 250,000 sea lions in 1975 to 50,000 in 2000 (**Figure 20.8**). Andrew Trites and C. P. Donnelly reviewed the available

TABLE 20.1

Production Efficiencies of Consumers

Consumer group	Production efficiency (%)
Endotherms	
Birds	1.3
Small mammals	1.5
Large mammals	3.1
Ectotherms	
Fishes and social insects	9.8
Nonsocial insects	40.7
Herbivores	38.8
Detritivores	47.0
Carnivores	55.6
Non-insect invertebrates	25.0
Herbivores	20.9
Detritivores	36.2
Carnivores	27.6

Source: Chapin et al. 2002; data from Humphreys 1979.

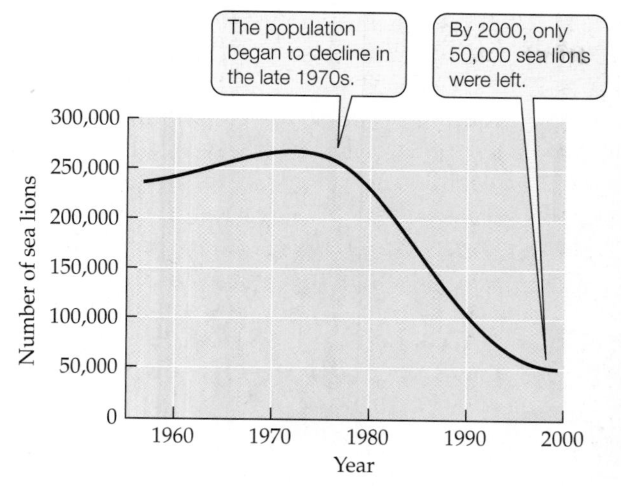

Eumetopias jubatus

FIGURE 20.8 Steller Sea Lion Population Decline in Alaska The population of sea lions in the Gulf of Alaska and the Aleutian Islands decreased by about 80% over 25 years. (After Trites and Donnelly 2003.)

information to try to determine possible causes for this decline (Trites and Donnelly 2003). They found that individual sea lions collected during the period of decline were smaller than individuals within the same age classes collected before the start of the decline. There was also a reduction in the number of pups born per female during this period, which resulted in a shift in the age structure toward older individuals. No evidence was found for outbreaks of disease or parasites. The lower body sizes and declining birth rates suggested that there were fewer prey available, or that the available prey were not providing sufficient nourishment to sustain the sea lions—in other words, that trophic efficiency had declined. Additional data indicated that the sea lions were obtaining prey—primarily fish—as regularly as they did before the decline. Nursing females in the declining population were actually spending less time hunting for the same amount of fish as nursing females in other populations that were not declining. Therefore, the availability of prey, or the sea lions' ability to capture it, did not appear to be limiting growth and reproduction.

Trites and Donnelly considered the possibility that changes in the species of prey fish available had contributed to the decline of the Steller sea lions. They and others suggested that the decline might be related to declining prey quality, an idea they referred to as the "junk food hypothesis." Prior to the decline, the diet of the sea lions was primarily herring, a fish that is relatively rich in fats, along with small amounts of pollock, cod, salmon, and squid. During the period of the population decline, the sea lion diet shifted away from herring toward a greater proportion of pollock and cod (**Table 20.2**). This change in diet reflected a shift toward cod dominance of the fish community from the 1970s through the 1990s. The causes of the change in fish community composition are uncertain, but may be associated with long-term climate change. The proportions of fat and energy per mass of pollock and cod are approximately half those of herring. Captive Steller sea lions raised on a diet of herring, and then switched to a diet of pollock, lose body mass and fat, even with an unlimited supply of pollock.

TABLE 20.2

Proportion of Steller Sea Lion Scats and Stomachs Containing Five Prey Categories

	Gadids (cod, pollock, hake)	Salmon	Small schooling fish (herring, capelin, eulachon, sand lance)	Cephalopods (squid)	Flatfish (flounder, sole)
1990–1993	85.2	18.5	18.5	11.1	13.0
1985–1986	60.0	20.0	20.0	20.0	5.0
1976–1978	32.1	17.9	60.7	0.0	0.0

Source: Trites and Donnelly 2003; data from Merrick et al. 1997.

Based on their review of the available information, Trites and Donnelly concluded that nutritional stress was the most likely cause of the decline in the Steller sea lion population. The amount of prey available to the sea lions did not appear to have changed, but changes in the quality of that prey, and associated changes in trophic efficiency, contributed to the decline in the population through their effects on individual growth rates and birth rates. Others have suggested that the decline in Steller sea lion numbers is linked to changes in trophic structure of the North Pacific (Springer et al. 2003). As described in the Case Study Revisited in Chapter 8 (p. 194), massive harvesting of great whales by humans in the mid-twentieth century may have forced their predators, killer whales, to hunt other prey, including Steller sea lions. As we describe in the next section, such "top-down" effects of predators on prey can have important consequences for energy flow in ecosystems.

CONCEPT 20.3 Changes in the abundances of organisms at one trophic level can influence energy flow at multiple trophic levels.

Trophic Cascades

There are two possible ways to look at the control of energy flow through ecosystems. First, the amount of energy that flows through trophic levels may be determined by how much energy enters an ecosystem via NPP, which in turn is related to the supply of resources (as we saw in Chapter 19). The greater the NPP entering the ecosystem, the more energy can be passed on to higher trophic levels. This view, which is often referred to as "bottom-up" control of energy flow, holds that the resources that limit NPP determine energy flow through an ecosystem (**Figure 20.9A**). Alternatively, energy flow may be governed by rates of consumption (as well as other, nonconsumptive interactions such as competition and facilitation, as discussed in Chapter 15) at the highest trophic levels, which influence abundances and species composition at multiple trophic levels below them. This view is often referred to as "top-down" control of energy flow (**Figure 20.9B**). In reality, both bottom-up and top-down controls are operating simultaneously in ecosystems, but the top-down view has important implications for the effects of trophic interactions on energy flow in ecosystems.

Trophic interactions can trickle down through multiple trophic levels

Changes in abundances or species composition at one trophic level can lead to important, somewhat unpredictable changes in the amount of biomass and species composition at other trophic levels. For example, increases in the rate of predation by a carnivore at the fourth trophic level on carnivores at the third trophic level would lead to a lower rate of consumption of herbivores at the second trophic level. More herbivory would result in lower rates of NPP, and possibly changes in abundances and species composition at the first trophic level. Nonconsumptive species interactions, such as competition, can have similar effects on abundances and species composition at lower trophic levels, as we'll see shortly. Such a series of changes in biomass and species composition is referred to as a **trophic cascade**.

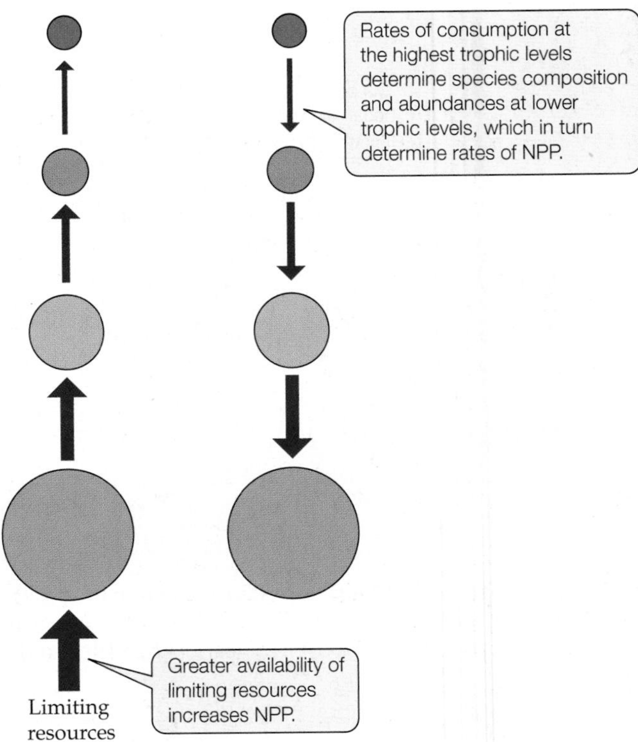

FIGURE 20.9 Bottom-Up and Top-Down Control of NPP Production in an ecosystem can be viewed as being controlled (A) by limiting resources or (B) by controls exerted on the species composition and abundances of autotrophs by consumption at higher trophic levels.

Our understanding of trophic cascades comes primarily from aquatic ecosystems, although there are examples from terrestrial ecosystems as well. Trophic cascades are most often associated with changes in the abundance of top specialist predators. Omnivory in food webs may act to buffer the effects of trophic cascades through the consumption of prey at multiple trophic levels. Finally, trophic cascades have been hypothesized to be most important in relatively simple, species-poor ecosystems. However, several recent experiments have demonstrated trophic cascades in ecosystems with relatively high species diversity.

An aquatic trophic cascade Many examples of trophic cascades come from unintended experiments associated with introductions of non-native species or near-extinctions of native species. A classic example of the latter type is the interaction among sea otters (*Enhydra lutris*), sea urchins, and killer whales on the west coast of North America, which was discussed in the Case Study Revisited in Chapter 8 (p. 194). Unfortunately, there is no shortage of examples of trophic cascades associated with the intentional or unintentional introduction of non-native species. One such example resulted from the release of brown trout (*Salmo trutta*), a popular sport fish, into streams and lakes of New Zealand. The stocking of Kiwi waters by European settlers began in the 1860s, and by 1920 an estimated 60 million fish had been released throughout New Zealand. Native fish populations have declined as a result, and some species have disappeared from streams now dominated by trout.

Alexander Flecker and Colin Townsend (1994) investigated the influence of brown trout on the species composition of its prey (primarily stream insects) and associated effects on primary production in the Shag River. Brown trout were originally released into the Shag River in 1869 by the "Otago Acclimatisation Society" to make settlers feel more at home. The Shag River is one of a small number of streams in New Zealand that still holds both native fish and trout in the same sections. The native fish include the common river galaxias (*Galaxias vulgaris*). The morphology and feeding behavior of galaxias are similar to those of trout, as indicated by their common name, Maori trout.

Flecker and Townsend compared the effects of brown trout and galaxias on stream invertebrate species composition and abundance as well as on primary production by algae. To manipulate fish presence and absence, they constructed artificial stream channels adjacent to the natural channel, made of 5 m lengths of half cylinders of PVC pipe. The PVC channels had mesh on the ends that kept fish in or out, but allowed free movement of stream invertebrates and algae. The researchers placed clean gravel and stone cobbles in the bottoms of the channels to provide a substrate for the invertebrates and algae. The channels were allowed to accumulate algae and invertebrates for 10 days before the fish were added. Three treatments were initiated: channels with introduced brown trout, channels with galaxias, and channels with no fish (controls). Eight fish of similar size and mass were used for both fish species additions. The experiment was run for 10 days, after which samples were collected to determine invertebrate species composition and abundance and algal biomass.

Flecker and Townsend had expected brown trout to decrease invertebrate diversity more than the native galaxias, but the effect of fish on invertebrate diversity was relatively small and did not differ between the two fish species. The brown trout, however, reduced total invertebrate density by approximately 40% relative to the control channels, while galaxias resulted in a smaller reduction (**Figure 20.10A**). The abundance of algae increased with both fish, but the effect was greater in the channels with trout (**Figure 20.10B**). Flecker and Townsend suggested that the effect on algal biomass was the result of a trophic cascade in which fish predation not only reduced the density of stream invertebrates, but also caused them to spend more time in refugia on the stream bottom rather than feeding on algae. The trout had a much greater effect on invertebrate density, and thus on primary production, than the native galaxias. These results suggested that trophic cascades associated with the stocking of non-native fish for sport may have consequences not just for native biodiversity, but for the functioning of stream ecosystems as well.

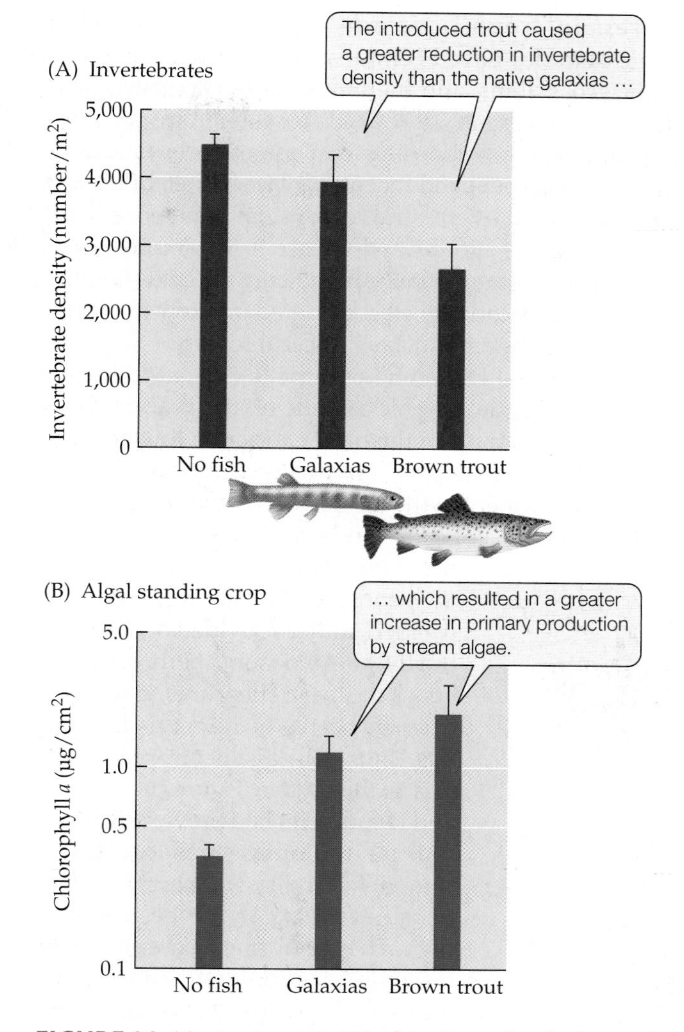

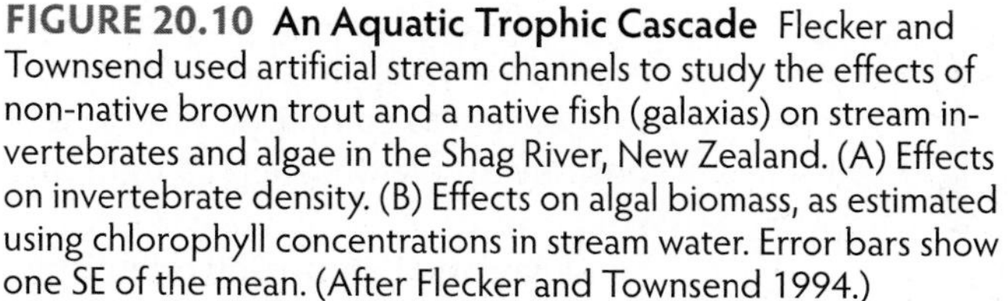

FIGURE 20.10 An Aquatic Trophic Cascade Flecker and Townsend used artificial stream channels to study the effects of non-native brown trout and a native fish (galaxias) on stream invertebrates and algae in the Shag River, New Zealand. (A) Effects on invertebrate density. (B) Effects on algal biomass, as estimated using chlorophyll concentrations in stream water. Error bars show one SE of the mean. (After Flecker and Townsend 1994.)

Q What factor other than overall consumption rate might explain why the presence of brown trout results in a larger increase in primary production than the presence of native galaxias?

A terrestrial trophic cascade As mentioned earlier, trophic cascades have been most commonly observed in aquatic ecosystems, and their existence in terrestrial ecosystems is less certain. Terrestrial ecosystems are generally thought to be more complex than aquatic ecosystems. In addition, it was believed that a decrease in the abundance of one species in a terrestrial ecosystem was more likely to be compensated for by an increase in the abundance of similar species that were not being consumed as heavily. Thus, trophic cascades were considered unlikely in diverse terrestrial ecosystems such as tropical forests.

Lee Dyer and Deborah Letourneau (1999a) tested the effects of a potential trophic cascade on the production of *Piper cenocladum* trees in the understory of a lowland wet tropical forest in Costa Rica. *Piper cenocladum* is a relatively common component of the understory in these forests and is eaten by dozens of different herbivore species. Ants of the genus *Pheidole* live in chambers of the petioles of the *Piper* trees. The ants eat food bodies provided by the trees, but they also consume herbivores that attack the trees. These ants, in turn, are eaten by beetles of the genus *Tarsobaenus*. Thus, four distinct trophic levels exist in this system (**Figure 20.11**). Dyer and Letourneau had previously noted that plant biomass was lower, and rates of herbivory were higher, when densities of *Tarsobaenus* beetles were high. They performed experiments to test whether a top-down trophic cascade involving the beetles, ants, and herbivores influenced the production of the *Piper* trees, and how strong that influence was compared with that of bottom-up factors such as light and soil fertility.

Dyer and Letourneau established experimental plots in the understory containing uniform-sized cuttings of *Piper* trees. They treated two groups of plots with an insecticide to kill any ants present, then added *Tarsobaenus* beetle larvae to one of those groups of plots. This procedure established three groups of treatment plots: one group of insecticide-treated plots with beetles and two control groups, one group of insecticide-treated plots without beetles, and one group of untreated plots. In the plots with beetles, the insecticide treatment improved the establishment of the beetles by preventing ant attacks on the beetle larvae. In addition, half of the plots were on a relatively fertile soil type, and the other half were on a relatively infertile soil type. Natural light levels in the plots were also varied such that half of the plots were assigned to a high-light treatment and half to a low-light treatment. Dyer and Letourneau maintained these treatments for 18 months and measured herbivory and leaf production within each of the plots.

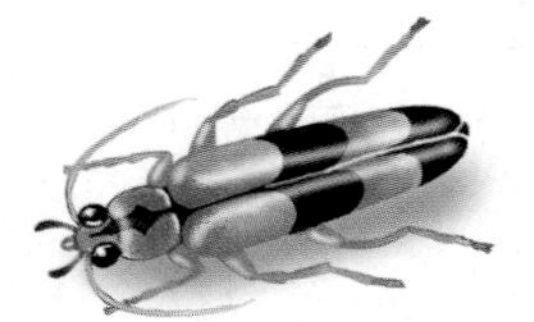

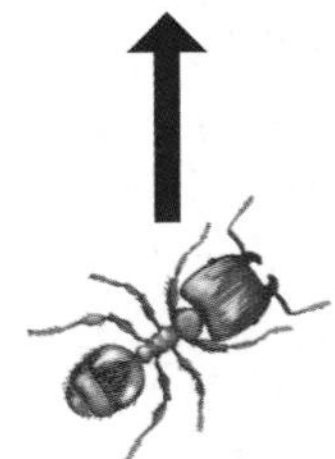

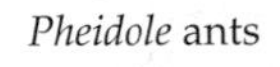

FIGURE 20.11 A Terrestrial Trophic Cascade Trophic interactions in the understory ecosystem of a lowland wet tropical rainforest in Costa Rica are indicated by the arrows. *Piper cenocladum* trees are consumed by herbivores, but provide shelter for *Pheidole* ants, which consume herbivores attacking the trees. *Pheidole* ants are consumed by *Tarsobaenus* beetles. Both ants and beetles also consume food bodies produced by the trees. (After Dyer and Letourneau 1999a.)

If the production of the *Piper* trees was limited primarily by resource supply, then the addition of the *Tarsobaenus* beetles would be expected to have little effect on *Piper* leaf production. Soil fertility and light levels would be expected to have greater effects on leaf production if these bottom-up effects were more important than the influence of the trophic cascade associated with beetles, ants, and herbivores. Dyer and Letourneau found, however, that the trophic cascade was the only significant influence on leaf production. The addition of the predatory beetles decreased ant abundance fivefold, increased rates of herbivory threefold, and decreased leaf area per tree to half that in the control plots (**Figure 20.12**). This experiment provided convincing evidence of a top-down trophic cascade affecting the production of the *Piper* trees. However, it should be noted that the lack of an effect of soil fertility and light in the control treatments, which had low rates of herbivory, indicate that the resource(s) that actually limit production may not have been manipulated in this experiment. An additional experiment that used more controlled manipulation of light levels and soil nutrients, rather than relying on variation in natural levels, found significant effects of these resources on *Piper* production, but also found a continued strong effect of herbivory (Dyer and Letourneau 1999b). Thus, it is clear that top-down trophic cascades do occur in diverse terrestrial ecosystems, although they may require strong interactions between specialist predators and their prey.

What determines the number of trophic levels?

What determines the variation among ecosystems in their numbers of trophic levels, and why do so few ecosystems have five or more trophic levels? This question is not simply an academic one. Through trophic cascades, the number of trophic levels in an ecosystem can influence movements of energy and nutrients as well as the potential for toxins in the environment to become concentrated at higher trophic levels, as we will see in this chapter's Case Study Revisited (p. 448). Variation in the number of trophic levels may be due to the addition or loss of a predator at the top of the

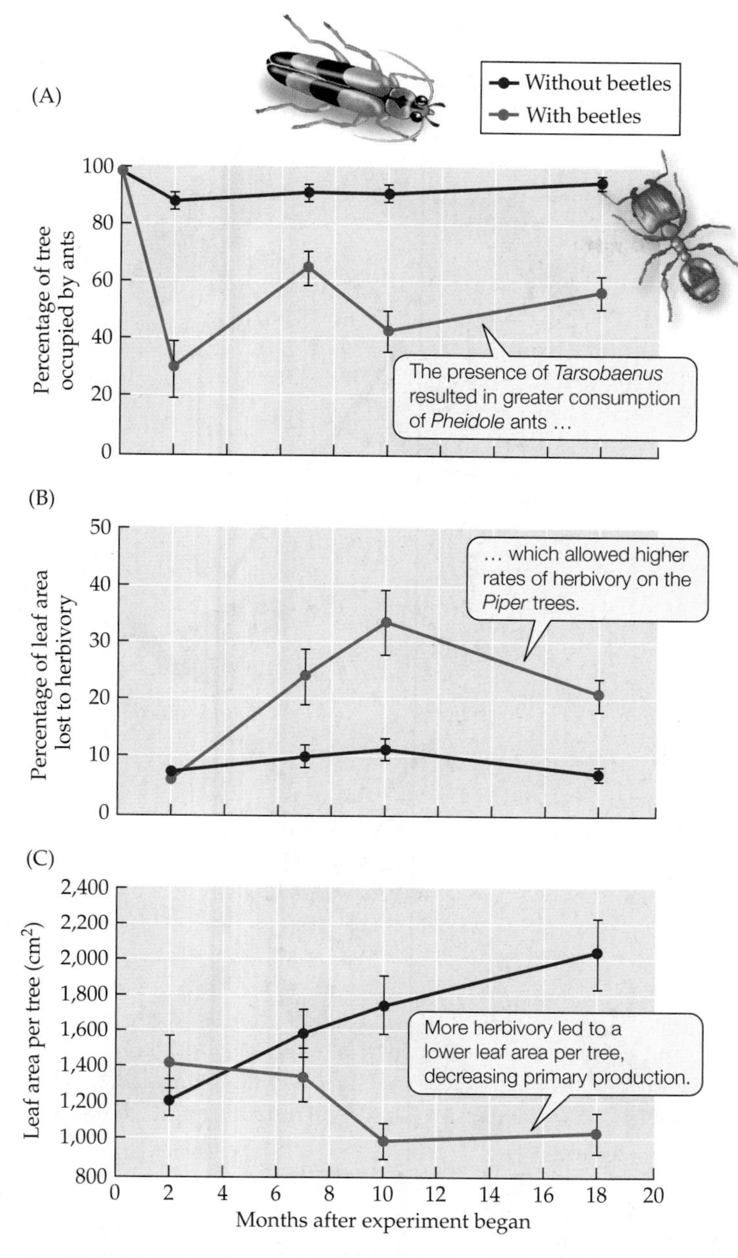

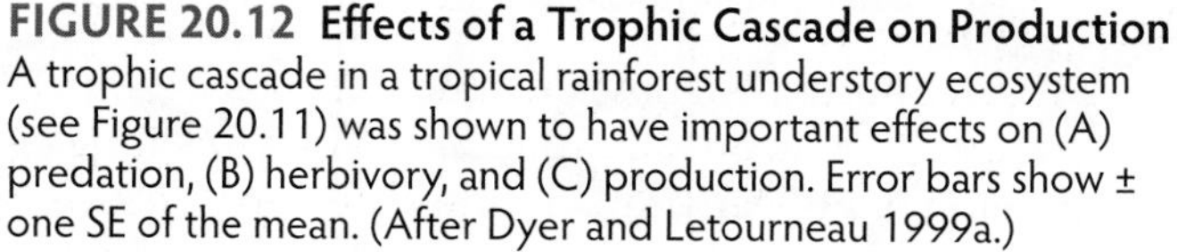

FIGURE 20.12 Effects of a Trophic Cascade on Production A trophic cascade in a tropical rainforest understory ecosystem (see Figure 20.11) was shown to have important effects on (A) predation, (B) herbivory, and (C) production. Error bars show ± one SE of the mean. (After Dyer and Letourneau 1999a.)

food chain, the insertion or loss of a predator in the middle of the food chain, or changes in omnivore feeding preference for prey at different trophic levels (**Figure 20.13**).

Several interacting ecological factors can control the number of trophic levels in ecosystems (Post 2002b). First, the amount of energy entering an ecosystem through primary production has been proposed as a determinant of the number of trophic levels. Because a relatively large amount of energy is lost in the transfer from one trophic level to the next, the more energy entering a system, the more is potentially available to support viable populations of higher-level predators. However, this explanation appears to be important primarily in ecosystems with low resource availability. Second, the frequency of disturbances or other agents of change, such as disease outbreaks, can determine whether populations of higher-level predators can be sustained. Because lower trophic levels are required to sustain higher trophic levels, there is a longer time lag for the reestablishment of the highest trophic levels following a disturbance. If disturbances occur frequently, then higher trophic levels may never become established, no matter how much energy is entering the system (Pimm and Lawton 1977). While some support for this hypothesis exists, the ability of some organisms to adapt to frequent disturbances and the potential for rapid colonization of disturbed sites (see Chapter 16) results in a smaller effect of disturbance on trophic level number than expected.

Finally, the physical size of an ecosystem can influence the number of trophic levels. Larger ecosystems support larger population sizes and have more habitat heterogeneity, and thus tend to have higher species diversity. Support for the effect of ecosystem size on the number of trophic levels is derived primarily from studies of lakes and oceanic islands, ecosystems with discrete boundaries. For example, Gaku Takimoto and colleagues (2008) tested the relative effects of disturbance and island size on the number of trophic levels on 36 islands in the Bahamas. The effect of disturbance was tested by examining 33 of the smaller islands that were either exposed to (19 islands) or protected from (14 islands) storm surges. The number of trophic levels was estimated using isotopic ratios of carbon and nitrogen (as described under Concept 19.4) in tissues from the top predators, spiders

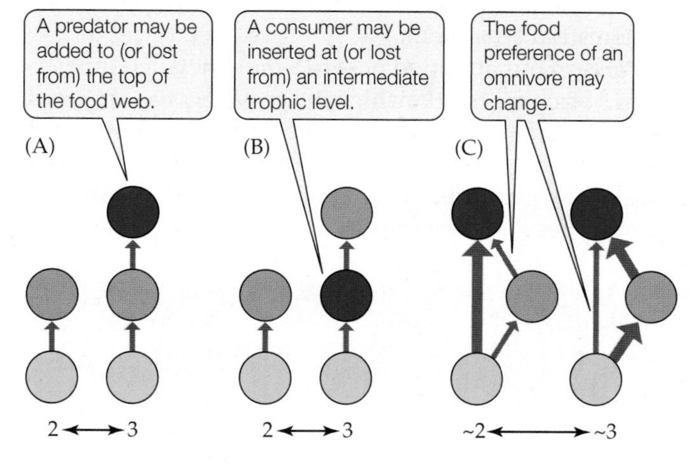

FIGURE 20.13 Changes in the Number of Trophic Levels Circles represent species at different trophic levels, and the thickness of the arrows represents the amount of energy flowing between species pairs. Differences among ecosystems in the number of trophic levels may occur due to (A) the addition or loss of a consumer at the top level, (B) the insertion of or loss of a consumer at an intermediate level, or (C) a change in the preferred feeding level of an omnivore. (After Post and Takimoto 2007.)

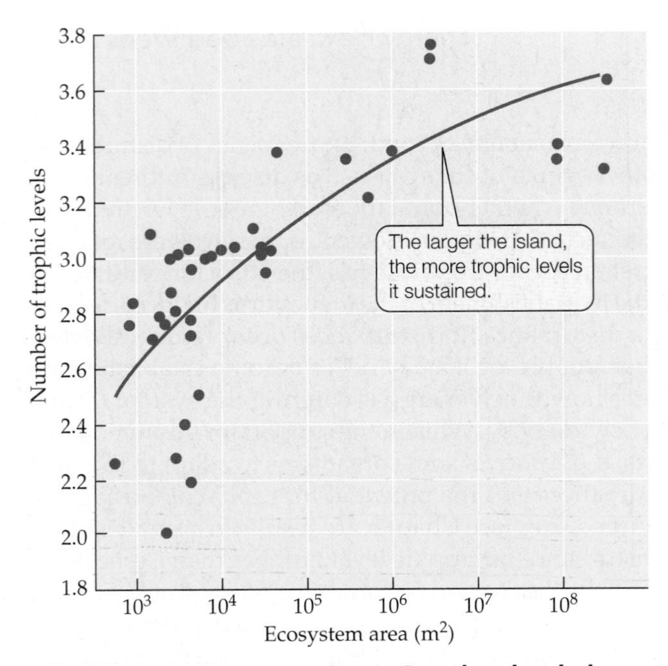

FIGURE 20.14 Ecosystem Size Is Correlated with the Number of Trophic Levels On islands in the Bahamas, Takimoto and colleagues found that as island size increased, the number of trophic levels also increased. (After Takimoto et al. 2008.)

and lizards. Takimoto and colleagues found that exposure to storm surges had no effect on the number of trophic levels. However, disturbance did influence the identity of the top predators: orb spiders were more frequently the top predators on exposed islands, and *Anolis* lizards were at the apex of the food web on protected islands. Island size, however, was strongly correlated with the number of trophic levels (**Figure 20.14**), providing evidence that ecosystem size can influence the number of trophic levels in a terrestrial ecosystem.

We turn our attention next to a more detailed investigation of trophic relationships in ecosystems as we cross the disciplinary boundaries of ecosystem ecology and community ecology (the topic of Unit 4) to examine how energy flow can influence the diversity and stability of communities and ecosystems.

CONCEPT 20.4 Food webs are conceptual models of the trophic interactions of organisms in an ecosystem.

Food Webs

Ever since Charles Darwin, in *The Origin of Species* (1859), described "a tangled bank, clothed with many plants of many kinds, with birds singing on the bushes, with various

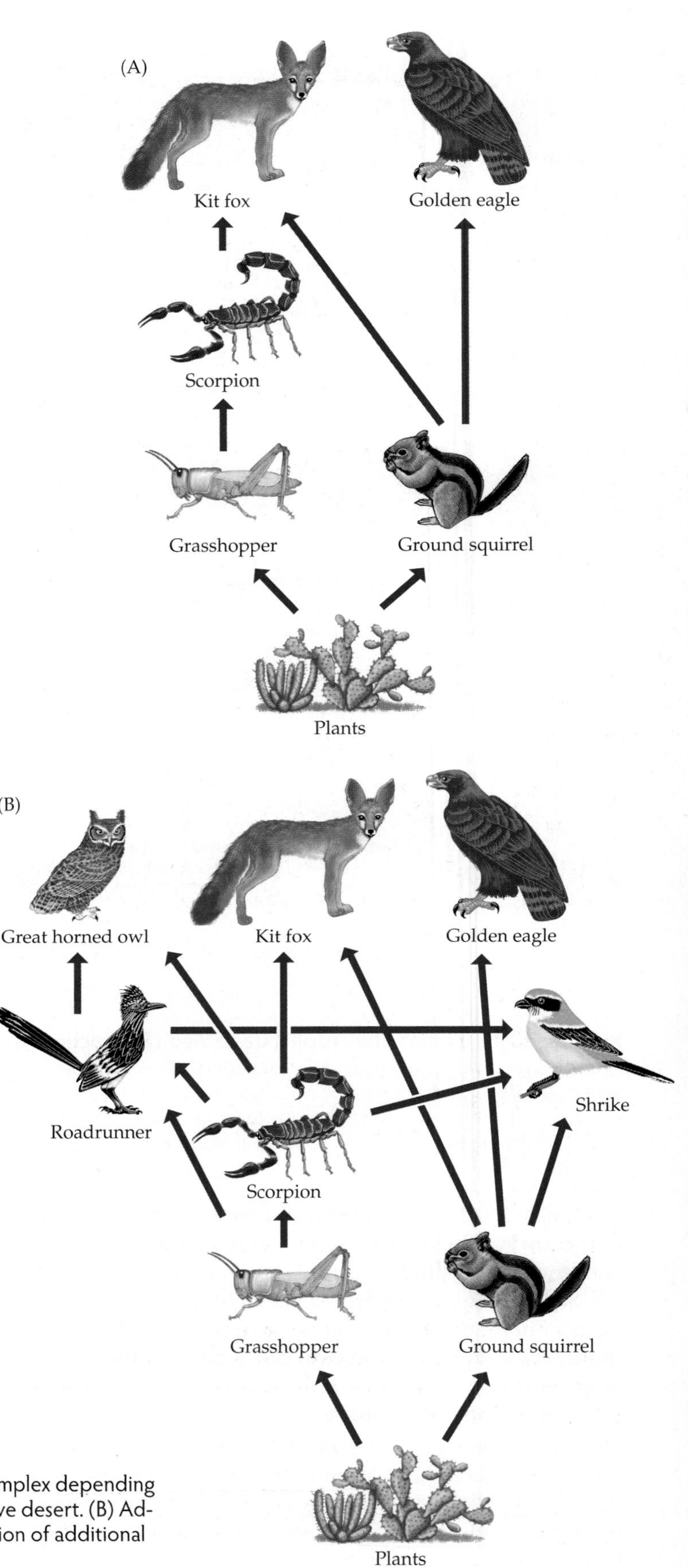

FIGURE 20.15 Desert Food Webs Food webs may be simple or complex depending on their purpose. (A) A simple six-member food web for a representative desert. (B) Addition of more participants to the food web adds realism, but the inclusion of additional species adds complexity.

insects flitting about, ... dependent upon each other in so complex a manner," the interdependence of species has been a central concept in ecology. When we examine these links among species with a focus on feeding relations, they can be described by a **food web**, a diagram showing the connections between organisms and the food they consume. For the desert ecosystem we considered at the start of this chapter, we can construct a simplified food web showing that plants are consumed by insects and ground squirrels, and that these herbivores are food for scorpions, eagles, and foxes (**Figure 20.15A**). In this way, we can begin to understand qualitatively how energy flows from one component of this ecosystem to another, and how that energy flow may influence changes in population sizes and in the species composition of communities.

Food webs are complex

The desert food web in Figure 20.15A is far from complete. Depending on our purposes, we may wish to add other organisms and links to the food web, providing additional complexity. For example, the scorpion consumes insects such as the grasshopper, but like the grasshopper, it may be food for birds such as shrikes and owls (**Figure 20.15B**). As we continue to add more and more organisms to the food web, we add complexity, such that the food web may take on the appearance of a "spaghetti diagram" (**Figure 20.16**). In order to add greater realism, it is important to recognize that the feeding relationships of animals can span multiple trophic levels (omnivory) and may even include cannibalism (circular arrows in Figure 20.16) (Polis 1991).

Although food webs are useful conceptual tools, even a simplified food web is a static description of energy flow and trophic interactions in a temporally dynamic ecosystem. Actual trophic interactions can change over time (Wilbur 1997). Some organisms alter their feeding patterns as they age; maturing frogs, for example, make the transition from omnivorous aquatic tadpoles to carnivorous adults. Some animals, such as migratory birds, are relatively mobile and are thus components of multiple food webs. Furthermore, most food webs fail to account for additional biological interactions among organisms that influence population and community dynamics, such as pollination mutualisms. (In community studies, this problem may be addressed by the use of interaction webs, as described in Chapter 15.)

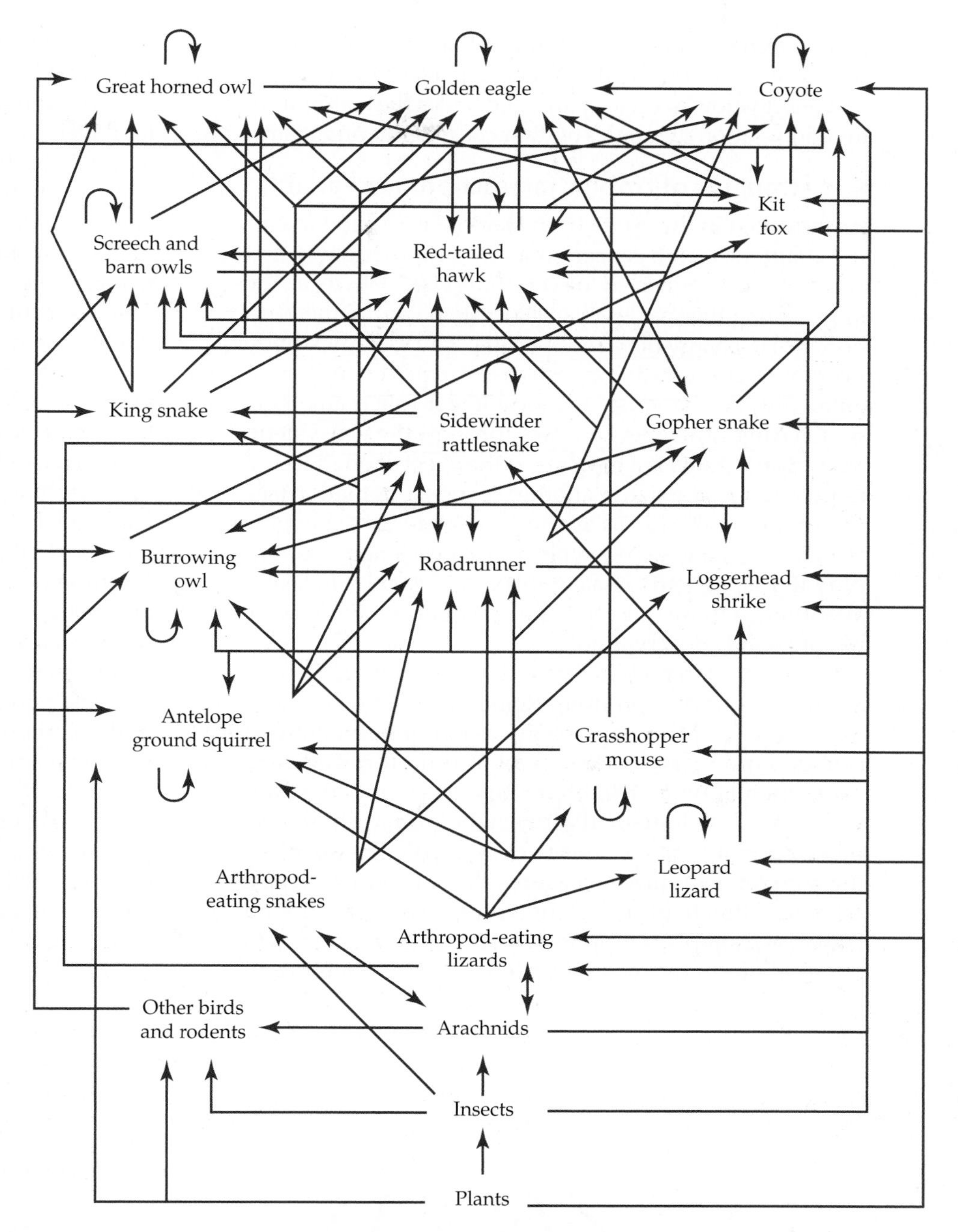

FIGURE 20.16 Food Webs Can Be Complex In this desert food web, complexity overwhelms any interpretation of interactions among the members. Even this food web, however, lacks the majority of the trophic interactions in the ecosystem. (From Polis 1991.)

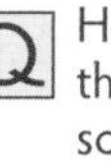
How many of the organisms or feeding groups depicted in this food web consume both plants and animals as food sources? What does this suggest about the frequency of omnivory in this food web?

The critically important roles of microorganisms are often ignored as well, despite their processing of a substantial amount of the energy moving through ecosystems. What are food webs good for, then? Despite these apparent shortcomings, food webs are important conceptual tools for understanding the dynamics of species interactions and energy flow in ecosystems, and hence the community and population dynamics of their component organisms.

The strengths of trophic interactions are variable

As indicated in the quote from Darwin above and in earlier chapters, a core concept of ecological thought is that "everything is connected to everything else." However, the links among the species in an ecosystem vary in their importance to energy flow and species population dynamics; in other words, not all connections are equally important. Some trophic relationships play a larger role than others in dictating how energy flows through the ecosystem. *Interaction strength* is a measure of the effect of one species' population on the size of another species' population (see Ecological Toolkit 15.1). Determining interaction strengths is an important goal of ecologists because it helps us simplify the "spaghetti" in a complex food web by focusing attention on those links that are most important for basic research and conservation.

How are interaction strengths determined? Several approaches have been used. Removal experiments, like those described in Chapter 15 to determine competition or facilitation, can be employed, but performing such experimental removals to quantify every link in a food web would be logistically overwhelming. Therefore, much current ecological research is devoted to discovering simpler, less direct measures that can still give us a reliable estimate of the relative importance of different links. For example, simple food webs can be coupled with observations of the feeding preferences of predators and of changes in the population sizes of predators and prey over time to provide an estimate of which interactions are the strongest. Similarly, comparisons of two or more food webs in which a predator or prey species is present in some but absent in others may provide evidence for the relative importance of links. Predator and prey body sizes have been used to predict the strengths of predator–prey interactions because feeding rate is known to be related to metabolic rate, which in turn is governed by body size. The best estimates of interaction strengths in food webs often come from a combination of these approaches.

A series of classic studies examining interaction strengths in food webs was performed in rocky intertidal zones of the Pacific Northwest by Robert Paine. Paine (1966) had observed that the diversity of organisms in rocky intertidal zones declined as the number of predators decreased. He reasoned that some of those consumers might be playing a greater role than others in controlling the diversity of these communities. One of Paine's critical observations was that one mussel species (*Mytilus californianus*) had the ability to overgrow and smother many of the other sessile invertebrate species that compete with it for space. Paine hypothesized that predators might play a key role in maintaining diversity in this community by consuming these mussels and preventing them from competitively excluding other species.

To test these hypotheses, Paine conducted an experiment in Washington State, in which he removed the top predator in the system, the sea star *Pisaster ochraceus*, from experimental plots. *Pisaster* feeds primarily on bivalves and barnacles, and to a lesser extent on other mollusks, including chitons, limpets, and a predatory whelk (genus *Nucella*) (**Figure 20.17**). Following the continuous manual

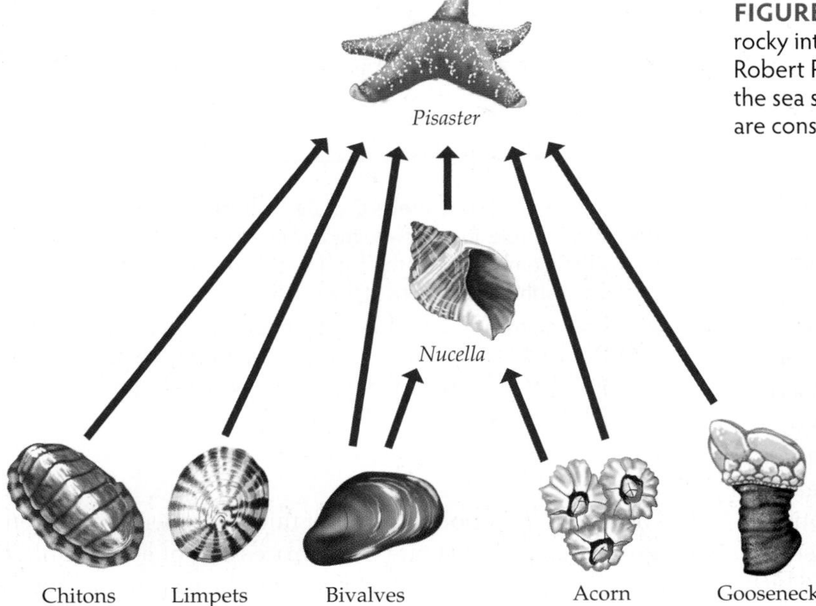

FIGURE 20.17 An Intertidal Food Web This food web from the rocky intertidal zone of Mukkaw Bay, Washington State, was used by Robert Paine to investigate the strength of the interaction between the sea star *Pisaster ochraceus* and its prey. The sea stars in the photo are consuming shellfish on a rocky shore at Strawberry Hill, Oregon.

removal of *Pisaster* from 16 m^2 plots, acorn barnacles (*Balanus glandula*) became more abundant, but with time, they were crowded out by mussels (*Mytilus*) and gooseneck barnacles (genus *Pollicipes*). After 2½ years, the number of species in the community had decreased from 15 to 8. Even 5 years after experiment began, when sea stars were no longer being removed, dominance by the mussels continued, as individual mussels had grown to sizes that prevented predation by sea stars, and diversity remained lower in the experimental plots than in adjacent control plots (Paine et al. 1985). Experimental removals of higher-level predators in other intertidal zones, including one in New Zealand, which shares no species with the intertidal zone of the Pacific Northwest, resulted in similar reductions in diversity. Predators in these intertidal ecosystems are thus key to maintaining species diversity by preventing competitive exclusion. Such species are more important in food webs than their numbers would indicate.

The experimental research of Paine and others was an encouraging advance in ecology because it demonstrated that, despite the potential complexity of trophic interactions among species, patterns of energy flow and community structure might be governed by a small subset of those species. Paine called animals such as *Pisaster keystone species*, defining them as species that have a greater influence on energy flow and community composition than their abundance or biomass would predict (see Figure 15.17). The keystone species concept has become an important focus in ecology and conservation biology because it implies that protecting such species may be critical for protection of the many other species that depend on it (as we'll see in Chapter 22). Many keystone species are predators at higher trophic levels, which tend to have large effects on prey populations relative to their own abundance.

Some species act as keystone species in only part of their geographic range, suggesting that interaction strengths are dependent on the environmental context. Several studies, including those described in Figure 15.20 and Ecological Toolkit 15.1, have found context-dependent variation in the degree to which species behave as keystone species. Thus, while the keystone species concept is intuitively simple, predicting when and where a particular species will behave as a keystone species remains a challenge.

Direct and indirect effects determine net interaction strength

One reason it remains difficult to predict the strength of trophic interactions is that the ecological importance of a keystone predator such as *Pisaster* manifests itself not only through one strong link, such as that between *Pisaster* and mussels, but also through strong indirect effects (see Figure 15.12), such as the effects *Pisaster* has on other species by reducing the abundance of mussels. If *Pisaster* consumed only the species that are inferior competitors for space (such as barnacles), it would not play a keystone role in the rocky intertidal community. Thus, predicting the effects of species losses on the remaining community requires an understanding not only of the strengths of individual links, but also the strengths of chains of indirect effects.

Removal experiments can provide estimates of the *net* effect of an interactor species on a target species (see Ecological Toolkit 15.1). This net effect includes the sum of the interactor's direct effect and all of its possible indirect effects mediated through the other species present in the community (**Figure 20.18**). The net effect of a predator on its prey, for example, includes not only the direct effect of prey consumption, but also the effects the predator has on other species that compete with, facilitate, or modify the environment of the target prey species. For example, *Pisaster* has a negative direct effect on barnacles by consuming them, but the positive indirect effect it has on barnacles by consuming mussels (and thus freeing the barnacles from competition for space) is much stronger; thus, *Pisaster* has a net positive effect on this inferior competitor. If the negative direct effect of a predator on the abundance of a target prey species is offset by its positive indirect effects on that target prey species, the net effect may appear weak. Alternatively, the net effect of a predator on a target prey

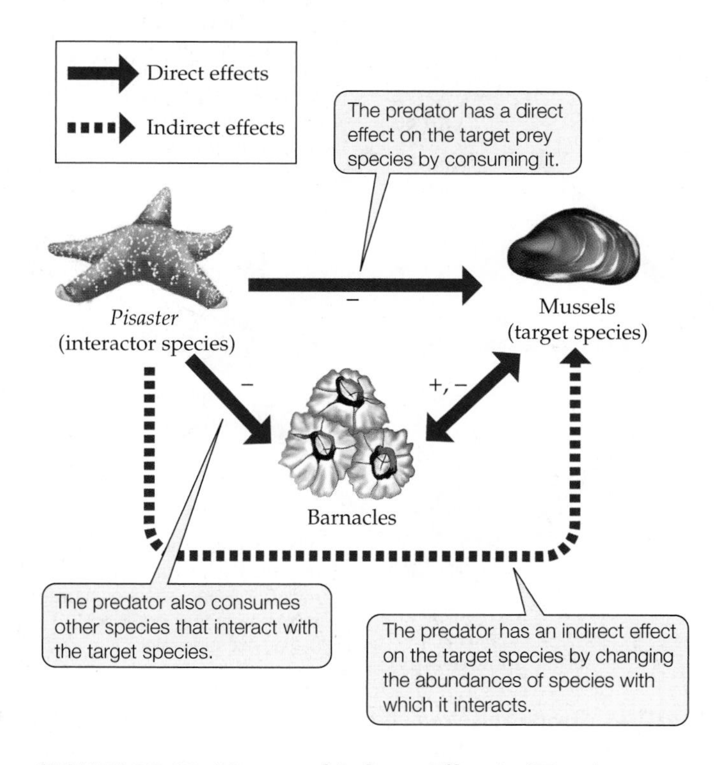

FIGURE 20.18 Direct and Indirect Effects of Trophic Interactions The net effect of a predator on a target prey species includes all possible indirect effects of the predator on other species in the community that interact with the target prey species as well as the direct effect of consumption.

species may appear very strong if both the direct and indirect effects are in the same direction.

Eric Berlow (1999) hypothesized that the potential for indirect effects to offset or reinforce the direct effect of a predator should be greatest when the direct effect is weak. Berlow tested his hypothesis by manipulating the direct and indirect effects of predatory whelks (snails of the genus *Nucella*) on a target prey species, the mussel *Mytilus trossulus*. The whelks also prey on acorn barnacles (*Balanus glandula*), which compete with the mussels for space. While whelks have a negative direct effect on mussels (by eating them), they can have either a positive or a negative indirect effect on mussels by eating barnacles. Barnacles generally facilitate mussels by providing safe "nooks and crannies" for mussel larvae to settle in. If barnacles settle on a rock in very dense clusters, however, they grow thin and are not well attached to the rock. When that happens, the barnacles, and any mussels that have settled on them, are more easily knocked off the rocks by waves. Therefore, when barnacles are at low densities, whelk predation on barnacles has a negative indirect effect on mussels by removing the mussels' preferred settlement substrate. However, when barnacles are at high densities, some thinning of the barnacles by whelks has a positive indirect effect on mussels by providing them with a more stable settlement substrate (**Figure 20.19A**).

Berlow manipulated the strength of the direct effect of whelk predation by using high and low densities of whelks. He then measured the effect of whelks on mussel settlement rates in the presence (with indirect effects) and absence (no indirect effects) of barnacles. Whelk predation, without the indirect effects mediated by barnacles, had a consistent negative direct effect on the settlement rate of mussels, regardless of whelk density (**Figure 20.19B**). In the presence of barnacles, however, the effect of whelk density on mussel settlement rates changed. When whelks were at low densities (i.e., when their direct effect was "weak"), whether their net effect on mussel settlement was positive or negative depended on the density of the barnacles, as described above. At high whelk densities (i.e., when the direct effect of whelks was "strong"), the whelks had a consistently negative net effect on mussel settlement, regardless of the densities of barnacles.

The results of Berlow's experiment supported his hypothesis, showing that weak direct effects could show up

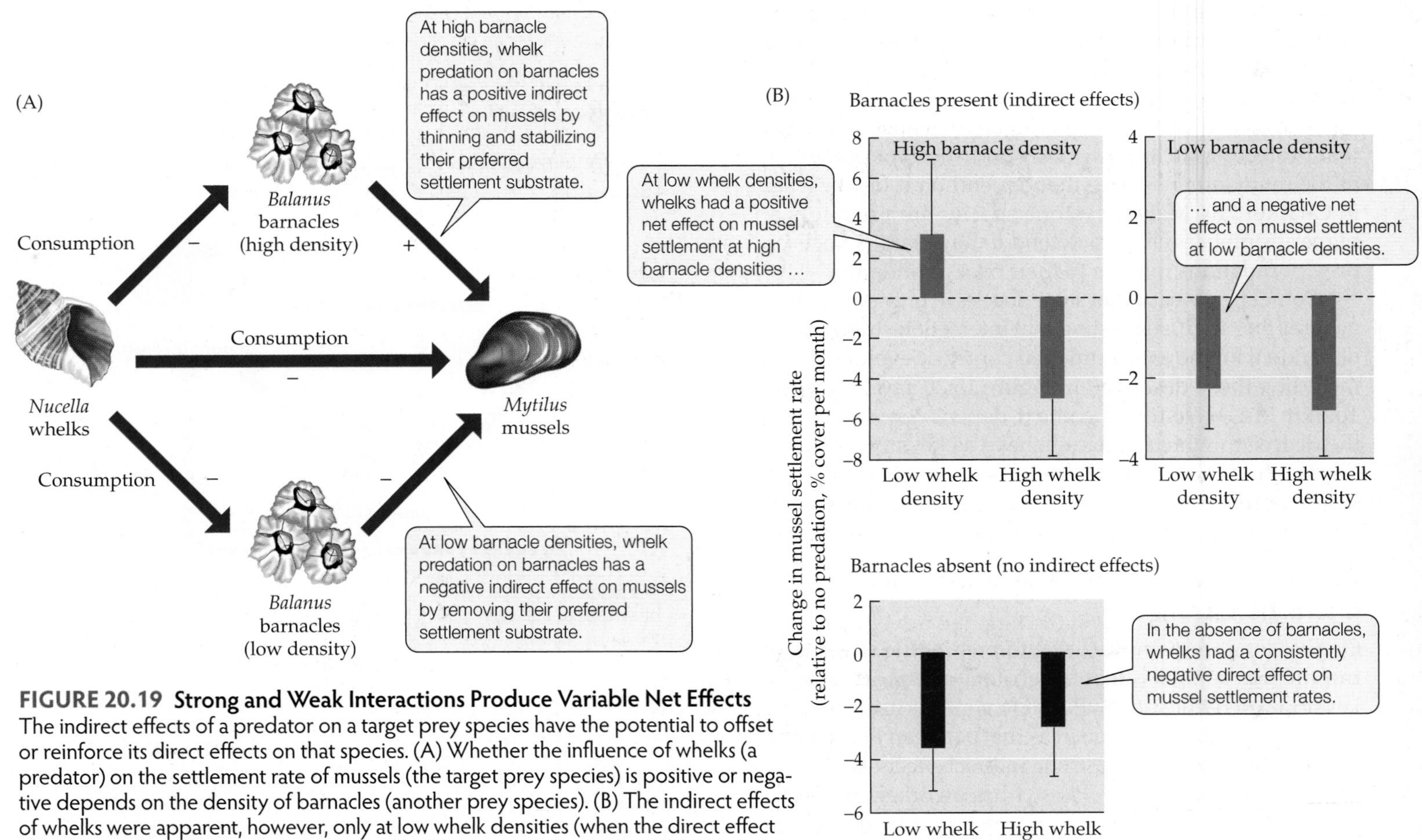

FIGURE 20.19 Strong and Weak Interactions Produce Variable Net Effects The indirect effects of a predator on a target prey species have the potential to offset or reinforce its direct effects on that species. (A) Whether the influence of whelks (a predator) on the settlement rate of mussels (the target prey species) is positive or negative depends on the density of barnacles (another prey species). (B) The indirect effects of whelks were apparent, however, only at low whelk densities (when the direct effect of whelks was weak). Error bars show one SE of the mean. (After Berlow 1999.)

as variable net effects under the influence of indirect effects. This variation reflects a combination of the direct effect of predation with indirect effects of predation, mediated by other prey species, which may offset, or in some cases reinforce, the direct effect. A mix of positive and negative indirect effects contributes to variation in the outcomes of species removal experiments when the direct effect is too weak to overcome the indirect effects. Why is this variation important? If a predator has varying (positive or negative) effects on a target prey species depending on the presence or absence of another species, the potential for the predator to eliminate that prey species throughout its range is lower. The variation associated with weak interactions may therefore promote coexistence of multiple prey species because different prey are facilitated in different places.

Does complexity enhance stability in food webs?

Ecologists have pondered whether more complex food webs—those with more species and more links among them—are more stable than simpler food webs with lower diversity and fewer links. Stability, in this context, is usually gauged by the magnitude of changes in the population sizes of the organisms in the food web over time. As we saw in Chapter 10, large oscillations in population size over time increase the susceptibility of species to local extinction. Thus, a less stable food web means a greater potential for extinction of its component species. The question of stability is taking on ever greater importance with increasing rates of biodiversity loss and non-native species invasions worldwide. How an ecosystem responds to species loss or gain is strongly related to the stability of its food webs.

Early proponents of the idea that food web complexity increases stability based their arguments on observations of real trophic interactions as well as on intuition. Ecologists such as Charles Elton and Eugene Odum argued that simpler, less diverse food webs should be more easily perturbed, experience larger changes in species population densities, and experience greater species losses as a result. More rigorous mathematical analyses of food webs, however, provided a contrary view. Robert May (1973) used food webs made up of random assemblages of organisms to demonstrate that food webs with higher diversity are less stable than those with lower diversity. The instability in May's models resulted from accentuation of population fluctuations by strong trophic interactions: the more interacting species there were, the more likely that their population fluctuations would reinforce one another, leading to the extinction of one or more of the species.

May's work overturned the notion that more complex systems are inherently more stable than simpler ones. Yet anyone visiting a tropical rainforest or a coral reef can attest to the fact that highly diverse and complex communities do persist in nature. Therefore, much ecological research has been devoted to discovering the factors that allow naturally complex food webs to be stable. More recent models, for example, have incorporated distributions of interaction strengths more closely resembling those observed in nature. In addition, there is a greater realization that weak interactions can stabilize trophic relationships, as demonstrated by the results of Eric Berlow's work described above. These models and experiments suggest that, while more complex systems are not necessarily more stable, some natural food webs may have a particular structure or organization that allows increased species diversity to have a stabilizing effect.

Sharon Lawler tested the relationship between species diversity and food web stability in an experiment using laboratory microcosms (small closed-system experimental containers) containing different numbers of protozoan species (Lawler 1993). She established microcosms in glass jars containing two, four, or eight protozoan species, all with equal numbers of prey and predator species. The prey species consumed bacteria, which in turn fed on a nutritive medium that was added to the microcosms. Lawler evaluated the stability of each of the component protozoan species by its persistence (whether its populations went extinct), the variation in its population size over time, and its mean population size.

Increasing the number of species in Lawler's microcosms resulted in more population extinctions (**Figure 20.20**), but no changes in variation in population sizes over time. The increase in extinctions was correlated with decreased population sizes, although the total abundance of individuals of all species differed little among the microcosms. Lawler suggested that the decrease in population sizes with increasing diversity may have increased the susceptibility of the populations to extinction.

Does Lawler's experiment indicate that diverse food webs are inherently unstable? No, such a conclusion would be too simplistic. Lawler cautioned that the species composition of the food web was also an important

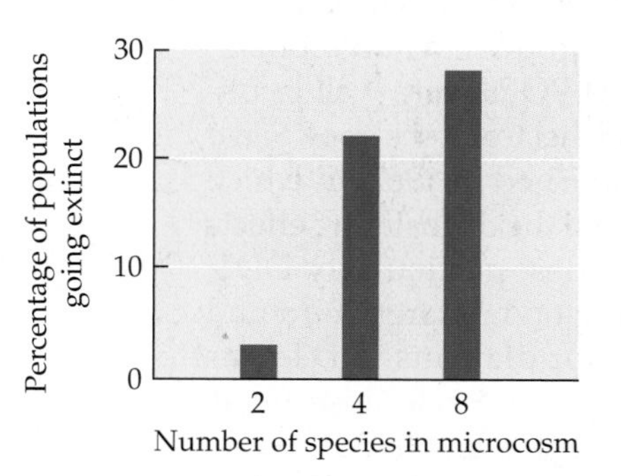

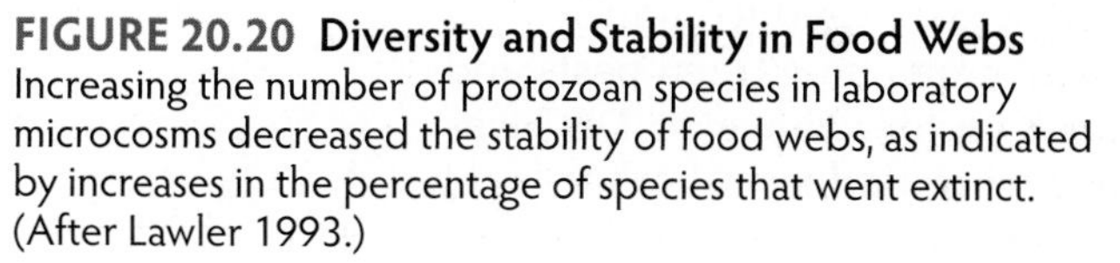
FIGURE 20.20 Diversity and Stability in Food Webs Increasing the number of protozoan species in laboratory microcosms decreased the stability of food webs, as indicated by increases in the percentage of species that went extinct. (After Lawler 1993.)

influence in her experiment. In other words, the *identity* of the species that made up the food web was important in determining its stability. In Lawler's experiment, some species were more likely than others to become extinct, and some showed greater variation in their population sizes in the presence of particular species. Furthermore, some predators were more likely to cause the extinction of prey species than others. Thus, both species diversity and species composition appeared to be important in determining the stability of the protozoan food webs. Additional studies suggest that the buffering influence of weak interactions (McCann et al. 1998, Neutel et al. 2002) and behavioral or evolutionary changes in prey choice (Kondoh 2003) can help to reduce the population fluctuations associated with complex food webs.

A CASE STUDY REVISITED
Toxins in Remote Places

Knowledge of how energy flows through the trophic levels of ecosystems is key to understanding the environmental effects of persistent organic pollutants like those described in this chapter's Case Study (see p. 430). Some chemical compounds taken up by organisms, either directly from the environment or by consumption with their food, can become concentrated in their tissues. For a variety of reasons, these compounds are not metabolized or excreted, so they become progressively more concentrated in the body over the organism's lifetime, a process known as **bioaccumulation**. Bioaccumulation can lead to increasing tissue concentrations of these compounds in animals at successively higher trophic levels as animals at each trophic level consume prey with higher concentrations of the compounds. This process is known as **biomagnification** (**Figure 20.21**). The POPs we discussed at the beginning of this chapter are particularly susceptible to these processes.

The potential dangers associated with bioaccumulation and biomagnification of POPs were well publicized by Rachel Carson's book *Silent Spring*, published in 1962, in which she described the devastating effects that pesticides, particularly DDT, were having on nontarget bird and mammal populations. DDT was thought of as a "miracle" insecticide during the 1940s and 1950s, when it was widely used to control a variety of crop and garden pests and disease vectors. However, DDT was also accumulating in higher-level predators as a result of biomagnification, and it contributed to the near-extinction of some birds of prey, including the peregrine falcon and the bald eagle. In *Silent Spring*, Carson described the persistence of DDT in the environment, its accumulation in the tissues of consumers, including humans, and its health hazards. Because of Carson's careful documentation and her ability to convey her message in a well-crafted manner that could be appreciated by the general public, *Silent Spring* led to increased scrutiny of the use of chemical pesticides, which eventually resulted in a ban on the manufacture and use of DDT in the United States.

The concept of biomagnification led researchers to suspect that the high concentrations of POPs found in the Inuit resulted from their position at the highest trophic levels of the Arctic ecosystem. This suspicion was reinforced by comparisons of the concentrations of toxins among different Inuit communities. The highest concentrations of toxins were found in communities that consumed marine mammals such as whales, seals, and walruses—animals that occupy the third, fourth, or fifth trophic levels. Inhabitants of communities where herbivorous caribou (at the second trophic level) were a more important part of the diet had lower concentrations of toxins. The Inuit preference for foods rich in fatty tissues, such as whale blubber (muktuk), poses a problem as well because many POPs are preferentially stored in the fatty tissues of animals.

Although emissions of some POPs and other pollutants are declining globally as awareness of their effects increases and regulations are put in place, the potential for long-term storage of these compounds in the Arctic environment means that their effects are unlikely to disappear any time soon (Pearce 1997). Concentrations of PCBs and DDT in Arctic lake sediments have continued to

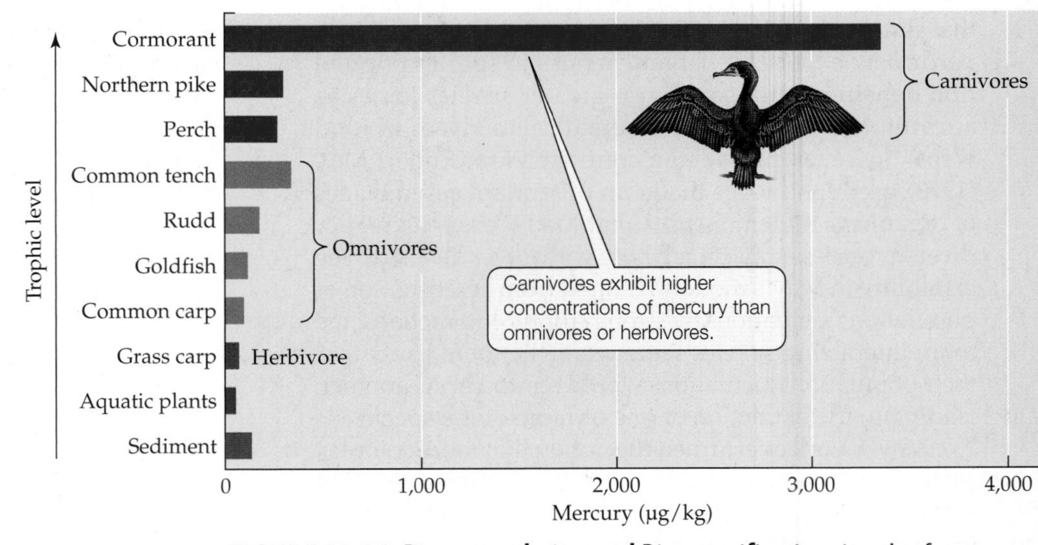

FIGURE 20.21 Bioaccumulation and Biomagnification Levels of mercury (a toxic heavy metal) show bioaccumulation and biomagnification in a Czech pond ecosystem. (After Houserová et al. 2007.)

increase over time, while concentrations of these POPs in lake sediments in lower latitudes have tended to decrease. The cold temperatures and relatively low light levels in the Arctic limit the chemical breakdown of POPs. Unfortunately, the long-term effects of POPs on Inuit populations are uncertain. While switching to alternative food sources might seem to be a potential solution to the problem, the cultural identity of the Inuit is strongly associated with their hunting traditions and their diet, and they would be unlikely to make such a switch easily.

CONNECTIONS IN NATURE

Biological Transport of Pollutants

Pollutants have been reported in almost all environments on Earth—even Antarctic ice holds trace amounts of DDT and lead emitted from the burning of leaded gasoline. Animals in many remote areas have high concentrations of industrial and agricultural toxins in their tissues. Fish in isolated alpine lakes of the Canadian Rockies, for example, contain high concentrations of POPs, which have been associated with condensation of these compounds in snowfields and glaciers above the lakes (Blais et al. 1998). As suggested above, the concentrations of these pollutants are related to the trophic positions of the animals: consumers at the highest trophic levels, such as polar bears, seals, and birds of prey, contain the highest concentrations. The widespread nature of this problem underscores the notion that ecosystems are connected by the movements of energy and materials among them. Ecological processes in one ecosystem can have effects on other ecosystems through these movements (Polis et al. 2004).

The movement of POPs and other human-made toxins is usually associated with atmospheric transport from low to high latitudes. However, the behaviors of animals can also influence the movement of POPs. Salmon, for example, have been shown to transport nutrients from marine to freshwater and terrestrial ecosystems during their spawning runs. At reproductive maturity, salmon leave the ocean and move up rivers in large numbers, as described in the Case Study in Chapter 2 (p. 22). From the rivers, they move into freshwater lakes and streams, where they spawn and then die. The potential exists for salmon to move toxins, as well as nutrients, from the oceans to freshwater ecosystems via this spawning behavior.

E. M. Krümmel and colleagues studied the potential for spawning sockeye salmon (*Oncorhynchus nerka*) to act as a "fish pump" for pollutants by moving PCBs from the ocean to remote lakes in Alaska (Krümmel et al. 2003). Salmon occupy the fourth trophic level, and thus, through bioaccumulation and biomagnification, they accumulate PCBs in their body fat at concentrations more than 2,500 times higher than those found in seawater. Krümmel and colleagues collected sediment cores from eight lakes in southwest Alaska that had different densities of spawning salmon (ranging from 0 to 40,000 spawners/km^2) and measured PCBs in the sediments. They found that the concentrations of PCBs were strongly correlated with the density of spawners (**Figure 20.22**). Lakes that did not have visits from spawning fish had concentrations of PCBs similar to expectations based on atmospheric transport alone. The lake with the highest density of spawning fish (40,000 per km^2) had PCB concentrations that were six times higher than the levels associated with atmospheric transport. A similar study found that DDT, other POPs, and mercury are transported by northern fulmars (*Fulmarus glacialis*, pelagic fish-eating seabirds) from the ocean to small ponds near their nesting colonies (Blais et al. 2005). These examples demonstrate how the behaviors of some species (spawning in fish, colonial nesting in birds) can exacerbate problems of pollution associated with biomagnification in ecosystems.

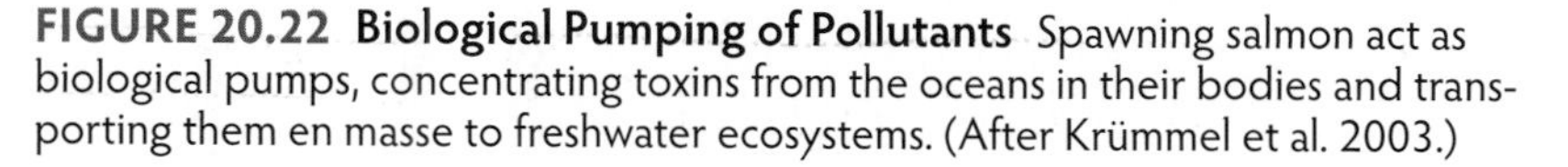

FIGURE 20.22 Biological Pumping of Pollutants Spawning salmon act as biological pumps, concentrating toxins from the oceans in their bodies and transporting them en masse to freshwater ecosystems. (After Krümmel et al. 2003.)

SUMMARY

CONCEPT 20.1 Trophic levels describe the feeding positions of groups of organisms in ecosystems.

- An organism's trophic level is determined by the number of feeding steps by which it is removed from the first trophic level, which contains autotrophs and detritus.
- Omnivores feed at multiple trophic levels, although their diets can be partitioned to reflect their consumption at each level.
- All organisms eventually end up as food for other organisms or as detritus.

CONCEPT 20.2 The amount of energy transferred from one trophic level to the next depends on food quality and consumer abundance and physiology.

- Energy and biomass pyramids portray the relative amounts of energy or biomass at different trophic levels.
- The high turnover of autotroph biomass in aquatic ecosystems can result in biomass pyramids that are inverted relative to energy pyramids.
- The proportion of autotroph biomass consumed in terrestrial ecosystems tends to be lower than that in aquatic ecosystems.
- The efficiency of energy transfer from one trophic level to the next is determined by food quality and the physiology of consumers.

CONCEPT 20.3 Changes in the abundances of organisms at one trophic level can influence energy flow at multiple trophic levels.

- Changes in the numbers and types of consumers at higher trophic levels can influence primary production through influences on the consumption of herbivores.
- Trophic cascades tend to be more apparent in aquatic ecosystems than in terrestrial ecosystems, but they have been demonstrated in complex terrestrial ecosystems as well.
- The number of trophic levels that can be sustained in an ecosystem is determined by the amount of energy entering the ecosystem through primary production, the frequency of disturbance, and the physical size of the ecosystem.

CONCEPT 20.4 Food webs are conceptual models of the trophic interactions of organisms in an ecosystem.

- Food webs are diagrams that portray the diverse trophic interactions among species in an ecosystem.
- Although trophic interactions are extremely complex, food webs can be simplified by focusing on the strongest interactions among the component organisms.
- Keystone species have greater effects on energy flow and community composition than their abundance or biomass would predict.
- Indirect effects of a predator on a target prey species, including its effects on other species that compete with, facilitate, or modify the environment of the target species, can offset or reinforce the direct effects of predation. These indirect effects may have stabilizing effects on inherently unstable food webs.

REVIEW QUESTIONS

1. Suppose one population of coyotes (population A) demonstrates a greater degree of omnivory than another population (population B). Population A relies on a diet that includes road-killed animal carcasses, plants, and rotten food from dumpsters, while population B has a steady diet of small rodents. Which population should have a higher assimilation efficiency, and why?
2. Mammals in temperate terrestrial and temperate marine ecosystems occupying similar trophic levels may have different production efficiencies. Assuming similar food quality, food abundance, and food capture rates, explain why the production efficiencies of these mammals would differ between a marine ecosystem and a terrestrial ecosystem. (Hint: Consider how the mammals maintain their body heat, as well as the temperature variation of their environments as described in Chapter 2.)
3. Which ecosystem would you expect to have a greater total amount of energy passing through its trophic levels: a lake or a forest adjacent to the lake? Which of these ecosystems would have a higher *proportion* of NPP moving through all of its trophic levels, the forest or the lake?
4. Generalist herbivores consume a greater number of plant species than specialist herbivores do. If a trophic cascade resulted in a reduction in the consumption of the predators of an herbivore, would there be a greater or a lesser effect on primary production if the herbivore were a specialist?

ON THE COMPANION WEBSITE
sites.sinauer.com/ecology2e

The website includes Chapter Outlines, Online Quizzes, Flashcards & Key Terms, Suggested Readings, a complete Glossary, and the Web Stats Review. In addition, the following resources are available for this chapter:

HANDS-ON PROBLEM SOLVING

Trophic Efficiency in a Coral Reef System

This Web exercise explores energy flow and efficiency of energy transfer in a coral reef community. You will read a recent paper that quantifies energy flows through multiple trophic levels in a community. Using data from the paper, you will calculate efficiencies of various steps in this system, and discuss the effects of trophic level on energy flow.

CHAPTER 21

Nutrient Supply and Cycling

KEY CONCEPTS

- **CONCEPT 21.1** Nutrients enter ecosystems through the chemical breakdown of minerals in rocks or through fixation of atmospheric gases.
- **CONCEPT 21.2** Chemical and biological transformations in ecosystems alter the chemical form and supply of nutrients.
- **CONCEPT 21.3** Nutrients cycle repeatedly through the components of ecosystems.
- **CONCEPT 21.4** Freshwater and marine ecosystems receive nutrient inputs from terrestrial ecosystems.

A Fragile Crust: A Case Study

The Colorado Plateau in western North America includes vast expanses of isolated mountains, intricately folded sandstone formations, and deeply cut, multicolored canyons. One of the most unusual features found in this rugged and beautiful region, however, occurs at a very small scale: its patchy cover of dark, bumpy soil (**Figure 21.1**). On closer examination, the soil looks like a miniature landscape of hills and valleys, covered with black, dark green, and white splotches resembling lichens. The comparison is apt, because this crust on the soil surface, known simply as a **biological crust** (or *cryptobiotic* crust), is composed of a mix of hundreds of species of cyanobacteria, lichens, and mosses (Belnap 2003). Approximately 70% of the soils on the Colorado Plateau, covering parts of Utah, Arizona, Colorado, and New Mexico, have some biological crust development. Similar crusts, containing a surprisingly similar suite of species, are found in many other arid and semiarid regions throughout the world. The crusty nature of the soil is largely the work of filamentous cyanobacteria, which create a sheath of mucilaginous material as they move through the soil after a rain. When the soil dries out, the cyanobacteria withdraw to deeper soil layers, leaving behind the sheathing material, which helps bind the coarse soil particles together (**Figure 21.2**).

The soils of the Colorado Plateau are exposed to tremendous climatic variation and strong erosive forces (Belnap 2003). Surface temperatures can range from –20°C (–4°F) in winter to 70°C (158°F) in summer. High evapotranspiration rates often dry out the soils, and the sparseness of the vegetation allows the strong surface winds to carry away fine soil particles. Precipitation in spring and summer often occurs as brief, intense thunderstorms. Biological crusts are critical for anchoring the soil in place in the face of high winds and torrential rains.

Although the Colorado Plateau is sparsely populated, humans have had a large and lasting effect on its landscape. Livestock grazing has been an important use of public lands in the region since cattle were introduced there in the 1880s. Most of the land has been affected to some degree by grazing, which has resulted in the trampling of biological crusts and overgrazing of vegetation. Until recently, grazing was the most important human-associated disturbance in the region. Recently, however, a proliferation of off-road vehicles has invaded the region. During the 2005 Moab Jeep Safari, for example, an estimated 30,000–40,000 participants descended on a town with a year-round population of 5,000. All-terrain vehicle use is also increasing dramati-

FIGURE 21.1 Biological Crust on the Colorado Plateau Biological crusts are a common feature in the deserts of the Colorado Plateau. The surface topography and coloration of the crust are clearly visible in this photo.

(A)

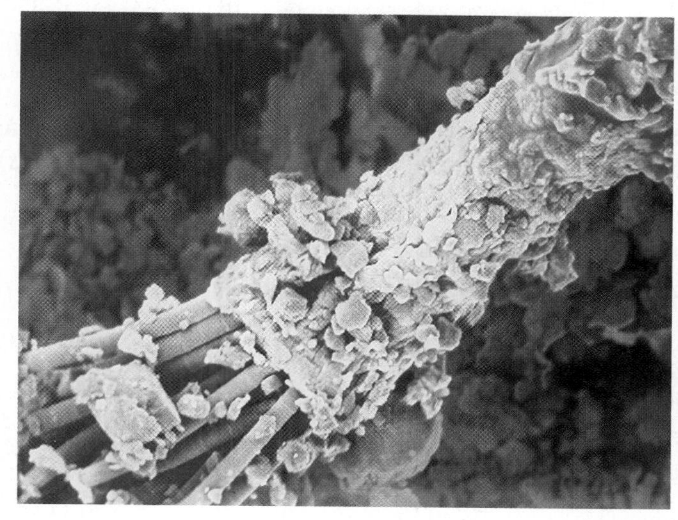

(B)

FIGURE 21.2 Cyanobacterial Sheaths Bind Soil into Crusts (A) Cyanobacterial strands surround themselves with a sheath of mucilaginous material as they move through the soil. (B) The sheaths left behind by the cyanobacteria help to bind soil particles together and protect soils from erosional loss.

cally, joining the motorcycle, mountain bike, and hiking traffic in the wilds. The majority of these users of the desert backcountry obey federal and local laws, staying on designated trails and roads. However, a minority of users drive their vehicles off designated roads and across soils covered with biological crusts.

While the spatial extent of soil surface disturbance associated with off-road vehicle use and livestock grazing has not been well quantified, it is clear that a large part of the landscape has been disturbed to some degree during the past 150 years, and that the rate of disturbance is increasing. The recovery of biological crusts following disturbance is extremely slow in arid environments: decades are required for the reestablishment of the cyanobacteria and up to centuries for recolonization by lichens and mosses (Belnap and Eldridge 2001).

What are the implications of the loss of biological crusts for the functioning of desert ecosystems? How important are they to the supply of nutrients in those ecosystems? Given the long-term nature of disturbances associated with livestock grazing across the Colorado Plateau, can we still find areas that can serve as controls for studies of the disturbance that has already occurred?

Introduction

In addition to energy, all organisms require specific chemical elements to meet their fundamental biochemical requirements for metabolism and growth. Organisms get these elements by absorbing them from the environment or by consuming other organisms, living or dead. Iron, for example, is needed by all organisms for several important metabolic functions, but how those organisms get their iron and where it comes from vary substantially. Phytoplankton in the Atlantic Ocean may take up iron that came from dust that blew in from the Sahara. Lions on an African savanna get their iron from the prey they kill and consume. Aphids get their iron in the sap they suck from a plant, whereas the plant takes up water containing dissolved iron from the soil. The ultimate source of all of this iron, however, is solid minerals in Earth's crust, which are subjected to chemical transformations as they move through the different physical and biological components of ecosystems.

The study of the physical, chemical, and biological factors that influence the movements and transformations of elements is known as **biogeochemistry**. An understanding of biogeochemistry is important for determining the availability of **nutrients**, which are defined as the chemical elements an organism requires for its metabolism and growth. Nutrients must be present in certain chemical forms to be available for uptake by organisms. Thus, the rate at which physical and chemical transformations occur determines the supply of nutrients. Biogeochemistry also encompasses the study of non-nutrient elements that can serve as tracers in ecosystems and of pollutant compounds that cause environmental damage. Biogeochemistry is a discipline that integrates contributions from soil science, hydrology, and atmospheric science as well as ecology.

In this chapter, we will consider the biological, chemical, and physical factors that control the supply and availability of nutrients in ecosystems. We will emphasize the roles of autotrophs (photosynthetic and chemosynthetic plants and bacteria) because they are the principal source of nutrients for heterotrophs. We will discover which nutrients are important for organismal functioning, describe the sources of those nutrients and how they enter ecosystems, and review some of the important chemical and biological transformations that constitute the cycling of nutrients in ecosystems. In Chapter 24, we will consider the global-scale cycling of some of these elements.

CONCEPT 21.1 Nutrients enter ecosystems through the chemical breakdown of minerals in rocks or through fixation of atmospheric gases.

Nutrient Requirements and Sources

All organisms, from bacteria to blue whales, share similar nutrient requirements. How those nutrients are obtained, the chemical forms of those nutrients that are taken up, and the relative amounts of those nutrients that are required vary greatly among organisms. All of these nutrients, however, come from a common source: inorganic mineral forms that are present in Earth's crust or as gases in the atmosphere.

Organisms have specific nutrient requirements

An organism's nutrient requirements are related to its physiology. The amounts and specific nutrients needed therefore vary according to the organism's mode of energy acquisition (autotrophs versus heterotrophs), mobility, and thermal physiology (ectotherms versus endotherms). Mobile animals, for example, generally have higher rates of metabolic activity than plants or bacteria, and they therefore have higher requirements for nutrients such as nitrogen (N) and phosphorus (P) to support the biochemical reactions associated with movement. Differences in nutrient requirements are reflected in the chemical composition of organisms (**Table 21.1**). Carbon is often associated with structural compounds in plant cells and tissues, while nitrogen is largely tied up in enzymes. Accordingly, the ratios of carbon to nitrogen (C:N) in organisms can indicate the relative concentrations of biochemical machinery in cells. Animals and microorganisms typically have lower C:N ratios than plants: for example, humans and bacteria have C:N ratios of 6.0 and 3.0, respectively, whereas those of plants ranges from 10 to 40. This difference is one reason why herbivores must consume more food than carnivores to acquire enough nutrients to meet their nutritional demands.

The nutrients essential for all plants, and the functions associated with them, are presented in **Table 21.2**. Some plant species have specific requirements for other nutrients not found in Table 21.2. For example, many, but not all, C_4 and CAM plants (see Chapter 5 for discussion of these photosynthetic pathways) require sodium. In contrast, sodium is an essential nutrient for all animals,

TABLE 21.1

Elemental Composition of Organisms (as Percentage of Dry Mass)

Element (symbol)	Bacteria (in general)	Plant (corn, *Zea mays*)	Animal (human, *Homo sapiens*)
Oxygen (O)	20	44.43	14.62
Carbon (C)	50	43.57	55.99
Hydrogen (H)	8	6.24	7.46
Nitrogen (N)	10	1.46	9.33
Silicon (Si)		1.17	0.005
Potassium (K)	1–4.5	0.92	1.09
Calcium (Ca)	0.01–1.1	0.23	4.67
Phosphorus (P)	2.0–3.0	0.20	3.11
Magnesium (Mg)	0.1–0.5	0.18	0.16
Sulfur (S)	0.2–1.0	0.17	0.78
Chlorine (Cl)		0.14	0.47
Iron (Fe)	0.02–0.2	0.08	0.012
Manganese (Mn)	0.001–0.01	0.04	—
Sodium (Na)	1.3	—	0.47
Zinc (Zn)		—	0.01
Rubidium (Rb)		—	0.005

Sources: Aiba et al. 1973; Epstein and Bloom 2005.

Note: Dashes indicate a negligible amount of an element; blank spaces indicate that the element has not been measured.

critical for maintaining pH and osmotic balances. Cobalt is required by some plants that host nitrogen-fixing symbionts (discussed later in this section). Selenium is toxic to most plants, but a small number of plants growing on soils rich in selenium may require it (in contrast, selenium is an essential nutrient for animals and bacteria).

Plants and microorganisms usually take up nutrients from their environment in relatively simple, soluble chemical forms, from which they synthesize the larger molecules needed for their metabolism and growth. Animals, on the other hand, typically take up their nutrients through the consumption of living organisms or detritus, obtaining their nutrients in larger, more complex chemical compounds. Animals break down some of these compounds and resynthesize new molecules; others are absorbed intact and used directly in biosynthesis. For example, 9 of the 20 amino acids that are essential for metabolism in humans and other mammals must be absorbed intact, since we cannot synthesize them ourselves.

Minerals and atmospheric gases are the ultimate sources of nutrients

All nutrients are ultimately derived from two abiotic sources: minerals in rocks and gases in the atmosphere. Over time, as they are taken up and incorporated by or-

TABLE 21.2

Plant Nutrients and Their Principal Functions

Nutrients	Principal functions
Carbon, hydrogen, oxygen	Components of organic molecules
Nitrogen	Component of amino acids, proteins, chlorophyll, nucleic acids
Phosphorus	Component of ATP, NADP, nucleic acids, phospholipids
Potassium	Ionic/osmotic balance, pH regulation, regulation of guard cell turgor
Calcium	Cell wall strengthening and functioning, ionic balance, membrane permeability
Magnesium	Component of chlorophyll, enzyme activation
Sulfur	Component of amino acids, proteins
Iron	Component of proteins (e.g., heme groups), oxidation–reduction reactions
Copper	Component of enzymes
Manganese	Component of enzymes, activation of enzymes
Zinc	Component of enzymes, activation of enzymes, component of ribosomes, maintenance of membrane integrity
Nickel	Component of enzymes
Molybdenum	Component of enzymes
Boron	Cell wall synthesis, membrane function
Chlorine	Photosynthesis (water splitting), ionic and electrochemical balance

Sources: Salisbury and Ross 1992; Marschner 1995.

ganisms, they accumulate in ecosystems in organic forms (i.e., in association with carbon and hydrogen molecules). Nutrients may be cycled within an ecosystem, repeatedly passing through organisms and the soil or water in which they live. They may even be cycled internally within an organism, stored or mobilized for use as its needs for specific nutrients change. Here we describe the inputs of nutrients into ecosystems from minerals and the atmosphere. In the following sections, we will complete the steps that constitute nutrient cycling within an ecosystem.

Mineral sources of nutrients The breakdown of minerals in rock supplies ecosystems with nutrients such as potassium, calcium, magnesium, and phosphorus. *Minerals* are solid substances with characteristic chemical properties, derived from a multitude of geologic processes. *Rocks* are collections of different minerals. Nutrients and other elements are released from minerals in a two-step process known as **weathering**. The first step, **mechanical weathering**, is the physical breakdown of rocks. Expansion and contraction processes, such as freezing–thawing and drying–rewetting cycles, act to break rocks into progressively smaller particles. Gravitational mechanisms (such as landslides) and the growth of plant roots also contribute to mechanical weathering. Mechanical weathering exposes greater amounts of surface area of mineral particles to **chemical weathering**, in which the minerals are subjected to chemical reactions that release soluble forms of nutrients.

Weathering is one of the processes involved in soil development. **Soil** is formally defined as a mix of mineral particles; solid organic matter (detritus, primarily decomposing plant matter); water containing dissolved organic matter, minerals, and gases (the *soil solution*); and organisms. Soils have several important properties that influence the delivery of nutrients to plants and microorganisms. One property is their texture, which is defined by the sizes of the particles that make up the soil. The coarsest soil particles (0.05–2 mm) are referred to as **sand**. Intermediate-sized particles (0.05 to 0.002 mm) are called **silt**. Fine soil particles (<0.002 mm), known as **clays**, have a semicrystalline structure and weak negative charges on their surfaces that can hold onto cations and exchange them with the soil solution. As a result, clay particles serve as a reservoir of nutrient cations such as Ca^{2+}, K^+, and Mg^{2+}. A soil's ability to hold these cations and exchange them with the soil solution, referred to as its **cation exchange capacity**, is determined by the amounts and types of clay the soil contains. Soil texture also influences the soil's water-holding capacity and thus the movement of nutrients in the soil solution. Soils with a high proportion of sand have a large volume of spaces between particles. These spaces (called *macropores*) allow water to drain through the soil and limit the amount of water it can hold.

The **parent material** of a soil is the rock or mineral material that was broken down by weathering to form that soil. It is usually the underlying bedrock, but may also include thick layers of sediment deposited by glaciers (known as **till**), by wind (**loess**), or by water. The chemistry and structure of the parent material are important determinants of the rate of weathering and the amount and types of nutrients released, and thus influence the fertility of the soil. Limestone, for example, is high in the nutrient cations Ca^{2+} and Mg^{2+}. Soils derived from more acidic parent material, such as granite, have lower concentrations of these elements. In addition, the higher acidity (lower pH) of soils derived from granite lowers the availability of nitrogen and phosphorus to plants.

The chemistry and pH of the parent material exerts an important influence on the abundance, growth, and diversity of plants in ecosystems. For example, Laura Gough and colleagues (2000) demonstrated that variation in the acidity of the parent material is associated with differences in plant species richness among Arctic ecosystems in Alaska. They surveyed Arctic vegetation across natural gradients in soil acidity associated with the differential distribution of calcium-rich loess, which has lower acidity

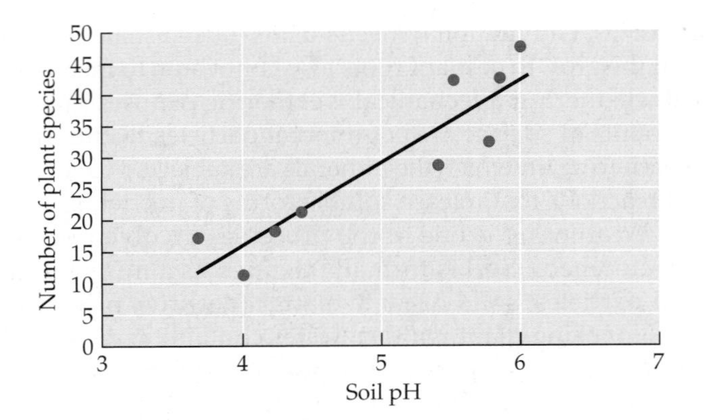

FIGURE 21.3 Plant Species Richness Increases with Decreasing Soil Acidity Vascular plant species richness in Alaskan Arctic tundra varies with soil acidity. The gradient in soil acidity is primarily due to differences in parent material: less acidic soils (with higher pH) are associated with greater loess deposits. (After Gough et al. 2000.)

than other parent materials. They found that the number of plant species increased as acidity decreased (**Figure 21.3**). This variation in diversity was attributed to the negative effects of soil acidity on nutrient availability as well as its inhibitory effects on plant establishment.

Over time, soils undergo changes associated with weathering, accumulation and chemical alteration of organic matter, and **leaching**: the movement of dissolved organic matter and fine mineral particles from upper to lower layers. These processes form **horizons**, layers of soil distinguished by their color, texture, and permeability (**Figure 21.4**). Variations in soil horizons are used by soil scientists to characterize different soil types.

Climate influences the rates of many of the processes associated with soil development, including weathering, biological activity (such as the input of organic matter and its decomposition in the soil), and leaching. In general, these processes occur most rapidly under warm, wet conditions. Thus, the soils of lowland tropical forest ecosystems, which have experienced high rates of weathering and leaching for a long time, are poor in mineral-derived nutrients such as calcium and magnesium. A high proportion of the nutrients in lowland tropical forest ecosystems are found in the living biomass of trees, in contrast to most other terrestrial ecosystems, in which the proportion of nutrients located in the soil is greater. When lowland tropical forests are cleared and burned to make way for pastures or cropland, most of the nutrients are lost in smoke and ash and through soil erosion following the fires. As a result, these ecosystems may become severely nutrient-impoverished, and it may take them centuries to return to their previous state. Soils in higher-latitude ecosystems have lower leaching rates and are usually richer in mineral-derived nutrients.

Organisms—primarily plants, bacteria, and fungi—influence soil development by contributing organic matter, which is an important reservoir of nutrients such as nitrogen and phosphorus. Organisms also increase rates of chemical weathering through the release of organic acids (from plants and detritus) and of respiratory CO_2. Thus, rates of biological activity have a strong influence on the development of soils.

Atmospheric sources of nutrients The atmosphere is composed of 78% nitrogen (as dinitrogen gas, N_2), 21% oxygen, 0.9% argon, increasing amounts of carbon dioxide (0.039%, or 388 parts per million, in 2010), and other trace gases—some natural, others pollutants derived from hu-

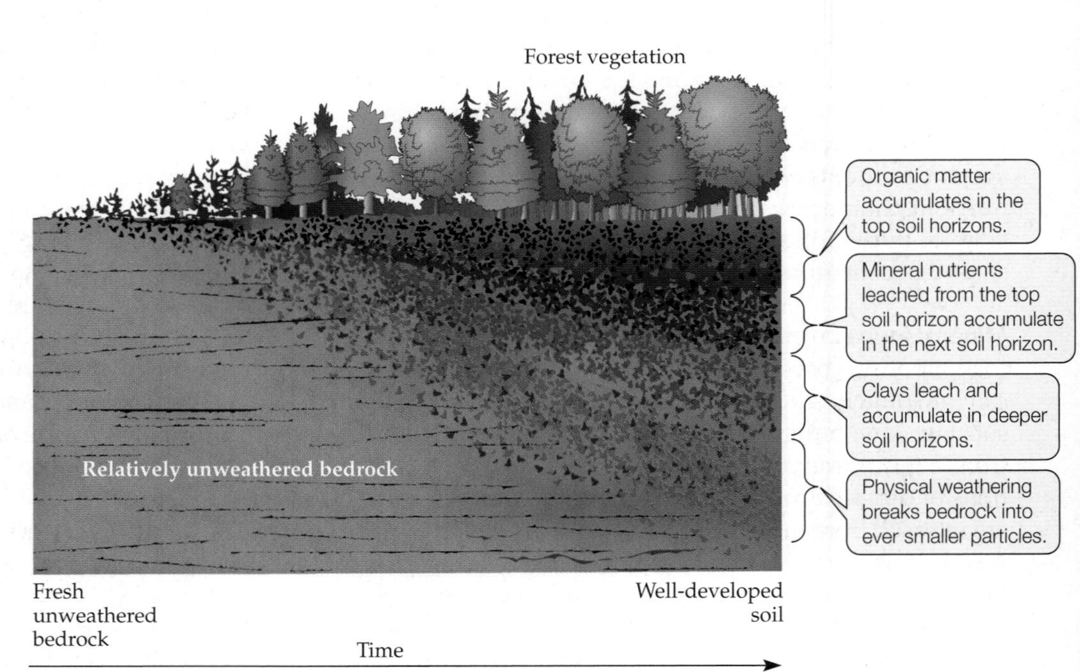

FIGURE 21.4 Development of Soil Horizons Soils develop over time as parent material is weathered and broken up into ever finer soil particles, increasing amounts of organic matter accumulate in the soil, and materials are leached and deposited in deeper soil layers. The rate of soil development is dependent on the climate, the parent material, and the organisms associated with the soil.

Q Given what you've learned about primary production in Chapter 19, and about the climatic factors that determine weathering and soil development in this chapter, what do you think the horizons of a desert soil would look like?

man activities. The atmosphere is the ultimate source of carbon and nitrogen for ecosystems. These nutrients become biologically available when they are taken up from the atmosphere and chemically transformed, or *fixed*, by organisms. They may then be transferred from organism to organism before returning to the atmosphere.

Carbon is taken up by autotrophs as CO_2 through photosynthesis. (The process of photosynthesis was described in Chapter 5, and the global cycling of carbon is discussed in Chapter 24.) Carbon compounds store energy in their chemical bonds, and they are important structural components of autotrophs (e.g., cellulose) as well.

Although the atmosphere is a huge reservoir of nitrogen, it is in a chemically inert form (N_2) that cannot be used by most organisms because of the high energy required to break the triple bond between the two molecules. The process of taking up N_2 and converting it into chemically available forms is known as **nitrogen fixation** (see Connections in Nature in Chapter 16). Biological nitrogen fixation is accomplished with the aid of the enzyme *nitrogenase*, which is found only in certain bacteria. Some of these nitrogen-fixing bacteria are free-living; others are partners in mutualistic symbiotic relationships (see Chapter 14). Nitrogen-fixing symbioses include associations between plant roots and soil bacteria, most notably between legumes and bacteria in the family Rhizobiaceae. Legumes "host" rhizobia in special root structures called nodules and supply them with carbon compounds as an energy source to meet the high energy demands of nitrogen fixation (**Figure 21.5**; see also Figure 16.21). In return for supplying the rhizobia with room and board, the plant gets nitrogen fixed by the bacteria. Other examples of nitrogen-fixing symbioses include associations between woody plants such as alders and bacteria in the genus *Frankia* (called actinorhizal associations), associations between the fern *Azolla* and cyanobacteria, lichens that include fungal and nitrogen-fixing symbionts, and termites with nitrogen-fixing bacteria in their guts. Humans also fix atmospheric nitrogen when they manufacture synthetic fertilizers using the Haber–Bosch process, in which ammonia is produced from atmospheric nitrogen and hydrogen under high pressures and temperatures using an iron catalyst. The Haber-Bosch process requires substantial energy input in the form of fossil fuels.

Natural nitrogen fixation requires a large amount of energy. It consumes as much as 25% of the photosynthetic energy obtained by plants with nitrogen-fixing symbiotic partners. Thus, nitrogen fixation provides these plants with a source of nitrogen, but represents a trade-off with other energy-demanding processes such as growth and reproduction. Allocation of energy to nitrogen fixation rather than to growth lowers the ability of nitrogen-fixing plants to compete for resources other than nitrogen. Nitrogen fixation is particularly important during the early stages of primary succession, as we saw in Chapter 16.

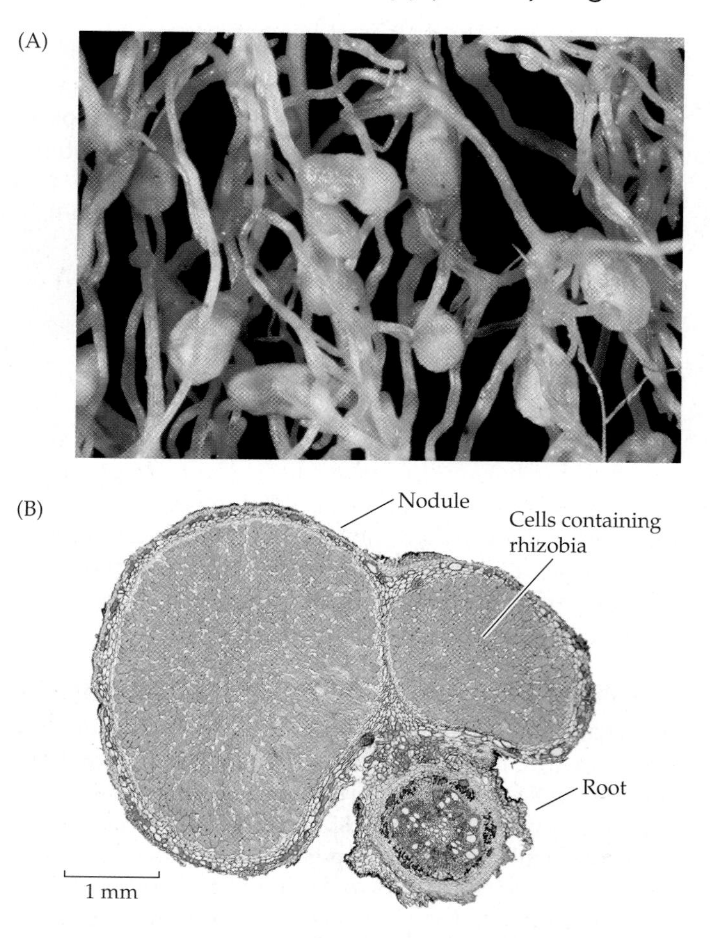

FIGURE 21.5 Legumes Form Nitrogen-Fixing Nodules (A) These swollen nodules on the roots of a red clover (*Trifolium pratense*) plant contain nitrogen-fixing bacteria. (B) Cells inside this soybean root nodule (yellow in this micrograph) are filled with rhizobia.

In addition to carbon and nitrogen, the atmosphere contains fine soil particles (dust) and a collection of suspended solid, liquid, and gaseous particles known as **aerosols**. Some of this particulate matter enters ecosystems when it falls from the atmosphere due to gravity or in precipitation, a process known as **atmospheric deposition**. Atmospheric deposition represents an important natural source of nutrients for some ecosystems. Aerosols containing cations derived from sea spray, for example, may be an important source of nutrients in coastal areas. Atmospheric deposition of dust originating in the Sahara is an important input of iron into the Atlantic Ocean and of phosphorus into the Amazon Basin. On the other hand, some ecosystems have been negatively affected by atmospheric deposition associated with human industrial and agricultural activities. Acid rain, for example, is an atmospheric deposition process that has been associated with declines in forest ecosystems in the eastern United States and Europe (as we will see in Chapter 24).

Now that we've seen how nutrients enter ecosystems, let's follow their movements within ecosystems as they are taken up and transformed. The next two sections will focus on terrestrial ecosystems; we will take a closer look at nutrient cycling in aquatic ecosystems in the final section.

CONCEPT 21.2 Chemical and biological transformations in ecosystems alter the chemical form and supply of nutrients.

Nutrient Transformations

Once they have entered an ecosystem, nutrients are subjected to further modifications as a result of uptake by organisms and other chemical reactions that alter their form and influence their movement and retention within the ecosystem. Foremost among these transformations is the decomposition of organic matter, which releases nutrients back into the ecosystem.

Decomposition is a key nutrient recycling process

As detritus (dead plants, animals, and microorganisms and egested waste products) builds up in an ecosystem, it becomes an increasingly important source of nutrients, particularly nitrogen and phosphorus. Those nutrients are made available by **decomposition**, the process by which detritivores break down detritus to obtain energy and nutrients (**Figure 21.6**). Decomposition releases nutrients as simple, soluble organic and inorganic compounds that can be taken up by other organisms.

Organic matter in soil is derived primarily from plant matter, which comes from above and below the soil surface. Fresh, undecomposed organic matter on the soil surface is known as **litter** and is typically the most abundant substrate for decomposition. The litter is used by animals, protists, bacteria, and fungi in the soil as a source of energy and nutrients. As animals such as earthworms, termites, and nematodes consume the litter, they break it up into progressively finer particles. This physical fragmentation enhances the chemical breakdown of the litter by increasing its surface area.

The chemical conversion of organic matter into inorganic nutrients (i.e., nutrients that are not associated with carbon molecules) is known as **mineralization**. It is the result of the breakdown of organic macromolecules in the soil by enzymes released by heterotrophic microorganisms. Because plants often rely on inorganic nutrients, ecologists use measurements of mineralization to estimate rates of nutrient supply. An understanding of the abiotic and biotic controls on decomposition and mineralization is key to understanding nutrient availability to autotrophs.

Rates of decomposition and mineralization are greatly influenced by climate. Decomposition, like other biologically mediated processes, proceeds most rapidly at warm temperatures. Soil moisture also controls rates of decomposition by influencing the availability of water and oxygen to detritivores. Dry soils may not provide enough water for these organisms, and wet soils have low oxygen concentrations, which lower aerobic respiration and the rate of biological activity. Therefore, the activity of detritivores is highest at intermediate soil moistures and warm temperatures (**Figure 21.7**).

FIGURE 21.6 Decomposition Decomposition of organic matter in the soil provides an important input of nutrients into terrestrial ecosystems. Similar steps occur in freshwater and marine ecosystems.

Q How would the use of a nonselective pesticide (i.e., one that does not target any specific animals) to control insect herbivores affect the rate of decomposition in a lawn ecosystem?

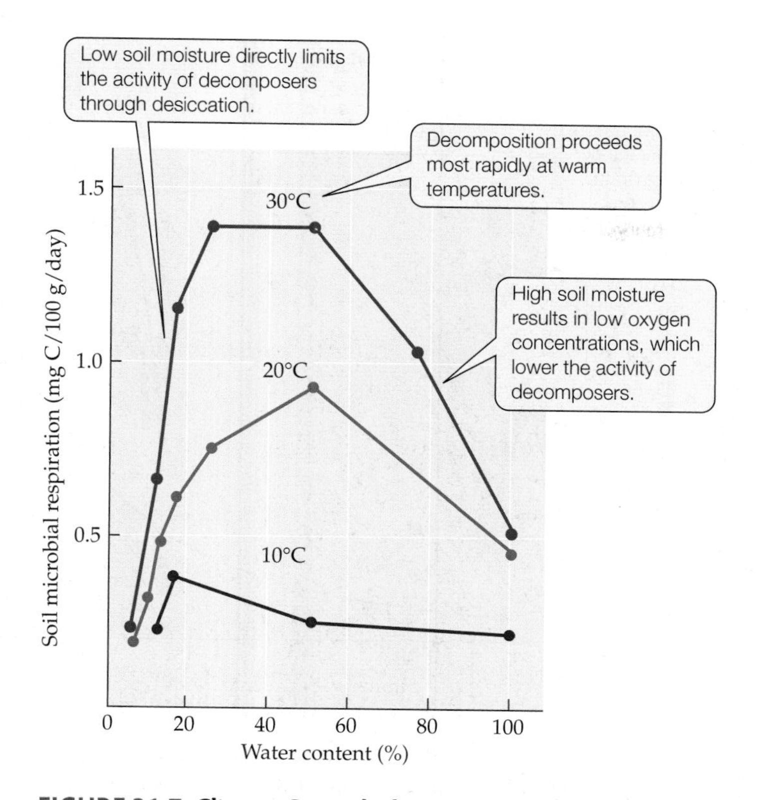

FIGURE 21.7 Climate Controls the Activity of Decomposers Changes in soil microbial respiration, used as an estimate of decomposition, are plotted as a function of soil moisture at different temperatures. (After Paul and Clark 1996.)

Some nutrients are consumed by detritivores during decomposition, so not all of the nutrients released during mineralization become available for uptake by autotrophs. The amounts of nutrients that are released from organic matter during its decomposition depend on the nutrient requirements of the decomposers and the amount of energy the organic matter contains. These factors can be approximated by the ratio of carbon (representing energy) to nitrogen (since nitrogen is the nutrient most often in short supply for detritivores) in the organic matter. A high C:N ratio in organic matter will result in a low net release of nutrients during decomposition, since heterotrophic microbial growth is more limited by nitrogen supply than by energy. For example, most heterotrophic microorganisms require approximately 10 molecules of carbon for every molecule of nitrogen they take up. About 60% of the carbon they take up is lost through respiration. Therefore, the optimal C:N ratio of organic matter for microbial growth is about 25:1; after a 60% loss of carbon to respiration, such material would yield a C:N ratio of 10:1, exactly what the microbes require. Organic matter with a C:N ratio greater than 25:1 would result in all of the nitrogen being taken up by the microbes during decomposition. Decomposition of organic matter with a C:N ratio of less than 25:1 would result in some nitrogen being released into the soil.

Not all of the carbon in litter is equally available as an energy source for decomposers: the chemistry of that carbon determines how rapidly the material can be decomposed. **Lignin**, a structural carbon compound that strengthens plant cell walls, is difficult for soil microorganisms to break down and thus decomposes very slowly (**Figure 21.8**). The rate of nutrient release from plant material containing high lignin concentrations, such as oak or pine leaves, is lower than that from material with low lignin concentrations, such as maple and aspen leaves. In addition, plant litter may contain secondary compounds, chemical compounds not used directly for growth (examples include those described in Chapters 5 and 12 associated with defense against herbivores and excess light), that can lower nutrient release during decomposition. Secondary compounds slow decomposition by inhibiting the activity of heterotrophic microorganisms and the enzymes they release into the soil or, in some cases, by stimulating their growth, leading to greater microbial uptake of nutrients.

By varying the chemistry of their litter, as well as the amount of litter they produce, plants can influence decomposition rates in the soil. Lowering decomposition rates lowers the fertility of the soil. What is the consequence for a plant of decreasing its own nutrient supply? For plants that have inherently slow growth rates, lowering soil fertility may protect them from competitive exclusion by neighbors with higher growth and resource uptake rates. Low soil nutrient concentrations can therefore be perpetuated through plant chemistry in a way that benefits the plants themselves (Van Breemen and Finzi 1998).

Microorganisms modify the chemical form of nutrients

Microorganisms in soil (as well as in freshwater and marine ecosystems) transform some of the inorganic nutri-

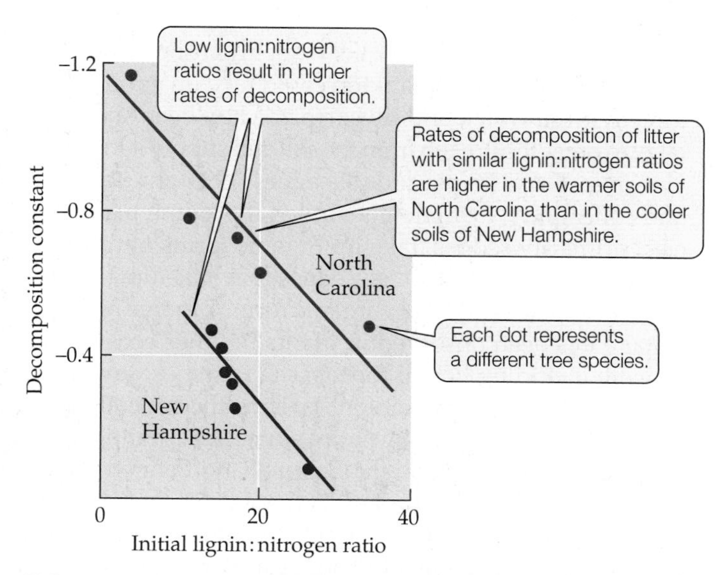

FIGURE 21.8 Lignin Decreases the Rate of Decomposition The rate of decomposition of leaf litter, expressed as the decomposition constant (lower—more negative—numbers mean higher decomposition rates), decreases as the ratio of lignin to nitrogen in the litter increases. This ratio varies among forest tree species. Note, however, that climate also has an important influence on decomposition rates. (After Melillo et al. 1982.)

ents released during the process of mineralization. These transformations are particularly important in the case of nitrogen, since they can determine its availability to plants and the rate at which it is lost from the ecosystem (see Figure 21.11 below). Certain chemoautotrophic bacteria, known as *nitrifying bacteria*, convert ammonia (NH_3) and ammonium (NH_4^+) released by mineralization into nitrate (NO_3^-) by a process called **nitrification**. Nitrification occurs under aerobic conditions, so it is limited primarily to terrestrial environments. Under hypoxic conditions, some bacteria use nitrate as an electron acceptor, converting it into N_2 and nitrous oxide (N_2O, a potent greenhouse gas) by a process known as **denitrification**. These gaseous forms of nitrogen are lost to the atmosphere and thus represent a loss of nitrogen from ecosystems.

Plant ecologists and physiologists once believed that nitrogen availability to plants was dependent solely on the supply of inorganic nitrogen—nitrate and ammonium. Therefore, soil fertility has traditionally been estimated using measurements of these inorganic forms of nitrogen.

During the 1990s, much effort was invested in understanding what controls nitrogen mineralization rates, particularly in ecosystems where fertilization experiments had indicated that nitrogen availability limits primary production and influences community diversity. Measurements of inorganic nitrogen production in forest and grassland soils generally came close to estimates of the amount taken up by plants. However, rates of inorganic nitrogen supply in Arctic and alpine ecosystems were substantially lower than what plants were actually taking up. These apparent shortfalls in nitrogen supply led to the realization that some plants were using organic forms of nitrogen to meet their nutritional requirements. Earlier work in marine ecosystems had shown that phytoplankton could take up amino acids directly from water, and mycorrhizae had been shown to take up organic nitrogen from the soil and supply it to plants. However, Terry Chapin and colleagues (1993) and Ted Raab and colleagues (1996) demonstrated that some plant species, primarily sedges, take up organic forms of nitrogen without mycorrhizae. Arctic sedges may take up as much as 60% of their nitrogen in organic form. Organic nitrogen uptake has been observed in plants in other ecosystems as well, including boreal forests, salt marshes, savannas, grasslands, deserts, and rainforests. Thus, the mineralization step in decomposition may not be as necessary for plants as has been commonly thought (Schimel and Bennett 2004).

The use of soluble organic nitrogen by plants has important implications for competition among plants and between plants and soil microorganisms. There is evidence to support the hypothesis that plants in some Arctic and alpine communities avoid competition through the preferential uptake of specific forms of nitrogen—an example of resource partitioning (described in Chapter 11). Robert McKane and colleagues (2002) examined the forms of nitrogen taken up by several plant species growing together in the Arctic tundra of northern Alaska. They measured each species' uptake of inorganic and organic forms of nitrogen, as well as the depth in the soil at which nitrogen was taken up and the time of year when it was taken up. They found that all three factors (form of nitrogen, depth of uptake, and timing of uptake) differed among species. Furthermore, the researchers found that the dominant plants in the community tended to be the species that used the form of nitrogen that was most abundant in the soil (**Figure 21.9**). Thus, a species' ability to dominate a community where nitrogen limits growth may be determined in part by its ability to take up a specific form of nitrogen.

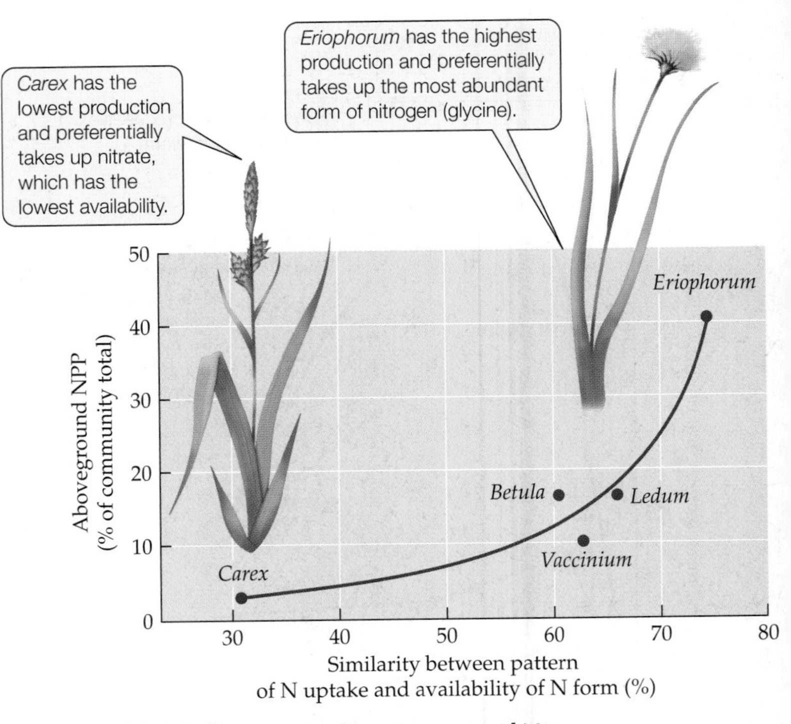

FIGURE 21.9 Community Dominance and Nitrogen Uptake Dominance of a plant community in the Alaskan Arctic tundra (measured by proportional contribution to the community's total NPP) is related to the similarity between a species' preferred form of nitrogen (ammonium, nitrate, or glycine, a small amino acid) and the availability of that form in the soil. (After McKane et al. 2002.)

Plants can recycle nutrients internally

Leaves, fine roots, and flowers, which are the metabolic powerhouses of plants, contain the highest nutrient concentrations of any plant organ. During seasonal leaf senescence, nutrients and nonstructural carbon compounds (such as starch and carbohydrates) in perennial plants are broken down into simpler, more soluble chemical forms and moved into stems and roots, where they are stored. This phenomenon is most obvious in mid- to high-latitude ecosystems as chlorophyll molecules in the leaves of deciduous species are broken down to recover their nitrogen and other nutrients, while other pigments, such as carotenoids, xanthophylls, and anthocyanins, remain, providing the autumnal splendor we humans enjoy. (Some of the fall coloration is due to an increase in pigment production, possibly to protect the leaves from high light levels or from herbivores.) When growth resumes in spring, the nutrients are transported to growing tissues for use in biosynthesis. Plants may resorb as much as 60%–70% of the nitrogen and 40%–50% of the phosphorus in their leaves before they fall. This recycling effort reduces their need to take up "new" nutrients in the following growing season.

As we've traced the chemical transformations of nutrients in terrestrial ecosystems, we've seen that they move through various components of those ecosystems as they are transformed. In the next section, we'll look at those movements in more detail and trace the fates of the nutrients as they move through an ecosystem.

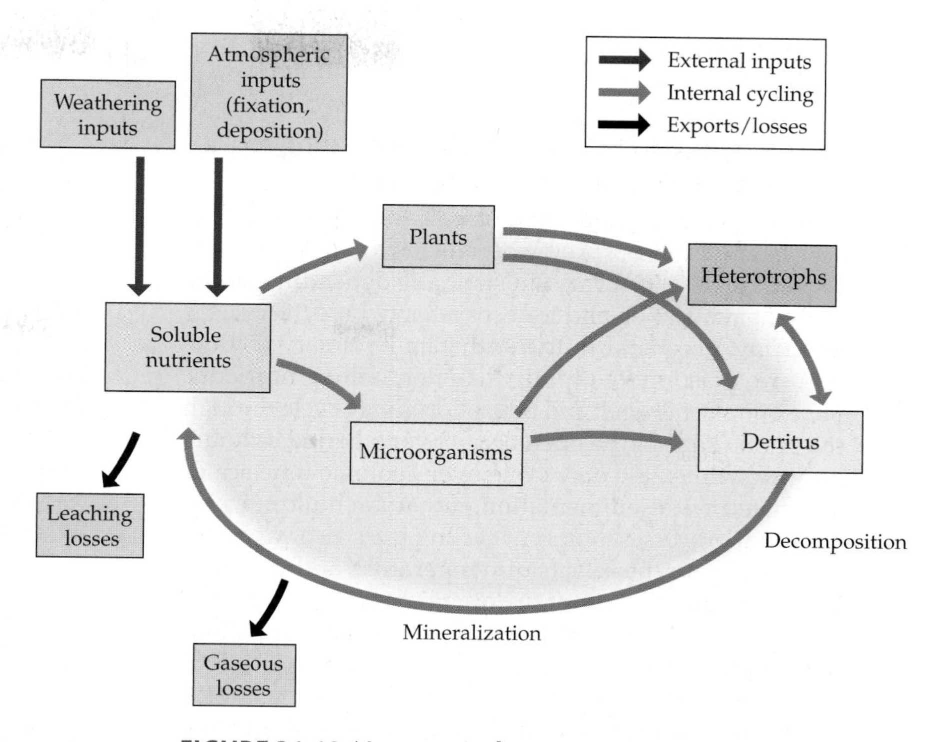

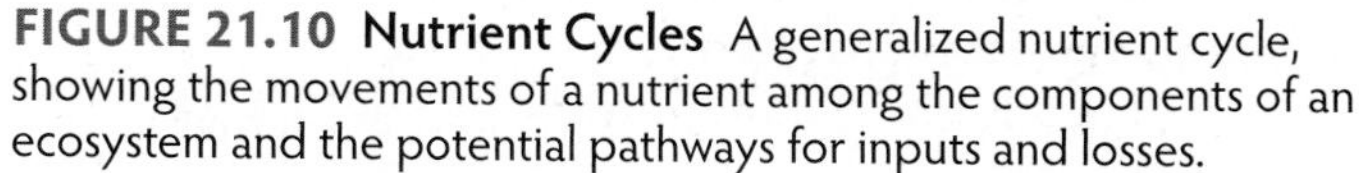

FIGURE 21.10 Nutrient Cycles A generalized nutrient cycle, showing the movements of a nutrient among the components of an ecosystem and the potential pathways for inputs and losses.

CONCEPT 21.3 Nutrients cycle repeatedly through the components of ecosystems.

Nutrient Cycles and Losses

In the previous section, we saw how nutrients undergo biological, chemical, and physical transformations as they are taken up by organisms and released through decomposition, ultimately returning to their original inorganic form, or a similar form, over time. This movement of nutrients within ecosystems is known as **nutrient cycling** (**Figure 21.10**). For example, we've traced the path of nitrogen into and through an ecosystem, starting with nitrogen-fixing microorganisms, as it is converted into chemical forms that can be used by plants. The plants incorporate the nitrogen into organic compounds (e.g., proteins and enzymes), which may end up being consumed by heterotrophs. Eventually plants, heterotrophs, and microorganisms all end up as detritus. Inorganic and organic nitrogen released from the detritus by decomposition is taken up again by plants and microorganisms, thereby completing the nitrogen cycle (**Figure 21.11**).

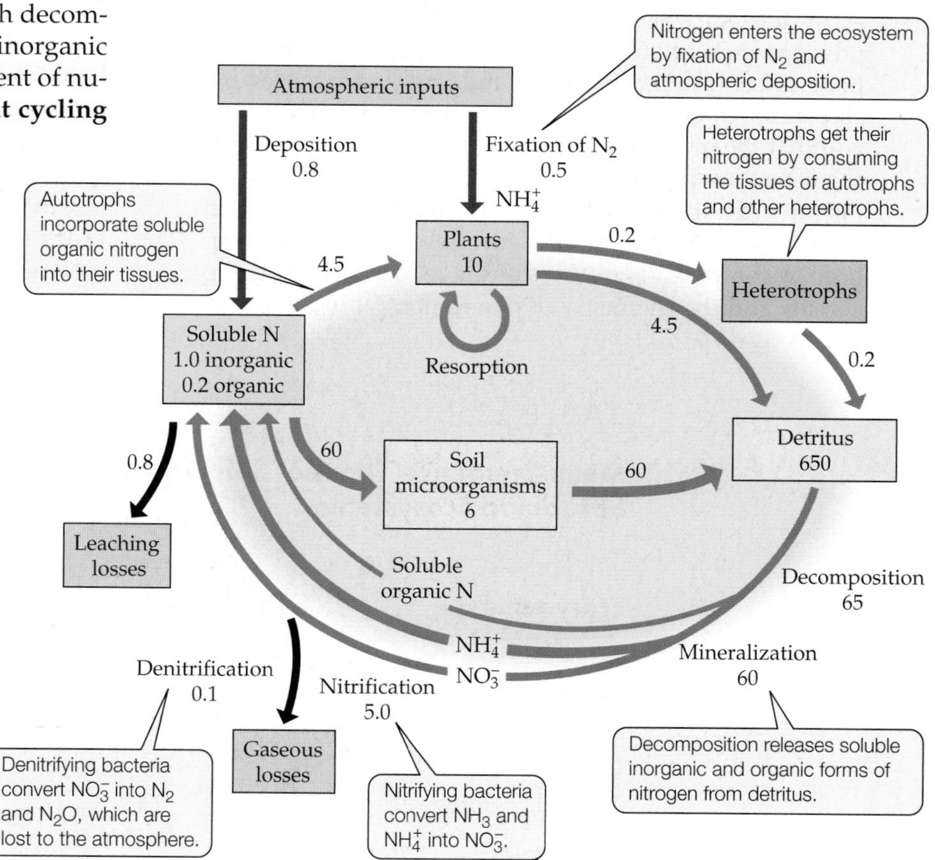

FIGURE 21.11 Nitrogen Cycle for an Alpine Ecosystem, Niwot Ridge, Colorado Boxes represent pools of nitrogen, measured in grams per square meter; arrows represent flows of nitrogen, measured in grams per square meter per year. Note the large amount of nitrogen passing through soil microorganisms, which indicates a high turnover rate for nitrogen in this relatively small pool. (Data from Bowman and Seastedt 2001.)

Nutrients cycle at different rates according to element identity and ecosystem type

The time it takes a nutrient molecule to cycle through an ecosystem, from uptake by organisms to release to subsequent uptake, can vary substantially depending on the element in question and the ecosystem where the cycle is occurring. In general, nutrients that limit primary production are cycled more rapidly than nonlimiting nutrients. For example, nitrogen and phosphorus may cycle through the photic zone of the open ocean over a period of hours or days, while zinc may cycle over geologic time scales associated with sedimentation, mountain building, and erosional processes. Nutrient cycling rates also vary with climate due to the effects of temperature and moisture on the metabolic rates of the organisms associated with production, decomposition, and chemical transformations of nutrients.

Biogeochemists measure rates of nutrient cycling by estimating the **mean residence times** of elements in some component of an ecosystem:

$$\text{mean residence time} = \frac{\text{total pool of element}}{\text{rate of input}}$$

In other words, the mean residence time is the amount of time an average molecule of an element spends in a pool before leaving it. The **pool** of an element is the total amount found within a physical or biological component of the ecosystem, such as soil or biomass. The inputs include all possible sources of the element for that ecosystem component. This approach to estimating mean residence time assumes that pools of nutrients do not change over time and that the mean residence time reflects the overall rate of nutrient cycling. It is most commonly used for estimating rates of nutrient turnover in soil organic matter, which reflect rates of nutrient input and subsequent decomposition. Decomposition rates, as we have seen, are related to climate and the chemistry of plant litter.

Given that both inputs of plant litter and decomposition rates control the mean residence times of nutrients in soil, and that both are subject to climatic control, what differences would we expect to see among ecosystems with similar plant growth forms (e.g., forests) in different climates? Tropical forests have higher net primary productivity, and therefore higher litter input rates, than boreal forests. Does this difference result in differences in the mean residence times of nutrients? A comparison of mean residence times for organic matter and for several nutrients indicates that nutrient pools in the soils of tropical forests are much smaller than those in boreal forests (**Table 21.3**). The turnover rates of nitrogen and phosphorus are more than 100 times faster in tropical forest soils than in boreal forest soils. Temperate forests and chaparral have turnover rates that fall in between, but are closer to those in the tropics.

The main reason for this trend in mean residence times is that the influence of climate on rates of decomposition is greater than its influence on primary productivity. Boreal forest soils often have permafrost layers underneath, which cool the soils and lower rates of biological activity. The permafrost also blocks the percolation of water through the soil, creating wet, anoxic soil conditions. Furthermore, the litter produced by boreal forest trees is rich in secondary compounds that slow rates of decomposition in the soil.

The variation in mean residence times among specific nutrients is related to their chemical properties (e.g., solubility). Some nutrients, such as potassium, occur in more soluble forms, and thus are lost from soil organic matter more quickly, than others, such as nitrogen, which is found as insoluble organic compounds.

In Chapter 24, we will return to nutrient cycling at a much larger spatial scale as we consider global cycles of carbon, nitrogen, phosphorus, and sulfur in the context of human alterations of these cycles.

TABLE 21.3

Mean Residence Times of Soil Organic Matter and Nutrients in Forest and Shrubland Ecosystems

	Mean residence time (yrs)					
Ecosystem type	**Soil organic matter**	**N**	**P**	**K**	**Ca**	**Mg**
Boreal forest	353	230	324	94	149	455
Temperate coniferous forest	17	18	15	2	6	13
Temperate deciduous forest	4	5	6	1	3	3
Chaparral	4	4	4	1	5	3
Tropical rainforest	0.4	2	2	1	1.5	1

Source: Schlesinger 1997.

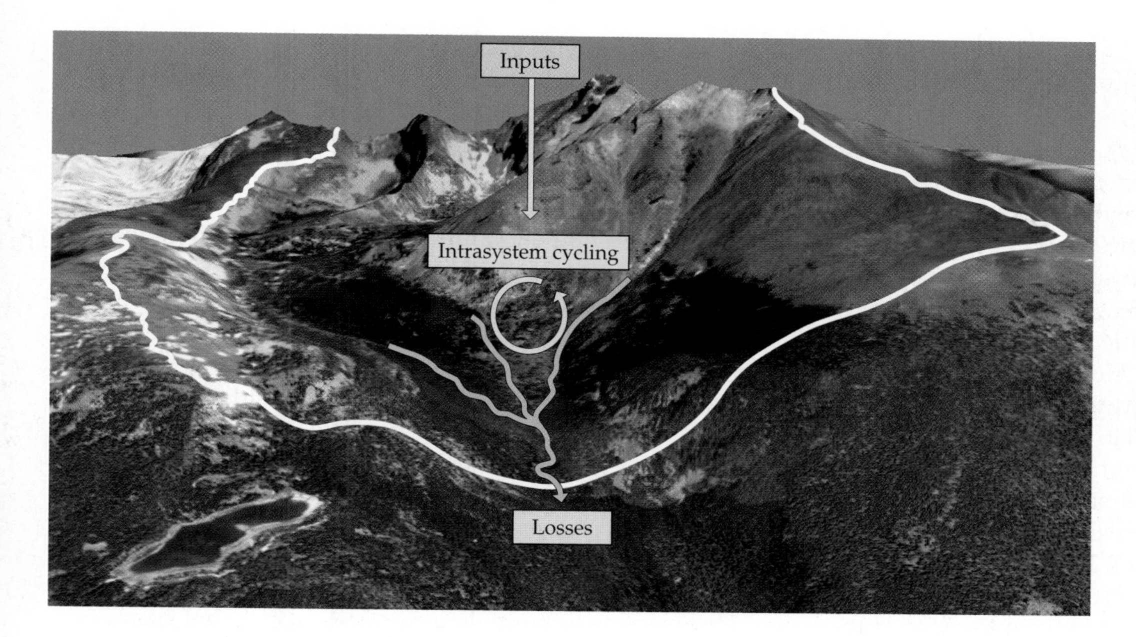

FIGURE 21.12 Catchments Are Common Units of Ecosystem Study A drainage basin (known as a catchment or watershed) associated with a single stream system (blue lines), with boundaries determined by topographic divides (outlined in white), is a unit commonly used in terrestrial ecosystem studies to measure inputs and outputs of nutrients. This catchment is the upper Hunters Creek basin, draining the south side of Longs Peak in Rocky Mountain National Park.

Q What assumptions are made in this simple input–output model of a catchment that may not be realistic? (Hint: Compare this figure with Figure 21.13).

Catchment studies measure losses of nutrients from ecosystems

What determines how long nutrients remain in an ecosystem? The retention of nutrients within an ecosystem is related to their uptake into its biological and physical pools and to the stability of their forms. Nitrogen, for example, is more stable as part of an insoluble organic molecule, such as a protein, than as nitrate, which is more easily leached from the soil. Nutrients are lost from an ecosystem when they move below the rooting zone by leaching, and from there into groundwater and streams. Nutrients are also lost to the atmosphere as gases and by conversion into chemical forms that cannot be used by organisms.

In our consideration of nutrient inputs into and losses from ecosystems, we have been referring to ecosystems as if they were definitive spatial units, but what defines the boundaries of an ecosystem? Ecologists studying terrestrial ecosystems commonly focus on a single drainage basin. This unit of study, which is called a **catchment** or **watershed**, includes the terrestrial area that is drained by a single stream (**Figure 21.12**). By measuring the inputs and outputs of elements in a catchment and calculating the balance between them, ecologists can make inferences about the use of nutrients in the ecosystem and their importance to ecosystem processes such as primary production.

Figure 21.13 presents a conceptual model of a catchment. Nutrient inputs into the catchment include atmospheric deposition and fixation. Nutrients that enter the catchment may be stored in the soil (on cation exchange sites or in the soil solution) or taken up by organisms.

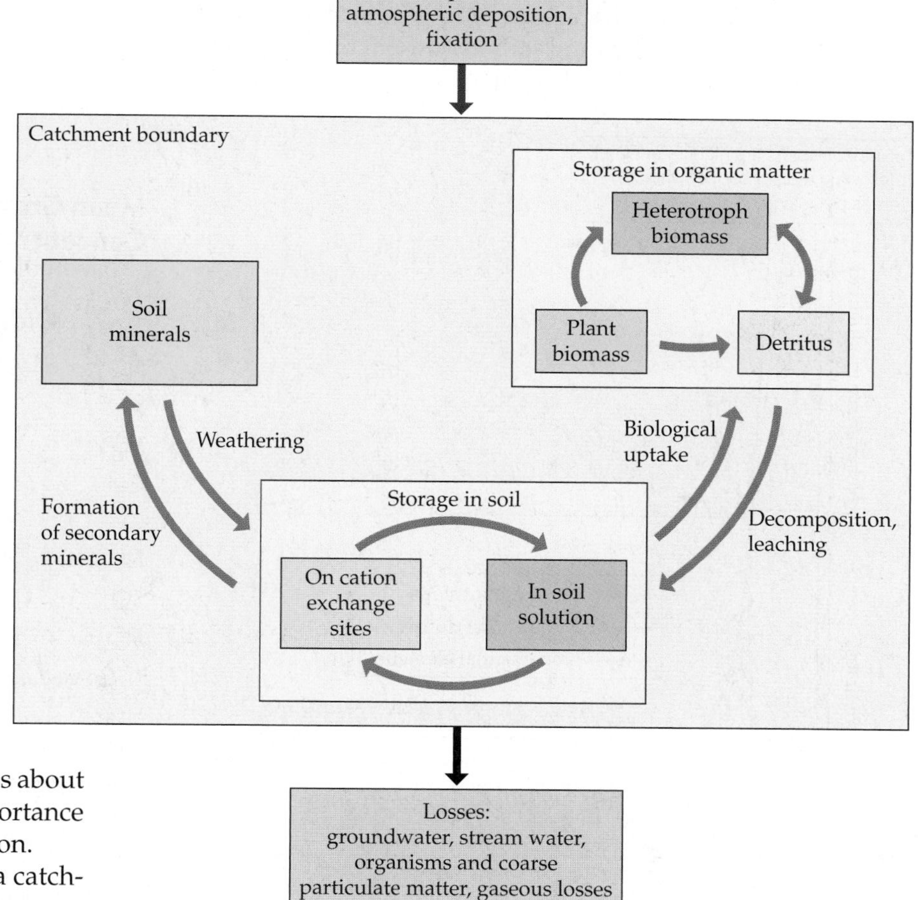

FIGURE 21.13 Biogeochemistry of a Catchment This conceptual model depicts the major pathways of nutrient movement into, through, and out of a catchment. (After Likens and Bormann 1995.)

They are transferred within and between these ecosystem components by herbivory and predation, decomposition, and weathering processes. Nutrients are assumed to be lost from the catchment primarily in stream water, so measurements of dissolved and particulate matter in streams draining the catchment are often used to quantify these losses. In reality, the situation is often more complicated, as nutrients are also lost to the atmosphere in gaseous forms (e.g., N_2 and N_2O from denitrification) and as coarse particulate matter, usually fragmented litter, and organisms moving out of the ecosystem. However, measurement of the input–output balance of different nutrients, using methods such as those described in **Ecological Toolkit 21.1**, is instructive for determining their biological importance.

The best-known catchment studies have been performed at the Hubbard Brook Experimental Forest in New Hampshire (Likens and Bormann 1995), which is considered to be representative of the northern deciduous forests of the United States. Continuous monitoring of the Hubbard Brook catchment began in 1963 under the direction of Herb Bormann and Gene Likens, whose studies have served as models for a number of other catchment-level studies. These studies are providing information about the roles of organisms and soils in nutrient retention, how ecosystems respond to disturbances such as logging and fire, and long-term trends in nutrient flows associated with acid rain and climate change.

An excellent example of the utility of catchment studies comes from an examination of the effect of disturbance on nutrient retention in ecosystems. How does a disturbance and subsequent recovery (succession, described in Chapter 16) influence nutrient cycling and nutrient losses in an ecosystem? Peter Vitousek (1977) used a catchment approach to study nutrient retention by spruce–fir forests in the White Mountains of New Hampshire at different stages of secondary succession following logging. Vitousek proposed that retention of nutrients would be related to the rate of forest growth. He predicted that the high rates of primary production usually observed during the intermediate stages of succession would result in the highest retention of nutrients, and that those nutrients most limiting to primary production would be retained more tightly than nonlimiting nutrients. His hypothesis was based on observations of multiple watersheds at different stages of succession. Vitousek's results showed that losses of nitrogen (a limiting nutrient) as nitrate in stream water from forests at mid-successional stages were much less than those from old-growth forests, while losses of nonlimiting nutrients, such as potassium, magnesium, and calcium, showed less sensitivity to forest successional stage (**Figure 21.14**).

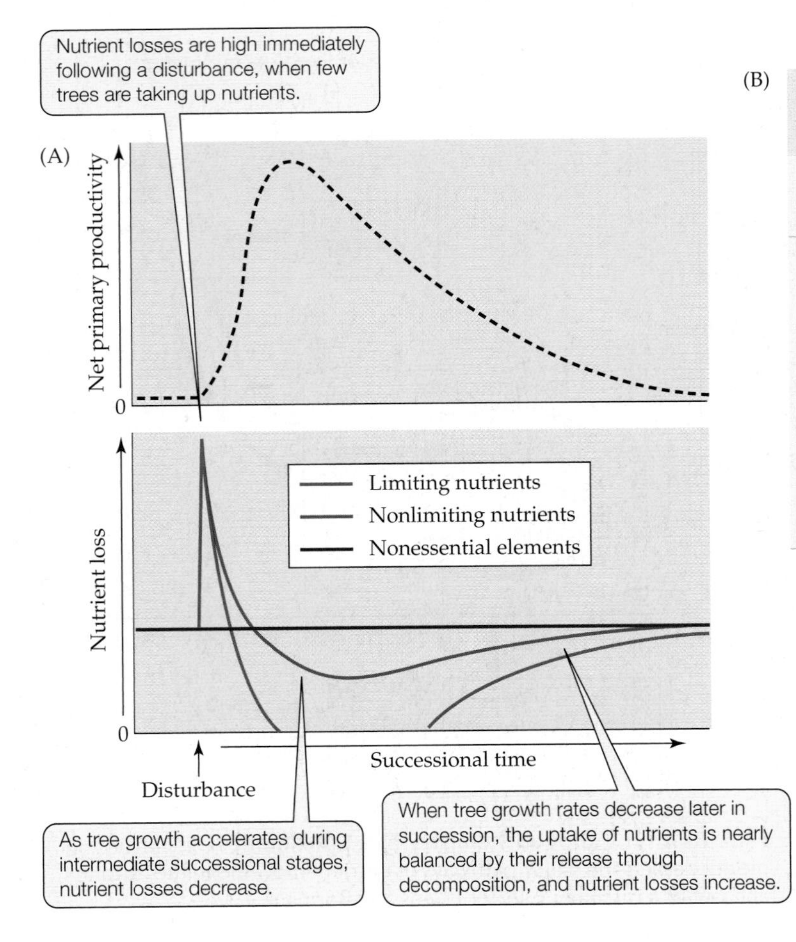

(B)

Mean Growing Season Stream Water Concentrations (μeq/L)

	Old-growth	Successional	Ratio of concentrations
NO_3^-	53 (5)	8 (1.3)	6.52
K^+	13 (1)	7 (0.5)	1.81
Mg^{2+}	40 (4.9)	24 (1.6)	1.66
Ca^{2+}	56 (4.5)	36 (2.5)	1.56
Cl^-	15 (0.3)	13 (0.3)	1.16
Na^+	29 (2.6)	28 (0.9)	1.03
SO_4^{2-}	119 (4.6)	123 (6.5)	0.97
Si	75 (7)	86 (5)	0.87

Note: Standard error in parentheses.

FIGURE 21.14 Retention of Nutrients Is Highest at Intermediate Stages of Forest Succession (A) Losses of nutrients from a spruce–fir forest catchment are hypothesized to vary over the course of succession as indicated in the graph, with limiting nutrients retained more tightly than nonlimiting nutrients at all stages. (B) The table shows changes in stream water outputs from old-growth stands and from younger successional stands of spruce–fir forests that support this hypothesis, with larger changes (higher ratios) in output of limiting nutrients (N, K) relative to nonlimiting nutrients (Ca, Cl, S). (Data from Vitousek 1977.)

21.1 Instrumenting Catchments

Measuring the inputs of nutrients into catchments via atmospheric deposition, as well as their losses in stream water, requires knowing the concentrations of the elements in water as well as the volume of water entering and leaving the catchment (i.e., the amount of precipitation and stream flow). The product of the two, concentration times volume, gives the total amount of the element entering or leaving the catchment. These values are usually averaged over periods ranging from a week to a year to provide input–output balances of specific elements.

Atmospheric deposition includes (1) elements captured in precipitation when it falls to the surface (wet deposition) and (2) particles, including aerosols and fine dust, that are transferred to the surface by gravitational fallout or air movement (dry deposition). Total atmospheric deposition can be sampled by placing buckets above the surrounding vegetation to collect the deposited material. However, buckets make good perches for birds, which may deposit their own contribution to ecosystem nutrient input inside the bucket, albeit at much higher concentrations than those found in most other parts of the catchment. This problem can be avoided by placing spiky projections around the edge of the bucket to prevent birds from landing on it. Another problem is that open buckets lose water to evaporation, increasing the concentration of the elements inside. Furthermore, in windy, cold climates, buckets are not good collectors or holders of snow due to their aerodynamics.

Wet deposition collectors have been developed that open to the atmosphere only during precipitation events and then close to prevent evaporation (**Figure A**). A moisture-sensitive surface controls a switch that opens and closes the collector. Where snow and wind occur together, windscreens help to prevent loss of snow from the bucket and enhance the capture of the deposition. Separate precipitation gauges may also be used to more accurately estimate the volume of precipitation entering the ecosystem. At regular intervals, the precipitation in the bucket or collector is analyzed for the elements of interest using chemical analyses that typically meet some government standard (e.g., in the United States, the Environmental Protection Agency provides these standards). In many developed nations, networks of wet deposition samplers have been established to provide spatial estimates of atmospheric deposition (e.g., the National Atmospheric Deposition Program in the United States: http://nadp.sws.uiuc.edu/; see Figure 24.19).

Dry deposition measurements are more complex, usually involving collection of atmospheric samples to measure the sizes of atmospheric particles and their chemical composition. These measurements are combined with wind speed and direction measurements to estimate movements of elements to the surface. Due to the greater difficulty of the sampling and the larger uncertainties, dry deposition is measured less frequently than wet or bulk (total) deposition. In some areas, however, such as deserts or Mediterranean-type ecosystems, dry deposition is the largest component of total deposition.

Measuring nutrient losses in stream flow is straightforward. The chemical composition of stream water leaving the catchment is measured by periodically collecting water samples and analyzing their chemistry. The volume of stream water is often estimated by constructing a *weir*, a small, usually V-shaped overflow dam made of concrete or wood and metal to control the size of the channel, and placing a depth gauge to calculate the volume of water passing through it (**Figure B**). The depth of the water can be measured with an automated system to give continuous volume estimates.

Figure A Measuring Precipitation A wet deposition collector is serviced on Niwot Ridge, Colorado. The bucket on the right is covered except during precipitation events.

Figure B Measuring Water Flow A weir on Fool Creek in the Fraser Experimental Forest, Colorado.

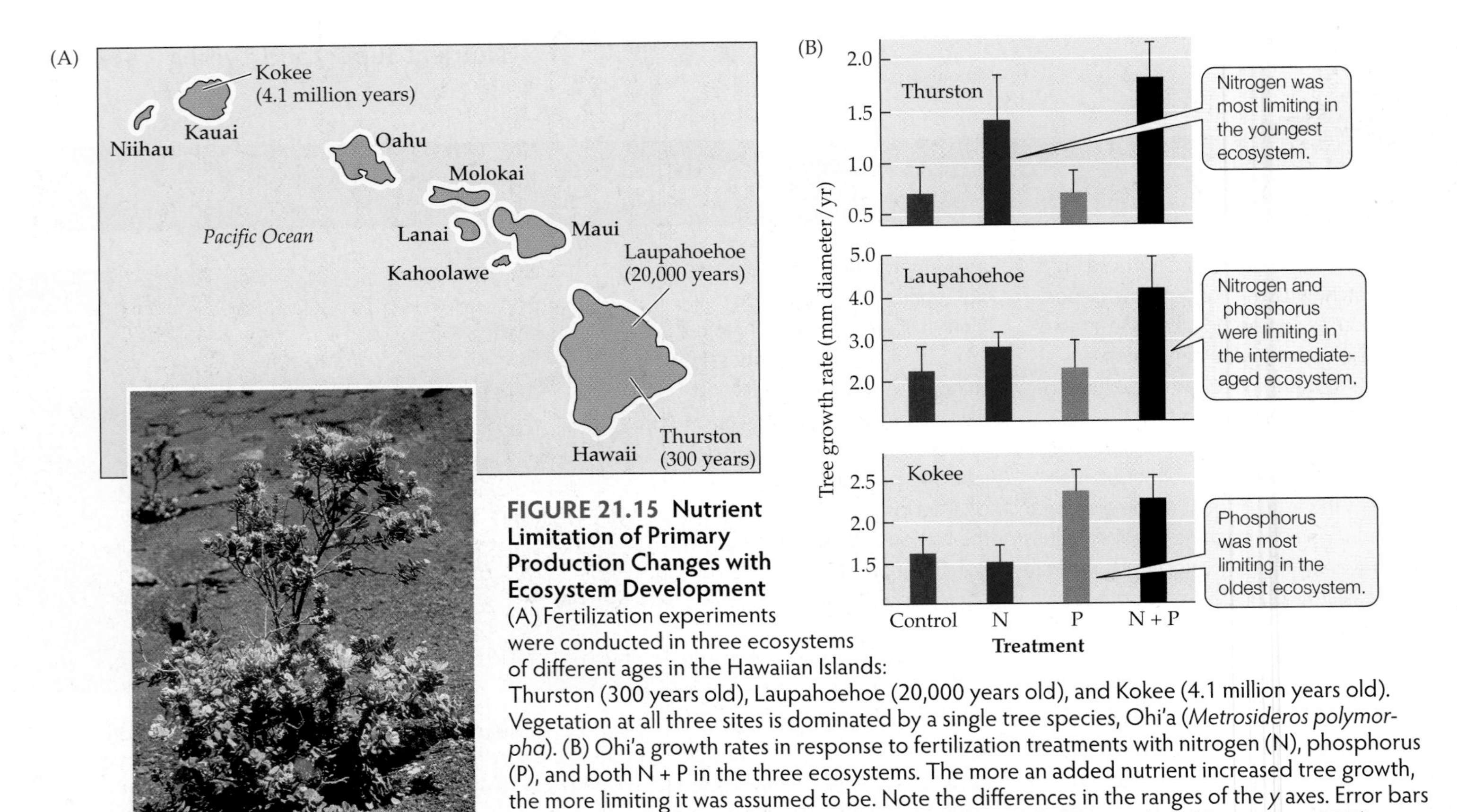

FIGURE 21.15 Nutrient Limitation of Primary Production Changes with Ecosystem Development (A) Fertilization experiments were conducted in three ecosystems of different ages in the Hawaiian Islands: Thurston (300 years old), Laupahoehoe (20,000 years old), and Kokee (4.1 million years old). Vegetation at all three sites is dominated by a single tree species, Ohi'a (*Metrosideros polymorpha*). (B) Ohi'a growth rates in response to fertilization treatments with nitrogen (N), phosphorus (P), and both N + P in the three ecosystems. The more an added nutrient increased tree growth, the more limiting it was assumed to be. Note the differences in the ranges of the *y* axes. Error bars show one SE of the mean. (A after Crews et al. 1995; B after Vitousek and Farrington 1997.)

Long-term ecosystem development affects nutrient cycling and constraints on primary production

As terrestrial ecosystems develop on new substrates (e.g., in primary succession on new volcanic flows), soil weathering, nitrogen fixation, and the buildup of organic matter in the soil determine the supply of nutrients available to plants. Early in ecosystem development, there is little organic matter in the soil, so supplies of nitrogen derived from decomposition are low. Supplies of mineral nutrients derived from weathering are also low, but higher than the supply of nitrogen. Accordingly, nitrogen availability should be an important constraint on primary production and plant community composition early in primary succession (see Chapter 16). As the pool of nitrogen in soil organic matter increases, its limitation of primary production should decrease.

Phosphorus enters ecosystems through the weathering of a single rock mineral (apatite), and its supply is high relative to that of nitrogen early in succession. As the supply of phosphorus from weathering is exhausted over time, however, decomposition becomes increasingly important in supplying phosphorus to plants. In addition, soluble phosphorus may combine with iron, calcium, or aluminum to form secondary minerals that are unavailable as nutrients, a process known as **occlusion**. The amount of phosphorus in occluded forms increases over time, further reducing its availability. As a result, phosphorus should become more limiting to primary production during later stages of succession (Walker and Syers 1976).

These observations of changes in nutrient cycling during ecosystem development thus provide a hypothetical framework for considering how those changes should influence the specific nutrients that limit primary production. Nitrogen should be most important in determining rates of primary production early in succession, nitrogen and phosphorus should both be important at intermediate stages of succession, and phosphorus should be most important late in succession. This hypothesis was tested in the Hawaiian Islands by Peter Vitousek and colleagues. The movement of the Pacific tectonic plate over millions of years has given rise to the chain of volcanoes that form these islands. The oldest islands are in the northwestern part of the chain, the youngest in the southwest (**Figure 21.15A**). Vitousek's group studied Hawaiian ecosystems on soils with ages ranging from 300 years to over 4 million years to determine which nutrients were most limiting to primary production. Their study was aided by the similarity of the vegetation and climate at each of the study sites. Vitousek and colleagues added nitrogen, phosphorus, and both nitrogen and phosphorus to plots in three ecosystems of different ages and measured the effects of these treatments on the growth of the dominant tree, Ohi'a (*Metrosideros polymorpha*). Consistent with their hypothesis, nitrogen was most limiting to tree growth in the youngest ecosystem, while phosphorus was most important in the oldest ecosystem (Vitousek and Farrington 1997) (**Figure 21.15B**). Nitrogen and phosphorus in combination increased tree growth in the intermediate-aged ecosystem. In contrast to these tropical soils, the soils of

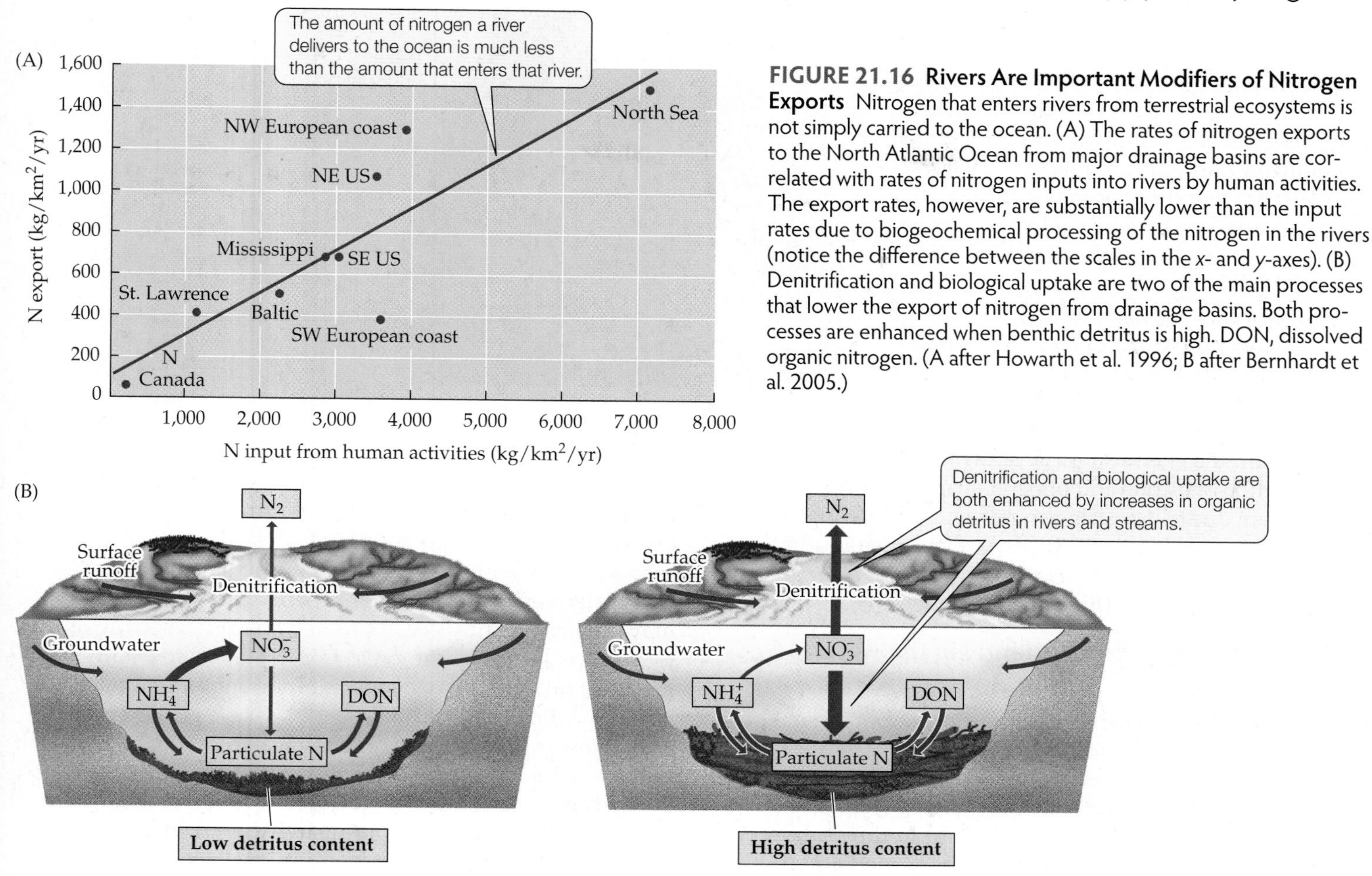

FIGURE 21.16 Rivers Are Important Modifiers of Nitrogen Exports Nitrogen that enters rivers from terrestrial ecosystems is not simply carried to the ocean. (A) The rates of nitrogen exports to the North Atlantic Ocean from major drainage basins are correlated with rates of nitrogen inputs into rivers by human activities. The export rates, however, are substantially lower than the input rates due to biogeochemical processing of the nitrogen in the rivers (notice the difference between the scales in the *x*- and *y*-axes). (B) Denitrification and biological uptake are two of the main processes that lower the export of nitrogen from drainage basins. Both processes are enhanced when benthic detritus is high. DON, dissolved organic nitrogen. (A after Howarth et al. 1996; B after Bernhardt et al. 2005.)

ecosystems in temperate, high-latitude, and high-elevation zones are often subjected to major disturbances (e.g., large-scale glaciation, landslides) and are less likely to reach ages at which phosphorus becomes limiting.

Nutrients lost from terrestrial ecosystems often end up in streams, lakes, and oceans. They are a critical source of nutrients for those aquatic ecosystems, but they can have negative effects as well, as we'll see in the next section.

CONCEPT 21.4 Freshwater and marine ecosystems receive nutrient inputs from terrestrial ecosystems.

Nutrients in Aquatic Ecosystems

In freshwater and marine ecosystems, nutrient transformations and transfers have the added complexity of occurring in a moving aqueous medium. Inputs of nutrients from outside the ecosystem are much more important than in terrestrial ecosystems. Furthermore, oxygen concentrations are often lower than in terrestrial ecosystems, constraining biological activity and the biogeochemical processes associated with it.

Nutrients in streams and rivers cycle in spirals

Nutrient supplies in streams and rivers are highly dependent on external inputs from terrestrial ecosystems. Terrestrial inputs of organic matter, dissolved nutrients derived from chemical weathering and decomposition in surrounding soils, and particulate minerals are the primary sources of nutrients for riverine organisms. Rivers and streams carry all of these materials to the ocean, but they are not just conduits for the movement of material between terrestrial and marine ecosystems. Biogeochemical processing in moving stream water can change the forms and concentrations of the elements it contains. For example, denitrification and biological uptake in streams and rivers may result in significant losses of nitrogen during transport in stream water. These processes may explain why rivers export less nitrate from regions receiving high amounts of nitrogen pollution than would be expected (**Figure 21.16A**). Both processes are enhanced when detritus is abundant on the stream bottom (**Figure 21.16B**).

Nutrients in rivers and streams are cycled repeatedly as the water flows downstream. Dissolved inorganic forms of nutrients are taken up by organisms, including fungi, bacteria, and phytoplankton, which incorporate them into organic molecules. These organisms may be consumed by

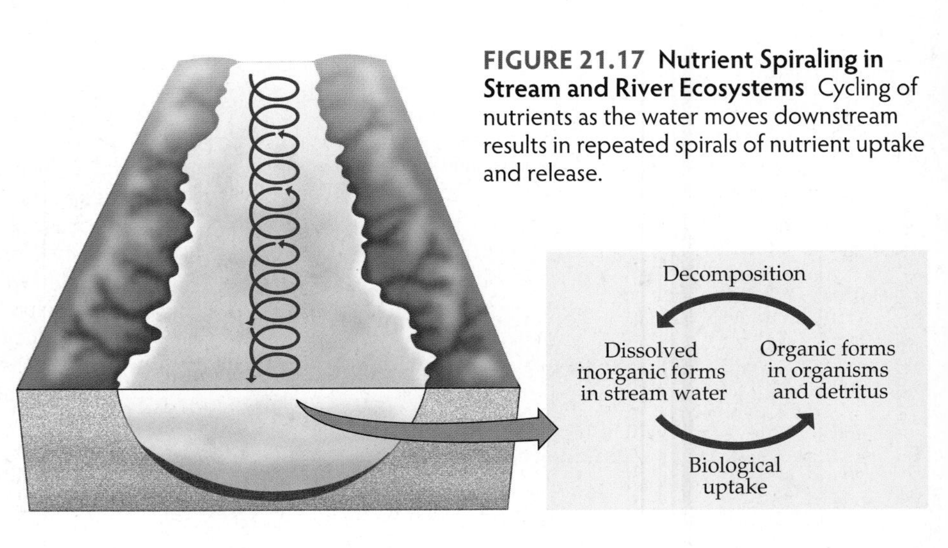

FIGURE 21.17 Nutrient Spiraling in Stream and River Ecosystems Cycling of nutrients as the water moves downstream results in repeated spirals of nutrient uptake and release.

others and pass through a food web, eventually entering the pool of stream detritus. Following decomposition of the detritus, the mineralized nutrients are released back into the water in dissolved inorganic forms. This repeated uptake and release in association with the movement of water can be thought of as nutrient "spiraling" (Newbold et al. 1983) (**Figure 21.17**). The time it takes for a full nutrient spiral to occur (i.e., from uptake and incorporation into organic forms to release in inorganic forms) is related to the amount of biological activity in the stream, the water velocity, and the chemical form of the nutrient. Retention of nitrate and phosphate in rivers increases downstream due to increasing spiral lengths; thus, higher-order streams (see Figure 3.13) are particularly important in buffering the effects of nutrient pollutants on estuarine and marine ecosystems (Ensign and Doyle 2006).

Nutrients in lakes cycle efficiently in the water column

Lake ecosystems receive inputs of nutrients from streams, by atmospheric deposition, and as litter falling from adjacent terrestrial ecosystems. Biological demand for nutrients is highest in the photic zone, where phytoplankton are suspended in the water column, and in the shallow zones at the margins of the lake, where rooted aquatic plants are found. Phosphorus commonly limits primary production in lakes, although nitrogen may also be limiting in some lakes. Nutrient transfers between trophic levels, like energy transfers (see Figure 20.5C), are very efficient in lakes. Some detritus is decomposed and mineralized in the water column and in sediments in the shallow zones, providing an internal input of nutrients. Nitrogen fixation by cyanobacteria occurs in the photic zone, particularly when demand for nitrogen by organisms is greater than for phosphorus. Rates of nitrogen fixation in lake ecosystems are similar to those in terrestrial ecosystems.

Over time, nutrients are progressively lost from the photic zone of a lake. Dead organisms sink through the water column and are deposited in the sediments of the benthic zone. These sediments are characterized by hypoxic conditions that limit biological activity, including decomposition, and by a reducing chemical environment that may change the chemical form of some nutrients. Iron, for example, is often reduced from Fe^{3+} to Fe^{2+}, contributing to the dark color of lake sediments. Denitrification is also promoted by the low oxygen concentrations in the sediments, and bacteria may reduce sulfate (SO_4^{2-}) to hydrogen sulfide (H_2S).

Decomposition in the benthic sediments cannot provide nutrients to the photic zone unless there is mixing of the water column. In stratified temperate-zone lakes, as we saw in the discussion of Concept 2.5, this mixing occurs in fall and spring, when the lake's water from top to bottom becomes isothermal and wind facilitates its turnover. This seasonal turnover brings dissolved nutrients from the bottom water to the surface layers, along with detritus that may be subsequently decomposed by bacteria. Mixing of water layers is less common in tropical lakes, so external inputs of nutrients may be more important for maintaining production in those lakes.

Lake ecosystems are often classified according to their nutrient status. Nutrient-poor waters with low primary productivity are referred to as **oligotrophic**, while nutrient-rich waters with high primary productivity are referred to as **eutrophic**. **Mesotrophic** waters are intermediate in nutrient status between oligotrophic and eutrophic waters. The nutrient status of a lake is the result of natural processes associated with climate and with lake size and shape. For example, lakes in high mountain areas are typically oligotrophic due to their short growing season, low temperatures, and tendency to be deep with a low surface area-to-volume ratio, which decreases the rate of nutrient input by atmospheric deposition. In contrast, shallow lakes at lower elevations or in the tropics tend to be eutrophic due to their warmer temperatures and higher nutrient availability.

The nutrient status of a lake tends to shift naturally from oligotrophic to eutrophic over time. This process, known as **eutrophication**, occurs as sediments accumulate on the lake bottom (**Figure 21.18**). As the lake becomes shallower, its summer temperatures become warmer, more decomposition occurs, nutrient pools and the amount of mixing increase, and the lake becomes more productive. Human activities have accelerated the process of eutrophication in many lakes through discharges of sewage, agricultural fertilizers, and industrial wastes containing high concentrations of nitrogen and phosphorus. For example, the water of Lake Tahoe, on the border between Nevada and California, has lost much of its clarity due to increased inputs of phosphorus and nitrogen from streams, groundwater, and surface runoff from neighboring communities. Water clarity, which is used as an indicator of a lake's nutrient status, is primarily

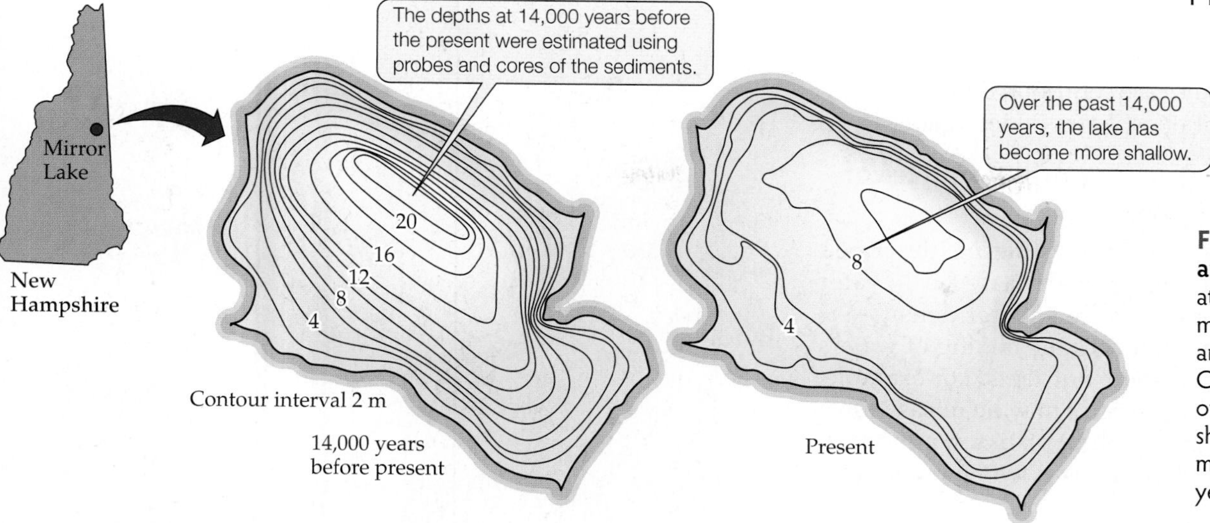

FIGURE 21.18 Lake Sediments and Depth Sediments accumulate at the bottom of a lake over time, making it progressively shallower and leading to eutrophication. Changes in the depth contours of Mirror Lake in New Hampshire show the accumulation of sediments there over the past 14,000 years. (After Davis et al. 1985.)

determined by the density of plankton in the water column. It can be measured using a Secchi disk, a black-and-white circular plate that is lowered gradually into the water; the maximum depth at which the disk can be seen is referred to as the *depth of clarity*. Over the past three decades, the average depth of clarity in Lake Tahoe has risen by 10 m (Murphy and Knopp 2000). The rate at which water clarity has been decreasing has declined since 2000, due in part to lower amounts of precipitation.

Anthropogenic eutrophication can be reversed if the discharge of wastes into surface waters is decreased. A classic example of such a reversal occurred in the 1960s and 1970s in Lake Washington, near Seattle. Treated sewage, containing high concentrations of phosphorus, was released into Lake Washington beginning in the late 1940s as neighborhoods and accompanying sewage treatment plants were built near the shore of the lake. Decreases in water clarity were noted during the 1950s, corresponding to increases in phytoplankton densities and blooms of cyanobacteria. Public concern grew, and local governments debated what action to take. A prominent local limnologist, W. T. Edmondson, believed that the problem was associated with phosphorus inputs from the treated sewage, which included wastewater from washing machines containing phosphorus-laden detergents. Based on Edmondson's advice, Seattle quit dumping its sewage into Lake Washington. Between 1963 and 1968, the amount of sewage discharged into the lake was progressively dropped to zero. Increases in water clarity were soon noted, and by 1975, the lake was considered recovered from eutrophication (**Figure 21.19**). Edmondson's recommendation was crucial to

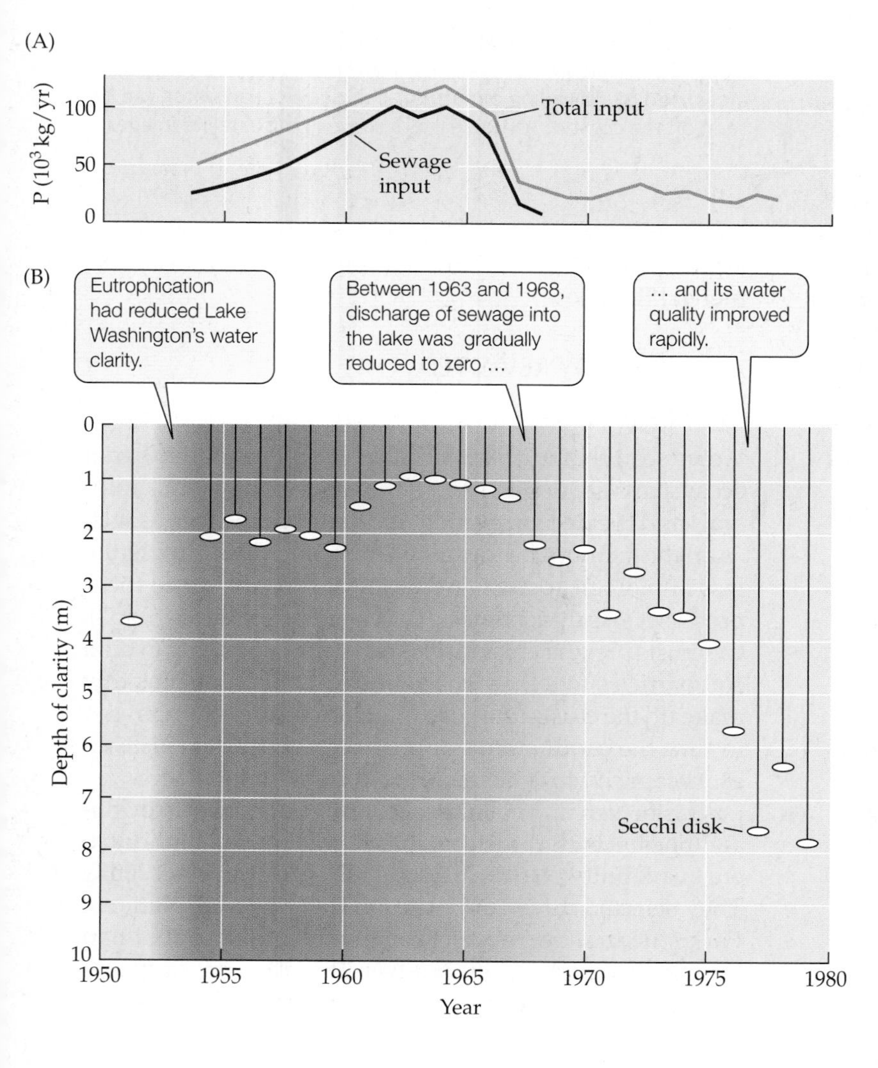

FIGURE 21.19 Lake Washington: Reversal of Fortune Inputs of treated sewage between the 1940s and the 1960s caused eutrophication in Lake Washington; cessation of sewage inputs between 1963 and 1968 increased lake clarity. (A) Phosphorus inputs. (B) Measurements of water clarity made with a Secchi disk. (A after Edmondson and Litt 1982.)

While the story of Lake Washington seems to be a clear "experimental" demonstration of pollution influencing the nutrient status of a lake, what would make it an even more convincing example?

the lake's recovery, and the case contributed to the current restrictions on the use of phosphates in detergents.

Imports and upwelling are important sources of nutrients in marine ecosystems

Rivers join marine ecosystems in estuaries (described under Concept 3.3). In these zones where fresh water meets seawater, salinity—and thus water density—is variable. This variation influences the mixing of waters and the chemical forms of some nutrients. For example, phosphorus bound to soil particles may be released in a form more easily available to phytoplankton as a result of changes in pH and water chemistry when river water mixes with seawater.

As the velocity of water flow decreases toward the mouth of a river, suspended sediments begin to settle out of the water. These sediments are resuspended by the upwelling of saltier, denser seawater into the fresher, less dense river water, providing detritus for detritivores and nutrients for phytoplankton in the estuary. Estuaries are often associated with salt marshes, which are rich in nutrients because they trap both riverine and ocean sediments. Like benthic sediments in lakes, estuarine and salt marsh sediments have low oxygen concentrations that limit decomposition.

As described under Concept 19.2, primary production in the open ocean is limited by several nutrients, including nitrogen, phosphorus, and in some areas, iron and silica. Seawater has relatively high concentrations of magnesium, calcium, potassium, chloride, and sulfur. Sources of nitrogen in marine ecosystems include inputs from rivers and atmospheric deposition as well as tight internal cycling through decomposition. Rates of nitrogen fixation by cyanobacteria in the oceans are lower than those in freshwater lakes, possibly because these organisms are limited by molybdenum, which is a component of the nitrogenase enzyme. Phosphorus, iron, and silica enter the marine ecosystem primarily in dissolved and particulate form in rivers; a smaller but important contribution comes from atmospheric deposition of dust. Inputs from both of these terrestrial sources are increasing as a result of human activities, including large-scale desertification and deforestation.

Deep deposits of sediments (up to 10 km, or 6 miles!) have accumulated in the benthic zones of the open ocean. These deposits, which consist of a mix of ocean-derived detritus and terrestrial erosional sediments, are important potential sources of nutrients. Sulfate reduction and denitrification are important processes in these anoxic sediments, but some decomposition and mineralization of organic matter occurs there. Bacteria have been found as deep as 500 m in these sediments. Mixing of deep, nutrient-rich waters with nutrient-poor surface waters occurs in zones of upwelling, where ocean currents bring deep waters to the surface (**Figure 21.20**). These zones of upwelling are highly productive and thus are important areas for commercial fisheries.

FIGURE 21.20 Zones of Upwelling Enhance Nutrient Supply for Marine Ecosystems Phytoplankton blooms (green areas), fed by upwelling of nutrient-rich deep ocean water, can be seen off the coast of Africa's eastern shore in this satellite image.

A CASE STUDY REVISITED

A Fragile Crust

We've seen that nutrient supplies for plants in terrestrial ecosystems are dependent on the weathering of rock minerals and the decomposition of detritus in the soil, as well as on the fixation of atmospheric nitrogen. How might the loss of biological crusts from desert soils influence these processes? As this chapter's Case Study explained (p. 452), the crusts prevent erosional losses of soil by helping to bind soil particles together. The activity of the organisms that make up the crusts may also influence nutrient inputs, and in turn the productivity of the desert ecosystem, as well as its capacity to withstand the desert climate.

Jason Neff and colleagues conducted a study to evaluate the effects of cattle grazing on soil erosion and nutrient availability on the Colorado Plateau (Neff et al. 2005). They selected three study sites in Canyonlands National Park: one that had never been grazed and two that had been grazed historically, but were closed to grazing after 1974 (30 years of recovery). Cattle grazing in the park was

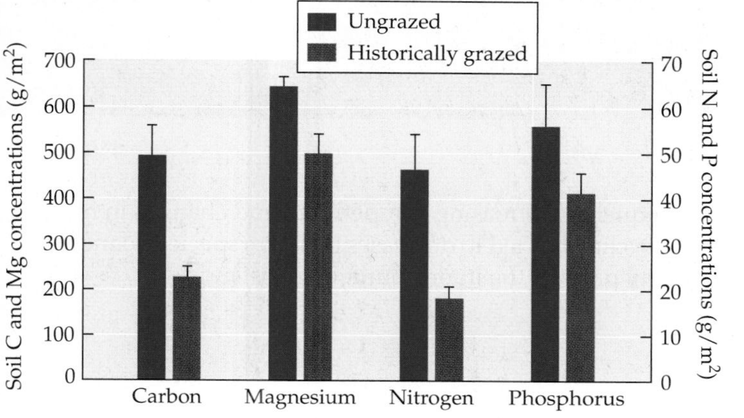

FIGURE 21.21 Loss of Biological Crusts Results in Smaller Nutrient Supplies Historically grazed soils in Canyonlands National Park contained less carbon, magnesium, nitrogen, and phosphorus than soils that had never been grazed. Error bars show one SE of the mean. (After Neff et al. 2005.)

initiated in the 1880s, and most of its soil surface has been affected. The ungrazed study site was surrounded by rock formations that prevented the movement of cattle into the area. The study sites all had the same parent material, had similar plant communities, and were located within 10 km of one another. Biological crusts were present at all three sites, although those at the historically grazed sites had clearly been damaged, as they appeared less well developed than those at the site that had never been grazed.

Samples of soil and bedrock were collected from each of the sites, and the textures and nutrient contents of the soils were compared. In addition, the retention of fine dust from the atmosphere was estimated by measuring the magnetic properties of the soil. Dust blown in from distant areas contains higher amounts of iron oxides than the native soil, so the more dust present, the stronger the magnetic signal. Retention of this dust is important because it is a source of mineral nutrients; in addition, loss of this dust indicates the potential for erosional loss of the native soil as well.

Neff and colleagues found that the historically grazed soils had less fine-textured soil, and substantially less magnesium and phosphorus, than the ungrazed soils (**Figure 21.21**). They attributed these differences to greater retention of dust and lower rates of erosion in the soils with better-developed biological crusts. The crusts may also enhance rates of weathering by altering pH, by increasing the rates of chemical reactions that release mineral nutrients, and by increasing water retention in the soil. Soils in the historically grazed sites also contained 60%–70% less carbon (from organic matter) and nitrogen than those in the ungrazed sites. These differences were also related to biological crusts. Although a crust had begun to recover at the historically grazed sites, comparison with the ungrazed site showed that the cumulative loss of carbon and nitrogen from the soils during the period of grazing was high. The cyanobacteria in biological crusts fix atmospheric N_2 (Belnap 2003), which represents an important input of a nutrient that may limit plant growth in the absence of water limitation during the spring growing season. In addition, crust-covered soils absorb more solar radiation and retain more water than soils without crusts, creating conditions more conducive to decomposition and mineralization.

CONNECTIONS IN NATURE
Nutrients, Disturbance, and Invasive Species

By increasing nutrient supplies and stabilizing soils, biological crusts enhance primary production. Plants growing in association with the crusts have higher growth rates, and contain more nutrients, than plants growing in soils without crusts. Plant cover also increases in the presence of biological crusts. Furthermore, biological crusts have been shown to lower the germination and survival rates of invasive plants (Mack and Thompson 1982; see Chapter 22). Thus, the destruction of crusts by cattle grazing has had multiple ecological effects.

Are the negative effects of cattle grazing on soil stability and nutrient availability that Neff and colleagues observed in Canyonlands National Park common in other areas? The answer lies in part with the long-term history of grazing and climate in North America. Prior to Euro-American settlement, soils in much of the intermountain West did not experience the amount of grazing by native animals that occurred in other areas, such as the Great Plains, where large herds of bison roamed [see the Case Study in Chapter 3 (p. 49) and the discussion under Concept 12.2 (p. 266)]. A combination of aridity and long-term development of biological crusts may have given the soils of the Colorado Plateau an especially low tolerance for heavy grazing.

In the grasslands of the intermountain West, the combination of soil disturbance and loss of biological crust has created a situation conducive to the spread of non-native species—most notably cheatgrass (*Bromus tectorum*;

FIGURE 21.22 Scourge of the Intermountain West Large areas of the intermountain West of North America are now dominated by cheatgrass (*Bromus tectorum*), an invasive species that increases fire frequencies, outcompetes native plants for resources, and spreads rapidly across the landscape.

Figure 21.22), a native of Eurasia. Cheatgrass has had profound effects on the ecology of much of western North America. Cheatgrass is a spring annual that sets seed, dies, and dries out by early summer. This life history increases the amount of dry, combustible vegetation that is present during the summer. As a result, cheatgrass has increased the frequency of fires, which now occur about every 3–5 years, compared with more natural fire frequencies of 60–100 years. Native grasses and shrubs cannot recover from such frequent fires, so cheatgrass increases its dominance under these conditions. Cheatgrass is an effective competitor for soil resources, and it also lowers rates of nitrogen cycling by producing litter with a C:N ratio higher than those of native species (Evans et al. 2001). This combination of increasing fire frequency, increasing competition, and changes in nutrient cycling has led to decreases in native species richness in many parts of the intermountain grasslands.

SUMMARY

CONCEPT 21.1 Nutrients enter ecosystems through the chemical breakdown of minerals in rocks or through fixation of atmospheric gases.

- The nutrient requirements of organisms are specific to their physiology and thus differ between autotrophs and heterotrophs.
- Autotrophs absorb nutrients in simple, soluble forms from their environment, while heterotrophs obtain them in more complex forms by consuming prey or detritus.
- The physical and chemical breakdown of minerals (weathering) releases soluble nutrients.
- Soils are made up of mineral particles, detritus, dissolved organic matter, water containing dissolved minerals and gases, and organisms.
- Carbon and nitrogen enter ecosystems through fixation of atmospheric gases by autotrophs and by bacteria, respectively.

CONCEPT 21.2 Chemical and biological transformations in ecosystems alter the chemical form and supply of nutrients.

- Decomposition of organic matter releases the nutrients it contains in soluble forms that can be reused by plants and microorganisms.
- Modification of the chemical forms of nutrients, particularly nitrogen, by microorganisms influences their availability to organisms or loss from the ecosystem.
- Plants recycle nutrients by reabsorbing them from senescing tissues and remobilizing them when growth commences again.

CONCEPT 21.3 Nutrients cycle repeatedly through the components of ecosystems.

- Nutrient cycling rates are controlled primarily by the rate of decomposition, which in turn is controlled by climate and the chemistry of plant litter.
- Losses of nutrients from terrestrial ecosystems can be estimated by measuring nutrient outputs in stream water.
- Changes in the relative amounts of nutrients supplied by weathering and decomposition determine the specific nutrients that limit primary production at different stages of ecosystem development.

CONCEPT 21.4 Freshwater and marine ecosystems receive nutrient inputs from terrestrial ecosystems.

- The cycling of nutrients in streams and rivers can be thought of as a spiral of repeated biological uptake and incorporation into organic forms followed by release in inorganic forms.
- In lakes, nutrients are cycled between the water column and the benthic sediments.
- Imports of nutrients from rivers and terrestrial ecosystems support production in marine ecosystems.

REVIEW QUESTIONS

1. Describe the processes involved in the transformation of solid minerals in rock into soluble nutrients in soil. What biological factors can influence the rate of this transformation?
2. Why is nitrogen often in short supply relative to other nutrients required by plants, despite being the most abundant element in the atmosphere? How does the supply of nitrogen change during terrestrial ecosystem development?
3. Which factor is more important in controlling the mean residence times and pools of nutrients in soil organic matter in terrestrial ecosystems, the rate of input (i.e., primary productivity) or the rate of decomposition? Would you expect to find larger nutrient pools in the soils of a tropical forest than in those of a boreal forest, given that primary productivity is higher in the tropics?
4. Why might you expect nutrient input from terrestrial and stream ecosystems to be more important in tropical lakes than in temperate-zone lakes?

ON THE COMPANION WEBSITE
sites.sinauer.com/ecology2e

The website includes Chapter Outlines, Online Quizzes, Flashcards & Key Terms, Suggested Readings, a complete Glossary, and the Web Stats Review. In addition, the following resources are available for this chapter:

HANDS-ON PROBLEM SOLVING

Dry Decomposition

This Web exercise explores how plant litter decomposes in a dry climate. You will interpret data from a recent paper investigating the factors responsible for litter decomposition in a semiarid ecosystem in Patagonia.

UNIT 6
Applied and Large-Scale Ecology

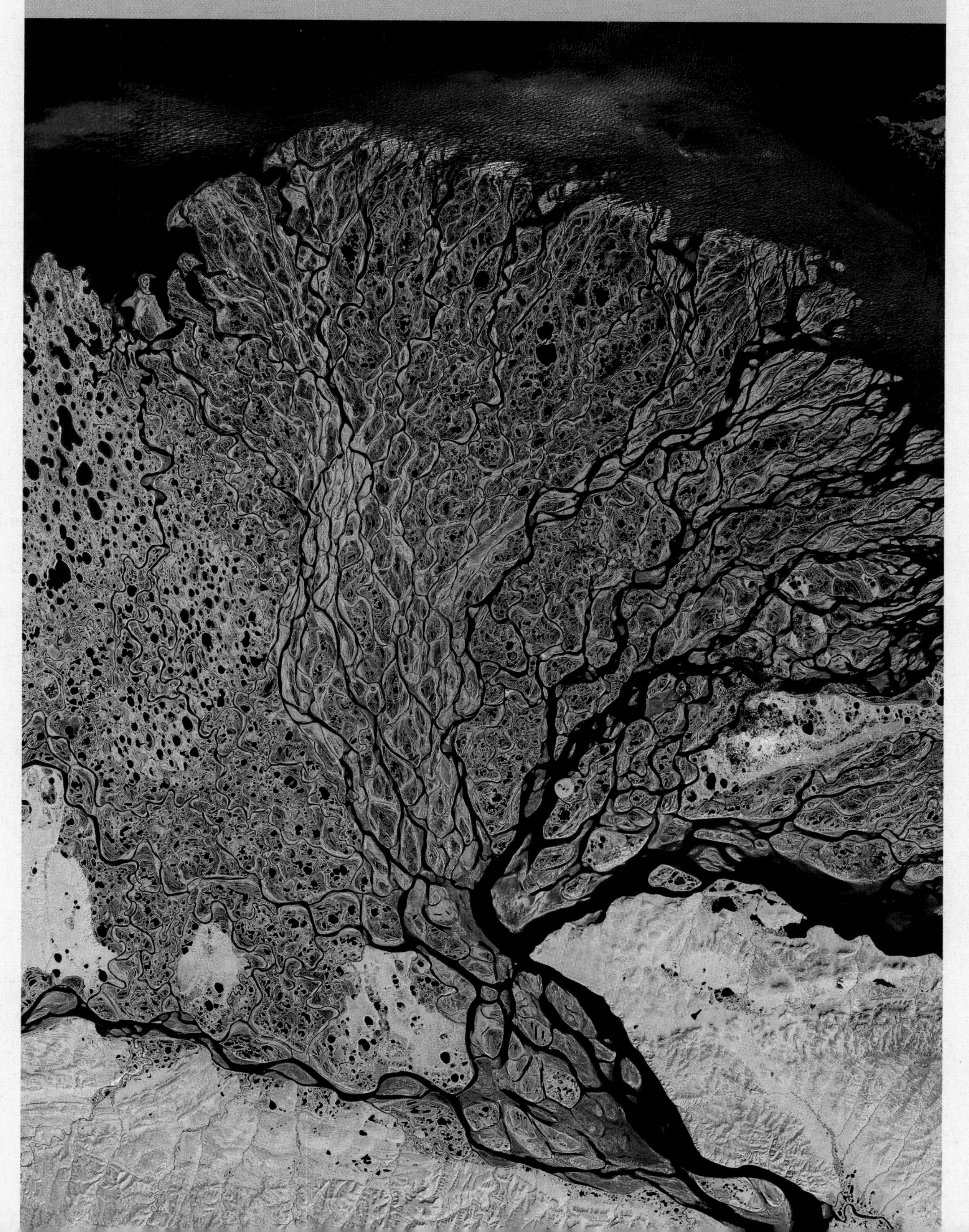

CHAPTER

22 Conservation Biology

KEY CONCEPTS

- **CONCEPT 22.1** Conservation biology is an integrative discipline that applies the principles of ecology to the protection of biodiversity.
- **CONCEPT 22.2** Biodiversity is declining globally.
- **CONCEPT 22.3** Primary threats to biodiversity include habitat loss, invasive species, overexploitation, pollution, disease, and climate change.
- **CONCEPT 22.4** Conservation biologists use many tools and work at multiple scales to manage declining populations.
- **CONCEPT 22.5** Prioritizing species helps maximize the biodiversity that can be protected with limited resources.

Can Birds and Bombs Coexist?: A Case Study

Could it be that using land as a bombing range has been the secret to conservation success? Although it may seem strange, decades of bombing on the Fort Bragg military base in the North Carolina Sandhills have inadvertently protected thousands of acres of longleaf pine savanna, aiding efforts to save the endangered red-cockaded woodpecker (**Figure 22.1**).

For 90 years, the forests of Fort Bragg have been used for military training exercises, degraded by off-road vehicles and earth-moving equipment, and set on fire by explosives. These destructive activities take place in the midst of a vibrant but now uncommon ecosystem—one that, ironically, survives in large part as a result of the military presence. How can this be? First, pine savanna depends on fire for its persistence, so the fires that result from explosions benefit rather than harm the ecosystem. Second, the designation of large blocks of forest land for military use has kept them from being converted to farmland, forest plantations, and residential uses.

While some longleaf pine savanna has been preserved at Fort Bragg and other military bases, overall, this ecosystem has been reduced to only 3% of the more than 35 million hectares (>86 million acres) it once covered (**Figure 22.2**). Various factors have contributed to this decline, including rapid growth of the human population, the clearing of land for large plantations where other tree species, such as loblolly pine, are grown, and fire suppression. With the decline of the longleaf pine ecosystem, several plant, insect, and vertebrate species that depended on it have also undergone substantial declines.

One of these species is the red-cockaded woodpecker (*Picoides borealis*), a small insectivorous bird that is well adapted to large tracts of open pine savanna. Once far more abundant, the species currently stands at about 6,100 breeding pairs and their associated helpers. (Red-cockaded woodpeckers are *cooperative breeders*, in which two to four males born to the breeding pair in previous years typically help their parents raise young.)

Whereas other woodpeckers nest in dead snags, red-cockaded woodpeckers require mature, living pine trees, especially the longleaf pine (*Pinus palustris*), for their nesting cavities. Periodic fires histori-

FIGURE 22.1 The Red-Cockaded Woodpecker: An Endangered Species A female red-cockaded woodpecker (*Picoides borealis*) approaching her nest hole. This species was once abundant throughout the pine savannas (communities dominated by grasses intermixed with pine trees) of the United States, but has been severely reduced in numbers due to the loss of its required habitat. Among the various types of pine savanna, longleaf pine savanna is the red-cockaded woodpecker's preferred habitat.

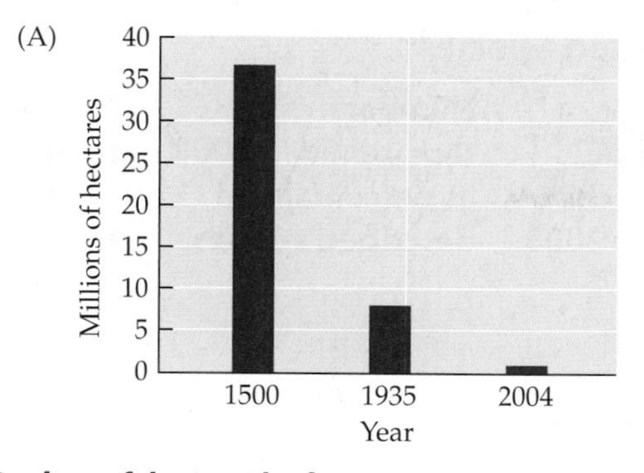

FIGURE 22.2 Decline of the Longleaf Pine Savanna Ecosystem (A) The estimated area covered by longleaf pine savanna at different times. (B) As seen in this photograph from the southeastern United States, longleaf pine (*Pinus palustris*) savanna consists of open forest with a grass understory. (A after Van Lear et al. 2005.)

Q Estimate the hectares of longleaf pine savanna that existed in 1500, 1935, and 2004. Was the annual loss of longleaf pine savanna greater from 1500 to 1935, or from 1935 to 2004?

cally helped to maintain longleaf pine savanna. Without those fires, the longleaf pine community soon undergoes succession. As an understory of young oaks and other hardwoods grows up, red-cockaded woodpeckers abandon their nesting cavities, apparently due to a decrease in food resources. In the past, the birds would move to parts of the forest that had been more recently burned, but as the area of suitably mature longleaf pines declines, there are fewer and fewer places for the birds to go. This loss of habitat has reduced the woodpeckers' populations, making them vulnerable to the problems of small, isolated populations that we discussed under Concept 10.3. There is evidence of inbreeding among the birds, and in 1989, Hurricane Hugo killed 70% of the birds in one population. West Nile virus has not yet been detected in this species, but it is highly lethal to other birds, so if it emerged, the results could be devastating.

The recent history of the red-cockaded woodpecker reflects that of thousands of other imperiled species around the world that have gone from being vital components of an extensive ecosystem through a gradual decline with loss of habitat to critically low numbers. When species rely on a specific habitat that is destroyed by human activities, they are found in diminishing numbers until, in some cases, they vanish. What can be done to protect species such as the red-cockaded woodpecker? Do we have a responsibility to protect existing biodiversity and to restore some of what has been lost? If so, how can we best allocate our limited resources to be most effective in our conservation efforts?

Introduction

Over the last few centuries, as the human population has grown and increased its use of resources, the species that evolved in the ecosystems around us have lost their habitats through their outright destruction or through changes in their biological or physical properties. These changes have given rise to a biodiversity crisis. The most recent Red List of Threatened Species, compiled by the International Union for Conservation of Nature (IUCN), lists 16,913 species as threatened with extinction—about 1% of all species worldwide (**Table 22.1**). This number is certainly an underestimate, as only the best-studied taxonomic groups have been assessed for their conservation status.

Ecologists play an important role in observing, measuring, and communicating the changes in species abundances, distributions, and biological traits that have resulted from the loss of habitat and other effects humans are having on the planet's ecosystems. As we'll see in this chapter and the next, ecologists are also one component of a diverse team working to find ways of reversing the decline of species and their habitats. We'll begin by introducing you to the field of biology dedicated to reversing those declines: conservation biology.

CONCEPT 22.1 Conservation biology is an integrative discipline that applies the principles of ecology to the protection of biodiversity.

Conservation Biology

The preservation of longleaf pine savanna at the Fort Bragg military base (described in the Case Study on p. 476) and on other federal and state lands, coupled with legal protection and extraordinary human effort, has led to stabilization and slow recovery of the numbers of red-cockaded woodpeckers (U.S. Fish and Wildlife Service 2003). This slow recovery has required expertise from biological disciplines such as demography, genetics, and pathology as well as contributions from arenas outside biology, including law, economics, political science, communications, and sociology. It has also required working with farmers, landowners,

TABLE 22.1

Global Summary of the Number of Documented Imperiled Species

Group	Number of described species	Number of species evaluated by 2008	Number of threatened species in 2008	Number threatened in 2008, as percentage of described species	Number threatened in 2008, as percentage of species evaluated
VERTEBRATES					
Mammals	5,488	5,488	1,141	21	21
Birds	9,990	9,990	1,222	12	12
Reptiles	8,734	1,385	423	5	31
Amphibians	6,347	6,260	1,905	30	30
Fishes	30,700	3,481	1,275	4	37
Subtotal	*61,259*	*26,604*	*5,966*	*10*	*22*
INVERTEBRATES					
Insects	950,000	1,259	626	0.07	50
Mollusks	81,000	2,212	978	1.2	44
Crustaceans	40,000	1,735	606	1.5	35
Others	161,384	955	286	0.2	30
Subtotal	*1,232,384*	*6,161*	*2,496*	*0.2*	*41*
PLANTS					
Mosses	16,000	95	82	1	86
Ferns and allies	12,838	211	139	1	66
Gymnosperms	980	910	323	33	35
Angiosperms	258,650	10,779	7,904	3	73
Subtotal	*288,468*	*11,995*	*8,448*	*3*	*70*
OTHERS					
Lichens	17,000	2	2	0.01	100
Mushrooms	30,000	1	1	<0.01	100
Subtotal	*47,000*	*3*	*3*	*<0.01*	*100*
TOTAL	1,629,111	44,763	16,913	1	38

Source: Data from the 2008 IUCN Red List of Threatened Species, as reported in Vié et al. (2009).

Note: "Imperiled" includes the IUCN Red List categories "critically endangered," "endangered" and "vulnerable." Some groups have been completely evaluated (mammals, birds) for conservation status, but for many groups, only a small percentage of described species have been evaluated. For those groups, there may be a bias toward completing assessments of imperiled species and making assessments of more common species a lower priority. That only 1% of described species are shown as imperiled is an artifact of incomplete evaluation, as the percentage is believed to be much higher.

the U.S. military, and the business community. Arriving at a successful management approach required not only data collection and analysis, but also creativity and the ability to work with a wide variety of stakeholders with varying interests in the ecosystem. Such an integrative approach is characteristic of conservation biology.

Conservation biology is the scientific study of phenomena that affect the maintenance, loss, and restoration of biodiversity. It applies many of the ecological principles and tools that you have studied in this book to the halting or reversal of biodiversity declines. Later in this chapter, we will look at the reasons why biodiversity is declining and at the tools conservation biologists use to address conservation problems. But first let's consider why it is so important to prevent and reverse declines in biodiversity.

Protecting biodiversity is important for both practical and moral reasons

People rely on nature's diversity. In addition to the hundreds of domesticated species that sustain us, we make abundant use of wild species for food, fuel, and fiber. We harvest wild species for medicines, building materials, spices, and decorative items. Many people rely on these natural resources for their livelihoods. As discussed under

Concept 18.4, the natural functioning of biological communities provides valuable services to humans. All of us are dependent on a wide range of these **ecosystem services**, such as water purification, generation and maintenance of soils, pollination of crops, climate regulation, and flood control (Ehrlich and Wilson 1991). These life-sustaining functions are themselves dependent on the integrity of natural communities and ecosystems. Furthermore, for our emotional health, most of us require time spent surrounded by nature's beauty and complexity. Spiritually, we go to natural ecosystems for solace, wonder, and insight.

But beyond our physical dependence on biodiversity, do we have some moral obligation to the other species that inhabit Earth? For many people, biodiversity has inherent value and warrants protection simply for that reason. For others, religious or spiritual beliefs lead to a sense of stewardship, or to the view that other species have a right to exist just as we do. Still others, however, do not share these views and see natural resources primarily as commodities awaiting human extraction.

The field of conservation biology arose in response to global biodiversity losses

Scientists have long been aware that human activity negatively affects the abundances and distributions of organisms. In the nineteenth century, Alfred Russel Wallace, the "father of biogeography" whose work we described in Chapter 17, foresaw the current biodiversity crisis, warning in 1869 that humanity was at risk of obscuring the record of past evolution by bringing about extinctions. In the United States, there was a rising public outcry over the rapid decline of bison in the West, the stunning harvest to extinction of the passenger pigeon (**Figure 22.3**), the extensive use of bird feathers in ladies' hats, and other assaults on animal populations.

Throughout the early history of ecology in the United States, ecologists were divided over how strongly they could advocate for the preservation of nature while still maintaining scientific objectivity (Kinchy 2006). Before 1945, the Ecological Society of America frequently lobbied Congress for the establishment of national parks or for better management of existing parks. In 1948, however, the Ecological Society of America decided to separate "pure" science from advocacy, and the Ecologists' Union branched off as an independent entity focused on the preservation of nature. In 1950, this offshoot organization changed its name to The Nature Conservancy, rising in prominence as a nonprofit organization that integrates science with advocacy and on-the-ground conservation work (Burgess 1977).

Conservation biology emerged as a scientific discipline in the early 1980s as ecologists and other scientists saw the need to apply their knowledge to the preservation of species and ecosystems. The Society for Conservation Biology, founded in 1985, arose in response to the biodiversity crisis.

FIGURE 22.3 The Passenger Pigeon: From Great Abundance to Extinction The passenger pigeon (*Ectopistes migratorius*), once one of the most abundant birds in North America, was subject to massive hunts in the nineteenth century. The last passenger pigeon died in the Cincinnati Zoo in 1914. The ecological effects of its extinction on the eastern deciduous forest, coincident with the loss of the American chestnut (see p. 294), are difficult to estimate, but are presumed to be considerable.

The emergence of professional journals dedicated to conservation biology during the 1980s and 1990s, and an ongoing increase in the number of academic programs for the training of graduate students and professionals, indicate the growing acceptance of and need for this specialized discipline.

Conservation biology is a value-based discipline

The methods of science call for objectivity—an assurance that the collection and interpretation of data are unbiased by preconceived ideas. Yet science is not free of human values, and it inevitably takes place within a larger social context. Conservation biologists have had to come to terms with the implicit and explicit values that are part of their work. From the founding of the Society for Conservation Biology, the designation of the discipline as "mission-driven" (Soulé and Wilcox 1980; Meine et al. 2006) and "crisis-oriented" (Soulé 1985) explicitly revealed the values behind the science.

Many ecologists have chosen to speak up or refocus their research programs, as they have come to understand the biological consequences of the changes taking place on the planet. In 1986, Dan Janzen, a tropical biologist who had largely committed himself until then to studying tropical plant–insect interactions, wrote that "if biologists want a tropics in which to biologize, they are going to have to buy it with care, energy, effort, strategy, tactics, time, and cash." Edward O. Wilson, who had distinguished himself in ant biology, island biogeography, and sociobiology, began to write in the mid-1980s about biodiversity loss and has rallied others to the conservation cause ever since. Such motivation does not necessarily detract from the objectivity of the scientific studies done by conservation biologists, as they understand that conserving biodiversity will require decisions based on sound and credible analyses. Furthermore, those analyses are subjected to rigorous scientific review by other scientists, who may challenge or even refute their conclusions. As we described in our discussion of Concept 1.3, such a critical examination of ideas and results is a standard and healthy feature of the scientific process.

In the next section, we'll meet one ecologist who put the values of conservation biology into practice. Then we'll examine the extent and causes of the current declines in biodiversity.

CONCEPT 22.2 Biodiversity is declining globally.

Declining Biodiversity

The tropical botanist Alwyn Gentry devoted his life to identifying, classifying, and mapping the immense diversity of plants found in Central and South America. He also became an eyewitness to plant species extinctions as the region underwent rapid deforestation. It was not uncommon for him to identify a new endemic plant species (that is, a plant species that occurs in a particular geographic region and nowhere else) during an expedition into Ecuador or Peru, only to return to the same spot a few years later to find the forest cleared and the species gone (Dodson and Gentry 1991) (**Figure 22.4**). Gentry worked with a growing sense of urgency to identify rare species in order to protect them from this fate. His death in a plane crash in the Ecuadorian forest in 1993, while doing an aerial survey of land proposed for conservation, cut this work short and was an enormous loss to tropical botany.

Gentry was just one of many ecologists who have been finding and describing species on the one hand and watching their destruction on the other. Extinctions of barely known tropical plant species (and most likely of other species that we have yet to discover) continue throughout the tropics despite our decades-long recognition of the problem. Through greater efforts to explore Earth's ecosystems, ecologists are gaining knowledge of the world's biota and tabulating new species at a faster rate, but threats to those species are keeping pace with such gains in our knowledge about them.

The rate at which Earth is losing species is accelerating

How rapidly are species are being lost? That is a difficult question to answer, in part because we do not know how many species exist that remain unknown to us. Most studies have estimated that there are about 5–10 million eukaryotic species on Earth, but there may be as few as 3 million or as many as 100 million.

Despite this uncertainty, extinction rates can be estimated using several indirect measures (May et al. 1995). For example, estimates of extinction rates from the fossil record can be used to establish a "background" extinction rate with which current rates can be compared. For the best-known taxonomic groups, the mammals and birds,

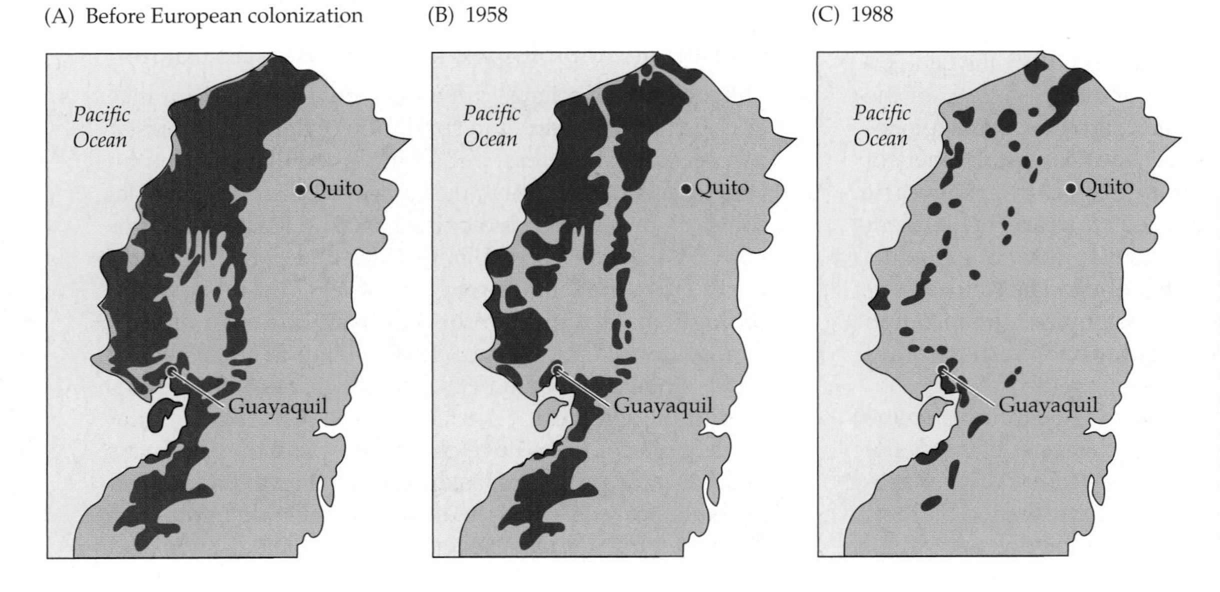

FIGURE 22.4 Loss of Forest Cover in Western Ecuador Between 1958 and 1988, a growing human population and government policies that served to stimulate rapid economic development led to rapid deforestation in western Ecuador. Green indicates forest cover. The extensive loss of forest habitat in this region is estimated to have resulted in the loss of more than 1,000 endemic species (After Brooks et al. 2002.)

paleontologists have estimated that the background extinction rate is on the order of one extinction every 200 years, which is equivalent to an average species life span of 1 million to 10 million years. By contrast, there was about one extinction per year among the mammals and birds over the twentieth century, which is equivalent to an average species life span of only 10,000 years. Thus, overall, the rate of extinction in the twentieth century was 100 to 1,000 times higher than the background rate estimated from the fossil record (May 2006).

A second method for estimating extinction rates uses the species–area relationship discussed under Concept 17.3. In particular, the relationship between number of endemic species and area is used to estimate the number of species that would be driven to extinction by a given amount of habitat loss (Kinzig and Harte 2000). In a third approach, biologists have used changes over time in the assessed conservation statuses of species (for instance, a shift from endangered to critically endangered) to forecast rates of extinction (Smith et al. 1993). Finally, a fourth approach is based on the rates of population decline or range contraction of common species (Balmford et al. 2003). All of these methods are fraught with uncertainty, yet they are the best ways we have devised to document losses of biodiversity.

It can also be difficult to ascertain when a species is definitely extinct. Many species are known from a single specimen or location, and the logistics of relocating them can be daunting. Even an exhaustive hunt for a very rare species can fail to detect some remnant populations. Declaring a species extinct, however, has been known to stimulate biologists' search efforts. Since the publication of a flora of Hawaiian plants in 1990, for example, 35 species listed there as extinct have been relocated, though only a few individuals were found. The joy of their rediscovery is compromised by the realization that these extremely small populations cannot serve the same ecological functions as more substantial populations, and that 8% of Hawaii's native flora of 1,342 species is now considered extinct (Wagner et al. 1999).

Although humanity's growing ecological footprint (see Chapter 9's Connections in Nature on p. 217) has accelerated the rates of biodiversity loss over the last century, people have had substantial effects on Earth's biota for millennia. Steadman (1995) describes how bones found on Pacific islands revealed the prehistoric extinction of up to 8,000 species of birds (of which perhaps 2,000 species were endemic flightless rails) after these islands were colonized by Polynesians. Most of these species were island endemics, and in some cases the extinctions encompassed entire ecological guilds (**Figure 22.5**). Ecologists can only speculate about the roles the lost frugivores and nectarivores played in maintaining endemic tree populations. Steadman's findings remind us that extinctions do

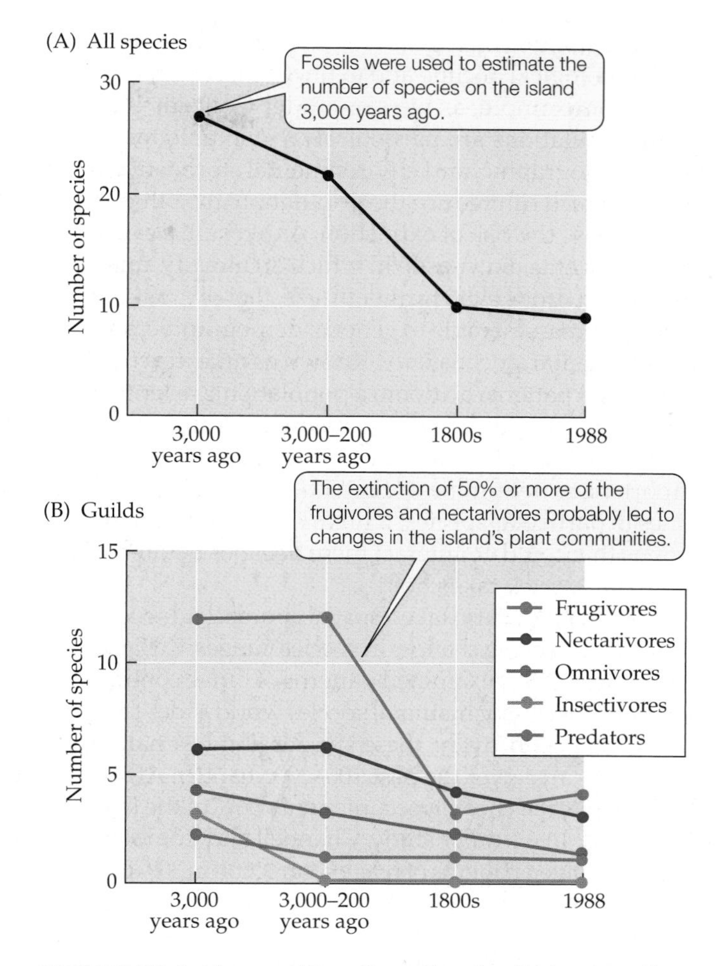

FIGURE 22.5 Humans Have Been Causing Extinctions for Millennia Trends over time in (A) the total number of bird species and (B) the number of species classified by feeding guild found on the Pacific island nation of Eua (Tonga). Prehistoric extinctions (3,000–200 years ago) occurred on many Pacific islands as a result of hunting and the introduction of rats, dogs, and pigs. (After Steadman 1995.)

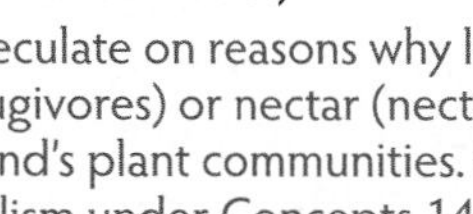

Q Speculate on reasons why losses of birds that feed on fruit (frugivores) or nectar (nectarivores) may have affected the island's plant communities. (Hint: See the discussion of mutualism under Concepts 14.1 and 14.2.)

not only eliminate individual species, but can also cause large changes in ecological communities.

Extinction is the end point of incremental biological decline

In 1954, Andrewartha and Birch wrote that "there is no fundamental distinction to be made between the extinction of a local population and the extinction of a species other than this: that the species becomes extinct with the extinction of the last local population." Sometimes the populations of a species gradually erode away, and sometimes they vanish in a spectacular collapse, as in the case of the passenger pigeon.

Conservation biologists have approached the process of biological decline and extinction in numerous ways. For example, as we saw under Concept 10.3, small populations are particularly vulnerable to genetic, demographic, and environmental stochasticity, each of which can reduce the population growth rate and increase the risk of extinction. As a result, a cyclic chain of events may ensue in which an already small population drops even further in size, thereby making it even more vulnerable to genetic, demographic, and environmental stochasticity. Known as an **extinction vortex**, this pattern can doom a population to eventual extinction once its size drops below a certain point. With this in mind, Caughley (1994) argued that it is important to determine the causes of population declines in particular species, with the aim of identifying actions that could counteract these declines before the extinction vortex takes hold.

Ecologists can also take a spatial approach to species declines by tracking changes in species' ranges. Ceballos and Ehrlich (2002) examined patterns of range contraction in 173 declining mammal species worldwide. They found that, collectively, these species had lost half of their range area over the past 100–200 years. In Africa, for example, the cheetah occupies only 56% of the land it once did. In a similar study, Channell and Lomolino (2000) examined patterns of range contraction in 309 declining species. They found that a decline often moves through the species' historic range like a wave, from one end to the other; this could occur, for example, if an invasive species entered the range at one edge and then spread through the range, eliminating the declining species population by population. Such a pattern contrasts with a retreat from all edges of the range into its center, which would probably occur if small population effects were prevailing.

Whatever its cause, the decline of a species does not take place in a vacuum. When populations are lost from an ecological community, there are consequences not only for the declining species, but also for that species' predators, prey, or mutualistic partners. The resulting changes at the community level may bring about secondary extinctions and ultimately affect ecosystem processes (Christianou and Ebenman 2005). Examples from earlier chapters include the local extinctions and other changes caused by the loss or removal of such species as the invertebrate *Corophium volutator* (Chapter 13), the marsh plant *Juncus gerardii* (Chapter 15), and the sea star *Pisaster ochraceus* (Chapter 20). In general, results from models indicate that while food webs can be resilient to species removal, the deletion of certain species can trigger a cascade of secondary extinctions. As might be expected, the stronger a species' interactions in the food web, the greater the effect of its removal (Solé and Montoya 2001). Similarly, models of plant–pollinator interaction webs in which pollinators were selectively

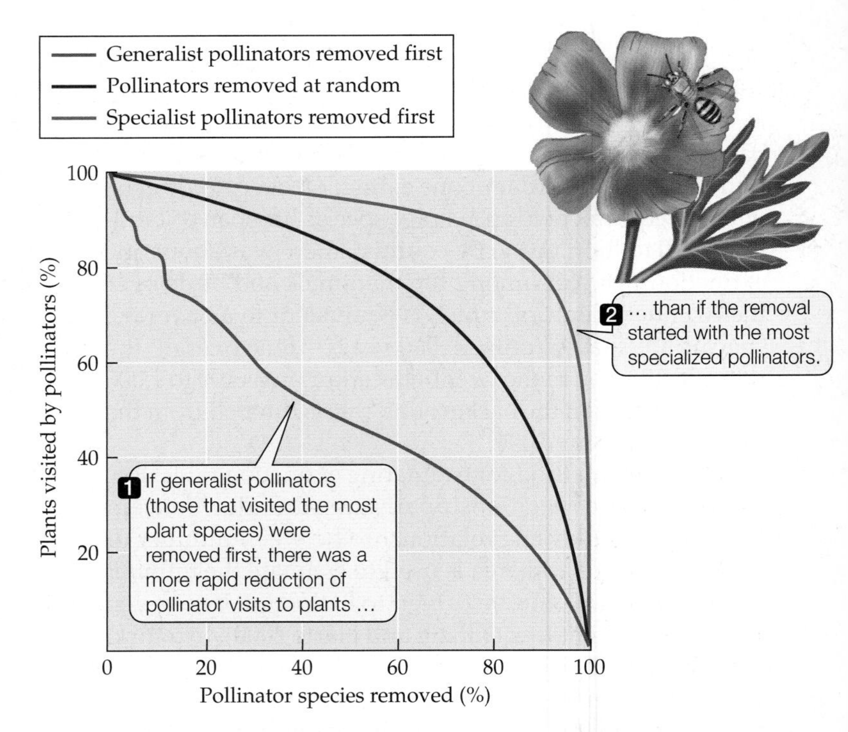

FIGURE 22.6 Effects of Pollinator Losses on Plant Species Depend on Pollinator Specialization In a model of a plant–pollinator interaction web from the Rocky Mountains (Pike's Peak, Colorado, USA), pollinator species were removed one by one and the effects of their losses on predicted plant visitation rates were recorded. (After Memmott et al. 2004.)

removed found that the effect on plant species depended on whether the removed pollinators were specialists or generalists (Memmott et al. 2004) (**Figure 22.6**). Both the food web and plant–pollinator models demonstrate the importance of considering the broader ecological consequences of incremental species loss.

Earth's biota is becoming increasingly homogenized

Organisms move about. They always have, and they always will. Over the last century, however, people have moved over Earth's surface at an unprecedented rate, carrying organisms with them and greatly enhancing rates of introductions of new species to all parts of the globe (**Figure 22.7**). These range expansions coincide with the range contractions of many native species whose numbers are declining due to habitat loss and other factors. Typically, the greatest "losers" among the native species tend to be specialists—those with morphological, physiological, or behavioral adaptations to a particular habitat—while the "winners" tend to be generalists with less stringent habitat requirements. The spread of introduced species and native generalists, coupled with declining abundances and distributions of native specialists, are part of a growing **taxonomic homogenization** of Earth's biota (Olden et al. 2004).

Island biotas are particularly vulnerable to both extinction and invasion. The decline of island endemics is often

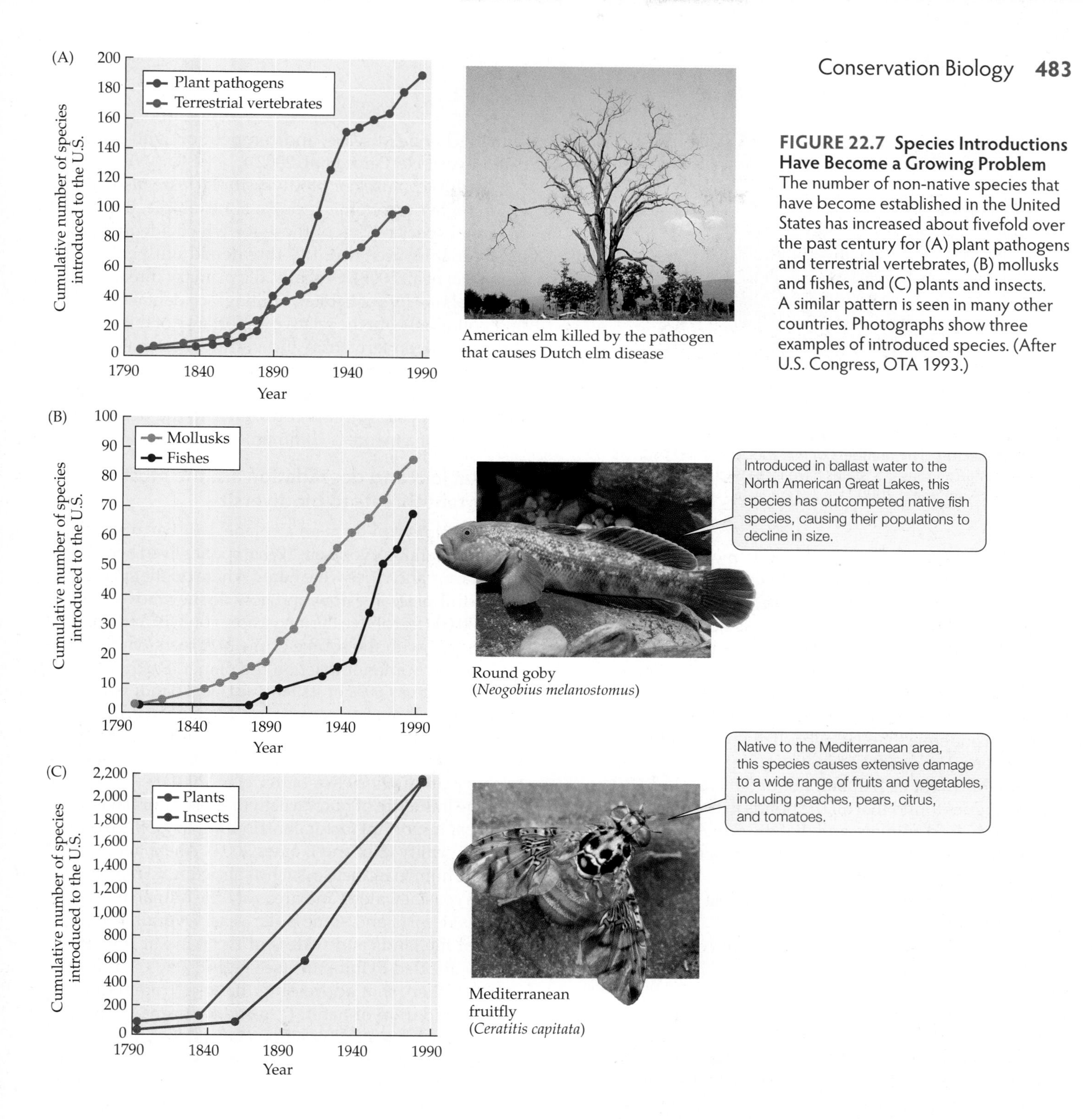

FIGURE 22.7 Species Introductions Have Become a Growing Problem The number of non-native species that have become established in the United States has increased about fivefold over the past century for (A) plant pathogens and terrestrial vertebrates, (B) mollusks and fishes, and (C) plants and insects. A similar pattern is seen in many other countries. Photographs show three examples of introduced species. (After U.S. Congress, OTA 1993.)

accelerated by the introduction of more cosmopolitan species. In a survey of American Samoa, Robert Cowie (2001) found just 19 of the 42 species of land snails historically known from that island group, plus 5 species not previously found there, but which he presumed were native. He also found that there were 12 non-native species present on the islands. These non-natives occurred in high abundances, representing about 40% of the individuals collected (there was also one abundant native species). Cowie concluded that most native species were declining in abundance, while many non-natives were increasing. Furthermore, the predators contributing to the declines of native land snail species were also non-natives, such as the predatory snail *Euglandina rosea* and the house mouse (*Mus musculus*). Cowie has found this trend toward homogenization of land snail faunas to be widespread among Pacific islands.

Homogenization has also been observed among the freshwater fishes of the United States, largely as the result of widespread introductions of game fishes. Rahel (2000) quantified the homogenization of U.S. fish faunas by examining the change in the number of species shared between all possible pairs of the 48 conterminous states. He found

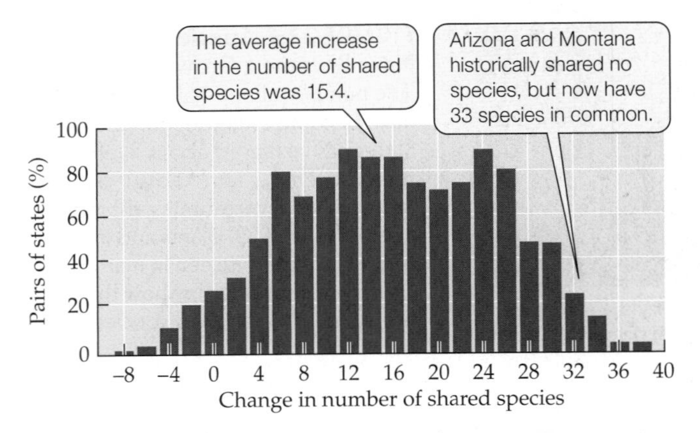

FIGURE 22.8 U.S. Fish Faunas Are Undergoing Taxonomic Homogenization The numbers of fish species shared by pairs of the 48 conterminous U.S. states have increased since European settlement. (After Rahel 2000.)

that, on average, pairs of states shared 15 more species than they did at the time of European colonization (**Figure 22.8**).

In addition to taxonomic homogenization across geographic regions, homogenization is occurring at the genetic level (Olden et al. 2004). Interspecific genetic homogenization is occurring through hybridization between native and non-native species. An example is the California tiger salamander (*Ambystoma californiense*), a threatened California endemic that has hybridized with another species of tiger salamander (*Ambystoma tigrinum*) introduced from the Midwest as fish bait about 50 years ago. Riley and colleagues found that all the ponds in their study area contained hybrid animals, and that the rare *A. californiense* was under threat of extinction through genetic swamping by *A. tigrinum* (Riley et al. 2003).

It is clear that biodiversity is being lost as a result of humanity's impact on the planet. Let's look in more detail at the reasons for these losses, and then consider what steps can be take to counteract them.

CONCEPT 22.3 Primary threats to biodiversity include habitat loss, invasive species, overexploitation, pollution, disease, and climate change.

Threats to Biodiversity

Understanding the causes of biodiversity losses is a first step toward reversing them. Multiple factors are likely to contribute to any particular species' decline and eventual extinction. For example, the last Pyrenean ibex (*Capra pyrenaica pyrenaica*) was killed in 2000 by a falling tree. This mountain goat, which was endemic to the Pyrenees in Spain and France, was abundant in the fourteenth century, but its numbers declined gradually due to hunting, climate change, disease, and competition with domesticated livestock (Perez et al. 2002).

Multiple causes of biodiversity loss are also apparent in higher taxonomic groups. For example, over 1,100 mammal species (25% of those for which adequate data are available) are currently threatened with extinction (Schipper et al. 2008). Globally, the primary threats facing mammals are loss of habitat, hunting, accidental mortality, and pollution—but the relative importance of these factors differs between terrestrial and marine mammals (**Figure 22.9**). Some mammals are threatened by additional factors, such as disease. As we'll see, this scenario, in which multiple types of threats contribute to the decline and extinction of a taxon, is common.

Habitat loss and degradation are the most important threats to biodiversity

The next time you fly in an airplane over Earth's surface, look down and ask yourself, "What species lived here before these farms and cities were here? Where do the species native to this place live now, and how do they move about?" From 30,000 feet above the landscape, you will find yourself face to face with the source of the biodiversity crisis: the scale of the human impact on the planet. Earth has been modified across 60% of its land surface (Sanderson et al. 2002), and all marine ecosystems have been affected by humans (Halpern et al. 2008). One species, *Homo sapiens*, is now appropriating between 10% and 55% of Earth's primary production (Rojstaczer et al. 2001).

The influence of human activities on natural habitat is the most important factor contributing to global declines in biodiversity (Sax and Gaines 2003). There are areas of extreme human influence, such as agricultural regions and certain coastal waters, and areas of little human influence, such as deserts and some polar seas. Overall, however, most of the lands and waters of Earth are at least moderately affected by humans (see Figure 3.5B,C). It is little wonder, then, that addressing the loss, fragmentation, and degradation of habitat caused by human activities is central to conservation work. **Habitat loss** refers to the outright conversion of habitat to another use, such as urban development or agriculture, while **habitat fragmentation** refers to the breaking up of once continuous habitat into a series of habitat patches amid a human-dominated landscape. Chapter 23 will address habitat fragmentation and its effects in detail. **Habitat degradation** refers to changes that reduce the quality of the habitat for many, but not all, species.

On a continental scale, the extent of habitat loss from some ecosystems is staggering (see Figure 23.12). Similar losses can be observed on more local scales, as in the Atlantic Forest of Brazil. This moist tropical forest has many endemic species, perhaps because it has been isolated from the Amazon rainforest for millions of years.

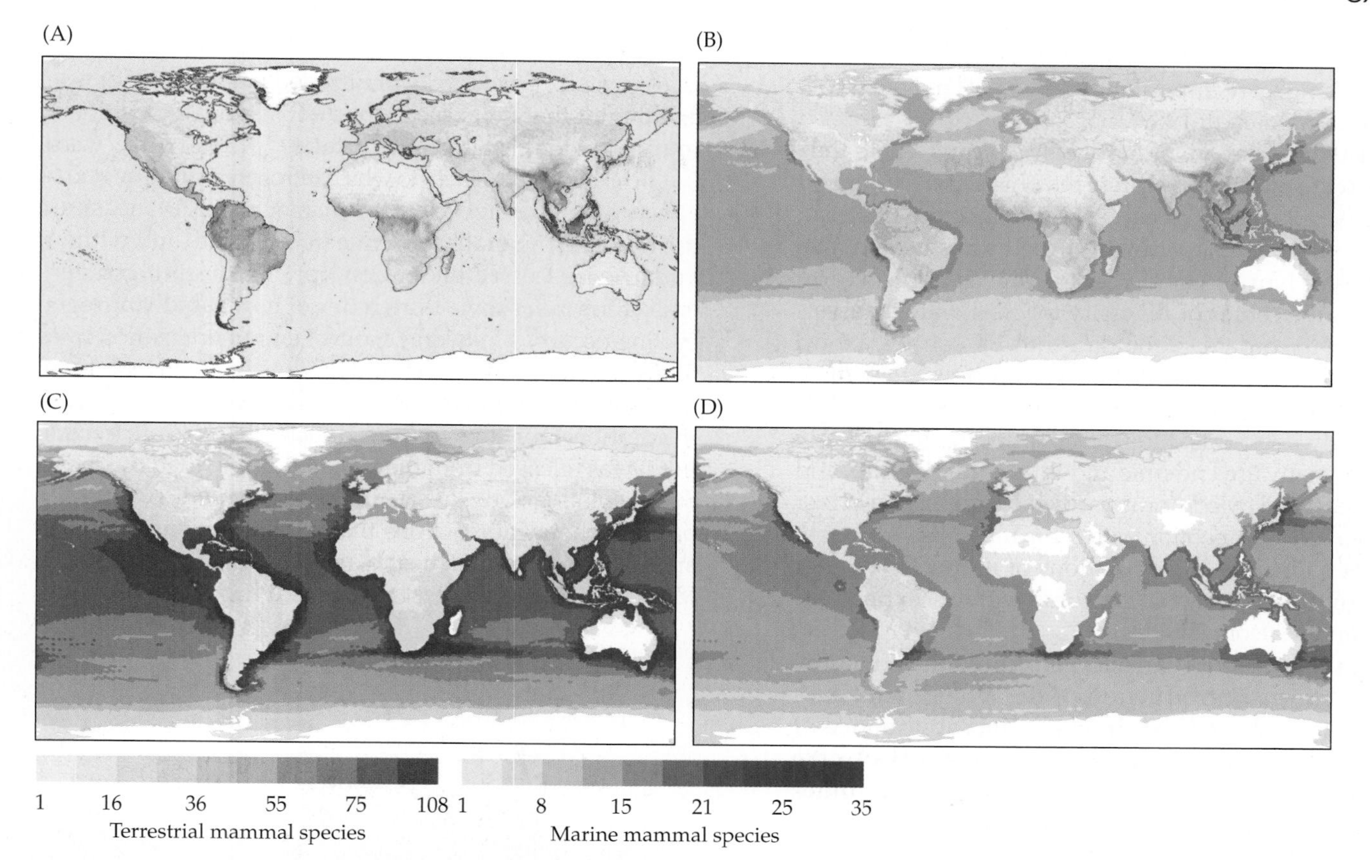

FIGURE 22.9 Threats to Mammal Species Globally, 25% of mammal species are threatened by extinction. These maps show the numbers of terrestrial and marine mammal species in various parts of the globe that are negatively affected by (A) habitat loss; (B) overharvesting; (C) accidental mortality; and (D) pollution. (After Schipper et al. 2008.)

Q Contrast the threats to land mammals and marine mammals.

Of South America's 904 mammal species, 73 are endemic to this forest, and 25 of those endemics are threatened with extinction. The forest's location also coincides with 70% of Brazil's human population. As a result, more than 92% of this habitat has been cleared to make room for agriculture and urban development, and what remains has been highly fragmented, pushing many species to endangerment. There is an ongoing call for conservation investment in this region to preserve what remains of the Atlantic Forest habitat and the species it contains (Fonseca 1985; Ranta et al. 1998).

How has the loss of Atlantic Forest habitat affected biodiversity? Brooks and colleagues (1999) asked why there have been no reports of extinctions among birds of this region. They offered three possible explanations, which might apply to patterns of biological decline in other ecosystems as well. First, the birds might be adjusting to living in forest fragments. Second, the most vulnerable species might have gone extinct before they were known to biologists. Their third explanation, which they see as the most plausible, is that the time lag between deforestation and extinction has not yet played out. While there may have been no extinctions yet, populations have been reduced to such an extent that the birds may no longer be capable of maintaining their populations. Unless drastic measures are taken, such species are doomed to extinction.

Habitat degradation is extremely widespread, and its effects have only begun to be clarified by ecological studies. It has diverse causes and takes many forms. In the Sinai Peninsula of Egypt, for example, Omar Attum and his colleagues compared undisturbed sand dune habitats with habitats that had been degraded through grazing, vegetation harvesting, and small-scale agriculture. The degradation had lowered the percentage of plant cover and the height of the vegetation. Lizard communities were also affected: there were fewer individual lizards as well as a lower diversity of lizard species in the degraded habitats (Attum et al. 2006).

Let's take a closer look at one way in which habitat can be degraded: through the presence of invasive species.

Invasive species can displace native species and alter ecosystem properties

Declines in biodiversity can result from the arrival of **invasive species**: non-native, introduced species that sustain growing populations and have large effects on communities. Worldwide, 20% of endangered vertebrates, especially those on islands, are imperiled as a result of invasive species (MacDonald et al. 1989).

Invasive species are of particular concern where they compete with, prey on, or change the physical environment of endangered native species. The effect of the Eurasian zebra mussel (*Dreissena polymorpha*) on the freshwater mussel species of North America is a prime example. North America is the center of diversity for freshwater mussels (bivalves of the order Unionoida), with 297 species, a third of those in the world. Prior to the invasion of the zebra mussel in the late 1980s, North American freshwater mussels were already in trouble. Most of these species are globally imperiled, many are endemic and thus naturally rare, and all are threatened by compromised water quality and river channelization. Competition with zebra mussels has brought about steep declines in populations of native freshwater mussels (60%–90%), including some regional extinctions (Strayer and Malcom 2007).

Invasive predators can also contribute to extinction. In Lake Victoria, introduction of the Nile perch (*Lates niloticus*) has reduced the diversity and abundance of the native cichlid fishes, a family that is a textbook case of adaptive radiation (a phenomenon discussed under Concept 6.4). Historically, about 600 species of cichlids had been recorded, most of which were endemic to the lake. The Nile perch is a large predator, and its introduction into the lake in the early 1960s has contributed to the extinction of roughly 200 cichlid species. Before the introduction, the cichlids made up 80% of the biomass of fish in the lake; the Nile perch now accounts for 80% of the biomass. As in most cases, more than one factor is driving the cichlids' decline: pollution and overfishing augment the negative effect of predation by the Nile perch (Seehausen et al. 1997).

In many ecosystems, habitat fragmentation and degradation have increased vulnerability to invasion by nonnative species, which in turn may lead to consequences that further degrade the ecosystem. The tropical dry forest of Hawaii, for example, harbors more than 25% of Hawaii's threatened plant species. The area of tropical dry forest has been reduced by 90% since human settlement. The arrival of an invasive species of fountain grass (*Pennisetum setaceum*) has made a bad situation worse. In addition to outcompeting and displacing local plants, fountain grass is an excellent source of fuel for brush fires. As a result, the frequency of fires has increased, furthering the decline of tropical dry forests, but favoring the spread of fountain grass—a fire-adapted species.

Some invasive species can alter properties of ecosystems such as nitrogen cycling (see Figure 21.11). One such species is kudzu (*Pueraria montana*), an invasive vine that covers more than 3 million ha (7.4 million acres) in the southeastern United States. This species disrupts communities by outcompeting other plants for light (see Figure 11.6). In addition, kudzu can fix up to 235 kg of nitrogen per hectare per year, an amount that far exceeds the atmospheric deposition of nitrogen in the eastern United States (7–13 kg N/ha/year).

To examine the extent to which nitrogen fixation by kudzu affects the nitrogen cycle, Hickman et al. (2010) measured the nitrogen mineralization rate in plots with and without kudzu (as discussed under Concept 21.2, the nitrogen mineralization rate provides an index of how rapidly nitrogen cycling occurs). On average, nitrogen mineralization rates increased more than sevenfold in plots invaded by kudzu, indicating a large effect on the nitrogen cycle (**Figure 22.10**). In addition, more than twice as much of the gas nitric oxide (NO) was released from the soil in plots invaded by kudzu than in plots lacking kudzu. In the atmosphere, NO participates in chemical reactions that produce ozone, a pollutant that affects human health and agricultural production. Modeling results suggest that kudzu has the potential to

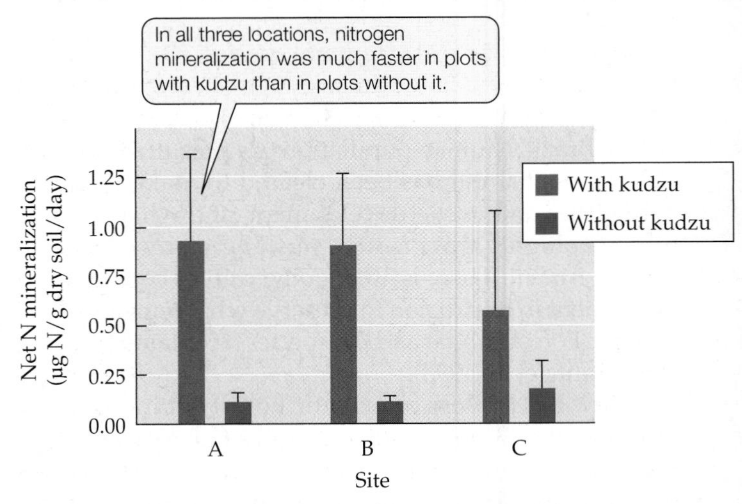

FIGURE 22.10 Invasive Species Can Alter the Nitrogen Cycle At three locations in Georgia (sites A, B, and C), net nitrogen mineralization rates (which provides an index of how rapidly nitrogen cycling occurs in an ecosystem) were much higher in soils supporting kudzu than in soils with native vegetation. Error bars show one SE of the mean. (After Hickman et al. 2010.)

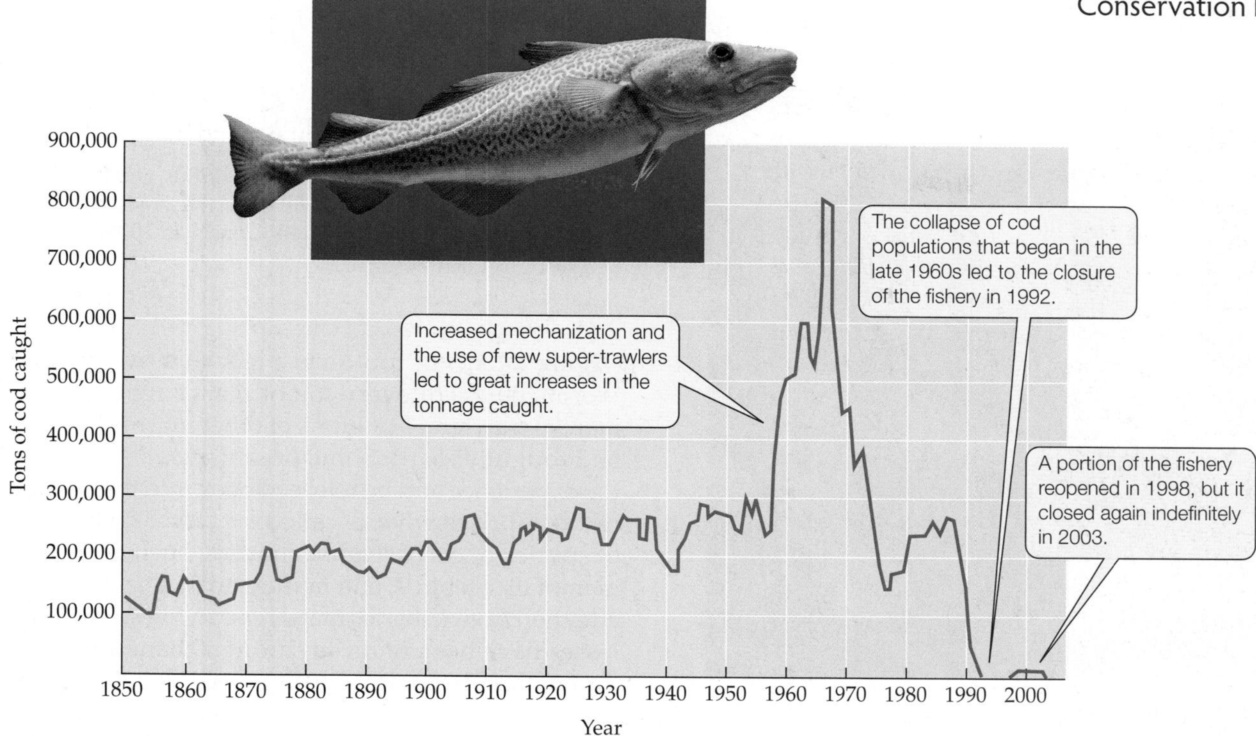

FIGURE 22.11 The Collapse of the Cod Fishery Changes over time in the tonnage of cod caught off the coast of Newfoundland, Canada. Overharvesting led to the collapse of cod populations, which still have not recovered. (After Millennium Ecosystem Assessment 2005.)

Q Based on data prior to 1950, roughly how many tons of cod could have been harvested in a sustainable manner? Explain.

increase the number of high-ozone event days by as many as 7 days per summer across broad regions of the southeastern United States (Hickman et al. 2010).

As we saw in the Case Study of the invasive alga *Caulerpa* in Chapter 15 (pp. 324–325), control or eradication of invasive species is difficult, labor-intensive, and expensive, but at times may be warranted in the interest of protecting economically or culturally valuable native species or natural resources. The best strategy for combating invasive species is to prevent their arrival through careful screening of biological materials at international borders. But once potentially invasive species are present, control measures are best implemented immediately; constant vigilance and quick action are key to minimizing their effects (Simberloff 2003).

Overexploitation of species has large effects on ecological communities

For many of the world's people, food comes, at least in part, directly from a natural ecosystem. The problem is that as the human population increases and natural habitat shrinks, the harvesting of many species from the wild has become unsustainable. Globally, overexploitation is contributing to the imperilment of many species, including many fishes, mammals, birds, reptiles, and plants. Overexploitation has been the cause of the probable extinction of at least one primate, Miss Waldron's red colobus monkey (*Procolobus badius waldroni*), a subspecies endemic to Ghana and Côte d'Ivoire whose last confirmed sighting was in 1978 (Oates et al. 2000; McGraw 2005).

The effects of overhunting on tropical forests have been substantial, resulting in what Kent Redford (1992) has called an "empty forest." This phrase refers to forests that look healthy in satellite images, but in which the abundances and diversity of large vertebrates have decreased. The increased accessibility of forests as roads are built through them facilitates this overharvesting of wildlife, as does the widespread availability of guns. The enormous quantity of "bushmeat" being taken from tropical forests is sobering. Redford has calculated that 13 million mammals are killed each year in the Amazon rainforests of Brazil by rural hunters, and it is estimated that in western and central Africa, 1 million tons of forest animals are taken annually for food (Wilkie and Carpenter 1999). Vast numbers of animals are also captured from tropical forests, coral reefs, and other ecosystems and then imported legally to other countries. For example, from 2000 to 2006, government records indicate that 1.5 billion animals, most of which were for the pet trade, were imported to the United States alone (Smith et al. 2009).

In the oceans, rapid and steep declines have taken place in both the abundance (**Figure 22.11**) and size

FIGURE 22.12 Overharvesting Has Led To a Decline in the Sizes of Top Marine Predators Photographs of trophy fish caught on charter fishing boats based in Key West, Florida, in (A) 1957, (B) the early 1980s, and (C) 2007. In commercial and recreational fisheries, the largest fish are often the preferred prey. (From McClenachan 2009.)

(**Figure 22.12**) of top-level predators (Myers and Worm 2003). For every ton of fish caught by commercial trawlers, 1 to 4 tons of other marine life may be brought aboard. Some organisms may survive the experience and be released back into the sea; the rest comprises what is called *bycatch*. The bycatch of species of conservation concern, such as marine mammals, seabirds, and marine turtles, has received attention from fisheries managers, and in some cases, losses have been reduced through changes in gear design (see Ecological Toolkit 9.1). But concern has been raised about the ecological effects of this unnecessary mortality on marine food webs (Lewison et al. 2004). In addition, repeated trawling on the coastal sea bottom has affected benthic species such as corals and sponges and has thereby degraded benthic habitat for many other species. Studies indicate that habitat recovery following trawling is very slow (National Research Council 2002).

Selective overharvesting of other resources from natural ecosystems, such as wood, fiber, medicines, and oils, has brought many species to a threatened status. All three species of mahogany (*Swietenia* spp.) have become imperiled by overharvesting as well as by habitat loss. Many of the wild palm species used for making rattan furniture are declining. The global growth in sales of herbal medicines has threatened wild populations of many medicinal plants. The premier wood for making violin bows comes from a tree (*Caesalpinia echinata*) that is globally endangered.

Whenever a species has market value, it is likely to be overharvested. And, in an unfortunate confluence between human behavior (i.e., greed) and declining animal and plant populations, when threatened species have monetary value, an "anthropogenic Allee effect" (see Figure 10.15) can occur in which their increasing economic value can lead to more aggressive search and collection missions. Many scientists and policymakers argue that the best approach to protecting overexploited species is to determine the levels of harvest that will be sustainable and to establish regulatory mechanisms to permit only those levels to be taken. In one example of how this could be done, Bradshaw and Brook (2007) describe management options that provide revenue from meat and trophy hunting of wild banteng (*Bos javanicus*), a member of the cattle genus, yet do not jeopardize the prospects for the recovery of this rare species.

Pollution, disease, and climate change erode the viability of populations

More insidious effects of human activities, such as air and water pollution and climate change, are chipping away at populations of many species. We are also seeing the emergence of new diseases and the crossing over of diseases from domesticated animals into wildlife. The effects of all these factors exacerbate declines in species already reduced by habitat loss, invasive species, or overexploitation.

Pollutants released by human activities are omnipresent in air and water—demonstrating one of the ecological maxims introduced in Table 1.1: Everything goes somewhere. These pollutants become a contributor to habitat degradation and biodiversity loss where they are present at levels that cause physiological stress. We will see in Chapter 24 how some of these pollutants degrade habitats, reduce populations, and threaten the persistence of species.

One emerging pollution threat is the growing concentration of persistent endocrine-disrupting contaminants (EDCs), particularly in the marine environment. As we saw in the Case Study in Chapter 20 (p. 430), persistent organic pollutants such as DDT, PCBs, flame retardants, and organophosphates from agricultural pesticides, some of which are EDCs, end up in marine food webs, where they are bioaccumulated and biomagnified, particularly in top predators. The number of chemicals found in marine mammals, the number of individuals affected, and the concentrations found have risen markedly in the last 40 years (Tanabe 2002). Peter Ross refers to the orcas of British Columbia as "fireproof killer whales" due to the extremely high levels of flame-retardant chemicals (polybrominated diphenyl ethers, or PBDEs) found in their bodies (**Figure 22.13**). These EDCs have been observed to interfere with reproduction, neurological development, and immune function in mammals (Ross 2006). EDCs have also interfered with reproduction—basically by turning males into females—in many other species, including a population of the endangered pallid sturgeon (*Scaphirhynchus albus*) in the Mississippi River downriver from Saint Louis. Such problems for species already at low numbers do not improve the outlook for their future.

Disease has also contributed to the decline of many endangered species. In a striking example, an emerging disease caused by the fungus *Batrachochytrium dendrobatidis* has decimated amphibian populations around the globe (Skerrat et al. 2007; see also the Case Study Revisited in Chapter 1 on p. 17). In the 1930s, the final decline to extinction of the thylacine, or Tasmanian wolf (*Thylacinus cynocephalus*), was hastened by an undetermined disease, and now the Tasmanian devil (*Sarcophilus harrisii*) appears to be similarly threatened due to the spread of a facial tumor disease (Hawkins et al. 2006). In the North American prairie, the threatened status of the black-footed ferret (*Mustela nigripes*) was exacerbated by canine distemper (Woodroffe 1999).

Finally, although hundreds of species have shifted their distributions to higher latitudes or elevations in response to global warming (Parmesan 2006), only a few cases are known in which species are imperiled directly by climate change. The decline of some amphibian species may have been caused in part by climate change

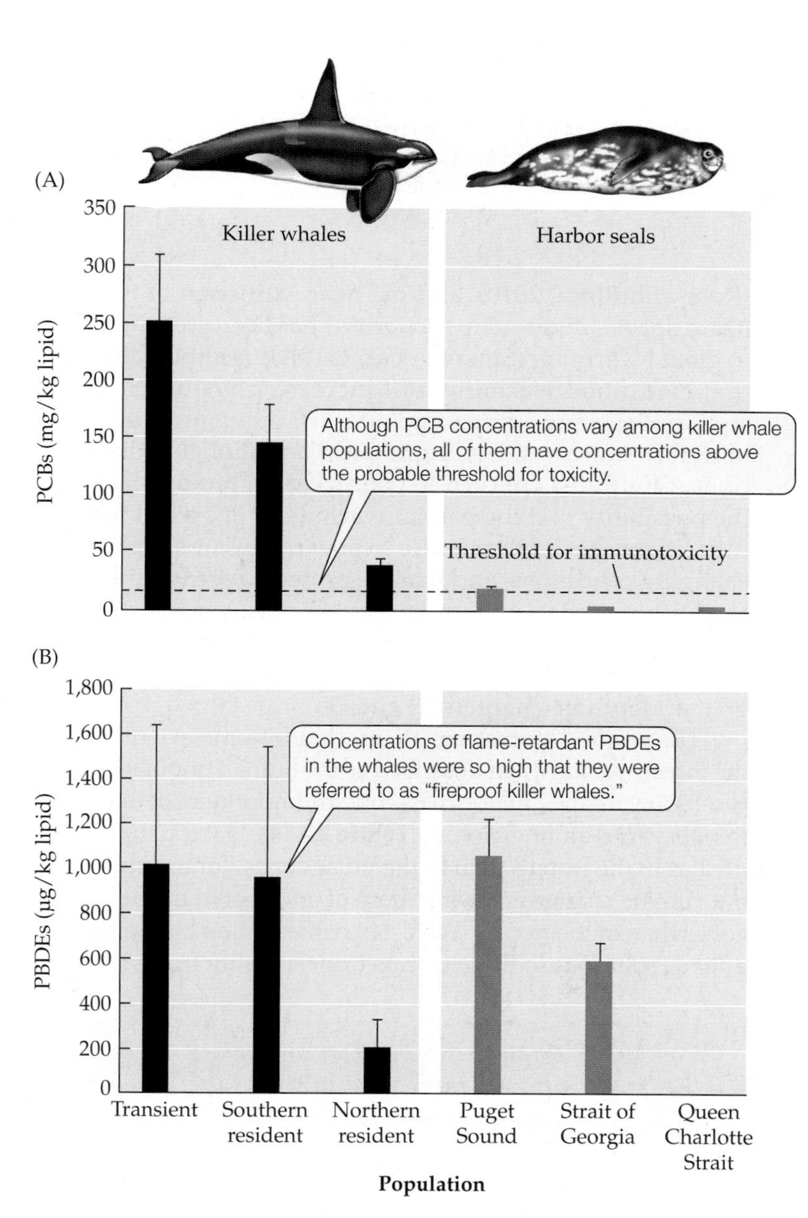

FIGURE 22.13 Persistent Synthetic Chemicals That Disrupt the Endocrine System Are a Growing Threat to Marine Mammals In British Columbia, the concentrations of PCBs (A) and PBDEs (B) found in killer whales (*Orcinus orca*) and harbor seals (*Phoca vitulina*) are very high. Error bars show one SE of the mean. (After Ross 2006.)

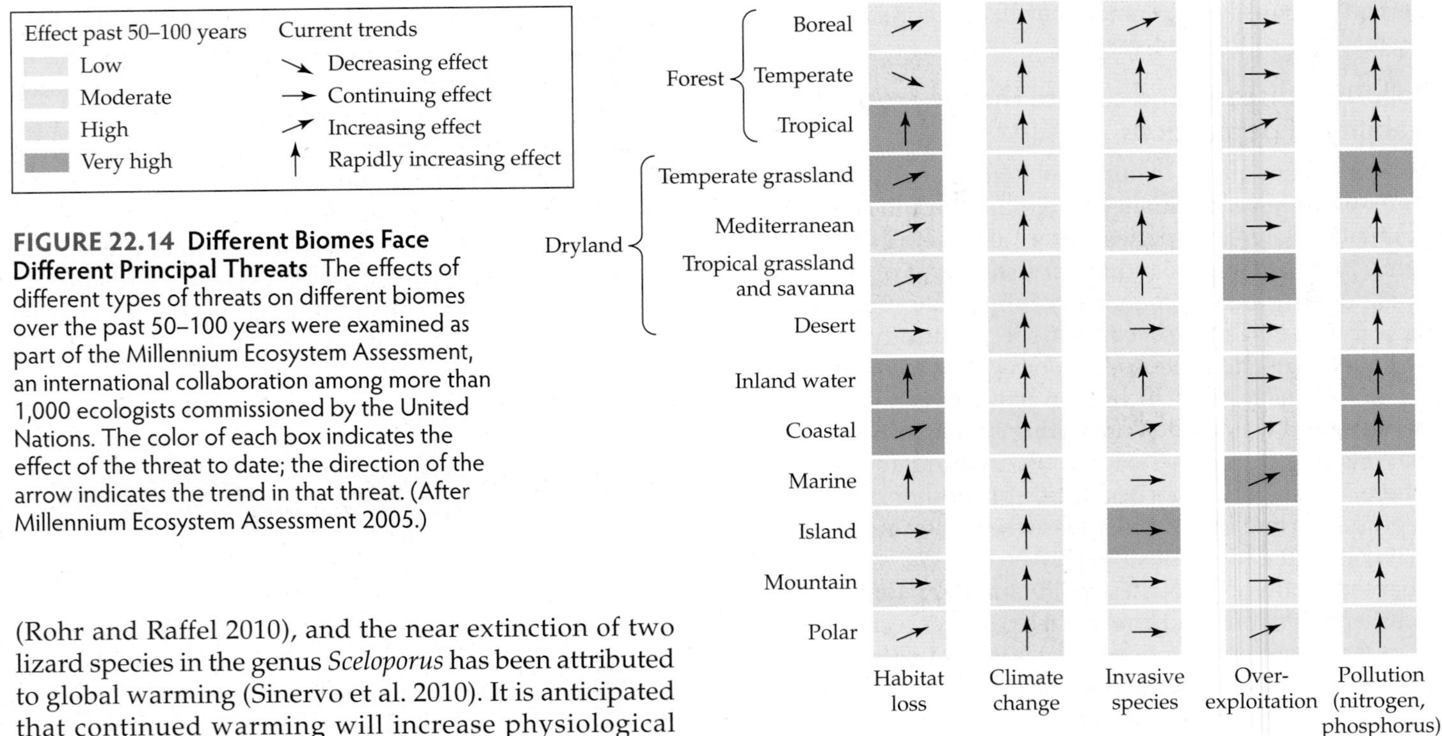

FIGURE 22.14 Different Biomes Face Different Principal Threats The effects of different types of threats on different biomes over the past 50–100 years were examined as part of the Millennium Ecosystem Assessment, an international collaboration among more than 1,000 ecologists commissioned by the United Nations. The color of each box indicates the effect of the threat to date; the direction of the arrow indicates the trend in that threat. (After Millennium Ecosystem Assessment 2005.)

(Rohr and Raffel 2010), and the near extinction of two lizard species in the genus *Sceloporus* has been attributed to global warming (Sinervo et al. 2010). It is anticipated that continued warming will increase physiological stresses for many species and alter the outcome of ecological interactions, leading to local and global extinctions. Of utmost concern to conservation biologists is the possibility that the pace of warming will exceed the capacity of species to migrate to new ranges or adapt to changing conditions, and that the protected areas we are establishing today will prove less effective over time as their physical environments become less suitable for the species living there. We will explore climate change in greater depth in Chapters 23 and 24.

The world's biomes are all affected to some extent by the threats we have just described as the human population climbs toward the 7 billion mark, but the importance of these threats varies among biomes (**Figure 22.14**). Habitat loss is greater in the tropics than in the polar zones, for example, but climate change is having more of an effect in the polar zones than in the tropics. What can conservation biologists offer as solutions to these threats from so many fronts?

CONCEPT 22.4 Conservation biologists use many tools and work at multiple scales to manage declining populations.

Approaches to Conservation

Species or habitat? Where should we put our focus? Conservation biologists have debated this question and have generally concluded that protecting habitat is of primary importance, but that understanding species is also important. There is no real dichotomy here, as we must understand a threatened species' biology in order to identify and preserve its habitat. The U.S. Endangered Species Act functions through the listing of particular species, but for each of those species, it mandates the identification and protection of critical habitat. Worldwide, many other laws protecting biodiversity take a similar approach.

These contrasting emphases have been labeled **fine-filter** (genes/populations/species) and **coarse-filter** (ecosystem/habitat/landscape) approaches to conservation. Chapter 23 will describe how the principles of ecology are applied to protecting habitat and how conservation biologists work to manage ecosystems and landscapes. In this section, we will look at the variety of ways in which conservation biologists work to understand biodiversity at the level of genes, populations, and species and how conservation solutions are generated using a fine-filter approach.

Genetic analyses are an important conservation tool

As we saw in Chapters 6 and 10, small populations are particularly vulnerable to the effects of genetic drift and inbreeding, which can result in a decrease in genetic variation and an increase in the frequency of deleterious alleles. A decrease in genetic variation can limit the extent to which a population can evolve in response to environmental change, potentially placing the population at a greater risk of extinction. An increase in the frequency of deleterious alleles is also of concern because it can cause birth or survival rates to drop, thereby decreasing the population growth rate—again increasing the risk of extinction.

By increasing the risk of extinction in these ways, genetic problems resulting from small population sizes can ruin efforts to conserve a species. In some cases, conservation biologists have addressed this threat head-on by attempting the "genetic rescue" of populations that otherwise would appear doomed to extinction. Consider the Florida panther (*Puma concolor coryi*), a subspecies of puma (pumas are also called panthers, cougars, and mountain lions). By the early 1990s, the number of panthers in Florida had decreased to fewer than 25 individuals. Compared with other puma populations, the Florida panther population had low genetic diversity and a high frequency of problems such as heart defects, kinked tails, poor sperm quality, and adult males in which one or both testes failed to descend properly. Models similar to those discussed in under Concept 10.3 indicated a 95% chance that the population would become extinct within 20 years.

In 1995, to rescue the Florida panther from genetic decline and likely extinction, biologists captured eight female pumas from populations in Texas and released them in southern Florida. They selected females from Texas because historically, gene flow occurred between the Florida and Texas puma populations. The results were striking (Johnson et al. 2010). Panther numbers tripled, levels of genetic variation doubled, and the frequency of genetic abnormalities decreased substantially (**Figure 22.15**). Increases in panther numbers no doubt were aided by other conservation efforts, including habitat protection and the construction of highway underpasses to reduce mortality from vehicle strikes, but it is clear that genetic restoration has contributed to the recovery of the Florida panther. Other examples of successful genetic rescue include the case of the greater prairie chicken, described under Concept 6.2.

As the Florida panther example suggests, genetic analyses can inform conservation decisions by revealing the genetic diversity present in a species and, in extreme cases, by guiding efforts to rescue a population or species from problems stemming from genetic decline. Genetic studies are also being used to determine the appropriate targets of management within species (e.g., subspecies or populations) through the identification of **evolutionarily significant units** or *management units* (Waples 1998). In the Pacific Northwest, for example, genetic studies of salmonid fishes (salmon and their relatives; *Oncorhynchus* spp.) have had both pragmatic aims—to distinguish populations warranting protection under the Endangered Species Act—and theoretical aims—to understand the evolutionary history of the Pacific salmonids.

Molecular genetic techniques can also be used in forensic applications related to conservation biology. For example, genetic analyses permitted the identification of illegally harvested whale species in meat that was sold in Japan and labeled as either dolphin or (Southern Hemisphere) minke whale, both of which are legal to hunt (Baker et al. 2002). Cycads, too, have been genetically "fingerprinted," allowing tracking of these highly valuable and frequently

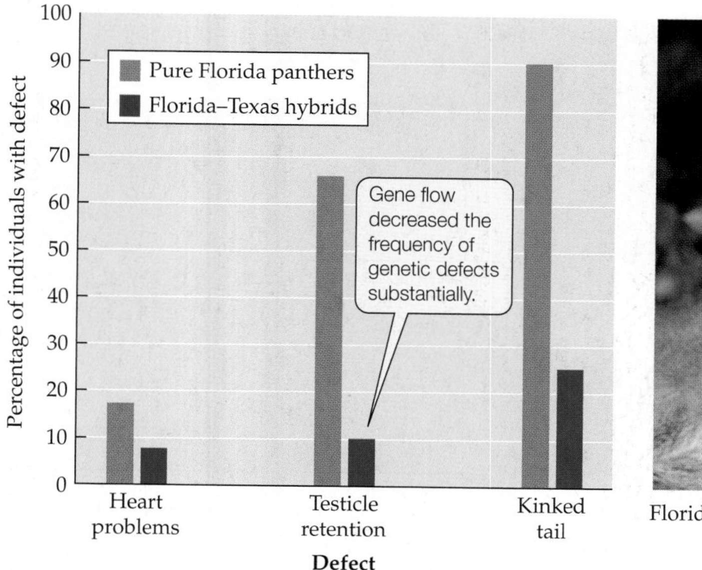

Florida panther (*Puma concolor coryi*)

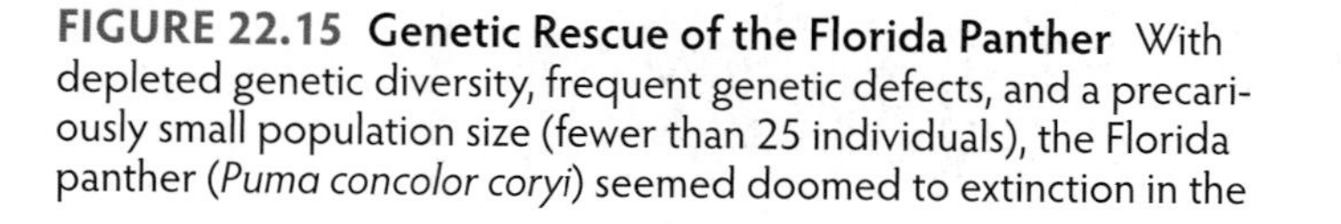

FIGURE 22.15 Genetic Rescue of the Florida Panther With depleted genetic diversity, frequent genetic defects, and a precariously small population size (fewer than 25 individuals), the Florida panther (*Puma concolor coryi*) seemed doomed to extinction in the early 1990s. The gene flow that resulted from the translocation of eight females from *P. concolor* populations in Texas helped to reverse these trends. (After Johnson et al. 2010.)

ECOLOGICAL TOOLKIT 22.1 Forensics in Conservation Biology

As we saw in our discussion of Concept 22.3, overharvesting of wildlife can lead to population declines across entire continents and throughout the world's oceans. In some cases, conservation biologists or wildlife authorities may know that individuals from protected populations have been captured or killed, but without further information they cannot determine the extent or source of such illegal harvests. This lack of information can make laws that protect threatened species difficult to enforce. Fortunately, in some species, molecular genetic techniques can be used to monitor the extent of illegal harvesting or trace the source of illegally harvested wildlife products.

As an example, consider the trade in ivory. High demand for ivory led to the widespread slaughter of African elephants (*Loxodonta africana*), causing their numbers to drop from 1.3 million to 600,000 individuals between 1979 and 1987. As a response to this problem, an international ban on ivory trade was established in 1989. Initially the ban was successful, but soon an illegal ivory trade sprang up, leading to further declines in elephant populations.

The illegal trade in ivory proved hard to combat because even if a shipment was intercepted, it could be difficult to identify where the tusks had come from. In June 2002, more than 5,900 kg (>13,000 pounds) of ivory were confiscated in Singapore—the largest seizure of ivory since the 1989 ban (**Figure A**). Law enforcement officials suspected that these tusks came from elephants killed in multiple regions of Africa. Were they correct?

As in some human forensic cases, DNA evidence was used to answer this question. First, DNA was obtained from tusks seized in the June 2002 raid. As you may recall from your introductory biology class, the polymerase chain reaction (PCR) can be used amplify (that is, produce many copies of) specific regions of DNA that often differ from one individual to another. Such highly variable DNA segments can then be visualized in a computer scan, as shown in **Figure B**. By amplifying several of these highly variable segments, researchers can create a "DNA profile" that characterizes an individual's genetic makeup.

Figure A Ivory from the 2002 Seizure in Singapore

To locate the source of the confiscated ivory, Samuel Wasser and colleagues amplified seven highly variable DNA segments and used them to produce a DNA profile for each of 37 of the confiscated tusks. The place of origin of each tusk was then estimated by comparing its DNA profile with those in a reference database of elephant DNA collected from known geographic locations (Wasser et al. 2007). Contrary to what law enforcement officials had originally suspected, the results indicated that all of the tusks came from a relatively small region in southern Africa, centered on Zambia (**Figure C**). These findings enabled wildlife authorities to focus their investigation on a smaller area and fewer trade routes, and they led the Zambian government to improve its anti-poaching efforts. More broadly, the approach described by Wasser and colleagues shows promise in forensic applications designed to limit illegal trade in a wide range of threatened animal and plant species.

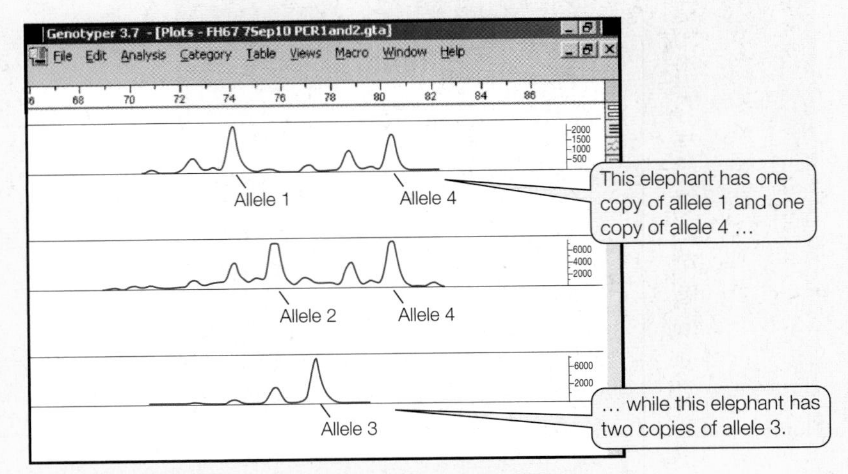

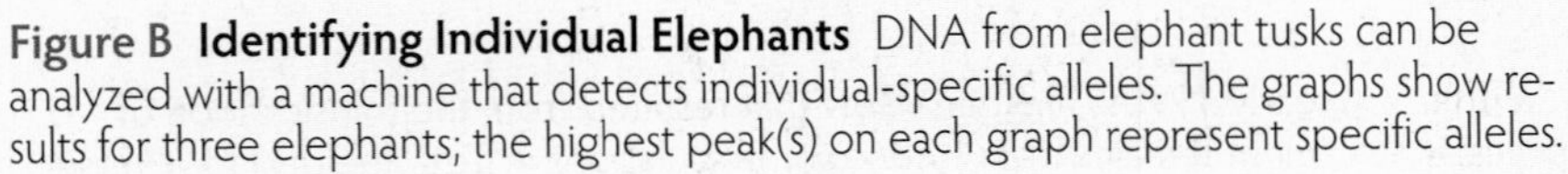

Figure B Identifying Individual Elephants DNA from elephant tusks can be analyzed with a machine that detects individual-specific alleles. The graphs show results for three elephants; the highest peak(s) on each graph represent specific alleles.

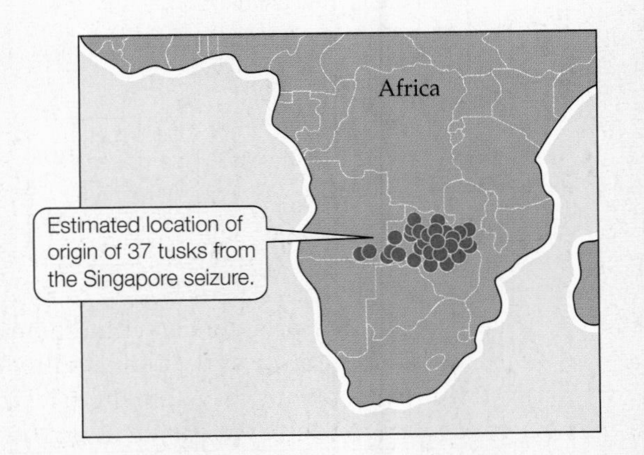

Figure C Tracking Contraband Ivory DNA methods indicated the ivory came from a relatively small geographic region—a finding that differed from what law enforcement officials had originally suspected. (After Wasser et al. 2007.)

poached plants (Little and Stevenson 2007). In **Ecological Toolkit 22.1**, we explore how such "forensic conservation biology" is done and how it was used to track the source of a large shipment of contraband ivory.

The availability of molecular genetic tools has enhanced our ability to understand the genetic problems faced by small populations and has helped us to address some of those problems. Let's turn next to some of the ways we can approach conservation at the population level.

Demographic models can guide management decisions

Is the growth rate of the Yellowstone grizzly bear population high enough to allow it to persist? At what life stages are loggerhead sea turtles most vulnerable to predation, and what management decisions would be most expedient to ensure their continued viability? How much old-growth forest habitat must be preserved to ensure the persistence of the northern spotted owl? Such questions arise with nearly any species of conservation concern, and demographic models offer approaches to answering them.

There are hundreds of quantitative demographic models in use, tailored to the specific biological traits of particular species. The quantitative approach most widely used for projecting the potential future status of populations is referred to as **population viability analysis (PVA)**. This approach allows ecologists to assess extinction risks and evaluate management options for populations of rare or threatened species (Morris and Doak 2002). PVA is a process by which biologists can calculate the likelihood that a population will persist for a certain amount of time under various scenarios. It encompasses a whole suite of models, ranging from relatively simple stage- or age-based demographic models like those described in Chapter 9 to more complex, spatially explicit models that can take actual landscape features and dispersal of individuals from multiple populations into account.

PVA provides conservation biologists with the probabilities that certain outcomes may occur, given certain assumptions about future conditions (e.g., changes in threats or in management efforts). Thus, PVA is a tool with which ecologists can synthesize data collected in the field, assess the risk of extinction of a population or populations, identify particularly vulnerable age or stage classes, determine how many animals to release or how many plants to propagate to ensure the establishment of a new population, or determine what might be a safe number of animals to harvest (Beissinger and Westphal 1998).

Ecological Toolkit 9.1 described the use of a demographic model for loggerhead sea turtles, from which ecologists concluded that protection of mature individuals would be key to slowing declines in populations. PVA has been used to make a wide variety of decisions about how best to manage rare species. In Florida, the fire regime that would best serve population growth in a rare plant (*Chamaecrista keyensis*) was determined through PVA simulations of burns at different times of year and at different intervals (Liu et al. 2005). In Australia, the forest cutting practices that would best serve the persistence of two endangered arboreal marsupial species, the greater glider and Leadbeater's possum, were determined through extensive PVA modeling coupled with long-term monitoring to verify the accuracy of the data going into the model (Lindenmayer and McCarthy 2006). Such analyses have played a critical role in management decisions for a number of species.

Some conservation biologists, however, caution against excessive reliance on conclusions based on the results of PVA. They point to the high level of uncertainty in the dynamics of small populations, the paucity of demographic and environmental data for many endangered species, and the high probability that a model will leave critical factors unaccounted for. To be used effectively, such models need to be constantly refined and revisited by different researchers to check their validity against field observations, just as management strategies must be checked and adjusted for effectiveness (Beissinger and Westphal 1998).

Ex situ conservation is a last-resort measure to rescue species on the brink of extinction

When remaining populations of a species fall below a certain number, direct, hands-on action may be called for. Such actions can include the introduction of individuals to threatened populations (as in the Florida panther) or extensive habitat manipulations intended to improve the chance that individuals can reproduce successfully (as in the red-cockaded woodpecker, as described in the Case Study Revisited on p. 497). In some cases, however, the only hope for a species' survival may be to take the remaining individuals out of their habitat—ex situ—and allow them to multiply in sheltered conditions under human care with the hope of later returning some individuals to the wild.

The rescue of the California condor (*Gymnogyps californianus*) is a leading example of this strategy (**Figure 22.16**). In the late Pleistocene, this great bird ranged throughout much of North America, and by the nineteenth century, it was still distributed from British Columbia to Baja California. The condor population declined steeply between the 1960s and 1980s, however, reaching a low of 22 birds by 1982. The species became extinct in the wild in 1987, when the last birds were captured and brought to an ex situ facility in California for breeding (Ralls and Ballou 2004).

The species now numbers more than 200 birds, some released into the wild and some remaining in captivity. Increasing the population to this point has required care-

FIGURE 22.16 Ex Situ Conservation Efforts Can Rescue Species from the Brink of Extinction Ex situ efforts to save the California condor (*Gymnogyps californianus*) involve multiple steps. (A) To reduce inbreeding and increase the number of eggs that hatch successfully, a USFWS biologist removes eggs from the wild (to be taken to an ex situ breeding facility) and replaces them with one egg from the San Diego Zoo. (B) At the San Diego Zoo, condor chick "Hoy" is being fed by a condor-feeding puppet to avoid its becoming acclimated to humans. (C) Two condors at the time of their release (spring 2000). The instrument in the right foreground is a scale from which condor weight can be read by telescope when a bird perches on it. (D) This adult, with a wingspan of 9 feet, was bred in captivity and later released.

ful genetic analysis, hand rearing of some chicks, and wide cooperation among zoos, managers of natural areas, hunters, and ranchers. Principal barriers to the condor's recovery include lead poisoning from ammunition found in the carrion condors feed on, the negative health effects of ingesting plastic and other trash, West Nile virus, and genetic drift. Given all of these risks and costs, is the recovery of the California condor worth all the effort that has gone into making it happen? Without that effort, the species would now be extinct.

Ex situ conservation programs are taking place in zoos, special breeding facilities, botanical gardens, and aquariums all over the world. Such programs have allowed many species at risk of extinction to build their numbers sufficiently to permit reintroduction into the wild. While ex situ programs play important roles in keeping our most threatened species from extinction, as well as in publicizing the plight of those species, they are expensive, and they can introduce a host of problems, such as exposure to disease, genetic adaptation to captivity, and behavioral changes (Snyder et al. 1996). Furthermore, they do not have a very high rate of success in terms of restoration of self-sustaining populations in the wild. Could the funds dedicated to ex situ efforts be better spent on managing species in the wild or on securing land for the establishment of new protected areas—that is, for in situ conservation? Sometimes the answer is no, usually when populations have been reduced to critical levels or when not enough suitable habitat is available. But the question must always be asked.

Legal and policy measures support biological methods of protecting species and habitat

Conservation biologists seek to gain the best scientific information possible to inform decisions that will benefit the welfare of species or ecosystems. The process of making many of these decisions, however, is more of a societal than a scientific one and falls into the realm of public policy and communications. The outcome of a society's collective decisions on conservation issues can be seen in national and state laws, the policies set by natural resource agencies, and the work and policies of nongovernmental organizations. At times, these decisions are guided by global processes in the form of international treaties, agreements, and conventions. At other times, they are driven by grassroots efforts. The interplay between science and this composite of human decision-making entities is complex, but it is an integral part of any successful conservation effort.

In the United States, the most prominent legislation protecting species, the Endangered Species Act (ESA), has played a vital role in protecting many of the country's most threatened species. It was passed by Congress in 1973 to "provide a means whereby the ecosystems on which endangered and threatened species depend may be conserved, and to provide a program for the conservation of these species." The U.S. Fish and Wildlife Service and the National Marine Fisheries Service are charged with listing federally threatened and endangered species, identifying critical habitat for each species, drafting recovery plans, and carrying out actions necessary to increase abundances to target numbers.

The ESA currently protects more than 1,300 species native to the United States and another 570 from other countries. The ESA extends its influence beyond U.S. borders by regulating trade in endangered species as a result of an international treaty called the Convention on International Trade in Endangered Species of Wild Fauna and Flora (CITES). This treaty, which has been in place for more than 35 years, regulates international trade in listed organisms and their parts. It mandates a virtual prohibition on trade in some species recognized to be endangered, while others, deemed to be less threatened, must be monitored in their home countries for indications that trade should be restricted. Currently, 167 countries have agreed to adhere to CITES regulations, and about 33,000 species receive some protection. While enforcement of CITES regulation remains a difficult task for governments, the treaty has been a key instrument in protecting species worldwide.

The Convention on Biological Diversity, which most nations have signed (the most notable exception being the United States), is the outcome of the Earth Summit held in Rio de Janeiro in 1992. This agreement acknowledges declining biodiversity as a problem shared by all the world's people and establishes goals for actions to counteract it. For example, the agreement urges nations to document the biodiversity contained within their borders, identifies the rights of nations to benefit from patents derived from their biodiversity, and calls for mechanisms to protect biodiversity from genetically modified organisms. It has provided a fruitful structure for nations developing their own conservation plans and has served as a framework for bringing nations together to address global biodiversity losses.

The protection of biodiversity is also dependent on a wide array of national, state, and local regulations and policies that set restrictions on land development, call for environmental review prior to land disturbances, and manage harvest levels. There is a limit to the percentage of the landscape we will be able to place under protected status, so much of the world's biodiversity will continue to reside on private lands and working landscapes. Therefore, it is vital that we maintain a legal framework that will serve to protect the most critical components of biodiversity in those places, and that those laws and policies be crafted with the best available science to support them.

CONCEPT 22.5 Prioritizing species helps maximize the biodiversity that can be protected with limited resources.

Ranking Species for Protection

How do we allocate the limited resources that are available for species conservation? Do we protect those species that are most threatened, or do we focus on those that play a substantial ecological role? How do conservation biologists and policymakers decide which areas are the most critical to protect? If a species is rare, is it necessarily at risk of extinction? Do we determine how threatened an insect or a lichen is in the same way that we do for a bird, a mammal, or a conifer?

The rarest and the most rapidly declining species are priorities for protection

Many species have become rare only recently as a result of the threats we outlined earlier in this chapter. Other species may have always been rare. In either case, having a measure of how threatened a species is permits us to focus our efforts on those species that are most threatened: the rarest and the most rapidly declining. We may be able to postpone attending to species that are naturally low in abundance but not particularly threatened.

What do we mean by rarity, and how do we determine just how rare something is? To clarify the different concepts of rarity, we can use a matrix that sorts out whether a species has a wide or a narrow geographic range, whether it is broad or restricted in its habitat specificity, and whether

TABLE 22.2

Seven Forms of Rarity

Local population size	Geographic distribution: WIDE		Geographic distribution: NARROW	
	Habitat specificity: BROAD	Habit specificity: RESTRICTED	Habitat specificity: BROAD	Habitat specificity: RESTRICTED
SOMEWHERE LARGE	Common	**RARE:** Widely distributed in an uncommon specific habitat	**RARE:** Narrow endemic, but with a broad ecological tolerance and locally abundant	**RARE:** Endemic, with narrow ecological requirements, but locally abundant
EVERYWHERE SMALL	**RARE:** Never abundant, but distributed over a wide geographic and habitat range	**RARE:** Small populations in a specific habitat, but over a wide geographic area	**RARE:** Endemic, broad ecological tolerance, small populations	**RARE:** Endemic, narrow ecological tolerance, small populations

Source: Rabinowitz et al. 1986.

local populations tend to be small or large (**Table 22.2**). There are some rare species, for example, that exist over a wide geographic area and are relatively broad in their habitat requirements, yet tend to occur in very small populations. Other rare species inhabit specific habitats within a narrow geographic range, but may have large populations in those specific locations (Rabinowitz et al. 1986). Conservation of these different types of rare species requires different approaches. Some species require small reserves to protect well-established populations; others require management practices that create habitat conditions suitable for a rare but geographically widespread species.

Objective, scientific assessment of the conservation status of species began in 1963 with the IUCN Red List (see Table 22.1). A parallel effort was developed in the United States by The Nature Conservancy, which established the Natural Heritage Program (now NatureServe) in the early 1970s in order to assess the conservation status of U.S. species. Both organizations have developed a ranking structure that indicates how threatened a species is and an assessment protocol to determine a species' rank. The assessment protocol takes into account not only numbers of populations or individuals, but also the total geographic area that the species occupies, the rate of its decline, and the threats it faces. Because of the challenge of creating a system that can be applied equally well to a skipper butterfly, a cycad, or a cheetah, and because the information available on rare species is often incomplete, both systems allow assessors to choose among different sets of criteria to decide whether a species is critically endangered, endangered, vulnerable, or under some lesser level of threat. The criteria must be flexible while still permitting consistent conclusions by dif-

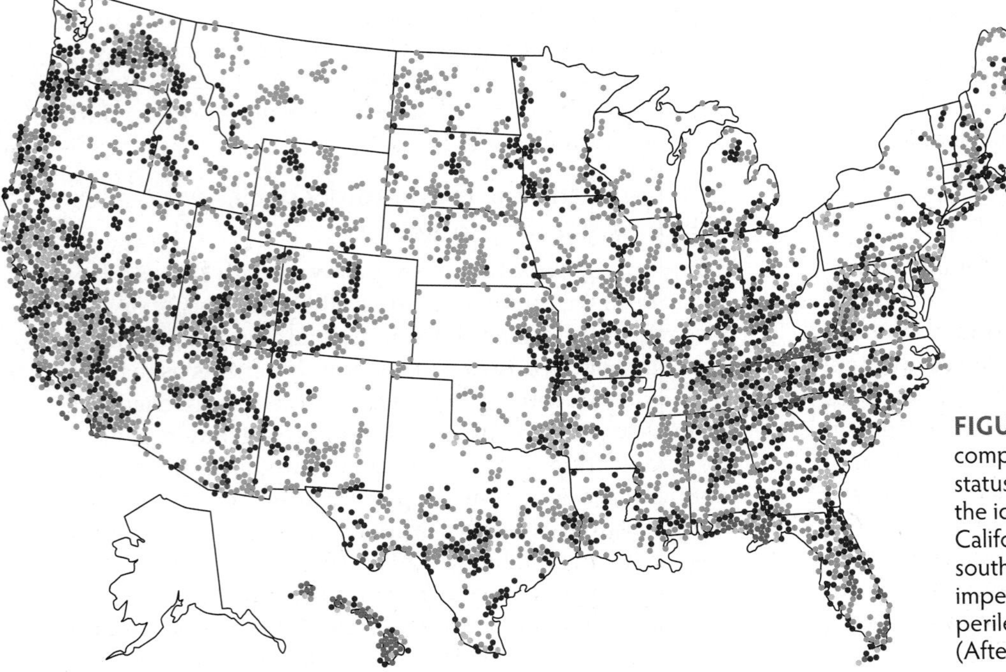

FIGURE 22.17 Hot Spots of Imperilment The compilation of NatureServe data on the conservation status of species in the United States has permitted the identification of the most critical areas to protect. California, Hawaii, the Florida Panhandle, and the southern Appalachian Mountains are "hot spots" of imperilment—they have high concentrations of imperiled species due to their high rates of endemism. (After Stein et al. 2000.)

FIGURE 22.18 A Flagship Species The giant panda (*Ailuropoda melanoleuca*), a native of China, is endangered, primarily due to habitat loss. About 2,000 pandas remain in the wild; another 300 pandas live in zoos or breeding centers, mostly in China.

ferent evaluators. Both systems come up with a global ranking, but are also applicable on a regional level. The effort to evaluate species is large and ongoing, but most taxonomic groups have had only a small proportion of their members evaluated (de Grammont and Cuarón 2006).

Such assessments of conservation status can be used to locate clusters of threatened species and thus identify areas that are critical to protect (**Figure 22.17**). They are frequently consulted when development projects are planned, and they are important for keeping the public aware of the degree of threat faced by Earth's biota. These databases are dynamic in that they can change as scientific information is updated: the conservation status assigned to a species can be downgraded if its numbers increase, or upgraded if its numbers decline.

Protection of surrogate species can provide protection for other species with similar habitat requirements

If we protect the habitat necessary for the red-cockaded woodpecker, as described in the Case Study on pp. 476–477, will we simultaneously provide protection for the gopher tortoise, Bachman's sparrow, Michaux's sumac, and other rare species that are dependent on the longleaf pine savanna ecosystem? Species may become conservation priorities not only because of their own conservation status, but also because of their capacity to serve as **surrogate species** whose conservation will serve to protect many other species with overlapping habitat requirements. We may choose surrogate species as a shortcut when we lack sufficient information on the distributions and abundances of many species of conservation concern in a region. We may be interested in a species that will help us garner public support for a conservation project; examples of such **flagship species** include charismatic animals such as the giant panda (**Figure 22.18**). An **umbrella species**, by contrast, is one that we select with the assumption that protection of its habitat will serve as an "umbrella" to protect many other species with similar habitat requirements. Umbrella species are typically species with large area requirements, such as grizzly bears, or habitat specialists, such as the red-cockaded woodpecker. But they may also include animals that are relatively easy to count, such as butterflies (Fleishman et al. 2000). Some researchers prefer to choose not just one species, but several **focal species**, selected for their different ecological requirements or susceptibility to different threats, with the realization that by thus casting a broader net, we improve our chances of covering regional biodiversity with protection.

Methods have been devised and criteria established to allow for strategic selection of the one or several surrogate species that will best serve conservation aims (Favreau et al. 2006). Conservation biologists recognize, however, that surrogate species approaches are not without problems, and that the distribution or habitat requirements of any one species cannot capture all the conservation targets we may have.

A CASE STUDY REVISITED

Can Birds and Bombs Coexist?

As the longleaf pine ecosystem lost 97% of its area over the last several hundred years, the biological traits of the red-cockaded woodpecker that had worked well in the extensive pine savannas of the past turned out to be detrimental in its changing environment. Prime woodpecker habitat became fragmented, consisting of islands of usable habitat in an unsuitable landscape. As a result, the woodpeckers' unusual habit of excavating cavities in living trees—a process that usually takes a year or more to complete—made the availability of cavities a limiting factor for woodpecker populations.

Jeff Walters and his colleagues tested the hypothesis that a lack of high-quality habitat was limiting the woodpecker's population growth by constructing artificial nest cavities, placing them in clusters, and observing woodpecker behavior. They tried this strategy for several reasons. First, each bird in a cooperative breeding group must have its own cavity for roosting; this explains why cavities are found in clusters. Second, cavity clusters are typically abandoned after several years' use, primarily due to cavity entrance enlargement by other species or mortality of cavity trees, so there is a continual demand for cavity clusters (Harding and Walters 2002). The cavity clusters constructed by the researchers were rapidly colonized, mostly by helper birds from the vicinity and young dispersing birds (Copeyon et al. 1991; Walters et al. 1992).

These results suggested that people could help the red-cockaded woodpecker increase its numbers by going out with drill, wood, wire, and glue and installing clusters

(A)

(B)

(C)

FIGURE 22.19 Construction and Installation of Artificial Nest Cavities Has Allowed Populations of Red-Cockaded Woodpeckers to Increase (A) An artificial nest cavity built for a red-cockaded woodpecker. (B) Cutting a hole for the artificial nest. (C) Installing the artificial nest.

of cavities within living longleaf pines (**Figure 22.19**). This realization has proved a boon to woodpecker recovery. Aided by the construction of artificial cavities, the population of red-cockaded woodpeckers at Fort Bragg increased from 238 breeding groups in 1992 to 368 breeding groups in 2006. Cavity construction also contributed to increased abundances of red-cockaded woodpeckers at other military bases, including Eglin Air Force Base (Florida), Fort Benning (Georgia), Fort Polk (Louisiana), Fort Stewart (Georgia), and Marine Corps Base Camp Lejeune (North Carolina). Similar successes have occurred at sites other than military bases. For example, when Hurricane Hugo hit the South Carolina coast in 1989, the population of red-cockaded woodpeckers in Francis Marion National Forest, previously home to 344 breeding groups, was severely reduced. The hurricane killed 63% of the birds, and another 18% died the following winter (Hooper et al. 2004). Within 2 years of the storm, however, National Forest workers had installed 443 artificial cavities. This strategy averted a severe population decline; by 1992, the population had recovered to 332 breeding groups.

Now that managers have identified cavity construction and maintenance as a critical factor for the species' recovery, they are obliged by the Endangered Species Act to continue doing it. This strategy is labor-intensive and expensive, but for now it is necessary for the red-cockaded woodpecker's continued existence. How long can we sustain this effort? Will we reach a point at which there is enough longleaf pine savanna that the woodpeckers will be able to maintain their own numbers without human assistance? What effect will artificial nest cavity construction have on the bird's behavior and evolution? We just don't know the answers to these questions.

In the decades that Walters and others have been researching the red-cockaded woodpecker, they have used many of the tools described in this chapter. Models of population dynamics have facilitated the identification of vulnerable stages in the woodpecker's life cycle. Genetic studies and modeling have focused attention on the threat of inbreeding. Field studies have demonstrated the need for prescribed burning to maintain the community structure required by the woodpeckers. Economic and sociological analyses have led to the development of a "safe-harbor" program that makes endangered species management more palatable to private landowners. And managers are reaching into a literal toolbox to build nesting cavities. Much of this work has been dictated by the creation more than 30 years ago of the U.S. Endangered Species Act.

CONNECTIONS IN NATURE

Some Burning Questions

As we saw in Chapter 3, recurrent fires promote the establishment of savanna. Hence, to maintain red-cockaded woodpecker populations and the longleaf pine savannas on which they depend, fire is key—whether it is ignited naturally or intentionally set under controlled conditions. Fire affects ecosystems at multiple scales, from the cellular and biochemical to the atmospheric. As with other regular forms of disturbance (see the discussion under Concept 8.2), differences in the frequency of fires can affect the distributions and abundances of species, and those changes, in turn, can affect the cycling of nutrients (see Chapter 21) and water. Because fire affects communities at so many levels, prescribed burning is used as a management tool for conserving species in numerous ecosystems where fire has been a regular natural disturbance (**Figure 22.20**).

But the use of fire as a management tool can have unintended and undesirable ecological outcomes where non-native invasive species are present. In some longleaf pine savannas in Florida, openings resulting from burning have provided favorable habitat for the establishment of cogongrass (*Imperata cylindrica*), an invasive plant from Asia. The presence of this grass, in turn, causes fires to burn hotter, higher, and more evenly on a horizontal plane. The consequences of these hotter fires are increased mortality

FIGURE 22.20 Prescribed Burning Is a Vital Management Tool in Some Ecosystems In the southeastern United States, regular burning is used to maintain the high plant biodiversity characteristic of the understory in pine savanna ecosystems. Many threatened species, including the red-cockaded woodpecker, rely on regular burning for their persistence. Here, USFWS firefighters monitor a prescribed burn intended to preserve habitat for the endangered Florida panther.

of longleaf pine seedlings and native wiregrass, favorable conditions for further infiltration of cogongrass, and a resulting threat to the high levels of native plant diversity found in the understory of the longleaf pine savanna (Lippincott 2000). Land managers in this situation are faced with a dilemma: to burn or not to burn? The right question is more likely to be when to burn, and how often.

Adding people to the burning landscape further complicates matters. Throughout the southeastern United States, prescribed burns are taking place in a complex landscape where patches of forest are adjacent to peoples' homes and businesses. Convincing the public that these fires are necessary has required considerable outreach and public education. In the North Carolina Sandhills, the days for prescribed burns are chosen not only for safe conditions, but also with regard to wind direction so as to minimize the amount of smoke in population centers.

Here, as elsewhere, recognition of people as an integral component of the landscapes that must harbor all of nature's diversity has been a vital piece of the conservation picture. Establishing protected natural areas as sanctuaries for wildlife is an important part of the solution to the biodiversity crisis, but we must also do what we can to ensure that the vast majority of Earth's surface outside of protected areas is able to sustain both people's livelihoods and habitat for other species. This is a difficult challenge that will involve constant education, negotiation, legislation, and many creative approaches.

SUMMARY

CONCEPT 22.1 Conservation biology is an integrative discipline that applies the principles of ecology to the protection of biodiversity.

- Conservation biology is the scientific study of phenomena that affect the maintenance, loss, and restoration of biodiversity.
- Biodiversity is important to human society because of our reliance on natural resources and ecosystem services that depend on the integrity of natural communities and ecosystems.
- With the growing awareness of accelerating losses of global biodiversity, ecologists saw a need for a separate discipline that would apply the principles of ecology to the preservation of species and ecosystems.
- Conservation biology is a scientific discipline instilled with the value of biodiversity.

CONCEPT 22.2 Biodiversity is declining globally.

- Earth is losing species at an accelerating rate, largely due to humanity's growing footprint on the planet.
- Extinction is the end point of incremental biological decline as species lose individuals and populations and become increasingly vulnerable to the problems of small populations.
- Earth's biota is becoming increasingly homogenized due to a rise in generalist species and a decline in specialist species, as well as losses of genetic diversity within some taxa.

CONCEPT 22.3 Primary threats to biodiversity include habitat loss, invasive species, overexploitation, pollution, disease, and climate change.

- Habitat degradation, fragmentation, and loss are the most important threats to biodiversity.
- Invasive species degrade local habitats by preying on or competing with native species and by altering ecosystem properties.
- Overexploitation of selected species has large effects on communities and ecosystems.
- Other factors that erode the viability of populations and contribute to losses of biodiversity include air and water pollution, diseases, and global climate change.

CONCEPT 22.4 Conservation biologists use many tools and work at multiple scales to manage declining populations.

- Genetic analyses have been used to understand and manage genetic diversity within rare species,

SUMMARY (continued)

to identify appropriate management units, and in forensic analyses of illegally harvested organisms.

- Population viability analysis (PVA) is an approach that uses demographic models to assess extinction risks and evaluate proposed management actions.
- Ex situ conservation, which involves taking organisms from the wild into human care, is a last-resort measure to rescue species on the brink of extinction.
- Laws, policies, and international treaties are vital supplements to biological methods of protecting species and habitat.

CONCEPT 22.5 Prioritizing species helps maximize the biodiversity that can be protected with limited resources.

- Conservation biologists identify those species of the highest priority for protection—the rarest and the most rapidly declining species—by assessing numbers of individuals and populations, total geographic area occupied, rates of decline, and the degree of threat faced.
- Identification of surrogate species can provide protection for other species with similar habitat requirements.

REVIEW QUESTIONS

1. Think about what motivates your own interest in ecology, and contemplate your own feelings as you witness destruction of natural places. Is your interest in part motivated by a desire to do something about it? How would you separate out your values as you consider conducting research in conservation biology? Are your research questions "value-laden"?
2. Describe two components of biodiversity that are being homogenized due to human actions.
3. What are the principal threats to biodiversity? Describe some examples in which multiple threats have contributed to a species' decline.
4. What is the difference between a species determined to be endangered by the Natural Heritage/NatureServe program and one that is listed as endangered under the U.S. Endangered Species Act? What are the consequences for management of each?
5. Identify five imperiled species that live in your region, including a plant, a mammal, a bird, a fish, and an invertebrate. Are any of these species endemic to your region? For each species you have identified, try to find out whether it was rare prior to human settlement of the region. What threats does this species face today? What is being done to protect this species? Based on the ecological knowledge you have gained, what questions do you think should be researched to aid in the species' recovery? (Much of this information is available at http://www.natureserve.org/.)

ON THE COMPANION WEBSITE
sites.sinauer.com/ecology2e

The website includes Chapter Outlines, Online Quizzes, Flashcards & Key Terms, Suggested Readings, a complete Glossary, and the Web Stats Review. In addition, the following resources are available for this chapter:

HANDS-ON PROBLEM SOLVING

Population Augmentation and Recovery of Endangered Species

This Web exercise explores the consequences of augmenting populations of endangered species with captive-raised individuals. You will read a recent paper on population augmentation in an endangered butterfly. Then, using a transition matrix model, you will explore the relative costs and benefits of population augmentation and habitat enhancement for an endangered fish, the June sucker.

CHAPTER

23 Landscape Ecology and Ecosystem Management

KEY CONCEPTS

- **CONCEPT 23.1** Landscape ecology examines spatial patterns and their relationship to ecological processes and changes.
- **CONCEPT 23.2** Habitat loss and fragmentation decreases habitat area, isolates populations, and alters conditions at habitat edges.
- **CONCEPT 23.3** Biodiversity can best be sustained by large reserves connected across the landscape and buffered from areas of intense human use.
- **CONCEPT 23.4** Ecosystem management is a collaborative process with the maintenance of long-term ecological integrity as its core value.

Wolves in the Yellowstone Landscape: A Case Study

Imagine that you have trekked by snowshoe to a wildlife blind in the Northern Range of Yellowstone National Park. You have your spotting scope trained on a herd of elk as they move in a loose cluster across an open icy meadow, scraping away the snow with their hooves to reach clumps of bunchgrass and forbs. As you watch, you notice that some animals are lifting their heads and casting their attention to the east. You look up and see an approaching pack of wolves. The herd draws together, and the elk begin to move, fleeing down the meadow. A young male in poor health drops behind, is separated from the herd, and is surrounded with wolves leaping at his hindquarters and neck. He falls, and is quickly killed.

In Yellowstone, such predation is an ancient scene, and it is now once again a regular occurrence as wolves, reintroduced to the northern Rockies in 1995 and 1996 after 70 years of absence, hunt among a smorgasbord of ungulates and other prey (**Figure 23.1**). The reintroduction of wolves was the culmination of years of research effort and hotly contested policy debate, with vociferous objection from some residents of the region. Fifteen years later, its ecological consequences have proved to be multifaceted and profound, and public opinion has become generally more favorable.

The Greater Yellowstone Ecosystem (GYE) is an area that both symbolizes the soul of the American wilderness and encapsulates the challenges of managing public lands. But how "wild" is it? Larger in area than the state of West Virginia, the GYE includes two national parks and seven national forests as well as other public and private lands (**Figure 23.2**). The region is actively managed by more than 25 different state and federal agencies as well as private corporations, nongovernmental organizations, and private landowners. Decisions about the use of its land and natural resources are complex and often uncoordinated, yet when considered together, these decisions determine which species will or will not be sustained by the ecosystem (Parmenter et al. 2003).

Despite its fragmented management, the GYE is often perceived as one of the most biologically intact ecosystems in North America. It sustains seven species of native ungulates and five large

FIGURE 23.1 A Top Predator Returns Gray wolves (*Canis lupus*) with an elk (*Cervus canadensis*) kill. Wolves were reintroduced to Yellowstone in 1995 after nearly 70 years of absence, where they are now the main predators of elk.

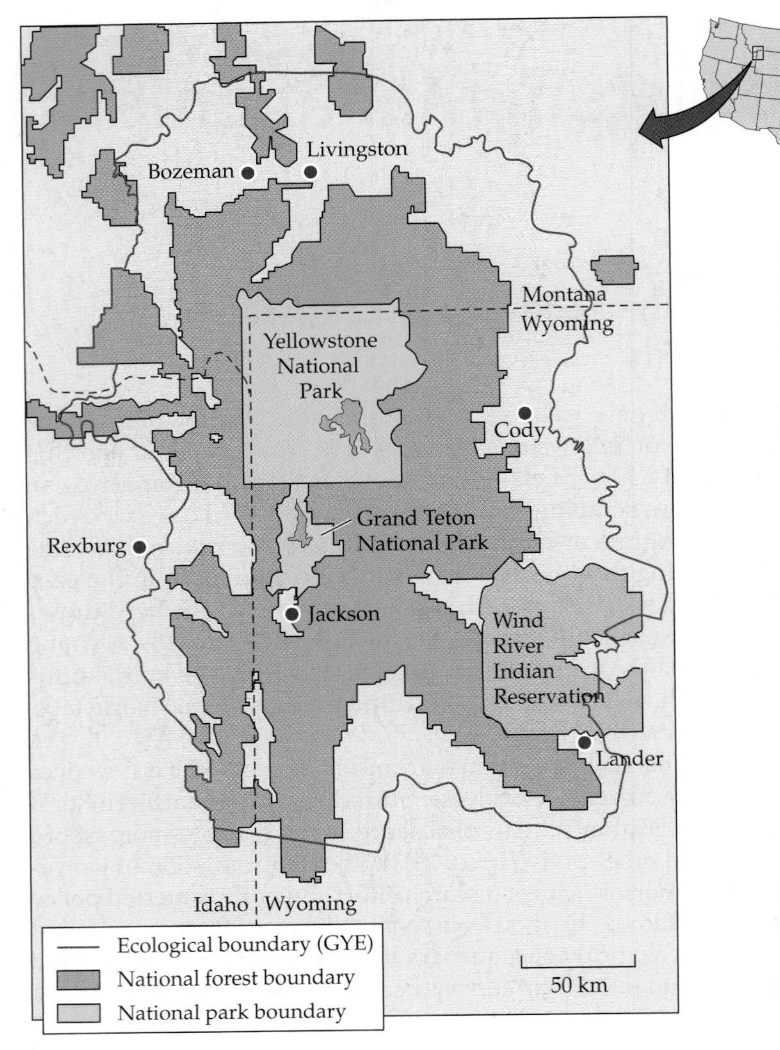

FIGURE 23.2 The Greater Yellowstone Ecosystem The Greater Yellowstone Ecosystem falls under a diverse array of management agencies. It contains Yellowstone and Grand Teton national parks, seven different national forests, and land managed by the Bureau of Land Management, as well as private lands. (After Parmenter et al. 2003.)

carnivore species. Understanding how these predator and prey populations interact, and how their relative numbers affect the whole ecosystem, has been a persistent challenge to ecologists who study the GYE, particularly in light of a century of management of wildlife populations. After wolves were eradicated in the mid-1920s, there were concerns that elk were overgrazing meadows in the northern part of the park. The elk population was regulated from the 1920s to the late 1960s by exporting animals to elk farms and by culling. In 1968, a new policy of "natural regulation" was implemented, and the elk population nearly quadrupled over a 30-year period, with subsequent suppression of the plants they feed on. The reintroduction of wolves has not only reduced the elk population, but has also affected the populations of many other species. How?

To start to answer that question, let's go back to the 1950s, when ecologists noticed that beavers had become scarce in Yellowstone National Park. Gradually, it became clear that the cause was elevated elk herbivory on the beavers' preferred food plants, willow and aspen. But a whole suite of other species depend on beaver ponds for their own persistence, and their abundances had declined along with the beavers'. The decision to eradicate wolves did not anticipate these ecological changes to the Yellowstone ecosystem. How can ecologists of today help managers of nature reserves make decisions that will take future consequences into account?

Introduction

In this chapter, we will take a step back, expand the scope of our view, and look at ecology from a landscape perspective. The emergence of this perspective has been accompanied by, or perhaps driven by, a powerful assemblage of tools that permits us to monitor the environment in multiple dimensions and at many scales. For example, the emergence of aerial photography gave ecologists a ready means of looking at "the big picture." More recently, our access to space has vastly expanded our ability to acquire images of Earth through remote sensing and has permitted the interpretation of many large-scale ecological patterns, such as global patterns of net primary production. The use of geographic information systems (GIS) has become standard for landscape planning efforts, whether for urban development or for conservation (**Ecological Toolkit 23.1**). In the field, handheld global positioning systems (GPS) have permitted ecologists to document precise locations and integrate them with other landscape variables through GIS. Radiotelemetry has greatly enhanced our ability to follow animal movements and migration patterns, again with the help of GIS. And our ability to analyze all this information is constantly growing, thanks to better computers and new statistical methods of spatial analysis.

We saw in Chapter 22 that habitat loss, fragmentation, and degradation are primary causes of the current declines in biodiversity. In this chapter, we'll see how the tools and methods of landscape ecology are used to address this problem at the landscape and ecosystem scales. Because protected natural areas are at the heart of conservation strategies, we will also consider how conservation biologists identify and design them to maximize their effectiveness. Finally, we'll examine how ecosystem management integrates ecological principles with social and economic information to help guide decisions about land and water use.

ECOLOGICAL TOOLKIT 23.1 Geographic Information Systems (GIS)

Geographic information systems (**GIS**) are computer-based systems that allow the storage, analysis, and display of data pertaining to specific geographic areas. The data used in GIS are derived from multiple sources, including aerial photographs, satellite imagery, and ground-based field studies (**Figure A**). Examples of such data include rainfall, elevation, and vegetation cover at specific locations. Each of these and many other variables could be used in a particular application of GIS—but whatever variables are used, the data are keyed to or referenced by spatial or geographic coordinates, so that they can be assembled into a multi-layered map.

Layers of mapped data can be put together in ways that help to address particular questions. We'll illustrate this process with an approach often used in conservation biology, called *gap analysis*. The acronym GAP refers to the Gap Analysis Program, a U.S. Geological Survey program whose mission is to help prevent biodiversity decline by identifying species and communities that are not adequately represented on existing conservation lands.

The lark bunting (*Calamospiza melanocorys*) is one such species. It depends on prairie habitat for its breeding grounds, but much of this habitat has been destroyed by humans. As a result, populations of the lark bunting have been declining by an average of 1.6% per year over the past 40 years, making it a species of conservation concern (U.S. Fish and Wildlife Service 2008).

For the lark bunting and any other species, gap analysis is a two-step process. First, input data on vegetation cover (see the top GIS layer in Figure A) and other environmental conditions required or preferred by the lark bunting is used to predict its geographic distribution (the second GIS layer in Figure A). Next, the lark bunting's distribution is compared with a third GIS layer showing the locations of conservation lands. By combining these two layers, we can calculate that only a small percentage of the bird's distribution is protected (**Figure B**).

Such information is critical to decisions about what lands should be protected to prevent future losses of biodiversity. (See **Web Extension 23.1** for a second example of GIS use in conservation biology.)

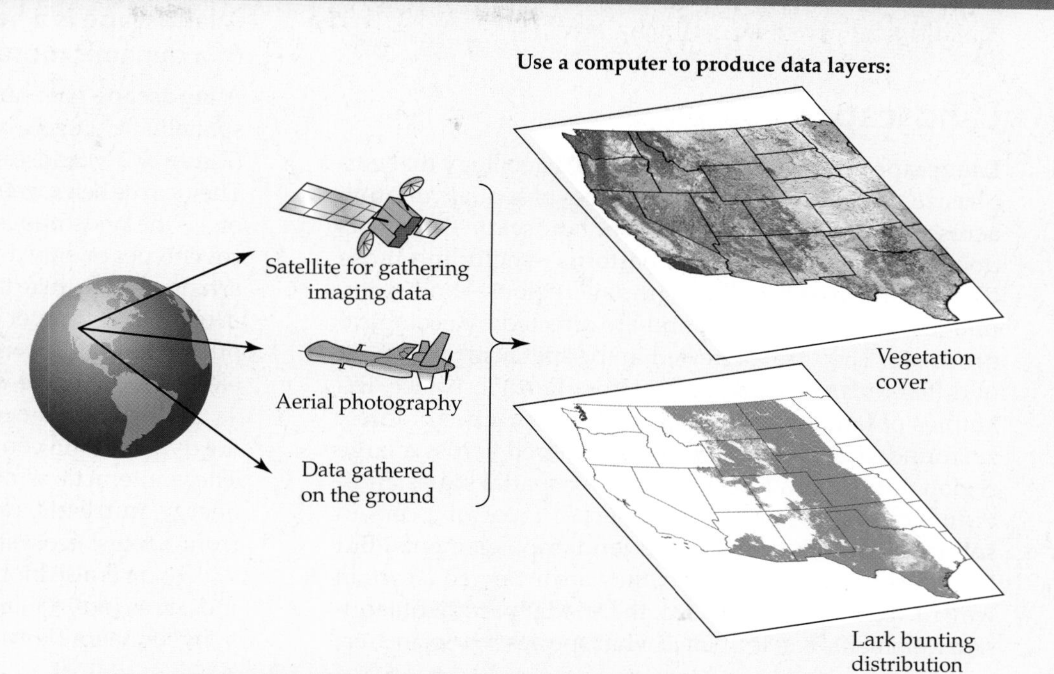

Figure A **GIS Integrates Spatial Data from Multiple Sources**

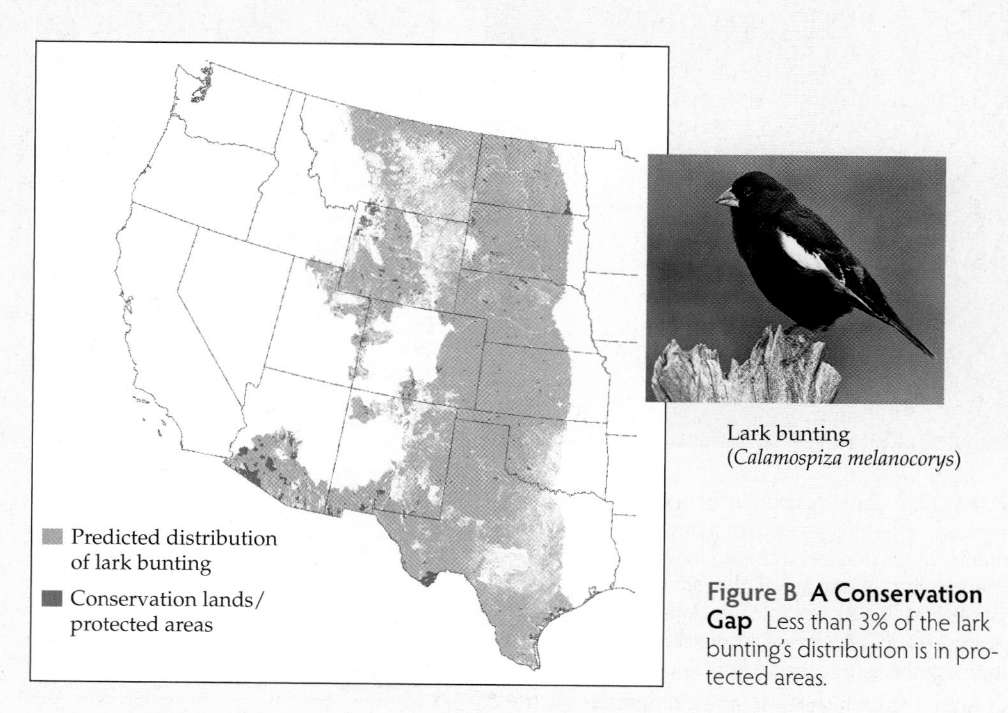

Figure B A Conservation Gap Less than 3% of the lark bunting's distribution is in protected areas.

CONCEPT 23.1 Landscape ecology examines spatial patterns and their relationship to ecological processes and changes.

Landscape Ecology

Landscape ecology is a subdiscipline of ecology that emphasizes the causes and consequences of spatial variation across a range of scales. As such, landscape ecologists document observed spatial patterns—including those that occur across broad geographic regions—and study how those patterns affect and are affected by ecological processes. They are interested in the spatial arrangement of different *landscape elements* across Earth's surface. Examples of landscape elements include patches of forest surrounded by pasture or lakes scattered across a large region of northern forest. At smaller spatial scales, individual creosote bushes in a desert or areas of a certain soil type could be considered landscape elements. But whatever these elements are, they are arranged a certain way in space. As we will see, the spatial pattern of landscape elements can influence what species live in an area, as well as the dynamics of ecological processes such as disturbance and dispersal.

A landscape is a heterogeneous area composed of a dynamic mosaic of interacting ecosystems

A **landscape** is an area in which at least one element is spatially heterogeneous (varies from one place to another) (**Figure 23.3**). Landscapes often include multiple ecosystems. They can be heterogeneous either in what they are composed of—is the landscape composed of twelve different vegetative cover types or only three?—or in the way their elements are arranged—are there many small patches arranged regularly over the landscape, or are there a few large patches? Ecologists often refer to this composite (or pattern) of heterogeneous elements that make up a landscape as a **mosaic**.

The different ecosystems that make up a landscape are dynamic and continually interacting with one another. These interactions may occur through the flow of water, energy, nutrients, or pollutants between ecosystems, as from a forest ecosystem into an adjacent lake.

There is also biotic flow between habitat patches in the mosaic as individuals or their gametes (e.g., pollen) move between them (Forman 1995). For such movement to oc-

(A)

(B)

(C)

FIGURE 23.3 Landscape Heterogeneity Landscapes can be heterogeneous in many different kinds of elements, which may be arranged in ways independent of one another. (A) An aerial photograph of Michigan's Upper Peninsula. (B) A map of six different soil types in the same area. (C) A map of seven different landscape elements in the same area. (After Delcourt 2002.)

Q In part (B), which landscape element covers the least area?

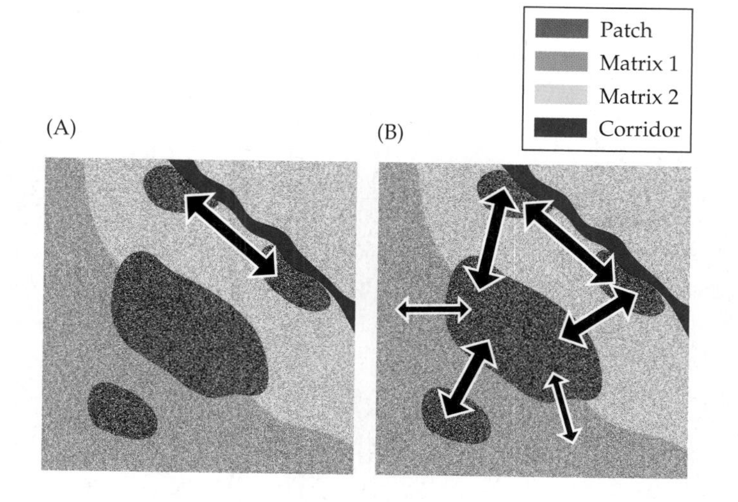

FIGURE 23.4 Movements Across the Landscape Movements between adjacent landscape elements may occur frequently (thicker arrows) or rarely (thinner arrows). (A) Exchange between patches of the same type occurs frequently if a corridor that allows movement connects the patches. (B) Exchange between patches of the same type occurs frequently, but exchange with the matrix occurs only rarely. (After Hersperger 2006.)

Do organisms move more freely across the matrix in (A) or (B)? Explain.

cur, patches of the same habitat type must be connected to one another, or the surrounding habitat (the *matrix*) must be of a type through which dispersal is possible (**Figure 23.4**). In Australia, for example, rats regularly leave patches of forest habitat to forage in adjacent macadamia nut plantations (a part of the surrounding matrix). As a result, nut losses along plantation edges adjacent to forests are greater than along edges adjacent to grasslands or agricultural fields (White et al. 1997).

Next, let's focus in more detail on two aspects of landscape heterogeneity: how it is described, and the scale at which it is studied.

Describing landscape heterogeneity The heterogeneity that we see in landscapes can be described in terms of composition and structure. **Landscape composition** refers to the kinds of elements or patches in a landscape, as well as to how much of each kind is present. These elements are defined by the investigator and are influenced by the source of the data used. In an example from Yellowstone National Park, researchers designated five different age classes of lodgepole pine forest using ground-based fieldwork, aerial photographs, and GIS (Tinker et al. 2003). The composition of the landscape in **Figure 23.5** can thus be quantified by counting the kinds of elements in the mapped area (five in this case), by calculating the proportion of the mapped area covered by each element, or by measuring the diversity and dominance of the different landscape elements much as one does for species, using a measure such as the Shannon index (described under Concept 15.2).

If we note that one portion of a landscape is more fragmented than another, we are comparing **landscape structure**: the physical configuration of the different elements that compose the landscape. In Figure 23.5, we can see that some parts of the landscape contain large contiguous blocks of older forest, while other parts are more fragmented and contain smaller patches of forest with a variety of different ages. How do landscape ecologists quantify these differences? Hundreds of different quantitative measures have been developed for analyzing and interpreting landscape patterns, many involving complex analysis. Overall, they address whether the landscape is characterized by large or small patches, how aggregated or dispersed the patches are, whether the patches are simple or complicated in their shape, and how fragmented the landscape is (Turner et al. 2001). Tinker and colleagues were able to use the measures of landscape structure that they derived for Yellowstone to compare the natural, fire-caused fragmentation within the park with fragmentation caused by clear-cutting in adjacent national forests (typically, trees cannot be harvested from national parks but can be harvested from national forests). As in this example, quantitative analyses of landscape struc-

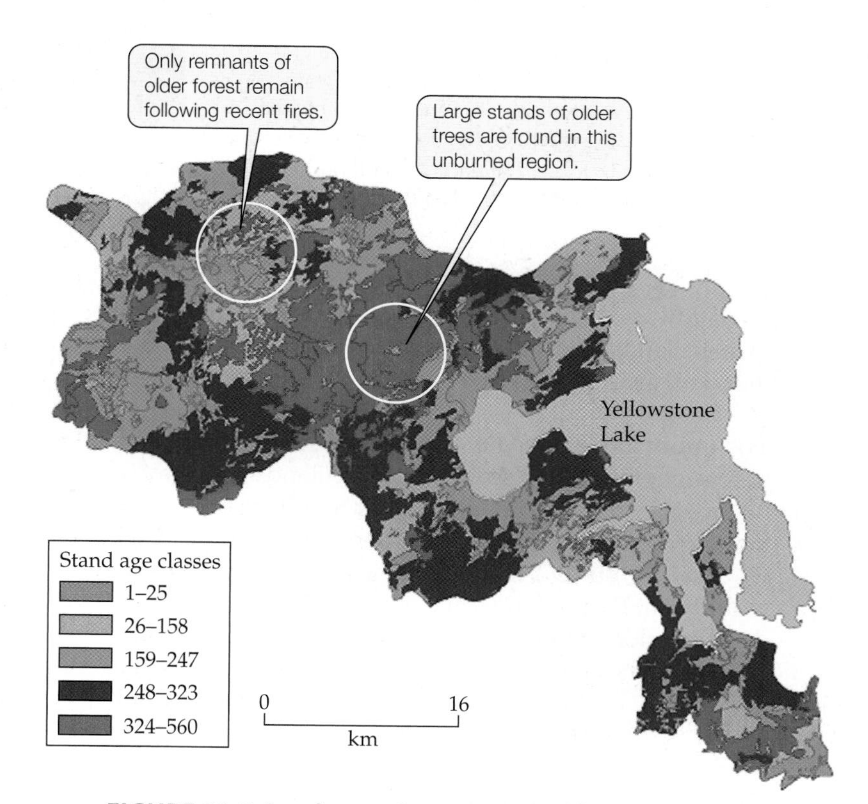

FIGURE 23.5 Landscape Composition and Structure In this 1985 map of lodgepole pine (*Pinus contorta* var. *latifolia*) forest in Yellowstone National Park, we can see that the landscape is composed of five different age classes of forest. Structural complexity varies across the landscape, as seen in the varying degree of natural fragmentation. (After Tinker et al. 2003.)

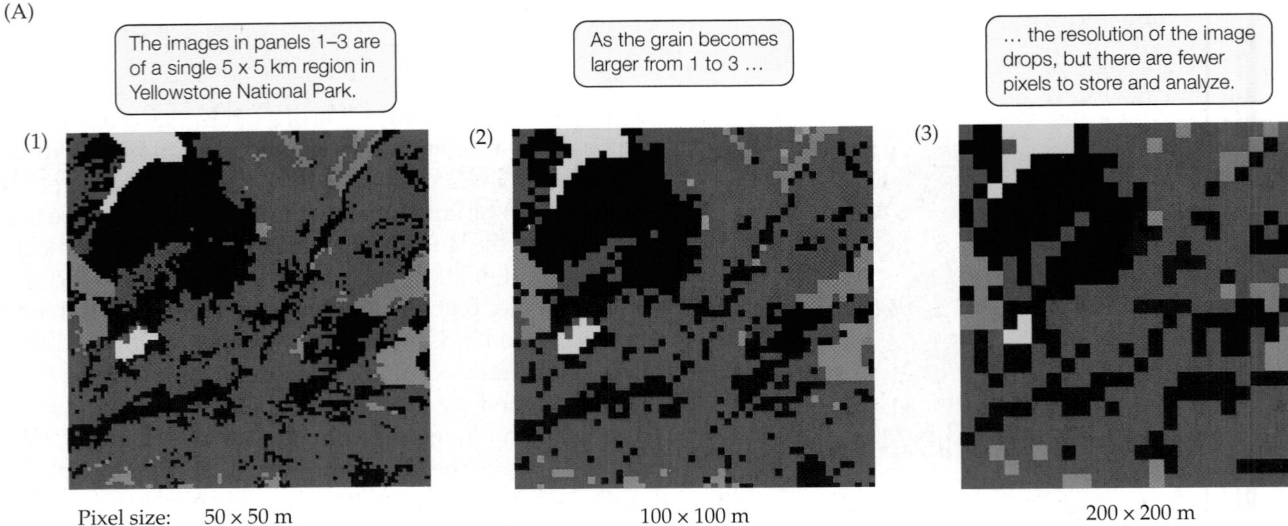

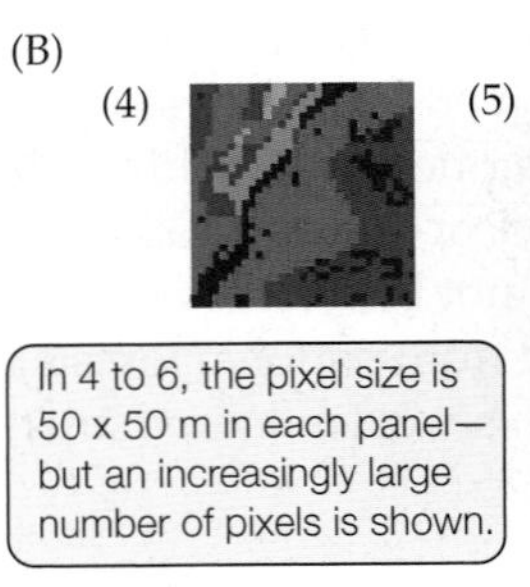

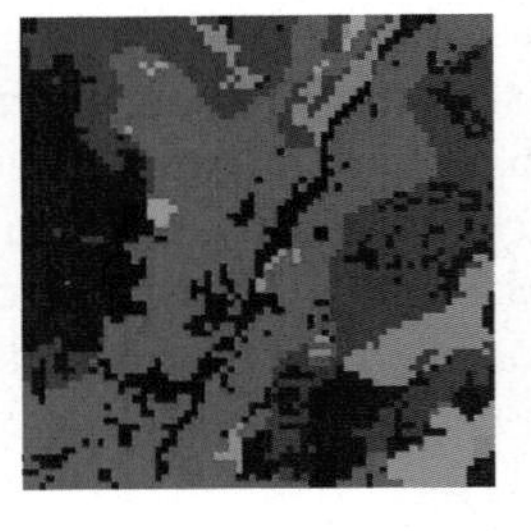

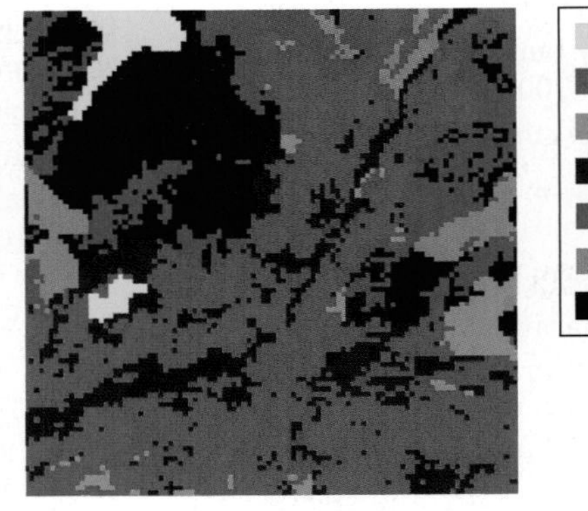

FIGURE 23.6 Effects of Grain and Extent (A) Panels 1–3 show the effect of increasing grain, measured here as pixel size. (B) Panels 4–6 show the effect of increasing extent. (After Turner et al. 2001.)

Q The grain in panel 1 of part (A) is identical to the grain in which of the panels of part (B)?

ture allow us to compare one landscape with another and to relate landscape patterns to ecological processes and to the dynamics of landscape change.

The importance of scale Considerations of scale cannot be ignored in landscape ecology. A landscape may be heterogeneous at a scale important to a tiger beetle, but homogeneous to a warbler or moose. The scale at which we choose to study a landscape determines the results we will obtain. Part of landscape ecology, therefore, is dedicated to understanding the implications of scale.

Scale, the spatial or temporal dimension of an object or process, is characterized by both grain and extent. **Grain** is the size of the smallest homogeneous unit of study (such as a pixel in a digital image) and determines the resolution at which we view the landscape. The selection of grain will affect the quantity of data that must be manipulated in analysis: using a large-grained approach may be appropriate when one is looking at patterns at a regional to continental scale (**Figure 23.6A**). **Extent** refers to the area or time period encompassed by a study. Consider how differently we might describe the composition of a landscape depending on how we define its spatial extent. Panel 4 of **Figure 23.6B**, for example, shows little late successional whitebark pine, while panel 6 contains a considerable area of it (Turner et al. 2001). There may be natural or human-created boundaries that determine the extent of a study, or they may be defined by the researcher.

In examining questions of scale, ecosystem and landscape studies must also determine how processes scale up or down. For example, a researcher studying carbon exchange at the landscape level needs to know how leaf-based measurements of CO_2 exchange scale up to the whole plant, the ecosystem, and ultimately the mosaic of ecosystems that make up the landscape. As in this example, it is often necessary to connect processes across different scales; as a result, ecologists have developed methods to analyze how patterns and phenomena at one scale affect those occurring at either larger or smaller scales (see Levin 1992).

Landscape patterns affect ecological processes

Landscape structure plays an important role in ecological dynamics. For example, it can affect whether and how animals move, and can therefore influence rates of pollination, dispersal, or consumption. In French Guiana,

FIGURE 23.7 The Bog Fritillary Butterfly The travel patterns of these butterflies (*Proclossiana eunomia*) are influenced by features of the surrounding landscape. Butterflies will hesitate to leave the patches they inhabit if there is not another suitable habitat patch nearby, but will traverse a matrix of unsuitable habitat when the next patch is close.

Mickaël Henry and his associates studied the movements of a fruit-eating bat (*Rhinophylla pumilio*) in a tropical forest that had been fragmented by the construction of a reservoir. By using landscape metrics that quantified the degree of patch connectivity at their sampling sites, they found that more isolated forest fragments were less likely to be visited by bats, even if they contained abundant food resources (Henry et al. 2007). Thus, the landscape structure affected bat foraging behavior. Furthermore, because frugivorous bats disperse plant seeds, it is also likely that landscape structure affected the dispersal of the plants that the bats fed on.

Landscape structure also influences biogeochemical cycling. Ecosystem ecologists have identified biogeochemical "hot spots" where chemical reaction rates are higher than in the surrounding landscape. Many such hot spots are found at the interfaces between terrestrial and aquatic ecosystems (McClain et al. 2003), but other factors may also play a part. For example, Kathleen Weathers and her colleagues found that inputs of sulfur, calcium, and nitrogen from atmospheric deposition were higher at forest edges than in forest interiors, primarily as a result of greater interception of airborne particles by the denser and more complex vegetation typically found at a forest edge. The fragmented forests typically surrounding urban areas may therefore be substantially influenced by atmospheric inputs of pollutants and nutrients—a finding that has implications for soil microbial dynamics, herbaceous vegetation growth, and animal communities in the edges of these fragments (Weathers et al. 2001). (We will discuss other such "edge effects" in the following section.)

Habitat patches typically vary in both their quality and their resource availability. This variation can affect the population densities of species inhabiting each patch, the time animals spend foraging in a patch, and the movement of organisms between patches. Patch boundaries, connections between patches, and the matrix between patches can also affect population dynamics, both within and among patches. For example, Schtickzelle and Baguette studied the movement patterns of the bog fritillary butterfly (*Proclossiana eunomia*) across fragmented landscapes in Belgium (**Figure 23.7**). Where patches of suitable habitat were aggregated, female butterflies crossed readily from patch to patch. However, where the habitat was more fragmented and there was a wider distance of matrix to cross, the butterflies were more hesitant to leave a patch (Schtickzelle and Baguette 2003).

The shape and orientation of landscape elements can also be important in physically intercepting organisms and may thus be instrumental in determining the species composition of the community. Gutzwiller and Anderson found that northward-migrating birds nesting in forest patches in the Wyoming grasslands were most likely to nest in patches that were oriented along an east–west axis. These habitat patch effectively served as a net, intercepting birds as they migrated north. The researchers found no such pattern for resident bird species (Gutzwiller and Anderson 1992).

Whereas ecological processes are influenced by landscape patterns, as we have just seen, landscape patterns are in turn influenced by ecological processes. Large grazing mammals, for example, often shape the landscapes they inhabit. The effects of moose (*Alces alces*) on Isle Royale in Lake Superior have been studied through the use of exclosures that have been in place since the 1940s. These studies have shown that high rates of browsing by moose depress net primary production, not just directly through the removal of biomass, but also indirectly by decreasing nitrogen mineralization rates and litter decomposition rates. Moose browsing also shifts the tree species composition toward spruce, and the predominance of spruce, in turn, feeds back to determine rates of biogeochemical processes (Pastor et al. 1988). The moose are thus both responding to and shaping the landscape. At a broader scale, landscape patterns interact with larger-scale disturbances, as we will see next.

Disturbance both creates and responds to landscape heterogeneity

Landscapes are dynamic. In nature, change sometimes comes to ecosystems suddenly in the form of large disturbances—forests and prairies burn over large areas, or floods bring sudden inputs of sediment into river ecosystems. (Although not our focus here, change can also come more slowly, as a result of shifting climates and moving continents.) We saw in Chapter 16 that disturbance is an important factor in determining community composition. Landscape ecologists have asked, in turn, whether particular landscape patterns slow or accelerate the spread of disturbances or increase or decrease an ecosystem's vulnerability to disturbances.

Consider, for example, the 1988 forest fires that burned nearly one-third of the 898,000 hectares (2.2 million acres) of Yellowstone National Park. These fires occurred in a summer of extreme drought and high winds. Similarly extensive fires are thought to have occurred in the northern Rockies at 100–500-year intervals over the past 10,000 years. The 1988 fires burned through forest stands of different ages and species compositions, leaving a complex mosaic of patches that were burned at different intensities (**Figure 23.8**). The type and arrangement of these patches will probably dictate the landscape composition for decades, if not centuries, to come (Turner et al. 2003). Here, a disturbance—fire—was a primary force shaping the landscape pattern of the future. At the same time, it was also responding to the existing landscape structure. This reciprocal interaction between landscape pattern and disturbance is a common one.

FIGURE 23.8 Disturbances Can Shape Landscape Patterns The fires that burned through nearly one-third of Yellowstone National Park in the summer of 1988 resulted in a complex mosaic of burned and unburned patches. Areas that appear black in this aerial view of Madison Canyon were burned by intense crown fires, and brown patches were burned by severe ground fires, both of which killed most or all of the vegetation.

Human actions have greatly altered the nature and extent of landscape-level disturbance. Some places have been more subject to human disturbance than others. People first settled and cleared the areas with the most fertile soils, and they naturally gravitated to good port locations, thus subjecting ecosystems in those places to the earliest human disturbance. Areas close to human settlements were converted to agriculture or subjected to logging and hunting earlier than outlying areas. These disturbance patterns can be detected in ecological communities even centuries after people have left the land and it has reverted to forest (Butzer 1992).

Such *landscape legacies* shape communities in ways that are just starting to be understood. In central France, Etienne Dambrine and his colleagues found that forest plant communities on the sites of Roman farming settlements still bore the mark of those disturbances 1,600 years later. These researchers studied plant diversity in the forest at various distances from sites of recently uncovered Roman ruins. The forest area they studied had not changed substantially since 1665, and it was probably maintained as forest for centuries before that. Dambrine and colleagues found that plant species richness increased in the vicinity of the ruins. An examination of soil properties revealed that this increase was primarily a consequence of higher soil pH, which was thought to result from remnants of the lime mortar used in Roman buildings and from Roman agricultural practices. Soil phosphorus levels were also higher closer to the settlement sites (Dambrine et al. 2007) (**Figure 23.9**). How many other ecosystems on Earth might display the signatures of human activities long since abandoned in their current community structure?

Disturbance, whether natural or human-caused, is an important factor shaping the landscape. Some current human activities are creating disturbances with far-reaching ecological effects, as we'll see in the next section.

CONCEPT 23.2 Habitat loss and fragmentation decreases habitat area, isolates populations, and alters conditions at habitat edges.

Habitat Loss and Fragmentation

In 1986, a massive hydroelectric project in the Caroni River valley of Venezuela inundated a large area of uneven terrain to create a reservoir known as Lago Guri (**Figure 23.10**). The result was the formation of scores of islands of tropical dry forest surrounded by water. This change in the landscape presented an opportunity for John Terborgh and his students and colleagues to study the effects of fragmentation in a tropical dry forest ecosystem. They found that small and medium-sized islands

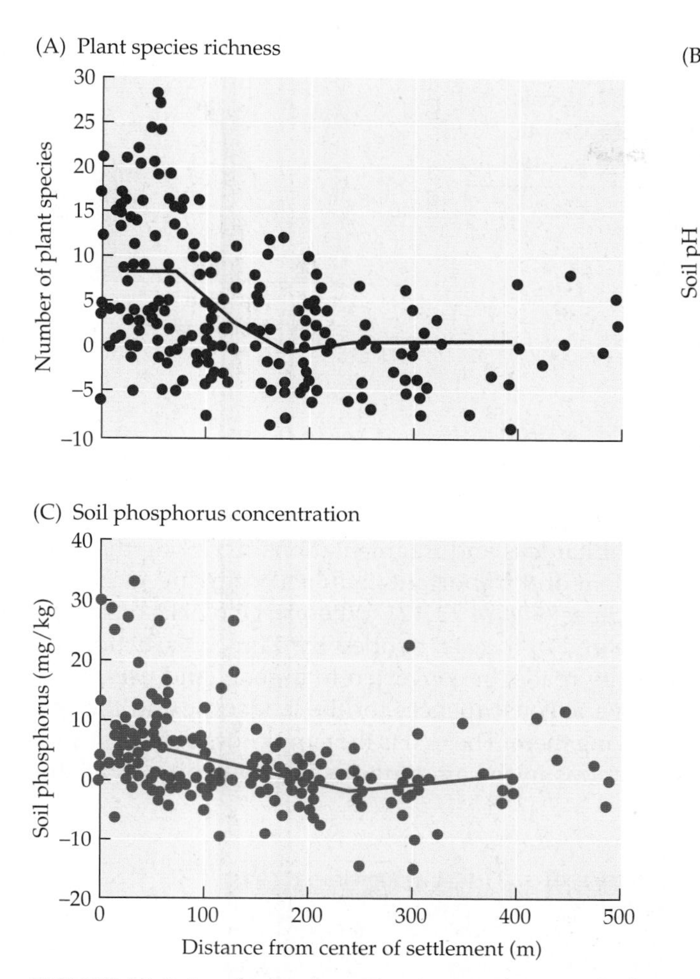

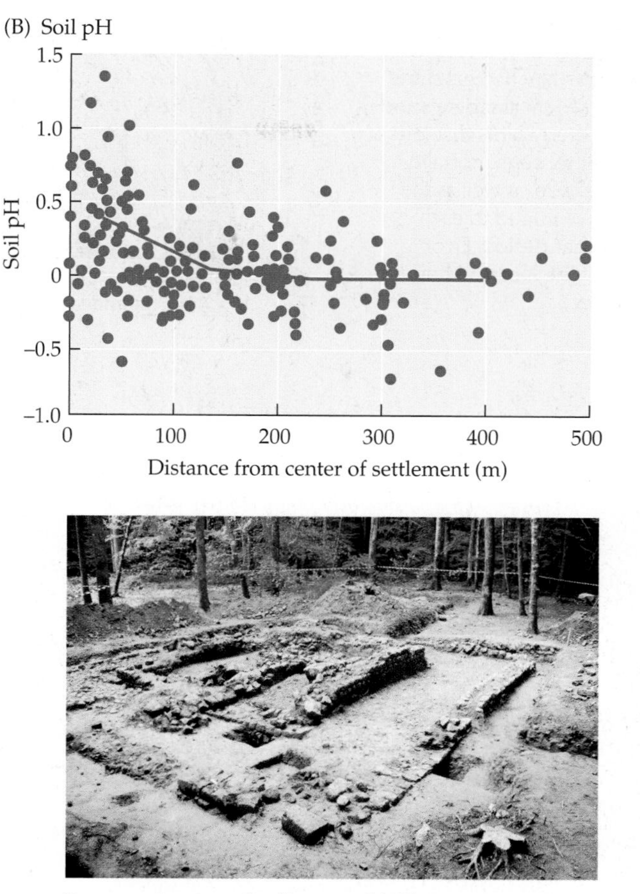

Roman ruins recently discovered in France

FIGURE 23.9 Landscape Legacies In central France, the legacy of Roman farming settlements, abandoned for nearly two millennia, is still reflected in plant species richness in the forest that replaced them. (A) More plant species were found closer to the center of settlement sites, including more species that prefer a higher soil pH. (B) Soil pH and (C) soil phosphorus were also higher closer to the settlement sites. The *y* axis on each graph represents departure from the mean calculated for plots 100–500 m from the settlement. (After Dambrine et al. 2007.)

were lacking the top predators found on the mainland, primarily wild cats (ocelots, jaguars, and pumas), raptors, and large snakes (Terborgh et al. 2006). As a result, generalist herbivores, seed predators, and predators of invertebrates were 10 to 100 times more abundant on the islands than in the remaining intact forest. Species that increased in abundance included leaf-cutter ants, birds, rodents, frogs, spiders, howler monkeys, porcupines, tortoises, and lizards. The increased abundances of these species had a dramatic effect on the vegetation of these islands as tree recruitment decreased and tree mortality increased due to high rates of herbivory, primarily by

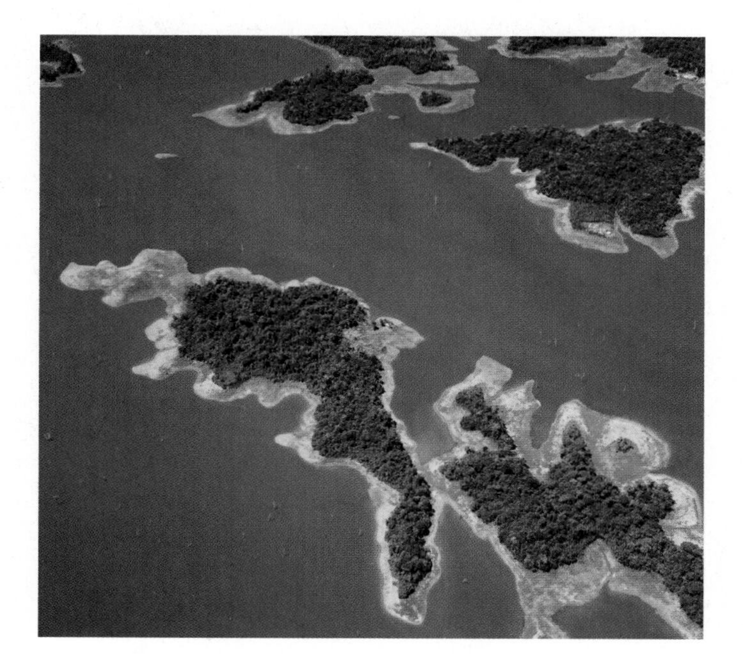

FIGURE 23.10 The Islands of Lago Guri An aerial view of Lago Guri, Venezuela. This lake was formed when 4,300 km^2 (1.1 million acres) of forested land were inundated by a hydroelectric dam, leaving isolated islands of tropical forest.

FIGURE 23.11 Effects of Habitat Fragmentation by Lago Guri The high abundances of herbivores on small and medium-sized islands in Lago Guri caused a dramatic decline in sapling establishment and survival. The bars show the percentages of (A) small saplings and (B) large saplings in study plots that left their size class either through mortality or growth to a larger size, as well as the number of saplings recruited to each size class, over a 5-year period. Error bars show one SE of the mean. (After Terborgh et al. 2006.)

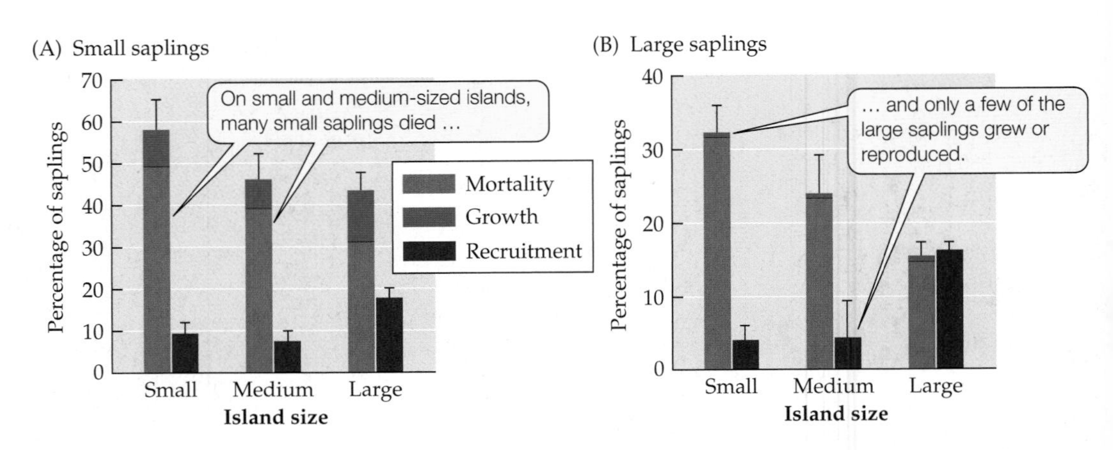

leaf-cutter ants (**Figure 23.11**). What lessons can we take from this "experiment" that apply to other fragmented ecosystems?

Habitat loss and fragmentation are among the most prevalent and important changes occurring in Earth's landscapes (**Figure 23.12**). When large blocks of habitat are cleared of forests, flooded by dam construction, divided by roads, or converted to human land uses, there are several consequences for the landscape and the species living there. The first is the simple loss of habitat area. Reductions in the amount of suitable habitat available

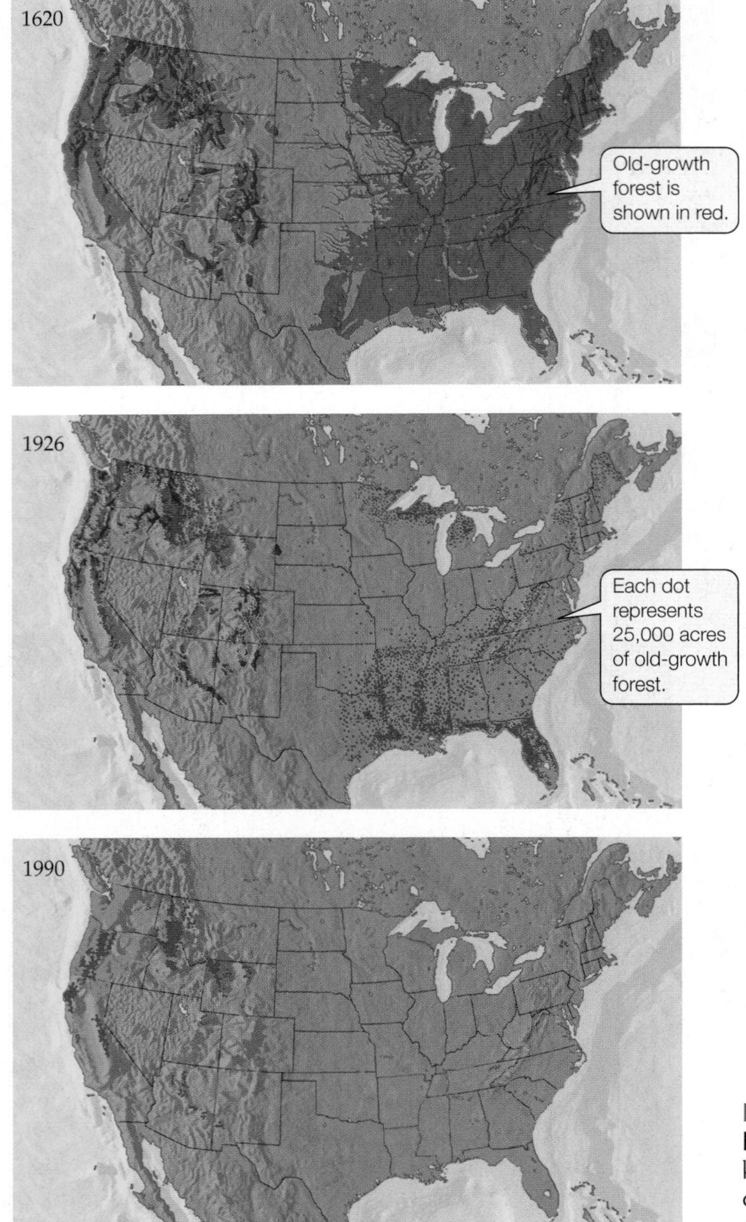

Tulip tree in old-growth forest, Great Smokey Mountains National Park

FIGURE 23.12 Loss and Fragmentation of U.S. Old-Growth Forests Beginning in 1620, vast regions of old-growth forest (also known as ancient or virgin forest) in the United States were cut down to provide lumber and to make room for agriculture, housing, and other forms of development.

FIGURE 23.13 The Process of Habitat Loss and Fragmentation Historically intact habitats are gradually reduced with increased human presence. These contemporaneous photographs illustrate a process that typically takes decades to complete. (A) An intact eucalyptus forest in Western Australia. (B) Areas within the forest have been cleared for grazing. (C) The forest has become further fragmented over time. (D) Only a few remnants of forest remain.

have contributed to the declines of thousands of species, including the red-cockaded woodpecker (see the Case Study in Chapter 22, p. 476). Second, as the remaining habitat becomes divided into smaller and smaller patches, it is increasingly degraded and influenced by edge effects, as we saw in the Case Study of the Biological Dynamics of Forest Fragments Project in Chapter 17 (p. 364). Third, fragmentation results in the spatial isolation of populations, making them vulnerable to the problems of small populations described under Chapter Concept 10.3.

The process of habitat loss and fragmentation may take place over many decades. A typical pattern begins with a clearing in a forest, which is then widened bit by bit until only isolated habitat fragments remain (**Figure 23.13**). Roads are often catalysts of habitat conversion (see Figure 3.6), though human access along rivers can also serve to accelerate deforestation. The principal drivers of habitat fragmentation are conversion of land for agriculture and urban expansion.

Habitat fragmentation is a reversible process. The northeastern United States, for example, has more forest cover than it did a century ago—but it will take centuries before these young forests contain as many species as were found in the old-growth forests that once covered the region. Furthermore, the global trend is toward net loss of forests (FAO, 2005) and toward increasingly fragmented forest, grassland, and riverine ecosystems. What are the ecological and evolutionary consequences of this fragmentation?

Fragmented habitats are biologically impoverished relative to intact habitats

When habitat is fragmented, some species go locally extinct within many of the fragments. There are a host of reasons why this occurs. There may be inadequate food resources, shelter, or nesting sites in the fragments. Animals may need to forage over larger areas than their conspecifics in intact habitat, using multiple fragments. Mutualisms may be disrupted as pollinators go missing or as mycorrhizal fungi fail to persist in a particular fragment. Some fragments may lack the microenvironments needed for seed germination. But local extinction is not inevitable; some species flourish under the changed conditions that follow fragmentation.

Fragmentation often leads to losses of top predators, giving rise to cascading effects, sometimes with dramatic consequences for the remaining community, as we saw at Lago Guri. An example of this phenomenon with implications for human health is the growing risk of Lyme disease as a result of forest fragmentation in the northeastern United States. Brian Allan, Felicia Keesing, and Richard Ostfeld found that in the Hudson River valley of New York, forest fragments of less than 2 ha (5 acres) contained very high populations of white-footed mice (*Peromyscus leucopus*). Fragments of that size did not support substantial predator populations, and the mice had few competitors there. White-footed mice are the most important reservoir of *Borrelia burgdorferi*, the spirochete bacterium that causes

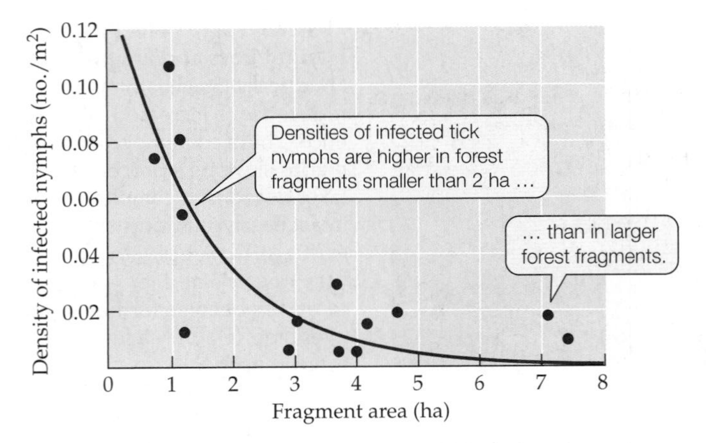

FIGURE 23.14 Habitat Fragmentation Can Have Consequences for Human Health The loss of predators from small forest fragments in New York State has led to elevated populations of white-footed mice in those fragments. As a result, densities of tick nymphs infected with the spirochete bacterium that causes Lyme disease are higher than in larger forest areas. (After Allan et al. 2003.)

Lyme disease. Ticks are the vector of this disease. Tick nymphs collected in these small fragments were significantly more likely to carry the disease, and occurred at higher densities, than in larger fragments (**Figure 23.14**). The outcome—an increased risk of human infection with Lyme disease—is ultimately a result of the biological impoverishment of habitat fragments (Allan et al. 2003).

The matrix between habitat fragments varies in permeability

Models of fragmented landscapes, which were initially derived from the theory of island biogeography, depict habitat fragments as islands isolated in a "sea" of unsuitable matrix, just as the islands of Lago Guri literally are. But do those models truly fit? For some species, such as the eastern wallaroo (*Macropus robustus*) of Australia, it appears that habitat fragments function as islands surrounded by a matrix that individuals will occasionally cross, as described in **Web Extension 23.1**. In other cases, however, fragmented landscapes have proved to be more complex than island models would suggest. The matrix may be *permeable* to some extent, and may form a mosaic of different patch types, of which some are more permeable than others.

In an example from South America, Traci Castellón and Kathryn Sieving studied the dispersal of a small insectivorous understory bird, the chucao tapaculo (*Scelorchilus rubecula*). They moved individual birds to habitat fragments located in different landscape contexts and followed their subsequent movements. They found that birds translocated to fragments surrounded by pasture were much more reluctant to leave the fragments to get to larger forest blocks than were birds that either had a shrubby habitat to cross or that were in fragments connected to larger forest blocks by a forested corridor (Castellón and Sieving 2006). Similar observations were made in a study of rodents in the Atlantic Forest of Brazil, in which some species moved readily through the matrix, while other species were hesitant to cross into unfamiliar patch types (Pardini 2004). As this study showed, the permeability of the matrix is species-dependent.

Edge effects change abiotic conditions and species abundances

As intact habitat is fragmented, an abrupt boundary between two dissimilar patch types is created. The total length of habitat boundary, or *edge*, increases as fragmentation increases. **Edge effects** are the diverse abiotic and biotic changes that are associated with such boundaries. The effect of edge formation is a change in the physical environment over a certain distance into the remaining fragment. As a result, biological interactions and ecological processes can change as well (**Figure 23.15**). The course of these changes

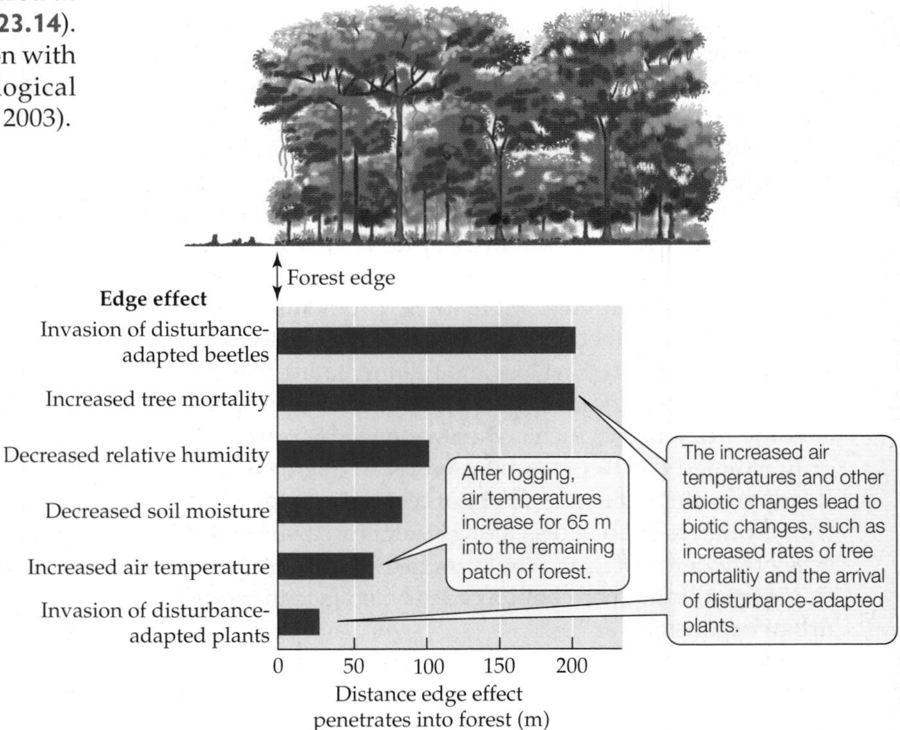

FIGURE 23.15 Edge Effects When an intact forest is first fragmented, abiotic conditions change near the edge of the patch of forest that remains, giving rise to biotic changes. The graph shows changes measured in Amazon rainforest fragments. (After Laurance et al. 2002.)

Q Based on this diagram, how far away from the edge must a tree be located if it is not to experience a drop in relative humidity?

plays out over time, so we can separate the immediate responses to fragmentation and edge formation from the responses that develop later (see Figure 17.24).

The Case Study Revisited in Chapter 17 (p. 384) described edge effects seen in large-scale experiments in Brazil. The effects of abiotic changes at a forest edge are also illustrated by a study of microclimates 10–15 years after the clear-cutting of an old-growth Douglas fir forest in the Pacific Northwest (Chen et al. 1995). Edges were generally characterized by higher temperatures, higher wind speeds, and more light penetration. Daily temperature extremes were also greater at the edges because heat was radiated from the forest edge at night. These abiotic edge effects varied in how far they extended into the forest interior. The biotic consequences of these abiotic changes included higher rates of decomposition, more windthrown trees and thus more woody debris on the forest floor, and differential seedling survival of some tree species (Pacific fir) over others (Douglas fir and western hemlock).

Habitat edges can serve either as barriers to or as facilitators of dispersal. Novel species interactions may take place at the junctions of two ecosystems. Some species may benefit from foraging in one habitat and reproducing in another. Invasive species are commonly more abundant in habitat edges, and population dynamics may be altered for many native species (Fagan et al. 1999).

For example, birds adapted to the forest interior often have lower breeding success when their nests are close to habitat edges; this can result from higher rates of egg predation by raccoons, crows, and other predators as well as higher rates of nest parasitism, especially by cowbirds. In the tallgrass prairie of Wisconsin, Johnson and Temple (1990) studied the reproductive success of five species of ground-nesting birds. They found that the closer nests were to a wooded edge of the prairie habitat, the greater the probability of nest predation by medium-sized predators and of nest parasitism by cowbirds, and the lower the rate of reproductive success. Similar patterns have been observed in other prairies, in Scandinavian forests, in eastern deciduous forests, and in the tropics (Paton 1994). Some biologists have characterized edges as "biological traps" as a result of the increased risks that some species face there (Battin 2004).

Species living near habitat edges also face increased risks from human activities. Species that are vulnerable to hunting, to selective logging, or to harvesting for other purposes have been found to decline following the creation of an edge. Domesticated grazing animals may wander into habitat fragments, causing degradation, and other domesticated species, especially cats and dogs, may prey on populations of wild species. The net outcome of these altered biological dynamics in fragment edges is a shift in species abundances within the fragment, with potential long-term consequences for species persistence.

Fragmentation alters evolutionary processes

In the time since G. Evelyn Hutchinson's 1965 depiction of "the ecological theatre and the evolutionary play," the stage set has been substantially rearranged by human actions. The "evolutionary play" will indeed go on, but in altered ways that we are only now trying to understand. What are the evolutionary consequences when populations of all species are split into smaller and more isolated populations and thrown together in new communities that lack historical precedent?

You have already read in Chapters 10 and 22 about the genetic and demographic problems of small, isolated populations. Marcel Goverde and his colleagues studied the evolutionary consequences of fragmentation by watching bumblebee behavior in the Jura Mountains of Switzerland (Goverde et al. 2002). Their experimental plots included meadow fragments of different sizes (created by mowing) and control plots in unfragmented meadow habitat. The researchers studied the foraging behavior of bumblebees as they visited the flowers of wood betony (*Stachys officinalis*), which were common in both experimental fragments and control plots. The bees visited fragments less frequently than they visited control plots, and once there, they tended to stay longer in the fragments. Ultimately, these two changes in bumblebee behavior resulted in a lower probability of pollination, and an increased likelihood of inbreeding, for the wood betony in the fragments—resulting in an altered evolutionary trajectory for those plants.

In many other cases, habitat fragmentation has been shown to increase rates of inbreeding and genetic drift and alter selection pressures for those species confined to fragments. For example, Keller and Largiadèr (2003) found significant genetic divergence between populations of a flightless ground beetle (*Carabus violaceus*) that had been isolated by roads, particularly by a highway. Where plant populations become small and isolated, their chances of encountering their pollinators, their pathogens, their herbivores, their dispersers, and their competitors all may be reduced, with subsequent evolutionary consequences. Similar effects have been observed in animals, whose breeding systems and survival patterns can be altered in small fragments. For example, in New Hampshire, New England cottontails in small habitat fragments had male-skewed sex ratios and higher mortality rates than rabbits in large habitat blocks, both factors that would influence natural selection (Barbour and Litvaitis 1993).

We have only begun to study the evolutionary implications of habitat fragmentation, and we still have much to learn. As we'll see in the next section, however, such evolutionary information is only one part of what must be considered in designing nature reserves that will work well to maintain biodiversity in landscapes increasingly modified by humans.

CONCEPT 23.3 Biodiversity can best be sustained by large reserves connected across the landscape and buffered from areas of intense human use.

Designing Nature Reserves

You may have a favorite national park, such as Everglades National Park in Florida, Grand Canyon National Park in Arizona, Bialowieski National Park in Poland, or Torres del Paine National Park in Chile. How did these places get to be national parks? What were they before they were parks? Are they the best possible sites for maintaining biodiversity in their regions? Now consider how well the land around you is functioning to sustain native species. Your view is undoubtedly shaped by where you are right now, by what the human history of your area is, and by how effective past conservation work there has been. We turn now to an examination of the ways in which people can work to improve the likelihood of the persistence of species native to their region.

To counteract habitat loss, conservation planners worldwide are working to locate and design protected areas where species can persist. The identification and preservation of core natural areas, buffer zones surrounding them, and habitat corridors connecting them is key to maintaining and allowing the growth of populations. In some cases, as we'll see, degraded ecosystems can be restored as viable habitat for wild species.

Core natural areas should be large and compact

The principles of landscape ecology and conservation biology have come together to guide biologists in selecting the most vital lands for conservation. The design of new nature reserves focuses on **core natural areas**, where the conservation of biodiversity and ecological integrity take precedence over other values or uses, and "where nature can operate in its own way in its own time" (Noss et al. 1999). Populations that are able to maintain themselves in core areas may serve as sources of individuals for populations outside the protected area. Ideally, core areas also provide enough land to meet the large habitat area requirements of top predators.

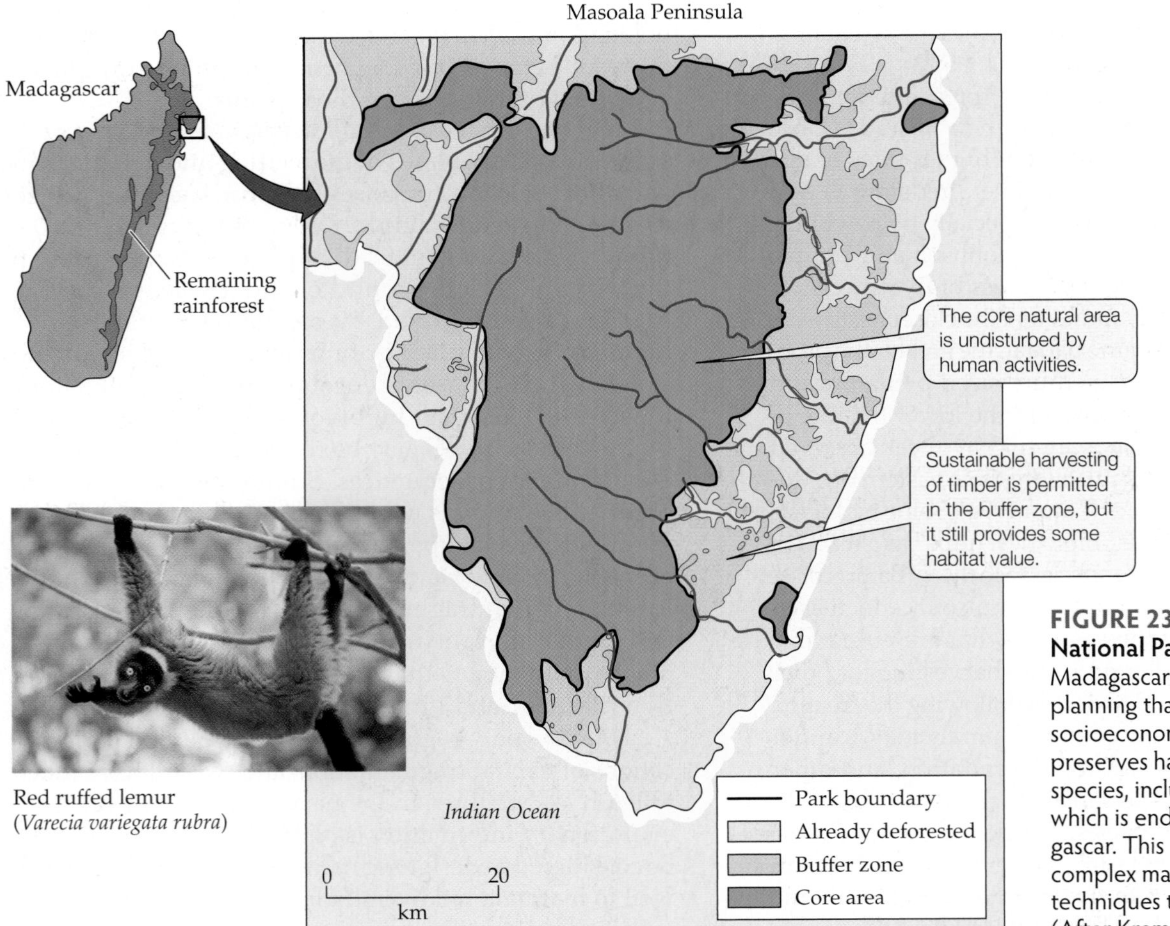

FIGURE 23.16 Designing Masoala National Park Masoala National Park in Madagascar was established after careful planning that took both ecological and socioeconomic concerns into account. It preserves habitat for many threatened species, including the red ruffed lemur, which is endemic to this region of Madagascar. This map was simplified from more complex maps generated by using GIS techniques to analyze satellite imagery. (After Kremen et al. 1999.)

Madagascar is a large island that is a global priority for conservation. It has a rich biota and many endemic species, including more than 70 species of lemurs, a group of primates found only on Madagascar. The biota of Madagascar is seriously imperiled, as only 15% of the island's original forest remains. Efforts are under way to put more of its land into conservation. In designing a new national park in northeastern Madagascar, Claire Kremen and her colleagues examined both the biological and the socioeconomic circumstances of the region. Their design (**Figure 23.16**) was based on a core natural area that extended across several elevational and precipitation zones, encompassing a range of vegetation types. The proposed core area encompassed habitat for all of the region's rare species of butterflies, birds, and primates, and had as yet been little affected by deforestation. The researchers excluded areas close to villages that had already been fragmented and where hunting had negatively affected animal populations (Kremen et al. 1999). The Masoala National Park, which opened in 1997, is now the largest national park in Madagascar at 211,230 ha (over 521,000 acres). With proper management, the park will give the unique biodiversity of this region an improved chance of being maintained in perpetuity.

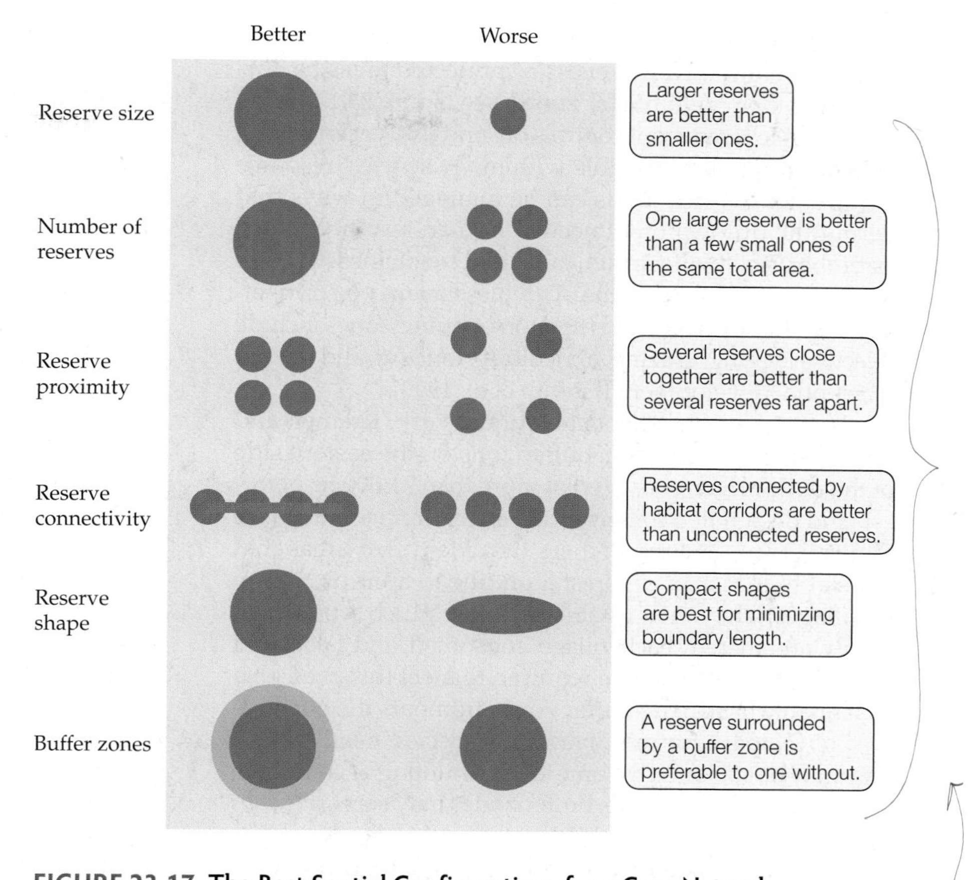

FIGURE 23.17 The Best Spatial Configurations for a Core Natural Area Some spatial configurations are usually better than others for fostering biodiversity. (After Diamond 1975 and Williams et al. 2005.)

Q For panels 1–5 (Reserve size; Number of reserves; Reserve proximity; Reserve connectivity; Reserve shape), explain the underlying reasons why the design on the left is better than the one on the right.

Ideally, core natural areas must be large and uncut by roads, or even by trails. Thus, not all protected areas qualify as core natural areas. Many do not fully serve the purpose of protecting the whole biota from human interference. Most national parks in the United States were not established with the conservation of biodiversity as their primary mission, but rather to preserve scenery, often on land that was not useful for anything else. Conservation planners recognize that many countries do not have the luxury of carving out thousands of hectares of land to be solely dedicated to biodiversity conservation. Therefore, the cores of many reserves do not meet all the criteria for core natural areas.

In the design of nature reserves, some spatial configurations are better than others for fostering the persistence of biodiversity (**Figure 23.17**). Overall, large, compact, and connected reserves are ideal, but there may be times when smaller or disconnected reserves may be more desirable; for example, diseases may spread less easily between isolated smaller reserves than within a large reserve. The primary biological objectives of reserve configuration are the maintenance of the largest possible populations of organisms, the provision of habitat for species throughout their area of distribution, and the provision of adequate area for maintenance of natural disturbance regimes.

Realistically, in many settings where conservation is being accomplished, either the landscape or the social context does not permit adhering to these principles (Williams et al. 2005). There are many smaller reserves that have been established with the conservation of a single species or ecological community as their main objective. Such **biological reserves**, which may be only a few hectares in size, are nevertheless an important part of our conservation efforts. Particularly where human population density is high and large reserves are unfeasible, critically situated smaller reserves may be the best available option.

Core natural areas should be buffered by compatible land uses

Due to many constraints, only relatively small areas of land can be designated as core natural areas. However, if we are to conserve the majority of the world's species, much larger areas of land will have to be able to provide adequate habitat for biodiversity persistence (Soulé and Sanjayan 1998). We can augment the effectiveness of protected areas by surrounding them with **buffer zones** (see Figure 23.17), large areas with less stringent controls on land use, yet which are at least partially compatible with many species' resource requirements. Such lands can be managed in ways that permit the production of needed human resources, such as timber, fiber, wild fruits, nuts, and medicines, but still maintain some habitat value. Activities that may be compatible with the conservation function of buffer zones include selective logging, grazing, agriculture, tourism, and limited residential development (Groom et al. 1999).

In the plan for Masoala National Park, Kremen and her colleagues included a buffer zone on the eastern side of the park, which consisted of more than 71,000 ha of forest land designated for sustainable timber harvesting (see Figure 23.16). The researchers first identified areas that were at high risk of deforestation due to their proximity to villages. They then established how much wood each family, and thereby each village, consumed, and calculated how much area would be required to meet this need on a sustainable basis. The buffer zone augments the effective area of the park for many lowland species, even though they may be subjected to some level of hunting or collection.

On a cautionary note, buffer zones may serve as *population sinks* (areas where death rates are higher than birth rates) for some species, as animals that stray out of core areas and into buffer zones become vulnerable to hunting, roadkill, or other sources of mortality. In Peru, for example, where slash-and-burn agriculture is commonly practiced just outside nature reserves, wild animals such as agoutis, armadillos, or tapirs often damage farmers' crops. As a result, these animals are targeted by hunters, and such hunting has altered the relative abundances of mammals in the forest (Naughton-Treves et al. 2003). In other cases, however, buffer zones do not appear to act as population sinks. A recent analysis of data from 785 animal species found that buffer zones can allow populations to persist in habitat fragments that might otherwise be too small or too isolated to support viable populations (Prugh et al. 2008). The key to success boils down to simple demography: if a buffer zone provides a threatened species with habitat in which birth rates are higher than death rates, it can aid conservation goals.

If we can succeed in establishing core areas for protection surrounded by sparsely inhabited buffer zones, have we done all that is necessary for conservation? Recall that landscape connectivity is another important consideration in reserve design.

Corridors can help maintain biodiversity in a fragmented landscape

Habitat corridors—linear patches that connect blocks of habitat—have become a staple of urban, suburban, and rural planning (**Figure 23.18**). It was clear to many ecologists early on that if habitat fragmentation is the problem, then connectivity might be the solution: if populations are at risk of isolation, then we should ensure that there are corridors of habitat that link them together. This solution made intuitive sense.

When designing Masoala National Park, Kremen and her colleagues looked at the larger landscape and anticipated connections that would be important in the future. Many of Masoala's target species are also found in areas northwest of the park, in lands that lie between Masoala and two important protected areas to the north. The park plan included three narrow corridors to provide connections to those protected areas. The researchers developed this part of the plan by examining maps, but out of expediency, they did not actually do studies of animal movements (Kremen et al. 1999).

The intended function of habitat corridors is to prevent the isolation of populations in fragments. But do we know that corridors actually help to overcome this isolation? And do corridors work at all scales, for beetles as well as for wolves? Is a stream corridor in the suburbs providing necessary landscape connectivity for some species? At the continental scale, could we link the Greater Yellowstone

FIGURE 23.18 A Habitat Corridor Grizzlies and other wildlife can cross this highway overpass in Banff National Park, Canada.

Ecosystem to the Yukon through habitat corridors, as some have proposed? Experimental and observational studies of corridors' utility have shown mixed results.

Nick Haddad and his colleagues established a test of the utility of corridors at the Savannah River Ecological Laboratory in South Carolina. They set up patches of early successional habitat in a matrix of pine forest, some of them connected by corridors, and observed the movements of organisms between patches (**Figure 23.19**). Their results showed that the corridors did indeed serve to facilitate the movement of butterflies, pollen, and bird-dispersed fruits (Tewksbury et al. 2002).

Other studies, however, have found no benefits of corridors, and still others have found negative effects. For example, in the same experimental system at the Savannah River Ecological Laboratory, predation on indigo bunting (*Passerina cyanea*) nests was higher in patches connected by corridors (Weldon 2006). There are also concerns that corridors could facilitate the movement of pathogens (Hess 1994) or invasive species (Simberloff and Cox 1987).

Ecological restoration can increase biodiversity in degraded landscapes

What if habitat corridors are lacking and organisms' ability to move is impaired by a hostile matrix of degraded habitat? This was the case in Guanacaste Province on the Pacific coast of Costa Rica, where Santa Rosa National Park, in a lowland area of tropical dry forest, was largely separated by 35 km of cattle pasture and forest fragments from the upland forest habitat of the nearby mountains.

Tropical ecologist Dan Janzen knew that many insects, birds, and mammals needed to migrate between these lowland and upland forests. He also saw that the tropical dry forest that he had spent his career studying was fast disappearing. Janzen's effort to reverse this trend became one of the largest and most ambitious ecological restoration projects ever undertaken in the Neotropics. Now covering some 120,000 ha of land and 70,000 acres of marine reserve, the Area de Conservación Guanacaste (ACG) includes three national parks, a protected corridor linking them, and the surrounding agricultural areas. The ACG is home to some 230,000 species, or 65% of the species in Costa Rica (Daily and Ellison 2002).

Within the ACG, cattle ranches have occupied much of the land between the three parks for decades. Janzen has launched an effort to restore 75,000 ha of these pasturelands to the original forest types. His strategies include planting trees, suppressing fires, and limiting hunting (to maintain mammalian and avian seed dispersers). Fire suppression is necessary to halt fires that burn readily in pastures covered in jaragua grass (*Hyparremia rufa*), an invasive introduced from Africa. Grazing will be maintained for some time in some areas to suppress the jaragua grass; cows and horses have also been found to help in tree seed dispersal.

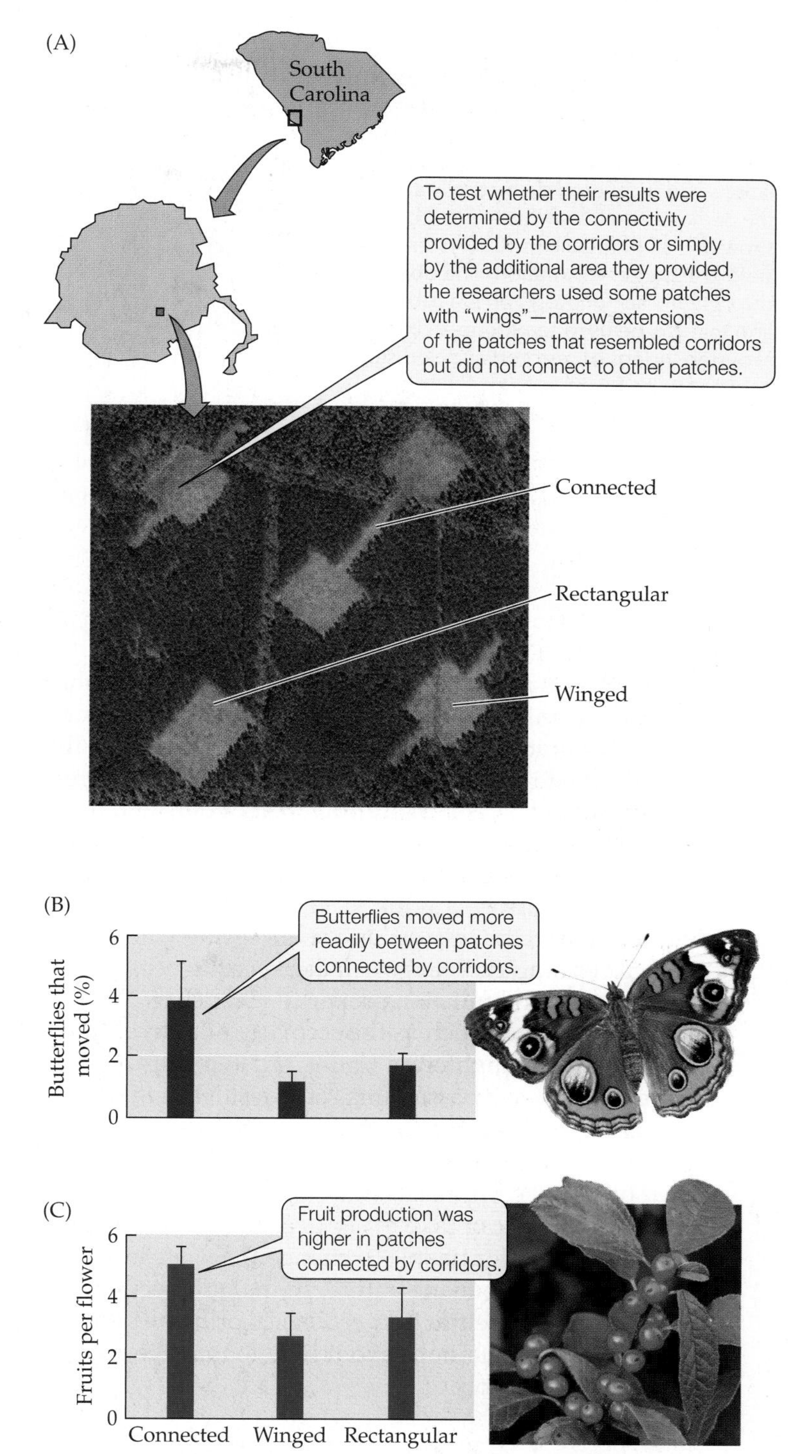

FIGURE 23.19 How Effective Are Habitat Corridors? (A) Nick Haddad and his colleagues tested the effectiveness of habitat corridors by creating experimental patches of early successional habitat within a pine forest and creating corridors between some of the patches. They then observed (B) movements of the common buckeye butterfly (*Junonia coenia*) between patches and (C) fruit production (which provides evidence of pollination) in winterberry (*Ilex verticillata*). Error bars in (B) and (C) show one SE of the mean. (After Tewksbury et al. 2002.)

FIGURE 23.20 Dramatic Effects of an Ecological Restoration Project Native oyster populations have collapsed worldwide as a result of habitat loss and overharvesting. (A) In an ecological restoration experiment that began in 2004, oyster reefs were constructed in nine protected areas along the Great Wicomico River in Virginia, USA. Three years later, native oyster populations had recovered dramatically across the 35 ha restoration project. Error bars show one SE of the mean. (B) Oyster habitat before and after restoration. The object on the right in each photograph is a robotic arm that can be used to pick up an individual oyster. Videos of restored and unrestored habitat can be found in **Web Extension 23.2**. (After Schulte et al. 2009.)

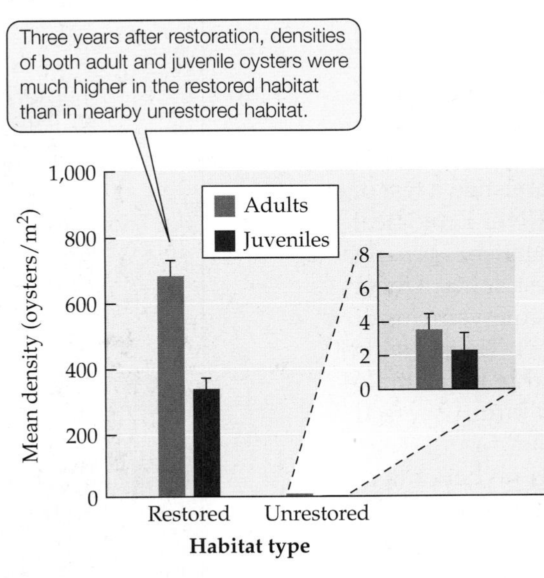

Before restoration

After restoration

Ecological restoration is being applied in many other ecosystems with varying degrees of success. To be successful, restoration ecologists must correctly diagnose the ecological state of an area, decide what the goals of the restoration should be, and then apply their understanding of ecological processes to recreate the desired type of ecosystem. Anthony Bradshaw, a founder of restoration ecology, referred to this process as the "acid test" of ecology: "Each time we undertake restoration we are seeing whether, in the light of our knowledge, we can recreate ecosystems that function, and function properly" (Bradshaw 1987).

In some cases, such as the recovery of native oyster populations highlighted in **Figure 23.20**, results quickly suggest that we've passed this "acid test." But in others, such as Janzen's efforts to restore tropical dry forests in Guanacaste, the process is likely to be a long and slow one. That is not surprising: large-scale changes to ecological communities can take many decades, and it can also take a long time for people to change the ways in which we relate to and manage nature. In the next section, we will look more closely at how ecological principles are applied in making decisions about how to manage natural resources sustainably.

CONCEPT 23.4 Ecosystem management is a collaborative process with the maintenance of long-term ecological integrity as its core value.

Ecosystem Management

The year was 1989. The setting was a packed public hearing in Olympia, Washington. A convoy of logging trucks and about 300 loggers had made the journey to the state capital to defend their jobs. There was talk that the northern spotted owl (*Strix occidentalis caurina*; see Figure 10.18) could be listed as a threatened species under the U.S. Endangered Species Act, which would place its old-growth forest habitat off-limits to logging. Tempers were flaring among loggers and others supported by the timber industry. "When it comes to choosing between owls and the welfare of families, the hell with the owl as far as I'm concerned," said a state politician. At times, some of the testimony was drowned out by the honking of truck horns. The contentious debate about the logging of forests in the Pacific Northwest was reduced to "owls versus jobs," and resulted in bumper sticker and T-shirt slogans, vandalism by both sides, and the exchange of many angry words.

Some people recognized that there might be a better way to make decisions about the use of public lands. The conflict in the Pacific Northwest was in part the outcome of a long history of top-down management of natural resources with a focus on resource production and extraction. In 1995, a federal Interagency Ecosystem Management Task Force was formed to develop alternatives to this approach (DellaSala and Williams 2006).

Approaches to managing natural resources have become more collaborative over time

Through most of the twentieth century, management of natural resources on U.S. public lands was focused on maintaining individual resources of economic value, whether timber, deer, ducks, or scenery for visitors. This focus remained at the core of many land management policies until Congress passed the Multiple Use Sustained Yield Act of 1960. By the late 1980s, natural resource agencies had gradually expanded their missions to include "multiple use," in recognition that different people had different interests and that it was possible to manage public lands to meet diverse, and at times competing, demands. This was frequently done through spatial compartmentalization of uses, as when different blocks of land were designated as timber extraction zones, recreation zones, or wilderness areas.

Since the 1980s, with our greater awareness of the necessity of preserving biodiversity, our goals for land management have shifted again. Ecosystem management has emerged as a way to expand the scope of management to include protection of all native species and ecosystems while focusing on the sustainability of the whole system, not just the sustainability of resources of interest.

What is ecosystem management? Most simply stated, it is "managing ecosystems so as to assure their sustainability" (Franklin 1996). A committee of the Ecological Society of America arrived at a less simple but more comprehensive consensus definition in 1996: "**Ecosystem management** is management driven by explicit goals, executed by policies, protocols, and practices, and made adaptable by monitoring and research based on our best understanding of the ecological interactions and processes necessary to sustain ecosystem structure and function" (Christensen et al. 1996). This definition emphasizes sustainability, but also recognizes the need for setting goals and using science to evaluate and adjust management practices over time.

The conflict in the late 1980s over old-growth forests in the Pacific Northwest was a stimulus to ecologists, forest service staff, industry, and citizens to seek a less confrontational way to make decisions. Since that time, more collaborative decision making has been combined with better incorporation of science to arrive at management plans that not only sustain both biodiversity and people's livelihoods, but are also responsive to changing conditions. In ecosystem management, the focus is on a particular biophysical ecosystem, or *ecoregion*, delineated by natural boundaries rather than political boundaries: a watershed, a mountain range, a stretch of coastline. The full range of *stakeholders*—people with some interest in the project—becomes involved in decision making for the ecoregion, joined together by the common goal of maintaining its ecological integrity and economic viability.

Ecosystem management sets sustainable goals, implements policies, monitors effectiveness, and adjusts as necessary

Ecosystem management is a process, one that may be implemented in different ways for different projects. Most ecosystem management projects begin with the gathering of scientific data to define the nature of the problems in the ecosystem. That information is then used to set sustainable goals. To meet those goals, a set of actions is needed, many of which may require adapting new policies. Once a new policy is implemented, the ecosystem is monitored to gauge whether that action brings about the desired result.

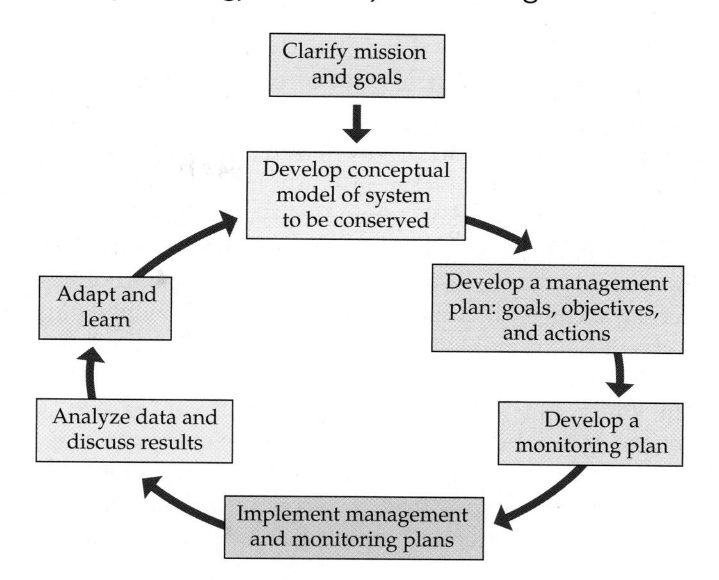

FIGURE 23.21 Adaptive Management Is a Vital Component of Ecosystem Management Adaptive management is a systematic way of learning from past management actions and adjusting future decisions accordingly. (After Margoluis and Salafsky 1998.)

Adjustments to the policies are then made as needed. In this iterative process, known as **adaptive management** (**Figure 23.21**), management actions are seen as experiments and future management decisions are determined by the outcome of present decisions.

Monitoring is a vital component of adaptive management. For example, Mark Boyce developed a model predicting elk and wolf population dynamics in Yellowstone following the reintroduction of wolves described in the Case Study on p. 501. He and his colleague Nathan Varley have taken an adaptive management approach by adjusting this model based on demographic data for the first 10 years of wolf presence. Since their original model estimated elk numbers well, but underestimated wolf numbers, they knew that some of the model's assumptions needed adjustment (Varley and Boyce 2006). This approach will be important for determining acceptable hunting levels for elk and for future adjustments in response to changing circumstances.

Although it is extremely useful, ecosystem management has limitations and drawbacks. One drawback is that it may take a long time to reach a consensus decision—yet averting an environmental crisis may require that preventative actions be taken immediately. There is also potential for continued conflict generated by those who simply want to disrupt the process, even when extensive efforts at stakeholder involvement have been made. In some instances, a lack of unbiased information, a struggle for power among different government agencies, the presence of corruption, or the unmet needs of the people in local communities can produce situations that may not lend themselves to participatory governance.

Humans are an integral part of ecosystems

Human actions affect natural ecosystems, and human economies are affected by supplies of natural resources. Ecosystem managers must not only manage natural resources and biodiversity across large landscapes, but also devise plans that protect both natural ecosystems and human economies. Ecosystem management incorporates human social and economic factors as a fundamental part of the decision-making process, along with legal requirements and, of course, ecological integrity (**Figure 23.22**). The integration of these different components is seen as necessary to achieve a successful management outcome.

As we have seen, people need natural ecosystems for many reasons, ranging from the economic to the spiritual. Ecosystem management incorporates education of the public about their reliance on ecosystem services as part of its mission. It also engages the public in helping to solve those problems that degrade the ecosystem services that they rely on.

Any conservation plan that excludes the human component will not be accepted, ultimately, by the stakeholders. The plan for Masoala National Park took the needs of the people living around the park into consideration. Conservation planners not only calculated their wood needs and provided for them in a buffer zone designated for managed forestry, but also surveyed the region for tree species that would have value in an export market and included them in an economic plan for future use. The idea was to remove economic pressure from park resources by identifying ways that people could support themselves and increase their incomes using resources outside the park. In addition, Kremen's team worked in conjunction with local people and with the Malagasy government to develop the plan, recognizing the importance of local acceptance of any proposal they made. In the end, the park plan provided for the economic needs of the people, by identifying forest resources that could be used to enrich the region, as well as for the habitat requirements of all the taxa included in the planners' analysis. Whether the park can achieve these goals will be ascertained only over time (Kremen et al. 1999).

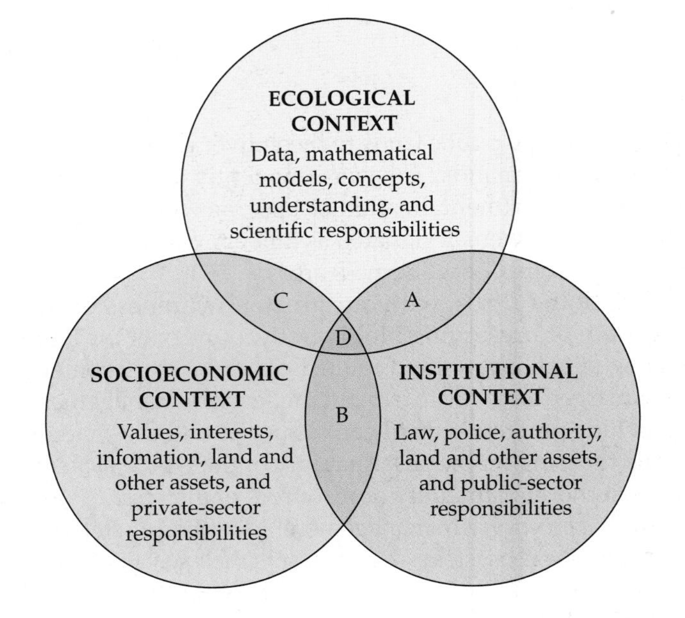

FIGURE 23.22 Humans Are an Integral Part of Ecosystem Management Ecosystem management integrates interests derived from ecological, institutional, and socioeconomic contexts. The letters represent the overlap of the three contexts: A, zone of regulatory or management authority; B, zone of social obligations; C, zone of informal decisions (as opposed to legal requirements); D, zone of win–win–win partnerships.

A CASE STUDY REVISITED

Wolves in the Yellowstone Landscape

The reintroduction of wolves into the Greater Yellowstone Ecosystem (GYE; see the Case Study on p. 501) in 1995 reflected the shift to an ecosystem management approach to decision making. It was a bold step that followed years of study and preparation. That it happened at all reflects a quantum shift in human attitudes toward nature over the last century. In the late 1800s and early 1900s, wolves were feared and reviled. They were perceived as a threat to people and livestock—an accurate perception as far as livestock were concerned. Wolves were hunted to extinction in the area of Yellowstone National Park by the late 1930s and throughout virtually all of the conterminous United States not long thereafter.

The removal of a top predator can alter the landscape substantially, in part because herbivores whose populations were once controlled by the predator may increase in number and negatively affect vegetation dynamics. In Yellowstone, the growth and reproduction of riparian tree species, such as cottonwoods, aspens, and willows, declined after wolves were removed (Ripple and Beschta 2007). A possible reason was that the trees experienced heavy browsing by herbivores such as elk, which roamed freely along rivers and streams once the wolves were gone. How strong is the support for this explanation?

Many observations are consistent with this idea. The reintroduction of wolves began in the winter of 1995–1996, when 31 wolves captured in Canada were released into the park. Their numbers increased rapidly; by 2004, there were about 250 wolves in the park. Following the reintroduction, populations of elk, the wolves' principal prey, have declined by 50%. Elk were initially naive and very vulnerable to wolf predation, but they have since modified their behavior, showing a preference for foraging in places that provide high visibility. Furthermore, cottonwoods, aspens, and willows have begun to recover in some areas. In some cases, the early signs of recovery appeared to be concentrated in areas where elk face a high risk of predation, such as locations where visibility is poor, escape routes are lacking, or ambush sites are common. Thus,

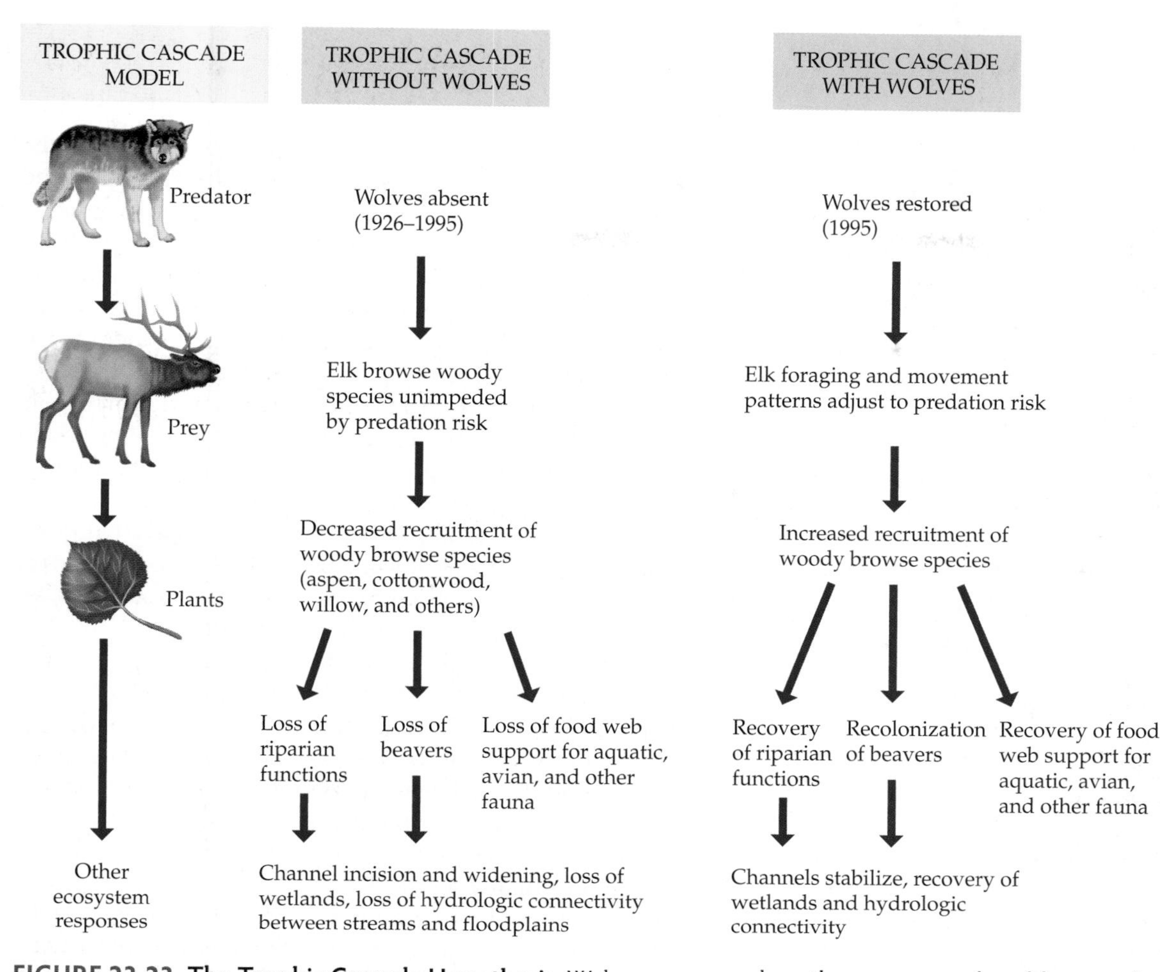

FIGURE 23.23 The Trophic Cascade Hypothesis Wolves are top predators, and their reintroduction to the Greater Yellowstone Ecosystem (GYE) has the potential to cause cascading trophic effects. According to the hypothesis shown here, elk now avoid those sites where they are most vulnerable to predation, and trees and shrubs are now returning to those sites after decades of suppression by elk. Researchers are actively testing this and other hypotheses about effects of wolves in the GYE. (After Ripple and Beschta 2004.)

elk may be avoiding areas where they are most vulnerable to attack by wolves, allowing trees in those areas to recover—and possibly leading to a series of other, cascading effects (**Figure 23.23**).

However, some studies have questioned whether a trophic cascade like that diagrammed in Figure 23.23 is occurring. In an experimental test of the hypothesis that elk forage less in areas with wolves, leading to the recovery of woody species in those areas, Kauffman et al. (2010) found that aspen survival was not affected by the presence of wolves. Similarly, Creel and Christianson (2009) found that willow consumption by elk was more strongly affected by snow conditions than by the presence of wolves; in fact, contrary to expectation, willow consumption actually increased when wolves were present! While the reintroduction of wolves may have affected willow and aspen, it may be because predation by wolves has decreased the size of the elk population—not because fear of predation has led to changes in elk foraging behavior. Whatever the outcome of this debate, the reintroduction of wolves provides a wonderful opportunity to test hypotheses about how ecosystems work in a large landscape.

CONNECTIONS IN NATURE

Future Changes in the Yellowstone Landscape

If riparian trees continue to recover in the GYE, a series of cascading effects (like those described under Concepts 15.3 and 20.3) may ensue. In some locations, increased numbers of willows have slowed stream flow and increased sedimentation rates (Beschta and Ripple 2006). The increased growth of riparian tree species is also expected to provide shade and habitat for trout, which prefer shade-cooled waters, and for migratory birds. More riparian bird species have been observed in a similarly recuperating ecosystem in Alberta (Hebblewhite et al. 2005). As populations of willow—a preferred food for beavers—have increased, new beaver colonies have appeared. In turn, the dams built by the beavers have changed patterns of water flow, creating marshlands that may favor the return of otters, ducks, muskrats, and mink.

But other, even more fundamental changes may be taking place in the Yellowstone ecosystem. Recall from Chapters 2–4 that climate is the single most important determinant of where species live. With rising concentrations

(A) Western red cedar (*Thuja plicata*)

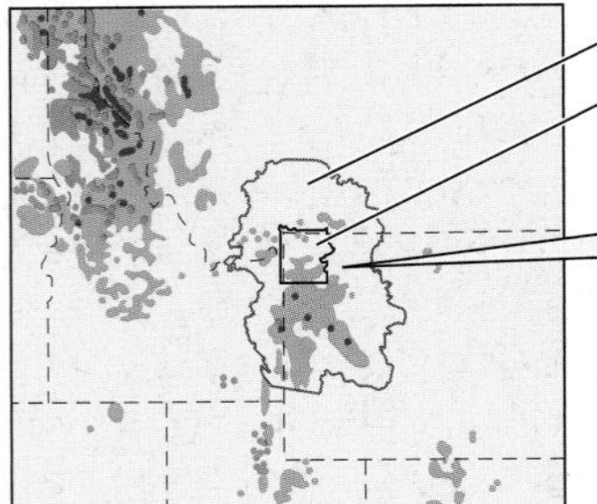

(B) Ponderosa pine (*Pinus ponderosa*)

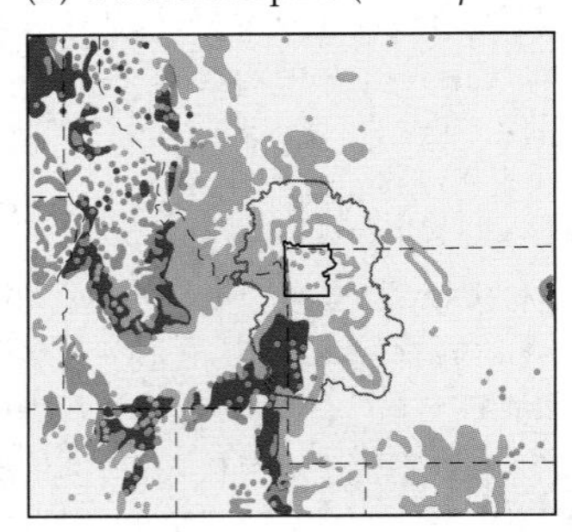

(C) Whitebark pine (*Pinus albicaulis*)

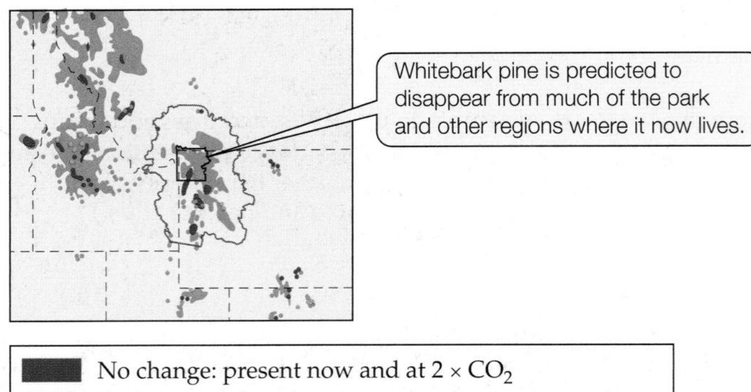

FIGURE 23.24 Projected Effects of Climate Change in the Northern Rockies Shifts in the distributions of some principal tree species in the northern Rocky Mountains are projected by a model of a future climate driven by twice the current atmospheric CO_2 concentrations. Note the increased distribution of some species currently uncommon in the region, including (A) western red cedar, and (B) ponderosa pine, and the near disappearance of (C) whitebark pine. (After Bartlein et al. 1997.)

of greenhouse gases in the atmosphere, climatic warming is anticipated in the coming century (see Chapter 24). Will Yellowstone be able to maintain its current biological diversity in the face of global climate change?

A modeling study showed what the vegetation of the region surrounding Yellowstone National Park might look like under a doubling of current atmospheric CO_2 concentrations, which could happen within a century (**Figure 23.24**). Generally, the projections are for higher temperatures, no increase in precipitation, and more frequent fires. Based on these projected changes to the physical environment, the model predicts upslope and northward migrations of many species. These migrations will cause shifts in forest communities, with some species declining within the park and others increasing their range to include the park. Species currently rare in or absent from the GYE that may increase substantially there include gambel oak, western red cedar, and ponderosa pine. A near-elimination of whitebark pine is predicted as suitable habitat for that species shifts to the north (Bartlein et al. 1997).

The loss of whitebark pine would likely trigger a number of other ecological shifts. This tree is a keystone species that produces large, fatty, and nutritious nuts, an important food source for Clark's nutcracker as well as for black and grizzly bears. Clark's nutcracker, in turn, is the primary disperser of the whitebark pine's seeds (Tomback 1982). One consequence of warmer winters during the past few decades has been an expansion of the range of the mountain pine beetle (*Dendroctonus ponderosae*) to high-elevation pine forests, including those with whitebark pine (Logan and Powell 2001). This beetle has devastating effects on whitebark pine (**Figure 23.25**). Whitebark pine is also being attacked throughout much of its North American range by the fungus *Cronartium ribicola*, an introduced pathogen that causes white pine blister rust (Tomback and Achuff 2010). The combined effects of mountain pine

FIGURE 23.25 Warm Winters Have Promoted a Devastating Insect Outbreak Once excluded from whitebark pine forests by cold winter temperatures, the mountain pine beetle has expanded its range as temperatures have warmed in recent decades. These beetles have contributed to the death of millions of whitebark pines, which turn red when they die (as in this Wyoming forest). In July 2010, the U.S. Fish and Wildlife Service announced that it is considering whether to list whitebark pine as endangered or threatened under the Endangered Species Act.

beetle and blister rust have caused an extensive die-off of whitebark pine, and preliminary data indicate that this has reduced the occurrence of Clark's nutcracker in some areas (McKinney et al. 2009). Loss of whitebark pine also means loss of a food source for grizzly bears. Thus, it appears that climate change and introduced disease are having a major influence on whitebark pine populations, and that these effects have the potential to be transferred to wildlife such as grizzly bears. (See **Climate Change Connection 23.1** for more information on how climate change is affecting biodiversity in forests and other ecosystems.)

As we've seen in this chapter, landscape ecology and the use of tools such as remote sensing and GIS can elucidate current patterns of biodiversity and help us to predict future ones. Over the past 30 years, we have put much effort into selecting, establishing, and undertaking management of new protected areas, but now we need to ask how well those areas will harbor their species in a warmer world. If losses are projected under climate change, are there steps we can take now that can improve habitat connectivity, create or improve buffer zones around core natural areas, or restore degraded areas to greater ecological integrity? Or will we need to move species to new areas of suitable habitat, especially if they cannot migrate quickly enough to keep up with climate change?

With a growing human population and growing demands on ecosystems, these challenges will be considerable. Ecologists will have the critical role of providing the scientific information needed to make decisions about how we proceed as a society. The future of untold numbers of species relies on how effective we can be at this task.

SUMMARY

CONCEPT 23.1 Landscape ecology examines spatial patterns and their relationship to ecological processes and changes.

- A landscape is a heterogeneous area made up of a dynamic mosaic of different components that interact through the exchange of materials, energy, and organisms.
- Landscapes are characterized by their composition—the elements that constitute them—as well as by their structure—how those elements are arranged on the landscape.
- Landscape patterns influence ecological processes by determining how easily organisms can move among elements as well as by influencing ecosystem properties such as rates of biogeochemical cycling.
- Landscape patterns both shape and are shaped by disturbances.

CONCEPT 23.2 Habitat loss and fragmentation decreases habitat area, isolates populations, and alters conditions at habitat edges.

- Habitat fragments are biologically impoverished compared with the intact habitat from which they were derived.
- Once a patch of habitat is isolated by fragmentation, organisms reliant on that habitat may be isolated or may be able to cross the intervening matrix to some extent.
- The edges of habitat fragments have different abiotic conditions, and thus have different population dynamics, than interior habitats.
- The isolation of populations and the shifts in ecological communities that result from habitat fragmentation alter the evolutionary process.

CONCEPT 23.3 Biodiversity can best be sustained by large reserves connected across the landscape and buffered from areas of intense human use.

- The ideal spatial configuration for a core natural area is large, compact, and connected to or close to other protected natural areas.
- Core natural areas should be surrounded by buffer zones where human economic uses that are compatible with biodiversity conservation are allowed.
- Habitat corridors are instrumental in facilitating the movements of organisms between natural areas.
- Ecological restoration allows areas that have been degraded to support native species and ecosystem processes once again.

CONCEPT 23.4 Ecosystem management is a collaborative process with the maintenance of long-term ecological integrity as its core value.

- Collaboration among all stakeholders is key to arriving at effective management plans.
- Ecosystem management is a process of setting sustainable goals, developing and implementing land use management policies, monitoring the effectiveness of prior decisions, and adapting plans accordingly.
- Humans are an integral part of ecosystems. Conservation plans that include the economic and social well-being of local human populations are more likely to succeed over the long term.

REVIEW QUESTIONS

1. How are islands of terrestrial habitat that result from habitat fragmentation like actual islands surrounded by water? How are they different? How do the principles of island biogeography apply to habitat islands in a fragmented landscape?
2. Are habitat corridors just long, skinny habitat patches? Describe how a corridor is like the habitat blocks it is meant to join and how it is different, and outline the implications for organisms using the corridor. Do you think corridors are beneficial? Do you think they are necessary? Why?
3. The western boundary of Yellowstone National Park, where it borders a national forest, is visible from space (check it out with Google Earth at 44°21′45″ N, 111°05′50″ W). Explain this observation in terms of the contrasting missions of a national park and that of a national forest. What are the ecological implications of these contrasting institutional functions for (a) biodiversity and (b) ecosystem management?
4. Describe the protected areas in the vicinity of your ecology class. Rank them by size, describe who has jurisdiction over each of them, and evaluate how well you think they may be doing to protect native biodiversity. Consider them as a network. How well connected are they? Is there any way landscape connectivity could be improved in your region? What is the potential and the need for increasing the area of conserved land in your region?
5. Give an example of a conservation effort in your region that has an ecosystem management approach. What is the ecosystem that is being protected? What are some of the target species or biological communities this effort seeks to protect? Who are the stakeholders involved in decision making? How well does this effort adhere to the core value of seeking to maintain ecological integrity?

ON THE COMPANION WEBSITE
sites.sinauer.com/ecology2e

The website includes Chapter Outlines, Online Quizzes, Flashcards & Key Terms, Suggested Readings, a complete Glossary, and the Web Stats Review. In addition, the following resources are available for this chapter:

HANDS-ON PROBLEM SOLVING

Patch Movement: Crickets versus Cybercrickets

This Web exercise explores how organisms move across patches in the landscape. You will interpret data from a recent paper that simulated the movement of virtual organisms between patches across landscapes with different levels of connectivity, and then compared the simulation results with results from manipulation studies. Do real crickets move like cybercrickets?

WEB EXTENSIONS

23.1 Habitat Islands and the Eastern Wallaroo

23.2 Effects of an Ecological Restoration Experiment

CLIMATE CHANGE CONNECTIONS

23.1 Effects of Climate Change on Biodiversity

CHAPTER

24 Global Ecology

KEY CONCEPTS

CONCEPT 24.1 Elements move among geologic, atmospheric, oceanic, and biological pools at a global scale.

CONCEPT 24.2 Earth is warming due to anthropogenic emissions of greenhouse gases.

CONCEPT 24.3 Anthropogenic emissions of sulfur and nitrogen cause acid deposition, alter soil chemistry, and affect the health of ecosystems.

CONCEPT 24.4 Losses of ozone in the stratosphere and increases in ozone in the troposphere each pose risks to organisms.

Dust Storms of Epic Proportions: A Case Study

Dust is usually a subtle nuisance for most city dwellers, a reminder of neglect and lax housekeeping. Living in islands of asphalt and concrete, most urbanites see little bare soil, let alone clouds of blowing dust in the sky. Yet in late spring of 1934, a massive dust storm shrouded the cities of Chicago and New York in a dark haze never seen before by their residents. People choked on the dust, and it burned their eyes. Twelve million tons of dust fell on Chicago, 4 pounds for each resident, and an estimated 350 million tons of dust were carried by the storm to the Atlantic Ocean. As frightening as this event was to city dwellers, farmers in the southern Great Plains of the United States had suffered through multiple years of frequent severe dust storms throughout the 1930s (**Figure 24.1**). During this period, many people in that region, known as the Dust Bowl, suffered from dust-induced pneumonia, similar to miner's black lung disease, which was often fatal.

Beijing, China, experienced comparable dust storms following the mid-1990s, associated with more widespread storms that affected China, South Korea, and Japan. Buildings, streets, sidewalks, and cars in Beijing were blanketed with a coat of orange dust during these events. Pedestrian footprints were visible in the dust of the sidewalks, and traffic was slowed by the lowered visibility. In an April 2006 storm, more than 300,000 tons of dust fell on the city. Residents were encouraged to stay indoors to avoid inhaling the dust and getting it in their eyes. Many of those brave enough to venture out wore surgical face masks to protect their lungs. Some residents lined their windows and doors with rags in an attempt to keep the dust out of their houses and apartments.

Large dust storms in urban areas are perceived as rare events, potentially linked to unsustainable land use practices such as overgrazing or farming on marginal lands. In both examples mentioned above, farming and grazing in arid areas had increased prior to the dust storms. There is evidence, however, that massive dust storms occur at regular, but infrequent, intervals irrespective of human activities, moving large amounts of soil across whole continents. Over the past century, these events have been associated with prolonged droughts. The urban dust storms in the United States during the 1930s were associated with a decade-long drought in

FIGURE 24.1 Massive Dust Storms A wall of dust approaches the town of Stratford, Texas, in April of 1935, one of several "black dusters" that swept through the Dust Bowl during the 1930s.

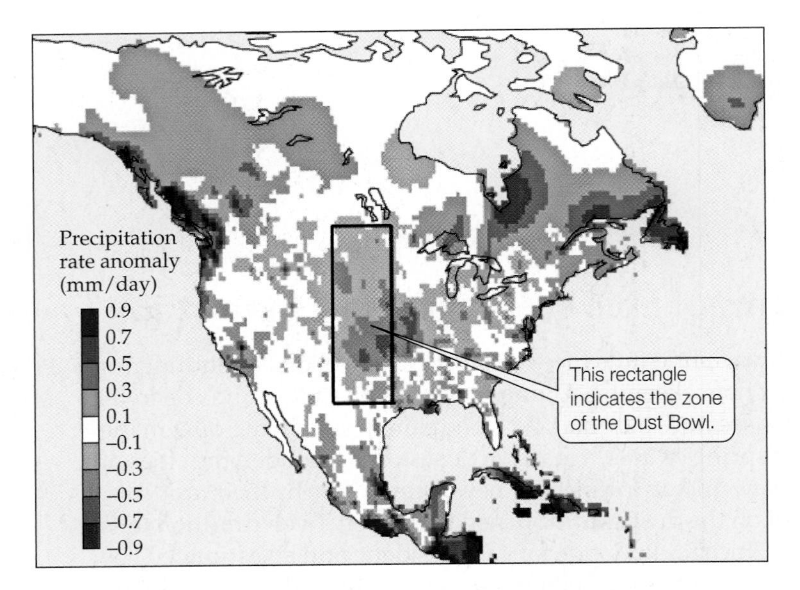

FIGURE 24.2 Drought in the Southern Plains During the 1930s, the southern Great Plains of the United States experienced the driest weather on record. The drought, in combination with loss of vegetation cover, created conditions conducive to dust input into the atmosphere. Values are anomalies (differences between averages for the period 1932–1939 and long-term averages). (After Cook et al. 2009.)

the Dust Bowl (**Figure 24.2**). Similarly, the Beijing dust storms of the past 2 decades were associated with drought in nearby Mongolia.

Dust in the atmosphere is made up of soil particles blown from regions lacking vegetative cover to protect their soils from the wind. As discussed in Chapters 4 and 21, soils are important as sources of nutrients, determinants of terrestrial moisture availability, and habitat for organisms. Therefore, the redistribution of soils from one area to another has the potential to cause ecological change. How widespread are these ecological effects? What role did humans play in the dust storms of the past century? As we will see in this chapter, the movement of dust is important component in the movement of elements at the global scale.

Introduction

In Chapter 21, we reviewed the cycling of nutrients within ecosystems associated with biological uptake and decomposition. The movements of these biologically important elements are linked at a global scale that transcends ecological boundaries at the ecosystem and biome scale. Ecological processes at the ecosystem scale (e.g., net primary production, decomposition) influence global phenomena (e.g., greenhouse gases emissions and uptake). In addition, the realization that humans are increasingly changing the physical and chemical environment at a global scale has fostered a greater awareness of ecology at these larger spatial scales. Emissions of pollutants, dust, and greenhouse gases into the atmosphere have caused widespread environmental problems, including climate change, acid precipitation, eutrophication, and loss of stratospheric ozone. A major focus of global ecology is therefore the study of the extensive environmental effects of human activities.

The first part of this chapter will cover the global-scale cycles, which are related to, but distinct from, the ecosystem-scale cycles covered in Chapter 21. Knowledge of these cycles is important for understanding global environmental change. Humans have had profound effects on these element cycles, and the environmental changes associated with these effects will be discussed in the final sections.

CONCEPT 24.1 Elements move among geologic, atmospheric, oceanic, and biological pools at a global scale.

Global Biogeochemical Cycles

In this section, we will follow the cycling of carbon, nitrogen, phosphorus, and sulfur at the global scale, incorporating the geologic and atmospheric components of those elements' cycles. These particular elements are emphasized both because of their importance to biological activity and because of their roles as pollutants in the global environment. The cycles are discussed in terms of *pools*, or reservoirs—the amounts of elements within components of the biosphere—and *fluxes*, or rates of movement, between pools. For example, terrestrial plants constitute a pool of carbon, while photosynthesis represents a flux—in this case, the movement of carbon from the atmospheric pool to the terrestrial plant pool.

The global carbon cycle is closely associated with energy

Carbon (C) is critically important for biological activity because of its role in energy transfer and the construction of biomass. At a global scale, C that is actively cycling is relatively dynamic, moving between atmospheric, terrestrial, and oceanic pools relatively quickly (over weeks to decades). It is important that we understand the global C cycle because changes in the fluxes of C among these pools are influencing Earth's climate system. Carbon in the atmosphere occurs primarily as carbon dioxide (CO_2) and methane (CH_4). As we saw in Chapter 2, both of these greenhouse gases influence atmospheric absorption of infrared radiation and its reradiation from Earth's surface. Thus, any changes in the atmospheric concentrations of these gases can have profound effects on the global climate, as we will see later in this chapter.

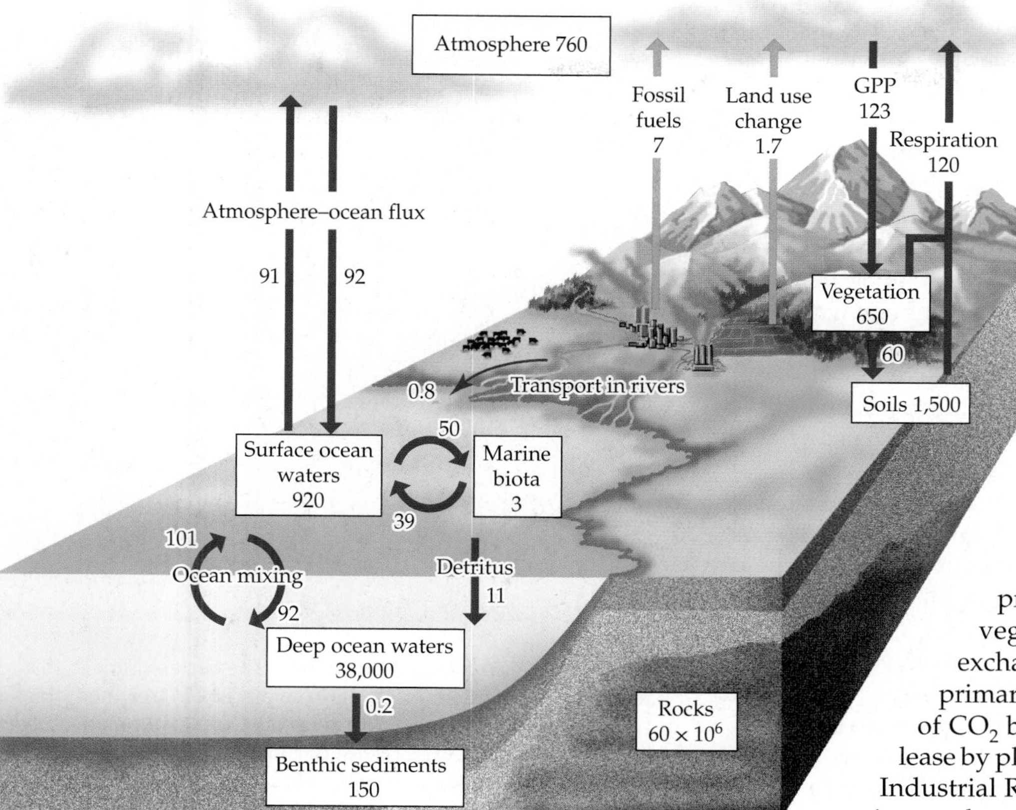

FIGURE 24.3 The Global Carbon Cycle Boxes represent major pools of C, measured in petagrams (1 Pg = 10^{15} g); arrows represent major fluxes of C, measured in Pg per year; anthropogenic fluxes are shown in orange. Note that the largest fluxes are for terrestrial GPP and respiration. (Additional data from IPCC 2007.)

Q How would deforestation influence the magnitude of these fluxes?

There are four major global pools of C: atmosphere, oceans, land surface (including soils and vegetation), and sediments and rock (**Figure 24.3**) (Schlesinger 1997). The largest of these pools is the combination of sediments and rock, which contain 99% of global C. The C in this pool is found primarily in the form of carbonate minerals and organic compounds. It is the most stable of the major pools, taking up and releasing C on geologic time scales.

The oceanic pool consists of two main components: surface waters (to depths of 75–200 m), where most marine biological activity occurs, and deeper, colder waters. Carbon dioxide dissolves in ocean water due to a concentration gradient between the atmosphere (higher concentration) and the ocean (lower concentration); thus, there is a net uptake of C from the atmosphere at the ocean surface. There is relatively little mixing between the ocean surface and the deeper waters, although C is transferred between them by the sinking of detritus and carbonate shells of marine organisms, and by the downwelling of polar ocean currents described under Concept 2.2. Most of the oceanic carbon (>90%) is in the deeper waters. Some flux from this deep ocean pool occurs when upwelling brings carbon-rich water to the surface, releasing CO_2 into the atmosphere.

The terrestrial pool, which includes vegetated and nonvegetated land surfaces and their associated soils, is the largest pool of biologically active C. The soil pool contains approximately twice as much C as the vegetation pool. The terrestrial pool exchanges C with the atmospheric pool primarily through photosynthetic uptake of CO_2 by plants and respiratory CO_2 release by plants and heterotrophs. Prior to the Industrial Revolution that began in the early nineteenth century, the exchanges between these two pools were roughly equal, with no net change in atmospheric CO_2.

As a result of the rapid growth of the human population over the past 160 years and associated industrial and agricultural development, there has been an increase in the release of C to the atmosphere from the terrestrial pool. This **anthropogenic** (human-associated) release of C is the result of land use change—mainly in the form of forest clearing for agricultural development—and the burning of fossil fuels. Prior to the mid-nineteenth century, deforestation was the largest contributor to anthropogenic C release to the atmosphere. Removing the forest canopy warms the soil surface, increasing rates of decomposition and heterotrophic respiration. Burning of the trees also releases CO_2, as well as small amounts carbon monoxide (CO) and methane (CH_4), into the atmosphere. During the last half of the twentieth century, deforestation for agricultural development shifted from the mid-latitudes of the Northern Hemisphere to the tropics.

The rate of anthropogenic emission of C into the atmosphere has continued to increase in recent decades. In 1970, anthropogenic CO_2 emissions added C to the atmosphere at a rate of 4.1 petagrams per year; by 2005, this rate had doubled (IPCC 2007). Today, burning of fossil fuels accounts for approximately 80% of the anthropogenic C flux to the atmosphere; the remaining 20% is associated with deforestation. Approximately half of these anthropogenic CO_2

emissions are taken up by the oceans and terrestrial biota. However, this proportion will decrease with time, as the uptake of CO_2 by terrestrial and marine ecosystems will not keep pace with the rate of emissions to the atmosphere (IPCC 2007).

Emissions of CH_4 to the atmosphere from the terrestrial pool have also increased as a result of human activities. Although atmospheric concentrations of CH_4 are much lower than those of CO_2, even small increases in CH_4 could influence the global climate because it is 25 times more effective as a greenhouse gas per molecule than CO_2. Methane is emitted naturally by anaerobic methanogenic archaea that live in wetlands and shallow marine sediments. Methanogenic archaea in the rumens of ruminant animals are also a source of atmospheric CH_4. Anthropogenic emissions of CH_4 have doubled since the early nineteenth century as a result of the processing and burning of fossil fuels, agricultural development (primarily that of rice, which is grown in flooded fields), burning of forests and crops, and livestock production (IPCC 2007). As a result, atmospheric CH_4 concentrations have more than doubled over the past two centuries.

The process of photosynthesis is sensitive to the concentration of CO_2 in the atmosphere. As a result, we might expect increases in rates of photosynthesis as anthropogenic CO_2 emissions increase, primarily in plants with the C_3 photosynthetic pathway (see Chapter 5). Experiments have shown, however, that for some herbaceous plants, these increases may be short-term because the plants may become acclimatized to elevated CO_2 concentrations. For other plants, such as forest trees, increases in photosynthetic rates may be more sustained. It is extremely important that we understand the response of forest ecosystems to elevated CO_2 concentrations. Because so much of terrestrial NPP, and thus C uptake, occurs in these ecosystems, their response will have a profound effect on the fate of anthropogenic CO_2 emissions. However, it is difficult to manipulate atmospheric CO_2 concentrations experimentally in an intact forest. In one successful approach, called free-air CO_2 enrichment, or FACE, researchers inject CO_2 into the air through vertical pipes surrounding stands of trees while monitoring the atmospheric concentration of CO_2 within the experimental stands. The rate of CO_2 injection is controlled to maintain a relatively constant elevated level.

A FACE experiment was used to investigate the ecosystem effects of elevated CO_2 concentrations in a young loblolly pine (*Pinus taeda*) forest in North Carolina by Evan DeLucia and his colleagues (DeLucia et al. 1999) (**Figure 24.4**). The experiment was initiated in 1997, when three plots exposed to elevated CO_2 concentrations and three control plots exposed to ambient CO_2 concentrations were established. The researchers used measurements of tree basal area to estimate aboveground NPP and repeated collections of soil cores to estimate fine root growth and belowground NPP. DeLucia and his colleagues found that the elevated CO_2 concentrations increased the overall NPP of the forest by 25%. Input of C into the soil, both from aboveground litter and from belowground fine root turnover, also increased. The results of this experiment indicated that forests may be an important sink (reservoir) for anthropogenic CO_2. However, DeLucia and colleagues suggested that their forest stand represents the upper limit of potential CO_2 uptake, and that older forests, and forests with lower water and nutrient supplies, may not have as great a capacity to take up CO_2. Results from other FACE experiments in forest ecosystems have found similar responses to elevated CO_2 concentrations (an average increase in NPP of 23%; Norby et al. 2005).

FIGURE 24.4 A FACE Experiment The circles visible in this aerial photo are free-air CO_2 enrichment (FACE) treatment rings in a loblolly pine (*Pinus taeda*) forest at the Duke Experimental Forest in North Carolina. CO_2 is released from plastic pipes surrounding treatment plots at a rate calculated to raise the CO_2 concentration to 200 ppm above ambient atmospheric CO_2 concentrations.

Atmospheric CO_2 concentrations directly affect the acidity (pH) of the oceans by affecting the rate at which CO_2 diffuses into seawater. Greater diffusion of CO_2 into seawater enhances the formation of carbonic acid, which lowers the pH of the seawater:

$$CO_2 + H_2O \leftrightarrows H_2CO_3 \text{ (carbonic acid)} \leftrightarrows H^+ + HCO_3^- \text{ (bicarbonate)} \leftrightarrows 2\,H^+ + CO_3^{2-} \text{ (carbonate)}$$

During the past century, ocean acidity has increased by about 30%. Further increases have been forecast using

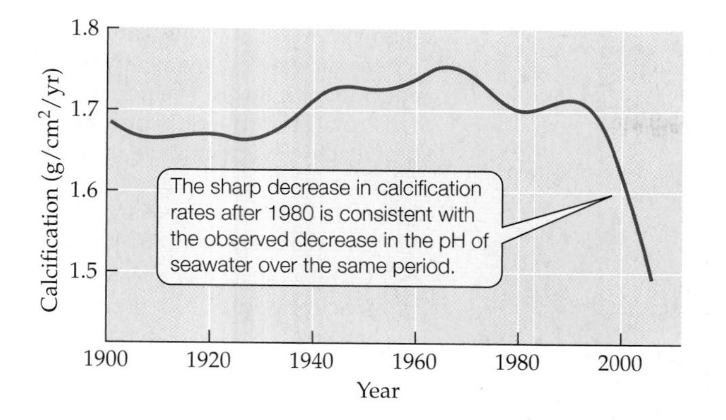

FIGURE 24.5 Rates of Calcification of Corals on Australia's Great Barrier Reef, 1900–2005 The sharp decline after 1980 is consistent with laboratory studies of the effect of decreasing pH on calcification rates due to increases in atmospheric CO_2, as well as increases in ocean water temperature. (After De'ath et al. 2009.)

model simulations incorporating the expected increases in anthropogenic CO_2 emissions over the twenty-first century (Caleira and Wickett 2003). The predicted increases will have two negative effects on marine organisms that form their protective external shells from calcium carbonate, including corals, mollusks, and many plankton. First, the increase in acidity will dissolve the existing shells of the organisms. Second, lower concentrations of carbonate in seawater will decrease the organisms' ability to synthesize shells (Feely et al. 2004; Orr et al. 2005). Between 1990 and 2009, the rate of formation of calcium carbonate by corals on Australia's Great Barrier Reef declined by 14%, an amount consistent with observed decreases in the pH of seawater (**Figure 24.5**) (De'ath et al. 2009). Both effects will increase mortality and lower the abundances of marine organisms that rely on calcium carbonate, altering the diversity and function of marine ecosystems.

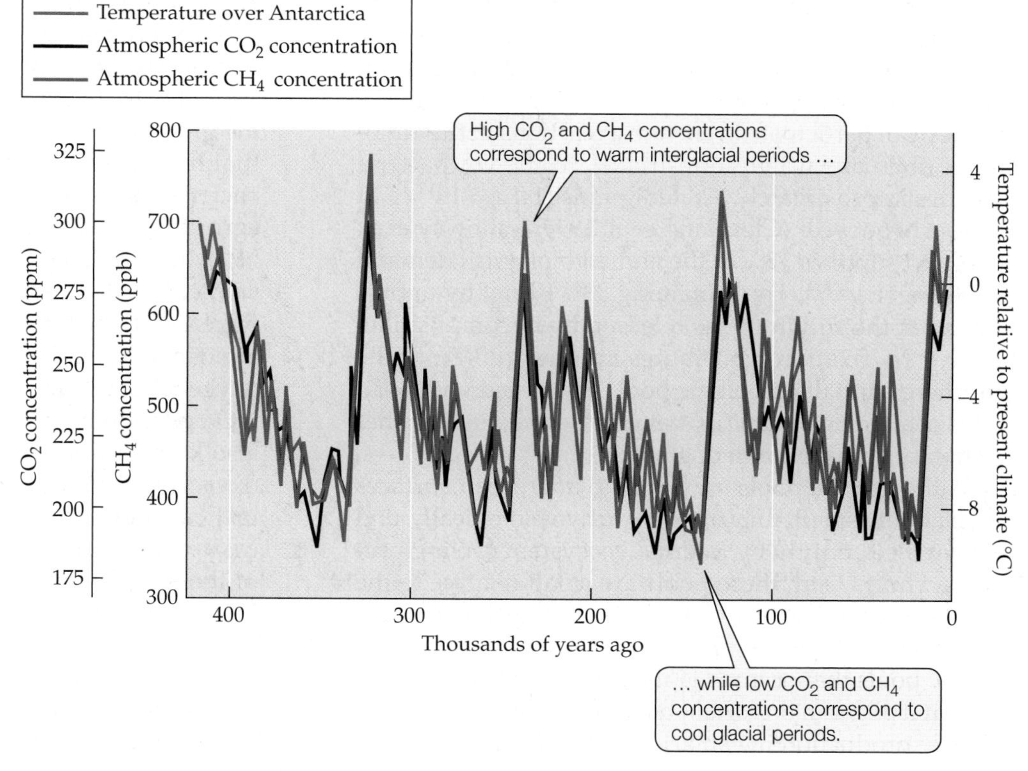

FIGURE 24.6 Changes in Atmospheric CO_2 and CH_4 Concentrations over Time Atmospheric CO_2 and CH_4 concentrations have varied with temperature over the past 400,000 years. These gas concentrations were measured in bubbles trapped in Antarctic ice; temperatures were estimated using oxygen isotopic analyses (see Ecological Toolkit 5.1). (Data from IPCC 2007.)

Atmospheric concentrations of C have changed dynamically throughout Earth's history in association with geologic and climate changes. Concentrations of CO_2 have ranged from greater than 3,000 parts per million (ppm) 60 million years ago to less than 200 ppm 140,000 years ago. Over the past 400,000 years, variations in the concentrations of CO_2 and CH_4, as measured in tiny bubbles preserved in polar ice, have followed glacial–interglacial cycles (see Chapter 2). The lowest CO_2 concentrations during this time were associated with glacial periods (**Figure 24.6**). Over most of the past 12,000 years, atmospheric CO_2 concentrations remained relatively stable, varying between 260 and 280 ppm. Since the mid-nineteenth century, however, CO_2 concentrations have increased at a rate faster than at any other time over the past 400,000 years (IPCC 2007), reaching values of 390 ppm in 2010. Even if we dramatically decreased our CO_2 emissions starting today, atmospheric CO_2 concentrations would remain elevated for a long time to come due to a time lag (decades to centuries) in oceanic uptake. The influence of CO_2 and CH_4 on climate change will be discussed later in this chapter.

Biological fluxes dominate the global nitrogen cycle

Nitrogen (N) plays a key role in biological processes as a constituent of proteins and enzymes, and it is one of the resources that most commonly limits primary production, as we saw under Concept 19.2. Thus, cycles of N and C are tightly coupled through the processes of photosynthesis and decomposition.

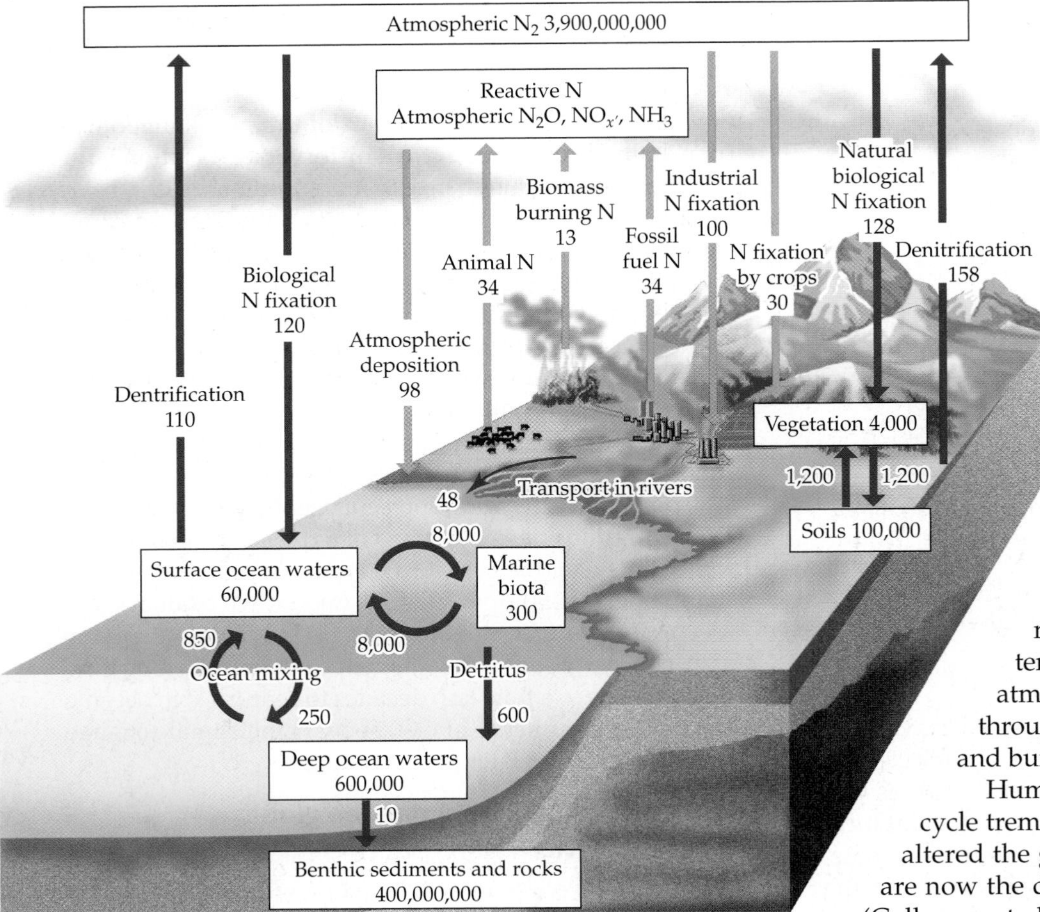

FIGURE 24.7 The Global Nitrogen Cycle Boxes represent major pools of N, measured in teragrams (1 Tg = 10^{12} g); arrows represent major fluxes of N, measured in Tg per year, with orange signifying anthropogenic fluxes. The percentage of the total atmospheric N pool made up of reactive N is minuscule (it is also difficult to quantify because it is very dynamic). (Data from Cleveland et al. 1999 and Galloway et al. 2004.)

Q Given its small size, why is the reactive pool of such great interest?

The largest pool of N (>90%) is atmospheric dinitrogen gas (N_2) (**Figure 24.7**). This form of N is very stable chemically and cannot be used by most organisms, with the important exception of nitrogen-fixing bacteria, which are able to convert it to more chemically usable forms, as described under Concept 21.1. These fixed chemical compounds are referred to as *reactive* N because, unlike N_2, they can participate in chemical reactions in the atmosphere, soils, and water. Terrestrial N_2 fixation by bacteria provides approximately 128 teragrams (1 Tg = 10^{12} g) of reactive N per year (Cleveland et al. 1999; Galloway et al. 2004) and supplies 12% of the annual biological demand (Schlesinger 1997). The remaining 88% is met by uptake of N from the soil in forms released by decomposition. Oceanic N_2 fixation contributes another 120 Tg to the biosphere annually. Geologic pools associated with sediments containing organic matter represent a much smaller fraction of global N than of global C.

Although the pools of N at land and ocean surfaces are relatively small, they are very active biologically, and they are held tightly by internal ecosystem cycling processes. Fluxes from these pools are small relative to the rates of internal cycling, usually less than 10% (Chapin et al. 2002). The natural flux of N between terrestrial and oceanic pools that occurs via movement of N in rivers is tiny, but it plays an important biological role by enhancing primary production in estuaries and salt marshes. Denitrification, a microbial process that occurs in anoxic soils and in the ocean (described under Concept 21.1), results in movement of N (as N_2) from terrestrial and marine ecosystems into the atmosphere. These ecosystems also lose N through burial of organic matter in sediments and burning of biomass.

Human activities have altered the global N cycle tremendously—even more than they have altered the global C cycle. Anthropogenic fluxes are now the dominant components of the N cycle (Galloway et al. 2004) (**Figure 24.8**). The rate of fixation of atmospheric N_2 by humans now exceeds the rate of natural terrestrial biological fixation. Emissions of N associated with industrial and agricultural activities are causing widespread environmental changes, including acid precipitation, discussed under Concept 24.3. Three major processes account for these anthropogenic effects. The first is the manufacture of agricultural fertilizers by the Haber–Bosch process, described in Chapter 21. Second, the growing of crops such as soybeans, alfalfa, and peas that have symbiotic relationships with nitrogen-fixing bacteria has increased biological N_2 fixation. Flooding of agricultural fields for other crops, such as rice, has increased N_2 fixation by cyanobacteria. Finally, anthropogenic emissions of certain gaseous forms of nitrogen have greatly increased the concentrations of these compounds in the atmosphere. Unlike N_2, these compounds, which include oxygenated nitrogen compounds (NO, NO_2, HNO_3, and NO_3^-; collectively referred to as NO_x), nitrous oxide (N_2O, also known as laughing gas), ammonia (NH_3), and peroxyacetyl nitrate (PAN), can undergo chemical reactions in the atmosphere and are potentially available for biological uptake. Fossil fuel combustion is the primary source of these nitrogenous gas emissions. Other contributors include biomass burning associated with deforestation, denitrification and volatilization (conversion to gaseous form) of fertilizers, and emissions from livestock feedlots and human sewage treatment plants. All of these reactive forms of N are returned to terrestrial and marine ecosystems

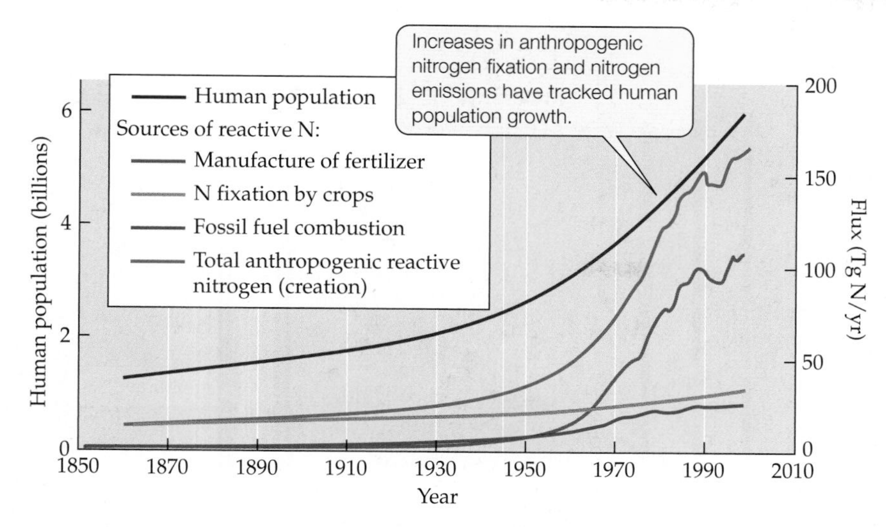

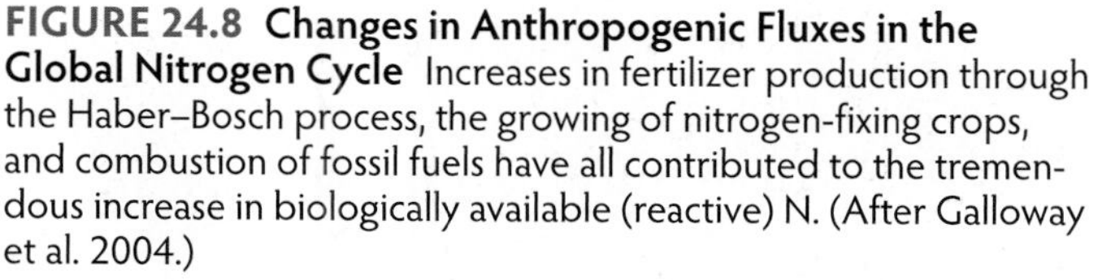

FIGURE 24.8 Changes in Anthropogenic Fluxes in the Global Nitrogen Cycle Increases in fertilizer production through the Haber–Bosch process, the growing of nitrogen-fixing crops, and combustion of fossil fuels have all contributed to the tremendous increase in biologically available (reactive) N. (After Galloway et al. 2004.)

through the process of atmospheric deposition (described under Concept 21.1).

The global phosphorus cycle is dominated by geochemical fluxes

Phosphorus (P) limits primary production in some terrestrial ecosystems—particularly those with old, well-weathered soils, such as tropical lowland forests—and in many freshwater and some marine ecosystems. Phosphorus is added to crops as a fertilizer globally. Phosphorus availability can also control the rate of biological N_2 fixation due to its high metabolic demand for P. Consequently, the C, N, and P cycles are linked to one another through photosynthesis and NPP, decomposition, and N_2 fixation.

Unlike C and N, P has essentially no atmospheric pool, with the exception of dust (**Figure 24.9**). Gaseous forms of P are extremely rare. The largest pools of P are in terrestrial soils and marine sediments. Phosphorus is released from sedimentary rocks in biologically available forms by weathering. The largest fluxes of P occur in internal ecosystem cycles, which form a tight recycling loop between biological uptake by plants and microorganisms and release by decomposition. Typically, very little of the P cycling through terrestrial and aquatic ecosystems is lost. In terrestrial ecosystems, most P loss is associated with the process of occlusion (described under Concept 21.3). Movement of P from terrestrial to aquatic ecosystems occurs primarily through erosion and movement of particulate organic matter—mainly bits and pieces of plants—into streams. Much of the P transported from terrestrial to marine ecosystems (about 90%) is lost when it is deposited in deep ocean sediments. Ultimately, P in sediments in both marine and terrestrial ecosystems is cycled again in association with tectonic uplift and weathering of rocks, which occurs on a scale of hundreds of millions of years.

Anthropogenic effects on the global P cycle are associated with use of agricultural fertilizers, discharges of sewage and industrial wastes, and increases in terrestrial surface erosion. Phosphorus fertilizers are usually derived from the mining of uplifted, ancient marine sediment deposits. As such, P is a nonrenewable resource, subject to depletion. Mining releases four times more P annually than is liberated through natural weathering of rock. Globally, P is applied as fertilizer in an amount equivalent

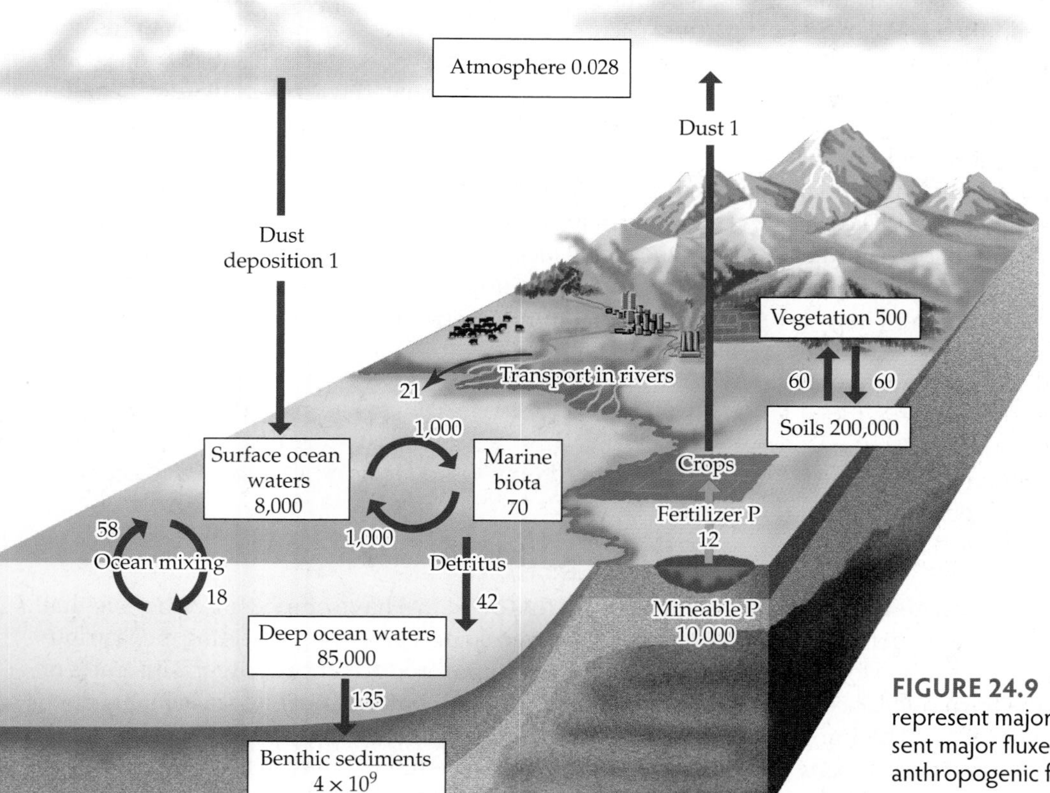

FIGURE 24.9 The Global Phosphorus Cycle Boxes represent major pools of P, measured in Tg; arrows represent major fluxes of P, measured in Tg per year. The major anthropogenic flux (P fertilization of crops) is indicated in orange.

to approximately 20% of the P that cycles naturally through terrestrial ecosystems (Schlesinger 1997). While occlusion of P in the soil minimizes the flux of anthropogenic P from terrestrial to aquatic ecosystems, that flux still has great potential for negative environmental effects. One such effect is eutrophication in lakes, as described in the case of Lake Washington (see p. 467).

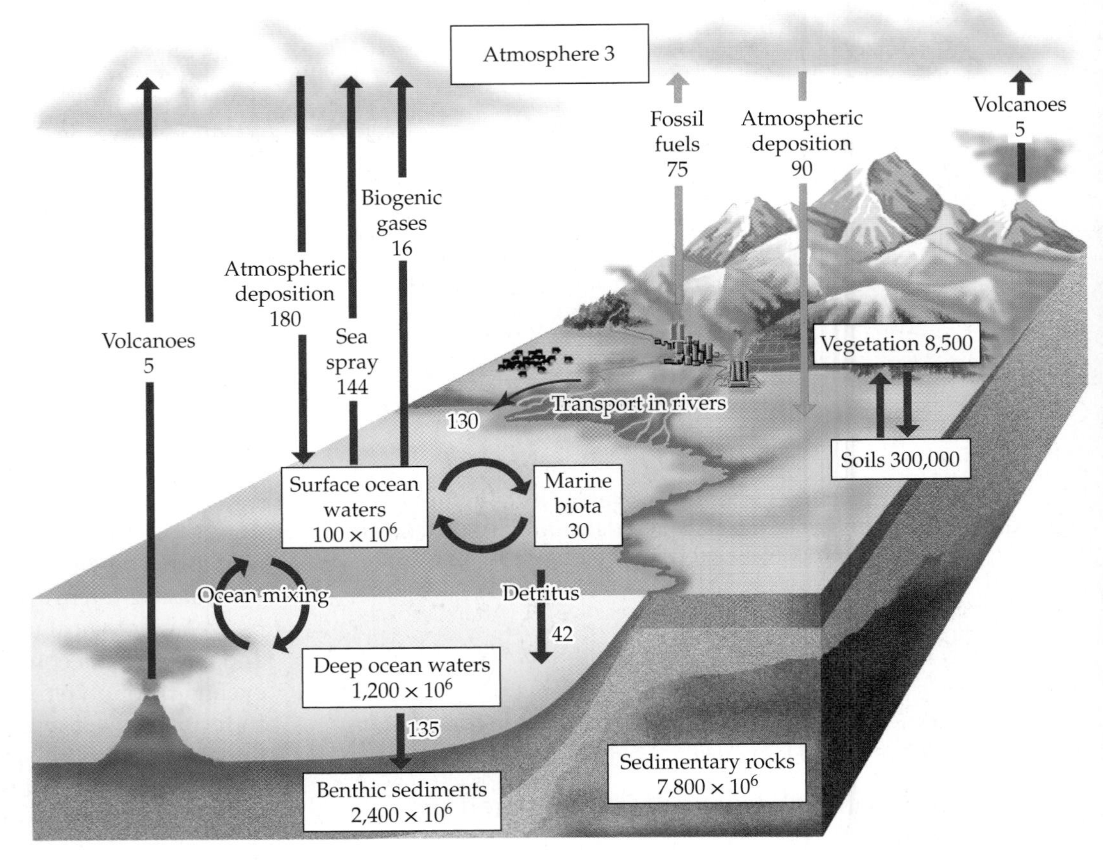

FIGURE 24.10 The Global Sulfur Cycle Boxes represent major pools of S, measured in Tg; arrows represent major fluxes of S, measured in Tg per year. Anthropogenic fluxes are shown in orange.

Biological and geochemical fluxes both determine the global sulfur cycle

Sulfur (S) is a constituent of some amino acids, but it is probably never in short supply for organismal growth. Sulfur plays important roles in atmospheric chemistry. As with the C, N, and P cycles, anthropogenic changes to the global S cycle have important negative environmental consequences, primarily through the generation of acid precipitation.

The major global pools of S are in rocks, sediments, and the ocean, which contains a large pool of dissolved sulfate (SO_4^{2-}) (**Figure 24.10**). Fluxes of S among these global pools can occur in gaseous, dissolved, or solid forms. Weathering of S-containing minerals, mainly sedimentary pyrite, releases soluble forms of S that may enter the atmosphere or oceans. There is a net movement of S from the terrestrial pool to the oceanic pool, associated with transport in rivers and in atmospheric dust. Volcanic eruptions emit substantial amounts of sulfur dioxide (SO_2) into the atmosphere. Because they are episodic events, however, the amount of S emitted to the atmosphere by volcanic eruptions, on a time scale of centuries, is approximately the same as the amount blown into the atmosphere as dust from bare soils. Oceans release S to the atmosphere as small particles of windborne ocean spray and as gaseous emissions associated with microbial activity. Bacteria and archaea in anaerobic soils also emit sulfur-containing gases such as hydrogen sulfide (H_2S). Most gaseous S compounds in the atmosphere undergo oxidation to SO_4^{2-} and H_2SO_4 (sulfuric acid), which are removed relatively quickly by precipitation.

Anthropogenic emissions of S to the atmosphere, which includes gaseous and particulate forms (e.g., dust, aerosols), have quadrupled since the Industrial Revolution. Most of these emissions are associated with the burning of S-containing coal and oil and the smelting of metal-containing ores. What goes up comes down in the form of atmospheric deposition, usually within the same region from which it was emitted, but not always. Long-distance transport of fine dust occurs episodically in association with droughts and major wind events, as described in the Case Study on p. 525. Increases in erosion associated with clearing of vegetation and overgrazing have contributed to anthropogenic input of S into the atmosphere as dust. Transport of S in rivers has doubled over the past 200 years (Schlesinger 1997).

Human activities have resulted in changes in all four of the global biogeochemical cycles we have just described, and as we have noted, some of those changes have had important environmental effects. Let's turn our attention to those effects next.

CONCEPT 24.2 Earth is warming due to anthropogenic emissions of greenhouse gases.

Global Climate Change

Throughout this book, we have emphasized the role that climate plays in ecological processes, including the distributions and physiological performance of organisms, rates of resource supply, and the outcomes of biological interactions such as competition. Thus, changes in climate—particularly changes in the frequency of extreme events such as extensive droughts, violent storms, or extreme high and low temperatures—have profound effects on ecological patterns and processes. Because they are disturbances that result in significant mortality within populations, these extreme events are often critical in determining species' geographic ranges.

In Chapter 2, we learned the difference between weather and climate. *Weather* refers to the current state of the atmosphere around us at any given time. *Climate* is the long-term description of weather, including both average conditions and the full range of variation. *Climatic variation* occurs at a multitude of time scales, from the daily changes associated with daytime solar heating and nighttime cooling, to seasonal changes associated with the tilt of Earth's axis, to decadal changes associated with interactions between ocean currents and the atmosphere (such as the Pacific Decadal Oscillation, described in Chapter 2's Case Study Revisited on p. 45). **Climate change**, on the other hand, refers to *directional* change in climate over a period of several decades. Earth is currently undergoing climate change.

Evidence of climate change is substantial

Climate change is distinguished from climatic variation by the presence of significant trends lasting at least three decades. Based on analyses of records from numerous climate monitoring stations, atmospheric scientists have determined that Earth is currently experiencing significant climate change (IPCC 2007) (**Figure 24.11A**). During the twentieth century, the average global surface temperature increased 0.6 ± 0.2°C (1.1 ± 0.4°F), with the greatest change occurring in the past 50 years. This rapid rise in global temperature is unprecedented in the past 10,000 years, although temperature changes at similar rates may have occurred at the onset and end of

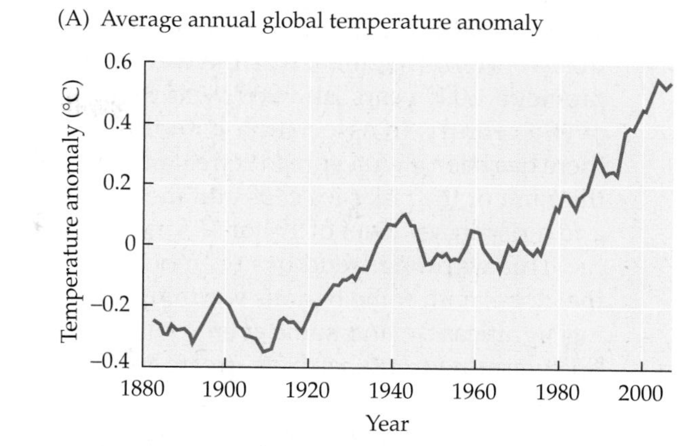

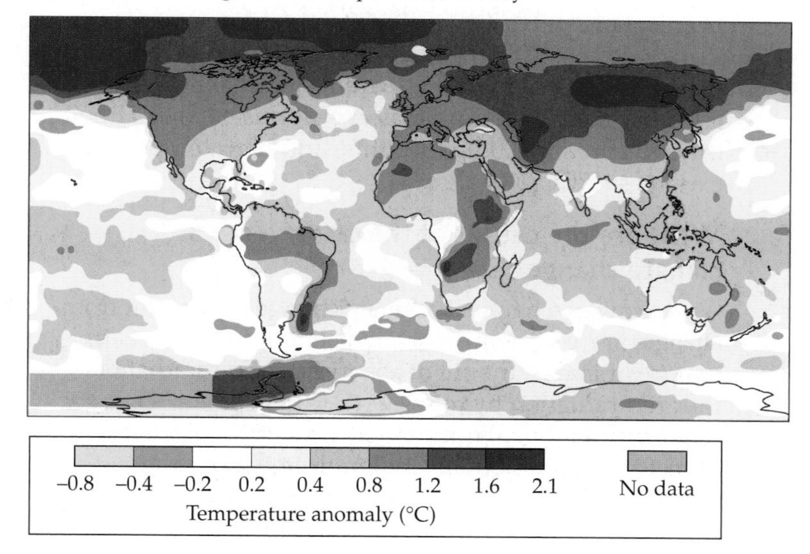

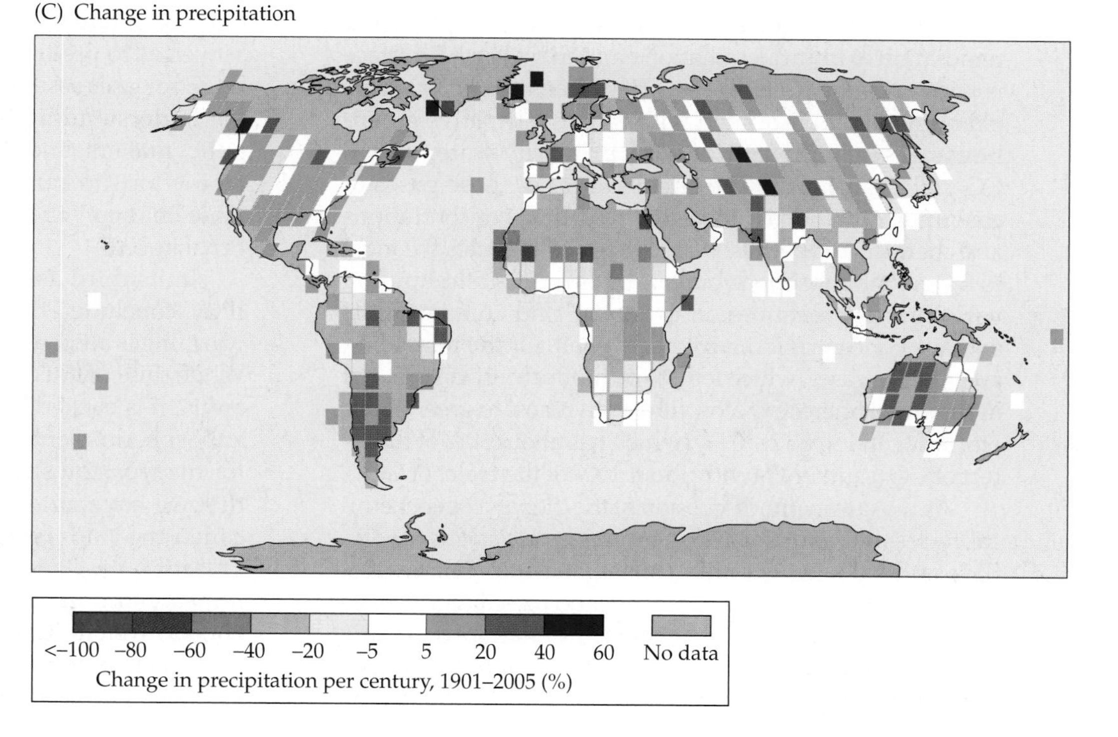

FIGURE 24.11 Changes in Global Temperature and Precipitation (A) Average annual global temperature anomalies between 1880 and 2005, averaged from air and sea surface temperature records from a number of meteorological stations and normalized to sea level. (B) Regional trends in average annual temperatures for 2001–2005. The values for parts (A) and (B) are expressed as differences (anomalies) from the average global temperature between 1951 and 1980. (C) Trends in global precipitation from 1901 to 2005, expressed as percentage change over that period. (A, B after Hansen et al. 2006; C after IPCC 2007.)

some glacial cycles (see Figure 24.6). The first decade of the twenty-first century was the warmest decade of the previous 1,000 years, and 2009 was the warmest year in over a century. In association with this warming trend, there has been a widespread retreat of mountain glaciers, thinning of the polar ice caps and melting of permafrost, and a rise in sea level of 19 cm (7.5 inches) since 1900.

This warming trend has been heterogeneous across the globe, with some regions warming, others not changing significantly, and some even cooling (**Figure 24.11B**). The warming trend has been greatest in the mid- to high latitudes of the Northern Hemisphere. Changes in precipitation have also occurred, with more precipitation in terrestrial portions of the high latitudes of the Northern Hemisphere and drier weather in the subtropics and tropics (**Figure 24.11C**). There has also been a tendency toward greater frequencies of some extreme weather events, such as hurricanes (including massive storms such as Hurricane Katrina in 2005), droughts, and high temperature extremes (IPCC 2007).

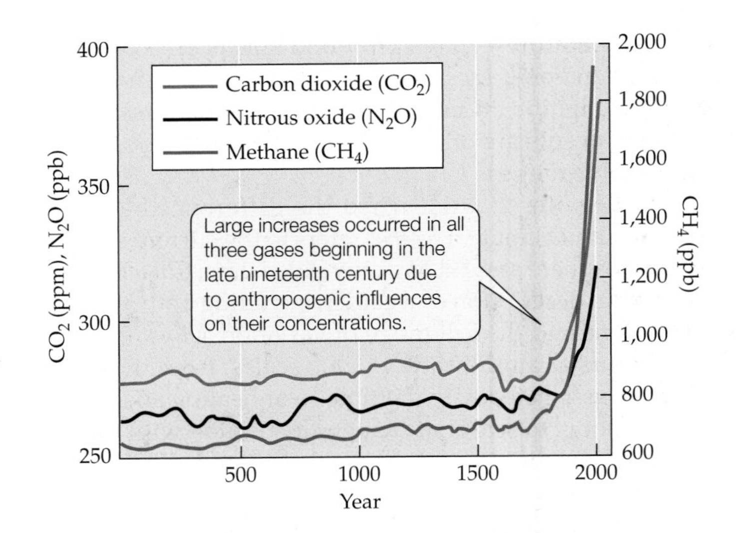

FIGURE 24.12 Increases in Greenhouse Gases The graph shows atmospheric concentrations of three greenhouse gases—CO_2, CH_4, and N_2O—over the past 2,000 years. Concentrations prior to 1958 were determined from ice cores; concentrations since 1958 have been measured directly. (After IPCC 2007.)

What are the causes of the observed climate change?

As we saw in Chapter 2, climate change may result from changes in the amount of solar radiation absorbed by Earth's surface or in the amount of absorption and reradiation of infrared radiation by gases in the atmosphere. Changes in absorption of solar radiation may be associated with variation in the amount of radiation emitted by the sun, in Earth's position relative to the sun, or in the reflection of solar radiation by clouds or surfaces with high reflectivity (albedo), such as snow and ice.

The warming of Earth by atmospheric absorption and reradiation of infrared radiation emitted by Earth's surface is known as the **greenhouse effect** (see Figure 2.4). This phenomenon is associated with radiatively active **greenhouse gases** in the atmosphere, primarily water vapor, CO_2, CH_4, and N_2O. The effectiveness of these gases in absorbing radiation is associated with their concentrations and chemical properties. Water vapor contributes the most to the greenhouse effect, but its atmospheric concentration varies greatly from region to region, and changes in its average concentration have been small. Of the remaining greenhouse gases, which tend to be more evenly distributed in the atmosphere, CO_2 contributes the most to greenhouse warming, followed by CH_4 (which has about 30% of the effect of CO_2) and N_2O (with about 10% of the effect of CO_2).

As we saw in our discussion of the global biogeochemical cycles of C and N, atmospheric concentrations of CO_2, CH_4, and N_2O are increasing substantially, primarily as a result of fossil fuel combustion and land use change (**Figure 24.12**). Are increases in anthropogenic emissions of these greenhouse gases responsible for global climate change? To evaluate the underlying causes of climate change, its potential effect on ecological and socioeconomic systems, and our options for limiting climate change associated with human activities, the World Meteorological Organization and the United Nations Environment Programme established the Intergovernmental Panel on Climate Change (IPCC) in 1988. The IPCC convenes panels of experts in atmospheric and climatic science to evaluate climatic trends and the probable causes for any changes observed. These experts use a combination of sophisticated modeling and analysis of data from the scientific literature to evaluate potential underlying causes of observed climate change, as well as to predict future climate change scenarios. The IPCC releases assessment reports periodically to enhance the understanding of climate change among scientists, policymakers, and the general public. In recognition of their efforts to spread "knowledge about man-made climate change," the IPCC was awarded the Nobel Peace Prize in 2007.

In its third assessment report, released in 2001, the IPCC concluded that the majority of the observed global warming is attributable to human activities (**Figure 24.13**). While this conclusion is still being debated in the political arena, it is backed by the majority of the world's leading atmospheric scientists. Paul Crutzen, a Nobel Prize winner for his work on stratospheric ozone loss, has suggested that we have entered a new geologic period, which he called the *Anthropocene epoch* (*anthropo*, "human"; *cene*, "recent"; *epoch*, geologic age) to indicate the extensive impact of humans on our environment, particularly the climate system (Crutzen and Stoermer 2000).

Will the climate continue to grow warmer? The IPCC's models project an additional increase in average global

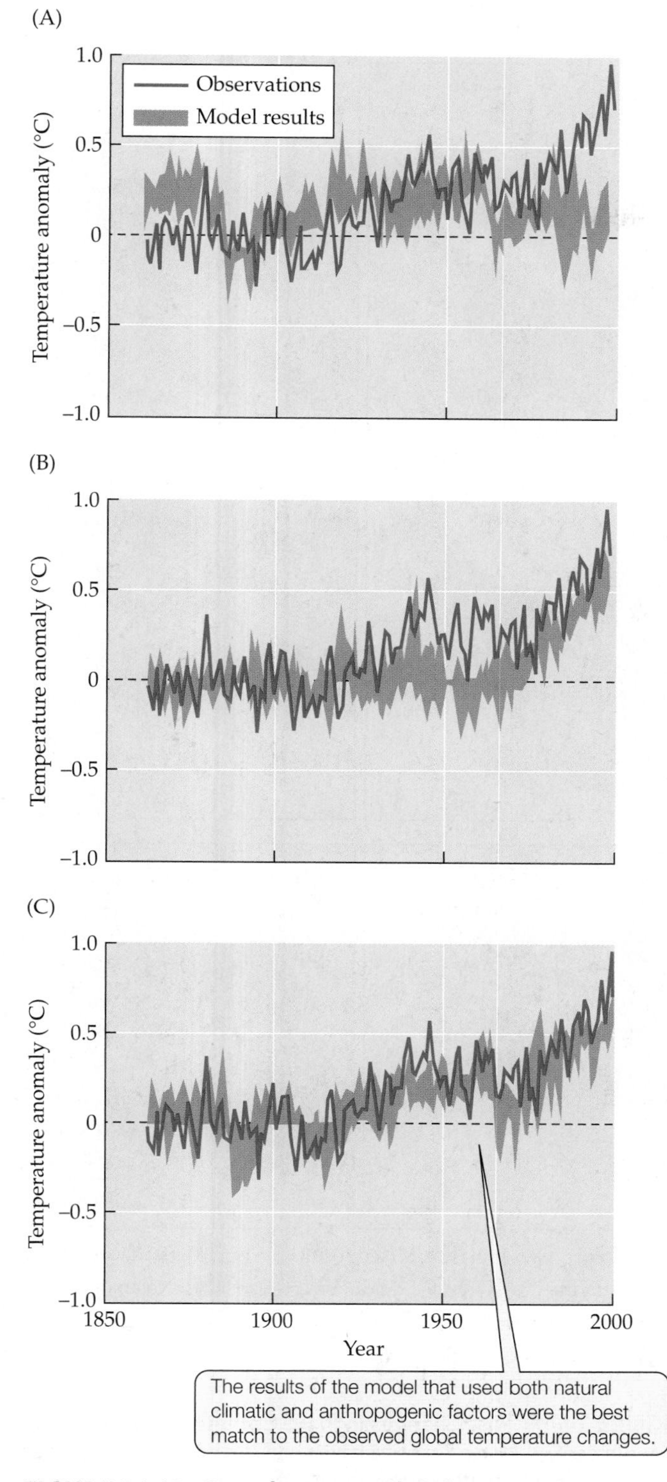

FIGURE 24.13 Contributors to Global Temperature Change IPCC scientists compared observed global temperature changes between 1850 and 2000 with the results of computer models. The models predicted the temperature changes that would be expected in that period based on (A) natural climatological factors only, including variation in solar radiation and in atmospheric concentrations of aerosols from volcanic eruptions, (B) anthropogenic factors only, including emissions of greenhouse gases and sulfate aerosols, and (C) a combination of natural and anthropogenic factors. The values are expressed as anomalies relative to the average temperature between 1901 and 1950. These results indicate that anthropogenic factors have played a large role in the observed warming. (After IPCC 2007.)

temperature of 1.8°C–4.0°C over the twenty-first century (IPCC 2007). The range of variation in this estimate is associated with uncertainties about future rates of anthropogenic greenhouse gas emissions and about the behavior of the terrestrial–atmospheric–oceanic system. Model simulations incorporating different economic development scenarios have predicted vastly different future rates of emissions. Aerosols in the atmosphere represent another source of uncertainty in the models' predictions. Aerosols, which reflect solar radiation, have a cooling effect on global temperatures; for example, emissions of large amounts of aerosols associated with major volcanic eruptions have had notable cooling effects at a global scale, as described in **Web Extension 24.1**. While some aerosols have been increasing in the atmosphere (e.g., dust, in association with land use change and desertification), others have decreased (e.g., SO_4^{2-}, due to decreasing anthropogenic SO_2 emissions). Water in the atmosphere may play contradictory roles: clouds may have a cooling effect, while water vapor, which may increase due to greater evapotranspiration, may increase greenhouse warming. Despite these uncertainties in predicting the magnitude of future climatic warming, there is a high probability that global temperatures will continue to rise.

Climate change will have ecological consequences

What does a 1.8°C–4.0°C change in average global temperature mean for biological communities? We can get a sense of what such a temperature change might mean by comparing it with the climatic variation associated with elevation in mountains. A median value for the projected temperature change (2.9°C) would correspond to a 500 m (1,600 foot) shift in elevation. In the Rocky Mountains, this change in climate would correspond approximately to a full change in vegetation zone, from subalpine forest (dominated by spruce and fir) to montane forest (dominated by ponderosa pine) (see Figure 3.11). Thus, if we assume perfect tracking of climate change by the current vegetation, climate change during the twenty-first century would result in an elevational shift in vegetation zones of 200–860 m. Similar predictions for latitudinal climatic shifts suggest movement of biological communities 500–1,000 km toward the poles.

Climate–biome correlations, such as those described under Concepts 3.1 and 4.1, are useful as a demonstration of what could happen with climate change, but it would be naive to use them to predict what will actually happen to biological communities. We know that community structure is influenced by a multitude of factors, including climate—particularly climatic extremes—as well as species interactions, the dynamics of succession, dispersal ability, and barriers to dispersal (as described in Unit 4). Because the coming climate change will be rapid relative to the climate changes that have shaped current biological communities, it is unlikely that the same assemblages of organisms will form the communities of the future.

Paleoecological records reinforce the suggestion that novel communities may emerge with climate change by showing that some plant communities of the past were quite different from modern plant communities. Jonathan Overpeck and colleagues used pollen records to reconstruct large-scale vegetation changes since the most recent glacial maximum in eastern North America (18,000 years ago) (Overpeck et al. 1992). They found not only that community types had made latitudinal shifts as the climate warmed, but also that community types without modern analogs existed under climatic regimes that were unique and no longer present (**Figure 24.14**). Overpeck and his colleagues concluded that future vegetation assemblages would follow similar trends, given the predicted rapid rate of global warming and the potential for the emergence of unique climatic patterns with no current analogs.

Because climate change will be continue to be relatively rapid, it is likely that evolutionary responses will not be possible for many organisms, and thus dispersal may be the only way for them to avoid extinction. Their rates of dispersal, and barriers to their dispersal associated with anthropogenic habitat fragmentation, will be important constraints on their responses to climate change. For most animals, mobility is not a problem, but their habitat and food requirements are intimately associated with the presence of specific vegetation types.

Plant dispersal rates are, on average, much slower than the predicted rate of climate change. In order to track the projected change in climate over the next century, plant species populations will need to move 5–10 km per year. Plant species that have animal-dispersed seeds, and which can establish viable populations and grow to reproductive maturity in a relatively short time, may be able to disperse rapidly enough to keep pace with climate change. However, this kind of dispersal strategy is common mainly in ruderal (weedy) herbaceous plants. Shrubs and trees have much slower rates of dispersal; as a result, there may be significant time lags in their response to climate change. In addition, barriers to dispersal may prevent organisms of all kinds from migrating in response to climate change. Dams, for example, may prevent fish from moving to water with more suitable temperatures. Fragmentation of habitat by human development may pose significant barriers to dispersal for some species (see Chapter 23). Without habitat corridors through which they can disperse, species face a greater probability of local extinction in the face of climate change.

In addition to affecting species' geographic ranges, climate change will affect ecosystem processes, such as net primary production (NPP), decomposition, and nutrient cycling and retention. Both photosynthesis and respiration are sensitive to temperature, and because their balance determines NPP, the direct effects of climatic warming on

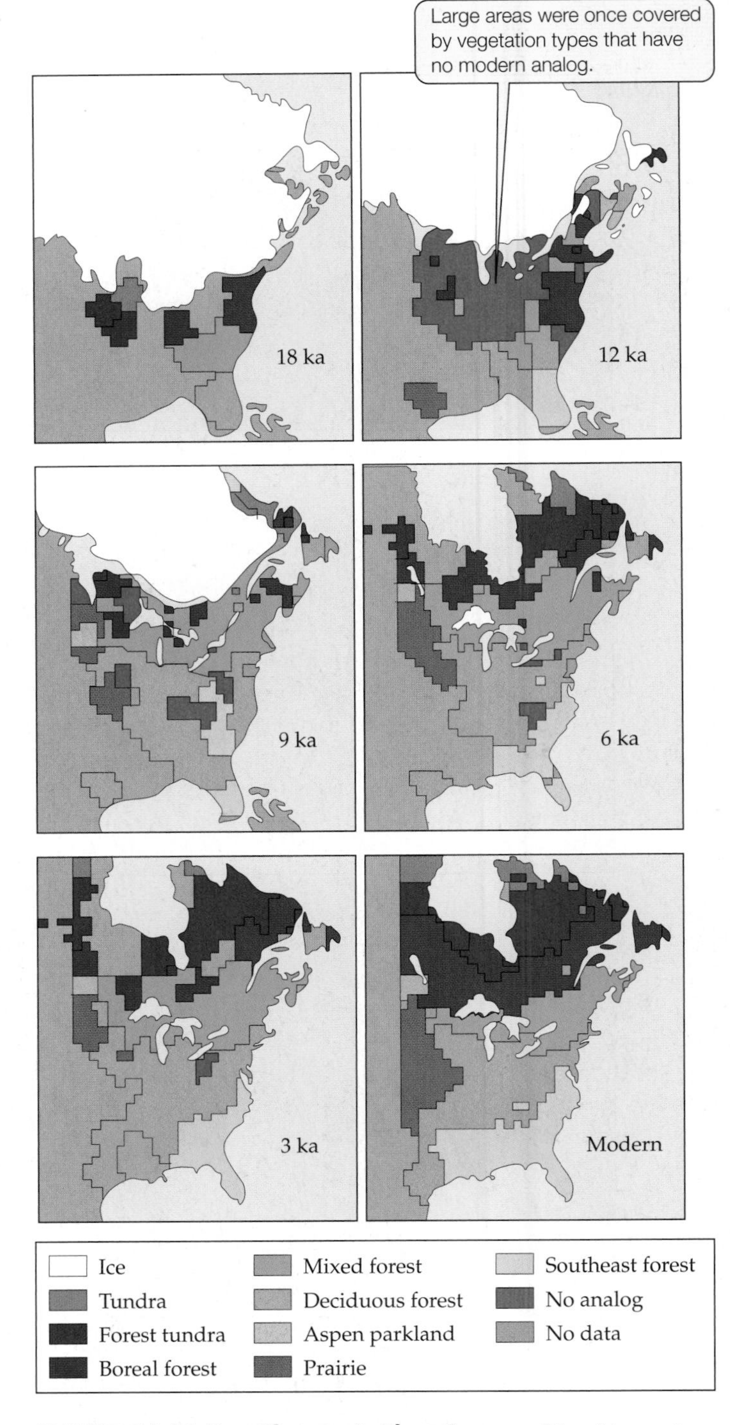

FIGURE 24.14 Past Changes in Plant Communities Vegetation types in eastern North America have changed since the last glacial maximum, 18,000 years ago (ka, thousands of years before present). Vegetation composition was determined from preserved pollen deposits. (After Overpeck et al. 1992.)

What factors may have led to the development of vegetation types different from those found today following retreat of the continental glacier?

NPP may be relatively minor. As indicated in Chapter 19, however, variation in NPP is related to water and nutrient availability and vegetation type, all of which may be affected by climate change. Changes in precipitation patterns and evapotranspiration rates resulting from climate change may strongly influence both water and nutrient availability. Because of the heterogeneity of climate change and of the resulting changes in vegetation types, both increases and decreases in NPP may occur. Thus, the effect of climate change on NPP will probably not be uniform. The effect of warming on nutrient supplies will be most pronounced in mid- to high-latitude terrestrial ecosystems, where low temperatures constrain rates of nutrient cycling and soils have large pools of nutrients. As a result, climate change may lead to increases in NPP in some temperate and boreal forest ecosystems.

Ecological responses to climate change are occurring

As noted earlier, global warming of 0.6°C has occurred over the past century. Several physical environmental changes have occurred over the same period, including the retreat of glaciers, increased melting of sea ice, and a rise in sea level. Have biological systems also responded to this climate change? Numerous reports of biological changes are consistent with recent global warming (Walther et al. 2002). These changes include earlier migration of birds, local extinction of amphibian and reptile populations, and earlier spring greening of vegetation.

Although they are more difficult to link directly to climate change, there have been changes in the geographic ranges of species that are consistent with warming. For example, Georg Grabherr and colleagues studied the vascular plant communities found on summits of mountains in the European Alps. They compared the current species richnesses of these communities with records dating back to the eighteenth and early nineteenth centuries (Grabherr et al. 1994). They found a consistent trend of upward movement of species from lower elevations onto the summits of the mountains (**Figure 24.15**). Similarly, Camille Parmesan and colleagues recorded a northward shift in the ranges of European nonmigratory butterfly species (Parmesan et al. 1999). Of the 35 species examined, 63% had shifted their ranges northward, while only 3% had shifted their ranges southward. An estimated 279 plant and animal species have shown geographic range shifts that are consistent with recent climate changes (Parmesan and Yohe 2003).

Climate change may be causing some species populations' to go extinct. Barry Sinervo and colleagues (2010) found that 12% of Mexico's *Sceloporus* lizard populations had gone extinct between 1975 and 2009. Recall from our discussion of Concept 22.2 that population extinctions are potentially the initial steps toward the extinction of a species. The extinctions of the lizard populations corresponded more closely to increases in temperature than to losses of habitat. Surprisingly, warming in the spring was better correlated with the extinctions than extreme temperatures during the summer. Sinervo and colleagues concluded that the warmer spring temperatures limited the lizards' foraging time during the breeding season. Ectothermic lizards must move into the shade and remain there to avoid overheating when temperatures become too warm (Chapter 4), and during that time they cannot seek out food. Consistent with this explanation, was the observation that the probability of extinction was greatest at low-elevation and low-latitude sites, where the animals were most likely to be at the limits of their thermal tolerance.

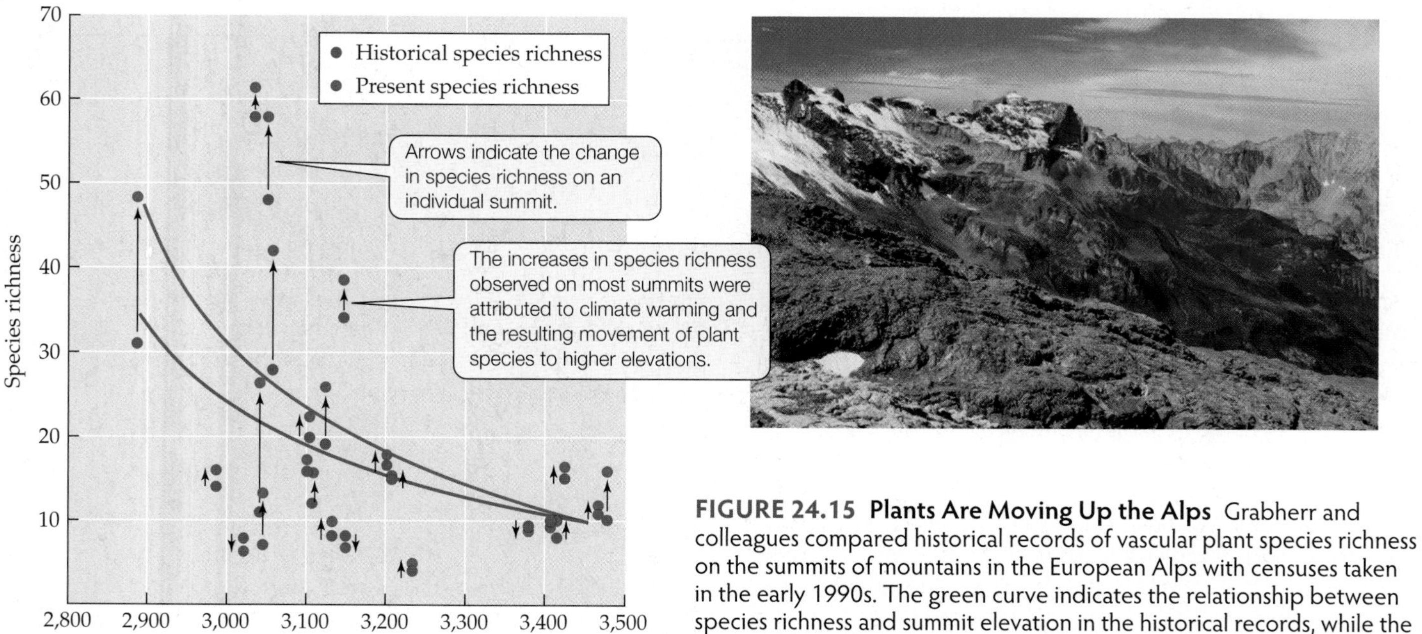

FIGURE 24.15 Plants Are Moving Up the Alps Grabherr and colleagues compared historical records of vascular plant species richness on the summits of mountains in the European Alps with censuses taken in the early 1990s. The green curve indicates the relationship between species richness and summit elevation in the historical records, while the red curve indicates the present relationship. (After Grabherr et al. 1994.)

Sinervo and colleagues also used a model of lizard thermal physiology to evaluate current and future worldwide effects of climate change on lizard populations. They estimated that climate change has already resulted in extinction of 4% of lizard populations worldwide. Using projections of future climate change, they suggested that 39% of the world's lizard populations, and 20% of its lizard species, may go extinct by 2050.

Migratory animals may also be adversely affected by climate change (Root et al. 2003). For example, migrating marine species, including whales and fish, may need to make longer journeys due to substantial changes in the distributions of their prey species as ocean temperatures warm. Some migratory birds that breed in England and North America have been arriving at nest sites as much as 3 weeks earlier than they did 30 years ago due to warmer spring temperatures and faster snowmelt. However, plants and invertebrate prey species have responded faster to climate change than the migrating birds, resulting in a mismatch between bird arrival and prey availability. On the other hand, longer breeding seasons may increase the number of offspring produced by some bird species, particularly in high-latitude ecosystems.

Changes in community composition may also be indicators of climate change. These effects may be particularly apparent in some sessile marine species. Chapters 3 and 16 have described the effects of rising water temperatures on corals and the resulting changes in coral reef communities. Changes in the abundances of marine foraminiferans—a type of zooplankton—also reflect global climatic trends during the past century (Field et al. 2006). Foraminiferan species have characteristic shells that allow them to be identified in marine sediments. Cores collected from benthic sediments can be examined to determine changes in the species composition of foraminiferans over time. Because the environmental tolerances of different species are known, these changes provide a means of reconstructing marine environments of the past. Following the mid-1970s, an increase in tropical and subtropical foraminiferan species, and a decrease in temperate and polar species, occurred in the eastern North Pacific Ocean, indicating a warming of ocean waters there.

Changes in global NPP also indicate biological responses to climate change. Ramakrishna Nemani and colleagues used remote sensing data to examine global patterns of NPP over an 18-year period (1982–1999) (Nemani et al. 2003). They found that global NPP increased 6% during the study period, or 0.3% per year (**Figure 24.16**). Tropical ecosystems exhibited the largest increase in NPP, which was associated with increases in solar radiation due to less cloud cover in the tropics during the study period. During the first decade of the twenty-first century, however, the trend toward increasing NPP was reversed, with lower global NPP. The decrease in global NPP was attributed to major droughts, particularly in the Southern Hemisphere (Zhao and Running 2010).

There has been a notable decrease in NPP at high northern latitudes during the past 30 years, which has coincided with some areas of the Arctic switching from a net uptake of CO_2 from the atmosphere (acting as a *sink*) to a net export of CO_2 (acting as a *source*) (Oechel et al. 1993). Large amounts of C are stored in the soils of boreal and

FIGURE 24.16 Changes in Terrestrial NPP Changes in net primary production (NPP) observed between 1982 and 1999, expressed as percentage change per year. (After Nemani et al. 2003.)

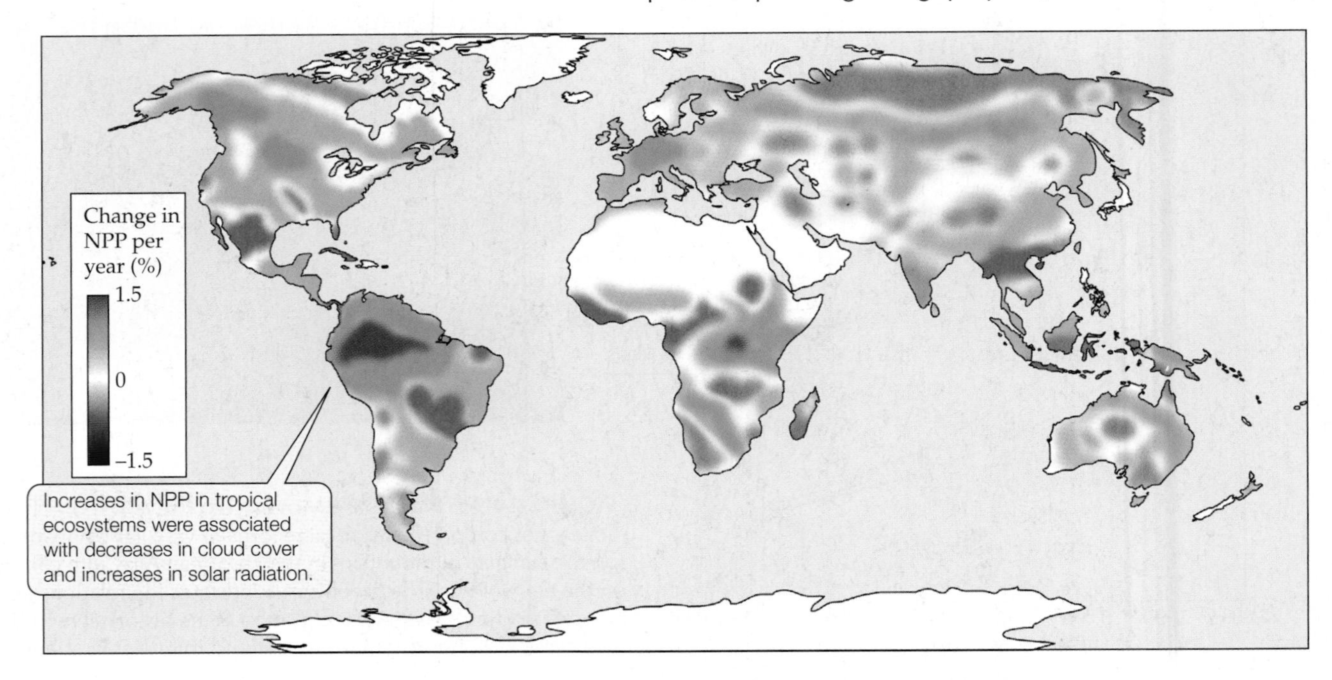

tundra ecosystems as a result of their low temperatures which inhibit decomposition, and the long-term buildup of carbon since the last glacial maximum. Warming during the twentieth century, however, increased the rate of CO_2 export from Arctic soils, such that losses now exceed gains from NPP. Warming of these high-latitude terrestrial ecosystems could provide a positive feedback to climate change by enhancing their emissions of CO_2 and CH_4. However, the rates of CO_2 loss from Arctic ecosystems have decreased since the early 1990s, possibly due to changes in rates of nutrient cycling and to physiological and compositional changes in the plants (Oechel et al. 2000).

Biological indicators of global climate change are diverse, and they are increasing with time. Experiments, modeling, and comparisons with historical and paleoecological records have all given us clues to how Earth's biota will respond to climate change. However, substantial uncertainties in predicting the effects of climate change still exist, many of which are associated with other environmental changes that are occurring at the same time. In the next section, we'll look at two such anthropogenic changes that are having profound effects on ecosystems: emissions of sulfur and nitrogen into the atmosphere.

CONCEPT 24.3 Anthropogenic emissions of sulfur and nitrogen cause acid deposition, alter soil chemistry, and affect the health of ecosystems.

Acid and Nitrogen Deposition

The negative effects of air pollution have been known since at least the time of the ancient Greeks, when laws protected the quality of air, as indicated by its odor (Jacobson 2002). Since the Industrial Revolution, air pollution has mainly been associated with urban industrial centers, power plants, and oil and gas refineries. These stationary sources of atmospheric pollutants mainly affect the areas immediately adjacent to them and are usually considered regional rather than global problems. During the twentieth century, however, effective emissions dispersal strategies (e.g., tall smokestacks), widespread industrial development, and greater emissions of pollutants from mobile sources, such as automobiles, have increased the spatial extent of air pollution tremendously.

Fossil fuel combustion, agriculture, and urban and suburban development have influenced fluxes of N and S to an even greater degree than fluxes of C. Emissions of N and S into the atmosphere have resulted in two related environmental issues: acid precipitation and N deposition. Emissions of N and S are only a subset of the multiple types of air pollution, but they are among the most far-reaching. Sites affected by acid precipitation and N deposition now include national parks and wilderness areas (**Figure 24.17**).

Acid precipitation causes nutrient imbalances and aluminum toxicity

The detrimental effects of air pollution on nearby vegetation, buildings, and human health have been known for several centuries, although the causes were not well understood. In England during the mid-nineteenth century, industrial processes that released acidic compounds into the atmosphere, primarily hydrochloric acid, were implicated as a major source of harmful pollution (Jacobson 2002). Legislation was enacted in 1863 to reduce these acidic emissions. The term "acid rain" was first coined by Robert Angus Smith, an inspector charged with enforcing the new regulations. Despite such legislation, acid precipitation continued to be a problem throughout the nineteenth and twentieth centuries in large industrialized urban centers. During the 1960s, awareness of the widespread effects of acid precipitation, including its effects on nearby "pristine" ecosystems and agriculture, increased. In particular, damage to forests and mortality among aquatic organisms in northern Europe, parts of Asia, and northeastern North America prompted greater attention to acid precipitation.

FIGURE 24.17 Air Quality Monitoring in Grand Canyon National Park Visibility, which serves as an index of air quality, is evaluated by the visual range: the maximum distance this Web camera can resolve (up to 225 miles). Air quality in national parks and wilderness areas, such as the Grand Canyon, has been compromised by emissions of pollutants, including sulfate aerosols. These pollutants not only lower visibility, but also pose a health hazard to the organisms that come into contact with them, including humans.

Sulfuric acid (H_2SO_4) and nitric acid (HNO_3) are the main acidic compounds found in the atmosphere. As we saw earlier in this chapter, sulfuric acid forms in the atmosphere from the oxidation of gaseous sulfur compounds, and nitric acid originates from the oxidation of NO_x compounds. Sulfuric and nitric acids can dissolve in water vapor and fall from the sky with precipitation (wet deposition). Naturally occurring precipitation has a pH of 5.0 to 5.6, due to the natural dissolution of CO_2 and formation of carbonic acid. Acid precipitation has a pH range from 5.0 to 2.0. Acidic compounds may also be deposited on Earth's surface when they form aerosols too large to be suspended, or when they attach to the surfaces of dust particles (dry deposition).

Research has focused on determining the causes of the environmental degradation associated with acid precipitation. Initially, the acidity was considered the main culprit. In most cases, however, rainfall and surface waters did not have a low enough pH to cause the observed biological responses. An exception is found in regions at high latitudes or high elevations that develop a seasonal snowpack. During winter, acidic compounds accumulate in the snow. When temperatures increase in spring, water percolates through the snowpack, leaching out all of the accumulated soluble compounds. The first meltwater of spring is therefore more acidic than the precipitation that fell during winter. This *acid pulse* has the potential to be toxic to sensitive organisms in soils and streams, including microorganisms, invertebrates, amphibians, and fish.

The vulnerability of organisms in soils, streams, and lakes to inputs of acid precipitation is determined by the ability of their chemical environment to counteract the acidity, known as its **acid neutralizing capacity**. The acid neutralizing capacity of soils and water is usually associated with their concentrations of base cations, including Ca^{2+}, Mg^{2+}, and K^+. Soils derived from parent material with high concentrations of these cations, such as limestone, are better able to neutralize acid precipitation than those derived from more acidic parent material, such as granite.

The detrimental effects of acid precipitation on plants and aquatic organisms are associated with biogeochemical reactions in the soil that decrease nutrient supplies and increase concentrations of toxic metals. As H^+ percolates through the soil, it replaces Ca^{2+}, Mg^{2+}, and K^+ at cation exchange sites on the surfaces of clay particles (see the description of cation exchange under Concept 21.3). These cations are released into the soil solution and can then leach out of the rooting zone of plants. The loss of base cations leads to a decrease in soil pH, or *soil acidification*. Deficiencies in Ca and Mg, sometimes in combination with other stresses, were associated with large-scale mortality of trees in European forests during the 1970s and 1980s (**Figure 24.18**). In advanced stages of soil acidification, aluminum (Al^{3+}) is released into the soil from cation exchange sites. Aluminum is toxic to plant roots, soil invertebrates, and aquatic organisms, including fish. The combination of increasing acidity in precipitation and increasing aluminum concentrations in terrestrial runoff has been linked to fish die-offs in lakes and streams in northern Europe and eastern North America.

FIGURE 24.18 Air Pollution Has Damaged European Forests The high tree mortality seen in this forest in southeastern Germany is associated with acid precipitation and the resulting nutrient imbalance, particularly losses of nutrient cations. Extensive forest decline occurred in Germany and northern Czechoslovakia (now part of the Czech Republic) in the 1970s and 1980s.

The realization that acid precipitation was negatively affecting the biota of forest and lake ecosystems prompted enhanced monitoring of atmospheric deposition and, eventually, laws to limit acidic emissions. Restrictions on emissions of S in North America and Europe have resulted in significant reductions in the acidity of precipitation (**Figure 24.19**). Forests are recovering from the effects of acid precipitation in central Europe, thanks to legislation limiting S emissions as well as decreased industrial activity in the former Soviet Union. Stream chemistry measurements also reflect the reduced acidity of precipitation and the recovery of aquatic ecosystems. Acid precipitation remains a problem, however, in some countries that have experienced rapid industrial development, such as China and India.

N deposition: too much of a good thing can be bad

Anthropogenic emissions of reactive nitrogen into the atmosphere have greatly altered global N cycles. Reactive N can fall back to Earth (via dry and wet deposition) after being transported some distance in the atmosphere. Globally, anthropogenic emissions and deposition of reactive N compounds are more than three times greater now than they were in 1860 (Galloway et al. 2004) (**Figure 24.20**). Emissions

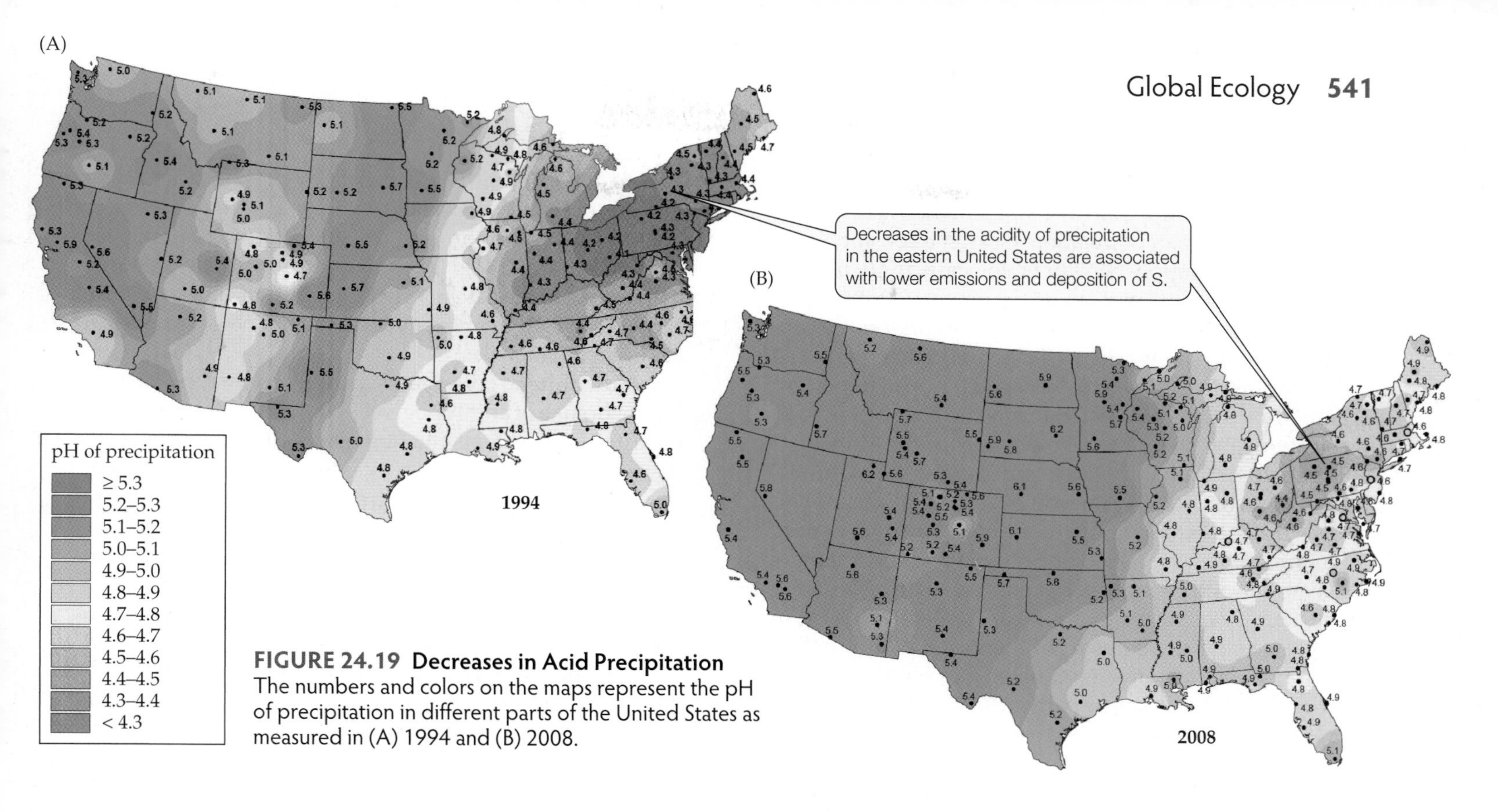

FIGURE 24.19 Decreases in Acid Precipitation The numbers and colors on the maps represent the pH of precipitation in different parts of the United States as measured in (A) 1994 and (B) 2008.

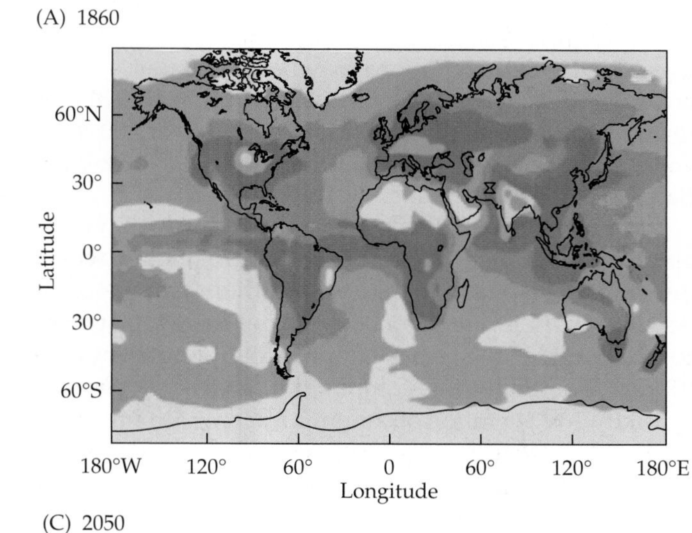

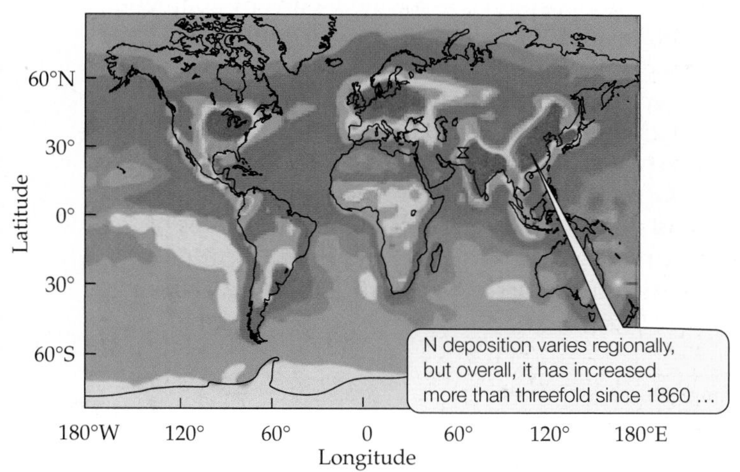

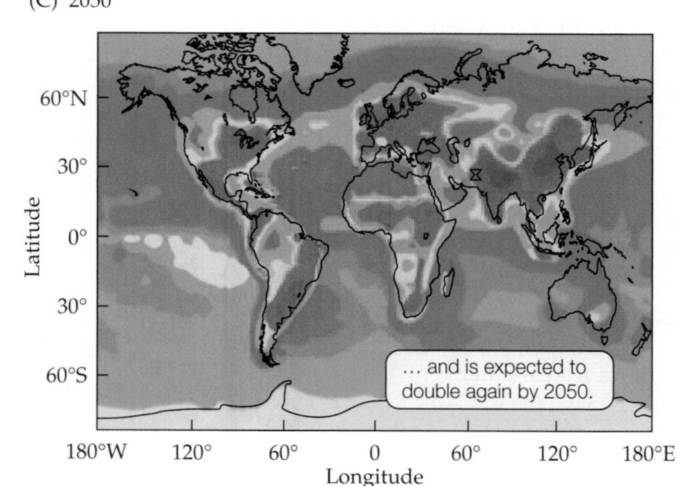

Deposition of inorganic N (mg N/m²/yr)

5,000
2,000
1,000
750
500
250
100
50
25
5

FIGURE 24.20 Historic and Projected Changes in Nitrogen Deposition (A) Estimated rates of deposition of inorganic N compounds (NH_4^+ and NO_3^-) in 1860. (B) Measured rates for the early 1990s. (C) Projected rates for 2050. (After Galloway et al. 2004.)

and deposition of reactive N are expected to double between 2000 and 2050 as industrial development increases to keep pace with the human population. Greater deposition of N will increase the supply of N for biological activity, but this abundance will come with an environmental cost.

The role of N as a determinant of rates of primary production in ecosystems ranging from tundra to open ocean was described in Chapter 19. Nitrogen plays an important role in photosynthesis, which forms the base of the food webs providing energy to all other organisms. Considerable benefit to humanity has accrued from the manufacture of N fertilizers and their widespread application to crops since the early twentieth century. We might expect, therefore, that an increased supply of N would facilitate plant growth and greater overall production in a N-limited ecosystem. Primary production has indeed increased in some ecosystems as a result of increased N deposition (e.g., forests in Scandinavia; Binkley and Högberg 1997). Nitrogen deposition may be partly responsible for a greater uptake of atmospheric CO_2 by terrestrial ecosystems in the Northern Hemisphere (Thomas et al. 2010).

Although primary production is increasing in some ecosystems due to N deposition, there is also strong evidence that N deposition is associated with environmental degradation, loss of biodiversity, and acidification. While N limits primary production in many terrestrial ecosystems, the capacity of vegetation, soils, and soil microbes to take up greater N inputs can be exceeded; this condition is known as *nitrogen saturation* (Aber et al. 1998) (**Figure 24.21**). Greater concentrations of inorganic N compounds NH_4^+ and NO_3^-) in the soil lead to enhanced rates of microbial processes (nitrification and denitrification) that release nitrous oxide (N_2O), a potent greenhouse gas. Nitrate (NO_3^-) is easily leached from soils and can move into groundwater, eventually entering aquatic ecosystems. When NO_3^- moves through the soil, it carries cations, including K^+, Ca^{2+}, Mg^{2+}, in solution to maintain a charge balance. As in the case of acid precipitation, losses of these cations can lead to nutrient deficiencies and eventually to acidification of soils.

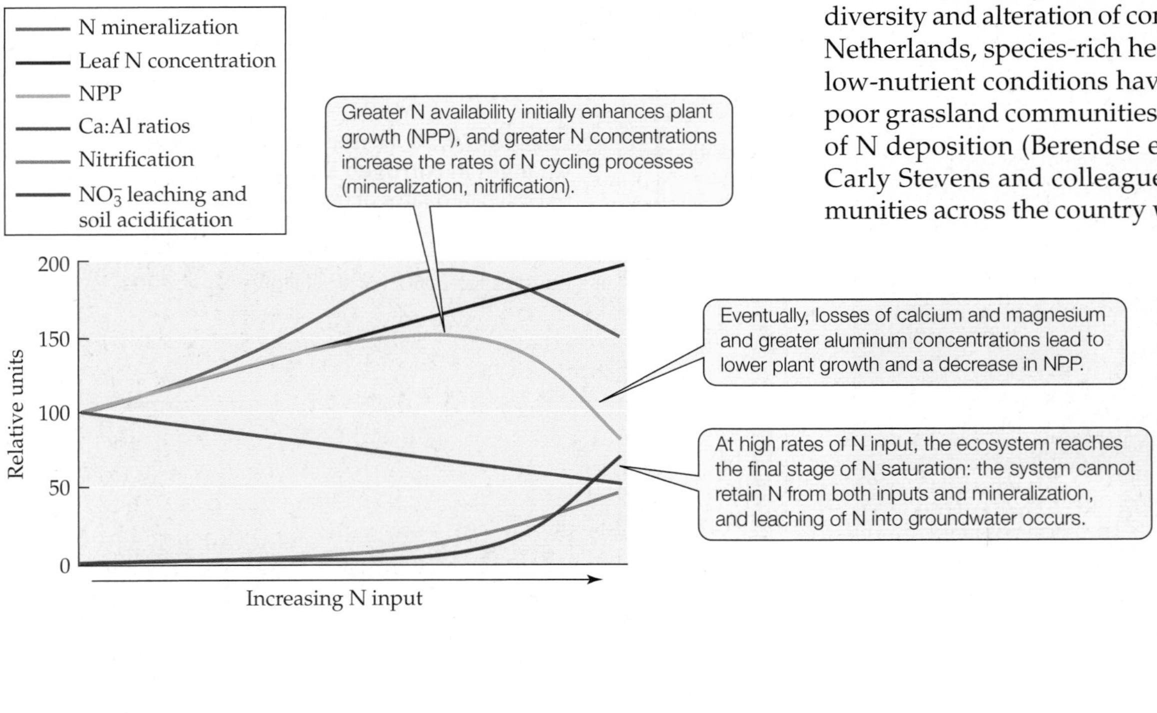

Most aquatic ecosystems are limited by P, so the biological uptake of anthropogenic NO_3^- that enters them from terrestrial ecosystems may be relatively small (although there is greater biological processing of N than expected; see Figure 21.16). Riverine transport of N to nearshore marine ecosystems has increased as inputs of N fertilizer have increased (Howarth et al. 1996). Primary production in estuarine and marsh communities is often limited by N, and thus the influx of N from terrestrial sources into these ecosystems has resulted in eutrophication (described under Concept 21.4). Eutrophication results in heavy algal growth, which can create hypoxic conditions in the bottom waters of nearshore ecosystems. High inputs of organic matter due to increased algal production lead to high rates of decomposition by microorganisms, which consume most of the available oxygen. The resulting hypoxic conditions are lethal for most marine life, including fish. Hypoxic conditions may occur over large areas, creating "dead zones." Dead zones of up to 18,000 km^2 form annually in the Gulf of Mexico, and over 400 dead zones form in locations around the world, including the Baltic Sea, the Black Sea, and Chesapeake Bay.

In nutrient-poor ecosystems, many plants have adaptations that lower their nutrient requirements, which lowers their capacity to take up additional inputs of N. As a result, N inputs may cause faster-growing species to outcompete the species adapted to low-nutrient conditions. Eventually, this competition can lead to losses of diversity and alteration of community composition. In the Netherlands, species-rich heath communities adapted to low-nutrient conditions have been replaced by species-poor grassland communities as a result of very high rates of N deposition (Berendse et al. 1993). In Great Britain, Carly Stevens and colleagues surveyed grassland communities across the country with a range of N deposition

FIGURE 24.21 Effects of Nitrogen Saturation Aber and colleagues devised a conceptual model of the response of forest ecosystems to increasing inputs of inorganic N resulting in nitrogen saturation. (After Aber et al. 1998.)

(A)

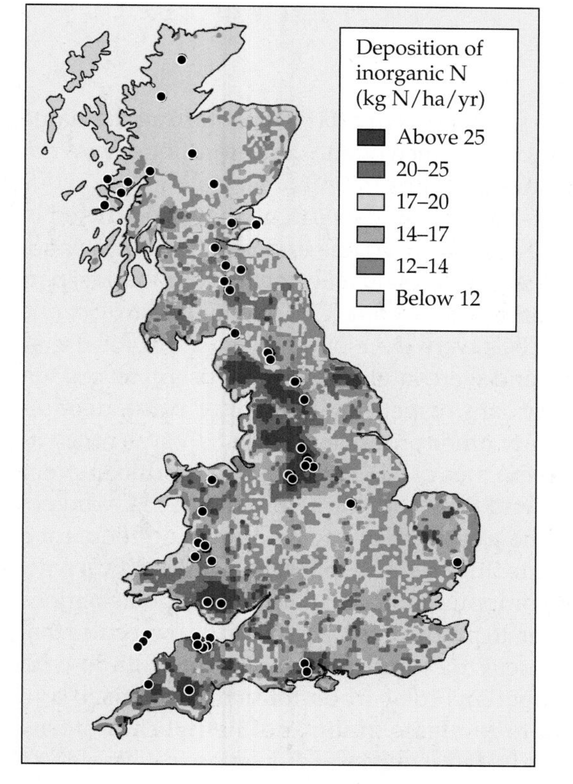

(B)

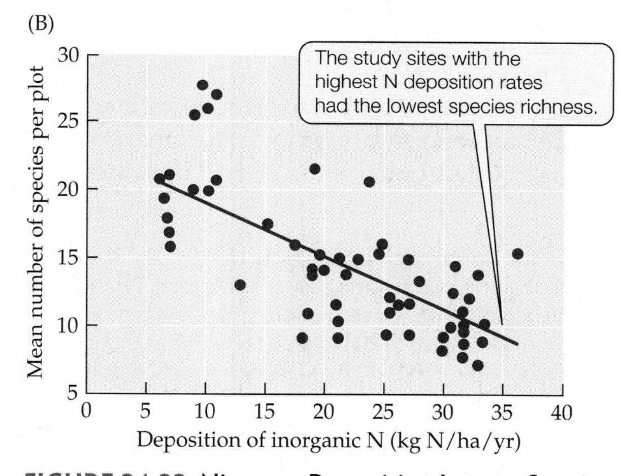

FIGURE 24.22 Nitrogen Deposition Lowers Species Diversity (A) Inorganic N deposition in Great Britain. Dots on the map indicate the study sites where plant species richness in grassland ecosystems was measured. (B) Correlation between rates of inorganic N deposition and plant species richness. (After Stevens et al. 2004.)

rates (**Figure 24.22A**). At 68 sites, they measured the mean plant species richness in multiple study plots, along with several environmental variables, to try to explain the variation in plant diversity among the sites. The environmental variables included nine soil chemical factors, nine physical environmental variables, grazing intensity, and the presence or absence of grazing enclosures (Stevens et al. 2004). Of the 20 possible factors that may have influenced differences in species richness among the study sites, the amount of N deposition explained the greatest amount of variation (55%): higher inputs of N were associated with lower species richness (**Figure 24.22B**). While this study was correlative in nature, its results are supported by a multitude of experimental studies in which adding N to experimental plots decreased species richness, often resulting in the loss of rare species (Suding et al. 2005). Stevens and colleagues suggest that a general trend of decreasing plant diversity has occurred in Great Britain, and will continue to occur with increasing anthropogenic N deposition. High rates of N deposition also facilitate the successful spread of some invasive plant species at the expense of native species (Dukes and Mooney 1999).

The ecological effects of sulfur and nitrogen result when atmospheric deposition returns anthropogenic emissions to Earth's surface. In the next section, we'll describe some anthropogenic compounds that exert negative effects while remaining in the atmosphere.

CONCEPT 24.4 Losses of ozone in the stratosphere and increases in ozone in the troposphere each pose risks to organisms.

Atmospheric Ozone

Ozone is good for biological systems, but only when it is not in close contact with them. In the upper atmosphere (the *stratosphere*), ozone provides a shield that protects Earth from harmful radiation. When in contact with organisms in the lower atmosphere (the *troposphere*), however, ozone can harm them. Detrimental changes in ozone concentrations have occurred in both the stratosphere and the troposphere as a result of anthropogenic emissions of air pollutants.

Loss of stratospheric ozone increases transmission of harmful radiation

About 2.3 billion years ago, when prokaryotes first evolved the capacity to carry out photosynthesis, oxygen began to accumulate in Earth's atmosphere, leading to a series of changes that facilitated the evolution of greater physiological and biological diversity. The increase in atmospheric oxygen (in the form of O_2) also led to the formation of a layer of ozone (O_3) in the stratosphere (at 10–50 km altitude). This ozone layer acts as a shield protecting Earth's surface from high-energy ultraviolet-B (UVB) radiation (0.25–0.32 μm). UVB radiation is harmful to all organisms, causing damage to DNA and photosynthetic pigments in plants and bacteria, impairment of immune responses, and cancerous skin tumors in animals, including humans.

Stratospheric ozone concentrations change seasonally as a result of changes in atmospheric circulation patterns, particularly in the polar zones, where they decline in spring (Jacobson 2002). British scientists measuring ozone concentrations in the Antarctic were the first to record an unusually large decrease in springtime stratospheric ozone concentrations starting in 1980. Since then, there has been a progressively larger decrease in stratospheric ozone concentrations in the Antarctic each spring, lasting about 2 months. Ozone concentrations have decreased by as much as 70% relative to those recorded in the same season prior to 1980 (**Figure 24.23**). There has also been an increase in the spatial extent of the Antarctic region experiencing a decrease in ozone, called the ozone hole. An **ozone hole** is defined as an area with an ozone concentration of less than 220 Dobson units (= 2.7×10^{16} molecules of ozone) per square centimeter; prior to 1980, ozone concentrations had never been recorded below this level. Ozone decreases have been recorded from 25° S to the South Pole. Similar reductions in ozone have been recorded in the Arctic (from 50° N to the North Pole), although the magnitude of the decrease has not been as great (thus conferring the name **Arctic ozone dent**, since ozone concentrations have not dropped below 220 Dobson units).

The decrease in stratospheric ozone was predicted in the mid-1970s by Mario Molina and Sherwood Rowland, who discovered that certain chlorinated compounds, particularly chlorofluorocarbons (CFCs), could destroy ozone molecules. CFCs were developed in the 1930s for use as refrigerants and were later found to be useful as propellants in spray cans dispensing hair spray, paint, deodorants, and many other products. By the 1970s, as much as a million metric tons of CFCs were being produced every year. Molina and Rowland (1974) found that CFCs did not degrade in the troposphere and could remain there for a very long time. In the stratosphere, however, CFCs react with other compounds, particularly in the polar regions during winter, to produce reactive chlorine molecules that destroy ozone. Other anthropogenic compounds with the same effect include carbon tetrachloride, used as a solvent and to fumigate grain, and methyl chloroform, used as an industrial solvent and degreaser. A single free chlorine atom has the potential to destroy 10^5 ozone molecules. Thus, the potential danger posed by chlorinated compounds to the stratospheric ozone layer was clear to Molina and Rowland.

Increases in UVB radiation at Earth's surface have been recorded as concentrations of stratospheric ozone have decreased (Madronich et al. 1998). These increases

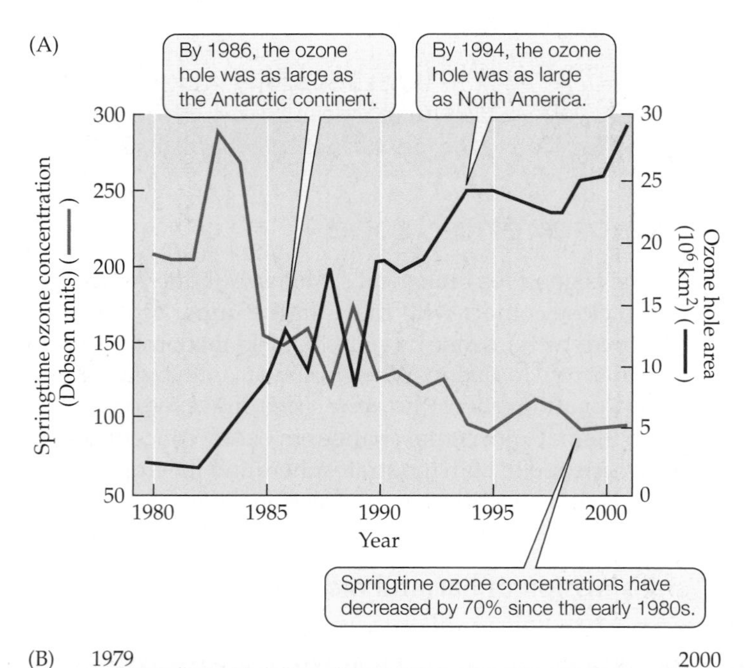

(B) 1979 2000

FIGURE 24.23 The Antarctic Ozone Hole (A) Since 1980, there have been dramatic decreases in springtime ozone concentrations over the Antarctic region, with concentrations dropping below the threshold for ozone hole status (220 Dobson units) for a large proportion of the region after 1984. (B) Average ozone concentrations over Antarctica for the months of September 1979 and September 2000, with the lowest ozone levels shown in purple, demonstrate the dramatic decrease in concentration that occurred during this period.

have been most striking in the Antarctic region, which has experienced as much as a 130% increase in UVB radiation during spring. Increases have also been recorded in the Northern Hemisphere, including a 22% increase at mid-latitudes during spring.

These increases in UVB radiation at Earth's surface have coincided with an increasing incidence of skin cancer in humans. The occurrence of skin cancer is now approximately ten times higher than it was in the 1950s. UVB radiation had an important role in the evolution of pigmentation in humans (Jablonski 2004). The production of melanin, a protective skin pigment, was selected for in humans living at low latitudes, where ozone levels are naturally lowest and the highest levels of UVB radiation reach Earth's surface. As humans migrated away from equatorial Africa into colder climates with less sunlight, however, high amounts of melanin in the skin limited production of vitamin D, resulting in selection for lower melanin production in peoples of higher latitudes. As these lighter-skinned humans have subsequently migrated into environments with higher UVB radiation, to which their complexions are not adapted, they have increased their risk of skin cancers. This has become particularly true for populations at high latitudes in the Southern Hemisphere, including Australia, New Zealand, Chile, Argentina, and South Africa, where exposure to UVB is enhanced by stratospheric ozone loss. Concern is particularly great in Australia, where nearly 30% of the population is diagnosed with some form of skin cancer.

Substantial evidence exists to indicate that increasing UVB radiation has important ecological effects (Caldwell et al. 1998; Paul and Gwynn-Jones 2003). Sensitivity to UVB radiation varies among the species within a community, and as a result, changes in community composition are likely to result from increased UVB radiation. The potential for detrimental UVB effects due to stratospheric ozone loss is greatest at high latitudes and at high elevations (>3,000 m; 9,800 feet), due to lower atmospheric filtering of UV radiation.

The realization of the rapid decreases in stratospheric ozone concentrations, and of their probable anthropogenic cause, resulted in several international conferences on ozone destruction in the 1980s. At these conferences, the Montreal Protocol, an international agreement calling for a reduction and eventual end of the production and use of CFCs and other ozone-degrading chemicals, was developed. The Montreal Protocol has been signed by more than 150 countries. Concentrations of CFCs have remained the same or, in most cases, declined since the Montreal Protocol went into effect in 1989 (**Figure 24.24**). A progressive recovery of the ozone layer is expected to occur over several decades, since the slow mixing of the troposphere and stratosphere, with the long-lived CFCs they still contain, will result in a time lag before stratospheric ozone concentrations rise. Full recovery of the ozone layer is expected by 2050.

Tropospheric ozone is harmful to organisms

Ninety percent of Earth's ozone is found in the stratosphere. The remaining 10% occurs in the troposphere. Tropospheric ozone is generated by a series of reactions involving sunlight, NO_x, and volatile organic compounds such as hydrocarbons, carbon monoxide, and methane. In some regions, natural vegetation can be an important source of volatile organic compounds, which include terpenes (which give pines their characteristic odor) and isoprene. Under natural atmospheric conditions, the amount of ozone produced in the troposphere is very small, but anthropogenic emissions of ozone precursor molecules have greatly increased its production. Air pollutants that produce ozone can travel long distances, and thus tropospheric ozone production is a widespread concern.

Tropospheric ozone is environmentally damaging for two main reasons. First, ozone is a strong oxidant; that is, the oxygen in it reacts easily with other compounds. Ozone causes respiratory damage and is an eye irritant in humans and other animals. Increases in the incidence of childhood asthma have been linked to exposure to ozone. Ozone damages the membranes of plants and can decrease their photosynthetic rates and growth. Ozone also increases the susceptibility of plants to other stresses, such as low water availability. Decreases in crop yields have been associated with exposure to ozone. Characteristic symptoms of ozone pollution have been found in native plants near urban areas since the 1940s and 1950s (e.g., in the San Gabriel Mountains near Los Angeles and in the northern Alps in Italy), but more recently, symptoms have been noted in national parks and wilderness areas farther from sources of pollution. For example, plants in the Sierra Nevada of California are negatively affected by ozone generated in the Central Valley and the San Francisco and Los Angeles urban areas (Bytnerowicz et al. 2003). Growth rates of trees in forests of the eastern United States are as much as 10% lower than they would be in the absence of ozone (Chappelka and Samuelson 1998).

Second, ozone is a greenhouse gas that can contribute to global climate change. Ozone has a short life span in the atmosphere relative to other greenhouse gases, however, and its concentration can vary greatly from place to place. Thus, the effect of anthropogenic ozone on climate change is difficult to estimate.

Strategies to limit tropospheric ozone production have focused on lowering anthropogenic emissions of NO_x and volatile organic compounds. In most developed countries, efforts to lower emissions of ozone-producing compounds have met with success. In the United States, for example, emissions of volatile organic compounds dropped by 50% between 1970 and 2004, emissions of NO_x dropped by 30%

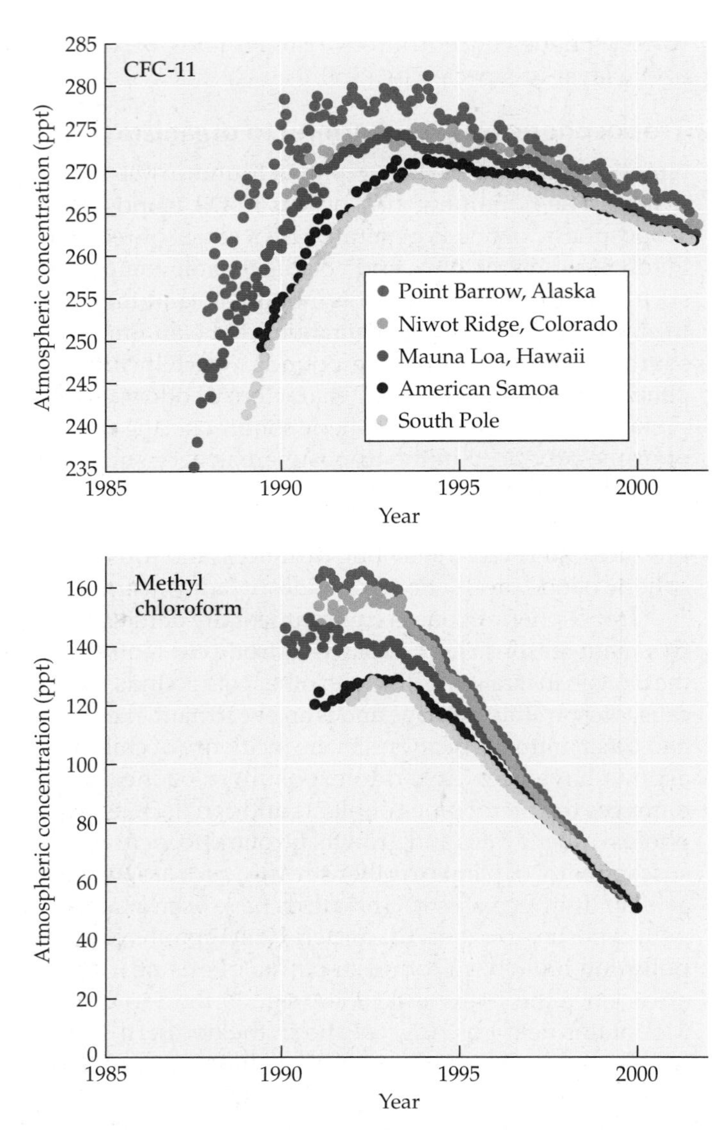

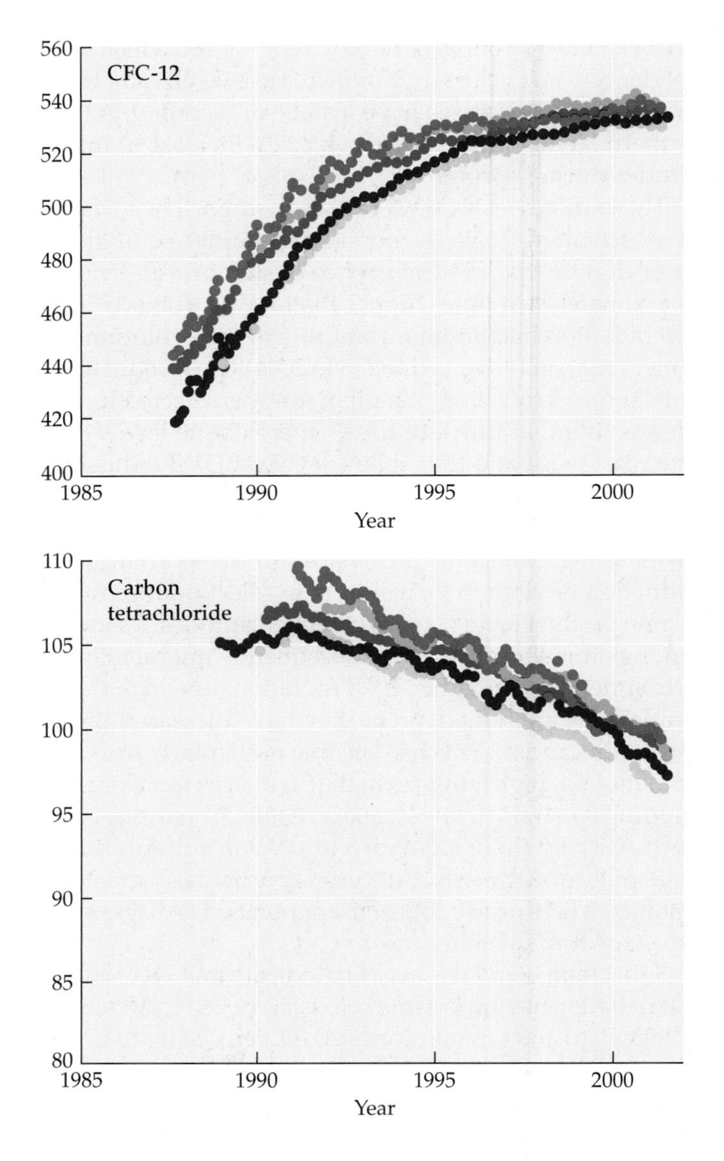

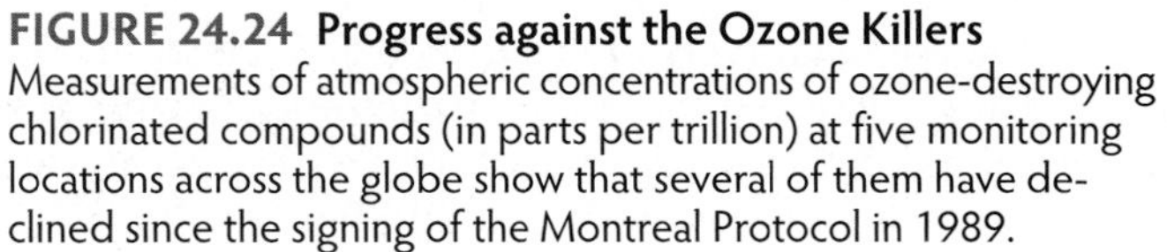

FIGURE 24.24 Progress against the Ozone Killers
Measurements of atmospheric concentrations of ozone-destroying chlorinated compounds (in parts per trillion) at five monitoring locations across the globe show that several of them have declined since the signing of the Montreal Protocol in 1989.

(U.S. EPA 2005), and tropospheric ozone concentrations are decreasing near large urban areas (Fiore et al. 1998). Regulation of emissions of ozone-producing compounds has not been as strict in some developing countries, however. Ozone is a serious air pollutant in urban and agricultural regions of China and India, but stricter environmental regulations are now being put in place.

A CASE STUDY REVISITED

Dust Storms of Epic Proportions

We've seen throughout this chapter that many aspects of global ecology—such as greenhouse gases and climate change, emissions and deposition of N and S, and stratospheric destruction and tropospheric production of ozone— involve transport and chemical processes in the atmosphere. The movements of dust described in this chapter's Case Study (p. 525) are also influenced by atmospheric processes, including rainfall patterns and wind. We've also seen that humans impact the environment at a global scale through emissions of greenhouse gases and pollutants into the atmosphere. Land use change, which alters the amount and type of vegetation cover primarily through deforestation for agricultural and pastoral development, as described in Chapter 3, generally influences the environment at a more local scale. However, land use change in arid zones that are subject to periodic severe droughts can have global-scale effects by enhancing the amount and spread of dust into the atmosphere.

During the early part of the twentieth century, the southwestern part of the U.S. Great Plains was opened up for agricultural development. The natural vegetation of the region consisted of drought- and grazing-tolerant grasses. Bison, which had grazed the land for centuries, were replaced by cattle in the late nineteenth century. Economic demand for wheat, due to losses of agricultural lands in Europe during World War I, and the recent population expansion into the southern Great Plains encouraged the development of agriculture. Although this area was known to experience periodic droughts, farmers, encouraged by the notion that "rain follows the plow" and by recent technological developments in farming, cultivated large areas of land, plowing under the native prairie grasses and replacing them with wheat. For a while, the weather was conducive to agriculture, and the farmers prospered. However, the 1930s brought prolonged severe drought. Fields dried up, and with no protective network of roots to hold the soil together, it began to blow away. Major dust storms carried the soil across the North American continent and all the way to the Atlantic Ocean. The Dust Bowl event is still considered the worst environmental disaster the United States has ever experienced (Egan 2006). Similar circumstances in Asia enhanced the severity of dust storms there. Deforestation, the development of agriculture in marginal zones, overgrazing, and the drainage of Aral Sea for irrigation have all been implicated in the increased severity of dust storms following the mid-1990s (Wang et al. 2004).

While dust storms in urban areas are a rarity, large-scale dust storms originate in desert regions regularly (**Figure 24.25**). However, both the Dust Bowl and Asian examples suggest that while dust storms are a natural phenomenon, a combination of agricultural development of marginal lands and severe drought exacerbates these events (Cook et al. 2009). At a global scale, extreme droughts and land use change contribute one-third to one-half of the inputs of dust into the atmosphere (Tegen and Fung 1995). Desert regions, such as the Gobi and Sahara–Sahel regions, have expanded at their margins due to land use change since the 1970s, increasing the global impact of dust storms. For example, Asian dust has been detected in the European

(A)

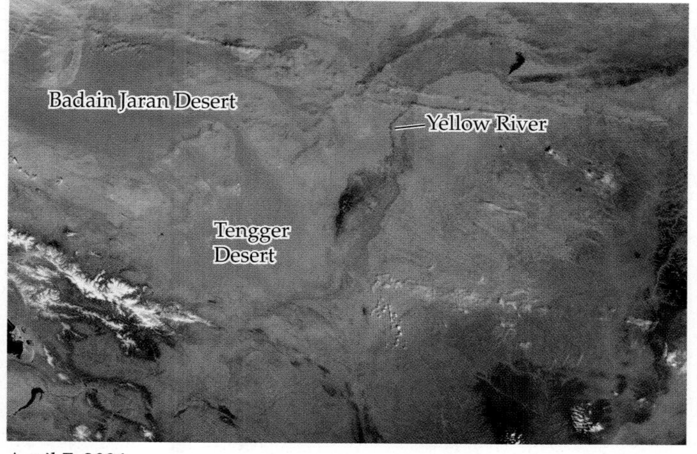

April 7, 2006

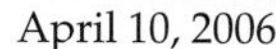
April 10, 2006

(B)

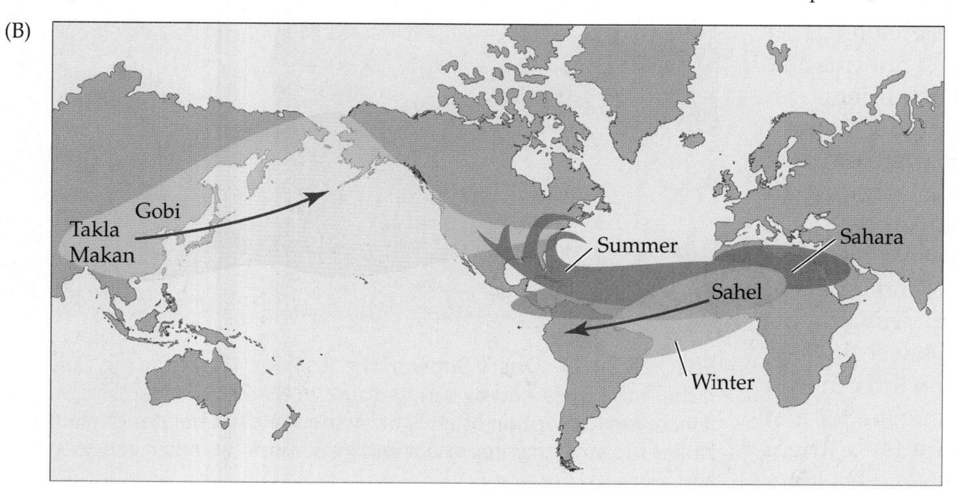

FIGURE 24.25 Desert Origins of Global Dust Storms Deserts are sources of dust that may travel large distances and have important ecological impacts in distant regions. (A) The photo on the left is a satellite image of the Gobi Desert in early April, 2006. The photo on the right shows the same region 3 days later, obscured by a massive dust storm. (B) Sources of the dust deposited in the Caribbean region include the deserts of North America and Asia. The main directions of dust flow are indicated by arrows. (After Garrison et al. 2003.)

(A)

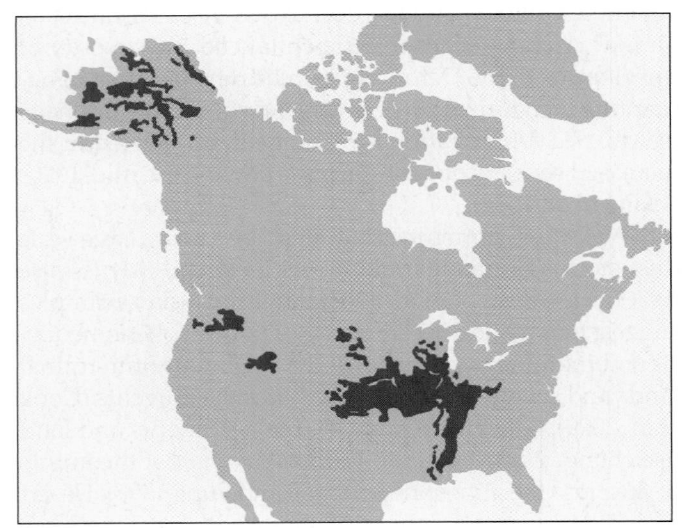

(B)

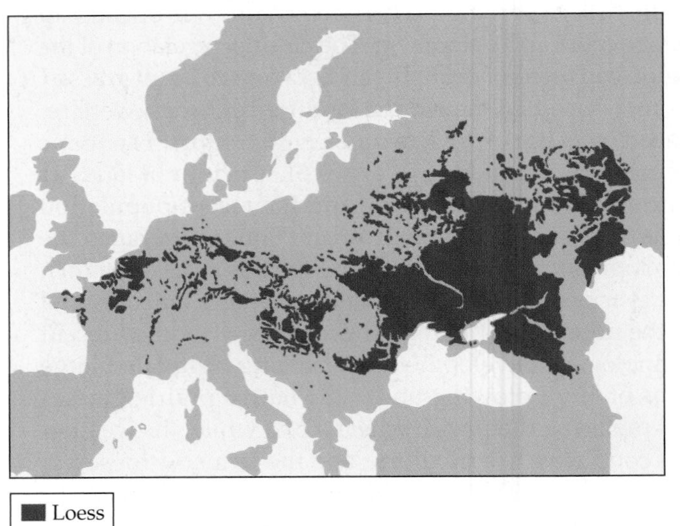

FIGURE 24.26 **Distribution of Loess Soils** As continental glaciers receded following the most recent glacial maximum, wind carried substantial amounts of loose soil from the exposed areas. Large areas of (A) North America and (B) Europe were covered with deep layers of this material, which developed into loess soils.

Alps, traveling two-thirds of the way around the globe in approximately a week (Grousset et al. 2003). On a geologic time scale, major periods of dust redistribution occur in association with the recession of large ice sheets during interglacial periods (see the discussion of glacial cycles under Concept 2.5), as evidenced by the distribution of loess soils across North America and Europe (**Figure 24.26**), some hundreds of meters thick.

CONNECTIONS IN NATURE

Dust as a Vector of Ecological Impacts

The ecological effects of dust removal and deposition are not fully understood, but one of the best-studied effects is the movement of nutrients (as described in Chapter 21) at spatial scales ranging from a few meters to continents and oceans (Field et al. 2010). Dust deposition of nutrients can have important consequences for primary production and the global carbon cycle. The supply of iron (Fe) from dust deposition is important for oceanic primary production (Mahowald et al. 2005), as we saw under Concept 19.2. Dust from the Asian storms described earlier has been associated with algal blooms in the Pacific, and inputs of cations from African dust are important to primary production in tropical forests in the Amazon (Okin et al. 2004). In contrast, the removal of surface soils by wind can lead to lower production due to losses of organic matter and fine mineral particles, which are important for nutrient supply and retention. Dust may also be important in long-distance transport of pathogens (Garrison et al. 2003) and pollutants (Jaffe et al. 2003) and could influence disease dynamics (described under Concept 13.5).

The ecological effects of dust movement can be both direct and indirect. Nutrient input and loss are examples of its direct effects. An example of an indirect effect occurs in the southwestern United States when dust transported from the Colorado Plateau falls in the Rocky Mountains and alters the timing of snowmelt. As noted in the Case

FIGURE 24.27 **Dusty Snow in the Rockies** Dust from the Colorado Plateau is carried by spring storms to the Rocky Mountains, where it increases absorption of sunlight by snow and accelerates its melting. Earlier snowmelt has important implications for mountain ecosystems and regional hydrology.

Study in Chapter 21 (p. 452), grazing and recreational vehicle use has disturbed biological crusts in arid lands of the Colorado Plateau, increasing their erodibility and dust input into the atmosphere. Most of the dust is swept away in spring storms, and some ends up deposited in snow on the Rockies (**Figure 24.27**). The dust increases the amount of sunlight absorbed by the land surface, warming the snow and causing accelerated melt. Earlier snowmelt has the potential to increase the length of the growing season for plants growing in areas with deep snow cover. However, rather than stimulating earlier growth of plants in areas that melt sooner, accelerated snowmelt delays the initiation of growth and flowering of alpine plants, which wait to initiate growth until air temperatures are suitable for growth. This delay results in greater synchrony of greening up of alpine plants, possibly leading to greater competition (Steltzer et al. 2009). In contrast, earlier snowmelt in lower-elevation subalpine meadows triggers some plants to initiate growth immediately, exposing them to potentially killing frosts (Inouye 2008). The surrounding subalpine forests may experience water shortages when snowmelt occurs earlier, which may lower their NPP (Hu et al. 2010). The ecological impacts of dust, both direct and indirect, remind us that ecological phenomena occur at a global scale, have widespread importance, and testify to the role of humans in intensifying their effects.

SUMMARY

CONCEPT 24.1 Elements move among geologic, atmospheric, oceanic, and biological pools at a global scale.

- The global carbon cycle includes large exchanges of CO_2 between the atmosphere and Earth's land surface associated with photosynthesis and respiration and, within the last 160 years, anthropogenic emissions of CO_2 and CH_4.
- Atmospheric concentrations of CO_2 and CH_4 are increasing due to burning of fossil fuels, deforestation, and agricultural development.
- Elevated atmospheric CO_2 concentrations may increase terrestrial plant growth and the acidity of the oceans, causing ecological changes.
- Global fluxes of nitrogen are associated with biological uptake and chemical transformations. Anthropogenic nitrogen fixation and emissions now dominate the global nitrogen cycle.
- The global cycles of phosphorus and sulfur include both geochemical and biological fluxes.
- Anthropogenic fluxes of phosphorus associated with mining and industrial emissions of sulfur far exceed natural fluxes associated with weathering.

CONCEPT 24.2 Earth is warming due to anthropogenic emissions of greenhouse gases.

- Elevated levels of CO_2, CH_4, N_2O, and other greenhouse gases in the atmosphere have warmed Earth, particularly since the 1950s. This warming trend is expected to continue throughout the twenty-first century.
- Large changes in species distributions, community composition, and ecosystem processes are expected as a result of global climate change.
- Recent changes in species' geographic ranges and in carbon source-sink relationships have been attributed to climate change.

CONCEPT 24.3 Anthropogenic emissions of sulfur and nitrogen cause acid deposition, alter soil chemistry, and affect the health of ecosystems.

- Sulfuric and nitric acids form in the atmosphere from compounds emitted by human activities, mainly by the burning of fossil fuels. These compounds are subsequently deposited on Earth's surface as acid precipitation.
- Acid precipitation causes nutrient imbalances and aluminum toxicity in soils.
- Atmospheric deposition of reactive nitrogen compounds emitted from anthropogenic sources can increase productivity in some ecosystems, but may also lead to soil acidification, dead zones in near-shore aquatic ecosystems, losses of species diversity, and increases in invasive species.

CONCEPT 24.4 Losses of ozone in the stratosphere and increases in ozone in the troposphere each pose risks to organisms.

- Anthropogenic emissions of chlorinated compounds have led to a loss of stratospheric ozone since the 1980s, particularly at high latitudes, and thus to an increase in the levels of harmful ultraviolet-B radiation reaching Earth's surface.
- Reactions involving volatile organic compounds, many of which are of anthropogenic origin, generate ozone in the troposphere, where it can harm organisms.

REVIEW QUESTIONS

1. What are the major biological influences on the global carbon cycle? How have human influences during the past two centuries affected the fluxes of CO_2 associated with these biological influences (i.e., other than fossil fuel burning) and subsequently increased atmospheric CO_2 concentrations?
2. Global climate change has occurred at an accelerated rate since the latter part of the twentieth century, and this trend is expected to continue into the twenty-first century. As the climate changes, terrestrial animals are capable of migrating to regions where the climate is optimal for their function. Despite their mobility, ecologists are still predicting that many animal species will experience local extinctions. Explain why animal responses to climate change will involve factors other than physiological tolerances and dispersal rates.
3. Describe how acid precipitation and nitrogen deposition are linked and how they differ.
4. How can ozone in the atmosphere be both good and bad for organisms?

ON THE COMPANION WEBSITE
sites.sinauer.com/ecology2e

The website includes Chapter Outlines, Online Quizzes, Flashcards & Key Terms, Suggested Readings, a complete Glossary, and the Web Stats Review. In addition, the following resources are available for this chapter:

HANDS-ON PROBLEM SOLVING

Nitrogen Cycle: Too Much or Too Little?

This Web exercise explores global flows in reactive nitrogen from anthropogenic sources. You will read a recent paper on anthropogenic transformation of the global nitrogen cycle. You will then calculate gains and losses of nitrogen on a continental scale and discuss the potential effects on humans and the natural environment.

WEB EXTENSION

24.1 Climate Models, Volcanoes, and Climate Change

Appendix
Some Metric Measurements Used in Ecology

MEASURES OF	UNIT	EQUIVALENTS	METRIC → ENGLISH CONVERSION
Length	meter (m)	base unit	1 m = 39.37 inches = 3.28 feet
	kilometer (km)	1 km = 1000 (10^3) m	1 km = 0.62 miles
	centimeter (cm)	1 cm = 0.01 (10^{-2}) m	1 cm = 0.39 inches
	millimeter (mm)	1 mm = 0.1 cm = 10^{-3} m	1 mm = 0.039 inches
	micrometer (µm)	1 µm = 0.001 mm = 10^{-6} m	
	nanometer (nm)	1 nm = 0.001 µm = 10^{-9} m	
Area	square meter (m^2)	base unit	1 m^2 = 1.196 square yards
	hectare (ha)	1 ha = 10,000 m^2	1 ha = 2.47 acres
Volume	liter (L)	base unit	1 L = 1.06 quarts
	milliliter (ml)	1 ml = 0.001 L = 10^{-3} L	1 ml = 0.034 fluid ounces
	microliter (µl)	1 µl = 0.001 ml = 10^{-6} L	
Mass	gram (g)	base unit	1 g = 0.035 ounces
	kilogram (kg)	1 kg = 10^3 g	1 kg = 2.20 pounds
	teragram (Tg)	1 Tg = 10^{12} g	
	petagram (Pg)	1 Pg = 10^{15} g	
	milligram (mg)	1 mg = 10^{-3} g	
	microgram (µg)	1 µg = 10^{-6} g	
	picogram (pg)	1 pg = 10^{-12} g	
Temperature	degree Celsius (°C)	base unit	°C = $\frac{5}{9}$(°F – 32)
			0°C = 32°F (water freezes)
			100°C = 212°F (water boils)
			20°C = 68°F ("room temperature")
Pressure	Megapascal (MPa)		1 MPa = 145 psi (pounds per square inch)
Energy	joule (J)		1 J ≈ 0.24 calorie = 0.00024 kilocalorie*

*A *calorie* is the amount of heat necessary to raise the temperature of 1 gram of water 1°C. The *kilocalorie*, or nutritionist's calorie, is what we commonly think of as a calorie in terms of food.

Answers to Figure Legend and Review Questions

CHAPTER 1

Answers to Figure Legend Questions

Figure 1.5 The results for cages from which *Ribeiroia* was excluded show that pesticides acting alone do not cause frog deformities. The results for cages exposed to *Ribeiroia* show that pesticides do affect frogs, since the percentage of frogs with deformities was higher in ponds where pesticides were present. However, the results do not indicate how pesticides caused that effect.

Figure 1.6 By comparing results from the controls with results from treatments in which pesticides were added, the investigator could test whether addition of a pesticide affected either the immune system response (number of eosinophils) of the tadpoles or the number of *Ribeiroia* cysts per tadpole. The intent of the "solvent control" was to check for possible effects of the solvent in which the pesticide was dissolved.

Figure 1.11 Producers absorb nutrients such as nitrogen from the environment and use them for growth (step 1). The nitrogen in the producer's body may then be transferred to a series of consumers: to an herbivore that eats the plant, a carnivore that eats the herbivore, a second carnivore that eats the first, and so on (step 2). Eventually, however, the nitrogen is returned to the physical environment when the dead body of the organism containing it is broken down by decomposers (step 3).

Figure 1.13 If the tanks exposed to UV light had been left uncovered, the environmental conditions in those tanks might have differed in many ways from the environmental conditions in tanks covered with plastic that blocked UV light. By covering both UV-exposed and UV-blocked tanks with clear plastic, the investigators could ensure that the main difference between the tanks was whether or not the tadpoles were exposed to UV light.

Answers to Review Questions

1. The phrase "connections in nature" is meant to evoke the fact that interactions among organisms and between organisms and their environment cause events in nature to be interconnected. As a result of such connections, an action that directly affects one part of an ecological community may cause unanticipated side effects in another part of the community. Various examples related to amphibian deformities and population declines illustrate such connections and their side effects. For example, it appears that the addition of fertilizers to ponds has led to the following chain of events: the fertilizer stimulates increased algal growth, which then leads to increased snail abundance, increased *Ribeiroia* abundance, and hence more frequent amphibian deformities.
2. Ecology is the scientific study of interactions between organisms and their environment. The scope of ecology is broad, and it may address virtually any level of biological organization (from molecules to the biosphere). Most ecological studies, however, emphasize on one or more of the following levels: individuals, populations, communities, or ecosystems. Thus, if ecologists studied the effects of a particular gene, they probably would emphasize how the gene affected interactions in nature—they might, for example, study how a gene affected the ability of an organism to cope with its environment, or how a gene affected interactions among species. Compared with a geneticist or cell biologist, an ecologist would be less likely to emphasize either the gene itself or its effects on the workings of a cell, and more likely to study how the gene affected interactions in nature that occur at the individual, population, community, or ecosystem levels.
3. The scientific method summarizes the process of scientific inquiry. The four key steps in this inquiry process are: (1) observe nature and ask a question about those observations; (2) use previous knowledge or intuition to develop hypotheses (possible answers) to those questions; (3) evaluate different hypotheses by performing experiments, collecting new observations, or analyzing results from quantitative models; and (4) use the results from the approaches taken in (3) to modify the hypotheses, pose new questions, or draw conclusions about the natural world. An essential feature of many scientific investigations is a controlled experiment in which results from an experimental group (that has the factor being tested) are compared with results from a control group (that lacks the factor being tested).

CHAPTER 2

Answers to Figure Legend Questions

Figure 2.4 An increase in atmospheric greenhouse gases would increase the flux of infrared radiation back to Earth's surface and would have a warming effect on Earth's climate. Atmospheric aerosols reflect incoming solar radiation, so an increase in these particles would have a cooling effect on Earth's climate.

Figure 2.15 The larger a continent, the greater the seasonal temperature changes there. Because water has a higher heat capacity than land, seasonal temperature changes increase with distance from the ocean. Higher latitudes experience greater seasonal changes in radiation, for reasons we will explore in Concept 2.5.

Figure 2.19 Changing a light-colored vegetation surface (grassland) to a dark green vegetation surface (crops) would decrease the albedo, increasing absorption of solar radiation. There would be greater latent heat loss due to greater transpiration of water to the atmosphere.

Figure 2.22 Seasonal changes in lake stratification would be unlikely in tropical lakes because seasonal changes in air temperature, and therefore water temperature, would be small.

Figure 2.26 Glacial periods would be promoted by (1) an elliptical orbit, taking Earth farther from the sun during the aphelion; (2) a maximum tilt in Earth's axis, lowering the amount of solar radiation received during winter, and (3) having Earth's axis tilted such that winter in the Northern Hemisphere, where the majority of the land mass is found, occurs during the aphelion, when Earth is farthest from the sun.

Figure 2.29 The PDO was in its warm phase in 1987, so the best fishing would have been in the waters off Alaska.

Answers to Review Questions

1. Extreme environmental conditions, such as high and low temperatures or droughts, are important determinants of mortality in organisms. As a result, species' distributions often reflect extreme environmental conditions more than average conditions. The timing of changes in the physical environment is also important, as exemplified by the response of vegetation to the timing of precipitation, which is not reflected in average annual conditions.
2. Differences in the intensity of solar radiation across Earth's surface establish latitudinal gradients of surface heating. Greater heating in the tropics results in rising air currents, which establish large-scale atmospheric circulation cells, called Hadley cells. The warm rising air also promotes high amounts of precipitation on the tropics. Polar cells form where cold, dense air descends at the poles. Between the Hadley and polar cells are the Ferrell cells, driven by the movement of the Hadley and polar cells and the exchange of energy between equatorial and polar air masses. The temperate zone is found at mid-latitudes in association with the Ferrell cells.
3. a. Mountains influence climate by steering the movements of large air masses. Mountains also force air masses passing over them to rise, cooling the air. Precipitation increases on their windward slopes, since the cooler air has a lower capacity to hold water vapor. When the air mass descends on the leeward slopes, it is drier and warmer, and thus the climate there is also drier and warmer, a phenomenon known as the "rain-shadow effect." Mountains can also create local upslope and downslope winds due to differential heating and cooling of their slopes.
   b. Ocean water can absorb more energy without a change in its own temperature than land can. As a result, land areas adjacent to oceans experience smaller seasonal temperature changes than inland areas. Oceans may also provide moisture to air masses passing over adjacent land surfaces, enhancing precipitation.
4. Salinization is a progressive increase in soil salinity due to surface evapotranspiration of water. Desert areas have high rates of evapotranspiration and little precipitation to leach salts to deeper soil layers. Some desert soils also have impervious soil layers underlying the surface layer that impede leaching, increasing the potential for salinization.

CHAPTER 3

Answers to Figure Legend Questions

Figure 3.4 Grasslands and shrublands might occur in areas with combinations of precipitation and temperature usually associated with forests or savannas due to disturbances such as fire or deforestation by humans or an outbreak of herbivory. These factors would limit successful establishment of trees, which would normally crowd out grasses and shrubs.

Figure 3.5 A comparison of Figures 3.5A and B shows that the greatest human impacts have occurred in grassland and deciduous forest biomes of North America and Eurasia (principally due to cropland development). Note that in the Indian subcontinent and in South America, human impacts have occurred primarily in the tropical seasonal forest biome.

Figure 3.11 Both east- and west-facing slopes would have distinct biological zonation associated with gradients of temperature and precipitation, but precipitation would be lower on the east-facing slope due to the rain-shadow effect. As a result, a forest community on the west-facing slope might be replaced by a shrub or grassland community at the same elevation on the east-facing slope.

Figure 3.14 Oxygen levels would be highest where the stream velocity is the fastest, in the main channel. This is where organisms with the highest oxygen demands, typically fish, are found. The lowest oxygen levels are found in the benthic and hyporheic zones, where organisms must be able to tolerate hypoxic conditions.

Answers to Review Questions

1. Plant growth forms are good indicators of the physical environment, particularly climatic and soil conditions. Because plants are immobile as adults (seeds can move), they have evolved morphological features that allow them to cope with their physical environment, including its extremes. Leaf life span (evergreen versus deciduous leaves), for example, reflects the fertility of the soil. Some biomes, such as grasslands, can also be indicators of disturbances such as grazing or fire. Animals can be important features of and controls on biome distribution, but their mobility renders them less useful as indicators of biomes.
2. Biomes are associated with the major climatic zones described in Chapter 2. Tropical rain forests are associated with a tropical climate characterized by high annual precipitation with only slight seasonal variations in the amount of precipitation. As the seasonality of rainfall becomes more pronounced further north and south from the tropics, regular dry periods occur, giving rise to the seasonal tropical forest biome. High pressure zones associated with Hadley cells create extremely dry zones that promote the desert biome. Seasonality of both temperature (cool winters, warm summers) and precipitation in the temperate climatic zone give rise to grassland (wet summers, dry winters) and shrubland (wet winters, dry summers) biomes. Temperate deciduous forests occur where seasonal temperature changes are moderate, and both summer and winter are moist. Moving toward the polar climatic zone, winter temperatures and precipitation decrease, the period of subfreezing winter temperature increases, marking the transition to the Boreal and tundra biomes.
3. According to the river continuum concept, water velocity, stream bed particle size, and input of detritus from riparian vegetation all decrease as rivers move downstream. As a result, the importance of the surrounding terrestrial ecosystems as sources of energy for stream organisms tends to decrease downstream. Stream insects include more shredders near the source of a stream and more collectors in the lower portions. Attached plants and free-floating algae become more abundant downstream.
4. Light penetration varies according to the depth and clarity of the water. Where there is enough light for photosynthesis (the photic zone), photosynthetic organisms provide food for consumers, increasing the abundance of those organisms. The stability of the substrate determines whether organisms can anchor themselves or bury themselves in sand. Nearshore zones with rocky substrata tend to have the most abundant organisms and the most diverse communities. Photosynthetic organisms are more sparse in nearshore zones with sandy bottoms and below the photic zone in the open ocean.

CHAPTER 4

Answers to Figure Legend Questions

Figure 4.4 The southern limit of aspen's range tends to be associated with survival of drought conditions, which are becoming more frequent in the center of the continent. As a result, the southern range limit of aspen may move to the north. At the northern limit of aspen, the effects of low temperatures on its survival and reproduction tend to limit its distribution. Climate warming may offset this effect, and aspen may move northward in the future.

Figure 4.9 Cooling is important in any biome where leaf temperatures may rise to levels that are stressful, including many temperate and tropical biomes. However, a steady supply of water is needed to support transpirational cooling, which would be the case in tropical biomes and subtropical biomes during the rainy season.

Figure 4.10 Cooling mechanisms that do not use water, such as leaf pubescence or increasing convective heat loss, may be more important to cooling in deserts than in moister habitats such as the tropics, where the water supply is sufficient water for transpirational cooling.

Figure 4.15 Moving between sun and shade influences the energy balance of the lizard. The lizard gains energy, particularly by solar radiation, when it moves to a sunny location. Moving into the shade results in net energy loss to the surrounding environment (losses > gains). If the rock on which the lizard basks is warmer than its body, then the lizard gains heat energy from the rock via conduction. A cooler rock in the shade will receive heat energy by conduction from the lizard's body.

Figure 4.21 Closing stomates during midday lowers transpiration by increasing the resistance to water loss. Opening the stomates later in the afternoon when the air is cooler exposes the leaf to a concentration gradient of water from the plant to the air that is lower than at midday. As a result, transpirational water loss is less than it would be during the hotter part of the day.

Figure 4.25 The rate of water loss for each animal is given by slope of the line. If the external environment (light, temperature, humidity) is kept relatively constant, then the gradient of water potential from the animal to the air is the same, and the resistance modifies the actual water loss. Differences in the slopes therefore reflect differences in resistance to water loss.

Answers to Review Questions

1. Plants as a group exhibit slightly greater tolerances of temperature extremes than ectotherms (see Figure 4.7), and both of these groups have tolerances much greater than those of endothermic animals. Plants and ectotherms, most of which do not generate heat internally, are more reliant on tolerance as a strategy for adapting to body temperature variation, while endotherms rely on avoidance of temperature extremes through internal heat generation and behavior, such as migration. Plants can exhibit avoidance of temperature extremes through leaf deciduousness.
2. The seasonal change in the thickness of the arctic fox's fur represents acclimatization to cold winter temperatures. Arctic foxes grow thicker fur in winter, increasing their insulation and decreasing their lower critical temperature. This seasonal acclimatization allows the arctic fox to stay warmer in winter and invest less energy in metabolic heat production than it would need to without the thicker fur.

   The thick fur of the arctic fox and its ability to alter its fur thickness in response to environmental temperatures represent adaptations that its relative, the bat-eared fox, does not share. The bat-eared fox has not been exposed to cold conditions that would favor the growth of thicker fur. In the distant past, species intermediate between the arctic and bat-eared foxes may have moved to colder, more seasonal climates where natural selection favored individuals with the capacity to grow thicker fur in winter.
3. a. Transpiration is an evaporative cooling mechanism that allows the plant to lower its leaf temperature below the air temperature. However, transpiration also results in water loss from the plant. If the water is not replaced, because the soil is too dry or the water loss is too rapid, the plant will experience water stress, and the rates of its physiological processes, such as photosynthesis, will decrease.

   b. Dark-colored animals may be able to warm themselves more effectively, but they may also be more visible to their predators or prey. In many cases, it appears that camouflage is more important than the ability to absorb sunlight effectively.
4. The principal ways in which plants determine their resistance to water loss are by adjusting the degree of opening of their stomates and by the thickness of the outer cuticle. Arthropods have cuticles that are extremely resistant to water loss. Similarly, skin thickness in amphibians, birds, and mammals affects their resistance to water loss. Reptiles have particularly thick skin, often overlain by scales, that provides a very effective barrier to water movement into the atmosphere. Note, however that increasing the resistance of a barrier to water loss requires trade-offs with evaporative cooling as well as gas exchange.

CHAPTER 5

Answers to Figure Legend Questions

Figure 5.7 The light saturation level would be lower than the maximum light level the plant experiences because the energy invested in achieving a higher light saturation level might not pay off. The plant experiences the maximum light level for only short periods of time, and the increase in CO_2 taken up during those short periods might not pay for the additional machinery (e.g., chlorophyll, enzymes) needed to increase the light saturation level.

Figure 5.13 Extrapolation of the line used to fit the data to the *x* axis indicates that the proportion of the grass flora that is C_4 drops to zero when the growing season minimum temperature is around 4–5°C. This would correspond to an average growing season temperature of 9–10°C, which is at or above the growing season temperatures for boreal forests and tundra shown in the climate diagrams. This result agrees well with the observed lack of C_4 plants in these biomes.

Ecological Toolkit 5.1 CAM plants exhibit a wider range of $\delta^{13}C$ values because some are facultative CAM plants. At some times they use C_3 photosynthesis, but during drier periods they use CAM photosynthesis. The $\delta^{13}C$ of their tissues would reflect a mixing of C taken up using both of these photosynthetic pathways.

Figure 5.23 The rate of energy gain with both long and short distances between patches declines if the quality or abundance of the prey is low. As a result, the giving up times come sooner.

Answers to Review Questions

1. Autotrophy is the use of sunlight (photosynthesis) or inorganic chemicals (chemosynthesis) to fix CO_2 and synthesize energy storage compounds containing carbon–carbon bonds. Photosynthesis occurs in archaea, bacteria, protists, algae, and plants. Heterotrophy is the consumption of organic matter to obtain energy. The organic matter includes both living and dead organisms. Living organisms vary in their mobility, and

their consumers (predators) have adapted ways to improve their efficiency in capturing their food (prey). Dead organic matter can be eaten and digested internally by multicellular heterotrophs or externally broken down by enzymes excreted into the environment and then absorbed by archaea, bacteria, and fungi.

2. C_4 plants separate CO_2 fixation from the Calvin cycle spatially. They have a biochemical pump that takes up CO_2 in the mesophyll tissue and concentrates it at the site of the Calvin cycle, in the bundle sheath tissue. The bundle sheath is surrounded by a waxy coating that keeps CO_2 from diffusing out. The higher CO_2 concentrations in the bundle sheath essentially eliminate the energy-robbing process of photorespiration in C_4 plants. Although additional energy is required to operate the C_4 biochemical pump, a higher rate of photosynthesis more than compensates for that energy expenditure, and the photosynthetic rates of C_4 plants are often higher than those of C_3 plants. C_4 plants also maintain greater water use efficiency than C_3 plants.
3. CAM plants open their stomates to take up CO_2 at night, when the humidity of the air is higher than it is during the day. They store CO_2 in the form of a four-carbon organic acid, then release it to the Calvin cycle during the day. The storage of CO_2 allows the stomates to be closed during the day, when the potential for transpirational water loss is greater.
4. Live animals are a higher-quality food source, but they are rarer and thus harder to find, and they may have defense mechanisms that require expenditure of energy to overcome. Plant detritus is abundant in many ecosystems, so little energy needs to be expended in locating it, but its food quality is low.
5. The greater expenditure of energy required by species B to fly between patches would dictate that it needs to spend longer in each patch in order to meet the assumptions of the marginal value theorem. Because its overall rate of energy gain in the habitat is lower, due to greater amount of energy it expends in traveling between patches, species B should deplete each patch to a greater degree before leaving it than species A.

CHAPTER 6

Answers to Figure Legend Questions

Figure 6.6 The "Before selection" and "After selection" data show that nearly all fly larvae in galls less than 17 mm in diameter were killed by wasps. A much greater proportion of larvae in the largest galls survived, suggesting that wasps provide a stronger source of selection than do birds.

Figure 6.7 When the simulation began, each population had 9 *A* alleles and 9 *a* alleles. At generation 20, 8 populations still had both alleles. Eventually, it is likely that the *A* allele would either reach fixation (a frequency of 100%) or be lost from each of those 8 populations.

Figure 6.14 No. The added risk of mortality due to reproduction is represented by the difference between the blue curve (females that reproduced) and the red curve (females that did not reproduce). That added risk decreases for females 3–7 years old, then rises for females 8–13 years old (and remains roughly constant thereafter).

Answers to Review Questions

1. Natural selection acts as a sorting process, favoring individuals with some heritable traits over individuals with other heritable traits. As a result, the frequency of the favored traits in a population may increase over time. When this occurs, the frequencies of alleles that determine the favored traits also increase over time, causing the population to evolve. But the individuals in the population do not evolve—each individual either has the trait favored by selection or it does not.
2. By consistently favoring individuals with one heritable trait over individuals with other heritable traits, natural selection can lead to a steady increase in the frequency of alleles that determine the favored trait. Although gene flow and genetic drift can also cause the frequency of alleles that determine an advantageous trait to increase over time, each of these processes can also do the reverse—that is, they can promote an increase in the frequency of disadvantageous alleles. Gene flow, for example, can transfer disadvantageous alleles to a population, thereby impeding adaptive evolution. Similarly, the random fluctuations in allele frequencies that result from genetic drift can promote an increase in the frequency of a disadvantageous allele. Hence, natural selection is the only evolutionary mechanism that consistently causes adaptive evolution.
3. Patterns of evolution over long time scales result from large-scale processes such as speciation, mass extinction, and adaptive radiation. The fossil record shows us that life on Earth has changed greatly over time, as seen in the rise and fall of different groups of organisms (for example, the rise of the amphibians and their later fall as reptiles became the dominant group of terrestrial vertebrates). Such changes in the diversity of life are due in part to speciation, the process by which one species splits to form two or more species. The rise and fall of different groups of organisms is also determined by mass extinctions and adaptive radiations. By removing large proportions of the species on Earth, a mass extinction forever changes the course of evolution and hence alters the patterns of evolution observed after the extinction event. Similarly, by pro-

moting an increase in the number of species in a group of organisms, an adaptive radiation shapes the patterns of evolution observed over long time scales.

4. Evolution occurs as organisms interact with one another and with their environment. Hence, evolution occurs partly in response to ecological interactions, and those interactions help to determine the course of evolution. The reverse is also true: as the species in a biological community evolve, the ecological interactions among those species change. Thus, ecology and evolution have joint effects because they both depend on how organisms interact with one another and their environments.

CHAPTER 7

Answers to Figure Legend Questions

Figure 7.2 Starting with the fish on the top left and proceeding clockwise, the genders are male, smallest nonbreeder, female, and largest nonbreeder. We can be confident of these predictions because the largest fish is female, the next largest a male, and the rest are nonbreeders.

Figure 7.4 A 5 m tall tree growing in a cool, moist climate is estimated to have a trunk diameter between 10 and 20 cm (the log scale makes it difficult to provide a precise estimate, but it is probably close to 15 cm), while a 5 m tall tree growing in a desert climate is estimated to have a trunk diameter between 20 and 30 cm (probably close to 22 cm). To illustrate how these estimates are obtained: if you follow the line that moves horizontally to the right from the 5 m mark on the y axis, that line intersects the blue curve (the regression line for a cool, moist climate) at a point whose trunk diameter is about 15 cm.

Figure 7.6 The larva would be genetically identical to the polyp because both result from the same zygote (which in turn was produced when a sperm cell fertilized an egg cell). Two different larvae, however, would not be genetically identical because each resulted from a different fertilization event.

Figure 7.7 In Generation 3 there are 8 sexual and 16 asexual individuals, while in Generation 4 there will be 16 sexual and 64 asexual individuals. The number of sexual individuals is increasing half as rapidly as the number of asexual individuals because half of the offspring produced by sexual females are males (and males do not give birth to offspring).

Figure 7.13 No. When $c > 1$, the average age of sexual maturity is greater than the average life span. For this occur, the majority of individuals must die before they are old enough to reproduce.

Answers to Review Questions

1. In many plants and marine invertebrate animals, dispersal is negatively correlated with propagule size: smaller propagules can disperse farther than larger ones. In invertebrate animals, smaller egg size is also correlated with longer development times and increased reliance on food (rather than yolk provided in the egg) to complete development. However, in some vertebrates (for example, Sinervo's fence lizards), smaller egg sizes actually lead to more rapid development to hatching. In both cases, the correlation between egg size and development time is striking, and the pattern that is favored varies with environmental conditions (e.g., temperature, rates of predation on larvae, etc.). An important reason why species that live in the same habitats may still exhibit different reproductive patterns is that different strategies may be favored in different years, depending on the particular environmental conditions. For example, in years with abundant food availability, a small-egg strategy may be favored, as offspring can acquire resources readily from the environment. However, in years when food is limited, a large-egg strategy may be advantageous due to its decreased reliance on external energy sources.
2. Asexual reproduction allows even a single individual to quickly increase the population size and allows a single highly successful genotype to dominate the population. The primary benefit of sex is the recombination of genetic material through the merging of unique genotypes. This recombination allows deleterious mutations to be purged from the population and allows potentially beneficial new combinations of genes to be introduced. The maintenance of both sexual and asexual reproduction allows rotifers to (1) quickly increase the size of the reproductive population under beneficial environmental conditions while (2) retaining the capacity to purge any deleterious mutations that may build up after prolonged periods of asexual reproduction.
3. Removal of small to medium-sized fish might produce selection for rapid growth through the size ranges that are favored by the fishery. This might lead to reproduction at older ages and larger sizes if there is a trade-off between growth and reproduction. Fish that are selected to grow quickly would allocate fewer resources to reproduction at smaller sizes so that they could allocate more resources to growth. Unfortunately, this is not the only effect of the Nassau grouper fishery. Because of heavy overfishing for both small and large fish and methods that target fish when they come together in large groups to spawn, Nassau grouper populations have declined precipitously.

CHAPTER 8

Answers to Figure Legend Questions

Figure 8.3 There was considerable variation in abundance from one field site to another in many of the years. In 1984 and 1989, for example, abundance was high at Hector but low at the other two locations.

Figure 8.4 There were 7 habitat patches in 1759 and about 86 patches in 1978. Thus, in 1759, the average patch size was 300 $km^2/7 = 42.9$ km^2. Patch sizes were much smaller in 1978: the average at that time was 60 $km^2/86 = 0.7$ km^2.

Figure 8.6 In clones that form by budding or apomixis, identification of groups of genetically identical individuals may require the use of genetic analyses. In clones that form by horizontal spread, groups of individuals that are still connected to one another could be marked; however, to tell whether members of two such groups were in fact genetically identical would again require genetic analyses.

Figure 8.9 Because it competes poorly with other barnacle species in relatively warm waters, *S. balanoides* is currently excluded from the region shaded purple on the map. Thus, by warming northern waters, global warming will probably decrease the geographic range of *S. balanoides*.

Figure 8.12 Both. Each curve increases as the density of offspring increases, indicating that wing production increases as offspring density increases. In addition, at all but the lowest offspring densities, the percentage of aphids that develop wings is higher for offspring whose mothers were reared at high densities than it is for offspring whose mothers were reared at low densities. This observation shows that the density experienced by the mother also influences whether offspring develop wings.

Answers to Review Questions

1. Complicating factors discussed in the text include (1) limited knowledge about the dispersal capabilities of the organism under study, (2) the fact that populations may have a patchy structure, and (3) the fact that individuals may be hard to define. The first two factors—limited information about dispersal and patchy populations—can make it difficult to determine the area within which individuals interact, and hence what constitutes a population. The third factor—difficulty in defining individuals—applies to the many organisms that reproduce asexually to form clones. In such organisms, it can be hard to determine what an individual is, thus making it difficult to estimate abundance.
2. The simplest reason that no species is found everywhere is that much of Earth does not provide suitable habitat. There can, in turn, be many reasons why portions of Earth are not suitable for a particular species. For example, the abiotic or biotic conditions of an environment may limit the growth, survival, or reproduction of the species, as may disturbance or the interaction between abiotic and biotic conditions. Furthermore, a species may be absent from environments where we would expect it to thrive because of dispersal limitation or historical factors (including evolutionary history and continental drift).
3. A niche model is a tool that predicts the environmental conditions occupied by a species based on the conditions at where the species has been found. Niche models can be used to predict the future distribution of an introduced species by collecting as much information as possible about environments where the species currently is found. Those data are then used to construct a niche model, which in turn is used to identify currently unoccupied locations that are likely to provide suitable habitat for the species. For such predictions to accurately reflect the future spread of the organism, information also must be gathered about its dispersal capabilities.
4. For a conservative estimate, assume there are 20 otters per square kilometer, each of which eats 20% of its body weight in food each day. Since urchins, on average, weigh 0.55 kg each, a kilogram of urchins consists of roughly $1/0.55 = 1.82$ urchins. Thus, the number of urchins per square kilometer that an otter population would be expected to eat each year is:

   $(20 \text{ otters/km}^2) \times (0.2 \times 23 \text{ kg/otter/day}) \times (365 \text{ days/year}) \times (1.82 \text{ urchins/kg}) = 61{,}116 \text{ urchins/km}^2\text{/year}$

CHAPTER 9

Answers to Figure Legend Questions

Figure 9.4 About 47% of Gambians born in the hungry season live to age 45; a similar percentage (44%) of U.S. females live to be 85 years old.

Figure 9.6 100 sheep survive to age 11; thus 10% (100/1,000) of sheep survive from birth to age 11.

Figure 9.8 The year-to-year population growth rate (λ) from year 4 to year 5 for age class 2 is the number of individuals in age class 2 at year 5 divided by the number in age class 2 at year 4. Filling in those numbers from (A), we find that $\lambda = 38/19 = 2$.

Figure 9.15 Since there were about 35 breeding females in 1975, results from previous years suggest that roughly 4 young per female should have survived to independence. In fact, less than 1.5 young per female survived to independence, suggesting that conditions on the island were different in 1975 than in other years (there could have been a drought or a disease outbreak, among many other possibilities).

Figure 9.18 As N becomes increasingly close to K, the term $(1 - N/K)$ becomes increasingly close to zero; this causes the population growth rate, dN/dt, to become increasingly close to zero. A population with a growth rate of zero does not increase in size; hence, as N approaches K, the population stops increasing in size.

Figure 9.21 The graph shows that the human population is projected to have an annual growth rate of roughly 0.4% in 2050. This rate is greater than zero, so the human population will still be increasing in size in 2050.

Figure 9.22 The best-estimate curve indicates there will be 9 billion people in 2050, and Figure 9.21 indicates that our annual growth rate will be about 0.5% at that time. Hence, from 2050 to 2051, we would expect to add about 45 million (9,000,000,000 × 0.005) to our population. Thus, the human population size in 2051 would be about 9,045,000,000.

Answers to Review Questions

1. a.

Age (x)	N_x	S_x	l_x
0	100	0.4	1.0
1	40	0.375	0.4
2	15	0.333	0.15
3	5	0	0.05
4	0		0

b. In a cohort life table, the fate of a group of individuals born during the same time period (a cohort) is followed from birth to death. This type of life table is often used for sessile or relatively immobile organisms that do not have long life spans, but is less useful for organisms that are highly mobile or long-lived. For those organisms, a static life table may be used, in which the survival and fecundity of individuals of different ages are observed during a single time period.

2. a.

Year	N_0	N_1	N_2
0	50	50	50
1	100	16.7	25
2	116.7	33.3	8.3
3	150	38.9	16.7
4	188.9	50	19.4
5	238.9	63	25
6	301.9	79.6	31.5

From these results we estimate:

$\lambda = 31.5/25 = 1.26$

The stable age distribution is 73% N_0; 19% N_1; 8% N_2.

b.

Year	N_0	N_1	N_2
0	80	50	20
1	130	26.7	25
2	156.7	43.3	13.3
3	200	52.2	21.7
4	252.2	66.7	26.1
5	318.9	84.1	33.3
6	403	106.3	42

From these results we estimate:

$\lambda = 42/33.3 = 1.26$

The stable age distribution is 73% N_0; 19% N_1; 8% N_2.

c. Based on the results in parts (a) and (b), we estimate:

$\lambda = 1.26$

The stable age distribution is 73% N_0; 19% N_1; 8% N_2.

3. a. 3,240.

b. Substituting the values $N_0 = 40$, $\lambda = 3$, and $t = 27$, we have

$N_t = N_0\lambda^t = 40 \times 3^{27}$

c. In this case, we have the values $N_0 = 100$, $\lambda = 0.75$, and $t = 3$, which we plug into the relation

$N_t = N_0\lambda^t = 100 \times (0.75)^3 = 42.19$

4. Factors that regulate population size are density-dependent: when N (the number of individuals in a population) is below some level, they cause the population size to increase, whereas when N goes above some level, they cause the population size to decrease. Even if density-independent factors, such as year-to-year variations in temperature or rainfall, are the primary cause of year-to-year changes in abundance, those factors do not regulate population size.

5. Each student will calculate their own answer.

CHAPTER 10

Answers to Figure Legend Questions

Figure 10.5 The carrying capacity that results from the second death rate curve, labeled K_2 in the drawing, is lower than the original carrying capacity, K.

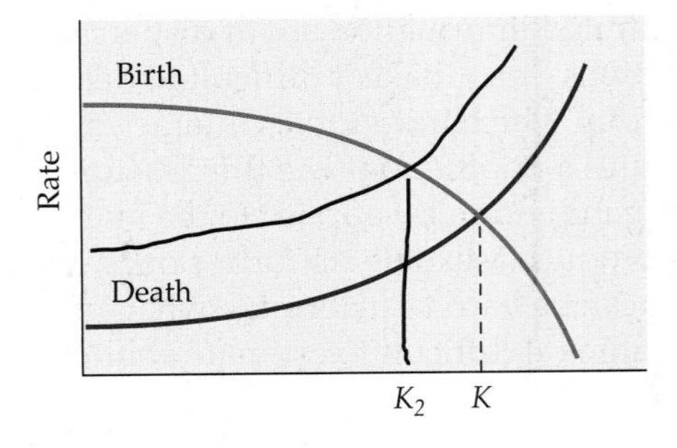

Figure 10.9 From 1988 to 2000, the collared lemming population exhibited regular cycles, reaching peak abundance every 4 years. Because abundances peaked at about 10 lemmings per hectare in 1990, 1994, and 1998, we would have expected the next peak to occur in 2002, again at about 10 lemmings per hectare. However, the actual abundance in 2002 was less than 1 lemming per hectare.

Figure 10.11 In (A), abundance rises and falls in a regular manner, reaching a peak about every 40 days; thus, this curve shows regular population cycles. In (B), to the left of the dotted vertical line, the results are again consistent with a regular population cycle that reaches peak abundance every 40 days. After food for adults is limited, however, the regular population cycle no longer occurs. Instead, abundance rises and then fluctuates around a roughly stable population size. This pattern that can be viewed as illustrating either population fluctuations or logistic growth (with fluctuations).

Figure 10.19 The chance of colonization is between 50% and 90%.

Figure 10.21 From 1952 to 1957, the abundance of predatory fish increased while the abundance of planktivorous fish showed little change. In the 1970s, predatory fish abundance dropped, planktivorous fish abundance increased, zooplankton abundance dropped, and phytoplankton abundance increased. Overall, the chain of feeding relationships for the Black Sea in the 1970s is more similar to that in Alaska pre-1990 than to that in Alaska in the late 1990s. In both cases, the organisms at the base of the food chain (phytoplankton in the Black Sea, kelp in Alaska) were only weakly controlled by their grazers (zooplankton in the Black Sea, urchins in Alaska), which in turn were strongly controlled by the organisms that ate them (planktivorous fish in the Black Sea, otters in Alaska).

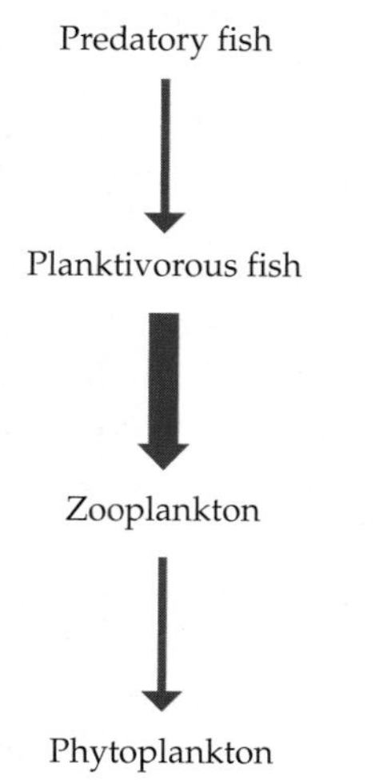

Answers to Review Questions

1. There are many built-in time lags in the responses of populations to changes in density. For example, the amount of available food may increase or decrease between the time the parent generation feeds and the time its offspring are born. In such a situation, the number of offspring produced may be more closely related to the previous conditions than to the conditions at the time of their birth. As a result of such time lags, the population may experience delayed density dependence, which may cause it to fluctuate in abundance over time.
2. Small populations can be threatened by chance events associated with genetic factors, demographic stochasticity, environmental stochasticity, and natural catastrophes. Genetic factors that increase the risk of extinction in small populations include genetic drift and inbreeding, both of which can increase the frequencies of harmful alleles. Demographic stochasticity results from chance events related to the reproduction and survival of individuals; such events can cause population growth rates to drop, as might occur if considerably more females than males happened to die in a small population, leaving few females to produce the next generation of offspring. Environmental stochasticity refers to unpredictable variation in environmental conditions; such variation can cause population growth rates to vary dramatically from year to year, increasing the chance of extinction in small populations. Finally, natural catastrophes can cause sudden reductions in population size, subjecting a population to increased risks from genetic factors, demographic stochasticity, and environmental stochasticity.
3. a. Yes, as illustrated by the two generations of parents and offspring in this diagram:

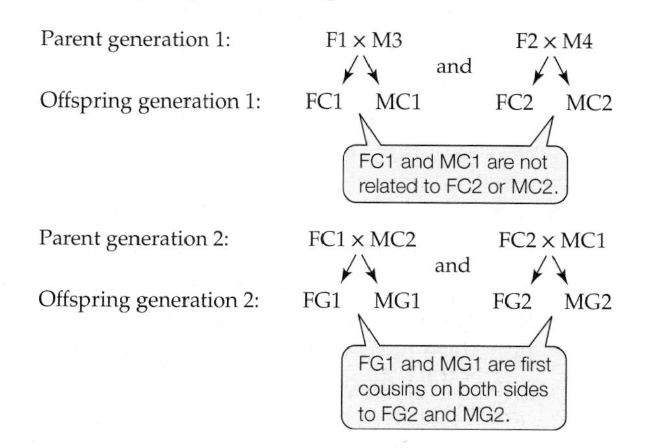

KEY: FC = female child; MC = male child; FG = female grandchild, MG = male grandchild

b. No, all of the individuals in the second generation of offspring are related to one another. As illustrated by this example, inbreeding is likely to be common in small populations.

4. a. As the proportion of the habitat that is suitable for the species drops, the colonization rate may decrease (because the distance between remaining populations increases) and the extinction rate may increase (because the loss of habitat may cause the remaining populations to become smaller, making them more prone to extinction). Once the extinction rate exceeds the colonization rate, the metapopulation will decline to extinction because existing populations will become extinct more rapidly than new populations are established.

   b. A large habitat patch is likely to have a larger population than a small habitat patch. As a result, the populations in the two large habitat patches are less likely to go extinct than the others. Thus, the large habitat patches could serve as source populations from which individuals disperse to small patches, thereby reducing the extinction rate of the small populations (the rescue effect). Positioning the large habitat patches far from each other and far from any of the small habitat patches would make it less likely that they could have a rescue effect, thus making it more likely that the metapopulation would not persist.

CHAPTER 11

Answers to Figure Legend Questions

Figure 11.2 When pitchers were deprived of prey (blue bars), the relative growth rate was about –0.05 when neighbors were intact and 0.1 when neighbors were reduced; hence, the relative growth rate increased by about 0.15 when competition was reduced. When pitchers had access to prey (orange bars), the relative growth rate was about –0.1 when neighbors were intact and 0.4 when neighbors were reduced; hence, the relative growth rate increased by about 0.5 when competition was reduced. These results show that when pitchers were deprived of prey, the effects of competition did not increase. Instead, the effects of competition appear to have decreased—the opposite of what was expected.

Figure 11.4 It is likely that *Asterionella* would drive the third diatom species to extinction. *Asterionella* reduces the concentration of silica to about 1 μmol/L when grown alone (see part A of this figure). This concentration is much lower than the concentration of silica (5 μmol/L) that results when the third diatom species is grown alone—suggesting that the third diatom species would not have enough silica and hence could not survive if it was grown in competition with *Asterionella*.

Figure 11.10 *Paramecium aurelia* feeds mainly on floating bacteria, while *P. bursaria* feeds mainly on yeast cells. Because they rely on different food sources, it is likely that both species would persist if they were grown together.

Figure 11.14 The density of species 2 would decrease because its abundance would be above its carrying capacity.

Figure 11.16 Two years. The observed replacement curve indicates that if a population begins with 100 individuals (in "year 0"), it will have about 22 individuals in the next year (year 1). A population that has 22 individuals in year 1 will have fewer than 10 individuals in year 2.

Answers to Review Questions

1. a. The key difference is that a resource can be depleted, but a physical factor cannot.

   b. Immediately after application of the fertilizer, it is likely that the intensity of competition for nitrogen will drop because soil nitrogen levels will not be as limited in supply. However, as the added nitrogen is used up by the plants (and leached from the soil by rainfall), soil nitrogen levels will decrease, and the intensity of competition for nitrogen will increase.

2. Four general features of competition and an example of each: (1) Competition can be direct (as in allelopathy) or indirect (as in both pitcher plants and bedstraws). (2) Competition is often asymmetrical (as in Tilman's diatoms). (3) Competition can occur between closely (as in bedstraws) or distantly (as in ants and rodents) related species. (4) Competition can determine the distribution and abundance of the competing species (as in Connell's barnacles or *Neotamias* chipmunks).

3. a. Results from laboratory experiments, field observations, and mathematical models all suggest that competing species are more likely to coexist when they use resources in different ways. For example, in Gause's experiments with *Paramecium*, *P. caudatum* coexisted with *P. bursaria*, most likely because one species fed primarily on bacteria, the other on yeast. Likewise, in the case of four species of *Anolis* lizards that lived together on Jamaica and ate similar food, Schoener's field observations indicated that these species used space in different ways (an example of resource partitioning). Finally, graphical analysis of the Lotka–Volterra competition model indicates that competing species can coexist when the inequality shown in Equation 11.2 holds. That inequality is more likely to hold when competing species use resources in very different ways (e.g., when α and β are not close to 1).

   b. Because $\beta = 1.6$ and there are 140 individuals of species 1, it would take $1.6 \times 140 = 224$ individuals of species 2 to reduce its own growth rate by the same amount that the 140 individuals of species 1 do.

Therefore, because there are 230 individuals of species 2 present, species 2 is having a slightly greater effect on its own growth rate than is species 1.

c. The statement is not correct. For example, if $\alpha = 0.5$ and $\beta = 1$, Equation 11.2 predicts that both species will persist when $0.5 < K_1/K_2 < 1$. Thus, for example, if $K_1 = 100$ and $K_2 = 150$, both species should persist when $\alpha = 0.5$ and $\beta = 1$. (The statement can be shown to be false in many other ways; for example, in Figure 11.14B, values for α, β, K_1, and K_2 can be selected such that species 2 always drives species 1 to extinction, even though $\alpha < \beta$.)

4. a. Possible reasons why these meadows might harbor one or the other (or both) of these two plant species: (1) Both species could persist at all locations, but one species (or the other) has yet to disperse to some meadows. (2) The physical conditions of the meadows differ such that in some meadows species 1 is favored, while in others species 2 is favored, and in still others, the species can partition resources such that both persist. (3) The abundances of herbivores or pathogens that feed on species 1 or 2 may vary between the meadows, causing the outcome of competition to differ from meadow to meadow. (4) The rates of a periodic disturbance such as fire may differ among meadows (if one of the species is an inferior competitor but is more tolerant of fire). (5) Evolutionary change may have occurred in some meadows, but not others, causing a range of possible outcomes, such as a competitive reversal or character displacement.

b. Addition and removal experiments would help us to evaluate these possible explanations for the observed distributions of the species. For example, in meadows where only species 1 was found, individuals of species 2 could be planted next to some individuals of species 1, but not others. Similarly, in meadows where both species are found, removal experiments could be performed in which individuals of species 1 could be removed from the vicinity of some species 2 individuals, but not others (and vice versa).

CHAPTER 12

Answers to Figure Legend Questions

Figure 12.2 The peak abundance of lynx usually occurs after the peak abundance of hares. One reason this might occur is that as hare abundance rises, the increased availability of food enables the lynx to produce more offspring; however, these offspring are not born immediately, so the rise in lynx abundance lags behind the rise in hare abundance.

Figure 12.7 To answer this question, we must use the data in the graph to determine the total number of agromyzid fly species and the number of agromyzid fly species that feed on fewer than five host plant species. We can do this using the scale on the *y* axis, which indicates that a bar that is 2.15 cm in height represents 50 fly species. Measuring all 13 bars on the graph, we find that their heights sum to 12.05 cm; this indicates that in total, there were about 280 fly species ($280 \approx \frac{12.05 \text{ cm}}{2.15 \text{ cm}} \times 50$). Similarly, the heights of the four bars representing fly species that feed on fewer than five host plant species sum to 10.4 cm, indicating that about 242 fly species feed on fewer than five host plant species. Thus, about 86% of agromyzid fly species feed on fewer than five host plant species.

Figure 12.10 On average (based on the height of the bar graph), the control plants produced about 11–12 fruits per plant. This indicates that a plant that compensated fully for clipping would also produce 11–12 fruits.

Figure 12.15 The density of other plants in the community would probably increase after herbivory by *Chrysolina* reduced the density of Klamath weed. Because Klamath weed was originally a dominant member of the community, it is likely that the community would change considerably after introduction of the beetle.

Figure 12.19 In the absence of snails, wetlands had phosphorus concentrations of less than 100 µg/L. When snails were present, phosphorus concentrations were usually much greater 100 µg/L; for example, in the seven wetlands with snail densities greater than 10 snails per square meter, the average phosphorus concentration was close to 1,000 µg/L. Thus, the presence of snails is associated with an increase in the phosphorus concentration of these wetlands.

Answers to Review Questions

1. Most predators have a broad diet in that they eat a wide range of prey species. Although there are a substantial number of herbivores that can eat many different plant species, the majority of herbivores are insects, most of which feed on just one or a few plant species.

2. A prey individual that cannot evade a predator is killed and eaten. While herbivores do not typically kill their food plants, they do have powerful negative effects on the plants on which they feed. As a result of this strong selection pressure that predators and herbivores exert on their food organisms, prey species and plants have evolved a wide range of defensive mechanisms that increase the chance that they will not be eaten. Predators and herbivores must eat if they are to survive, so there is also strong selection pressure on predators and herbivores to overcome the defenses of their food organisms.

These effects are pervasive because all organisms must obtain food—setting in motion the conflicts described in the previous sentences. The effects are pronounced because there is such strong selection for both defensive and counterdefensive mechanisms.

3. In this chapter, we claim that exploitative interactions such as predation and herbivory can have pronounced effects on ecological communities.
   a. Evidence described in this and preceding chapters indicates that exploitative interactions such as predation and herbivory can have a powerful effect on the abundances and distributions of populations of the interacting species. Ecological communities are composed of sets of interacting populations. Thus, this claim is likely to be true because if predation or herbivory affect the populations involved in those interactions, then they will also have an effect on other populations that interact with those populations.
   b. The scientific evidence strongly supports this claim. In this chapter, we described several cases in which the effects of predation or herbivory were so pronounced that they altered ecological communities greatly, in some cases causing a shift from one community type to another. For example, lesser snow geese feeding on marsh grasses, arctic foxes feeding on seabirds, and aquatic snails feeding on large aquatic plants had such effects.
4. The prey population changes in size according to the equation

$$\frac{dN}{dt} = rN - aNP$$

where N is the prey density and P is the predator density. The prey population does not change in size when $dN/dt = 0$, or in this case, when

$$\frac{dN}{dt} = rN - aNP = 0$$

From this we can calculate that the prey population does not change in size when $rN = aNP$, or $P = r/a$. This implies that if there are exactly $P = r/a$ predators in the population, the prey population does not change in size. Thus, we can conclude that if there are more than r/a predators, the prey population should decrease in size (and if there are fewer than r/a predators, the prey population should increase in size).

CHAPTER 13

Answers to Figure Legend Questions

Figure 13.4 Averaging across the six groups, there are about 21 parasite species per host. This average would probably not be close to the number of parasite species found in a previously unstudied host from of one of the six groups of organisms. A reason for this is that in five of the groups (all but the trees), the average number of parasites per host is less than 12, while in the trees, the average is 95. Thus, we might expect that 95 parasite species would be found in another tree, 7 parasite species would be found in another wasp, etc.—but we would not expect to find 21 parasite species in a host from any of the six groups.

Figure 13.9 The gamete-producing cells enable the parasite to disperse from a human host to a mosquito.

Figure 13.11 No. For example, with an infection rate of 70%, the Lake Wahapo snails are very poorly defended against parasites from their own lake, but they are reasonably well defended against parasites from both other lakes. Similarly, Lake Paringa snails are poorly defended against parasites from their own lake (infection rate = 51%), but they are well defended against parasites from Lake Mapourika (infection rate = 11%).

Figure 13.16 If the cycles stopped completely, we would not expect the numbers in both of the treated populations to drop in 1989 and again in 1993—the same years that the control populations were predicted to drop based on long-term data on population cycles in red grouse.

Figure 13.23 The percentage of unparasitized isopods under cover rises from about 70% to more than 90% when the number of predators increases from 0 to 2, indicating a change in their behavior when predators are present. Parasitized isopods did not respond in a consistent way to the presence of predators—when one predator was present, they spent less time under cover than they did in the absence of predators, but when two predators were present, this effect was not observed.

Answers to Review Questions

1. Ectoparasites live on the surface of their host, whereas endoparasites live inside the body of their host. Examples of ectoparasites include plants such as dodder and fungi such as rusts and smuts; examples of endoparasites include tapeworms and bacterial pathogens such as *Mycobacterium tuberculosis*. Ectoparasites can disperse more easily from one host individual to the next than can endoparasites; however, ectoparasites are at greater risk from natural enemies than are endoparasites.
2. Parasites can greatly reduce the growth, survival, or reproduction of host individuals, thereby reducing the growth rate of host populations. As a result, we would expect that parasites could also alter both the outcomes of species interactions and the composition of ecological communities. For example, if two plant species compete for resources and one typically outcompetes the other, a parasite that reduces the performance of the superior competitor may cause a competitive reversal in which the inferior competitor becomes the superior competitor. Such changes in the outcome of

species interactions can cause changes in the relative abundances of the interacting species, thus altering the ecological community.

3. a. A given disease will become established and spread in a given host population only if the density of susceptible hosts exceeds a critical threshold density (ST). The concept of a threshold density has considerable medical and ecological importance because it indicates that a disease will *not* spread if the density of susceptible hosts can be held below the threshold density.

   b. A disease can become established and spread only if the density of infected individuals (I) increases over time. This means that a disease will become established and spread if the rate of change of the density of infected individuals is greater than zero, which can be written mathematically as $dI/dt > 0$. Based on Equation 13.1, we can therefore calculate that a disease will become established and spread if $dI/dt = \beta SI - mI > 0$.

   This equation implies that the disease will spread if $\beta SI > mI$, which can be rewritten as $\beta S > m$ or $S > m/\beta$.

   Thus, we can conclude that the disease will become established and spread if the density of susceptible individuals (S) exceeds m/β; hence, $S_T = m/\beta$ is the critical, threshold density.

4. a. Host organisms have a wide range of defensive mechanisms that include a protective outer covering, an immune system that kills or limits the effectiveness of the parasite, and biochemical conditions inside the host's body that reduce the ability of the parasite to grow or reproduce.

   b. The statement could be true if the plant populations in Australia possessed specific defensive features that limited the ability of the parasite to grow or reproduce, yet the populations in Europe lacked such adaptations. Among many other possible examples, plants in the Australian populations might possess a specific allele that enabled them to kill or disable the parasite—hence causing the parasite to have mild effects there—whereas plants in the European populations might lack this allele, making them more vulnerable to parasite attack.

CHAPTER 14

Answers to Figure Legend Questions

Figure 14.3 The regions colored light green are similar to the regions in which tropical rainforests are found. Thus, the plants in this mycorrhizal association are likely to be tropical rainforest trees and other plants found in the rainforest biome.

Figure 14.4 It is likely to be an arbuscular mycorrhizal association because the hyphae of arbuscular mycorrhizal fungi can extend considerable distances into the soil.

Figure 14.8 No. In the absence of *Typha*, there is little difference between the dissolved oxygen content of soils that are 11°C–12°C and soils that are 18°C–20°C (compare the red curve in part A with the red curve in part B).

Figure 14.20 These results would suggest that although ants increase their frequency of weeding when parasites are present, they do not discriminate among parasites.

Answers to Review Questions

1. Commensalism and mutualism share a number of characteristics: they are both very common, they can evolve in many ways, and they can cease to be beneficial if conditions change such that the costs of the interaction exceed its benefits. In addition, some evidence indicates that positive interactions may be particularly common in stressful environments. Positive interactions can also differ from one another in that they can range from species-specific, obligate, and coevolved relationships to those that show none of these three characteristics.
2. When a species in a mutualistic interaction provides its partner with a benefit, that action comes at a cost to the species providing the benefit. If circumstances change such that the costs of the interaction are greater than the benefits to one of the species, that species may cease to provide benefits to its partner, or it may penalize its partner. The fact that mutualists may stop providing benefits to their partners when it is not advantageous for them to do so has convinced researchers that mutualism is not an altruistic interaction.
3. Initially, we could expect a decrease in the growth or reproduction of the coral species that are most sensitive to high water temperatures. If high temperatures continued long enough to cause repeated bleaching, it is likely that these more sensitive species would begin to suffer heavy mortality. As a result of the decreased growth, reproduction, and survival of the these sensitive species, the species composition of the reef would change: those coral species that were better able to tolerate high water temperatures would constitute an increasingly high percentage of the corals found in the reef. Such changes in the composition of the coral reef community might also affect other species; for example, a fish that depended on an increasingly rare coral for shelter or food might also decline in abundance. As water temperatures continued to rise, other, less sensitive corals might also experience negative effects. Eventually, if temperatures continued to rise, the abundance of all corals in the reef might decline, as would the abundance of the many species that depend on the reef.

CHAPTER 15 ANSWERS

Answers to Figure Legend Questions

Figure 15.2 Based on the dates of *Caulerpa* sightings, it is likely the seaweed spread from Monaco west to Spain and east to Sicily, Italy, at about the same time (1992, 1993). It was restricted to these locations for 2 years, but then traveled to the eastern coast of Italy, and from there to the island of Hvar (1995), from which it spread to the northern islands of Croatia (1996). Finally, *Caulerpa* was sighted much later in Tunisia (2000), even though Tunisia is closer to Sicily than is Croatia. This may have been because there is less boat traffic to Tunisia than to Croatia, thus lowering the chance of invasion, or it may have been due to a lack of recognition until 2000 that the seaweed was present.

Figure 15.3 The desert and hot spring communities are defined by physical attributes of their environment, while the tropical rainforest and coral reef communities are defined by biological attributes of their environment, particularly by the presence and importance of abundant species.

Figure 15.11 The tropical soil bacterial community requires much more sampling because each sample contains new species, thus producing a linear species accumulation curve. The sampling in the temperate forest plant and tropical bird communities was sufficient to identify a large majority of the species in these communities, and thus more sampling would not be needed. This is clear from the leveling off of the species accumulation curve once all the samples were analyzed. Finally, although the human oral bacterial and tropical moth communities showed some leveling off of their species accumulation curves, new species were being found even once all the samples were analyzed. Thus, they also need more sampling to adequately capture their species richness.

Figure 15.19 Beavers act as ecosystem engineers by damming streams with cut trees and woody debris. This behavior creates a flooded area, which accumulates sediment and eventually becomes dominated by marsh vegetation. At a landscape scale, by creating a mosaic of wetlands within a larger forest community, the beavers' actions enhance regional species diversity. Thus, beavers can also be classified as keystone species because they have such a large effect on diversity relative to their abundance.

Answers to Review Questions

1. A community is a group of interacting species that exist together at the same place and time. Interactions among species act in a synergistic fashion to shape the structure of the community over time.
2. Species richness is the number of species in a community, but that measure tells us nothing about the relative abundances of those species. If two communities had a similar number of species, but great differences in species evenness (as in Figure 15.6), species richness would not reflect this difference, but species diversity indices would. Rank abundance curves (as in Figure 15.8) allow hypotheses to be generated about how those species are interacting in the community based on their abundances.
3. Dominant species have a large effect on other species due to their high abundance or biomass. For example, kelp and trees have a large influence on species diversity by virtue of providing their communities with habitat, food, and other services that are directly related to their size. Keystone species have a large effect not because of their abundance or biomass, but because of the role they play in their communities. For example, sea otters have large effects on their communities by preying on herbivores (sea urchins), which, in turn, eat primary producers (kelp). This indirect interaction can allow primary producers to have higher abundances. Finally, ecosystem engineers are able to create, modify, or maintain physical habitat for themselves and other species. Trees and kelp are examples of ecosystem engineers that are dominant species, and beavers are an example of a keystone species that is also an ecosystem engineer.
4. a. Direct effect 1: Removing seagulls will increase the abundance of the gooseneck barnacle due to decreased predation by seagulls.

   Direct effect 2: Removing seagulls will increase the abundance of the limpet *Lottia digitalis* due to decreased predation by seagulls.

   b. Indirect effect 1: Removing seagulls will decrease or not change the abundance of the mussel due to increased competition with the gooseneck barnacle.

   Indirect effect 2: Removing seagulls will decrease or not change the abundance of the limpet *Lottia strigatella* due to increased competition with the limpet *Lottia digitalis*.

   Indirect effect 3: Removing seagulls will decrease or not change the abundance of phytoplankton due to increased predation by the gooseneck barnacle.

   Indirect effect 4: Removing seagulls will decrease or not change the abundance of seaweeds due to increased predation by the limpet *Lottia digitalis*.

CHAPTER 16

Answers to Figure Legend Questions

Figure 16.2 The most destruction occurred immediately below the mountain, where a huge magma-filled bulge exploded and released rock and mud down the north side of the mountain. An area later known as the Pumice Plain, formed by the hot, pelting pumice rock, experienced the most destruction. The massive wave of debris from the explosion was funneled down the North Fork Toutle River, removing most life along the way. Spirit Lake was also completely destroyed due to its location within the path of the avalanche. Other areas, such as the south side of the mountain and the locations farther from the explosion (mudflow zone and blowdown zone), experienced blowdown of all trees, but some life remained, especially underground. Finally, the least amount of destruction occurred in the scorch zone, where trees were denuded but remained standing.

Figure 16.4 Whether a disturbance is intense or frequent will depend on the susceptibility of the organisms involved and their ability to respond to the disturbance. The intensity and frequency of disturbance for an insect population will be quantitatively different than that for an elephant population. The same disturbance—let's say, a tree falling in a forest—could cause major destruction for the insect population living on that tree while having little effect on the elephant population, even if an elephant were struck by the tree. Of course, the insect population would recover much faster than an elephant population might.

Figure 16.9 The oldest communities are located in the areas that have been exposed the longest since glacial retreat, such as the mouth of the bay. Here, succession has been able to proceed for over 200 years and has allowed the formation of mature spruce forests. As the glacial retreat becomes more recent, the communities become younger, such that the youngest, pioneer community is located closest to the glacier.

Figure 16.17 The fish preferred to eat the tunicate *Styela* because, when the tiles were protected from fish predation, it was the species that dominated. When fish predation was allowed, the bryozoan *Schizoporella* dominated, suggesting that it was unpalatable to the fish. This experiment suggests that *Styela* is the dominant competitor over *Schizoporella* in the absence of predation.

Answers to Review Questions

1. Abiotic and biotic agents of change include those listed in Table 16.1. Intense disturbances such as hurricanes, tsunamis, fires, and volcanic eruptions can cause major damage, but are relatively infrequent. Other agents of change, such as sea level rise, competition, or parasitism, may not cause major damage, but may be frequent or constant. Still others, such as predation, may be relatively frequent, but not very intensive, thus forming patches of available resources.
2. Primary succession involves the colonization of habitats devoid of life. Species colonizing these habitats must deal with stressful conditions and transform their habitats to create soils, nutrients, and food. Secondary succession involves the reestablishment of a community in which most, but not all, of the organisms have been destroyed. Under these conditions, colonizing species benefit from the biological legacy of the preexisting species, but are likely to face more competition for resources.
3. A hypothetical community might be a newly cleared vacant lot in an unnamed city. The facilitation model would be supported if the first species to arrive were stress-tolerant and had the ability to modify their habitat in positive ways. In this case, those early species would facilitate the growth of later species, which would be better competitors but less stress-tolerant. Over time, these later species would dominate as they outcompeted the facilitating species. The tolerance model assumes that the earliest species modify the environment, but in ways that neither help nor hinder later species. Later species are merely those that live longer and tolerate stressful conditions longer than early species. Finally, the inhibition model would be supported if the early species created conditions that benefited themselves, but inhibited later species. Only through the removal of those inhibitory early species—for example, via disturbance or stress—would later species be able to displace them.
4. It is hard to know whether a community is stable because stability depends on the spatial and temporal scale at which the community is observed. All communities fluctuate and change over time, but how long must we wait for a community to return to some original state before we assume it is stable? There is no single answer. Although Sutherland did observe the formation of alternative communities on his tiles when predators were manipulated, did he follow the communities long enough, and at a large enough spatial scale, to show stability? Again, it depends on how you define "stability," leaving us with an unresolved question.

CHAPTER 17

Answers to Figure Legend Questions

Figure 17.2 The goal of the study was to look at the effect of fragmentation on species diversity in the *remaining* forest fragments rather than considering the direct effects of deforestation itself.

Figure 17.6 No, there could never be more local species than would be contained within a region because the spatial scale of the region is larger than that of the local community.

Figure 17.11 One would expect speciation to increase as

land masses separate because species would become reproductively isolated from one another, thus increasing the chance that they would follow different evolutionary trajectories. The separation of species in this way is known as vicariance.

Figure 17.12 Cassowaries and emus come from Australia while the moas lived on New Zealand. Based on Figure 17.11B, Australia and New Zealand separated 80 million years ago, so the correct answer is 80 million years.

Answers to Review Questions

1. The largest spatial scale is the global scale, which covers the entire world, over which there are major differences in species diversity and composition with latitude and longitude. These patterns are controlled by speciation, extinction, and dispersal. The next scale down is the regional scale, defined by areas of uniform climate and by species that are bound by dispersal limitation to the region. Within a region, species diversity and composition depend on dispersal and extinction rates across the landscape. The regional species pool (also called gamma diversity) has an important influence on the species present at the next scale down, the local scale (also called alpha diversity). The relationship between regional and local species richness can help us to determine the extent to which the regional species pool or the local effects of species interactions and physical conditions determine local species richness.
2. Six terrestrial biogeographic regions, which represent distinct biotas that vary in species diversity and composition, have been identified. These biogeographic regions reflect the evolutionary isolation of species due to the movements of the continents. Thus, the ancestors of many modern species may have occurred together in the evolutionary past, but since Pangaea began breaking up into the continents we know today, they have evolved separately. There are also impediments to dispersal within oceans, such as currents, thermal gradients, differences in water depth, and the continents themselves, so it is assumed that the oceans could be divided into biogeographic regions, but that effort has received considerably less attention.
3. The three main hypotheses focus on (1) species diversification rate, (2) species diversification time, and (3) productivity. The first hypothesis proposes that both the large geographic land area and the thermal stability of the tropics might promote higher speciation rates and lower extinction rates, thereby increasing species' population sizes and geographic ranges. Speciation rates should increase because larger geographic ranges should lead to greater reproductive isolation. Extinction rates should decrease because larger population sizes should lower the risk of extinction due to chance events and larger species ranges should spread extinction risk over a larger area.

   The second hypothesis suggests that the tropics have had a longer evolutionary history than the temperate or polar zones because of their greater climatic stability. This stability may have allowed more species to evolve without the interruption of severe climatic conditions that would have hindered speciation and increased extinctions in the temperate and polar zones.

   The third hypothesis suggests that the high productivity of the tropics increases species diversity by promoting larger population sizes, which should lead to lower extinction rates and overall higher species richness.
4. Design 3 would be the best choice because it has the largest and closest fragments of rare habitat. The equilibrium theory of island biogeography predicts that the largest fragments will have the lowest extinction rates because species population sizes will be larger than in smaller fragments. In addition, the theory predicts that fragments closest to sources of species (e.g., mainland areas, or in this case, other, larger fragments of the rare habitat) will have the highest species diversity because immigration rates will be higher than in more distant fragments.

CHAPTER 18

Answers to Figure Legend Questions

Figure 18.4 No, it does not make sense that the fish and frog species should be present in the local community given in the figure because that community contains terrestrial species. The abiotic filter should have excluded any aquatic species from this terrestrial community.

Figure 18.7 B shows the most resource partitioning (least overlap in resource use). A and C show the least resource partitioning (most overlap in resource use).

Figure 18.15 The lowest species richness occurred on the small boulders (their maximum richness was 4), which rolled over more frequently, and thus experienced more disturbance, than the other boulders.

Answers to Review Questions

1. Yes. Community membership is dependent on dispersal, abiotic factors, and biological interactions. Given all the introductions of non-native species that have occurred worldwide, it is clear that "getting there" has been an important constraint on species entrance into communities. In this particular case, the seeds on the ecologist's shoes are physically and biologically adapted to prairie grassland communities and are thus prime candidates for successful introduction into New Zealand grassland communities.
2. Resource partitioning is the idea that coexistence among species is possible if the species in a community

use its resources in slightly different ways. Other models, such as the intermediate disturbance hypothesis, rely on population fluctuations due to disturbance, stress, or predation as the mechanism of coexistence. As long as species' populations never reach their carrying capacities, competitive exclusion will not occur, and coexistence will be possible. Lottery models assume that resources made available by disturbance, stress, or predation are captured at random by recruits from a larger pool of colonists, all of which have an equal chance to do so.

3. The lottery model best supports the tropical rainforest data set. The lottery model assumes that resources made available by the deaths of individuals are captured at random by recruits from a larger pool of colonists such that no one species has an advantage, and that species diversity is maintained as a result.
4. Species diversity–community function relationships can differ depending on two variables: the degree of overlap in the ecological functions of species and variation in the strength of ecological functions of species. Graph A is best described by the complementarity hypothesis, which proposes that as species richness increases, there will be a linear increase in community function. This linear relationship occurs because each species added to the community has a unique and equally incremental effect on community function. Graph B is best described by the redundancy hypothesis, in which there is an upper limit on the effect of species richness on community function. This curvilinear relationship occurs because the unique functional contributions of species reach a threshold due to the overlap of species' functions. Graph C best describes the driver and passenger hypothesis, which suggests that the strengths of the effects of species' functions vary dramatically. "Driver" species have a large effect on community function, while "passenger" species have a minimal effect. The addition of driver and passenger species to a community has unequal effects on community function, producing a "stair step" curve. Graph D best describes a variation on the driver and passenger hypothesis in which the curve levels off at high species richness as the functions of the species become redundant.

CHAPTER 19

Answers to Figure Legend Questions

Figure 19.5 Greater allocation of NPP to belowground tissues can be an adaptation to disturbances, such as fire, or to herbivory. Allocation of NPP to storage compounds allows more rapid recovery and higher survival rates following disturbance or loss of tissues to herbivory.

Figure 19.8 Cacti are CAM plants (see Chapter 5), which open their stomates and take up CO_2 during the night when air temperatures are cooler and humidities are higher. The daily pattern of atmospheric CO_2 concentrations would be reversed from what is shown for the boreal forest, with lower concentrations at night and higher concentrations during the day.

Figure 19.9 Estuaries also have high NPP due to the inputs of nutrients brought in by rivers. These nutrient subsidies include organic matter from both terrestrial and aquatic ecosystems as well as agricultural runoff.

Figure 19.12 The proportional allocation to belowground NPP would be greater in the more nutrient-poor community, the dry meadow. Greater allocation to roots enhances the uptake of the resources that most limit NPP, whereas light is more likely to be limiting in the more nutrient-rich wet meadow. Allocation to belowground NPP would decrease in response to fertilization.

Answers to Review Questions

1. Primary production is the source of the energy entering an ecosystem, and it therefore determines the amount of energy available to support that ecosystem. Primary production also results in the exchange of carbon between the atmosphere and the biosphere and thus is important in determining the atmospheric concentration of CO_2, an important greenhouse gas. Finally, primary production is a measure of the functioning of an ecosystem and provides a biological indicator of the ecosystem's response to stress.
2. As NPP increased in a terrestrial ecosystem, the leaf area index would increase along with overall plant biomass. The amount of shading would increase as the leaf area index increased, and light would become increasingly limiting to growth. To compensate, plants would allocate more energy to stems and less to roots so as to increase their height and overtop neighbors in order to acquire more light.
3. The researchers found a correlation between NPP and soil temperature, and they assumed that the causal link was through the effect of soil temperature on root growth. While this assumption may be correct, the researchers failed to show the causal link conclusively, which would require careful experimentation, or at least more thorough measurements of the effect of soil temperature on the factors that can influence plant growth. For example, soil temperature can affect the rate of decomposition of organic matter in the soil, and thus the availability of nutrients, which may influence growth rates.
4. Harvest techniques are simple and don't require high-tech equipment. However, harvesting can be labor-

intensive, may fail to account for production that is lost to herbivores or decomposition, and is impractical at large scales. Remote sensing provides estimates of NPP at larger spatial scales and can be used at frequent intervals. However, remote sensing is expensive and requires handling of massive amounts of data. Because it is based on absorption of light by chlorophyll, remote sensing can potentially overestimate NPP if a plant canopy is physiologically inactive.

5. The ^{15}N ratio in the tissues of a heterotroph reflects that in the sources of food it is eating. If an animal population relies on a single food source, its ^{15}N ratio would tend to vary little among individuals, particularly if its food source was plants. An omnivore, by definition, consumes both plants and animals. A population of omnivores should therefore show greater variation among individuals in ^{15}N ratios as a result of their varying degrees of reliance on plant and animal foods.

CHAPTER 20

Answers to Figure Legend Questions

Figure 20.7 In Figure 20.4, consumption efficiency increases as the percentage of NPP that ends up as detritus decreases and as the percentage of NPP consumed by herbivores increases. Figure 20.6 shows that overall consumption efficiency in aquatic ecosystems is higher than in terrestrial ecosystems, as the line fitting the aquatic ecosystem data lies above the line fitting the terrestrial ecosystem data, indicating that a greater percentage of the NPP is being consumed.

Figure 20.10 Brown trout might preferentially feed on predators that are more effective in controlling insect herbivores than are the predators that galaxias feed on. As a result, the effect of the brown trout on algal abundance would be greater than the effect of the galaxias.

Figure 20.16 Eight of the 21 species or feeding groups (38%) eat both plants and animals, and most of the others eat at more than one trophic level, indicating that omnivory is very common in this desert food web.

Answers to Review Questions

1. Population B should have a higher assimilation efficiency due to the higher food quality of its diet. The garbage and plant component of population A's diet is higher in materials that are difficult to digest, and its C:N ratio is also lower than that of population A's rodent diet. Thus, the amount of food assimilated would be greater in population B.

2. The seasonal and diurnal temperature variations in these animals' environments are different and should result in different production efficiencies. The marine environment is more thermally stable, and thus the marine mammals should need to invest less energy in coping with temperature changes than the mammals in the terrestrial ecosystem. As a result, the marine mammals should be able to invest more energy in growth and reproduction.

3. The forest would have a greater total amount of energy flowing through its trophic levels because a greater amount of energy would enter that ecosystem at the first trophic level. However, a larger proportion of the energy entering the lake ecosystem would pass through its higher trophic levels due to its higher consumption and production efficiencies.

4. The effect of the trophic cascade on primary production (plant growth) would be less if the herbivore were a specialist. Increases in the abundance of a specialist herbivore would result in greater consumption of one or a few plant species, potentially releasing other plant species from the effects of competition with the species being consumed.

CHAPTER 21

Answers to Figure Legend Questions

Figure 21.4 Primary production is low and plants are sparse in desert ecosystems, so the amount of soil organic matter should also be low. Wetting–drying events should enhance mechanical weathering of soils, producing a range of soil particle sizes. However, without a protective covering, winds may remove some of the finest particles, as we describe in the Case Study Revisited in Chapter 21. The low amount of precipitation and plant growth should limit the development and depth of distinct soil horizons.

Figure 21.6 Pesticides applied to plants can wash into the organic surface layers of soils, where they can kill both herbivorous animals and soil detritivores. The loss of these animals would effectively lower the rate of decomposition and would thereby decrease soil fertility.

Figure 21.12 The simple input–output model depicted in the figure assumes that elements enter the ecosystem primarily through deposition and leave it in stream water. As noted in Figure 21.13, other modes of input and output occur, including inputs through N_2 fixation, outputs in groundwater, and gaseous losses (e.g., denitrification).

Figure 21.19 The study of eutrophication in Lake Washington is very convincing, but it lacks an appropriate control. Therefore, it is correlational; that is, it shows a quantitative link between depth of clarity and phosphorus inputs, but that link isn't necessarily causal. Appropriate controls might have included another lake that didn't have sewage inputs, or a lake that continued to have inputs of phosphorus-laden sewage during the time sewage inputs to Lake Washington were halted. (Experiments with appropriate controls have demonstrated beyond a doubt that inputs of phosphorus in sewage entering lakes do cause eutrophication.)

Answers to Review Questions

1. The transformation of minerals in rock involves both the physical breakdown (mechanical weathering) and chemical alteration (chemical weathering) of the minerals. Mechanical weathering occurs through expansion and contraction of solid materials due to freezing–thawing or drying–rewetting cycles, gravitational forces such as landslides, and pressure exerted by plant roots. Mechanical weathering exposes the surfaces of mineral particles to chemical weathering. Weathering is a soil-building process, leading to the development of ever finer mineral particles and greater release of the nutrients in the minerals. The release of CO_2 and organic acids into the soil from organisms and detritus enhances the rate of chemical weathering.
2. The original source of nitrogen for plants is dinitrogen gas (N_2) in the atmosphere, but they cannot use it unless it is converted to other forms by the process of nitrogen fixation. Only bacteria can carry out nitrogen fixation, which is an energetically expensive process. Some plants, such as legumes, have symbiotic relationships with nitrogen-fixing bacteria. As ecosystems develop, nitrogen builds up in the pool of detritus and is converted into soluble organic and inorganic forms through decomposition. Some of the nitrogen released by decomposition is consumed by microorganisms, lowering the supply available to plants.
3. While both primary production and decomposition influence the buildup of organic matter and associated nutrients in the soil, decomposition is more sensitive to climatic controls than is primary production. The mean residence time of nutrients is therefore more strongly controlled by decomposition. Low soil temperatures in boreal forests result in very long mean residence times. High rates of decomposition limit the buildup of soil organic matter in tropical forests, and the mean residence times of nutrients such as nitrogen and phosphorus are two orders of magnitude lower than those in boreal forests.
4. Nutrient transfers between trophic levels are efficient in both tropical and temperate-zone lakes, but organic matter is progressively lost from the surface layers in both systems, falling into the sediments in the benthic zone, where oxygen concentrations, and thus decomposition rates, are low. In the temperate zone, some of these nutrient-rich sediments are brought back to the surface layers during seasonal turnover of water, where they decompose, providing nutrients to support production. Turnover is largely absent in tropical lakes, which are therefore more dependent on external inputs of nutrients from streams and terrestrial ecosystems.

CHAPTER 22

Answers to Figure Legend Questions

Figure 22.2 The bar graphs indicate there were about 36 million ha in 1500, 8 million ha in 1935, and 1 million ha in 2004. The annual rate of loss appears to have been greater from 1935 to 2004 (7 million ha lost over 69 years, or approximately 100,000 ha lost per year) than from 1500 to 1935 (28 million ha lost over 435 years, or approximately 64,000 ha lost per year).

Figure 22.5 As discussed in Chapter 14, the seeds of many plant species are dispersed by animals that eat their fruit; hence the extinction of many frugivores may have reduced the ability of such plant species to disperse their seeds. Likewise, as also discussed in Chapter 14, many plants are pollinated by animals that visit flowers to collect nectar. Hence, the loss of nectarivores may have reduced the reproductive success of some plant species.

Figure 22.9 Habitat loss is the most important factor affecting terrestrial mammals; overharvesting is also an important threat to them. In contrast, accidental mortality and pollution are the most important threats affecting marine mammals.

Figure 22.11 Individual answers may vary but should include a line of reasoning similar to the following: Although there was year-to-year fluctuation in the cod harvest, overall the catch increased from roughly 100,000 tons caught in 1850 to roughly 300,000 tons caught in 1950. Because the harvest was maintained at these levels for 100 years, this suggests that at about 200,000 tons could have been caught in a sustainable manner.

Answers to Review Questions

1. This is a thought exercise, and answers will be personal, but should reflect that conservation biologists are scientists, motivated by a personal sense that loss of biodiversity is a problem—in part an ethical problem—that society needs to address. Answers should reflect that science must be rigorous and can be applied to solving conservation problems even if there are values behind the science.
2. Two components of biodiversity undergoing increased homogenization are species diversity and genetic diversity. Range expansions of invasive species, the increase in generalist species, and the loss of more specialized species through range contractions or extinction are leading to taxonomic homogenization. Hybridization between native and introduced species contributes to genetic homogenization.

3. The principal threats to biodiversity are habitat loss, degradation, and fragmentation; the spread of invasive species; overharvesting; and climate change. For some species, disease poses a threat, and for others, particularly aquatic species, pollution is a particular threat. Many freshwater mussel species of North America are threatened both by pollution and by the invasion of the zebra mussel. The Pyrenean ibex was driven extinct by hunting, climate change, disease, and competition with domesticated species. Many other examples are possible.
4. The classification system set up by Natural Heritage/NatureServe documents each species' conservation status from a biological perspective, while a listing under the U.S. Endangered Species Act is a legal designation. While federally endangered species would generally also be considered globally rare by Natural Heritage/NatureServe, the reverse does not necessarily hold true: many extremely rare or threatened species are not on the federal endangered species list. The Endangered Species Act (ESA) provides legal protection for listed species, and it requires the designation of critical habitat and the development and implementation of a recovery plan for those species. In contrast, Natural Heritage/NatureServe can only recommend the protection of species.
5. Answers to this question will depend on where students are located and what species they identify. The object of this question is to make students aware of species of conservation concern, threats to biodiversity, and efforts that are under way to protect species in their own region. It also invites them to identify research needs and to think about scientific approaches to conservation.

CHAPTER 23

Answers to Figure Legend Questions

Figure 23.3 Wet calcareous loam.

Figure 23.4 Organisms move more freely across the matrix in (B). We can infer this because exchange occurs between habitat patches separated by matrix in (B) whereas it does not occur in (A) (unless patches are connected to one another by a corridor).

Figure 23.6 It is identical to the grain in all three panels of part (B)—they each have a pixel size of 50 × 50 m.

Figure 23.15 About 100 meters.

Figure 23.17 *Reserve size*: A reserve that covers a small area typically harbors small populations—and small populations are at greater risk than larger ones from genetic factors (genetic drift and inbreeding), demographic stochasticity, environmental stochasticity, and natural catastrophes (see Chapter 10). In addition, a smaller proportion of the area is exposed to edge effects in a large reserve than in a small reserve; in a very small reserve, the entire area may be exposed to edge effects. *Number of reserves*: Although the total protected area is the same for both designs, in the design on the right each reserve is small in area and hence is likely to be at risk from problems associated with small populations. *Reserve proximity*: When several reserves are close to one another, individuals can move more freely between them. These movements help to prevent each reserve from experiencing problems associated with small population sizes. *Reserve connectivity*: Habitat corridors enable organisms to cross boundaries or landscape elements that otherwise might isolate each reserve from the other reserves (thereby exposing each reserve to problems associated with small population sizes). *Reserve shape*: When two reserves of equal area are compared, the reserve with a more compact shape (the best possible shape being a circle) will have proportionately less of its area exposed to edge effects.

Answers to Review Questions

1. Habitat islands resemble actual islands in the way that they spatially isolate populations of some species from one another, with potential demographic and genetic consequences. They differ from islands, however, in that the matrix between habitat fragments may be more or less permeable to some species, so that movement between habitat fragments may be constrained, but may still occur with some frequency. As we saw in Chapter 17, the principles of island biogeography apply to habitat islands in that there is immigration to fragments, extinction within fragments, and some equilibrium level of species diversity. Larger habitat islands can sustain greater species diversity than smaller fragments.
2. In a sense, corridors are long, skinny habitat patches. Animals may nest in them, plants will germinate in them if conditions are right, and predation and competition occur in them. But they are likely to be biologically impoverished relative to larger habitat blocks because of the effects of their narrow dimensions on their abiotic and biotic properties. They are likely to resemble edge habitat in experiencing more light, more rapid biogeochemical cycling, and more predation than larger habitat blocks. They may be more vulnerable to invasive species, and they may permit movement of diseases between habitat blocks. Nevertheless, they are generally beneficial, at least for some species, in allowing movement of organisms across a fragmented landscape.
3. National forests and national parks have different management objectives. The difference in the resulting land uses is visible from space, in the form of a clear line separating clear-cut patches of the Targhee National Forest from the uncut forests of Yellowstone National Park. National forests permit the harvesting

of timber, which is generally not permitted in national parks. Timber harvesting makes for a patchy forest of different-aged stands, which may support a different group of species than is found in a national park, and may favor early successional species over old-growth-associated species. While both national parks and national forests have a mandate to protect biodiversity, national parks must balance these aims with recreation and visitor needs, while national forests must include timber production needs in their mission as well. Under an ecosystem management approach, the emphasis would be regional, and so the national forest and national park administrations would be working together to achieve conservation goals set by consensus.

4. Answers to this question will be place-specific and will depend on where students are located. The object of this question is to get students to evaluate the size, connectivity, and effectiveness of protected areas in their region. They may also find themselves wondering what kinds of lands serve as protected areas (i.e., state and federal lands, local public lands, lands held by land trusts and private conservation organizations such as the Audubon Society or The Nature Conservancy) and see the challenge of keeping track of where all these lands are. They should consider the role that private lands play in harboring local species, and they should consider how lands change over time depending on their ownership. In considering connectivity between protected areas, they should evaluate the protection status of corridors and whether those corridors are vulnerable to future disruption. They may consider possibilities for habitat restoration that could improve the landscape-level picture.
5. Again, the response here will be place-specific. This question invites students to identify and examine regional ecosystem management projects. They may choose a watershed protection project, a regional effort to preserve a forest, or a coordinated effort to protect habitat for a particular species. The key will be to identify a diverse partnership of stakeholders coming together to address land use in a region in a manner that serves both human needs and regional biodiversity. Students are asked to identify the successes of the program, and they may also identify the societal compromises and actions that slow progress in land protection or in resolving problems causing habitat degradation.

CHAPTER 24

Answers to Figure Legend Questions

Figure 24.3 Deforestation would immediately lower the flux of carbon from the atmosphere to the land surface due to photosynthesis, but would increase the flux from the land surface to the atmosphere due to respiration. In other words, the deforested land would change from a sink to a source of atmospheric CO_2. Cutting the trees removes the most important autotrophs in the system. It also supplies carbon (from roots and woody debris) to soil heterotrophs and warms up the soil, both of which increase respiratory C emissions to the atmosphere.

Figure 24.7 Reactive N is chemically and biologically active, as the name infers. As a result, the pool of reactive N is a potential source of nutrients for organisms. In addition, it can influence soil chemistry and the health of organisms, as we will see later in the chapter. N_2, on the other hand, is chemically inert and must be converted to other chemical forms by nitrogen fixation to be used by organisms.

Figure 24.14 In Chapters 3, 15, and 16 we discussed several factors that determine the makeup of vegetation assemblages. These factors include physiological tolerances, biotic interactions such as competition and herbivory, and dispersal ability. Following deglaciation, combinations of temperature and precipitation different from any found today occurred in parts of North America, which resulted in unique combinations of plants relative to those that occur today. In addition, by differentially consuming specific plant species, particular species of herbivores can have an effect on vegetation types. As noted in the Case Study in Chapter 3 (p. 49), the animals that occurred at this time were quite different from those found today, including sloths, mastodons, and camels. Finally, the rates at which different species dispersed into the newly exposed substrate would have influenced the composition of the vegetation.

Answers to Review Questions

1. The two major biological influences on the global carbon cycle are photosynthesis, which takes up CO_2 from the atmosphere, and respiration, which releases CO_2 back to the atmosphere. Prior to the Industrial Revolution, uptake by photosynthesis and release by respiration were roughly equal at a global scale, and thus there was no net flux associated with Earth's biota. However, increasing human population growth rates resulted in increasing deforestation and agricultural development, which in turn resulted in greater decomposition and heterotrophic respiration due to warming of the soil surface. As a result, atmospheric CO_2 concentrations increased. Deforestation was the primary reason for increasing atmospheric CO_2 concentrations until the early part of the twentieth century.
2. While animals can respond to climate change by moving, their habitats cannot. Animals are dependent on plants to provide their food (or food for their prey). Climate change will be so rapid that evolutionary responses will not be possible for most species of plants,

and the dispersal rates of most plant species are too slow to track the predicted climate changes. Dispersal may be inhibited by fragmentation of dispersal corridors due to land-use change. Loss of habitat will therefore result in decreased population growth for some animals. Additionally, migrating animals may respond to climate change more slowly than nonmigratory species. As a result, prey species may be less abundant or absent when these animals arrive at their destination.

3. One of the components of acid precipitation is nitric acid (HNO_3), which is a reactive form of nitrogen (i.e., it is biologically available). Both acid deposition and nitrogen deposition can cause acidification of soils and surface waters, loss of nutrient cations, and increases in aluminum concentrations. Nitrogen deposition can increase plant growth when it occurs at low rates. Over long periods, or at high rates of N deposition, N saturation occurs, leading to increased plant susceptibility to stress and, as noted above, soil acidification.
4. The effect of atmospheric ozone on organisms depends on where in the atmosphere it is found. Ozone in the stratosphere acts as a shield against high-energy ultraviolet-B radiation, which is harmful to organisms. In contrast, ozone in the troposphere damages organisms that come in direct contact with it. Ozone in the troposphere also acts as a greenhouse gas, contributing to global climate change.

Glossary

Numbers in brackets refer to the chapter(s) where the term is introduced.

A

abiotic Of or referring to the physical or nonliving environment. *Compare* biotic. [1]

absolute population size The actual number of individuals in a population. *Compare* relative population size. [8]

abundance The number of individuals in a species that are found in a given area; abundance is often measured by population size or population density. [8]

acclimatization An organism's adjustment of its physiology, morphology, or behavior to lessen the effect of an environmental change and minimize the associated stress. [4]

acid neutralizing capacity The ability of the chemical environment to counteract acidity, usually associated with concentrations of base cations, including Ca^{2+}, Mg^{2+}, and K^{+}. [24]

acidity A measure of the ability of a solution to behave as an acid, a compound that releases protons (H^{+}) to the water in which it is dissolved. *Compare* alkalinity. [2]

adaptation A physiological, morphological, or behavioral trait with an underlying genetic basis that enhances the survival and reproduction of its bearers in their environment. [1, 4]

adaptive evolution A process of evolutionary change in which traits that confer survival or reproductive advantages tend to increase in frequency in a population over time. [6]

adaptive management A component of ecosystem management in which management actions are seen as experiments and future management decisions are determined by the outcome of present decisions. [23]

adaptive radiation An event in which a group of organisms gives rise to many new species that expand into new habitats or new ecological roles in a relatively short time. [6]

aerosols Solid or liquid particles suspended in the atmosphere. [21]

age structure The proportions of a population in each age class. [9]

albedo The amount of solar radiation reflected by a surface, usually expressed as a percentage of the incident solar radiation. [2]

alkalinity A measure of the ability of a solution to behave as a base, a compound that takes up protons (H^{+}) or releases hydroxide ions (OH^{-}). *Compare* acidity. [2]

Allee effect A decrease in the population growth rate (r or λ) as the population density decreases. [10]

allele One of two or more forms of a gene that result in the production of different versions of the protein that the gene encodes. [6]

allochthonous Produced outside the ecosystem. *Compare* autochthonous. [20]

allometry Differential growth of body parts that results in a change of shape or proportion with size. [7]

alpha diversity Species diversity at the local or community scale. *Compare* beta diversity, gamma diversity. [17]

alternation of generations A complex life cycle, found in many algae and all plants, in which there is both a multicellular diploid form, the sporophyte, and a multicellular haploid form, the gametophyte. [7]

alternative stable states Different community development scenarios that are possible at the same location under similar environmental conditions. [16]

anisogamy Production of two types of gametes of different sizes. *Compare* isogamy. [7]

anoxia The absence of oxygen. [2]

anthropogenic Of, relating to, or caused by humans or their activities. [24]

aposematic coloration *See* warning coloration.

arbuscular mycorrhizae Mycorrhizae in which the fungal partner grows into the soil, extending some distance away from the plant root, and also grows between some root cells while penetrating others. *Compare* ectomycorrhizae. [14]

Arctic ozone dent An area of the stratosphere over the Arctic region where ozone concentrations are low, but have not dropped below 220 Dobson units. [24]

assimilation efficiency The proportion of ingested food that is assimilated by an organism. [20]

atmospheric deposition The movement of particulate and dissolved matter from the atmosphere to Earth's surface by gravity or in precipitation. [21]

atmospheric pressure The pressure exerted on a surface due to the mass of the atmosphere above it. [2]

autochthonous Produced within the ecosystem. *Compare* allochthonous. [20]

autotroph An organism that converts energy from sunlight or from inorganic chemical compounds in the environment into chemical energy stored in the carbon–carbon bonds of organic compounds. *Compare* heterotroph. [5]

avoidance A response to stressful environmental conditions that lessens their effect through some behavior or physiological activity that minimizes an organism's exposure to the stress. *Compare* tolerance. [4]

B

benthic zone The bottom of a body of water including the surface and shallow subsurface layers of sediment. [3]

beta diversity The change in species number and composition, or turnover of species, as one moves from one community to another. *Compare* alpha diversity, gamma diversity. [17]

bioaccumulation A progressive increase in the concentration of a substance in an organism's body over its lifetime. [20]

biodiversity The diversity of important ecological entities that span multiple spatial scales, from genes to species to communities. [15]

biofuel A liquid or gas fuel made from plant material (biomass). [18]

biogeochemistry The study of the physical, chemical, and biological factors that influence the movements and transformations of chemical elements. [21]

biogeographic region A portion of Earth containing a distinct biota that differs markedly from the biotas of other biogeographic regions in its species composition and diversity. [17]

biogeography The study of variation in species composition and diversity among geographic locations. [17]

biological crust A crust on the soil surface composed of a mix of species of cyanobacteria, lichens, and mosses; also called a cryptobiotic or cryptogamic crust. [21]

biological reserve An often small nature reserve established with the conservation of a single species or ecological community as the main conservation objective. [23]

biomagnification A progressive increase in the tissue concentrations of a substance in animals at successively higher trophic levels that results as animals at each trophic level consume prey with higher concentrations of the substance due to bioaccumulation. [20]

biomass The mass of living organisms, usually expressed per unit of area. [19]

biome A large-scale terrestrial biological community shaped by the regional climate, soil, and disturbance patterns where it is found, usually classified by the growth form of the dominant plants. [3]

biosphere The highest level of biological organization, consisting of all living organisms on Earth plus the environments in which they live; located between the lithosphere and the troposphere. [1, 3]

biotic Of or referring to the living components of an environment. *Compare* abiotic. [1]

biotic resistance Interactions of the native species in a community with non-native species that exclude or slow the growth of those non-native species. [18]

bottom-up control Limitation of the abundance of a population by nutrient supply or by the availability of food. *Compare* top-down control. [10]

boundary layer A zone close to a surface where a flow of fluid, usually air, encounters resistance and becomes turbulent. [4]

buffer zone A portion of a nature reserve surrounding a core natural area where controls on land use are less stringent than in the core natural area, yet land uses are at least partially compatible with many species' resource requirements. *Compare* core natural area. [23]

C

C_3 photosynthetic pathway A biochemical pathway involving the uptake of CO_2 by the enzyme ribulose 1,5 bisphosphate carboxylase/oxygenase (rubisco) and synthesis of sugars by the Calvin cycle. *Compare* C_4 photosynthetic pathway, crassulacean acid metabolism. [5]

C_4 photosynthetic pathway A biochemical pathway involving the daytime uptake of CO_2 by the enzyme phosphoenol pyruvate carboxylase (PEPcase) in mesophyll cells; the carbon is then transferred as a four-carbon acid to the bundle sheath cells, where CO_2 is released to the Calvin cycle for sugar synthesis. *Compare* C_3 photosynthetic pathway, crassulacean acid metabolism. [5]

Calvin cycle The biochemical pathway used by photosynthetic and chemosynthetic organisms to fix carbon and synthesize sugars. [5]

carbon neutral Of or relating to fuels that produce an amount of CO_2 when burned that is equal to or less than the amount taken up by the plants from which they are made. [18]

carrying capacity The maximum population size that can be supported indefinitely by the environment, represented by the term *K* in the logistic equation. [9]

catchment The area in a terrestrial ecosystem that is drained by a single stream; a common unit of study in terrestrial ecosystem studies; also called a watershed. [21]

cation exchange capacity A soil's ability to hold nutrient cations such as Ca^{2+}, K^+, and Mg^{2+} and exchange them with the soil solution, determined by the clay content of the soil. [21]

character displacement A process in which competition causes the phenotypes of competing species to evolve to become more different over time, thereby causing the species to become more different where they live together than where they live apart. [11]

cheater In a mutualism, an individual that increases its production of offspring by overexploiting its mutualistic partner. [14]

chemical weathering The chemical breakdown of soil minerals leading to the release of soluble forms of nutrients and other elements. *Compare* mechanical weathering. [21]

chemolithotrophy *See* chemosynthesis.

chemosynthesis The use of energy from inorganic chemical compounds to fix CO_2 and produce carbohydrates using the Calvin Cycle; also called chemolithotrophy. [5]

clay Fine soil particles (<2 μm) that have a semicrystalline structure and weak negative charges on their surfaces that can hold onto cations and exchange them with the soil solution. [21]

climate The long-term description of weather, based on averages and variation measured over decades. *Compare* weather. [2]

climate change Directional change in climate over a period of three decades or longer. [1, 24]

cline A pattern of gradual change in a characteristic of an organism over a geographic region. [6]

clone A genetically identical copy of an individual. [7]

clumped dispersion A dispersion pattern in which individuals are grouped together. *Compare* random dispersion, regular dispersion. [8]

coarse-filter Of or referring to an approach to conservation biology that focuses on habitats, landscapes, and ecosystems. *Compare* fine-filter. [22]

coevolution The evolution of two interacting species, each in response to selection pressure imposed by the other. [13]

cohort life table A life table in which the fate of a group of individuals born during the same time period (a cohort) is followed from birth to death. [9]

commensalisms An interaction between two species in which individuals of one species benefit while individuals of the other species do not benefit and are not harmed. [14]

community A group of interacting species that occur together at the same place and time. [1,15]

community structure The set of characteristics that shape a community, including the number, composition, and abundance of species. [15]

compensation An adaptive growth response of plants to herbivory in which removal of plant tissues stimulates the plant to produce new tissues. [12]

competition An interaction between individuals of two species in which each is harmed by their shared use of a resource that limits their ability to grow, survive, or reproduce (a –/– relationship). [11]

competition coefficient A constant used in the Lotka–Volterra competition model to describe the extent to which an individual of one competing species decreases the per capita growth rate of the other species. [11]

competitive exclusion principle The principle that two species that use a limiting resource in the same way cannot coexist indefinitely. [11]

competitive networks Sets of competitive interactions involving multiple species in which every species negatively interacts with every other species thus promoting species coexistence. [15]

competitive plants In Grime's triangular model, plants that are superior competitors under conditions of low stress and low disturbance. *Compare* ruderals, stress-tolerant plants. [7]

complementarity hypothesis A hypothesis proposing that as the species richness of a community increases, there is a linear increase in the positive effects of those species on community function. *Compare* redundancy hypothesis. [18]

complex life cycle A life cycle in which there are at least two distinct stages that differ in their habitat, physiology, or morphology. [7]

conduction The transfer of sensible heat through the exchange of kinetic energy between molecules due to a temperature gradient. *Compare* convection. [2]

conservation biology The scientific study of phenomena that affect the maintenance, loss, and restoration of biodiversity. [22]

consumer An organism that obtains its energy by eating other organisms or their remains. *Compare* producer. [1]

consumption efficiency The proportion of the biomass available in an ecosystem that is ingested. [20]

continental climate The climate typical of terrestrial areas in the middle of large continental land masses at high latitudes, characterized by high variation in seasonal temperatures. *Compare* maritime climate. [2]

continental drift The slow movement of tectonic plates (sections of Earth's crust) across Earth's surface. [17]

controlled experiment A standard scientific approach in which an experimental group (that has the factor being tested) is compared with a control group (that lacks the factor being tested). [1]

convection The transfer of sensible heat through the exchange of air and water molecules as they move from one area to another. *Compare* conduction. [2]

convergence The evolution of similar growth forms among distantly related species in response to similar selection pressures. [3]

cooperative breeding A behavioral pattern in which young animals postpone breeding and instead help their parents raise offspring. [8]

core natural area A portion of a nature reserve where the conservation of biodiversity and ecological integrity takes precedence over other values or uses. *Compare* buffer zone. [23]

Coriolis effect The apparent deflection of air or water currents when viewed from a rotating reference such as Earth's surface. [2]

crassulacean acid metabolism (CAM) A photosynthetic pathway in which CO_2 is fixed and stored as an organic acid at night, and then released to the Calvin cycle during the day. *Compare* C_3 photosynthetic pathway, C_4 photosynthetic pathway. [5]

crypsis A defense against predators in which prey species have a shape or coloration that provides camouflage and allows them to avoid detection. [12]

cryptobiotic crust *See* biological crust.

D

damped oscillations A pattern of population fluctuations where the extent to which the population rises and falls in abundance gradually become smaller over time. [10]

decomposition The physical and chemical breakdown of detritus by detritivores, leading to the release of nutrients as simple, soluble organic and inorganic compounds that can be taken up by other organisms. [21]

delayed density dependence Delays in the effect of population density on population size that can contribute to population fluctuations. [10]

demographic stochasticity Chance events associated with whether individuals survive or reproduce. [10]

denitrification A process by which certain bacteria convert nitrate (NO_3^-) into nitrogen gas (N_2) and nitrous oxide (N_2O) under hypoxic conditions. [21]

density-dependent Of or referring to a factor that causes birth rates, death rates, or dispersal rates to change as the density of the population changes. *Compare* density-independent. [9]

density-independent Of or referring to a factor whose effects on birth and death rates are independent of population density. *Compare* density-dependent. [9]

desertification Degradation of formerly productive land in arid regions resulting in loss of plant cover and acceleration of soil erosion. [3, 24]

detritus Freshly dead or partially decomposed remains of organisms. [3, 8]

direct development A simple life cycle that goes directly from fertilized egg to juvenile without passing through a free-living larval stage. [7]

direct interactions Interactions between two species, including competition, exploitation, and positive interactions. *Compare* indirect interactions. [15]

directional selection Selection that favors individuals with one extreme of a heritable phenotypic trait. *Compare* disruptive selection, stabilizing selection. [6]

dispersal The movement of organisms or propagules from their birthplace. [7]

dispersal limitation A situation in which a species' limited capacity for dispersal prevents it from reaching areas of suitable habitat. [8]

dispersion The spatial arrangement of individuals within a population. [8]

disruptive selection Selection that favors individuals with a phenotype at either extreme over those with an intermediate phenotype. *Compare* directional selection, stabilizing selection. [6]

distribution The geographic area where individuals of a species are present. [8]

disturbance An abiotic event that kills or damages some individuals and thereby creates opportunities for other individuals to grow and reproduce. [8, 16]

dominant species A species that has large, community-wide effects by virtue of its size or abundance, its strong competitive ability, or its provision of habitat or food for other species, also called a foundation species. [15]

dormancy A state in which little or no metabolic activity occurs. [4]

doubling time (t_d) The number of years it takes a population to double in size. [9]

driver and passenger hypothesis A hypothesis proposing that the strengths of the effects of species' ecological functions on their communities vary dramatically, such that "driver" species have a large effect on community function, while "passenger" species have a minimal effect. [18]

dynamic equilibrium model An elaboration of the intermediate disturbance hypothesis proposing that species diversity is maximized when the level of disturbance and the rate of competitive displacement are roughly equivalent. [18]

E

ecological footprint The total area of productive ecosystems required to support a population. [9]

ecological niche The physical and biological conditions that a species needs to grow, survive, and reproduce. [8]

ecology The scientific study of interactions between organisms and their environment. [1]

ecosystem All the organisms in a given area as well as the physical environment in which they live; an ecosystem can include one or more communities. [1]

ecosystem engineer A species that influences its community by creating, modifying, or maintaining physical habitat for itself and other species. [15]

ecosystem management An approach to habitat management in which scientifically based policies and practices guide decisions on how best to meet an overarching goal of sustaining ecosystem structure and function for long periods. [23]

ecosystem services Natural processes that sustain human life and which depend on the functional integrity of natural communities and ecosystems. [22]

ecotype A population with adaptations to unique local environmental conditions. [4]

ectomycorrhizae Mycorrhizae in which the fungal partner typically grows between plant root cells and forms a mantle around the exterior of the root. *Compare* arbuscular mycorrhizae. [14]

ectoparasite A parasite that lives on the surface of another organism. *Compare* endoparasite. [13]

ectotherm An animal that regulates its body temperature primarily through energy exchange with its external environment. *Compare* endotherm. [4]

edge effects Abiotic and biotic changes that are associated with an abrupt habitat boundary such as that created by habitat fragmentation. [23]

El Niño Southern Oscillation (ENSO) An oscillation of pressure cells and sea surface temperatures in the equatorial Pacific Ocean causing widespread climate variation and changes in upwelling currents. *See also* La Niña. [2]

endemic Occurring in a particular geographic location and nowhere else on Earth. [17]

endoparasite A parasite that lives inside the body of its host organism. *Compare* ectoparasite. [13]

endosperm Nutrient-rich material in a seed that sustains the developing embryo and often the young seedling. [7]

endotherm An animal that regulates its body temperature primarily through internal metabolic heat generation. *Compare* ectotherm. [4]

environmental science An interdisciplinary field of study that incorporates concepts from the natural sciences (including ecology) and the social sciences (e.g., politics, economics, ethics), focused on how people affect the environment and how we can address environmental problems. [1]

environmental stochasticity Erratic or unpredictable changes in the environment. [10]

epilimnion The warm surface layer of water in a lake, lying above the thermocline, that forms during the summer in some lakes of temperate and polar regions. *Compare* hypolimnion. [2]

equilibrium theory of island biogeography A theory proposing that the number of species on an island or in an island-like habitat results from a dynamic balance between immigration rates and extinction rates. [17]

eutrophic Nutrient-rich; characterized by high primary productivity. *Compare* oligotrophic, mesotrophic. [21]

eutrophication A change in the nutrient status of an ecosystem from nutrient-poor to nutrient-rich; such changes occur naturally in some lakes due to the accumulation of sediments, but they may also be caused by nutrient inputs that result from human activities. [21]

evapotranspiration The sum of water loss through evaporation and transpiration. [2]

evolution (1) Change in allele frequencies in a population over time. (2) Descent with modification; the process by which organisms gradually accumulate differences from their ancestors. [1, 6]

evolutionarily significant unit An appropriate target of management within a species (e.g., a subspecies or population) as determined by genetic analyses; also called a management unit. [22]

exploitation A relationship in which one organism benefits by feeding on, and thus directly harming, another. [12]

exploitation competition An interaction in which species compete indirectly through their mutual effects on the availability of a shared resource. *Compare* interference competition. [11]

exponential growth Change in the size of a population of a species with continuous reproduction by a constant proportion at each instant in time. *Compare* geometric growth. [9]

exponential population growth rate (r) A constant proportion by which a population of a species with continuous reproduction changes in size at each instant in time; also called the intrinsic rate of increase. *Compare* geometric population growth rate. [9]

extent In landscape ecology, the area or time period encompassed by a study; together with grain, extent characterizes the scale at which a landscape is studied. *Compare* grain. [23]

extinction vortex A pattern in which a small population that drops below a certain size may decrease even further in size, perhaps spiraling toward extinction. [22]

F

fecundity The average number of offspring produced by a female while she is of age x (denoted F_x in a life table). [9]

Ferrell cell A large-scale, three-dimensional pattern of atmospheric circulation in each hemisphere, located at midlatitudes between the Hadley and polar cells. [2]

fine-filter Of or referring to an approach to conservation biology that focuses on genes, populations, and species. *Compare* coarse-filter. [22]

finite rate of increase *See* geometric population growth rate.

fitness The genetic contribution of an organism's descendants to future generations. [7]

fixation (1) The uptake of the gaseous form of a compound, including CO_2 in photosynthesis and N_2 in nitrogen fixation, by organisms for use in metabolic functions. [5] (2) With respect to the genetic composition of a population, an allele frequency of 100%. [6]

flagship species A charismatic species that may be emphasized in conservation efforts because it helps to garner public support for a conservation project. [22]

focal species One of a group of species selected as a priority for conservation efforts, chosen because its ecological requirements differ from those of other species in the group, thereby helping to ensure that as many different species as possible receive protection. [22]

food web A diagram showing the connections between organisms and the food they consume. [20]

foundation species *See* dominant species.

functional group A subset of a community that includes species that function in similar ways, but do not necessarily use the same resources. *Compare* guild. [15]

G

gamma diversity Species diversity at the regional scale; the regional species pool. *Compare* alpha diversity, beta diversity. [17]

gene flow The transfer of alleles from one population to another via the movement of individuals or gametes. [6]

gene-for-gene interaction A specific defensive response that makes particular plant genotypes resistant to particular parasite genotypes. [13]

genet A genetic individual, resulting from a single fertilization event; in organisms that can reproduce asexually, a genet may consist of multiple, genetically identical parts, each of which has the potential to function as an independent physiological unit. *Compare* ramet. [8]

genetic drift A process in which chance events determine which alleles are passed from one generation to the next, thereby causing allele frequencies to fluctuate randomly over time; the effects of genetic drift are most pronounced in small populations. [6, 10]

genotype The genetic makeup of an individual. [6]

geographic information systems (GIS) Computer-based systems that allow the storage, analysis, and display of data pertaining to specific geographic areas. [23]

geographic range The entire geographic region over which a species is found. [8]

geometric growth Change in the size of a population of a species with discrete reproduction by a constant proportion from one discrete time period to the next. *Compare* exponential growth. [9]

geometric population growth rate (λ) A constant proportion by which a population of a species with discrete reproduction changes in size from one discrete time period to the next; also called the finite rate of increase. *Compare* exponential population growth rate. [9]

grain In landscape ecology, the size of the smallest homogeneous unit of study (such as a pixel in a digital image), which determines the resolution at which a landscape is observed; together with extent, grain characterizes the scale at which a landscape is studied. *Compare* extent. [23]

gravitational potential The energy associated with gravity. [4]

greenhouse effect The warming of Earth by gases in the atmosphere that absorb and reradiate infrared energy emitted by Earth's surface. [2, 24]

greenhouse gases Atmospheric gases that absorb and reradiate the infrared radiation emitted by Earth's surface, including water vapor (H_2O), carbon dioxide (CO_2), methane (CH_4), and nitrous oxide (N_2O). [2, 24]

gross primary production (GPP) The amount of energy that autotrophs capture by photosynthesis and chemosynthesis per unit time. *Compare* net primary production. [19]

guild A subset of a community that includes species that use the same resources, whether or not they are taxonomically related. *Compare* functional group. [15]

H

habitat corridor A relatively narrow patch that connects blocks of habitat and often facilitates the movement of species between those blocks. [23]

habitat degradation Anthropogenic change that reduces the quality of habitat for many, but not all, species. [22]

habitat fragmentation The breaking up of once continuous habitat into a complex pattern of spatially isolated habitat patches amid a matrix of human-dominated landscape. [10, 22]

habitat loss The outright conversion of an ecosystem to another use by human activities. [22]

habitat mutualism A mutualism in which one partner provides the other with shelter, a place to live, or favorable habitat. [14]

Hadley cell A large-scale, three-dimensional pattern of atmospheric circulation in each hemisphere in which air is uplifted at the equator and subsides at about 30° N and S. [2]

heat capacity The amount of energy required to raise the temperature of a substance. [2]

herbivore An organism that eats the tissues or internal fluids of living plants or algae. [12]

heterotroph An organism that obtains energy by consuming energy-rich organic compounds from other organisms. *Compare* autotroph. [5]

hibernation Torpor lasting several weeks during the winter; a strategy that is possible only for animals that have access to enough food and can store enough energy reserves. [4]

horizons Layers of soil distinguished by their color, texture, and permeability. [21]

horizontal interactions Non-trophic interactions, such as competition and some positive interactions, that occur within a trophic level. [15]

host An organism on or within which a parasite or other symbiont lives. [13]

hypolimnion The densest, coldest water layer in a lake, lying below the thermocline. *Compare* epilimnion. [2]

hyporheic zone The portion of the substrate below and adjacent to a stream bed where water movement still occurs, either from the stream or from groundwater moving into the stream. [3]

hypothesis A possible answer to a question developed using previous knowledge or intuition. *See also* scientific method. [1]

hypoxia The condition of oxygen depletion, usually below a level that can sustain most animals. [2]

hysteresis The inability of a community that has undergone change to shift back to the original community type, even when the original conditions are restored. [16]

I

inbreeding Mating between related individuals. [10]

indirect interactions Interactions in which the relationship between two species is mediated by a third (or more) species. *Compare* direct interactions. [15]

induced defense In plant–herbivore interactions, a defense against herbivory, such as production of a secondary compound, that is stimulated by herbivore attack. [12]

interaction strength A measure of the effect of one species (the interactor) on the abundance of another species (the target species). [15]

interaction web A concept that describes both the trophic (vertical) and non-trophic (horizontal) interactions among the species in a traditional food web. [15]

interference competition An interaction in which species compete directly by performing antagonistic actions that interfere with the ability of their competitors to use a resource that both require, such as food or space. *Compare* exploitation competition. [11]

intermediate disturbance hypothesis A hypothesis proposing that species diversity in communities should be greatest at intermediate levels of disturbance (or stress or predation) because competitive exclusion at low levels of disturbance and mortality at high levels of disturbance should reduce species diversity. [18]

Intertropical Convergence Zone (ITCZ) The zone of maximum solar radiation, atmospheric uplift, and precipitation within the tropical zone. [2]

intrinsic rate of increase *See* exponential population growth rate.

invasive species An introduced species that survives and reproduces in its new environment, sustains a growing population, and has large effects on the native community. [22]

island biogeography *See* equilibrium theory of island biogeography.

isogamy The production of equal-sized gametes. *Compare* anisogamy. [7]

isolation by distance A metapopulation pattern in which habitat patches located far away from occupied patches are less likely to be colonized than are nearby patches. [10]

iteroparous Having the capacity to reproduce multiple times in a lifetime. *Compare* semelparous. [7]

J

jump dispersal A long-distance dispersal event by which a species colonizes a new geographic region. [10]

K

K *See* carrying capacity.

***K*-selection** In the *r*- and *K*-selection continuum used for classifying life history strategies, the selection pressure for slower rates of increase faced by organisms that live in environments where population densities are high (at or near the carrying capacity, *K*). *Compare* *r*-selection. [7]

keystone species A strong interactor that has an effect on energy flow and community structure that is disproportionate to its abundance or biomass. [15, 20]

L

La Niña A stage of the El Niño Southern Oscillation consisting of a stronger-than-average phase of the normal climatic pattern, with high atmospheric pressure off the coast of South America and low pressure in the western equatorial Pacific. [2]

land use change The alteration of terrestrial surface, including vegetation and land forms, by human activities such as agriculture, forestry, and mining. [3]

landscape An area that often includes multiple ecosystems and that is spatially heterogeneous in one or more features of the environment, such as the number or arrangement of different habitat types. [1, 23]

landscape composition In landscape ecology, the kinds of elements or patches comprised by a landscape and how much of each kind is present. *Compare* landscape structure. [23]

landscape ecology The study of landscape patterns and the effects of those patterns on ecological processes. [23]

landscape structure In landscape ecology, the physical configuration of the different compositional elements of a landscape. *Compare* landscape composition. [23]

lapse rate The rate at which atmospheric temperature decreases with increasing distance from the ground. [2]

latent heat flux Heat transfer associated with the phase change of water, such as evaporation, sublimation, or condensation. [2]

leaching The vertical movement of dissolved matter and fine mineral particles from upper to lower layers of soil. [21]

leaf area index The area of leaves per unit of ground area (a dimensionless number, since it is an area divided by an area). [19]

lentic Of or referring to still water. *Compare* lotic. [3]

life history A record of major events relating to an organism's growth, development, reproduction, and survival; these events include the age and size of first reproduction, the amount and timing of reproduction, and longevity. [7]

life history strategy The overall pattern in the timing and nature of life history events, averaged across all the individuals of a species. [7]

life table A summary of how survival and reproductive rates in a population vary with the age of individuals; in species for which age is not informative or is difficult to measure, life tables may be based on the size or life history stage of individuals. [9]

lignin A structural compound that strengthens plant tissues. [21]

litter Fresh, undecomposed organic matter on the soil surface. [21]

littoral zone The nearshore zone of a lake where the photic zone reaches to the bottom. [3]

local scale A spatial scale that is essentially equivalent to a community. [17]

loess Sediment deposited by wind. [21]

logistic growth Change in the size of a population that is rapid at first, then decreases as the population approaches the carrying capacity of its environment. [9]

lotic Of or relating to flowing water. *Compare* lentic. [3]

Lotka–Volterra competition model A modified form of the logistic equation used to model competition. [11]

lottery model A hypothesis proposing that species diversity in communities is maintained by a "lottery" in which resources made available by the effects of disturbance, stress, or predation are captured at random by recruits from a larger pool of potential colonists. [18]

lower critical temperature The environmental temperature at which the heat loss of an endotherm triggers an increase in metabolic heat generation. [4]

M

macroparasites Relatively large parasite species, such as arthropods and worms. *Compare* microparasites. [13]

macrophyte A rooted or floating aquatic vascular plant. [3]

marginal value theorem A conceptual optimal foraging model proposing that an animal should stay in a food patch until the rate of energy gain in that patch has declined to the average rate for the habitat, then depart for another patch. [5]

maritime climate The climate typical of coastal terrestrial regions that are influenced by an adjacent ocean, characterized by low daily and seasonal variation in temperature. *Compare* continental climate. [2]

mass extinction An event in which a large proportion of Earth's species are driven to extinction worldwide in a relatively short time. [6]

matric potential The energy associated with attractive forces on the surfaces of large molecules inside cells or on the surfaces of soil particles. [4]

mean residence time The amount of time an average molecule of an element spends in a pool before leaving it. [21]

mechanical weathering The physical breakdown of rocks into progressively smaller particles without chemical change. *Compare* chemical weathering. [21]

mesotrophic Having a nutrient status that is intermediate between oligotrophic and eutrophic, usually used in reference to lakes. *Compare* eutrophic, oligotrophic. [21]

metamorphosis An abrupt transition from a larval to a juvenile life cycle stage that is sometimes accompanied by a change in habitat. [7]

metapopulation A set of spatially isolated populations linked to one another by dispersal. [10]

microparasites Parasite species too small to be seen with the naked eye, such as bacteria, protists, and fungi. *Compare* macroparasites. [13]

Milankovitch cycles Cycles of regular change over thousands of years in the shape of Earth's orbit, in the angle of tilt of its axis, and in its orientation toward other celestial bodies, that change the intensity of solar radiation received by Earth. [2]

mimicry A defense against predators in which prey species resemble less palatable organisms or physical features of their environment, causing potential predators to mistake them for something less desirable to eat. [12]

mineralization The chemical conversion of organic matter into inorganic compounds. [21]

morphs Discrete phenotypes with few or no intermediate forms. [7]

mosaic A composite or pattern of different habitat types or other heterogeneous features of the environment in a landscape. [23]

mutation Change in the DNA of a gene. [6]

mutualism A mutually beneficial interaction between individuals of two species (a +/+ relationship). [14]

mycorrhizae Symbiotic associations between plant roots and various types of fungi that are usually mutualistic. [14]

N

natural catastrophe An extreme environmental event such as a flood, severe windstorm, or outbreak of disease that can eliminate or drastically reduce the sizes of populations. [10]

natural selection The process by which individuals with certain heritable characteristics tend to survive and reproduce more successfully than other individuals because of those characteristics. [1, 6]

nekton Swimming organisms capable of overcoming water currents. *Compare* plankton. [3]

net ecosystem exchange (NEE) The combined fluxes of CO_2 into and out of an ecosystem principally by net primary production and autotrophic and heterotrophic respiration. [19]

net primary production (NPP) The amount of energy (per unit time) that autotrophs capture by photosynthesis and chemosynthesis, minus the amount they use in cellular respiration. *Compare* gross primary production. [1, 19]

net reproductive rate (R_0) The mean number of offspring produced by an individual in a population during its lifetime. [9]

net secondary production The balance between heterotroph energy gains through ingestion and heterotroph energy losses by cellular respiration and egestion. [19]

niche model A predictive tool that models the ecological niche occupied by a species based on the conditions at locations the species is known to occupy. [8]

nitrification A process by which certain chemoautotrophic bacteria, known as nitrifying bacteria, convert ammonia (NH_3) and ammonium (NH_4^+) into nitrate (NO_3^-) under aerobic conditions. [21]

nitrogen fixation The process of taking up nitrogen gas (N_2) and converting it into chemical forms that are more chemically available to organisms. [21]

North Atlantic Oscillation An oscillation in atmospheric pressures and ocean currents in the North Atlantic Ocean that affects climatic variation in Europe, in northern Asia, and on the eastern coast of North America. [2]

nutrient A chemical element required by an organism for its metabolism and growth. [21]

nutrient cycling The cyclic movement of nutrients between organisms and the physical environment. [1, 21]

O

occlusion A process by which soluble phosphorus combines with iron, calcium, and aluminum to form insoluble compounds (secondary minerals) that are unavailable to organisms as nutrients. [21]

oligotrophic Nutrient-poor, characterized by low primary productivity. *Compare* eutrophic, mesotrophic. [21]

omnivore (1) In trophic studies, an organism that feeds on more than one trophic level. [15] (2) An organism that feeds on both plants and animals. [20]

optimal foraging A theory proposing that animals will maximize their rate of energy gain. [5]

osmotic adjustment An acclimatization response to changing water availability or salinity in terrestrial and aquatic environments that involves changing the solute concentration, and thus the osmotic potential, of the cell. [4]

osmotic potential The energy associated with dissolved solutes. [4]

ozone hole An area of the stratosphere with an ozone concentration of less than 220 Dobson units (= 2.7×10^{16} molecules of ozone) per square centimeter; found primarily over the Antarctic region. [24]

P

Pacific Decadal Oscillation (PDO) A long-term oscillation in sea surface temperatures and atmospheric pressures in the North Pacific Ocean that has widespread climatic effects. [2]

paedomorphic Resulting from a delay of a developmental event relative to sexual maturation. [7]

parasite An organism that lives in or on a host organism and feeds on its tissues or body fluids. [13]

parasitoid An insect that lays one or a few eggs on or in a host organism (itself usually an insect), which the resulting larvae remain with, eat, and almost always kill. [12]

parent material The rock or sediments that are broken down by weathering to form mineral particles in soil. [21]

pathogen A parasite that causes disease. [13]

pelagic zone The open water column of a lake or ocean. [3]

permafrost A subsurface soil layer that remains frozen year-round for at least 3 years. [3]

phenotype The observable characteristics of an organism. [6]

phenotypic plasticity The ability of a single genotype to produce different phenotypes under different environmental conditions. [7]

photic zone The surface layer of a lake or ocean where enough light penetrates to allow photosynthesis. [3]

photorespiration A chemical reaction in photosynthetic organisms in which the enzyme rubisco takes up O_2, leading to the breakdown of sugars, the release of CO_2, and a net loss of energy. [5]

photosynthesis A process that uses sunlight to provide the energy needed to take up CO_2 and synthesize sugars. [5]

physical factor A feature of the environment that affects organism function and population growth rates but is not consumed or depleted. *Compare* resource. [11]

phytoplankton Photosynthetic plankton. *Compare* zooplankton. [3]

pioneer stage The first stage of primary succession. [16]

plankton Small, often microscopic organisms that live suspended in water; although many plankton are mobile, none can swim strongly enough to overcome water currents. *Compare* nekton. [3]

polar cell A large-scale, three-dimensional pattern of atmospheric circulation in which air subsides at the poles and moves toward the equator when it reaches Earth's surface, and is replaced by air moving through the upper atmosphere from lower latitudes. [2]

polar zone The major climatic zone above 60° N and S. [2]

pool (1) The total amount of a nutrient or other element found within a component of an ecosystem. [21] (2) A relatively deep and slowly flowing portion of a stream. *Compare* riffle. [3]

population A group of individuals of the same species that live within a particular area and interact with one another. [1, 8]

population cycles A pattern of population fluctuations in which alternating periods of high and low abundance occur after nearly constant intervals of time. [10]

population density The number of individuals per unit of area. [8]

population fluctuations The most common pattern of population growth, in which population size rises and falls over time. [10]

population outbreak An extremely rapid increase in the number of individuals in a population. [10]

population regulation A pattern of population growth in which one or more density-dependent factors increase population size when numbers are low and decrease population size when numbers are high. [9]

population size The number of individuals in a population. [8]

population viability analysis (PVA) Projection of the potential future status of a population through use of demographic models; a PVA approach is often used to estimate the likelihood that a population will persist for a certain amount of time in different habitats or under different management scenarios. [22]

positive interactions Interactions between species in which one or both species benefit and neither is harmed. [14]

predator An organism that kills and eats other organisms, referred to as its prey. [12]

prey An organism eaten by a predator. [12]

pressure potential The energy associated with the exertion of pressure; has a positive value if pressure is exerted on the system and a negative value if the system is under tension. [4]

primary producer *See* producer.

primary production The rate that chemical energy in an ecosystem is generated by autotrophs, derived from the fixation of carbon during photosynthesis and chemosynthesis. *Compare* secondary production. *See also* gross primary production, net primary production. [19]

primary succession Succession that involves the colonization of habitats devoid of life. *Compare* secondary succession. [16]

producer An organism that can produce its own food by photosynthesis or chemosynthesis; also called a primary producer or autotroph. *Compare* consumer. [1]

production efficiency The proportion of assimilated food that is used to produce new consumer biomass. [20]

pubescence The presence of hairs on the surface of an organism. [4]

Q

quadrat A sampling area (or volume) of any size or shape. [8]

R

***r*-selection** In the *r*- and *K*-selection continuum used for classifying life history strategies, the selection pressure for high population growth rates faced by organisms that live in environments where population densities are usually low. *Compare* *K*-selection. [7]

radiatively active gases *See* greenhouse gases.

rain-shadow effect The effect a mountain range has on regional climates by forcing moving air upward, causing it to cool and release precipitation on the windward slopes, resulting in lower levels of precipitation and soil moisture on the leeward slope. [2]

ramet An actually or potentially physiologically independent member of a genet that may compete with other members for resources. *Compare* genet. [8]

random dispersion A dispersion pattern that is similar to what would occur if individuals were positioned at locations selected at random. *Compare* clumped dispersion, regular dispersion. [8]

rank abundance curve A graph that plots the proportional abundance of each species in a community relative to the others in rank order, from most abundant to least abundant. [15]

recombination Rearrangements of genetic material during sexual reproduction that result in the production of offspring that have combinations of alleles that differ from those in either of their parents. [6]

redundancy hypothesis A hypothesis that assumes an upper limit on the positive effect of species richness on community function because once species richness reaches some threshold, the functions of species in the community will overlap. *Compare* complementarity hypothesis. [18]

regional scale A spatial scale that encompasses a geographic area where the climate is roughly uniform and the species contained therein are often restricted to that region by dispersal limitation. [17]

regional species pool All the species contained within a region; sometimes called gamma diversity. [17]

regular dispersion A dispersion pattern in which individuals are relatively evenly spaced throughout their habitat. *Compare* clumped dispersion, random dispersion. [8]

relative population size An estimate of population size based on data that are related in an unknown way to the absolute population size, but can be compared from one time period or place to another. *Compare* absolute population size. [8]

remote sensing A technique that measures the reflection of solar radiation off surfaces to estimate the density and composition of objects on Earth's surface and atmosphere. [19]

replicate To perform each treatment of a controlled experiment, including the control, more than once. [1]

reproductive barrier A barrier that prevents members of a population from breeding freely with members of its parental species and which may result in speciation. [6]

rescue effect A tendency for high rates of immigration to protect a population from extinction. [10]

resistance (1) Any force that impedes the movement of compounds such as water or CO_2 across an energy or concentration gradient; its opposite is conductance. [4] (2) The ability of a community to resist change from outside influences such as disturbance. [18]

resource A feature of the environment that is required for growth, survival, or reproduction and which can be consumed or otherwise used to the point of depletion. *Compare* physical factor. [11]

resource partitioning The use of limiting resources by different species in a community in different ways. [11, 18]

resource ratio hypothesis A hypothesis proposing that species can coexist in a community by using the same resources, but in differing proportions. [18]

riffle A fast-moving portion of a stream. *Compare* pool. [3]

ruderals In Grime's triangular model, plants that are adapted to environments with high levels of disturbance and low levels of stress. *Compare* competitive plants, stress-tolerant plants. [7]

S

salinity The concentration of dissolved salts in water. [2]

salinization A process by which high rates of evapotranspiration in arid regions result in a progressive buildup of salts at the soil surface. [2]

sand The coarsest soil particles (0.05–2 mm). [21]

savanna A vegetation type dominated by grasses with intermixed trees and shrubs. [3]

scale The spatial or temporal dimension at which ecological observations are collected. [1, 23]

scientific method An iterative and self-correcting process by which scientists learn about the natural world, consisting of four steps: (1) observe nature and ask a question about those observations; (2) develop possible answers to that question (hypotheses); (3) evaluate competing hypotheses with experiments, observations, or quantitative models; (4) use the results of those experiments, observations, or models to modify the hypotheses, pose new questions, or draw conclusions. [1]

secondary compound A chemical compound in plants not used directly in growth, and often used in such functions as defense against herbivores or protection from harmful radiation. [12]

secondary production Energy in an ecosystem that is derived from the consumption of organic compounds produced by other organisms. *Compare* primary production. [19]

secondary succession Succession that involves the reestablishment of a community in which some, but not all, of the organisms have been destroyed. *Compare* primary succession. [16]

semelparous Reproducing only once in a lifetime. *Compare* iteroparous. [7]

senescence A decline in the fitness of an organism with age as a result of physiological deterioration. [7]

sensible heat flux The transfer of heat through the exchange of energy by conduction or convection. [2]

sequential hemaphroditism A change or changes in the sex of an organism during the course of its life cycle. [7]

service mutualism A mutualism in which one partner performs an ecological service for the other. [14]

Shannon index The index most commonly used to describe species diversity quantitatively. [15]

silt Intermediate-sized soil particles, often ranging in size between 0.05 and 0.002 mm. [21]

sit-and-wait predator A predator that hunts by remaining in one place and attacking prey that move within striking distance. [12]

soil A mix of mineral particles, detritus, dissolved organic matter, water containing dissolved minerals and gases (the soil solution), and organisms that develops in terrestrial ecosystems. [21]

speciation The process by which one species splits into two or more species. [6]

species accumulation curve A graph that plots species richness as a function of the total number of individuals that have accumulated with each additional sample. [15]

species–area relationship The relationship between species richness and area sampled. [17]

species composition The identity of the species present in a community. [15]

species diversity A measure that combines the number of species (species richness) in a community and their relative abundances compared with one another (species evenness). [15]

species evenness The relative abundances of species in a community compared with one another. [15]

species richness The number of species in a community. [15]

stability The tendency of a community to remain the same in structure and function. [18]

stabilizing selection Selection that favors individuals with an intermediate phenotype. *Compare* directional selection, disruptive selection. [6]

stable Returning to the original state after some perturbation. [16]

stable age distribution A population age structure that does not change from one year to the next. [9]

stable isotope A rare form of an element containing a different number of neutrons than its most common form, that does not decay over time. [5]

stable limit cycle A pattern of population fluctuations in which abundance cycles indefinitely. [10]

static life table A life table that records the survival and reproduction of individuals of different ages during a single time period. [9]

stomata A pore in plant tissues, usually leaves, surrounded by specialized guard cells that control its opening and closing. [4]

stratification The layering of water in oceans and lakes due to differences in water density and temperature with depth. [2]

stress An abiotic factor that results in a decrease in the rate of an important physiological process, thereby lowering the potential for an organism's survival, growth, or reproduction; the condition caused by such a factor. [4, 7, 16]

stress-tolerant plants In Grime's triangular model, plants that are adapted to conditions of high stress and low disturbance. *Compare* competitive plants, ruderals. [7]

subsidence A sinking (downward) movement of air in the atmosphere, usually over a broad area, leading to the development of a high-pressure cell. *Compare* uplift. [2]

succession The process of change in the species composition of a community over time as a result of abiotic and biotic agents of change. [16]

surrogate species A species selected as a priority for conservation with the assumption that its conservation will serve to protect many other species with overlapping habitat requirements. [22]

survival rate The proportion of individuals of age x that survive to be age $x + 1$ (denoted S_x in a life table). [9]

survivorship The proportion of individuals that survive from birth (age 0) to age x (denoted l_x in a life table). [9]

survivorship curve A graph based on survivorship data (l_x) that plots the number of individuals from a hypothetical cohort (typically, of 1,000 individuals) that will survive to reach different ages. [9]

symbiont An organism that lives in or on an organism of another species, referred to as its host; a symbiont is the smaller member of a symbiosis. *See also* host, symbiosis. [13]

symbiosis A relationship in which two species live in close physiological contact with each other. *See also* host, symbiont. [14]

T

taxonomic homogenization The worldwide reduction of biodiversity resulting from the spread of non-native and native generalists coupled with declining abundances and distributions of native specialists and endemics. [22]

temperate zone The major climatic zone between 30° and 60° N and S. [2]

thermocline The zone of rapid temperature change in a lake beneath the epilimnion and above the hypolimnion. [2]

thermoneutral zone The range of environmental temperatures over which endotherms maintain a constant basal metabolic rate. [4]

threshold density The minimum number of individuals susceptible to a disease that must be present in a population for the disease to become established and spread. [13]

tides Patterns of rising and falling of ocean water generated by the gravitational attraction between Earth and the moon and sun. [3]

till Layers of sediment deposited by glaciers. [21]

tolerance The ability to survive stressful environmental conditions. *Compare* avoidance. [4]

top-down control Limitation of the abundance of a population by consumers. *Compare* bottom-up control. [10]

torpor A state of dormancy in which endotherms drop their lower critical temperature and associated metabolic rate. [4]

trade-off An organism's allocation of its limited energy or other resources to one structure or function at the expense of another. [6]

transpiration The evaporation of water from the inside of a plant, typically through stomatal pores. [2]

trophic cascade A change in the rate of consumption at one trophic level that results in a series of changes in species abundance or composition at lower trophic levels. [15, 20]

trophic efficiency A measure of the transfer of energy between trophic levels, consisting of the amount of energy at one trophic level divided by the amount of energy at the trophic level immediately below it. [20]

trophic facilitation An interaction in which a consumer is indirectly facilitated by a positive interaction between its prey or food plant and another species. [15]

trophic level A group of species that obtain energy in similar ways, classified by the number of feeding steps by which the group is removed from primary producers, which are the first trophic level. [15, 20]

trophic mutualism A mutualism in which one or both of the mutualists receives energy or nutrients from its partner. [14]

trophic pyramid A common approach to conceptualizing trophic relationships in an ecosystem in which a stack of rectangles is constructed, each of which represents the amount of energy or biomass within one trophic level. [20]

tropical zone The major climatic zone between 25° N and S, encompassing the equator. [2]

turgor pressure Hydrostatic pressure that develops in a plant cell when water moves into it, following a gradient in water potential. [4]

turnover (1) The mixing of the entire water column in a stratified lake when all the layers of water reach the same temperature and density. [2] (2) The replacement of one species with another over time or space. [17]

type I survivorship curve A survivorship curve in which newborns, juveniles, and young adults all have high survival rates and death rates do not begin to increase greatly until old age. [9]

type II survivorship curve A survivorship curve in which individuals experience a constant chance of surviving from one age to the next throughout their lives. [9]

type III survivorship curve A survivorship curve in which individuals die at very high rates when they are young, but those that reach adulthood survive well later in life. [9]

U

umbrella species A surrogate species selected with the assumption that protection of its habitat will serve as an "umbrella" to protect many other species; often a species with large or specialized habitat requirements or one that is easy to count. [22]

uplift The rising of warm, less dense air in the atmosphere due to surface heating. *Compare* subsidence. [2]

upwelling The rising of deep ocean waters to the surface. [2]

V

vicariance The evolutionary separation of species due to a barrier such as continental drift that result in the geographic isolation of species that once were connected to one another. [17]

W

warning coloration A defense against predators in which prey species that contain powerful toxins advertise those toxins with bright coloration; also called aposematic coloration. [12]

water potential The overall energy status of water in a system; the sum of osmotic potential, gravitational potential, turgor pressure, and matric potential. [4]

watershed *See* catchment.

weather The current temperature, humidity, precipitation, wind, and cloudiness. *Compare* climate. [2]

weathering The physical and chemical processes by which rock minerals are broken down, eventually releasing soluble nutrients and other elements. [21]

Z

zero population growth isoclines Lines derived from the Lotka–Volterra competition model marking the conditions under which a population does not increase or decrease in size. [11]

zooplankton Nonphotosynthetic plankton. *Compare* phytoplankton. [3]

Illustration Credits

The following figures use elements originally rendered for Sadava et al., *Life: The Science of Biology* (Sinauer Associates and W.H. Freeman, 2011): Figures 2.24, 5.11, 7.10, 13.3, 13.9, and 17.10.

The following figures use elements originally rendered for Cain et al., *Discover Biology* (Sinauer Associates and W.W. Norton, 2002): Figures 1.11, 8.9, 8.17, 11.8, 11.19, and 23.12.

UNIT OPENER PHOTOS

Unit 1 A bullfrog (*Rana catesbeiana*) rests among yellow pond lilies (*Nuphar variegatum*), Horseshoe Lake, Ontario, Canada. © Andrew McLachlan/All Canada Photos/Photolibrary.com.

Unit 2 Frailejones (*Espletia hartwegiana*), El Angel Nature Reserve, Ecuador. © Kevin Schafer/Peter Arnold Images/Photolibrary.com.

Unit 3 A red-billed oxpecker (*Buphagus erythrorhynchus*) removes ticks from a Cape buffalo (*Syncerus caffer*), Hluhluwe Game Reserve, South Africa. © James Hager/Robert Harding Travel/Photolibrary.com.

Unit 4 Giant green sea anemones (*Anthopleura xanthogrammica*), goose barnacles (*Lepas anserifera*) and ochre sea stars (*Pisaster ochraceus*) at low tide, Olympic National Park, Washington. © Konrad Wothe/Minden Pictures.

Unit 5 Elephants (*Loxodonta africana*), common wildebeest (*Connochaetes taurinus*), and plains zebra (*Equus quagga*) by the Tarangire River, Tanzania. © Sean Russell/FStop/Photolibrary.com.

Unit 6 The Lena Delta Reserve in Siberia is the most extensive protected wilderness area in Russia. Courtesy of the USGS EROS Data Center Satellite Systems Branch.

Chapter 1 1.1: © Suzanne L. & Joseph T. Collins/Photo Researchers, Inc. 1.2: Data from IUCN 2000, AmphibiaWeb (www.amphibiaweb.org), and Hero and Shoo 2003. 1.5: © Joseph M. Kiesecker. 1.6A: © Phototake Inc./Alamy. 1.6B: © Joseph M. Kiesecker. 1.8 *organism*: © Kanwarjit Singh Boparai/Shutterstock. 1.8 *population*: © David Davis/Shutterstock. 1.8 *community*: © T. & S. Allofs/Bios/Photolibrary.com. 1.8 *landscape*: Courtesy of the U.S. Geological Survey. 1.8 *biosphere*: NASA images by Reto Stöckli, based on data from NASA and NOAA. 1.9A: © LMR Group/Alamy. 1.9B: Courtesy of Andrew Sinauer. 1.9C: © Jeremy Woodhouse/Photodisc Green/Getty Images. 1.9D: © WaterFrame/Alamy. 1.12A: Courtesy of Tim Cooper and Richard Lenski, Michigan State University. 1.12B: Courtesy of Simone DesRoches and Dolph Schluter. 1.12C: Courtesy of the U.S. Forest Service. Box 1.1: Courtesy of Walter Carson.

Chapter 2 2.1: © Christopher S. Miller/Alaska Stock LLC/Alamy. 2.3: Courtesy of Craig Allen. 2.5, 2.6, 2.8, 2.10, 2.11, 2.12: After Ahrens 2005. 2.13: After IPCC 2005. 2.14: © Robert A. Rohde/Global Warming Art. 2.15: After Strahler and Strahler 2005. 2.16: Courtesy of the Center for Sustainability and the Global Environment (SAGE) through their *Atlas of the Biosphere*, www.sage.wisc.edu/atlas. 2.17: Data from www.globalbioclimatics.org. 2.18B *west*: © W. Cody/Flirt Collection/Photolibrary.com. 2.18B *east*: © Claver Carroll/Ticket/Photolibrary.com. 2.18B *bottom*: NASA images by Reto Stöckli, based on data from NASA and NOAA. 2.20: After Ahrens 2005. 2.22: After Dodson 2005. 2.23: Data from the NOAA Tropical Atmosphere Ocean Project. 2.25: Courtesy of NOAA. 2.26: After Lomolino et al. 2006. 2.27: Courtesy of NOAA.

Chapter 3 3.1: © Michel & Christine Denis-Huot/Photolibrary.com. 3.2: © Mark Hallett Paleoart/Photo Researchers, Inc. 3.3 *sclerophyll*: © Andy Jackson/Alamy. 3.3 *deciduous*: © Arco Images/Alamy. 3.3 *cactus*: © Dan Eckert/istock. 3.3 *needle-leaved*: © David Robertson/Alamy. 3.3 *grass*: © maurice joseph/Alamy. 3.3 *broad-leaved*: © Kevin Schafer/Alamy. 3.3 *forb*: Courtesy of William Bowman. 3.4: After Whittaker 1975. *Tropical rainforest, left*: © Juan Carlos Muñoz/AGE Fotostock. *Tropical rainforest, right*: Courtesy of William Bowman. 3.6: Courtesy of NASA/Goddard Space Flight Center Scientific Visualization Studio. *Tropical seasonal forest, left*: Courtesy of Andrew Sinauer. *Tropical seasonal forest, right*: © Pete Oxford/Naturepl.com. *Hot subtropical desert, left*: © Purestock/Alamy. *Hot subtropical desert, right*: © The Africa Image Library/Alamy. 3.7A: © Paroli Galperti/CuboImages srl/Alamy. 3.7B: © Hervé Lenain/Bios/Photolibrary.com. *Temperate grasslands, left*: © Tom Bean/Alamy. *Temperate grasslands, right*: © Pavel Filatov/Alamy. *Temperate shrublands, right*: © Oxford Scientific Films/Photolibrary.com. *Temperate shrublands, left*: © Bill Bachman/Alamy. *Temperate deciduous forests, left*: © Tom Till/Alamy. *Temperate deciduous forests, right*: © Digital Archive Japan/Alamy. 3.8: © David Noton/Naturepl.com. *Temperate evergreen forests, left*: © José Enrique Molina/AGE Fotostock. *Temperate evergreen forests, right*: © James Brunker/Alamy. *Boreal forests, left*: © John R Delapp/AlaskaStock/Photolibrary.com. *Boreal forests, right*: © Wild Wonders of Europe/Widstrand/Naturepl.com. 3.9: Courtesy of U.S. Forest Service Research. 3.10: © Fred Bruemmer/Peter Arnold Images/Photolibrary.com. *Tundra, left*: © Juan-Carlos Muñoz/Bios/Photolibrary.com. *Tundra, right*: Courtesy of William Bowman. 3.11 *grassland, montane, alpine*: Courtesy of William Bowman. 3.11 *subalpine*: © mediacolor's/Alamy. 3.12: © Kevin Schafer/Peter Arnold Images/Photolibrary.com. 3.15A: © Phototake Inc./Alamy. 3.15B: © Laguna Design/Science Photo Library/Photo Researchers, Inc. 3.17: © Mark Brownlow/Naturepl.com. 3.18: David McIntyre. 3.19: © Ron Giling/Lineair/Photolibrary.com. 3.20: © Dennis Frates/Alamy. 3.22: © Georgette Douwma/Naturepl.com. 3.23: Courtesy of Earth Sciences and Image Analysis Laboratory, NASA Johnson Space Center. 3.24: © Ralph A Clevenger/Flirt Collection/Photolibrary.com. 3.25A: © D. P. Wilson/Minden Pictures. 3.25B: © Harold Taylor/OSF/Photolibrary.com. 3.26: © Norbert Wu/Minden Pictures. 3.27: From Halpern et al. 2008. 3.28: © North Wind Picture Archives/Alamy. 3.30A: © Bryon Palmer/Shutterstock. 3.30B, C: Courtesy of the Konza Prairie LTER.

Chapter 4 4.1: Courtesy of J. M. Storey. 4.2 *sylvatica*: David McIntyre. 4.2 *maculata*: © Robin Arnold/istock. 4.4 *aspens*: © Brian Balster/Shutterstock. 4.7: After Willmer et al. 2005. 4.8: After Nobel 1983. 4.9A:

© Cheryl Power/Photo Researchers, Inc. 4.9B: Courtesy of G. H. Holroyd and A. M. Hetherington. 4.10: Courtesy of James Ehleringer. 4.12: Courtesy of Wayne Law, New York Botanical Garden. 4.13A: © Alastair MacEwen/OSF/Photolibrary.com. 4.13B: © Takahashi, via Creative Commons. 4.14: After Willmer et al. 2005 and Schmidt-Nelson 1997. 4.15: Courtesy of William Bowman. 4.17: After Armitage et al. 2003. 4.17 *inset*: © Wildlife GmbH/Alamy. 4.18: After Kramer 1983. 4.21: After Slatyer 1967. 4.23: After Willmer et al. 2005 and Edney 1980. 4.24, 4.25: After Schmidt-Nelson 1997. 4.27A, B: © Simon Colmer/Naturepl.com. 4.27B: © Simon Colmer/Naturepl.com. 4.27C: © Andrew Syred/Photo Researchers, Inc. 4.27D: © Manfred Kage/Peter Arnold Images/Photolibrary.com.

Chapter 5 5.1: © Stan Osolinski/OSF/Photolibrary.com. 5.2: © Jean-Paul Ferrero/Minden Pictures. 5.3A: David McIntyre. 5.3C: Courtesy of Brytten Steed, USDA Forest Service, Bugwood.org. 5.4: © Stephen Frink/Nomad/Photolibrary.com. 5.5: © Outdoor-Archiv/Alamy. 5.8: Courtesy of T. Vogelmann. 5.11A: © George Clerk/istock. 5.11B: © Elburg Botanic Media/AGE Fotostock. 5.16A: David McIntyre. 5.16B: © Organica/Alamy. 5.16C: © Michel de Nijs/istock. 5.17 *beetle, deerfly*: David McIntyre. 5.17 *moth*: © Tom Stack/Alamy. 5.17 *housefly*: © John Kimbler/Tom Stack & Associates/Alamy. 5.18 *eagle*: © Gerry Ellis/DigitalVision. 5.18 *finch*: © Eric Isselée/Shutterstock. 5.18 *avocet*: © David P. Smith/ShutterStock. 5.18 *toucan*: © Shutterstock. 5.18 *kingfisher*: Courtesy of Andrew Sinauer. 5.19A: © Tim Zurowski/All Canada Photos/AGE Fotostock. 5.26: © Rolf Nussbaumer Photography/Alamy. 5.28: © Ron Toft. 5.29: © National Academy of Sciences, Janet Mann/AP Photo.

Chapter 6 6.1: © Wildlife/Alamy. 6.3A: Courtesy of Matthew P. Travis. 6.3B, C, D: Courtesy of Peter J. Park. 6.5: © Emmanuel Lattes/Alamy. 6.7: After Futuyma 2005. 6.8A *inset*: © Enrique R. Aguirre/AGE Fotostock. 6.10A, B: © Kim Taylor/Naturepl.com. 6.10C: © Ross Armstrong/AGE Fotostock. 6.10D: Courtesy of J. Schmidt/National Park Service. 6.17A: © Stanley M. Awramik/Biological Photo Service. 6.17B *left*: Courtesy of A. H. Knoll. 6.17B *right*: © Shuhai Xiao. 6.17C: © Barbara J. Miller/Biological Photo Service. 6.17D: © David M. Dennis/OSF/Photolibrary.com. 6.17E: © Ted Daeschler/Academy of Natural Sciences/VIREO. 6.20 *wasp*: Courtesy of Andrew Forbes. 6.20 *fly*: Courtesy of Joseph Berger, Bugwood.org. 6.20 *apples*: © Rambleon/Shutterstock. 6.21 *H. annuus*: © Noella Ballenger/Alamy. 6.21 *H. petiolaris*: Dave Powell, USDA Forest Service, Bugwood.org. 6.21 *H. anomalus*: © Jack Dykinga/Naturepl.com. 6.22: From Dietl et al. 2004, courtesy of Gregory Dietl. 6.24 *inset*: © Steven Kazlowski/Alaskastock/Photolibrary.com.

Chapter 7 7.1: © Gerald Lacz/Peter Arnold Images/Photolibrary.com. 7.2: © Images&Stories/Alamy. 7.4 *inset*: © Morales/AGE Fotostock. 7.5: Courtesy of David Pfennig. 7.8A: © Dr. Peter Siver/Visuals Unlimited, Inc. 7.8B: © David M. Phillips/Photo Researchers, Inc. 7.9A *larva*: © Joël Héras/Bios/Photolibrary.com. 7.9A *adult*: © André Simon/Bios/Photolibrary.com. 7.9B *larva*: © Wim van Egmond/Visuals Unlimited, Inc. 7.9B *adult*: © WaterFrame/Alamy. 7.10 *inset*: © David Forster/Alamy. 7.11: © Frans Lanting Studio/Alamy. 7.17 *inset*: © blickwinkel/Alamy. 7.18: Courtesy of the Otorohanga Zoological Society. 7.20: © FLPA/D. P. Wilson/AGE Fotostock. 7.21A: © Barry Mansell/Naturepl.com. 7.21B: © Michael Redmer/Visuals Unlimited, Inc. 7.22: David McIntyre.

Chapter 8 8.1: © Accent Alaska.com/Alamy. 8.3 *beetle*: © Beatriz Moisset. 8.3 *goldenrod*: © John Martin/Alamy. 8.4 *inset*: © Bob Gibbons/OSF/Photolibrary.com. 8.5: Courtesy of William W. Dunmire/National Park Service. 8.7 *bush*: Courtesy of Sue in AZ/Wikipedia. 8.7 *cactus*: © Dan Eckert/istock. 8.8: Reproduced with permission of the Department of Natural Resources, Queensland, Australia. 8.9 *inset*: © Arterra Picture Library/Alamy. 8.10: From Lomolino et al. 2006. 8.12 *left*: © BugsLife/Alamy. 8.12 *right*: © Geoff du Feu/Alamy. 8.13: © Purestock/Alamy. 8.13 *inset*: © Tom McHugh/Photo Researchers, Inc. 8.14 *top*: David McIntyre. 8.14 *right*: © Alan Marsh/First Light Associated Photographers/Photolibrary.com. 8.15: Courtesy of James C. Trager. 8.16 *inset*: © blickwinkel/Alamy. 8.17A: © Jack Dykinga/Naturepl.com. 8.17B: © Imagebroker RF/Photolibrary.com. 8.20: © Ariadne Van Zandbergen/OSF/Photolibrary.com. Box 8.1A: © WaterFrame/Alamy. Box 8.1B: © AfriPics.com/Alamy. Box 8.1C: © Visual&Written SL/Alamy.

Chapter 9 9.1: Courtesy of NASA. 9.3: © Wild Wonders of Europe/Zankl/Naturepl.com. 9.9: © Jason O. Watson/Alamy. 9.12: © Don Johnston/Alamy. Box 9.1A: © Guillen Photography/Alamy. Box 9.1B: Courtesy of NOAA Fisheries, Office of Protected Resources.

Chapter 10 10.1: © Hiroya Minakuchi/Minden Pictures. 10.3: © Inge Schepers/istock. 10.4: © Margaret Walton/Ticket/Photolibrary.com. 10.7: © Chip Clark, Museum of Natural History, Smithsonian. 10.8: © Tom Neversely/All Canada Photos/Photolibrary.com. 10.14 *lion*: Courtesy of Andrew Sinauer. 10.14 *sperm*: Courtesy of David Wildt. 10.15B: © Wild Wonders of Europe/Zankl/Naturepl.com. 10.15C: © Robin Bush/OSF/Photolibrary.com. 10.15D: © CuboImages srl/Alamy. 10.16: After Morris and Doak 2002. 10.17B: Courtesy of DigitalGlobe. 10.18: © Jared Hobbs/All Canada Photos/Photolibrary.com. 10.20: © David Wrobel/Visuals Unlimited, Inc.

Chapter 11 11.1: © Photolibrary Group. 11.3: © Ilja Dubovskis/Alamy. 11.6: Courtesy of Randy Cyr, Greentree, Bugwood.org. 11.9: After Lomolino et al. 2006. 11.16: © Doug Sokell/Visuals Unlimited, Inc. 11.17: Courtesy of David Pimentel. 11.21: Courtesy of David Ruple/Grand Bay National Estuarine Research Reserve. 11.21 *inset*: © blickwinkel/Alamy. Box 11.1 *inset*: © Bar Aviad.

Chapter 12 12.1: © White/Photolibrary.com. 12.3A: © Andrew Darrington/Alamy. 12.3B: © Guy Van Langenhove/Bios/Photolibrary.com. 12.3C: © Jonathan Bird/Peter Arnold Images/Photolibrary.com. 12.4: © Alex Wild/Visuals Unlimited, Inc. 12.4 *inset*: © Peter J. Bryant/Biological Photo Service. 12.8A: © Leonardo Díaz Romero/AGE Fotostock. 12.8B: © imagebroker/Alamy. 12.8C: © Robert Oelman/OSF/Photolibrary.com. 12.8D: © Ingo Arndt/Naturepl.com. 12.8E: © Wayne Lynch/All Canada Photos/Photolibrary.com. 12.9A: © Robert Thompson/Naturepl.com. 12.9B: © Philip Smith/Alamy. 12.12B: © Don Johnston/AGE Fotostock. 12.13 *left*: © John Sullivan/Alamy. 12.13 *right*: © Jack Goldfarb/AGE Fotostock. 12.14: Courtesy of Judith Becerra and D. Lawrence Venable. 12.15A: Courtesy of USDA ARS, European Biological Control Laboratory/Bugwood.org. 12.16: Courtesy of Jonathan Losos. 12.17B: Courtesy of R. L. Jefferies (University of Toronto), a member of the Hudson Bay Project partly funded by NSERC, Canada. 12.18 *inset*: David McIntyre. 12.21A: From Huffaker 1958. 12.23 *inset*: Courtesy of Takehito Yoshida and R. O. Wayne. 12.24A: Photo by Tim Karels courtesy of Charles Krebs.

Chapter 13 13.1: © Pascal Goetgheluck/SPL/Photo Researchers, Inc. 13.2 *inset*: © David Maitland. 13.4: After Stiling 2002. 13.5A: © Dayton Wild/Visuals Unlimited, Inc. 13.5B: © blickwinkel/Alamy. 13.6A, B: © Eye of Science/Photo Researchers, Inc. 13.6C: © Nigel Cattlin/Alamy. 13.8: Courtesy of Michael A. Huffman. 13.10: © Renee Morris/Alamy. 13.11: © David Wall/Alamy. 13.13 *inset*: Courtesy of CSIRO.

13.15B: Courtesy of the Forest History Society, Durham, NC. 13.16: After Turchin 2003. 13.18: From Mouritsen et al. 1998. 13.21B *woodrat*: © Rick & Nora Bowers/Alamy. 13.21B *leishmaniasis*: © Andy Crump, TDR, World Health Organization/Photo Researchers, Inc. 13.22: From Eberhard 2001, courtesy of William Eberhard.

Chapter 14 14.1: © Bence Mate/Naturepl.com. 14.2A: After Wirth et al. 2003. 14.2B: © Mark Moffett/Minden Pictures. 14.4A: After Rovira et al. 1983. 14.4A *SEM*: © R. L. Peterson/Biological Photo Service. 14.4B: After Mauseth 1988. 14.5: Courtesy of Kevin Carpenter and Patrick Keeling. 14.5 *inset*: Courtesy of Patrick Keeling. 14.7: © Les Howard/istock. 14.13: © Dr. Morley Read/Photo Researchers, Inc. 14.14: From Pellmyr 2003, courtesy of Olle Pellmyr. 14.16 *insets*: Courtesy of John Jaenike. 14.17A: © Alex Wild/Visuals Unlimited, Inc. 14.17B: © Mark Moffett/Minden Pictures. 14.17C: © Nicholas Smythe/Photo Researchers, Inc. 14.18A: © Splashdown Direct/OSF/Photolibrary.com.

Chapter 15 15.1: © Laurent Piechegut/Bios/Photolibrary.com. 15.2: After Meinesz 2001. 15.3A: Courtesy of Joan Gemme. 15.3B: Courtesy of Jim Peaco/NPS. 15.3C: © Garry DeLong/Alamy. 15.3D: © Shutterstock. 15.7: © David Forman/Ticket/Photolibrary.com. 15.16: Courtesy of Robert S. Steneck. 15.19: © Pete Ryan/National Geographic Stock. 15.21: © Juan Carlos Calvin/AGE Fotostock. Box 15.1: Courtesy of Sally Hacker.

Chapter 16 16.1: Courtesy of Jim Vallance/USGS. 16.2 *debris, pumice*: Courtesy of the U.S. Forest Service. 16.2 *blowdown, scorch*: Courtesy of Jerry Franklin. 16.6: © Brad Wieland/istock. 16.7A: Courtesy of Charles Elton. 16.7B: From Elton 1927. 16.9: © Beth Davidow/Visuals Unlimited, Inc. 16.13: Courtesy of Sally Hacker. 16.20 *left*: From Crisafulli et al. 2005, courtesy of Charlie M. Crisafulli. 16.20 *right*: © Wayne Lynch/All Canada Photos/Photolibrary.com. 16.21A: © Jonathan Titus, courtesy of John Bishop.

Chapter 17 17.1: © Michael DeFreitas/Robert Harding Travel/Photolibrary.com. 17.2B: © R. O. Bierregaard. 17.3A: © SMuller/Wildlife/Photolibrary.com. 17.3B: © Phil Schermeister/National Geographic Stock. 17.3C: © Inger Hogstrom/AGE Fotostock. 17.3D: © Gunter Marx/Alamy. 17.4A *top, bottom*: Courtesy of Sally Hacker. 17.4A *middle*: © Craig Lovell/Eagle Visions Photography/Alamy. 17.4B *top*: © Simon Lane/Alamy. 17.4B *bottom*: Courtesy of Sally Hacker. 17.8B: © The Natural History Museum, London. 17.11: After Pielou 1979. 17.14B *puffin*: © Jerome Whittingham/istock. 17.14B *pelican*: © Irina Tischenko/Shutterstock. 17.14B *penguin*: © Jacqueline Abromeit/Shutterstock. 17.22B: © Philippe Crochet/Photononstop/Photolibrary.com. 17.23A: Courtesy of DigitalGlobe. 17.23B, C: Courtesy of E. O. Wilson.

Chapter 18 18.1: © Larry Michael/Naturepl.com. 18.2: Courtesy of G. David Tilman. 18.3: © Robert Glusic/White/Photolibrary.com. 18.5A: © Jinny Goodman/Alamy. 18.5B: Courtesy of NOAA. 18.5B *inset*: © Wolfgang Pölzer/Alamy. 18.6: © Arco Images GmbH/Alamy. 18.7: After Ricklefs 2001. 18.13: © Science and Scenics/Alamy. 18.20: © Kelvin Aitken/Peter Arnold Images/Photolibrary.com.

Chapter 19 19.1: © Stan Wayman/Photo Researchers, Inc. 19.2: © Scripps Institution of Oceanography/OSF/Photolibrary.com. 19.4: © Juan Carlos Muñoz/AGE Fotostock. 19.6A: Courtesy of Stephen Ausmus/USDA ARS. 19.6B: Courtesy of Dylan Fischer. 19.8A: Courtesy of Sean Burns. 19.9: Image by Robert Simmon, NASA GSFC Earth Observatory, based on data provided by Watson Gregg, NASA GSFC. 19.12A: Courtesy of William Bowman. 19.14: Courtesy of Wayne Wurtsbaugh and the American Association of Limnology and Oceanography. 19.15: Courtesy of Fisheries and Oceans Canada, Experimental Lakes Area. Reproduced with the permission of Her Majesty the Queen in Right of Canada, 2007. 19.16B: © Ken Johnson/Monterey Bay Aquarium Research Institute. 19.20: From Shank et al. 1998. Box 19.1B: Courtesy of NASA.

Chapter 20 20.1: © Chlaus Lotscher/Peter Arnold Images/Photolibrary.com. 20.2: © Mia & Klaus Matthes/Superstock/Photolibrary.com. 20.4A *left*: David McIntyre. 20.4A *right*: © blickwinkel/AGE Fotostock. 20.8: © Michael S. Nolan/AGE Fotostock. 20.17: Courtesy of Sally Hacker. 20.22: © Bill Stevenson/Photolibrary.com.

Chapter 21 21.1: Courtesy of William Bowman. 21.2: Courtesy of Jayne Belnap/USGS. 21.4: After Brady and Weil 2001. 21.5A: David McIntyre. 21.5B: © P&R Fotos/AGE Fotostock. 21.7: After Paul and Clark 1996. 21.12: Courtesy of DigitalGlobe. 21.15: © T. C. Middleton/OSF/Photolibrary.com. 21.19B: After Dodson 2004. 21.20: Courtesy of the SeaWiFS Project, NASA/Goddard Space Flight Center, and ORBIMAGE. 21.21: Courtesy of Jason C. Neff. 21.22: Courtesy of Steve Dewey, Utah State University, Bugwood.org. 21.22 *inset*: Courtesy of Tom Heutte, USDA Forest Service, Bugwood.org. Box 21.1A: Courtesy of Mark Losleben. Box 21.1B: Courtesy of William Bowman.

Chapter 22 22.1: © Rolf Nussbaumer Photography/Alamy. 22.2B: © Raymond Gehman/Nomad/Photolibrary.com. 22.3: © Mary Evans Picture Library/Alamy. 22.7A: Courtesy of Roland J. Stipes, Virginia Polytechnic Institute and State University, Bugwood.org. 22.7B: © blickwinkel/Alamy. 22.7C: Courtesy of Scott Bauer/USDA ARS. 22.9: From Schipper et al. 2008. 22.10: Courtesy of John D. Byrd, Mississippi State University, Bugwood.org. 22.11 *inset*: © imagebroker/Alamy. 22.15: © Mark Conlin/Alamy. 22.16A, B, C: Courtesy of the U.S. Fish and Wildlife Service. 22.16D: © John Cancalosi/Alamy. 22.18: © J-L. Klein & M-L. Hubert/Bios/Photolibrary.com. 22.19, 22.20: Courtesy of the U.S. Fish and Wildlife Service. Box 22.1A, B: Courtesy of the Center for Conservation Biology, University of Washington.

Chapter 23 23.1: © William Ervin/Photo Researchers, Inc. 23.3A: Courtesy of NASA. 23.7: © imagebroker/Alamy. 23.8: Courtesy of Jim Peaco/NPS Photo. 23.9: Courtesy of Laure Laüt. 23.10: © Porky Pies Photography/Alamy. 23.12: © danilo donadoni/Marka/AGE Fotostock. 23.13: From McIntyre and Hobbs 1999, courtesy of Richard Hobbs. 23.16: © Rainer Raffalski/Alamy. 23.18: © Joel Sartore/National Geographic Stock. 23.19A: Courtesy of DigitalGlobe. 23.19B *inset*: © Millard H. Sharp/Photo Researchers, Inc. 23.19C *inset*: © Bruce Coleman, Inc./Alamy. 23.20: From Schulte et al. 2009. 23.25: © Kurt Repanshek/National Parks Traveler. Box 23.1A, B: Courtesy of the USGS Gap Analysis Program at the University of Idaho, 530 S. Asbury Street, Suite 1, Moscow, ID 83843. Box 23.1B *inset*: © William Leaman/Alamy.

Chapter 24 24.1A: Courtesy of NOAA George E. Marsh Album. 24.3: After Chapin et al. 2002. 24.4: Courtesy of Evan DeLucia. 24.7, 24.9, 24.10: After Chapin et al. 2002. 24.15: Courtesy of William Bowman. 24.17: Courtesy of Julie Winchester/Cooperative Institute for Research in the Atmosphere. 24.18: © Angelo Gandolfi/Naturepl.com. 24.19: Courtesy of the National Atmospheric Deposition Program (NRSP-3). 24.23A: After Jacobson 2002. 24.23B: Courtesy of NASA. 24.24: Courtesy of NOAA. 24.25A: Courtesy of the MODIS Rapid Response Team at NASA GSFC. 24.26A: After *Eolian history of North America*, esp.cr.usgs.gov/info/eolian/. 24.26B: After Haase et al. 2007. 24.27: Courtesy of the Center for Snow and Avalanche Studies, Silverton, CO.

Literature Cited

Numbers in brackets refer to the chapter(s) where this reference is cited.

A

Aber, J., W. and 9 others. 1998. Nitrogen saturation in temperate forest ecosystems: Hypotheses revisited. *BioScience* 48: 921–934. [24]

Abramsky, Z., M. L. Rosenzweig and B. Pinshow. 1991. The shape of a gerbil isocline measured using principles of optimal habitat selection. *Ecology* 72: 329–340. [11]

Adey, W. H. and R. S. Steneck. 2001. Thermogeography over time creates biogeographic regions: A temperature/space/time-integrated model and an abundance-weighted test for benthic marine algae. *Journal of Phycology* 37: 677–698. [17]

Agrawal, A. A. and M. Fishbein. 2006. Plant defense syndromes. *Ecology* 87: S132–S149. [12]

Aiba, S., A. E. Humphrey and N. F. Mills. 1973. *Biochemical Engineering,* 2nd ed. Academic Press, New York. [21]

Allan, B. F., F. Keesing and R. S. Ostfeld. 2003. Effect of forest fragmentation on Lyme disease risk. *Conservation Biology* 17: 267–272. [23]

Allan, B. F. and 15 others. 2009. Ecological correlates of risk and incidence of West Nile virus in the United States. *Oecologia* 158: 699–708. [1]

Allen, E. B., P. J. Temple, A. Bytnerowicz, M. J. Arbaugh, A. G. Sirulnik and L. E. Rao. 2007. Patterns of understory diversity in mixed coniferous forests of Southern California impacted by air pollution. *Scientific World Journal* 7: 247–263. [17]

Alliende, M. C. and J. L. Harper. 1989. Demographic studies of a dioecious tree. I. Colonization, sex, and age structure of a population of *Salix cinerea. Journal of Ecology* 77: 1029–1047. [9]

Allin, E. F. and J. A. Hopson. 1992. Evolution of the auditory system in Synapsida ("mammal-like reptiles" and primitive mammals) as seen in the fossil record. In *The Evolutionary Biology of Hearing,* D. B. Webster, R. R. Fay and A. N. Popper (eds.), 587–614. Springer, New York. [6]

Allison, G. 2004. The influence of species diversity and stress intensity of community resistance and resilience. *Ecological Monographs* 74: 117–134. [18]

Alpert, P. 2006. Constraints of tolerance: Why are desiccation tolerant organisms so small or rare? *Journal of Experimental Biology* 209: 1575–1584. [4]

Andersen, S. B. and 7 others. 2009. The life of a dead ant: The expression of an adaptive extended phenotype. *American Naturalist* 174: 424–433. [13]

Andrewartha, H. G. and L. C. Birch. 1954. *The Distribution and Abundance of Animals.* University of Chicago Press, Chicago. [22]

Anthony, P. A., J. A. M. Holtum and B. R. Jackes. 2002. Shade acclimation of rainforest leaves to colonization by lichens. *Functional Ecology* 16: 808–816. [14]

Arcese, P. and J. N. M Smith. 1988. Effects of population density and supplemental food on reproduction in song sparrows. *Journal of Animal Ecology* 57: 119–136. [9]

Arias, E., B. L. Rostron and B. Tejada-Vera. 2010. *United States Life Tables, 2005.* National Vital Statistics Reports, vol. 58, no. 10. National Center for Heath Statistics, Hyattsville, MD. http://www.cdc.gov/nchs/data/nvsr/nvsr58/nvsr58_10.pdf. [9]

Armitage, K. B., D. T. Blumstein and B. C. Woods. 2003. Energetics in hibernating yellow belly marmots (*Marmota flaviventris*). *Comparative Biochemistry and Physiology* A 134: 101–114. [4]

Arnold, G. W., D. E. Steven, J. R. Weeldenburg and E. A. Smith. 1993. Influences of remnant size, spacing pattern and connectivity on population boundaries and demography in euros *Macropus robustus* living in a fragmented landscape. *Biological Conservation* 64: 219–230. [23]

Asner, G. P., T. K. Rudel, T. M. Aide, R. Defries and R. Emerson. 2009. A contemporary assessment of change in humid tropical forests. *Conservation Biology* 6: 1386–1395. [3]

Attum, O., P. Eason, G. Cobbs and S. M. B. El Din. 2006. Response of a desert lizard community to habitat degradation: Do ideas about habitat specialists/generalists hold? *Biological Conservation* 133: 52–62. [22]

B

Baker, C. S., M. L. Dalebout, G. M. Lento and N. Funahashi. 2002. Gray whale products sold in commercial markets along the Pacific coast of Japan. *Marine Mammal Science* 18: 295–300. [22]

Balanyá, J., J. M. Oller, R. B. Huey, G. W. Gilchrist and L. Serra. 2006. Global genetic change tracks global climate warming in *Drosophila subobscura. Science* 313: 1773–1775. [6]

Balmford, A., R. E. Green and M. Jenkins. 2003. Measuring the changing state of nature. *Trends in Ecology and Evolution* 18: 326–330. [22]

Barbier, E. B., S. D. Hacker, C. Kennedy, E. W. Koch, A. C. Stier and B. R. Silliman. In press. The value of estuarine and coastal ecosystem services. *Ecological Monographs.* [3]

Barbour, M. S. and J. A. Litvaitis. 1993. Niche dimensions of New England cottontails in relation to habitat patch size. *Oecologia* 95: 321–327. [23]

Barnosky, A. D., P. L. Koch, R. S. Feranec, S. L. Wing and A. B. Shabel. 2004. Assessing the causes of Late Pleistocene extinctions on the continents. *Science* 306: 70–75. [3]

Bartlein, P., J. C. Whitlock and S. L. Shafer. 1997. Future climate in the Yellowstone National Park region and its potential impact on vegetation. *Conservation Biology* 11: 782–792. [23]

Battin, J. 2004. When good animals love bad habitats: Ecological traps and the conservation of animal populations. *Conservation Biology* 18: 1482–1491. [23]

Beall, C. M. 2007. Two routes to functional adaptation: Tibetan and Andean high-altitude natives. *Proceedings of the National Academy of Sciences USA* 104: 8655–8660. [4]

Becerra, J. X. 2003. Synchronous coadaptation in an ancient case of herbivory. *Proceedings of the National Academy of Sciences USA* 100: 12804–12807. [12]

Becerra, J. X. 2007. The impact of herbivore–plant coevolution on plant community structure. *Proceedings of the National Academy of Sciences USA* 104: 7483–7488. [12]

Beck, B. B. 1980. *Animal Tool Behavior: The Use and Manufacture of Tools by Animals.* Garland STPM Press, New York. [5]

Begon, M., M. Mortimer and D. J. Thompson. 1996. *Population Ecology: A Unified Study of Animals and Plants.* 3rd ed., Blackwell Science Ltd., Oxford. [9]

Beisner, B. E., D. T. Haydon and K. Cuddington. 2003. Alternative stable states in ecology. *Frontiers in Ecology and the Environment* 1: 376–382. [16]

Beissinger, S. R. and M. I. Westphal. 1998. On the use of demographic models of population viability in endangered species management. *Journal of Wildlife Management* 62: 821–841. [22]

Bell, M. A., M. P. Travis and D. M. Blouw. 2006. Inferring natural selection in a fossil threespine stickleback. *Paleobiology* 32: 562–577. [6]

Bellows, T. S. Jr. 1981. The descriptive properties of some models for density dependence. *Journal of Animal Ecology* 50: 139–156. [9]

Belnap, J. 2003. The world at your feet: Desert biological crusts. *Frontiers in Ecology and Environment* 1: 181–189. [21]

Belnap, J. and D. Eldridge. 2001. Disturbance and recovery of biological soil crusts. In *Biological Soil Crusts: Structure, Function, and Management*, J. Belnap and O. L. Lange (eds.), 363–383. Berlin: Springer-Verlag. [21]

Belt, T. 1874. *The Naturalist in Nicaragua.* Murray, London. [14]

Belz, R. G. 2007. Allelopathy in crop/weed interactions—an update. *Pest Management Science* 63: 308–326. [11]

Benkman, C. W. 1993. Adaptation to single resources and the evolution of crossbill (*Loxia*) diversity. *Ecological Monographs* 63: 305–325. [5]

Benkman, C. W. 2003. Divergent selection drives the adaptive radiation of crossbills. *Evolution* 57: 1176–1181. [5]

Benton, M. J. and B. C. Emerson. 2007. How did life become so diverse? The dynamics of diversification according to the fossil record and molecular phylogenetics. *Palaeontology* 50: 23–40. [6]

Berendse, F., R. Aerts and R. Bobbink. 1993. Atmospheric nitrogen deposition and its impact on terrestrial ecosystems. In *Landscape Ecology of a Stressed Environment*, C. C. Vos and P. Opdam (eds.), 104–121. Chapman and Hall, London. [24]

Berlow, E. L. 1997. From canalization to contingency: Historical effects in a successional rocky intertidal. *Ecological Monographs* 676: 435–460. [16]

Berlow, E. L. 1999. Strong effects of weak interactions in ecological communities. *Nature* 398: 330–334. [20]

Bernhardt, E. S. and 13 others. 2005. Can't see the forest for the stream? In-stream processing and terrestrial nitrogen exports. *BioScience* 55: 219–230. [21]

Bertness, M. D. 1989. Interspecific competition and facilitation in a northern acorn barnacle population. *Ecology* 70: 257–268. [14]

Bertness, M. D. and R. Callaway. 1994. Positive interactions in communities. *Trends in Ecology and Evolution* 9: 191–193. [18]

Bertness, M. D. and S. D. Hacker. 1994. Physical stress and positive associations among marsh plants. *American Naturalist* 144: 363–372. [18]

Bertness, M. D. and G. H. Leonard. 1997. The role of positive interactions in communities: Lessons from intertidal habitats. *Ecology* 78: 1976–1989. [14, 18]

Bertness, M. D. and S. W. Shumway. 1993. Competition and facilitation in marsh plants. *American Naturalist* 142: 718–724. [16]

Beschta, R. L. and W. J. Ripple. 2006. River channel dynamics following extirpation of wolves in northwestern Yellowstone National Park, USA. *Earth Surface Processes and Landforms* 31: 1525–1539. [23]

Bierregaard, R. O. Jr., C. Gascon, T. E. Lovejoy and R. C. G. Mesquita. 2001. *Lessons from Amazonia: The Ecology and Conservation of a Fragmented Forest.* Yale University Press, New Haven, CT. [17]

Bigler, C., D. G. Gavin, C. Gunning and T. T. Veblen. 2007. Drought induces lagged tree mortality in a subalpine forest in the Rocky Mountains. *Oikos* 116: 1983–1994. [9]

Binkley, D. and P. Högberg. 1997. Does atmospheric deposition of acidity and nitrogen threaten Swedish forests? *Forest Ecology and Management* 92: 119–152. [24]

Birkeland, C. 1997. Symbiosis, fisheries and economic development on coral reefs. *Trends in Ecology and Evolution* 12: 364–367. [3]

Bishop, J., W. F. Fagan, J. D. Schade and C. M. Crisafulli. 2005. Causes and consequences of herbivory on Prairie lupine (*Lupinus lepidus*) in early primary succession. In *Ecological Responses to the 1980 Eruption of Mount St. Helens*, V. H. Dale, F. J. Swanson and C. M. Crisafulli (eds.), 151–161. Springer, New York. [16]

Bjorkman, O. 1981. Responses to different quantum flux densities. In *Physiological Plant Ecology* I: *Encyclopedia of Plant Physiology*, O. L. Lange, P. S. Nobel, C. B. Osmond and H. Ziegler (eds.), 57–101. Springer-Verlag, Berlin. [5]

Blais, J. M., D. W. Schindler, D. C. G. Muir, L. E. Kimpe, D. B. Donald and B. Rosenberg. 1998. Accumulation of persistent organochlorine compounds in mountains of western Canada. *Nature* 395: 585–588. [20]

Blais, J. M., L. E. Kimpe, D. McMahon, B. E. Keatley, M. L. Mattory, M. S. V. Douglas and J. P. Smol. 2005. Arctic seabirds transport marine-derived contaminants. *Science* 309: 445–445. [20]

Blaustein, A. R. and P. T. J. Johnson. 2003. Explaining frog deformities. *Scientific American* 288: 60–65. [1]

Boonstra, R., D. Hik, G. R. Singleton and A. Tinnikov. 1998. The impact of predator-induced stress on the snowshoe hare cycle. *Ecological Monographs* 79: 371–394. [12]

Booth, M. G. and J. D. Hoeksema. 2010. Mycorrhizal networks counteract competitive effects of canopy trees on seedling survival. *Ecology* 91: 2294–2302. [14]

Borland, A. M., H. Griffiths, C. Maxwell, M. S. J. Broadmeadow, N. M. Griffiths and J. D. Barnes. 1992. On the ecophysiology of the Clusiaceae in Trinidad: Expression of CAM in *Clusia minor* L. during the transition from wet to dry season and characterization of three endemic species. *New Phytologist* 122: 349–357. [5]

Bouzat, J. L., H. A. Lewin and K. N. Paige. 1998. The ghost of genetic diversity past: Historical DNA analysis of the greater prairie chicken. *American Naturalist* 152: 1–6. [6]

Bowman, W. D. and C. J. Bilbrough. 2001. Influence of a pulsed nitrogen supply on growth and nitrogen uptake in alpine graminoids. *Plant and Soil* 233: 283–290. [19]

Bowman, W. D. and T. R. Seastedt. 2001. *Structure and Function of an Alpine Ecosystem, Niwot Ridge, Colorado.* Oxford University Press, New York. [21]

Bowman, W. D., T. A. Theodose, J. C. Schardt and R. T. Conant. 1993. Constraints of nutrient availability on primary production in two alpine communities. *Ecology* 74: 2085–2098. [19]

Braby, M. F. 2002. Life history strategies and habitat templets of tropical butterflies in north-eastern Australia. *Evolutionary Ecology* 16: 399–413. [7]

Bradshaw, A. D. 1987. Restoration: An acid test for ecology. In *Restoration Ecology—a Synthetic Approach to Ecological Restoration,* W. R. Jordan, M. E. Gilpin and J. D. Aber (eds.), 23–30. Cambridge University Press, Cambridge. [23]

Bradshaw, C. J. A. and B. W. Brook. 2007. Ecological–economic models of sustainable harvest for an endangered but exotic megaherbivore in northern Australia. *Natural Resource Modeling* 20: 129–156. [22]

Bradshaw, W. E. and C. M. Holzapfel. 2001. Genetic shift in photoperiodic response correlated with global warming. *Proceedings of the National Academy of Sciences USA* 98: 14509–14511. [6]

Breshears, D. D. and 12 others. 2005. Regional vegetation die-off in response to global-change-type drought. *Proceedings of the National Academy of Sciences USA* 10: 15144–15148. [2]

Brewer, J. S. 2003. Why don't carnivorous pitcher plants compete with non-carnivorous plants for nutrients? *Ecology* 84: 451–462. [11]

Briggs, J. C. 2006. Proximate sources of marine biodiversity. *Journal of Biogeography* 33: 1–10. [17]

Brommer, J. E., H. Pietiainen, K. Ahola, P. Karell, T. Karstinen and H. Kolunen. 2010. The return of the vole cycle in southern Finland refutes the generality of the loss of cycles through "climatic forcing." *Global Change Biology* 16: 577–586. [10]

Bronstein, J. L. 1992. Seed predators as mutualists: Ecology and evolution of the fig/pollinator interaction. In *Insect-Plant Interactions* (Vol. 4), E. A. Bernays (ed.), p. 1–44. CRC Press, Boca Raton, FL. [14]

Bronstein, J. L. 1994. Conditional outcomes in mutualistic interactions. *Trends in Ecology and Evolution* 9: 214–217. [14]

Brooks, T., J. Tobias and A. Balmford. 1999. Deforestation and bird extinctions in the Atlantic forest. *Animal Conservation* 2: 211–222. [22]

Brooks, T. M. and 9 others. 2002. Habitat loss and extinction in the hotspots of biodiversity. *Conservation Biology* 16: 909–923. [22]

Brower, L. P. 1996. Monarch butterfly orientation: Missing pieces of a magnificent puzzle. *The Journal of Experimental Biology* 199: 93–103. [8]

Brown, J. H. and D. W. Davidson. 1977. Competition between seed-eating rodents and ants in desert ecosystems. *Science* 196: 880–882. [11]

Brown, J. H. and A. Kodric-Brown. 1977. Turnover rates in insular biogeography: Effect of immigration on extinction. *Ecology* 58: 445–449. [10]

Bruno, J. F., J. J. Stachowicz and M. D. Bertness. 2003. Inclusion of facilitation into ecological theory. *Trends in Ecology and Evolution* 18: 119–125. [18]

Bruno, J. F. and 7 others. 2007. Thermal stress and coral cover as drivers of coral disease outbreaks. *PLoS Biology* 5: 1220–1227. [13]

Bullock, S., H. Mooney and E. Medina, eds. 1995. Seasonally dry tropical forests. Cambridge University Press, Cambridge. [3]

Bunn, R., Y. Lekberg and C. Zabinski. 2009. Arbuscular mycorrhizal fungi ameliorate temperature stress in thermophilic plants. *Ecology* 90: 1378–1388. [14]

Burgess, R. L. 1977. The Ecological Society of America: Historical data and preliminary analyses. In *History of American Ecology,* F. N. Egerton (ed.), 1–24. Arno Press, New York. [22]

Buss, L. W. and J. B. C. Jackson. 1979. Competitive networks: Nontransitive competitive relationships in cryptic coral reef environments. *American Naturalist* 113: 223–234. [15]

Buston, P. M. 2003a. Size and growth modification in clownfish. *Nature* 424: 145–146. [7]

Buston, P. M. 2003b. Forcible eviction and prevention of recruitment in the clown anemonefish. *Behavioral Ecology* 14: 576–582. [7]

Buston, P. M. 2004. Territory inheritance in clownfish. *Proceedings of the Royal Society of London* B (suppl.) 271: S252–S254. [7]

Butzer, K. W. 1992. The Americas before and after 1492: An introduction to current geographical research. *Annals of the Association of American Geographers* 82: 345–368. [23]

Bytnerowicz, A., M. J. Arbaugh and R. Alonso. 2003. *Ozone Pollution in the Sierra Nevada: Distribution and Effects on Forests.* Elsevier, Amsterdam. [24]

C

Cain, M. L., B. G. Milligan and A. E. Strand. 2000. Long-distance seed dispersal in plant populations. *American Journal of Botany* 87: 1217–1227. [8]

Caldwell, M. M., L. O. Björn, J. F. Bornman, S. D. Flint, G. Kulandaivelu, A. H. Teramura and M. Tevini. 1998. Effects of increased solar ultraviolet radiation on terrestrial ecosystems. *Journal of Photochemistry and Photobiology* B: *Biology* 46: 40–52. [24]

Caleira, K. and M. E. Wickett. 2003. Anthropogenic carbon and ocean pH. *Nature* 425: 365. [24]

Callaway, R. M. and L. King. 1996. Temperature-driven variation in substrate oxygenation and the balance of competition and facilitation. *Ecology* 77: 1189–1195. [14]

Callaway, R. M., E. H. DeLucia and W. H. Schlesinger. 1994. Biomass allocation of montane and desert ponderosa pine: An analog for response to climate change. *Ecology* 75: 1474–1481. [7]

Callaway, R. M. and 12 others. 2002. Positive interactions among alpine plants increase with stress. *Nature* 417: 844–848. [14, 18]

Carlsson, J. O. L., C. Brönmark and L.-A. Hansson. 2004. Invading herbivory: The golden apple snail alters ecosystem functioning in Asian wetlands. *Ecology* 85: 1575–1580. [12]

Carlton, J. T. and J. B. Geller. 1993. Ecological roulette: The global transport and invasion of non-indigenous marine organisms. *Science* 261: 78–82. [18]

Carroll, S. P. and C. Boyd. 1992. Host race radiation in the soapberry bug: Natural history with the history. *Evolution* 46: 1052–1069. [6]

Carroll, S. P., H. Dingle and S. P. Klassen. 1997. Genetic differentiation of fitness-associated traits among rapidly evolving populations of the soapberry bug. *Evolution* 51: 1182–1188. [6]

Carson, R. 1962. *Silent Spring.* Houghton Mifflin, Boston. [20]

Carson, W. P. and R. B. Root. 2000. Herbivory and plant species coexistence: Community regulation by an outbreaking phytophagous insect. *Ecological Monographs* 70: 73–99. [1]

Cary, J. R and L. B. Keith. 1979. Reproductive change in the ten-year cycle of snowshoe hares. *Canadian Journal of Zoology* 57: 375–390. [12]

Castellón, T. D. and K. E. Sieving. 2006. An experimental test of matrix permeability and corridor use by an endemic understory bird. *Conservation Biology* 20: 135–145. [23]

Caswell, H. 2001. *Matrix Population Models: Construction, Analysis, and Interpretation*. 2nd ed. Sinauer Associates, Sunderland, MA. [9]

Caughley, G. 1994. Directions in conservation biology. *Journal of Animal Ecology* 63: 215–244. [22]

Caughley, G., N. Shepherd and J. Short. 1987. *Kangaroos: Their Ecology and Management in the Sheep Rangelands of Australia*. Cambridge University Press, Cambridge. [8]

Ceballos, G. and P. R. Ehrlich. 2002. Mammal population losses and the extinction crisis. *Science* 296: 904–907. [22]

Cebrian, J. and J. Lartigue. 2004. Patterns of herbivory and decomposition in aquatic and terrestrial ecosystems. *Ecological Monographs* 74: 237–259. [20]

Cerling, T. E., J. M. Harris, B. J. MacFadden, M. G. Leakey, J. Quade, V. Eisenmann and J. R. Ehleringer. 1997. Global vegetation change through the Miocene/Pliocene boundary. *Nature* 389: 153–158. [5]

Channell, R. and M. V. Lomolino. 2000. Trajectories to extinction: Spatial dynamics of the contraction of geographical ranges. *Journal of Biogeography* 27: 169–179. [22]

Chapin, F. S. III, L. Moilanen and K. Kielland. 1993. Preferential use of organic nitrogen for growth by a non-mycorrhizal Arctic sedge. *Nature* 361: 150–153. [21]

Chapin, F. S. III, L. R. Walker, C. L. Fastie and L. C. Sharman. 1994. Mechanisms of primary succession following deglaciation at Glacier Bay, Alaska. *Ecological Monographs* 64: 149–175. [16]

Chapin, F. S. III, P. A. Matson and H. A. Mooney. 2002. *Principles of Terrestrial Ecosystem Ecology*. Springer, New York. [19, 20, 24]

Chappelka, A. and L. Samuelson. 1998. Ambient ozone effects on forest trees of the eastern United States: A review. *New Phytologist* 139: 91–108. [24]

Charnov, E. L. 1976. Optimal foraging, the marginal value theorem. *Theoretical Population Biology* 9: 129–136. [5]

Charnov, E. L. 1993. *Life History Invariants*. Oxford University Press, Oxford. [7]

Charnov, E. L. and D. Berrigan. 1990. Dimensionless numbers and life history evolution: Age of maturity versus the adult lifespan. *Evolutionary Ecology* 4: 273–275. [7]

Chen, J. Q., J. F. Franklin and T. A. Spies. 1995. Growing-season microclimatic gradients from clear-cut edges into old-growth douglas-fir forests. *Ecological Applications* 5: 74–86. [23]

Chen, J.-Y. and 10 others. 2009. Complex embryos displaying bilaterian characters from Precambrian Doushantuo phosphate deposits, Weng'an, Guizhou, China. *Proceedings of the National Academy of Sciences USA* 106: 19056–19060. [6]

Chen, M., M. Schliep, R. D. Willows, Z. Cai, B. A. Neilan and H. Scheer. 2010. A red-shifted chlorophyll. *Science* 329: 1318–1319. [5]

Chesson, P. L. and R. R. Warner. 1981. Environmental variability promotes coexistence in lottery competitive systems. *American Naturalist* 117: 923–943. [18]

Christensen, N. L. and 12 others. 1996. The report of the Ecological Society of America Committee on the Scientific Basis for Ecosystem Management. *Ecological Applications* 6: 665–691. [23]

Christianou, M. and B. Ebenman. 2005. Keystone species and vulnerable species in ecological communities: Strong or weak interactors? *Journal of Theoretical Biology* 235: 95–103. [22]

Clements, F. E. 1916. *Plant Succession: An Analysis of the Development of Vegetation*. Publication 42. Carnegie Institution of Washington, Washington, DC. [16]

Cleveland, C. C. and 10 others. 1999. Global patterns of terrestrial biological nitrogen (N_2) fixation in natural ecosystems. *Global Biogeochemical Cycles* 13: 623–645. [24]

Clutton-Brock, T. H., F. E. Guinness and S. D. Albon. 1983. The costs of reproduction to red deer hinds. *Journal of Animal Ecology* 52: 367–383. [6]

Cohen, J. E. 1995. *How Many People Can the Earth Support?* W. W. Norton, New York. [9]

Coltman, D. W., P. O'Donoghue, J. T. Jorgenson, J. T. Hogg, C. Strobeck and M. Festa-Bianchet. 2003. Undesirable evolutionary consequences of trophy hunting. *Nature* 426: 655–658. [6]

Connell, J. H. 1961a. The effects of competition, predation by *Thais lapillus*, and other factors on natural populations of the barnacle, *Balanus balanoides*. *Ecological Monographs* 31: 61–104. [11, 18]

Connell, J. H. 1961b. The influence of interspecific competition and other factors on the distribution of the barnacle *Chthamalus stellatus*. *Ecology* 42: 710–723. [11]

Connell, J. H. 1978. Diversity in tropical rain forests and coral reefs. *Science* 199: 1302–1310. [18]

Connell, J. H. 1983. On the prevalence and relative importance of interspecific competition: Evidence from field experiments. *American Naturalist* 122: 661–696. [11]

Connell, J. H. and R. O. Slatyer. 1977. Mechanisms of succession in natural communities and their role in community stability and organization. *American Naturalist* 111: 1119–1144. [16]

Connell, J. H. and W. P. Sousa. 1983. On the evidence needed to judge ecological stability or persistence. *American Naturalist* 121: 789–824. [16]

Cook, B. I., R. L. Miller and R. Seager. 2009. Amplification of the North American "Dust Bowl" drought through human-induced land degradation. *Proceedings of the National Academy of Sciences USA* 106: 4997–5001. [24]

Cooper, W. S. 1923a. The recent ecological history of Glacier Bay, Alaska: The interglacial forests of Glacier Bay. *Ecology* 4: 93–128. [16]

Cooper, W. S. 1923b. The recent ecological history of Glacier Bay, Alaska: Permanent quadrats at Glacier Bay: An initial report upon a long-period study. *Ecology* 4: 355–365. [16]

Copeyon, C. K., J. R. Walters and J. H. Carter III. 1991. Induction of red-cockaded woodpecker group formation by artificial cavity construction. *Journal of Wildlife Management* 55: 549–556. [22]

Costanzo, J. P., R. E. Lee Jr., A. L. DeVries, T. Wang and J. R. Layne Jr. 1995. Survival mechanisms of vertebrate ectotherms at subfreezing temperatures: Applications in cryomedicine. *FASEB Journal* 9: 351–358. [4]

Cotton, P. A., S. D. Rundle and K. E. Smith. 2004. Trait compensation in marine gastropods: Shell shape, avoidance behavior, and susceptibility to predation. *Ecology* 85: 1581–1584. [12]

Courchamp, F., T. Clutton-Brock and B. Grenfell. 1999. Inverse density dependence and the Allee effect. *Trends in Ecology and Evolution* 14: 405–410. [10]

Cowie, R. H. 2001. Decline and homogenization of Pacific faunas: The land snails of American Samoa. *Biological Conservation* 99: 207–222. [22]

Cowie, R. J. 1977. Optimal foraging in great tits (*Parus major*). *Nature* 268: 137–139. [5]

Cowles, H. C. 1899. The ecological relations of the vegetation on the sand dunes of Lake Michigan. *Botanical Gazette* 27: 95–117, 167–202, 281–308, 361–391. [16]

Creel, S. and D. Christianson. 2009. Wolf presence and increased willow consumption by Yellowstone elk: Implications for trophic cascades. *Ecology* 90: 2454–2466. [23]

Crews, T. E., K. Kitayama, J. H. Fownes, D. A. Herbert, R. H. Riley, D. Mueller-Dombois and P. M. Vitousek. 1995. Changes in soil phosphorus fractions and ecosystem dynamics across a long chronosequence in Hawaii. *Ecology* 76: 1407–1424. [21]

Crisafulli, C. M., J. A. MacMahon and R. R. Parmenter. 2005. Small-mammal survival and colonization on the Mount St. Helens volcano: 1980–2005. In *Ecological Responses to the 1980 Eruption of Mount St. Helens*, V. H. Dale, F. J. Swanson and C. M. Crisafulli (eds.), 183–197. Springer, New York. [16]

Croll, D. A., J. L. Maron, J. A. Estes, E. M. Danner and G. V. Byrd. 2005. Introduced predators transform subarctic islands from grassland to tundra. *Science* 307: 1959–1961. [12]

Crouse, D. T., L. B. Crowder and H. Caswell. 1987. A stage-based population model for loggerhead sea turtles and implications for conservation. *Ecology* 68: 1412–1423. [9]

Crowder, L. B., D. T. Crouse, S. S. Heppell and T. H. Martin. 1994. Predicting the impact of Turtle Excluder Devices on loggerhead sea turtle populations. *Ecological Applications* 4: 437–445. [9]

Crutzen, P. J. and E. F. Stoermer. 2000. The "Anthropocene." *Global Change Newsletter* 41: 12–13. [24]

Currie, C. R. and A. E. Stuart. 2001. Weeding and grooming of pathogens in agriculture by ants. *Proceedings of the Royal Society of London* B 268: 1033–1039. [14]

Currie, C. R., U. G. Mueller and D. Malloch. 1999a. The agricultural pathology of ant fungus gardens. *Proceedings of the National Academy of Sciences USA* 96: 7998–8002. [14]

Currie, C. R., J. A. Scott, R. C. Summerbell and D. Malloch. 1999b. Fungus-growing ants use antibiotic-producing bacteria to control garden parasites. *Nature* 398: 701–704. [14]

Currie, C. R., M. Poulsen, J. Mendenhall, J. J. Boomsma and J. Billen. 2006. Coevolved crypts and exocrine glands support mutualistic bacteria in fungus-growing ants. *Science* 311: 81–83. [14]

D

Dahlgren, C. P. and D. B. Eggleston. 2000. Ecological processes underlying ontogenetic habitat shifts in a coral reef. *Ecology* 81: 2227–2240. [7]

Daily, G. C. and K. Ellison. 2002. *The New Economy of Nature: The Quest to Make Conservation Profitable*. Island Press, Washington, DC. [23]

Dale, V. H., F. J. Swanson and C. M. Crisafulli, eds. 2005. *Ecological Responses to the 1980 Eruption of Mount St. Helens*. Springer, New York. [16]

Dambrine, E., J. L. Dupouey, L. Laut, L. Humbert, M. Thinon, M. T. Beaufils and H. Richard. 2007. Present forest biodiversity patterns in France related to former Roman agriculture. *Ecology* 88: 1430–1439. [23]

Damman, H. and M. L. Cain. 1998. Population growth and viability analyses of the clonal woodland herb, *Asarum canadense*. *Journal of Ecology* 86: 13–26. [9]

Darimont, C. T., S. M. Carlson, M. T. Kinnison, P. C. Paquet, T. E. Reimchen and C. C. Wilmers. 2009. Human predators outpace other agents of trait change in the wild. *Proceedings of the National Academy of Sciences USA* 106: 952–954. [6]

Darwin, C. 1859. *On the Origin of Species*. J. Murray, London. [6, 11, 12, 15, 17, 18, 20]

Darwin, C. 1875. *Insectivorous Plants*. Appleton and Company, New York. [11]

Daskalov, G. M., A. N. Grishin, S. Rodionov and V. Mihneva. 2007. Trophic cascades triggered by overfishing reveal possible mechanisms of ecosystem regime shifts. *Proceedings of the National Academy of Sciences USA* 104: 10518–10523. [10]

Davidson, D. W., S. C. Cook, R. R. Snelling and T. H. Chua. 2002. Explaining the abundance of ants in lowland tropical rainforest canopies. *Science* 300: 969–972. [19]

Davidson, J. 1938. On the growth of the sheep population in Tasmania. *Transactions of the Royal Society of South Australia* 62: 342–346. [10]

Davidson, J. and H. G. Andrewartha. 1948. The influence of rainfall, evaporation and atmospheric temperature of fluctuations in the size of a natural population of *Thrips imaginis* (Thysanoptera). *Journal of Animal Ecology* 17: 200–222. [9]

Davis, C. C. 1964. Evidence for the eutrophication of Lake Erie from phytoplankton records. *Limnology and Oceanography* 9: 275–283. [10]

Davis, M. B., J. Ford and R. E. Moeller. 1985. Paleolimnology. In *An Ecosystem Approach to Aquatic Ecology: Mirror Lake and Its Environment*, G. E. Likens (ed.), 345–366. Springer-Verlag, New York. [21]

Dawson, J. and R. Lucas. 2000. *Nature Guide to the New Zealand Forest.* A Godwit Book. Random House New Zealand, Auckland, NZ. [17]

Dayton, P. K. 1971. Competition, disturbance and community organization: The provision and subsequent utilization of space in a rocky intertidal community. *Ecological Monographs* 41: 351–389. [15]

De'ath, G., J. M. Lough and K. E. Fabricus. 2009. Declining coral calcification in the Great Barrier Reef. *Science* 323: 116–119. [24]

Deevey, E. S. 1947. Life tables for natural populations. *Quarterly Review of Biology* 22: 283–314. [9]

de Grammont, P. C. and A. D. Cuarón. 2006. An evaluation of threatened species categorization systems used on the American continent. *Conservation Biology* 20: 14–27. [22]

Delcourt, H. R. 2002. Creating landscape pattern. In *Learning Landscape Ecology*, S. E. Gergel and M. G. Turner (eds.), 62–82. Springer Verlag, New York. [23]

Delcourt, P. A., H. R. Delcourt, C. R. Ison, W. E. Sharp and K. J. Gremillion. 1998. Prehistoric human use of fire, the eastern agricultural complex, and Appalachian oak–chestnut forests: Paleoecology of Cliff Palace Pond, Kentucky. *American Antiquity* 63: 263–278. [3]

DellaSala, D. A. and J. E. Williams. 2006. The northwest forest plan: A global model of forest management in contentious times. *Conservation Biology* 20: 274–276. [23]

Del Moral, R., D. M. Wood and J. H. Titus. 2005. Proximity, microsites, and biotic interactions during early succession. In *Ecological Responses to the 1980 Eruption of Mount St. Helens*, V. H. Dale, F. J. Swanson and C. M. Crisafulli (eds.), 93–107. Springer, New York. [16]

DeLucia, E. H. and 10 others. 1999. Net primary production of a forest ecosystem under experimental CO_2 enrichment. *Science* 284: 1177–1179. [24]

DeMaster, D. P., A. W. Trites, P. Clapham, S. Mizroch, P. Wade, R. J. Small and J. Ver Hoef. 2006. The sequential megafaunal collapse hypothesis: Testing with existing data. *Progress in Oceanography* 68: 329–342. [8]

Denslow, J. S. 1987. Tropical rainforest gaps and tree species diversity. *Annual Review of Ecology and Systematics* 18: 431–451. [7]

Department of Economic and Social Affairs, Population Division. 2007. *World Population Prospects: The 2006 Revision. Highlights.* United Nations Secretariat, New York. [9]

Dewailly, E., P. Ayotte, S. Bruneau, C. Laliberte, D. C. G. Muir and R. J. Norstrom. 1993. Inuit exposure to organochlorines through the aquatic food-chain in Arctic Quebec. *Environmental Health Perspectives* 101: 618–620. [20]

Diamond, J. M. 1975. The island dilemma: Lessons of modern biogeographic studies for the design of natural reserves. *Biological Conservation* 7: 129–146. [23]

Diamond, J. M. 1997. *Guns, Germs, and Steel: The Fates of Human Societies*. W. W. Norton, New York. [13]

Dietl, G. P., G. S. Herbert and G. J. Vermeij. 2004. Reduced competition and altered feeding behavior among marine snails after a mass extinction. *Science* 306: 2229–2231. [6]

Dirzo, R. and P. H. Raven. 2003. Global state of biodiversity and loss. *Annual Review of Environment and Resources* 28: 137–167. [3]

Dixon, A. F. G. 1971. The role of aphids in wood formation. II. The effect of the lime aphid, *Eucallipterus tiliae* L. (Aphididae), on the growth of lime *Tilia* × *vulgaris* Hayne. *Journal of Applied Ecology* 8: 393–399. [12]

Dobson, A. and M. Meagher. 1996. The population dynamics of brucellosis in the Yellowstone National Park. *Ecology* 77: 1026–1036. [13]

Dodd, D. M. B. 1989. Reproductive isolation as a consequence of adaptive divergence in *Drosophila pseudoobscura*. *Evolution* 43: 1308–1311. [6]

Dodson, C. H. and A. H. Gentry. 1991. Biological extinction in Western Ecuador. *Annals of the Missouri Botanical Garden* 78: 273–295. [22]

Duffy, J. E., B. J. Cardinale, K. E. France, P. B. McIntyre, E. Thebault and M. Loreau. 2007. The functional role of biodiversity in ecosystems: Incorporating trophic complexity. *Ecology Letters* 10: 522–538. [18]

Duggins, D. O. 1980. Kelp beds and sea otters: An experimental approach. *Ecology* 61: 447–453. [8]

Duggins, D. O., C. A. Simenstad and J. A. Estes. 1989. Magnification of secondary production by kelp detritus in coastal marine ecosystems. *Science* 245: 170–173. [8]

Dukes, J. S. and H. A. Mooney. 1999. Does global change increase the success of biological invaders? *Trends in Ecology and Evolution* 14: 135–139. [24]

Dybdahl, M. F. and C. M. Lively. 1998. Host–parasite coevolution: Evidence for rare advantage and time-lagged selection in a natural population. *Evolution* 52: 1057–1066. [13]

Dyer, L. A. and K. K. Letourneau. 1999a. Trophic cascades in a complex terrestrial community. *Proceedings of the National Academy of Sciences USA* 96: 5072–5076. [20]

Dyer, L. A. and K. K. Letourneau. 1999b. Relative strengths of top-down and bottom-up forces in a tropical forest community. *Oecologia* 119: 265–274. [20]

E

Eberhard, W. G. 2001. Under the influence: Webs and building behavior of *Plesiometa argyra* (Araneae, Tetragnathidae) when parasitized by *Hymenoepimecis argyraphaga* (Hymenoptera, Ichneumonidae). *Journal of Arachnology* 29: 354–366. [13]

Edmondson, W. T. and A. H. Litt. 1982. *Daphnia* in Lake Washington. *Limnology and Oceanography* 27: 272–293. [21]

Egan, T. 2006. *The Worst Hard Time: The Untold Story of Those Who Survived the Great American Dust Bowl.* Houghton Mifflin, Boston. [24]

Ehleringer, J. R. and C. S. Cook. 1990. Characteristics of *Encelia* species differing in leaf reflectance and transpiration rate under common garden conditions. *Oecologia* 82: 484–489. [4]

Ehleringer, J. R., T. E. Cerling and B. R. Helliker. 1997. C_4 photosynthesis, atmospheric CO_2, and climate. *Oecologia* 112: 285–299. [5]

Ehrlich, P. R. and E. O. Wilson. 1991. Biodiversity studies—Science and policy. *Science* 253: 758–762. [22]

Elser, J. J. and 11 others. 2000. Nutritional constraints in terrestrial and freshwater food webs. *Nature* 408: 578–580. [20]

Elser, J. J. and 9 others. 2007. Global analysis of nitrogen and phosphorus limitation of primary producers in freshwater, marine and terrestrial ecosystems. *Ecology Letters* 10: 1135–1142. [19]

Elton, C. S. 1927. *Animal Ecology*. Sidgwick and Jackson, London. [16]

Elton, C. S. 1942. *Voles, Mice and Lemmings: Problems in Population Dynamics*. Oxford University Press, Oxford. [6]

Elton, C. S. 1958. *The Ecology of Invasions by Animals and Plants.* Chapman and Hall, London. [18]

Endler, J. A. 1986. *Natural Selection in the Wild*. Princeton University Press, Princeton, NJ. [6]

Ensign, S. H. and M. W. Doyle. 2006. Nutrient spiraling in streams and river networks. *Journal of Geophysical Research* 111: G04009, doi:10.1029/2005JG000114. [21]

Epstein, E. and A. J. Bloom. 2005. *Mineral Nutrition of Plants: Principles and Perspectives*. 2nd ed. Sinauer Associates, Sunderland, MA. [21]

Epstein, P. R. 2000. Is global warming harmful to health? *Scientific American* 283: 50–57. [13]

Erickson, R. O. 1945. The *Clematis fremontii* var. *riehlii* population in the Ozarks. *Annals of the Missouri Botanical Garden* 32: 413–460. [8]

Estes, J. A. and D. O. Duggins. 1995. Sea otters and kelp forests in Alaska: Generality and variation in a community ecology paradigm. *Ecological Monographs* 65: 75–100. [8]

Estes, J. A., M. T. Tinker, T. M. Williams and D. F. Doak. 1998. Killer whale predation on sea otters linking oceanic and nearshore ecosystems. *Science* 282: 473–476. [8]

Evans, R. D., R. Rimer, L. Sperry and J. Belnap. 2001. Exotic plant invasion alters nitrogen dynamics in an arid grassland. *Ecological Applications* 11: 1301–1310. [21]

F

Fagan, W. F., R. S. Cantrell and C. Cosner. 1999. How habitat edges change species interactions. *American Naturalist* 153: 165–182. [23]

Faithfull, I. 1997. *Etiella behrii* (Zeller) (Lepidoptera: Pyralidae) bred from pods of gorse, *Ulex europaeus* L. Fabaceae. *Victorian Entomology* 27: 34. [18]

FAO. Global Forest Resources Assessment 2005. Food and Agriculture Organization of the United Nations. www.fao.org/forestry/site/24690/en. [23]

Farrell, B. D. 1998. "Inordinate fondness" explained: Why are there so many beetles? *Science* 281: 555–559. [6]

Farrell, T. M. 1991. Models and mechanisms of succession: An example from a rocky intertidal community. *Ecological Monographs* 61: 95–113. [16]

Favreau, J. M., C. A. Drew, G. R. Hess, M. J. Rubino, F. H. Koch and K. A. Eschelbach. 2006. Recommendations for assessing the effectiveness of surrogate species approaches. *Biodiversity and Conservation* 15: 3949–3969. [22]

Feder, J. L. 1998. The apple maggot fly, *Rhagoletis pomonella*: Flies in the face of conventional wisdom about speciation? In *Endless Forms: Species and Speciation*, D. J. Howard and S. H. Berlocher (eds.), 130–144. Oxford University Press, Oxford. [6]

Fedonkin, M. A., A. Simonetta and A. Y. Ivantsov. 2007. New data on *Kimberella*, the Vendian mollusc-like organism (White Sea region, Russia): Palaeoecological and evolutionary implications. *Geological Society, London, Special Publications* 286: 157–179. [6]

Feely, R. A., C. L. Sabine, K. Lee, W. Berrelson, J. Kleypas, V. J. Fabry and F. J. Millero. 2004. Impact of anthropogenic CO_2 on the $CaCO_3$ system in the oceans. *Science* 305: 362–366. [24]

Fenner, F. and F. N. Ratcliffe. 1965. *Myxomatosis*. Cambridge University Press, Cambridge. [13]

Fernández-Marín, H., J. K. Zimmerman, D. R. Nash, J. J. Boomsma and W. T. Wcislo. 2009. Reduced biological control and enhanced chemical pest management in the evolution of fungus farming in ants. *Proceedings of the Royal Society of London* B 276: 2263–2269. [14]

Ferraz, G., G. J. Russell, P. C. Stouffer, R. O. Bierregaard Jr., S. L. Pimm and T. E. Lovejoy. 2003. Rates of species loss from Amazonian forest fragments. *Proceedings of the National Academy of Sciences USA* 100: 14069–14073. [17]

Field, C. B., M. J. Behrenfeld, J. T. Randerson and P. Falkowski. 1998. Primary productivity of the biosphere: Integrating terrestrial and oceanic components. *Science* 281: 237–240. [19]

Field, D. B., T. R. Baumgartner, C. D. Charles, V. Ferreira-Bartrina and M. D. Ohman. 2006. Planktonic foraminifera of the California current reflect 20th-century warming. *Science* 311: 63–66. [24]

Field, J. P. and 9 others. 2010. The ecology of dust. *Frontiers in Ecology and the Environment* 8: 423–430. [24]

Fine, P. V. A. and R. H. Ree. 2006. Evidence for a time-integrated species–area effect on the latitudinal gradient in tree diversity. *American Naturalist* 168: 796–804. [17]

Fiore, A. M., D. J. Jacob, J. A. Logan and J. H. Yin. 1998. Long-term trends in ground level ozone over the contiguous United States, 1980–1995. *Journal of Geophysical Research* 103 D1: 1471–1480. [24]

Fisher, S. G. and G. E. Likens. 1973. Energy budget in Bear Brook, New Hampshire: An integrative approach to stream ecosystem metabolism. *Ecological Monographs* 43: 421–439. [20]

Fitter, A. H. 2005. Darkness visible: Reflections on underground ecology. *Journal of Ecology* 93: 231–243. [14]

Flecker, A. S. and C. R. Townsend. 1994. Community-wide consequences of trout introduction in New Zealand streams. *Ecological Applications* 4: 798–807. [20]

Fleishman, E., D. D. Murphy and P. E. Brussard. 2000. A new method for selection of umbrella species for conservation planning. *Ecological Applications* 10: 569–579. [22]

Flessa, K. W. 1975. Area, continental drift and mammalian diversity. *Paleobiology* 1: 189–194. [17]

Foley, J. A., M. H. Costa, C. Delire, N. Ramankutty and P. Snyder. 2003. Green surprise? How terrestrial ecosystems could affect Earth's climate. *Frontiers in Ecology and Environment* 1: 38–44. [2]

Fonseca, G. A. B. da. 1985. The vanishing Brazilian Atlantic forest. *Biological Conservation* 34: 17–34. [22]

Forbes, A. A., T. H. Q. Powell, L. L. Stelinski, J. J. Smith and J. L. Feder. 2009. Sequential sympatric speciation across trophic levels. *Science* 323: 776–779. [6]

Forman, R. T. T. 1995. Some general principles of landscape and regional ecology. *Landscape Ecology* 10: 133–142. [23]

Food and Agriculture Organization of the United Nations. *See* FAO.

Frank, P. W., C. D. Boll and R. W. Kelly. 1957. Vital statistics of laboratory cultures of *Daphnia pulex* De Geer as related to density. *Physiological Zoology* 30: 287–305. [9]

Franklin, J. F. 1996. Ecosystem management: An overview. In *Ecosystem Management: Applications for Sustainable Forest and Wildlife Resources*, M. S. Boyce and A. Haney (eds.), 21–53. Yale University Press, New Haven, CT. [23]

Franklin, J. F. and C. T. Dyrness. 1988. *Natural Vegetation of Oregon and Washington*. Oregon State University Press, Corvallis. [17]

Franks, S. J., S. Sim and A. E. Weis. 2007. Rapid evolution of flowering time by an annual plant in response to a climate fluctuation. *Proceedings of the National Academy of Sciences USA* 104: 1278–1282. [6]

Freeman, S. and J. C. Herron. 2007. *Evolutionary Analysis*. 4th ed. Pearson Prentice Hall, Upper Saddle River, NJ. [6]

Fry, B. 2007. *Stable isotope ecology*. Springer, New York. [5]

Funk, D. J., P. Nosil and W. J. Etges. 2006. Ecological divergence exhibits consistently positive associations with reproductive isolation across disparate taxa. *Proceedings of the National Academy of Sciences USA* 103: 3209–3213. [6]

Futuyma, D. J. 2009. *Evolution*. 2nd ed. Sinauer Associates, Sunderland, MA. [6]

G

Galloway, J. N. and 14 others. 2004. Nitrogen cycles: Past, present, and future. *Biogeochemistry* 70: 153–226. [24]

Garrison, V. H. and 9 others. 2003. African and Asian dust: From desert soils to coral reefs. *BioScience* 53: 469–480. [24]

Gascon, C., G. B. Williamson and G. A. B. da Fonseca. 2000. Receding forest edges and vanishing reserves. *Science* 288: 1356–1358. [17]

Gasol, J. M., P. A. del Giorio and C. M. Duarte. 1997. Biomass distribution in marine planktonic communities. *Limnology and Oceanography* 42: 1353–1363. [20]

Gaston, K. J. 2003. *The Structure and Dynamics of Geographic Ranges*. Oxford University Press, Oxford. [8]

Gaston, K. J., P. H. Williams, P. Eggleton and C. J. Humphries. 1995. Large scale patterns of biodiversity: Spatial variation in family richness. *Proceedings of the Royal Society of London* B 260: 149–154. [17]

Gause, G. F. 1934a. Experimental analysis of Vito Volterra's mathematical theory of the struggle for existence. *Science* 79: 16–17. [11, 18]

Gause, G. F. 1934b. *The Struggle for Existence*. Williams & Wilkins, Baltimore, MD. [11]

Geffeney, S. L., E. Fujimoto, E. D. Brodie III, E. D. Brodie Jr. and P. C. Ruben. 2005. Evolutionary diversification of TTX-resistant sodium channels in a predator–prey interaction. *Nature* 434: 759–763. [12]

Gilg, O., I. Hanski and B. Sittler. 2003. Cyclic dynamics in a simple vertebrate predator–prey community. *Science* 302: 866–868. [10, 12]

Gilg, O., B. Sittler and I. Hanski. 2009. Climate change and cyclic predator–prey population dynamics in the high Arctic. *Global Change Biology* 15: 2634–2652. [10]

Gillooly, F. J., A. P. Allen, G. B. West and J. H. Brown. 2005. The rate of DNA evolution: Effects of body size and temperature on the molecular clock. *Proceedings of the National Academy of Sciences USA* 102: 140–145. [17]

Gleason, H. A. 1917. The structure and development of the plant association. *Bulletin of the Torrey Botanical Club* 44: 463–481. [16]

González, C., O. Wang, S. E. Strutz, C. González-Salazar, V. Sánchez-Cordero and S. Sarkar. 2010. Climate change and risk of leishmaniasis in North America: Predictions from ecological niche models of vector and reservoir species. *PLoS Neglected Tropical Diseases* 4: 1–16. [13]

Gotelli, N. J. and A. M. Ellison. 2004. *A Primer of Ecological Statistics*. Sinauer Associates, Sunderland, MA. [1]

Gough, L., G. R. Shaver, J. Carroll, D. L. Royer and J. A. Laundre. 2000. Vascular plant species richness in Alaskan arctic tundra: The importance of soil pH. *Journal of Ecology* 88: 54–66. [21]

Goverde, M., K. Schweizer, B. Baur and A. Erhardt. 2002. Small-scale habitat fragmentation effects on pollinator behaviour: Experimental evidence from the bumblebee *Bombus veteranus* on calcareous grasslands. *Biological Conservation* 104: 293–299. [23]

Grabherr, G., M. Gottfried and H. Pauli. 1994. Climate effects on mountain plants. *Nature* 369: 448. [24]

Graham, I. M. and X. Lambin. 2002. The impact of weasel predation on cyclic field-vole survival: The specialist predator hypothesis contradicted. *Journal of Animal Ecology* 71: 946–956. [10]

Grant, B. R. and P. R. Grant. 2003. What Darwin's finches can teach us about the evolutionary origin and regulation of biodiversity. *BioScience* 53: 965–975. [6]

Grant, P. R. and B. R. Grant. 2006. Evolution of character displacement in Darwin's finches. *Science* 313: 224–226. [11]

Grime, J. P. 1977. Evidence for the existence of three primary strategies in plants and its relevance to ecological and evolutionary theory. *American Naturalist* 111: 1169–1194. [7]

Grime, J. P. 1997. Biodiversity and ecosystem function: The debate deepens. *Science* 277: 1260–1261. [18]

Groom, M. J. and 7 others. 1999. Buffer zones: Benefits and dangers of compatible stewardship. In *Continental Conservation: Scientific Foundations of Regional Reserve Networks*, M. E. Soulé and J. Terborgh (eds.), 171–197. Island Press, Washington, DC. [23]

Grousset, F. E., P. Ginoux, A. Bory and P. E. Biscaye. 2003. Case study of a Chinese dust plume reaching the French Alps. *Geophysical Research Letters* 30: 1277, doi:10.1029/2002GL016833. [24]

Grutter, A. S., J. M. Murphy and J. H. Choat. 2003. Cleaner fish drives local fish diversity on coral reefs. *Current Biology* 13: 64–67. [14]

Gurevitch, J., L. L. Morrow, A. Wallace and J. S. Walsh. 1992. A meta-analysis of competition in field experiments. *American Naturalist* 140: 539–572. [11]

Gutzwiller, K. J. and S. H. Anderson. 1992. Interception of moving organisms: Influences of patch shape, size, and orientation on community structure. *Landscape Ecology* 6: 293–303. [23]

H

Haase, D. and 7 others. 2007. Loess in Europe—its spatial distribution based on a European Loess Map, scale 1:2,500,000. *Quaternary Science Review*. 26: 1301–1312. [24]

Hacker, S. D. and M. D. Bertness. 1996. Trophic consequences of a positive plant interaction. *American Naturalist* 148: 559–575. [15]

Hacker, S. D. and M. D. Bertness. 1999. Experimental evidence for the factors maintaining plant species diversity in a New England salt marsh. *Ecology* 80: 2064–2073. [18]

Hacker, S. D. and S. D. Gaines. 1997. Some implications of direct positive interactions for community species diversity. *Ecology* 78: 1990–2003. [18]

Haddrath, O. and A. J. Baker. 2001. Complete mitochondrial DNA genome sequences of extinct birds: Ratite phylogenetics and the vicariance biogeography hypothesis. *Proceedings of the Royal Society of London* B 268: 939–945. [17]

Haines, B. L. 1978. Element and energy flows through colonies of the leaf-cutting ants, *Atta colombica*, in Panama. *Biotropica* 10: 270–277. [14]

Hairston, N., F. Smith and L. Slobodkin. 1960. Community structure, population control and competition. *American Naturalist* 94: 421–425. [20]

Halpern, B. S. and 18 others. 2008. A global map of human impact on marine ecosystems. *Science* 319: 948–952. [3, 22]

Hansen, J., M. Sato, R. Ruedy, K. Lo, D. W. Lea and M. Medina-Elizade. 2006. Global temperature change. *Proceedings of the National Academy of Sciences USA* 103: 14288–14293. [24]

Hansen, T. A. 1978. Larval dispersal and species longevity in lower Tertiary gastropods. *Science* 199: 885–887. [7]

Hanski, I. 1999. *Metapopulation Ecology*. Oxford University Press, New York. [10]

Harding, S. R. and J. R. Walters. 2002. Processes regulating the population dynamics of red-cockaded woodpecker cavities. *Journal of Wildlife Management* 66: 1083–1095. [22]

Hare, S. R. and R. C. Francis. 1994. Climate change and salmon production in the Northeast Pacific Ocean. In *Climate Change and Northern Fish Populations*, R. J. Beamish (ed.), 357–372. Canadian Special Publication of Fisheries and Aquatic Sciences, 121. [2]

Harrison, P. 1987. *A Field Guide to Seabirds of the World*. Stephen Greene Press, Lexington, MA. [17]

Harrison, P. and F. Pearce. 2001. *AAAS Atlas of Population and the Environment*. University of California Press, Berkeley. [3]

Harrison, R. G. 1980. Dispersal polymorphisms in insects. *Annual Review of Ecology and Systematics* 11: 95–118. [8]

Hart, M. W. and R. R. Strathmann. 1994. Functional consequences of phenotypic plasticity in echinoid larvae. *Biological Bulletin* 186: 291–299. [7]

Hartl, D. L. and A. G. Clark. 2007. *Principles of Population Genetics*. 4th ed. Sinauer Associates, Sunderland MA. [6]

Hartnett, D. C. and G. W. T. Wilson. 1999. Mycorrhizae influence plant community structure and diversity in tallgrass prairie. *Ecology* 80: 1187–1195. [14]

Harvell, C. D., C. E. Mitchell, J. R. Ward, S. Altizer, A. P. Dobson, R. S. Ostfeld and M. D. Samuel. 2002. Climate warming and disease risks for terrestrial and marine biota. *Science* 296: 2158–2162. [13]

Harvell, C. D., S. Altizer, I. M. Cattadori, L. Harrington and E. Weil. 2009. Climate change and wildlife diseases: When does the host matter the most? *Ecology* 90: 912–920. [13]

Hatch, A. C. and A. R. Blaustein. 2003. Combined effects of UV-B radiation and nitrate fertilizer on larval amphibians. *Ecological Applications* 13: 1083–1093. [1]

Hawkins, C. E. and 11 others. 2006. Emerging disease and population decline of an island endemic, the Tasmanian devil *Sarcophilus harrisii*. *Biological Conservation* 131: 307–324. [22]

Hays, J. D., J. Imbrie and N. J. Shackleton. 1976. Variations in Earths orbit—Pacemaker of ice ages. *Science* 194: 1121–1132. [2]

Hearne, S. 1911. *A Journey from Prince of Wales's Fort in Hudson's Bay to the Northern Ocean in the years 1769, 1770, 1771 and 1772*. The Champlain Society, Toronto. [4]

Hebblewhite, M. and 7 others. 2005. Human activity mediates a trophic cascade caused by wolves. *Ecology* 86: 2135–2144. [23]

Hechtel, L. J., C. L. Johnson and S. A. Juliano. 1993. Modification of antipredator behavior of *Caecidotea intermedius* by its parasite *Acanthocephalus dirus*. *Ecology* 74: 710–713. [13]

Hector, A. and 33 others. 1999. Plant diversity and productivity experiments in European grasslands. *Science* 286: 1123–1127. [18]

Hegazy, A. K. 1990. Population ecology and implications for conservation of *Cleome droserifolia*: A threatened xerophyte. *Journal of Arid Environments* 19: 269–282. [9]

Henderson, S., P. Hattersley, S. von Caemmerer and C. B. Osmond. 1995. Are C_4 pathway plants threatened by global climatic change? In *Ecophysiology of Photosynthesis*, E.-D. Schulze and M. M. Caldwell (eds.), 529–549. Springer, New York. [5]

Henry, M., J. M. Pons and J. F. Cosson. 2007. Foraging behaviour of a frugivorous bat helps bridge landscape connectivity and ecological processes in a fragmented rainforest. *Journal of Animal Ecology* 76: 801–813. [23]

Hersperger, A. M. 2006. Spatial adjacencies and interactions: Neighborhood mosaics for landscape ecological planning. *Landscape and Urban Planning* 77: 227–239. [23]

Hess, G. R. 1994. Conservation corridors and contagious disease—A cautionary note. *Conservation Biology* 8: 256–262. [23]

Hetrick, B. A. D., G. W. T. Wilson and D. C. Hartnett. 1989. Relationship between mycorrhizal dependence and competitive ability of two tallgrass prairie grasses. *Canadian Journal of Botany* 67: 2608–2615. [14]

Hickman, J. E., S. Wu, L. J. Mickley and M. T. Lerdau. 2010. Kudzu (*Pueraria montana*) invasion doubles emissions of nitric oxide and increases ozone pollution. *Proceedings of the National Academy of Sciences USA* 107: 10115–10119. [22]

Hik, D. S. and R. L. Jefferies. 1990. Increases in the net above-ground primary production of a salt-marsh forage grass: A test of the predictions of the herbivore-optimization model. *Journal of Ecology* 78: 180–195. [12]

Hiller, N. L., S. Bhattacharjee, C. van Ooij, K. Liolios, T. Harrison, C. Lopez-Estraño and K. Haldar. 2004. A host-targeting signal in virulence proteins reveals a secretome in malarial infection. *Science* 306: 1934–1937. [13]

Himmel, M. E., S.-Y. Ding, D. K. Johnson, W. S. Adney, M. R. Nimlos, J. W. Brady and T. D. Foust. 2007. Biomass recalcitrance: Engineering plants and enzymes for biofuels production. *Science* 315: 804–807. [18]

Hochochka, P. W. and G. N. Somero. 2002. *Biochemical Adaptation, Mechanism and Process in Physiological Evolution*. Oxford University Press, Oxford. [4]

Hollinger, D. Y. and 14 others. 1998. Forest–atmosphere carbon dioxide exchange in eastern Siberia. *Agricultural and Forest Meteorology* 90: 291–306. [19]

Hooper, R. G., W. E. Taylor and S. C. Loeb. 2004. Long-term efficacy of artificial cavities for red-cockaded woodpeckers: Lessons learned from Hurricane Hugo. In *Red-Cockaded Woodpecker: Road to Recovery*, R. Costa and S. J. Daniels (eds.), 430–438. Hancock House Publishers, Blaine, WA. [22]

Houserová, P., V. Kubánˇ, S. Krácˇmar and J. Sitko. 2007. Total mercury and mercury species in birds and fish in an aquatic ecosystem in the Czech Republic. *Environmental Pollution* 145: 185–194. [20]

Howarth, R. W. and 14 others. 1996. Regional nitrogen budgets and riverine N & P fluxes for the drainages to the North Atlantic Ocean: Natural and human influences. *Biogeochemistry* 35: 75–139. [21, 24]

Hu, J., D. J. P. Moore, S. P. Burns and R. K. Monson. 2010. Longer growing seasons lead to less carbon sequestration by a subalpine forest. *Global Change Biology* 16: 771–783. [24]

Hudson, P. J., A. P. Dobson and D. Newborn. 1998. Prevention of population cycles by parasite removal. *Science* 282: 2256–2258. [13]

Huffaker, C. B. 1958. Experimental studies on predation: Dispersion factors and predator–prey oscillations. *Hilgardia* 27: 343–383. [12]

Huffaker, C. B. and C. E. Kennett. 1957. A ten-year study of vegetational changes associated with biological control of Klamath weed. *Journal of Range Management* 12: 69–82. [12]

Huffman, M. A. 1997. Current evidence for self-medication in primates: A multidisciplinary perspective. *Yearbook of Physical Anthropology* 40: 171–200. [13]

Hughes, J. B., J. J. Hellman, T. H. Ricketts and B. J. M. Bohannan. 2001. Counting the uncountable: Statistical approaches to estimating microbial diversity. *Applied and Environmental Microbiology* 67: 4399–4406. [15]

Humphreys, W. F. 1979. Production and respiration in animal populations. *Journal of Animal Ecology* 48: 427–454. [20]

Hunt, G. R. 1996. Manufacture and use of hook-tools by New Caledonian crows. *Nature* 379: 249–251. [5]

Hunt, G. R. and R. D. Gray. 2003. Diversification and cumulative evolution in New Caledonian crow tool manufacture. *Proceedings of the Royal Society of London* B 270: 867–874. [5]

Hunziker, J. H., R. A. Palacios, A. G. de Valesi and L. Poggio. 1972. Species disjunctions in *Larrea*: Evidence from morphology, cytogenetics, phenolic compounds and seed albumins. *Annals of Missouri Botanical Garden* 59: 224–233. [8]

Huston, M. 1979. A general hypothesis of species diversity. *American Naturalist* 113: 81–101. [18]

Hutchings, J. A. 2005. Life history consequences of overexploitation to population recovery in Northwest Atlantic cod (*Gadus morhua*). *Canadian Journal of Fisheries and Aquatic Sciences* 62: 824–832. [7]

Hutchinson, G. E. 1959. Homage to Santa Rosalia or why are there so many kinds of animals? *American Naturalist* 93: 145–159. [17]

Hutchinson, G. E. 1961. The paradox of the plankton. *American Naturalist* 95: 137–145. [18]

Hutchinson, G. E. 1965. *The Ecological Theater and the Evolutionary Play.* Yale University Press, New Haven, CT. [23]

I

Inouye, D. W. 2008. Effects of climate change on phenology, frost damage, and floral abundance of montane wildflowers. *Ecology* 89: 353–362. [24]

IPCC. 2007. *Climate Change 2007*. Cambridge University Press, Cambridge; also available at the IPCC Web site: http://www.ipcc.ch/. [17, 18]

J

Jablonski, D. 1995. Extinctions in the fossil record. In *Extinction Rates*, J. H. Lawton and R. M. May (eds.), 25–44. Oxford University Press, Oxford. [6]

Jablonski, D., K. Roy and J. W. Valentine. 2006. Out of the tropics: Evolutionary dynamics of the latitudinal diversity gradient. *Science* 314: 102–106. [17]

Jablonski, N. G. 2004. The evolution of human skin and skin color. *Annual Review of Anthropology* 33: 585–623. [24]

Jablonski, N. G. 2006. *Skin: A Natural History*. University of California Press, Berkeley. [4]

Jacobson, M. Z. 2002. *Atmospheric Pollution, History, Science, and Regulation*. Cambridge University Press, Cambridge. [24]

Jaenike, J., R. Unckless, S. N. Cockburn, L. M. Boelio and S. J. Perlman. 2010. Adaptation via symbiosis: Recent spread of a *Drosophila* defensive symbiont. *Science* 329: 212–215. [14]

Jaffe, D., I. McKendry, T. Anderson and H. Price. 2003. Six "new" episodes of trans-Pacific transport of air pollutants. *Atmospheric Environment* 37: 391–404. [24]

Janzen, D. H. 1966. Coevolution of mutualism between ants and acacias in Central America. *Evolution* 20: 249–275. [14]

Janzen, D. H. 1967. Why mountain passes are higher in the tropics. *American Naturalist* 101: 233–249. [17]

Janzen, D. H. 1986. The future of tropical ecology. *Annual Review of Ecology and Systematics* 17: 305–324. [22]

Jefferies, R. L., R. F. Rockwell and K. F. Abraham. 2003. The embarrassment of riches: Agricultural food subsidies, high goose numbers, and loss of Arctic wetlands—a continuing saga. *Environmental Reviews* 11: 193–232. [12]

Jentsch, A., J. Kreyling and C. Beierkuhnlein. 2007. A new generation of climate change experiments: Events, not trends. *Frontiers in Ecology and the Environment* 5: 315–324. [2]

Jeon, K. W. 1972. Development of cellular dependence on infective organisms: Microsurgical studies in amoebas. *Science* 176: 1122–1123. [14]

Jeschke, J. M. 2007. When carnivores are "full and lazy." *Oecologia* 152: 357–364. [5]

Johnson, P. T. J., K. B. Lunde, E. G. Ritchie and A. E. Launer. 1999. The effect of trematode infection on amphibian limb development and survivorship. *Science* 284: 802–804. [1]

Johnson, P. T. J. and 7 others. 2007. Aquatic eutrophication promotes pathogenic infection in amphibians. *Proceedings of the National Academy of Sciences USA* 104: 15781–15786. [1]

Johnson, R. G. and S. A. Temple. 1990. Nest predation and brood parasitism of tallgrass prairie birds. *Journal of Wildlife Management* 54: 106–111. [23]

Johnson, W. E. and 15 others. 2010. Genetic restoration of the Florida panther. *Science* 329: 1641–1645. [22]

Jones, C. G., J. H. Lawton and M. Shachak. 1994. Organisms as ecosystem engineers. *Oikos* 69: 373–386. [15]

Jones, C. G., J. H. Lawton and M. Shachak. 1997. Positive and negative effects of organisms as physical ecosystem engineers. *Ecology* 78: 1946–1957. [15]

Jones, H. L. and J. M. Diamond. 1976. Short-time-base studies of turnover in breeding bird populations on the California Channel Islands. *Condor* 78: 526–549. [10]

Jordan, M. J. and G. E. Likens. 1975. An organic carbon budget for an oligotrophic lake in New Hampshire. *Verhandlungen Internationale Vereinigung fur Theoretische and Angewandte Limnologie* 19: 994–1003. [20]

Jousson, O. and 8 others. 2000. Invasive alga reaches California. *Nature* 408: 157–158. [15]

K

Karban, R. and G. English-Loeb. 1997. Tachinid parasitoids affect host plant choice by caterpillars to increase caterpillar survival. *Ecology* 78: 603–611. [13]

Kauffman, M. J., J. F. Brodie and E. S. Jules. 2010. Are wolves saving Yellowstone's aspen? A landscape-level test of a behaviorally mediated trophic cascade. *Ecology* 91: 2742–2755. [23]

Kausrud, K. L. and 10 others. 2008. Linking climate change to lemming cycles. *Nature* 456: 93–97. [10]

Keever, K. 1953. Present composition of some stands of the former oak–chestnut forest in the southern Blue Ridge mountains. *Ecology* 34: 44–54. [13]

Keller, I. and C. R. Largiadèr. 2003. Recent habitat fragmentation caused by major roads leads to reduction of gene flow and loss of genetic variability in ground beetles. *Proceedings of the Royal Society of London* B 270: 417–423. [23]

Kenward, B., A. A. S. Weir, C. Rutz and A. Kacelnik. 2005. Tool manufacture by naive juvenile crows. *Nature* 433: 121. [5]

Kessler, A. and I. T. Baldwin. 2001. Defensive function of herbivore-induced plant volatile emissions in nature. *Science* 291: 2141–2144. [12]

Kessler, A., R. Halitschke and I. T. Baldwin. 2004. Silencing the jasmonate cascade: Induced plant defenses and insect populations. *Science* 305: 665–668. [12]

Kideys, A. E. 2002. Fall and rise of the Black Sea ecosystem. *Science* 297: 1482–1484. [10]

Kiehl, J. T. and K. E. Trenberth. 1997. Earth's annual global mean energy budget. *Bulletin of the American Meteorological Society* 78: 197–208. [2]

Kiesecker, J. M. 2002. Synergism between trematode infection and pesticide exposure: A link to amphibian limb deformities in nature? *Proceedings of the National Academy of Sciences USA* 99: 9900–9904. [1]

Kinchy, A. J. 2006. On the borders of post-war ecology: Struggles over the Ecological Society of America's preservation committee, 1917–1946. *Science as Culture* 15: 23–44. [22]

King, A. A. and W. M. Schaffer. 2001. The geometry of a population cycle: A mechanistic model of snowshoe hare demography. *Ecology* 82: 814–830. [12]

Kingsland, S. E. 1991. Defining ecology as a science. In *Foundations of Ecology: Classic Papers with Commentaries*, L. A. Real and J. H. Brown (eds.), 1–13. University of Chicago Press, Chicago. [16]

Kinnison, M. T. and A. P. Hendry. 2001. The pace of modern life II: From rates of contemporary evolution to pattern and process. *Genetica* 112–113: 145–164. [6]

Kinzig, A. P. and J. Harte. 2000. Implications of endemics–area relationships for estimates of species extinctions. *Ecology* 81: 3305–3311. [22]

Kitzes, J. and M. Wackernagel. 2009. Answers to common questions in Ecological Footprint accounting. *Ecological Indicators* 9: 812–817. [9]

Knapp, A. K. and 9 others. 2002. Rainfall variability, carbon cycling and plant species diversity in a mesic grassland. *Science* 298: 2202–2205. [3]

Kodric-Brown, A. and J. M. Brown. 1993. Highly structured fish communities in Australian desert springs. *Ecology* 74: 1847–1855. [17]

Kohler, S. L. and M. J. Wiley. 1997. Pathogen outbreaks reveal large-scale effects of competition in stream communities. *Ecology* 78: 2164–2176. [13]

Köhler, W. 1927. *The Mentality of Apes*. Routledge & Kegan Paul, London. [5]

Komdeur, J. 1992. Importance of habitat saturation and territory quality for evolution of cooperative breeding in the Seychelles warbler. *Nature* 358: 493–495. [8]

Kondoh, M. 2003. Foraging adaptation and the relationship between food-web complexity and stability. *Science* 299: 1388–1391. [20]

Korpimäki, E. and K. Norrdahl. 1998. Experimental reduction of predators reverses the crash phase of small-rodent cycles. *Ecology* 79: 2448–2455. [10]

Koskela, T., S. Puustinen, V. Salonen and P. Mutikainen. 2002. Resistance and tolerance in a host plant–holoparasitic plant interaction: Genetic variation and costs. *Evolution* 56: 899–908. [5]

Kozaki, A. and G. Takeba. 1996. Photorespiration protects C_3 plants from photooxidation. *Nature* 384: 557–580. [5]

Kraaijeveld, A. R., K. A. Hutcheson, E. C. Limentani and H. C. J. Godfray. 2001. Costs of counterdefenses to host resistance in a parasitoid of *Drosophila*. *Evolution* 55: 1815–1821. [13]

Krebs, C. J. 1999. *Ecological Methodology*. Addison Wesley, Menlo Park, CA. [8]

Krebs, C. J. and 7 others. 1995. Impact of food and predation on the snowshoe hare cycle. *Science* 269: 1112–1115. [12]

Krebs, J. B., J. T. Eriksen, M. T. Webber and E. L. Charnov. 1977. Optimal prey selection in the Great Tit (*Parus major*). *Animal Behavior* 25: 30–38. [5]

Kremen, C., V. Razafimahatratra, R. P. Guillery, J. Rakotomalala, A. Weiss and J. S. Ratsisompatrarivo. 1999. Designing the Masoala National Park in Madagascar based on biological and socioeconomic data. *Conservation Biology* 13: 1055–1068. [23]

Kricher, J. 1998. *A Field Guide to Eastern Forests*. Illustrated by G. Morrison. Houghton Mifflin, New York. [17]

Krümmel, E. M. and 7 others. 2003. Delivery of pollutants by spawning salmon. *Nature* 425: 255. [20]

Krützen, M., J. Mann, M. R. Heithaus, R. C. Connor, L. Bejder and W. B. Sherwin. 2005. Cultural transmission of tool use in bottlenose dolphins. *Proceedings of the National Academy of Sciences USA* 102: 8939–8943. [5]

Kuhnlein, H. V., O. Receveur, D. C. G. Muir, H. M. Chan and R. Soueida. 1995. Arctic indigenous women consume greater than acceptable levels of organochlorines. *Journal of Nutrition* 125: 2501–2510. [20]

Kurz, W. A. and 7 others. 2008. Mountain pine beetle and forest carbon feedback to climate change. *Nature* 452: 987–990. [10]

L

Lack, D. 1947. The significance of clutch size. *Ibis* 89: 302–352. [7]

Lafferty, K. D. 2009. The ecology of climate change and infectious diseases. *Ecology* 90: 888–900. [13]

Lambers, H., F. S. Chapin III and T. L. Pons. 1998. *Plant Physiological Ecology*. Springer, New York. [4, 5]

Lande, R. 1988. Demographic models of the northern spotted owl (*Strix occidentalis caurina*). *Oecologia* 75: 601–607. [10]

Landesman, W. J., B. F. Allan, R. B. Langerhans, T. M. Knight and J. M. Chase. 2007. Inter-annual associations between precipitation and human incidence of West Nile virus in the United States. *Vector-Borne and Zoonotic Diseases* 7: 337–343. [1]

Langerhans, R. B., M. E. Gifford and E. O. Joseph. 2007. Ecological speciation in *Gambusia* fishes. *Evolution* 61: 2056–2074. [6]

Larcher, W. 1980. *Physiological Plant Ecology*. Springer-Verlag, Berlin. [19]

Lauenroth, W. K. and O. E. Sala. 1992. Long-term forage production of North American shortgrass steppe. *Ecological Applications* 2: 397–403. [19]

Laurance, W. F. 2001. The hyper-diverse flora of the Central Amazon. In *Lessons from Amazonia: The Ecology and Conservation of a Fragmented Forest*, R. O. Bierregaard Jr., C. Gascon, T. E. Lovejoy and R. C. G. Mesquita (eds.), 47–53. Yale University Press, New Haven, CT. [17]

Laurance, W. F. and 10 others. 2002. Ecosystem decay of Amazonian forest fragments: A 22-year investigation. *Conservation Biology* 16: 605–618. [17, 23]

Law, R. 1975. Colonization and the evolution of life histories in *Poa annua*. Unpublished PhD thesis, University of Liverpool. [9]

Law, R., A. D. Bradshaw and P. D. Putwain. 1977. Life-history variation in *Poa annua*. *Evolution* 31: 233–246. [7]

Lawler, S. P. 1993. Species richness, species composition and population dynamics of protists in experimental microcosms. *Journal of Animal Ecology* 62: 711–719. [20]

Lawton, J. H. 2000. *Community Ecology in a Changing World*. Excellence in Ecology, O. Kinne (ed.), no. 11. Ecology Institute, Luhe, Germany. [18]

Layne, J. R. and R. E. Lee. 1995. Adaptations of frogs to survive freezing. *Climate Research* 5: 53–59. [4]

Lennartsson, T., P. Nilsson and J. Tuomi. 1998. Induction of overcompensation in the field gentian, *Gentianella campestris*. *Ecology* 79: 1061–1072. [12]

Levin, S. A. 1992. The problem of pattern and scale in ecology. *Ecology* 73: 1943–1967. [23]

Levine, J. M. 2000. Species diversity and biological invasions: Relating local process to community pattern. *Science* 288: 852–854. [18]

Levins, R. 1969. Some demographic and genetic consequences of environmental heterogeneity for biological control. *Bulletin of the Entomological Society of America* 15: 237–240. [10]

Levins, R. 1970. Extinction. In *Some Mathematical Problems in Biology*, M. Gerstenhaber (ed.), 75–107. American Mathematical Society, Providence, RI. [10]

Lewison, R. L., L. B. Crowder, A. J. Read and S. A. Freeman. 2004. Understanding impacts of fisheries bycatch on marine megafauna. *Trends in Ecology and Evolution* 19: 598–604. [22]

Lewontin, R. C. 1969. The meaning of stability. In *Diversity and Stability in Ecological Systems*, 13–24. Brookhaven Symposia in Biology, no. 22. Brookhaven National Laboratories, Brookhaven, NY. [16]

Likens, G. E. and F. H. Bormann. 1995. *Biogeochemistry of a Forested Ecosystem*. Springer-Verlag, New York. [21]

Lindeman, R. L. 1942. The trophic–dynamic aspect of ecology. *Ecology* 23: 399–418. [19]

Lindenmayer, D. B. and M. A. McCarthy. 2006. Evaluation of PVA models of arboreal marsupials: Coupling models with long-term monitoring data. *Biodiversity and Conservation* 15: 4079–4096. [22]

Lindqvist, C. and 13 others. 2010. Complete mitochondrial genome of a Pleistocene jawbone unveils the origin of polar bear. *Proceedings of the National Academy of Sciences USA* 107: 5053–5057. [8]

Ling, S. D. 2008. Range expansion of a habitat-modifying species leads to loss of taxonomic diversity: A new and impoverished reef state. *Oecologia* 156: 883–894. [8]

Ling, S. D., C. R. Johnson, S. D. Frusher and K. R. Ridgway. 2009. Over-fishing reduces resilience of kelp beds to climate-driven catastrophic phase shift. *Proceedings of the National Academy of Sciences USA* 106: 22341–22345. [8]

Lippincott, C. L. 2000. Effects of *Imperata cylindrica* (L.) Beauv. (cogongrass) invasion on fire regime in Florida sandhill. *Natural Areas Journal* 20: 140–149. [22]

Little, D. P. and D. W. Stevenson. 2007. A comparison of algorithms for the identification of specimens using DNA barcodes: Examples from gymnosperms. *Cladistics* 23: 1–21. [22]

Liu, H., E. S. Menges and P. F. Quintana-Ascencio. 2005. Population viability analyses of *Chamaecrista keyensis*: Effects of fire season and frequency. *Ecological Applications* 15: 210–221. [22]

Lively, C. M. 1989. Adaptation by a parasitic trematode to local populations of its snail host. *Evolution* 43: 1663–1671. [13]

Logan, J. A., and J. A. Powell. 2001. Ghost forest, global warming, and the mountain pine beetle (Coleoptera: Scolytidae). *American Entomologist* 47: 160–172. [23]

Lomolino, M. V., J. H. Brown and R. Davis. 1989. Island biogeography of montane forest mammals in the American Southwest. *Ecology* 70: 180–194. [17]

Loreau, M. and 11 others. 2001. Biodiversity and ecosystem functioning: Current knowledge and future challenges. *Science* 294: 804–808. [18]

Lotka, A. J. 1932. The growth of mixed populations: Two species competing for a common food supply. *Journal of the Washington Academy of Sciences* 22: 461–469. [11]

Louda, S. M., D. Kendall, J. Conner and D. Simberloff. 1997. Ecological effects of an insect introduced for the biological control of weeds. *Science* 277: 1088–1090. [8]

Lubchenco, J. 1978. Plant species diversity in a marine intertidal community: Importance of herbivore food preference and algal competitive abilities. *American Naturalist* 112: 23–39. [18]

M

MacArthur, R. H. 1955. Fluctuations in animal populations and a measure of community stability. *Ecology* 36: 533–536. [18]

MacArthur, R. H. 1958. Population ecology of some warblers of Northeastern coniferous forests. *Ecology* 39: 599–619. [18]

MacArthur, R. H. and J. W. MacArthur. 1961. On bird species diversity. *Ecology* 42: 594–598. [18]

MacArthur, R. H. and E. O. Wilson. 1963. An equilibrium theory of insular zoogeography. *Evolution* 17: 373–387. [17]

MacArthur, R. H. and E. O. Wilson. 1967. *The Theory of Island Biogeography*. Princeton University Press, Princeton, NJ. [7, 17]

MacDonald, I. A. W., L. L. Loope, M. B. Usher and O. Hamann. 1989. Wildlife conservation and the invasion of nature reserves by introduced species: A global perspective. In *Biological Invasions: A Global Perspective*, J. A. Drake, H. A. Mooney, F. di Castri, R. H. Groves, F. J. Kruger, M. Rejmanek and M. Williamson (eds.), 215–256. SCOPE (Scientific Committee on Problems of the Environment), 37. John Wiley & Sons, Chichester. [22]

Mack, R. N. and J. N. Thompson. 1982. Evolution in steppe with few large, hooved mammals. *American Naturalist* 119: 757–773. [21]

Macnair, M. R. and P. Christie. 1983. Reproductive isolation as a pleiotropic effect of copper tolerance in *Mimulus guttatus*. *Heredity* 50: 295–302. [6]

MacPhee, R. D. E. and P. A. Marx. 1997. Humans, hyperdisease, and first-contact extinctions. In *Natural Change and Human Impact in Madagascar*, S. M. Goodman and B. D. Patterson (eds.), 169–217. Smithsonian Institution Press, Washington, DC. [3]

Madigan, M. T. and J. M. Martinko. 2005. *Brock Biology of Microorganisms*. Prentice Hall, Upper Saddle River, NJ. [5]

Madronich, S., R. L. McKenzie, L. O. Björn and M. M. Caldwell. 1998. Changes in biologically active ultraviolet radiation reaching the Earth's surface. *Journal of Photochemistry and Photobiology* B: *Biology* 46: 5–19. [24]

Magurran, A. E. 2004. *Measuring Biological Diversity*. Blackwell Science Publishing, Malden, MA. [15]

Mahowald, N. and 8 others. 2005. The atmospheric global dust cycle and iron inputs to the ocean. *Global Biogeochemical Cycles* 19: GB4025 10.1029/2004GB002402. [24]

Maitland, D. P. 1994. A parasitic fungus infecting yellow dungflies manipulates host perching behaviour. *Proceedings of the Royal Society of London* B 258: 187–193. [13]

Mantua, N. J. 2001. The Pacific Decadal Oscillation. In *The Encyclopedia of Global Environmental Change*, vol. 1, *The Earth System: Physical and Chemical Dimensions of Global Environmental Change*, M. C. McCracken and J. S. Perry (eds.), 592–594. Wiley, New York. [2]

Mantua, N. J. and S. R. Hare. 2002. The Pacific Decadal Oscillation. *Journal of Oceanography* 58: 35–44. [2]

Mantua, N. J., S. R. Hare, Y. Zhang, J. M. Wallace and R. C. Francis. 1997. A Pacific interdecadal climate oscillation with impacts on salmon production. *Bulletin of the American Meteorological Society* 78: 1069–1079. [2]

Margoluis, R. and N. Salafsky. 1998. *Measures of Success: Designing, Managing, and Monitoring Conservation and Development Projects*. Island Press, Washington, DC. [23]

Maron, J. L. and M. Vila. 2001. When do herbivores affect plant invasion? Evidence for the natural enemies and biotic resistance hypotheses. *Oikos* 95: 361–373. [18]

Marr, J. W. 1967. *Ecosystems of the East Slope of the Front Range of Colorado*. University of Colorado Studies, Series in Biology, no. 8. University of Colorado Press, Boulder. [3]

Marschner, H. 1995. *Mineral Nutrition of Higher Plants*. Academic Press, San Diego, CA. [21]

Marti, M., R. T. Good, M. Rug, E. Knuepfer and A. F. Cowman. 2004. Targeting malaria virulence and remodeling proteins to the host erythrocyte. *Science* 306: 1930–1933. [13]

Martin, A. P. and S. R. Palumbi. 1993. Body size, metabolic rate, generation time, and the molecular clock. *Proceedings of the National Academy of Sciences USA* 90: 4087–4091. [17]

Martin, J. A., B. E. Hamilton, P. D. Sutton, S. J. Ventura, F. Menacker, S. Kirmeyer and T. J. Mathews. 2009. Births: Final data for 2006. National Vital Statistics Reports, Vol. 57 no. 7. Hyattsville, Maryland: National Center for Heath Statistics. http://www.cdc.gov/nchs/data/nvsr/nvsr57/nvsr57_07.pdf. [9]

Martin, J. H. and 43 others. 1994. Testing the iron hypothesis in ecosystems of the equatorial Pacific Ocean. *Nature* 371: 123–129. [19]

Martin, P. S. 1984. Prehistoric overkill: The global model. In *Quaternary Extinctions: A Prehistoric Revolution*, P. S. Martin and R. G. Klein (eds.), 354–403. University of Arizona Press, Tucson. [3]

Martin, P. S. 2005. *Twilight of the Mammoths: Ice Age Extinctions and the Rewilding of America*. University of California Press, Berkeley. [3]

Maslin, M. A. and E. Thomas. 2003. Balancing the deglacial global carbon budget: The hydrate factor. *Quaternary Science Reviews* 22: 1729–1736. [5]

Mattson, W. J. Jr. 1980. Herbivory in relation to plant nitrogen content. *Annual Review of Ecology and Systematics* 11: 119–161. [12]

Mauseth, J. D. 1988. *Plant Anatomy*. Benjamin/Cummings, Menlo Park, CA. [14]

May, R. M. 1973. *Stability and Complexity in Model Ecosystems*. Princeton University Press, Princeton, NJ. [20]

May, R. M. 1976. Models for single populations. In *Theoretical Ecology: Principles and Applications*, R. M. May (ed.), 4–25. W. B. Saunders, Philadelphia. [10]

May, R. M. 2006. Threats to tomorrow's world. *Notes and Records of the Royal Society* 60: 109–130. [22]

May, R. M. and R. M. Anderson. 1983. Parasite–host coevolution. In *Coevolution*, D. J. Futuyma and M. Slatkin (eds.), 186–206. Sinauer Associates, Sunderland, MA. [13]

May, R. M., J. H. Lawton and N. E. Stork. 1995. Assessing extinction rates. In *Extinction Rates*, J. H. Lawton and R. M. May (eds.), 1–24. Oxford University Press, Oxford. [22]

McCaig, A. E., L. A. Glover and J. I. Prosser. 1999. Molecular analysis of bacterial community structure and diversity in unimproved and improved upland grass pastures. *Applied and Environmental Microbiology* 65: 1721–1730. [15]

McCann, K., A. Hastings and G. R. Huxel. 1998. Weak trophic interactions and the balance of nature. *Nature* 395: 794–798. [20]

McClain, M. E. and 11 others. 2003. Biogeochemical hot spots and hot moments at the interface of terrestrial and aquatic ecosystems. *Ecosystems* 6: 301–312. [23]

McClenachan, L. 2009. Documenting the loss of large trophy fish from the Florida Keys with historical photographs. *Conservation Biology* 23: 636–643. [22]

McCracken, K. G. and 9 others. 2009. Parallel evolution in the major haemoglobin genes of eight species of Andean waterfowl. *Molecular Ecology* 18: 3992–4005. [6]

McEvoy, P., C. Cox and E. Coombs. 1991. Successful biological control of ragwort, *Senecio jacobaea*, by introduced insects in Oregon. *Ecological Applications* 1: 430–442. [11]

McGraw, W. S. 2005. Update on the search for Miss Waldron's red colobus monkey. *International Journal of Primatology* 26: 605–619. [22]

McKane, R. B. and 10 others. 2002. Resource-based niches provide a basis for plant species diversity and dominance in Arctic tundra. *Nature* 415: 68–71. [21]

McKinney, S. T., C. E. Fiedler, and D. F. Tomback. 2009. Invasive pathogen threatens bird-pine mutualism: implications for sustaining a high-elevation ecosystem. *Ecological Applications* 19: 597–607. [23]

McMahon, T. E. and J. C. Tash. 1988. Experimental analysis of the role of emigration in population regulation of desert pupfish. *Ecology* 69: 1871–1883. [8]

McNeill, W. H. 1976. *Plagues and Peoples*. Anchor Press/Doubleday, Garden City, New York. [13]

McNeilly, T. 1968. Evolution in closely adjacent plant populations. III. *Agrostis tenuis* on a small copper mine. *Heredity* 23: 99–108. [6]

Meine, C., M. Soulé and R. F. Noss. 2006. "A mission-driven discipline": The growth of conservation biology. *Conservation Biology* 20: 631–651. [22]

Meinesz, A. 2001. *Killer Algae*. University of Chicago Press, Chicago. [15]

Meire, P. M. and A. Ervynck. 1986. Are oystercatchers (*Haematopus ostralegus*) selecting the most profitable mussels (*Mytilus edulis*)? *Animal Behaviour* 34: 1427–1435. [5]

Melillo, J. M., J. D. Aber and J. F. Muratore. 1982. Nitrogen and lignin control of hardwood leaf litter decomposition dynamics. *Ecology* 63: 621–626. [21]

Mellars, P. 1989. Major issues in the emergence of modern humans. *Current Anthropology* 30: 349–385. [5]

Memmott, J., N. M. Waser and M. V. Price. 2004. Tolerance of pollination networks to species extinctions. *Proceedings of the Royal Society of London* B 271: 2605–2611. [22]

Menge, B. A. 1995. Indirect effects in marine rocky intertidal interaction webs: Patterns and importance. *Ecological Monographs* 65: 21–74. [15]

Menge, B. A. and J. P. Sutherland. 1987. Community regulation: Variation in disturbance, competition, and predation in relation to environmental stress and recruitment. *American Naturalist* 130: 730–757. [18]

Menge, B. A., E. L. Berlow, C. A. Blanchette, S. A. Navarette and S. B. Yamada. 1994. The keystone concept—Variation in interaction strength in a rocky intertidal habitat. *Ecological Monographs* 64: 249–286. [20]

Menge, B. A., B. Daley and P. A. Wheeler. 1996. Control of interaction strength in marine benthic communities. In *Food Webs: Integration of Patterns and Dynamics*, G. A. Polis and K. O. Winemiller (eds.), 258–274. Chapman and Hall, New York. [15]

Merrick, R. L., T. R. Loughlin and D. G. Calkins. 1987. Decline in abundance of the northern sea lion, *Eumetopias jubatus*, in Alaska, 1956–86. *Fishery Bulletin* 85: 351–365. [20]

Miller, G. T. 2007. *Living in the Environment: Principles, Connections, and Solutions.* Thomson Brooks/Cole, Belmont, CA. [9]

Millennium Ecosystem Assessment. 2005. *Ecosystems and Human Well-being: Biodiversity Synthesis.* World Resources Institute, Washington, DC. [17, 18, 22]

Minorsky, P. V. 2002. Allelopathy and grain crop production. *Plant Physiology* 130: 1745–1746. [11]

Mittelbach, G. G. and 21 others. 2007. Evolution and the latitudinal diversity gradient: Speciation, extinction and biogeography. *Ecology Letters* 10: 315–331. [17]

Mokany, K., R. J. Raison and A. S. Prokushkin. 2006. Critical analysis of root:shoot ratios in terrestrial biomes. *Global Change Biology* 12: 84–96. [4]

Molina, M. J. and F. S. Rowland. 1974. Stratospheric sink for chlorofluoromethanes: Chlorine atom-catalysed destruction of ozone. *Nature* 249: 810–812. [24]

Molino, J.-F. and D. Sabatier. 2001. Tree diversity in tropical rain forests: A validation of the intermediate disturbance hypothesis. *Science* 294: 1702–1704. [18]

Moore, J. A. 1957. An embryologist's view of the species concept. In *The Species Problem,* E. Mayr (ed.), 325–388. American Association for the Advancement of Science, Washington DC. [6]

Moore, S. E., T. J. Cole, E. M. E. Poskitt, B. J. Sonko, R. G. Whitehead, I. A. McGregor and A. M. Prentice. 1997. Season of birth predicts mortality in rural Gambia. *Nature* 388: 434. [9]

Morin, X., C. Augspurger and I. Chuine. 2007. Process-based modeling of species' distributions: What limits temperate tree species' range boundaries? *Ecology* 88: 2280–2291. [4]

Morris, W. F. and D. F. Doak. 2002. *Quantitative Conservation Biology: Theory and Practice of Population Viability.* Sinauer Associates, Sunderland, MA. [10, 22]

Mouritsen, K. N. and R. Poulin. 2002. Parasitism, community structure and biodiversity in intertidal ecosystems. *Parasitology* 124: S101–S117. [13]

Mouritsen, K. N., L. T. Mouritsen and K. T. Jensen. 1998. Changes of topography and sediment characteristics on an intertidal mud-flat following mass-mortality of the amphipod *Corophium volutator. Journal of the Marine Biological Association of the United Kingdom* 78: 1167–1180. [13]

Moutinho, P., C. D. Napstad and E. A. Davidson. 2003. Influence of leaf-cutting ant nests on secondary forest growth and soil properties in Amazonia. *Ecology* 84: 1265–1276. [14]

Muir, J. 1915. *Travels in Alaska.* Houghton Mifflin, Boston. [16]

Munger, J. 1984. Optimal foraging? Patch use by horned lizards (Iguanidae: *Phrynosoma*). *American Naturalist* 123: 654–680. [5]

Murdoch, W. W., S. Avery and M. E. B Smith. 1975. Switching in a predatory fish. *Ecology* 56: 1094–1105. [12]

Murphy, D. D. and C. M. Knopp. 2000. *Lake Tahoe Watershed Assessment.* Vol. 1. General Technical Report PSW-GTR-175. Pacific Southwest Research Station, Forest Service, U.S. Department of Agriculture, Albany, CA. [21]

Myers, J. H. and D. Bazely. 1991. Thorns, spines, prickles, and hairs: Are they stimulated by herbivory and do they deter herbivores? In *Phytochemical Induction by Herbivores,* D. W. Tallamy and M. J. Raupp (eds.), 325–344. Wiley, New York. [12]

Myers, R. A. and B. Worm. 2003. Rapid worldwide depletion of predatory fish communities. *Nature* 423: 280–283. [22]

N

Nabhan, G. P. and A. R. Holdsworth. 1998. *State of the Sonoran Desert Biome: Uniqueness, Biodiversity, Threats and the Adequacy of Protection in the Sonoran Bioregion.* The Wildlands Project, Tucson, AZ. [3]

Naeem, S. L., J. Thompson, S. P. Lawler, J. H. Lawton and R. M. Woodfin. 1994. Declining biodiversity can alter the performance of ecosystems. *Nature* 368: 734–736. [18]

Nager, R. G., P. Monaghan and D. Houston. 2000. Within-clutch trade-offs between the number and quality of eggs: Experimental manipulations in gulls. *Ecology* 81: 1339–1350. [7]

Naiman, R. J., C. A. Johnston and J. C. Kelley. 1988. Alterations of North American streams by beaver. *BioScience* 38: 753–762. [15]

National Research Council. 2002. Effects of trawling and dredging on seafloor habitat. Committee on Ecosystem Effects of Fishing: Phase I—Effects of Bottom Trawling on Seafloor Habitats, Ocean Studies Board, Division on Earth and Life Studies, National Research Council. National Academy of Sciences Press, Washington, DC. [22]

Naughton-Treves, L., J. L. Mena, A. Treves, N. Alvarez and V. C. Radeloff. 2003. Wildlife survival beyond park boundaries: The impact of slash-and-burn agriculture and hunting on mammals in Tambopata, Peru. *Conservation Biology* 17: 1106–1117. [23]

Nee, S., N. Colegrave, S. A. West and A. Grafen. 2005. The illusion of invariant quantities in life histories. *Nature* 309: 1236–1239. [7]

Neff, J. C., R. L. Reynolds, J. Belnap and P. Lamothe. 2005. Multi-decadal impacts of grazing on soil physical and biogeochemical properties in southeastern Utah. *Ecological Applications* 15: 87–95. [21]

Nemani, R. R. and 7 others. 2003. Climate-driven increases in global terrestrial net primary production from 1982 to 1999. *Science* 300: 1560–1563. [24]

Neutel, A.-M., J. A. P. Heesterbeek and P. C. de Ruiter. 2002. Stability in real food webs: Weak links in long loops. *Science* 296: 1120–1123. [20]

Newbold, J. D., J. W. Elwood, R. V. O'Neill and A. L. Sheldon. 1983. Phosphorus dynamics in a woodland stream ecosystem: A study of nutrient spiralling. *Ecology* 64: 1249–1265. [21]

Nicholson, A. J. 1957. The self-adjustment of populations to change. *Cold Spring Harbor Symposia on Quantitative Biology* 22: 153–173. [10]

Norby, R. J. and 13 others. 2005. Forest response to elevated CO_2 is conserved across a broad range of productivity. *Proceedings of the National Academy of Sciences USA* 102: 18052–18056. [24]

Noss, R., E. Dinerstein, B. Gilbert, M. Gilpin, B. J. Miller, J. Terborgh and S. Trombulak. 1999. Core areas: Where Nature reigns. In *Continental Conservation: Scientific Foundations of Regional Reserve Networks,* M. E. Soulé and J. Terborgh (eds.), 99–128. Island Press, Washington, DC. [23]

O

Oates, J. F., M. Abedi-Lartey, W. S. McGraw, T. T. Struhsaker; and G. H. Whitesides. 2000. Extinction of a West African red colobus monkey. *Conservation Biology* 14: 1526–1532. [22]

O'Brien, E. L., A. E. Burger and R. D. Dawson. 2005. Foraging decision rules and prey species preferences of Northwestern Crows (*Corvus caurinus*). *Ethology* 111: 77–87. [5]

Odum, H. T. and C. F. Jordan. 1970. Metabolism and evapotranspiration of the lower forest in a giant plastic cylinder. In *A Tropical Rain Forest,* H. T. Odum and R. F. Pigeon (eds.), 165–

189. Division of Technical Information, U.S. Atomic Energy Commission. [19]

Oechel, W. C., S. J. Hastings, G. Vourlitis, M. Jenkins, G. Riechers and N. Grulke. 1993. Recent change of Arctic tundra ecosystems from a net carbon dioxide sink to a source. *Nature* 361: 520–523. [24]

Oechel, W. C., G. L. Vourlitis, S. J. Hastings, R. C. Zulueta, L. Hinzman and D. Kane. 2000. Acclimation of ecosystem CO_2 exchange in the Alaskan Arctic in response to decadal climate warming. *Nature* 406: 978–981. [24]

Ogren, W. L. 1984. Photorespiration: Pathways, regulation, and modification. *Annual Review of Plant Physiology* 35: 415–442. [5]

Okin, G. S., N. Mahowald, O. A. Chadwick and P. Artaxo. 2004. Impact of desert dust on the biogeochemistry of phosphorus in terrestrial ecosystems. *Global Biogeochemical Cycles* 18: GB2005. [24]

Olden, J. D., N. L. Poff, M. R. Douglas, M. E. Douglas and K. D. Fausch. 2004. Ecological and evolutionary consequences of biotic homogenization. *Trends in Ecology and Evolution* 19: 18–24. [22]

Olsen, E. M., M. Heino, G. R. Lilly, M. J. Morgan, J. Brattey, B. Ernande and U. Dieckmann. 2004. Maturation trends indicative of rapid evolution preceded the collapse of northern cod. *Nature* 428: 932–935. [7]

Orr, J. C. and 26 others. 2005. Anthropogenic ocean acidification over the twenty-first century and its impact on calcifying organisms. *Nature* 437: 681–686. [2, 3, 24]

Osmond, C. B., K. Winter and H. Ziegler. 1982. Functional significance of different pathways of photosynthesis. In *Physiological plant ecology*, vol. 2, O. L. Lange, P. S. Nobel, C. B. Osmond and H. Ziegler (eds.), 479–547. Encyclopedia of Plant Physiology, new series, vol. 12B. Springer Verlag, New York. [5]

Ostfeld, R. S. 2009. Climate change and the distribution and intensity of infectious diseases. *Ecology* 90: 903–905. [13]

Overpeck, J. T., R. S. Webb and T. Webb. 1992. Mapping Eastern North American vegetation change of the past 18,000 years: No-analogs and the future. *Geology* 20: 1071–1074. [24]

Owen-Smith, N. 1987. Pleistocene extinctions: The pivotal role of megaherbivores. *Paleobiology* 13: 351–362. [3]

P

Pace, M. L. and 8 others. 2004. Whole-lake carbon-13 additions reveal terrestrial support of aquatic food webs. *Nature* 427: 240–243. [20]

Packer, C., A. E. Pusey, H. Rowley, D. A. Gilbert, J. Martenson and S. J. O'Brien. 1991. Case study of a population bottleneck: Lions of the Ngorongoro Crater. *Conservation Biology* 5: 219–230. [10]

Paine, R. T. 1966. Food web complexity and species diversity. *American Naturalist* 100: 65–75. [18, 20]

Paine, R. T. 1974. Intertidal community structure: Experimental studies on the relations between a dominant competitor and its principal predator. *Oecologia* 15: 93–120. [12]

Paine, R. T. 1979. Disaster, catastrophe, and local persistence of the sea palm, *Postelsia palmaeformis*. *Science* 205: 685–687. [11]

Paine, R. T., J. C. Castillo and J. Cancino. 1985. Perturbation and recovery patterns of starfish-dominated intertidal assemblages in Chile, New Zealand, and Washington State. *American Naturalist* 125: 679–691. [20]

Palumbi, S. R. 2001. Humans as the world's greatest evolutionary force. *Science* 293: 1786–1790. [6]

Pardini, R. 2004. Effects of forest fragmentation on small mammals in an Atlantic Forest landscape. *Biodiversity and Conservation* 13: 2567–2586. [23]

Park, T. 1948. Experimental studies of interspecies competition. I. Competition between populations of the flour beetles, *Tribolium confusum* Duvall and *Tribolium castaneum* Herbst. *Ecological Monographs* 18: 267–307. [13]

Parmenter, A. W. and 8 others. 2003. Land use and land cover change in the Greater Yellowstone Ecosystem: 1975–1995. *Ecological Applications* 13: 687–703. [23]

Parmesan, C. 2006. Ecological and evolutionary responses to recent climate change. *Annual Review of Ecology, Evolution, and Systematics* 37: 637–669. [1, 6, 22]

Parmesan, C. and G. Yohe. 2003. A globally coherent fingerprint of climate change impacts across natural systems. *Nature* 421: 37–42. [8, 24]

Parmesan, C. and 12 others. 1999. Poleward shifts in geographical ranges of butterfly species associated with regional warming. *Nature* 399: 579–583. [24]

Pastor, J., R. J. Naiman, B. Dewey and P. Mcinnes. 1988. Moose, microbes, and the boreal forest. *BioScience* 38: 770–777. [23]

Paton, P. W. C. 1994. The effect of edge on avian nest success: How strong is the evidence? *Conservation Biology* 8: 17–26. [23]

Patterson, B. D. 1980. Montane mammalian biogeography in New Mexico. *Southwestern Naturalist* 25: 33–40. [11]

Patterson, B. D. 1981. Morphological shifts of some isolated populations of *Eutamias* (Rodentia: Sciuridae) in different congeneric assemblages. *Evolution* 35: 53–66. [11]

Paul, E. A. and F. E. Clark. 1996. *Soil Microbiology and Biochemistry*. Academic Press, San Diego, CA. [21]

Paul, N. D. and D. Gwynn-Jones. 2003. Ecological roles of solar UV radiation: Towards an integrated approach. *Trends in Ecology and Evolution* 18: 48–55. [24]

Paulay, G. 1997. Diversity and distribution of reef organisms. In *Life and Death of Coral Reefs*, C. Birkeland (ed.), 298–353. Chapman & Hall, New York. [3]

Pearce, F. 1997. Why is the apparently pristine Arctic full of toxic chemicals that started off thousands of kilometres away? *New Scientist* 154: 24–27. [20]

Pearcy, R. W. 1977. Acclimation of photosynthetic and respiratory carbon dioxide exchange to growth temperature in *Atriplex lentiformis* (Torr) Wats. *Plant Physiology* 59: 795–799. [5]

Pearl, R. and L. J. Reed. 1920. On the rate of growth of the population of the United States since 1790 and its mathematical representation. *Proceedings of the National Academy of Sciences USA* 6: 275–288. [9]

Peek, A. S., R. A. Feldman, R. A. Lutz and R. C. Vrijenhoek. 1998. Cospeciation of chemoautotrophic bacteria and deep sea clams. *Proceedings of the National Academy of Sciences USA* 95: 9962–9966. [19]

Pellmyr, O. and C. J. Huth. 1994. Evolutionary stability of mutualism between yuccas and yucca moths. *Nature* 372: 257–260. [14]

Pellmyr, O., C. C. Labandeira and C. M. Herrera, eds. 2002. *Plant–Animal Interactions: An Evolutionary Approach*. Blackwell Science Publishing, Malden, MA. [12]

Perez, J. M., J. E. Granados, R. C. Soriguer, P. Fandos, F. J. Marquez and J. P. Crampe. 2002. Distribution, status and conservation problems of the Spanish Ibex, *Capra pyrenaica* (Mammalia: Artiodactyla). *Mammal Review* 32: 26–39. [22]

Peterson, A. T. 2009. Shifting suitability for malaria vectors across Africa with warming climates. *BMC Infectious Diseases* 9: 59. [13]

Peterson, B. J. and B. Fry. 1987. Stable isotopes in ecological studies. *Annual Review of Ecology and Systematics* 18: 293–320. [19]

Peterson, G., C. R. Allen and C. S. Holling. 1998. Ecological resilience, biodiversity, and scale. *Ecosystems* 1: 6–8. [18]

Pfennig, D. W. 1992. Proximate and functional causes of polyphenism in an anuran tadpole. *Functional Ecology* 6: 167–174. [7]

Pfennig, D. W., A. M. Rice and R. A. Martin. 2007. Field and experimental evidence for competition's role in phenotypic divergence. *Evolution* 61: 257–271. [11]

Pickett, S. T. A. 1989. Space-for-time substitution as an alternative to long-term studies. In *Long-term Studies in Ecology: Approaches and Alternatives*, G. E. Likens (ed.), 110–135. Springer-Verlag, New York. [16]

Pimentel, D., E. H. Feinberg, P. W. Wood and J. T. Hayes. 1965. Selection, spatial distribution, and the coexistence of competing fly species. *American Naturalist* 99: 97–109. [11]

Pimentel, D., L. Lach, R. Zuniga and D. Morrison. 2000. Environmental and economic costs of nonindigenous species in the United States. *BioScience* 50: 53–65. [15]

Pimm, S. L. 2002. *Food Webs*. University of Chicago Press, Chicago. [20]

Pimm, S. L. and J. H. Lawton. 1977. The number of trophic levels in ecological communities. *Nature* 275: 542–544. [20]

Pimm, S. L., H. L. Jones and J. Diamond. 1988. On the risk of extinction. *American Naturalist* 132: 757–785. [10]

Pinder, A. W., K. B. Storey and G. R. Ultsch. 1992. Estivation and hibernation. In *Environmental Physiology of the Amphibia*, M. E. Feder and W. W. Burggren (eds.), 250–274. University of Chicago Press, Chicago. [4]

Pinto-Tomás, A. A. and 7 others. 2009. Symbiotic nitrogen fixation in the fungus gardens of leaf-cutter ants. *Science* 326: 1120–1123. [14]

Poethke, H. J., W. W. Weisser and T. Hovestadt. 2010. Predator-induced dispersal and the evolution of conditional dispersal in correlated environments. *American Naturalist* 175: 577–586. [8]

Polis, G. A. 1991. Complex trophic interactions in deserts: An empirical critique of food web theory. *American Naturalist* 138: 123–155. [20]

Polis, G. A., M. E. Power and G. R. Huxel. 2004. *Food Webs at the Landscape Level*. University of Chicago Press, Chicago. [20]

Pollock, M. M., R. J. Naiman and T. A. Hanley. 1998. Plant species richness in riparian wetlands–a test of biodiversity theory. *Ecology* 79: 94–105. [18]

Post, D. M. 2002a. Using stable isotopes to estimate trophic position: Models, methods, and assumptions. *Ecology* 83: 703–718. [20]

Post, D. M. 2002b. The long and short of food-chain length. *Trends in Ecology and Evolution* 17: 269–277. [20]

Post, D. M. and G. Takimoto. 2007. Proximate structural mechanisms for variation in food-chain length. *Oikos* 116: 775–782. [20]

Pounds, J. A. and 13 others. 2006. Widespread amphibian extinctions from epidemic disease driven by global warming. *Nature* 439: 161–167. [1]

Power, M. E., M. S. Parker and J. T. Wootton. 1996a. Disturbance and food chain length in rivers. In *Food Webs: Integration of Patterns and Dynamics*, G. A. Polis and K. O. Winemiller (eds.), 286–297. Chapman and Hall, New York. [15]

Power, M. E. and 9 others. 1996b. Challenges in the quest for keystones. *BioScience* 46: 609–620. [15]

Power, M. E., M. S. Parker and W. E. Dietrich. 2008. Seasonal reassembly of a river food web: Floods, droughts, and impacts on fish. *Ecological Monograph* 78: 263–282. [15]

Primack, R. B. and S. L. Miao. 1992. Dispersal can limit local plant distribution. *Conservation Biology* 6: 513–519. [8]

Prugh, L. R., K. E. Hodges, A. R. E. Sinclair and J. S. Brashares. 2008. Effect of habitat area and isolation on fragmented animal populations. *Proceedings of the National Academy of Sciences USA* 105: 20770–20775. [23]

Pyke, G. H., H. R. Pulliam and E. L. Charnov. 1977. Optimal foraging: A selective review of theory and tests. *Quarterly Review of Biology* 52: 137–154. [5]

R

Raab, T. K., D. A. Lipson and R. K. Monson. 1996. Non-mycorrhizal uptake of amino acids by roots of the alpine sedge *Kobresia myosuroides*: Implications for the alpine nitrogen cycle. *Oecologia* 108: 488–494. [21]

Rabinowitz, D., S. Cairns and T. Dillon. 1986. Seven forms of rarity and their frequency in the flora of the British Isles. In *Conservation Biology: The Science of Scarcity and Diversity*, M. E. Soulé (ed.), 182–204. Sinauer Associates, Sunderland, MA. [22]

Rahel, F. J. 2000. Homogenization of fish faunas across the United States. *Science* 288: 854–856. [22]

Ralls, K. and J. D. Ballou. 2004. Genetic status and management of California Condors. *Condor* 106: 215–228. [22]

Ranta, P., T. Blom, J. Niemela, E. Joensuu and M. Siitonen. 1998. The fragmented Atlantic rain forest of Brazil: Size, shape and distribution of forest fragments. *Biodiversity and Conservation* 7: 385–403. [22]

Raxworthy, C. J., E. Martinez-Meyer, N. Horning, R. A. Nussbaum, G. E. Schneider, M. A. Ortega-Huerta and A. T. Peterson. 2003. Predicting distributions of known and unknown reptile species in Madagascar. *Nature* 426: 837–841. [8]

Raymond, M., C. Chevillon, T. Guillemaud, T. Lenormand and N. Pasteur. 1998. An overview of the evolution of overproduced esterases in the mosquito *Culex pipiens*. *Philosophical Transactions of the Royal Society of London* B 353: 1707–1711. [6]

Réale, D., A. G. McAdam, S. Boutin and D. Berteaux. 2003. Genetic and plastic responses of a northern mammal to climate change. *Proceedings of the Royal Society of London* B 270: 591–596. [6]

Redford, K. H. 1992. The empty forest. *BioScience* 42: 412–422. [22]

Redman, R. S., K. B. Sheehan, R. G. Stout, R. J. Rodriguez and J. M. Henson. 2002. Thermotolerance generated by plant/fungal symbiosis. *Science* 298: 1581. [14]

Rees, W. E. 1992. Ecological footprints and appropriated carrying capacity: What urban economics leaves out. *Environment and Urbanization* 4: 121–130. [9]

Reiners, W. A., I. A. Worley and D. B. Lawrence. 1971. Plant diversity in a chronosequence at Glacier Bay, Alaska. *Ecology* 52: 55–69. [16]

Reisen, W. K., Y. Fang and V. M. Martinez. 2006. Effects of temperature on the transmission of West Nile virus by *Culex tarsalis* (Diptera; Culicidae). *Journal of Medical Entomology* 43: 309–317. [1]

Relyea, R. A. 2003. Predator cues and pesticides: A double dose of danger for amphibians. *Ecological Applications* 13: 1515–1521. [1]

Reusch, T. B. H., M. A. Häberli, P. B. Aeschlimann and M. Milinski. 2001. Female sticklebacks count alleles in a strategy of sexual selection explaining MHC polymorphism. *Nature* 414: 300–302. [13]

Reznick, D. N., M. J. Bryant, D. Roff, C. K. Ghalambor and D. E. Ghalambor. 2004. Effect of mortality on the evolution of senescence in guppies. *Nature* 431: 1095–1099. [7]

Richardson, H. and N. A. M. Verbeek. 1986. Diet selection and optimization by northwestern crows feeding on Japanese littleneck clams. *Ecology* 67: 1219–1226. [5]

Rieseberg, L. H. and 9 others. 2003. Major ecological transitions in wild sunflowers facilitated by hybridization. *Science* 301: 1211–1216. [6]

Riley, S. P. D., H. B. Shaffer, S. R. Voss and B. M. Fitzpatrick. 2003. Hybridization between a rare, native tiger salamander and its introduced congener. *Ecological Applications* 13: 1263–1275. [22]

Ripple, W. J. and R. L. Beschta. 2004. Wolves and the ecology of fear: Can predation risk structure ecosystems? *BioScience* 54: 755–766. [23]

Ripple, W. J. and R. L. Beschta. 2007. Restoring Yellowstone's aspen with wolves. *Biological Conservation* 138: 514–519. [23]

Rizki, R. M. and T. M. Rizki. 1990. Parasitoid virus-like particles destroy *Drosophila* cellular immunity. *Proceedings of the National Academy of Sciences USA* 87: 8388–8392. [13]

Robertson, G. P., M. A. Huston, F. C. Evans and J. M. Tiedje. 1988. Spatial variability in a successional plant community: Patterns of nitrogen availability. *Ecology* 69: 1517–1524. [18]

Rodriguez, R. J., J. F. White Jr., A. E. Arnold and R. S. Redman. 2009. Fungal endophytes: Diversity and functional roles. *New Phytologist* 182: 314–330. [14]

Rohde, K. 1978. Latitudinal gradients in species-diversity and their causes. I. A review of the hypotheses explaining the gradients. *Biologisches Zentralblatt* 97: 393–403. [17]

Rohr, J. R. and T. R. Raffel. 2010. Linking global climate and temperature variability to widespread amphibian declines putatively caused by disease. *Proceedings of the National Academy of Sciences USA* 107: 8269–8274. [1, 22]

Rohr, J. R. and 11 others. 2008. Agrochemicals increase trematode infections in a declining amphibian species. *Nature* 455: 1235–1239. [1]

Rojstaczer, S., S. M. Sterling and N. J. Moore. 2001. Human appropriation of photosynthesis products. *Science* 294: 2549–2552. [22]

Root, R. B. and N. Cappuccino. 1992. Patterns in population change and the organization of the insect community associated with goldenrod. *Ecological Monographs* 62: 393–420. [8]

Root, T. L., J. T. Price, K. R. Hall, S. H. Schneider, C. Rosenzweig and J. A. Pounds. 2003. Fingerprints of global warming on wild animals and plants. *Nature* 421: 57–60. [24]

Rosenzweig, M. L. 1992. Species diversity gradients: We know more and less than we thought. *Journal of Mammology* 73: 715–730. [17]

Rosenzweig, M. L. and R. H. MacArthur. 1963. Graphical representation and stability conditions of predator–prey interactions. *American Naturalist* 97: 209–223. [12]

Ross, P. S. 2006. Fireproof killer whales (*Orcinus orca*): Flame-retardant chemicals and the conservation imperative in the charismatic icon of British Columbia, Canada. *Canadian Journal of Fisheries and Aquatic Sciences* 63: 224–234. [22]

Rovira, A. D., C. D. Bowen and R. C. Foster. 1983. The significance of rhizosphere microflora and mycorrhizas in plant nutrition. In *Inorganic Plant Nutrition* (*Encyclopedia of Plant Physiology*, new series, Vol. 15B), A. Läuchli and R. L. Bieleski (eds.), 61–93. Springer-Verlag, Berlin. [14]

Rutz, C. and 7 others. 2010. The ecological significance of tool use in New Caledonian crows. *Science* 329: 1523–1526. [5]

S

Sale, P. F. 1977. Maintenance of high diversity in coral reef fish communities. *American Naturalist* 111: 337–359. [18]

Sale, P. F. 1979. Recruitment, loss and coexistence in a guild of territorial coral reef fishes. *Oecologia* 42: 159–177. [18]

Salisbury, F. B. and C. Ross. 1992. *Plant Physiology*. 4th ed. Wadsworth, Belmont, CA. [21]

Sanderson, E. W., M. Jaiteh, M. A. Levy, K. H. Redford, A. V. Wannebo and G. Woolmer. 2002. The human footprint and the last of the wild. *BioScience* 52: 891–904. [3, 22]

Sandquist, D. R. and J. R. Ehleringer. 2003. Population- and family-level variation of brittlebush (*Encelia farinosa*, Asteraceae) pubescence: Its relation to drought and implications for selection in variable environments. *American Journal of Botany* 90: 1481–1486. [4]

Saugier, B., J. Roy and H. A. Mooney. 2001. Estimations of global terrestrial productivity: Converging toward a single number? In *Terrestrial Global Productivity*, J. Roy, B. Saugier and H. A. Mooney (eds.), 543–557. Academic Press, San Diego. [19]

Sax, D. F. and S. D. Gaines. 2003. Species diversity: From global decreases to local increases. *Trends in Ecology and Evolution* 18: 561–566. [22]

Schall, J. J. 1992. Parasite-mediated competition in *Anolis* lizards. *Oecologia* 92: 58–64. [13]

Schimel, J. P. and J. Bennett. 2004. Nitrogen mineralization: Challenges of a changing paradigm. *Ecology* 85: 591–602. [21]

Schindler, D. W. 1974. Eutrophication and recovery in experimental lakes: Implications for lake management. *Science* 184: 897–899. [19]

Schipper, J. and 132 others. 2008. The status of the world's land and marine mammals: Diversity, threats, and knowledge. *Science* 322: 225–230. [22]

Schlesinger, W. H. 1997. *Biogeochemistry: An Analysis of Global Change*. 2nd ed. Academic Press, San Diego, CA. [21, 24]

Schluter, D. 1994. Experimental evidence that competition promotes divergence in adaptive radiation. *Science* 266: 798–801. [11]

Schluter, D. 1998. Ecological causes of speciation. In *Endless Forms: Species and Speciation*, D. J. Howard and S. H. Berlocher (eds.), 114–129. Oxford University Press, Oxford. [6]

Schluter, D., T. D. Price and P. R. Grant. 1985. Ecological character displacement in Darwin's finches. *Science* 227: 1056–1059. [11]

Schmidt-Nielsen, B. and K. Schmidt-Nielsen. 1951. A complete account of the water metabolism in kangaroo rats and an experimental verification. *Journal of Cellular and Comparative Physiology* 38: 165–182. [4]

Schmidt-Nielsen, K. 1964. *Desert Animals: Physiological Problems of Heat and Water*. Clarendon Press, Oxford. [4]

Schmidt-Nielsen, K. 1997. *Animal Physiology, Adaptation and Environment*. Cambridge University Press, Cambridge. [4]

Schnee, C., T. G. Kollner, M. Held, T. C. J. Turlings, J. Gershenzon and J. Degenhardt. 2006. The products of a single maize sesquiterpene synthase form a volatile defense signal that attracts natural enemies of maize herbivores. *Proceedings of the National Academy of Sciences USA* 103: 1129–1134. [12]

Schoener, T. W. 1971. Theory of feeding strategies. *Annual Review of Ecology and Systematics* 2: 370–404. [5]

Schoener, T. W. 1974. Resource partitioning in ecological communities. *Science* 185: 27–39. [11]

Schoener, T. W. 1983. Field experiments on interspecific competition. *American Naturalist* 122: 240–285. [11]

Schoener, T. W. and D. A. Spiller. 1996. Devastation of prey diversity by experimentally introduced predators in the field. *Nature* 381: 691–694. [12]

Scholander, P. F., V. Walters, R. Hock and L. Irving. 1950. Body insulation of some arctic and tropical mammals and birds. *Biological Bulletin* 99: 225–236. [4]

Schtickzelle, N. and M. Baguette. 2003. Behavioural responses to habitat patch boundaries restrict dispersal and generate emigration–patch area relationships in fragmented landscapes. *Journal of Animal Ecology* 72: 533–545. [23]

Schulte, D. M., R. P. Burke and R. N. Lipcius. 2009. Unprecedented restoration of a native oyster metapopulation. *Science* 325: 1124–1128. [23]

Schuur, E. A. G. 2003. Productivity and global climate revisited: The sensitivity of tropical forest growth to precipitation. *Ecology* 84: 1165–1170. [19]

Schwarz, C. J. and G. A. F. Seber. 1999. Estimating animal abundance: Review III. *Statistical Science* 14: 427–456. [8]

Seehausen, O., J. J. M. van Alphen and F. Witte. 1997. Cichlid fish diversity threatened by eutrophication that curbs sexual selection. *Science* 277: 1808–1811. [22]

Selosse, M.-A. and F. Le Tacon. 1998. The land flora: A phototroph–fungus partnership? *Trends in Ecology and Evolution* 13: 15–20. [14]

Sessions, S. K. and S. B. Ruth. 1990. Explanation for naturally occurring supernumary limbs in amphibians. *Journal of Experimental Zoology* 254: 38–47. [1]

Shaw, M. J. P. 1970. Effects of population density on alienicolae of *Aphis fabae* Scop. I. The effect of crowding on the production of alatae in the laboratory. *Annals of Applied Biology* 65: 191–196. [8]

Sheriff, M. J., C. J. Krebs and R. Boonstra. 2009. The sensitive hare: Sublethal effects of predator stress on reproduction in snowshoe hares. *Journal of Animal Ecology* 78: 1249–1258. [12]

Shmida, A. and M. V. Wilson. 1985. Biological determinants of species diversity. *Journal of Biogeography* 12: 1–20. [17]

Sidor, C. A. 2003. Evolutionary trends and the origin of the mammalian lower jaw. *Paleobiology* 29: 605–640. [6]

Sih, A. and B. Christensen. 2001. Optimal diet theory: When does it work, and when and why does it fail? *Animal Behavior* 61: 379–390. [5]

Simberloff, D. 2003. Eradication: Preventing invasions at the outset. *Weed Science* 51: 247–253. [22]

Simberloff, D. and J. Cox. 1987. Consequences and costs of conservation corridors. *Conservation Biology* 1: 63–71. [23]

Simberloff, D. S. and E. O. Wilson. 1969. Experimental zoogeography of islands: The colonization of empty islands. *Ecology* 50: 278–296. [17]

Simenstad, C. A., J. A. Estes and K. W. Kenyon. 1978. Aleuts, sea otters, and alternative communities. *Science* 200: 403–411. [15]

Sinervo, B. 1990. The evolution of maternal investment in lizards: An experimental and comparative analysis of egg size and its effects on offspring performance. *Evolution* 44: 279–294. [7]

Sinervo, B. and 25 others. 2010. Erosion of lizard diversity by climate change and altered thermal niches. *Science* 328: 894–899. [22, 24]

Sjögren-Gulve, P. 1994. Distribution and extinction patterns within a northern metapopulation of the pool frog, *Rana lessonae*. *Ecology* 75: 1357–1367. [10]

Skerratt, L. F. and 7 others. 2007. Spread of chytridiomycosis has caused the rapid global decline and extinction of frogs. *EcoHealth* 4: 125–134. [1, 22]

Smith, F. D. M., R. M. May, R. Pellew, T. H. Johnson and K. R. Walter. 1993. How much do we know about the current extinction rate? *Trends in Ecology and Evolution* 8: 375–378. [22]

Smith, F. E. 1961. Density-dependence in the Australian thrips. *Ecology* 42: 403–407. [9]

Smith, K. F., M. Behrens, L. M. Schloegel, N. Marano, S. Burgiel and P. Daszak. 2009. Reducing the risks of the wildlife trade. *Science* 324: 594–595. [22]

Smith, R. L. 1974. *Ecology and Field Biology*. 2nd ed. Harper & Row, New York, NY. [10]

Smith, S. E. and D. J. Read. 2008. *Mycorrhizal Symbiosis*. 3rd ed. Academic Press, San Diego, CA. [14]

Smith, T. B. 1993. Disruptive selection and the genetic basis of bill size polymorphism in the African finch *Pyrenestes*. *Nature* 363: 618–620. [6]

Snyder, N. F. R., S. R. Derrickson, S. R. Beissinger, J. W. Wiley, T. B. Smith, W. D. Toone and B. Miller. 1996. Limitations of captive breeding in endangered species recovery. *Conservation Biology* 10: 338–348. [22]

Sokal, R. R. and F. J. Rohlf. 1995. *Biometry*. 3rd ed. W. H. Freeman, New York. [1]

Solé, R. V. and J. Montoya. 2001. Complexity and fragility in ecological networks. *Proceedings of the Royal Society of London* B 268: 2039–2045. [22]

Soulé, M. E. 1985. What is conservation biology? *BioScience* 35: 727–734. [22]

Soulé, M. E. and M. A. Sanjayan. 1998. Conservation targets: Do they help? *Science* 279: 2060–2061. [23]

Soulé, M. E. and B. A. Wilcox, eds. 1980. *Conservation Biology: An Evolutionary–Ecological Perspective*. Sinauer Associates, Sunderland, MA. [22]

Sousa, W. P. 1979a. Disturbance in marine intertidal boulder fields: The non-equilibrium maintenance of species diversity. *Ecology* 60: 1225–1239. [18]

Sousa, W. P. 1979b. Experimental investigations of disturbance and ecological succession in a rocky intertidal algal community. *Ecological Monographs* 49: 227–254. [16, 18]

Sousa, W. P. 2001. Natural disturbance and the dynamics of marine benthic communities. In *Marine Community Ecology*,

M. D. Bertness, S. D. Gaines and M. E. Hay (eds.), 85–130. Sinauer Associates, Sunderland, MA. [16]

Spencer, K. A. 1972. Handbooks for the identification of British insects, *Diptera, Agromyzidae* (Vol. X, Part 5 g). Royal Entomological Society, London. [12]

Springer, A. M. and 7 others. 2003. Sequential megafaunal collapse in the North Pacific Ocean: An ongoing legacy of industrial whaling? *Proceedings of the National Academy of Sciences USA* 100: 12223–12228. [8, 20]

Stachowicz, J. J., J. R. Terwin, R. B. Whitlatch and R. W. Osman. 2002. Linking climate change and biological invasions: Ocean warming facilitates nonindigenous species invasions. *Proceedings of the National Academy of Sciences USA* 99: 15497–15500. [18]

Steadman, D. W. 1995. Prehistoric extinctions of Pacific island birds—Biodiversity meets zooarchaeology. *Science* 267: 1123–1131. [22]

Stebbins, G. L. 1974. *Flowering Plants: Evolution Above The Species Level.* Belknap, Cambridge, MA. [17]

Stein, B. A., L. S. Kutner and J. S. Adams. 2000. *Precious Heritage.* Oxford University Press, Oxford. [22]

Steltzer, H., C. Landry, T. H. Painter, J. Anderson and E. Ayers. 2009. Biological consequences of earlier snowmelt from desert dust deposition in alpine landscapes. *Proceedings of the National Academy of Sciences USA* 106: 11629–11634. [24]

Steneck, R. S., S. D. Hacker and M. N. Dethier. 1991. Mechanisms of competitive dominance between crustose coralline algae: An herbivore-mediated competitive reversal. *Ecology* 72: 938–950. [11]

Stevens, C. J., N. B. Dise, J. O. Mountford and D. J. Gowling. 2004. Impact of nitrogen deposition on the species richness of grasslands. *Science* 303: 1876–1879. [24]

Stevens, O. A. 1932. The number and weight of seeds produced by weeds. *American Journal of Botany* 19: 784–794. [7]

Stinner, J. N. and V. H. Shoemaker. 1987. Cutaneous gas exchange and low evaporative water loss in the frogs *Phyllomedusa sauvagei* and *Chiromantis xerampelina*. *Journal of Comparative Physiology* 157: 423–427. [4]

Stomp, M. and 8 others. 2004. Adaptive divergence in pigment composition promotes phytoplankton biodiversity. *Nature* 432: 104–107. [11]

Stomp, M., J. Huisman, L. Vörös, F. R. Pick, M. Laamanen, T. Haverkamp and L. J. Stal. 2007. Colourful coexistence of red and green picocyanobacteria in lakes and seas. *Ecology Letters* 10: 290–298. [11]

Storey, K. B. 1990. Life in a frozen state: Adaptive strategies for natural freeze tolerance in amphibians and reptiles. *American Journal of Physiology* 258: R559–R568. [4]

Strayer, D. L. and H. M. Malcom. 2007. Effects of zebra mussels (*Dreissena polymorpha*) on native bivalves: The beginning of the end or the end of the beginning? *Journal of the North American Benthological Society* 26: 111–122. [22]

Strebel, P. M. and S. L. Cochi. 2001. Waving goodbye to measles. *Nature* 414: 695–696. [13]

Stuart, S. N., J. S. Chanson, N. A. Cox, B. E. Young, A. S. L. Rodrigues, D. L. Fischman and R. W. Waller. 2004. Status and trends of amphibian declines and extinctions worldwide. *Science* 306: 1783–1786. [1]

Suding, K. and 7 others. 2005. Functional and abundance-based mechanisms explain diversity loss due to nitrogen fertilization. *Proceedings of the National Academy of Sciences USA* 102: 4387–4392. [24]

Sutherland, J. P. 1974. Multiple stable points in natural communities. *American Naturalist* 108: 859–873. [16]

T

Takimoto, G., D. A. Spiller and D. M. Post. 2008. Ecosystem size, but not disturbance, determines food chain length on islands of the Bahamas. *Ecology* 89: 3001–3007. [20]

Tanabe, S. 2002. Contamination and toxic effects of persistent endocrine disrupters in marine mammals and birds. *Marine Pollution Bulletin* 45: 69–77. [22]

Tansley, A. G. 1917. On competition between *Galium saxatile* L. (*G. hercynicum* Weig.) and *Galium sylvestre* Poll. (*G. asperum* Schreb.) on different types of soil. *Journal of Ecology* 5: 173–179. [11]

Tansley, A. G. 1935. The use and abuse of vegetational concepts and terms. *Ecology* 16: 284–307. [19]

Tebbich, S., M. Taborsky, B. Fessl and D. Blomqvist. 2001. Do woodpecker finches acquire tool-use by social learning? *Proceedings of the Royal Society of London* B 268: 2189–2193. [5]

Tegen, I. and I. Fung. 1995. Contribution to the atmospheric mineral aerosol load from land surface modification. *Journal of Geophysical Research–Atmosphere* 100: 18707–18726. [24]

Terborgh, J. 1973. On the notion of favorableness in plant ecology. *American Naturalist* 107: 481–501. [17]

Terborgh, J., K. Feeley, M. Silman, P. Nunez and B. Balukjian. 2006. Vegetation dynamics of predator-free land-bridge islands. *Journal of Ecology* 94: 253–263. [23]

Tewksbury, J. J. and 9 others. 2002. Corridors affect plants, animals, and their interactions in fragmented landscapes. *Proceedings of the National Academy of Sciences USA* 99: 12923–12926. [23]

Thomas, C. D. and T. M. Jones. 1993. Partial recovery of a skipper butterfly (*Hesperia comma*) from population refuges: Lessons for conservation in a fragmented landscape. *Journal of Animal Ecology* 62: 472–481. [10]

Thomas, F., A. Schmidt-Rhaesa, G. Martin, C. Manu, P. Durand and F. Renaud. 2002. Do hairworms (Nematomorpha) manipulate the water seeking behaviour of their terrestrial hosts? *Journal of Evolutionary Biology* 15: 356–361. [13]

Thomas, F. and 7 others. 2003. Biochemical and histological changes in the brain of the cricket *Nemobius sylvestris* infected by the manipulative parasite *Paragordius tricuspidatus* (Nematomorpha). *International Journal for Parasitology* 33: 435–443. [13]

Thomas, R. Q., C. D. Canham, K. C. Weathers and C. L. Goodale. 2010. Increased tree carbon storage in response to nitrogen deposition in the U.S. *Nature Geoscience* 3: 13–17. [24]

Thompson, J. N. 1998. Rapid evolution as an ecological process. *Trends in Ecology and Evolution* 13: 329–332. [6]

Thrall, P. H. and J. J. Burdon. 2003. Evolution of virulence in a plant host–pathogen metapopulation. *Science* 299: 1735–1737. [13]

Tilman, D. 1977. Resource competition between plankton algae: An experimental and theoretical approach. *Ecology* 58: 338–348. [18]

Tilman, D. 1997. Community invisibility, recruitment limitation, and grassland biodiversity. *Ecology* 78: 81–92. [18]

Tilman, D. and J. A. Downing. 1994. Biodiversity and stability in grasslands. *Nature* 367: 363–365. [18]

Tilman, D., M. Mattson and S. Langer. 1981. Competition and nutrient kinetics along a temperature gradient: An experimental test of a mechanistic approach to niche theory. *Limnology and Oceanography* 26: 1020–1033. [11, 18]

Tilman, D., D. Wedin and J. Knops. 1996. Productivity and sustainability influenced by biodiversity in grassland ecosystems. *Nature* 379: 718–720. [18]

Tilman, D., J. Hill and C. Lehman. 2006. Carbon-negative biofuels from low-input high-diversity grassland biomass. *Science* 314: 1598. [18]

Tinker, D. B., W. H. Romme and D. G. Despain. 2003. Historic range of variability in landscape structure in subalpine forests of the Greater Yellowstone Area, USA. *Landscape Ecology* 18: 427–439. [23]

Tomback, D. F. 1982. Dispersal of whitebark pine seeds by Clark's nutcracker: A mutualism hypothesis. *Journal of Animal Ecology* 51: 451–467. [23]

Tomback, D. F. and P. Achuff. 2010. Blister rust and western forest biodiversity: Ecology, values and outlook for white pines. *Forest Pathology* 40: 186–225. [23]

Townsend, C. R., M. R. Scarsbrook and S. Doledec. 1997. The intermediate disturbance hypothesis, refugia, and biodiversity in streams. *Limnology and Oceanography* 42: 938–949. [18]

Tracy, R. L. and G. E. Walsberg. 2002. Kangaroo rats revisited: Re-evaluating a classic case of desert survival. *Oecologia* 133: 449–457. [4]

Trites, A. W. and C. P. Donnelly. 2003. The decline of Steller seal lions *Eumetopias jubatus* in Alaska: A review of the nutritional stress hypothesis. *Mammal Review* 33: 3–28. [20]

Tsukaya, H., K. Fujikawa and S. Wu. 2002. Thermal insulation and accumulation of heat in the downy inflorescences of *Saussurea medusa* (Asteraceae) at high elevation in Yunnan, China. *Journal of Plant Research* 115: 263–268. [4]

Turchin, P. 2003. *Complex Population Dynamics: A Theoretical/ Empirical Synthesis*. Princeton University Press, Princeton, NJ. [12]

Turner, M. G., R. H. Gardner and R. V. O'Neill. 2001. *Landscape Ecology in Theory and Practice: Pattern and Process.* Springer Verlag, New York. [23]

Turner, M. G., W. H. Romme and D. B. Tinker. 2003. Surprises and lessons from the 1988 Yellowstone fires. *Frontiers in Ecology and the Environment* 1: 351–358. [23Tyerman, J. G., M. Bertrand, C. C. Spencer and M. Doebeli. 2008. Experimental demonstration of ecological character displacement. *BMC Evolutionary Biology* 8: ARTN 34. [11]

U

Uddenberg, N., T. Fagerström and A. Jeffner. 1995. Kontakten med urkällan: Naturen, människan och religionen. In *Det Stora Sammanhanget*, N. Uddenberg (ed.), 141–186. Doxa, Nora, Sweden. [1]

Umina, P. A., A. R. Weeks, M. R. Kearney, S. W. McKechnie and A. A. Hoffmann. 2005. A rapid shift in a classic clinal pattern in *Drosophila* reflecting climate change. *Science* 308: 691–693. [6]

United Nations, Department of Economic and Social Affairs, Population Division. 2009. *World Population Prospects: The 2008 Revision, Highlights.* Working Paper No. ESA/P/WP.210. [9]

U.S. Congress, Office of Technology Assessment. 1993. *Harmful Non-Indigenous Species in the United States.* OTA-F-565. Washington, DC: U.S. Government Printing Office. [22]

U.S. Environmental Protection Agency. 2005. *Evaluating Ozone Control Programs in the Eastern United States: Focus on the NO_x Budget Trading Program, 2004*. EPA454-K-05–001. http://www.epa.gov/airtrends/2005/ozonenbp.pdf. [24]

U.S. Fish and Wildlife Service. 2003. Recovery plan for the red-cockaded woodpecker (*Picoides borealis*): Second revision. U.S. Fish and Wildlife Service, Atlanta, GA. [22]

U.S. Fish and Wildlife Service. 2008. Birds of conservation concern 2008. U.S. Fish and Wildlife Service, Division of Migratory Bird Management, Arlington, VA.

V

Valverde, T. and J. Silvertown. 1997. A metapopulation model for *Primula vulgaris*, a temperate forest understorey herb. *Journal of Ecology* 85: 193–210. [10]

Van Breemen, N. and A. C. Finzi. 1998. Plant-soil interactions: Ecological aspects and evolutionary implications. *Biogeochemistry* 42: 1–19. [21]

Van den Bosch, F., R. Hengeveld and J. A. J. Metz. 1992. Analysing the velocity of animal range expansion. *Journal of Biogeography* 19: 135–150. [10]

Van der Heijden, M. G. A. and 7 others. 1998. Mycorrhizal fungal diversity determines plant biodiversity, ecosystem variability and productivity. *Nature* 396: 69–72. [14]

Van der Veken, S., J. Rogister, K. Verheyen, M. Hermy and R. Nathan. 2007. Over the (range) edge: A 45-year transplant experiment with the perennial forest herb *Hyacinthoides non-scripta*. *Journal of Ecology* 95: 343–351. [8]

Van Dover, C. L. 2000. *The Ecology of Deep-Sea Hydrothermal Vents*. Princeton University Press, Princeton, NJ. [19]

Van Lear, D. H., W. D. Carroll, P. R. Kapeluck and R. Johnson. 2005. History and restoration of the longleaf pine–grassland ecosystem: Implications for species at risk. *Forest Ecology and Management* 211: 150–165. [22]

Van Mantgem, P. J. and 10 others. 2009. Widespread increase of tree mortality rates in the western United States. *Science* 323: 521–524. [9]

Vannote, R. L., G. W. Minshall, K. W. Cummins, K. R. Sedell and C. E. Cushing. 1980. The River Continuum Concept. *Canadian Journal of Fisheries and Aquatic Sciences* 37: 130–137. [3]

Van Tyne, J. and A. J. Berger. 1959. *Fundamentals of Ornithology*. Wiley, New York. [17]

Varley, N. and M. S. Boyce. 2006. Adaptive management for reintroductions: Updating a wolf recovery model for Yellowstone National Park. *Ecological Modelling* 193: 315–339. [23]

Veblen, T. T., T. Kitzberger and J. Donnegan. 2000. Climatic and human influences on fire regimes in ponderosa pine forests in the Colorado Front Range. *Ecological Applications* 10: 1178–1195. [2]

Vellend, M., J. A. Myers, S. Gardescu and P. L. Marks. 2003. Dispersal of *Trillium* seeds by deer: Implications for long-distance migration of forest herbs. *Ecology* 84: 1067–1072. [14]

Vié, J.-C., C. Hilton-Taylor and S. N. Stuart, eds. 2009. *Wildlife in a Changing World—An Analysis of the 2008 IUCN Red List of Threatened Species*. Gland, Switzerland: IUCN. [1, 22]

Vilà, C. and 9 others. 2003. Rescue of a severely bottlenecked wolf (*Canis lupus*) population by a single immigrant. *Proceedings of the Royal Society of London* B 270: 91–97. [22]

Vitousek, P. M. 1977. The regulation of element concentrations in mountain streams in the northeastern United States. *Ecological Monographs* 47: 65–87. [21]

Vitousek, P. M. and H. Farrington. 1997. Nutrient limitation and soil development: Experimental test of a biogeochemical theory. *Biogeochemistry* 37: 63–75. [21]

Volterra, V. 1926. Variazioni e fluttuazioni del numero d'individui in specie animali conviventi. *Memorie della R. Accademia dei Lincei* 2: 31–113. Reprinted 1931 as "Variations and fluctuations of the number of individuals in animal species living together," in R. N. Chapman, *Animal Ecology*, McGraw-Hill, New York. [11]

W

Wackernagel, M. and 7 others. 1999. National natural capital accounting with the ecological footprint concept. *Ecological Economics* 29: 375–390. [9]

Wagner, W. L., M. M. Bruegmann, D. R. Herbst and J. Q. C. Lau. 1999. Hawaiian vascular plants at risk: 1999. *Bishop Museum Occasional Papers* 60: 1–58. [22]

Walker, T. W. and J. K. Syers. 1976. The fate of phosphorus during pedogenesis. *Geoderma* 15: 1–19. [21]

Wallace, A. R. 1855. On the law which has regulated the introduction of new species. *Annals and Magazine of Natural History* 16. [17]

Wallace, A. R. 1858. On the tendency of varieties to depart indefinitely from the original type. *Journal of the Proceedings of the Linnean Society, Zoology* 3: 53–62. [17]

Wallace, A. R. 1860. On the zoological geography of the Malay Archipelago. *Journal of the Linnean Society of London* 4: 172–184. [8]

Wallace, A. R. 1869. *The Malay Archipelago, The Land of The Orangutan And The Bird of Paradise: A Narrative of Travel With Studies of Man And Nature*. Macmillan, London. [17, 22]

Wallace, A. R. 1876. *The Geographic Distribution of Animals*. 2 vols. Harper and Brothers, New York. [17]

Wallace, A. R. 1878. *Tropical Nature and Other Essays*. Macmillan, London. [17]

Walters, C. 1995. *Fish on the Line: The Future of Pacific Fisheries*. A report to the David Suzuki Foundation, Fisheries Project Phase I, 219–2211 West Fourth Avenue, Vancouver, BC, V6K 4S2, Canada. [2]

Walters, J. R., C. K. Copeyon and J. H. Carter. 1992. Test of the ecological basis of cooperative breeding in red-cockaded woodpeckers. *Auk* 109: 90–97. [22]

Walther, G.-R. and 8 others. 2002. Ecological responses to recent climate change. *Nature* 416: 389–395. [24]

Wang, X., Z. Dong, J. Zhang and L. Liu. 2004. Modern dust storms in China: An overview. *Journal of Arid Environments* 58: 559–574. [24]

Waples, R. S. 1998. Evolutionarily significant units, distinct population segments, and the Endangered Species Act: Reply to Pennock and Dimmick. *Conservation Biology* 12: 718–721. [22]

Ward, M. P., J. S. Milledge and J. B. West. 1995. *High Altitude Medicine and Physiology*. Chapman and Hall, London. [4]

Weathers, K. C., M. L. Cadenasso and S. T. A. Pickett. 2001. Forest edges as nutrient and pollutant concentrators. *Conservation Biology* 15: 1506–1514. [23]

Webb, N. R. and L. E. Haskins. 1980. An ecological survey of heathlands in the Poole Basin, Dorset, England, in 1978. *Biological Conservation* 17: 281–296. [8]

Webberley, K. M. and 8 others. 2004. Host reproduction and a sexually transmitted disease: Causes and consequences of *Coccipolipus hippodamiae* distribution on coccinellid beetles. *Journal of Animal Ecology* 73: 1–10. [13]

Weber, N. A. 1966. Fungus-growing ants. *Science* 153: 587–604. [14]

Weber, R. E. 2007. High-altitude adaptations in vertebrate hemoglobins. *Respiratory Physiology and Neurobiology* 158: 132–142. [6]

Weeks, A. R., M. Turelli, W. R. Harcombe, K. T. Reynolds and A. A. Hoffmann. 2007. From parasite to mutualist: Rapid evolution of *Wolbachia* in natural populations of *Drosophila*. *PLoS Biology* 5: 997–1005. [14]

Weir, A. A. S., J. Chappell and A. Kacelnik. 2002. Shaping of hooks in New Caledonian crows. *Science* 297: 981. [5]

Weis, A. E. and W. G. Abrahamson. 1986. Evolution of host-plant manipulation by gall makers: Ecology and genetic factors in the *Solidago–Eurosta* system. *American Naturalist* 127: 681–695. [6]

Weiss, R. A. and A. J. McMichael. 2004. Social and environmental risk factors in the emergence of infectious disease. *Nature Medicine* 10: S70–S76. [1]

Weldon, A. J. 2006. How corridors reduce Indigo Bunting nest success. *Conservation Biology* 20: 1300–1305. [23]

Werner, E. E. 1988. Size scaling and the evolution of complex life cycles. In *Size-Structured Populations*, B. Ebenman and L. Persson (eds.), 61–81. Springer-Verlag, Berlin. [7]

Westemeier, R. L. and 8 others. 1998. Tracking the long-term decline and recovery of an isolated population. *Science* 282: 1695–1698. [6]

Westoby, M. 1997. What does "ecology" mean? *Trends in Ecology and Evolution* 12: 166. [1]

White, G. 1789. *The Natural History of Selborne*. Reprinted 1977, Penguin, New York. [10]

White, J., J. Wilson and K. Horskins. 1997. The role of adjacent habitats in rodent damage levels in Australian macadamia orchard systems. *Crop Protection* 16: 727–732. [23]

Whittaker, R. H., K. J. Willis and R. Field. 2001. Towards a general, hierarchical theory of species diversity. *Journal of Biogeography* 28: 453–470. [17]

Wilbur, H. M. 1997. Experimental ecology of food webs: Complex systems in temporary ponds. The Robert H. MacArthur Award Lecture. *Ecology* 78: 2279–2302. [20]

Wilkie, D. S. and J. F. Carpenter. 1999. Bushmeat hunting in the Congo Basin: An assessment of impacts and options for mitigation. *Biodiversity and Conservation* 8: 927–955. [22]

Williams, B. K., J. D. Nichols and M. J. Conroy. 2002. *Analysis and Management of Animal Populations*. Academic Press, San Diego, CA. [8]

Williams, C. B. 1943. Area and number of species. *Nature* 152: 264–267. [17]

Williams, C. B. 1964. *Patterns in the Balance of Nature*. Academic Press, London. [17]

Williams, J. C., C. S. ReVelle and S. A. Levin. 2005. Spatial attributes and reserve design models: A review. *Environmental Modeling and Assessment* 10: 163–181. [23]

Willig, M. R., D. M. Kaufman and R. D. Stevens. 2003. Latitudinal gradients of biodiversity: Pattern, process, scale, and synthesis. *Annual Review of Ecology and Systematics* 34: 273–309. [17]

Willmer, P., G. Stone and I. Johnston. 2005. *Environmental Physiology of Animals*. Blackwell Publishing, Malden, MA. [4]

Wilson, E. O. 1994. *Naturalist*. Island Press, Washington, DC. [17]

Wilson, E. O. and D. S. Simberloff. 1969. Experimental zoogeography of islands: Defaunation and monitoring techniques. *Ecology* 50: 267–278. [17]

Wilson, S. D. and D. Tilman. 1993. Plant competition and resource availability in response to disturbance and fertilization. *Ecology* 74: 599–611. [11]

Windsor, D. A. 1998. Most of the species on Earth are parasites. *International Journal for Parasitology* 28: 1939–1941. [13]

Winterhalder, B. P. 1980. Canadian fur bearer cycles and Cree-Ojibwa hunting and trapping practices. *American Naturalist* 115: 870–879. [12]

Witman, J. D., R. J. Etter and F. Smith. 2004. The relationship between regional and local species diversity in marine benthic communities: A global perspective. *Proceedings of the National Academy of Sciences USA* 101: 15664–15669. [17]

Woodroffe, R. 1999. Managing disease threats to wild mammals. *Animal Conservation* 2: 185–193. [22]

Wright, S. J. 1981. Intra-archipelago vertebrate distributions: The slope of the species–area relation. *American Naturalist* 118: 726–748. [17]

Wright, S. J. 2005. Tropical forests in a changing environment. *Trends in Ecology and Evolution* 10: 553–560. [2, 3]

Y

Yates, T. L. and 15 others. 2002. The ecology and evolutionary history of an emergent disease: Hantavirus pulmonary syndrome. *BioScience* 52: 989–998. [8]

Yoda, J. A., T. Kira, H. Ogawa and K. Hozumi. 1963. Self-thinning in overcrowded pure stands under cultivated and natural conditions. *Journal of Biology, Osaka City University* 14: 107–120. [9]

Yoshida, T., L. E. Jones, S. P. Ellner, G. F. Fussmann and N. G. Hairston Jr. 2003. Rapid evolution drives ecological dynamics in a predator–prey system. *Nature* 424: 303–306. [12]

Z

Zar, J. H. 2006. *Biostatistical Analysis*. 5th ed. Prentice Hall, Upper Saddle River, NJ. [1]

Zhao, M. and S. W. Running. 2010. Drought-induced reduction in global terrestrial net primary production from 2000 through 2009. *Science* 329: 940–943. [24]

Zhao, M., S. W. Running and R. R. Nemani. 2006. Sensitivity of Moderate Resolution Imaging Spectroradiometer (MODIS) terrestrial primary production to the accuracy of meteorological reanalyses. *Journal of Geophysical Research–Biogeosciences* 111: G01002. [19]

Index

Page numbers in *italic* type indicate the information will be found in an illustration. The designation "n" following a page number indicates the information will be found in a footnote.

B

C

F

M

N

O